2020 22nd European Conference on Power Electronics and Applications (EPE'20 ECCE Europe)

Lyon, France
7-11 September 2020

Pages 2743-3427

IEEE Catalog Number: CFP20850-POD
ISBN: 978-1-7281-9807-1

Copyright © 2020, EPE Association
All Rights Reserved

*** *This is a print representation of what appears in the IEEE Digital Library. Some format issues inherent in the e-media version may also appear in this print version.*

IEEE Catalog Number: CFP20850-POD
ISBN (Print-On-Demand): 978-1-7281-9807-1
ISBN (Online):   978-9-0758-1536-8

Additional Copies of This Publication Are Available From:

Curran Associates, Inc
57 Morehouse Lane
Red Hook, NY 12571 USA
Phone: (845) 758-0400
Fax: (845) 758-2633
E-mail: curran@proceedings.com
Web: www.proceedings.com

2020 22nd European Conference on Power Electronics and Applications (EPE'20 ECCE Europe)

Lyon, France
7-11 September 2020

Pages 2743-3427

IEEE Catalog Number: CFP20850-POD
ISBN: 978-1-7281-9807-1

TABLE OF CONTENTS

VALIDATION OF THERMAL STRESS MODELING IN PV INVERTERS UNDER MISSION PROFILE OPERATION 1
Ariya Sangwongwanich, Huai Wang, Frede Blaabjerg

ON THE LIMITATIONS OF USING A LTI MODELLING APPROACH FOR CONTROL TUNING OF VSC-HVDC SYSTEMS 9
Pablo Briff, Julián Freytes, Guillaume De-Preville, Jiaqi Li, Omar Jasim

A VOLTAGE CONTROL METHOD FOR POWER DISTRIBUTION LINES UTILIZING DISPERSED CUSTOMER RESOURCES 19
Hiroki Ishihara, Kaho Nada, Miwako Tanaka, Sadayuki Inoue, Akiko Kuwata, Tomihiro Takano

PERFORMANCE COMPARISON BETWEEN SIC AND SI INVERTER MODULES IN AN ELECTRICAL VARIABLE TRANSMISSION APPLICATION 27
Mauricio Dalla Vecchia, Simon Ravyts, Florian Verbelen, Jeroen Tant, Peter Sergeant, Johan Driesen

SEAMLESS INTEGRATION OF FEEDFORWARD AND FEEDBACK CONTROL OF BALANCE OF ARM CAPACITOR VOLTAGES IN STATCOMS BASED ON CHAIN LINKS OF H BRIDGE MODULES 37
D. Basic, N. Lapassat

ASYNCHRONIZED ELECTROMECHANICAL CONVERTER IN THE ELECTRICAL SUPPLY SYSTEM OF POWERFUL ENERGY CONSUMERS 47
Aleksey G. Vorontsov, Mikhail V. Pronin, Anastasiia D. Stotckaia, Vasiliy V. Glushakov, Pavel V. Sokur

SYMMETRIC AND ASYMMETRIC OPERATING MODES OF HYBRID CASCADE FREQUENCY CONVERTERS 56
Aleksey G. Vorontsov, Vasiliy V. Glushakov, Mikhail V. Pronin, Anastasiia D. Stotckaia

SYSTEM FREQUENCY DYNAMIC RESPONSE OF A NOVEL, SELF-SYNCHRONIZING INVERTER IN A HIGH RENEWABLE PENETRATION GRID 65
Christian Perenyi, Moath Alqatamin, Thibaut Harzig, Michael McIntyre, Brandon M. Grainger

ROTOR POSITION ESTIMATION WITH HALL-EFFECT SENSORS IN BEARINGLESS DRIVES 75
Patricio Peralta, Jacopo Leo, Yves Perriard

NON-UNIT ROCOV SCHEME FOR PROTECTION OF MULTI-TERMINAL HVDC SYSTEMS 85
María José Pérez-Molina, Pablo Eguia, Marene Larruskain, Garikoitz Buigues, Esther Torres

MODELLING OF CONVERTER SYSTEMS PARALLELED VIA INTERPHASE TRANSFORMERS IN CYCLIC CASCADE TOPOLOGY AND OPTIMIZATION OF PWM CARRIER SHIFTS 95
D. Basic, H. Baërd, S. Siala

MEASUREMENT AND CALCULATION METHOD OF WIRELESS POWER TRANSFER COIL EQUIVALENT SERIES RESISTANCE UNDER THE VEHICLE....................................... 105
Norihito Kimura, Hiroaki Yuasa

DESIGN OF A CIRCUMSCRIBING POLYGON WIDE BANDGAP BASED INTEGRATED MODULAR MOTOR DRIVE TOPOLOGY WITH THERMALLY DECOUPLED WINDINGS AND POWER CONVERTERS 115

Abdalla Hussein Mohamed, Hendrik Vansompel, Peter Sergeant

LIMITS OF ENHANCED DESATURATION DETECTION METHOD WITH ADAPTIVE BLANKING FOR GAN HEMTS 124

Jan Schmitz, Markus Meißner, Steffen Bernet

CURRENT CONTROL OF A GRID-CONNECTED SINGLE-PHASE VOLTAGE-SOURCE INVERTER WITH LCL FILTER 134

Alfonso Parreño Torres, Fco. Javier López-Alcolea, Pedro Roncero-Sánchez, Javier Vázquez, Emilio J. Molina-Martínez, Felix García-Torres

FOUR SWITCH BUCK/BOOST CONVERTER FOR DC MICROGRID APPLICATIONS 143

Matthias Schulz, Nico Schleippmann, Kilian Gosses, Bernd Wunder, Martin März

STABILITY INVESTIGATION OF THREE-PHASE GRID-TIED PV INVERTERS WITH IMPEDANCE-BASED METHOD 153

Zhiqing Yang, Wanchao Gou, Xian Luo, Chirag Shah, Nurhan Rizqy Averous, Rik W. De Doncker

STABILITY INVESTIGATION OF LARGE-SCALE PV PARKS WITH EIGENVALUE-BASED METHOD 163

Zhiqing Yang, Christian Bendfeld, Jin Qiang, Benedict Mortimer, Rik W. De Doncker

COMPACT CORE LOSS MODEL BASED ON AN EFFECTIVE FREQUENCY FOR ARBITRARY CORE EXCITATIONS INCLUDING DC-BIAS 173

Erika Stenglein, Manfred Albach, Thomas Dürbaum

ASSESSMENT OF AGING AND PERFORMANCE DEGRADATION OF SUPERCAPACITORS INTEGRATED INTO A MODULAR MULTILEVEL CONVERTER 183

F. Errigo, L. Chédot, F. Morel, P. Venet, A. Sari, A. Hijazi, R. A. Peña

SEPARATION OF MAGNETIC FLUX DENSITY TRAJECTORIES INTO SUBLOOPS FOR THE PREDICTION OF HYSTERESIS LOSS 193

Erika Stenglein, Manfred Albach, Thomas Dürbaum

INFLUENCE OF GENERALIZED DISCONTINUOUS PULSE WIDTH MODULATION (GDPWM) ON THE DC-LINK CURRENT AND VOLTAGE RIPPLE IN BATTERY-FED PWM INVERTER SYSTEMS 203

Panagiotis Mantzanas, Alexander Bucher, Daniel Kuebrich, Alexander Pawellek, Christian Hasenohr, Harald Hofmann, Thomas Duerbaum

AUTOMATED DESIGN METHOD FOR SINE WAVE FILTERS IN MOTOR DRIVE APPLICATIONS WITH SIC-INVERTERS 213

Thorben Schobre, Regine Mallwitz

A SYMMETRICAL BOOST CONVERTER WITH REDUCED COMMON-MODE LEAKAGE CURRENTS FOR EV APPLICATIONS 223

Caniggia Viana, Netan Yakop, Damien Frost, Peter Lehn

MODELING AND ANALYSIS OF CONDUCTED EMI ON FLYBACK CONVERTER USING POWER MANAGEMENT IC WITH CHAOTIC SUPPRESSION EMI 231

Diao Jiaqi, Yang Ru, Liu Zuolian, Yang Hong, Jie Hai

HIGH PERFORMANCE DRIVE INVERTER FOR AN ELECTRIC TURBO COMPRESSOR IN FUEL CELL APPLICATIONS 241

N. Langmaack, G. Tareilus, R. Mallwitz

DEVELOPMENT OF AN ALGORITHM FOR THE AUTOMATION OF THE MODELLING PROCESS OF POWER CONVERTERS 251

Jon Anzola, Iosu Aizpuru, Asier Arruti

A NOVEL FULLY DISTRIBUTED COST OPTIMAL CONTROL METHOD FOR DC MICROGRID 260

Qingping Xia, Hua Han, Yao Liu, Zhangjie Liu, Yao Sun, Mei Su

MEASUREMENT OF DYNAMIC ON-STATE RESISTANCE OF HIGH-VOLTAGE GAN-HEMTS UNDER REAL APPLICATION CONDITIONS 266

Benedikt Kohlhepp, Carsten Kuring, Stefan Peller, Daniel Kübrich

ANALYSIS OF DC-SIDE FAULT RESPONSE OF MMCS WITH CONTROLLED FAULT BLOCKING CAPABILITY FOR DIFFERENT TRANSMISSION LINE TYPES 276

Willem Leterme, Paul D. Judge, Tim C. Green

A HYBRID SERIES-PARALLEL MICROGRID AND ITS LOW-DEPENDENT COMMUNICATION CONTROL 285

Lang Li, Yao Sun, Hua Han, Mei Su

ADAPTIVE VOLTAGE CONTROL OF ISLANDED RES-BASED RESIDENTIAL MICROGRID WITH INTEGRATED FLYWHEEL/BATTERY HYBRID ENERGY STORAGE SYSTEM 292

Linda Barelli, Gianni Bidini, Ermanno Cardelli, Dana-Alexandra Ciupageanu, Andrea Ottaviano, Dario Pelosi, Simone Castellini, Gheorghe Lazaroiu

AN IMPROVED λ -CONSENSUS CONTROL METHOD FOR DC MICROGRIDS 302

Siqi Fu, Yao Sun, Zhangjie Liu, Hua Han, Mei Su

DECREASE OF POWER ELECTRONIC SWITCHING LOSSES USING VARIABLE SWITCHING EVENTS 307

Hannes Ramm, Michael Homann, Torben A. Schulze, Faical Turki, Heiko Rabba

OPTIMIZATION OF MEDIUM-FREQUENCY TRANSFORMERS WITH LARGE CAPACITY AND HIGH INSULATION REQUIREMENT 317

Xuan Guo, Chi Li, Zedong Zheng, Yongdong Li

IMPROVED SOC BALANCING AND ACTIVE POWER SHARING CONTROL METHOD IN HIGHLY RESISTIVE LINE MICROGRID 326

Yuanhao Zhu, Hua Han, Guangze Shi, Zhangjie Liu, Yao Sun, Mei Su

TECHNO-ECONOMIC ANALYSIS OF SECOND-LIFE LITHIUM-ION BATTERIES INTEGRATION IN MICROGRIDS 332

Camille Birou, Xavier Roboam, Hugo Radet, Fabien Lacressonnière

DESIGN, MODELLING, AND TEST OF A SOLID-STATE MAIN BREAKER FOR HYBRID DC CIRCUIT BREAKER 342

Jiawen Xi, Xiaoze Pei, Xianwu Zeng, Liyong Niu

MODEL PREDICTIVE CONTROL FOR THREE-PHASE SPLIT-SOURCE INVERTER 352

Youssuf Elthokaby, Islam Mohamed, Naser Abdel-Rahim

HARDWARE IMPLEMENTATION STUDY OF VARIABLE SPEED WIND-TURBINE-DFIG IN STAND-ALONE MODE 362
Fayssal Amrane, Bruno Francois, Azeddine Chaiba

INFLUENCE OF WIRE-BONDING LAYOUT ON RELIABILITY IN IGBT MODULE 370
Lubin Han, Lin Liang, Wei Xin, Fang Luo

RAIL POTENTIAL CALCULATION MODEL FOR DC RAILWAY POWER SUPPLY EQUIPPED WITH VOLTAGE LIMITING DEVICE 377
Shota Kimura, Tsutomu Miyauchi, Kenji Oguma, Hirotaka Takahashi, Keiko Teramura

HOMOGENIZATION OF CURRENT DISTRIBUTION IN PARALLEL CONNECTION OF INTERLEAVED WINDING LAYERS OF HIGH-FREQUENCY TRANSFORMERS BY OPTIMIZING DISTANCE BETWEEN WINDING LAYERS 386
Ryo Murata, Tomohide Shirakawa, Kazuhiro Umetani, Eiji Hiraki, Hiroto Mizutani, Takaaki Takahara, Osamu Mori

REAL-TIME PARAMETERS IDENTIFICATION OF LITHIUM-ION BATTERIES MODEL TO IMPROVE THE HIERARCHICAL MODEL PREDICTIVE CONTROL OF BUILDING MICROGRIDS 396
Daniela Yassuda Yamashita, Ionel Vechiu, Jean-Paul Gaubert

IMPACT OF DC FAULT BLOCKING CAPABILITY ON THE SIZING OF THE DC-DC MODULAR MULTILEVEL CONVERTER 406
J. D. Paez, F. Morel, S. Bacha, Piotr Dworakowski, D. Frey

OPTIMIZATION OF HIGH FREQUENCY MAGNETIC DEVICES WITH CONSIDERATION OF THE EFFECTS OF THE MAGNETIC MATERIAL, THE CORE GEOMETRY AND THE SWITCHING FREQUENCY 416
Sobhi Barg, Muhammad Farhan Alam, Kent Bertilsson

REAL TIME CONTROL HARDWARE IN THE LOOP TEST OF A NOVEL MVDC SOLID-STATE BREAKER 424
Alessio Clerici, Riccardo Chiumeo, Chiara Gandolfi

IGBT LIFETIME ESTIMATION IN A MODULAR MULTILEVEL CONVERTER FOR BIDIRECTIONAL POINT-TO-POINT HVDC APPLICATION 433
Diego Velazco, Guy Clerc, Emmanuel Boutleux, François Wallart, Laurent Chédot

OPTIMIZATION DESIGN FOR SIC DRIFT STEP RECOVERY DIODE (DSRD) 443
Xiaoxue Yan, Lin Liang, Ziyue Wang, Guoqiang Tan

DISCRETE SUPER-TWISTING SLIDING MODE CURRENT CONTROLLER FOR INDUCTION MOTOR DRIVES 450
Tianqing Wang, Bo Wang, Yong Yu, Yangming Zhu, Dianguo Xu

NEW GRID-CONNECTED MULTILEVEL BOOST CONVERTER TOPOLOGY WITH INHERENT CAPACITORS VOLTAGE BALANCING USING MODEL PREDICTIVE CONTROLLER 460
Rasoul Shalchi Alishah, Kent Bertilsson, Frede Blaabjerg, Mohd. Ali Jagabar Sathik, Ali Yahya Rezaee

DCM OPERATION OF SINGLE-SWITCH HIGH STEP-UP DC-DC CONVERTER WITH THREE-WINDING COUPLED INDUCTOR 467
Masataka Minami, Genki Hase

POWER LOSSES CALCULATION FOR MEDIUM VOLTAGE DC/DC CURRENT-FED SOLID STATE TRANSFORMER FOR BATTERY GRID-CONNECTED 471

E. K. Hussain, Mohammad Abusara, S. M. Sharkh

MODELLING AND EXPERIMENTAL VALIDATION OF A POLE-TO-GROUND PROTECTION DEVICE IN LOW VOLTAGE DC MICROGRIDS 480

L. Hallemans, G. Govaerts, G. Van Den Broeck, S. Ravyts, M. M. Alam, P. Van Tichelen, J. Driesen

DESIGN OF A DUAL ACTIVE BRIDGE CONVERTER FOR ON-BOARD VEHICLE CHARGERS USING GAN AND INTO TRANSFORMER INTEGRATED SERIES INDUCTANCE 490

K. Siebke, M. Giacomazzo, R. Mallwitz

AN EXPERIMENTAL ANALYSIS OF CIRCULATING CURRENT CONTROL CIRCUIT FOR OUTPUT POWER FROM VIBRATION GENERATOR FOR VIBRATION INCLUDING THE THIRD HARMONICS 498

Masataka Minami, Akito Nakagaki, Genki Hase

IMPLEMENTATION OF CONTROL STRATEGY FOR STEP-DOWN DC-DC CONVERTER BASED ON PIEZOELECTRIC RESONATOR 503

Mustapha Touhami, Ghislain Despesse, François Costa, Benjamin Pollet

THERMAL IMPEDANCES AND TEMPERATURE SENSORS: A COMBINED APPROACH FOR A NOVEL THERMAL MODEL OF POWER SEMICONDUCTORS 512

Maria De Lauretis, Jonas Millinger, Erik Baker, Martin Karlsson, Diane -Perle Sandik

A 3A LOW VOLTAGE LASER DIODE DRIVER IC IN A CMOS TECHNOLOGY FOR AN ITOF-BASED 3D IMAGE SENSOR 522

Romain David, Bruno Allard, Xavier Branca, Charles Joubert

COMPARISON OF DECOUPLING TECHNIQUES VIA DISCRETE LUENBERGER STYLE OBSERVER FOR VOLTAGE ORIENTED CONTROL 532

Gyanendra Kumar Sah, Michael Schütt, Hans-Günter Eckel

VARIABLE SWITCHING POINT PARALLEL PREDICTIVE CURRENT CONTROL (VSP3CC) FOR INDUCTION MOTOR 542

Qing Chen, Ralph Kennel

OPERATION OF AN EXTERNALLY EXCITED SYNCHRONOUS MACHINE WITH A HYBRID MULTILEVEL INVERTER 551

C. Terbrack, J. Stöttner, C. Endisch

A FACILITY FOR MIXED FLOWING GAS TESTING OF AND EXPERIMENTATION WITH POWER ELECTRONIC COMPONENTS AND SYSTEMS 563

Juuso Rautio, Janne Jäppinen, Tommi J. Kärkkäinen, Markku Niemelä, Pertti Silventoinen, Mika Kiviniemi, Joonas Leppänen, Jonny Ingman

IMPACT OF IMPLEMENTATION OF AUXILIARY BIAS-WINDINGS ON CONTROLLABLE INDUCTORS FOR POWER ELECTRONIC CONVERTERS 571

Jonas Pfeiffer, Pierre Küster, Yeliz Erenler, Ziyad H. S. Qashlan, Peter Zacharias

APPROXIMATED SLIDING-MODE CONTROL OF PARALLEL-CONNECTED GRID INVERTERS 581

Albrecht Gensior

EQUIVALENT MODEL AND CONTROL OF A NEUTRAL POINT SUPPLY SYNRM DRIVE 590
Xiaokang Zhang, Jean-Yves Gauthier, Xuefang Lin-Shi

IMPROVEMENTS ON SIGNAL-TO-NOISE RATIO IN FEEDBACK MEASUREMENT IN
DC/DC CONVERTERS 598
Fernando Davalos Hernandez, Federico Ibanez, Sebastian Gutierrez, Wilmar Martinez

APPROACH OF AN ACTIVE DEVICE PROTECTION FOR DRIVE INVERTERS AGAINST
SHORT CIRCUIT FAULTS IN AN OPEN INDUSTRIAL DC GRID.. 608
Simon Puls, Urs Obernolte, Martin Ehlich, Holger Borcherding

A NEW DESIGN OF AN AIR CORE TRANSFORMER FOR ELECTRIC VEHICLE ON-
BOARD CHARGER 618
Valentin Rigot, Tanguy Phulpin, Daniel Sadarnac, Jihen Sakly

ENABLING FOIL WINDINGS OF MEDIUM-FREQUENCY TRANSFORMERS FOR HIGH
CURRENTS 627
Thomas B. Gradinger, Uwe Drofenik, Filip Grecki

A HIGH-EFFICIENCY WIRELESS POWER TRANSFER SYSTEM FOR UNMANNED
AERIAL VEHICLE CONSIDERING CARBON FIBER BODY 637
Kai Song, Peng Zhang, Zhengxin Chen, Guang Yang, Jinhai Jiang, Chunbo Zhu

ANALYTICAL COMPUTATION OF NORMAL AND FAULT-TOLERANT ACTIVE SHORT
CIRCUIT OPERATION OF ANISOTROPIC SYNCHRONOUS DOUBLE STAR MACHINES 644
Michael Gleissner, Johannes Häring, Wolfgang Wondrak, Mark-M. Bakran

FULL-SILICON 98.7% EFFICIENT THREE-PHASE FIVE-LEVEL 3-PORT UPS
ARCHITECTURE WITH WIDE VOLTAGE RANGE BATTERY BASED ON MULTIPLEXED
TOPOLOGY 654
Kepa Odriozola, Thierry A. Meynard, Alain Lacarnoy

ON-GRID/OFF-GRID DC MICROGRID OPTIMIZATION AND DEMAND RESPONSE
MANAGEMENT 667
Wenshuai Bai, Manuela Sechilariu, Fabrice Locment

SHEDDING AND RESTORATION ALGORITHMS FOR AN EV CHARGING STATION TO
MAXIMIZE AVAILABLE POWER 677
Dian Wang, Fabrice Locment, Manuela Sechilariu

EFFICIENCY AND COST COMPARISON OF B6 AND HYBRID ANPC CONVERTERS FOR
TRACTION DRIVES 686
Johannes Häring, Michael Gleissner, Wolfgang Wondrak, Mark-M. Bakran

DESIGN AND CONTROL OF A KE (KINETIC ENERGY) - COMPENSATED
GRAVITATIONAL ENERGY STORAGE SYSTEM 696
Alfred Rufer

A NOVEL POWER FLOW CONTROL STRATEGY FOR HETEROGENEOUS BATTERY
ENERGY STORAGE SYSTEMS BASED ON PROGNOSTIC ALGORITHMS FOR
BATTERIES 707
Markus Muehlbauer, Samantha Klier, Herbert Palm, Oliver Bohlen, Michael A. Danzer

AN IGCT-BASED MULTI-FUNCTIONAL MMC SYSTEM WITH COMMUTATION AND
SWITCHING... 718
Chaoqun Xu, Mingzhu Guo, Biao Zhao, Bojin Tang, Zhanqing Yu, Dongling Zhai, Chunpin Ren

COMMON-MODE NOISE MODELLING AND RESONANT ESTIMATION IN A THREE-PHASE MOTOR DRIVE SYSTEM: 9-150 KHZ FREQUENCY RANGE 726
Hansika Rathnayake, Amir Ganjavi, Firuz Zare, Dinesh Kumar, Pooya Davari

POLYNOMIAL MULTI-VARIABLE CONTROL STRATEGY FOR FLUX BALANCING IN DUAL ACTIVE BRIDGE CONVERTER 736
Pierre-Baptiste Steckler, Jean-Yves Gauthier, Xuefang Lin-Shi, François Wallart

ENHANCED POWER SYSTEM DAMPING ESTIMATION VIA OPTIMAL PROBING SIGNAL DESIGN 745
S. Boersma, X. Bombois, L. Vanfretti, V. Peric, J-C. Gonzalez-Torres, R. Segur, A. Benchaib

IMPROVED HIGH STEP-UP BOOST-BASED DC/DC CONVERTER WITH BUILT-IN TRANSFORMER AND ACTIVE CLAMP FOR DC MICROGRIDS 755
Konstantinos Zaoskoufis, Emmanuel C. Tatakis

ELIMINATION/MITIGATION OF OUTPUT VOLTAGE HARMONICS FOR MULTILEVEL CONVERTERS OPERATED AT FUNDAMENTAL SWITCHING FREQUENCY USING MATLAB'S GENETIC ALGORITHM OPTIMIZATION 765
Anton Kersten, Manuel Kuder, Arthur Singer, Weiji Han, Torbjörn Thiringer, Thomas Weyh, Richard Eckerle

EVALUATION OF DRIVE TOPOLOGIES FOR MACRO SCALE SYNCHRONOUS ELECTROSTATIC MACHINES 777
Peter Killeen, Daniel C. Ludois

DECENTRALIZED VOLTAGE REGULATION IN ISLANDED DC MICROGRIDS IN THE PRESENCE OF DISPATCHABLE AND NON-DISPATCHABLE DC SOURCES 787
Mohammadreza Nabatirad, Reza Razzaghi, Behrooz Bahrani

AN ULTRA-FAST GATE DRIVER WITH OVER CURRENT PROTECTION FOR GAN POWER TRANSISTORS 797
Qingqing Nie, Han Peng, Yong Kang

A NEW GAN HYBRID RESONANT-CLAMPING GATE DRIVER FOR HIGH FREQUENCY SIC MOSFETS 804
Ziyue Dang, Han Peng, Hao Peng, Yong Kang, Yu Chen, Xudan Liu, Maojun He

MAINTENANCE SCHEDULING IN POWER ELECTRONIC CONVERTERS CONSIDERING WEAR-OUT FAILURES 810
Saeed Peyghami, Frede Blaabjerg, Jose Rueda Torres, Peter Palensky

AC/DC DYNAMIC INTERACTIONS OF MMC-HVDC IN GRID-FORMING FOR WIND-FARM INTEGRATION IN AC SYSTEMS 820
Rayane Mourouvin, Kosei Shinoda, Jing Dai, Abdelkrim Benchaib, Seddik Bacha, Didier Georges

A DESIGN OF SOLID STATE POWER CONTROLLER FOR A BIDIRECTIONAL DC-DC CONVERTER IN AN AERONAUTIC CONTEXT 829
Hassan Cheaito, Bruno Allard, Guy Clerc, Joris Pallier, Pascal Pommier-Petit

A NEW APPROACH OF RESONANT CONVERTER USING LARGE AIR GAP TRANSFORMER 835
Michael Finkenzeller, Monika Poebl, Thomas Komma

REDUCED CAPACITOR SIZE AND ON-STATE LOSSES IN ADVANCED MMC
SUBMODULE TOPOLOGIES.. 843
Christopher Dahmen, Rainer Marquardt

STABILITY AND ROBUSTNESS ANALYSIS OF FRACTIONAL PROPORTIONAL
RESONANT CONTROLLERS IN CURRENT-CONTROLLED VOLTAGE-SOURCE-
INVERTERS... 853
Daniel Heredero-Peris, Cristian Chillón-Antón, Daniel Montesinos-Miracle

EMPLOYING VIRTUAL SYNCHRONOUS GENERATOR WITH A NEW CONTROL
TECHNIQUE FOR GRID FREQUENCY STABILIZATION................................. 863
*Meysam Saeedian, Bahman Eskandari, Kumars Rouzbehi, Shamsodin Taheri, Edris
Pouresmaeil*

A HYBRID PULSE WIDTH MODULATION TECHNIQUE WITH TEMPERATURE
CONTROL FOR MODULAR MULTILEVEL CONVERTERS 871
Ara Bissal, Waqas Ali, Rob Leedham, Mark Snook, Ibrahim Elsabrouty, Ilknur Colak

DESIGN FLOW OF A COMPACT HIGH-FREQUENCY DC/DC CONVERTER WITH
OPTIMUM AVERAGE EFFICIENCY IN A WIDE OPERATION RANGE.................. 880
Maximilian Nitzsche, Matthias Zehelein, Julian Weimer, Dominik Koch, Jörg Roth-Stielow

ANALYSIS OF THE TRANSFORMER MODULARIZATION FOR HIGH FREQUENCY
ISOLATED IIIGII VOLTAGE GENERATOR WITH THE SILICON CARBIDE DEVICES..................... 892
Saijun Mao, Popovic Jelena, Jan Abraham Ferreira

IMPROVED DIRECT-MODEL PREDICTIVE CONTROL WITH A SIMPLE DISTURBANCE
OBSERVER FOR DFIGS... 900
Mohamed Abdelrahem, Christoph Hackl, José Rodríguez, Ralph Kennel

MODELING OF SIC-MOSFET CONVERTER LEG INCLUDING PARASITICS OF PRINTED
CIRCUIT BOARD LAYOUT AND DEVICE PACKAGING 909
M. Pulvirenti, L. Salvo, A. G. Sciacca, G. Scelba, M. Cacciato

PERFORMANCE ANALYSIS OF RL DAMPER IN GAN-BASED HIGH-FREQUENCY
BOOST CONVERTER... 919
A. Gutierrez, E. Marcault, C. Alonso, D. Tremouilles

RAPID IMPEDANCE ESTIMATION ALGORITHM FOR MITIGATION OF
SYNCHRONIZATION INSTABILITY OF PARALLELED CONVERTERS UNDER GRID
FAULTS... 927
Mads Graungaard Taul, Robert Eric Betz, Frede Blaabjerg

ADAPTIVE THERMAL CONTROL FOR MOSFET-BASED MODULAR MULTILEVEL
CONVERTER.. 937
Tianxiang Yin, Lei Lin, Chen Xu

ELECTRIC IMPULSE TECHNOLOGY – BREAKING ROCK 944
Matthias Voigt, Erik Anders, Franziska Lehmann, Margarita Mezzetti, Frank Will

IMPACT OF COMBINED THERMO-MECHANICAL AND ELECTRO-CHEMICAL STRESS
ON THE LIFETIME OF POWER ELECTRONIC DEVICES................................. 954
Felix Hoffmann, Stefan Schmitt, Nando Kaminski

CURRENT CONTROL AND FPGA–BASED REAL–TIME SIMULATION OF GRID–TIED
INVERTERS.. 962
Sabin Carpiuc, Matthias Schiesser, Carlos Villegas

IMPACT OF CONTROL LOOPS ON THE LOW-FREQUENCY PASSIVITY PROPERTIES OF GRID-FORMING CONVERTERS 969

Mebtu Beza, Massimo Bongiorno, Anant Narula

GRID IMPEDANCE ESTIMATION WITH OVERSAMPLING FOR GRID-CONNECTED CONVERTERS 979

Niklas Himker, Robin Strunk, Axel Mertens

LOW SPEED SENSORLESS CURRENT CONTROL FOR PMSM WITH SEARCH-BASED OBSERVER (SBO) 989

K. Scicluna, C. Spiteri Staines, R. Raute

INSIGHT INTO THE PECULIARITIES OF OPTIMIZED PULSE PATTERNS FOR PERMANENT-MAGNET SYNCHRONOUS MACHINES 998

Georgios Darivianakis, Ioannis Tsoumas

INVESTIGATING THE EFFECT OF DIFFERENT PARAMETERS ON HARMONICS AND EMI EMISSIONS AT THE FREQUENCY RANGE OF 0–9 KHZ 1006

Amir Ganjavi, Hansika Rathnayake, Firuz Zare, Dinesh Kumar, Amin Abbosh, Pooya Davari

FIVE-LEVEL NESTED INVERTER WITH NEUTRAL POINT CONNECTION 1016

Juhamatti Korhonen, Aleksi Mattsson, Heikki Järvisalo, Pertti Silventoinen, William Giewont, Dan Isaksson

ELECTRIC SPRING-BASED SMART WATER HEATER FOR LOW VOLTAGE MICROGRIDS 1025

Alexander Micallef, Racquel Ellul, John Licari

ENERGY-BALANCING OF A MODULAR MULTILEVEL CONVERTER USING AN ONLINE TRAJECTORY PLANNING ALGORITHM 1030

Qiuye Gui, Jan Lasse Gnärig, Hendrik Fehr, Albrecht Gensior

CAPACITOR SIZE COMPARISON ON HIGH-POWER DC-DC CONVERTERS WITH DIFFERENT TRANSFORMER WINDING CONFIGURATIONS ON THE AC-LINK 1040

Babak Khanzadeh, Torbjörn Thiringer, Yuhei Okazaki

DYNAMIC CHARACTERISTICS VERIFICATION OF LINEAR INDUCTION MOTOR BY SIMULTANEOUS PROPULSION AND LEVITATION CONTROL 1047

Shota Nakatani, Daichi Okamori, Toshimitsu Morizane, Hideki Omori

'IG,VGS' MONITORING FOR FAST AND ROBUST SIC MOSFET SHORT-CIRCUIT PROTECTION WITH HIGH INTEGRATION CAPABILITY 1057

Yazan Barazi, François Boige, Nicolas Rouger, Jean-Marc Blaquiere, Frédéric Richardeau

FAULT-TOLERANT CONTROL OF SERIES CONNECTABLE MODULAR FULL-BRIDGE INVERTER MITIGATING OPEN SWITCH FAULTS 1067

Juris Arrozy, Darian V. Retianza, Jorge L. Duarte, Henk Huisman

DESIGN AND CONTROL OF A MODULAR POWER ELECTRONIC BACK-TO-BACK CONVERTER FOR WAVE ENERGY HARVESTING APPLICATIONS 1076

Mattia Mantellini, Riccardo Morici, Marcos Blanco, Marcos Lafoz, Gustavo Navarro, Jorge Torres, Jorge Najera, Miguel Santos

INTELLIGENT HIGH CURRENT SENSOR FOR VARIOUS FREQUENCY 1086

Bohumil Skala, Vladimir Kindl, Pavel Turjanica, Ales Vobornik, Libor Polacek, Josef Stengl, Vladimir Pavlicek, Jiri Fort

FAIL-SAFE SWITCHING-CELLS ARCHITECTURES BASED ON MONOLITHIC ON-CHIP FUSE 1096

Amirouche Oumaziz, Emmanuel Sarraute, Frédéric Richardeau, Abdelhakim Bourennane

HOW GOOD ARE THE DESIGN TOOLS IN POWER ELECTRONICS? 1106

Thomas Lagier, Piotr Dworakowski, Laurent Chédot, François Wallart, Bruno Lefebvre, Jose Maneiro, Juan Páez, Philippe Ladoux, Cyril Buttay

ANALYSIS OF THE IMPACT OF MANUFACTURING DISSYMMETRY ON CURRENT DISTRIBUTION FOR MAGNETICALLY COUPLED INTERLEAVED INVERTERS 1118

Rita Mattar, Mickael Petit, Eric Monmasson, Stéphane Lefebvre, Christelle Saber, Cyrille Gautier, Marwan Ali

POWER FLOW CONTROL USING A BIDIRECTIONAL Z-SOURCE INVERTER–BASED STATIC SYNCHRONOUS SERIES COMPENSATOR 1128

Xuejiao Pan, Han Huang, Li Zhang

INVESTIGATION OF HARMONICS CONTENT IN PWM NATURAL AND REGULAR SAMPLING INCLUDING DEAD TIME AND LOAD CURRENT PHASE 1138

Tonny Wederberg Rasmussen, Anushruti Vashishtha, Ankit Jotwani

USING A WEB SCRAPING ALGORITHM FOR COMPONENT MODEL GENERATION IN MULTIOBJECTIVE OPTIMIZATION OF POWER ELECTRONIC APPLICATIONS 1148

Marcel Gladen, Volker Staudt

IMPACT ON THE ELECTRICAL CHARACTERISTICS, WAVEFORMS AND LOSSES OF THE ZERO-SEQUENCE INJECTION ON THE MODULAR MULTILEVEL CONVERTER 1158

Francois Gruson, Pierre Vermeersch, Philippe Delarue, Philippe Le Moigne, Frédéric Colas, Haibo Zhang, Moez Belhaouane, Xavier Guillaud

WIDE BANDWIDTH CURRENT SENSOR FOR COMMUTATION CURRENT MEASUREMENT IN FAST SWITCHING POWER ELECTRONICS 1168

Philipp Ziegler, Nathan Tröster, Dimitri Schmidt, Johannes Ruthardt, Manuel Fischer, Jörg Roth-Stielow

A SERIES–PARALLEL-TYPE RESONANT CIRCUIT WIRELESS POWER TRANSFER SYSTEM WITH A DUAL ACTIVE BRIDGE DC–DC CONVERTER 1177

Kohei Sugiyama, Taishi Kitamura, Shuto Uwai, Takahiro Yano, Yoshitaka Kawabata

STRAY VOLTAGE CAPTURE FOR ROBUST AND ULTRA-FAST SHORT CIRCUIT DETECTION IN POWER ELECTRONICS WITH HALF-BRIDGE STRUCTURE: THE LIMITATION AND IMPLEMENTATION 1186

Darian Verdy Retianza, Jeroen Van Duivenbode, Henk Huisman

ON THE INFLUENCE OF THE STATOR WINDING TOPOLOGY ON THE ELECTROMAGNETIC EMISSIONS OF FRACTIONAL HORSEPOWER BLDC MOTORS 1196

Felix Krall, Annette Muetze

IMPACT OF SILICON CARBIDE DEVICES IN 2 MW DFIG BASED WIND ENERGY SYSTEM 1205

Antxon Arrizabalaga, Aitor Idarreta, Mikel Mazuela, Iosu Aizpuru, Unai Iraola, José Luis Rodriguez, Daniel Labiano, Ibrahim Alisar

SMALL-SIGNAL STABILITY OF HVDC SYSTEM COMPRISING DC REACTORS 1215

Kosei Shinoda, Abdelkrim Benchaib, Jing Dai

MODEL PREDICTIVE CONTROL FOR THE REDUCTION OF DC-LINK CURRENT RIPPLE IN TWO-LEVEL THREE-PHASE VOLTAGE SOURCE INVERTERS 1224

Junzhong Xu, Fei Gao, Thiago Batista Soeiro, Linglin Chen, Luca Tarisciotti, Houjun Tang, Pavol Bauer

CARRIER-BASED MODULATED MODEL PREDICTIVE CONTROL FOR VIENNA RECTIFIERS 1233

Junzhong Xu, Fei Gao, Thiago Batista Soeiro, Linglin Chen, Luca Tarisciotti, Houjun Tang, Pavol Bauer

NEW HIGH-EFFICIENCY POWER GENERATION USING POSITION SENSOR-LESS PERMANENT MAGNET SYNCHRONOUS GENERATOR 1243

Somi Takeuchi, Hiroyuki Takahashi, Shota Yamada, Yoshitaka Kawabata

ACTIVE CLAMPING METHOD FOR SIC MOSFET HIGH POWER MODULES - BENEFITS AND LIMITS 1252

Robert W. Maier, Mark-M. Bakran

PREDICTIVE TORQUE CONTROL OF INDUCTION MACHINE WITH AN ADAPTIVE OBSERVER FOR TRAJECTORY PLANNING OF SERVO PRESS 1262

Qi Li, Jianbo Gao, Qiwu Wang, Ralph Kennel

FUTURE GRID STABILITY, A COST COMPARISON OF GRID-FORMING AND SYNCHRONOUS CONDENSER BASED SOLUTIONS 1270

Thibault Prevost, Guillaume Denis, Clementine Coujard

DEMONSTRATION OF THE SHORT-CIRCUIT RUGGEDNESS OF A 10 KV SILICON CARBIDE BIPOLAR JUNCTION TRANSISTOR 1279

Besar Asllani, Hervé Morel, Pascal Bevilacqua, Dominique Planson

LOSS MINIMIZATION OF TRACTION SYSTEMS IN BATTERY ELECTRIC VEHICLES USING VARIABLE DC-LINK VOLTAGE TECHNIQUE — EXPERIMENTAL STUDY 1289

Libo Liu, Boyang Li, Gunther Götting, Yusheng Xiang, Qusay Salem, Muhammad Hamid, Jian Xie

DIRECT MULTIVARIABLE CONTROL FOR MMC: DIGITAL SIGNAL PROCESSING AND EXPERIMENTAL RESULTS 1297

Daniel Dinkel, Claus Hillermeier, Rainer Marquardt

STATE OF CHARGE CONTROL FOR A FREQUENCY-SUPPORTING STORAGE SYSTEM BASED ON AN AUTO-REGRESSIVE FREQUENCY FORECAST 1306

A. Bolzoni, R. Todd, Q. Zhu, A. J. Forsyth

DESIGN OF A WIDE INPUT VOLTAGE RANGE CURRENT-FED DC/DC CONVERTER WITHIN A REDUCED DUTY-CYCLE RANGE 1316

Michael Gerstner, Martin Maerz, Armin Dietz

AN IMPROVED CONTROL STRATEGY FOR RENEWABLE ENERGY SOURCES (RES) BASED DC MICROGRID WITH ENHANCED SYSTEM STABILITY AND CONTROL PERFORMANCE 1326

Muhammad Adnan Mumtaz, Zheng Yan

TRANSIENT VOLTAGE DIP MITIGATION SYSTEM BASED ON HYBRID MODULAR MULTILEVEL CONVERTERS 1336

Manuel Colmenero, Francisco R. Blanquez, Karsten Kahle

A LOSS-COMPENSATED CONTROL SCHEME FOR SIC-BASED DUAL ACTIVE BRIDGE CONVERTER 1346

Ishan Pendharkar, Tobias Strittmatter, Paula Diaz Reigosa, Nicola Schulz

EXPERIMENTAL HYBRID AC/DC-MICROGRID PROTOTYPE FOR LABORATORY RESEARCH 1354

Enrique Espina, Claudio Burgos-Mellado, Juan S. Gomez, Jacqueline Llanos, Erwin Rute, Alex Navas F., Manuel Martínez-Gómez, Roberto Cárdenas, Doris Sáez

EXPERIMENTAL AND NUMERICAL CHARACTERIZATION OF PCB-EMBEDDED POWER DIES USING SOLDERLESS PRESSED METAL FOAM 1363

S. Bensebaa, M. Berkani, S. Lefebvre, M. Petit, N. Schmitt

FEASIBILITY STUDY OF A SUPERCONDUCTING POWER FILTER FOR HVDC GRIDS 1373

Loïc Quéval, Olivier Despouys, Frédéric Trillaud, Bruno Douine

POWER DECOUPLING METHOD OF DC TO SINGLE-PHASE AC CONVERTER USING FLYING CAPACITOR DC/DC CONVERTER WITH BOUNDARY CURRENT MODE 1380

Hiroki Watanabe, Keisuke Kusaka, Jun-Ichi Itoh

AN ARCHITECTURE FOR LEVEL-3 EV BATTERY CHARGER STATIONS USING INTEGRATED SOLID STATE TRANSFORMER (I-SST) 1390

Erick I. Pool-Mazun, Prasad Enjeti, Gerardo Escobar, Ira Pitel

LQR AND H-INFINITY CONTROL OF VOLTAGE SOURCE INVERTERS FOR AC MICROGRIDS 1400

Tenorio Jorge, Jose Miguel Ramirez Scarpetta, Fabio Andrade

FAMILY OF SPLITTING CURRENT SINGLE-LOOP CONTROL FOR *LCL*- TYPE GRID-CONNECTED INVERTER 1410

Yuying He, Xuehua Wang, Xinbo Ruan, Guoxing Su, Fuxin Liu

ANALYSIS AND DESIGN OF HIGH-POWER SINGLE-STAGE THREE-PHASE DIFFERENTIAL-BASED FLYBACK INVERTER FOR PHOTOVOLTAIC APPLICATIONS 1417

Ahmed Ismail M. Ali, Mahmoud A. Sayed, Takaharu Takeshita

INVESTIGATION OF IMPROVEMENT OF MODELING PRECISION FOR CONDUCTED NOISE ON ISOLATED AC/DC CONVERTER USING SIC DEVICES 1425

Kazuki Kuwana, Kohei Mitani, Wataru Kitagawa, Takaharu Takeshita

PASSIVITY-BASED DESIGN FOR THE PLUG-AND-PLAY SINGLE-LOOP CONTROLLED LCL-FILTERED INVERTER 1435

Yuying He, Xuehua Wang, Xinbo Ruan, Yixiao Ma, Fuxin Liu

CHARACTERISTICS OF AN INTEGRATED MOTOR CONTROLLED INDEPENDENTLY BY MULTI-INVERTERS TO ACHIEVE HIGH EFFICIENCY AND A WIDE SPEED RANGE 1442

Kazuto Sakai, Yano Hideaki

AN ISOLATED MEDIUM-VOLTAGE AC-DC CONVERTER USING LEVEL-SHIFTED PWM CONTROL OF A MODULAR MATRIX CONVERTER 1450

Kohei Budo, Takaharu Takeshita

DETAILED SIMULATION MODEL OF AN ASYMMETRICAL HALF-BRIDGE PWM CONVERTER WITH SYNCHRONOUS RECTIFICATION INCLUDING PARASITIC ELEMENTS 1460

Benedikt Kohlhepp, Valentin Zeller, Markus Barwig, Thomas Dürbaum

ELECTRICAL PROPERTY VARIABILITY OF GAN TRANSISTORS IN PARALLEL AND THEIR IMPACT ON FAST SWITCHING OPERATIONS 1470

Thilini Wickramasinghc, Bruno Allard, Réne Escofficr, Marc Plissonnicr

A COMPARISON BETWEEN DIFFERENT MODELS OF THE MODULAR MULTILEVEL CONVERTER 1479

Rafael Coelho-Medeiros, Bogdan Džonlaga, Jean-Claude Vannier, Jing Dai, Loic Queval, Philippe Egrot

PACKAGING TECHNOLOGY FOR THE IMPROVEMENT OF POWER CYCLING CAPABILITY OF HVIGBTS 1489

Kenji Hatori, Keiichi Nakamura, Nobuhiko Tanaka, Yasuhiro Sakai, Norikazu Sakai, Kenji Ota, Takeshi Higashihata, Eckhard Thal, Nils Soltau

A BIDIRECTIONAL DAB-LLC DCX TO ACHIEVE VOLTAGE REGULATION AND WIDE ZVS RANGE CAPABILITY 1498

Yuefeng Liao, Tao Peng, Mei Su, Yao Sun, Weijing Xiong, Guo Xu

SALIENCY SELECTION FOR SEARCH-BASED AC MACHINE LOW AND ZERO SPEED ESTIMATION METHODS 1506

K. Scicluna, C. Spiteri Staines, R. Raute

GENETIC ALGORITHM BASED MULTI OBJECTIVE OPTIMIZATION FOR INDUCTOR DESIGN 1515

Thorben Schobre, Raquel González Aríztegui, Regine Mallwitz

DIGITAL SMART DRIVER FOR SIC MOSFETS... 1524

Nerea Arandia, José Ignacio Garate, Jon Mabe, Ander Ordoño

FASTER SWITCHING WITH LESS OVERVOLTAGE - OPERATING A SIC-MOSFET AT ITS SPEED LIMIT.. 1533

Pablo Rodriguez De Mora, Mark-M. Bakran

THE ENERGY RING TO SUPPLY THE EXPOELECTRIC'18 SHOW WITH RENEWABLE ENERGY SOURCES AND ELECTRIC VEHICLES 1542

Cristian Chillón-Antón, Daniel Heredero-Peris, Francesc Girbau-Llistuella, Paula González-Fontderubinat, Marc Llonch-Masachs, Daniel Montesinos-Miracle, Oriol Gomis-Bellmunt

IMPEDANCE-BASED MODELING OF A THREE-LEVEL CONVERTER UNDER BALANCED AND UNBALANCED CONDITION FOR THE STABILITY ANALYSIS OF BIPOLAR LVDC GRIDS 1551

T. Roose, G. Van Den Broeck, M. M. Alam, J. Beerten

LCL FILTER DESIGN FOR THREE PHASE AC-DC CONVERTERS CONSIDERING SEMICONDUCTOR MODULES AND MAGNETICS COMPONENTS PERFORMANCE...................... 1561

Marco Stecca, Thiago Batista Soeiro, Laura Ramirez Elizondo, Pavol Bauer, Peter Palensky

SWITCHING BEHAVIOR AND COMPARISON OF 600V SMD WIDE BANDGAP POWER DEVICES 1569

Markus Meißner, Jan Schmitz, Steffen Bernet

ANALYSIS OF THE COUPLING BETWEEN THE OUTER AND INNER CONTROL LOOPS OF A GRID-FORMING VOLTAGE SOURCE CONVERTER 1579

T. Qoria, F. Gruson, F. Colas, X. Kestelyn, X. Guillaud

INFLUENCE OF DIFFERENT PULSE-WIDTH MODULATION METHODS ON MAGNET LOSSES IN PERMANENT MAGNET SYNCHRONOUS MACHINES 1589
Narciso G. Marmolejo, Xiaohu Tang, Martin Doppelbauer

RESONANT DC/DC CONVERTER WITH CLASS ϕ_2 INVERTER AND CLASS DE RECTIFIER BASED ON GAN HEMT 1599
Cai Si-Yuan, He Jun-Ping, Li Zi-Fan

FOUR-LEVEL INVERTER WITH VARIABLE VOLTAGE LEVELS FOR HARDWARE-IN-THE-LOOP EMULATION OF THREE-PHASE MACHINES 1605
Manuel Fischer, Johannes Ruthardt, Vasken Ketchedjian, Philipp Ziegler, Maximilian Nitzsche, Jörg Roth-Stielow

POWDER INJECTION MOLDING IN THE FABRICATION OF SOFT FERRITE MATERIAL FOR POWER ELECTRONICS 1613
J-S Ngoua-Teu, U. Soupremanien, P. Sallot, G. Delette, M. Bohnke

MODULATION SCHEME WITH COMMON MODE AND DIFFERENTIAL MODE VOLTAGE ELIMINATION FOR A FIVE LEVEL INVERTER FED OPEN END WINDING INDUCTION MOTOR DRIVE 1619
Greeshma Nadh, Durga Nair S., Arun Rahul S.

A FAST AND ROBUST MODEL OF DUAL-ACTIVE BRIDGE CONVERTERS IN REAL-TIME SIMULATION 1627
Ming Jia, Philipp Joebges, Rik W. De Doncker

DUAL INTERLEAVED 3.6 KW LLC CONVERTER OPERATING IN HALF-BRIDGE, FULL-BRIDGE AND PHASE-SHIFT MODE AS A SINGLE-STAGE ARCHITECTURE OF AN AUTOMOTIVE ON-BOARD DC-DC CONVERTER 1638
Philipp Rehlaender, Sergey Tikhonov, Frank Schafmeister, Joachim Bocker

SWITCHING LOSS ESTIMATION USING A VALIDATED MODEL OF 650 V GAN HEMTS 1648
Joao Oliveira, Florent Loiselay, Hervè Morel, Dominique Planson

REDUCTION OF CONDUCTION LOSSES IN RESONANT CONVERTERS BY CONNECTING THREE SINGLE-PHASE INVERTERS TO A COMMON GENERATOR 1658
Sergio Tárraga, John Paul Mayorga, Esther De Jódar, José Villarejo

COMPARISON OF DIFFERENT LOW VOLTAGE MULTILEVEL CONVERTER TOPOLOGIES FOR DISTRIBUTED POWER GENERATION 1666
Ingmar Kaiser, Hans-Günter Eckel

LOSS DISTRIBUTION COMPARISON OF VARIABLE AND FIXED INDUCTOR DAB CONVERTERS 1675
Erik Smailus, Gerd Griepentrog, Markus Pfeifer, Marcel Lutze

DESIGN BY OPTIMIZATION OF MULTIPHASE INVERTER FOR ELECTRIC VEHICLE DRIVE 1685
Nasreddine Kesbia, Jean-Luc Schanen, Hadi Alawieh, Lauric Garbuio, Yvan Avenas

OPTIMAL TORQUE/SPEED CHARACTERISTICS OF A FIVE-PHASE SYNCHRONOUS MACHINE UNDER PEAK OR RMS CURRENT CONTROL STRATEGIES 1693
Tiago José Dos Santos Moraes, Hailong Wu, Eric Semail, Ngac Ky Nguyen, Duc Tan Vu

COMPARATIVE STUDY OF TWO CONTROL TECHNIQUES OF REGENERATIVE BRAKING POWER RECOVERING INVERTER BASED DC RAILWAY SUBSTATION 1700
Youssef Krim, Khaled Almaksour, Hervé Caron, Tony Letrouvé, Christophe Saudemont, Bruno Francois, Benoit Robyns

JUNCTION TEMPERATURE CONTROL STRATEGY FOR LIFETIME EXTENSION OF POWER SEMICONDUCTOR DEVICES 1709
Johannes Ruthardt, Hendrik Schulte, Philipp Ziegler, Manuel Fischer, Maximilian Nitzsche, Jörg Roth-Stielow

HIGH DYNAMIC POWER BALANCING FOR DUAL TWO-LEVEL INVERTERS DURING HIGH-SPEED MACHINE OPERATION 1718
Johannes Büdel, Johannes Teigelkötter, Alexander Stock, Christian Herkommer, Kai Kuhlmann

CHARGING HIGH VOLTAGE CAPACITORS IN PULSED POWER APPLICATIONS WITH A CAPACITOR DIODE VOLTAGE MULTIPLIER OF REDUCED SIZE AND LOWER RIPPLE CURRENTS 1727
Tristan Weinert, Wolfgang Oberschelp, Günter Schröder

REVIEW OF OPTIMIZATION METHODS FOR THE DESIGN OF POWER ELECTRONICS SYSTEMS 1737
Mylène Delhommais

A FLEXIBLE POWER CROSSBAR-BASED ARCHITECTURE FOR SOFTWARE-DEFINED POWER DOMAINS 1747
Francesco Di Gregorio, Gilles Sassatelli, Abdoulaye Gamatié, Arnaud Castelltort

IMPACT OF GRID-FORMING CONTROL ON THE INTERNAL ENERGY OF A MODULAR MULTILEVEL CONVERTER 1756
Ebrahim Rokrok, Taoufik Qoria, Antoine Bruyere, Bruno Francois, Haibo Zhang, Moez Belhaouane, Xavier Guillaud

COMBINING MULTIPLE TEMPERATURE-SENSITIVE ELECTRICAL PARAMETERS USING ARTIFICIAL NEURAL NETWORKS 1766
Daniel Herwig, Torben Brockhage, Axel Mertens

SINGLE-PHASE MEASUREMENT OF THE OUTPUT IMPEDANCE OF THE FOUR-QUADRANT CASCADED H-BRIDGE CONVERTER CELL USING WIDEBAND SIGNALS 1776
Marko Petkovic, Dražen Dujic

A NOVEL THREE-PHASE PFC DIODE RECTIFIER BY LC NETWORK CIRCUITS FOR HIGH FREQUENCY GENERATOR 1786
Shin-Ichi Motegi, Yasuyuki Nishida

FREQUENCY-DOMAIN SIMULATION OF POWER ELECTRONIC SYSTEMS BASED ON MULTI-TOPOLOGY EQUIVALENT SOURCES MODELLING METHOD 1793
Stephane Vienot, Arnaud Videt, Nadir Idir, Lamine Kone, Sébastien Weiss, Frederic Lafon

MODULAR MULTILEVEL CONVERTER WITH DISTRIBUTED GALVANIC ISOLATION: A DECENTRALIZED VOLTAGE BALANCING ALGORITHM WITH SMART GATE DRIVERS 1803
Darbas Corentin, Ginot Nicolas, Olivier Jean-Christophe, Poitiers Frédèric

COMPARISON AND OPTIMIZATION OF MAGNETICALLY COUPLED AND NON-COUPLED MAGNETIC DEVICES IN INTERLEAVED OPERATION 1813
Peter Zacharias, Alejandro Aganza-Torres

EXPERIMENTAL TUNING AND DESIGN GUIDELINES OF A DYNAMICALLY
RECONFIGURED WEIGHTING FACTOR FOR THE PREDICTIVE TORQUE CONTROL OF
AN INDUCTION MOTOR.. 1823
  Ilker Sahin, Ozan Keysan, Eric Monmasson

COMPENSATION OF TEMPERATURE DEPENDENCE IN A MODULE PARASITIC BASED
CURRENT MEASUREMENT SYSTEM 1831
  Frank Lautner, Mark-M. Bakran

DEVELOPMENT AND IMPLEMENTATION OF A LOW-COST RESEARCH PLATFORM
FOR CONTROL APPLICATIONS FOR INVERTER-BASED GENERATORS 1841
  Jesus D. Vasquez Plaza, Juan F. Patarroyo-Montenegro, Fabio Andrade

CONTROL OF PARALLEL CONNECTED VOLTAGE SOURCE INVERTERS IN A
MICROGRID FOR EXPERIMENTAL TESTING 1850
  *Jesus D. Vasquez-Plaza, Jorge Tenorio, J. M. Ramírez-Scarpetta, Jose Alex Restrepo, Fabio
  Andrade*

OPTIMIZATION STRATEGY FOR THE SIZING OF PASSIVE MAGNETIC COMPONENTS 1858
  *Guillaume Devos, Maya Hage-Hassan, Philippe Dessante, Cyrille Gautier, Adrien Mercier,
  Eric Labouré*

EXPLOITING A MULTI-PORT TRANSFORMER FOR MINIMAL DC-LINK CAPACITANCE
FOR AN AUTOMOTIVE ONBOARD CHARGER 1866
  Franz Vollmaier, Alexander Connaughton, Thomas Langbauer, Klaus Krischan

DESIGN AND OPTIMIZATION OF HIGH-EFFICIENCY 1W 500V-12V ISOLATED LOW-
COST DC/DC CONVERTER... 1874
  Etienne Foray, Christian Martin, Bruno Allard

CHALLENGES IN CALIBRATING AN UNCONVENTIONAL PARTIAL DISCHARGE
MEASUREMENT SYSTEM FOR PULSED VOLTAGES 1885
  Markus Fürst, Mark-M. Bakran

ELECTROTHERMAL MODELING OF GAN POWER TRANSISTOR FOR HIGH
FREQUENCY POWER CONVERTER DESIGN... 1895
  *Loris Pace, Florian Chevalier, Arnaud Videt, Nicolas Defrance, Nadir Idir, Jean-Claude De
  Jaeger*

MODELING AND FAULT DETECTION IN PHOTOVOLTAIC SYSTEMS USING THE I-V
SIGNATURE 1905
  *Abdelhadi Benzagmout, Thierry Talbert, Olivier Fruchier, Thierry Martire, Philippe
  Alexandre, Carolina Penin*

EFFICIENCY REQUIREMENTS FOR PASSIVELY COOLED CONVERTERS WITH
THERMAL MEASUREMENT BASED 3D-FEM SIMULATION 1915
  Julian Weimer, Dominik Koch, Maximilian Nitzsche, Matthias Zehelein, Ingmar Kallfass

GENERIC CONTROL LAW FOR DC AND AC MACHINES... 1923
  Pierre-Philippe Robet, Maxime Gautier, Yannick Aoustin

A HIGH PERFORMANCE 48-TO-8 V MULTI-RESONANT SWITCHED-CAPACITOR
CONVERTER FOR DATA CENTER APPLICATIONS... 1934
  Rose A. Abramson, Zichao Ye, Robert C. N. Pilawa-Podgurski

SISO CONTROL STRATEGY OF RESONANT DUAL ACTIVE BRIDGE WITH A TUNED CLC NETWORK 1944
Meiqi Wang, Bo Yang, Lie Xu, Jing Li, David Gerada, Chunyang Gu, He Zhang, Chris Gerada, Yongdong Li

IMPACT OF STEADY-STATE GRID-FREQUENCY DEVIATIONS ON THE PERFORMANCE OF GRID-FORMING CONVERTER CONTROL STRATEGIES 1952
Anant Narula, Massimo Bongiorno, Mebtu Beza, Jan R Svensson, Xavier Guillaud, Lennart Harnefors

A GENERAL METHOD TO DAMP WIND TURBINE SSR WITH DIFFERENT TRANSMISSION SYSTEMS 1962
Ignacio Vieto, Jian Sun

A TEST SCHEME FOR THE COMPREHENSIVE QUALIFICATION OF MMC SUBMODULE BASED ON 10 KV SIC MOSFETS UNDER HIGH DV/DT 1972
Xingxuan Huang, Shiqi Ji, Dingrui Li, Cheng Nie, William Giewont, Leon M. Tolbert, Fred Wang

PWM GAIN LINEARIZATION ALGORITHM FOR MEDIUM VOLTAGE SOURCE INVERTER 1982
Hamza El Jihad, Sami Siala, Elise Savarit

AUTO-COMMISSIONING OF ACOUSTIC CONTROL OF IM DRIVE USING BAYESIAN OPTIMIZATION 1992
Michal Kroneisl, Václav Šmídl

EXPERIMENTAL EMI STUDY OF A 3-PHASE 100KW 1200V DUAL ACTIVE BRIDGE CONVERTER USING SIC MOSFETS 2000
Hadiseh Geramirad, Florent Morel, Piotr Dworakowski, Philippe Camail, Bruno Lefebvre, Thomas Lagier, Christian Vollaire

MODELING OF A DAB UNDER PHASE-SHIFT MODULATION FOR DESIGN AND DM INPUT CURRENT FILTER OPTIMIZATION 2010
Glauber De Freitas Lima, Yves Lembeye, Fabien Ndagijimana, Jean-Christophe Crebier

ACTIVE CURRENT AND ENERGY CONTROL FOR THE QUASI-THREE-LEVEL OPERATION MODE OF AN EXTENDED MODULAR MULTILEVEL CONVERTER TOPOLOGY 2020
Malte Lorenz, Jakub Kucka, Axel Mertens

TORQUE RIPPLE REDUCTION TECHNIQUE FOR A SWITCHED RELUCTANCE MOTOR 2029
Krzysztof Jackiewicz, Arkadiusz Kaszewski, Andrzej Stras, Bartlomiej Ufnalski, Tomasz Balkowiec

EXPERIMENTAL VALIDATION OF THE PERFORMANCES OF AN INVERTER SIZED WITH OPTIMIZATION METHODS 2039
Adrien Voldoire, Jean-Luc Schanen, Jean-Paul Ferrieux, Alexis Derbey, Cyrille Gautier, Marwan Ali

INFLUENCE OF SYSTEM PARAMETERS IN VARIABLE SPEED AC-INDUCTION MOTOR DRIVES ON PARASITIC ELECTRIC BEARING CURRENTS 2049
Martin Weicker, Guilherme Bello, Dennis Kampen, Andreas Binder

PLASMA IMPACT ON OVERVOLTAGE SHORT-CIRCUIT FAILURES IN ANPC CONVERTERS 2059
David Hammes, Sidney Gierschner, Dietmar Krug, Hans-Günter Eckel

NOVEL SOFT-SWITCHING INTERLEAVED BOOST CONVERTERS FOR RENEWABLE ENERGY CONVERSION SYSTEMS 2068

Madhuchandra Popuri, V. V. Subrahmanya Kumar Bhajana, Pavel Drabek, Manoj Kumar Maharana

POWER DENSITY OF PLANAR TRANSFORMERS DESIGNED WITH COMMERCIAL STANDARD CORES 2078

Reda Bakri, Xavier Margueron, Jean Sylvio Ngoua Teu Magambo, Philippe Le Moigne, Nadir Idir

EFFECTS OF PV PANEL AND BATTERY DEGRADATION ON PV-BATTERY SYSTEM PERFORMANCE AND ECONOMIC PROFITABILITY 2088

Monika Sandelic, Ariya Sangwongwanich, Frede Blaabjerg

FULL SENSORLESS OPERATION OF INDUCTION MACHINES BASED ON ONLINE IDENTIFICATION OF SALIENCIES USING HARMONIC COMPENSATION LUTS IN TRACTION APPLICATIONS 2098

E. Rodriguez Montero, M. Vogelsberger, T. Wolbank

MITIGATING DRAIN SOURCE VOLTAGE OSCILLATION WITH LOW SWITCHING LOSSES FOR SIC POWER MOSFETS USING FPGA-CONTROLLED ACTIVE GATE DRIVER 2106

Zheming Li, Robert W. Maier, Mark-M. Bakran

ONLINE TRAJECTORY PLANNING DURING LOW-VOLTAGE FRT OF A MODULAR MULTILEVEL CONVERTER 2116

Hendrik Fehr, Albrecht Gensior

EVALUATING FREQUENCY STABILITY WITH CONSIDERATION OF LOAD TYPE IN DIFFERENT SHARE OF RENEWABLES AND EMULATED INERTIA IN CASE OF SYSTEM SPLIT 2126

Nastaran Fazli, Sidney Gierschner, Hans-Günter Eckel

DISCRETE-TIME DIRECT POLE PLACEMENT FOR STABILITY ENHANCEMENT OF LCL-FILTERED INVERTERS IN THE SYNCHRONOUS-REFERENCE FRAME 2135

Pei Cai, Xiaohua Wu, Yongheng Yang, Wenli Yao, Weilin Li, Frede Blaabjerg

ON THE SWITCHING-INDUCED DC-LINK VOLTAGE RIPPLE IN THREE-LEVEL CONVERTERS WITH A NEUTRAL POINT 2145

Ioannis Tsoumas, Tobias Geyer

EFFECT OF PASSIVE INVERTER OUTPUT MOTOR FILTERS ON DRIVE SYSTEMS 2153

Dennis Kampen, Martin Weicker

IMPACT OF THE NEUTRAL POINT POTENTIAL RIPPLE ON THE GRID SIDE HARMONICS OF A 3LNPC BACK-TO-BACK CONVERTER EMPLOYED IN A MEDIUM VOLTAGE WECS 2163

Ioannis Tsoumas

TWO-LAYER GENETIC ALGORITHM FOR THE CHARGE SCHEDULING OF ELECTRIC VEHICLES 2172

Nikolaos T. Milas, Dimitris A. Mourtzis, Panagiotis I. Giotakos, Emmanuel C. Tatakis

SIX-PHASE PMSM DRIVE INVERTER TESTING ON A HIGH PERFORMANCE POWER HARDWARE-IN-THE-LOOP TESTBED 2182

Yasser Rahmoun, Patrick Winzer, Alexander Schmitt, Horst Hammerer

AN IMPROVED BIDIRECTIONAL HYBRID SWITCHED INDUCTOR CONVERTER.........................2192

Dan Hulea, Mihaita Gireada, Danut Vitan, Octavian Cornea, Nicolae Muntean

HYBRID MULTIPLE CHOPPER CELLS OF PWM AND SQUARE-WAVE OPERATION FOR SOLID-STATE TRANSFORMER 2200

Naoto Kikuchi, Jun-Ichi Itoh, Keisuke Kusaka, Hoai Nam Le

A NEW ZVS ZONE IDENTIFICATION FOR DUAL ACTIVE BRIDGE WITH A GENERAL MODULATION OBJECTIVE.........................2210

Suman Maharana, Dipankar De, Alberto Castellazzi

SINGLE-STAGE BOOST MODULAR MULTILEVEL CONVERTER (BMMC) FOR ENERGY STORAGE INTERFACE.........................2220

Ahmed Abdelhakim, Frede Blaabjerg, Hans-Peter Nee

LOW VOLTAGE GAN-BASED GATE DRIVER TO INCREASE SWITCHING SPEED OF PARALLELED 650 V E-MODE GAN HEMTS 2230

Raffael Risch, Jürgen Biela

GATE STRESSES AND THRESHOLD VOLTAGE INSTABILITY IN NORMALLY-OFF GAN HEMTS 2241

Jose Ortiz Gonzalez, Burhan Etoz, Olayiwola Alatise

NEW ENERGY MANAGEMENT ALGORITHM BASED ON FILTERING FOR ELECTRICAL LOSSES MINIMIZATION IN BATTERY-ULTRACAPACITOR ELECTRIC VEHICLES 2251

Bakou Traoré, Moustapha Doumiati, Cristina Morel, Jean-Christophe Olivier, Ousmane Soumaoro

MECHANISTIC POWER MODULE DEGRADATION MODELLING CONCEPT WITH FEEDBACK.........................2258

Martin Bendix Fogsgaard, Paula Diaz Reigosa, Francesco Iannuzzo, Michael Hartmann

EXPERIMENTAL VALIDATION AND COMPARISON OF A SIC MOSFET BASED 100 KW 1.2 KV 20 KHZ THREE-PHASE DUAL ACTIVE BRIDGE CONVERTER USING TWO VECTOR GROUPS 2265

Thomas Lagier, Piotr Dworakowski, Cyril Buttay, Philippe Ladoux, Andrzej Wilk, Philippe Camail, Elissa Cresenta Anak Justin

IMPEDANCE ANALYSIS OF AN AUTOMOTIVE DC BUS.........................2274

Michael Schlüter, Marius Gentejohann, Sibylle Dieckerhoff

A NEW DUAL-MODE MPPT ALGORITHM APPLIED TO A QUADRATIC CONVERTER IN A SOLAR ENERGY SYSTEM 2284

Ahmad Ghamrawi, Jean-Paul Gaubert, Driss Mehdi

THERMAL MODEL DEVELOPMENT FOR SIC MOSFETS ROBUSTNESS ANALYSIS UNDER REPETITIVE SHORT CIRCUIT TESTS 2293

M. Pulvirenti, D. Cavallaro, N. Bentivegna, S. Cascino, E. Zanetti, M. Saggio

COMPENSATION OF THE RADIAL AND CIRCUMFERENTIAL MODE 0 VIBRATION OF A PERMANENT MAGNET ELECTRIC MACHINE BASED ON AN EXPERIMENTAL CHARACTERISATION.........................2303

Jan Andresen, Stephan Vip, Axel Mertens, Sebastian Paulus

MEASUREMENT BASED MODEL FOR THE CALCULATION OF CURRENT DISTRIBUTIONS BETWEEN PARALLELED POWER SEMICONDUCTORS DURING HIGH CURRENT OPERATION 2312
  Julian Da Cunha

DUAL-LOOP CONTROL SCHEME WITH OPTIMIZED TYPE-III CONTROLLER BASED ON GENETIC ALGORITHM FOR 6-PHASE INTERLEAVED CONVERTER IN ELECTRIC VEHICLE DRIVETRAINS 2320
  Dai-Duong Tran, Sajib Chakraborty, Thomas Geury, Joeri Van Mierlo, Mohamed El Baghdadi, Omar Hegazy

HIGH SENSITIVITY CURRENT TRANSFORMER WITH LOW SETTLING TIME, FOR MAGNIFIED AC CURRENT MEASUREMENTS IN PULSED APPLICATIONS 2331
  Georgios Tsolaridis, Pascal Seiler, Juergen Biela

LOSS SEPARATION IN HARD- AND SOFT-SWITCHING GAN HEMTS OPERATED IN A 10 KW ISOLATED DC/DC CONVERTER 2341
  Jan Böcker, Sören Heucke, Sibylle Dieckerhoff

A SWITCHED-MODE POWER AMPLIFIER FOR ION ENERGY CONTROL IN PLASMA ETCHING 2350
  Qihao Yu, Erik Lemmen, Korneel Wijnands, Bas Vermulst

EXPLORING THE BOUNDARIES AND EFFECTS OF THE DISCONTINUOUS CONDUCTION MODE IN H-BRIDGE INVERTER WITH DEAD-TIME 2358
  Qihao Yu, Erik Lemmen, Korneel Wijnands, Bas Vermulst

FIGURES-OF-MERIT AND CURRENT METRIC FOR THE COMPARISON OF IGCTS AND IGBTS IN MODULAR MULTILEVEL CONVERTERS 2366
  Arthur Boutry, Cyril Buttay, Dong Dong, Rolando Burgos, Bruno Lefebvre, Florent Morel, Colin Davidson

ZERO-CURRENT SWITCHING WITH LC RESONANT TANK CIRCUIT AND CAPACITOR ISOLATION DC-DC CONVERTER 2376
  Hideki Jonokuchi, Osamu Nakashima, Daichi Hiwatari, Hiroshi Hirayama

A FULL STATE-VARIABLE PREDICTIVE CONTROL OF BI-DIRECTIONAL BOOST CONVERTERS WITH GUARANTEED STABILITY 2386
  Yu Li, Zhenbin Zhang, Ralph Kennel

SYSTEM-LEVEL RELIABILITY ANALYSIS OF A REPAIRABLE POWER ELECTRONIC-BASED POWER SYSTEM CONSIDERING NON-CONSTANT FAILURE RATES 2393
  Amirali Davoodi, Yongheng Yang, Tomislav Dragicevic, Frede Blaabjerg

AN EFFICIENCY ANALYSIS OF A FERRITE MAGNET ASSISTED SYNCHRONOUS RELUCTANCE MACHINE FOR LOW POWER DRIVES INCLUDING FLUX WEAKENING 2403
  Matthias Hofer, Mario Nikowitz, Thomas Kirowitz, Manfred Schrödl

HIGH PERFORMANCE LQR CONTROL OF MODULAR MULTILEVEL CONVERTERS WITH SIMPLE CONTROL STRUCTURE AND IMPLEMENTATION 2409
  Min Jeong, Simon Fuchs, Jürgen Biela

FAULT DETECTION AND CLASSIFICATION BASED ON DEEP LEARNING IN LVDC OFF-GRID SYSTEM 2419
  Iurii Demidov, Antti Pinomaa, Andrey Lana, Olli Pyrhönen

AN INPUT-SERIES OUTPUT-INDEPENDENT FULL-BRIDGE DUAL ACTIVE BRIDGE CONVERTER WITH SOFT-SWITCHING CHARACTERISTICS FOR CHARGING AND BALANCING ELECTRIC VEHICLE BATTERY STACKS 2429
Alex V. Mirtchev, Emmanuel C. Tatakis

A METHOD TO SEARCH GLOBAL MAXIMA BY PERMANENT MONITORING OF VOLTAGE AND CURRENT OF EACH PV PANEL 2439
Shailendra Rajput, Moshe Averbukh

SURVEY AND COMPARISON OF 1D/2D ANALYTICAL MODELS OF HF LOSSES IN LITZ WIRE 2446
Qingchao Meng, Jürgen Biela

HIGH-FREQUENCY SIC-BASED MEDIUM VOLTAGE QUASI-2-LEVEL FLYING CAPACITOR DC/DC CONVERTER WITH ZERO VOLTAGE SWITCHING............................ 2457
Rafal Kopacz, Przemyslaw Trochimiuk, Grzegorz Wrona, Jacek Rabkowski

SMART FUEL CELL MODULE (6.5 KW) FOR A RANGE EXTENDER APPLICATION 2467
Pascal Bazin, Bruno Beranger, Jacques Ecrabey, Laurent Garnier, Sylvain Mercier

IMPACT OF THE INITIAL TRANSIENT INTERRUPTION VOLTAGE (ITIV) ON THE DESIGN AND OPERATION OF HYBRID CURRENT-INJECTION DC CIRCUIT BREAKERS............ 2475
Andreas Jehle, Jürgen Biela

FOUR QUADRANT BUS-TIE SWITCH FOR PROTECTION OF SHIPBOARD POWER SYSTEMS 2486
Gabriele Ulissi, Seong-Yong Lee, Drazen Dujic

ESTIMATION OF AN UNBALANCED GRID IMPEDANCE USING A THREE-PHASE POWER CONVERTER 2495
Jarno Kukkola, Ville Pirsto, Mikko Routimo, Marko Hinkkanen

FAULT DIAGNOSIS OF HVDC TRANSMISSION SYSTEM USING WAVELET ENERGY ENTROPY AND THE WAVELET NEURAL NETWORK 2505
Cuicui Liu, Feng Wang, Fang Zhuo, Ziqian Zhang

REDUCING THE ENERGY STORAGE REQUIREMENTS OF MODULAR MULTILEVEL CONVERTERS WITH OPTIMAL CAPACITOR VOLTAGE TRAJECTORY SHAPING 2513
Simon Fuchs, Min Jeong, Jürgen Biela

LEAKAGE INDUCTANCE MODELLING OF TRANSFORMERS: ACCURATE AND FAST MODELS TO SCALE THE LEAKAGE INDUCTANCE PER UNIT LENGTH.. 2524
Richard Schlesinger, Jürgen Biela

A GAN-BASED DC/DC CONVERTER FOR E-VEHICLES APPLICATIONS 2535
Eduardo F. De Oliveira, Sebastian Sprunck, Jonas Pfeiffer, Peter Zacharias

THEORY OF INFLUENCING THE BREATHING MODE AND TORQUE PULSATIONS OF PERMANENT MAGNET ELECTRIC MACHINES WITH HARMONIC CURRENTS 2545
Jan Andresen, Stephan Vip, Axel Mertens, Sebastian Paulus

POWER HARDWARE IN THE LOOP SYSTEM BASED ON INTERLEAVED CONVERTER AND FPGA - APPLICATION TO DC AND AC SIDE EMULATION FOR PHOTOVOLTAIC INVERTER TESTING.. 2554
R. Kadri, R. Bakri, A. Omrane, F. Colas, F. Delpech

IMPLEMENTATION OF TAPIR SWITCHING CELLS WITH INTEGRATED DIRECT AIR-COOLING FOR SIC POWER DEVICES 2564
Wendpanga Fadel Bikinga, Kouceila Alkama, Bachir Mezrag, Jean Michel Guichon, Yvan Avenas

EFFECT OF UNIPOLAR AND BIPOLAR SPWM ON THE LIFETIME OF DC-LINK CAPACITORS IN SINGLE-PHASE VOLTAGE SOURCE INVERTERS 2573
Silpa Baburajan, Saeed Peyghami, Dinesh Kumar, Frede Blaabjerg, Pooya Davari

TRANSIENT THERMAL MODELS OF CAPACITORS AND INDUCTORS FOR SYSTEM OPTIMIZATION 2583
Vasilios Karaventzas, Juergen Biela, Felix Rodriguez Mateos

ENERGY MANAGEMENT FOR ISOLATED RENEWABLE-POWERED MICROGRIDS USING REINFORCEMENT LEARNING AND GAME THEORY 2594
Rui Hu, Alexis Kwasinski

ALL-GAN BIDIRECTIONAL ANPC-BASED RESONANT DC-DC CONVERTER 2603
Tino Kahl, Laurenz Wernicke, Sibylle Dieckerhoff, Christopher Fromme, Marvin Tannhäuser, Ag Siemens

LIFETIME ESTIMATION AND DIMENSIONING OF THE MACHINE-SIDE CONVERTER FOR PUMPING-CYCLE AIRBORNE WIND ENERGY SYSTEM 2613
Bakr Bagaber, Patrick Junge, Axel Mertens

A DESIGN OF HIGH-POWER INVERTER CIRCUIT INCLUDING GAN POWER DEVICES 2623
Takashi Sawada, Hiroshi Tadano, Koji Shiozaki

SPEED SENSORLESS COMMISSIONING OF RESONATING MECHANICAL SYSTEM IN ELECTRIC DRIVES... 2630
A. Putkonen, N. Nevaranta, O. Liukkonen, M. Niemelä, O. Pyrhönen

CONTROL OF A TWO-STAGE, SINGLE-PHASE GRID-TIED, GAN BASED SOLAR MICRO-INVERTER 2638
Anthony Bier, Van Sang Nguyen, Stéphane Catellani, Jérémy Martin

A DC/DC BUCK-BOOST CONVERTER CONTROL USING SLIDING SURFACE MODE CONTROLLER AND ADAPTIVE PID CONTROLLER... 2648
Bassem Saleh, Ahmed Teirelbar, Amr Wasfi

SENSORLESS NEUTRAL POINT VOLTAGE STABILIZATION IN THREE-PHASE FOUR-WIRE CONVERTERS... 2656
Xinwei Xu, Gabriel Tibola, Jorge L. Duarte

BIDIRECTIONAL ISOLATED RIPPLE CANCEL TRIPLE ACTIVE BRIDGE DC-DC CONVERTER.. 2666
Takahiro Ohta, Pin-Yu Huang, Yuichi Kado

DESIGN OF THE SPEED SENSORLESS FIELD ORIENTED CONTROL SYSTEM FOR INDUCTION MOTORS CONSIDERING SUDDEN CHANGE OF THE ROTOR SPEED 2675
Yoshiki Sakurazawa, Osamu Yamazaki, Kazuaki Yuki, Yosuke Nakazawa, Kenji Natori, Keiichiro Kondo

EFFICIENCY POTENTIAL OF SOLID-STATE PULSE MODULATORS USING SIC DEVICES 2684
Spyridon Stathis, Michael Jaritz, Sebastian Blume, Jürgen Biela

EFFICIENT AND SCALABLE POWER CONTROL IN MULTI-PORT ACTIVE-BRIDGE CONVERTERS 2695
Soleiman Galeshi, David Frey, Yves Lembeye

COMPARISON OF PRESS-PACK AND WIRE-BONDING TECHNOLOGIES FOR SIC MOSFETS UNDER SHORT-CIRCUIT CONDITIONS 2704
Ran Yao, Francesco Iannuzzo, Amir Sajjad Bahman, Hui Li

ERROR INDUCED BY THE OPTICAL PATH OF A HIGH ACCURACY AND HIGH BANDWIDTH OPTICAL CURRENT MEASUREMENT SYSTEM 2712
Stefan Rietmann, Jürgen Biela

ANALYSIS OF THE RMS CURRENT STRESS ON THE DC LINK CAPACITORS OF THE FOUR PHASE 3-LEVEL T-TYPE VOLTAGE SOURCE CONVERTER 2723
Zoran Miletic, Werner Tremmel, Roland Bründlinger, Johannes Stöckl, Petar J. Grbovic

AN ADAPTIVE DROOP CONTROL METHOD FOR INTERLINK CONVERTER IN HYBRID AC/DC MICROGRIDS 2733
Mohammad S. Golsorkhi, Rasool Heydari, Mehdi Savaghebi

SIMPLIFIED CALCULATION OF PARASITIC ELEMENTS AND MUTUAL COUPLINGS OF WIDE-BANDGAP POWER SEMICONDUCTOR MODULES 2743
Mohammad Ali, Jens Friebe, Axel Mertens

VARIABLE-SPEED-DRIVE-BASED SENSORLESS ESTIMATION OF PUMP SYSTEM RESERVOIR FLUID LEVEL 2753
Santeri Pöyhönen, Aleksi Simola, Jero Ahola

ANALYSIS OF SWITCHING PERFORMANCE AND EMI EMISSION OF SIC INVERTERS UNDER THE INFLUENCE OF PARASITIC ELEMENTS AND MUTUAL COUPLINGS OF THE POWER MODULES 2763
Mohammad Ali, Jan-Kaspar Müller, Jens Friebe, Axel Mertens

WIRE-WOUND MULTI-PHASE STATOR BASED EMEH WITH MPPT SELF-POWERED ENERGY MANAGEMENT SYSTEM 2773
Mahmoud Shousha, Dragan Dinulovic, Talha Zafar, Michael Brooks, Martin Haug

COMPARISON OF OPTIMIZED MOTOR-INVERTER SYSTEMS USING A STACKED POLYPHASE BRIDGE CONVERTER COMBINED WITH A 3-, 6-, 9-, OR 12-PHASE PMSM 2780
Thilo Bringezu, Jürgen Biela

DESIGN OF A PULSE MODULATOR BASED ON TRANSMISSION LINES FOR GENERATING FAST CURRENT PULSES FOR PLASMA DRILLING 2791
Oliver Keel, Melissa Artiglia, Juergen Biela

ANALYSIS OF CURRENT IN PULSATING DC LINK CONVERTER WITH ZERO VOLTAGE TRANSITION 2802
Daniele Marciano, Giovanni Busatto, Carmine Abbate, Annunziata Sanseverino, Davide Tedesco, Francesco Velardi

SIGNAL INJECTION FOR SENSORLESS CURRENT SHARING WITH EXPERIMENTAL VERIFICATION ON 1 MHZ GAN PROTOTYPE 2812
N. Boškovic, J. Duarte, E. A. Lomonova

MODELLING AND ANALYSIS OF SENSORLESS CURRENT SHARING APPROACH 2820
N. Boškovic, J. Duarte

PWM-INDUCED HARMONIC POWER IN 75 KW IM DRIVE SYSTEM 2829
Lassi Aarniovuori, Hannu Kärkkäinen, Markku Niemelä, Juha Pyrhönen

PROPOSAL OF BOOST CONVERTER WITHOUT REACTOR USING OPEN-ENDED
WINDING PMSM FOR PHOTOVOLTAIC PUMP SYSTEM.. 2838
Akihiro Okazaki, Sari Maekawa

THE PROPOSAL OF DISCRIMINATING STABLE CONTROL BANDWIDTH USING ANN IN
SENSORLESS SPEED CONTROL SYSTEM FOR PMSM.. 2844
Ami Tanaka, Sari Maekawa

COST FUNCTION DESIGN FOR STABILITY ASSESSMENT OF MODULATED MODEL
PREDICTIVE CONTROL 2851
*Jordan P. Zucuni, Fernanda Carnielutti, Humberto Pinheiro, Margarita Norambuena, Jose
Rodriguez*

A ROBUST FUZZY-BASED CONTROL TECHNIQUE FOR WIND FARM TRANSIENT
VOLTAGE STABILITY USING SVC AND STATCOM: COMPARISON STUDY 2860
*Reza Ebrahimi, Vahid Eslampanah, Hossein Madadi Kojabadi, Mohammadreza Azizian,
Naser Nourani Esfetanaj, Dao Zhou*

TEMPERATURE EVOLUTION AS AN EFFECT OF WIRE-BOND FAILURES IN A MULTI-
CHIP IGBT POWER MODULE.. 2865
N. Degrenne, R. Delamea, S. Mollov

COST OF ENERGY ASSESSMENT OF WIND TURBINE CONFIGURATIONS 2873
Catalin Dincan, Philip Kjær, Lars Helle

ENERGY MANAGEMENT IN A MULTI-SOURCE SYSTEM USING ISOLATED DC-DC
RESONANT CONVERTERS... 2881
M. Arazi, A. Payman, M. B. Camara, B. Dakyo

LONG-TERM CLIMATE IMPACT ON IGBT LIFETIME... 2888
Martin Vang Kjaer, Yongheng Yang, Huai Wang, Frede Blaabjerg

COMMUNICATION-FREE SECONDARY FREQUENCY AND VOLTAGE CONTROL OF
VSC-BASED MICROGRIDS: A HIGH-BANDWIDTH APPROACH 2898
*Rasool Heydari, Mohammad S. Golsorkhi, Mehdi Savaghebi, Tomislav Dragicevic, Frede
Blaabjerg*

OFFSHORE WIND FARM LAYOUT OPTIMIZATION CONSIDERING WAKE EFFECTS 2907
Asma Dabbabi, Salvy Bourguet, Rodica Loisel, Mohamed Machmoum

SMALL-SIGNAL STABILITY ANALYSIS OF SMART GRIDS CONSIDERING HIGH
PENETRATION OF POWER ELECTRONICS CONVERTERS AND ENERGY MARKETS 2917
Javiera Meneses, Patricio Mendoza-Araya

COMPONENT-LEVEL RELIABILITY ASSESSMENT OF A DIRECT-DRIVE PMSG WIND
POWER CONVERTER CONSIDERING LONG-TERM AND SHORT-TERM THERMAL
CYCLES.. 2928
Shuaichen Ye, Dao Zhou, Frede Blaabjerg

A SUBMODULE IMPLEMENTATION FOR PARALLEL CONDUCTION OF DIODES IN
MODULAR MULTILEVEL CONVERTERS.. 2938
Martin Geske, Duro Basic, Christian Keller, Thomas Brückner

EVALUATION OF THE I_{MAX}-F_{SW}-DV/DT TRADE-OFF OF HIGH VOLTAGE SIC MOSFETS BASED ON AN ANALYTICAL SWITCHING LOSS MODEL 2946

Anliang Hu, Jürgen Biela

PROTECTION MEASURES FOR MODULAR MULTILEVEL CONVERTERS IN CASE OF DC SHORT-CIRCUIT FAULTS 2957

Martin Geske, Duro Basic, Roland Jakob, Christian Keller, Thomas Brückner

INVESTIGATION ON PARALLEL OPERATION OF TWO MMC-HVDC LINKS IN GRID FORMING CONNECTED TO AN EXISTING NETWORK 2967

H. Saad, P. Rault, S. Dennetière

MODELLING AND EXPERIMENTAL VALIDATION OF A LABORATORY-SCALED HVDC CABLE EMULATOR TESTED IN AN MMC-BASED PLATFORM 2977

Enric Sánchez-Sánchez, Adrià Junyent-Ferré, Eduardo Prieto-Araujo, Oriol Gomis-Bellmunt, Tim Green

DAISY CHAIN PN CELL FOR MULTILEVEL CONVERTER USING GAN FOR HIGH POWER DENSITY 2987

Faheem Ahmad, Asger Bjørn Jørgensen, Szymon Michal Beczkowski, Stig Munk-Nielsen

GRID-FREQUENCY VIENNA RECTIFIER AND ISOLATED CURRENT-SOURCE DC-DC CONVERTERS FOR EFFICIENT OFF-BOARD CHARGING OF ELECTRIC VEHICLES 2996

Jacek Rabkowski, Andrei Blinov, Denys Zinchenko, Grzegorz Wrona, Mariusz Zdanowski

UNIDIRECTIONAL THYRISTOR-BASED DC-DC CONVERTER FOR HVDC CONNECTION OF OFFSHORE WIND FARMS 3006

Pierre Le Métayer, Piotr Dworakowski, Jose Maneiro

INDUCTOR SIZE EVALUATION OF AN ELECTROMAGNETIC INTERFERENCE FILTER FOR A TWO-LEVEL POWER FACTOR CORRECTION RECTIFIER USING DIFFERENT MODULATION TECHNIQUES 3015

Mohammad Najjar, Alireza Kouchaki, Morten Nymand

EVALUATION OF MMCS FOR HIGH-POWER LOW-VOLTAGE DC-APPLICATIONS IN COMBINATION WITH THE MODULE LLC-DESIGN 3024

Roland Unruh, Frank Schafmeister, Joachim Böcker

IRON LOSS CHARACTERISTICS OF MNZN FERRITES UNDER GAN INVERTER EXCITATION IN THE MHZ ORDER 3034

Wilmar Martinez, Camilo Suarez, Federico Ibanez

VIBRATION SUPPRESSION AND CONTROL PARAMETER DESIGN OF A SENSORLESS PMSM ROTARY COMPRESSOR DRIVE 3044

Tao Li, Chaohui Liang

3D PCB PACKAGE FOR GAN INVERTER LEG WITH LOW EMC FEATURE 3054

Pawel B. Derkacz, Jean-Luc Schanen, Pierre-Olivier Jeannin, Piotr Musznicki, Piotr J. Chrzan, Mickael Petit

ESTIMATION OF THE WINDING LOSSES OF MEDIUM FREQUENCY TRANSFORMERS WITH LITZ WIRE USING AN EQUIVALENT PERMEABILITY AND CONDUCTIVITY METHOD 3064

Mohammad Kharezy, Morteza Eslamian, Torbjörn Thiringer

IMPROVEMENT OF DRIVING EFFICIENCY OF PMSM BY USING MODIFIED TRAPEZOIDAL MODULATING SIGNAL 3071
Kento Betto, Satoshi Joryo, Toshimitsu Morizane

DESIGN AND CONTROL OF A VIRTUAL DC-LINK FOR A FULL GAN-BASED SINGLE PHASE CONVERTER WITH HIGH POWER DENSITY 3081
Yugandhara H. Wankhede, Leon Fauth, Jens Friebe

USING BOTH THE CIRCULATING CURRENTS AND THE COMMON-MODE VOLTAGE FOR THE BRANCH ENERGY CONTROL OF MODULAR MULTILEVEL CONVERTERS 3091
Rebecca Dierks, Jakub Kucka, Axel Mertens

ANALYTICAL HARMONIC CURRENT MODEL FOR A PERMANENT MAGNET ASSISTED SYNCHRONOUS RELUCTANCE MOTOR (PMA-SYNRM) FED BY PWM INVERTER 3101
Jessica Neumann, Carole Hénaux, Maurice Fadel, Etienne Founier, Dany Prieto, Mathias Tientcheu Yamdeu

GENERALIZED SMALL-SIGNAL AVERAGED SWITCH MODEL ANALYSIS OF A WBG-BASED INTERLEAVED DC/DC BUCK CONVERTER FOR ELECTRIC VEHICLE DRIVETRAINS 3111
Sajib Chakraborty, Dai-Duong Tran, Joeri Van Mierlo, Omar Hegazy

ADAPTIVE PREDICTIVE-DPC FOR LCL-FILTERED GRID CONNECTED VSC WITH REDUCED NUMBER OF SENSORS 3119
Hosein Gholami-Khesht, Pooya Davari, Frede Blaabjerg

FPGA IMPLEMENTATION OF MODIFIED SPACE VECTOR MODULATION (SVM) FOR HIGH-FREQUENCY HYBRID ACTIVE NEUTRAL-POINT-CLAMPED (NPC) POWER FACTOR CORRECTION RECTIFIER 3129
Mohammad Najjar, Alireza Kouchaki, Morten Nymand

ENHANCED FLUX CONTROL INCLUDING A CLOSED LOOP VOLTAGE CONTROLLER TO OPTIMIZE THE VOLTAGE USAGE AND THE TORQUE COMPUTATION FOR A 48V IPMSM 3137
Felix Bertele, Ulrich Ammann, Christoph Cheshire, Tobias Röser

EXTENDED BOOST PV INVERTER TOPOLOGY FOR THE REDUCTION OF COMMON-MODE LEAKAGE CURRENT IN THREE-PHASE APPLICATIONS 3146
Georgios I. Orfanoudakis, Eftychios Koutroulis, Michael A. Yuratich, Suleiman M. Sharkh

A ROBUST CONTROL DESIGN TO REAL-TIME CONDITIONS AND MODELLING OF A MICROGRID 3156
Iréna Horvatic, Delphine Riu, Moataz Elsied, Sébastien Benjamin

DESIGN OF MODULAR LOW-PROFILE FREQUENCY CONVERTER FOR MULTI-MOTOR MANIPULATORS 3166
Tomas Glasberger, Zdenek Kehl, Tomas Kosan, Jan Molnar

STUDY OF THE CONTROL OF A NEW AC VOLTAGE STABILIZER USING LINEAR CONTROLLER WITH REFERENCE FRAME TRANSFORMATION 3172
Bunthern Kim, Etienne Boulaud, Emile Boisaubert, Sokchea Am, Phok Chrin

HYBRID ENERGY STORAGE SYSTEM FOR MVDC-GRIDS 3179
Florian Mahr, Johann Jaeger, Stefan Henninger, Hubert Rubenbauer

A COMBINED MODEL FOR OPTIMAL POWER FLOW APPLIED TO MT-HVDC SYSTEMS 3189
Fernando Torres, Javier Muñoz, Fredy Muñoz, Claudio Roa

CHARACTERIZATION OF LITHIUM ION SUPERCAPACITORS 3198
Zeyang Geng, Felix Mannerhagen, Torbjöm Thiringer

GREY WOLF OPTIMIZER BASED PREDICTIVE TORQUE CONTROL FOR ELECTRIC
VEHICLE APPLICATIONS 3205
Ali Djerioui, Azeddine Houari, Mohamed Machmoum, Malek Ghanes, Tedjani Mesbahi,
Mohamed Fouad Benkhoris

OPERATION PRINCIPLE AND PERSPECTIVE PERFORMANCES OF METAL OXIDE
VACUUM FIELD EFFECT TRANSISTOR - MOVFET 3210
Davide Patti, G. Busatto, G. Golluccio, D. Marciano, A. Sanseverino, F. Velardi

IMPROVED METHODOLOGY FOR PREDICTING CORRELATED COLOR TEMPERATURE
IN MIXED LED LIGHTING SOURCES 3217
Thais E. Bolzan, Bruno F. Almeida, Renan R. Duarte, Vitor C. Bender, Rafael A. Pinto

DC MICROGRID CONCEPT FOR MINE ENVIRONMENT 3227
Jooa Pursiainen, Jenni Rekola, Raimo Juntunen, Mikko Valtee, Pasi Peltoniemi

A COMPARISON OF TWO-STAGE INVERTER AND QUASI-Z-SOURCE INVERTER FOR
HYBRID ENERGY STORAGE APPLICATIONS 3237
V. Castiglia, R. Miceli, F. Blaabjerg, Y. Yang

STATE ESTIMATION FOR MEDIUM AND LOW VOLTAGE DISTRIBUTION GRIDS
BASED ON NEAR REAL-TIME GRID MEASUREMENTS AND DELAYED SMART
METERS DATA 3247
Mohammad Rayati, Thomas Pidancier, Mauro Carpita, Mokhtar Bozorg

GROUND FAULT ACTIVE COMPENSATION IN EMULATED DISTRIBUTION GRID OF 10
KV 3257
Tomáš Komrska, Antonín Glac, Jakub Talla, Bohumil Skala, Jan Štepánek, Lubeš Streit,
Zdenek Peroutka

MODELING OF A POWER TRANSFORMER INCLUDING HIGHER ORDER RESONANCES 3263
Lukas Reißenweber, Alexander Stadler

A COMPARISON OF TWO STATE-SPACE MODELS OF AN INDUCTION MACHINE
CONSIDERING DIFFERENT SETS OF WINDING DISTRIBUTION HARMONICS 3272
Julien Cordier, Stefan Klass, Ralph Kennel

PERFORMANCE IMPROVEMENT FOR PLUG-IN REVERSE CONDUCTING IGBTS
THROUGH GATE-VOLTAGE OBSERVATION 3282
Daniel Lexow, Hans-Günter Eckel

DIFFERENTIAL FLATNESS FOR SMOOTH TRANSITION BETWEEN GRID-CONNECTED
AND STANDALONE MODE OF THREE-PHASE INVERTER 3289
Abdelhakim Saim, Azeddine Houari, Mourad Ait-Ahmed, Mohamed Machmoum, Josep. M
Guerrero

DIFFERENTIAL MODEL EMI FILTER ANALYSIS FOR INTERLEAVED BOOST PFC
CONVERTERS CONSIDERING OPTIMAL PHASE SHIFTING 3295
Naser Nourani Esfetanaj, Yamen Saad, Omar Ahmed Sakaria, Huai Wang, Pooya Davari

MODULAR HYBRID DC BREAKER-BASED ADAPTIVE AUTO-RECLOSING METHOD
FOR MMC-HVDC SYSTEMS 3305
Hossein Iman-Eini, M. Langwasser, L. Camurca, Marco Liserre

MULTISTEP MPC OF DUAL INVERTER FOR SWITCHING LOSSES OPTIMIZATION 3314
Martin Votava, Tomas Glasberger, Zdenek Peroutka

A HIGH-EFFICIENCY CONTROL OF A DOUBLE-INPUT CONVERTER FOR RENEWABLE
ENERGIES AND HYBRID VEHICLES 3321
Mario Marchesoni, Massimiliano Passalacqua, Luis Vaccaro

DEAD-TIME INFLUENCE ON FAST SWITCHING PULSED POWER CONVERTERS
DESIGN - A HIGH CURRENT APPLICATION FOR ACCELERATOR'S MAGNETS 3330
*Ludovic Horrein, Jean-Marc Cravero, Philippe Delarue, Alain Bouscayrol, Davide Aguglia,
Carmen Ortega-Perez*

DYNAMIC CHARACTERIZATION OF A SIC-MOSFET HALF BRIDGE IN HARD- AND
SOFT-SWITCHING AND INVESTIGATION OF CURRENT SENSING TECHNOLOGIES 3340
Janine Ebersberger, Jan-Kaspar Müller, Axel Mertens

POWER SUPPLY DESIGN CONSIDERATIONS FOR 400HZ AIRCRAFT APPLICATIONS 3348
Bilal Ahmad, Jorma Kyyrä, Juha Mäkelä

DC CAPACITOR VOLTAGE FEEDBACK METHOD FOR A PEAK VOLTAGE
SUPPRESSION CONTROL WITH MULTIPLE LEG-SHORT-CIRCUITS USING SIC-
MOSFETS EMPLOYED IN POWER CONVERTERS 3358
Tomoyuki Mannen, Takanori Isobe, Keiji Wada

INVESTIGATION OF BOND WIRE LIFT-OFF BY ANALYZING THE CONTROLLER
OUTPUT VOLTAGE HARMONICS FOR THE PURPOSE OF CONDITION MONITORING 3366
Firat Yüce, Marc Hiller

FRUGAL INNOVATION FOR SUSTAINABLE RURAL ELECTRIFICATION 3376
Bunthern Kim, Phok Chrin, Maria Pietrzak-David, Pascal Maussion

A CURRENT-MODULUS DERIVATIVE-BASED PROTECTION METHOD IN A FLEXIBLE
DC GRID 3385
Jianquan Liao, Niancheng Zhou, Qianggang Wang

COMPARATIVE ASSESSMENT OF VOLTAGE MODULATION METHODS FOR
ASYMMETRIC SIX-PHASE MACHINES 3393
R. S. Kanchan, Omer Ikram Ul Haq, Luca Peretti

SIMULATION AND MEASUREMENT-BASED ANALYSIS OF EFFICIENCY
IMPROVEMENT OF SIC MOSFETS IN A SERIES-PRODUCTION READY 300 KW / 400 V
AUTOMOTIVE TRACTION INVERTER 3403
*A. Nisch, M. Heller, W. Wondrak, A. Bucher, C. Hasenohr, K. Kefer, B. Lunz, A. Pawellek, A.
Smit, M. Gärtner, N. Twardon, U. Kirchenberger*

VALIDITY OF POWER CYCLING LIFETIME MODELS FOR MODULES AND EXTENSION
TO LOW TEMPERATURE SWINGS 3413
Josef Lutz, Christian Schwabe, Guang Zeng, Lukas Hein

ROADMAP FOR DC 3422
Pavol Bauer

THE ROLE OF COLLABORATIVE RESEARCH TO SUPPORT INNOVATION FOR CLEAN
ENERGY TRANSITION 3424
Hubert De La Grandiere

THOMAS EDISON VINDICATED — THE RESURGENCE OF DC IN MV AND HV POWER GRIDS 3425

Colin Davidson

INTEGRATION OF ELECTRIC MOBILITY IN THE FRENCH PUBLIC ELECTRICITY DISTRIBUTION NETWORK 3426

Anne-Sophie Cochelin

A CRITICAL ROLE FOR R&I FOR CLEAN ENERGY FOR THE EU GREEN AND DIGITAL RECOVERY 3427

Hélène Chraye

Author Index

Simplified Calculation of Parasitic Elements and Mutual Couplings of Wide-bandgap Power Semiconductor Modules

Mohammad Ali, Jens Friebe, Axel Mertens
LEIBNIZ UNIVERSITY HANNOVER
Institute for Drive Systems and Power Electronics
Welfengarten 1
30167 Hannover, Germany
Tel: +49 / (0) - 511 762 3778
Fax: +49 / (0) - 511 762 3040
E-mail: mohammad.ali@ial.uni-hannover.de
URL: www.ial.uni-hannover.de

Acknowledgments

This work was supported by the German Ministry of Economics and Technology - 19236 N (FVA).

Keywords

≪Parasitic Capacitance≫, ≪Parasitic Inductance≫, ≪Magnetic Coupling≫, ≪3D Packaging≫, ≪SiC Module≫.

Abstract

This paper presents a simplified calculation of parasitic elements (LC) and mutual couplings between parasitics of wide-bandgap (WBG) power semiconductor modules, based on analytical equations and on 3D FEM. A simplified parallel plate capacitor is derived from stray fields of different plate surfaces. The simple structures e. g. two parallel round wires with different directions of current, are considered to calculate the parasitic inductance and the magnetic coupling. The analytical models are verified by ANSYS Q3D results. This method includes stray fields of capacitive and inductive parasitic structures based on a simplified geometric approach. The package of a SiC-MOSFET half-bridge power module is 3D-modeled and the parasitic elements are extracted. The analytical models are verified by numerical results. At last, the influence of parasitic elements and mutual couplings on the switching characteristics is analyzed.

I. Introduction

The parasitic inductances which are associated with the interconnections inside a package strongly affect the switching characteristics of the power semiconductor devices. They excite overshoots and oscillations in the switching waveforms [1], [2]. The overshoots and oscillations further contribute to increased EMI emissions [3], [4]. Also, the parasitic capacitance of the power semiconductor devices has a significant impact on common mode EMI [5], [6]. Looking at WBG semiconductors, their steeper voltage edges and higher switching frequencies cause much larger high-frequency excitations than Si devices and increasing the common mode currents through parasitic capacitances [7], [8]. All these effects contribute to an increased level of EMI emissions and other high-frequency effects, such as machine insulation stress and bearing currents in case of electric drives with WBG devices.

For a streamlined design and development process of WBG power electronics, it is important to evaluate expected EMI emissions in an early stage, when there is no experimental setup available. In order to do so, parasitic elements and mutual couplings of the interconnections inside a power semiconductor module need to be identified based on their geometric layout, so that switching waveforms and, further on, common mode and differential mode EMI noise can be accurately predicted.

In [2], [5], the parasitic inductance without mutual couplings among different interconnected structures in a package and the parasitic capacitance without considering stray fields are discussed in detail, using analytical and numerical (ANSYS Q3D) methods. In fact, most of the literature on parasitics in power electronics neglects electric stray fields and mutual magnetic couplings between elements.

Therefore, the goal of this paper is to identify the influence of stray fields on the parasitic capacitances and the influence of mutual couplings within a WBG semiconductor module. Mutual couplings affect the parasitic gate-to-source inductances as well as the power loop inductance.

In section II, the parasitic capacitance with stray fields of the different plate surfaces is analyzed. These analytical results are validated using numerical results. Next, general considerations on magnetic simple structures, such as two parallel round wires with different directions of current, are used to calculate the parasitic inductance and the magnetic coupling between them, again validating the results with ANSYS Q3D. Finally, the package of a SiC-MOSFET half-bridge power module is 3D-modeled and the high-frequency (HF) parasitic elements are extracted according to the geometry and material properties of the actual module using the analytical results to simplify the numerical procedures.

II. Parasitic Capacitance and Inductance

A. Simplified parallel plate capacitor

Capacitance stores energy in the electric field in and around a structure. It varies according to dimension, but the basic configuration is two conductors carrying equal but opposite charges, which is shown in Fig. 1. For parallel-plate capacitors, the influence of the distance between the plates on fringing electric fields is explained in [9]–[11]. The capacitance between two parallel plates including stray fields which are shown in Figs. 1b and 1c can be expressed as follows:

$$
\begin{aligned}
C_{total} &= C_1 + 2C_{2wl} + 2C_{3wl} + 4C_4, \\
&= \varepsilon_0 \varepsilon_r \frac{wl}{d} + 2 \cdot \varepsilon_0 \varepsilon_{r_1} \left[\left[\ln \frac{(d+2t)}{d} \right] \cdot \left[\frac{(l+w)}{\pi} \right] + \frac{2wl}{\pi} \left[\frac{1}{d+2t+w} + \frac{1}{d+2t+l} \right] \right]
\end{aligned}
\tag{1}
$$

Here, w, l denote the width and length of parallel plates, d indicates the height of the dielectric material and ε_0, ε_r are the free space and relative permittivity, respectively. C_1 denotes the capacitance between two parallel plates without stray fields and fringing effects, as shown in Fig. 1b (red arrows); C_{2wl} indicates the capacitance linked to stray fields at the edge surfaces of two plates at the dielectric material, shown in Figs. 1b and 1c (in magenta); C_{3wl} is the capacitance linked to a stray field through the outer surface of the plates, shown in Figs. 1b and 1c (in green) and C_4 represents the capacitance linked to a stray field of the quarter spherical shells at the edges of the plates, as shown in Figs. 1b and 1c (in blue). Since the capacitance C_4 is very small, it is neglected.

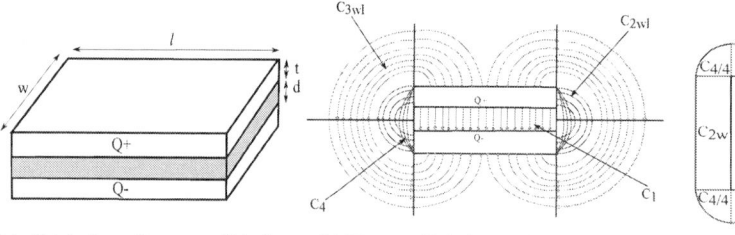

(a) Total view of two parallel plates (b) Two parallel plates with stray fields (c) Top view of stray fields

Fig. 1. Total capacitance between two parallel plates.

B. Capacitance in terms of static charge

In ANSYS Q3D, Maxwell capacitance matrix represents the relationship between charges on the conductors and voltages of the conductors. For example, Fig. 2 shows the capacitance between three conductors with the background object used as a grounded reference. The most important capacitances are $C_{12} = C_{21}$, $C_{13} = C_{31}$ and $C_{23} = C_{32}$. These capacitances are called the mutual capacitances. The ANSYS Q3D software simulates the Maxwell capacitance

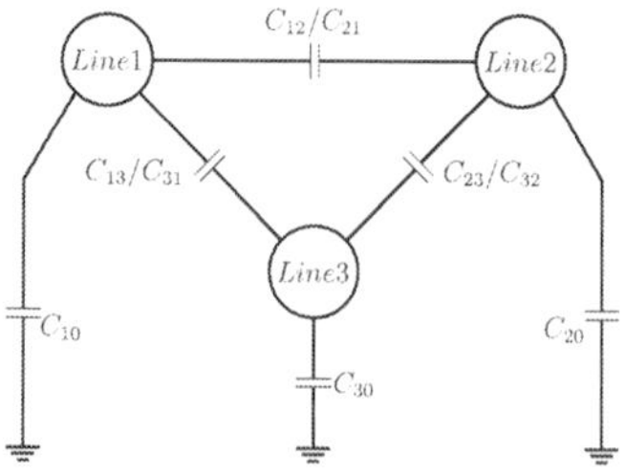

Fig. 2. Capacitances between three conductors.

matrix format. So, the charge on each conductor can be written as following relationships [14]:

$$
\begin{bmatrix} Q_1 \\ Q_2 \\ Q_3 \end{bmatrix} = \underbrace{\begin{bmatrix} C_{10} + C_{12} + C_{13} & -C_{12} & -C_{13} \\ -C_{21} & C_{20} + C_{21} + C_{23} & -C_{23} \\ -C_{31} & -C_{32} & C_{30} + C_{31} + C_{32} \end{bmatrix}}_{\textbf{Maxwell capacitance matrix}} \cdot \begin{bmatrix} V_1 \\ V_2 \\ V_3 \end{bmatrix} \tag{2}
$$

C. Validation between the analytical calculation and the simulation results

The parallel-plate capacitance determination using ANSYS Q3D is shown in Fig. 3a. Fig. 3b shows the comparison between the analytical and numerical results of the total capacitance with varying the plate lengths. Here, it can be seen that the analytical result is in a good agreement with the numerical result. Moreover, the difference between the total capacitance, C_{total} with stray fields and the capacitance C_1 without stray fields is seen to be important. Neglecting those would introduce large errors.

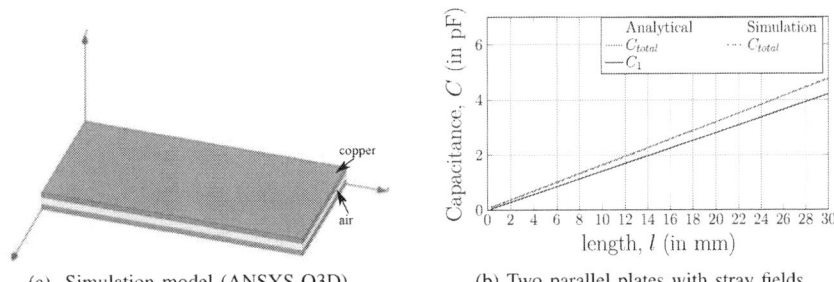

(a) Simulation model (ANSYS-Q3D) (b) Two parallel plates with stray fields

Fig. 3. Comparison between the analytical and simulation results of the Parallel-plate capacitance, where $t = 0.30\,\text{mm}$, $d = 0.38\,\text{mm}$ and $w = 6\,\text{mm}$.

D. Net and loop parasitic inductance of parallel wires

Fig. 5a and 5b show two parallel round wires with currents in the same and in the opposite direction. The net inductance and the loop inductance of the two parallel wires can be expressed as follows [12], [13],

$$
L_{pnet} = \frac{L_{p1} L_{p2} - L^2_{mp}}{L_{p1} + L_{p2} - 2L_{mp}}, \qquad L_{loop} = L_{p1} + L_{p2} - 2L_{mp} \tag{3}
$$

Here, L_{p1}, L_{p2} indicate the self-inductances of each of the parallel wires and L_{mp} is the mutual inductance between two parallel wires. The self-inductance of a round wire and the mutual partial inductance between two parallel round wires can be calculated by the following relationships respectively [12], [13]:

$$
\begin{aligned}
L_{cir} &= \frac{\mu}{2\pi} l_{cir} \left[\ln\left(\frac{l_{cir}}{r}\right) + \sqrt{\left(\frac{l_{cir}}{r}\right)^2 + 1} - \sqrt{1 + \left(\frac{r}{l_{cir}}\right)^2} + \frac{r}{l_{cir}} \right], \text{where} \quad l_{cir} \gg r, \\
&\cong \frac{\mu}{2\pi} l_{cir} \left[\ln\frac{2l_{cir}}{r} - 1 \right], \text{where} \quad l_{cir} \gg r, \text{[For AC]} \\
&\cong \frac{\mu}{2\pi} l_{cir} \left[\ln\frac{2l_{cir}}{r} - \frac{3}{4} \right], \text{where} \quad l_{cir} \gg r, \text{[For DC]}
\end{aligned} \tag{4}
$$

$$
\begin{aligned}
L_{mp} &= \frac{\mu_0}{2\pi} l_{cir} \left[\ln\left(\frac{l_{cir}}{s} + \sqrt{\left(\frac{l_{cir}}{s}\right)^2 + 1} \right) - \sqrt{1 + \left(\frac{s}{l_{cir}}\right)^2} + \frac{s}{l_{cir}} \right], \quad \text{where} \quad l_{cir} \gg s, \\
&\cong \frac{\mu_0}{2\pi} l_{cir} \left[\ln\frac{2l_{cir}}{s} - 1 \right], \quad \text{where} \quad l_{cir} \gg s, \quad \text{[For AC and DC]}
\end{aligned} \tag{5}
$$

Here, l_{cir}, r denote the length and radius of the wire, respectively, and s is the distance between two parallel wires.

E. Inductance in terms of flux linkage

In ANSYS Q3D, the inductance matrix represents the magnetic flux linkage within a group of conductors. Fig. 4 shows the group of conductors with a linkage flux. The relationship between induced flux and currents is given below [14]:

$$\begin{bmatrix} \Psi_1 \\ \Psi_2 \\ \Psi_3 \end{bmatrix} = \begin{bmatrix} L_{11} & L_{12} & L_{13} \\ L_{12} & L_{22} & L_{23} \\ L_{13} & L_{23} & L_{33} \end{bmatrix} \cdot \begin{bmatrix} I_1 \\ I_2 \\ I_3 \end{bmatrix} \quad (6)$$

Here, Ψ is the linkage flux. To validate the analytical calculations with simulation results, two parallel wires with currents in the same direction are shown in Figs. 5a and 5c. In Fig. 6, the per-length inductance calculation and simulation results for two round wires are in a good agreement. The DC inductance is higher than the AC

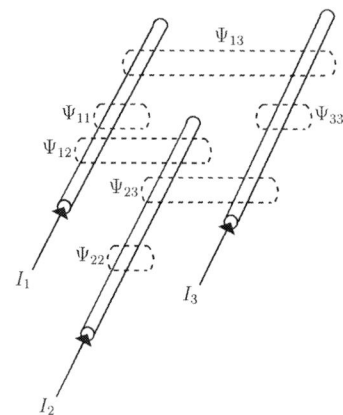

Fig. 4. Magnetic flux linkage within a group of conductors.

inductance, because in the DC case, eddy currents do not occur because the magnetic field created by the current through the conductor is static. In contrast, in the AC case, the oscillating magnetic field induces eddy currents in the conductor, which affect the inductance, which is explained in section II-F.

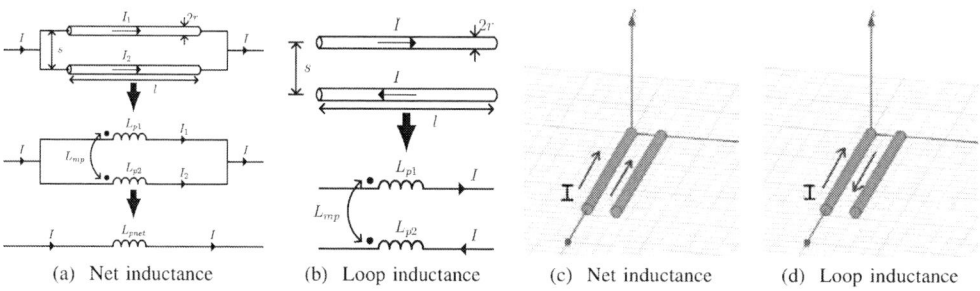

(a) Net inductance   (b) Loop inductance   (c) Net inductance   (d) Loop inductance

Fig. 5. Parasitic inductance and the simulation model of two parallel wires with current in the same and the opposite direction.

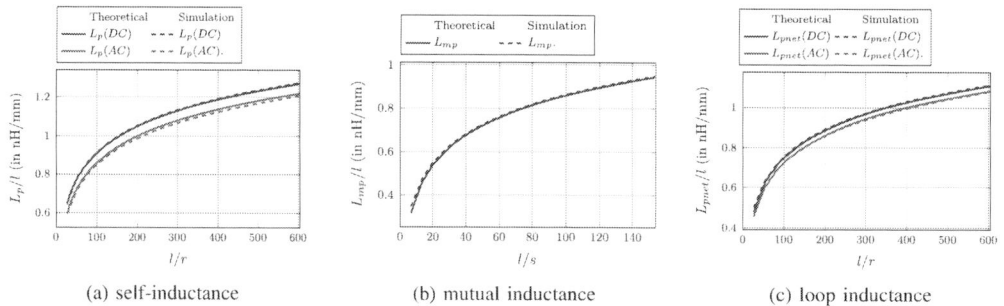

(a) self-inductance   (b) mutual inductance   (c) loop inductance

Fig. 6. Similarity between the simplified analytical and simulated results of the per-length parasitic inductance of two parallel round wires, where, (a) $l = 10, 20 \ldots 230$mm, $r = 0.381$mm (b) $l = 10, 20 \ldots 230$mm, $s = 1.5$mm, (c) $l = 10, 20 \ldots 230$mm, $r = 0.381$mm.

F. AC vs. DC inductance and resistance

ANSYS Q3D extractor can compute inductance and resistance matrices for AC and DC problems. In the software, the AC solvers calculate high-frequency asymptotes of inductance and resistance. The external inductance values do not increase according to the frequency. In contrast, the resistance values are unbounded and increase according to the square root of the frequency, which is shown in Figure 7a. Thus, the software needs the operating frequency to compute the finite resistance. The default operating frequency is $100\,\text{MHz}$. The software can calculate the skin depth according to the following relationship [14]:

$$\delta = \sqrt{\frac{2}{\omega \sigma \mu_0 \mu_r}},\tag{7}$$

where, ω denotes the angular frequency, σ indicates the conductivity of the conductor and μ_0, μ_r are the free space and relative permeability, respectively. Figure 7a illustrates the valid region of the AC inductance and resistance and the DC inductance and resistance in Ansys Q3D. The lower frequency bound for AC inductance and resistance depends on the skin depth of the conductor. The frequency must be calculated so that it produces a skin depth much smaller (at least 3 times smaller) than the thickness d of the conductor. According to the relationship, the frequency limit of the AC region can be written from equation (7) as follows,

$$f_{AC} \geq \frac{9}{\pi \sigma d^2 \mu_0 \mu_r}.\tag{8}$$

Similarly, the upper frequency bound for DC inductance and resistance depends on the skin depth of the conductor. The frequency must be calculated so that it produces a skin depth greater than the thickness of the conductor. According to the relationship, the frequency limit of the DC region can be written from equation (7) as follows,

$$f_{DC} \leq \frac{1}{\pi \sigma d^2 \mu_0 \mu_r}.\tag{9}$$

Here, d is the thickness of the conductor. In the DC region, the inductance and resistance are almost constant over the frequency. In the AC region, the inductance is nearly constant with the frequency. The self-inductance in the AC region is lower than the self-inductance in the DC region, because the skin effect reduces the magnetic field and the corresponding stored magnetic energy inside the conductors [14]. On the other hand, the AC resistance increases proportionally with the square root of the frequency, because the skin depth decreases with the frequency, reducing the effective cross-section for the current flow. For example, Fig. 7b illustrates R and L of one wire conductor with the radius, r=0.05 mm, and the length l=10 mm. It becomes clear that the frequency limit for the DC region is around $2\,\text{MHz}$ and the frequency limit for the AC region is around $50\,\text{MHz}$, as shown in Fig. 7b

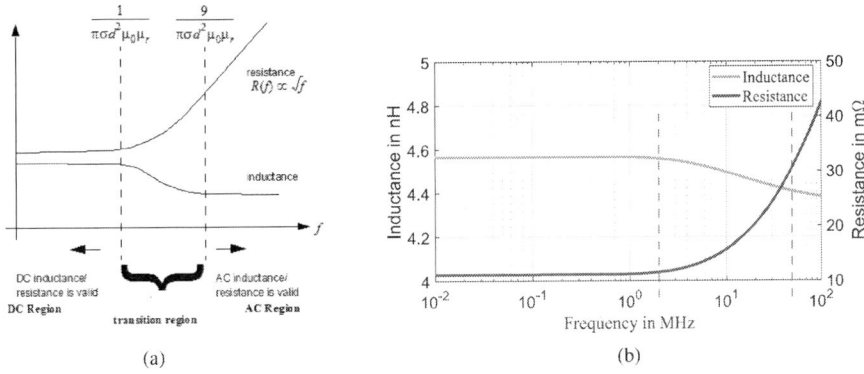

Fig. 7. (a) The valid region of the AC inductance and resistance and the DC inductance and resistance in Ansys Q3D [14]. (b) Inductance and resistance of a single conductor with skin effect.

III. Simulation Model of a Half-bridge SiC-MOSFET Module

In this section, the SiC-MOSFET half-bridge module (commercially available FF23MR12W1M1_B11 from Infineon) is taken to investigate the parasitic parameters and the mutual couplings. Figs. 8a and 8b show the practical module and the simulation model of the SiC-MOSFET half-bridge power module using ANSYS Q3D extractor. Components used in the SiC-MOSFET power module are shown in Table II. The AC parasitic elements such as inductances and capacitances are extracted according to the geometry and material properties of the actual module at $100\,\mathrm{MHz}$.

TABLE I
COMPONENTS USED IN SiC-MOSFET POWER MODULE

Parts	Part number/Thickness	Description
SiC-MOSFET	FF23MR12W1M1_B11	$1200\,\mathrm{V}/50\,\mathrm{A}$
Solder	0.08 mm	Connection between dies, pins and a top layer of DCB
Copper	0.30 mm	Top layer of DCB
Bond wire	0.381 mm (Diameter)	Connection wire between a die and a top layer of DCB
Al$_2$O$_3$	0.38 mm	Middle layer of DCB
Copper	0.301 mm	Bottom layer of DCB
Background material	Silica gel (SiO$_2$)	Surrounding of the power module

A. Extraction of parasitic capacitances

Fig. 8c shows the determination of parasitic capacitances between different conductors and the DCB ground plane. Alumina (Al$_2$O$_3$, $\varepsilon_r = 9.8$) is used as a dielectric material. Silica gel (SiO$_2$, $\varepsilon_r = 2.5$) is used as a surrounding material of the power module. The numerical and analytical results are shown in Tab. II and III. In Tab. II, the diagonal elements are the capacitance to the reference ground, called the self-capacitance. The reference ground is assumed to surround the power module in Q3D. The first column and the first row except the diagonal elements are capacitances to the bottom copper layer of DCB. The other elements are called the mutual parasitic capacitances. The numerical results for DC-to ground are a little than the analytical results because of the complexity of the dielectric material.

(a) Practical module

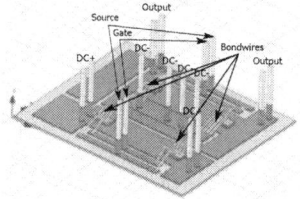
(b) ANSYS-Q3D FEM model

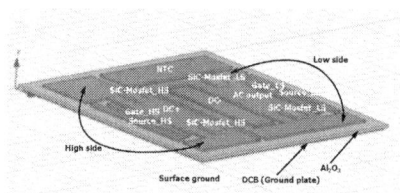
(c) Determination of parasitic capacitances

Fig. 8. Practical module and simulation model of the SiC-MOSFET half-bridge power module.

TABLE II
MUTUAL CAPACITANCES ACCORDING TO EQUATION (2), IN (PF) (NUMERICAL)

Conductor	DCB	AC output	DC+	DC-
DCB	–	71.65	61.89	16.63
AC output	71.65	–	–	–
DC+	61.89	–	–	–
DC-	16.63	–	–	–

TABLE III
ANALYTICAL RESULTS FROM EQUATION (1), IN (PF), WHERE (AL$_2$O$_3$, $\varepsilon_r = 9.8$) AND (SIO$_2$, $\varepsilon_r = 2.5$)

Conductor	DCB	AC output	DC+	DC-
DCB	–	72.52	62.83	17.92
AC output	72.52	–	–	–
DC+	62.83	–	–	–
DC-	17.92	–	–	–

B. Extraction of parasitic inductances of parallel bond wires

Nowadays, bond wires are used for interconnections between the top of the device and the direct-bonded-copper (DCB) substrate of semiconductor power modules. The bond wires inside the half-bridge SiC-MOSFET power module are visible in Figs. 8a and 8b. A number of parallel bond wires is used to increase the current capacity. Fig. 9a shows parameters of a single bond wire. The length of a single bond wires l_1, l_2 and l_3 are considered individually. So, the total self-inductance of a single bond wire can be written by

$$L_{ptotal} = L_{p1} + L_{p2} + L_{p3}. \qquad (10)$$

Here, the self-inductance L_p is calculated by equation (4). The equation (10) is not considering the partial mutual inductance of the wire section. Figure 9b represents two parallel bond wires. The total mutual inductance of two paralleled bond wires can be written by

$$L_{mptotal} = L_{mp1} + L_{mp2} + L_{mp3}, \tag{11}$$

where, the AC mutual inductances L_{mp} are calculated by equation (5). The total partial net inductance can be written from equation (3)

$$L_{pnet} = \frac{L_{ptotal1} L_{ptotal2} - L^2_{mptotal}}{L_{ptotal1} + L_{ptotal2} - 2L_{mptotal}}. \tag{12}$$

Table IV illustrates the validation of the theoretical calculation and the simulation results of two parallel bond wires, according to Figure 9b. Here, it can be seen that numerical results are higher than the analytical results, because the analytical results of the three sections of each bond wires are calculated without considering the partial mutual inductances between them.

TABLE IV

THEORETICAL CALCULATION AND SIMULATION RESULT FOR TWO PARALLELED BONDWIRES AT 100 MHz

self inductance L_p nH/mm		mutual inductance L_{mp} nH/mm		net inductance L_{pnet} nH/mm	
analytical (10)	simulation	analytical (11)	simulation	analytical (12)	simulation
5.34(AC)	5.81(AC)	2.76(AC)	3.43(AC)	4.05(AC)	4.62(AC)
deviation: approx. 10%		deviation: approx. 25%		deviation: approx. 15%	

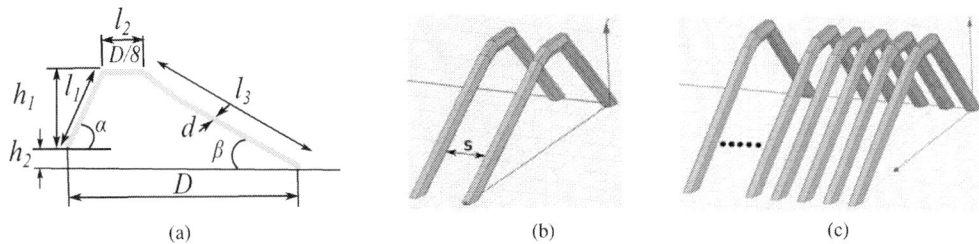

(a)   (b)   (c)

Fig. 9. Parasitic inductance calculation of a single bond wire and the simulation result of the net parasitic inductance of different numbers of bond wires. (a) parameters of a bond wire, where $h_1 = 2.5$ mm, $h_2 = 0$ mm, $D = 8$ mm, $d = 0.381$ mm, $s = 0.7$ mm, $\alpha = 45°$ and $\beta = 45°$. (b) two parallel bond wires. (c) parallel bond wires, where, the number of bond wires $= 1, 2, 3 \ldots n$.

Fig. 9c shows a number of parallel bond wires. The angles of α and β are the same for all bond wires. The net parasitic inductances of different numbers of parallel bond wires for the diameter of 0.381 mm and 0.254 mm are depicted in Fig. 10. The inductances are simulated at 100 MHz. The inductance is reduced significantly around 50% when taking five paralleled bond wires instead of a single bond wire. The reduction of inductance is around 15% less, when increasing the number of parallel bond wires from six to ten. So, the reduction will be less at higher numbers of bond wires. Besides, the diameter of the bond wire is important. A thicker bond wire is required with a high current carrying-capacity and thermomechanical robustness. Moreover, it has a lower inductance than thinner bond wires.

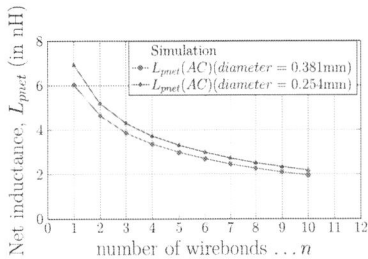

Fig. 10. Net partial inductance of different numbers of bond wires.

C. Extraction of parasitic inductances of the total module

The parasitic inductances of the power module are extracted using the current paths, as shown in Figure 11. These inductances store magnetic energy when current passes through them. Table V presents the parasitic inductance extracted from the power module using ANSYS Q3D.

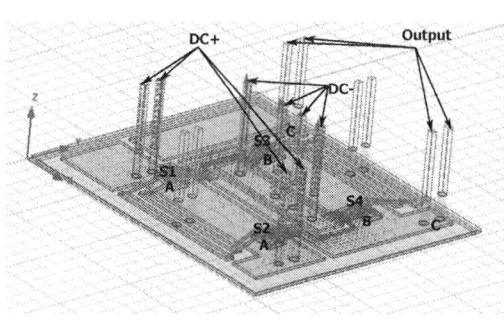 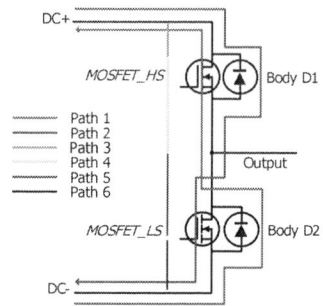

(a) Current density from $DC+$ to $DC-$ (b) Different current paths

Fig. 11. Determination of parasitic inductances using a commutation current path.

TABLE V
PARASITIC INDUCTANCE EXTRACTED FROM THE POWER MODULE (nH) AT $100\,\mathrm{MHz}$

Current path	Parasitic inductance (DC)	Parasitic inductance (AC)
DC+ to DC- (Path-1)	9.28	6.83
DC- to DC- (Path-2)	9.28	6.83
DC+ to A (Path-3)	3.18	2.68
A to C (via S1 and S2) (Path-4)	8.35	5.91
C to B (via S3 and S4) (Path-5)	2.05	1.40
B to DC- (Path-6)	4.47	3.61

Figure 12 represents a total overview of the power loop and net parasitic inductances with considering mutual inductance and the comparison among the analytical, numerical and the data sheet result from Infineon. Figure 12a shows the parasitic inductance for the MOSFET_HS (high side) and the MOSFET_LS (low side). For the MOSFET_HS, two current paths from DC+ to A (red line) and C to A (green line) are indicated. The inductances such as self- and mutual inductances are calculated using Q3D, which is shown in Fig. 12a (arrangement 1). Similarly, for the MOSFET_LS, two current paths from DC- to B (blue line) and C to B (black line) are indicated. Figure 12b depicts the loop parasitic inductance for switching (arrangement 2) and the net parasitic inductance for design purposes (arrangement 3). For switching purposes, two current paths from DC+ to C (red line) and DC- to C (blue line) are indicated. Fig. 12c gives the electric circuit diagrams for the three model arrangements (1, 2, 3) and the values of the self-inductances (including mutual inductances). For design purposes, the current path from DC+ to DC- is indicated. The net inductance is simulated using Q3D. It can be seen that the total calculated net inductance from DC+ to DC- is $L_{net} = 6.9287\,\mathrm{nH}$ from arrangement 2 using equation (3), which is in a good agreement with the numerical result. However, the stray inductance given by data sheet from Infineon [15] is larger than the numerical and analytical result.

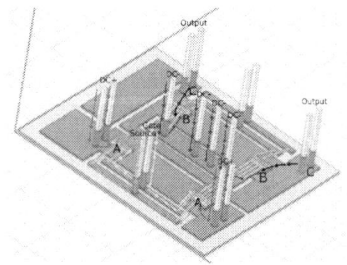 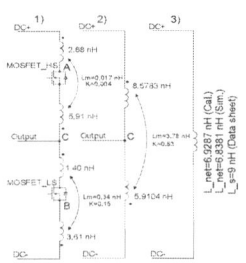

(a) parasitic inductances for the MOS-FET_HS and the MOSFET_LH (arrangement 1)   (b) parasitic inductances for switching (arrangement 2) and design purposes (arrangement 3)   (c) electrical circuit arrangement 1, 2, 3, respectively

Fig. 12. Total overview of the power loop and net parasitic inductances with considering mutual inductance.

Similarly, in Fig. 12a, two current paths from the source pin to A (cyan line) and gate pin to A (yellow line) are indicated. The parasitic inductance such as self- and mutual inductances are calculated using Q3D software. The results are given in Fig. 13. For the same and the opposite direction of the current, the total calculated net and loop inductance are $L_{net} = 7.03\,\text{nH}$ and $L_{loop} = 9.43\,\text{nH}$ from equation (3), which are in a good agreement with the numerical results, shown in Fig. 13b (blue text). At last, Fig. 13a and 13b depict the schematic of the electrical circuit using extracted parasitic elements without and with consideration of mutual inductances of the half bridge SiC-MOSFET, respectively.

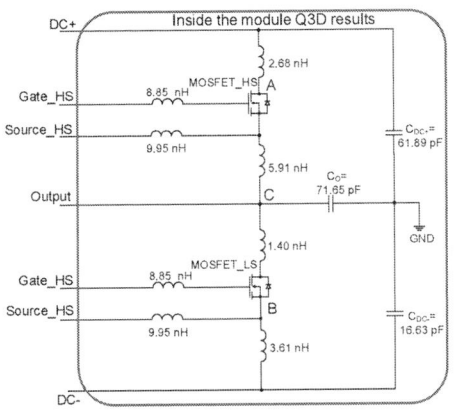 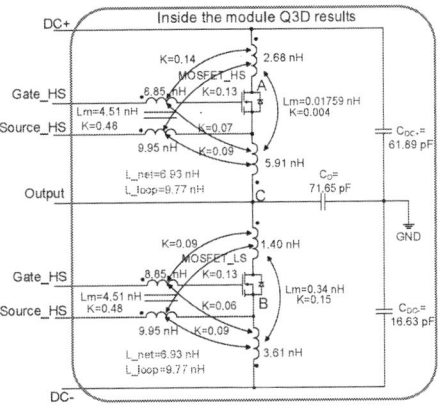

(a) Extracted parasitic elements without mutual inductances   (b) Extracted parasitic elements with mutual inductances

Fig. 13. Schematic representation of the electrical circuit of the half-bridge SiC-MOSFET with the parasitic AC elements.

D. Influence of parasitic elements and mutual couplings on switching characteristics

An ideal equivalent model of the SiC-MOSFET half-bridge module is built by ANSYS Twin Builder simulation software, which is shown in Fig. 14a. This software provides detailed characterization tools for MOSFETs, IGBTs and diodes [16]. The SiC-MOSFET half-bridge module is characterized according to data sheet information [15]. The specification of the half-bridge module is shown in Tab. VI. Comparing the three cases in Figs. 14b and 14c, it can be seen that the presence of parasitic elements and mutual couplings has a large influence on the switching waveforms during turn-on and turn-off transitions. Besides, a notable difference is observed in current and voltage overshoots.

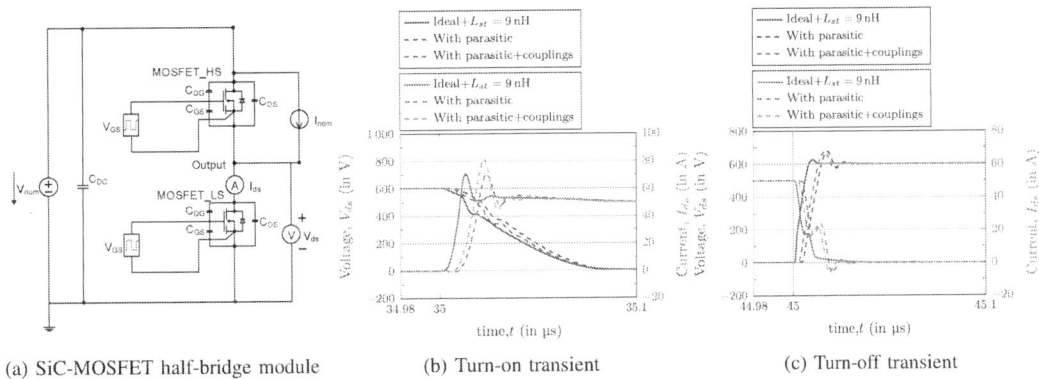

(a) SiC-MOSFET half-bridge module   (b) Turn-on transient   (c) Turn-off transient

Fig. 14. (a) An ideal equivalent model of the half-bridge SiC-MOSFET using ANSYS Twin Builder software, where the total stray inductance of the module is $L_{st} = 9\,\text{nH}$. (b, c) influence of internal parasitic elements and couplings on the switching characteristics of a low-side SiC-MOSFET half-bridge module.

TABLE VI
SPECIFICATION OF THE SiC-MOSFET HALF-BRIDGE MODULE

Drain-to-source voltage, $V_{nom} = 600\,\text{V}$	Drain current, $I_{nom} = 50\,\text{A}$
Gate-voltage-on, $V_{gon} = 15\,\text{V}$	Gate-voltage-off, $V_{goff} = -5\,\text{V}$
Gate-resistance-on, $R_{gon} = 1\,\Omega$	Gate-resistance-off, $R_{goff} = 1\,\Omega$
Switching frequency, $f = 50\,\text{kHz}$	DC-bus capacitor, $C_{DC} = 5\,\mu\text{H}$

IV. Conclusion

The parasitic capacitances of a SiC-MOSFET module, as well as the parasitic inductances and their mutual couplings, have been addressed in this paper. While those capacitances that are interesting for the circuit simulation (see Figs. 13a and 13b) can be calculated analytically with an accuracy in the range of a few percentage, the inductances can be better addressed using numerical results (for partial stray- and mutual inductances) and an analytic equation to combine them. This way, the common-mode inductances (current in the same direction) and loop inductances (currents in opposite direction) can be extracted using the same set of FEM results. The circuit simulation shows that the mutual couplings have an effect on the current and voltage overshoot and the switching times. It is expected that they will also make a difference in EMI simulation. This will be discussed in more detail in further research.

REFERENCES

[1] S. Li, L. M. Tolbert, F. Wang, and F. Peng, "Reduction of stray inductance in power electronic modules using basic switching cells," *in Energy Conversion Congress and Exposition (ECCE)*, 2010 IEEE, 2010, pp. 2686-2691.

[2] F. Yang, Z. Liang, Z. Wang, and F. Wang, "Parasitic inductance extraction and verification for 3D Planar Bond All Module," *in 2016 International Symposium on 3D Power Electronics Integration and Manufacturing (3D-PEIM)*, 2016, pp. 1-11.

[3] Di Han and Bulent Sarlioglu, "Study of the switching performance and EMI signature of SiC MOSFETs under the influence of parasitic inductance in an automotive DC-DC converter," *in IEEE Transportation Electrification Conference and Expo (ITEC)*, 2015.

[4] Di Han and Bulent Sarlioglu, "Comprehensive Study of the Performance of SiC MOSFETs Based Automotive DC-DC Converter under the Influence of Parasitic Inductance," *in IEEE Transac- tions on Industry Applications*, vol. 52, 2016.

[5] J. Z. Chen, L. Yang, D. Boroyevich and W. G. Odendaal, "Modeling and Measurements of Parasitic Parameters for Integrated Power Electronics Modules," *in Nineteenth Annual IEEE Applied Power Electronics Conference and Exposition*, vol.1, 2004.

[6] Liyu Yang,and Willem G. Hardus Odendaal, "Measurement-Based Method to Characterize Parasitic Parameters of the Integrated Power Electronics Modules," *in IEEE Transactions on Power Electronics*, vol.22, 2007.

[7] Andre Domurat-Linde, Eckart Hoene, "Investigation and PEEC Based Simulation of Radiated Emissions Produced by Power Electronic Converter," *in 6th International Conference on Integrated Power Electronics Systems (CIPS)*, March, 16-18, 2010,Nuremberg/Germany.

[8] Andre Domurat-Linde, Eckart Hoene, "Analysis and Reduction of Radiated EMI of Power Modules basic switching cells," *in 7th International Conference on Integrated Power Electronics Systems (CIPS)*, March, 16-18, 2010,Nuremberg/Germany.

[9] G. W. Parker, "Electric field outside a parallel plate capacitor," *in American Journal of Physics*, 2002.

[10] M. C. Hegg and A. V. Mamishev, "Influence of variable plate separation on fringing electric fields in parallel-plate capacitors," *in Conference Record of the 2004 IEEE International Symposium on Electrical Insulation*, 2004.

[11] Mehran Hosseini, Guchuan Zhu, and Yves Alain Peter, "A new model of fringing capacitance and its application to the control of parallel-plate electrostatic micro actuators," *in DTIP of MEMS &MOEMS*, Stresa, Italy, 26-28 April 2006.

[12] Paul CR (ed.), "Inductance Loop and Partial," Hoboken, NJ, USA: John Wiley & Sons, Inc; 2009. 2008.

[13] Ruehli AE, "Inductance Calculations in a Complex Integrated Circuit Environment," *IBM Journal of Research and Development*, 1972;16(5):470–81.

[14] "ANSYS Inc. Q3D Extractor Online Help," *http://www.ansys.com 2017*, Release 18.2.

[15] "Dual 1200 V, 23 m Ω half-bridge module with module with Cool SiC-MOSFET," *https://www.infineon.com/cms/en/product/power/mosfet/silicon-carbide/modules/ff23mr12w1m1_b11/*, 2018.

[16] "ANSYS Inc. Twin Builder Manual," *http://www.ansys.com 2019*, Release 18.2.

Variable-speed-drive-based sensorless estimation of pump system reservoir fluid level

Santeri Pöyhönen, Aleksi Simola, and Jero Ahola
LUT UNIVERSITY
School of Energy Systems
P.O. Box 20, FI-53851
Lappeenranta, Finland
Tel.: +358 44 06 089 12
E-mail: santeri.poyhonen@lut.fi, aleksi.simola@lut.fi, jero.ahola@lut.fi
URL: http://www.lut.fi

Keywords

«Adjustable-speed drive», «Estimation technique», «Intelligent drive», «Variable-speed drive»

Abstracts

This paper proposes a soft sensor for estimating the surface level of a reservoir in a centrifugal pumping system. The proposed soft sensor can be implemented in the control scheme of a variable-speed drive and requires no extra equipment to function.

Introduction

Pumps are a significant end-user of electricity, accounting for approximately 8 % of all the electric energy consumed globally [1]. Increasing legislative pressure towards energy efficiency [2] and discovered energy-saving potential [3] have led to the wide implementation of variable-speed drives (VSD) in the capacity control of pumps. With a VSD, the energy loss associated with throttling the flow with a valve can be avoided by instead adjusting the pump's rotational speed to meet varying process demands. In addition, some pump systems with less strictly defined demand for the pump's output may encompass degrees of freedom and control-wise flexibility that allow the use of variable-speed control strategies, which aim to minimize the overall energy consumption of the process [4]–[6].

The benefits of VSDs in pump systems are not limited to efficient control. They provide accurate estimates of the shaft power, rotational speed, and shaft torque of the electric motor running the pump [7]. These estimates can be used as inputs to computational models of the process beyond the motor shaft, which enables the development of methods for monitoring the pumping process from the points of view of energy efficiency and maintenance. Moreover, the computational capacity and programmability of a modern VSD makes it possible to execute the methods directly within the control scheme of the drive, making it an intelligent process diagnostics tool, which can simultaneously log data, carry out analyses, and generate signals of interest [8]. Previous research has yielded, for instance, VSD-based methods for the sensorless estimation of the centrifugal pump operating point [9]–[10], detection of cavitation [11], and detection of non-return valve failure [12]. Some studies have also explored the possibility of process variable estimation through algorithm-based control sequences embedded in the control scheme of the drive: in [13], an identification (ID) run sequence was proposed, with which the process characteristics of a pump system could be determined. VSD-based soft sensing and condition monitoring methods have also been developed for fan and compressor systems [14]–[16], and some have been implemented as added functionality in commercial VSDs [17]–[18].

The delivery of fluid to and from reservoirs and tanks (hereinafter, simply referred to as *reservoirs*) is a common process task handled by pumps. In many such applications, measurement or detection of the fluid surface level in the reservoir is required. Depending on the process, the level is either continuously measured with a continuous level sensor, or a level switch is used to detect when the level reaches certain points of elevation. The level can be measured and controlled to meet the demands of the rest of the process, or simply to prevent reservoir overflow. In applications where the pump draws water from a reservoir, level sensing helps to prevent the complete emptying of the reservoir, which could lead to the pump running dry. Measuring the level can also help pump cavitation avoidance: the suction-side reservoir level can be kept above a threshold, below which cavitation-inducing low pump suction pressure would occur.

This paper presents a VSD-based soft sensor for estimating the surface level of a reservoir to or from which a centrifugal pump delivers fluid. The proposed method requires no extra equipment or instrumentation to function and can be implemented within the control scheme of a modern VSD, making it virtually cost free. Systems with level switches can benefit from the soft sensor, since it enables estimation of the fluid level between the measuring points of the switch. In systems with level sensors, the proposed method improves the reliability of the level measurement by providing a redundant estimate. In the following section, the theory of the proposed soft sensor is explained. In the subsequent section, the results of the laboratory tests for verifying the feasibility of the soft sensor are presented.

Sensorless estimation of reservoir surface level

The method presented in this paper enables continuous sensorless estimation of the fluid level of a reservoir in a pump system. It is based on sensorless pump operating point estimation with a VSD and on an identification (ID) run sequence for determining the dynamic flow resistance of a pumping system. As such, the method only estimates the level when the pump is running and producing flow. Furthermore, the suitability of the method is limited to pump systems, in which the dynamic flow resistance of the system does not change over time. The method can be implemented directly in the control scheme of a programmable VSD. In the following, the principle of and more detailed restrictions and requirements related to the presented fluid level soft sensor are first discussed. Subsequently, the identification run sequence included in the presented method is introduced.

The operating point of a pump can be determined based on the pump's characteristic performance curves and a VSD's estimates for the pump's shaft power and rotational speed. The pump-specific flow rate vs. shaft power (QP) and flow rate vs. head (QH) curves at a certain rotational speed n_0 are available from the pump manufacturer. To estimate the operating point based on the curves with a pump that is run at varying rotational speeds, they must be converted into the instantaneous speed using the following pump affinity laws:

$$Q = \frac{n}{n_0} Q_0 \tag{1}$$

$$H = \left(\frac{n}{n_0}\right)^2 H_0 \tag{2}$$

$$P = \left(\frac{n}{n_0}\right)^3 P_0. \tag{3}$$

When the curves at the present speed have been defined, an estimate for the flow rate can be interpolated from the QP curve based on the pump shaft power. Subsequently, using the flow rate estimate, an estimate for the pump head can be interpolated from the QH curve. The VSD-based operating point estimation is illustrated in Fig. 1. Centrifugal pumps, in particular, are suitable for this kind estimation due to the high $|dP/dQ|$ of their QP curves and due to the curves' typical monotonic shape – an estimate for shaft power will lead to only one possible estimate for flow rate.

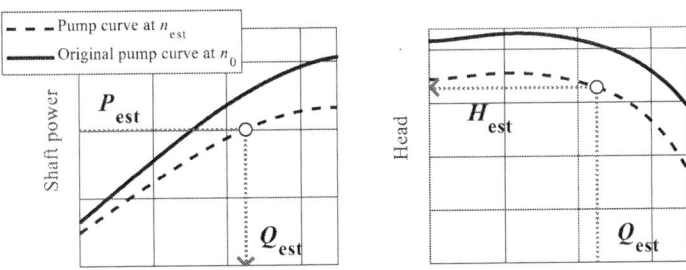

Flow rate

Fig. 1. Pump operating point estimation based on the pump's characteristic performance curves and a VSD's estimates for the shaft power and rotational speed of the pump's motor.

The total head produced by the pump consists of a static component H_{st} and a dynamic component H_{dyn}:

$$H = H_{st} + H_{dyn} = H_{st} + kQ^2, \tag{4}$$

where k is the dynamic flow resistance factor of the system the pump is connected to.

The operating point of a pump lies in the intersection of the QH curves of the pump and the system as shown in Fig. 2.

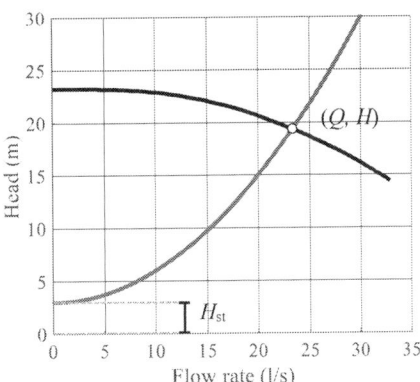

Fig. 2. Pump operating point at the intersection of the curves of the pump and the surrounding system.

With (4) and an estimate of the pump's operating point, the static head H_{st} can be determined with the equation:

$$H_{st} = H - kQ^2. \tag{5}$$

The method presented in this paper can be applied in pump systems, where a reservoir is connected to either the discharge or suction side of the pump in such a way that the surface level in the reservoir affects the static pressure difference over the pump. In such systems, the value of H_{st} depends on the surface level of the fluid in the reservoir, the elevation of the pumping system components, and the atmospheric pressure differential between the suction and discharge sides. If the atmospheric differential can be assumed to be negligible, i.e. the atmospheric pressure felt on the surface of the fluid in the reservoir can be considered equal to the atmospheric pressure where the fluid leaves the system on the other side of the pump, H_{st} depends on the fluid level as illustrated in Fig. 3. Thus, when the elevation from the bottom of the reservoir to the height to which the pump delivers fluid (Z) is known, the height of the fluid column (h) in the reservoir can be calculated with the equation:

$$h = Z - H_{st} = Z - H + kQ^2. \tag{6}$$

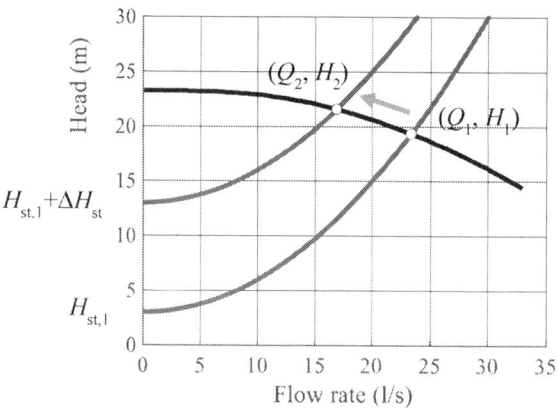

Fig. 3. Knowledge of the pump static head and the elevation from the bottom of the reservoir to the height to or from which the fluid is pumped can be used to determine the surface level in the reservoir.

To estimate h according to (6), the value of k must be known. It can be determined with an ID run sequence, in which the pump is run at a constant speed, and the operating point shifts due to the change in the reservoir's surface level, as illustrated in Fig. 4.

Fig. 4. System curves and the corresponding pump operating points at the start and end of the ID run.

For the pump head at the start and end points of the ID run (denoted 1 and 2, respectively) and the change in static head, the following equations apply:

$$H_1 = H_{st,1} + kQ_1^2 \tag{7}$$

$$H_2 = H_{st,1} + \Delta H_{st} + kQ_2^2 \tag{8}$$

$$\Delta H_{st} = \frac{\int_1^2 Q(t)\,dt}{A}, \tag{9}$$

where A is the cross-sectional area of the reservoir. Similarly to how the elevation parameter Z is determined, the value of A can be defined by inspecting and measuring the dimensions of the reservoir. Given that the value of k does not change during the ID run, it can be calculated by combining (7)–(9):

$$k = \frac{H_2 - H_1 - \frac{\int_1^2 Q(t)\,dt}{A}}{Q_2^2 - Q_1^2}. \tag{10}$$

Once k is defined, (6) can be constantly used to estimate the surface level in the reservoir when the pump is running.

Experimental evaluation of the proposed soft sensor

The feasibility of the presented estimator was evaluated with laboratory measurements. The method's ID run was conducted in a pump system with a virtual water reservoir. The static head caused by the water level was emulated with a flow-throttling valve, as proposed in [19].

Test setup

The laboratory pump system includes a Sulzer A22-80 centrifugal pump driven with an 11-kW induction motor and an ABB ACS880 VSD employing direct-torque control (DTC). The pump is connected to a closed loop system comprising piping, a water tank, and a control valve. No static head exists in the system. The test setup is illustrated in Fig. 5.

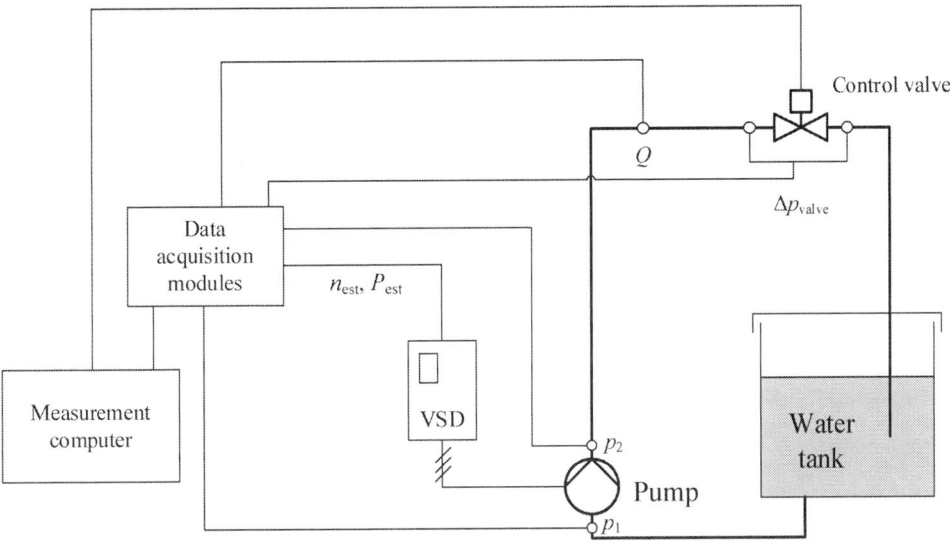

Fig. 5. The laboratory setup.

The flow rate (Q in Fig. 5) of the pump is measured with an ABB FEP321 electromagnetic flow meter, and pressure transducers are used to measure the pressure before and after the pump (p_1 and p_2) for the determination of the pump head as well as the pressure drop over the control valve (Δp_{valve}). The measuring instruments are connected to an analog data acquisition module, from which the data are read and recorded with a computer running a LabVIEW-based measurement program. Measurement data is logged by reading 250 samples per second and recording the mean of those values each second.

A virtual water tank was implemented in the measurement program. In the implementation, one can set the cross-sectional area of the virtual tank and the initial water level to desired values. The change in water level in the tank is then determined according to the tank's cross-sectional area and the volume of the water pumped into the tank, as defined in (9). The control valve is operated with a PI controller within the measurement program to maintain a pressure difference across the valve equal to the back pressure that would be caused by the virtual water level. With this setup, the ID run presented in this paper can be conducted.

Test results

First, the actual flow resistance factor k of the laboratory pump system was determined by recording the pump operating point at different speeds while maintaining a constant artificial static head with the control valve, and subsequently fitting the polynomial equation (4) on the operating points with the least-squares method. The value of k was determined at different emulated static heads to verify the feasibility of the method to create an artificial static head; correct operation of the method would require that the values of k at different emulated static heads take on similar values. Fig. 6 presents the measured system curves at the emulated static heads of 2 m, 5 m, 8 m, 11 m, and 14 m. The resulting small discrepancy between the values of k with different artificial static heads suggests that the static head emulation method can be used in the verification test for the ID run of the method presented in this paper. The mean value of these five measurements, $k_{meas} = 5.38\times10^{-3}$ m/(l/s)^2, will be used as the reference which the k values estimated by the presented method will be compared with.

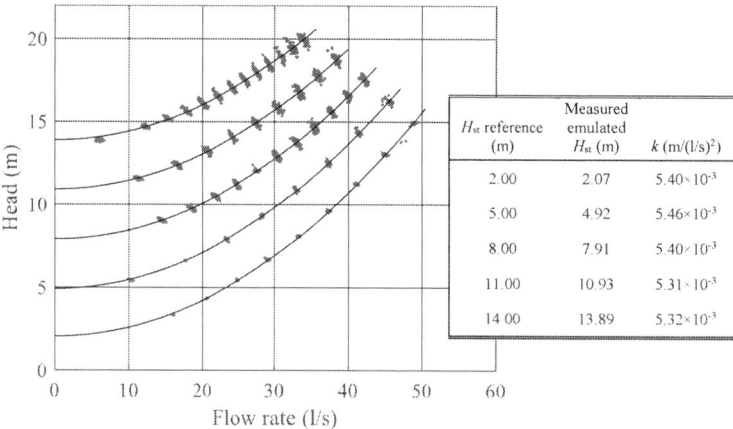

H_{st} reference (m)	Measured emulated H_{st} (m)	k (m/(l/s)^2)
2.00	2.07	5.40×10^{-3}
5.00	4.92	5.46×10^{-3}
8.00	7.91	5.40×10^{-3}
11.00	10.93	5.31×10^{-3}
14.00	13.89	5.32×10^{-3}

Fig. 6. System curves with five different emulated static heads.

The ID run of the presented method was conducted at constant rotational speeds of 1200 rpm and 1400 rpm. The cross-sectional area of the virtual water tank was set to 5 m^2, and the tank was filled from an initial virtual water level of 5 m to a level of 13 m. Then, (10) was used to estimate the value of k in the two following ways. First, to demonstrate the feasibility of the method in principle, k was determined using the measured pump operating points. Second, to study the practical sensorless use of the method, the VSD-based sensorless operating point estimation method was used to provide the operating point data to (10).

Fluctuation was observed in the position control of the control valve over the course of the filling of the virtual tank, which led to variation in measured variables that depend on the valve position: flow rate, head, and shaft power. Since the method involves using instantaneous values in (10), the fluctuation can cause uncertainty in the estimation of k. To reduce the error, a median filter was applied to the shaft power estimate and the flow rate and head measurements. The filter replaces each data point with the median of the signal within 60 seconds before and 60 seconds after the data point in question. The fluctuation and the filtered signal are shown in Fig. 7.

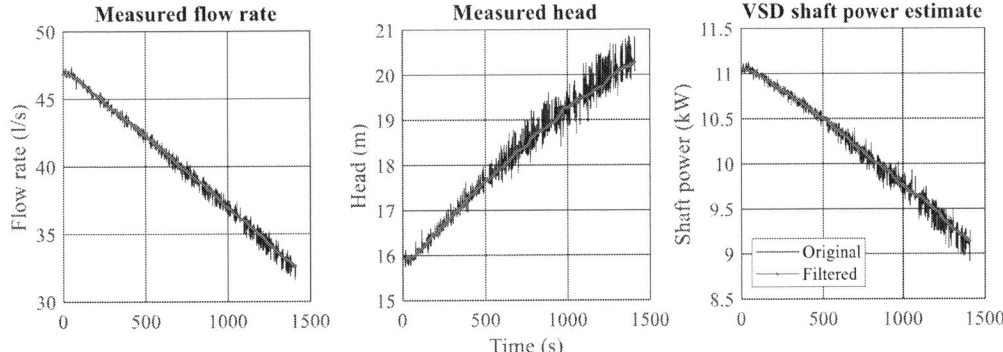

Fig. 7. Filtered and original signals.

To evaluate the performance of the method at different pump capacities, the calculation of k was done for multiple water level ranges within the executed 5–to–13 m fill-up. Table 1 presents the error of the estimates for k with respect to its actual measured value earlier, $k_{meas} = 5.38{\times}10^{-3}$ m/(l/s)^2. The error is calculated with

$$E = \frac{k_{est} - k_{meas}}{k_{meas}}\ 100\ \%, \tag{11}$$

where E is the relative estimation error and k_{est} is the estimated flow resistance factor. Results are shown for when the measured pump operating points are used in the calculation, and when the estimated operating point were used. Using the measured operating points, the estimate was within 13.6 % of the actual value, whereas the error using estimated operating points was at its largest -50.0 %. The latter significant error can most likely be attributed to mismatch between the performance curves provided by the manufacturer and the actual performance of the pump and uncertainty in the shaft power estimate of the VSD. Together they lead to erroneous pump operating point estimation. However, the results imply that given accurate knowledge of the pump operating point, the method presented in this paper can provide a reasonably accurate estimate of the pump system's flow resistance factor k.

Table 1. Estimation error of the flow resistance factor k. The largest absolute error value of each ID run is written in **bold**.

Estimation error (%) using measured operating points

1200 rpm		$H_{st,1}$ (m)						
		5	6	7	8	9	10	11
$H_{st,2}$ (m)	13	4.8	5.6	4.1	2.7	2.1	-1.2	-1.9
	12	4.3	5.2	3.3	1.4	0.3	-5.4	
	11	7.1	8.8	7.1	5.7	6.2		
	10	8.6	11.1	9.6	8.6			
	9	7.7	10.6	8.1				
	8	8.7	**13.6**					
	7	7.2						

1400 rpm		$H_{st,1}$ (m)						
		5	6	7	8	9	10	11
$H_{st,2}$ (m)	13	6.5	6.5	5.5	8.2	5.0	6.7	5.7
	12	3.5	3.2	1.3	3.7	-2.0	-2.9	
	11	6.7	6.9	5.4	10.0	4.2		
	10	6.3	6.4	4.2	**10.5**			
	9	8.0	8.8	6.5				
	8	3.4	2.2					
	7	9.6						

Estimation error (%) using operating points estimated based on manufacturer's pump curves

1200 rpm		$H_{st,1}$ (m)						
		5	6	7	8	9	10	11
$H_{st,2}$ (m)	13	-24.2	-20.5	-18.1	-15.2	-11.2	-4.5	-1.9
	12	-27.7	-24.0	-22.0	-19.5	-15.7	-8.2	
	11	-30.8	-27.2	-25.5	-23.4	-19.9		
	10	-34.5	-31.1	-30.2	-29.6			
	9	-35.8	-31.8	-30.8				
	8	-37.6	-32.6					
	7	**-40.5**						

1400 rpm		$H_{st,1}$ (m)						
		5	6	7	8	9	10	11
$H_{st,2}$ (m)	13	-36.2	-32.3	-31.2	-26.6	-23.8	-16.7	-14.3
	12	-41.7	-38.4	-38.4	-34.9	-34.1	-30.0	
	11	-42.4	-38.6	-38.6	-33.9	-32.2		
	10	-46.0	-42.3	-43.5	-39.5			
	9	-47.0	-42.5	-44.4				
	8	**-50.0**	-45.1					
	7	-49.5						

To analyze the practical accuracy of the presented method without the effect of the uncertainties in the manufacturer's pump curves and the VSD shaft power estimate, the pump's QP and QH curves were determined at 1200 rpm using the head and flow rate measurements and the VSD's shaft power estimate. The measured pump curves were used as basis for the operating point estimation and k was calculated again with (10). Table 2 presents the estimation error when the pump curves' and shaft power estimate's uncertainties have been alleviated in this manner. The estimation error was at most 12.3 %, and in most of the tested ID run water level ranges, within ±5 %.

Table 2. Estimation error of k using operating points estimated based on measured pump curves. The largest absolute error value of each ID run is written in **bold**.

1200 rpm		$H_{st,1}$ (m)						
		5	6	7	8	9	10	11
$H_{st,2}$ (m)	13	2.3	2.4	1.9	2.5	4.8	10.5	**12.3**
	12	0.2	0.1	-1.0	-0.9	0.9	7.3	
	11	-1.0	-1.4	-3.1	-3.7	-2.4		
	10	-2.5	-3.4	-6.2	-8.5			
	9	-0.3	-0.8	-3.8				
	8	1.8	2.0					
	7	3.3						

1400 rpm		$H_{st,1}$ (m)						
		5	6	7	8	9	10	11
$H_{st,2}$ (m)	13	1.7	1.7	-0.5	1.6	2.1	8.0	8.4
	12	-3.3	-4.1	-7.6	-7.0	**-9.1**	-6.7	
	11	-0.5	-0.9	-4.7	-2.7	-3.9		
	10	-2.0	-2.8	-8.3	-7.2			
	9	1.3	1.2	-5.4				
	8	1.8	2.0					
	7	8.5						

A comparison of the estimated and the actual emulated water levels during the 1200 rpm and 1400 rpm ID runs is shown in Fig. 8. Here, the k estimate is based on the 1200-rpm ID run from 5 m to 13 m. Two estimates of k are calculated: based on the re-measured pump curves and based on the manufacturer's curves. In this example, H_{st} is equal to the water level, and thus (5) is used to calculate the level instead of (6). To demonstrate the sensorless capability of the presented method, the VSD-based operating point estimates are used in place of Q and H in (5). Throughout the runs, the estimate remained within ± 7 % of the actual value when accurate pump curves were used.

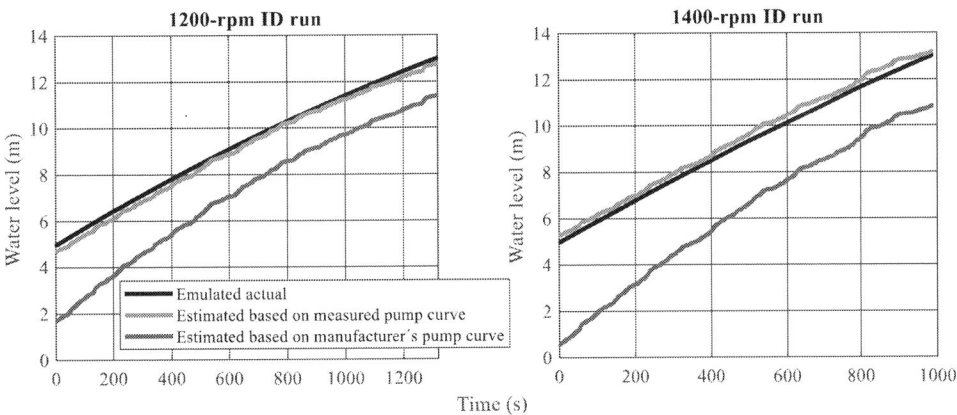

Fig. 8. Emulated and estimated water levels during the ID runs.

Conclusion

This paper presents a variable-speed-drive-based soft sensor for the surface level of a water reservoir that is filled or emptied with a centrifugal pump. The soft sensor employs the characteristic performance curves of a pump as parameters and the shaft power and rotational speed signals provided by a VSD as inputs and requires no extra instrumentation to function. The method includes an ID run to determine the flow resistance factor of the piping surrounding the pump. After the ID run, provided that the flow resistance factor does not change over time, and that the pump is being run, the static head of the system can be continuously estimated. Considering the physical elevation of the reservoirs and the destination of the pumped fluid, the water level can be calculated from the static head of the pump with a simple subtraction.

The accuracy of the presented soft sensor was tested with laboratory measurements. The suggested ID run was conducted in a laboratory pump system where the static head caused by the changing water level in a reservoir was emulated with a flow-throttling control valve. From the results, it is evident that the accuracy of the presented soft sensor depends highly on the accuracy of the pump operating point estimate, which in turn depends on the accuracies of the pump performance curves and the VSD's estimate for shaft power. In test runs with accurate pump curves and shaft power estimation, the error of the presented soft sensor was within ± 7 % of the actual value.

As such, the presented soft sensor can improve the reliability of existing surface level measurements by providing a redundant estimate. In the absence of pre-existing sensors, when for instance level switches are used, it enables the estimation of the level between the switches. Suggested future studies include further laboratory testing with an actual reservoir and evaluating the practical applicability of the soft sensor in a real industrial reservoir pumping application.

References

[1] P. Waide and C. Brunner, "Energy-Efficiency Policy Opportunities for Electric Motor-Driven Systems," International Energy Agency, 2011.

[2] A. de Almeida, J. Fong and H. Falkner, "New European Ecodesign Regulation for Electric Motors and Drives," in *9th International Conference on Energy Efficiency in Motor Driven Systems - EEMODS 2015*, Helsinki, 2015.

[3] T. Fleiter and W. Eichhammer, "Energy efficiency in electric motor systems: Technology, saving potentials and policy options for developing countries," United Nations Industrial Development Organization, Vienna, 2012.

[4] J. Viholainen, J. Tamminen, T. Ahonen, J. Ahola, E. Vakkilainen and R. Soukka, "Energy-efficient control strategy for variable speed-driven parallel pumping systems," *Energy Efficiency*, vol. 6, no. 3, pp. 495-509, 2013.

[5] M. Lindstedt and R. Karvinen, "Optimal control of pump rotational speed in filling and emptying a reservoir: minimum energy consumption with fixed time," *Energy Efficiency*, vol. 9, no. 6, pp. 1461-1474, 2016.

[6] T. Ahonen, S. Pöyhönen, J. Siimesjärvi and J. Tolvanen, "Graphic determination of available energy-saving potential in a reservoir pumping application with variable-speed operation," *Energy Efficiency*, vol. 12, no. 5, pp. 1041-1051, 2019.

[7] T. Ahonen, J. Tamminen, J. Ahola and M. Niemelä, "Accuracy study of frequency converter estimates used in the sensorless diagnostics of induction-motor-driven systems," in *Proceedings of the 14th European Conference on Power Electronics and Applications*, Birmingham, 2011.

[8] T. Ahonen, S. Pöyhönen, J. Ahola and J. Siimesjärvi, "Remote monitoring of fluid handling systems with variable-speed drive," in *19th European Conference on Power Electronics and Applications (EPE'17 ECCE Europe)*, Warsaw, 2017.

[9] J. Tamminen, J. Viholainen, T. Ahonen, J. Ahola, S. Hammo and E. Vakkilainen, "Comparison of model-based flow rate estimation methods in frequency-converter-driven pumps and fans," *Energy Efficiency*, vol. 7, no. 3, pp. 493-505, 2014.

[10] S. Pöyhönen, T. Ahonen, J. Ahola, P. Punnonen, S. Hammo and L. Nygren, "Specific speed-based pump flow rate estimator for large-scale and long-term energy efficiency auditing," *Energy Efficiency*, vol. 12, no. 5, pp. 1279-1291, 2019.

[11] T. Ahonen, J. A. J. Tamminen and J. Kestilä, "Novel method for detecting cavitation in centrifugal pump with frequency converter," *Insight - Non-Destructive Testing and Condition Monitoring*, vol. 53, no. 8, pp. 439-449, 2011.

[12] T. Ahonen, J. Tamminen, J. Ahola and J. Saukko, "Monitoring of non-return valve operation with a variable-speed drive," in *17th European Conference on Power Electronics and Applications (EPE'15 ECCE-Europe)*, Geneva, 2015.

[13] T. Ahonen, J. Tamminen, J. Ahola, L. Niinimäki and J. Tolvanen, " Sensorless estimation of the pumping process characteristics by a frequency converter," in *15th European Conference on Power Electronics and Applications (EPE)*, Lille, France, 2013.

[14] S. Pöyhönen, P. Punnonen, S. Hammo, M. Niemelä and J. Ahola, "Variable-Speed-Drive-Based Estimation of the Pressure Drop Caused by Filter Fouling in Fan Systems," in *European Conference On Power Electronics And Applications*, Riga, Latvia, 2018.

[15] S. Pöyhönen, T. Ahonen and J. Saukko, "Variable-speed-drive-based soft sensor for the twin-rotary screw-compressor output pressure," in *European Conference On Power Electronics And Applications*, Warsaw, Poland, 2017.

[16] S. Pöyhönen, J. Ahola, T. Ahonen, S. Hammo and M. Niemelä, "Variable-Speed-Drive-Based Estimation of the Leakage Rate in Compressed Air Systems," *IEEE Transactions on Industrial Electronics*, vol. 65, no. 11, pp. 8906-8914, 2018.

[17] ITT Corporation, "ITT Pro services," 2014. [Online]. Available: https://www.ittproservices.com/ittgp/medialibrary/ITTPROServices/website/Literature/Brochures/PRO%20Services/PSmartbulletin.pdf?ext=.pdf. [Accessed 22 December 2016].

[18] IEN Europe, "Variable Speed Drives for Sensorless Detection of Pump Cavitation," 15 June 2017. [Online]. Available: https://www.ien.eu/article/variable-speed-drives-for-sensorless-detection-of-pump-cavitation/. [Accessed 13 12 2019].

[19] A. Simola, S. Pöyhönen and J. Ahola, "Emulating pump system static head using PID-controlled flow-regulating valve," in *European Conference on Power Electronics and Applications*, Genova, 2019.

Analysis of Switching Performance and EMI Emission of SiC Inverters under the Influence of Parasitic Elements and Mutual Couplings of the Power Modules

Mohammad Ali, Jan-Kaspar Müller, Jens Friebe, Axel Mertens
LEIBNIZ UNIVERSITY HANNOVER
Institute for Drive Systems and Power Electronics
Welfengarten 1
30167 Hannover, Germany
Tel: +49 / (0) - 511 762 3778
Fax: +49 / (0) - 511 762 3040
E-mail: mohammad.ali@ial.uni-hannover.de
URL: www.ial.uni-hannover.de

Acknowledgments

This work was supported by the German Ministry of Economics and Technology - 19236 N (FVA).

Keywords

≪Inverter≫, ≪EMI≫, ≪Parasitic≫ ≪Inductance≫, ≪Mutual couplings≫, ≪Parasitic capacitance≫, ≪Silicon Carbide (SiC)≫.

Abstract

Parasitic elements and mutual couplings of SiC half-bridge modules strongly affect the switching characteristics of devices. They excite overshoots and oscillations that further contribute to increased EMI emissions. This paper will explain and analyze such effects on the switching performance and EMI emissions, based on 3D FEM models of the module. It can be said that knowledge about the effects of parasitic elements and mutual couplings on the switching behavior is an important basis for the design guidelines of fast switching wide-bandgap (WBG) power converters. A three-phase DC-AC inverter prototype with three SiC half-bridge MOSFET modules and an EMI measurement test setup are constructed for the experiments. The experiments results are validated with simulation results.

I. Introduction

In recent years, many power semiconductor manufacturers have expanded their portfolio with silicon carbide (SiC) wide bandgap (WBG) power semiconductors. In particular, SiC components in the form of MOSFETs are now also available on the market for higher power applications in the three-digit kilowatt range and represent a competitive technology to conventional silicon (Si) - power semiconductors (IGBT, MOSFET). WBG power semiconductors allow a much faster switching compared to conventional Si semiconductors, which significantly reduces the switching losses [1]–[5]. As a result, the efficiency increases and higher switching frequencies are enabled, whereby the size of the passive components in the converter system can be significantly reduced [6]. These properties make the technology particularly interesting for systems with high power density requirements, such as in electro mobility applications.

On the other hand, the mentioned advantages are also faced with disadvantages caused by the fast switching of WBG semiconductors: The edge steepness of the currents (di/dt) and voltages (du/dt) increases by about one order of magnitude. The circuit switching transients induce voltage spikes across parasitic inductances and further contribute to increased EMI emissions [7]–[9]. The steeper voltage edges and higher clock frequencies cause critical high-frequency excitations and increase the common mode currents through parasitic capacitances, which leads to an increased filtering effort for electromagnetic compatibility (EMC) [10], [11]. Hence, the influence of parasitic inductances and mutual couplings and parasitic capacitances on the switching performances of a WBG semiconductor module are significant. In [12]–[14], the impact of parasitic inductances on conventional Si-MOSFET is studied. However, only limited literature specially analyzes the influence of the parasitic inductances of a SiC-MOSFET power module on the switching performance. But they neglect the mutual couplings among the internal and external parasitic elements. Therefore, this paper aims to investigate the influence of mutual couplings among the internal parasitic elements of a WBG semiconductor module, the influence of mutual couplings between the external dc-link inductances and between external gate-to-source inductances on the switching behavior, including a consideration of their impact on the EMI emissions. Also, the influence of parasitic capacitances of a WBG semiconductor module on the CM EMI emissions is analyzed.

In this paper, the package of a SiC-MOSFET half-bridge power module is 3D-modeled and the high-frequency (HF) parasitic elements are extracted according to the geometry and material properties of the actual module. Then, the influences of parasitic elements and mutual couplings of a SiC-MOSFET half-bridge module on the switching characteristic, as well as their impact on the EMI emissions are presented. Finally, a three-phase DC-AC inverter prototype with three SiC half-bridge MOSFET modules and an EMI measurement test setup are constructed to validate the simulation by experimental results.

II. SiC-MOSFET DC–AC Inverter

The studied configuration of a DC-AC inverter of a drive system is shown in Fig. 1a. The drive system consists of an input filter, a DC-AC inverter, an output filter, a three-phase cable, and an electrical machine. The SiC-MOSFET half-bridge module (commercially available from Infineon FF23MR12W1M1_B11) is used to build the DC-AC inverter. An ideal equivalent model of the half-bridge SiC-MOSFET is built by ANSYS Twin Builder simulation software, which is shown in Fig. 1b. This software provides detailed characterization tools for MOSFETs, IGBTs, and diodes [15]. The SiC-MOSFET half-bridge module is characterized according to data sheet information [16]. The specification of the half-bridge module is shown in Tab. I.

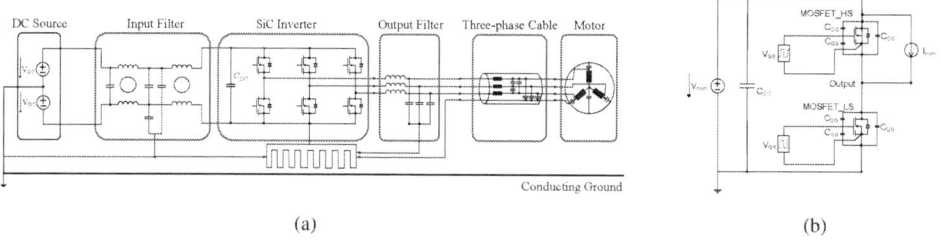

Fig. 1. (a) Circuit diagram of a drive system with a DC-AC inverter. (b) An ideal equivalent model of the SiC-MOSFET half-bridge module using ANSYS Twin Builder software, where the total stray inductance of the module is $L_{st} = 9\,\mathrm{nH}$.

TABLE I
SPECIFICATION OF THE SiC-MOSFET HALF-BRIDGE MODULE

Drain-to-source voltage, $V_{nom} = 600\,\mathrm{V}$	Drain current, $I_{nom} = 50\,\mathrm{A}$	Gate-voltage-on, $V_{gon} = 15\,\mathrm{V}$
Gate-voltage-off, $V_{goff} = -5\,\mathrm{V}$	Gate-resistance-on, $R_{gon} = 1\,\Omega$	Gate-resistance-off, $R_{goff} = 1\,\Omega$
Switching frequency, $f = 50\,\mathrm{kHz}$	DC-bus capacitor, $C_{DC} = 5\,\mu\mathrm{H}$	Load, $R_{load} = 40\,\Omega$

A. Extraction of parasitic capacitances and inductances of a SiC-MOSFET half-bridge module

A SiC-MOSFET half-bridge module is used to study the parasitic parameters, which is shown in Fig. 2a. Figs. 2b and 2c show the simulation model of the SiC-MOSFET half-bridge power module using the ANSYS Q3D extractor. The parasitic elements such as inductances and capacitances are extracted according to the geometry and material properties of the actual module at $100\,\mathrm{MHz}$. Fig. 2b shows the different conductors and the DCB ground plane. The main parasitic capacitances such as DC+ to DCB ground, DC- to DCB ground, and AC output to DCB ground are extracted from the geometry of the power module. These capacitances are shown in Fig. 3. The parasitic inductances of the power module are extracted using the current paths, as shown in Fig. 2c. These inductances store magnetic energy when current passes through them. In Fig. 2c, two current paths from source to A (cyan line) and gate to A (yellow line) are indicated. The parasitic inductance such as self- and mutual inductances are calculated using Q3D output, which is seen in Fig. 3. For the MOSFET_HS, two current paths from DC+ to A (red line) and C to A (green line) are indicated. Similarly, for the MOSFET_LS, two current paths from DC- to B (blue line) and C to B (black line) are indicated and the parasitic self- and mutual inductances are extracted, as shown in Fig. 3.

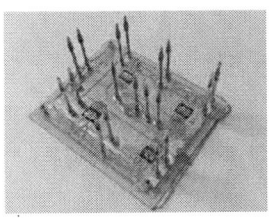

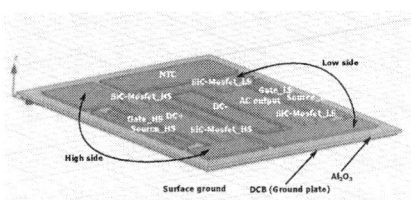

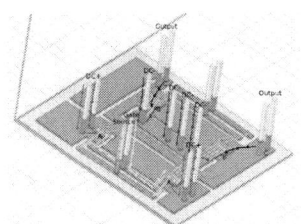

(a) Practical module (F23MR12W1M1_B11)

(b) Determination of parasitic capacitances

(c) Determination of parasitic inductances

Fig. 2. ANSYS-Q3D simulation model of the half-bridge SiC-MOSFET module.

III. Impact of Parasitic Elements and Mutual Couplings on a SiC-MOSFET Half-bridge Module

A. Influence of internal parasitic inductances and couplings

Figure 3a illustrates the differential mode (DM) equivalent simulation circuit of the SiC-MOSFET half-bridge module with extracted parasitic elements (Q3D results inside the module) and with the external gate-to-source, dc-link inductancs including mutual couplings. The DM current paths are shown in Fig. 3a by the blue line (switched on low-side MOSFET) or by the green line (switched off low-side MOSFET). To analyze the influence of parasitic elements on the switching performance inside the module, SPICE simulations were built with the circuits of Fig. 3 (a, b). The resulting waveforms are recorded as turn-on and turn-off transitions, as shown in Fig. 4. The drain-to-source voltage V_{ds} and drain-to-source current I_{ds} are obtained by the voltage and current probe at a low-side MOSFET of the SiC half-bridge module, as shown in Fig. 3a. Three cases e. g. ideal circuit with only a stray inductance in the commutation loop, considering the parasitics inside the module and considering parasitics with mutual couplings, are investigated.

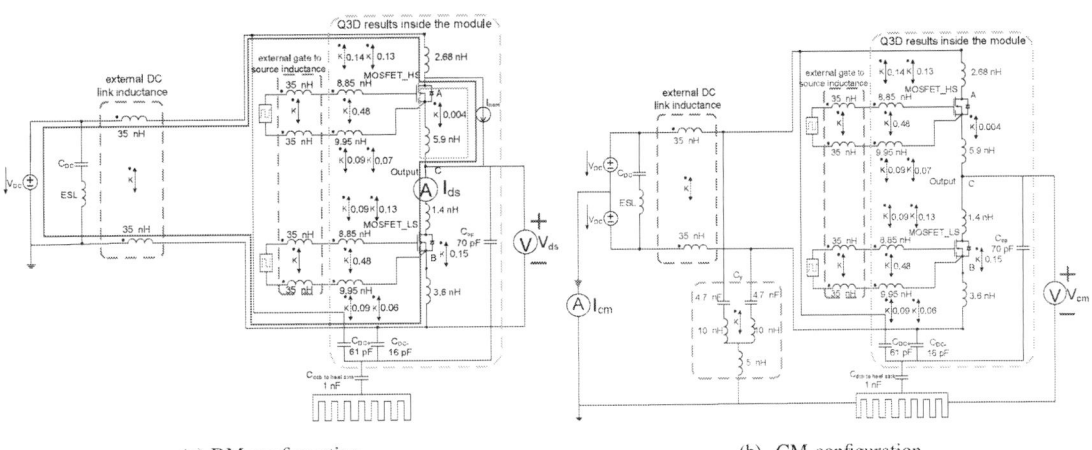

(a) DM configuration

(b) CM configuration

Fig. 3. DM and CM equivalent circuit of the half-bridge SiC-MOSFET, where the extracted parasitic elements (Q3D results inside the module) and an external dc-link and gate-to-source inductances with mutual couplings are used.

1) Drain terminal waveforms: Comparing the three cases in Figs. 4c and 4d, it can be seen that the presence of parasitic elements and mutual couplings has an influence on the current overshoots during the turn-on and turn-off. These also slightly decrease the current slew rate during turn-on transition. Similarly, the presence of parasitic elements and mutual couplings affects the voltage overshoots during turn-off over the ideal case and decrease the voltage slew rate.

EPE'20 ECCE Europe

Assigned jointly to the European Power Electronics and Drives Association & the Institute of Electrical and Electronics Engineers (IEEE)

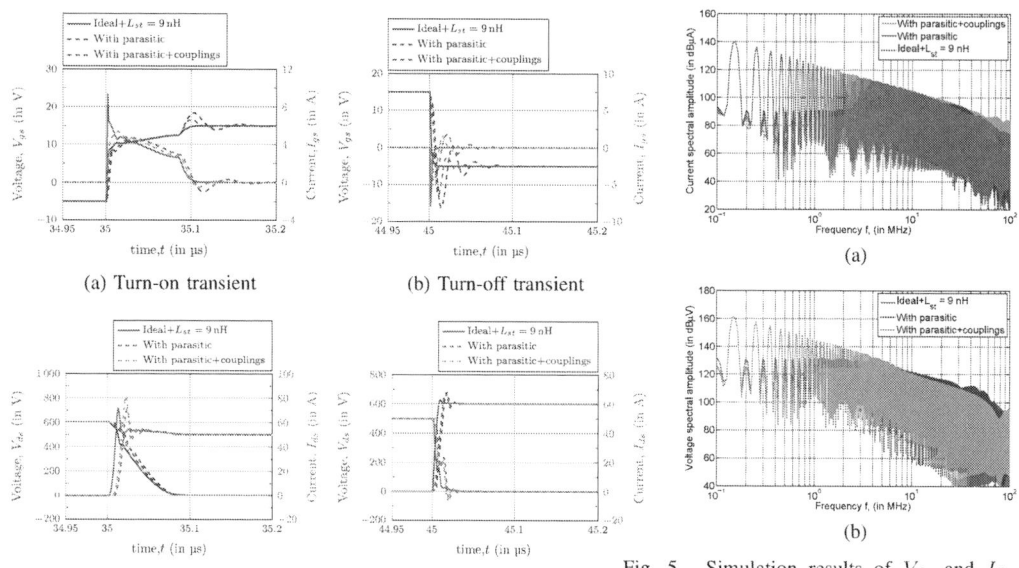

(a) Turn-on transient   (b) Turn-off transient

(c) Turn-on transient   (d) Turn-off transient

Fig. 4. The influence of internal parasitic elements and couplings on the switching characteristics of a low-side MOSFET of the SiC half-bridge module, which is shown in Fig. 3a by the orange dashed box.

Fig. 5. Simulation results of V_{ds} and I_{ds} spectral amplitudes of a low-side MOSFET of the SiC half-bridge module. (a, b) the influence of internal parasitics and couplings of the module on the current and voltage spectral amplitudes, which is shown in Fig. 3a by the orange dashed box.

2) Gate terminal waveforms: In Figs. 4a and 4b, it can be noted that the internal gate-to-source parasitics and couplings can induce overshoots and ringing on voltage and current at the gate terminal during the turn-on and turn-off transitions.

3) EMI emissions: In Fig. 5a, it can be observed that the presence of parasitic elements and mutual couplings affects the current spectral amplitudes in the frequency range of 20 MHz to 100 MHz. In this frequency range, the current amplitude increases about 5-12 dB, which is due to the high frequency ringing in the waveforms during the turn-on and turn-off transitions. Similarly, in Fig. 5b, the presence of parasitic elements and mutual couplings affects the voltage spectral amplitudes in the frequency range of 10 MHz to 100 MHz. In this frequency range, the voltage amplitude increases about 5 dB, which is due to the high frequency ringing in the waveforms during the turn-off transition, as shown in Fig. 5b, blue waveform. For the same frequency range, the voltage amplitude decreases about 5 dB which is due to the high frequency coupling effects inside the SiC-MOSFET half-bridge module during turn-on and turn-off transitions, as shown in Fig. 3.

B. Influence of additional single external parasitic inductances and couplings

The DM equivalent simulation circuit of the SiC-MOSFET half-bridge module with the external dc-link and gate-to-source inductances including the mutual couplings is shown in Fig. 3a. The chosen inductance value of 35 nH is consistent with the value usually encountered in a non-optimal PCB layout [8]. To monitor the influence of external mutual couplings between dc-link inductances and gate-to-source inductances on the switching performance, the resulting waveforms are recorded in Figs. 6 and 7, respectively.

4) Drain terminal waveforms: In Figs. 6c and 6d, it can be seen that the presence of external dc-link inductances and the mutual couplings has a large influence on the current overshoots and ringing during the turn-on transition. The magnetic coupling between the dc-link reduces the current ringing during the turn-on transition. This is because the loop inductance of the commutation loop is decreased with increasing coupling. Besides, the slew rate increases with increased mutual couplings between dc-link inductances. The presence of external dc-link inductances and the mutual couplings has a large influence on the voltage overshoots and ringing during the turn-off transition and the

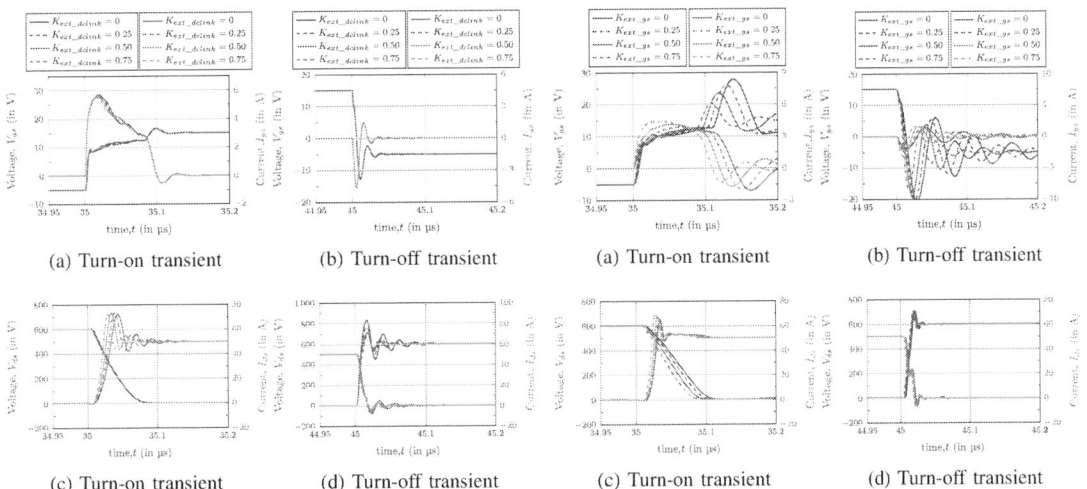

(a) Turn-on transient   (b) Turn-off transient   (a) Turn-on transient   (b) Turn-off transient

(c) Turn-on transient   (d) Turn-off transient   (c) Turn-on transient   (d) Turn-off transient

Fig. 6. The influence of external dc-link parasitic inductances and couplings, including the internal parasitics inside the module, on the switching characteristics of a low-side MOSFET of the SiC half-bridge module, where other external gate-to-source parasitics are zero.

Fig. 7. The influence of external gate-to-source parasitic inductances and couplings, including the internal parasitics inside the module, on the switching characteristics of a low-side MOSFET of the SiC half-bridge module, where other external dc-link parasitics are zero.

magnetic coupling between the dc-link inductances decreases the voltage overshoot and ringing during the turn-off transition.

In Figs. 7c, it can be seen that the presence of gate-to-source inductances and the mutual couplings has little influence on the current and voltage overshoots and ringing during the turn-on transition. The slew rate increases with increased mutual couplings between gate-to-source inductances. In Figs. 7d, it can be noted that the presence of gate-to-source inductances and the mutual couplings does not affect the current and voltage overshoots and ringing during the turn-off transition.

5) Gate terminal waveforms: In Figs. 6a and 6b, it can be noted that the presence of external drain-to-source inductances and couplings does not affect the voltage and current overshoots and ringing at the gate terminal during the turn-on and turn-off transitions compared to the ideal case.

Figs. 7a and 7b show that the presence of external gate-to-source inductances has a large influence on the current and voltage overshoots and ringing at the gate terminal during the turn-on and turn-off transitions. During the turn-on transition, the gate-to-source, V_{gs} is -5 to +15 V ideally. However, due to the influence of external inductance, the gate-to-source, V_{gs} is increased -5 to +27 V. Similarly, during the turn-off transition, the gate-to-source V_{gs} are +15 to -5 V in the ideal case. However, due to the influence of external inductance, the gate-to-source V_{gs} is increased +15 to -23 V. It can be seen that the gate-to-source voltage during turn-on and turn-off transitions decreases with increasing couplings, because this decreases the loop inductance in the gate-to-source circuit loop.

C. Common influence of all parasitic inductances and couplings

To monitor the influence of all external parasitics inductances and couplings, including the internal parasitics inside the module, on the switching performance, the resulting waveforms are recorded in Fig. 8.

6) Drain terminal waveforms: In Fig. 8a, it can be seen that the presence of external mutual couplings between dc-link inductances and the gate-to-source inductances has a large influence on the current overshoots and ringing during the turn-on transition. The slew rate increases with increased mutual couplings between dc-link inductances and gate-to-source inductances. The current ringing during the turn-on transition decreases with increased mutual couplings. In Fig. 8b, the presence of external mutual couplings between dc-link inductances and the gate-to-source inductances proves an influence on the voltage overshoots and ringing during the turn-off transition. The voltage overshoot and ringing during the turn-off transition decreases with increased mutual couplings.

EPE'20 ECCE Europe

Assigned jointly to the European Power Electronics and Drives Association & the Institute of Electrical and Electronics Engineers (IEEE)

7) EMI emissions: Fig. 9a shows that the presence of external mutual couplings affects the current spectral amplitudes in the frequency range of 45 MHz to 100 MHz. In this frequency range, the current amplitude decreases around 15-20 dB compared to the ideal case. The cause is that the inductance provides an impedance that reduces the noise currents. The amplitude decreases with increasing mutual couplings. In Fig. 9b, it can be observed that the presence of external mutual couplings affects the voltage spectral amplitudes in the frequency range of 15 MHz to 100 MHz. In this frequency range, the voltage amplitude increases about 3-7 dB, which is due to the high frequency overshoots and ringing in the waveforms during the turn-on and turn-off transitions. The amplitude decreases with increasing mutual couplings.

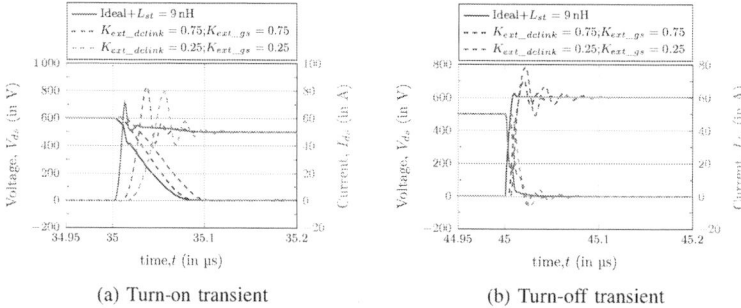

(a) Turn-on transient   (b) Turn-off transient

Fig. 8. The influence of all external parasitic inductances and couplings, including the internal parasitics inside the module, on the switching characteristics of a low-side MOSFET of the SiC half-bridge module, where the external dc-link and gate-to-source couplings are varied.

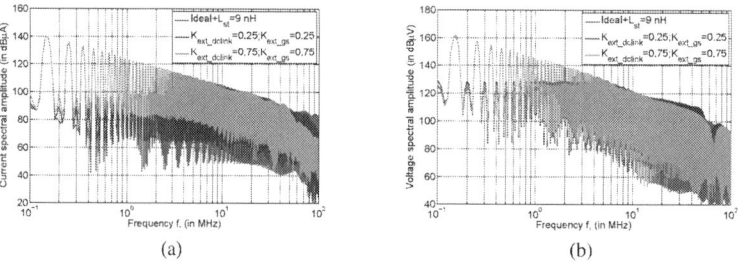

(a)   (b)

Fig. 9. Simulation results of V_{ds} and I_{ds} spectral amplitudes of a low-side MOSFET of the SiC half-bridge module. (a, b) the influence of all external parasitic inductances and couplings, including the internal parasitics inside the module, on the current and voltage spectral amplitudes, where the external dc-link and gate-to-source couplings are varied.

IV. Influence on Common Mode (CM) EMI

Fig. 3b illustrates the CM equivalent circuit of the half-bridge SiC-MOSFET with extracted parasitic elements (Q3D results inside the module) and with external gate-to-source, dc-link inductances including the mutual couplings. To investigate the effect of parasitic elements and the mutual couplings inside the module on the CM voltage and current spectral amplitude, the high-side MOSFET is considered in turn-on and turn-off transitions. The parasitic elements and the mutual couplings inside the module are shown in Fig. 3b by the orange dashed box. The CM current path is shown in Fig. 3b by the red arrow. The CM voltage is obtained by the voltage probe line-to-ground at the half-bridge module terminal. Figs. 10a and 10b show the influence of the internal parasitic elements and mutual couplings of the SiC half-bridge MOSFET on the CM current and voltage spectral amplitude. Here, it can be seen that the presence of the internal parasitic elements and mutual couplings increases the CM voltage spectral amplitudes by a few dB in the frequency range of 15 MHz to 100 MHz. Figs. 10c and 10d present the influence of the external parasitic elements and mutual couplings of the SiC half-bridge MOSFET on the CM current and voltage spectral amplitude. The presence of the external parasitic elements and mutual couplings increases the CM current and voltage spectral amplitudes around 3-8 dB in the frequency range of 15 MHz to 100 MHz.

To monitor the influence of parasitic capacitances inside the module on CM current, three different cases (without parasitic capacitances, with parasitic capacitances and 3 times more parasitic capacitances) are considered, which is shown in Fig. 10e. Here, it can be observed that the CM current spectral amplitude increases with increasing parasitic capacitances. The presence of parasitic capacitances affects the CM spectral amplitude in the frequency range of 25 MHz to 100 MHz. In this frequency range, the current spectral amplitude with parasitic capacitances increases around 3-8 dB compared to the case without parasitic capacitances. Similarly, the current spectral amplitude with 3 times more parasitic capacitances increases around 10-15 dB compared to the case without parasitic capacitances. So, it can be stated that the presence of parasitic capacitances of a half-bridge module has a significant impact on the CM current spectral amplitude.

The influence of C_y capacitors with parasitics including coupling on the CM current spectral is illustrated in Fig. 10f. Here, it can be seen that a resonance occurs at 11 MHz, which is due to the C_y capacitors. At this frequency, the CM current spectral is around 20 dB higher compared to the spectral amplitude without capacitors. The resonance is shifted from 11 MHz to 9 MHz due to the presence of parasitic inductances of C_y capacitors with coupling. Besides, it increases the noise level about 5-12 dB in the frequency range of 30 MHz to 100 MHz. On the other hand, the C_y capacitors decrease the CM spectral amplitude around 5-15 dB in the frequency range of 13 MHz to 100 MHz.

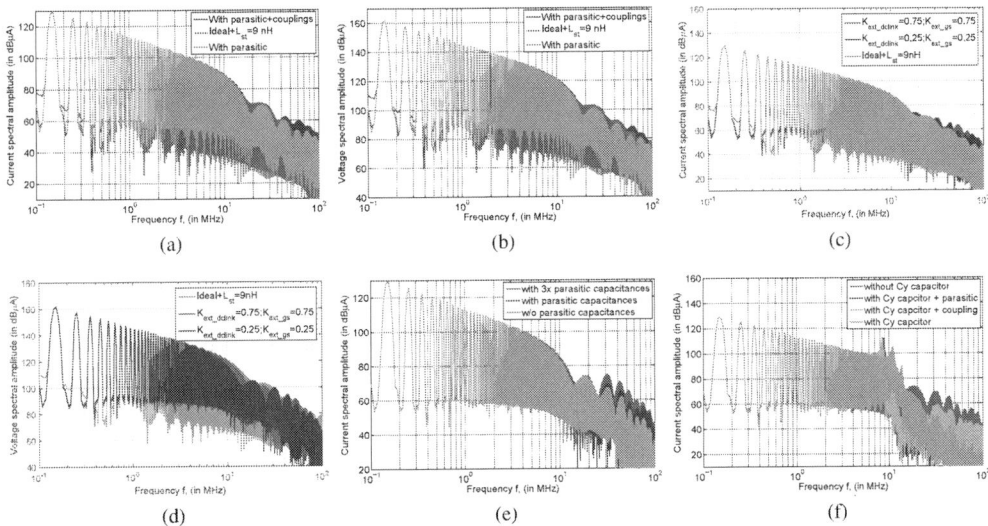

Fig. 10. Simulation results of CM spectral amplitudes, V_{cm} and I_{cm} of a SiC-MOSFET half-bridge module. (a, b) the influence of the internal parasitic elements and mutual couplings of a SiC half-bridge MOSFET on the common mode current and voltage spectral amplitude (I_{cm} and V_{cm}). (c, d) the influence of the external parasitic elements and mutual couplings of a SiC half-bridge MOSFET on the common mode current and voltage spectral amplitude (I_{cm} and V_{cm}). (e) Internal parasitic capacitances of a SiC-MOSFET half-bridge module and (f) C_y capacitance including parasitic inductances and coupling influence on the common mode current (I_{cm}).

V. Experimental Validation

To validate the simulation and experimental results, an experimental prototype of the SiC three-phase DC-AC inverter is built, which is illustrated in Fig. 11a. A schematic circuit diagram of the the SiC three-phase DC-AC inverter is shown in Fig. 11b. The prototype inverter is constructed using the three SiC half-bridge MOSFET modules (F23MR12W1M1_B11), which is shown in Fig. 2a. The three half-bridge SiC MOSFET modules are driven by three isolated gate drivers with +15/-5 V gate-to-source value. The bus capacitor is 5 μF. The gate resistor of 1 Ω is used. The value of the C_y capacitor is 4.7 nF.

A. Parasitic inductances and mutuals couplings of the main power board

Parasitic inductances and mutual couplings of the main power board are calculated using ANSYS Q3D. First, the PCB layout drawing is designed by Eagle software. Second, the layout drawing is exported as a DXF file, which is a computed-aided design file. Third, the DXF file is imported to the ANSYS Q3D extractor to extract the parasitic elements. Fig. 12 shows the imported PCB of the

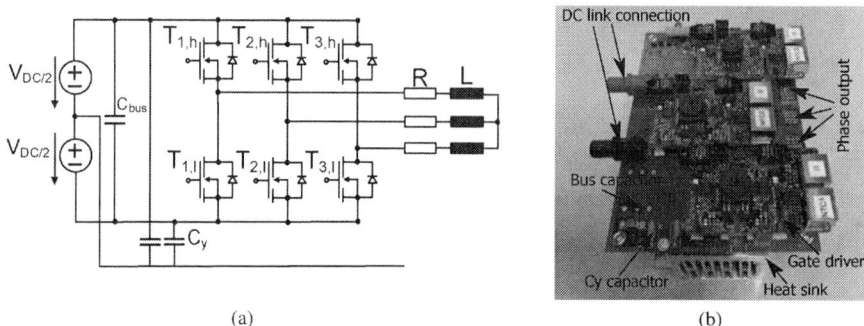

(a)   (b)

Fig. 11. (a) Schematic of the SiC three-phase DC-AC inverter.(b) Experimental prototype of the SiC three-phase DC-AC inverter using the three half-bridge SiC MOSFET modules.

main power board in ANSYS Q3D. To extract the parasitic inductances and mutual coupling between the top copper layer (V_{bus+}) and the bottom copper layer (V_{bus-}), the current paths from V_{bus+} to DC+ pins (red line) and from V_{bus-} to DC- pins (black line) are indicated, as shown in Fig. 12a and Fig. 12b, respectively. In ANSYS Q3D, the AC parasitic elements such as the self-inductance and mutual coupling are extracted at $100\,\mathrm{MHz}$. FR4 epoxy is used as a surrounding material of the main power board. From the simulation, the value of self-inductances are $L_{bus+} = 11.03\,\mathrm{nH}$ and $L_{bus-} = 16.10\,\mathrm{nH}$. The mutual coupling coefficient between two layers is $K = 0.19$. So, the loop and net inductance of the main power board are $L_{loop} = 22.06\,\mathrm{nH}$ and $L_{net} = 7.75\,\mathrm{nH}$, which are calculated from the relationships $L_{loop} = L_{p1} + L_{p2} - 2L_{mp}$ and $L_{pnet} = \frac{L_{p1}L_{p2} - L^2_{mp}}{L_{p1} + L_{p2} - 2L_{mp}}$, respectively. L_p and L_{mp} are the self- and mutual inductance. Similarly, to extract the parasitic inductances and mutual coupling between the gate trace (top layer) and the source trace (bottom layer), the current paths from V_{gate} to $V_{gmosfet}$ (blue line) and from V_{gate} to $V_{smosfet}$ (green line) are indicated, which is shown in Fig. 12a and Fig. 12b, respectively. From the simulation, the value of self-inductances are $L_g = 27.38\,\mathrm{nH}$ and $L_s = 27.38\,\mathrm{nH}$. The mutual coupling coefficient between two traces is $K = 0.52$. So, the loop and net inductance of the gate-to-source are $L_{loop} = 26.28\,\mathrm{nH}$ and $L_{net} = 20.80\,\mathrm{nH}$.

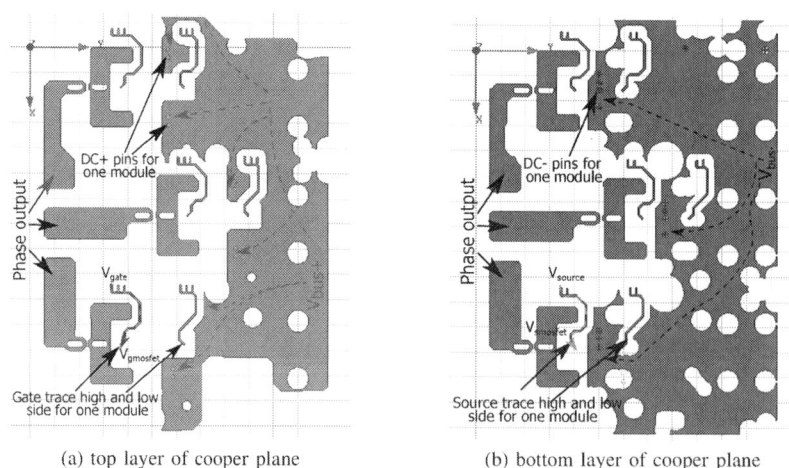

(a) top layer of cooper plane   (b) bottom layer of cooper plane

Fig. 12. Estimation of parasitic inductances and mutuals couplings of the main power board of the SiC three-phase DC-AC inverter using ANSYS Q3D.

B. Test setup for EMI measurement and experimental results

To validate the simulation and measurement results, an experimental test bench of the total drive system is built, which is shown in Fig. 13. The switching frequency of the inverter is set to $50\,\mathrm{kHz}$ and the switching duty ratio at a constant value. EMI measurements (conducted disturbances) at DC and AC side are measured in an EMI chamber. The frequency range of the measurement is $150\,\text{kHz}$ to $30\,\text{MHz}$. The DC voltage U_{DC} is $300\,\text{V}$. The frequency of the machine is $f\_out = 0$. The common voltage (V_{cm}) at the inverter output is measured by using a voltage probe, as shown in Fig. 13b.

(a) test bench

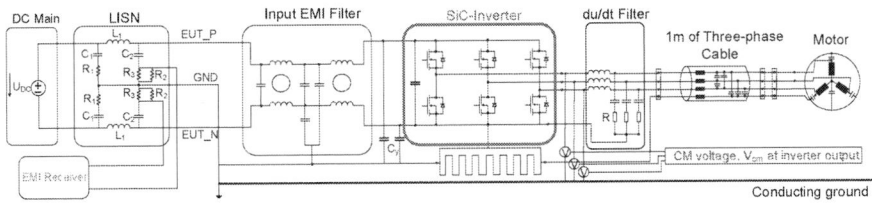

(b) schematic circuit diagram

Fig. 13. Total measurement overview.

The comparison of the simulated and measured common mode voltage amplitude V_{cm} at the inverter output during the turn-on and turn-off transitions is shown in Fig. 14a. It can be seen that the voltage overshoot and the ringing are roughly the same. The overshoot and ringing is caused by the internal and external parasitic elements and mutual couplings of the SiC half-bridge DC-AC inverter. Figure 14a shows the comparison of the simulated and measured common mode voltage spectral amplitude V_{cm} at the inverter output. It can be noted that the simulation result agrees well with the measurement result in the frequency range from $100\,\text{kHz}$ to $7\,\text{MHz}$. In the frequency range from $7\,\text{MHz}$ to $30\,\text{MHz}$, the measured voltage spectral is around 2-3 dB higher than the simulated result.

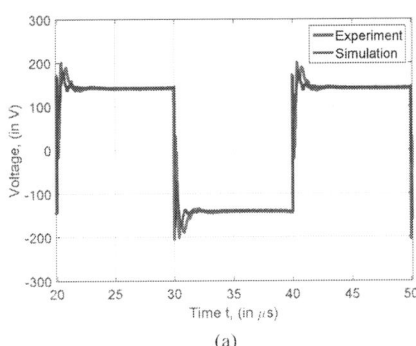
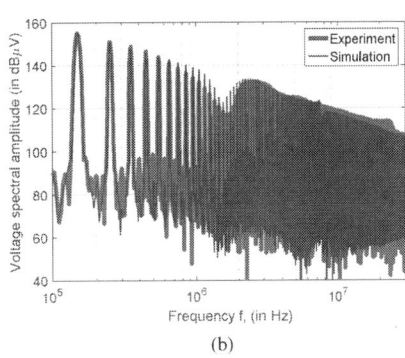

(a)   (b)

Fig. 14. Comparison of measured and simulated CM voltage and CM spectral amplitude for $50\,\text{kHz}$ switching frequency.

VI. Conclusion

The influence of parasitic elements and mutual couplings of a SiC-MOSFET half-bridge module in a DC-AC inverter on the switching characteristic, as well as their impact on the EMI emissions are presented. First, the package of a SiC-MOSFET half-bridge power module is 3D-modeled and the high frequency (HF) parasitic elements are extracted according to the geometry and material properties of the actual module. Second, simulations are performed to investigate the influence of parasitic elements and mutual couplings of a SiC-MOSFET half-bridge module on the switching characteristics, as well as the voltage and current spectral amplitude. It is observed that the presence of internal parasitic elements and mutual couplings of a module has an influence on the current and voltage overshoots and ringing during the turn-on and turn-off. These affect the current and voltage spectral amplitude above 20 MHz. Third, the presence of external mutual couplings between the dc-link inductances and gate-to-source inductances has a large influence on the current and voltage overshoots and ringing during turn-on and turn-off transitions. It is found that the slew rate increases with increased mutual couplings between dc-link inductances and gate-to-source inductances during turn-on transition. The voltage overshoot and ringing during the turn-off transition decreases with increasing mutual couplings. These affect the current and voltage spectral amplitude above 15 MHz. Fourth, the influence of parasitic capacitances of a SiC-MOSFET half-bridge module and the C_y capacitance with parasitics including mutual couplings on the CM EMI emissions are investigated. It is clarified that the presence of parasitic capacitances of a half-bridge module has a significant impact on the CM current spectral amplitude above 25 MHz. Besides, parasitic inductances of the C_y capacitor with coupling increase the CM current amplitude.

Finally, a prototype of the three-phase SiC DC-AC inverter was built. Parasitic inductances and mutual couplings of the main power board of the inverter are calculated using ANSYS Q3D. To validate the simulation and measurement results, an EMI measurement setup was built. It could be observed that the simulation result agrees well with the measurement result in terms of ringing frequency.

The analysis of the paper focused on the important influences of the parasitic elements and mutual couplings on the switching behavior of the SiC half-bridge MOSFET module, which is used to build the three-phase DC-AC inverter. To investigate the EMI emissions, the EMI measurement setup of a drive system was constructed. More studies such as conducted disturbances at DC and AC side, parasitic effects as well as an optimized EMI measurement setup will be discussed in future research.

REFERENCES

[1] A. Merkert, T. Krone and A. Mertens, "Characterization and Scalable Modeling of Power Semiconductors for Optimized Design of Traction Inverters with Si- and SiC-Devices," in *IEEE Transactions on Power Electronics*, vol. 29, no. 5, pp. 2238–2245, 2014.

[2] T. Köneke , A. Mertens, D. Domes and P. Kanschat, "Highly Efficient 12kVA Inverter with Natural Convection Cooling Using SiC Switches," in *PCIM Europe* , 2011, Nuremberg.

[3] J. Biela ,M. Schweizer, S. Waffler and J. W. Kolar, " SiC versus Si-Evaluation of Potentials for Performance Improvement of Inverter and DC–DC Converter Systems by SiC Power Semiconductors," in *IEEE Transactions on Industrial Electronics*, Vol. 58, No. 7, 2011.

[4] B. Wrzecionko, J. Biela and J. W. Kolar," SiC Power Semiconductors in HEVs: Influence of Junction Temperature on Power Density,Chip Utilization and Efficiency," in *Industrial Electronics IECON*, 2009.

[5] D. Bortis, B. Wrzecionko and J. W. Kolar," A 120°C ambient temperature forced air-cooled normally-off SiC JFET automotive inverter system," in *Applied Power Electronics Conference and Exposition (APEC)*, 2011.

[6] Di Han J. Noppakunkajorn and B. Sarlioglu, "Comprehensive efficiency, weight, and volume comparison of SiC and Si-based bidirectional DC-DC converters for hybrid electric vehicles," in *IEEE Trans. Veh. Technol*, vol. 63, no. 7, pp. 3001–3010, Sep. 2014.

[7] J. Noppakunkajorn, Di Han and B. Sarlioglu, "Analysis of high-speed PCB with SiC devices by investigating turn-off overvoltage and interconnection inductances influence," in *IEEE Trans. Transp. Electr.*, vol. 1, no. 2, pp. 118–125, Aug. 2015.

[8] Di Han, and Bulent Sarlioglu, "Comprehensive Study of the Performance of SiC MOSFETs Based Automotive DC-DC Converter under the Influence of Parasitic Inductance," in *IEEE Transactions on Industry Applications*, vol. 52, no. 6, pp. 5100–5111, Nov./Dec. 2016.

[9] Shengnan Li, L. M. Tolbert, F. Wang, and F. Peng, "Reduction of stray inductance in power electronic modules using basic switching cells," in *Energy Conversion Congress and Exposition (ECCE)*, pp. 2686-2691, 2010.

[10] Andre Domurat-Linde, Eckart Hoene, "Investigation and PEEC Based Simulation of Radiated Emissions Produced by Power Electronic Converter," in *6th International Conference on Integrated Power Electronics Systems (CIPS)*, March, 16-18, 2010,Nuremberg/Germany.

[11] Andre Domurat-Linde, Eckart Hoene, "Analysis and Reduction of Radiated EMI of Power Modules basic switching cells," in *7th International Conference on Integrated Power Electronics Systems (CIPS)*, March, 16-18, 2010,Nuremberg/Germany.

[12] Z. Chen,D. Boroyevich, and R. Burgos, "Experimental parametric study of the parasitic inductance influence on MOSFET switching characteristics," in *International Power Electronics Conference (ECCE)*, pp. 164-169, Jun. 2010.

[13] Saeed Safari, Alberto Castellazzi and Pat Wheeler, "Experimental Study of Parasitic Inductance Influence on SiC MOSFET Switching Performance in Matrix Converter," in *European Conference on Power Electronics and Applications (EPE)*, 2-6 Sept. 2013.

[14] J. Wang, H. Chung, and R. Li, "Characterization and experimental assessment of the effects of parasitic elements on the MOSFET switching performance," in *IEEE Trans. Power Electron*, vol. 28, no. 1, pp. 573–590, Jan. 2013. .

[15] "ANSYS Inc. Twin Builder Manual," *http://www.ansys.com 2019*, Release 18.2.

[16] "Dual 1200 V, 23 m Ω half-bridge module with module with Cool SiC-MOSFET," *https://www.infineon.com/cms/en/product/power/mosfet/silicon-carbide/modules/ff23mr12w1m1_b11/*, 2018.

Wire-Wound Multi-Phase Stator Based EMEH with MPPT Self-Powered Energy Management System

Mahmoud Shousha, Dragan Dinulovic, Talha Zafar, Michael Brooks, and Martin Haug

MagI3C R&D Division, Würth Elektronik eiSos GmbH & Co. KG Parkring 29 85748
Garching bei München, Germany

Email: mahmoud.shousha@we-online.de

URL: http://katalog.we-online.de/en/pm

Keywords

«Electromagnetic Energy Harvester (EMEH)», «Kinetic Energy», « Energy Management System (EMS)», « Maximum Power Point Tracking (MPPT)», « Power Management System (PMS)».

Abstract

This paper presents a cost-effective electromagnetic energy harvester (EMEH) based on wire-wound technology. It also shows the energy management system used to convert the EMEH's AC voltage to a DC one. The EH system is self-powered and does not require external sources or pre charge of any of its components and provides maximum power point tracking.

Introduction

Energy harvesting (EH) is the process of obtaining energy from the ambient and converting this energy into useful electrical energy. This harvested electrical energy can be used as a power source for miniaturized autonomous devices where a wired power is unavailable or cost-intensive. Multiple ambient energy sources exist in the environment, whose energy by applying various physical effects can be converted into electrical energy [1]-[4]. A multitude of EH applications has been developed recently such as health monitoring, smart home (buildings automation) applications, communication systems, transportation, robotics, sensor networks, various Internet of Things (IoT) applications, and wearable electronics [5]–[10].

EMEHs can be divided into three groups, namely, (*i*) oscillatory harvesters that are driven by vibrating energy; (*ii*) hybrid harvesters that transform vibrations into rotation; and (*iii*) rotational harvesters that are driven by rotational movements. The main advantage of the rotational EMEHs in comparison to vibrations-driven harvesters is the independence from the resonant frequency, expanding their potential applications. In addition, rotational harvesters can achieve higher energy density compared to other kinds of electromagnetic transducers [11]-[13]. However, the disadvantage of rotational harvesters is their complex design. In this paper, an electromagnetic energy harvesting system, namely energy harvesting device and energy management system, that converts kinetic energy to electrical energy is developed and tested. The EMEH is developed using wire-wound technology, unlike previous developments [14]-[18] and miniaturized to present a cost effective solution, helping in further deployment of the EH system.

Design of the Energy Harvester

The energy harvester system consists of a mechanism for movement conversion, electromagnetic EH transducer and energy management system (EMS) with RF transmitter. The first proof of concept design is described in [14], [18]. The disadvantages of latter design are big volume and high development cost associated with the big packaging and the multi-layer stator design. In the next itertion, the EMEH was miniaturized, as the volume of the harvester was reduced by about 70% of the original volume, cutting significantly the solution size and partially the development cost due to saving in the housing material but the design continues to feature the multi-layer stator design [15]. In this work, we will show the development and testing of the updated EH system with a new coil system as well as the modified power management system, compared to what was presented in [16]-[18].

The electromagnetic harvester belongs to the group of push button harvesters as it converts the kinetic movement energy of the button into electrical energy using electromagnetic working principle (Fig. 1). The electromagnetic working principle is based on inducing the electrical voltage during rotation of a coil in the static homogenous magnetic field. Therefore, the electromagnetic transducer consists of a movable part (rotor) and a stator. Both parts, the stator and the rotor, are mounted on axis forming a defined air gap in between. The rotor has 4 multi pole permanent magnet segments of NdFeB mounted on soft magnetic sheet metal. As a soft magnetic material, a Co-alloy sheet metal Vacoflux by Vacuumschmelze is used. Vacoflux shows a very high saturation flux density of 2.3 T, which is important to avoid the saturation of the magnetic system. The hard magnet poles are magnetized perpendicular to the surface of the poles. The stator features one coil system consisting of 4 phases. The previous coil system design with multilayer coils embedded into FR-4 material are now replaced with wire-wounded coils. Using embedded coil system, we had a restriction with respect to the number of layers and number of turns, as only 100 turns were achieved before. The induced voltage strongly depends on number of turns; therefore, the induced voltage was not high enough, creating a challenging task for the EMS designers. With the new wire-wound coils, the number of turns is increased to 726 turns. Figure 2 shows the phase design where it can be noted that the wire is wounded on a plastic frame. Electrical properties of one phase are shown in Fig. 3. The resistance of the coil is about 45 Ω. The resistance is increased 3 times in comparison to the embedded coil, but the inductance is increased more than 20 times. The inductance has a value of 2.6 mH (the previous system has only 124 µH). The inductance is stable until 200 kHz and maximum Q-factor is about 18 at frequency between 100 kHz and 300 kHz. The stator consists of 4 phases; two of them are connected in series and then connected in parallel with the other two series phases.

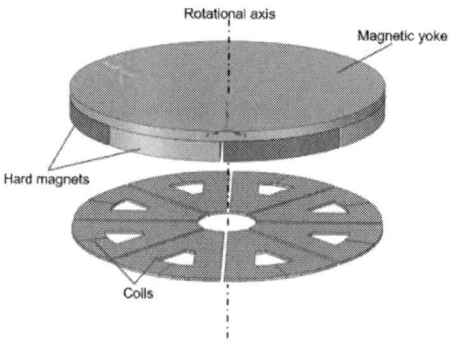

Fig. 1: Design of electromagnetic part of the harvester.

The linear movement of the button is first converted into rotation by using a mechanism for movement conversion. The mechanism for movement conversion is integrated in the harvester and allows the conversion of small linear motion of the button (about 10mm) into rotation with rotational speed higher than 1000 rpm.

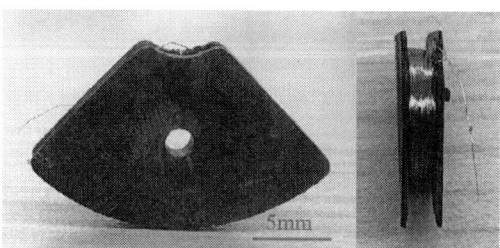

Figure 2: Design of a wire-wounded phase.

The movement conversion mechanism consists of a button, round gear rack, spring, and gear with one-way clutch. For this harvester type, it is important to generate a rotation only in one direction. The rotation in opposite direction generates a negative voltage signal and the sum of the induced voltage is zero. Therefore, the one-way clutch is applied to allow the rotation in only one direction. The gear is a standard mechanical part and the round gear rack is easy for fabrication by the mechanical machining. By using of standard and easy-to-fabricate parts, the cost of the EMEH is reduced.

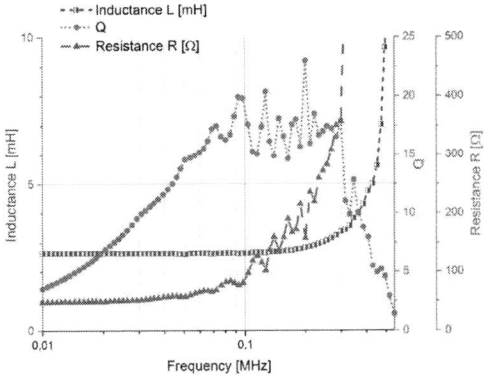

Figure 3: Electrical properties of one wire-wounded coil.

By pressing the button, the spring is loaded and the energy is stored in the spring. During releasing of the spring, the button and round gear rack are moving in the starting position back. The round gear rack moves the gear and the one–way clutch allows the uninhibited free rotation of the rotor with assembled permanent magnets. The EMEH produces around 1 mJ from a single pressing and an exponentially damped open circuit voltage of 1.2 V.

Energy Management System

The EMS presented in [18] requires retrofitting to make it capable of working with the presented energy harvester. Since the output energy is quite low compared to the previous harvester design, the PMS needs to be more energy efficient without compromising other key performance parameters like size or impedance matching (maximum power point tracking) [18]. The first design parameter that has been adjusted is the switching frequency (8 KHz vs 40 KHz) to reduce the switching losses, allowing more energy to be delivered to the output of the EMS. As a consequence, the inductance value needs to be adjusted to achieve the desired impedance matching but at the same time without increasing the inductor package that was used in [18]. The highest inductance value in the same package that can be found is 330 μH. It is worth mentioning that the 330 μH inductor has a higher DC resistance (180 $m\Omega$ vs 770 $m\Omega$) but due to the low current of the whole system, the conduction losses are less of a concern and hence the efficiency of the EMS is not significantly affected. In other words, the system is more sensitive to switching losses compared to conduction losses due to the continuous light load operation of the system. The duty cycle is adjusted to 0.33 compared to 0.5 in the previous implementation, which further improves efficiency [18]. Equation 1 describes the relationship between the converter input impedance and the circuit parameters.

$$Z_{in} = \frac{2 f_s L}{D^2}, \quad (1)$$

Where Z_{in} is the input impedance of the converter, L is the inductance value, f_s is the switching frequency and D is the duty cycle of the switching signal. The converter needs to be tuned to achieve input impedance of around 46 Ω to harvest the maximum available energy of the harvester, i.e. maximum power point tracking.

Another modification to further reduce the power losses of the control circuit is to reduce the resistors used with the PWM generation circuit (R_6 as 4.3 MΩ and R_5 as 3 MΩ, in Fig. 4, vs 1 MΩ in [18]) by reducing the current consumption of the comparators circuit. The number of the charge pump stages needs to be increased from two to four, as shown in the bias supply block in Fig. 4, since the input energy is lower compared to the previous implementation (1 mJ vs. 20 mJ). Another approach would be to decrease the charge pump capacitor to allow for a higher charge pump voltage but that causes higher ripples resulting in slower charging of supply capacitor and might cause failure of the start-up mechanism. For the PMS to have better cold start capability, better subthreshold MOSFETs have been used with threshold voltage $V_{Gs,th}$ of 1 V vs 2 V for the previous implementation [18] which enables the operation of the PMS from the first cycle of energy harvester voltage. Due to the improved power consumption of the control circuit and the bias supply scheme, the bleeding resistors used with the system presented in [18] was completely removed, further improving the harvested energy. Figure 4 shows the modified EMS used with the presented EMEH.

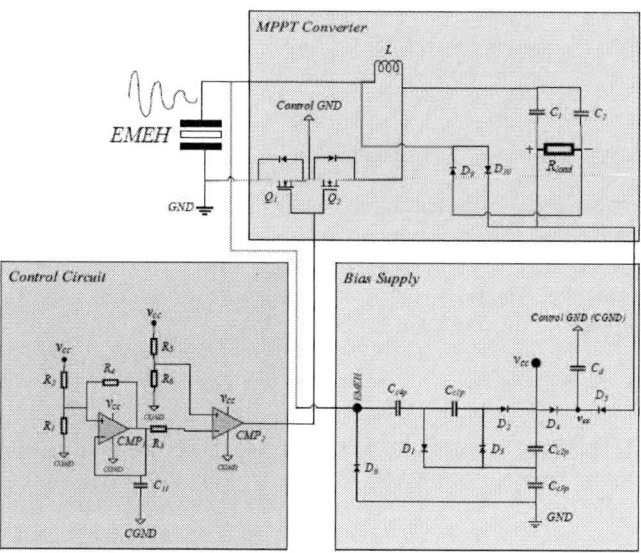

Figure 4: The modified EMS.

Design of Bias Supply Scheme for Cold Start

To drive the MOSFETs and provide adequate supply voltage to the control circuit, a Villiard charge pump is implemented to generate the required voltage for operation, as shown in the highlighted pink square in Fig. 4. Since the number of input cycles from the harvester are limited in terms of voltage level due to non continuous rotation of the rotary part inside the energy harvester, improving the transient response as well as the output voltage level of the bias supply is of a great importance. The voltage of harvester can be modelled as damped sinusoidal with peak voltage as 1 V with very low damping in the first few cycles. However, for the ease of analysis we consider the design of bias supply such that the input voltage equals 4 half cycles of sinusoid with no damping because only these 4 half cycles provide the significant input voltage for charge build up and hence the rest of cycles can be discarded.

The charge pump circuit rectifies input AC voltages to DC output whose steady state voltage is determined by the number of charge pump stages.

$$V_{dc} = N(\hat{v}_{ac} - V_d) \qquad (2)$$

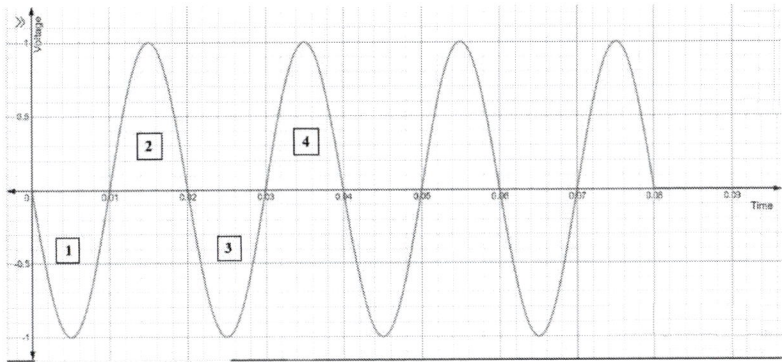

Figure 5: Modelled input waveform.

Where $\hat{v}_{ac}$ is the peak value of input voltage, V_{dc} is the steady state output voltage and V_d is the diode forward voltage drop with $V_d = 0.14$ V, and N the number of stages can be multiples of 2. The design goal for bias supply is to reach a value of 2 V with the modelled waveform in Fig. 5 for turning on MOSFETs as well as reduce the rise time of output voltage in transients. If N = 2, the steady value V_{dc} equals 1.72 V according to (2) which is the main reason that N needs to be at least 4 for the defined design goal.

The design goal is to reach 2 V at the positive peak of the fourth half cycle of the EMEH voltage. In the first cycle, the source charges the first stage and the next cycle charges the subsequent stage. The capacitances are optimized such that it allows a maximum voltage drop of 0.5 V across each capacitor while discharging and a voltage rise of 1 V while charging. Therefore, the maximum possible output voltage for the first charging cycle of each stage equals

$$V_o = N(\hat{v}_{ac} - V_d) - 0.5m \qquad (3)$$

Where m is the number of discharging capacitors in the cycle.

In the first cycle, only the first stage charges hence N = 2, and m = 1 and hence $V_o = 1.22$ V. In the second cycle, the first stage loads the second stage and now N = 4, m = 3 and hence $V_o = 1.94$ V.

It can be concluded that a four stage charge pump can enable the operation of PMS topology with only two input cycles given that capacitance of each stage are optimized using a similar approach to the one presented in [19]-[20].

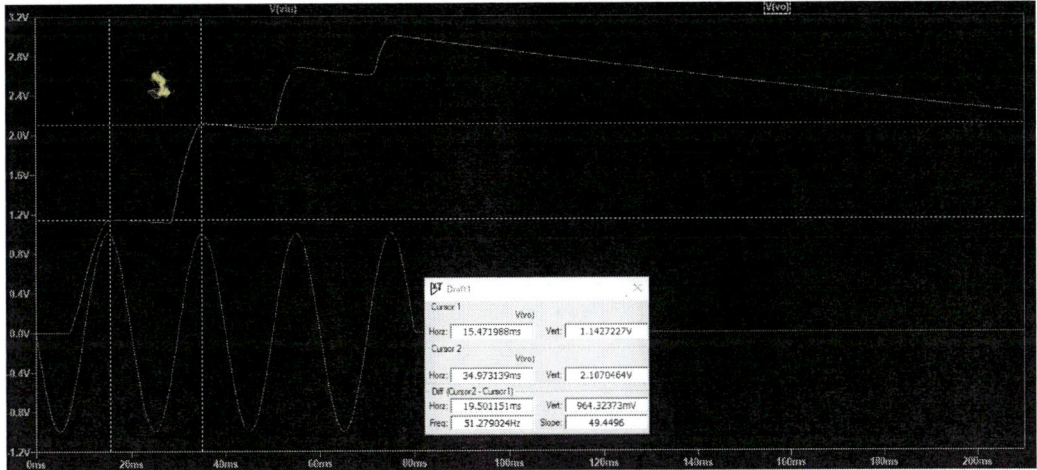

Figure 6: Spice simulation of four stage bias supply.

It can be inferred from Fig. 6, with the optimal capacitances, the bias supply reaches 2 V in the 2nd input cycle.

Intuitively, increasing the number of stages beyond four would not improve the rise time because as observed the each stage is activated subsequently after one complete cycle, which will increase the steady state voltage but would rather not improve the transients, which is of the interest here.

Experimental Results

Both the EMEH and the EMS were produced and tested. Figure 7 shows the impedance matching results. As it can be noted, both the harvester current and voltage has the same shape and phase shift, which means that the MPPT converter appears as a controlled resistive element to the EMEH, achieving the maximum power point operation. It can be noted also that the switching takes place after one cycle, which is the startup time needed to cold start the PMS as explained in the previous section. Figure 8 shows the input current and the output voltage under very light load conditions, i.e. with 500 KΩ. It can be seen from the figure that the voltage is stepped up from 1 V to around 12 V in a single step while achieving maximum power point tracking as can be noted from the input current waveform. That condition simulates the case where the energy produced from a single push button can be stored in output capacitor with very light load drawing current from that reservoir.

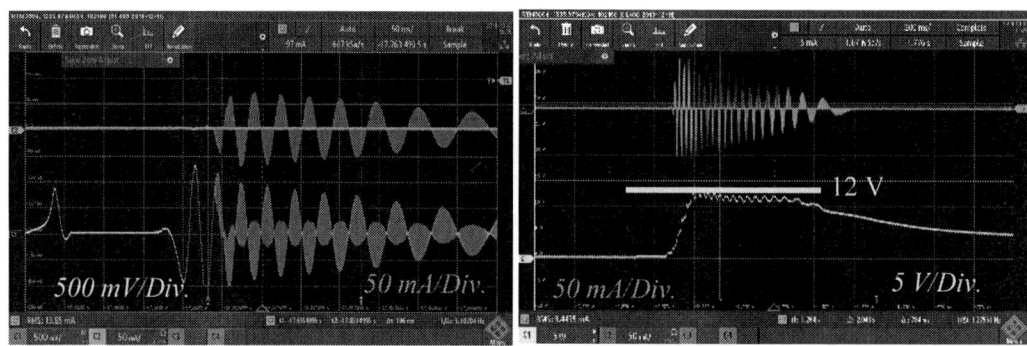

Figure 7: Input current and voltage of the EMEH.   Figure 8: Input current and output voltage.

Figure 9 shows the operation of the EH system when a high voltage LED is connected at the output of the EMS, a condition that is used to demonstrate the heavy load operation of the EH system. Figure 9 shows multiples pressing to show that the system can start without any pre charge even when the load is connected to the EMS. As it can be noted, that system can start if the output capacitors are completely depleted or if it has an arbitrary pre charge value.

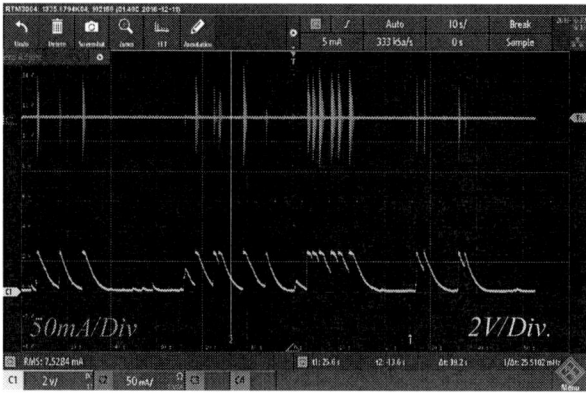

Figure 9: EH system operation with a high voltage LED.

Conclusions

A miniaturized and cost effective EMEH based on wire-wound technology is presented and shown in this paper. In addition, a modified self-powered maximum power point tracking based EMS is presented. The experimental results presented in this paper prove the effectiveness of the adopted approach.

References

[1] D. P. Arnold, "Review of Microscale Magnetic Power Generation," in *IEEE Trans. Mag.*, Vol. 43 no. 11, pp. 3940 – 3951, 2007.

[2] J. W. Matiko, N. J. Grabham, S. P. Beeby and M. J. Tudor, "Review of the application of energy harvesting in buildings," in *Measurement Science and Technology*, Vol. 25, no. 1, 2013.

[3] A. P-Vadean, P. P. Pop, T. Latinovic, C. Barz, and C. Lung, "Harvesting energy an sustainable power source, replace batteries for powering WSN and devices on the IoT," in *Proc. IOP Conf. Series: Materials Science and Engineering*, Vol. 200, pp. 1-10, 2017.

[4] B. Pozo, J. I. Garate, J. A. Araujo, and S. Ferreiro, "Energy Harvesting Technologies and Equivalent Electronic Structural Models – Review," in *Electronics*, Vol. 8 no. 5, 2019.

[5] Y. Kuang, T. Ruan, Z. J. Chew, and M. Zhu, "Energy harvesting during human walking to power a wireless sensor node," in *Sensors and Actuators A*, Vol. 254, pp. 69 – 77, 2017.

[6] J. Zhang, Z. Fang, C. Shu, J. Zhang, Q. Zhang, and C. Li, "A rotational piezoelectric energy harvester for efficient wind energy harvesting," in *Sensors and Actuators A*, Vol. 262, pp. 123 – 129, 2017.

[7] A.T.Papagiannakis, S.Dessouky, A. Montoya, and H. Roshani, "Energy Harvesting from Roadways," in *Procedia Computer Science*, Vol. 83, pp. 758 – 765, 2016.

[8] D. Mallick, P. Constantinou, P. Podder, and S. Roy, "Multi-frequency MEMS Electromagnetic Energy Harvesting," in *Sensors and Actuators A*, Vol. 264, pp. 247 – 259, 2017.

[9] M. Safaei, H. A. Sodano, and S. R. Anton, "A review of energy harvesting using piezoelectric materials: state-of-the-art a decade later (2008–2018)," in *Smart Materials and Structures*, Vol. 28, pp. 1-63, 2019.

[10] N. Thasni and K. Sebastian, "Review on Energy Harvesting and Data Collection," in *Proc. IOP Conf. Ser.: Mater. Sci. Eng.* Vol. 396, 2018.

[11] D. P. Arnold, F. Herrault, I. Zana, P. Galle, J-W. Park, S. Das, J. H. Lang and M. G. Allen, "Design optimization of an 8-Watt, microscale, axial-flux, permanent-magnet generator," in *J. Micromech. Microeng.*. Vol. 16 no. 9, pp. 290 – 296, 2006.

[12] H. Vocca and F. Cottone, "Kinetic Energy Harvesting, ICT - Energy - Concepts Towards Zero - Power Information and Communication Technology," *InTech*, pp. 25 – 48, 2014.

[13] M. Niroomand and H.R. Foroughi, "A rotary electromagnetic microgenerator for energy harvesting from human motions, " in *J. Appl. Research Techn.*, Vol. 14, pp. 259–267, 2016.

[14] D. Dinulovic, M. Brooks, M. Haug, and T. Petrovic, "Rotational Electromagnetic Energy Harvesting System," in *Physics Procedia*, vol. 75, pp. 1244 – 1251, 2015.

[15] D. Dinulovic, M. Shousha, M. Brooks, M. Haug and T. Petrovic, "Portable rotational electromagnetic energy harvester for IoT," *in Proc. IEEE Int. Mag. Conf.*, pp. 1-2, 2017.

[16] M. Shousha, D. Dinulovic, M. Haug, and A. Mahougb, "A Bias Supply Scheme for a Self-Powered EMS for Battery-less IoT Applications Powered by Electromagnetic Energy Harvesters," in Proc. 20th Annu. IEEE European Conf. Power Electron. Applicat., pp. P.1-P.6, 2018.

[17] M. Shousha, D. Dinulovic, M. Brooks, and M. Haug, "A miniaturized cost effective shared inductor based energy management system for ultra-low-voltage electromagnetic energy harvesters in battery powered applications," in Proc. Annu. *IEEE Appl. Power Electron. Conf. Expo.*, pp. 703-707, 2018.

[18] M. Shousha, D. Dinulovic, M. Haug, T. Petrovic and A. Mahgoub, "A Power Management System for Electromagnetic Energy Harvesters in Battery/Batteryless Applications," in *IEEE J. Emerg. Sel. Topics Power Electron.* Vol., no. , pp. 1-15, 2019.

[19] T. Tanzawa, "An Optimum Design for Integrated Switched-Capacitor Dickson Charge Pump Multipliers With Area Power Balance," in *IEEE Trans. Power Electron.*, vol. 29, no. 2, pp. 534-538, Feb. 2014.

[20] I. C. Kobougias and E. C. Tatakis, "Optimal design of a Half Wave Cockroft-Walton Voltage Multiplier with different capacitances per stage," *in Proc. Annu. Int. Power Electron. Motion Control*, pp. 1274-1279, 2008.

Comparison of optimized motor-inverter systems using a Stacked Polyphase Bridge Converter combined with a 3-, 6-, 9-, or 12-phase PMSM

Thilo Bringezu and Jürgen Biela
Laboratory for high power electronic systems (HPE), ETH Zürich
E-Mail: bringezu@hpe.ee.ethz.ch

Keywords

≪Modular converter≫, ≪Multiphase drive≫, ≪Permanent magnet motor≫, ≪System integration≫

Abstract

Stacked Polyphase Bridge Converters (SPB-C) are well suited for integrated motor drives, i.e. for systems that combine an electric motor with its supplying power electronic converter in one compact unit. A SPB-C requires a motor with multiple 3-phase winding systems. In the literature, the SPB-C is typically combined with motors whose winding systems are phase-aligned. This type of winding is magnetically equivalent to a 3-phase winding (with parallel branches). In this paper, the SPB-C with either 2, 3, or 4 modules is combined with a motor that has non-phase-aligned winding systems what results in a 6-, a 9-, or a 12-phase system, respectively. These multiphase systems are compared to 3-phase systems with the same number of modules with regard to efficiency, power density, and the possible degree of modularity and the benefits of the multiphase systems are evaluated.

1 Introduction

One solution to face the demand for more compact and efficient drive systems is to integrate the power electronic converter into the motor [1]. An example for such an integrated motor drive is given in Fig. 1 which consists of a tooth-coil wound PMSM for high-torque applications and a modular converter cooled by an end plate with cooling fins. Modular converter topologies, such as Stacked Polyphase Bridge Converters (SPB-C), are the preferred choice for this integration concept as the converter modularity ensures a very compact design. The SPB-C is introduced in [2] as a modular converter topology consisting of series con-

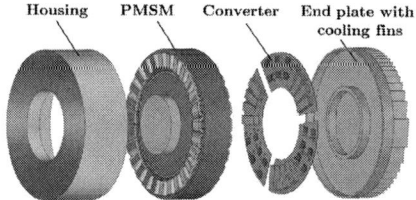

Figure 1: Basic concept of an integrated motor drive.

nected three-phase converters for hybrid electric vehicles. The topology of the SPB-C is shown in Fig. 2a for 2, 3, and 4 three-phase modules which are called SPB-modules in the following. Each SPB-module has a dc link capacitor and is connected to a 3-phase winding system of a motor in multi-star point connection. The main advantages of the SPB-C are: Low voltage semiconductor devices with low $R_{\text{ds,on}}$ can be used due to the series connection, fault-tolerance, and modularity ensuring a compact integrated system. The latter advantage is presented in different publications that analyze converter topologies suitable for an Integrated Modular Motor Drive (IMMD) [3], [4]. The IMMD-concept is introduced in [5] as a motor with segmented stator and axially integrated converter units leading to a compact modular system.

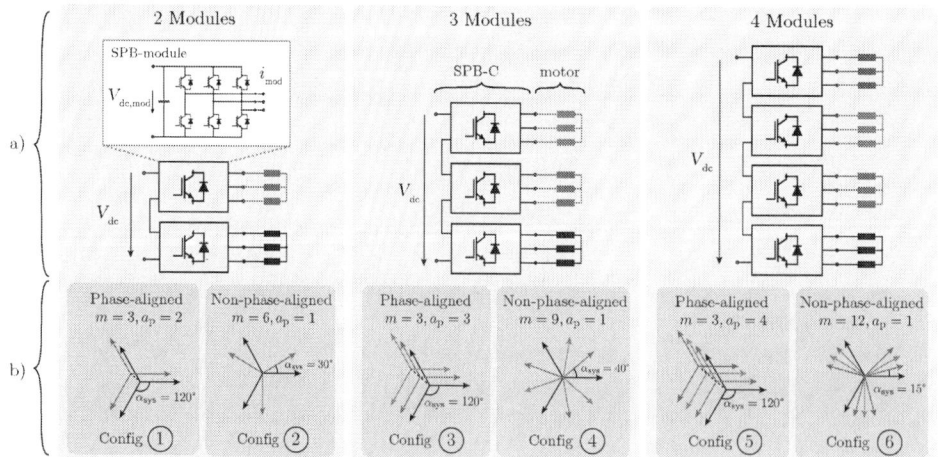

Figure 2: a) SPB-C with 2, 3, and 4 SPB-modules connected to motor windings [2]. b) For each SPB-C, two configurations of the magnetic winding axes are shown with m phases and a_p phase-aligned winding systems.

Publications [3], [6] consider the SPB-C in combination with modular three-phase motors, i.e. motors having as many phase-aligned three-phase winding systems as SPB-modules. However, the SPB-C is also suitable for motors with non-phase-aligned three-phase winding systems. In [4], the possibility of using either phase-aligned or non-phase-aligned winding systems is mentioned but the effect on the system performance is not analyzed. Therefore, this analysis is presented in this paper. Fig. 2b shows the magnetic winding axes for different possible alignments. The motor-converter configurations ①, ③, and ⑤ have phase-aligned three-phase winding systems resulting in 3-phase motors, whereas configurations ②, ④, and ⑥ have non-aligned three-phase winding systems, resulting in a 6-, a 9-, and a 12-phase motor, respectively.

The motivation for considering higher phase numbers is that higher phase numbers result in higher winding factors and consequently in systems with higher power density and/or efficiency as will be shown based on the design procedure described in section 2. The motor related models and design steps for the design procedure are presented in section 3. Thereafter, the converter related models and design steps are presented in section 4. Finally, the results of the design procedure, applied to all system configurations shown in Fig. 2, are presented and discussed for specifications given in Tab. IV-VI in section 5.

2 Design procedure

The main goal of this paper is to determine and to compare the system performance of the motor-inverter configurations ① to ⑥ (c.f. Fig. 2) in terms of the performance indicators efficiency, power density, and the possible degree of motor-converter modularity that is defined in section 3.2. A focus is put on the comparison of configurations ① & ②, ③ & ④, and of ⑤ & ⑥ because these pairs have the same amount of semiconductor devices at the same ratings and thus comparable costs.

The procedure used for an optimal system design in terms of the aforementioned performance indicators is shown in Fig. 3. First, the system specifications are defined which consist of the rated torque, the rated speed, the dc-link voltage, the selected semiconductor devices, and the considered system configuration ①/.../⑥ shown in Fig. 2. For these specifications, a motor is designed in the steps S1 and S2. In **S1**, the number of stator slots and the number of rotor pole pairs are selected based on the criteria described in 3.2. This selection already determines the possible motor-converter modularity as will be explained later. In **S2**, the motor dimensions are optimized and electrical motor parameters are calculated using the optimization routine presented in 3.3. In the next steps S3-S8, the switching frequency is optimized. For each switching frequency selected in **S3**, the motor/converter currents are calculated in **S4** using the equations for sinus-PWM given in [7] and the motor model equations given in 3.1, where the electrical motor parameters resulting from S2 are inserted. In **S5**, the dc link is dimensioned using the method described in 4.1. In **S6** and **S7**, the system losses and the system volume are calculated, respectively. The used volume models and the loss models are described in 3.4 for the motor and in 4.2 for the converter. The sum of all volumes and the sum of all losses are finally used to calculate the system efficiency and power density in **S8**.

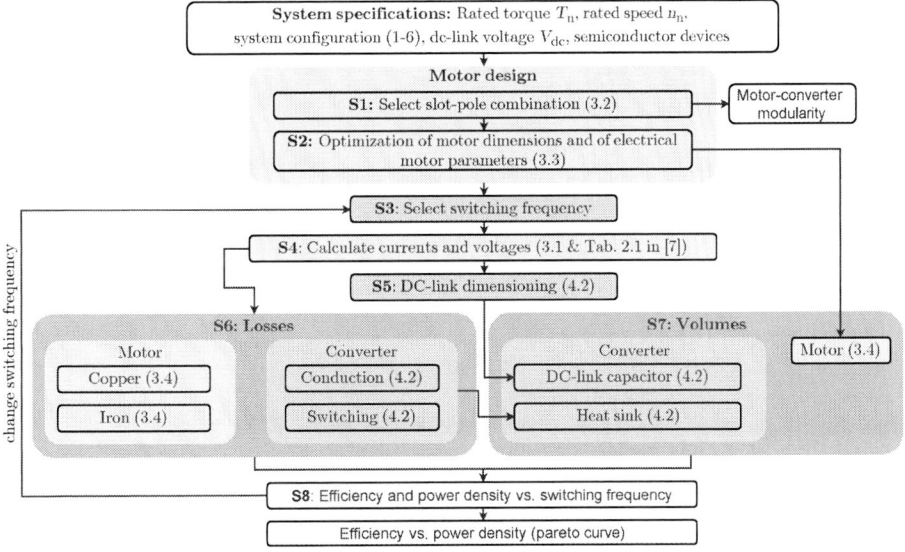

Figure 3: Design procedure for an optimized motor-converter system using the SPB-C and a tooth-coil wound PMSM for given system specifications. Optimization goals: high efficiency, high power density, and high degree of motor-converter modularity. Yellow: Motor modelling and design steps. Green: Converter modelling and design steps.

Table I: Used motor variables.

$A_{c,\text{skin}}(f)$	Effective conductor cross sectional area	m	Number of phases (distinct winding axes)
A_δ	Air gap area $(2\pi r_g l_m)$	μ_0	Vacuum permeability
α	Angular position on the stator	N	Number of stator slots
$\alpha_{N[-\pi,\pi]}$	El. angle between adjacent slots in $[-\pi,\pi]$	$N_i(\alpha)$	Winding function of the i-th phase
α_{sys}	Angle between winding axes (cf. Fig. 2)	N_l	Number of tooth-coils
α_z	Angle between phasors in the star of slots	n_l	Number of winding layers (1 or 2)
a_p	Number of phase-aligned winding axes	n_n	Rated speed
$\hat{B}_{\delta,1}$	Fundamental air gap flux density	n_t	Number of turns per tooth-coil
b_s	Slot opening	ω	Electrical angular frequency
D	Outer stator diameter	$p_{\text{Cu,loss}}$	Copper loss density (related to motor surface)
δ	Air gap length	PF	Power factor
h_M	Length of rotor magnets in radial direction	p	Number of pole pairs
$\vec{i}_{\text{dq}k}$	Motor current in the k-th dq-subspace	$\vec{\psi}_{\text{dq}k}$	Motor flux linkage in the k-th dq-subspace
J	Current density in the stator winding	$\vec{\psi}_{\text{PM}k}$	Magnet flux linkage in the k-th dq-subspace
k_{fill}	Copper fill factor	$R(f)$	AC phase resistance
k_p	Pole embrace of rotor magnets	r_g	Air gap radius
k_w	Winding factor	ρ_{Cu}	Copper conductivity
$L_{\text{dq}k}$	Motor inductance in the k-th dq-subspace	T	Electromagnetic torque
L_{ij}	(i,j)-th motor inductance matrix entry	T_n	Rated torque
l_m	Motor length	t	Winding periodicity number
$\bar{l}_t$	Mean length of one winding turn	$\vec{v}_{\text{dq}k}$	Motor voltage in the k-th dq-subspace
M	Modulation index		

3 Motor modelling and design

In this section, the motor model and the motor design steps of the system design procedure in Fig. 3 are explained. The used motor variables are summarized in Tab. I.

3.1 Electromechanical model for multiphase non-salient PMSMs

An electromechanical motor model comprising a voltage, a flux, and a torque equation in dq-coordinates is given in [8] in (5.9) for electrically excited salient multiphase synchronous machines. This model is adapted to non-salient PMSMs by replacing the product of the excitation current and of the main inductance with the magnet flux linkage $\psi_{\text{PM}k}$ and by setting the inductance in the d-axis equal to the inductance in the q-axis. These adaptations result in the following set of equations

$$\vec{v}_{\text{dq}k} = R\vec{i}_{\text{dq}k} + \text{j}k\omega\vec{\psi}_{\text{dq}k} + \frac{d\vec{\psi}_{\text{dq}k}}{dt} \tag{1}$$

$$\vec{\psi}_{\text{dq}k} = L_{\text{dq}k}\vec{i}_{\text{dq}k} + \vec{\psi}_{\text{PM}k} \tag{2}$$

$$T = \frac{m}{2}p\sum_{k=1,3,\dots}^{k_{\max}} k\psi_{\text{PM}k}i_{\text{q}k} \overset{!}{=} \frac{m}{2}p\psi_{\text{PM}1}i_{\text{q}1} \tag{3}$$

for $k = 1,3,\dots k_{\max}$ with $k_{\max} = m-2$, if m is odd and $k_{\max} = m-1$, if m is even. In multiphase machines ($m > 3$), different harmonics of phase voltages, currents and flux linkages are mapped into $k_{\max} > 1$ dq-subspaces [8], in contrast to 3-phase machines where $k_{\max} = 1$. This paper does not consider harmonic current injection ($i_{\text{q}k} \approx 0$ for $k > 1$) so that only the first summand in (3) remains.

3.2 Selection of a slot-pole combination

In this section, the criteria used for selecting a slot-pole combination, i.e. a combination of the number of stator slots N and of the number of rotor pole pairs p are described (step S1 in Fig. 3). There, tooth-coil windings (double-layer and single-layer) are considered because of their shorter end windings compared to distributed windings, which is important for high torque motors whose diameter is large compared to its length. The considered criteria are summarized in Tab. II and explained in the following. Criteria C1 to C5 are commonly used in motor design whereas criterion C6 is particularly added to determine how suitable a slot-pole combination is for an integrated modular motor-converter system.

C1: An m-phase, n_l-layer tooth-coil winding for a motor with N stator slots, p rotor pole pairs, and a_p phase-aligned winding systems is feasible if the two conditions C1a and C1b are fulfilled. **C1a** ensures that the number of stator slots can be equally shared among all phase windings and that α_{sys} matches an integer multiple of the electrical distance of slots along the stator circumference. **C1b** ensures that a_p phase-aligned winding systems are possible. The analytical expressions for C1a and C1b in (4) are valid for single and for double layer tooth-coil windings. The expressions are derived based on [9] and [10] by defining the number of layers n_l (1 for single layer,

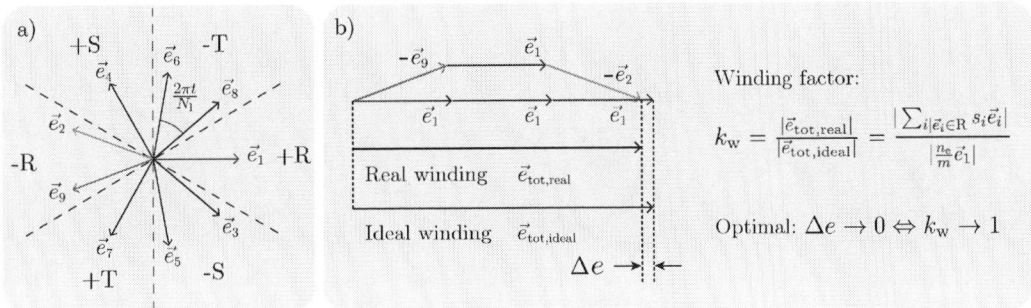

Figure 4: a) Star of slots for an example winding with $m = 3$, $n_l = 2$, $N = 27$ and $p = 12$. b) Derivation of the winding factor from the star of slots.

2 for double layer windings) and the number of tooth-coils $N_l := n_l N/2$.

$$\mathbf{C1a}: \frac{\alpha_{\text{sys}} N_l}{2\pi t} \in \mathbb{N} \quad \wedge \quad \mathbf{C1b}: \frac{a_{\text{p,max}}}{a_{\text{p}}} \in \mathbb{N} \quad \text{with } a_{\text{p,max}} = \begin{cases} 2t, & \text{if } \frac{N_l}{mt} \text{ is even} \\ t, & \text{if } \frac{N_l}{mt} \text{ is odd} \end{cases} \tag{4}$$

The winding periodicity number t in (4) is given by the greatest common divider (gcd) of N_l and p. This number indicates how often a so-called base winding is repeated along the stator circumference.

C2: The fundamental winding factor $k_{\text{w},1}$, further called winding factor k_{w}, can be defined as $\psi_{\text{PM1}}/w\phi_{\text{PM1}}$, where w is the number of turns in series per phase of the stator winding and ϕ_{PM1} is the fundamental air gap flux of one rotor magnet [11]. Therefore, the product $k_{\text{w}} \cdot w$ describes the effective number of turns linked to the complete rotor flux ϕ_{PM1}. In the considered design approach where torque, speed and available motor voltage are imposed, a higher winding factor can be used to reduce the number of turns or to reduce the flux by decreasing the motor size. These changes have an impact on efficiency and/or power density as will be shown later. Here, the winding factor is determined using the method of the star of slots [9], [10], illustrated in Fig. 4 for one example configuration. The "star" in Fig. 4a consists of so-called slot phasors indicating the voltages induced by the rotating excitation field in the coil sides of one layer of one base winding. The phasors are placed at an angular distance of $\alpha_z = 2\pi t/N_l$ resulting in a number of $n_e := N_l/t$ phasors. Each of the slot phasors, denoted by $\vec{e}_i$, is assigned to one phase with a positive or negative sign s_i according to the zones labeled +R, -T, etc. that are defined by the considered winding axes configuration (cf. Fig. 2b). Then, the winding factor is calculated as the ratio of the total induced phase voltage to the induced phase voltage of an ideal reference winding whose slot phasors lie on the same axis, as shown in Fig. 4b. There, the coil pitch and the winding distribution are inherently taken into account [9].

C3: Eddy currents in the rotor magnets and in the rotor iron due to parasitic stator magnetomotive force harmonics should be negligible. In this paper, a relative indicator $p_{\text{r,loss}}$ derived in [3], [12] is used to take this loss component into account. Slot-pole combinations with $p_{\text{r,loss}} > 100$ are discarded in the analysis.

C4 & C5: As a further criterion, $\gcd(N_l, 2p) > 1$ is applied in order to avoid unbalanced magnetic pull that can lead to motor vibration and noise [13]. Furthermore, a high value of the lowest common multiple (lcm) of N and $2p$ indicates low cogging torque [13]. Here, $\text{lcm}(N, 2p) > 50$ is chosen.

C6: To evaluate how suitable a winding is for an integrated modular motor-converter system, three degrees of modularity for motor-converter systems are defined (cf. Fig. 5). The three degrees A, B and C define to what extend the stator and the converter can be modularized into motor-converter modules. In Fig. 5, coils of the same color are connected to the same SPB-module. Modularity degree A means that only converter modules but no combined motor-converter modules are possible because the coils connected to one SPB-module (i.e. one color in Fig. 5) are spread around the stator. Degree B means that combined motor-converter modules are possible where each module consists of one stator segment and one half-bridge of the SPB-C. Degree C means that motor-converter modules are possible where each module consists of one stator segment and one complete SPB-module. The disadvantage of degree B compared to C is that half-bridges of the same three-phase SPB-module are spread around the stator, which requires wiring across converter modules. For each degree of modularity, conditions on

Table II: Criteria C1-C6 for selecting a slot-pole combination.

C1	Winding feasibility	C4	$\gcd(N_l, 2p) > 1$ to avoid unbalanced magnetic pull
C2	High winding factor k_{w}	C5	High $\text{lcm}(N, 2p)$ for low cogging torque
C3	Low rotor loss indicator $p_{\text{r,loss}}$	C6	High degree of modularity

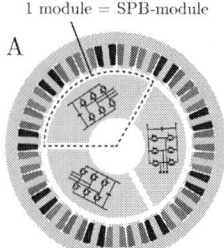
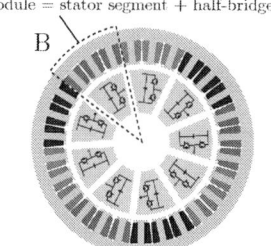
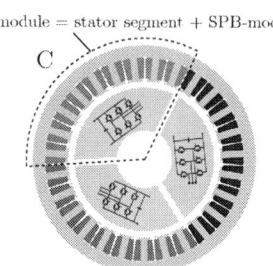

Figure 5: Defined degrees of modularity A, B, and C of motor-converter systems using the SPB-C and tooth-coil windings for exemplary motors with $N = 27$, $a_p = a_{p,max} = 3$, $n_l = 2$ and A: $p = 1$ & $m = 9$, B: $p = 13$ & $m = 9$, and C: $p = 12$ & $m = 3$. Windings of the same color (green/red/black) are connected to the same SPB-module as also indicated in Fig. 2.

the slot-pole combination are analyzed. For modularity degree A, no condition applies. For modularity degree B, the conditions in (5) are derived based on the method of the star of slots where $\alpha_{N[-\pi,\pi]} := 2\pi p/N_l$ is the electrical angle between two mechanically adjacent slots in case of a double layer winding and between every second slot in case of a single layer winding. For modularity degree C, it is found that (5) must be fulfilled and that the winding axes of phases connected to one SPB-module must be adjacent. Fig. 2 shows that this is only the case if $m = 3$. Hence, the considered multiphase-motors cannot reach the highest degree of modularity.

$$
\begin{cases}
a_p = a_{p,max} \wedge \left(\frac{N_l}{mt} = 2 \vee \alpha_{N[-\pi,\pi]} = \pm\alpha_z \vee \alpha_{N[-\pi,\pi]} = \pm(\frac{N_l}{2t} - 1)\alpha_z \right), & \text{if } \frac{N_l}{mt} \text{ is even} \\
a_p = a_{p,max} \wedge \left(\frac{N_l}{mt} = 1 \vee \alpha_{N[-\pi,\pi]} = \pm(\frac{N_l}{2t} - \frac{1}{2})\alpha_z \right), & \text{if } \frac{N_l}{mt} \text{ is odd}
\end{cases}
\tag{5}
$$

3.3 Optimization of motor dimensions and of electrical motor parameters

In the following, step S2 of the design procedure shown in Fig. 3 is explained. In this step, the motor length l_m and split ratio χ, as well as the electrical motor parameters R, L_{dqk}, and ψ_{PM1} are determined for each of the system configurations ①-⑥ (cf. Fig. 2). The split ratio is defined as the ratio of rotor diameter to outer stator diameter. The following approach is chosen to compare the configurations with regard to power density and efficiency: The three-phase motors (systems ①/③/⑤ in Fig. 2) are designed for maximum power density. For the multiphase motors (systems ②/④/⑥ in Fig. 2), two variants, "a" and "b", are defined. In variant "a" (②ₐ/④ₐ/⑥ₐ), the multiphase motors are designed for maximum power density. In variant "b" (②ᵦ/④ᵦ/⑥ᵦ), the multiphase motors are designed with the same outer dimensions as the three-phase motors in order to compare the achievable efficiencies when the same volume/power density is imposed.

The motor design routine for maximum power density is illustrated in Fig. 6. First, motor specifications are defined. The specifications that are used for all system configurations are summarized in Tab. IX. Therein, T_n, n_n, and D are selected based on application requirements. For the constraints (J_{max}, $p_{Cu,loss,max}$ & PF_{min}) and for the other specifications in Tab. IX, reasonable choices are made based on literature [14] in order to limit the calculation effort. For the configuration dependent specifications (dc link voltage and slot-pole combination) it is referred to Tab. V and Tab. VIII.

After defining the specifications, the design routine starts in **M1** (cf. Fig. 6) by selecting values for l_m and χ. In **M2**, $\hat{B}_{\delta,1}$, A_δ, and $n_t \cdot i_{q1}$ are calculated using the Carter factor concept [14] and (3), where sinusoidal currents and $B_{fe} \leq B_{fe,sat}$ (no saturation) are assumed. In **M3-M5**, the number of turns per tooth-coil n_t is determined in an internal iteration ensuring that the modulation index $M := 2\hat{V}_{dq1}/V_{dc,mod}$ matches the reference value M_{ref} in steady state, where $\hat{V}_{dq1}$ is calculated using (1). In each iteration step (**M4**), the electrical model parameters R, L_{dqk}, and ψ_{PM1} are recalculated using (6)-(8).

$$
R(f) = \frac{n_t N_l}{m a_p^2} \rho_{Cu} \frac{\bar{l}_t}{A_{c,skin}(f)}
\tag{6}
$$

$$
L_{ij} = \frac{\mu_0 r_g l_m}{\delta} \int_0^{2\pi} N_i(\alpha) N_j(\alpha) d\alpha + L_{ij,slot}
\tag{7}
$$

$$
\psi_{PM1} = k_w \frac{n_t N_l}{m a_p} l_m \frac{2 r_g}{p} \hat{B}_{\delta,1}
\tag{8}
$$

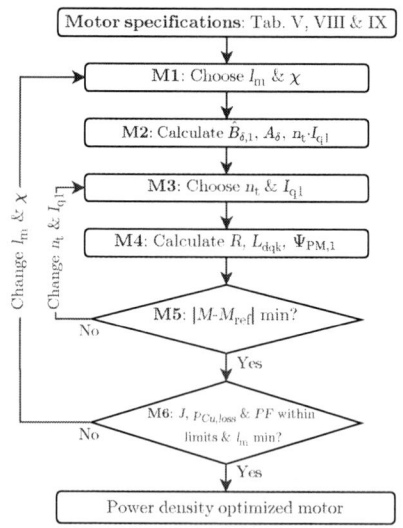

Figure 6: Optimization of the motor length and split ratio for a PMSM with maximum power density and specifications given in Tab. V, VIII & IX.

L_{ij} in (7) is the (i,j)-th entry of the winding inductance matrix. The integral term corresponds to the main and the leakage harmonic inductance [15]. The slot leakage inductance $L_{ij,\text{slot}}$ is calculated based on [16]. Inductance L_{dqk} is obtained from the k-th diagonal entry of the matrix $[T]_m [L_{ij}] [T]_m^{-1}$ where $[T]_m$ is the dq-transformation matrix for an m-phase system derived from [17]. In **M6**, the constraints are checked and the motor with minimum length among all valid results is returned. This motor is used for the further design steps S4-S8 (cf. Fig. 3).

3.4 Motor volume and losses

In steps S6 and S7 of the design procedure shown in Fig. 3, the motor volume and the motor losses are calculated. The motor volume is approximated by the cylindrical stator of length l_m and diameter D. Concerning losses, the copper losses are calculated using $\sum R(f)I(f)^2$. For the iron loss calculation in the stator, a reluctance model is used to estimate the magnetic flux density $\hat{B}$ in the stator teeth and in the yoke elements. Then, iron losses are calculated in the frequency domain using the Bertotti equation $p = K_{\text{hyst}}\hat{B}^2 f + K_{\text{ec}}(\hat{B}f)^2 + K_{\text{ex}}(\hat{B}f)^{1.5}$ [18] where the loss coefficients are fitted to loss curves provided by the data sheet of the used laminated steel.

4 Converter modelling and design

In this section, the converter model and the converter design steps of the system design procedure in Fig. 3 are explained.

4.1 DC link dimensioning

For the dc link, film capacitors are considered as they are well suited for the high temperatures occurring in integrated motor-inverter systems [4]. In general, dc link capacitors are selected based on the rated voltage, the capacitance required to limit the dc link voltage ripple, and the current rating. The minimum required voltage rating is determined by the module dc link voltage $V_{\text{dc,mod}}$. The capacitance required to keep the dc link voltage ripple below a predefined maximum $\Delta v_{\text{dc,max}}$ is separately determined for static load conditions ($C_{\text{dc,min,s}}$) and for dynamic load conditions ($C_{\text{dc,min,d}}$) with the methods described in sections 4.1.1 and 4.1.2, respectively. The current rating requirement is also transformed into a minimum capacitance ($C_{\text{dc,min,th}}$) as is explained in section 4.1.3. Hence, the overall condition on the capacitance rating results in $C_{\text{dc,min}} = \max\{C_{\text{dc,min,s}}, C_{\text{dc,min,d}}, C_{\text{dc,min,th}}\}$.

4.1.1 Capacitance required for steady state

In steady state, the voltage ripple is caused by the PWM. Analytical formulae for calculating $C_{\text{dc,min,s}}$ are derived in [7] for various PWM techniques and a current-voltage phase shift angle of $\phi = 0$. As inductive loads with $\phi > 0$ are considered here, a similar procedure is applied to derive (9), valid for arbitrary ϕ and sinus-PWM:

$$C_{\text{dc,min,s}} = \frac{I_{\text{mod}}}{\Delta v_{\text{dc,max}}f_s} \left| \frac{3}{\sqrt{8}}M\cos(\phi)(\delta_{000}(\alpha_{\text{crit}}) + \delta_{100}(\alpha_{\text{crit}})) - \sqrt{2}\cos(\alpha_{\text{crit}} - \phi)\delta_{100}(\alpha_{\text{crit}}) \right| \tag{9}$$

$$\delta_{000}(\alpha) = \frac{1}{2} - \frac{M}{2}\cos(\alpha), \qquad \delta_{100}(\alpha) = \frac{\sqrt{3}}{2}M\cos\left(\frac{\pi}{6} + \alpha\right) \tag{10}$$

I_{mod} is the phase rms-current of one SPB-module, f_s is the switching frequency, and $\delta_{000}(\alpha_{\text{crit}})$ and $\delta_{100}(\alpha_{\text{crit}})$ are the duty cycles of the switching states <000> and <100> at the angle α_{crit} of the reference voltage vector. The values for α_{crit} are determined numerically and are shown in Fig. 7. To derive an analytical expression, these angles are least squares fitted to a 2D polynomial, given in (11) with coefficients given in Tab. III. The duty cycle functions in (10) are already known from [7].

$$\alpha_{\text{crit}} = \begin{cases} p(M,\phi) = c_1\phi^3 + c_2\phi^2 + c_3M\phi + c_4\phi + c_5M + c_6, & \text{if } p(M,\phi) \in [0,60] \\ 0, & \text{if } p(M,\phi) \notin [0,60] \end{cases} \tag{11}$$

Fig. 8 shows the resulting normalized dc link capacitance $C_{\text{dc,min,n}} := C_{\text{dc,min,s}}/(I_{\text{mod}}/(\Delta v_{\text{dc,max}}f_s))$. The result shows that the minimum required dc link capacitance increases towards the maximum modulation index and towards the phase shift angle $\phi = 90\,\text{deg}$. Therefore, dimensioning the dc link capacitance for the angle $\alpha_{\text{crit}} = 0$ is not sufficient but the presented formula in (9) should be taken into account.

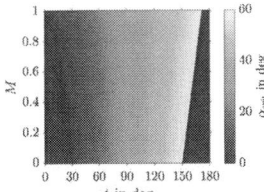 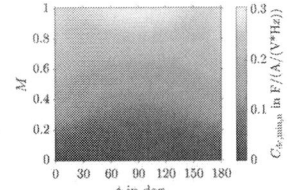

Table III: Coefficients of (11).

c_1	0.03408
c_2	-0.1606
c_3	-0.1852
c_4	0.7286
c_5	0.2910
c_6	-0.3567

Figure 7: Critical voltage angle α_{crit} vs. modulation index M and phase shift ϕ.

Figure 8: Normalized capacitance $C_{\text{dc,min,n}}$ vs. modulation index M and phase shift ϕ.

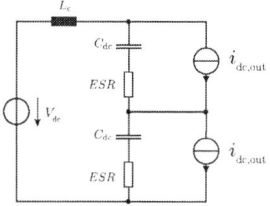
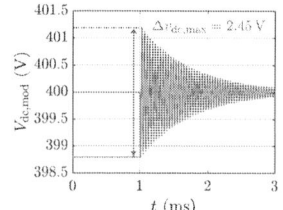
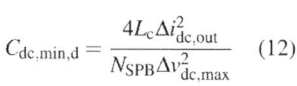

$$C_{\text{dc,min,d}} = \frac{4L_c \Delta i_{\text{dc,out}}^2}{N_{\text{SPB}} \Delta v_{\text{dc,max}}^2} \quad (12)$$

Figure 9: Model of a 2-module SPB-C with converter modules modelled as current sources.

Figure 10: Module dc link voltage $V_{\text{dc,mod}}$ for a load step from 0 A to 2 A, $L_c = 7.5\,\mu\text{H}$, $C_{\text{dc,mod}} = 10\,\mu\text{F}$, $ESR = 10\,\text{m}\Omega$.

4.1.2 Capacitance required for load steps

If cables are used to connect the motor-inverter system to the dc source, load steps lead to an additional ripple on the dc link voltage due to the cable inductance. This effect is analyzed using the model of the 2-module SPB-C shown in Fig. 9 where the SPB-modules are modelled with controlled current sources $i_{\text{dc,out}}$. Here, the PWM induced current ripple is neglected so that $i_{\text{dc,out}} = 3/4 M \hat{i}_{\text{mod}} \cos(\phi)$. In Fig. 10, the module dc link voltage is shown for a load step of 2 A and a cable inductance of 7.5 μH. The equation for the minimum required dc link capacitance $C_{\text{dc,min,d}}$ is given in (12) for the maximum voltage ripple $\Delta v_{\text{dc,max}}$ and load step $\Delta i_{\text{dc,out}}$. There, L_c is the cable inductance and N_{SPB} is the number of SPB-modules.

4.1.3 Capacitor current rating requirement

The current rating requirement is based on the thermal condition $T_c < T_{c,\text{max}}$ with $T_c \approx T_{\text{amb}} + ESR(f_s) I_c^2 R_{\text{th,c}}$ where T_c, $T_{c,\text{max}}$ and T_{amb} are the capacitor operating, maximum rated, and ambient temperature, ESR and $R_{\text{th,c}}$ are the equivalent series and thermal resistance of the capacitor, and I_c is the rms capacitor current. I_c is calculated analytically based on [19]. The thermal condition is translated into a required capacitance $C_{\text{dc,min,th}}(I_c, f_s, T_{\text{amb}}, T_{c,\text{max}})$ which is numerically determined based on data sheet values of a selected capacitor series.

4.2 Volume and loss estimation

Fig. 1 shows the considered motor-converter integration concept. The converter volume is approximated as the sum of the populated PCB volume and the heat sink volume where the populated PCB volume is defined here by the smallest possible cylindrical box around the board with all its components. The height of this box is imposed by the highest component on the PCB which is usually the dc link capacitor. For small dc link capacitors, a minimum height of the populated PCB $h_{\text{PCB,min}}$ is defined. The available dc link footprint area is estimated in Fig. 11 for a converter with 2, 3, or 4 SPB-modules. The capacitor height results from this area and the volume that is given by (13) [7]. The coefficients in (13) are given in Tab. X for the selected high energy density capacitor series.

$$Vol_c(C, V_c) = k_1 C V_c^2 + k_2 C V_c + k_3 V_c + k_4 \quad (13)$$

The heat sink shown in Fig. 1 is considered to be thermally isolated from the motor. The heat sink volume is estimated based on the semiconductor losses, their thermal resistances given in Tab. VI, and a cooling system performance index CSPI of $4.4\,\text{W}\,\text{dm}^{-3}\,\text{K}^{-1}$ that was already experimentally achieved for naturally convective cooled heat sinks [20]. The semiconductor losses (conduction and switching) are calculated using the equations from [7]. The transistor switching losses are derived based on data sheet values and the method described in [21].

5 Application and results

The system design procedure in Fig. 3 is applied to all system configurations ① to ⑥ (cf. Fig. 2). The system specifications are presented in section 5.1. The results of the design steps S1, S2, and S3-S8 (cf. Fig. 3) are presented in sections 5.2, 5.3, and 5.4, respectively.

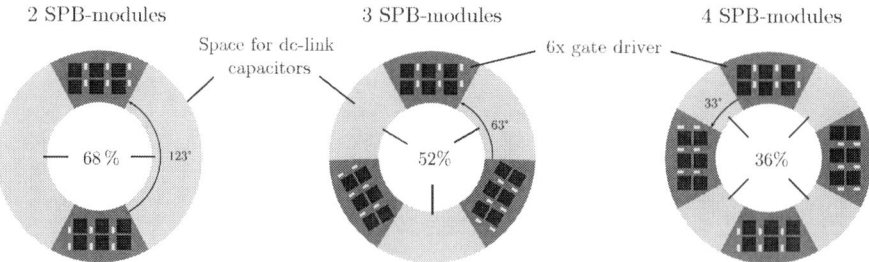

Figure 11: PCB-topside of a SPB-C with 2, 3, and 4 SPB-modules for estimating the PCB area available for dc-link capacitors. Transistors are mounted on the bottom-side. The used gate-driver dimensions result from gate-driver *1EDS5663H* (Infineon).

Table IV: System specifications used for all system configurations ①-⑥ (cf. Fig. 2) in the design procedure (Fig. 3).

Rated torque	T_n	(Nm)	50
Rated speed	n_n	(rpm)	300
Heat sink	$CSPI$	$(\mathrm{W\,K^{-1}\,dm^{-3}})$	4.4
Cable inductance	L_c	(µH)	7.5
Min. PCB height	$h_{PCB,min}$	(mm)	10
Max. DC link voltage ripple	$\Delta v_{dc,max}/V_{dc,mod}$	(-)	0.02
Max. load step	$\Delta i_{dc,out}/I_{mod}$	(-)	1
Max. junction temperature	$T_{jc,max}$	(°C)	125
Ambient temperature	T_{amb}	(°C)	55

Table V: Considered dc link voltages in V for each system configuration ①-⑥ (cf. Fig. 2) and for device blocking voltages of 200 V/600 V.

Config	$V_{ds,max}$	V_{dc}	$V_{dc,mod}$
① & ②	200	266	133
③ & ④		400	133
⑤ & ⑥		533	133
① & ②	600	800	400
③ & ④		1000	333
⑤ & ⑥		1000	250

Table VI: Specifications of the selected transistors (dimensions: $w \times h \times l$).

Transistor	$V_{ds,max}(\mathrm{V})$	$R_{ds,on}(\mathrm{m\Omega})$	$R_{th,jc}(\mathrm{K/W})$	$w(\mathrm{mm})$	$l(\mathrm{mm})$	$h(\mathrm{mm})$
IGOT60R070D1	600	70	1	15.9	14.2	3.5
EPC2034C	200	8	0.3	4.6	2.2	0.79

5.1 System specifications

The system specifications used for all system configurations ①-⑥ (cf. Fig. 2) are summarized in Tab. IV. GaN-HEMTs are used as semiconductor devices to achieve low losses. To evaluate the system performance at different dc link voltage levels, device blocking voltages of 200 V and 600 V are considered. Tab. VI lists the specifications of the selected GaN-HEMTs with these blocking voltages and lowest available $R_{ds,on}$ values. Tab. V summarizes the dc link voltage V_{dc} and the module dc link voltage $V_{dc,mod}$ for each configuration ① to ⑥. There, V_{dc} is limited to 1000 V to stay within low-voltage standards.

5.2 Selection of slot-pole combinations

Tab. VII shows the comparison of slot-pole combinations based on the criteria described in 3.2 for all configurations ①-⑥, slot numbers $N \in [1, 40]$, and pole pair numbers $p \in [1, 20]$, for single and for double layer windings ($n_l = 1$, 2). Columns of pole pair numbers p and rows of slot numbers N with only grey entries (not feasible or $k_w < 0.9$) are not shown for the sake of brevity. The rows are ordered in such a way that the results of the configuration pairs ① & ②, ③ & ④, and ⑤ & ⑥ lie next to each other to enable a direct comparison.

From Tab. VII, the following results: In all (N, p)-combinations where both configurations are feasible (marked with a star), the multiphase configuration (②/④/⑥) has a higher winding factor than the 3-phase configuration (①/③/⑤). On average, only considering the relevant/green combinations, the increase in k_w is 3.7%. As expected, only three-phase configurations can reach the highest modularity degree C. In all cases where the multiphase configurations ensure the modularity degree B, the slot-pole combination suffers from unbalanced magnetic pull (orange) or high rotor losses (red). Therefore, combined motor-converter modularity is practically not feasible for multiphase motors with the considered slot-pole combinations. Furthermore, there are significantly less feasible slot-pole combinations for single layer than for double layer windings. This is due to condition C1a given in (4) that requires an even number of stator slots N for single layer windings.

For the next design steps, only double layer windings are considered and one recommended (green) slot-pole combination is chosen for each configuration ①-⑥. The selected combinations are summarized in Tab. VIII.

5.3 Motor dimensions and electrical motor parameters

The motor length and the electrical motor parameters resulting from step S2 of the system design procedure (Fig. 3) are summarized in Tab. XI, for all system configurations ①-⑥$_{a/b}$ (cf. section 3.3). As explained in 3.3, the multiphase motors of variant "b" are derived by choosing the same dimensions as those of the three-phase motors, i.e. l_m and χ are fixed for system configurations ②$_b$, ④$_b$, and ⑥$_b$ in the optimization routine shown in Fig. 6. The motor parameters in Tab. XI are inserted into (1)-(3) in S4 (Fig. 3) of the design procedure. S4 is part of the switching frequency optimization (S3-S8 in Fig. 3), the results of which are presented in the following section.

Table VIII: Selected slot-pole combinations for system configurations ①-⑥ (cf. Fig. 2).

Configuration	①	②	③	④	⑤	⑥
N	36	36	27	27	24	24
p	17	17	12	12	10	11
m	3	6	3	9	3	12
k_w	0.953	0.986	0.945	0.985	0.933	0.991

Table VII: Winding factor k_w and degree of modularity (A/B/C) (cf. Fig. 5) of tooth-coil windings for slot-pole combinations with $N \in [1,40]$, $p \in [1,20]$, and system configurations ①-⑥ (cf. Fig. 2). Other criteria of Tab. II are color-coded.

Double layer windings

N	Config	4	5	7	8	10	11	12	13	14	15	16	17	19	20
Configurations ① & ②															
12	①		0.933-C	0.933-C⋆									0.933-C	0.933-C	
	②		0.966-A	0.966-A									0.966-A	0.966-A	
18	①			0.902-A	0.945-C	0.945-C	0.902-A								
	②														
24	①					0.933-A	0.949-C⋆		0.949-C	0.933-A⋆					
	②					0.966-A	0.983-A		0.983-A	0.966-A					
30	①								0.936-A	0.951-C		0.951-C	0.936-A		
	②														
36	①									0.902-A	0.933-A⋆	0.945-A	0.953-A	0.953-C⋆	0.945-A
	②										0.966-A		0.986-A	0.986-A	
Configurations ③ & ④															
9	③	0.985-B	0.985-B						0.985-B	0.985-B					
	④														
18	③			0.940-A	0.985-A	0.985-A	0.940-A								
	④														
27	③						0.945-C⋆				0.945-C⋆				
	④					0.914-A	0.954-A	0.985-A	0.994-B	0.994-B	0.985-A	0.954-A	0.914-A		
36	③										0.933-A				
	④								0.903-A	0.940-A		0.985-A	0.992-A	0.992-A	0.985-A
Configurations ⑤ & ⑥															
24	⑤					0.933-C				0.933-C					
	⑥						0.991-A		0.991-A						
36	⑤										0.902-A		0.945-C		0.945-C
	⑥														

Single layer windings

N	Config	4	5	7	8	10	11	12	13	14	15	16	17	19	20
Configurations ① & ②															
12	①		0.966-C	0.966-C									0.966-C	0.966-C	
	②														
24	①					0.966-A	0.958-C		0.958-C	0.966-A					
	②						0.991-A		0.991-A						
36	①									0.902-A	0.966-A	0.945-A	0.956-A	0.956-C	0.945-A
	②														
Configurations ③ & ④															
18	③			0.940-B	0.985-B	0.985-B	0.940-B								
	④														
36	③										0.966-A				
	④								0.906-A	0.940-A		0.985-A	0.996-A	0.996-A	0.985-A
Configurations ⑤ & ⑥															
24	⑤				0.966-C				0.966-C						
	⑥														

Colorlegend

C1-C5 fulfilled	gcd(N_l, $2p$) == 1	$p_{r,loss} > 100$	Not feasible	$k_w < 0.9$

Table IX: Motor specifications used for all system configurations ①-⑥$_{a/b}$ (cf. section 3.3) in the optimization routine shown in Fig. 6.

T_n	(Nm)	50	b_s	(mm)	3
n_n	(rpm)	300	k_{fill}	(-)	0.35
M_{ref}	(-)	0.9	J_{max}	(A mm^{-2})	7
D	(mm)	250	$p_{Cu,loss,max}$	(kW m^{-2})	1.1
δ	(mm)	1	PF_{min}	(-)	0.9
h_M	(mm)	4	Iron material	(-)	M330-35a
k_p	(-)	0.9	Magnet material	(-)	NdFeB

Table X: Coefficients of (13) for capacitor series EPCOS B3277x [7].

k_1	cm^3 F^{-1} V^{-2}	0.7
k_2	cm^3 F^{-1} V^{-1}	1700
k_3	cm^3 V^{-1}	0.0016
k_4	cm^3	3.6

Table XI: Electrical motor parameters (R, L_{dqk} & ψ_{PM1}), rated motor current for one SPB-module (I_{mod}), and motor length (l_m) resulting from the design routine in Fig. 6 for system configurations ①-⑥$_{a/b}$ (cf. section 3.3) and $V_{ds,max} = 600$ V.

Configuration		①	②$_a$	②$_b$	③	④$_a$	④$_b$	⑤	⑥$_a$	⑥$_b$
R	(Ω)	2.05	3.99	3.86	1.41	4.07	3.92	0.82	3.06	2.78
L_{dq1}	(mH)	20.4	41.2	40.3	20.0	60.6	62.0	12.5	49.9	51.0
L_{dq5}	(mH)	-	32.1	31.4	-	36.3	36.8	-	38.0	38.5
L_{dq7}	(mH)	-	-	-	-	23.4	23.4	-	29.2	29.3
L_{dq11}	(mH)	-	-	-	-	-	-	-	17.3	16.8
ψ_{PM1}	(mWb)	280	279	281	328	328	330	301	269	269
I_{mod}	(A)	2.47	2.48	2.47	2.00	2.00	1.99	1.96	1.99	1.99
l_m	(mm)	54	52	54	56	54	56	58	54	58

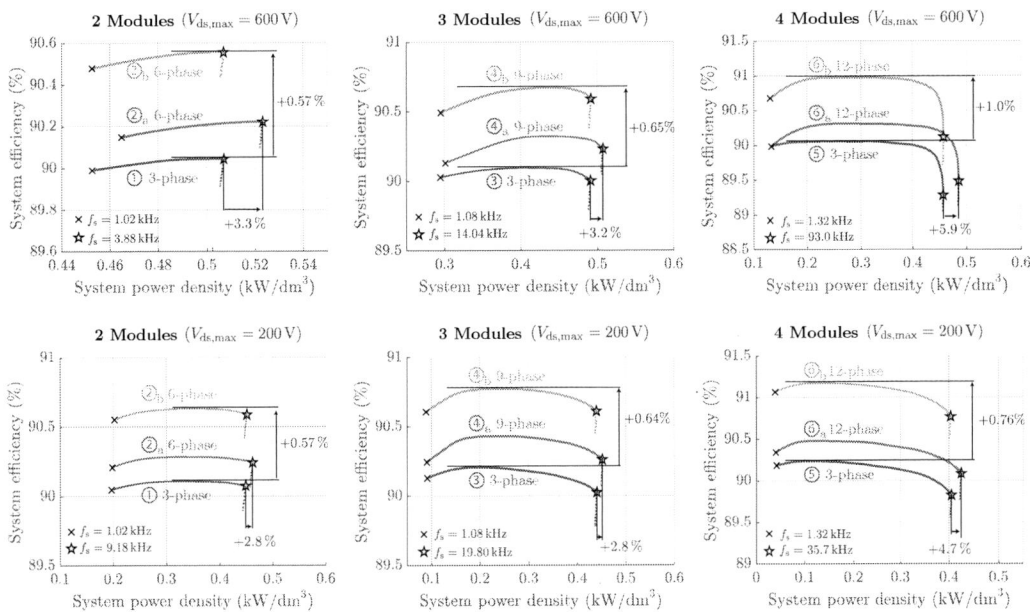

Figure 12: System efficiency and power density of the 3-phase and of the multiphase system configurations ①–⑥$_{a/b}$ (cf. section 3.3) for transistor blocking voltages $V_{ds,max} = 600\,\text{V}/200\,\text{V}$ and switching frequencies $f_s = 1..100\,\text{kHz}$. Considered losses: Motor copper and stator core losses, converter switching and conduction losses (cf. sections 3.4/4.2). Considered volumes: Motor, populated PCB and heat sink (cf. sections 3.4/4.2).

5.4 Efficiency vs. power density

Fig. 12 shows the resulting power densities and efficiencies of all considered system configurations for switching frequencies between 1 kHz and 100 kHz where each graph compares one of the 3-phase configurations ①, ③ or ⑤ to its corresponding multiphase configuration ②$_{a/b}$, ④$_{a/b}$ or ⑥$_{a/b}$ (cf. 3.3). As expected, all configurations show a similar dependency on the switching frequency. At low switching frequencies f_s, large capacitance values are required to limit the voltage ripple on the dc link, leading to non-optimal power densities. Also, the efficiency is not maximum for small values of f_s as could be expected from common pareto curves. This is due to the motor inductance that is held constant in the applied design procedure (Fig. 3) which leads to a high current ripple and high harmonic losses in the motor for low switching frequencies. Increasing the switching frequency increases the motor efficiency but decreases the converter efficiency due to the switching losses. Also, the required capacitor volume is reduced but the heat sink volume to dissipate more switching losses increases. The curve shape near the point of maximum power density (marked with a star) is relatively abrupt for the 2-module configurations and becomes smoother with increasing amount of SPB-modules. This is due to the capacitance limit $C_{dc,min,d} \sim 1/N_{SPB}$ (cf. (12)) that is reached earlier (i.e. for smaller f_s) for a small number of modules compared to a large number of modules. Overall, Fig. 12 shows that multiphase configurations on average enable an increase of 0.70 % in maximum efficiency or an increase of 3.8 % in maximum power density compared to 3-phase configurations.

6 Conclusion

In this paper, the SPB-C combined with a 3-, 6-, 9-, or 12-phase tooth-coil wound non-salient PMSM is analyzed and compared with respect to efficiency and power density based on a comprehensive system optimization. The results show that the winding factors of the proposed 6-, 9-, and 12-phase (multiphase) motors are on average 3.7 % higher than the winding factors of 3-phase motors with 2, 3, or 4 phase-aligned winding systems. The higher winding factor can be used to increase the maximum efficiency by 0.70 % or the maximum power density by 3.8 %. Furthermore, the possible motor-converter modularity is analyzed. As a result, 3-phase motors allow a higher degree of motor-converter modularity compared to multiphase motors.

References

[1] R. Abebe *et al.*, "Integrated motor drives: state of the art and future trends," *IET Electric Power Applications*, vol. 10, no. 8, pp. 757–771, Sep 2016.

[2] S. Norrga, L. Jin, O. Wallmark, A. Mayer, and K. Ilves, "A novel inverter topology for compact EV and HEV drive systems," in *Conf. of the Industrial Electronics Society (IECON)*, Nov 2013.

[3] H. Zhang, O. Wallmark, M. Leksell, S. Norrga, M. N. Harnefors, and L. Jin, "Machine design considerations for an MHF/SPB-converter based electric drive," in *Conf. of the Industrial Electronics Society (IECON)*, Oct 2014.

[4] J. Wang, Y. Li, and Y. Han, "Integrated modular motor drive design with gan power fets," *IEEE Trans. on Industry Applications*, vol. 51, no. 4, pp. 3198–3207, Jul 2015.

[5] N. R. Brown, T. M. Jahns, and R. D. Lorenz, "Power converter design for an integrated modular motor drive," in *IEEE Industry Applications Annual Meeting (IAS)*, Sep 2007.

[6] M. Ugur, H. Sarac, and O. Keysan, "Comparison of inverter topologies suited for integrated modular motor drive applications," in *International Power Electronics and Motion Control Conference (PEMC)*, Aug 2018.

[7] J. Wyss, "Multi-domain optimization for highly efficient, active rectifiers in drive systems," Ph.D. dissertation, ETH Zürich, 2018.

[8] A. A. Rockhill, "On the modelling and control of high phase order synchronous machines," Ph.D. dissertation, University of Wisconsin-Madison, 2012.

[9] B. P. Germar Müller, Karl Vogt, *Berechnung Elektrischer Maschinen*. WILEY-VCH, 2008.

[10] N. Bianchi and M. D. Pre, "Use of the star of slots in designing fractional-slot single-layer synchronous motors," *IEE Proceedings - Electric Power Applications*, vol. 153, no. 3, p. 459, 2006.

[11] G. Müller and B. Ponick, *Theorie Elektrischer Maschinen*. WILEY-VCH, 2009.

[12] B. Aslan, E. Semail, J. Korecki, and J. Legranger, "Slot/pole combinations choice for concentrated multiphase machines dedicated to mild-hybrid applications," in *Conf. of the Industrial Electronics Society (IECON)*, Nov 2011.

[13] F. Meier, "Permanent-magnet synchronous machines with non-overlapping concentrated windings for low-speed direct-drive applications," Ph.D. dissertation, KTH Stockholm, 2008.

[14] K. Hameyer, *Entwurf, Berechnung und Technologie elektrischer Maschinen (lecture notes)*. Institut für Elektrische Maschinen der RWTH Aachen, 2014.

[15] A. El-Refaie, T. Jahns, and D. Novotny, "Analysis of surface permanent magnet machines with fractional-slot concentrated windings," *IEEE Trans. on Energy Conversion*, vol. 21, no. 1, pp. 34–43, Mar 2006.

[16] R. Krall, "Permanentmagneterregte mehrphasen-synchronmaschine in zahnspulenausführung einschliesslich des phasendezimierten betriebs," Ph.D. dissertation, Montanuniverstität Leoben, 2015.

[17] A. A. Rockhill and T. A. Lipo, "A generalized transformation methodology for polyphase electric machines and networks," in *International Electric Machines & Drives Conference (IEMDC)*, May 2015.

[18] M. T. Kakhki, "Modeling of losses in a permanent magnet machine fed by a pwm supply," Ph.D. dissertation, University of Laval, 2016.

[19] J. Kolar and S. Round, "Analytical calculation of the RMS current stress on the DC-link capacitor of voltage-PWM converter systems," *IEE Proceedings - Electric Power Applications*, vol. 153, no. 4, p. 535, 2006.

[20] D. Christen, M. Stojadinovic, and J. Biela, "Energy efficient heat sink design: Natural versus forced convection cooling," *IEEE Trans. on Power Electronics*, 2017.

[21] D. Christen and J. Biela, "Analytical switching loss modeling based on datasheet parameters for mosfets in a half-bridge," *IEEE Trans. on Power Electronics*, 2019.

Design of a pulse modulator based on transmission lines for generating fast current pulses for plasma drilling

Oliver Keel, Melissa Artiglia and Juergen Biela
Laboratory for High Power Electronics, ETH Zurich
8092 Zurich, Switzerland
keel@hpe.ee.ethz.ch

Keywords

"Pulse current generator", "Plasma drilling"

Abstract

This paper presents the design procedure of a pulse modulator based on parallel connected transmission lines for plasma drilling. The procedure focuses on minimizing the parasitic capacitances/inductances of the modulator to achieve a fast current rise time, which is necessary for plasma drilling. Furthermore, a concept for a scaled prototype system is presented.

1 Introduction

For efficiently generating base load electrical power with geothermal energy sources, wells with a depth, of up to 10 km are required. In such depths the ambient temperature generally reaches up to $300\,°C$ and therefore enables an efficient electricity generation [1]. However, the cost of drilling deep holes with a mechanical grinding process grows exponentially with the depth. Therefore, conventional drilling is typically not attractive for holes deeper than 5 km [2].

New drilling concepts have been proposed for reducing the drilling cost. Examples are laser cutting and drilling [3], rock spallation with a hydrothermal flame [4] or electro pulse drilling [2]. Another promising concept is plasma drilling [5], which uses a pulsed plasma to generate fast temperature changes at the drill head to disintegrate the rock by thermal contraction. In the pulse breaks, the plasma is kept alive with a pilot current with an amplitude of typically 200-300 A so that no high voltage in the range of several $100\,kV$ is required for reigniting the plasma. During the pulse period a pulse current I_{pulse} at least 5-10 times higher than the pilot current flows through the plasma in order to heat it up (Fig. 1a). To achieve fast temperature changes and a efficient disintegration of the rock, the pulse current I_{pulse} requires a rise time t_{rise} in the range of maximal a few $100\,ns$ and a pulse length t_{pulse} of approximately $10\,\mu s$.

Possible pulse modulators which can fulfill these demanding specifications for example are pulse forming networks (PFN) [6], marx generators [7], and pulse forming lines [8] (PFL). All those topologies basically allow to generate pulses with a fast voltage rise time. In order to achieve a long lifetime of the pulse modulator, the goal is to minimize the complexity of the system, especially the amount of (active) switches in the bore hole with very high ambient temperatures should be minimized. Therefore, marx

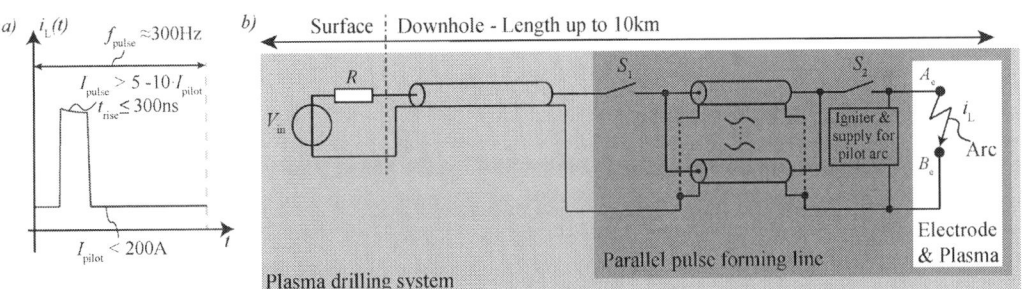

Figure 1: a) Pulse current specification for plasma drilling. b) Plasma drilling system with the concept parallel connected pulse forming line (P-PFL).

generators with its multitude of spark gaps and a relatively slow repetition rate is unfavorable. In contrast, PFLs allow to build robust and simple pulse modulators which require only two spark gaps and a transmission line cable. In many applications, the PFL is operated under very high voltages ($>500\,\text{kV}$) and pulse lengths of less than 1 µs as shown for example in [9], [10], and [8].

In contrast, this paper presents the concept and a design procedure of parallel connected pulse forming lines (P-PFL), which can operate with larger pulse currents at relatively low pulse voltage and low load impedance. Furthermore, the pulse lengths are much longer (Fig. 1).

In the section 2, first the basic model of a transmission line for a pulse current generator is introduced. Thereafter, the concept of a parallel pulse forming line (P-PFL) and its working principle is presented. For achieving a fast current rise time, it is important to model the parasitics of the complete discharge circuit and the load which is presented in section 6. To validate the model of the parasitics, section 7 shows a prototype system and the related simulation. In the final section, the simulation result of the full scale pulse modulator are presented.

2 Modelling of transmission lines for pulse current generators

In general, an accurate approach to model transmission lines is to use a infinite number of infinitesimally small transmission line sections of length dx [11, 12] resulting in the distributed model. A single segment is illustrated in figure 2, where R', L', C', G' represent, the series resistance, the series inductance, the shunt capacitance, and the shunt conductance per unit length.

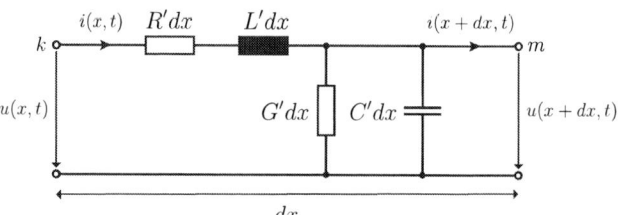

Figure 2: Schematic of a transmission line section.

For a single segment, the voltage and the current are calculated with the telegrapher's equations [13]:

$$\frac{\partial^2 u}{\partial x^2} = R'(\omega)G'(\omega)u + \left(R'(\omega)C'(\omega) + L'(\omega)G'(\omega) \right)\frac{\partial u}{\partial t} + L'(\omega)C'(\omega)\frac{\partial^2 u}{\partial t^2} \tag{1}$$

$$\frac{\partial^2 i}{\partial x^2} = R'(\omega)G'(\omega)i + \left(R'(\omega)C'(\omega) + L'(\omega)G'(\omega) \right)\frac{\partial i}{\partial t} + L'(\omega)C'(\omega)\frac{\partial^2 i}{\partial t^2} \tag{2}$$

The analytical solution to calculate the transients would require the solution of the telegrapher's equations in the frequency domain for each frequency component [11]. The time domain solution would be determined by summing all transformed time domain solutions, making the relatively approach unpractical for arbitrary signals. Moreover, the frequency dependence of the line parameters would require numerically unstable convolution integrals [14]. In the past, several simplifications have been introduced at the expense of the model accuracy to overcome the problem of unstable convolution integrals and to reduce the computational effort. Two examples of such simplifications are the PI and the Bergeron model. However, the PI-model uses only a single transmission line section and is limited to a single frequency representation of the signal. Furthermore, the PI-model is suitable only for steady state conditions and if the travel time of the line is not of interest [11, 15]. In contrast, the Bergeron-model is a distributed approach, based on traveling wave theory. It solves the telegrapher's equations directly in the time domain by assuming a lossless transmission line (i.e. $R' = G' = 0$) and constant line parameters evaluated at a single frequency [16]. Moreover, the Bergeron-model does not accurately represent transients and the lossless approximation implies that an amplitude wave reduction is neglected. Therefore, the following frequency dependent frequency dependent modal and phase model are used in this paper.

2.1 Frequency Dependent Modal and Phase model

José R. Martì introduced in [14] the frequency dependent modal model to calculate the current and voltage waveform for any signal in transmission lines with the help of an electromagnetic transient simulation program. The model decouples multi-phase lines by using a matrix manipulation and transforms them into single phase to neutral systems. The introduced transformation matrix is frequency dependent and needs to be recalculated for each frequency harmonic of the signal. This approach leads to a less time consuming evaluation of the convolutions integrals and to better numerical stability [17]. Additional adaptions are inroduced in [18] to further reduce the computational effort. This results in the frequency dependent phase model.

For both models, voltages and currents at the beginning and at the end of the line are related to each other by a propagation matrix H and a characteristic impedance matrix Y_c as shown in (3) and (4).

$$Y_c V_k - I_k = 2H^T I_{mr} \tag{3}$$

$$Y_c V_m - I_m = 2H^T I_{kr} \tag{4}$$

The indices k, m, and r indicate the nodes at the beginning and end of the line respectively the reflected quantity. The equations are solved in the frequency domain and the propagation H and the characteristic impedance matrices Y_c are defined in (5) and (6).

$$H = e^{-\sqrt{ZY}l} \tag{5}$$

$$Y_c = Z^{-1}\sqrt{ZY} \tag{6}$$

The variables Z, Y are the series impedance and the shunt admittance matrices and l is the cable length. Each entry of the matrices H and Y_c is approximated by rational fitting, which allows to find an equivalent Laplace transfer function of simpler form. The newly obtained Laplace function is then transformed back into the time domain, allowing to consider the full frequency character of the system. This technique leads to a lower computational effort [17].

Figure 3 shows a comparison of how the different models are derived from the exact distributed model. The PI-model represents a strong approximation since it is neglecting both delay time and the frequency dependent characteristic of the transmission line. An improvement is given by the Bergeron-model,

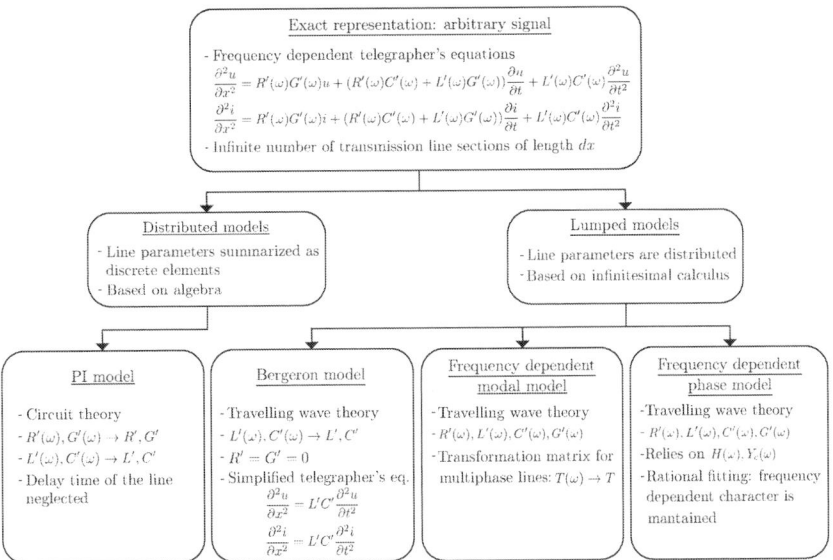

Figure 3: Flow diagram of the different transmission line models and their capabilities

which is based on the traveling wave theory but still suffers from the single frequency representation of the system. The frequency dependent modal model can represent the frequency dependency of the line parameters, although only for multi phase systems. By using rational fitting, the frequency dependent phase model takes into account the frequency and is computationally efficient what enables simulations of transients in transmission lines.

3 Influences of frequency dependent phase and analytical model on the pulse waveform calculation

The phase dependent models are implemented in PSCAD and the numerical results are compared to the results of an analytical model of a single transmission line segment. The analytical model is based on the characteristic impedance of a generic lossy line with impedance Z_o to represent the transmission line. The closing switch is modeled with by step voltage with the amplitude V_{in}. The load current I_L can be calculated with

$$I_L = \frac{V_{in}}{Z_L + Z_o} \tag{7}$$

where Z_L is the load impedance. For short lines, the simulation result given in Fig. 4 show only small differences caused by high frequent oscillations. Those are not modeled with the analytical model. Therefore, the result shows that for short transmission lines, the analytical model gives relatively accurate results.

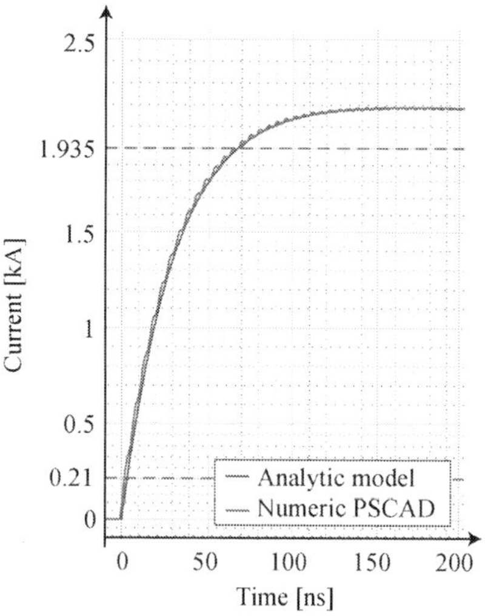

4 Calculation of coaxial transmission line parameters

Figure 4: Pulse current waveform with the analytical in blue and, frequency dependent phase model in red.

After the model has been presented, in the following section the methods to calculate the line parameters are presented. Electromagnetic transient simulation programs (EMTP) as for example PSCAD require the transmission line parameter R', L', C', G' to calculate the current and voltage waveforms. These parameters depend on the material of transmission line and physical dimensions of parameters r_1, r_2, and r_3 which are he core conductor radius, the outer radius of the insulation and the outer sheat radius (Fig. 5).

The series impedance of a transmission line is the sum of the line resistance and the line inductance in the frequency domain and is defined as $Z_{TML} = R + j\omega L$. The nonuniform current distribution due to he skin effect [19] results in a frequency dependent series impedance. The core conductor impedance Z_{core} is modeled as a cylindrical solid conductor whereas the sheath impedance Z_{sheath} as a hollow conductor. The influence of the stranded property of the conductor is corrected with factors presented in [17, 20].

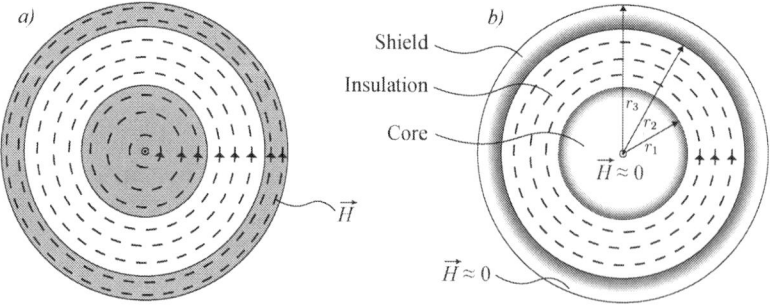

Figure 5: a) DC case: Current is uniformly distributed. b) High frequency: Non uniform current distribution. The magnetic field in the core and in the sheath is approximately canceled.

Based on [21] the impedance of the core is presented in (8).

$$Z'_{core}(\omega) = \frac{m}{2\pi r_1 \kappa} \cdot \frac{I_0(mr_1)}{I_1(mr_1)}, \qquad m = \sqrt{j\omega\mu\kappa} \tag{8}$$

The variable $m = \sqrt{j\omega\mu\kappa}$ is the intrinsic propagation constant of a solid metal, κ is the conductor conductivity, μ the conductor permeability, and I_0, I_1 the zero and first order modified Bessel's functions of the first kind. For high frequencies, the sheath impedance is given by (9), where K_0, K_1 are the zero and first order modified Bessel's functions of the second kind.

$$Z'_{sheath}(\omega) = \frac{m}{2\pi r_2 \kappa} \cdot \frac{I_0(mr_2)K_1(mr_3) + K_0(mr_2)I_1(mr_3)}{I_1(mr_3)K_1(mr_2) - I_1(mr_2)K_1(mr_3)} \tag{9}$$

The value of the resistance and the inductance are obtained by taking the real and the imaginary part of the complex impedance at the considered frequency [19, 21].

$$R'_{core}(\omega) = \Re Z'_{core}(\omega) \qquad\qquad \omega L'_{core}(\omega) = \Im Z'_{core}(\omega) \tag{10}$$
$$R'_{sheath}(\omega) = \Re Z'_{sheath}(\omega) \qquad\qquad \omega L'_{sheath}(\omega) = \Im Z'_{sheath}(\omega) \tag{11}$$

The total resistance per unit length is the sum of the core and the sheath resistance as shown in (12).

$$R'(\omega) = R'_{core}(\omega) + R'_{sheath}(\omega) \tag{12}$$

Due to the proximity effect, the current at high frequencies is confined at the core-insulation and at the insulation-sheath interfaces (Fig. 5). As a result, the magnetic field H is concentrated in the insulation and the total inductance per unit length L', which is the sum of core inductance L'_{core}, insulation inductance L'_{ins}, and sheath inductance L'_{sheath} can be simplified to $L' \approx L'_{ins}$ for high frequencies. The insulation inductance is defined by (13), where μ is the permeability of the insulation material [13].

$$L' \approx L'_{ins} = \frac{\mu}{2\pi} \cdot \ln\left(\frac{r_2}{r_1}\right) \tag{13}$$

The shunt admittance includes the line capacitance and conductance and is defined as $Y_{TML} = G + j\omega C$. The shunt conductance is assumed to be zero $(G = 0)$ and the shunt capacitance is approximated as a cylinder capacitor [19, 13]:

$$C' = \frac{2\pi\varepsilon}{\ln\left(\frac{r_2}{r_1}\right)} \tag{14}$$

The relative permittivity ε of the insulation material is assumed to be constant. Usually, transmission lines have multiple semiconductive layers, where influences are taken into account by computing an effective relative permittivity value for the main insulation [20, 17].

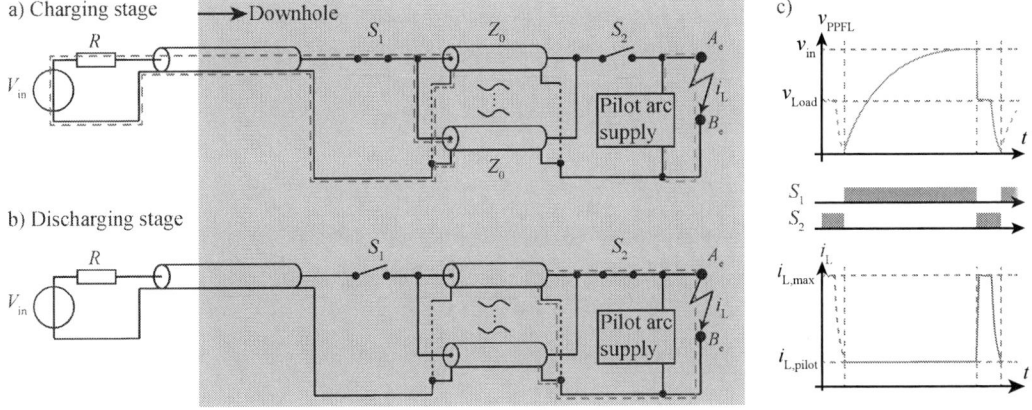

Figure 6: Two operation stages of the P-PFL modulator

5 Parallel pulse forming line modulator

The basic operating principle of the P-PFL is similar to the pulse forming line and can be divided in two operation stages as shown in figure 6. Switches S_1 and S_2 are two closing switches implemented with spark gaps. During the first stage (Fig. 6a), the P-PFL is charged up to the input voltage V_{in} by triggering S_1. In the following second stage (Fig. 6b), the P-PFL is discharged into the load after the parallel transmission lines have been charged up by triggering spark gap S_2. There, the pulse waveform depends on the length and the characteristic impedance Z_0 (15) of the parallel lines. The capacitance and the inductance of the transmission line cable do change diametrical, i.e. for a larger capacitance value a smaller inductance value results and vise versa.

$$Z_{0,s} = \sqrt{\frac{L'}{C'}} \tag{15}$$

The load voltage and current waveforms depend on the reflection coefficient ρ which describes the relation between the load Z_L and the single transmission line $Z_{0,s}$ impedance as shown in (16). In the ideal case, Z_L has the same value as Z_0 with the consequence that $\rho = 0$ and a single rectangular pulse with the voltage $V_{in}/2$ and pulse length of $t_p = 2l \cdot \sqrt{L'C'}$ is generated (Fig. 7).

$$\rho = \frac{Z_L - Z_0}{Z_L + Z_0} \tag{16}$$

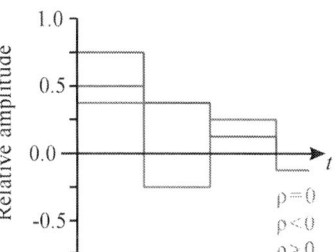

Figure 7: Resulting load voltage waveform for different ρ. The relative amplitude describes the relation of V_{in} to pulse voltage.

Usually, pulse forming lines are operated at high voltages [9] and therefore, the utilized transmission line requires a large distance between the core and the shield so that L' becomes larger and C' small. As a result, the characteristic impedance Z_0 is large.

However, in the considered case the load impedance Z_L is the plasma impedance Z_{Plasma}, which is highly dynamic and can reach minimum values of less than $0.5\,\Omega$. In order to avoid negative back reflections Z_{Plasma} must be larger than Z_0 during multiple consecutive pulse periods. Such a low cable impedance Z_0 can be achieved by paralleling the cables according to (17).

$$n_{parallel} = \left\lceil \frac{Z_0}{Z_{Plasma,min}} \right\rceil \tag{17}$$

5.1 Effects of manufacturing tolerances on parallel pulse forming line

Due to manufacturing tolerances, the parallel connected transmission line can have different line parameters and the capacitance as well as the inductance per unit length affect the characteristic impedance and the pulse length, which are crucial for the design of the parallel pulse forming line. With larger capacitances, Z_0 is decreasing and t_p increasing, whereas larger inductance values lead to an increasing Z_0 and an decreasing t_p. Furthermore, the variation of Z_0 can change the sign on the reflection coefficient ρ (16). However, the manufacturing tolerances are not influencing the current pulse rise time.

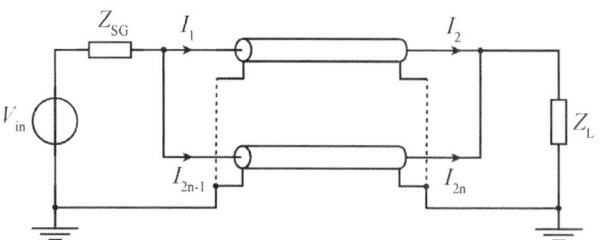

Figure 8: Simplified P-PFL with parallel transmission lines.

In order to investigate the effect of the tolerances on the current distribution between the parallel lines, the P-PFL is simplified to two parallel transmission lines. The network can be analyically solved according to the circuit in figure 8, where Z_1 and Z_2 are the characteristic impedance of the lines, Z_{SG} is the spark gap impedance and Z_L the load impedance. The peak of the currents I_3 and I_4 is calculated as in (18). The deviation of the characteristic impedance results in a higher share of the load current in the cable with the lower impedance.

$$I_3 = V_{in} \cdot \frac{Z_2}{Z_1 Z_2 + Z_L(Z_1 + Z_2)} \qquad\qquad I_4 = V_{in} \cdot \frac{Z_1}{Z_1 Z_2 + Z_L(Z_1 + Z_2)} \qquad (18)$$

For illustration purposes, it is assumed that for example a LEMO 121221 coaxial transmission line cable has $\pm 10\%$ manufacturing tolerances of the outer insulation layer radius r_2, which results in an deviation of $\pm 6\%$ on the parameter L' and C'. Therefore, the characteristic impedance Z_0 would deviate by about 3% which means that this transmission line would have to carry 3% more current.

6 P-PFL parasitic model

The parasitics model of the P-PFL is crucial for achieving the required current rise time for plasma drilling. In the simulation, the system is represented as a detailed network including parasitic components. The main components in the discharge circuit are the spark gap, the electrode, the plasma, the ground return, and the cable mount. Figure 10 shows the physical setup and the associated associated network.

The transmission line cables are feed to the cable mount (CM) which separates the shields from the core of the cable and connects them parallel. The cable mount is modeled by the parasitic inductance L_{CM} and the parasitic capacitance C_{CM}. The parasitic inductance L_{CM} is calculated with (19), which is the equation for the inductance of round conductors [22] divided by the number of parallel transmission line cables $n_{parallel}$. Variable l_{CM} is the length of the cable without shield and D the diameter of the cable core.

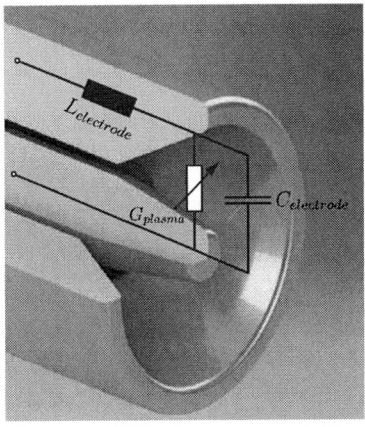

Figure 9: CAD of the plasma drilling head.

$$L_{CM} = \frac{2 l_{CM}}{n_{parallel}} \cdot \left\{ \ln\left[\left(\frac{2 l_{CM}}{D}\right)\left(1 + \sqrt{1 + \left(\frac{D}{2 l_{CM}}\right)^2}\right)\right] - \sqrt{1 + \left(\frac{D}{2 l_{CM}}\right)^2} + \frac{\mu}{4} + \left(\frac{D}{2 l_{CM}}\right) \right\} \quad (19)$$

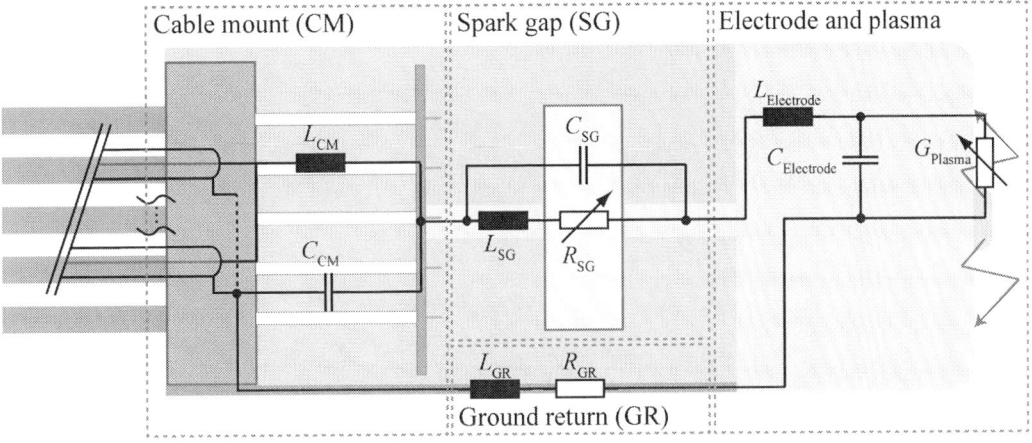

Figure 10: Equivalent P-PFL network with parasitic components.

The parasitic capacitance C_{CM} is calculated with (20) which is the equation for plate capacitors.

$$C_{CM} = \frac{\varepsilon A}{l_{CM}} \tag{20}$$

The spark gap (SG) switch is modeled by the parasitic inductance L_{SG}, the parasitic gap capacitance C_{SG}, and the spark gap arc, which is described as a variable resistor R_{SG}. The values for the spark gap switch parasitics are taken from the datasheet. The electrode is modeled with the parasitic inductance $L_{Electrode}$, the capacitance $C_{Electrode}$, and the plasma with the current dependent conductance model G_{Plasma} presented in [23]. The basic shape of the electrodes is shown in figure 9. The plasma burns between the inner and the outer electrode and is modeled by R_{Arc}, whereas $C_{Electrode}$ models the gap capacitance.

The values for $L_{Electrode}$ and $C_{Electrode}$ are determined with FEM simulations, where the plasma is modeled as a conductor in the space between the inner and the outer electrode. Since the pulse period is relatively short, it is assumed that the arc is not moving during the pulse period. To estimate the largest value (worst case) of $L_{Electrode}$ and $C_{Electrode}$, the electrode geometry and the position of the plasma are varied in the insulation. The inductance and capacitance values result from the energy in the magnetic and the electrostatic field. Since the exact path of the arc is not known, an estimated maximum value for $L_{Electrode}$ as a worst case scenario is chosen for the models.

The pulse current returns to the shields through the ground return path (GR) which is modeled by the parasitic inductance L_{GR} and the resistance R_{GR}. The parasitic inductance of the load PCB is modeled as a rectangular conductor and PCB track is calculated with (21) given in [22]. The variables w, t, and l are the conductor width, conductor thickness, and the length of the conductor.

$$L_{Load,\,PCB} = 2 \cdot 10^{-3} \cdot l \left[\ln\left(\frac{2l}{w+t}\right) + 0.5 + 0.2235\left(\frac{w+t}{l}\right) \right] \tag{21}$$

7 Model validation by scaled prototype system

Figure 11 shows a CAD of the components between the electrode and the transmission line cables. For the prototype, the electrode is replaced with a resistive load in order to enable an easier and more precise measurement of the pulse rise time. To avoid that the parasitics are significantly different to the real system with an arc, the current path is kept the same as with the arc. In an ideal case, the ground return

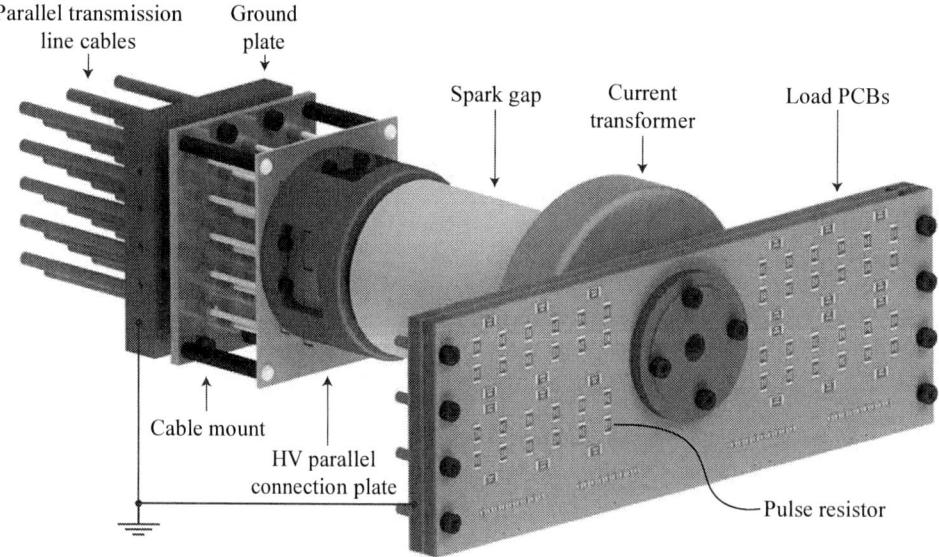

Figure 11: 3D CAD model of the scaled prototype system for measuring the achievable pulse rise time and for validating the models.

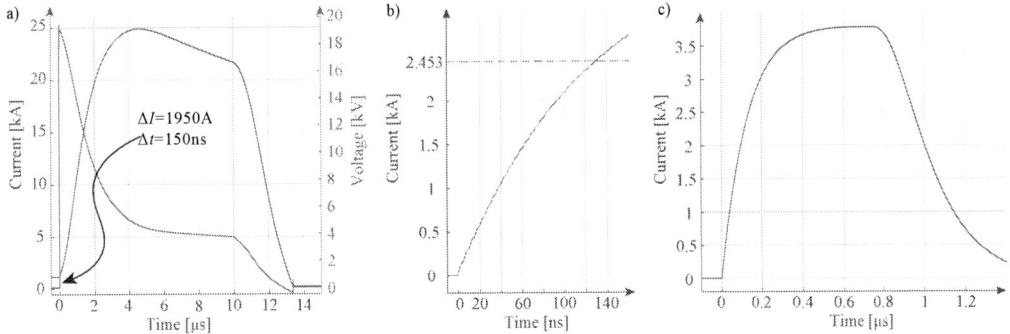

Figure 12: a) Plasma pulse current waveform b) Load current rise waveform of the prototype system c) Load current pulse waveform on the prototype system

path has a cylindrical shape like the electrode and forms a coaxial arrangement with the spark gap and cable mount. To make the assembly of the prototype system easier, the return paths are split in two "paths" on each side of the spark gap. The parasitics of the load are calculated with the inductance values from the data sheet of the resistor and analytical equations for the inductance of the PCB via (22) and track inductance (21).

$$L_{\mathrm{via}} = \frac{t_{\mathrm{PCB}}}{5} \cdot \left[1 + \ln\left(\frac{t_{\mathrm{PCB}}}{D_{\mathrm{via}}}\right) \right] \tag{22}$$

For desiging the load PCB, in a first step, the total number of required SMD resistor is calculated. Then, the resistors are arranged on the PCB to determine the track lengths and the number of vias from which the parasitic inductance can be calculated. The distances between the tracks are designed based on the required minimum creepage and clearance distance. Moreover, the resistors are placed on both sides of the PCB to reduce the size and the parasitic inductance of the PCB. The maximum pulse energy of the resistor is used to calculate the allowed pulse length at a chosen pulse power.

Figure 11 shows the 3D CAD model of the scaled prototype. The cable sheaths are separated from the insulation and connected to a conducting plate at ground potential. The inner conductors are soldered on a high voltage plate PCB and connected to the resistive load (PCB) through a triggered spark gap. For the setup, 25 coaxial cables (LEMO 121221) each of 25 m length are used and a GXG150L spark gap from e2v with 15 kV hold off voltage is chosen. For the resisitive load, 3.3 Ω resistors arranged in 19 series and 24 parallel connected branches on three stacked PCBs are used. The maximum pulse flattop length for this configuration is 250 ns at a maximum peak pulse power of 11.56 MW. The parasitics add up to 9.45 nH per PCB and are therefore smaller than the electrode and plasma parasitics. To achieve similar current rise times as the final full scale system with arc/plasma, additional inductive and capacitive components are added.

The simulated current pulse waveform in the load PCB of the scaled prototype is shown in Fig. 12b and c. The current pulse reaches a value of 2.453 kA after 130 ns faster then the required 150 ns. The total pulse length is 800 ns.

Table I: Input voltage and values of the parasitics

Parameter		
V_{in}	10	kV
R_{charge}	10	Ω
L_{CM}	1.53	nH
C_{CM}	1.55	pF
L_{SG}	18	nH
C_{SG}	7	pF
$R_{\mathrm{SG,on}}$	8	mΩ
$L_{\mathrm{Electrode}}$	60	nH
$C_{\mathrm{Electrode}}$	15	pF
R_{GR}	0.03	mΩ
L_{GR}	63.5	nH
C'	101	$\frac{\mathrm{pF}}{\mathrm{m}}$
L'	253	$\frac{\mathrm{nH}}{\mathrm{m}}$
Z_0	50	Ω

8 Simulation results of full scale system

This section presents the resulting current waveform for a full scale system. The values of the parasitics are given in table I. The transmission line cable used for the simulation is the 2012STJ from HiVolt. To match the low resistance of the plasma, 84 cables in parallel are required. Moreover, the length of the P-PFL is 646.5 m for a minimum pulse length of 10 µs. The current rises in 150 ns from 50 A pilot current to 1.93 kA pulse current. Then, the current increase further as the plasma resistance is further dropping, reaching a maximal value of 24.8 kA as shown in figure 12a. After 10 µs, the current starts to decrease with intermediate peaks because of the positive reflection coefficient.

9 Conclusion

This paper presents the model and the design procedure for a parallel pulse forming line (P-PFL) to be used as a pulse modulator for plasma drilling. The working principle of the P-PFL as well as the selection of the cable impedance and the length for the given waveform requirements are explained. Moreover, the model of the parasitics in the discharge circuit is introduced. The simulation results shows that the load current rises from 50 A to 1.93 kA in 150 ns for the full scale system. The scalled prototype system is being built and will be used to validate the introduced models.

10 Acknowledgment

This work would not have been possible without the financial support of the Swiss federal office of energy. Our gratitude goes especially to Mr. Siddiqi and Mrs. Weber.

References

[1] I. Kocis, T. Kristofic, M. Gebura, G. Horvath, M. Gajdos, and V. Stofanik, "Novel deep drilling technology based on electric plasma developed in Slovakia," in *Assembly and Scientific Symposium of the Union of Radio Science (URSI GASS)*, August 2017.

[2] H. O. Schiegg, A. Rodland, G. Zhu, and D. A. Yuen, "Electro-pulse-boring (epb): Novel super-deep drilling technology for low cost electricity," *Journal of Earth Science*, February 2015.

[3] M. S. Zediker, C. C. Rinzler, B. O. Faircloth, Y. Koblick, and J. F. Moxley, "High power laser downhole cutting tools and systems," Patent US2 014 060 930 (A1), March, 2014, cIB: E21B7/15.

[4] C. R. Augustine, "Hydrothermal spallation drilling and advanced energy conversion technologies for Engineered Geothermal Systems," Thesis, Massachusetts Institute of Technology, 2009.

[5] I. Kocis, I. Kocis, T. Kristofic, and D. Kocis, "Method of disintegrating rock by melting and by synergism of water streams," Patent EP2 809 867 (B1), December, 2016, cIB: E21B7/14.

[6] G.N. Glasoe & J.V. Lebacqz, *Pulse Generators*, 1965. [Online]. Available: http://archive.org/details/PulseGenerators

[7] W. J. Carey and J. R. Mayes, "Marx generator design and performance," in *International Power Modulator Symposium*, June 2002.

[8] G. Rim, E. Pavlov, H. Lee, and J. Kim, "Pulse forming lines for square pulse generators," June 2001.

[9] J. M. Lehr, *Foundations of pulsed power technology*. IEEE Press Wiley, 2017.

[10] J. D. Ivers and J. A. Nation, "Compact 1-gw pulse power source," *Review of Scientific Instruments*, vol. 54, no. 11, pp. 1509–1510, 1983.

[11] C. M. Franck and G. Hug, "Introduction to Electric Power Transmission: System and Technology," 2016.

[12] J. Luo, K. Zhang, T. Chen, G. Zhao, P. Wang, and S. Feng, "Distributed parameter circuit model for transmission line," in *2011 International Conference on Advanced Power System Automation and Protection*, vol. 2. IEEE, 2011, pp. 1529–1534.

[13] W. Bächtold, "Leitungen und Filter," 2010.

[14] J. R. Marti, "The problem of frequency dependence in transmission line modelling," Ph.D. dissertation, University of British Columbia, 1981.

[15] E. Haginomori, T. Koshiduka, H. Ikeda, and J. Arai, *Power System Transient Analysis: Theory and Practice Using Simulation Programs (ATP-EMTP)*. John Wiley & Sons, 2016.

[16] N. Watson and J. Arrillaga, *Power systems electromagnetic transients simulation*. IET, 2003, vol. 39.

[17] U. S. Gudmundsdottir, *Modelling of long High Voltage AC Cables in the Transmission System*. Department of Energy Technology, Aalborg University, 2010.

[18] A. Morched, B. Gustavsen, and M. Tartibi, "A universal model for accurate calculation of electromagnetic transients on overhead lines and underground cables," *IEEE Transactions on Power Delivery*, vol. 14, no. 3, pp. 1032–1038, 1999.

[19] C. M. Franck, "Technology of Electric Power Systems Components (lecture notes)," 2019.

[20] B. Gustavsen, "Panel session on data for modeling system transients insulated cables," in *2001 IEEE Power Engineering Society Winter Meeting. Conference Proceedings (Cat. No. 01CH37194)*, vol. 2. IEEE, 2001, pp. 718–723.

[21] S. A. Schelkunoff, "The electromagnetic theory of coaxial transmission lines and cylindrical shields," *Bell system technical journal*, vol. 13, no. 4, pp. 532–579, 1934.

[22] F. E. Terman, *Radio Engineers' Handbook*, 1st ed. McGraw-Hill Book Company, inc, 1943.

[23] O. Keel and J. Biela, "Pulse current generator with fast rise time based on transformers and single active switch for plasma drilling," in *European Conference on Power Electronics and Applications*, 2019.

Analysis of Current in Pulsating DC Link Converter with Zero Voltage Transition

Daniele Marciano, Giovanni Busatto, Carmine Abbate, Annunziata Sanseverino,
Davide Tedesco and Francesco Velardi
DIEI - UNIVERSITY OF CASSINO AND SOUTHERN LAZIO
Via G. Di Biasio 43
Cassino, Italy
Tel.: +39 0776/2993724
E-Mail: daniele.marciano@unicas.it
URL: https://www.unicas.it/diei/

Acknowledgements

This work has been supported partially by the project "HEROGRIDS – Holistic approach to EneRgy-efficient smart nanOGRIDS" funded by the MIUR Progetti di Ricerca di Rilevante Interesse Nazionale (PRIN) Bando 2017 - grant 2017WA5ZT3_001, and partially by MIUR in the mainframe of the measure n. 232 year 2016: "Departments of Excellence".

Keywords

« ZVS converters», « Reliability», « Industrial application», « Soft switching», « Voltage Source Inverters (VSI)» .

Abstract

An exhaustive analysis of the evolution of current in the input power stage of insulated multistage DC/AC power converter based on the Pulsating DC link principle, characterized by Zero Voltage Transition (ZVT) of the output stage, is provided in this work. This topology is featured by an input power stage which provides on the second stage a pulsed voltage characterized by zero phases (pulsating DC link) which are used to achieve the ZVT conditions for the commutations of this stage. The specific and innovative topology of this converter influences the evolution of the main electrical characteristics and requires more attention on the design. The parameters that affects the behavior of the current in this particular topology, especially in the crucial working phases, are identified and their influence are analyzed to define the constraints of this system. A complete study of the current has been accomplished by LTSPICE simulator for different combination of leakage parameters of the architecture and switching frequency of the two power stages. The obtained results are presented and discussed to provide the limits of the design of the main components that compose this particular power converter. For the experimental validation of the results of the simulation, a 3.5kW prototype of the power converter was assembled and tested at different load conditions.

Introduction

Nowadays power applications and market require that power converters are featured by high power density, compactness, low weight, high efficiency, robustness and reliability.
The demanded characteristics in terms of robustness and reliability are limited by failure mechanisms and instabilities of the power devices [1, 2, 3] which are in addition subjected to external factors like cosmic rays effects [4, 5]. From the reliability point of view, components like insulation transformers and electrolytic capacitors can affect the converter performances. These performances are also strongly influenced by the architecture of the converter and its modulation technique.
In the last years an innovative two stages isolated DC/AC converter based on pulsating DC link principle was proposed [6]. This converter is characterized by the absence of the inductive-capacitive filter between the two power stages and consequently the voltage on DC link has a pulsating

evolution. The absence of the inductor between the first stage and the DC link influences the modulation and control technique of this particular converter, requiring suited innovative control techniques. For this reason the attention in the literature was focused on the optimization of the modulation technique to achieve a very low THD of the output voltages, as it was done in [7] where two different modulation techniques and their results are discussed and compared. Lately, the use of pulsating DC link principle to decrease the power losses of the second power stage and increase overall efficiency, thanks to ZVT commutations, in high power and voltage applications was presented [8]. The papers about the Pulsating DC Link Converter (PDLC) presented so far, were focused on modulation techniques of this particular topology [9]-[11] and nobody has never investigated so far about the design and the limits of this topology. So in this paper we show how this innovative power converter works and which are the critical constraints in its design. In particular, we present how the current on the primary of high frequency transformer evolves during the powering phases of the first stage and which are the parameters that influence its evolution.

This paper is structured as follows: a complete description of the analyzed PDLC ZVT architecture is provided in the first section; the second section describes the evolution of the main electrical characteristics during the crucial working phases of power converter and which parameters influence them. Then the results of multi parameters simulation analysis of the power converter are presented and discussed. In the subsequent section the experimental characterization of a PDLC ZVT prototype is described. The obtained results are presented and evaluated. The conclusions are summarized in the final section.

Architecture and features of analyzed topology

The schematic of analyzed PDLC ZVT is supplied in Fig. 1, where each power stage is highlighted to simplify the analysis. It is composed of a Phase Shifted Bridge, PSB, which includes the switches Ap-An-Bp-Bn and is connected to an insulation transformer L1-L2. The secondary winding L2 is connected to a diode bridge rectifier. A three-phase 2-level Voltage Source Inverter (VSI) made of the switches Rp-Rn-Sp-Sn-Tp-Tn is fed by the pulsating DC link at the output rectifier. The active clamp Dac-Sac-Cac is used to limit the overvoltage at the turn off of the switches.

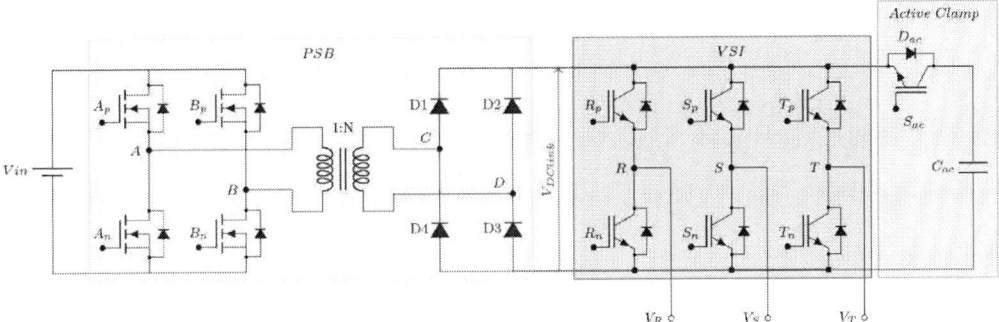

Fig.1: Schematic of Pulsating DC Link Converter with Zero Voltage Transition (PDLC ZVT).

The main advantages of this topology are:

- The absence of reactive elements between the two power stages results in:
  - Fast dynamic response: the topology does not require large and bulky components (inductor and capacitor) which are used as a filter to obtain the DC link voltage; consequently, the transfer function does not include two low frequency, often dominant, poles;
  - Better converter reliability thanks to the absence of the electrolytic capacitor which can seriously limit the overall reliability of the converter. In facts, these components are

nowadays becoming a bottle neck for the reliability issues because they are affected by degradation process during their lifetime related to their working conditions [12]-[13].

- The use of a pulsating DC link instead of fixed DC link, provides the possibility to increase the efficiency of the power converter thanks to the turn-on of the switches of VSI stage when the bus voltage is zero. In such a way the turn-on of VSI switches are Zero Voltage Transition (ZVT) with zero energy losses. So this feature can help to improve the efficiency of VSI by using a proper VSI modulation technique without using additional resonant circuits.

These features make PDLC ZVT a good architecture for all industrial applications where reliability, efficiency, reduced size and weight are strongly required.

The choice of the modulation logic for this particular power converter is fundamental to exploit its intrinsic characteristics and overcome the related issues. In particular, to achieve the ZVT turn on of the VSI switches, the PSB stage must be controlled in order to achieve a pulsating voltage on the DC link which is characterized by the presence of opportune zero voltage phases during which the VSI switches can be commutated. Consequently, the modulation logic must guaranty the synchronization between the gate signals of the PSB and VSI stages. Moreover, during every powering phase of the PSB, the switch Sac of the Active Clamp must be turned on to connect the clamping capacitor Cac to the DC bus and to transfer the energy due the inductive voltage overshoots back to the VSI stage. In addition, the implemented modulation technique provides a switching frequency of the PSB stage larger than the that of the VSI one as it will be discussed in the next section. The increase of the switching frequency of the PSB stage allows us to reduce weight and volume of the insulation transformer and to increase power density and compactness of the converter.

Analysis of the primary current of the transformer

The evolution of main electrical characteristics of the analyzed power converter are strongly conditioned by the topological features of the system like the absence of low-pass output filter of PSB stage and the necessity of the active clamp. So in this paragraph we describe how these features influence the evolution of the main electrical quantities of the analyzed power converter. In particular, the evolution of the current on the primary winding and the voltage across the capacitor of the active clamp are characterized by infrequent behaviors, especially during the powering phases of the PSB stage. For this reason their time evolution must be evaluated to find out the design limits of the main converter's parameters. First of all, the absence of the filter inductor at the output of PSB causes the current evolution of the PSB switches and transformer to be strongly dependent on the parameters of the transformer, the switches and the stray elements of the PSB stage.

To study the circuit behavior let us consider the equivalent circuit of Fig. 2 which refers to the powering phase when Ap - Bn switches and D1 - D3 diodes are in the on state. It is worth to note that Cac is connected to the DC link by Dac or Sac switch which is turned on during the powering phase. The current in Cac flows through Dac during the recirculating phase of the load/filter inductance and through Sac when this recirculating phase ends and Cac gives back power to the DC link. The circuit of Fig. 2 includes: winding resistance R$_{winding}$; magnetizing and leakage inductances L_{mag} and $L_{leakage}$ of the transformer, parasitic resistance of switches and diodes, R$_{onAp}$, R$_{onBn}$, R$_{onSacp}$, R$_{onD1}$, R$_{onD3}$, and R$_{onDac}$, respectively.

The current in the primary side of the transformer is given by the sum of the magnetizing current I_{Lmag} and the output current referred to the primary side I'_{out} which includes the current I'_{VSI} in the VSI and the current in I'_{Cac} in the active clamp.

The evolution of I_{Lmag} is determined by magnetizing inductance of the transformer, L_{mag}, and switching frequency of the PSB stage, f_{sPSB}. For them we can write the following equation:

$$V_{Lmag} = L_{mag} * \frac{dI_{Lmag}}{dt} \tag{1}$$

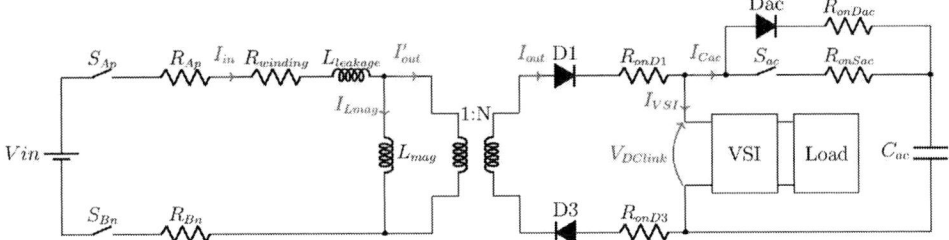

Fig. 2: Equivalent schematic circuit of the proposed PDLC ZVT during the powering phases of the PSB stage.

Discretizing and solving Eq. 1 for the current I_{Lmag}, we obtain:

$$I_{Lmag} = \frac{V_{Lmag}}{L_{mag} * f_{sPSB}} \qquad (2)$$

We can observe that the rise of I_{Lmag} can be limited by increasing the product $L_{mag} * f_{sPSB}$.

On the other side, the evolution of I'_{out} is influenced by the leakage inductance, the winding resistance of the transformer and the difference between the input voltage V_{IN} and the voltage V'_{DClink} on the DC-link referred to the primary. The evolution of V'_{DClink} is strictly related to the evolution of the voltage across the clamping capacitor Cac which is practically constant due to the large capacitance chosen for this capacitor. Using the Kirchhoff's voltage law on the loop of Fig. 2, we can obtain the following differential equation:

$$V_{IN} - Ron_{SAp} * I'_{out} - L_{leakage} * \frac{dI'_{out}}{dt} - R_{leakage} * I'_{out} - Ron_{SBn} * I'_{out} - V'_{DClink} = 0, \qquad (3)$$

which can be written as a function of the current I'_{out} as:

$$\frac{dI'_{out}}{dt} + \frac{R_{total}}{L_{leakage}} * I'_{out} = \frac{V_{IN} - V'_{DClink}}{L_{leakage}} \qquad (4)$$

where the resistance R_{total} includes all resistive components of the circuit in Fig. 2.
Eq. 4 indicates that the evolution of I'_{out} is strongly influenced by the evolution of the voltage across C'_{ac} and how quickly this voltage reaches the steady state value during the powering phases. The rise of the voltage across the capacitor also depends on the value of time constant $\frac{L_{leakage}}{R_{total}}$ and the way how the active Clamp is controlled by the modulation logic. So for very low values of this time constant the voltage across the capacitor reaches the steady state very quickly and permits to limit the value of $V_{IN} - V'_{DClink}$ during powering phases of PSB stage.

Simulation Results of the analyzed PDLC

To investigate about the relationship between the input current of the transformer, the voltage across Cac' and the leakage parameters of the system, a multiparameter analysis of the power converter has been performed and the results are reported in this chapter.

The behavior of the discussed PDLC architecture has been analyzed keeping in mind the application of the proposed topology to a DC/AC converter rated at 15kW with a supply voltage of 750V and a three-phase 400V/50Hz output voltage.

The modulation technique described in [8] has been modified to take advantage of the use of SiC power MOSFET in the PSB stage. In particular, the modulation techniques continues to guarantee the zero voltage portions on the V_{DClink} at the switching frequency of the VSI switches in order to achieve ZVT of these switches. During each powering phase the switches of the PSB are commutated several times in such a way to increase the actual switching frequency of this stage and reduce the size of the transformer. In particular, as it is done in a PSB, the switches of each leg are commutated with a 50% duty cycle and the phase shift between the two legs is set to apply several times the input voltage on the transformer for the total time necessary to transfer the required energy. In other words we subdivide each powering phase of VSI stage in several powering phases of the PSB stage. Each powering phase associated with the switches Ap-Bn is followed by the powering phase of the same duration of the switches Bp-An in order to avoid DC component of the current in the transformer. These two powering phases are separated by a free-wheeling phase whose minimum duration must be enough for achieving the complete turn-off of the involved switch. The active clamp is turned on during the first PSB powering phase at the beginning of the VSI powering phase and is turned-off during the last PSB powering phase before the end the VSI powering phase. In our application we use an IGBT module as the active clamp switch and the durations of the first and last powering phase of the PSB are larger than the other powering phases for permitting the complete turn-on and off of this switch.

To analyze the operation of the converter we report in Fig. 3 the waveforms of V_{DClink}, I_{in} and the currents in the switches of the S phase of VSI, i.e. I_{Stp} and I_{Stn}. The time base was chosen to highlight the switching periods of both the PSB and VSI stages.

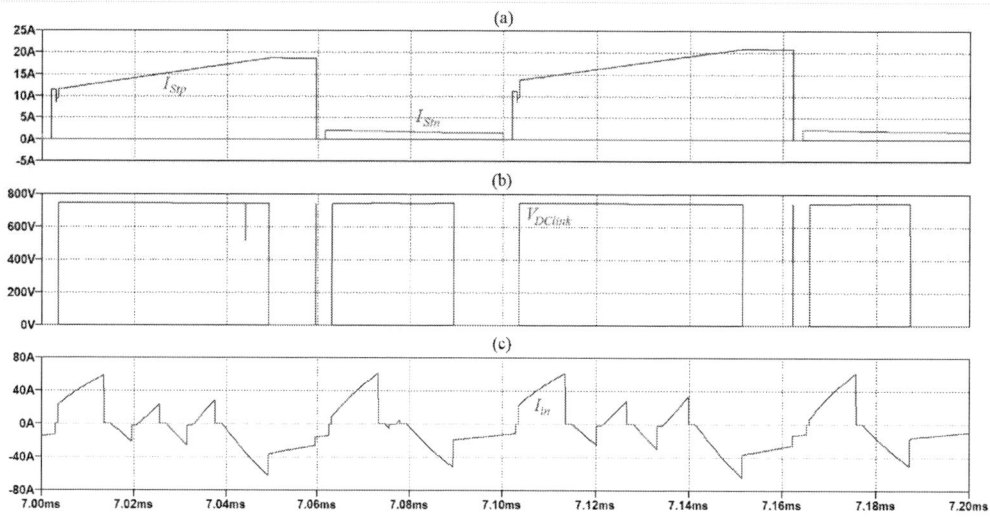

Fig. 3: Evolution of the current in the leg S of the VSI stage (a), pulsating DC link voltage (b) and primary transformer current (c) represented in very resolute time scale.

We can see that in the switching period of the VSI, $T_{sVSI} = 100\mu s$, both the switches are turned on and off during the zero voltage phase of the pulsating DC link. At the turn-offs of the Stp, which take place at about 7.06ms ($I_{Stp} = 19A$) and 7.16ms ($I_{Stp} = 21A$), respectively, a very short voltage spike is registered on V_{DClink} due to the recirculation of the energy stored in the inductance. Its amplitude is limited to about 750V by the active clamp. No spike is observed at the turn-offs of Stn which take place at much lower currents ($\approx 2A$) in the shown time interval.

The waveform of the current in the transformer (see Fig. 3c) shows that the several commutations of the PSB switches are performed during each powering phase of the VSI. As said before, the durations of the first and last PSB pulses during each powering phases of the VSI are quite large ($\approx 10\mu s$) to permit a complete turn-on and turn-off of the IGBT clamp switch. Because of this the current in the transformer reaches quite large current peaks ($\approx$60A) at the end of these powering phases. All the other current peaks are much smaller.

To study the performances of the PDLC architecture we performed several LTspice simulations in the following test conditions: DC supply voltage, V_{IN} = 750V, Switching frequency of the VSI stage, f_{sVSI} = 10kHz, Maximum output power, P_{OUTMax} = 15kW, Power Factor of the three phase load, $Cos\ \emptyset$ = 0.8, Capacitance of the Active Clamp, Cac = 220uF.

The simulations were done for different combinations of the following parameters which have a strong impact on the performances of the converter:

1. Switching frequency of the PSB stage, f_{sPSB} = 30, 40, 60 and 80kHz;
2. Leakage inductance of the transformer, $L_{leakage}$ = 0.5, 1 and 2µH;
3. Magnetizing inductance of the transformer, L_{mag} = 0.5, 1 and 2mH.

The attention of the analysis was first focused on the current's limits of transformer and switches of the PSB stage due to the different combinations of analysis parameters.

Fig. 4 (a), (b), (c), and (d) report the simulated waveforms of the current in the primary side of the transformer (bottom waveforms) and the difference between the input voltage V_{IN} and the DC-link voltage V_{DClink} (top waveforms) for f_{sPSB} = 30, 40, 60 and 80kHz, respectively. Fig. 4 clearly shows that the increase of f_{sPSB} results in the increase of the number of the powering phases of the PSB and the reduction of their durations. Contrary to what we would have expected, the increase of f_{sPSB} causes the increase of the peak current in the transformer. To understand the reasons of this behavior we have to consider the interaction between the stray inductance of the circuit, $L_{leakage}$, and clamping capacitor, C_{ac} (see Fig. 2). If we neglect the parasitic resistances of the circuit, during the powering phase C'_{ac} is in series with $L_{leakage}$. At increasing frequency the reactance of C'_{ac} decreases whereas the reactance of $L_{leakage}$ increases and, consequently, the voltage across C'_{ac} and C_{ac} decreases causing the increase of V_{IN} - V_{DClink} as it is shown in the top waveforms of Fig. 4. This voltage is

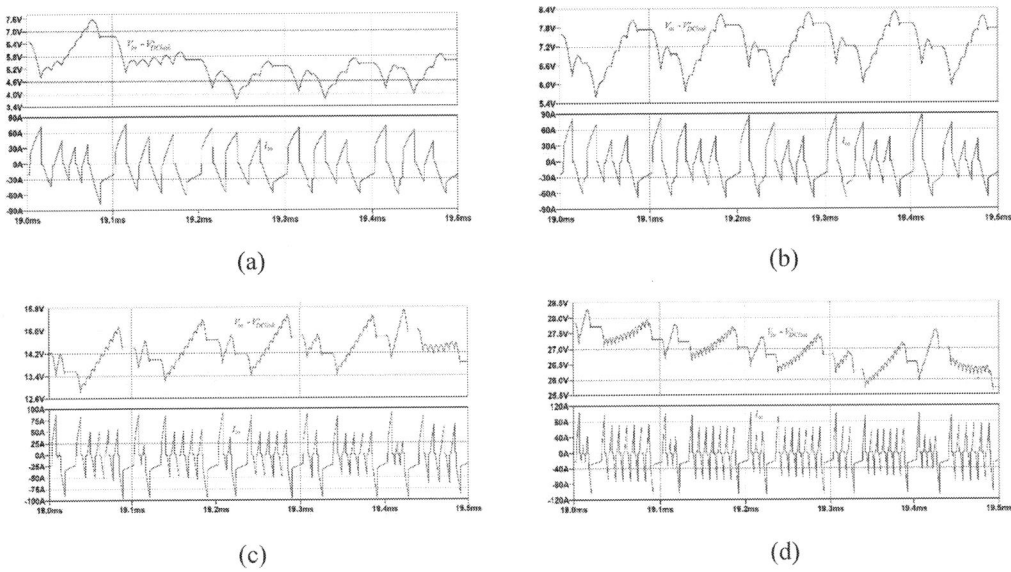

(a)

(b)

(c)

(d)

Fig. 4: The potential difference between the input voltage V_{IN} and the voltage V_{DClink} across the capacitor C'_{ac} (top waveform) and primary transformer current (bottom waveform) for a switching frequency of PSB stage of 30kHz (a), 40kHz (b), 60kHz (c) and 80kHz (d).

applied on $L_{leakage}$ so the slope of the current in it increases and with it the current peak even if the duration of each pulse is smaller at increasing frequency.

The simulated values of the maximum peak current in the transformer are reported in Fig. 5 as a function of PSB switching frequency in two cases: (a) $L_{leakage}$ = 1µH and different values of L_{mag} and (b) L_{mag} = 1mH and different values of $L_{leakage}$. The two plots confirm the significant increase of the peak current with the switching frequency. The maximum current peak is practically independent of L_{mag} whereas the current peak increases significantly with $L_{leakage}$ for the lower switching frequencies, i.e. 30kHz and 40kHz.

To analyze the effects of the switching frequency on the distortion, Fig. 6 reports the simulated waveforms of the three phase line to line output voltage for the four values of the PSB switching frequency described above. The other test conditions are: V_{IN} = 750V, f_{SVSI} = 10kHz, P_{OUTMax} = 15kW, $Cos\ \emptyset$ = 0.8, Cac = 220uF, L_{mag} = 2mH, and $L_{leakage}$ = 1µH.

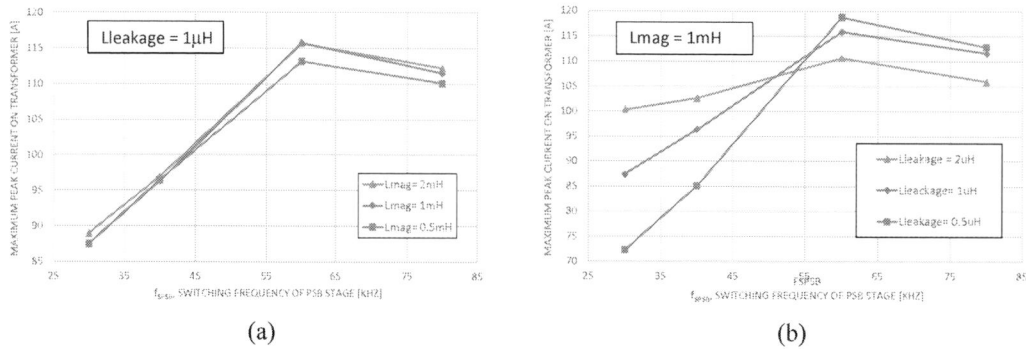

(a)   (b)

Fig. 5: Maximum peak current in the transformer vs. switching frequency of the PSB for Lleakage = 1µH and different values of Lmag (a) and for Lmag = 1mH and different values of Lleakage (b).

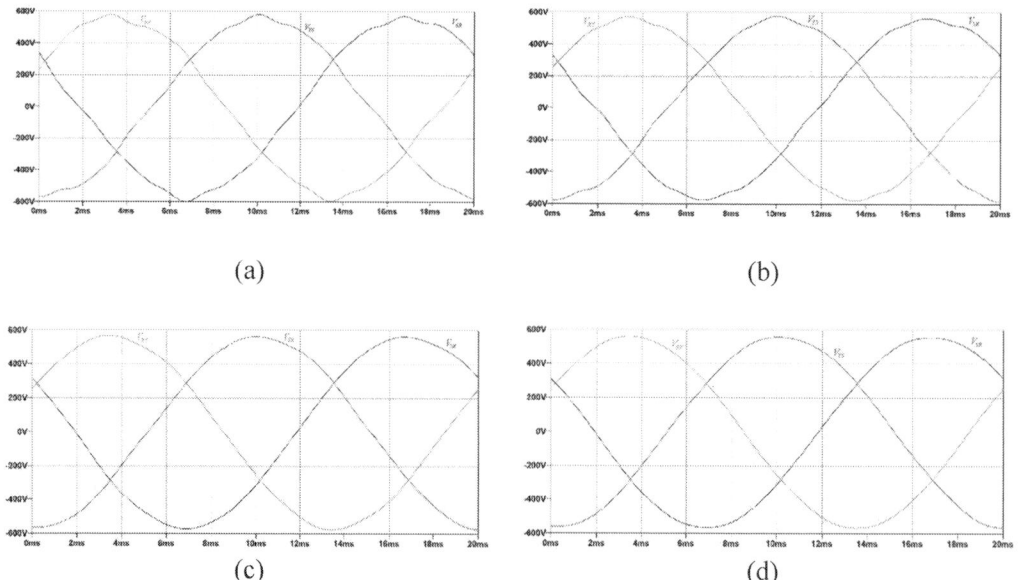

(a)   (b)

(c)   (d)

Fig. 5: Waveforms of the three phase line to line output voltage for four values of the PSB switching frequency: 30kHz (a), 40kHz (b), 60kHz (c) and 80kHz (d).

We can observe that in the cases with lower switching frequency, namely 30 and 40kHz, the line to line voltage waveforms show a significant ripple around the peaks. Lower distortion is observed on the output voltages at the higher switching frequencies for which the ripple practically disappears. This behavior is due to the implemented modulation logic of the converter which provides the best performances for high output power and high ratio between the switching frequencies, f_{sVSI}/f_{sPSB}.

As a conclusive remark of the section, we can say that the increase of PSB switching frequency results in the decrease of the distortion of the output voltages but it causes the increase of the current peaks in the PSB switches and the transformer. So we can use f_{sPSB} to get a trade-off between these two quantities.

From the above analysis we can achieve the following guideline for the design of the proposed PDLC ZVT architecture:

- To obtain output voltage with low distortion it is necessary to use a high value of the ratio between the switching frequency of the PSB stage and the VSI stage;
- To limit the peak and control the evolution of the transformer's current it is necessary to design the transformer with very low leakage inductance and find a tradeoff between the values of the switching frequency and the leakage inductance.
- The operation of the Active Clamp is very important in the choice of the switching frequency.

Experimental characterization of PDLC prototype

To validate the simulations' results and demonstrate the particular behavior of the analyzed PDLC ZVT architecture during the powering phases of the PSB stage, a light prototype of power converter has been realized. The prototype is characterized by 3.5kW nominal output power with input voltage range 500-800 VDC and 230 VAC-50Hz nominal output voltage. IGBT half bridge modules were used for the VSI stage and for the active clamp. Instead, the PSB stage was realized with all power devices, diodes and MOSFET, in Silicon Carbide (SiC) to achieve high switching frequencies and to reduce the power losses. To assure the galvanic insulation between the input DC supply and the AC load we used a high frequency transformer with a transform ratio of 6/7. Fig.7 reports two views of the power stages of the constructed PDLC ZVT prototype.

The switching frequency f_{sVSI} was fixed to 10kHz whereas different values of f_{sPSB} were used. A variable ohmic-inductive three phase load was used to characterize the power converter in several load conditions.

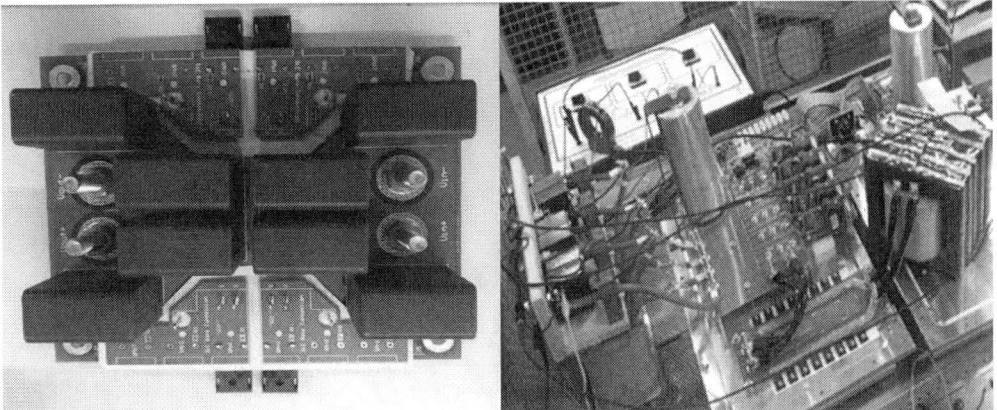

Fig. 6: Top view of the full bridge converter of the PSB stage (on left) and top view of the VSI, transformer and output rectifier of PSB (on right) of analyzed Pulsating DC Link Converter with Zero Voltage Transitions (PDLC ZVT)

Fig.8 (a) and (b) report the waveforms of line to line output voltage and current phase for $V_{in}=600V$ and $P_{OUT}=2.6kW$ pure resistive load measured for two different values of f_{sPSB}, 60kHz and 80kHz, respectively. Comparing the evolution of the electrical characteristic reported in the figure, we can observe that the line to line voltage is quite sinusoidal in both situation with very small deformation on the peaks. We can also note that the distortion of voltage is smaller for 80kHz. This result experimentally confirms that to reduce the distortion we have to increase the ratio between the switching frequency of PSB and VSI stages.

Fig.9 reports the measured pulsating voltage V_{DClink} at the output of PSB stage and the current in the primary winding of the high frequency transformer, for the same test condition of the previous figure. Comparing the voltage waveforms we can note that the evolution of the pulsating voltage is practically the same in the two cases because it does not depend on the value of f_{sPSB} but it depends only on the duration of the VSI powering phase necessary to obtain the requested output power. Regarding the behavior of the transformer's current, we can observe that the modulation logic allows us to provide the same quantity of energy to the VSI stage distributing it in a number of couples of powering phases proportional to the PSB switching frequency. It can also be noted that the rise of the switching frequency f_{sPSB}, from 60kHz to 80kHz, produces the increase of the slope and of the peak of current as predicted by the model presented in the previous section.

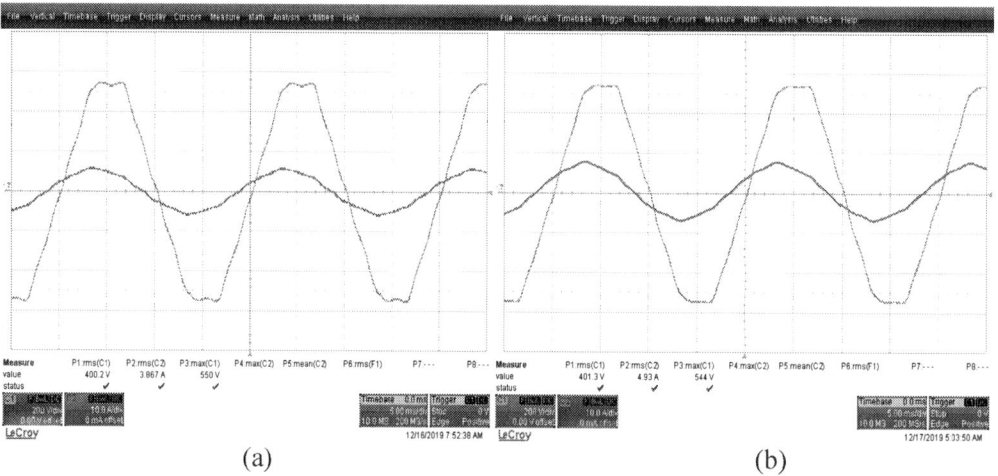

(a)   (b)

Fig. 7: Line to line voltage (yellow) and phase current (red) in the case $V_{in}=600V$, $P_{OUT}=2.6kW$ for a switching frequency of the PSB stage equal to 60kHz (a) and to 80kHz (b).

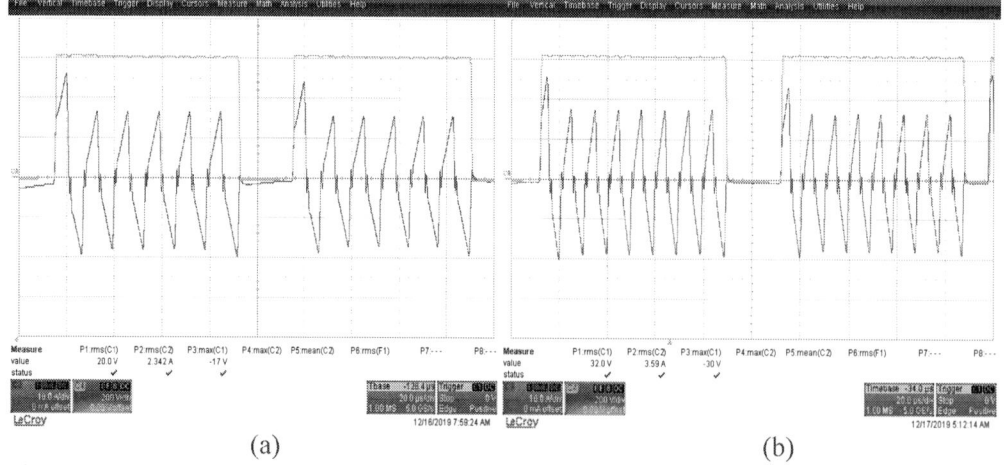

(a)   (b)

Fig. 8: Pulsating voltage (green) and primary transformer current (blue) in the case $V_{in}=600V$, $P_{OUT}=2.6kW$ for a switching frequency of the PSB stage equal to 60kHz (a) and to 80kHz (b).

Conclusions

The features of high efficiency, overall reliability and low volume has become in the last years the main characteristic of the power converters. For this reason the power converter architectures and the modulation logic have been reviewed to improve their performances and reach the requested features. In this paper we have shown how the proposed Pulsating DC Link Converter with ZVT conditions works, especially during the powering phases of the PSB stage. In particular, we have investigated about the main parameters which may influence the evolution of the current in the transformer and PSB switches. A simulation analysis of the current evolution was made for different combinations of influence parameters and the results are reported to show the limits of this innovative power converter topology and provide design constraints for this architecture. A low power PDLC ZVT prototype was constructed and tested for different switching frequencies of PSB stage and different load conditions to show the features of this topology and validate simulation results.

References

[1] B. Wang, J. Cai, X. Du and L. Zhou, "Review of power semiconductor device reliability for power converters," in CPSS Transactions on Power Electronics and Applications, vol. 2, no. 2, pp. 101-117, 2017, doi: 10.24295/CPSSTPEA.2017.00011.

[2] F. Iannuzzo, C. Abbate and G. Busatto, "Instabilities in Silicon Power Devices: A Review of Failure Mechanisms in Modern Power Devices," in IEEE Industrial Electronics Magazine, vol. 8, no. 3, pp. 28-39, Sept. 2014, doi: 10.1109/MIE.2014.2305758.

[3] C. Abbate, G. Busatto, A. Sanseverino, F. Velardi, C. Ronsisvalle, "Analysis of Low and High Frequency Oscillations in IGBTs during Turn on Short Circuit", Transaction on Electron Devices, September 2015, pp. 2952-2958

[4] H.R. Zeller, "Cosmic ray induced failures in high power semiconductor devices," Microelectronics Reliability, vol. 37, Issue 10, pp. 1711-1718, 1997.

[5] C. Abbate, G. Busatto, F. Iannuzzo, S. Mattiazzo, A. Sanseverino L. Silvestrin, D. Tedesco, F. Velardi: "Experimental Study of Single Event Effects Induced by Heavy Ion Irradiation in Enhancement Mode GaN Power HEMT" Microelectronics Reliability 2015, 55(9-10), pp. 1496-1500.

[6] Vitorino M. A., Alves L. F. S., da Silva Í. R. F. M. P., Corrêa M. B. R. and dos Santos G. G.: High-frequency pulsating DC-link three-phase inverter without electrolytic capacitor, 2017 IEEE Applied Power Electronics Conference and Exposition (APEC), Tampa, FL, 2017, pp. 3456-3461

[7] Diao L., Du H., Shu Z., Xue Y., Li M. and Sharkh S. M.: Vanderkeyn Ralf W.: A Comparative Study Between AI-HM and SPD-HM for Railway Auxiliary Inverter With Pulsating DC Link, in IEEE Transactions on Industrial Electronics, vol. 65, no. 7, pp. 5816-5825, July 2018

[8] C. Abbate, G. Busatto, F. Iannuzzo, D. Marciano and D. Tedesco, "Isolated DC/AC Converter with ZVT based on Pulsating DC Link," 2020 IEEE Applied Power Electronics Conference and Exposition (APEC), New Orleans, LA, USA, 2020, pp. 1162-1167, doi: 10.1109/APEC39645.2020.9124076.

[9] Diao L., Wang L., Du H., Wang L., Liu Z. and Sharkh S. M.: AI-HM Based Zero Portion Effects and Phase-Shift Optimization for Railway Auxiliary Inverter With Pulsating DC-Link, in IEEE Access, vol. 5, pp. 7444-7453, 2017

[10] Huang R. and Mazumder S. K.: A Soft-Switching Scheme for an Isolated DC/DC Converter With Pulsating DC Output for a Three-Phase High-Frequency-Link PWM Converter, in IEEE Transactions on Power Electronics, vol. 24, no. 10, pp. 2276-2288, Oct. 2009

[11] Rahnamaee A. and Mazumder S. K.: A Soft-Switched Hybrid-Modulation Scheme for a Capacitor-Less Three-Phase Pulsating-DC-Link Inverter, in IEEE Transactions on Power Electronics, vol. 29, no. 8, pp. 3893-3906, Aug. 2014

[12] Zhou D., Wang H., Blaabjerg F., Kær S. K. and Blom-Hansen D.: Degradation effect on reliability evaluation of aluminum electrolytic capacitor in backup power converter, Proc. IEEE 3rd Int. Future Energy Electron. Conf. and ECCE Asia, pp. 202-207, Jun. 2017.

[13] Wang H. and Blaabjerg F.: Reliability of Capacitors for DC-Link Applications in Power Electronic Converters—An Overview, in IEEE Transactions on Industry Applications, vol. 50, no. 5, pp. 3569-3578, Sept.-Oct. 2014

Signal Injection for Sensorless Current Sharing with Experimental Verification on 1 MHz GaN Prototype

N. Bošković, J. Duarte and E.A. Lomonova
Eindhoven University of Technology
Electromechanics & Power Electronics Group
Eindhoven, The Netherlands
Email: n.boskovic@tue.nl
URL: https://www.tue.nl

Keywords

≪Gallium Nitride (GaN)≫, ≪Parallel operation≫, ≪Multiphase converter≫, ≪Estimation technique≫, ≪Sensorless current sharing≫

Abstract

The drawback of estimation-based sensorless current sharing applied in multiphase converters is that it requires an injection of a lengthy perturbation signal into the system, which can influence the load current or stress the components. This paper experimentally demonstrates the effectiveness of a method to reduce the duration of the perturbation signal. By modifying the shape of the signal which is injected, the steady-state, needed for sensorless balancing, can be reached faster, although the duration of the injected signal is much shorter.

Introduction

The lasting challenge in power electronics is to reach higher power ratings of power converters. The limiting factor is insufficient current rating of switching power devices which are used to control the power flow. Transistors, such as Gallium Nitrides (GaN), have limited current capability which can be increased if transistors are paralleled directly [1]. Paralleling of GaNs, however, brings new challenges which are still not completely solved. During the transients, equal current distribution can not be guaranteed, and can substantially increase the temperature difference between the transistors, see [2] where three transistors are paralleled. This effect is even more severe for higher switching frequencies. Another issue when paralleling transistors is a gate-source voltage oscillation, see [3] where half-bridges are paralleled, which can lead to a breakdown of transistors. In order to mitigate these issues, a small inductor is connected to the switching node of each paralleled half-bridge. The added inductances limit the current transients, and enable current control. Furthermore, they allow to interleave the currents, leading to a smaller output voltage ripple. Although the current transients are limited, the steady-state currents still have to be controlled. Due to the differences between the branch (half-bridge plus inductor) resistances, the steady-state branch currents can be unequal. This leads to uneven loading of the branches and their incomplete utilization. Therefore, it is necessary to balance the branch currents.

Placing a current sensor in each paralleled branch of the converter will give exact information about the current distribution. It is then possible to balance the currents with active current control by using the measured branch currents, as in [4]. Other methods obtain information about current distribution by measuring the input voltage [5], the output voltage [6], or the voltages across the branch inductors [7]. In order to lower complexity, sensorless current sharing is explored. An estimation-based sensorless current sharing is an attractive solution for balancing currents in converters with many branches in parallel due to the lack of current sensors. In [8] – [10], perturbation signals are injected in each branch of the converter. By observing the output voltage steady-state response, it is possible to estimate the current distribution due to the branch resistance mismatch.

Since the injected perturbation signals have an impact on the temperature of the components of the converter, and have an impact on the load current, the modified perturbation signal is introduced in [11]. It enables reduced duration of the perturbation, making the impact smaller. This paper experimentally verifies the proposed method. By reducing the duration of the perturbation, it is possible to apply it in any moment during converters operation, without having a huge impact on the temperature of the components.

Modified injected signal for sensorless current sharing

In order to increase the current rating, branches are paralleled. Each branch consists of a half-bride and an inductor together with the decoupling capacitor and the gate driver, as shown in Fig. 1. A circuit of the converter, consisting of three paralleled branches, is shown in Fig. 2. The equivalent branch resistances R_1, R_2 and R_3, represent the on-state resistance of the transistors, where the on-state resistances of the upper and lower transistor in one half-bridge are considered equal, of the resistance of the PCB traces, and the equivalent series resistance of the inductor. This topology also allows interleaving of the branch currents.

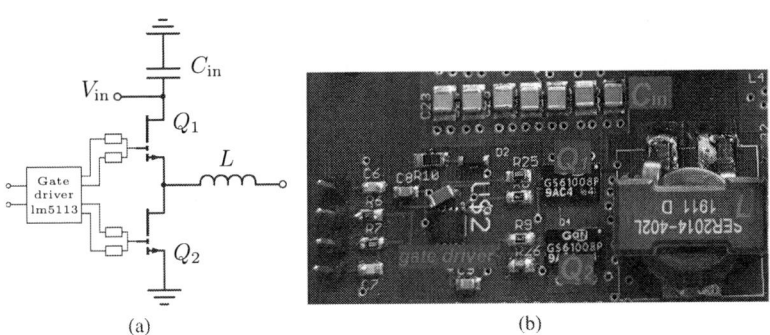

Fig. 1: Branch consisting of the half-bridge with GaN transistors (Q_1, Q_2), the inductor (L), the decoupling capacitor of the half-bridge (C_{in}) and the gate driver. (a) Schematic of the branch, and (b) photo of one branch in the prototype.

The converter with the paralleled branches can exhibit an uneven distribution of the branch currents if there is a difference between the branch resistances or the difference between the duty cycles of each half-bridge. The sensorless current balancing method compensates for the branch resistances mismatch. In this method, the perturbation signals are applied consecutively to each half-bridge. Each time, the quasi-steady-state response of the output voltage is measured. The quasi-steady-state represents the steady-state of the equivalent circuit in which the variables are averaged over one switching period. In other words, the switching behaviour is omitted. The measured output voltage response is then used to estimate the current sharing during the operation of the converter. Detail description and analysis is given in [8] and [11].

The standard perturbation signals required for the estimation have the square-wave shape with period T_{pert}, as represented with the dash-dotted trace in Fig. 3a. After the signals are injected, it is necessary to wait until the output voltage reaches it's quasi-steady-state. This waiting time is determined by the converter's parameters. If it is too long, the perturbation will influence the load current or increase the temperature of the components. The solid yellow trace in Fig. 3a shows the modified perturbation signal with period $T_{\text{pert,m}}$, introduced in [11], which enables shorter perturbation time while still reaching the same quasi-steady-state response. The response to the perturbation of the quasi-steady-state current is

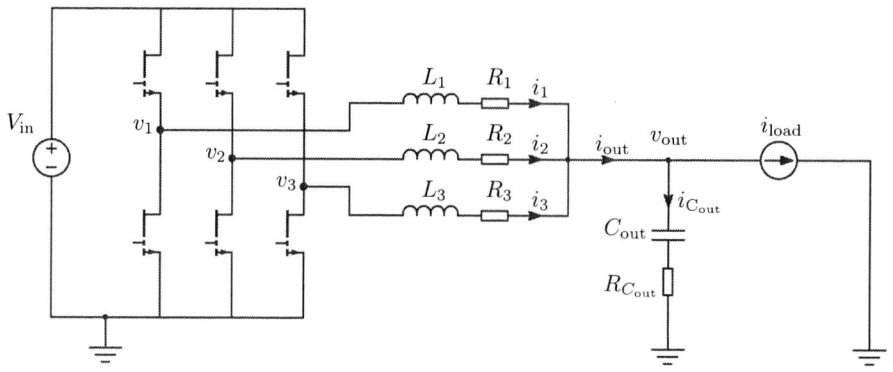

Fig. 2: Converter with the three paralleled branches.

shown in Fig. 3b. It can be seen that the current response is shorter when the modified perturbation is applied but still reaches the same steady-state value.

Parameters $t_{\text{init}1}$ and $t_{\text{init}2}$ are durations of the initial pulses, as shown in Fig. 3. The values of these parameters are defined in such a way that the current of the perturbed branch reaches the value equal to the quasi-steady-state current when the square-wave signal (dash-dotted trace) is applied [11]. After the initial pulse with amplitude $\Delta_{\text{pert,init}}$ has passed, the standard pulse with amplitude Δ_{pert} is applied until the end of period $T_{\text{pert,m}}$ to ensure that the branch current reaches the desired value. The parameters are defined as

$$t_{\text{init}1} = \frac{L_{\text{nom}}}{R_{\text{nom}}} \ln\left(\frac{\Delta_{\text{pert,init}}}{\Delta_{\text{pert,init}} - \Delta_{\text{pert}}}\right), \qquad t_{\text{init}2} = \frac{L_{\text{nom}}}{R_{\text{nom}}} \ln\left(\frac{\Delta_{\text{pert,init}}}{\Delta_{\text{pert,init}} - 2\Delta_{\text{pert}}}\right), \qquad (1)$$

where $\Delta_{\text{pert,init}}$ is the amplitude of the initial pulse, and Δ_{pert} is the amplitude of the square-wave signal. The values L_{nom} and R_{nom} represent the nominal values of the branch inductance and the branch resistance.

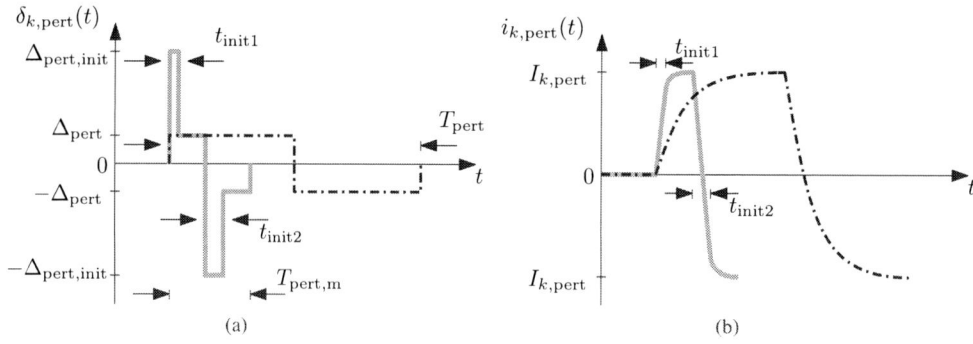

Fig. 3: Waveforms of (a) the perturbation signals, and (b) corresponding quasi-steady-state branch current responses due to the perturbation. The dash-dotted traces represent the standard square-wave perturbation signal and its current response, while the solid yellow traces represent the modified perturbation signal which enables shorter perturbation time and its current response.

Experiments

The converter under test is a low-voltage converter which consists of three branches, as shown in Fig. 4, where each branch consists of a half-bridge with Gallium Nitride transistors GS61008P switching at 1

MHz (period 1000 ns), and the inductor SER2014-402L with inductance value of 4 μH. The control of the converter is implemented in a STM32G474RE microcontroller (MCU), which is part of a NUCLEO-G474RE development board. The clock frequency of the MCU is 170 MHz, and the timer frequency is 170 MHz. However, the MCU has a high-precision timer (HRTIM) which gives a resolution of the PWM generator of $\pm$184 ps. The input and output voltages are measured through a voltage divider and a low-pass filter, and then sampled with the MCU's internal 12-bit analogue-to-digital converter (ADC) with a sampling rate of 4 MSPS (mega samples per seconds). The measured voltage signals have a ripple induced by the switching behaviour. A peak-to-peak voltage ripple can be large enough to produce a significant measurement error. To overcome this, an oversampling is applied. A measured signal is oversampled by a factor of 16, and a resulting value is shifted by 2 bits, so that the effective resolution is 14 bits. In that way, the moving-average value is measured, and the influence of the ripple at the switching frequency on the measurements is minimised.

Fig. 4: Photo of the three branch GaN converter.

The measurement results are shown in Fig. 5 – Fig. 9. These experiments are done at zero load current, while the load inductance is large enough to restrict the current to significantly change during the perturbation sequence. It is also worth noting that the scale of the x-axis is different among Figures. The y-axis is the same, except for Fig. 7 in which different perturbation amplitudes are applied. The branch currents are in red (i_1), blue (i_2) and green (i_3). Standard square-wave perturbation is applied in Fig. 5, and Fig. 6a. The branches are perturbed one by one, and the period of the injected signal (single perturbation) is $T_{\text{pert}} = 3000\ \mu$s, $1000\ \mu$s, and $300\ \mu$s, respectively. The time-gap between each perturbation is equal to the single perturbation period T_{pert}. The amplitude of the perturbation is $\Delta_{\text{pert}} = 10\%$ (a percentage of the maximum duty cycle, which is 1).

The amplitude of the injected signal Δ_{pert} is chosen such that the branch inductors do not saturate during the perturbation. Fig. 7 shows standard perturbation with the period of $3000\ \mu$s when the injected signal has an amplitude of 7.5%, 15%, and 20%, respectively. For comparison, the current ripple of current i_2 is 3 A, 11 A, and 17 A, respectively. A significant saturation, visible for the perturbation signal amplitudes of 15% and 20%, distorts the output voltage and decrease the measurement precision of the output voltage variation. Moreover, high currents, due to saturation, force the transistors to leave the Ohmic region, which distorts the switching node voltage. Therefore, the amplitude of the injected signals is restricted to 10%, so that the inductors do not saturate during the perturbations.

In order to decrease the duration of the perturbation signal, it is possible to apply the proposed modified perturbation signal to reach the quasi-steady-state. The responses of the branch currents due to the consecutive modified perturbations is shown in Fig. 6b. The period of the injected modified perturbation signal is $300\ \mu$s. Even though the modified perturbation signal is ten times shorter, the same quasi-steady-state is reached as with the standard square-wave perturbation signal with a period of $3000\ \mu$s.

The amplitude of the initial pulses $\Delta_{\text{pert,init}}$ is set to 40%. Based on this value, it is possible to calculate parameters t_{init1} and t_{init2} according to the expression (1). The nominal branch inductance is $L_{\text{nom}} = 4\ \mu$H, and the estimated nominal branch resistance is $R_{\text{nom}} = 200$ mΩ. The values of t_{init1} and t_{init2} are 5.8 μs, and 13.9 μs, respectively. For comparison, Fig. 8 shows the cases when $\Delta_{\text{pert,init}}$ is different from the assumed value when the parameters t_{init1} and t_{init2} were calculated. For $\Delta_{\text{pert,init}} = 30\%$, it is apparent that the currents in the branches two and three do not reach the quasi-steady-state. For $\Delta_{\text{pert,init}} = 50\%$, the amplitude of the response of the current in the branch one is much larger than for the case when the modified perturbation amplitude is 40%.

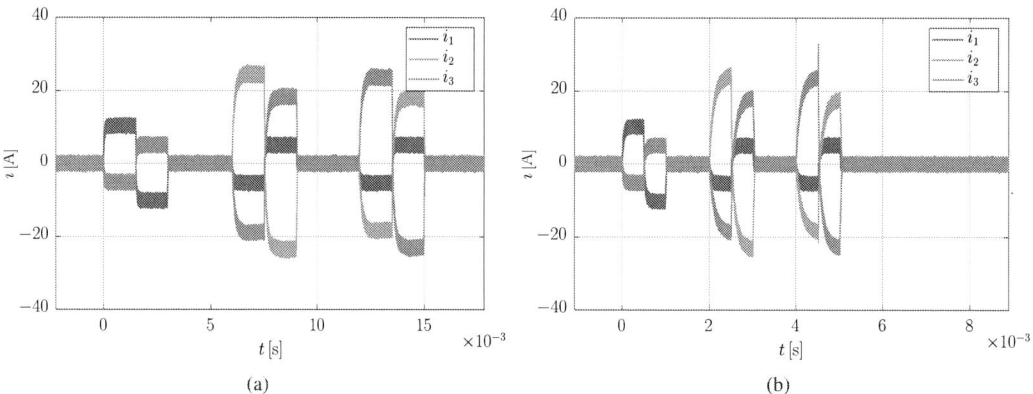

(a)　　　　　　　　　　　　　　　　(b)

Fig. 5: Perturbation of the branch currents (red, blue and green) with the standard square-wave signal with a period T_{pert} of 3000 μs (a), and 1000 μs (b).

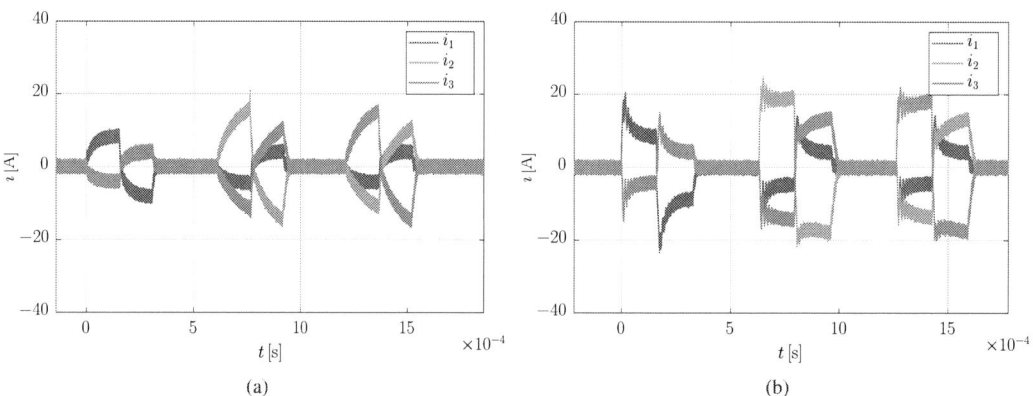

(a)　　　　　　　　　　　　　　　　(b)

Fig. 6: Perturbation of the branch currents (red, blue and green) with standard square-wave signal (a), and modified signal (b). Both periods, T_{pert} and $T_{\text{pert,m}}$, are 300 μs, but with the modified perturbation the branch currents are closer to reaching the quasi-steady-state, although the injected perturbations have the same final amplitude Δ_{pert}.

During the perturbation of each branch, the quasi-steady-state output voltage response is obtained. The obtained measurements represent the variation of the output voltage v_{out} when the perturbation is applied to one of the branches. It is denoted as ΔV_{out_k}, where $k = 1 \ldots 3$. Values of ΔV_{out_k} are obtained for the perturbations shown in Fig. 5-6, and given in Table I.

It is possible to express a branch resistance R_k as a function of ΔV_{out_k} [11], as given by

$$R_k = R_{\text{thev}} V_{\text{in}} 2\Delta_{\text{pert}} \frac{1}{\Delta V_{\text{out}_k}}, \tag{2}$$

where Δ_{pert} is the amplitude of the square-wave perturbation signal shown in Fig. 3a, and the param-

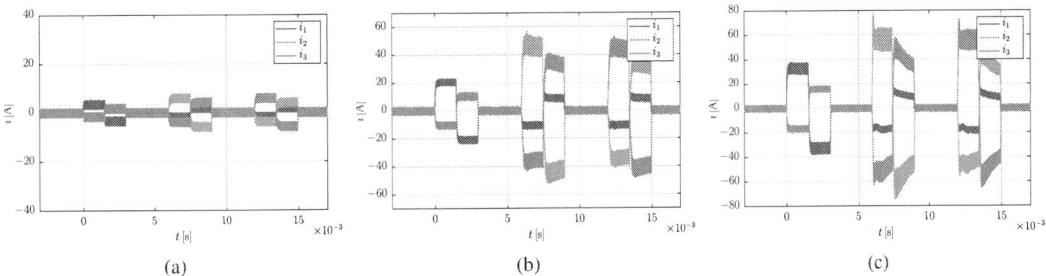

(a)   (b)   (c)

Fig. 7: Standard square-wave perturbation with the period $T_{\text{pert}} = 3000 \ \mu s$, with the amplitudes Δ_{pert} of 0.075 (a), 0.15 (b), and 0.2 (c). The large amplitude of the perturbation signal can lead to the saturation of the branch inductors.

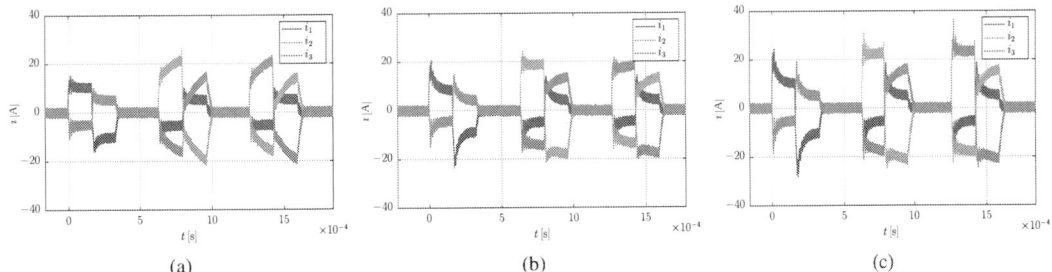

(a)   (b)   (c)

Fig. 8: Modified perturbation with the initial signal amplitude $\Delta_{\text{pert,init}}$ of 0.3 (a), 0.4 (b), and 0.5 (c).

eter R_{thev} represents the Thevenin equivalent of the branch resistances, in other words, an equivalent resistance of the three paralleled branch resistances.

If the previous expression (2) is incorporated in the expression for the branch DC current,

$$I_k = \frac{\delta_k V_{\text{in}} - V_{\text{out}}}{R_k}, \tag{3}$$

then the following expression is obtained

$$\hat{I}_k = \frac{\delta_k V_{\text{in}} - V_{\text{out}}}{R_{\text{thev}} V_{\text{in}} 2 \Delta_{\text{pert}}} \Delta V_{\text{out}_k}, \tag{4}$$

where $\hat{I}_k$ represents the estimated value of the current in k-th branch. During the modified perturbation, the initial pulses exist only to accelerate the rise or fall of the branch currents. The final amplitude of the perturbation signal is still Δ_{pert}. Therefore, the expression (4) is valid both for the standard and the modified perturbation.

The resistance R_{thev} is not known and, therefore, it is not possible to calculate $\hat{I}_k$ from (4). However, it is possible to determine the current distribution based on the known non-constant part of the (4), which is $(\delta_k V_{\text{in}} - V_{\text{out}}) \Delta V_{\text{out}_k}$. Both the input and the output voltage are measured, and the duty cycles δ_k are known since they are input commands. After estimation of the current distribution, the branch currents are balanced as described in [11].

The parameters ΔV_{out_k} are also calculated directly from (2) in order to compare them with the measured values. The branch resistances are estimated from $I_1 R_1 = I_2 R_2 = I_3 R_3$, while the branch currents are measured in the quasi-steady-state. The calculated values of ΔV_{out_k} are 1.144 V, 4.641 V, and 3.815 V, respectively. Compared to the results in Table I, the main difference is in the first parameter $\Delta V_{\text{out},1}$ which differs for about two volts.

Table I: Experimentally obtained values of ΔV_{out_k}, where $k = 1\dots3$, during the standard square-wave and modified perturbations.

	T_{pert}	ΔV_{out_1}	ΔV_{out_2}	ΔV_{out_3}
Standard	$3000\,\mu s$	3.204 V	4.426 V	4.577 V
Standard	$1000\,\mu s$	3.154 V	4.599 V	4.057 V
Standard	$300\,\mu s$	3.497 V	4.623 V	4.568 V
Modified	$300\,\mu s$	3.875 V	4.753 V	4.548 V

The existing difference is due to the dead time. When the positive part of the perturbation signal is applied to one branch, the other two have negative currents and the switching node voltage of the half-bridges in these branches has longer turn-on time than predicted. The resulting amplitude of the signal injected in the first branch is Δ_{pert}, and the resulting amplitude of the signal injected in the rest of the branches, which was assumed to be zero, is actually $-2t_{\text{dt}}/T_{\text{sw}}$, where $t_{\text{dt}} = 33$ ns is the dead time, and $T_{\text{sw}} = 1000$ ns is the switching period. When a negative perturbation is applied to one of the branches, the other two have a positive currents, and the resulting amplitude of the injected signal is lowered by $2t_{\text{dt}}/T_{\text{sw}}$, being $-\Delta_{\text{pert}} - 2t_{\text{dt}}/T_{\text{sw}}$.

The difference between the assumed and the actual perturbation amplitudes, existing due to the influence of the dead time, is the reason that the measured values of ΔV_{out} are different than the calculated. In order to compensate for the difference, the dead time is incorporated into the injected perturbation signal pattern such that the actual resulting amplitude of the signal is equal to Δ_{pert}. The dead time t_{td} is set in the firmware of the MCU. At the output of the MCU, two complementary signals with the dead time t_{dt} are generated for the each branch. A perturbation signal with a dead time compensation leads to the similar values of ΔV_{out_k} as the one directly calculated from (2). The new values of ΔV_{out_k} are given in Table II.

Table II: Obtained values of ΔV_{out_k}, where $k = 1\dots3$, during the standard square-wave and modified perturbations with the dead time compensation.

	T_{pert}	ΔV_{out_1}	ΔV_{out_2}	ΔV_{out_3}
Standard	$3000\,\mu s$	1.142 V	4.068 V	3.809 V
Standard	$1000\,\mu s$	0.708 V	3.682 V	2.754 V
Modified	$300\,\mu s$	0.903 V	3.042 V	2.556 V

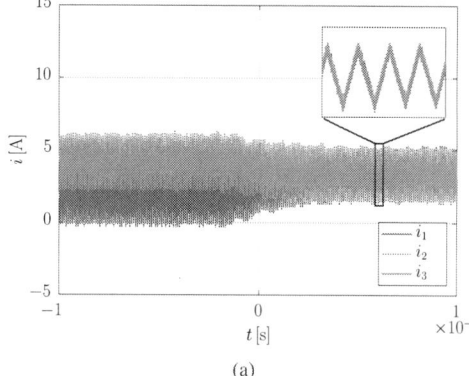
(a)

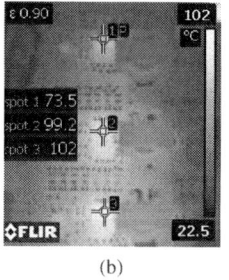
(b)

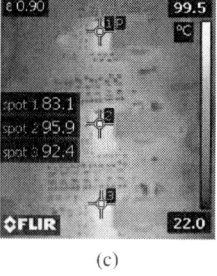
(c)

Fig. 9: Waveforms of the branch currents at the moment when balancing controller is activated (a). Thermal images of the prototype, including steady-state case temperatures of the GaN devices, before (b), and after (c) balancing the branch currents. The temperatures are closer to the average temperature after balancing (c), than before balancing (b).

The experimental demonstration of balancing is shown in Fig. 9a. The calculated values of ΔV_{out_k} are used for balancing the currents. The output current is 10.8 A in this case.

The benefit of the balancing is verified not only by showing the balanced branch currents in Fig. 9a, but also by measuring the steady-state case temperatures of the GaN transistors of each branch. Fig. 9b, and Fig. 9b show the thermal image of the prototype before and after the current balancing. The images are taken from the same perspective as the photo in Fig. 4. On each thermal image, three temperatures are detected. The converter was in operation for a longer period so that the steady-state temperatures are captured. These temperatures, on the images labelled as spots 1, 2 and 3, refer to temperatures of the transistors of the branches 1, 2 and 3 (red, blue and green), respectively. It can be seen that before the balancing, the temperature difference between the first and the third branch is 28.5 °C, and after the balancing it is reduced to 9.3 °C. The temperature of the second branch is also reduced, from 99.2 °C to 95.9 °C, approaching the average case temperature.

Conclusion

Current sharing in the paralleled converters can be guaranteed if the perturbation-based sensorless method is applied. The perturbation required to estimate the current distribution can be shortened to reduce its impact on the load current and the components. By modifying the perturbation pattern, it is possible to reach the needed quasi-steady-state faster. This paper verifies the proposed method experimentally on a Gallium Nitride based prototype with 1 MHz switching frequency. The obtained values of ΔV_{out_k} are compared for different periods of injected signal and different shape of the injected signal. Additionally, the influence of the dead time is compensated by further modifying the amplitude of the perturbation signal. As a consequence, the resulting output voltage variations ΔV_{out_k} correspond to the calculated ones, and the balancing of the branch currents is accomplished.

References

[1] J. L. Lu and D. Chen: Paralleling GaN E-HEMTs in 10kW–100kW systems, 2017 IEEE Applied Power Electronics Conference and Exposition (APEC), Tampa, FL, 2017, pp. 3049-3056.

[2] N. Sang, P. Jeannin and P. Lefranc: Analyses of the unbalanced paralleled GaN HEMT transistors, 2018 20th European Conference on Power Electronics and Applications (EPE'18 ECCE Europe), Riga, 2018, pp. P.1-P.9.

[3] J. Burkard and J. Biela: Paralleling GaN switches for low voltage high current half-bridges, 2019 21th European Conference on Power Electronics and Applications (EPE'19 ECCE Europe), Genova, 2019.

[4] J. Burkard, M. Pfister and J. Biela: Control Concept for Parallel Interleaved Three-Phase Converters with Decoupled Balancing Control, 2018 20th European Conference on Power Electronics and Applications (EPE'18 ECCE Europe), Riga, 2018, pp. P.1-P.9.

[5] G. Eirea and S. R. Sanders: Phase Current Unbalance Estimation in Multiphase Buck Converters, IEEE Transactions on Power Electronics, vol. 23, no. 1, pp. 137-143, Jan. 2008.

[6] S. Mariethoz, A. G. Beccuti and M. Morari: Model predictive control of multiphase interleaved DC-DC converters with sensorless current limitation and power balance, 2008 IEEE Power Electronics Specialists Conference, Rhodes, 2008, pp. 1069-1074.

[7] E. Dallago, M. Passoni and G. Sassone: Lossless current sensing in low-voltage high-current DC/DC modular supplies, IEEE Transactions on Industrial Electronics, vol. 47, no. 6, pp. 1249-1252, Dec. 2000.

[8] R. F. Foley: Sensorless current estimation and sharing in multiphase buck converters, IEEE Transactions on Power Electronics, vol 27, no 6, pp. 2936-2946, 2012.

[9] X. Zhang, L. Corradini and D. Maksimovic: Sensorless Current Sharing in Digitally Controlled Two-Phase Buck DC-DC Converters, 2009 Twenty-Fourth Annual IEEE Applied Power Electronics Conference and Exposition, Washington, DC, 2009, pp. 70-76.

[10] J. Gordillo and C. Aguilar: A Simple Sensorless Current Sharing Technique for Multiphase DC–DC Buck Converters, IEEE Transactions on Power Electronics, vol. 32, no. 5, pp. 3480-3489, May 2017.

[11] N. Boskovic, M. G. L. Roes, C. G. E. Wijnands and E. A. Lomonova: Improved current estimation in paralleled half-bridge converters, 2019 21st European Conference on Power Electronics and Applications (EPE '19 ECCE Europe), Genova, Italy, 2019, pp. P.1-P.9.

Modelling and Analysis of Sensorless Current Sharing Approach

N. Bošković, and J. Duarte
Eindhoven University of Technology
Electromechanics & Power Electronics Group
Eindhoven, The Netherlands
Email: n.boskovic@tue.nl
URL: https://www.tue.nl

Keywords

≪Sensorless current sharing≫, ≪Balancing controller≫, ≪Estimation technique≫, ≪Gallium Nitride (GaN)≫, ≪Parallel operation≫

Abstract

Sensorless current sharing enables even distribution of the currents in multiphase converters. This paper presents a detailed model of a proposed balancing controller based on sensorless current sharing, and a converter consisting of paralleled branches. Additionally, stability and sensitivity analysis are included. The model is verified with experiments by balancing the branch currents at different load currents.

Introduction

In order to increase a current rating of converters, it is a common practice to parallel half-bridges (HBs). If HBs are directly paralleled, then transient and steady-state current sharing become an issue [1]. When a small inductor is connected to the middle point of each paralleled HB, the problem of current sharing is reduced to the steady-state [2]. The impedances of the added inductors limit the dynamics of the circulating currents, which can occur between the paralleled branches. As a consequence, it is easier to balance these currents.

In order to avoid using current sensors to balance the steady-state branch currents, a sensorless current sharing is applied [3]. By measuring the response during the injected perturbation, it is possible to estimate the steady-state current distribution. A balancing controller uses this estimation to equalize the branch currents. It compensates for the steady-state current mismatch, which occurs due to the difference between the branch resistances.

Sensorless current sharing approach has been implemented in [3] – [5]. However, there is a lack of system-level model, which combines a balancing controller with a converter. Models of power converters and controllers can give an insight into the behaviour of the converters during all operating points. Large-signal models that are solved numerically can be complex and include converter nonlinearities. On the other hand, small-signal models based on analytical equations are linear and usually less complex. Analytical models can be easily used for stability and uncertainty analysis. Furthermore, these models can be used to optimise the design of the converter and the controller.

This manuscript presents an simplified analytical model of the parallel HBs converter and the balancing controller, along with the stability and sensitivity analysis. The previous papers use a structure of the balancing controller with a proportional-integral (PI) block. In this work, a more effective structure of the balancing controller is proposed. Before showing the proposed structure of the balancing controller, the model of the converter is first derived. Afterwards, the sensorless current sharing approach is described and a structure of a balancing controller is proposed, together with a model of the controller. The efficacy of the models is verified by means of experiments performed on Gallium Nitride based converter with three paralleled branches.

Converter model

The circuit of the converter with n paralleled branches is shown in Fig. 1. The branch resistance R_k, where $k = 1 \ldots n$, represents the sum of an equivalent on-state resistance of the transistors in the branch, the equivalent series resistance of the branch inductor, and the resistance of the PCB tracks in the branch. The on-state resistances of the bottom and the top transistors in the HB are considered equal, and constant even with a junction temperature change. The equivalent series resistance of the output capacitor C_{out} is denoted as $R_{C_{\text{out}}}$.

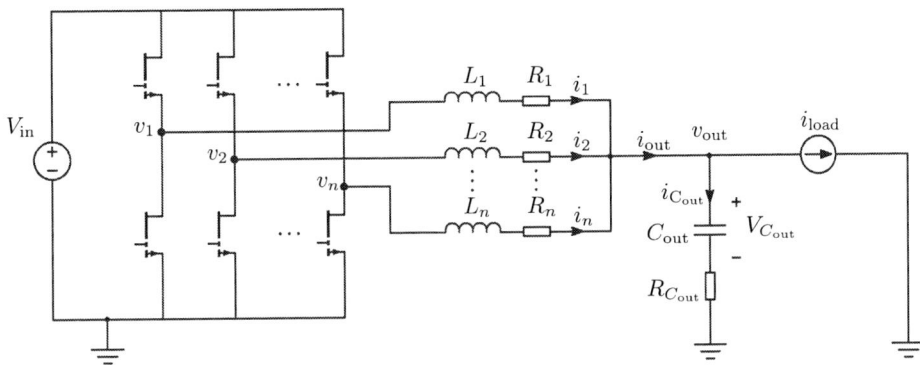

Fig. 1: Circuit of the converter with the paralleled branches.

The linear state-space model of the converter is based on the circuit shown in Fig. 1 and it is derived in a similar manner as in [6]. The middle-point voltage of the HBs, v_k, is averaged over one switching period. Therefore, the voltage is calculated as $v_k = \delta_k V_{\text{in}}$, where δ_k is the duty cycle of the k-th HB. The state-space model of the converter is given by

$$\dot{x} = \mathbf{A}x + \mathbf{B}u, \tag{1}$$

$$y = \mathbf{c}x + \mathbf{d}u, \tag{2}$$

where $x = [i_1, \cdots, i_n, v_{C_{\text{out}}}]^T \in \mathbb{R}^{(n+1) \times 1}$ is the state vector, $u = [\delta_1, \cdots, \delta_n, i_{\text{load}}]^T \in \mathbb{R}^{(n+1) \times 1}$ is a vector of the inputs, and $y = v_{\text{out}}$ is the output variable. A system matrix $\mathbf{A} \in \mathbb{R}^{(n+1) \times (n+1)}$, an input matrix $\mathbf{B} \in \mathbb{R}^{(n+1) \times (n+1)}$, an output matrix $\mathbf{c} \in \mathbb{R}^{1 \times (n+1)}$, and a feedthrough matrix $\mathbf{d} \in \mathbb{R}^{1 \times (n+1)}$ are given by

$$\mathbf{A} = \begin{bmatrix} -\frac{R_1 + R_{C_{\text{out}}}}{L_1} & \cdots & -\frac{R_{C_{\text{out}}}}{L_1} & -\frac{1}{L_1} \\ \vdots & \ddots & \vdots & \vdots \\ -\frac{R_{C_{\text{out}}}}{L_n} & \cdots & -\frac{R_n + R_{C_{\text{out}}}}{L_n} & -\frac{1}{L_n} \\ \frac{1}{C_{\text{out}}} & \cdots & \frac{1}{C_{\text{out}}} & 0 \end{bmatrix}, \quad \mathbf{B} = \begin{bmatrix} \frac{V_{\text{in}}}{L_1} & \cdots & 0 & \frac{R_{C_{\text{out}}}}{L_1} \\ \vdots & \ddots & \vdots & \vdots \\ 0 & \cdots & \frac{V_{\text{in}}}{L_n} & \frac{R_{C_{\text{out}}}}{L_n} \\ 0 & \cdots & 0 & -\frac{1}{C_{\text{out}}} \end{bmatrix}, \tag{3}$$

$$\mathbf{c} = \begin{bmatrix} R_{C_{\text{out}}} & \cdots & R_{C_{\text{out}}} & 1 \end{bmatrix}, \quad \mathbf{d} = \begin{bmatrix} 0 & \cdots & 0 & -R_{C_{\text{out}}} \end{bmatrix}. \tag{4}$$

Sensorless current sharing

Once the state-space model of the converter is derived, the balancing controller should be modelled as well. First, the main expressions describing sensorless current sharing approach are given. Then the balancing controller is introduced, and the model is derived.

As already mentioned, the sensorless current sharing requires an injection of the perturbation signal to obtain the output voltage response. The required information about the response is contained in measured quantities ΔV_{out_k}, where $k = 1 \ldots n$, which represent the output voltage response variations when the signal is accordingly injected in the duty cycle of each HB [5]. The expressions describing the current sharing from [3] are, using a similar approach, derived in [5]. It is shown that the absolute value of the estimated branch DC current is related to the variation ΔV_{out_k}, as

$$\hat{I}_k = \frac{\delta_k V_{\text{in}} - V_{\text{out}}}{R_{\text{thev}} V_{\text{in}} 2\Delta_{\text{pert}}} \Delta V_{\text{out}_k}, \tag{5}$$

where Δ_{pert} represents the amplitude of the injected perturbation signal, and R_{thev} is the equivalent Thévenin resistance of the branch resistances. Since the parameter R_{thev} is unknown, it is not possible to calculate the absolute value of the branch currents from (5). However, it is possible to make use of an expression proportional to (5), assuming that when applying the same Δ_{pert} for each branch, the ratios

$$\frac{\Delta V_{\text{out}_j}}{\Delta V_{\text{out}_k}} \tag{6}$$

with $j, k \in [1 \ldots n]$, are constants. In a practical setup, the values in (6) should be obtained through experimental results. By defining g_k as

$$g_k = (\delta_k V_{\text{in}} - V_{\text{out}}) \Delta V_{\text{out}_k}, \tag{7}$$

the expression (5) can now be rewritten as $\hat{I}_k = g_k / (R_{\text{thev}} V_{\text{in}} 2\Delta_{\text{pert}})$. From (7), the duty cycle is expressed as

$$\delta_k = \frac{g_k}{V_{\text{in}} \Delta V_{\text{out}_k}} + \frac{V_{\text{out}}}{V_{\text{in}}}. \tag{8}$$

What remains is to set an appropriate value for g_k in order to have the balanced branch currents. That is performed by calculating the average value of all g_k, as follows

$$g_{\text{avg}} = \frac{1}{n} \sum_{k=1}^{n} g_k = \frac{1}{n} V_{\text{in}} \sum_{k=1}^{n} (\delta_k \Delta V_{\text{out}_k}) - \frac{1}{n} V_{\text{out}} \sum_{k=1}^{n} \Delta V_{\text{out}_k}, \tag{9}$$

and setting it as the same value of all g_k. Therefore, $g_k = g_{\text{avg}}$, for $k = 1 \ldots n$. Based on the calculated average value g_{avg}, it is possible to calculate the correction factors, $\Delta \delta_k$, which are added to the nominal duty cycle $V_{\text{out}} / V_{\text{in}}$ in (8) in order to have the balanced branch currents. The correction factors are

$$\Delta \delta_k = \frac{g_{\text{avg}}}{V_{\text{in}} \Delta V_{\text{out}_k}}. \tag{10}$$

Therefore, the total duty cycle applied to the k-th HB is $\delta_k = \delta_{\text{nom}} + \Delta \delta_k$, where δ_{nom} denotes the common duty cycle applied to all HBs. If these duty cycles δ_k, where $k = 1 \ldots n$, are applied to the converter, a difference between the branch currents, that exists due to the branch resistance mismatch, is reduced.

Balancing controller

The balancing controller, based on the sensorless current sharing described in the last section, compensates for the resistance mismatch between the branches. In order to develop the balancing controller, it is required to know the relation between the correction factors $\Delta\delta_k$ and the duty cycles δ_k. This relation can be obtained if (9) and (10) are combined, resulting in

$$\Delta\delta_k = \frac{1}{n\Delta V_{\mathrm{out}_k}} \sum_{k=1}^{n} (\delta_k \Delta V_{\mathrm{out}_k}) - \frac{1}{n\Delta V_{\mathrm{out}_k}} \frac{V_{\mathrm{out}}}{V_{\mathrm{in}}} \sum_{k=1}^{n} \Delta V_{\mathrm{out}_k}, \tag{11}$$

which, when rewritten in the matrix form, results in

$$
\begin{aligned}
\Delta\delta = \begin{bmatrix} \Delta\delta_1 \\ \vdots \\ \Delta\delta_n \end{bmatrix} &= \begin{bmatrix} \frac{\Delta V_{\mathrm{out}_1}}{n\Delta V_{\mathrm{out}_1}} & \cdots & \frac{\Delta V_{\mathrm{out}_n}}{n\Delta V_{\mathrm{out}_1}} & -\frac{\sum_{k=1}^{n}\Delta V_{\mathrm{out}_k}}{nV_{\mathrm{in}}\Delta V_{\mathrm{out}_1}} \\ \vdots & & \vdots & \vdots \\ \frac{\Delta V_{\mathrm{out}_1}}{n\Delta V_{\mathrm{out}_n}} & \cdots & \frac{\Delta V_{\mathrm{out}_n}}{n\Delta V_{\mathrm{out}_n}} & -\frac{\sum_{k=1}^{n}\Delta V_{\mathrm{out}_k}}{nV_{\mathrm{in}}\Delta V_{\mathrm{out}_n}} \end{bmatrix} \begin{bmatrix} \delta_1 \\ \vdots \\ \delta_n \\ v_{\mathrm{out}} \end{bmatrix} \\
&= \begin{bmatrix} \frac{\Delta V_{\mathrm{out}_1}}{n\Delta V_{\mathrm{out}_1}} & \cdots & \frac{\Delta V_{\mathrm{out}_n}}{n\Delta V_{\mathrm{out}_1}} \\ \vdots & & \vdots \\ \frac{\Delta V_{\mathrm{out}_1}}{n\Delta V_{\mathrm{out}_n}} & \cdots & \frac{\Delta V_{\mathrm{out}_n}}{n\Delta V_{\mathrm{out}_n}} \end{bmatrix} \begin{bmatrix} \delta_1 \\ \vdots \\ \delta_n \end{bmatrix} + \begin{bmatrix} 0 & \cdots & 0 & -\frac{\sum_{k=1}^{n}\Delta V_{\mathrm{out}_k}}{nV_{\mathrm{in}}\Delta V_{\mathrm{out}_1}} \\ \vdots & & \vdots & \vdots \\ 0 & \cdots & 0 & -\frac{\sum_{k=1}^{n}\Delta V_{\mathrm{out}_k}}{nV_{\mathrm{in}}\Delta V_{\mathrm{out}_n}} \end{bmatrix} \begin{bmatrix} i_1 \\ \vdots \\ i_n \\ v_{\mathrm{out}} \end{bmatrix} \\
&= \mathbf{\Gamma} \begin{bmatrix} \delta_1 \\ \vdots \\ \delta_n \end{bmatrix} + \mathbf{\Phi} \begin{bmatrix} i_1 \\ \vdots \\ i_n \\ v_{\mathrm{out}} \end{bmatrix} = \mathbf{\Gamma}\delta + \mathbf{\Phi}x, \tag{12}
\end{aligned}
$$

where $\Delta\delta \in \mathbb{R}^{n\times 1}$ is a vector of the correction factors, and $\mathbf{\Gamma} \in \mathbb{R}^{n\times n}$ and $\mathbf{\Phi} \in \mathbb{R}^{n\times(n+1)}$ are matrices with coefficients assumed constant as described in (6). Although the branch currents have no impact in (12), they are included as a part of the vector x so that the state vector is the same as in (1) – (2).

The proposed control diagram for balancing the branch currents is shown in Fig. 2, where the block $\mathbf{P}$ represents the model of the converter described by (1) and (2). The blocks G_d represent time delays which are introduced to make the system stable.

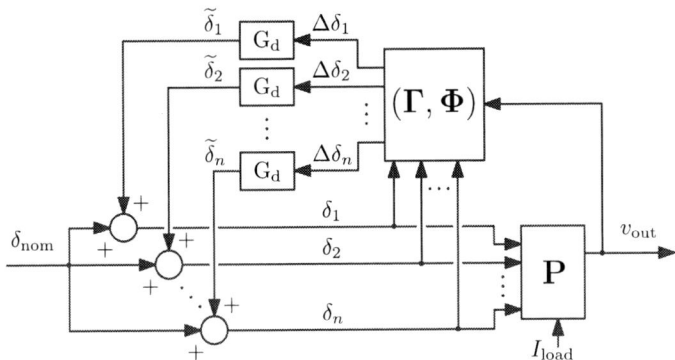

Fig. 2: Block diagram of the current balancing loop.

The model of the delays G_d in s domain is given by

$$s\widetilde{\boldsymbol{\delta}} = s \begin{bmatrix} \widetilde{\delta}_1 \\ \vdots \\ \widetilde{\delta}_n \end{bmatrix} = \begin{bmatrix} -\frac{1}{\tau} & 0 & \cdots & 0 \\ 0 & -\frac{1}{\tau} & \cdots & \vdots \\ \vdots & \vdots & \ddots & 0 \\ 0 & \cdots & 0 & -\frac{1}{\tau} \end{bmatrix} \begin{bmatrix} \widetilde{\delta}_1 \\ \vdots \\ \widetilde{\delta}_n \end{bmatrix} + \begin{bmatrix} \frac{1}{\tau} & 0 & \cdots & 0 \\ 0 & \frac{1}{\tau} & \cdots & \vdots \\ \vdots & \vdots & \ddots & 0 \\ 0 & \cdots & 0 & \frac{1}{\tau} \end{bmatrix} \begin{bmatrix} \Delta\delta_1 \\ \vdots \\ \Delta\delta_n \end{bmatrix} = \mathbf{M}\widetilde{\boldsymbol{\delta}} + \mathbf{N}\boldsymbol{\Delta\delta}, \quad (13)$$

where parameter τ is a time constant, matrices $\mathbf{M}, \mathbf{N} \in \mathbb{R}^{n \times n}$, and $\mathbf{M} = -\mathbf{N}$. The vector of the duty cycles is obtained by combining (12) and (13), resulting in

$$\begin{aligned} \boldsymbol{\delta} = \widetilde{\boldsymbol{\delta}} + \boldsymbol{\delta}_{\mathrm{nom}} &= (s\mathbf{I} - \mathbf{M})^{-1}\mathbf{N}\boldsymbol{\Delta\delta} + \boldsymbol{\delta}_{\mathrm{nom}} \\ &= (s\mathbf{I} - \mathbf{M})^{-1}\mathbf{N}(\boldsymbol{\Gamma\delta} + \boldsymbol{\Phi x}) + \boldsymbol{\delta}_{\mathrm{nom}} \\ &= \mathbf{D}(\boldsymbol{\Gamma\delta} + \boldsymbol{\Phi x}) + \boldsymbol{\delta}_{\mathrm{nom}}, \end{aligned} \quad (14)$$

where $\mathbf{I} \in \mathbb{R}^{n \times n}$ is the identity matrix. The common duty cycle $\boldsymbol{\delta}_{\mathrm{nom}}$ is generated by a standard feedback control loop responsible for controlling the load current. Here, the $\boldsymbol{\delta}_{\mathrm{nom}}$ is assumed to be an input.

Expression (14) can be rewritten as

$$\boldsymbol{\delta} = (\mathbf{I} - \mathbf{D}\boldsymbol{\Gamma})^{-1}\mathbf{D}\boldsymbol{\Phi x} + (\mathbf{I} - \mathbf{D}\boldsymbol{\Gamma})^{-1}\boldsymbol{\delta}_{\mathrm{nom}}. \quad (15)$$

The matrix $\mathbf{D} = (s\mathbf{I} - \mathbf{M})^{-1}\mathbf{N}$ describes the introduced delays G_d. The case without the delay, i.e., when τ equals zero, results in $\mathbf{D}$ being equal to the identity matrix, as calculated:

$$\begin{aligned} \mathbf{D} = (s\mathbf{I} - \mathbf{M})^{-1}\mathbf{N} &= \begin{bmatrix} s+\frac{1}{\tau} & 0 & \cdots & 0 \\ 0 & s+\frac{1}{\tau} & \cdots & \vdots \\ \vdots & \vdots & \ddots & 0 \\ 0 & \cdots & 0 & s+\frac{1}{\tau} \end{bmatrix}^{-1} \begin{bmatrix} \frac{1}{\tau} & 0 & \cdots & 0 \\ 0 & \frac{1}{\tau} & \cdots & \vdots \\ \vdots & \vdots & \ddots & 0 \\ 0 & \cdots & 0 & \frac{1}{\tau} \end{bmatrix} \\ &= \begin{bmatrix} \frac{1}{s\tau+1} & 0 & \cdots & 0 \\ 0 & \frac{1}{s\tau+1} & \cdots & \vdots \\ \vdots & \vdots & \ddots & 0 \\ 0 & \cdots & 0 & \frac{1}{s\tau+1} \end{bmatrix}\Bigg|_{\tau=0} = \mathbf{I}. \end{aligned}$$

If $\mathbf{D} = \mathbf{I}$, then $(\mathbf{I} - \mathbf{D}\boldsymbol{\Gamma})^{-1} = (\mathbf{I} - \boldsymbol{\Gamma})^{-1}$ from (15) is not solvable, because the determinant of the matrix

$$(\mathbf{I} - \boldsymbol{\Gamma})^{-1} = \left(\mathbf{I} - \begin{bmatrix} \frac{\Delta V_{\mathrm{out}_1}}{n\Delta V_{\mathrm{out}_1}} & \cdots & \frac{\Delta V_{\mathrm{out}_n}}{n\Delta V_{\mathrm{out}_1}} \\ \vdots & & \vdots \\ \frac{\Delta V_{\mathrm{out}_1}}{n\Delta V_{\mathrm{out}_n}} & \cdots & \frac{\Delta V_{\mathrm{out}_n}}{n\Delta V_{\mathrm{out}_n}} \end{bmatrix} \right)^{-1} = \begin{bmatrix} 1 - \frac{\Delta V_{\mathrm{out}_1}}{n\Delta V_{\mathrm{out}_1}} & \cdots & -\frac{\Delta V_{\mathrm{out}_n}}{n\Delta V_{\mathrm{out}_1}} \\ \vdots & & \vdots \\ -\frac{\Delta V_{\mathrm{out}_1}}{n\Delta V_{\mathrm{out}_n}} & \cdots & 1 - \frac{\Delta V_{\mathrm{out}_n}}{n\Delta V_{\mathrm{out}_n}} \end{bmatrix}^{-1}$$

is equal to zero, for any n. Therefore, the system is not guaranteed to be stable without the delay G_d.

By combining (1) and (15), the system including models of the converter and the balancing controller is reduced to the following equation

$$\boldsymbol{x} = \left[s\mathbf{I} - \mathbf{A} - \mathbf{B}(\mathbf{I} - \mathbf{D}\boldsymbol{\Gamma})^{-1}\mathbf{D}\boldsymbol{\Phi} \right]^{-1} \cdot \mathbf{B}(\mathbf{I} - \mathbf{D}\boldsymbol{\Gamma})^{-1}\boldsymbol{\delta}_{\mathrm{nom}}. \quad (16)$$

It should be noted that it is assumed that the load current i_{load} is considered constant when compared to the bandwidth of the balancing controller. For that reason, the vector of the duty cycles $\boldsymbol{\delta}$ is used instead of a vector of inputs $u = [\delta,\ i_{\text{load}}]^{\text{T}}$.

The stability of the current-balancing controller depends on the coefficient τ. In order to observe the stability of the system when parameter τ varies, the transfer function between δ_{nom} and v_{out} is analysed. The transfer function is easily obtained when (16) is combined with (2). The model is implemented in MATLAB, and the poles of the transfer function are obtained for different values of the parameter τ. For a converter with three paralleled branches and parameters given in the experiments section, the dependency of the real parts of the poles versus τ is shown in Fig. 3a. The system is stable when all real parts of the poles are negative, which for this particular case is true for τ larger than $56.4\ \mu s$. To be on the safe side, the τ is set to a value three times higher than the one that puts the system on boundary stability. The Nyquist plot of the system with this value of τ is shown in Fig. 3b. The resulting phase margin is $30.9°$.

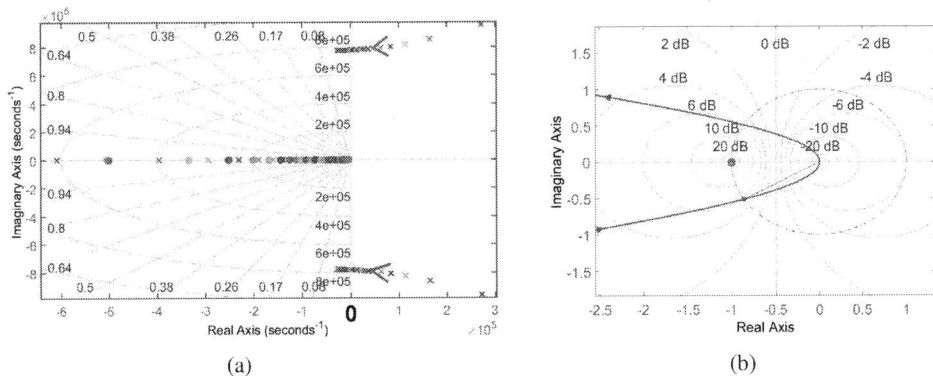

(a)   (b)

Fig. 3: (a) Pole movement when the parameter τ changes. Red arrows show the movement of the poles as τ increases. (b) Nyquist plot of the system when τ is chosen such that the system is stable.

Besides the time constant τ, the current balancing also depends on the obtained quantities ΔV_{out_k}, where $k = 1\ldots n$. An error between the actual and the measured value of this parameter leads to an uneven balancing of the branch currents. The error is defined as $e_k = \Delta V_{\text{out}_k}^{(\text{meas})}/\Delta V_{\text{out}_k}$, where $\Delta V_{\text{out}_k}^{(\text{meas})}$ is the measured value, and ΔV_{out_k} is the actual value. If the actual values are used for the balancing, then the branch currents will equalize. However, if the measured values are different than the actual ones, the branch current distribution will not be ideal. The difference between the values can be due to a measurement error, or due to a short perturbation injection signal which does not result in the complete steady-state of the output voltage.

In order to observe the influence of the error e_k on the branch current distribution, a following expression can be obtained from (5),

$$\frac{\hat{I}_k^{(\text{calc})}}{\hat{I}_k} = \frac{\Delta V_{\text{out}_k}^{(\text{meas})}}{\Delta V_{\text{out}_k}} = e_k, \tag{17}$$

where $\hat{I}_k$ is the k-th branch current estimate based on the actual value ΔV_{out_k}, and $\hat{I}_k^{(\text{calc})}$ is the k-th branch current estimate based on the measured value $\Delta V_{\text{out}_k}^{(\text{meas})}$. Since it is considered that the actual values ΔV_{out_k}, where $k = 1\ldots n$, lead to equal branch currents when used for the balancing, then the following applies, $\hat{I}_k = \hat{I}_j$, where $k, j \in [1\ldots n]$. According to this equality and (17), the following can be derived:

$$\hat{I}_k^{(\text{calc})} e_j = \hat{I}_j^{(\text{calc})} e_k, \tag{18}$$

where $k, j \in [1\ldots n]$.

The last expression shows that the branch current balancing is directly influenced by the error e_k.

Considering (18), and the fact that the load DC current is the sum of all branch DC currents, $I_{\text{load}} = \sum_{k=1}^{n} I_k$, it is possible to express a branch current through the load current and the errors. For example, the current of the first branch is given by

$$I_1 = \frac{I_{\text{load}}}{\frac{1}{e_1}\sum_{k=1}^{n} e_k}.$$

The difference between the actual current through the branch one when $\Delta V_{\text{out}_k}^{(\text{meas})}$ are used to balance the currents, where $k = 1 \ldots n$, and the average branch current is given by

$$\Delta I_1 = I_{\text{load}}\left(\frac{1}{\frac{1}{e_1}\sum_{k=1}^{n} e_k} - \frac{1}{n}\right),$$

and graphically shown in Fig. 4 for a different number of paralleled branches.

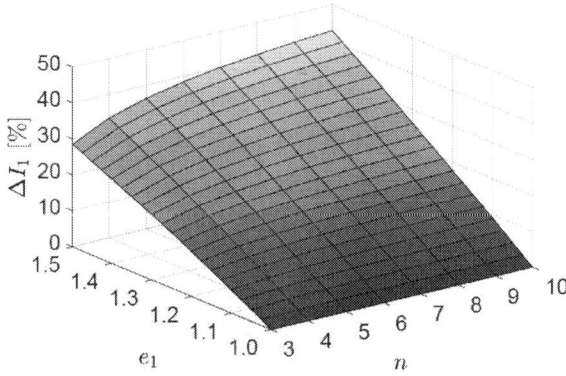

Fig. 4: Current deviation in the first branch given as a percentage of the average branch current, for a different error e_1, and a different number of paralleled branches. The values ΔV_{out_k} for all other branches are considered without an error ($e_k = 1$, for $k = 2 \ldots n$).

Experiments

A Gallium Nitride (GaN) based converter with three paralleled branches ($n = 3$) switching at 1 MHz is used for experimental verification. Each branch has an inductor with a value of $L = 4\ \mu H$, and a half-bridge consisting of two GS61008P transistors. The total output capacitance is $C_{\text{out}} = 3.96\ \mu F$. The measured value of the equivalent series resistance of the output capacitors is $R_{C_{\text{out}}} = 7.4\ m\Omega$.

The control is implemented in STM32G474RE microcontroller (MCU) with 170 MHz clock frequency. Additional feature of this MCU is that it contains high-resolution timer (HRTIM) which allows 184 ps duty cycle resolution. The MCU is part of NUCLEO-G474RE development board which is visible in the photo of the setup shown in Fig. 5.

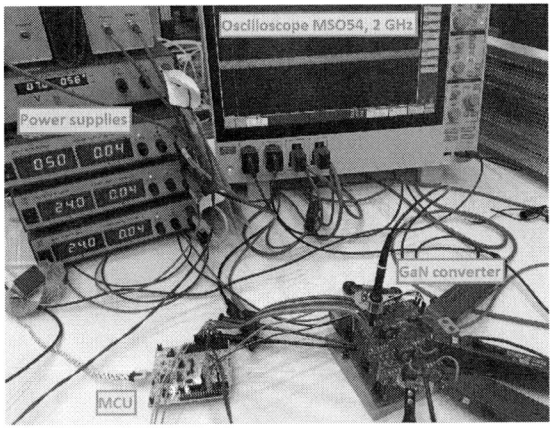

Fig. 5: Photo of the setup.

First, a perturbation signal is consequently injected in the duty cycle of each branch. The perturbation procedure is performed at zero load current. During the signal injection, the output voltage variations, $\Delta V_{\text{out},k}$, where $k = 1 \ldots 3$, are measured and further used in the balancing algorithm, as described in the previous sections. Obtained ratios are $\Delta V_{\text{out},1}/\Delta V_{\text{out},2} = 1/4.661$ and $\Delta V_{\text{out},1}/\Delta V_{\text{out},3} = 1/3.3838$. The parameter τ is set according to the calculation in the previous section.

The experimental measurement of the branch current balancing is shown in Fig. 6. The moment when the balancing controller is activated is captured. The input voltage is 48 V, and the load current is 10 A, 15 A, and 20 A, respectively. It can be seen that the sensorless current balancing results in even branch currents for the lower load current. However, as the load current increases, the difference between the branch currents, after the balancing, increases.

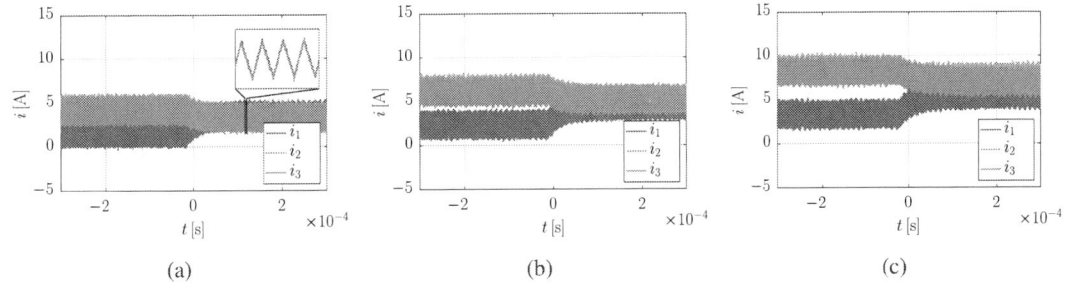

Fig. 6: Branch current balancing for a different DC load currents, 10 A, 15 A, and 20 A, respectively. A moment when balancing controller is activated is captured.

Conclusion

Sensorless current sharing approach is a promising method to balance the currents in the converter with the paralleled half-bridges. This paper proposes a structure of the balancing controller. Additionally, the model of the whole system, including the converter and the balancing controller, is derived and analysed. The analysis includes the stability and sensitivity study, where measurement errors are taken into account. The experimental verification is also included to verify the stability of the system in different operating points.

References

[1] J. L. Lu and D. Chen: Paralleling GaN E-HEMTs in 10kW-100kW systems, 2017 IEEE Applied Power Electronics Conference and Exposition (APEC), Tampa, FL, 2017, pp. 3049-3056.

[2] J. Burkard and J. Biela: Paralleling GaN switches for low voltage high current half-bridges, 2019 21th European Conference on Power Electronics and Applications (EPE'19 ECCE Europe), Genova, 2019.

[3] R. F. Foley: Sensorless current estimation and sharing in multiphase buck converters, IEEE Transactions on Power Electronics, vol 27, no 6, pp. 2936-2946, 2012.

[4] X. Zhang, L. Corradini and D. Maksimovic: Sensorless Current Sharing in Digitally Controlled Two-Phase Buck DC-DC Converters, 2009 Twenty-Fourth Annual IEEE Applied Power Electronics Conference and Exposition, Washington, DC, 2009, pp. 70-76.

[5] N. Boskovic, M. G. L. Roes, C. G. E. Wijnands and E. A. Lomonova: Improved current estimation in paralleled half-bridge converters, 2019 21st European Conference on Power Electronics and Applications (EPE '19 ECCE Europe), Genova, Italy, 2019, pp. P.1-P.9.

[6] J. Gordillo and C. Aguilar: A Simple Sensorless Current Sharing Technique for Multiphase DC-DC Buck Converters, IEEE Transactions on Power Electronics, vol. 32, no. 5, pp. 3480-3489, May 2017.

PWM-Induced Harmonic Power in 75 kW IM Drive System

Lassi Aarniovuori, Hannu Kärkkäinen, Markku Niemelä and Juha Pyrhönen
LUT-University
Yliopistonkatu 34, 53850
Lappeenranta, Finland
Tel.: +358 40 833 7984
lassi.aarniovuori@lut.fi
URL: www.lut.fi

Keywords

Experimental testing, Harmonic power, Measurement, Uncertainty, Variable Speed Drives.

Abstract

More and more commonly, the rotating field machine drives are equipped with a frequency converter. The frequency converters enable the control of the magnetization state, torque and speed of the machine according to the needs of applications and, in principle, save energy. In turn, the pulse-width-modulation (PWM) creates additional harmonic losses over the fundamental losses in the machine. The harmonic power induced by the PWM is examined here in a case of 75 kW high efficiency squirrel-cage induction motor. The examination is carried out using an experimental setup with three commercial frequency converters at 30 different operation points of the motor. In each of the measurement points, the three-phase voltage and current waveforms are recorded as well as the corresponding electric power, current and voltage values. In the analysis, the harmonic power is segregated from the total electric power and examined as a function of the frequency, load and switching frequency.

Introduction

The frequency converter-fed electrical machines are becoming more popular and the effect of the PWM on the motor losses has gained more interest [1]. Two-level voltage source converter (VSC) is the most used converter topology in the motor drives. A typical VSC has a passive rectifier bridge, line or DC – reactor, intermediate circuit and an inverter bridge. The desired amplitude and frequency AC voltage waveform is produced with the inverter bridge that utilizes pulse-width modulation. The pulse pattern of the inverter bridge is controlled using a microcontroller and/or ASIC-circuit. There are several modulation methods such as space-vector and different discontinuous modulation methods. In addition, some control systems create a pulse pattern with an integrated modulator such as direct torque control (DTC) or model predictive control (MPC). In motor applications, a higher number of switching instants consumes more energy in the frequency converter but at the same time, the voltage harmonics and the resulting harmonic losses in the motor are decreased. Similarly, a lower number of switching instants consumes less energy in the frequency converter but increases harmonic losses in the motor. Therefore, the switching frequency of a voltage source converter is a key design parameter when minimizing the losses of a converter-fed motor system.

A comparison between the sinusoidal and converter supply in case of a 15 kW squirrel-cage induction motor is presented in 16 operation points in [2] while in [3] the frequency converter is proposed as single power supply to determine the losses of the machine with sinusoidal or converter supply. The approach in [3] is based on the filtering settings of the power analyzer. An automated testing procedure for converter fed IMs is presented in [4] according to IEC/TS 60034-2-3 [5]. The above-mentioned technical specification has been a topic for many publications e.g. [6], [7]. The next edition of 60034-2-3 was launched as a standard on March 2020.

The frequency converter and the induction motor total losses and efficiencies are investigated in [8], while in this study, the losses of three commercial voltage source converters with the rated power of 90 kW supplying a 75 kW four-pole induction motor are analyzed. The utilized switching frequencies and the fundamental wave voltages are analyzed in all 30 operation points. The motor losses are segregated in the fundamental wave losses and in the harmonic losses.

Experimental setup

The measurement setup consisted of a 75 kW induction machine, a 200 kW line converter controlled IM to create the load torque and three different frequency converters, Fig. 1. The manufacturers of the frequency converters are all well-known globally active companies and the converters are intended to be used in dynamic industrial processes rather than in pump or fan applications. The continuous power ratings of the converters are 90 kW being therefore slightly over dimensioned. The 75 kW squirrel-cage induction motor is measured in 30 operating points below the rated speed and using these converters. The measurement points are set according to the motor operating points with torque values of 10%, 25%, 50%, 75% and 100% of the motor rated torque and supply frequency values of 10%, 25%, 50%, 75%, 90% and 100% of the rated motor frequency. In each of the operating points, the three-phase voltage and current waveforms are recorded as well as the corresponding electric power, current and voltage values. In addition, the torque and speed are measured and recorded using HBM T12 torque transducer, and Pt100 sensors are used to capture the motor winding and ambient temperatures.

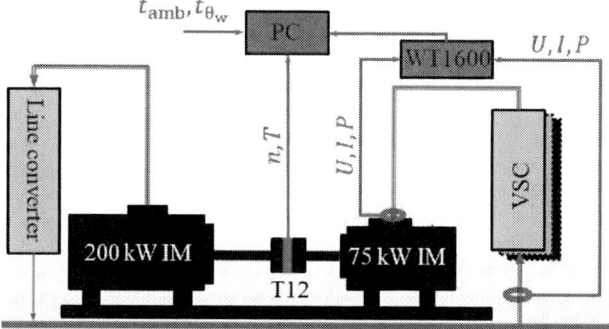

Fig 1: The schematic of the experimental setup and measurement data collection.

Switching frequency

The terms carrier frequency and switching frequency are used to denote the average number of switching instants during a specific time period, typically given as Hertz (Hz). The terms are used in commutative manner; however, a simple distinction can be made between the terms. The carrier frequency is the typical interval between the switching instants and the switching frequency is the number of the realized switching instants within some period. In some cases, such as when utilizing symmetrical three-phase modulation in the linear modulation region, the switching frequency is equal to the carrier frequency, but in practice, modern converters use different modulation algorithms at different operating points and also control and limit the switching frequency. The term 'carrier frequency' is based on analog devices' triangular carrier waveform that is compared to voltage references to create the binary switching commands. Nowadays, in the digital modulator implementation, the carrier is typically a counter but the term is still commonly used. The reference switching frequency depends typically on the size of the

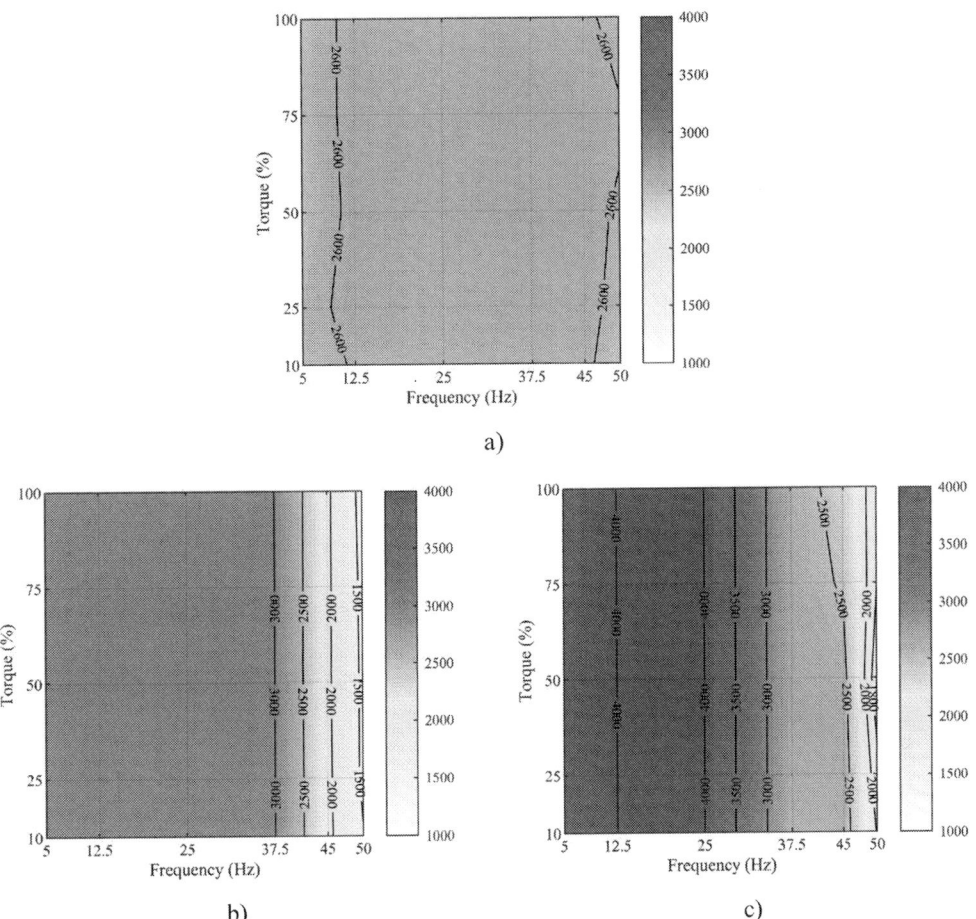

a)

b)   c)

Fig 2: The switching frequencies of the three frequency converters recorded in fundamental frequency – torque plane. The grid crossings show the actual measurement points that were used to determine the switching frequencies.

converter: the larger the converter the lower the switching frequency. In hysteresis-based modulation methods, the switching frequency can be controlled by adjusting the widths of the hysteresis bands. The switching frequency is a function of cooling conditions that depend on the power of the ventilation fan and ambient air temperature or on the cooling fluid temperature in the liquid cooling converters.

Since the harmonic loss in a motor is a function of the switching frequency, the actually used switching frequency was determined from the recorded three phase voltage waveforms. The voltage levels in each phase were compared and the position of the switches were determined. A one second sample with 1 µs sampling interval was used. The minimum pulse length for this size of a converter is few microseconds and therefore all pulses were easily detected. The vectors "zero" (000 all phases connected to zero potential) and "seven" (111 all phases connected to DC bus potential) were distinguished using the measured common mode voltage. The results are presented in Fig. 2. The isoline-figures are formed using all the 30 measurement points.

The results in Fig. 2 show three different approaches to control the switching frequency. Converter A is using the switching frequency around 2660 Hz in the whole frequency-torque plane. Only at very low output frequency and at the motor rated frequency the switching frequency is slightly decreased. This kind of switching frequency behaviour is typically a result of a feedback control.

Converter B is using the switching frequency of precisely 3 kHz at and below the 37.5 Hz points and then the switching frequency is decreased as the output frequency is increased. This can be a result of limited voltage at 45 Hz output frequency and above it. Depending on the location of the reference voltage vector, the second active voltage vector and zero vectors are too short, and they are not realized at all. Some of the switching sequences can contain only one active vector, and thus the actually realized switching frequency is decreasing.

Converter C utilizes two modulation methods, three-phase modulation at output frequencies between 5 Hz and 25 Hz and two-phase modulation (discontinuous) at output frequencies above 25 Hz. The same carrier frequency 4 kHz is used in the whole operating area that results in the switching frequency of 2/3 times the carrier frequency in case of discontinuous modulation.

Voltages and currents

The different switching frequencies and modulation methods create diverse voltage waveforms that form also slightly different currents. The recorded voltage and current waveforms are illustrated with all three converters at the 25 Hz operation point with 75% load, Fig. 3.

Harmonic power analysis

Using Discrete Fourier Transform the active power can be presented using the frequency components from n to infinity

$$P_{\text{tot}} = P_{\text{DC}} + \sum_{n=1}^{n=\infty} U(n)I(n)\cos(\varphi(n)) \tag{1}$$

where P_{DC} is the DC power, $U(n)$ and $I(n)$ are the RMS values of the nth components of voltage and current, and $\varphi(n)$ is the phase difference between $U(n)$ and $I(n)$. Each nth harmonic can be a harmonic, an interharmonic or a subharmonic. The total active AC-power can be split between the fundamental power and harmonic power as

$$P_{\text{AC,tot}} = P_{\text{fund}} + P_{\text{harm}}$$

$$= U(\text{fund})I(\text{fund})\cos(\varphi(\text{fund})) + \sum_{n=1}^{n=\text{fund}-1} U(n)I(n)\cos(\varphi(n)) + \sum_{n=\text{fund}+1}^{n=\infty} U(n)I(n)\cos(\varphi(n)), \tag{2}$$

where 'fund' is the index of the fundamental frequency power component. Hence, the total harmonic power is

$$P_{\text{harm}} = P_{\text{AC,tot}} - P_{\text{fund}}. \tag{3}$$

Therefore, the total harmonic power is simply the difference of the total active power $P_{\text{active,tot}}$ and the power at the fundamental wave frequency P_{fund}. Assuming that only P_{fund} produces the mechanical power of the shaft P_{shaft}, we can define the fundamental losses as

$$P_{\text{fund,losses}} = P_{\text{fund}} - P_{\text{shaft}}. \tag{4}$$

In addition, we can define the harmonic losses as

$$P_{\text{harm,losses}} = P_{\text{AC,tot}} - P_{\text{shaft}} - P_{\text{fund,losses}} \tag{5}$$

PWM-Induced Harmonic Power in 75 kW IM Drive System   AARNIOVUORI Lassi

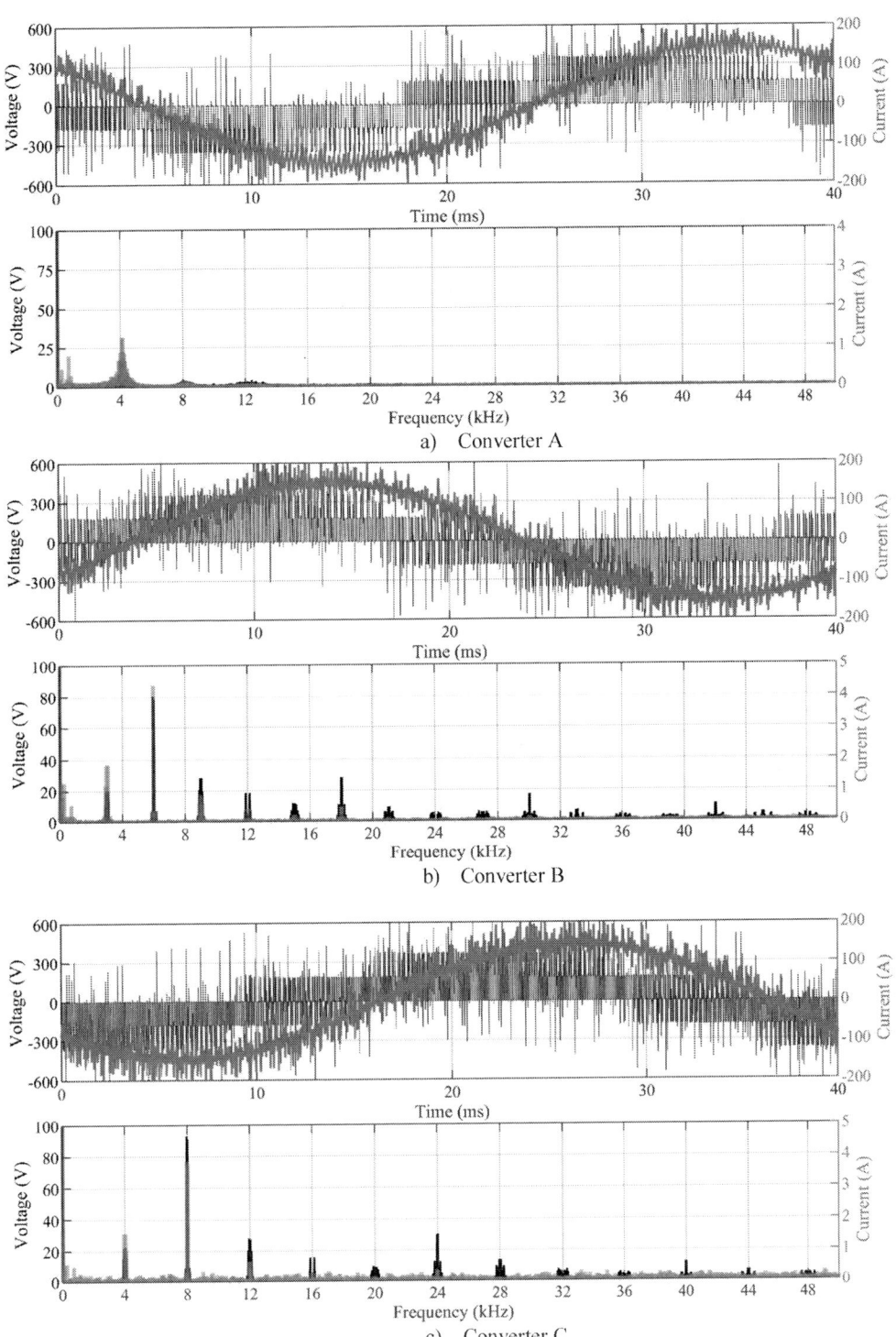

Fig. 3: The 75 kW IM voltage and current waveforms in the time and frequency domains using the three frequency converters. The fundamental wave amplitude has been scaled out of the frequency domain figures.

Combining (3) - (5), we get

$$P_{\text{harm,losses}} = P_{\text{harm}} \tag{6}$$

Based on (6), we can assume that the harmonic power of the PWM waveform is creating the additional harmonic losses in the electrical machine. This assumption is valid for all electrical machines and simplifies the analysis of the converter-fed machines. Therefore, the segregation of losses can be performed between the fundamental losses and the harmonic losses without any comparative measurements with a sinusoidal supply. The fundamental losses are the excitation frequency losses and the related phenomena. This is closely related to the losses that are obtained when the machine is fed by a sinusoidal supply. However, the state of the machine is never exactly the same in converter and sinusoidal supply, since a small portion of the fundamental wave power is used to overcome the effects of losses formed by the PWM harmonics. The harmonic power of the PWM waveforms with all three converters and in all operating points are shown in Fig. 3.

The results in Fig. 4 show that the IE3 induction motor harmonic loss is with Converter A from 150 W to 520 W, with Converter B from 230 W to 520 W and with Converter C from 270 W to 520 W. The distribution of the losses in the fundamental frequency – torque plane is similar with all converters even though the differences in the

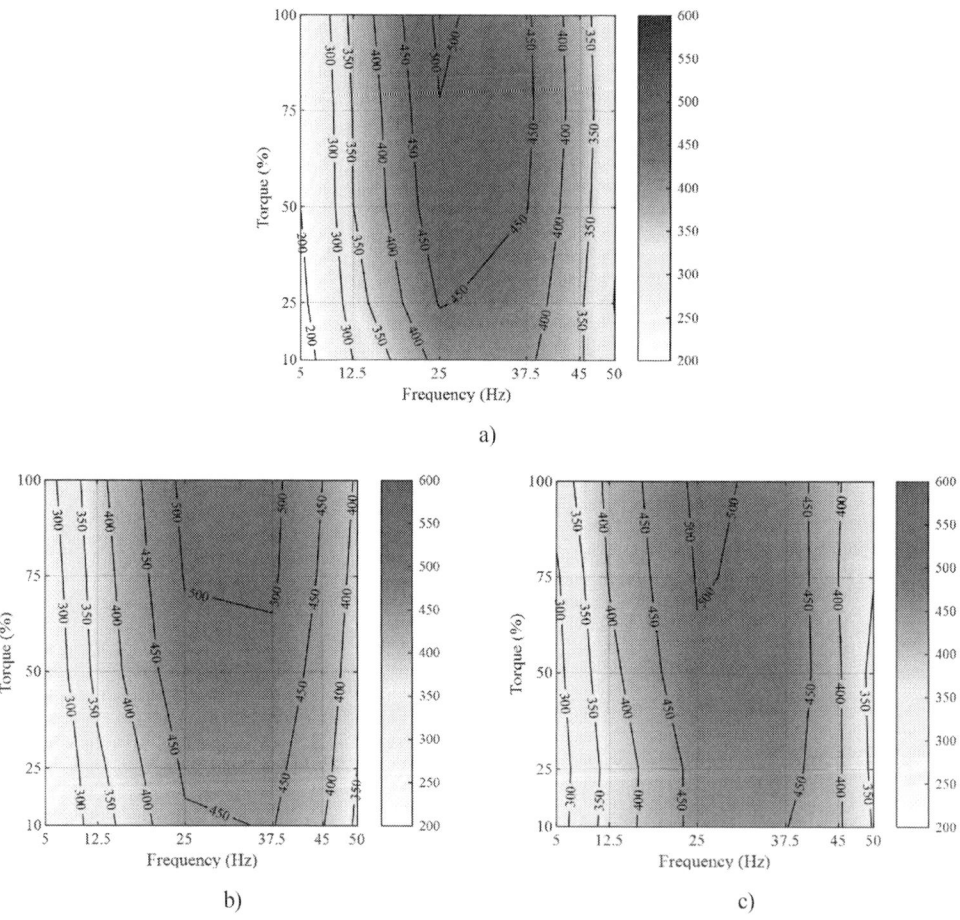

Fig. 4: The harmonic power of the system with converters A, B and C presented in frequency – torque plane.

switching frequencies in Fig. 2 and in the harmonic voltages and current distribution in the frequency domain were remarkable. The maximum harmonic losses are examined at 25 Hz output frequency and with the rated torque. It can be concluded from the results in Fig. 4, that the harmonic losses of the induction motor are a function of the fundamental wave frequency and load torque.

Uncertainty analysis

The measurements were made with Yokogawa PX8000 Precision Power Scope equipped with Hitec Zero-Flux current measurement system. The power measurement reading and range accuracies of the used power analyzer is presented in Table I. When the measurement uncertainty is analyzed, the range accuracy is applied only according to the most significant frequency band.

Table I: The reading and range power accuracies of the used power analyser.

Frequency (Hz)	Accuracy	
	of Reading (%)	of Range (%)
DC	0.2	0.4
$0.1 \leq f < 10$	0.2	0.2
$10 \leq f < 45$	0.2	0.1
$45 \leq f < 1000$	0.1	0.1
$1\,000 \leq f < 10\,000$	0.1	0.16
$10\,000 \leq f < 50\,000$	0.2	0.2
$50\,000 \leq f < 100\,000$	0.6	0.4
$100\,000 \leq f < 200\,000$	1.5	0.6
$200\,000 \leq f < 400\,000$	1.5	0.6
$400\,000 \leq f < 500\,000$	$0.1 + 0.006 \times f^*$	0.6
$500\,000 \leq f < 1\,000\,000$	$0.1 + 0.006 \times f^*$	6

*frequency unit is kHz

In this case, the most significant component is the motor excitation frequency (fundamental wave frequency). It can be examined in Table I that below 10 Hz the power measurement accuracy is already remarkably lower than with the optimum frequency band from 45 Hz to 1 kHz. Above this band, the power measurement accuracy is decreasing as a function of the frequency. As shown in [9], the total electric power measurement uncertainty can be analyzed using the fundamental wave accuracy and the harmonic power in the uncertainty analysis can be neglected. However, the uncertainty analysis can also be executed only for harmonic power assuming the uncertainty of the fundamental frequency power does not have an influence to harmonic power measurement accuracy. Still, the range error related to fundamental wave frequency must be taken into account. Since, the uncertainties of the different harmonic components definitely have correlation, the total measurement uncertainty of the harmonic power is

$$u_c(y) = \sum_{i=1}^{N} |u(y_i)|, \tag{7}$$

where $u(y_i)$ is the uncertainty of active power related to each frequency band given in Table I. Each of the uncertainty components in the sum can be analyzed separately. Figure 5 illustrates the harmonic power measurement uncertainty. Figure 5 has been created using Converter C data, but the uncertainty is almost identical for all three converters, since the dominant uncertainty source is the power analyzer range error that is a function of the voltage and current ranges and is fixed to a motor operating point.

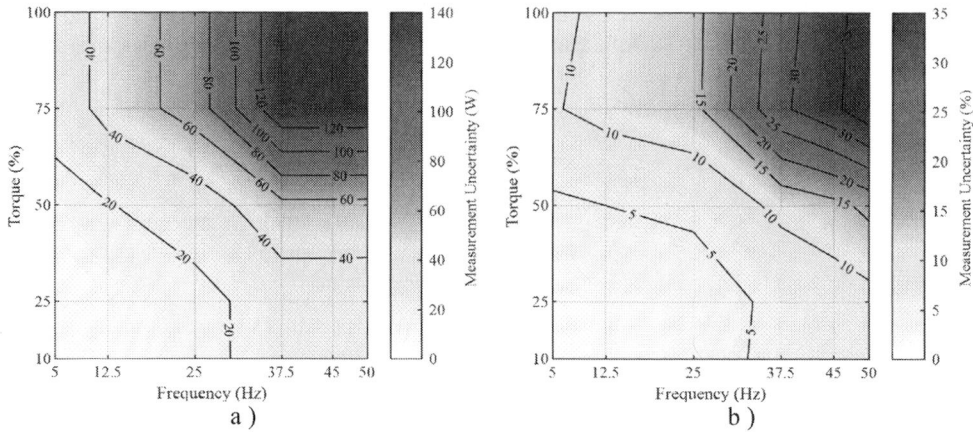

Fig. 5: The measurement uncertainty of the harmonic power presented as a) absolute values, b) relative to total harmonic power value.

The level of the measurement uncertainty of the harmonic power is acceptable. The measurement uncertainty is a function of total measured power. As shown in Fig. 4. the harmonic power value does not change significantly at different operation points and therefore the relative uncertainty gets its lowest values at low power operation points. Although, the measurement uncertainty is high, the data is valid for comparison purposes when the same measurement instruments have been used to obtain the measured values.

Conclusion

When measuring electric motors with modern power analyzers, the fundamental waveform power and the harmonic power can be easily separated. The harmonic power does not, in practice, convert to mechanical power can therefore be considered responsible for the additional harmonic losses in an electrical machine supplied by a converter. There is no need to use complex and troublesome comparisons by using also a sinusoidal supply to study this loss component. The active power can be further split in frequency bands to examine what frequencies are the most dominant in the harmonic loss generation. The results in this paper strengthens the hypothesis that the harmonic losses in the induction motors are a function of the frequency and load. The three converters used here utilize different control and modulation methods, but still, the harmonic losses are relatively similar. This indicates, that in the energy efficiency classification and in the related tests, the modulation pattern is not critical.

References

[1] E. B. Agamloh, "Power and Efficiency Measurement of Motor-Variable-Frequency Drive Systems," *IEEE Trans. Ind. Appl. Vol. 53. No. 1, 2017.*

[2] H. Kärkkäinen, L. Aarniovuori, M. Niemelä and J. Pyrhönen, "Converter-fed induction motor losses in different operating points," *2016 18th European Conference on Power Electronics and Applications (EPE'16 ECCE Europe),* Karlsruhe, 2016, pp. 1-8.

[3] E. Agamloh, A. Cavagnino and S. Vaschetto, "Induction machine efficiency measurement using a variable frequency drive source," *2017 IEEE Energy Conversion Congress and Exposition (ECCE),* Cincinnati, OH, 2017, pp. 768-775.

[4] J. Mushenya, M. A. Khan and P. S. Barendse, "Development of a Test Rig to Automate Efficiency Testing of Converter-Fed Induction Motors," in *IEEE Transactions on Industry Applications*, vol. 55, no. 6, pp. 5916-5924, Nov.-Dec. 2019.

[5] Rotating Electrical Machines - Part 2-3: Specific Test Methods for Determining Losses and Efficiency of Converter-Fed AC Induction Motors, IEC-TS 60034-2-3, Nov. 2013.

[6] A. Boglietti, A. Cavagnino, M. Cossale, A. Tenconi, and S. Vaschetto, "Efficiency determination of converter-fed induction motors: Waiting for the IEC 60034–2–3 standard," in Proc. IEEE Energy Convers. Congr.Expo., Denver, CO, USA, 2013, pp. 230–237.

[7] H. Kärkkäinen, L. Aarniovuori, M. Niemela and J. Pyrhönen, "Converter-Fed Induction Motor Efficiency: Practical Applicability of IEC Methods," in *IEEE Industrial Electronics Magazine*, vol. 11, no. 2, pp. 45-57, June 2017.

[8] L. Aarniovuori, H. Kärkkäinen, M. Niemelä, K. Cai, J. Pyrhönen and W. Cao, "Experimental Investigation of the Losses and Efficiency of 75 kW Induction Motor Drive System," *IECON 2019 - 45th Annual Conference of the IEEE Industrial Electronics Society*, Lisbon, Portugal, 2019, pp. 1052-1058.

[9] H. Kärkkäinen, L. Aarniovuori, M. Niemelä and J. Pyrhönen, "Advanced Uncertainty Calculation Method for Converter-Fed Motor Loss Determining," 2019 IEEE International Electric Machines & Drives Conference (IEMDC), San Diego, CA, USA, 2019, pp. 1551-1558.

Proposal of Boost converter without reactor using Open-ended Winding PMSM for Photovoltaic Pump System

Akihiro Okazaki, Sari Maekawa
SEIKEI UNIVERSITY
〒180-8633 3-3-1, Kichijoji Kitamachi,
Musashino-city, Tokyo-to, Japan
E-mail : sari1.maekawa@st.seikei.ac.jp
Tel : +81-422-37-3768

Keywords

«Open-ended winding permanent magnet synchronous motor», «Photovoltaic», «Boost converter», «Reactor»

Abstract

The way to use of clean energy is promoted actively, improving this system is important to solve environment issues. In photovoltaic pump system, voltage to work that system is generated by chopper circuit. However, in this system, there is problem that increase of cost and upsizing by chopper circuit for boost converter. In this paper, we propose the system to replace a reactor in chopper circuit with three phase windings of the open-ended winding PMSM.

Introduction

Recently, various clean energy is focused. photovoltaic is typical method and, popular system of them. The power generation system is familiar to have high penetration in the world clean energy consumption ratio and, researched actively in the world. The reason is useful technology to save earth's resource. However clean energy to have represented by photovoltaic have a problem that power generation capacity fluctuates by external factor such as weather.

There is a pump system in one of photovoltaic application, the conventional systems have a boost converter to adjust voltage. However, in order to construct chopper circuit, the large reactor and a lot of switching devices are needed. And these devices increase system cost and size [1]. On the other hand, there is an Open-ended winding motor system with two inverters and an Open-ended winding structure, which improves the voltage utilization rate [2]-[4]. In these conventional converters, there is the method on boosting the secondary inverter source using a capacitor, however the power factor decreases [2][3]. In association with, the system that is combined photovoltaic pump system and Open-ended winding PMSM is also proposed [5][6].

In this paper, we propose the method which makes suppress photovoltaic voltage without additional boost circuits. The proposed method constitutes boost converter without additional reactors to replace a reactor in a boost converter with three phase windings for photovoltaic open-ended winding PMSM. In the system, we verify that the proposed system can be operated.

Theory of proposal system

Fig.1 shows three phases open-ended winding PMSM system. This system controls rotation speed with d-q axis current, and calculates switching duty for generating PWM by feedback DC voltage Vin. In this paper, we propose to apply three phase windings of the open-ended winding PMSM for a reactor of boost converter, and boost or step-down operations. Fig.2 shows proposal converter.

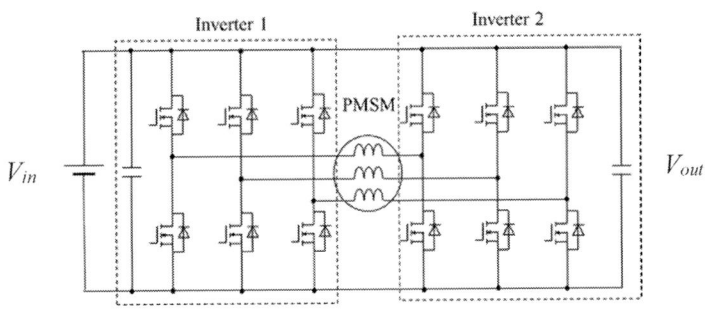

Fig.1 Open-ended winding PMSM drive system

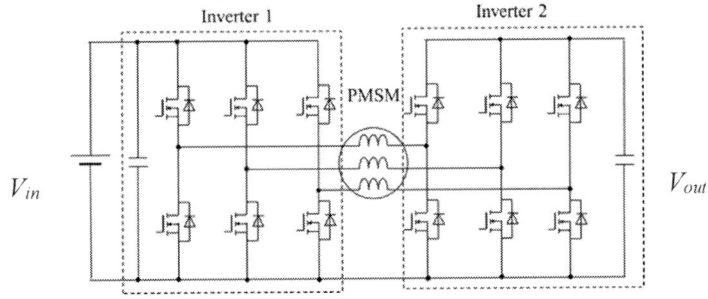

Fig.2 Proposal Open-ended winding PMSM drive system

The proposal converter uses three phase windings for reactor of boost converter to connect only DC negative electrode between inverter1(INV1) and inverter2(INV2). During the operation, INV1 side can replace to step-down chopper, INV2 side is boost chopper as well to consider of three phases as one. Fig.3 shows equivalent circuit of proposed converter.

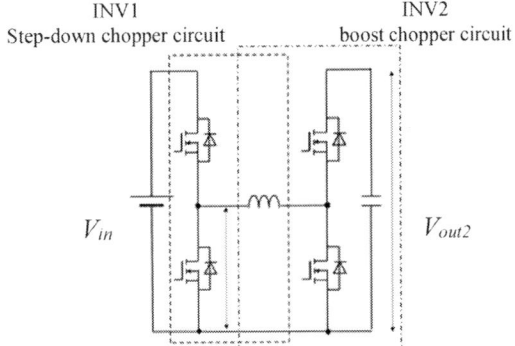

Fig.3 Equivalent circuit proposed converter

First, we describe the INV1 operation as a step-down chopper. In case, high side switching duty as D_1, output voltage as V_{out}, the circuit operates to follow in Eq. (1)

$$V_{out1}=D_1 V_{in}$$ …(1)

In INV2, a reactor saves energy during connecting to GND at OFF state and, can discharge energy to secondary side at ON state. Low side switching duty as D_2, INV2 side works to follow in Eq. (2)

$$V_{out2}=\frac{1}{1-D_2}V_{out1}$$ … (2)

In other words, the proposal converter connects these chopper operations in series, can boost V_{out2} with large D_1 and small D_2.

Control method

This system calculates rotation speed by motor rotation speed commend value and, d-q axis current by PI control. From the result value, it generates inverter's duty and PWM. The capacitor voltage controller calculates zero-sequence duty D_0 with capacitor voltage V_{out} voltage and, add to three phase duties which are calculated by d-q axis current controller. Fig.4 shows the control configuration of proposal converter. On the other hands, some other conventional open-ended winding PMSM system has the controller in order to suppress zero-sequence current I_0.

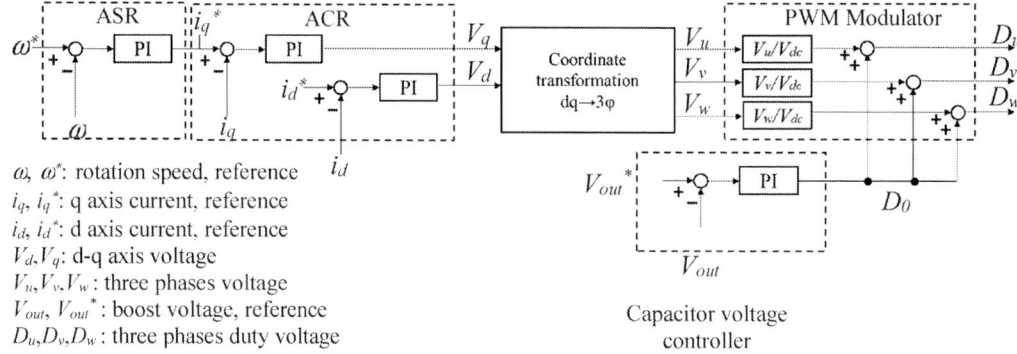

ω, ω^*: rotation speed, reference
i_q, i_q^*: q axis current, reference
i_d, i_d^*: d axis current, reference
V_d, V_q: d-q axis voltage
V_u, V_v, V_w: three phases voltage
V_{out}, V_{out}^*: boost voltage, reference
D_u, D_v, D_w: three phases duty voltage

Capacitor voltage controller

Fig.4 Control configuration of proposal converter

Simulation result

At first, we simulate proposal circuit to confirm when the motor is operated or not. From that result, we guess boost duty control affect to I0. Then we simulate other condition, these are conventional circuit, conventional circuit when the zero-sequence current controller is excluded. Table.1 shows simulation condition. And Fig.5 shows simulation result of proposal circuit when the motor is not operated. V_{in}, V_{out} and V_{out}^*, three phases switching duty, three phases current and zero-sequence current are shown in (a), (b), (c), (d).

Table.1 Simulation conditions

	Operate motor	Not operate motor
Fundamental frequency [Hz]	0	30
Number of polepair	2	2
d axis inductance [H]	0.02	0.02
q axis inductance [H]	0.03	0.03
Winding resistance [W]	1	1
Magnetic flux [Wb]	0.2	0.2
Input voltage [V]	100	100
Capacitor voltage reference [V]	200	200

In Fig.5, three phases winding of motor is able to replace to a reactor of chopper circuit. And we verify that proposal converter can be boost chopper. In other words, the converter boosts voltage following our theory. Fig.6 shows same signals as Fig.5 when motor is operated.

In Fig.6, we confirm that proposal circuit operates boost in the condition that the motor is operated as well. However, Harmonics depending on I0 increase when motor is operated. We consider this phenomenon is caused by V_{out} fluctuation because of V_{out} and I0 waveform synchronized. Fig.7 shows the phase current waveform of 0.75 to 0.8 seconds in Fig.6 (c). Fig.7 (a), (b), and (c) show each result of conventional circuit, conventional circuit when I0 control is excluded and proposal circuit. Fig.8 (a) and (b) show FFT analysis of I_u and I0.

In Fig.7, the proposal circuit increases I0 than conventional circuit that I0 control is excluded. And Fig.8 shows phase current waveform increase of distortion by I0 dependent harmonics. The phenomenon causes to generate vibration and audible noise.

From simulation results, proposal converter achieves boost operating and reach target voltage about 0.2 seconds after start control when motor is operating or not. However, V_{out} has an overshoot, we consider that this phenomenon is not solved by adjusting PI gain of the capacitor voltage controller. Therefore, to solve this problem need other method.

Proposal of Boost converter without reactor using Open-ended Winding PMSM for Photovoltaic Pump System OKAZAKI Akihiro

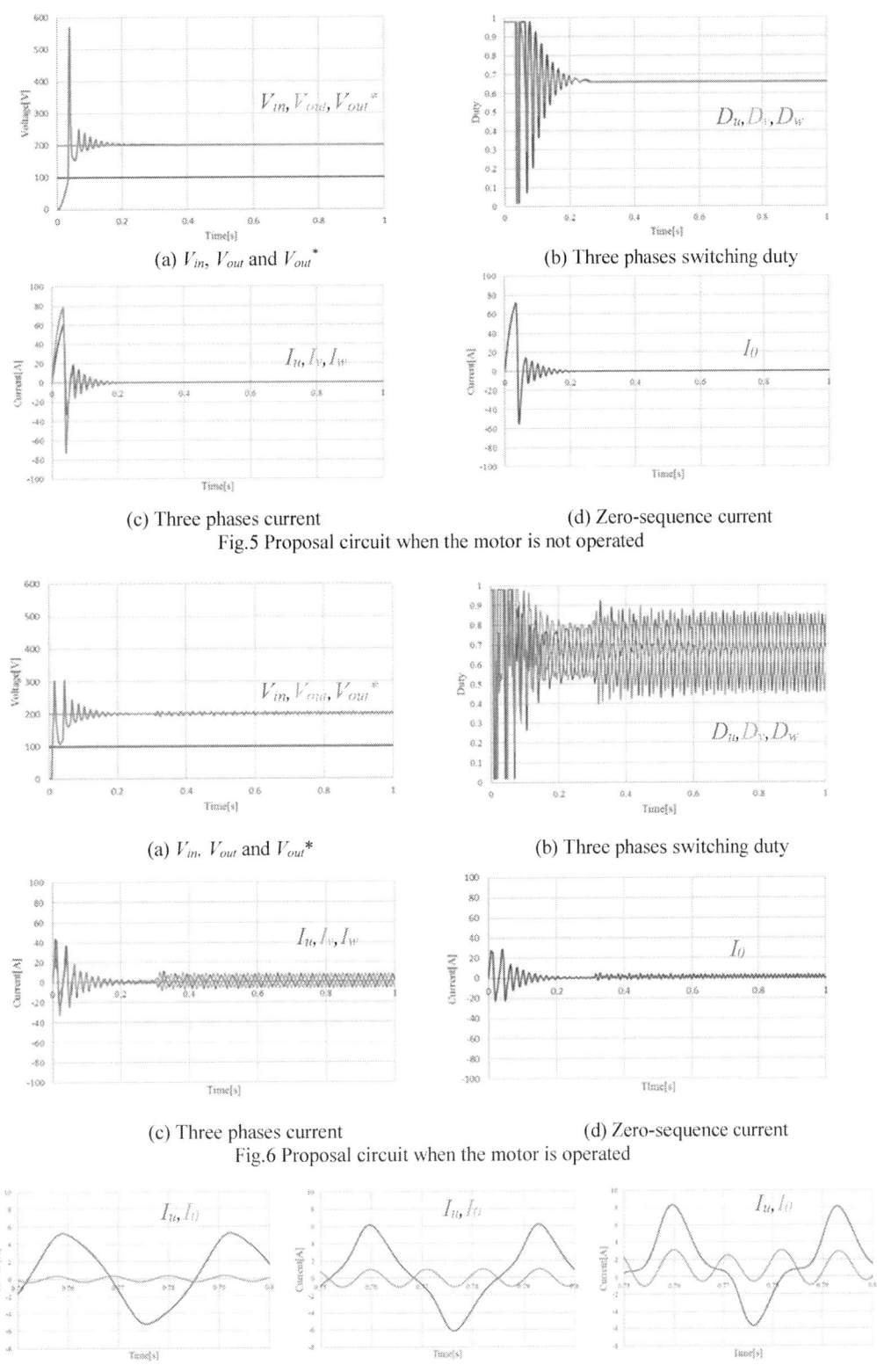

(a) V_{in}, V_{out} and V_{out}^*

(b) Three phases switching duty

(c) Three phases current

(d) Zero-sequence current

Fig.5 Proposal circuit when the motor is not operated

(a) V_{in}, V_{out} and V_{out}^*

(b) Three phases switching duty

(c) Three phases current

(d) Zero-sequence current

Fig.6 Proposal circuit when the motor is operated

(a) Conventional Circuit   (b) Conventional Circuit(I_0 control is excluded)   (c) Proposal Circuit

Fig.7 Comparison of waveform of the zero-sequence and phase current

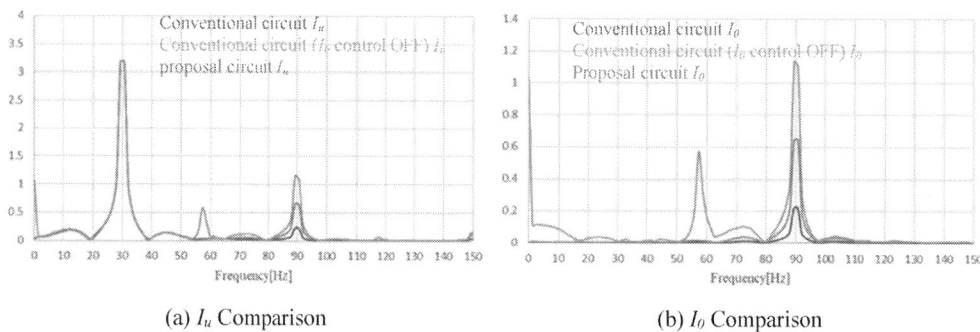

(a) I_u Comparison   (b) I_0 Comparison

Fig.8 FFT Analysis of the zero-sequence and phase current

In Fig.9 and 10 we confirm the upper limit of the rotation speed in the conventional and proposed circuits, and they show the three-phase current and the rotation speed against the reference value. In this simulation, when the duty reaches 1.0, the rotation speed does not follow the reference value. The simulation was performed under the condition of I_d=0.Table 2 shows rotation speed from the simulation results and the parameters related to the cost and size of each circuit. From Table 2, it can be seen that the proposed system is able to increase the maximum rotation speed without additional reactor.

Table.2 Comparison of the rotation speed and components

	1 inverter	1 inverter +3 leg interleaved boost converter	Conventional (DC link connected OEW)	proposal
Maximum rotation speed [rpm]	1290	2460	2310	2520
V_{in} [V]	100	200	100	100
V_{out} [V]	-	-	100	200
Switching Devices	6	12	12	12
Reactor	0	3	0	0

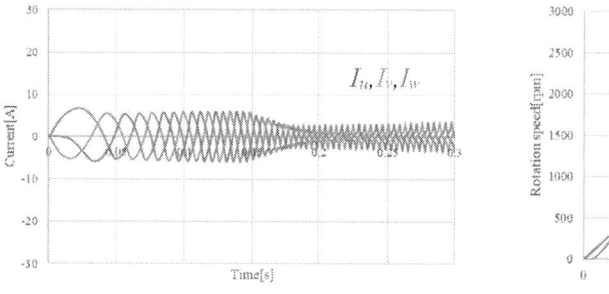 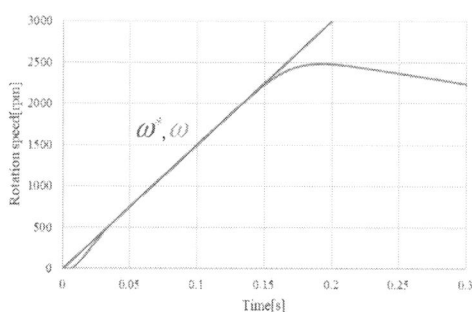

(a) Three phases current   (b) Rotation speed

Fig.9 Conventional circuit rotation speed

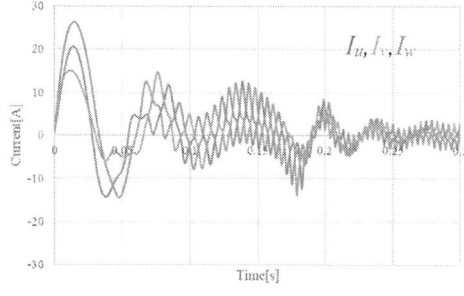 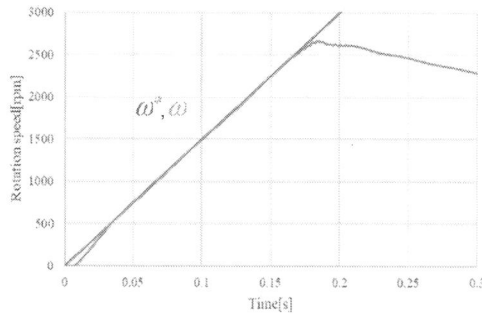

(a) Three phases current   (b)Rotation speed

Fig.10 Proposal circuit rotation speed

The proposed system is suitable for motors with a large number of windings because it can raise the drive voltage and keep the speed high. In addition, the proposed circuit can increase the efficiency of the system without additional a reactor, thus reducing the cost and space. Although the current that can flow through the semiconductor is limited due to the current increase, the proposed system can maintain its superiority in terms of the maximum rotation speed that can be driven.

Conclusion

In this paper, we propose a system that use three phases windings of a motor to realize chopper circuit in open-ended winding PMSM. Proposal converter can boost voltage however, additional duty control generates harmonics. Then, they affect to fluctuate boost voltage. We should consider other method of solving these problems in order to increase practicality of proposal circuit. The proposed circuit is expected to be more effective than other systems in increasing the maximum rotation speed at a lower cost and in a smaller space, mainly in motors with a large number of windings.

References

[1] Farhat Mayssa, Flah Aymen, and V. T. Somasekhar: Influence of photovoltaic DC bus voltage on the high speed PMSM drive", IECON 2012 - 38th Annual Conference on IEEE Industrial Electronics Society (2012)

[2] Jeffrey Ewanchuk, John Salmon, and Chris Chapelsky: "A Method for Supply Voltage Boosting in an Open-Ended Induction Machine Using a Dual Inverter System With a Floating Capacitor Bridge", IEEE Transactions on Power Electronics, Vol.28 No.3 pp.1348 - 1357 (2012)

[3] Wenzhi Zhou, Dan Sun, and Bin Lin: "A modified flux weakening direct torque control for open winding PMSM system fed by hybrid inverter", 2014 17th International Conference on Electrical Machines and Systems (ICEMS) (2014)

[4] Bo Wang, G. Localzo, G. El Murr, J. Wang, A. Griffo , C. Gerada, and T. Cox: "Overall assessments of dual inverter open winding drives", 2015 IEEE International Electric Machines & Drives Conference (IEMDC) (2015)

[5] Sachin Jain, Athiesh Kumar Thopukara, Ramsha Karampuri, and V. T. Somasekhar: "A Single-Stage Photovoltaic System for a Dual-Inverter-Fed Open-End Winding Induction Motor Drive for Pumping Applications", IEEE Transactions on Power Electronics, Vol.30 No.9 pp. 4809 - 4818 (2015)

[6] Ramsha Karampuri, Sachin Jain, and Chris Chapelsky: "A single-stage solar PV power fed Open-End Winding Induction Motor pump drive with MPPT", 2014 IEEE International Conference on Power Electronics, Drives and Energy Systems (PEDES) (2014) [15] S. Khaldoune, M. Pietrzak-David, K. Abdelaziz, F. Maurice, "Hardware in loop methodologies for the control of dual-PMSM connected in parallel: FPGA implementation and experimentation," 2015 17th European Conference on Power Electronics and Applications, pp.1-10 (2015)

[16] K. Sahri, M. Pietrzak-David, M. Fadel, A. Kheloui, "Sensorless Tolerant Fault Control for Dual Permanent Magnet Synchronous Motor Drive with Global FPGA Emulator," 2018 20th European Conference on Power Electronics and Applications, pp.1-8 (2018)

The Proposal of discriminating stable control bandwidth using ANN

in sensorless speed control system for PMSM

Ami Tanaka, Sari Maekawa
Seikei University
〒180-8633 3-3-1, Kichijoji Kitamachi,
Musashino-city, Tokyo-to, Japan
E-mail : sari1.maekawa@st.seikei.ac.jp
Tel : +81-422-37-3768

Keywords

«Sensorless control», «simulation», «Neural network», «Current control», «Speed control»

Abstract

In the sensorless control of a permanent magnet synchronous motor (PMSM), there is a method of determining stability by analyzing the pole placement of the closed-loop transfer function regarding the influence of the three control bandwidths of current control, speed control, and sensorless control on stability. However, due to various effects, there is a problem that the theoretical analysis and the range of the control bandwidth that can be stably driven in the actual machine do not match. In this paper, we propose a method using ANN (Artificial Neural Network) in order to determine the condition of the stable control band more accurately.

Introduction

Although a method for estimating rotor magnetic flux and induced voltage in the medium to high speed range is used in sensorless control of permanent magnet synchronous motor, various factors cause the problem of instability in sensorless control. Thus, the stability of control system as a whole considering there relationship between speed control and minor loops which are the frequency of current control and sensorless control is studied.

For example, as an approach based on linear analysis, there is a method of determining stability by analyzing the pole placement of the closed-loop transfer function with respect to the effects of the three control bands of current control, speed control, and sensorless control on stability [1][2]. However, due to various effects such as approximation of transfer function and estimated angle error, there is a problem that the theoretical analysis and the result of simulation do not match [3].

On the other hand, in the fields of power electronics and motor drive, the various method using neural networks such as ANN and SNN (Structure Neural Network) are studied [4]-[10]. In this paper, we propose the determination method which has a stability bandwidth about unknown PMSMs, after learning relationship between the parameters of many PMSMs and stability bandwidth..

Configuration of PMSM and Control System

A. PMSM model

The voltage equation and speed equation of the PMSM are linearized as minute fluctuations from the equilibrium point (ω_0, I_{d0}, I_{q0}) where each state variable is stable, and the state equation of equation (1) is derived.

$$\frac{d}{dt}\begin{pmatrix} \Delta\omega_m \\ \Delta I_d \\ \Delta I_q \end{pmatrix} = \begin{pmatrix} -\dfrac{D}{J} & \dfrac{PL_1 I_{q0}}{J} & -\dfrac{P(\phi_f + L_1 I_{d0})}{J} \\ -I_{q0} & -\dfrac{R}{L_d} & \omega\dfrac{L_q}{L_d} \\ -\dfrac{\phi_f}{L_q} & -\omega\dfrac{L_d}{L_q} & -\dfrac{R}{L_q} \end{pmatrix}\begin{pmatrix} \Delta\omega_m \\ \Delta I_d \\ \Delta I_q \end{pmatrix} + \begin{pmatrix} 0 & 0 \\ \dfrac{1}{L_d} & 0 \\ 0 & \dfrac{1}{L_q} \end{pmatrix}\begin{pmatrix} V_d \\ V_q \end{pmatrix} = \begin{pmatrix} \dfrac{1}{J} \\ 0 \\ 0 \end{pmatrix} T_l \qquad \cdots(1)$$

Where, $I_{d,q}$ are d-q-axes currents, ω is electrical frequency, $V_{d,q}$ are d-q-axes voltages, $L_{d,q}$ are d-q-axes inductances, R is winding (stator) resistance, and ϕ_f is linked flux of permanent magnet, P is pole pair number, J is moment of inertia, D is friction coefficient, and T_i is load torque. The electrical frequency ω can be expressed by the following equation.

$$\omega = P\omega_m \qquad \cdots(2)$$

Where, ω_m is the mechanical frequency.

B. Configuration of Control System

Fig. 1 shows the control configuration used in the analysis. The control method used is sensorless vector control, and utilizes speed control (speed regulator: SR) based on control of PMSM speed command by detecting the motor current and direct current voltage. As shown in Fig. 1, sensorless control utilizes a general phase locked loop (PLL) configuration whereby, based on the voltages V_{dc}, V_{qc}, and currents I_{dc}, I_{qc} along the estimated axes dc, qc, axial error $\Delta\theta$ is calculated as the difference between the actual position, and the estimated position, and the estimated speed $\omega\hat{}$ and position $\theta\hat{}$ are calculated using PI controllers.

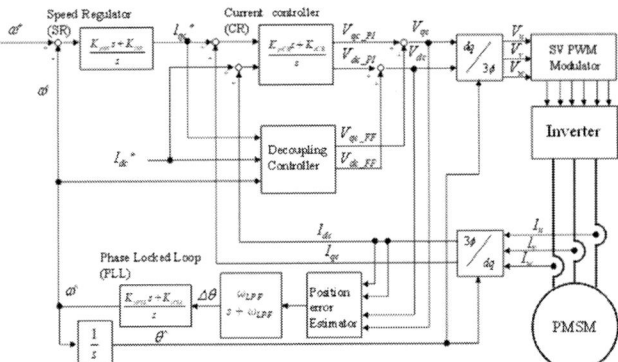

Fig.1 Control configuration

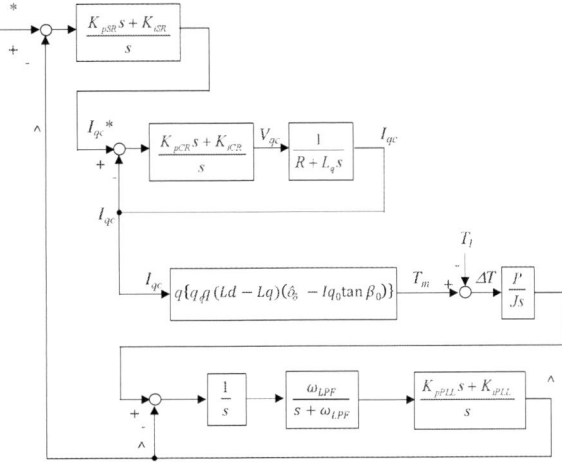

Fig.2 Block diagram of sensorless speed control

Fig. 2 shows the block diagram of the sensorless speed control.

The closed-loop transfer function $G_{SR^\wedge}(s)$ of estimated speed $\omega^\wedge$ corresponding to speed command value $\omega*$ including the current control, speed control, and sensorless control systems is given by equation (3).

$$G_{SR^\wedge}(s) = \frac{\omega^\wedge}{\omega^*} = \frac{G_1(s)G_3(s)}{1 + G_1(s)G_3(s)} \qquad \cdots(3)$$

Where,

$$G_2(s) = \frac{1}{s}G_{LPF}(s)C_{PLL}(s) \qquad \cdots(4)$$

$$G_3(s) = \frac{\omega^\wedge}{\omega} = \frac{G_2(s)}{1 + G_2(s)} \qquad \cdots(5)$$

Stability Judgement by Pole Placement and Natural angular frequency

A. Stability Judgement of Sensorless Speed Control by Pole Placement

The analysis of the pole placement of the closed-loop transfer function can determine the stability of the influence of the three control bands of current control, speed control, and sensorless control on the stability in the sensorless control system. We derive the closed-loop transfer function of the estimated speed $\omega^\wedge$ for the speed command value ω^* from the PMSM state equation model from (1) and the model of each control system shown in Fig. 1, and analyze it by pole placement. The characteristic equation of the closed-loop transfer function expressed by Equations (6) has 12 poles [3]

$$\begin{aligned}
D_{SR^\wedge}(s) = &s^6 + (\omega_{CR} + \omega_{LPF})s^5 + \omega_{LPF}(2\omega_{PLL} + \omega_{CR})s^4 + \omega_{LPF}(\omega_{PLL}^2 + 2\xi_{PLL}\omega_{CR}\omega_{PLL})s^3 \\
&+ \omega_{LPF}\omega_{CR}\omega_{PLL}(\omega_{PLL} + 4\xi_{SR}\xi_{PLL}\omega_{SR}P^4\varphi_f^2)s^2 \\
&+ 2\omega_{LPF}\omega_{CR}\omega_{SR}\omega_{PLL}P^4\varphi_f^2(\xi_{SR}\omega_{PLL} + 4\xi_{PLL}\omega_{SR})s + \omega_{LPF}\omega_{CR}\omega_{SR}^2\omega_{PLL}^2P^4\varphi_f^2
\end{aligned} \qquad \cdots(6)$$

Where, ω_{CR}, ω_{SR}, ω_{PLL} are current, speed, sensorless contol bandwidth respectively. ω_{LPF} is cut off frequency of the position error estimator. ξ_{SR}, ξ_{PLL} are the respective damping coefficients. We do not consider control delays when Equation (6) is calculated.

Fig.3 (a) and 3 (b) show the trajectories of the poles when the speed control bandwidth is increased. (a) is sensorless control, and (b) is sensored control. It can be seen that the sensorless control is more likely to become unstable because the poles A and B move to the right half plane than the control with the sensor

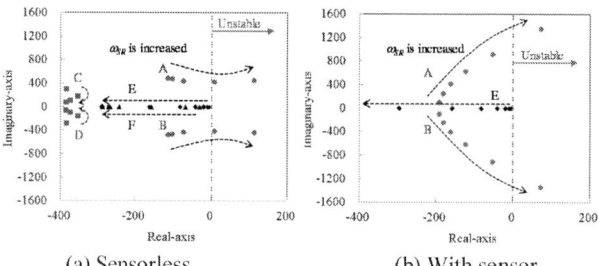

(a) Sensorless   (b) With sensor

Fig.3 Root locus when the frequency of SR is increased.

B. Comparison of Frequency Stability Verification and Transient Response Controls

Fig. 4(a) shows the results of stability discrimination when the speed control band and the sensorless control band are changed by pole analysis. Fig. 4(b) shows the results of repeated transient analysis using an inverter motor simulator. According to these analyses, the stable control bandwidth range which is analyzed with pole analysis is narrower than the result of simulation. Various factors such as approximation of a transfer function and the influence of an estimated angle error can be considered for this mismatch factor. Therefore, in order to determine the condition of the stable control band more accurately, this time we will consider a method using ANN.

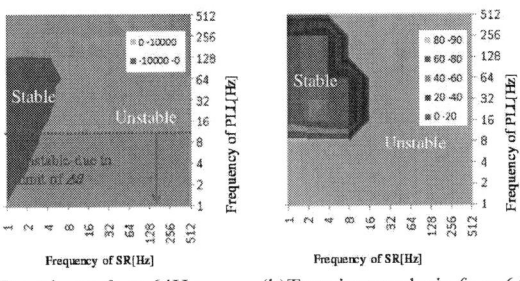

(a)Root locus,f_{CR} = 64Hz   (b)Transient analysis,f_{CR} = 64Hz

Fig.4 The comparison of root locus analysis and transient analysis when frequency of SR is changed.

Stable control band determination using of ANN

A neural network has a network structure in which neurons are arranged in layers. The neural network layers are classified into three layers: input layer, intermediate layer, and output layer. The following expression represents the output y to the next layer.

$$y = (f(\sum_{k=1}^{m} x_k \omega_{k1} + b_1), f(\sum_{k=1}^{m} x_k \omega_{k2} + b_2), \dots (f \sum_{k=1}^{m} x_k \omega_{kn} + b_n) \qquad \cdots(7)$$

Where, f is an activation function, x is a value of input to neuron, ω is weight, b: bias, y is a vector.

Fig. 5 and Table 1 show the input and output of theANN in order to estimate stable control bandwidth.

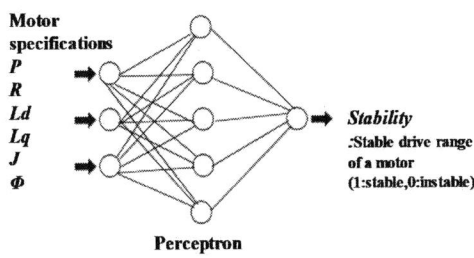

Fig.5 Configuration of neural network

Table1. Input and Output/Tuning data of ANN in order to determine of stable control

Division	Symbol	Parameters
Input	P	Number of pole pair
	R	One-phase winding resistance[Ω]
	Ld	D-axes inductances[H]
	Lq	Q-axes inductances[H]
	J	Moment of inertia[kgm2]
	Φ	Linked flux of permanent magnet[Wb]
Output	$Stability$	Stable drive range of a motor(1:stable,0:unstable)

The input is number of pole pair P, winding resistance R, d-axes inductances L_d, q-axes inductances L_q, moment of inertia J, linked flux of permanent magnet ϕ_f. The output is a range of a speed control bandwidth and a sensorless control bandwidth that can drive a PMSM with stable. When the output is "0", the sensorless control system becomes unstable. On the other hands, the "1" of output makes it stable. We recommend using a lookup table when we try to implement ANN.

Results of automatic adjustment of stable control bandwidth by ANN

Fig. 6 shows the flowchart of learning to the ANN in order to determine of stable control. First, we prepare 200 samples of the PMSM specifications shown in Table 1 and perform a simulation with transient analysis, and make the ANN learn the results of stable operation for the speed control bandwidth and sensorless control bandwidth for each PMSM. Because of simple study in this paper, we set to the current control bandwidth 100Hz.

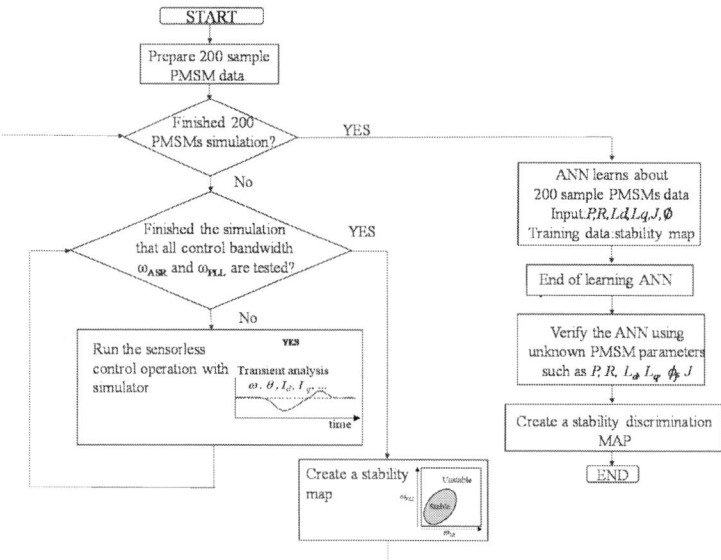

Fig.6 Process of learning with ANN

Then, we train them into the ANN and then determine the stability control bandwidth range of **u**nknown PMSMs that had not yet been trained by the ANN. Table 2 and Table 3 show two PMSM parameters which are used to test the ANN. Fig.7 and Fig.8 show the experimental results. Fig.7(a) and Fig.8(a) are the stability determination results by ANN, and Fig.7(a) and Fig.8(a) are the simulation results by transient analysis. Fig.7 and Fig.8 use the PMSMs with different specifications respectively.

Comparing the results of the two methods, the stable control bandwidth range in which the PMSM can be operated is almost the same, so we can consider that the stability determination method using ANN can determine the stable range even for unknown PMSMs.

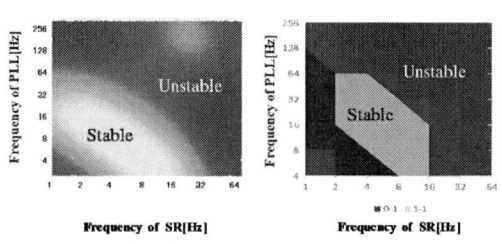

(a) ANN results of PMSM1   (b)Actual results of PMSM1

Fig.7 Results of discriminating stable control

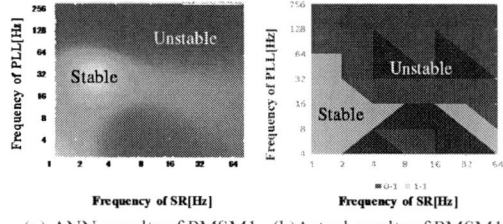

(a) ANN results of PMSM1   (b)Actual results of PMSM1

Fig.8 Results of discriminating stable control bandwidth
by ANN and simulation

Table.2 Parameters of PMSM1

Parameter	Parameter	Motor1
P	Polar logarithm	3.000
R	One-phase winding resistance[Ω]	1.624
Ld	D-axes inductances[H]	0.012
Lq	Q-axes inductances[H]	0.019
J	Moment of inertia[kgm2]	0.000
φf	Linked flux of permanent magnet[Wb]	0.145
A MAX	Maximum rated current	20
Vdc	Rated voltage	300
Torque	Rated torque	2
N-rpm	Rated rpm	1,800

Table.3 Parameters of PMSM2

Parameter	Parameters	Motor2
P	Polar logarithm	4
R	One-phase winding resistance[Ω]	10.717
Ld	D-axes inductances[H]	0.054
Lq	Q-axes inductances[H]	0.056
J	Moment of inertia[kgm2]	0.003
φf	Linked flux of permanent magnet[Wb]	0.109
A MAX	Maximum rated current	20
Vdc	Rated voltage	200
Torque	Rated torque	2
N-rpm	Rated rpm	300

Furthermore, it is confirmed by the same method whether ANN applies when a high-power PMSM is used. Table 4 shows the PMSM parameters used for ANN testing. Fig. 9(a) shows the stability system discrimination result by ANN, and Fig. 9(b) shows the simulation result by transient analysis. 9(a) and 9(b), the stability control bands are almost the same, so the stability determination method using ANN can determine the unknown PMSM stability range even when using a high-power motor.

Table.4 Parameters of PMSM3

Parameter	Parameters	Motor3
P	Polar logarithm	3
R	One-phase winding resistance[Ω]	0.812
Ld	D-axes inductances[H]	0.003
Lq	Q-axes inductances[H]	0.005
J	Moment of inertia[kgm2]	0.003
φf	Linked flux of permanent magnet[Wb]	0.290
A_MAX	Maximum rated current	80
Vdc	Rated voltage	600
$Torque$	Rated torque	10
$N\text{-}rpm$	Rated rpm	1,800

(a) ANN results of PMSM1　(b)Actual results of PMSM1
Fig.9 Results of discriminating stable control
bandwidth by ANN and simulation

Fig. 10 shows the result of setting the speed control bandwidth and sensorless control bandwidth determined by ANN that the PMSM drive stably.

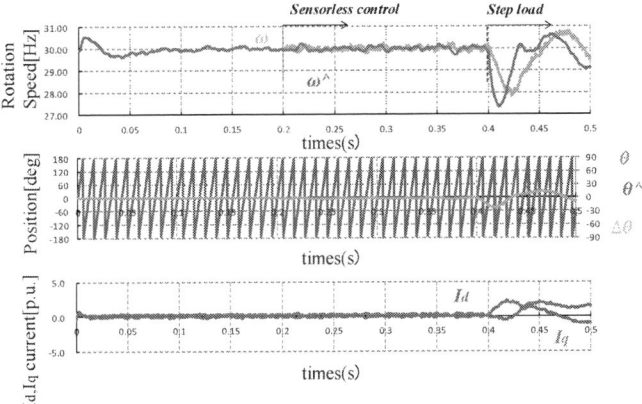

Fig.10 The result of operation waveform of the PMSM1, the condition of control bandwidth is selected with the point of A in the Fig.6(a) by ANN. (f_{CR}=100Hz, f_{SR}=8Hz, f_{PLL}=8Hz)

Conclusion

We propose the method that uses ANN to determine the stable control bandwidth range that can operate the stability of an unknown PMSM. As a result, the range almost coincided with the simulation result by transient analysis. The stability determination method using ANN can determine the unknown PMSM stability range even when using a high-power motor. In the future, we will verify whether the ANN functions effectively under non-linear conditions such as magnetic saturation, and synchronous PWM control during voltage saturation.

References

[1] K. Shinohara, T. Nagano, and K. Ohyama: "Stability Analysis of Vector Control of InductionMotor without Speed Sensor Taking into Account the Effects of Current Control Loop", IEEJ Trans. IA, Vol.116, No.3, pp.337-347 (1996).

[2] K. Ohyama, K. Shinohara, T. Nagano, and H. Arima: "Stability Analysis of the Direct Field Oriented Control System of the Induction Motor without a Speed Sensor using the Adaptive Rotor Flux Observer", IEEJ Trans. IA, Vol.119, No.3, pp.333-344 (1999).

[3] S. Maekawa, M. Sugimoto, K. Ishida, M. Nogi, and M. Kanamori: "Stability Analysis of Sensorless Speed Control for PMSM Considered Current Control System", IEEJ Journal of Industry Applications. Vol.8 No.4 pp.736-744 (2019)

[4] Karanayil, B, Rahman, M.F, Grantham, C: "Speed sensorless vector controlled induction motor drive with rotor time constant identification using artificial neural networks", Intelligent Control, 2002. Proceedings of the 2002 IEEE International Symposium, pp. 715- 720, 2002

[5] Garcia, Pablo, Reigosa, David, Briz, Fernando, Raca, Dejan, and Lorenz, Robert D: "Automatic Self-Commissioning for Secondary-Saliencies Decoupling in Sensorless-Controlled AC Machines Using Structured Neural Networks", Industrial Electronics, 2007. ISIE 2007. IEEE International Symposium, pp.2284-2289, 4-7 June 2007

[6] Gadoue, S.M, Giaouris, D, Finch, J.W: "An experimental assessment of a stator current MRAS based on neural networks for sensorless control of induction machines", Sensorless Control for Electrical Drives (SLED), 2011 Symposium, pp.102-106, 1-2 Sept. (2011)

[7] Rajesh Kumar, R. A. Gupta, Bhim Singh: "Intelligent Tuned PID Controllers for PMSM Drive - A Critical Analysis", 2006 IEEE International Conference on Industrial Technology, 2006

[8] Tomasz Pajchrowski, Krzysztof Zawirski, Krzysztof Nowopolski: "Neural Speed Controller Trained Online by Means of Modified RPROP Algorithm", IEEE Transactions on Industrial Informatics. Vol.11 No.2 (2015)

[9] Wang Tong-xu, Ma Hong-yan: "The research of PMSM RBF neural network PID parameters self-tuning in elevator", The 27th Chinese Control and Decision Conference (2015)

[10] Wangpeng An, Haoqian Wang, Qingyun Sun, Jun Xu, Qionghai Dai, Lei Zhang: "A PID Controller Approach for Stochastic Optimization of Deep Networks", 2018 IEEE/CVF Conference on Computer Vision and Pattern Recognition (2018)

Cost Function Design for Stability Assessment of Modulated Model Predictive Control

Jordan P. Zucuni, Fernanda
Carnielutti, Humberto Pinheiro
Power Electronics and Control
Research Group - GEPOC
Federal University of Santa Maria
Av. Roraima, 1000,
Santa Maria, RS, Brazil
[jzucuni, fernanda.carnielutti,
humberto.ctlab.ufsm.br]gmail.com

Margarita Norambuena
Universidad Tecnica
Federico Santa Maria
Avenida España 1680
Valparaíso, Chile
margarita.norambuena@gmail.com

Jose Rodriguez
Universidad Andres Bello
República 239, Santiago, Chile
jose.rodriguez@unab.cl

Acknowledgments

J. Rodriguez acknowledges the support of ANID through projects FB0008, ACT192013 and 1170167.

Keywords

≪Model Predictive Control≫, ≪Stability≫, ≪Power Electronics≫

Abstract

This paper proposes a quadratic cost function for Modulated Model Predictive Control, M^2PC, designed in a way such as to guarantee that the inverter under consideration will be stable for a desired steady-state operational condition, also considering parametric variations. The M^2PC results in better harmonic performance, less current ripple and less steady-state error when compared to the classical FCS-MPC. Hardware-in-the-Loop, HIL, results are presented comparing the performance of the implemented M^2PC and a classical FCS-MPC for a two-level inverter connected to an RL load, considering the same cost function. The HIL results have demonstrated the good performance of the proposed approach, showing the stability and performance despite the model of the implemented M^2PC.

Introduction

Model Predictive Control, MPC, is becoming a very attractive control solution for power electronics applications [4, 5, 6], such as machine drives, grid-tied inverters, renewable energy sources, microgrids, etc., as it takes advantage of the discrete nature of static power electronic converters. As power converters have a finite number of possible voltage vectors, i.e. system inputs, a Finite Control Set MPC approach, FCS-MPC, can be easily implemented [5, 16, 14, 17, 7]. In the classical FCS-MPC, a cost function is designed in a way such as to contemplate as many control objectives as desired and needed, such as reference current tracking, capacitor voltage regulation and/or balance for multilevel converters, minimization of switching transitions, etc. [5]. In each sampling instant k, the future system states are predicted and the cost function is evaluated for each one of the converter voltage vectors, and the one that minimizes the cost function is chosen and implemented by the converter. For an improved prediction of the system states, a large prediction horizon should be used; however, this leads to an increased computational burden. As a result, in practical control applications, usually a prediction horizon of one or two steps ahead is implemented.

A problem that may arise in the classical FCS-MPC is that the same converter voltage vector can be implemented during various consecutive sampling instants k, leading to variable switching frequency, large current ripples and a widespread harmonic content (including low-order harmonics), which, in turn, make the design of the output filter more complex. In order to solve these issues, a switching sequence can be implemented during each k instead of just one voltage vector, such as in a conventional Space Vector Modulation, SVM. This approach is known as Modulated Model Predictive Control, M^{2}PC [3, 9, 15, 12, 13, 24, 25, 26], resulting in fixed switching frequency, smaller current ripple and better harmonic performance when compared to the classical FCS-MPC.

Even though M^{2}PC and classical FCS-MPC are well-suited for controlling power converters, some issues still need to be studied in greater detail. One of them is the stability of MPC controllers [10, 21]. Some papers have already dealt with this issue in the literature. For example, in [22] a robust modified classical FCS-MPC is proposed for controlling a three-phase active-front-end rectifier. The design of the FCS-MPC is based on a Lyapunov function, such as to ensure stability, robustness, and fast dynamic response of the controller. As a FCS-MPC is used, only one vector is implemented during each sampling instant k. In [23] an improved FCS-MPC is presented for three-phase inverters with LC filter. A robust prediction is proposed and the stability-constrained finite states are selected by means of asymptotic stability conditions such as to reduce THD and steady-state errors. Kalman filter observers are used to improve the system robustness against parametric variations. On the other hand, closed-loop stability is investigated for Model Predictive Direct Current Control, MPDCC, in [20], and applied to a three-level Neutral Point clamped, NPC, converter with an active RL load. The MPDCC algorithm guarantees controller stability, and, by adding minor modifications to the classical FCS-MPC, robustness to parametric variations can also be achieved.

An interesting approach to deal with stability issues for MPC controllers was presented in [18, 19]. Here, the cost function design is also based on Lyapunov stability, such as to guarantee that the system states will converge to the desired steady-state operating point. The controller stability is demonstrated in the paper, and results are presented for a Buck DC-DC converter and a three-phase two-level inverter with an RL load. However, the developments of [18, 19] are presented for the classical FCS-MPC. In this context, this paper proposes to apply the cost function design of [18, 19] to an M^{2}PC algorithm, considering as an example also a three-phase two-level inverter with an RL load. The results are analysed and compared to the classical FCS-MPC, considering steady-state and transient responses, harmonic performance and parametric variations.

This paper is divided as follows: in Section II, the M^{2}PC is reviewed for a three-phase two-level inverter connected to an RL load; Section III presents the design of the cost function for stable performance of the inverter; in Section IV, Hardware-in-the-Loop, HIL, results are presented and compared for the proposed M^{2}PC and classical FCS-MPC; at last, Section V brings the conclusions of the paper.

Modulated Model Predictive Control - M^{2}PC

Before describing the proposed approach, let us first briefly review the M^{2}PC strategy presented in [3], taking as example a three-phase two-level inverter with an RL load, as shown in Fig.1. The converter voltage vectors are arranged in the same sectors in the SV diagram as in a conventional SVM. During each sampling instant k, the designed cost function is calculated for each individual switching vector, as for the classical FCS-MPC. Then, the cost of each sector is calculated based on the cost of its respective voltage vectors. The sector with the minimum cost is selected, and its associated switching sequence is implemented by the inverter during k.

As the M^{2}PC controller is implemented in discrete-time, let us write the dynamic equations of the system in the $\alpha\beta$ reference frame, considering Euler discretization, as:

$$\mathbf{i}[k+1] = \mathbf{i}[k] + \frac{T_s}{L}(-R\mathbf{i}[k] + \mathbf{u}[k]) \tag{1}$$

or, in state-space:

$$\mathbf{i}[k+1] = \mathbf{A}\mathbf{i}[k] + \mathbf{B}\mathbf{u}[k] \tag{2}$$

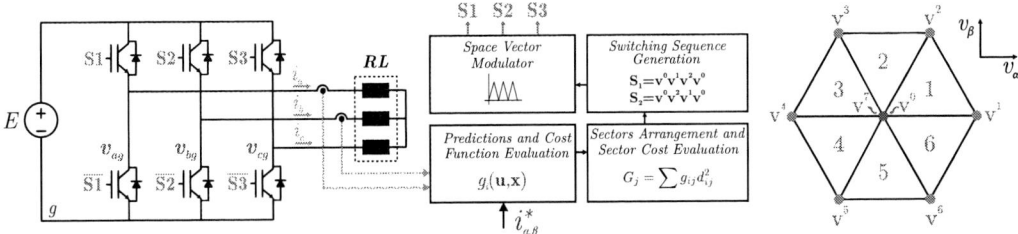

Fig. 1: Three-phase two-level inverter with RL load, block diagram of the proposed M²PC and SV diagram in αβ coordinates.

where T_s is the sampling period, R and L are the load resistance and inductance, $\mathbf{i}[k] = [i_\alpha \quad i_\beta]^T$ are the system states, i.e. the output currents, $\mathbf{u}[k] = [u_\alpha \quad u_\beta]^T$ are the control inputs, i.e. the inverter voltage vectors, and matrices $\mathbf{A}$ and $\mathbf{B}$ are given by:

$$\mathbf{A} = (1 - \frac{RT_s}{L}) \begin{bmatrix} 1 & 0 \\ 0 & 1 \end{bmatrix} \qquad \mathbf{B} = \frac{T_s}{L} \begin{bmatrix} 1 & 0 \\ 0 & 1 \end{bmatrix} \tag{3}$$

As the M²PC will be implemented in a Digital Signal Processor, DSP, the implementation delay must be accounted for [3]. To accomplish this, at each sampling instant k, the system states at $\mathbf{x}[k+2]$ and the current references $\mathbf{i}^*[k+2] = \mathbf{x}^*[k+2]$ are predicted. In this application, the only control objective is to minimize the current tracking error, so we can write a cost function such as:

$$g = (\mathbf{i}^*[k+2] - \mathbf{i}[k+2])^2. \tag{4}$$

In the M²PC approach, the cost function (4) is calculated for each individual switching vector, and then the cost of each sector is calculated based on the cost of its respective voltage vectors. The sector with the minimum cost is selected, and its associated switching sequence is implemented by the inverter during k. The cost G_j of each sector is calculated based on the cost of the its voltage vectors as:

$$G_j = \sum g_{ij} d_{ij}^2 \tag{5}$$

$$s.t. \sum g_{ij} d_{ij} = 1 \qquad 0 \le d_{ij} \le 1 \tag{6}$$

where d_{ij} and g_{ij} are, respectively, the duty cycle and the cost associated to the ith vector of the designed switching sequence for the jth sector. The duty cycles depend on the cost g_{ij} of its associated voltage vector, and are given by:

$$d_{ij} = \frac{Q_j}{d_{ij}} \qquad Q_j = \frac{1}{\sum g_{ij}^{-1}}. \tag{7}$$

As this solution satisfies the Karush-Kuhn-Tucker conditions and its a convex problem, the solution is optimal [3]. The sector with minimum cost is then implemented in k by means of its associated switching sequence. In this paper, the sequences are defined offline and stored in the DSP memory, and two sequences comprised of four voltage vectors are defined:

$$\mathbf{S}_1 = \mathbf{v}^0 \mathbf{v}^1 \mathbf{v}^2 \mathbf{v}^0 \qquad \mathbf{S}_2 = \mathbf{v}^0 \mathbf{v}^2 \mathbf{v}^1 \mathbf{v}^0 \tag{8}$$

where the first and second sequences are designed for the odd- and even-numbered sectors, respectively. Vector $\mathbf{v}^0$ represents the redundancies of the null vector, and $\mathbf{v}^1$ and $\mathbf{v}^2$ are the vectors of the SV diagram of Fig. 1 on the counterclockwise direction. These sequences were designed in order to minimize the switching commutations and to be implementable in a commercial DSP, with interruptions on the underflow and period match of the DSP carrier. As a result, the inverter output line-to-line voltages have a constant switching frequency with better harmonic performance than the classical FCS-MPC, as will be presented and discussed in Section IV.

Cost Function Design for Stability

For MPC controllers, the stability problem can be stated as how to design a proper cost function to guarantee that the controlled variable, or variables, will converge to the desired steady-state operating point. As mentioned in the Introduction of this paper, in [18, 19] a cost function for stable performance was defined as:

$$g(\mathbf{u},\mathbf{x}) = |\mathbf{x}[k] - \mathbf{x}^*[k]|_{\mathbf{Q}}^2 + |\mathbf{u}[k] - \mathbf{u}^*[k]|_{\mathbf{R}}^2 + |\mathbf{x}[k+1] - \mathbf{x}^*[k]|_{\mathbf{P}}^2 \tag{9}$$

where $\mathbf{u}^*$ is the steady-state input reference, matrices $\mathbf{Q}$ and $\mathbf{R}$ are semipositive definite, matrix $\mathbf{P}$ is positive definite and $|\mathbf{x}[k] - \mathbf{x}^*[k]|_{\mathbf{Q}}^2 = \mathbf{e}^T\mathbf{Q}\mathbf{e}$ is the quadratic norm of the error (the same applies to the other terms in (9)). To account for the implementation delay, (9) can be rewritten as:

$$g(\mathbf{u},\mathbf{x}) = |\mathbf{x}[k+2] - \mathbf{x}^*[k+2]|_{\mathbf{Q}}^2 + |\mathbf{u}[k+2] - \mathbf{u}^*[k+2]|_{\mathbf{R}}^2 + |\mathbf{x}[k+2] - \mathbf{x}^*[k+2]|_{\mathbf{P}}^2 \tag{10}$$

Matrices $\mathbf{Q}$ and $\mathbf{R}$ are defined by the user, and matrix $\mathbf{P}$ is the solution of the algebraic Riccati equation, that can be found by a DLQR algorithm. By assigning a large value to matrix $\mathbf{Q}$ and a small value to matrix $\mathbf{R}$, the controller gives more emphasis to minimizing the steady-state current tracking error. If, on the contrary, we have a a large value for matrix $\mathbf{R}$ and a small one to matrix $\mathbf{Q}$, emphasis is given to minimizing the control inputs of the system. In the M^2PC proposed in this paper, the cost function (4) is changed by (10) and is evaluated at each sampling instant for all inverter voltage vectors. The same procedure of [3] is used to calculate the vector duty cycles and sector costs. The switching sequence of the sector with the minimum cost is then implemented by the inverter.

In order to verify the theoretical analysis described here, in the next section, Hardware-in-the-Loop, HIL, results are presented, comparing the performance of proposed M^2PC to the classical FCS-MPC approach, where both controllers use the cost function (10).

Hardware-in-the-Loop Results

In this section, Hardware-in-the-Loop, HIL, results obtained with Typhoon HIL are presented in order to demonstrate the good performance of the proposed M^2PC with cost function designed for stability. The considered two-level inverter has the following parameters: nominal power $P_{non} = 100$kW, DC bus voltage $V_{DC} = 800$V, $R = 1.3\Omega$, $L = 2$mH (power factor 0.9 inductive), current $I = 238$A, fundamental frequency of 50Hz, and sampling frequency $f_{sw} = 10$kHz (interruptions on the underflow and period match of the DSP carrier). For comparison purposes, a classical FCS-MPC was also implemented for the same inverter and the same cost function (10). The matrices $\mathbf{R}$ and $\mathbf{Q}$ were chosen as $\mathbf{R} = 0.0001\mathbf{I}_{2\times2}$ and $\mathbf{Q} = \mathbf{I}_{2\times2}$. By means of a DLQR algorithm, matrix $\mathbf{P}$ is equal to $\mathbf{P} = 1.0035\mathbf{I}_{2\times2}$. In this design, by assigning a small value to the entries of matrix $\mathbf{R}$, the controller gives more emphasis to minimizing the steady-state tracking error given by matrix $\mathbf{Q}$, and not the input control action.

First, the results were obtained for the nominal operation conditions described in the previous paragraph. Fig. 2 shows the output phase currents i_a and i_b, reference phase currents i_a^* and i_b^* and the line-to-line voltage v_{ab} in abc coordinates at nominal load conditions for the proposed M^2PC and the classical FCS-MPC, respectively. It can be seen that both strategies can track the current reference; however, the M^2PC presents a much better steady-state performance, less ripple on the phase currents and better line-to-line voltage v_{ab}. The apparent lower switching frequency of the PWM pulses for the classical FCS-MPC is due to the fact that is operates with variable switching frequency, and the same voltage vector can be implemented by the inverter for multiple consecutive switching instants k, while the proposed M^2PC implements a switching sequence over k. Fig. 3 shows the current tracking errors e_a and e_b for phases a and b in abc coordinates for the proposed M^2PC and the classical FCS-MPC. It can be seen that the amplitude of the current tracking error is smaller for the M^2PC, and the error also presents less ripple than for the classical FCS-MPC. The FFTs of the output current i_a and of the line-to-line voltage v_{ab}

are shown, respectively, in Figs. 4 and 5, for both controllers. From both figures, we can see that the proposed M^2PC presents a much better harmonic performance than the classical FCS-MPC, specially regarding the low-order harmonics. It can be noted that the amplitude of the harmonics for the classical FCS-MPC is higher, specially for the currents. Also, the spectra of the current and voltage is more widespread for the classical FCS-MPC approach, resulting in higher THDs. The calculated current and voltage THDs are, respectively, 1.18% and 86.84% for the proposed M^2PC and 5.64% and 86.09% for the classical FCS-MPC. The superior performance of the proposed M^2PC can be better visualized when computing the Weighted Total Harmonic Distortions, WTHDs, for the currents and voltages. The calculated current and voltage WTHDs are, respectively, 0.070% and 0.524% for the proposed M^2PC and 0.328% and 2.525% for the classical FCS-MPC. This demonstrates that the proposed M^2PC has less low-order harmonics, and that their amplitudes are smaller, when compared to the classical FCS-MPC.

Finally, a step change was applied to the reference current, decreasing it by 50%, from 238A to 119A. Fig.6 shows the waveforms of the output phase currents i_a and i_b, reference phase currents i_a^* and i_b^* and the line-to-line voltage v_{ab} in *abc* coordinates at nominal load conditions and with the step change in the current reference, respectively, for the proposed M^2PC and the classical FCS-MPC. Both controllers have a very fast dynamic response, converging to the new desired steady-state operating point. However, we can see that, again, the proposed M^2PC has a better steady-state response than the classical FCS-MPC.

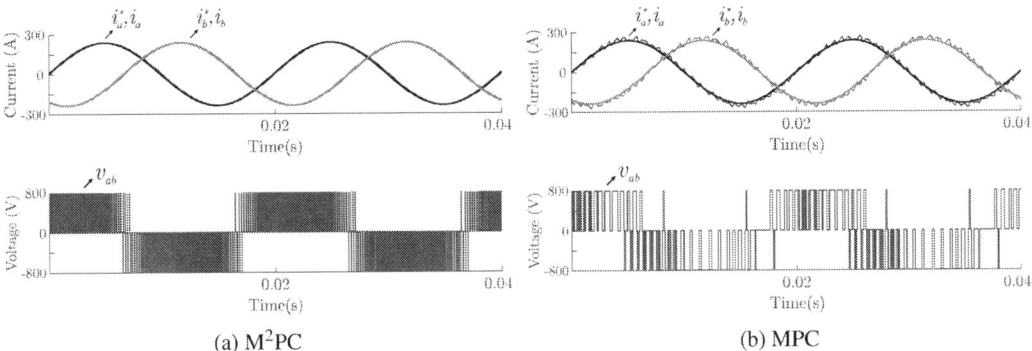

(a) M^2PC (b) MPC

Fig. 2: Reference and phase currents in *abc* coordinates, and line-to-line voltage v_{ab} at nominal conditions for a) Proposed M^2PC and b) classical FCS-MPC.

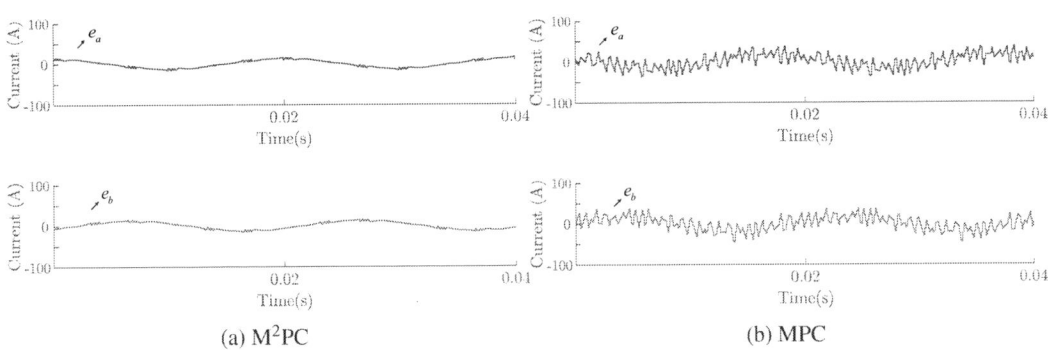

(a) M^2PC (b) MPC

Fig. 3: Current tracking error e_a and e_b for phases a and b in *abc* coordinates at nominal conditions, for a) Proposed M^2PC and b) classical FCS-MPC.

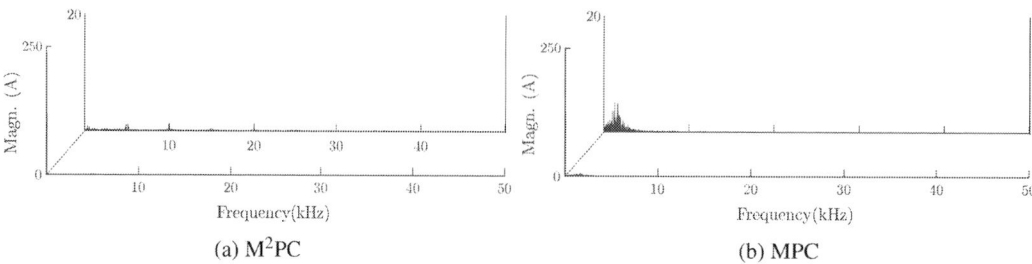

(a) M²PC   (b) MPC

Fig. 4: FFT of the output current i_a at nominal conditions for a) Proposed M²PC and b) classical FCS-MPC.

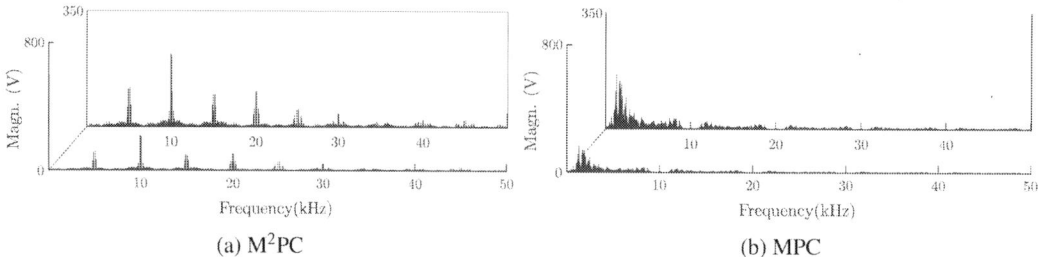

(a) M²PC   (b) MPC

Fig. 5: FFT of the output line-to-line voltage at nominal conditions v_{ab} for a) Proposed M²PC and b) classical FCS-MPC.

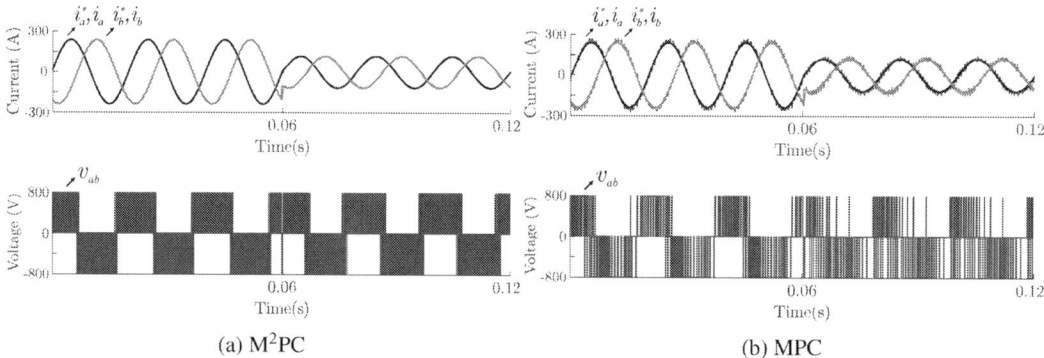

(a) M²PC   (b) MPC

Fig. 6: Reference and phase currents in *abc* coordinates and line-to-line voltage v_{ab} with a 50% decrease in the reference currents for nominal conditions for a) Proposed M²PC and b) classical FCS-MPC.

The previous results have shown that the proposed M²PC has good steady-state and transient performances for nominal conditions. Let us now perform the same tests for parametric variations on the plant. Now, the load inductance and resistance are decreased by 50%, to $R = 0.65\Omega$ and $L = 1$mH. The results are shown in Fig. 7, that depicts the output phase currents i_a and i_b, reference phase currents i_a^* and i_b^* and the line-to-line voltage v_{ab} in *abc* coordinates with parametric variations for the proposed M²PC and the classical FCS-MPC. The current tracking errors e_a and e_b for phases a and b in *abc* coordinates are shown in Fig.8 for the proposed M²PC and the classical FCS-MPC. It can be seen from these results that both strategies can track the current reference even with parametric variations; however, in this case, there is more ripple in the output currents and, consequently, the steady-state error is increased, specially for the classical FCS-MPC [11]. The output line-to-line voltages are also deteriorated. This is reflected in the harmonic content of the currents and voltages. The FFTs of the output current i_a and of the line-to-line voltage v_{ab} are shown, respectively, in Figs. 9 and 10, for both controllers. We can see that the spectra of the currents and voltages is worse than the ones presented for nominal conditions; however,

EPE'20 ECCE Europe

Assigned jointly to the European Power Electronics and Drives Association & the Institute of Electrical and Electronics Engineers (IEEE)

the proposed M^{2}PC still outperforms the classical FCS-MPC. The calculated current and voltage THDs are, respectively, 2.86% and 164.28% for the proposed M^{2}PC and 18.4% and 250.55% for the classical FCS-MPC. On the other hand, the current and voltage WTHDs are, respectively, 0.073% and 1.248% for the proposed M^{2}PC and 0.993% and 7.382% for the classical FCS-MPC.

As for the nominal conditions, a step change was applied to the reference current, decreasing it by 50%, from 238A to 119A, and with the same parametric variations. Fig. 11 shows the waveforms of the output phase currents i_a and i_b, reference phase currents i_a^* and i_b^* and the line-to-line voltage v_{ab} at nominal load conditions and with the step change in the current reference, respectively, for the proposed M^{2}PC and the classical FCS-MPC. Again, both controllers have a very fast dynamic response, also converging to the new desired steady-state operating point, and with the proposed M^{2}PC having a better steady-state response than the classical FCS-MPC.

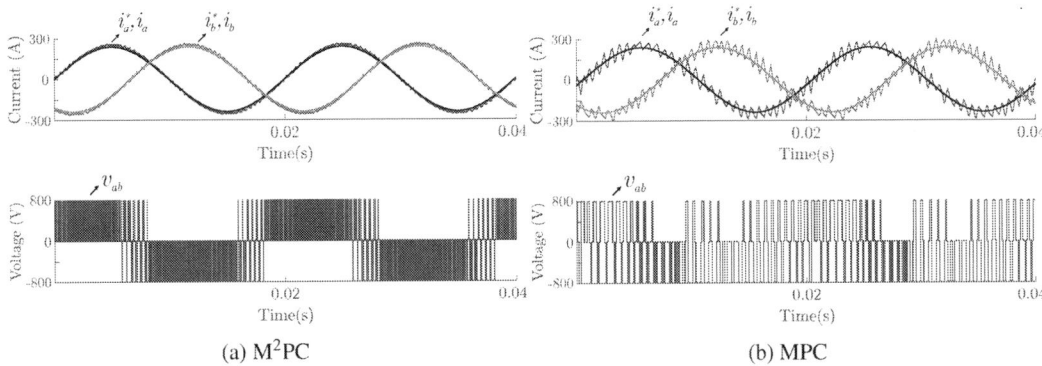

(a) M^{2}PC   (b) MPC

Fig. 7: Reference and phase currents in *abc* coordinates, and line-to-line voltage v_{ab} with 50% decrease in the load parameters for a) Proposed M^{2}PC and b) classical FCS-MPC.

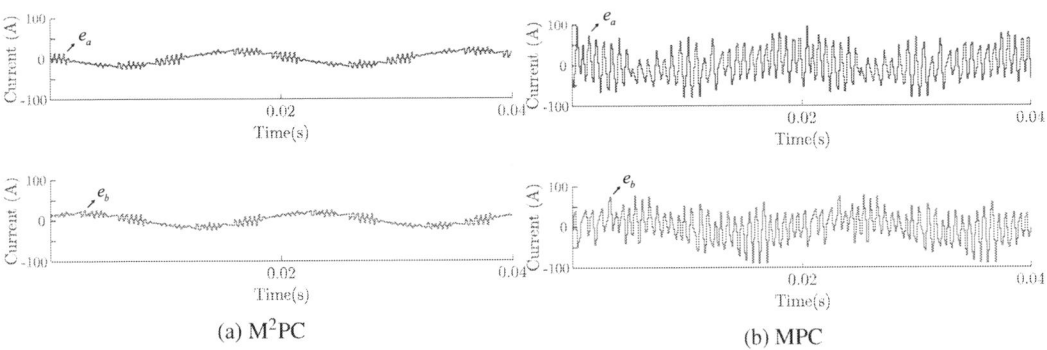

(a) M^{2}PC   (b) MPC

Fig. 8: Current tracking error e_a and e_b for phases a and b in *abc* coordinates with 50% decrease in the load parameters for a) Proposed M^{2}PC and b) classical FCS-MPC.

Conclusions

This paper presented a quadratic cost function for Modulated Model Predictive Control, M^{2}PC, designed such as to guarantee that the inverter under consideration will be stable for the desired steady-state operational conditions. Hardware-in-the-Loop, HIL, results were presented comparing the performance of the implemented M^{2}PC and a classical FCS-MPC for a two-level inverter with an RL load, considering the same cost function. Results were obtained for nominal operating conditions and for parametric variations on the load, both for steady-state and transients. The HIL results have demonstrated the good performance of the proposed approach, showing the stability and performance despite the model of the implemented M^{2}PC.

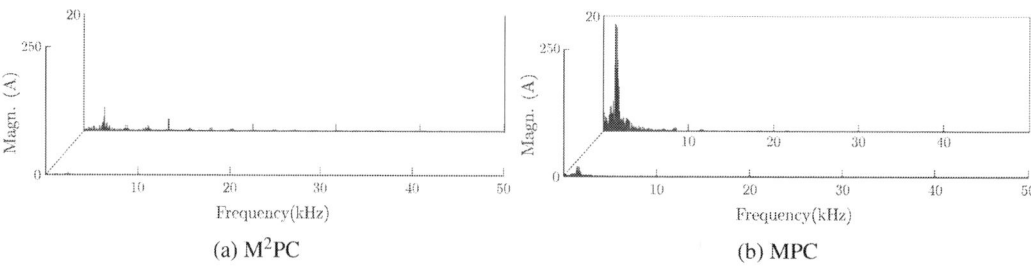

Fig. 9: FFT of the output current i_a with 50% decrease in the load parameters for a) Proposed M^2PC and b) classical FCS-MPC.

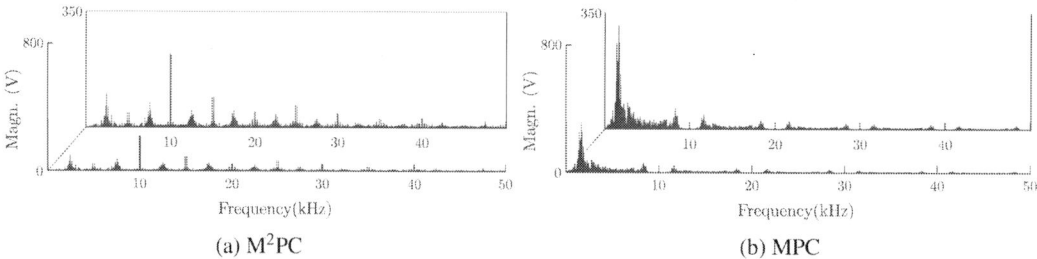

Fig. 10: FFT of the output line-to-line voltage v_{ab} with 50% decrease in the load parameters for a) Proposed M^2PC and b) classical FCS-MPC.

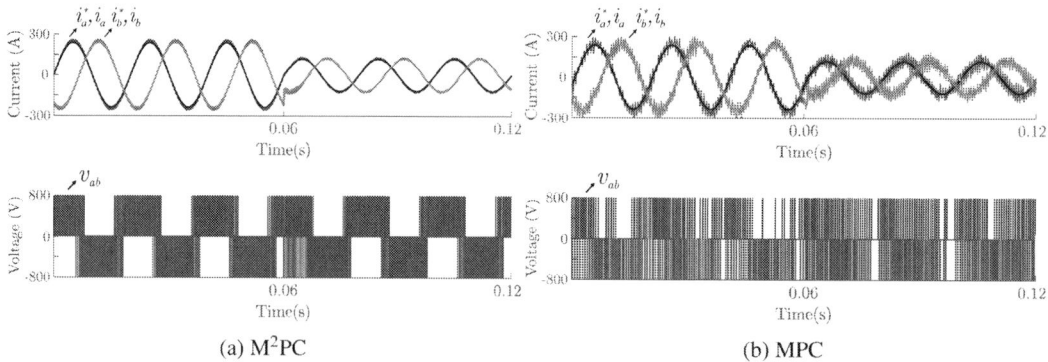

Fig. 11: Reference and phase currents in abc coordinates and line-to-line voltage v_{ab} with a 50% decrease in the reference currents for nominal conditions for a) Proposed M^2PC and b) classical FCS-MPC.

References

[1] D. G. Holmes, T. A. Lipo, B. P. McGrath and W. Y. Kong, "Optimized Design of Stationary Frame Three Phase AC Current Regulators," IEEE Transactions on Power Electronics, vol. 24, no. 11, pp. 2417-2426, Nov. 2009.

[2] Katsuhiko Ogata,"Discrete-Time Control Systems",1994,Prentice Hall.

[3] F. Donoso, A. Mora, R. Crdenas, A. Angulo, D. Sez and M. Rivera, "Finite-Set Model-Predictive Control Strategies for a 3L-NPC Inverter Operating With Fixed Switching Frequency," in IEEE Transactions on Industrial Electronics, vol. 65, no. 5, pp. 3954-3965, May 2018.

[4] J. Rodriguez et al., "State of the Art of Finite Control Set Model Predictive Control in Power Electronics," in IEEE Transactions on Industrial Informatics, vol. 9, no. 2, pp. 1003-1016, May 2013.

[5] S. Vazquez, J. Rodriguez, M. Rivera, L. G. Franquelo and M. Norambuena, "Model Predictive Control for Power Converters and Drives: Advances and Trends," in IEEE Transactions on Industrial Electronics, vol. 64, no. 2, pp. 935-947, Feb. 2017.

[6] R. E. Prez-Guzmn, M. Rivera and P. W. Wheeler, "Recent Advances of Predictive Control in Power Converters," 2020 IEEE International Conference on Industrial Technology (ICIT), Buenos Aires, Argentina, 2020, pp. 1100-1105, doi: 10.1109/ICIT45562.2020.9067169.

[7] T. Geyer and D. E. Quevedo, "Performance of Multistep Finite Control Set Model Predictive Control for Power Electronics," in IEEE Transactions on Power Electronics, vol. 30, no. 3, pp. 1633-1644, March 2015, doi: 10.1109/TPEL.2014.2316173.

[8] C. Zheng, T. Dragievi, B. Majmunovi and F. Blaabjerg, "Constrained Modulated Model-Predictive Control of an LC-Filtered Voltage-Source Converter," in IEEE Transactions on Power Electronics, vol. 35, no. 2, pp. 1967-1977, Feb. 2020.

[9] C. Zheng, T. Dragievi, B. Majmunovi and F. Blaabjerg, "Constrained Modulated Model-Predictive Control of an LC-Filtered Voltage-Source Converter," in IEEE Transactions on Power Electronics, vol. 35, no. 2, pp. 1967-1977, Feb. 2020.

[10] D.Q. Mayne and J. B. Rawlings and C.V. Rao and P.O.M Scokaert, "Constrained model predictive control: Stability and optimality", Automatica, vol. 36, no. 6, pp. 789-814, June 2000.

[11] H. A. Young, M. A. Perez and J. Rodriguez, "Analysis of Finite-Control-Set Model Predictive Current Control With Model Parameter Mismatch in a Three-Phase Inverter," in IEEE Transactions on Industrial Electronics, vol. 63, no. 5, pp. 3100-3107, May 2016.

[12] S. Vazquez et al., "Model Predictive Control for Single-Phase NPC Converters Based on Optimal Switching Sequences," in IEEE Transactions on Industrial Electronics, vol. 63, no. 12, pp. 7533-7541, Dec. 2016.

[13] S. Vazquez, P. Acuna, R. P. Aguilera, J. Pou, J. I. Leon and L. G. Franquelo, "DC-Link Voltage-Balancing Strategy Based on Optimal Switching Sequence Model Predictive Control for Single-Phase H-NPC Converters," in IEEE Transactions on Industrial Electronics, vol. 67, no. 9, pp. 7410-7420, Sept. 2020.

[14] R. P. Aguilera, P. Lezana and D. E. Quevedo, "Finite-Control-Set Model Predictive Control With Improved Steady-State Performance," in IEEE Transactions on Industrial Informatics, vol. 9, no. 2, pp. 658-667, May 2013.

[15] L. Tarisciotti, P. Zanchetta, A. Watson, J. C. Clare, M. Degano and S. Bifaretti, "Modulated Model Predictive Control for a Three-Phase Active Rectifier," in IEEE Transactions on Industry Applications, vol. 51, no. 2, pp. 1610-1620, March-April 2015, doi: 10.1109/TIA.2014.2339397.

[16] D. Quevedo and G. Goodwin and M. Rivera and J. de. Don, "Finite constraint set receding horizon quadratic control," International Journal of Robust and Nonlinear Control, vol. 14, no. 4, pp. 355-377, March. 2004.

[17] T. Geyer and D. E. Quevedo, "Multistep Finite Control Set Model Predictive Control for Power Electronics," in IEEE Transactions on Power Electronics, vol. 29, no. 12, pp. 6836-6846.

[18] R. P. Aguilera and D. E. Quevedo, "Stability Analysis of Quadratic MPC With a Discrete Input Alphabet," in IEEE Transactions on Automatic Control, vol. 58, no. 12, pp. 3190-3196, Dec. 2013.

[19] R. P. Aguilera and D. E. Quevedo, "Predictive Control of Power Converters: Designs With Guaranteed Performance," in IEEE Transactions on Industrial Informatics, vol. 11, no. 1, pp. 53-63, Feb. 2015.

[20] T. Geyer, R. P. Aguilera and D. E. Quevedo, "On the stability and robustness of model predictive direct current control," 2013 IEEE International Conference on Industrial Technology (ICIT), Cape Town, 2013, pp. 374-379.

[21] T. Geyer, R. P. Aguilera and D. E. Quevedo, "On the stability and robustness of model predictive direct current control," 2013 IEEE International Conference on Industrial Technology (ICIT), Cape Town, 2013, pp. 374-379.

[22] M. Parvez, S. Mekhilef, N. M. L. Tan and H. Akagi, "A robust modified model predictive control (MMPC) based on Lyapunov function for three-phase active-front-end (AFE) rectifier," 2016 IEEE Applied Power Electronics Conference and Exposition (APEC), Long Beach, CA, 2016, pp. 1163-1168.

[23] H. T. Nguyen, J. Kim and J. Jung, "Improved Model Predictive Control by Robust Prediction and Stability-Constrained Finite States for Three-Phase Inverters With an Output LC Filter," in IEEE Access, vol. 7, pp. 12673-12685, 2019.

[24] A. Mora, R. Crdenas-Dobson, R. P. Aguilera, A. Angulo, F. Donoso and J. Rodriguez, "Computationally Efficient Cascaded Optimal Switching Sequence MPC for Grid-Connected Three-Level NPC Converters," in IEEE Transactions on Power Electronics, vol. 34, no. 12, pp. 12464-12475, Dec. 2019.

[25] M. Aguirre, S. Kouro, C. A. Rojas, J. Rodriguez and J. I. Leon, "Switching Frequency Regulation for FCS-MPC Based on a Period Control Approach," in IEEE Transactions on Industrial Electronics, vol. 65, no. 7, pp. 5764-5773, July 2018.

[26] M. Aguirre, S. Kouro, C. A. Rojas and S. Vazquez, "Enhanced Switching Frequency Control in FCS-MPC for Power Converters," in IEEE Transactions on Industrial Electronics.

A Robust Fuzzy-based Control Technique for Wind Farm Transient Voltage Stability Using SVC and STATCOM: Comparison Study

Reza Ebrahimi
Sahand University of Technology
Tabriz, Iran
r_ebrahimi@sut.ac.ir

Vahid Eslampanah
Sahand University of Technology
Tabriz, Iran
v_eslampanah@sut.ac.ir

Hossein Madadi Kojabadi
Sahand University of Technology
Tabriz, Iran
hmadadi64@yahoo.ca

Mohammadreza Azizian
Sahand University of Technology
Tabriz, Iran
azizian@sut.ac.ir

Naser Nourani Esfetanaj
Aalborg University
Aalborg, Denmark
nne@et.aau.dk

Dao Zhou
Aalborg University
Aalborg, Denmark
zda@et.aau.dk

Keywords

«Wind Farm», «Transient Stability», «SVC», «STATCOM», «Fuzzy Logic».

Abstract

The wind farm is subjected to various fault conditions to observe voltage and power fluctuations. To improve the transient voltage stability, under various fault conditions, STATCOM and SVC are used which are controlled by fuzzy logic controller also be compared with PI controller.

Introduction

Wind energy is now widely used as a non-polluting energy source. However, its natural effects on power quality and network stability are undeniable. Many power grids use rotary generators that, under normal conditions of speed control and excitation field, can withstand the gradual changes in power supply and maintain frequency and voltage constant; However, if transient events such as sudden load arrivals, renewable energy outages, and system failures occur, the generators are subject to greater stress. Generators try to adjust the mechanical power of the input according to the output changes. The speed controller inside the generator, which is responsible for regulating the frequency of the system, tries to track the actual power changes quickly. However, some limitations due to mechanical motion cause the generator to delay in response to the instantaneous output power changes. Several instability fluctuations can occur and can be considered as one of the critical conditions that cause the whole system to be unstable. This instability is one of the problems of the system of mixed generation and renewable energy. The protection system may identify the abnormal conditions listed as a problem and remove existing resource components from the network [1].

Recently, flexible AC transmission system (FACTS) have been introduced to instantaneously stabilize system voltage [2]. In [3], a UPFC is investigated to improve the LVRT capability of a DFIG-equipped wind power system during grid-side voltage drop fault. The results show that UPFC significantly improves the LVRT capability of the wind energy conversion system. The effect of STATCOM on the damping of power oscillations and the increase in transmission power has been investigated in [4] and it has been shown that using STATCOM improves the stability of the wind turbine connected to the power system after various perturbations including wind farm mechanical power changes. In [5], a simple method for evaluating the rated power of STATCOM for regulating the voltage level of point of common coupling (PCC) is presented. Vector control is used as a robust control for the reactive power injection required by STATCOM. Reference [6] compared IPFC and UPFC controllers using a small signal model to improve the transient voltage stability of the power system. The damping capability of both controllers is also evaluated. Reference [7] introduces a new method for improving the dynamic behavior of power systems using SVC and variable power absorber. The SVC installed to improve power quality is combined with a power absorber that improves system performance.

In this paper, the effect of wind farm presence on power system is investigated by considering the wind shear effect and tower shadow. For this purpose, the power system is subjected to various fault conditions to observe voltage and power fluctuations. The fuzzy logic controller for STATCOM and SVC is used to improve the transient voltage stability under various fault conditions. Finally, the performance of STATCOM and SVC controlled with PI controller and fuzzy logic controller will be compared.

Wind turbine model

The mechanical power and aerodynamic torque delivered by the turbine are as follows:

$$P_w = \frac{\pi \rho r^2}{2} v_{wind}^3 C_p (\lambda, \beta) \tag{1}$$

Where P_w is the output mechanical power of the turbine in W, C_p is power coefficient, ρ is the air density in kg/m^3 and v_{wind} is the wind speed in m/s. C_p in addition to the aerodynamic shape of the blades is a function of both tip-speed-ratio λ and the blade pitch angle β. ω_m is the angular turbine speed in rad/s.

$$\lambda = \frac{\omega_m r}{v_{wind}} \tag{2}$$

Wind Shear effect

Wind speed changes with increasing and decreasing height are called wind shear effect. Because of periodic changes in wind speed due to height change, torque fluctuations and subsequently power fluctuations occur. It will also vary in torque and power due to each blade being exposed to different wind conditions during a full cycle [8]. The blade upward is exposed to higher wind speeds than the blade downward. During a rotation, the torque fluctuates three times as each blade is once exposed to the maximum wind speed. Therefore, modeling of torque oscillation due to wind shear effect is very important in studying wind turbine systems. The common model of wind shear effect extracted directly from the literature is as follows [9]:

$$V(z) = V_H \left(\frac{z}{H} \right)^\alpha \tag{3}$$

Equation 3 is written as a function of r (radial distance from the rotor axis) and θ (side angle):

$$V(r,\theta) = V_H \left(\frac{r\cos\theta + H}{H} \right)^\alpha = V_H \left[1 + W_s(r,\theta) \right] \tag{4}$$

Where V_H is the wind speed at the center of the rotor, W_s is the function of the shear shape of the wind, α is the experimental number of the shear effect, H is the height of the center of rotor and z is the elevation above ground level. The expression W_s is a disturbance in wind speed due to the shear effect of the wind which is added to the wind speed at the height of the center of the rotor.

Tower shadow effect

In the presence of the tower, the wind distribution varies. For rotors facing the wind, the wind shifts in front of the tower, thus reducing the blade torque facing the tower. This is called the shadow effect. Torque fluctuations caused by tower shadows become very important when the turbine blades are in the wind direction and the wind turns in the opposite direction [10]. The wind field which only the shadow effect of the tower is considered, defined as follows:

$$V(y,x) = V_H + v_{tower}(y,x) \tag{5}$$

The expression v_{tower} is the oscillation observed at the wind speed due to the shadow of the tower, which is added to V_H. In [11], tower disturbance is modeled using the potential flow theory for wind movements around the tower. The following relation is obtained using Figure 1(a).

$$v_{tower}(y,x) = V_0 a^2 \frac{y^2 - x^2}{\left(x^2 + y^2 \right)^2} \tag{6}$$

In the above relation, V_0 is the mean spatial wind speed, α is the radius of the tower, y is the lateral distance between the middle of the blade and the middle of the tower, and x is the distance between the blade and the middle of the tower.

STATCOM Modeling

STATCOM is a GTO and IGBT based voltage converter that is powered by batteries and is usually connected to a grid in parallel by a transformer [12]. The reactive power relationship in STATCOM is as follows:

$$Q_{sh} = \frac{|V_N|^2}{X_{sh}} - \frac{|V_N||V_{sh}|}{X_{sh}} \cos(\theta_N - \theta_{sh}) = \frac{|V_N|^2 - |V_N||V_{sh}|}{X_{sh}} \tag{7}$$

Where V_N network voltage, V_{sh} controller voltage, X_{sh} reactance between the controller and the network, θ_{sh}, θ_N are the angle of the network voltage and the controller voltage, respectively. If in the above relation $|V_N| > |V_{sh}|$, Q_{sh} is positive and STATCOM absorbs reactive power and if $|V_N| < |V_{sh}|$ is negative Q_{sh} and STATCOM injects reactive power into the network.

STATCOM is a static synchronous generator whose inductive or capacitive output current is controlled independently of the ac system voltage. STATCOM is a voltage source inverter that generates a three-phase ac voltage set and connects to the ac system voltage through a relatively small reactor. By controlling the output voltage amplitude, the voltage exchange between STATCOM and the power system can be control. STATCOM has three operational areas. The structure of the STATCOM and how it is connected to the power system are shown in figure 1(b) and its characteristic curve illustrated in figure 1(c).

1. If the STATCOM output voltage amplitude is greater than the ac system voltage amplitude, the current flows through the reactor to the ac system and STATCOM delivers reactive power to the system.
2. If the output voltage of the STATCOM is lower than the output voltage of the ac system, the reactive current flows from the ac to the STATCOM and the STATCOM absorbs reactive power.
3. If the output voltage of STATCOM is equal to the ac system voltage, the reactive power exchange is zero.

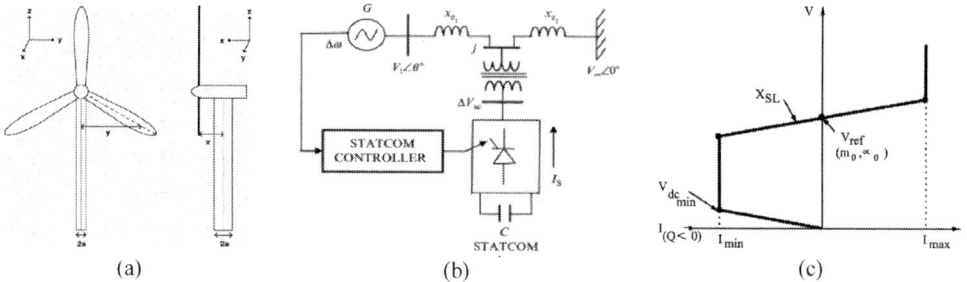

(a)   (b)   (c)

Figure 1. (a) Dimensions used in (6), (b) Structure of the STATCOM and how it is connected to the power system, (c) V-I characteristic curve.

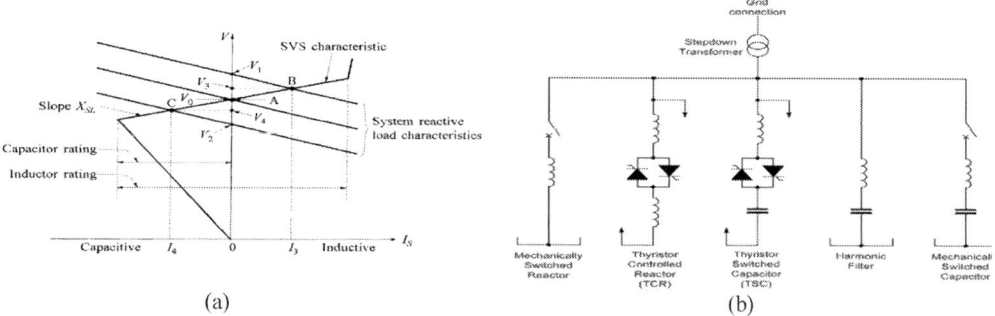

(a)   (b)

Figure 2. (a) characteristic of TCR / TCS, (b) SVC structure composed of TSC and TCR

SVC Modeling

The SVC consists of a thyristor controlled reactor (TCR) and a thyristor switching capacitor (TSC). The TSCs are group-switched and the TCR is controlled by a thyristor. The dynamic change of reactive power is controlled by the SVC of the shunt voltage to which the SVC is connected. Under normal condition, continuous reactive power control is performed by the TCR, which is powered by a reactor and two parallel thyristors. The reactive power absorbed by the TCR is calculated as follows:

$$Q_L = \frac{2\beta - \sin 2\beta}{\pi \omega L} V^2 \qquad (8)$$

Where β is the thyristor fire angle and must be $\left(\beta \in \left[0, \frac{\pi}{2} \right] \right)$ and ω is the base angular frequency. TSC is a combination of step capacitors that provide the reactive power of each step to the system as follows:

$$Q_C = \omega C V^2 \qquad (9)$$

Where C is the capacitance. From Equation (8) and (9) we can obtain the SVC reactive injection power to the system as follows:

$$Q_{SVC} = Q_C - Q_L = \left(\omega C - \frac{2\beta - \sin 2\beta}{\pi \omega L} \right) V^2 \qquad (10)$$

According to (10) the SVC composite characteristic of TCR / TCS will be plotted in Fig. 2(a). the SVC structure composed of TSC and TCR is presented in fig. 2(b). When the system voltage is higher than the rated voltage, the SVC absorbs the maximum reactive power from the system and works with pure self-propulsion. If the system voltage is less than the nominal value, the SVC injects the maximum reactive power into the system and works with a pure capacitive characteristic.

Fuzzy Logic Controller

Each fuzzy controller consist of four steps: 1. Fuzzification, 2. Rule base, 3. Inference mechanism, 4. Defuzzification. The block diagram of the control for STATCOM is shown in figure. 3(a), where it is called the Supplementary Secondary Control Ring, which controls the DC link capacitor voltage by adjusting the output voltage phase angle of the STATCOM. The control structure of the fuzzy-based controller for SVC is depicted in figure. 3(b), where the fuzzy logic controller inputs are the error between the reference voltage and the measured voltage and its derivative. The output of fuzzy controller is suspension dB.

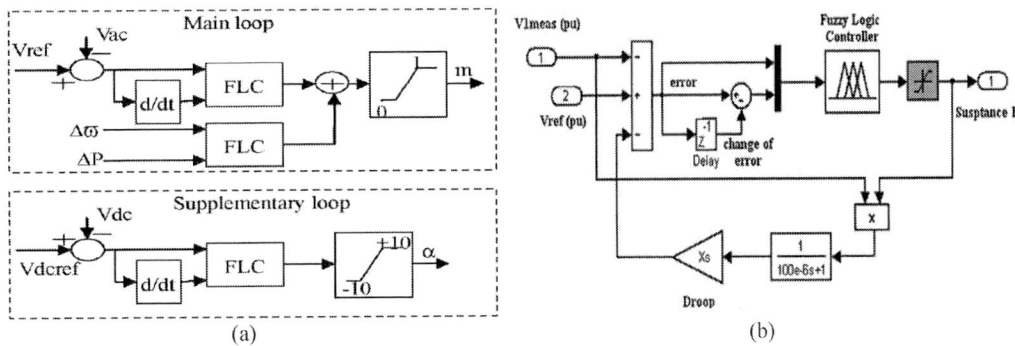

(a)   (b)

Figure 3. The block diagram of fuzzy controller for (a) STATCOM and (b) SVC.

Simulation Results

The simulation model consists of 6×1.5 MW wind turbines connected to a 25 kV distribution system that injects power through a 25 kV feeder into a 120 kV grid. The dynamic behavior of the studied system is investigated in two cases. In the first case, the effect of increasing and decreasing wind speed on transient voltage stability will be studied. In the second case, a three-phase error is applied to the B25 shin to evaluate the system behavior during the fault.

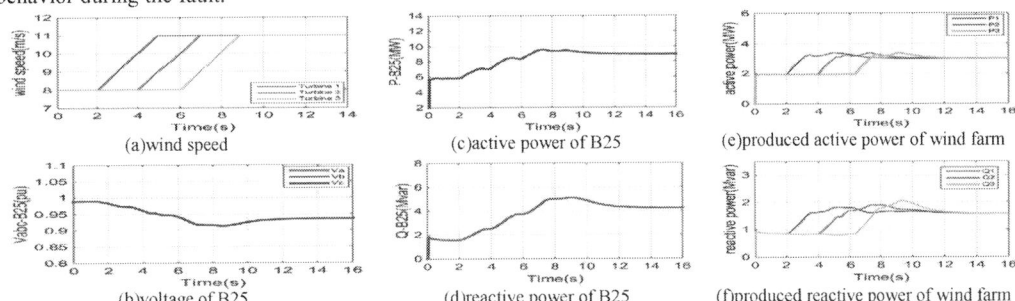

Figure 4. The changing system specifications for wind speed changes.

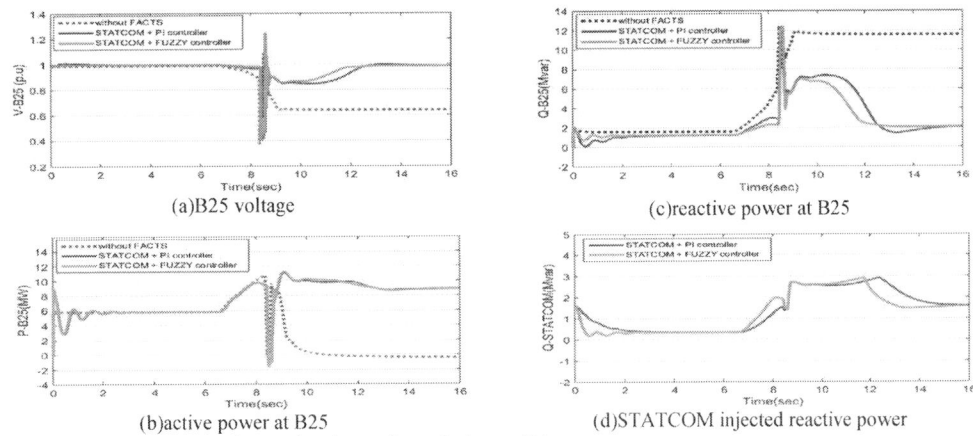

Figure 5. STATCOM operation under three-phase fault conditions

Whenever the reactive power absorbed by the induction generator is increased, the B25 voltage is reduced, so we use FACTS controllers to compensate for the reactive power consumed in the system. In order to investigate the effects of wind speed change in short-circuit conditions, a short-circuit fault is simulated at the high voltage terminal of the wind turbine transformer. Figures 5-6 shows the status of the system during short circuit. As shown in Figs. 5-6, after the three-phase error the voltage drops to 0.64 p.u, the rotor speed also increases rapidly. Therefore, it can be said that wind speed variations cause instability of the wind farm and its associated power system.

However, STATCOM and SVC prevents voltage instability and increases the voltage to an acceptable level (0.985 p.u) by injecting reactive power into the grid.

According to the results it can be seen that in the presence of STATCOM and SVC the wind farm is stable, so that after the fault, the voltage returns to the acceptable range (0.985 p.u). The results show that STATCOM and SVC keep the voltage in the acceptable range by injecting 1.6 and 1.8 MW of reactive power into B25, respectively. The results also show that the speed of STATCOM response is higher than that of SVC, due to the use of electronic power converter in STATCOM structure.

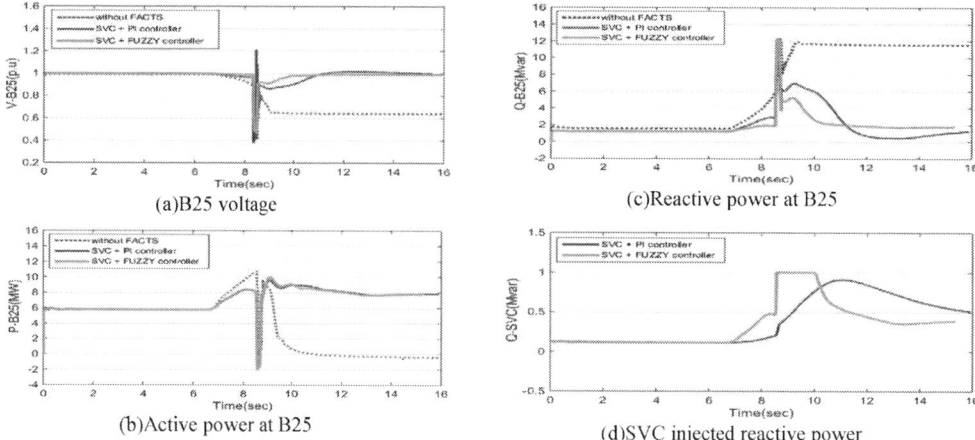

(a)B25 voltage

(c)Reactive power at B25

(b)Active power at B25

(d)SVC injected reactive power

Figure 6. SVC operation under three-phase fault conditions

Conclusion

Reactive power flow analysis for voltage stabilization clearly shows that when using STATCOM for stable performance, reactive power and settling time are low. The maximum capacitance generated by the SVC is proportional to the square of the system voltage (constant suspension), while the maximum capacitance generated by the STATCOM (constant current) decreases linearly with the voltage. Capacitive power output during error is a very important advantage of STATCOM comparing to SVC. In addition, STATCOM usually has a faster response than SVC because there is no delay proportional to thyristor firing (4 milliseconds in SVC) using voltage source converters in STATCOM. Therefore, STATCOM performs reactive power compensation for very fast voltage regulation in case of emergency.

References

[1]   P. Kundur, N. J. Balu, and M. G. Lauby, *Power system stability and control* vol. 7: McGraw-hill New York, 1994.

[2]   X.-P. Zhang, C. Rehtanz, and B. Pal, *Flexible AC transmission systems: modelling and control*: Springer Science & Business Media, 2012.

[3]   Y. M. Alharbi and A. Abu-Siada, "Application of UPFC to improve the low-voltage-ride-through capability of DFIG," in *2015 IEEE 24th International Symposium on Industrial Electronics (ISIE)*, 2015, pp. 665-668.

[4]   M. R. Nasiri, S. Farhangi, and J. Rodríguez, "Model predictive control of a multilevel CHB STATCOM in wind farm application using diophantine equations," *IEEE Transactions on Industrial Electronics,* vol. 66, pp. 1213-1223, 2018.

[5]   M. Mahfouz and M. A. El-Sayed, "Static synchronous compensator sizing for enhancement of fault ride-through capability and voltage stabilisation of fixed speed wind farms," *IET Renewable Power Generation,* vol. 8, pp. 1-9, 2014.

[6]   S. Jiang, A. M. Gole, U. D. Annakkage, and D. Jacobson, "Damping performance analysis of IPFC and UPFC controllers using validated small-signal models," *IEEE Transactions on Power Delivery,* vol. 26, pp. 446-454, 2010.

[7]   P. Srithorn and N. Theejanthuek, "The Enhanced Performance SVC for Transient Instability Oscillation Damping," *Energy Procedia,* vol. 56, pp. 510-517, 2014.

[8]   E. Mohammadi, R. F. Bahramjerdi, H. R. Naji, and G. Moschopoulos, "Investigation of Horizontal and Vertical Wind Shear Effects Using a Wind Turbine Emulator," *IEEE Transactions on Sustainable Energy,* 2018.

[9]   T. Thiringer and J.-A. Dahlberg, "Periodic pulsations from a three-bladed wind turbine," *IEEE Transactions on Energy Conversion,* vol. 16, pp. 128-133, 2001.

[10]   Z. Xie, Z. Xu, X. Zhang, S. Yang, and L. Wang, "Improved power pulsation suppression of DFIG for wind shear and tower shadow effects," *IEEE Transactions on Industrial Electronics,* vol. 64, pp. 3672-3683, 2017.

[11]   P. Sørensen, A. D. Hansen, and P. A. C. Rosas, "Wind models for simulation of power fluctuations from wind farms," *Journal of wind engineering and industrial aerodynamics,* vol. 90, pp. 1381-1402, 2002.

[12]   R. Fadaeinedjad, G. Moschopoulos, and M. Moallem, "Using STATCOM to mitigate voltage fluctuations due to aerodynamic aspects of wind turbines," in *2008 IEEE Power Electronics Specialists Conference*, 2008, pp. 3648-3654.

Temperature Evolution as an effect of Wire-bond Failures in a Multi-Chip IGBT Power Module

N. Degrenne, R. Delamea, S. Mollov
Mitsubishi Electric R&D Centre Europe
1, allée de Beaulieu
Rennes, France
n.degrenne@fr.merce.mee.com
https://www.mitsubishielectric-rce.eu/

Keywords

«IGBT», «Reliability», «Modelling», «Diagnostics».

Abstract

Multi-chip power switches consist of several dies in parallel. The temperature of these dies is not equal and evolves during ageing, in part as a consequence of wire-bond degradation. The temperature distribution may impact the condition monitoring and the overall reliability. It is thus necessary to understand how the temperature distribution is modified by wire-bond degradation. In this paper, wire-bond lift-offs are reproduced experimentally by sequentially sectioning the hottest wire of the hottest die of a multi-chip IGBT module. The results show that at the beginning of the degradation, hot spots move away from the more recent wire-bond lift-off, contributing to temperature equalization. However, this effect collapses as fewer bonds remain attached towards the end of the module life.

Introduction

In power cycling experiments, the temperature of switches needs to be estimated. When a switch is composed of several dies in parallel, the dies may not have the same temperature. In the absence of individual monitoring of each die, it is customary to use V_{ON} as a TSEPs (Thermo-Sensitive Electrical Parameters). However, the temperature of each individual die is unknown. The initial temperature unbalance is attributable to a number of factors, from die parameter dispersion, to manufacturing tolerances [1].

When the module ages due to cyclic thermo-mechanical stress, wire-bonds degrade and detach from the die, modifying the current distribution among the dies, and thus modifying the loss and temperature distribution. The main question addressed in this paper is: "How does the chip temperature distribution evolve in multi-chip power modules during power cycling, and in field ageing?"

In the prior art, some papers address the validity of temperature estimations of multi-chip power modules during ageing under switching conditions. The evolution of the estimated temperature (i.e. with V_{ON}) with wire degradation is assessed in the case of a single [2] and multi-chip [3, 4] power module respectively. The reference [3] observes the temperature over-estimation with V_{ON} at high current value. The reference [4] observes and models the temperature under-estimations with V_{ON} at low current value, with and without correction based on the electrical resistance increase estimated for example at the Zero Temperature Coefficient (ZTC) current value. In [4], the evolution of the temperature distribution as a consequence to wire degradation is modelled, and experimentally verified by cutting the wires of a die in a 3-die power module. This sequence has a low probability to occur, but offers the advantage to ease the calibration of the models.

While most papers concentrate on the validity of temperature estimation, there is a lack of literature on how the die temperatures evolve as a consequence to degradation during DC or switching power cycling.

In this paper, a lift-off sequence of wire-bonds is experimentally reproduced, and the consequences on the temperature distribution are observed and analysed. The sectioning adopts the assumption that the hottest wire will fail first and is performed until power module functional failure. The failure instant was captured optically and electrically, and is analysed.

Experimental protocol

An industrial six-pack 1200V/150A IGBT power module was used as a DUT (Device Under Test). Its gel was removed, and the surface was black painted. The 3 phases were connected in parallel, to produce a single half-bridge leg with 3 dies in parallel.

The DUT was used in a back-to-back water-cooled power converter as previously described in [5] in unipolar modulation mode operating at 600 V, 80 A, 15 kHz, and a duty-cycle of 0.5. This operating point was selected to generate a reasonably high (about 60K) temperature rise on the chips. Note that the operating current is below the ZTC current value that is at around 120A for this 3-phase module. This means that current will be naturally steered towards the hottest die, amplifying the temperature mismatches.

A thermal camera was used to capture the average temperature of the area of each die. After identification of the hottest die, the hottest wire-bond was identified and sectioned. This sectioning order was selected because the hottest wire-bonds are typically the most subject to CTE-induced shear stain and stress. In between each sectioning, large and short (150A/200μs) current pulses were generated at a constant heat-sink temperature, so as to capture in-situ the electrical resistance increase associated with the cuts. The experimental protocol is shown in Fig. 1.

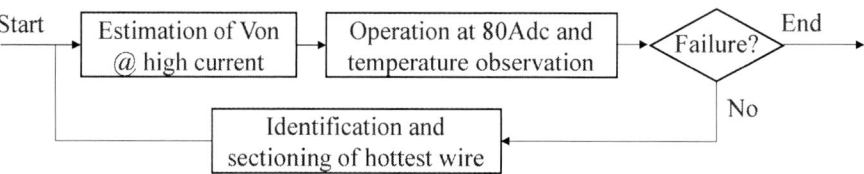

Fig. 1: Experimental protocol

Observation and analysis of temperature distribution

The temperature distributions are shown in Fig. 2, with infrared imaged on the right. The y-axis show the individual die temperatures normalised to their average and the x-axis shows the proportion of cut wire-bonds normalised to the total number for the 3-die switch. The cross indicates to which die the wire-bond being cut belongs to.

The graph can be divided in 3 regions. The first part is the **balancing region**. Initially, when 0% of the wire-bonds are cut (instance **a**), the die B is the hottest (since it is in the middle of the power module), and the die A the coldest. From 0% to 14% degradation, the wire-bonds of die B are cut, until the temperature of the die B and C equalises. From 14% to 30%, the dies C and B are cut, until their temperature equalize the one of the die A (instance **b**). This balancing phase can be understood with the insight of [4]. The wire sectioning increases the electrical resistance of the top connection to the concerned die. As a consequence, a portion of the current and of the associated conduction losses is redistributed to the other parallel dies, causing the die temperatures to converge.

The second part is the **balanced region**. From 30% to 56%, the die temperatures are well balanced (instances **b** to **c**).

The third part is the **diverging region**. At 56% of degradation, the die A is the less damaged, and probably the one passing the largest current, so as to compensate for its better thermal performances. It is subjected to the sectioning of its 4th wire-bond. Its resistivity increases significantly, and the current is redistributed. As a result, its temperature decreases below the one of dies B and C. The same phenomenon occurs several times (instances **e** to **k** in the graph, **k** being the instance of catastrophic failure) from 56% degradation to 100% degradation.

The infrared images illustrate the temperature distribution in the later stages of the power module life. Instant **e** indicates slight temperature divergence across the dies and also across their surface. As degradation progresses, the system nevertheless attempts to converge, for example between instances **g-h**. However, as fewer wire-bonds are available to redirect the current, successive cuts result in large temperature excursions (instances **f**, **g** and **j-k**). Instant **j** captures the state in which die B has no longer connected wire-bonds and dies A and C are hotter and exhibit large temperature gradients. The latter is attributed to the fact that very few (or singular) wire-bonds bring the current in a very localised manner.

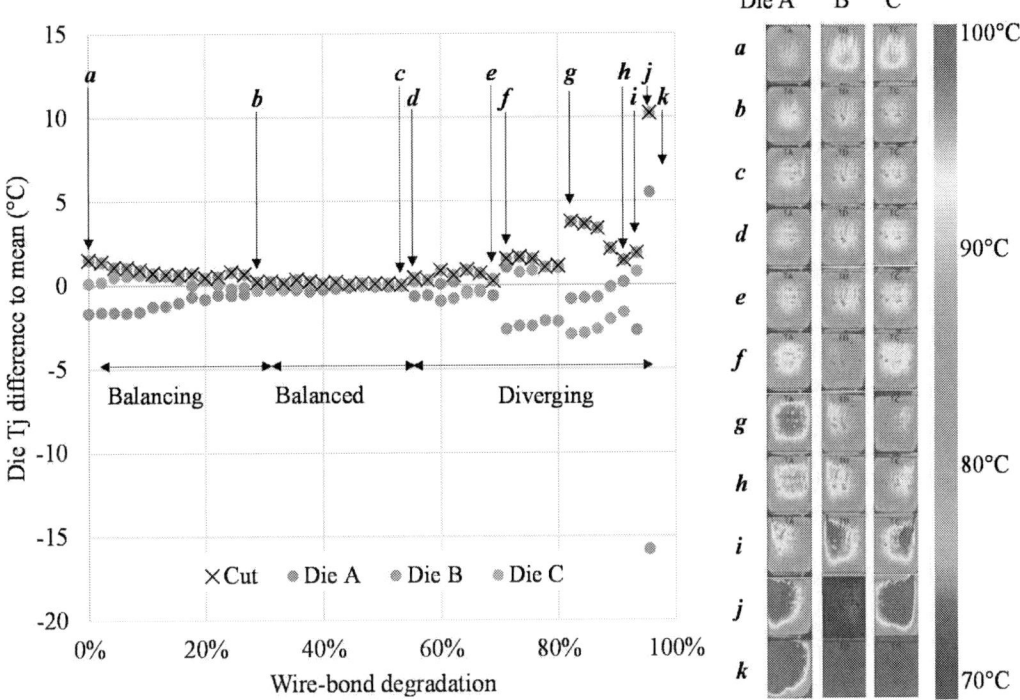

Fig. 2: Temperature distribution with wire-bond degradation and corresponding selection of images (obtained with the IR camera)

Figure 3 shows the average temperature obtained with the infra-red camera of the 3 dies with wire-bond degradation progression. Few data points have lower temperature of around 1°C. This is attributed to the application of the experimental protocol but has no influence on the results of Fig. 2. Otherwise, the mean temperature is rather constant, though slightly increasing towards the end of the experiment.

The temperature and electrical resistance increases are typically considered as failure precursors and their values can be extrapolated in order to estimate the end of life as in [6]. This experiment shows that the average temperature has low sensitivity to damage attributable to wire-bond lift-off. However, the temperature mismatch between dies in the unbalancing phase may allow to identify the end of life if individual die temperature can be assessed such as in [7].

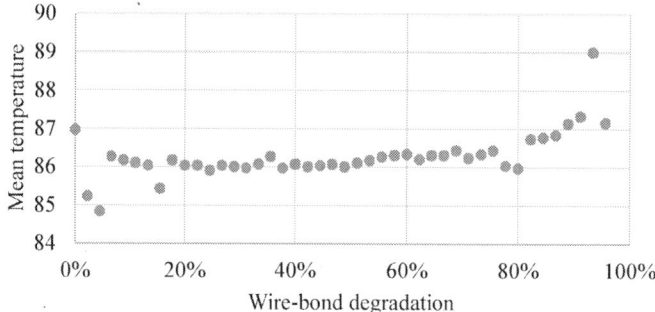

Fig. 3: Average temperature value of the 3 dies with wire-bond degradation (obtained with the IR camera)

Figure 4 represents the measured in-situ electrical resistance increase as the number of cuts increases. The three degradation regions observed in Fig. 3 can be observed as well on the electrical resistance increase. The large temperature deviations observed in the temperature unbalancing region, for example instances f and g, are accompanied by sharp electrical resistance increases. Such electrical resistance progression is also observed in traction power modules with 8 or more dies in parallel.

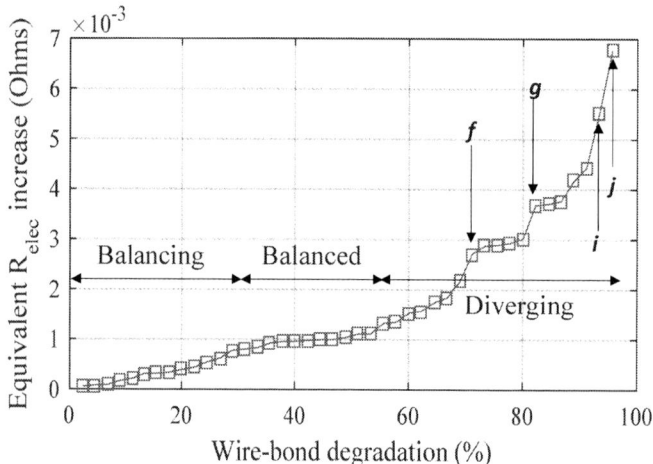

Fig. 4: Electrical resistance increase with wire-bond degradation

The applicability of these results to power cycling relies on two main hypotheses:

1. The wear-out concerns the wires only (i.e. no die-attach or baseplate degradation).
2. The hottest wire-bond is the next to fail.

In practice, the module failure is often a combination of several failure mechanisms. For example, the die-attach degradation cannot be reproduced artificially in the manner employed above. It will either have to be reproduced by accelerated ageing coupled with non-destructive analysis or by simulation. Another degradation mechanism not taken into account by our experiment is the increase of the resistivity of the top metallization, which will further impede the current distribution and enhance the temperature dispersion after each lift-off across the die area.

Observation and analysis of catastrophic failure

At the instant of failure, only one wire-bond was remaining on the die A conducting the entire load current, while the dies B and C with initial worst thermal performances are disconnected as a result to wire sectioning. Note that die A was initially the coldest die (Fig. 2).

The current was increased progressively, and failure occurred for a current of 70A. Figure 5-6 show the evolution of the surface temperature of the die just prior to the failure, captured by the high-speed infrared and optical camera respectively. The hot spot remains limited to the stich region and very little heat is transported to the rest of the dies surface. The last frame captures the metallic melt spray generated during the disintegration of the joint. The failure mode is the thermal runaway of the remaining wire-bond. The melting of aluminium and the ensuing arc lead to the complete failure of the power module in the form of a catastrophic event.

Figure 7 shows the electrical waveforms at the instant of the failure. The gate-to-emitter and collector-to-emitter voltage correspond to the left top-switch on which the wires were cut. The electrical waveforms reveal the following sequence of events illustrated in Fig. 8:

t1. Failure of the left top switch in short-circuit during its on-state
t2. Turn-on of the bottom switches leading to a type 1 arm short-circuit on the left leg
t3. Turn-off of the left bottom switch by the 'desat' protection leading to a wrong state of the left arm and to the increase of the load current through the top left and the bottom right switches
t4. Turn-on of the top switches by the gate drivers leading to a normal state of the arms (i.e. top right and left), across the load and to the stabilization of the current
t5. Second arm short-circuit eventually leading to an explosion

In this experiment, the traditional 'desat' protection was thus ineffective on the failed die since it is uncontrollable after its failure in short-circuit. Even if the complementary die was effectively turned-off, the current was still high and lead to a thermal runaway. It is also likely that a protection based on over-temperature detection would have been ineffective since the temperature over the 3 dies is low before explosion. On the other hand, estimation of the temperature of each individual die may be effective.

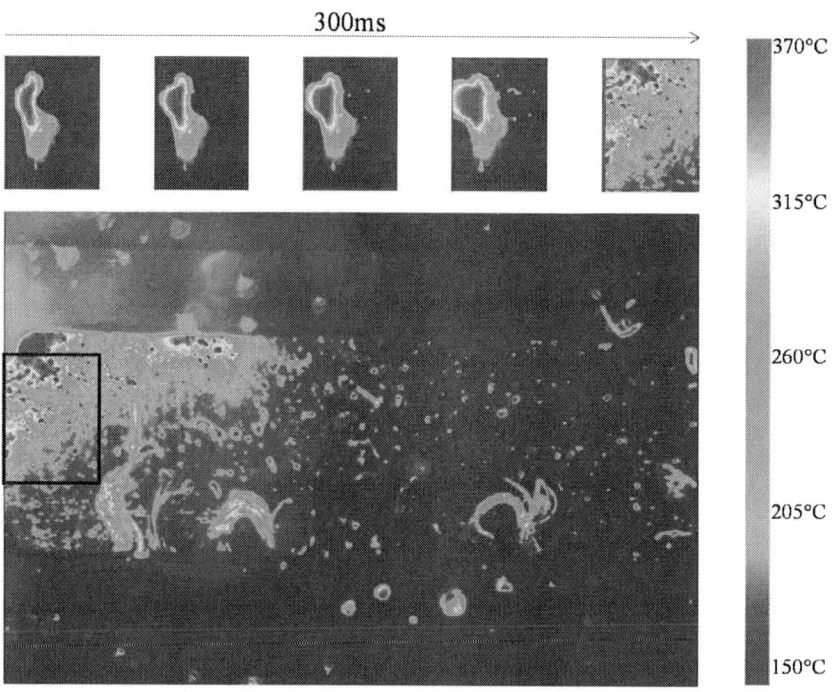

Fig. 5: Capturing the last instances of failure with an high-speed infra-red camera

Fig. 6: Capturing the last instances of failure with an optical camera

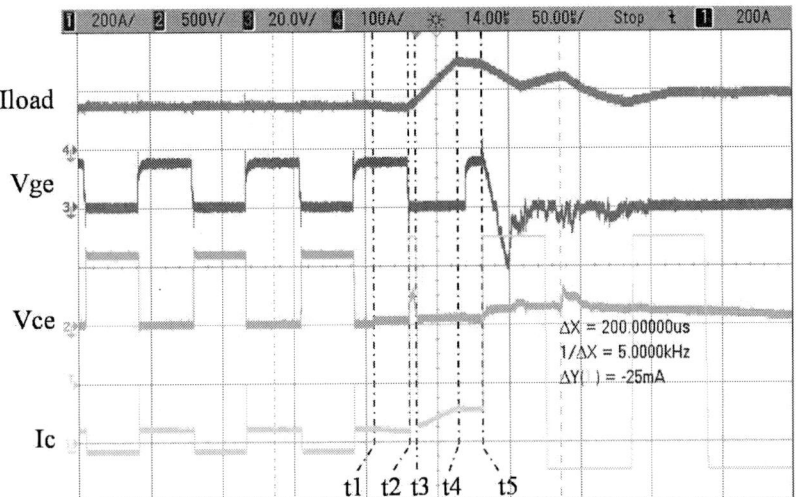

Fig. 7: Main waveforms acquired on the top left switch subject to sectioning at the end of life

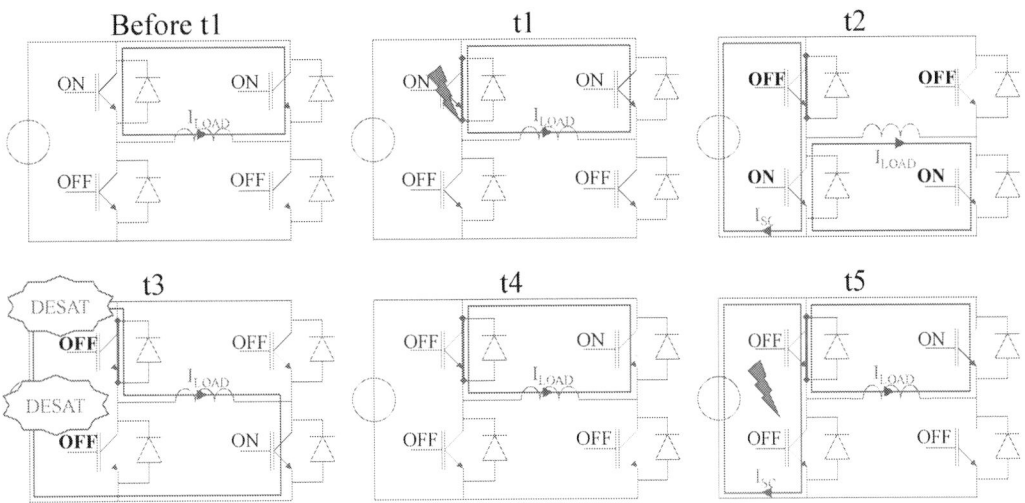

Fig. 8: Circuit states during the sequence of events acquired in Fig. 7

The generalization of these results should be considered with care since other experimental protocols would probably lead to a different failure mode:

1. A higher operating current would probably have led to a failure with the dies B and C still connected. This would probably have changed the failure location and the dynamics of the thermal runaway.

2. Manual wire cutting constitutes a frank disconnection, whereas during wire degradation, the lift-off can be of few tens of μm only, and momentary partial contact may occur even after wire lift-off. Instead of wire cutting, power cycling could be performed, with the following particularities: a) must be done with nominal voltage under switching conditions to represent

more faithfully the stress distribution b) thermal images require gel removal which may alter the module integrity, especially in a high-voltage switching test bench; c) the wire lift-offs are difficult to locate visually; d) power cycling may trigger other failure mechanisms and the interpretation of the results is more complex.

3. At time t3, and as shown in Fig. 8, the left arm is in a "wrong" state with a top switch failed in short-circuit and a bottom switch off due to its 'desat' protection. This state creates a current increase through the low inductive load, which leads to even higher thermal stress on the top left switch. According to the circuit topology and to the instant of failure, the consequence of a wrong arm state could have been different and led to different failure modes. On some other modules tested with the same set-up, the failure actually did not result in an explosion but a dark spot visible on the top-side of the die.

Conclusions

The results of this experiment shows that stress sharing between parallel dies in the same module structure occurs naturally, in the initial stages of degradation of the wire-bonds. When more than half of the wire-bonds are detached, the stress sharing diverges progressively. The evolution of the electrical resistance follows a shape that is related to the temperature imbalance between dies. This observation provides a hint to understand and model the evolution of the electrical resistance during power cycling, and to perform prognostics. Additionally, the temperature mismatch between dies could be used as a precursor to failure too, assuming individual chip temperature is accessible. This would eventually also allow to prevent catastrophic failures in field applications.

The unbalancing ultimately leads to the failure of one die. The catastrophic failure resulting from wire-bond sectioning leads to a short-circuit failure and is not prevented by the traditional 'desat' protection. As a result, a highly energetic event could be observed.

In the future, the combined influence of wire-bond and die-attach degradation will be modelled and simulated in order to reproduce the die temperature evolutions observed during power cycling. Thus, a relation between chip temperature and degradation can be established and used in an on-line prognostic and health management system.

References

[1] M. Piton, B. Chauchat, and J. F. Servière, "Microelectronics Reliability Implementation of direct Chip junction temperature measurement in high power IGBT module in operation — Railway traction converter," Microelectron. Reliab., 2018.

[2] N. Degrenne and S. Mollov, "Robust On-line Junction Temperature Estimation of IGBT Power Modules based on Von during PWM Power Cycling," 2019 IEEE Int. Work. Integr. Power Packag., pp. 107–116, 2019.

[3] F. Gonzalez-Hernando, J. San-Sebastian, A. Garcia-Bediaga, M. Arias, and A. Rujas, "Junction Temperature Model and Degradation Effect in IGBT Multichip Power Modules," pp. 2957–2962, 2019.

[4] N. Degrenne, R. Delamea, S. Mollov, "Electro-thermal modelling and Tj estimation of wire-bonded IGBT power module with multi-chip switches subject to wire-bond lift-off," AIMS Electronics and Electrical Engineering, ElectronEng-04-02-154, 2020.

[5] N. Degrenne and S. Mollov, "Robust On-line Junction Temperature Estimation of IGBT Power Modules based on Von during PWM Power Cycling," in 2019 IEEE International Workshop on Integrated Power Packaging (IWIPP), 2019, pp. 107–116.

[6] N. Degrenne and S. Mollov, "Experimentally-Validated Models of On-State Voltage for Remaining Useful Life Estimation and Design for Reliability of Power Modules," in CIPS 2018.

[7] J. Brandelero, J. Ewanchuk and S. Mollov, "Selective Gate Driving in Intelligent Power Modules," in *IEEE Transactions on Power Electronics*, doi: 10.1109/TPEL.2020.3002188.

Cost of energy assessment of wind turbine configurations

Catalin Dincan*, Philip Kjær, Lars Helle
Vestas Wind Systems A/S
Hedeager 42,
Aarhus N, Denmark
*Phone: +45 5086 6867, Email: cagdi@vestas.com

Keywords

«LCoE », «Wind energy systems», «medium voltage», «silicon carbide», «annual energy production ».

Abstract

This paper presents findings from a comparative study on LCoE (levelized cost of energy) impact from various configurations of electrical power conversion in on-shore wind turbines. All comparisons are made against a baseline configuration, based on proprietary models and input data, thus boundary conditions and assumptions apply, which do not necessarily translate to other manufacturers of wind turbines and electrical power conversion solutions. In our analysis, we asked the question: which power conversion topology and converter building block size can offer the best LCoE reduction across a broad range of wind turbine products employing uniform architecture? And which contributions to life-cycle cost dominate? The methodology is explained, and results indicate, among those analysed, that a scalable, modular medium-voltage converter platform should be employed in a turbine down-tower architecture.

Introduction

Levelized cost of energy for on-shore wind-energy has experienced a drastic decrease, as seen in Fig. 1 [1]. Prognosis predicts a far more moderate reduction in the next years, and it is relevant to search for contributions across turbine design and its life-cycles. This paper therefore evaluates which role the semiconductor technology and power conversion topology may have in a reduction of LCoE when tied to the choice of turbine architecture. A baseline turbine architecture is considered, where the converter (2-level, low-voltage, silicon-based) and transformer are located in the nacelle. If one could assign the converter and transformer zero cost, zero weight, zero physical volume and weight, zero failures and maintenance, a total of approximately 8.5% LCoE reduction is achieved from the model (see Fig. 2) This relatively modest ultimate reduction can be understood considering the example of on-shore wind turbine cost structure illustrated in Fig. 3.

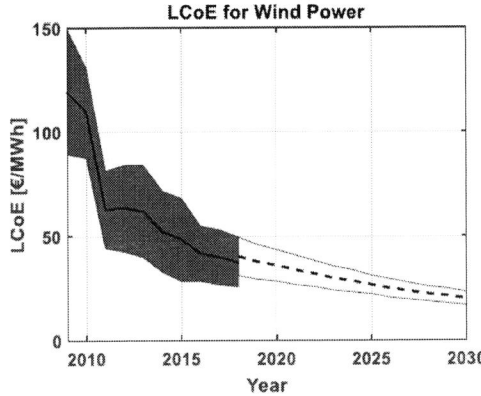

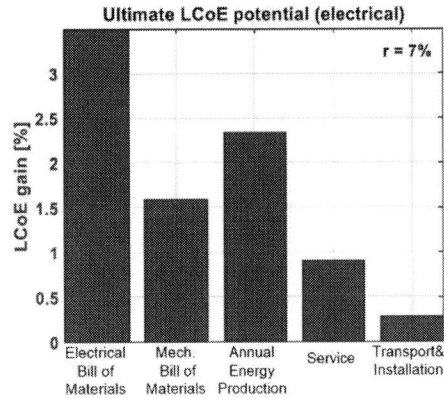

Fig. 1 Lazard's levelized cost of energy results [1]   Fig. 2 Electrical system LCOE potential

LCOE tool structure and methodology

The LCoE metric allows comparison of power plants with different generating technologies and cost structures. The LCoE results from the aggregation of all costs through the life-cycles of the power plant divided by the accumulated generated energy throughout the plants life [2].

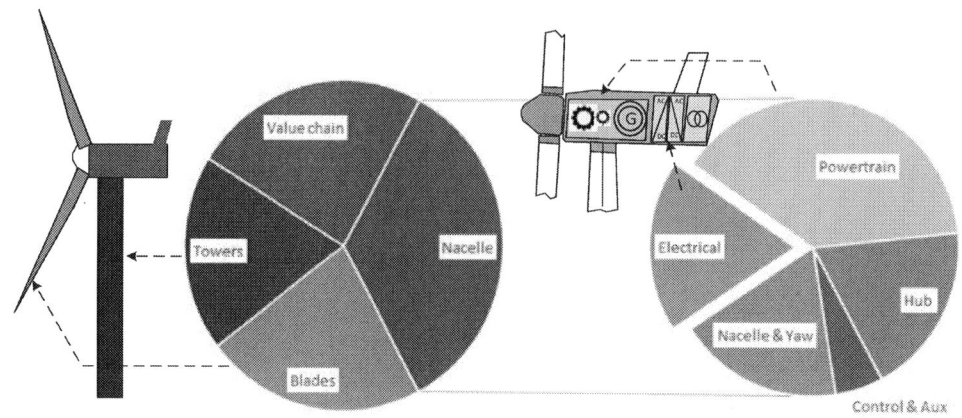

Fig. 3 Example of on-shore wind turbine cost structure

Fig. 4 Wind turbine product life-cycles considered in LCoE analysis

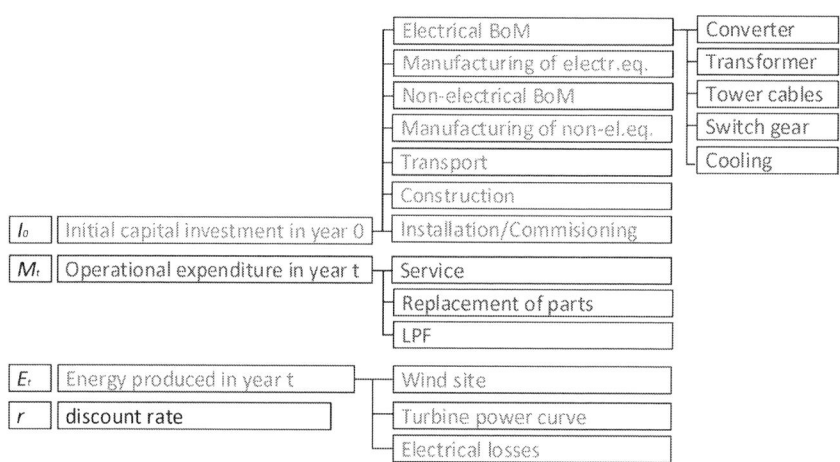

Fig. 5 LCoE tool structure

$$LCOE = \frac{I_0 + \Sigma_{t=1}^{n} \dfrac{M_t}{(1+r)^t}}{\Sigma_{t=1}^{n} \dfrac{E_t}{(1+r)^t}}$$

Eq. 1

LCOE reduction for the next generation wind turbines has been addressed before in ([3], [4], [5]), where impact of optimized blade designs, low induction rotor, drivetrain components, fixed and floating substructures was investigated. To the knowledge of the authors, no report has been made to date of development of any tool, and results from its use, that evaluates LCoE impact from power semiconductor technology, converter topology & its location across turbine rated power ranges.

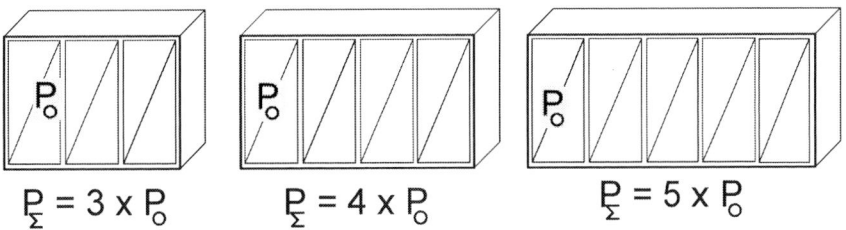

Fig. 6 Illustration of scalable, modular converter with building block rated power P_o

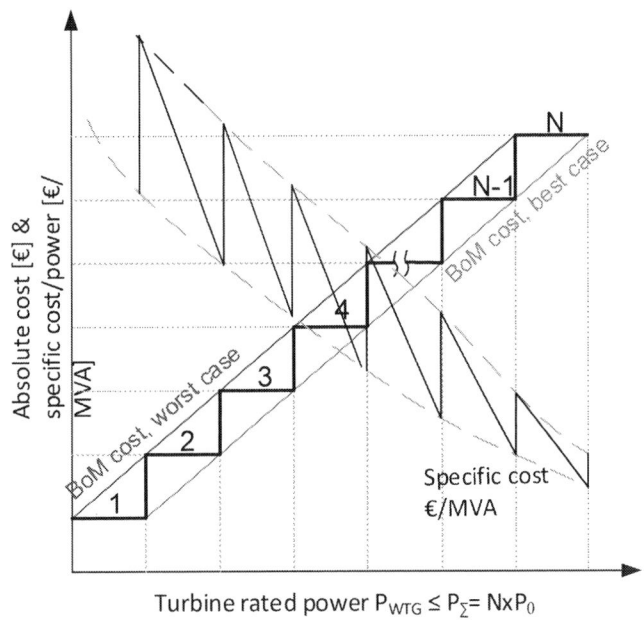

Fig. 7 Impact from integer no. of parallel power stacks on BoM cost of converter

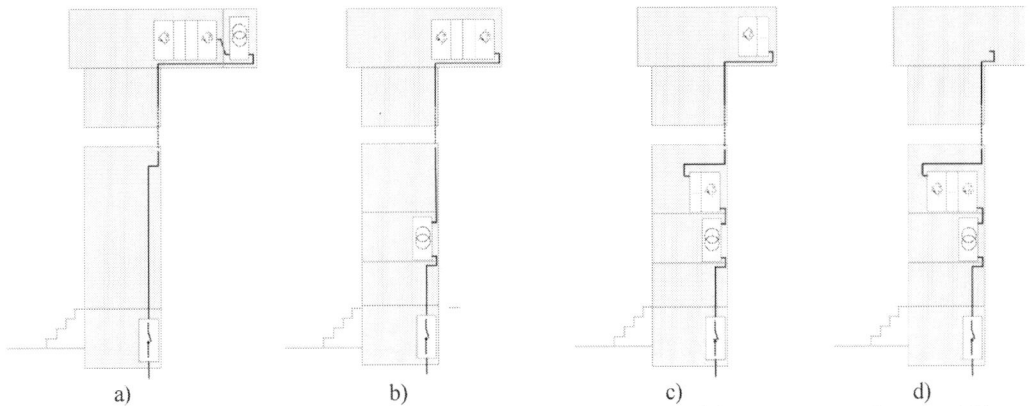

Fig. 8 Wind turbine configuration: a) Up-tower; b) Up-tower converter and down-tower transformer; c) Up-tower machine side converter and down tower line side converter and transformer; d) Down-tower electrical system. *Configurations with DFIG generator have been excluded from the study, while focus was on PMG with medium speed.*

As indicated in Fig. 5, the expression for LCoE in Eq. 1 include important value chain contributions in the life-cycles transport, installation, service, (see Fig. 4). It is known that value chain costs scale with turbine ratings due to higher costs in manufacturing, transportation and service and identifying an architecture that brings costs out in the particular areas is crucial.

Returning to the objective stated in the abstract: "to compare LCoE for various <u>scalable, modular converter topologies</u> – and underlying semiconductor technologies – for ranges of turbine rated power with select architectures", below we elaborate on how the converter's building block ratings and the turbine's architecture are included in the analysis.

All converter solutions considered employ series- or parallel-connection of identical so-called power stacks, ie. dc/ac converter building blocks containing semiconductors, dc-capacitors, magnetics where applicable, and more, see principle in Fig. 6.

(A) The turbine's architecture (converter location and connection voltage level), together with the power converter topology and its semiconductor type (all to be introduced later) define the power stack nominal power rating.
(B) Any set of values of turbine nominal power rating are chosen.
(C) To permit a range of converters across a power range, an integer number N identical power stacks may be connected to meet the turbine nominal power rating. The integer number of modules in a converter for a particular turbine power rating is best illustrated by the stair-case curve for absolute bill-of-material cost in Fig. 7. If turbine power ratings exploit converter ratings in the best possible way, the green curve ('best case') is obtained. Similarly, if turbine power rating just exceeds $(N-1) \times P_0$ it drives a converter configured with N modules, and 'worst case' red curve is obtained. These two lines define the cost band. Translated into specific cost (cost per MVA), the stepped hyperbola results. All other life-cycle cost will also scale depending on N (for example manufacturing, transport, operating power losses, failure rate).
(D) A baseline configuration is chosen (see overleaf), against which comparisons are made. Rather than calculate absolute cost, we estimate and report the differences (ΔLCoE) between concepts and baseline.
(E) Annual energy production is calculated for the specific turbine configuration.

Design drivers

The analysis is restricted to four turbine architectures:
(i)   up-tower converter and transformer (baseline)
(ii)   up-tower converter and down-tower transformer
(iii)   up-tower machine side converter and down-tower line-side converter and transformer
(iv)   down-tower converter and transformer

Further, four converter topologies and semiconductor are considered with the specifications illustrated in Table 1 and turbine power ratings ranging between 3.0 and 8.0MW.

(i)   LV Si = low-voltage converter (two-level) employing silicon semiconductors
(ii)   LV SiC = low-voltage converter (two-level) employing silicon-carbide semiconductors
(iii)   MV Si = medium-voltage converter (3-level, or MMC) employing silicon semiconductors
(iv)   MV SiC = medium-voltage converter (3-level, or MMC) employing silicon-carbide semiconductors

Table 1 Use cases analyzed

	LV Si	LV SiC	MV Si	MV SiC
Topology	2-level	2-level	a) 3-level NPC b) MMC1 c) MMC2	3-level NPC
Granularity	1.4MW	1.4MW	a) 3.0MW b) 1.1MW c) 1.3MW	3.0 MW
Line-side voltage level	$0.69kV_{ac}$	$0.69kV_{ac}$	a) 4.0kV b) 3.3-8.8kV c) 2.7-4.0kV	4.0 kV

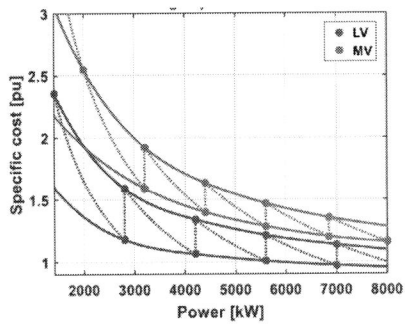

Fig. 9 Power stack rating impact on converter BoM

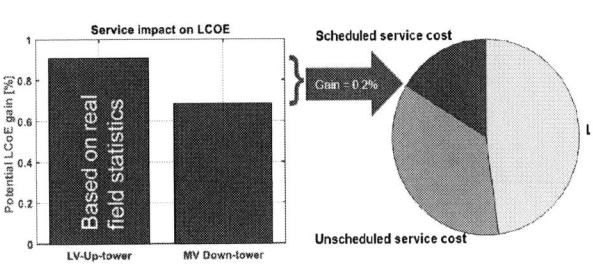

Fig. 10 Service impact on LCoE

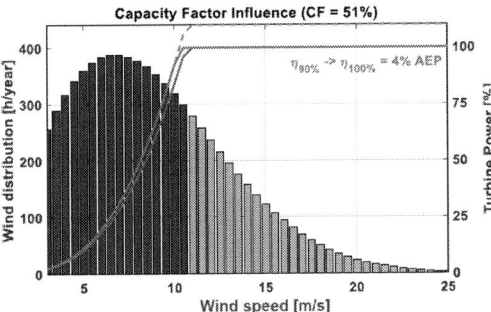

Fig. 11 Capacity factor influence at 51%

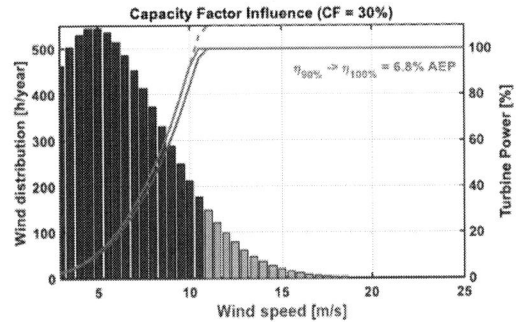

Fig. 12 Capacity factor influence at 30%

"LV Si" represents state-of-the-art in the wind turbine power conversion using fully rated converters [6]. It offers low BoM cost, and is scalable by paralleling of converter modules, until the aggregated line current becomes unmanageable.

"LV SiC"-based converters should offer lower losses and higher power density, while component maturity and cost remain challenges [7].

"MV Si" covers (a) 3-level neutral-point clamped topologies using semiconductors with MV blocking capability, and (b) modular multilevel converters using semiconductors with LV blocking capability. Two variants of the MMC have been analised: MMC1 with 8.8kV$_{ac}$ grid and MMC2 with 4.0kV$_{ac}$ grid.

"MV Si" power modules are utilized by different converter topologies: three level NPC or modular multilevel converters (MMC) [3]. They can offer building block standardization and are suitable for a modular turbine architecture, while paying a penalty on high costs.

In similar manner to LV SiC, MV SiC technology has also the potential for high power density, high efficiency, building block standardization, but time to market and costs are deficits. It can be employed on two or three level topologies. MV SiC modules have been proposed and evaluated extensively in different works ([8], [9]) and with measured efficiencies above 98%. In this study, same granularity and grid voltage as the MV Si counter part have been selected: 3.0MW per stack and 4.0kV output grid voltage.

Sensitivity to input data

The cost of energy model has many parameters and assumptions, where some are very approximate, but there are a few factors affecting the sensitivity of LCoE to input data, such as power stack modularity, service tasks duration, lost production factor and the influence of capacity factor on annual energy production (AEP). The impact of service tasks duration and lost production (LPF) during service on LCoE is presented in Fig. 10, with the assumption that equal failure rates are present in both LV and MV solutions. Another important factor in LCoE is the Annual Energy Production (AEP), which is further impacted by the converter losses and turbine capacity factor. Capacity factor influences proportionally also failure rates, LPF during service, and mechanical BoM. The capacity factor in this assessment was kept constant and across assessed turbine powers. Fig. 11 and Fig. 12 show the capacity factor influence at 51% and 30%.

Examples of findings from comparative analysis

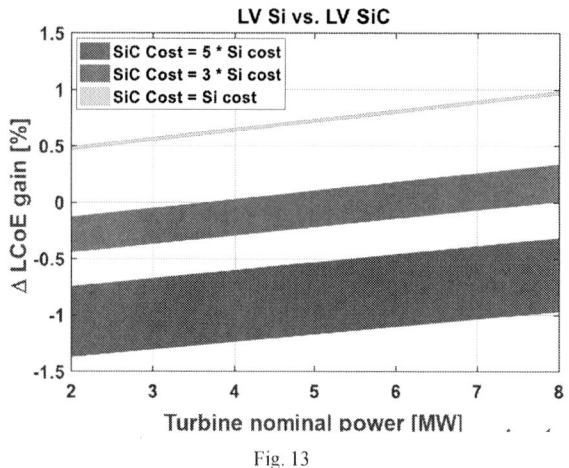

Fig. 13

Up-tower LV SiC

LV-Si vs. LV SiC results are shown in Fig. 13. For this comparison, the following assumptions were made: converter failure rate optimistically assumed unchanged, while only electrical BoM cost and converter efficiency contribute to ΔLCoE. The results permit the following conclusions: The semiconductor cost increase (Si→SiC) directly offsets the LCoE curves. The larger the SiC component cost, the broader the band between best/worst case number of modules. SiC converter efficiency gain (>50% decrease in total losses at Pn) cannot compensate a significant BoM cost increase. Assuming same cost between Si and SiC (highly unrealistic in immediate time frame), the ΔLCoE gain is between 0.5% and 1%.

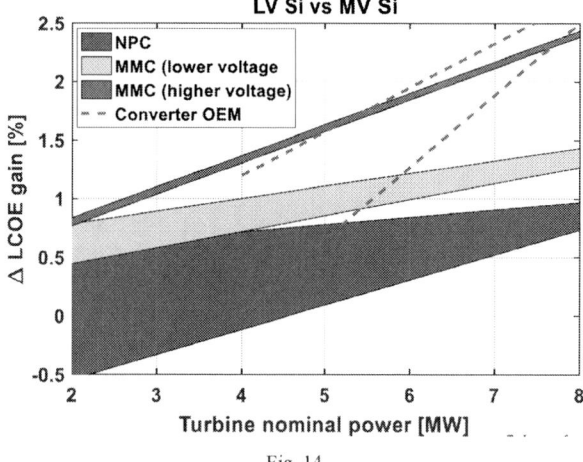

Fig. 14

Down-tower MV Si

Fig. 14 presents the ΔLCoE impact when comparing LV Si converters (up-tower location) to MV Si (down-tower). In this comparison, not only losses and BoM cost differ, but also transport, installation and service cost. Further, turbines based on down-tower MV converters benefit from smaller nacelle size and lower tower loads. In this specific study the 3-level NPC is "penalized" from a 3MW modularity (based on best semiconductor rating), significant filter losses, tower cable losses and cost. The MMC2 (4.0kV) is penalized from tower cable losses and tower cable cost. On the other hand, the MMC1(8.8kV) benefits from lower module rated power (1.1MW), lower tower cable loss and cost as compared to MMC2 . To further support our findings based on proprietary data, comparative numbers were supplied from a converter OEM's own analysis. In summary, the largest LCoE gain comes with the MMC1 topology, ranging from 0.8% up to 2.5%.

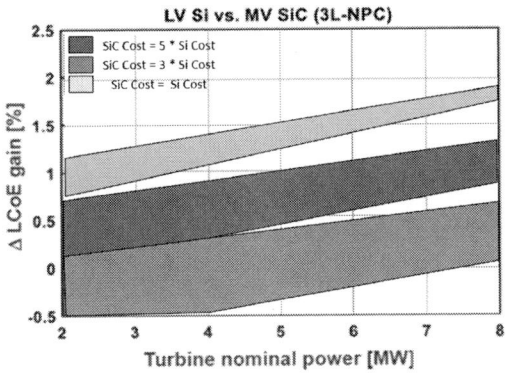

Fig. 15

Down-tower MV SiC

For the down-tower MV SiC, a three-level NPC version was investigated with results shown in Fig. 15. Similar to previous comparisons converter failure rate assumed unchanged, while full value chain impact contribute to LCoE impact. For the MV SiC, the electrical system benefits from higher efficiency, while the electrical system cost is uncertain as no commercial MV SiC modules are present at the moment. Ficticious 4.5kV SiC modules (extrapolated from existing 3.3kV SiC modules) where employed in AEP estimation. The turbine benefits from lower nacelle size and tower weights, while electrical system benefits from lower service and maintenance time. Similar to the LV SiC counterpart, even if total losses are significantly decreased, the LCoE gains are not significantly larger as compared to the down-tower MV Si counterpart. Again, if cost parity is assumed between LV Si and MV SiC, the LCOE gains will range between 1% to 1.8% versus the base line.

Discussions

Fig 16. compares the electrical losses in the equipment from terminals of the generator through converter, transformer, tower cable (also in different sequence) to the turbine switchgear for all configurations. Interestingly, this model shows the up-tower, low-voltage two-level SiC-based converter offers the larger reduction in losses, yet the down-tower medium-voltage MMC yields the larger LCoE reduction. The explanation is in the fact that it is not only losses that play a role, but also the capacity factor and the number of hours a turbine operates in a year at lower speeds. Meaning the impact in percentage [%] on annual energy production will have a decisive role, as observed in Fig.17 , where the MV MMC based topologies will have larger impact on AEP, followed by SiC based topologies (both LV and MV) and MV NPC with lowest impact. Further on, due to down-tower turbine architecture benefits in the value chain savings, the larger the power rating of the turbine, the larger drop will be in direct product cost (as compared to the baseline). On the other hand, there are no significant value chain savings with 2L SiC up-tower architecture. Thus, it is not enough to simply decrease the electrical system losses, but selecting a turbine architecture that has impact on all the life cycle stages (manufacturing, transportation, installation, service) for an entire product family should be the path forward.

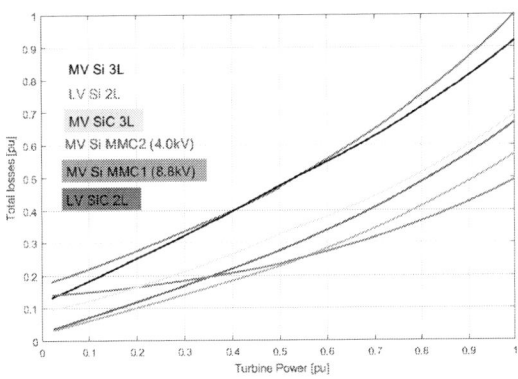

Fig. 16 Total losses for different architecture

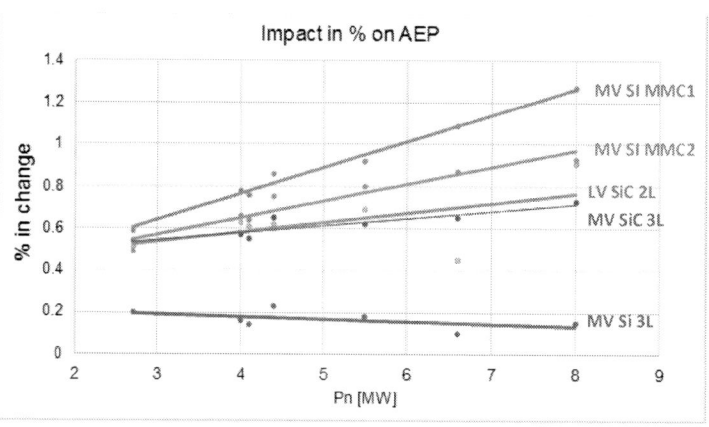

Fig. 17 Impact in percentage on annual energy production

Conclusion

This paper has evaluated the role of semiconductor technology for wind-energy in driving the LCoE in the years to come and the potential of it. The interest has been to illustrate the LCoE change across a large range of turbine powers, where a uniform building block is used in a scalable, modular converter. Alternative configurations (up-tower/down-tower, LV/MV, Si/SiC) have been analyzed. It turns out that the largest benefits will come from a down-tower/MV, Si based architecture, offering LCoE gains up to 2.5%, versus an up-tower, LV Si baseline configuration. The reduction of LCoE is explained due to higher AEP and value chain savings (with largest costs out coming from nacelle re-design, transportation, installation and service.). Transitioning to a LV SiC architecture will have the benefits of increased AEP, but due to cost penalty and no significant value chain savings, the configuration is not that attractive at this point and even assuming a cost parity with Si counterparts, the LCOE gains will range from 0.5% to 1%. In the end, it is re-emphasized that the results of this paper are built based on one company's cost model, with its assumptions and the transformation of an entire turbine product range is always tied to where the investments need to go and the time to market.

References

[1] Lazard. et.al, "Levelized cost of energy analysis ver.12.0.0," 2018.

[2] Ke. Ma. Frede Blaabjerg, "Wind energy systems," *Proceedings of the IEEE*, pp. 2116-2131, 2017.

[3] S. Bhattacharya, "High MegaWatt MV Drives," FREEDM systems center, NC State University, 2018.

[4] F. Rasmussen, "Emerging wind energy technologies," DTU, 2014.

[5] D. Rothmund, et.al, "Design and experimental analysis of a 10 kV SiC MOSFET Based 50 kHz Soft-switching Single-Phase 3.8kV AC/ 400V DC Solid-state transformer," in *IEEE Energy Conversion Congress and Expo, ECCE*, Portland, 2018.

[6] S. Madhusoodhanan, et.al, "Solid state transformer and MV grid tie applications enabled by 15kV SiC IGBTs and 10 kV SiC mosfets based multilevel converters," *IEEE Trans. on Ind. Applic.*, vol. 51, no. 4, pp. 3343-3360, 2015.

[7] C. Kost, "Levelized cost of electricity renewable energy technologies," Fraunhofer Institute for Solar Energy Systems ISE, 2018.

[8] P. English, Bruce Valpy, "Future renewable energy costs: onshore wind," KIC InnoEnergy, 2014.

[9] P.H.Jensen, et. al., "LCOE reduction for the next generation offshore wind turbines," INNWIND.eu, 2017.

Energy management in a Multi-source System using isolated DC-DC resonant converters

M.Arazi; A.Payman; M.B.Camara, *Member IEEE*; and B.Dakyo, *Member IEEE*

GREAH Laboratory, University of Le Havre Normandie, 75 Rue Bellot, 76600 Le havre, France

mouncif.arazi@univ-lehavre.fr; paymana@univ-lehavre.fr ; camaram@univ-lehavre.fr; brayima.dakyo@univ-lehavre.fr.

Keywords

«DC-DC resonant converter», «soft switching», «energy management», «Fuel cell», «Supercapacitor».

Abstract

This paper presents an energy management strategy for a multi-source system composed of a fuel cell (FC) as the primary source and supercapacitors (SC) as secondary source. Isolated LLC resonant converter is used to connect the fuel cell to the DC-link, while the supercapacitors are connected through a bidirectional resonant converter. The resonant operation ensures high efficiency and high power density by means of soft switching and high frequency operation. The aim of this paper is the control of the DC-DC resonant converters to share the requested power between the SC and FC according to their dynamic response. So, the Fuel cell supplies the nominal power, while the supercapacitors ensure the mitigation of the fluctuations from the load power. The performances of the proposed topology are evaluated through some simulations.

1. INTRODUCTION

The Fuel cells can be a good alternative of recurrent sources of electric power in order to decrease the CO_2 emissions. They are characterized by a high specific energy, but their main weak point is the slow dynamic response [1]. In order to overcome this issue, the hybridization of the fuel cell with a high power density source can be a better solution for the applications which presents dynamic variations such as electric vehicles. Supercapacitors can play this role since they are characterized by a high dynamic response. In fact fuel cell and SC are considered as complimentary sources [2]. In this paper, the dynamic properties of the two sources are considered to propose an adequate energy management strategy. In a hybrid electric system with DC-link configuration, DC-DC converters that connect the different sources to the DC-bus are the main element of the system. However, the key part in this association is to find the best configuration and control strategy. In [3] an association of FC, SC and the load via one buck-boost converter was proposed. In this configuration, the voltage of the DC link is not controlled. In [4] two non-isolated buck-boost converter are used to control DC-bus voltage and the power flow. In fact, galvanic isolation is highly recommended for applications with high voltage ratio. Some isolated topologies based on Dual-active Bridge (DAB) converter [5] were proposed for electrical hybrid systems. In [6] [7] triple active bridge converter with different control strategies is proposed to associate FC and SC. However, these topologies suffer from the complexity of the power flow control and they have limited soft switching. In this paper an association of the FC and the SC based on two DC-DC resonant converters is proposed. These topologies of the DC-DC converters allows soft switching for the whole load range and guarantee a good efficiency and a high power density. The Fuel cell is coupled to the DC-bus through unidirectional LLC resonant converter [8], while the supercapacitors are connected via a symmetrical bidirectional resonant converter [9]. Fig.1 shows the configuration of the system.

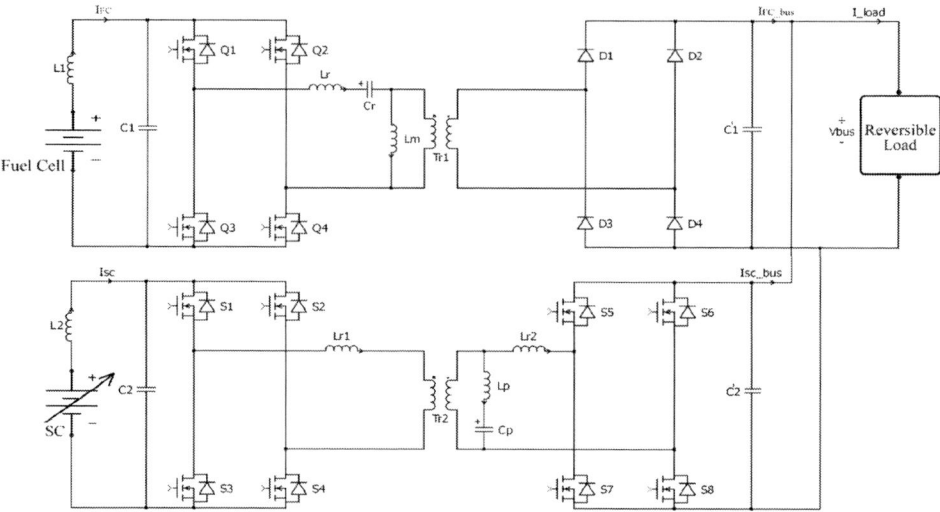

Fig. 1: Electric hybrid system configuration.

2. DC/DC RESONANT CONVERTER TOPOLOGIES AND MODELS OF THE SOURCES

A. Supercapacitors model

The dynamic model of the supercapacitors is presented in (1) ,where R_w is the wiring resistance. R_{sc} is the SC series resistance, N_s is the number of series cells, N_p is the number of parallel series, $\beta.V(t)$ is the variable capacitor component, and C_0 is the constant capacitor and V_{t0} presents the initial voltage of the SC at $t = 0$, [10]. This model describes the SC behavior during the charge and discharge operations.

$$\begin{cases} V_{sc} = V_{t0} - \int_0^t \frac{I_{sc}}{C_{eq}} \cdot dt - R_{eq} \cdot I_{sc} \\ C_{eq} = \frac{N_p}{N_s}.(2 \cdot \beta \cdot V(t) + C_0) \\ R_{eq} = \frac{N_s}{N_p}.R_{sc} + \frac{N_s - 1}{N_p}.R_w \end{cases} \quad (1)$$

B. Fuel cell model

The fuel cell is used as main energy source. The voltage of one fuel cell can be described by (2) and the Fuel cell stack voltage is given by (3), where E_N is the thermodynamic potential of the cell, V_c is the capacitor voltage , V_{ohm} is the ohmic voltage loss, V_{conc} is the concentration voltage drop, and N_s is the is the number of cells in series. More details about Fuel cell modeling can be found in [11].

$$V_{cell} = E_N - V_c - V_{ohm} - V_{conc} \quad (2)$$

$$V_{FC} = N_s.V_{cell} \quad (3)$$

C. LLC Resonant converter topology

LLC resonant converter is proposed to connect the fuel cell to the DC-bus. As shown in Fig.2, the topology of this converter is composed of an active full bridge. a high frequency transformer. a diode rectifier and a resonant tank. This last one includes a series inductance L_r, a series capacitor C_r and a parallel inductance L_m. The frequency modulation with a fixed duty cycle is proposed to control the converter, thus zero voltage switching (ZVS) for primary bridge switches and Zero Current Switching (ZCS) for secondary rectifier diodes can be reached. Based on the first harmonic approximation, the voltage gain of the converter versus the switching frequency is described in (4), where Q is the factor of the quality, and f_n is the normalised frequency.

$$H = \frac{mV_o}{V_i} = \frac{1}{\left[1+\gamma-\frac{1}{fn^2}-\frac{\gamma}{fn^2}\right]+j\left[Q.(\frac{\gamma.fn}{\gamma+1}-\frac{1}{fn})\right]} \qquad (4)$$

$$\gamma = \frac{L_r}{L_m} = \frac{1}{k}; \qquad Q = \frac{1}{R}\sqrt{\frac{L_r}{C_r}} \ ; \qquad f_n = \frac{f_s}{f_r} \ ; \qquad f_r = \frac{1}{2\pi\sqrt{L_r\,C_r}} \qquad (5)$$

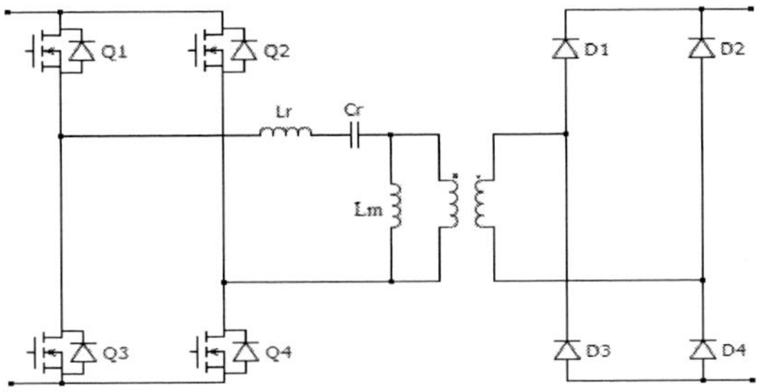

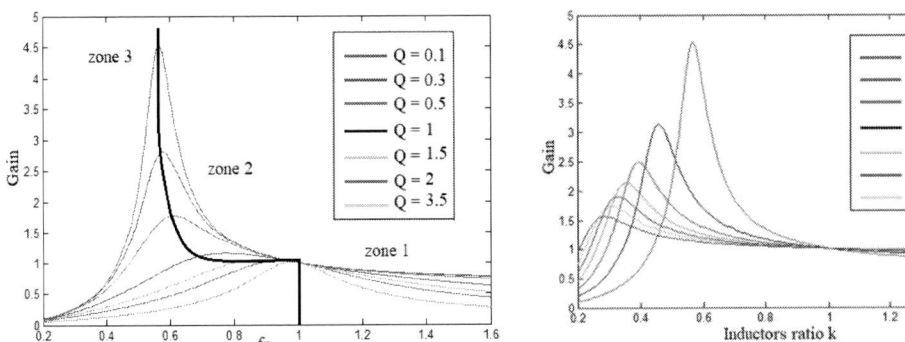

Fig. 2 : Topology of LLC resonant converter

The voltage gain versus the normalized frequency for different values of Q and for different values of the inductance ratio k are presented in Fig.3 and Fig.4. In this study, the LLC converter operates in boost mode and its operation area is located in zone 2 of the Fig.3.

Fig. 3: Voltage gain of LLC resonant converter for different values of Q.

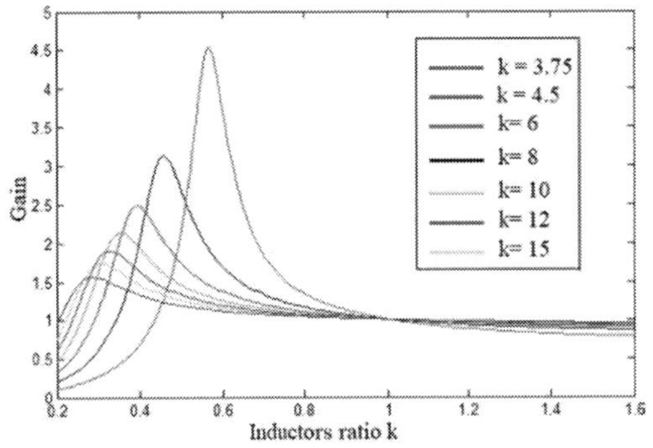

Fig. 4 : Voltage gain of LLC resonant converter for different values of k.

The design of the LLC converter requires the definition of the nominal power and the input and output voltages. Based on the characteristics of the studied system, the required voltage gain is from 2.7 to 4.5. According to Fig.3 and Fig.4 we choose Q_{max} =0.1 and k=3.75. Then, resonant tank component can be calculated using (6) .

$$L_r = \frac{Q_{max}.R_{min}}{2\pi.f_{r1}} \ ; \quad L_m = k.L_r \ ; \quad C_r = \frac{1}{L_r.(2\pi.f_{r1})^2} \qquad (6)$$

The parameters of the LLC converter are given in Table I.

Table I: LLC resonant converter parameters

NAMES	PARAMETERS	VALUES
Fuel Cell voltage range	V_{bat}	60-100 V
DC-link voltage	V_{bus}	270 V
Transformer turn ratio	η_{LLC}	4
Magnetizing inductance	L_m	20μH
Series resonant inductance	L_r	4 μH
Series resonant capacitor	C_r	0.9 μF
Series resonant frequency	F_{r1}	80kHz
Maximum power	P_{LLC}	3kW

D. Bidirectional resonant converter topology

Supercapacitors are connected to the DC-bus via a bidirectional resonant converter to ensure charging and discharging of the SC module according to the applied energy management strategy. The proposed topology is illustrated in Fig.5. Using first harmonic approximation, the voltage gain in forward mode is given in (7). Forward voltage gain versus normalized frequency for different values of Q is shown in Fig.6. Note that the resonant tank is symmetrical, so forward and backward modes have same behavior and characteristics. The parameters of the proposed bidirectional converter according to the studied system are given in Table II.

$$\frac{V_{bus}}{V_{bat}} = \left\lVert \frac{m}{\left[1+\frac{kf_n^2}{f_n^2-1}\right]+\left[jf_nQ(2k+\frac{k^2}{(1-f_n^2)})\right]} \right\rVert \tag{7}$$

$$Q = \frac{1}{R'_e}\sqrt{\frac{L'_p}{C'_p}} \;;\; f_n = \frac{f}{f_r} \;;\; f_r = \frac{1}{2\pi\sqrt{L_pC_p}} \;;\; k = \frac{L_{r1}}{L'_p} = \frac{L_{r2}}{L_p} \;;\; L'_{r2} = L_{r2}/m^2;\; L'_p = L_p/m^2;\; C'_p = C_pm^2 \tag{8}$$

Table II: bidirectional resonant converter parameters

NAMES	PARAMETERS	VALUES
Supercapacitors voltage range	V_{Sc}	60-80 V
Transformer turn ratio	m	4
Primary Series inductance	L_{r1}	$4\mu H$
Secondary Series inductance	L_{r2}	$60\ \mu H$
Parallel resonant capacitor	C_p	$25\ nF$
Parallel resonant frequency	L_p	$150\mu H$

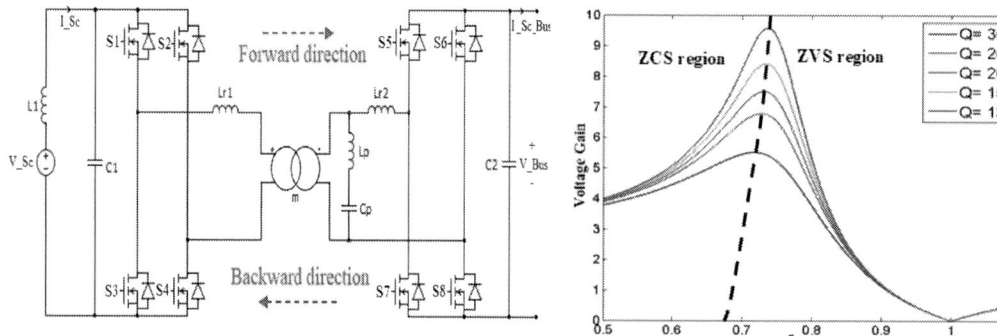

Fig. 5 : Proposed bidirectional resonant DC-DC converter to link the SC

Fig. 6 : Proposed bidirectional resonant DC-DC converter to link the SC

3. ENERGY MANAGEMENT STRATEGY

The energy management strategy based on the dynamics of the SC and FC is proposed. The adopted strategy proposes to supply the load nominal power by the FC, while the SCs will ensure the load variations due to their fast dynamic response. Three modes of operation are considered:

- Mode 1: $P_{Load} > P_{FC}$, in this mode, the FC and SC supply the load power demand. The bidirectional resonant converter operates in forward mode.

- Mode 2: $0 < P_{Load} < P_{FC}$, in this mode, the FC supply the load power requirement and ensure the charging of the supercapacitors.

- Mode 3: $P_{Load} < 0$, in this mode, the SC are charged by the power delivered by the FC and the regenerative power from the load.

For Mode 2 and 3, the bidirectional resonant converter operates in backward mode. The LLC resonant converter is controlled via frequency variation with 50% of duty cycle to control the DC-bus voltage. The control loop is illustrated in Fig.7. The reference of the power delivered by the supercapacitors is calculated using (9), and the reference of supercapacitor's current is given in (10). The supercapacitor's current control loop is illustrated in Fig.8.

$$P_{sc\_ref} = P_{Load} - P_{FC\_Nom} \tag{9}$$

$$I_{sc\_ref} = \frac{P_{sc\_ref}}{V_{sc}} \tag{10}$$

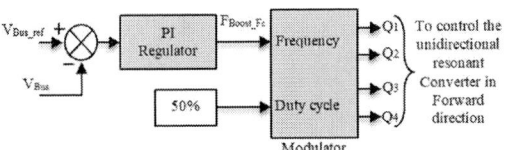 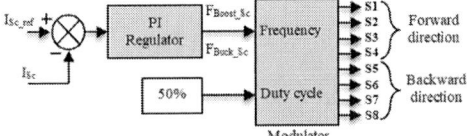

Fig. 7 : DC-bus voltage control loop.

Fig. 8 : Control loop of the supercapacitor's current.

4. SIMULATION RESULTS

To validate the performances of the proposed resonant converters, an energy management strategy using Matlab/Simulink coupled to Plecs software are done. Fig.9 shows the load's power profile and the contribution of each source. This figure shows that the power delivered by the FC still constant at the nominal power, while the load's power variations are compensated by the SC. Fig.10 shows the DC-bus voltage control result, where the measured DC-bus voltage follows the reference. ZVS for active switches is highly recommended to eliminate the switching losses and enhance the efficiency of the converter. On the other hand, ZCS for body diodes is preferred to decrease the reverse recovery losses. Fig.11 shows the SC's voltage, we can see that the evolution of the SC's voltage matches well with the variations of the load. Fig.12 shows the simulation waveforms of LLC converter switches. It can be seen that primary MOSFET work with ZVS and rectifier diodes reach ZCS. The bidirectional converter is controlled in forward or backward mode according to the load variations, and should ensure ZVS for active switches and ZCS for rectifier diodes for all load regions. Fig.13 and Fig.14 show simulation waveforms in forward mode for Psc=1.5kW and in backward mode for Psc= -3kW, respectively. As we see in Fig.13 and Fig.14, ZVS for primary active switches and ZCS for secondary rectifier diodes are reached independently of the energy transfer direction.

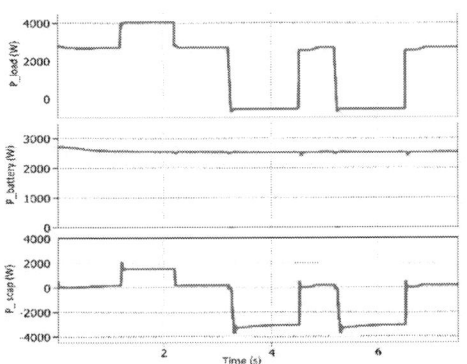 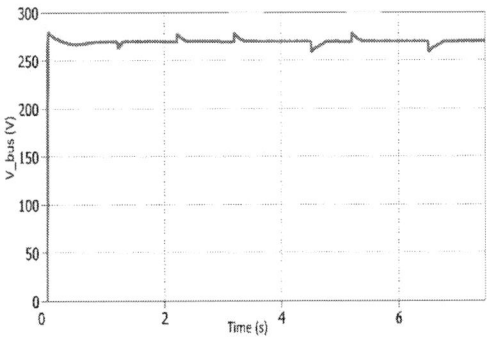

Fig. 9 : Load's power, FC's power and SC's power.

Fig. 10 : Evolution of DC-bus Voltage .

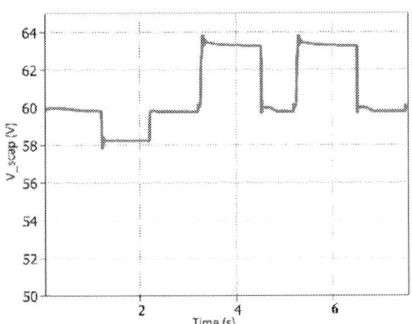 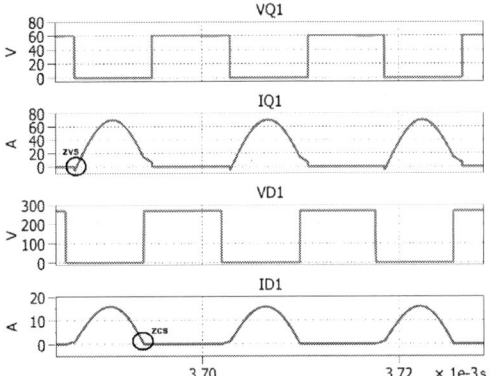

Fig. 11 : Evolution of Supercapacitor's voltage.

Fig. 12 : Switching waveforms of the SC converter in forward mode at P_{SC} =2.5kW.

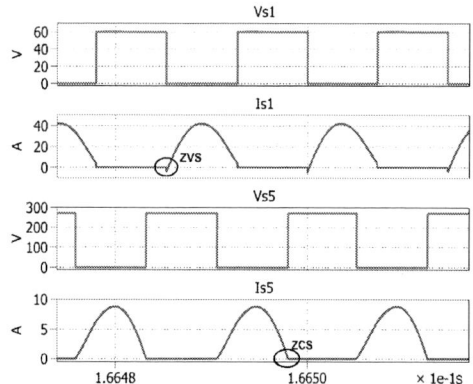

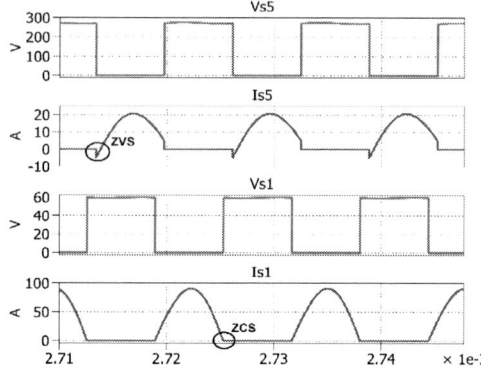

Fig. 13 : Switching waveforms of the SC converter in forward mode at Psc =1.5kW.

Fig. 14 : Switching waveforms of the SC converter in backward mode at Psc =-3.5kW.

5. CONCLUSION

This paper proposes an energy management strategy for a hybrid electric system using DC-DC resonant converters. Models of the fuel cell and the supercapacitors are described. Therefore, two topologies of isolated resonant DC/DC converters for FC and SC association are proposed. Finally, energy management strategy based on the dynamics of the sources is presented. The simulation results confirm the benefits of using the isolated resonant converters in term of reduction of switching losses, also the benefits of the proposed energy management strategy, especially for protecting the battery against the fast variations of the load.

References

[1] J.Snoussi; S. Ben Elghali ; M.Benbouzid ; M.F.Mimouni; "Optimal Sizing of Energy Storage Systems Using Frequency-Separation-Based Energy Management for Fuel Cell Hybrid Electric Vehicles"; IEEE Transactions on Vehicular Technology; Volume: 67 , Issue: 10, pp 9337 - 9346 , Oct. 2018.

[2] Clément Dépature ; Walter Lhomme ; Pierre Sicard ; Alain Bouscayrol ; Loïc Boulon, " Real-Time Backstepping Control for Fuel Cell Vehicle Using Supercapacitors", IEEE Transactions on Vehicular Technology, Volume: 67 , pp 306 - 314, Issue: 1 , Jan. 2018

[3] Qian Xun, Yujing Liu, Elna Holmberg. "A comparative study of fuel cell electric vehicle hybridization with battery or supercapacitor," International Symposium on Power Electronics, Electrical Drives, Automation and Emotion, Italy, 2018, pp. 390-395.

[4] A. Tani, M.B. Camara, B. Dakyo,Y.Azzouz, "DC/DC and DC/AC Converters Control for Hybrid Electric Vehicles Energy Management-Ultracapacitors and Fuel Cell ", IEEE Trans. on Industrial Informatics, ISSN: 1551-3203, Vol. 9, No.2, pp. 686 - 696, May 2013.

[5] B. Zhao, Q. Song, W. Liu, and Y.Sun,Overview of dual-active bridge isolated bidirectional dc dc converter for high-frequency-link power conversion system, IEEE Trans. on P.E, Vol. 29, pp. 4091–4106, 2014.

[6] M.Phattanasak ; R. G.Ghoachani ; J.P.Martin ; B.Nahid-Mobarakeh ; S.Pierfederici ; B.Davat; "Control of a Hybrid Energy Source Comprising a Fuel Cell and Two Storage Devices Using Isolated Three-Port Bidirectional DC–DC Converters", IEEE Transactions on Industry Applications, Vol.: 51 ,pp 491 - 497, Iss.: 1 , Feb. 2015.

[7] Ching-Ming Lai ; Yu-Huei Cheng ; Yun-Hsiu Li ; Hsin-Yu Chen; "A novel multiport converter with an auxiliary voltage pumping circuit for fuel-cell/battery hybrid energy sources"; 2017 IEEE 12th International Conference on Power Electronics and Drive Systems (PEDS); 12-15 Dec. 2017.

[8] Z. Fang, T. Cai, S. Duan, and C. Chen, "Optimal Design Methodology for LLC Resonant Converter in Battery Charging Applications Based on Time-Weighted Average Efficiency," IEEE Trans. Power Electron., vol. 30, no. 10, pp. 5469-5483, Oct. 2015.

[9] Mouncif Arazi ; Alireza Payman ; Mamadou Bailo Camara ; Brayima Dakyo; "Analysis of a Bidirectional Resonant Converter for Wide Battery Voltage Range in Electric Vehicles Application" ; 2017 IEEE Vehicle Power and Propulsion Conference (VPPC); 11-14 Dec. 2017.

[10] Kosseila Bellache; Mamadou Baïlo Camara ; Brayima Dakyo ; "Supercapacitors Characterization and Modeling Using Combined Electro-Thermal Stress Approach Batteries", IEEE Transactions on Industry Applications; Volume: 55, pp 1817 – 1827; Year: 2019.

[11] Ismail Oukkacha ; Mamadou Baïlo Camara ; Brayima Dakyo ; ''Energy Management in Electric Vehicle based on Frequency sharing approach, using Fuel cells, Lithium batteries and Supercapacitors'', 2018 7th International Conference on Renewable Energy Research and Applications (ICRERA), 14-17 Oct. 2018.

Long-Term Climate Impact On IGBT Lifetime

Martin Vang Kjaer, Yongheng Yang, Huai Wang and Frede Blaabjerg

Department of Energy Technology, Aalborg University
Pontoppidanstæde 111, 9220 Aalborg East, Denmark
Email:{mkj, yoy, hwa, fbl}@et.aau.dk

Abstract

Considerable efforts have been made to estimate the lifetime of power devices, e.g., IGBTs, when they are subjected to a specific loading profile, which is affected by real field mission profiles. In those cases, a yearly mission profile, e.g., ambient temperatures, wind speeds and solar irradiance levels, is adopted. However, the prior art assumes that the same accumulated damage is caused in each year during the operation of the solutions. In practice the mission profile will vary, making the assumption invalid. This paper thus examines the assumption of equally distributed damage accumulation throughout the lifetime, by analyzing multiple years of mission profiles.

Introduction

Power electronics are becoming are essential in power systems, concurrently as increased use of renewable generation. One of the key challenges associated with the growth of renewable generation, is the demand of reliable power conversion, as power electronic converters are known to constitute a potential bottleneck of the entire system in terms of reliability [1]. Moreover, IGBTs under the exposure of cyclic thermal stress are identified as being critical, which are prone to fail [2]. The reliability of power converters in renewable generation is highly affected by its operating condition, i.e. the mission profile and the lifetime will vary accordingly [3]. All prior mission profile-based lifetime prediction of IGBT modules is based on the assumption, that the annual damage is the same from year-to-year throughout the service life of the device, however, this may not hold true in practice. This paper investigates the variation of the annual accumulated damage (A_D) from year-to-year by considering a mission profile data span of multiple years to gain precision when estimating the lifetime. A case study of a 3-phase 12-kVA converter is given in order to provide an overview of the complete procedure of the method used for predicting the lifetime. Each individual annual mission profile results in corresponding annual accumulated damage and initially the lifetime from each profile is compared in order to examine if there is any significant difference from each other. Then, each profile is subjected to uncertainties and the failure percentage of the sample distribution from each annual profile, at the desired lifetime is compared, to provide a more understandable reliability measure. Finally, it is shown how much statistical uncertainty is needed to be included in the stress variables in order to account for the climate variation when compared to the analysis based on a single annual loading profile. This paper will further provide insights into the effectiveness of making the assumption of constant annual damage when predicting the lifetime of IGBT power modules.

Proposed Analysis Method

A three-phase 2-level voltage source PV-inverter (2L-VSI) is chosen as the case study to illustrate the proposed method as shown in Fig. 1. The component under investigation in this paper is the IGBT module [4], which realizes the power inversion stage. The parameters of the power module are summarised in Table I. A state-of-the-art reliability analysis is conducted, which is essential to clarify how much annual damage is caused by the thermal loading due to yearly weather conditions.

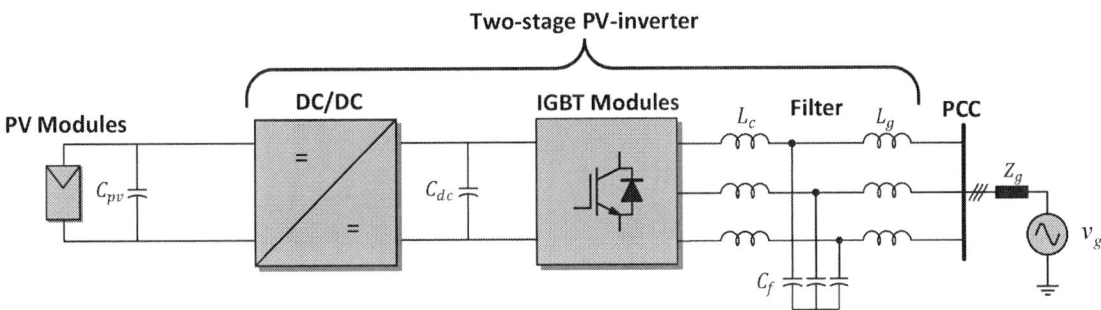

Fig. 1: Grid-Connected PV-system: C_{pv} is the PV input capacitor, DC/DC is the boost stage, C_{dc} is the DC-link capacitor and PCC is the point of common connection.

Table I: Data of the Power Module Under Study [4].

IGBT Based Inverter Power Module	
Parametrics	FS50R12KT4_B11
Configuration	Sixpack
Rated collector-emitter voltage V_{CE}	1200 V
Rated continuous collector current $I_{C,nom}$	50 A
Rated repetitive peak collector current I_{CRM}	100 A for $t_p = 1$ ms
Rated gate-emitter peak voltage V_{GE}	±20 V

Translation of Mission Profiles

The stress variables, which are used to evaluate the lifetime, are contained in the thermal loading profile. The environmental conditions are therefore needed to be translated into a thermal loading profile [5]. Initially, the PV array is sized in a manner, which suits the typical climate conditions of the operational location. The sizing of the PV array will influence the amount of the output power and therefore also the stresses applied to the IGBTs. A common approach is to oversize the PV array in respect to the converter rating, in order to obtain a wide-scale utilization of such systems, as the PV array rarely operates at its rated condition due to insufficient available solar radiance. The amount of surplus loading depends on the installation site, which also affects the converter lifetime [6]. Due to the relatively low available solar irradiance during winter time in Denmark, the PV array is chosen to be oversized by 20% in respect to the converter rating. Each annual profile, which contains solar radiance and ambient temperature, is the analysis input. The systems operation conditions are defined in terms of different solar irradiance levels and ambient temperatures, which amounts to 13 different operating conditions for the ambient temperature, T_a, and 11 different operating conditions for the solar irradiance, SI. Each entry in the operational interval maps a corresponding output power of the PV array, which is correlated by [7]

$$i = I_{ph}(G,T) - I_o(T)\left(e^{\frac{v+R_s i}{nN_s V_{th}(T)}} - 1\right) - \frac{v+R_s i}{R_p} \tag{1}$$

where I_{ph} is the photo-generated current, I_o is the dark saturation current of the PV module, R_s is the series resistance, n is the ideality factor, N_s is the number of series connected PV cells, V_{th} is the thermal voltage of a single cell and R_p is the shunt resistance. The output power lookup table is used to map the complete time series of the annual mission profiles into the annual output power profile of the PV-array, as elaborated in Fig. 2. The peturb and observe (PO) algorithm is used to control the boost converter, for maximum power point tracking (MPPT) as shown in Fig. 3.

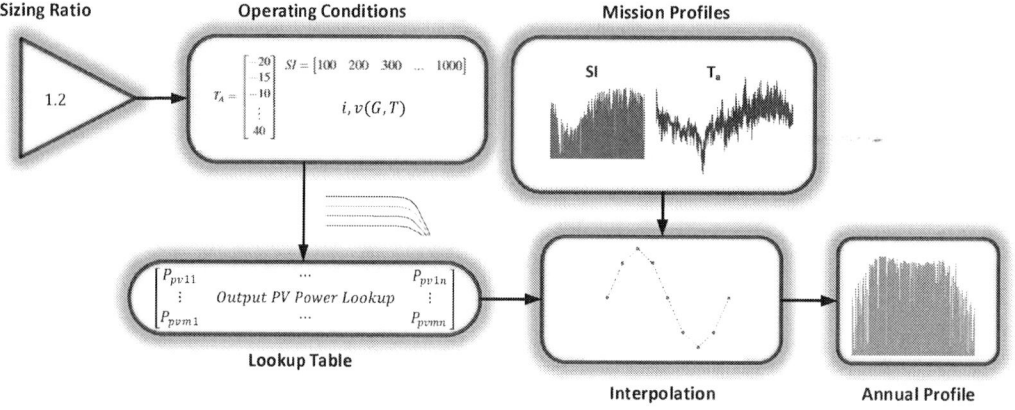

Fig. 2: Methodology for obtaining the annual profiles.

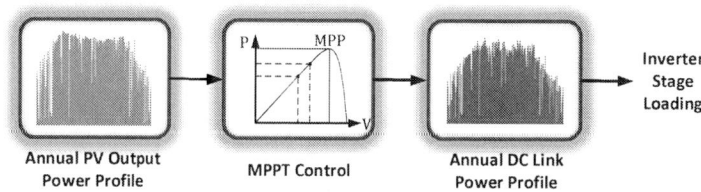

Fig. 3: Methodology for obtaining the inverter loading.

The extracted power is the annual loading profile supplied to the inversion stage. Applying the annual input DC-link power to the electrical circuit model of the inversion stage, will result in the decomposed loading profiles, i.e., the specific electrical loading of the components of interest. When the knowledge of the decomposed loading profiles is obtained, they can be applied to the loss and thermal model of the IGBTs, which willl result in an annual thermal loading. The switching and conduction losses are obtained by utilizing the built-in loss model in the simulation tool, i.e., PLECS. Instead of determining the switching losses of the IGBTs from current and voltage transients, the simulation tool measures the operation condition before and after each switching event.

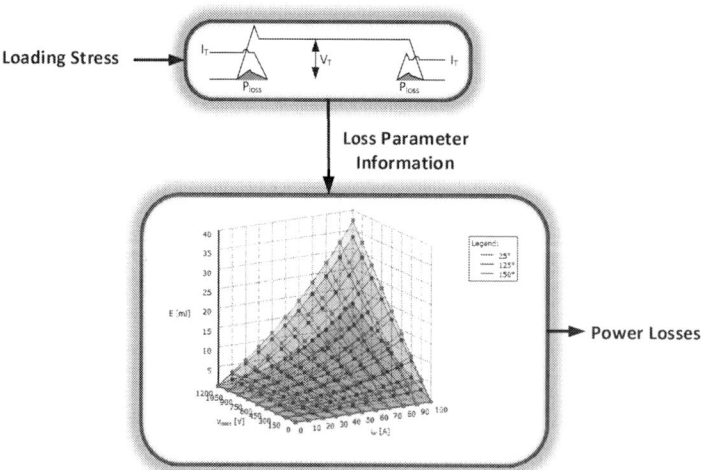

Fig. 4: Methodology for obtaining the IGBT power losses.

Based on these conditions, the dissipated losses are calculated by means of multi-dimensional look-up tables, which are constructed on the basis of the extracted datasheet information provided by the manufacturer [4]. Look-up tables for the turn-on and turn-off energies and the conduction losses are constructed, as shown in Fig. 4. The annual loss profile is subjected to the thermal model of the IGBTs, which consists of a thermal model describing the internal thermal characteristics of the power module and a thermal model describing the external cooling network. The internal and external parts are modelled by different types of networks, and the different layers of the module connecting the power chips to the case are modeled as a Foster network. Each layer is represented by a thermal resistance connected in parallel with a thermal capacitance, which describes the transient temperature characteristics, as shown in Fig. 5. The calculation of the total impedance from junction to case is analogous to those of a first-order electrical circuit as [8], [9]

$$Z_{th(j-c)}(t) = \sum_{i}^{n} R_i \left(1 - e^{-\frac{t}{\tau_i}} \right) \tag{2}$$

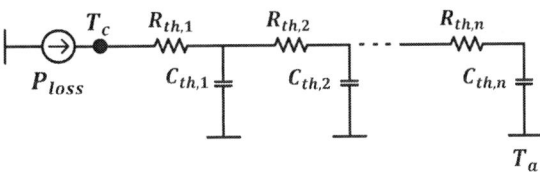

Fig. 5: Foster network, which is used to model the temperature difference from junction to case.

The thermal impedance representing the internal four layers of the power module is provided by the manufacturer datasheet [4]. As opposed to the internal layers of the power module, the cooling components are modeled by means of the Cauer model, whose parameters are based on actual physical properties. The parameters of each layer are obtained by means of finite element simulations and the parameters obtained for each layer are used in each of the RC lumps, as shown in Fig. 6 [8].

Fig. 6: Cauer network used to model the external cooling components

Unfortunately, it is not possible to combine the non-physical Foster model with an external Cauer network to model the case to ambient, without altering the thermal tendency of the junction to case temperature. A solution is to divide the thermal network into two, as shown in Fig. 7: one part, which will provide the right amount of power to the cooling network by means of a low pass filter based on Foster to Cauer transformation, as desribed in [10]. The adaption of this solution will then provide the right case temperature to the foster network, which will then provide the correct junction temperature.

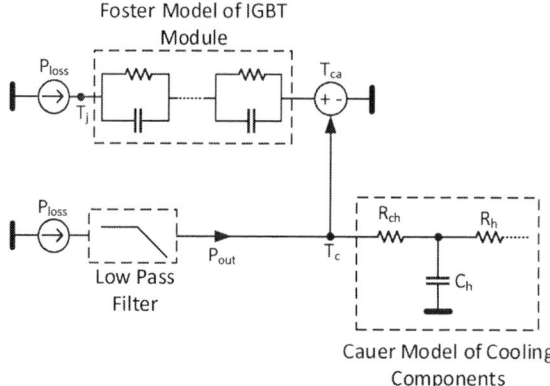

Fig. 7: Two circuit thermal network used as solution to combine the Foster network with an external Cauer network

With the use of the electro-thermal model, the thermal loading caused by each of the annaul mission profiles can be obtained, which is realised by obtaining the thermal junction temperature for the set of operating conditions defined previously and as shown in Fig. 2. The corresponding annual loading profiles are shown in Fig. 8.

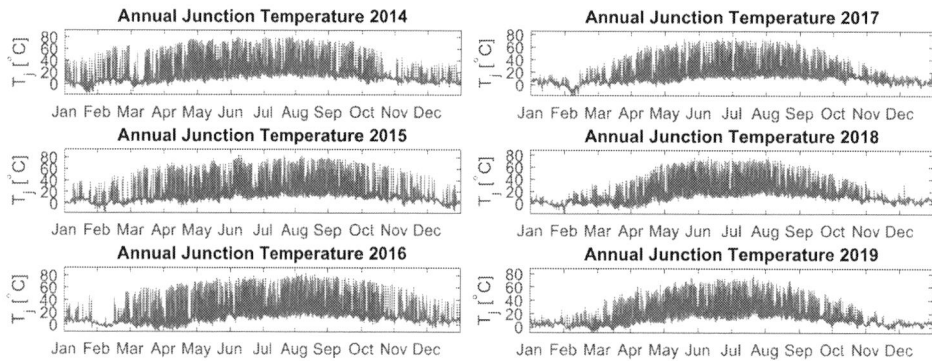

Fig. 8: Annual thermal loading profiles obtained from conduction the mission profile translation analysis on the annual mission profiles.

Wear-out Analysis

Due to the variation in the amount of cycles and the magnitude in each of these cycles contained in the thermal loading of each consecutive year, it will result in different of numbers of cycles to failure. However, due to the dynamical nature of the obtained annual thermal loading profiles, the profiles are not applicable with the used lifetime model. A counting method is therefore required, in order to decompose the cycles contained in the profile into categorized discrete values and then apply the categorized cycles to a damage model [11]. A rainflow algorithm, which is a widely used counting method in the stress analysis related to thermal cycling, is applied to the annual thermal loading profiles to extract the thermal stress variables at load reversals, as shown in Fig. 9.

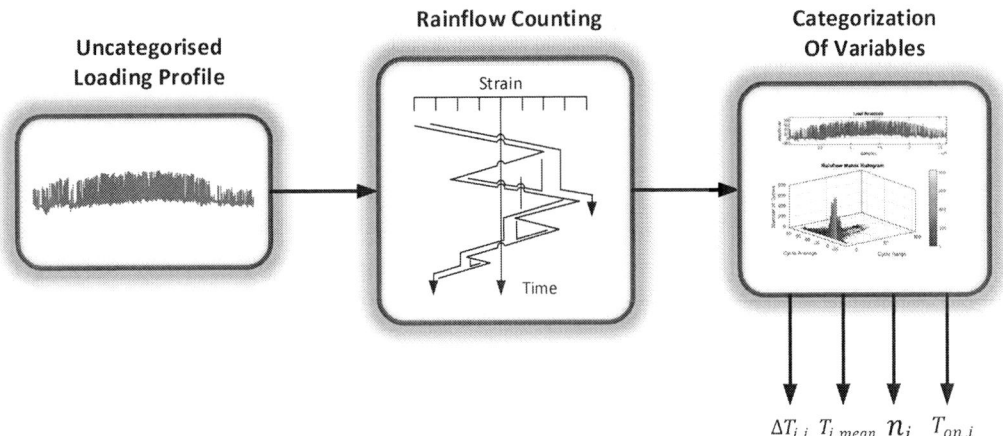

Fig. 9: Methodology to obtain categorized, discrete variables, which are directly applicable with the used lifetime model.

The extracted discrete stress variables are then applied to the lifetime model used for estimating the lifetime of the IGBTs, which is expressed as [12]

$$N_f = L \cdot (\Delta T_j)^{\beta_1} \cdot e^{\left(\frac{-\beta_2}{T_{j,mean}}\right)} \cdot \left(\frac{t_{on}}{1.5}\right)^{-\beta_3} \tag{3}$$

The model in (3) is based on the Coffin-Manson law, which relates the temperature variation amplitude, ΔT_j, with the number of cycles to failure. Additionally, the cycle period, t_{on}, and the exponential Arrhenius term is added to take into account the lifetime degrading effect of the mean junction temperature. The parameters of the lifetime model, L, β_1, β_2 and β_3, is determined with the aid of curve fitting the experimental data, as presented in [13]. The annual damage of each profile is calculated using the Miner's rule, with the assumption of linear damage accumulation and it is expressed as [14]

$$A_D = \sum_{i=1}^{n} \frac{n_i}{N_{f,i}} \tag{4}$$

where n_i is the amount of cycles extracted from the rainflow counting and $N_{f,i}$ is the number of cycles until the end-of-life obtained from (3). The corresponding lifetime from each individual year is presented in Fig. 10.

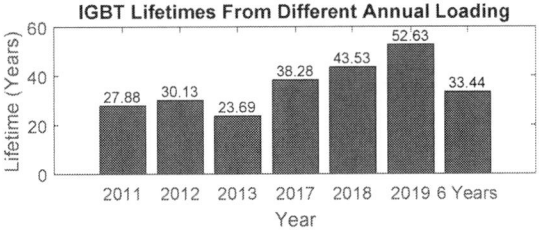

Fig. 10: The lifetimes obtained from each of the individual loading profiles.

As it can be observed in the Fig. 10, there is a significant amount of difference in the resulting lifetime. The far right column represents the lifetime based on all 6 years of mission profiles, which is the correct measure to use for estimating the lifetime. As for the lifetime based on single years, they deviate as much as up to 57% with respect to the lifetime obtained when using all 6 years. Still, the fixed values of the lifetime evaluation is not a useful measure, as these values only occur at one certain case, where all power devices fail at the same instant. The lifetime estimation can be improved by including statistical uncertainty, which takes into account the difference in operating conditions, which vary from those during the experimental testing by including tolerance in the lifetime model parameters. Additionally the electrical parameters of the components can differ due to variations in the manufacturing process. Each parameter is modelled as a probability density function, with each having carefully chosen variance [15], [16]. The tolerances used to model the lifetime model parameters are based on [17], which conducted some regression models to determine the tolerances. An exception is the multiplicative L parameter, which is given a tolerance of 10% of its mean value. Each of the parameters respective tolerances are summarised in Table II.

Table II: Tolerances used when modeling the lifetime model parameters as distribution functions [17]

Tolerances Used for Distributions Functions	
Multiplicative Temperature Variation Parameter A	$1.34 \cdot 10^{24}$
Lifetime Model Exponential Temperature Variation Parameter β_1	0.281
Lifetime Model Exponential Mean Temperature Parameter β_2 (E_a/k_b)	83.33
Lifetime Model Exponential Cycle On-time Parameter β_3	0.1

As for the case of the electrical parameter variation, the collector-emitter voltage tolerance is taken into consideration with a tolerance in accordance with the datasheet information [4]. The variation of the collector emitter voltage will have an indirect influence on the junction temperature by causing variance in the conduction losses. Finally, the variation of all the parameters are executed, in order to obtain an overall lifetime distribution of the IGBTs. 10000 random samples are taken from each of the model parameters distributions, which are subjected to variation within their predefined tolerances as outlined in Fig. 11.

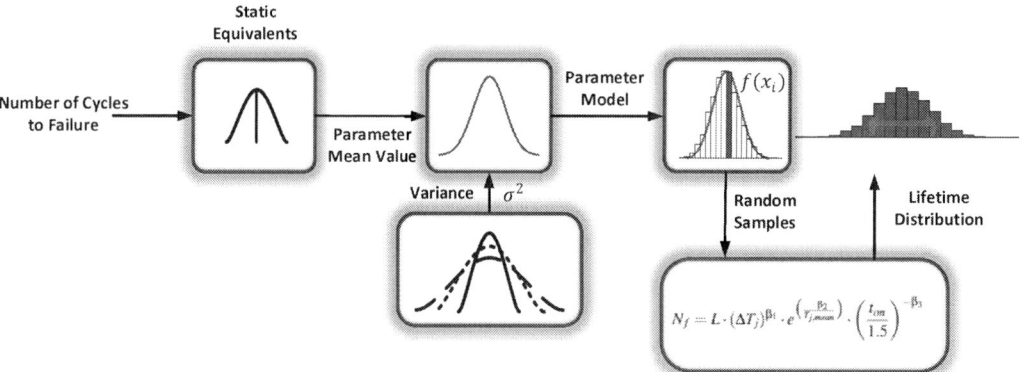

Fig. 11: Used method to obtain the lifetime distributions

The obtained lifetime yield for each of the 10000 samples is fitted by use of the Weibull distribution function, which can be expressed as [18]

$$f(t) = \frac{\beta}{\eta^\beta} t^{\beta-1} \exp\left[-\left(\frac{t}{\eta}\right)^\beta\right] \tag{5}$$

where β is the shape parameter, η is the scale parameter, corresponding to the time when 63.2% of a population has failed. From the obtained lifetime distributions, the reliability of the power modules can be expressed in terms of how big a percentage of the entire population that have failed at the desired operational lifetime, which is, opposite the fixed lifetime value, a useful measure. In order to evaluate this measure of reliability, the Weibull cumulative distribution function, commonly referred to as the failure probability function, needs to be considered

$$F(t) = \int_0^t f(t)dt = 1 - \exp\left[-(\frac{t}{\eta})^\beta\right] \tag{6}$$

By means of the probability function (6), the overall failed population at the desired lifetime can be executed. The failure probability functions of the IGBT power module for each annual profile are shown in Fig. 12.

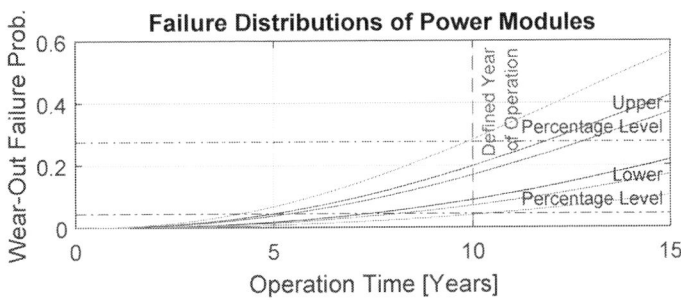

Fig. 12: The failure probability functions of the IGBT power module for each annual profile.

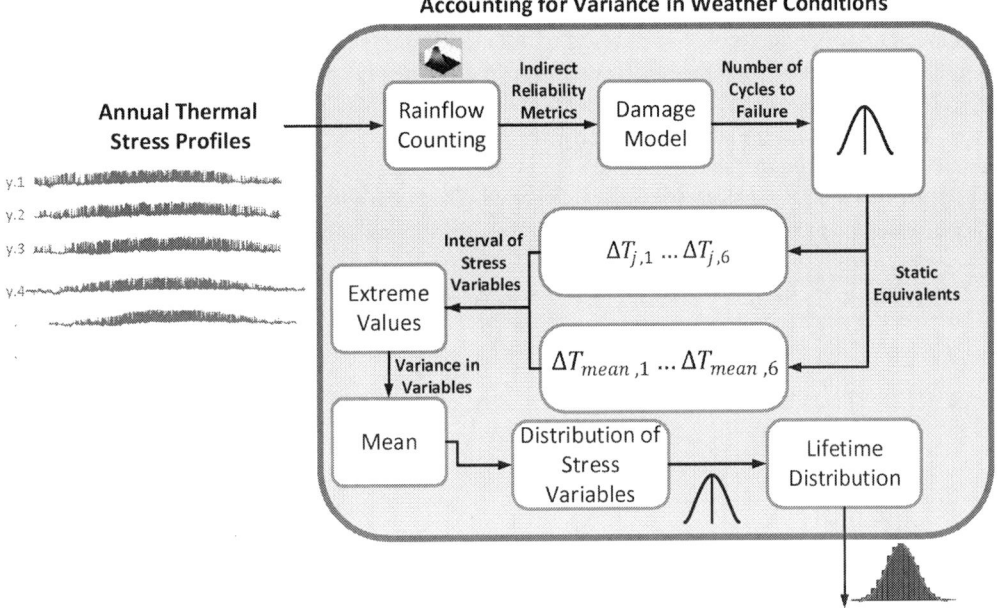

Fig. 13: Workflow needed to include the statistical uncertainty caused by difference in climate conditions.

The upper and lower percentage level of failure at 10 years of operation is 27.6% and 4.42%, respectively. The difference in the failure proportion is a clear indication of the lifetime of the IGBTs being significantly dependent on the environmental weather variation on a year-to-year basis, which therefore needs to be taken into account when estimating the lifetime.

Annual Climate Variation Effect on Stress Variables

The lifetime of IGBTs can still be validly estimated by means of a single year mission profile by including the statistical uncertainty that is a result of having different annual climate conditions. Each of the applied mission profiles results in a specific amount of number of cycles to failure, when evaluating the entire stress profile by iteration.

Table III: Parameters used to include statistical uncertainty in the stress variables to take into account the difference in annual climate conditions.

Uncertainty Parameters				
Year	N_f (Obtained from Iteration)	$\Delta T_{j,static}$	$T_{j,mean,static}$	$T_{on,static}$
2011	362020	79.53	41.78	8.67
2012	391880	79.63	42.57	9.64
2013	310840	81.13	43.93	8.82
2017	542700	68.61	38.12	8.36
2018	600040	68.94	38.04	8
2019	754780	63.12	36.56	9.95

Considering each term in the lifetime model, one specific static value of the stress variables corresponds to the same amount of cycles to failure, as the ones obtained when the iteration of the entire loading profiles was performed as shown in the following expression

$$N_{f,i} = N_{f,static} = A \cdot \left(\Delta T_{j,static}\right)^{\beta_1} \cdot e^{\left(\frac{E_a}{k_b \cdot T_{j,mean,static}}\right)} \cdot \left(\frac{T_{on,static}}{1.5}\right)^{\beta_2} \tag{7}$$

Each of the applied mission profiles will result in different static stress variable values, which will constitute the variance and mean of the of stress variable distribution by considering the extreme values and the overall mean values as shown in Fig. 13. From each thermal loading profile, the corresponding parameters, which represent the statistical uncertainty for inclusion in the stress variables, are summarised in Table III. As observed in Table III, the extreme values are mostly contained in the annual profiles of 2013 and 2019 with the exception of the on cycle period. Using the obtained extreme values to constitute the variance and the mean values of the stress variables, the models shown in Fig 14 are obtained.

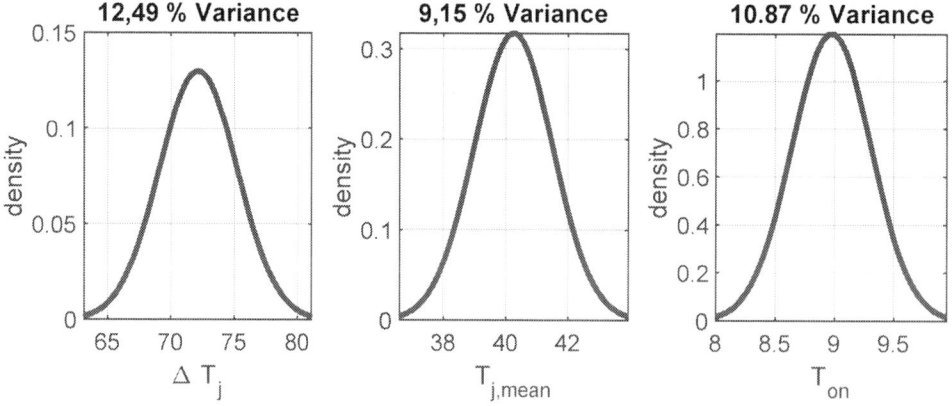

Fig. 14: Model of stress variables, which accounts for the annual diversity in climate conditions.

Including the variances stated in the Fig. 14, the lifetime can be estimated on the basis of only one single year of the applied mission profile.

Conclusion

This paper examines how much influence the year-to-year climate variation has on the lifetime prediction of IGBT power modules used in three-phase PV inverters. State-of-the-art reliability modelling is used to obtain fixed value lifetime for each of the individual mission profiles. Special attention is put on presenting a useful reliability measure in terms of failure percentage under realistic operational lifetime. A considerable difference in failure percentage of more than 20% indicates that it is not valid to perform the lifetime estimation only based on a single year mission profile. Finally, a method is presented to ensure the effectiveness of single year profile based estimations, by including the statistical uncertainty in the stress variables, which will take into account the year-to-year climate variation.

References

[1] H. Wang, F. Blaabjerg, K. Ma, and R. Wu, "Design for reliability in power electronics in renewable energy systems – status and future," in 4th International Conference on Power Engineering, Energy and Electrical Drives, Istanbul, 2013, pp. 1846-1851.

[2] S. Yang, A. Bryant, P. Mawby, D. Xiang, L. Ran and P. Tavner, "An industry-based survey of reliability in power electronic converters," in 2009 IEEE Energy Conversion Congress and Exposition, San Jose, CA, 2009, pp. 3151-3157.

[3] A. Sangwongwanich, Y. Yang, D. Sera and F. Blaabjerg, "Mission Profile-Oriented Control for Reliability and Lifetime of Photovoltaic Inverters," IEEE Transactions on Industry Applications, vol. 56, no. 1, pp. 601-610, Jan.-Feb. 2020.

[4] Infineon. Technical Information FS50R12KT4B11, 2013.: 21052020.

[5] P. D. Reigosa, H. Wang, Y. Yang and F. Blaabjerg, "Prediction of Bond Wire Fatigue of IGBTs in a PV Inverter Under a Long-Term Operation," IEEE Transactions on Power Electronics, vol. 31, no. 10, pp. 7171-7182, Oct. 2016.

[6] A. Sangwongwanich, Y. Yang, D. Sera and F. Blaabjerg, "Impacts of PV array sizing on PV inverter lifetime and reliability," in 2017 IEEE Energy Conversion Congress and Exposition (ECCE), Cincinnati, OH, 2017, pp. 3830-3837.

[7] D. Sera, R. Teodorescu, and P. Rodriguez, "PV panel model based on datasheet values," 2007 IEEE International Symposium on Industrial Electronics, Vigo, 2007, pp. 2392-2396.

[8] M. Marz. Thermal Modeling of Power-electronic Systems, Fraunhofer Institute, 2013.

[9] F. Blaabjerg and K. Ma, "Future on Power Electronics for Wind Turbine Systems," IEEE Journal of Emerging and Selected Topics in Power Electronics, vol. 1, no. 3, pp. 139-152, Sept. 2013.

[10] K. Ma, N. He, M. Liserre and F. Blaabjerg, "Frequency-Domain Thermal Modeling and Characterization of Power Semiconductor Devices," IEEE Transactions on Power Electronics, vol. 31, no. 10, pp. 7183-7193, Oct. 2016.

[11] H. Huang and P. A. Mawby, "A Lifetime Estimation Technique for Voltage Source Inverters," IEEE Transactions on Power Electronics, vol. 28, no. 8, pp. 4113-4119, Aug. 2013.

[12] U. Scheuermann, R. Schmidt and P. Newman, "Power cycling testing with different load pulse durations," in 7th IET International Conference on Power Electronics, Machines and Drives (PEMD 2014), Manchester, 2014, pp. 1-6.

[13] Infineon, Technical Information IGBT Modules Use of Power Cycling curves for IGBT 4, Warstein, 2010.

[14] G. Zhang, D. Zhou, F. Blaabjerg and J. Yang, "Mission profile resolution effects on lifetime estimation of doubly-fed induction generator power converter," in 2017 IEEE Southern Power Electronics Conference (SPEC), Puerto Varas, 2017, pp. 1-6.

[15] A. Sangwongwanich, Y. Yang, D. Sera and F. Blaabjerg, "Lifetime Evaluation of Grid-Connected PV Inverters Considering Panel Degradation Rates and Installation Sites," IEEE Transactions on Power Electronics, vol. 33, no. 2, pp. 1225-1236, Feb. 2018.

[16] Y. Shen, H. Wang, Y. Yang, P. D. Reigosa and F. Blaabjerg, "Mission profile based sizing of IGBT chip area for PV inverter applications," in 2016 IEEE 7th International Symposium on Power Electronics for Distributed Generation Systems (PEDG), Vancouver, BC, 2016, pp. 1-8.

[17] R. Bayerer, T. Herrmann, T. Licht, J. Lutz, and M. Feller. Model for power cyclinglifetime of igbt modules - various factors influencing lifetime. pages 1 – 6, 04 2008.

[18] H. Chung, H. Wang, F. Blaabjerg, and M. Pecht. Reliability of Power ElectronicConverter Systems. Number ISBN-978: 1-84919-901-8 in IET Power and EnergySeries. The Institution of Engineering and Technology, London, United Kingdom,2015.

Communication-Free Secondary Frequency and Voltage Control of VSC-Based Microgrids: A High-Bandwidth Approach

Rasool Heydari[1], Mohammad S. Golsorkhi[2], Mehdi Savaghebi[1], Tomislav Dragicevic[3], and Frede Blaabjerg[4]

[1] Electrical Engineering Section, The Mads Clausen Institute, University of Southern Denmark, Odense, Denmark

[2] Department of Electrical and Computer Engineering, Isfahan University of Technology, Isfahan, Iran

[3] Department of Electrical Engineering, Technical University of Denmark (DTU), Copenhagen, Denmark

[4] Department of Energy Technology, Aalborg University, Aalborg, Denmark

rah@mci.sdu.dk, golsorkhi@iut.ac.ir, mesa@mci.sdu.dk, tomdr@elektro.dtu.dk, fbl@et.aau.dk

Keywords

≪Communication-free control≫, ≪Decentralized control≫, ≪Microgrids≫, ≪Secondary frequency control≫, ≪voltage source converter (VSC)≫

Abstract

In this paper, a decentralized secondary control strategy for microgrids, with fast dynamic response is proposed. This high bandwidth approach is realized by applying a finite control set, model predictive control (FCS-MPC) at the primary control level of the voltage source converters (VSCs) control. At the upper control level, a novel decentralized secondary control structure is proposed to regulate the islanded microgrid voltage and frequency subsequent to load change, with no need of any communication infrastructure. The proposed control strategy, restores the microgrid frequency and voltage to the nominal value while maintaining accurate power-sharing of the droop mechanism. Experimental results are also provided to verify the effectiveness of the proposed approach.

Introduction

Microgrid (MG) is introduced as a promising solution for integrating renewable energy sources (RES) in the modern power grid. However, a high penetration of the converter-based RES, and non-synchronous solid state generators (NSGs) introduce new challenges in terms of frequency and voltage control in the power grid. Due to the inherent low-inertia characteristic of NSGs, voltage and frequency control have become a major challenge in the MG operation and control. Therefore, a proper and robust control strategy, suitable for the low inertia-grids, is highly demanded.

Power converters are the backbone of an MG, which serve as the interface between the dc and ac sides and enable the RES to inject power into the MG system. Several control structures have been presented in the literature to control the output voltage and frequency of paralleled power converters in the MG. The major roles of the control strategy are as follows [1, 2]:

- MG frequency and voltage amplitude regulation

- Accurate RES coordination and load sharing

- Control of power flow between the main grid and the MG

- MG synchronization with the main grid

- Minimizing the MG operational cost

- Emergency control

These control objectives are very different in terms of timescale and functionalities. Therefore, the hierarchical control structure for the MG is presented and reviewed in [3] in order to perform a stable operation either in islanding mode and connected to the main grid [4]. The hierarchical control structure presented in the literature comprises cascading inner voltage and current control loops at the primary control level (PCL) [5, 6]. On top of the PCL, a secondary control level (SCL) is utilized, which consists of more slow frequency and voltage controllers as well as low bandwidth communication link (LBCL). Finally, the tertiary control level manages the optimal operation of the MG, economically dispatching and control of the power flow between the MG and the main power grid. The tertiary controllers are out of the scope of this study. Although this control structure is widely accepted in the literature, it suffers from several limitations. Firstly, multi-loop cascaded control structures have an inherent slow dynamic response. Furthermore, from a practical point of view, this control structure can be affected by the communication network (CN) uncertainties and also threated by cyber-attacks. Therefore, a fast and robust control structure is one of the key requirements of a reliable MG control.

An alternative solution is to use a new control structure that reacts more quickly compared to conventional control strategies [7, 8]. To increase the dynamic of the voltage and frequency restorations, the PCL should be designed for a higher frequency bandwidth, and consequently the SCL can be operated at higher bandwidths. Such a system, however, should at the same time be sufficiently robust against communication uncertainties.

In this regard, the SCL architectures can be categorized into three main structures based on the required communication link, i.e., centralized secondary control (CSC), distributed secondary control (DISC), and decentralized secondary control (DESC) [2]. Fig. 1 shows the three secondary control architecture diagrams. Conventionally, the secondary control comprised a centralized control unit and low bandwidth communication links to share the required data among distributed generation units (DGUs). This structure relies on central controller and communication links. Therefore, any failure in the main control unit or communication infrastructure degrades the SCL performance and consequently may lead to the MG frequency and voltage instability. In order to address the aforementioned problem, distributed SCL is presented in [9, 10, 11, 12]. The distributed SCL architecture is divided into three main categories, i.e., averaging, consensus and event-triggered structures. Although these structures don't need a central controller, they still need a communication infrastructure to share the required data [13, 14].

In the DESC structure, communication links are not required for frequency and voltage restoration, and each DGU regulates the voltage amplitude and frequency locally. Fig. 2 shows a general schematic diagram of the proposed DESC structure. In this structure, firstly an finite control set, model predictive control (FCS-MPC) is utilized to enhance dynamic response of the MG in the inner control loop of PCL, then at the upper control level, a droop control and virtual impedance are employed to guarantee accurate power sharing. The proposed communication-free voltage and frequency restoration are embedded into the droop control, which is presented in the following section.

Secondary Voltage and Frequency Control Architectures

In the CSC structure, DGUs send the required data, e.g., voltage amplitude and the frequency, to a central controller through a communication infrastructure. The central controller also sends appropriate signals to each DGU to restore the voltage and frequency in the MG.

As it can be seen from Fig. 1(a), CSC structure totally relies on the communication links. Consequently, any failure, either in communication infrastructure or the main central controller, may affect the SCL performance and system stability. As the DGUs in the MG can be spatially distributed and heterogeneous, the distributed control structure is a promising approach to improve the MG stability, scalability, and performance. In this control structure shown in Fig.1(b), the DGUs cooperate together to fulfill a

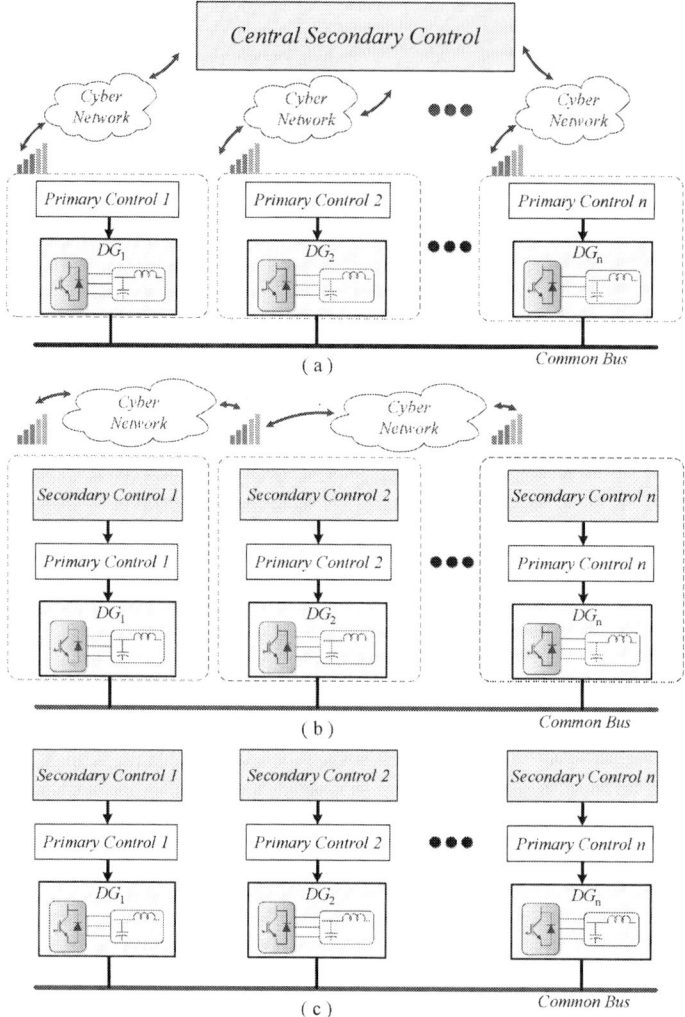

Fig. 1: Microgrid secondary control architectures with and without communication: a) centralized secondary control (CSC), b) distributed secondary control (DISC) , and c) decentralized secondary control (DESC) [2]. DG: Distributed generation.

set of objectives. According to the communication link topologies and transmission algorithms, the distributed secondary control (DISC) can be classified into three main categories, i.e., distributed averaging, distributed consensus and event-triggered structures [2, 15].

Finally, in a decentralized secondary control (DESC) architecture, communication links are not required and the voltage and frequency regulations are carried out locally, using the local measurements.

Proposed High Bandwidth DESC structure

At the inner loop of PCL, a finite control set model predictive control (FCS-MPC) of the power converter is employed. The FCS-MPC takes advantage of discrete nature of the power converter and a known model of its output filter to explicitly find the converter switch configuration, which minimizes a certain cost function at every sampling instant [16].

A general two-level three-phase power converter comprises three legs (a, b and c). Each leg consists of upper and lower switches, which are controlled by external gating signals (S_a, S_b, and S_c). Only one of the switches in each leg is turned on at a given time in order to avoid shoot-through of the leg.

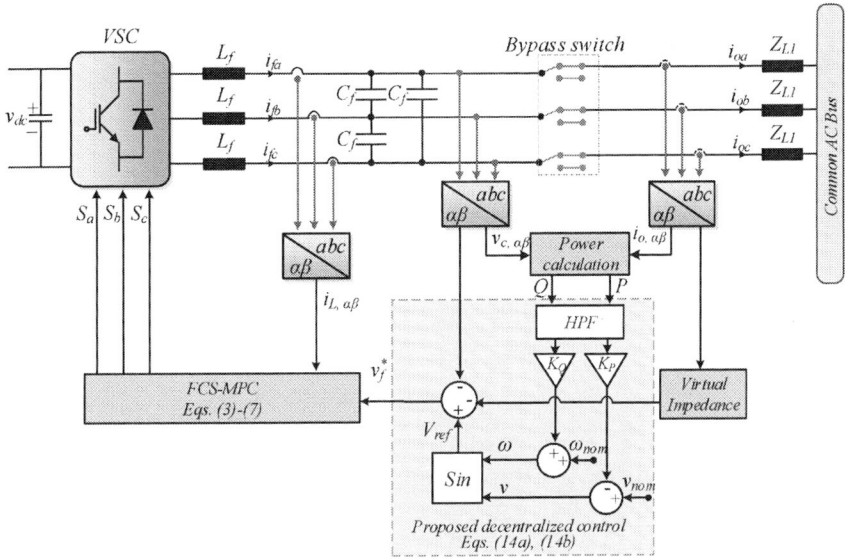

Fig. 2: Proposed communication-free control structure employing a high bandwidth inner finite control set, model predictive control (FCS-MPS).

The controller predicts the behavior of the controlled variables for all feasible switching states. Then, it evaluates predefined cost function (CF) for each prediction. Finally, the switching state that minimizes the CF is selected.

The main objective of the control approach is to successively select the voltage input vectors so that the output voltage accurately tracks the reference voltage trajectory with minimum error [17]. To get a discrete-time model, the Euler forward method is employed in the discretization as follows:

$$\frac{dx}{dt} = \frac{x(k+1) - x(k)}{\bar{t}_c}, \tag{1}$$

where, $\bar{t}_c$ is the sampling time. Conventionally, an Euclidean distance based cost function is used in a finite control set MPC to minimize the error between the reference voltage and predicted output voltage of the VSC as follows:

$$g_{conv}(t_k) = \left[v^*_{c\alpha}(t_k) - v_{c\alpha}(t_{k+1}) \right]^2 + \left[v^*_{c\beta}(t_k) - v_{c\beta}(t_{k+1}) \right]^2, \tag{2}$$

In order to decrease the total harmonic distortion (THD) a derivative error term (G_d) is added to the cost function (CF) as follows [7]:

$$CF : \|\bar{v}_e(i)\|^2 + \xi_{lim}(i) + \zeta_w SW^2(i)) + G_d, \tag{3}$$

$$\bar{v}_e(i) = \bar{v}^*_f(i) - \bar{v}_f(i), \tag{4}$$

$$\xi_{lim}(i) = \begin{cases} 0, & \text{if } |i_f(i)| \le i_{max} \\ \infty, & \text{if } |i_f(i)| > i_{max} \end{cases}, \tag{5}$$

$$SW(i) = \sum |u(i) - u(i-1)|, \tag{6}$$

$$G_d = \left(\frac{d\bar{v}_f^*(t)}{dt} - \frac{d\bar{v}_f(t)}{dt} \right) = \\ (C_f \omega_{ref} v_{f\beta}^* - i_{f\alpha} + i_{o\alpha})^2 + (C_f \omega_{ref} v_{f\alpha}^* - i_{f\beta} + i_{o\beta})^2, \tag{7}$$

where, $v_e(i)$ is the output prediction error, $\bar{v}_f^*(i)$ represents the reference voltage and $\bar{v}_f(i)$ is the predicted output voltage in α-β reference frame. $\xi_{lim}(i)$ is the current constraint, $SW(i)$ is the switching effort with a weighting factor ζ_w and G_d exposes the derivative error. The reference voltage is determined through droop control and virtual impedance at the upper level.

Since the pulse-width modulation (PWM) delay and inner cascaded control loops are eliminated in the FCS-MPC algorithm, the bandwidth of the MG is limited by only the sampling time. By replacing the inner control loops in the PCL with an FCS-MPC, the dynamic response time of the MG can be reduced in the order of magnitude, and it is much faster than the conventional control structures. However, the reference set-points are determined by the droop control at upper control level, which has inherent steady state error.

Frequency and Voltage Restoration

The conventional droop control has steady state error, and also the line impedance (resistive or inductive) affects on the droop characteristic and control performance [18]. By applying virtual impedance, the output impedance, seen by the VSC, can be enforced to be either purely inductive or resistive. Therefore, for the resistive virtual impedance, the active and reactive power exchange can be obtained as follows:

$$Q = \frac{-V_{MG}V_c}{R_{L_i}} \delta_i , \tag{8a}$$

$$P = \frac{V_{MG}}{R_{L_i}} (V_c - V_{MG}) , \tag{8b}$$

where, P and Q are the exchange active and reactive power respectively. Moreover, V_{MG} and V_c stand for the MG voltage and the VSC voltage amplitude, respectively. Finally, δ_i is power angle of the VSC. Therefore, the droop characteristic can be achieved as follows:

$$\omega_i = \omega_{nom} + k_Q Q , \tag{9a}$$

$$V_i = V_{nom} - k_P P , \tag{9b}$$

where, k_Q and k_P are the droop coefficients. And ω_{nom}, and V_{nom} are the reference frequency and nominal voltage amplitude, respectively.

Taking a derivative from the droop equation with respect to time, and defining $\Delta\omega_i = \omega_i - \omega_{nom}$, $\Delta P_i = P_i - P_{nom}$ and $\Delta v_i = v_i - v_{nom}$, the following equations are achieved:

$$\frac{d}{dt}\Delta\omega - k_Q\frac{d}{dt}Q = 0 \ , \tag{10a}$$

$$\frac{d}{dt}\Delta v + k_P\frac{d}{dt}P = 0 \ . \tag{10b}$$

The above equations show the power sharing accuracy in steady state, however they cannot restore the MG frequency and voltage. It can be accomplished by adding compensation terms as follows:

$$\frac{d}{dt}\Delta\omega - k_Q\frac{d}{dt}Q + m_Q\Delta\omega = 0 \ , \tag{11a}$$

$$\frac{d}{dt}\Delta v + k_P\frac{d}{dt}P + m_P\Delta v = 0 \ . \tag{11b}$$

Note that the derivative terms can be considered zero in steady state. Thus, equations (11) enforce the voltage and frequency deviations (i.e. Δv and $\Delta\omega$) to be zero in steady state. Therefore, if the MG is stable, then by applying the final value theorem to the equations (11), one can show that ω and v converge to their reference values [2]. By applying Laplace transform to (11) the transfer function can be obtained as follows:

$$s\Delta\omega(s) - k_Q s Q(s) + m_Q\Delta\omega(s) = 0 \ , \tag{12a}$$

$$s\Delta v(s) + k_P s P(s) + m_P\Delta v(s) = 0 \ , \tag{12b}$$

which can be rewritten as follows:

$$\frac{\Delta\omega(s)}{Q(s)} = \frac{k_Q s}{s + m_Q} \ , \tag{13a}$$

$$\frac{\Delta v(s)}{P(s)} = -\frac{k_p s}{s + m_P} \ . \tag{13b}$$

and in the more common droop form, it can be rewritten as:

$$\omega = \omega_{nom} + k_Q\left(\frac{s}{s + m_Q}\right)Q \ , \tag{14a}$$

$$v = v_{nom} - k_P\left(\frac{s}{s + m_P}\right)P \ . \tag{14b}$$

As it can be seen, the power-sharing, frequency and voltage restoration are realized by applying a first-order high-pass filter (HPF), which passes the transient component of the signals. This washout-based filter can be tuned based on the restoration time of voltage and frequency. Therefore, the proposed approach maintained the reference frequency and voltage subsequent to the load change in the MG with no need for a communication network infrastructure.

Experimental Setup and Results

In order to evaluate the performance of the proposed control structure, a three-DGU MG setup as shown in Fig. 3 is implemented. Each 15 kVA power converter is connected to a common ac bus through an LC filter, and a three-phase RL load is supplied in the MG. The rated voltage amplitude and frequency are 200 V, and 50 Hz, respectively. The system parameters can be obtained from Table I. In order to have an

Table I: Parameters of the Test System

Electrical Parameters		
Parameters	Symbol	Value
Output voltage of rectifier	V_{DC}	550 V
Nominal voltage magnitude	V_{nom}	200 V
Nominal Frequency	ω_{nom}	50 Hz
Sampling time	T_s	25 μ s
Capacitance of LC filter	C_f	25 μ F
Inductance of LC filter	L_f	1.8 mH
Inner loop coefficients and other control Parameters		
Control Parameters	DGU: 2	DGU: 1 and 3
$P-v$ droop coefficient	0.001 V/W	0.002 V/W
$Q-\omega$ droop coefficient	0.005 rad/VAr.s	0.01 rad/VAr.s

interface between the controller and the power hardware in the loop (PHiL), a dSPACE MicroLabBox board is utilized.

The main results achieved by using the proposed control strategy are shown in Fig. 4. In the first scenario, the proposed FCS-MPC is activated, but the secondary control is deactivated. The virtual impedance, as well as conventional droop control are also implemented to share the power among DGUs and control the frequency and voltage of the system.

Fig. 4 (a) shows the frequency deviations during a load change. As it can be seen from Fig. 4 (a), the primary control and droop controller work with high bandwidth and react very fast to load change. Compared to the conventional control structures presented in the literature, where the restoration time is in the order of seconds, the proposed model restores the frequency deviations in the order of milliseconds (see Fig. 16 in [9]). This is a significant achievement since both systems use similar physical parameters. However, as it can be seen from Fig. 4 (a), the conventional droop controller suffers from steady-state error. Thus, a secondary controller and communication links are required in the conventional structure to restore the MG frequency to the nominal values.

It is worth to note that a resistive virtual impedance is employed, hence, based on the (9) the load connection leads to frequency increase. In the second scenario, the proposed control structure is activated. The virtual resistive impedance is also implemented. Fig. 4 (b) shows the frequency control performance of the proposed control structure. As can be seen, after a load change,

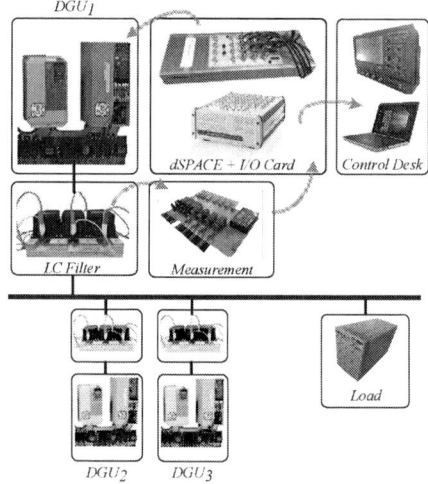

Fig. 3: Experimental setup with three converters, LC filters, measurements and dSPACE system for control.

the frequency deviates from the nominal value. However, the proposed controller can quickly restore the MG frequency to the nominal value. Furthermore, there is no need for communication links to share the desired data among DGUs and the controllers. Fig. 4 (c) shows the voltage deviations during a load change in the first scenario when the conventional control method is implemented. As there is no secondary controller, and the conventional droop control has a steady-state error, the voltage cannot reg-

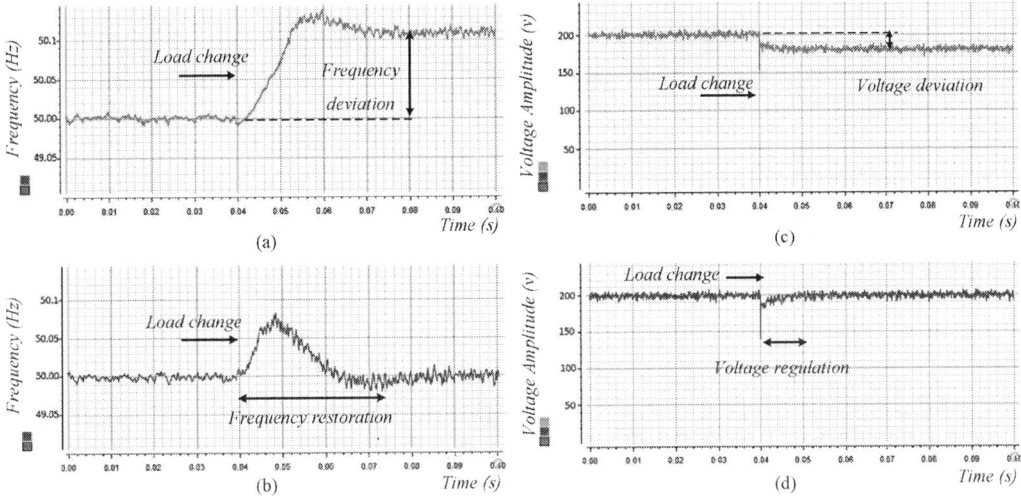

Fig. 4: Experimental verification and results, (a) conventional droop control performance for frequency restoration during a load change, (b) the proposed frequency control performance, (c) performance of voltage control employing conventional droop control, (d) the proposed voltage control performance.

ulate to the nominal value. However, it still has a fast transient response. When several voltage source converters are connected in parallel in the MG, additional voltage drops are also introduced because of the inherent droop control laws. Hence, the ability to regulate the drops with high bandwidth is of instrumental importance. In Fig. 4 (d), the proposed control structure is applied. As can be seen, the voltage restores very fast to the nominal values with no need for communication links and data exchange. Thereby it also is robust in cyber security sense.

Conclusion

In this paper, a high-bandwidth control structure is presented for parallel voltage source converters forming an islanded microgrid. This approach is realized by employing a finite control set, model predictive controller (FCS-MPC) at the inner control loop of the power converters. Droop control and resistive virtual impedance are also employed to maintain accurate power sharing among the DGUs. Compared to the conventional control structures, which have a steady-state error and need a secondary controller, in the proposed control structure, there is no need for communication links and secondary controller at the upper control level of hierarchical control structure. These communication-free voltage and frequency restoration method are realized by modifying the droop control to integrate the secondary control. Experimental results validate the performance of the proposed control structure.

References

[1] A. Bidram and A. Davoudi, "Hierarchical structure of microgrids control system," *IEEE Transactions on Smart Grid*, vol. 3, no. 4, pp. 1963–1976, 2012.

[2] Y. Khayat, Q. Shafiee, R. Heydari, M. Naderi, T. Dragicevic, J. W. Simpson-Porco, F. Dorfler, M. Fathi, F. Blaabjerg, J. M. Guerrero, and H. Bevrani, "On the secondary control architectures of ac microgrids: An overview," *IEEE Trans. Power Electrons.*, 2019.

[3] J. M. Guerrero, J. C. Vasquez, J. Matas, L. G. De Vicuña, and M. Castilla, "Hierarchical control of droop-controlled ac and dc microgridsa general approach toward standardization," *IEEE Trans. Ind. Electron.*, vol. 58, no. 1, pp. 158–172, 2011.

[4] M. Chen and X. Xiao, "Hierarchical frequency control strategy of hybrid droop/vsg-based islanded microgrids," *Electric Power Systems Research*, vol. 155, pp. 131–143, 2018.

[5] M. Savaghebi, A. Jalilian, J. C. Vasquez, and J. M. Guerrero, "Secondary control for voltage quality enhancement in microgrids," *IEEE Trans. Smart Grid*, vol. 3, no. 4, pp. 1893–1902, 2012.

[6] M. Najjar, A. Moeini, M. K. Bakhshizadeh, F. Blaabjerg, and S. Farhangi, "Optimal selective harmonic

mitigation technique on variable dc link cascaded h-bridge converter to meet power quality standards," *IEEE Journal of Emerging and Selected Topics in Power Electronics*, vol. 4, no. 3, pp. 1107–1116, 2016.

[7] R. Heydari, T. Dragicevic, and F. Blaabjerg, "High-bandwidth secondary voltage and frequency control of vsc-based ac microgrid," *IEEE Trans. Power Electrons.*, vol. 34, no. 11, pp. 11 320–11 331, 2019.

[8] R. Heydari, M. Gheisarnejad, M. H. Khooban, T. Dragicevic, and F. Blaabjerg, "Robust and fast voltage-source-converter (vsc) control for naval shipboard microgrids," *IEEE Trans. Power Electrons.*, vol. 34, no. 9, pp. 8299–8303, 2019.

[9] Q. Shafiee, Č. Stefanović, T. Dragičević, P. Popovski, J. C. Vasquez, and J. M. Guerrero, "Robust networked control scheme for distributed secondary control of islanded microgrids," *IEEE Trans. Ind. Electron.*, vol. 61, no. 10, pp. 5363–5374, 2014.

[10] J. Hu, Y. Li, T. Yong, J. Cao, J. Yu, and W. Mao, "Distributed cooperative regulation for multiagent systems and its applications to power systems: A survey," *The Scientific World Journal*, pp. 1–12, 2014.

[11] R. Heyderi, M. Alhasheem, T. Dragicevic, and F. Blaabjerg, "Model predictive control approach for distributed hierarchical control of vsc-based microgrids," in *2018 20th European Conference on Power Electronics and Applications (EPE'18 ECCE Europe)*. IEEE, 2018, pp. P–1.

[12] J. Schiffer, F. Dörfler, and E. Fridman, "Robustness of distributed averaging control in power systems: Time delays & dynamic communication topology," *Automatica*, vol. 80, pp. 261–271, 2017.

[13] Y. Khayat, R. Heydari, M. Naderi, T. Dragicevic, Q. Shafiee, M. Fathi, H. Bevrani, and F. Blaabjerg, "Estimation-based consensus approach for decentralized frequency control of ac microgrids," in *2019 21st European Conference on Power Electronics and Applications (EPE '19 ECCE Europe)*, 2019, pp. 1–8.

[14] R. Heydari, Y. Khayat, M. Naderi, A. Anvari-Moghaddam, T. Dragicevic, and F. Blaabjerg, "A decentralized adaptive control method for frequency regulation and power sharing in autonomous microgrids," in *2019 IEEE 28th International Symposium on Industrial Electronics (ISIE)*, 2019, pp. 2427–2432.

[15] D. Shi, T. Chen, and L. Shi, "An event-triggered approach to state estimation with multiple point-and set-valued measurements," *Automatica*, vol. 50, no. 6, pp. 1641–1648, 2014.

[16] J. Rodriguez, J. Pontt, C. A. Silva, P. Correa, P. Lezana, P. Cortes, and U. Ammann, "Predictive Current Control of a Voltage Source Inverter," *IEEE Trans. on Ind. Electron.*, vol. 54, no. 1, pp. 495–503, Feb 2007.

[17] J. Rodriguez, M. P. Kazmierkowski, J. R. Espinoza, P. Zanchetta, H. Abu-Rub, H. A. Young, and C. A. Rojas, "State of the art of finite control set model predictive control in power electronics," *IEEE Trans. Ind. Informat.*, vol. 9, no. 2, pp. 1003–1016, 2012.

[18] H. Bevrani, *Robust power system frequency control*. Springer, 2014.

Offshore wind farm layout optimization considering wake effects

Asma DABBABI · Salvy BOURGUET · Rodica LOISEL · Mohamed MACHMOUM
University of Nantes,
IREENA, Institut de Recherche en Energie Electrique de Nantes Atlantique,
CRTT, 37 Bd de l'université, BP406, 44602 Saint Nazaire, Cedex, France
Tel.: +33 / 2 49 14 20 61
E-Mails: Asma.Dabbabi@univ-nantes.fr
Salvy.Bourguet@univ-nantes.fr
Rodica.Loisel@univ-nantes.fr
Mohamed.Machmoum@univ-nantes.fr
URL: http://www.ireena.univ-nantes.fr/

Keywords

« Wind farm optimization », « mixed AC-HVDC topology », « load flow », « wake effects », « LCOE».

Abstract

With the serious concern for environmental problems, the exploitation of renewable energies is becoming a need and no longer a choice, offshore wind energy is one of the trends in the renewable energy market. The exploitation of this energy demands important investment costs that's why optimizing the wind farm layout is primordial to make a cost effectiveness during the lifetime of the park. Otherwise, minimizing the Levelized Cost Of Energy (LCOE) is the key to have the tradeoff between wind farm cost and the annual energy delivered to the terrestrial network. For a realistic framework, the wind farm losses must take into account the wake losses between turbines so a calculation tool of wake effect is developed in this paper.

Introduction

Compared to onshore wind farm, offshore wind farm has many advantages. It mainly focuses on a higher extracted wind speed with fewer fluctuations due to favorable wind conditions off the coast, a lower energy loss thanks to the possibility to extract maximum energy so as to improve the utilization of the installed wind energy capacity compared to onshore sites. Environmental influence is also under consideration since the offshore wind farms are less noisy and have less impact on the landscape.

The exploitation of this energy requires investments in the transmission of energy to the terrestrial network. Nevertheless, it is important to consider cheaper transmission solutions with less energy loss. There are two main technologies of energy transmission: HVAC and HVDC. In fact, HVAC is the most economic connection solution for short transmission distances and HVDC transmission method provides more advantages for distances exceed 100-150km [1]. Furthermore, a cost-effective wind farm is impacted by the annual production capacity that takes into account the overall energy losses. The topology of the park and the placement of its turbines affect meaningfully the losses calculation since the wake effect between turbines can induce a reduction of wind speed at the downstream wind turbine so the reduction of its capacity factor. The large spacing between the turbines can reduce the wake effect impact but it's driving up the cost of offshore connections so the best solution is to find the tradeoff between energy yields and investment costs. Several research works have been done for the optimization of wind farms with the incorporation of the wake effect calculation. In fact, Wu et al. [2] and Hou et al. [3] respectively in 2014 and 2015, considered wake effects in the optimization of offshore wind farms, thus allowing the evaluation of energy loss in the system that promotes the optimization of offshore wind farms under more realistic conditions. Wu et al. [2] applied two optimization algorithms. Indeed, the genetic algorithm was used to find the optimal positions of the turbines, then the ant colony algorithm (AC) was applied to optimize the connection of the MV collection network.

Hou et al. [3] used a standard PSO with an inertial weight and assumed a dispersed topology of wind turbines (non-regular wind farm) with a constant distance between turbines of different rows.

In 2016, Hou et al. [4] brought back further studies by proposing a new way of positioning wind turbines for a regularly shaped wind farm, taking into account the optimization of the direction of placement of the wind farm and the spacing between the turbines, as well as the impact of the variation of the pitch angle assigned to each turbine on the energy yields of the entire structure.

Rodriguez et al. [5] proposed an optimization technique based on "Covariance Matrix Adaptation" where the turbines (floating turbines) move in the optimal wind direction to minimize the wake effect. In general, four analytical wake effect models can be considered: Lissaman's wake model, Larsen's wake model, Jensen's wake model and Ainslie's wake model [6], these models can be called low fidelity engineering models because it describes the wake losses with a simple mathematical way with a rapid computation time and an accurate way to estimate the velocity deficit even for far wake zones. Other models are more accurate such as the Dynamic Wake Meandering model [7] and the Computational Fluid Dynamics [8] but they take a long time to evaluate the wake effect calculation. The most common model used is the Jensen model [9] because it offers a rapid computation time to calculate multiple wake effects, it's a model that can estimate the partial or the multiple wake effect.

Mikel de Prada in his research [10], studies the impact of the wake effect on overall losses using the Jensen's wake model, the proposed idea was trying to operate the turbines at their non optimum points in the aim to reduce their wake effects. Hou et al., Wu et al. and Rodriguez et al. integrate also the Jensen's model to calculate the wake losses in the optimization of the wind farm layout design.

For this work, we combined two approaches to optimize offshore wind farms: the genetic algorithm and the Prim algorithm. Indeed, the GA provides the first topologies of connections such as the connection between wind turbines and the connection of offshore substations as well as their positions. Then, the Prim algorithm is used to complete the connection between each group of wind turbines and the nearest substation (search for the shortest path) [11]. The positions of wind turbines are fixed but the substation positions and numbers are determined by the heuristic optimization GA, both AC distribution and HVDC transport networks are built with the optimization algorithm while minimizing the losses and cable costs. The ultimate goal of this article is to optimize a mixed AC/HVDC connection architecture with the integration of wake losses by the Jensen's model.

The paper is organized as follows. The first part is dedicated to the problem statement that presents the mathematic wake model, the wind farm load flow modelling and the optimization flowchart. The second part shows a study that is carried out on an offshore wind farm composed of 40 turbines located at 130 km from the coast with a MVAC distribution network and a HVDC transmission network. The optimized layouts with and without wake effect consideration are shown with a comparison between both cases. Conclusions are discussed in the final part.

Problem statement

In this part, the overall problem statement is shown. First of all, the Jensen's wake model is presented with its formulation and calculation methodology. After that, the steps of a mixed AC/DC load flow calculation are introduced in the aim to calculate the total system losses taking into account the wake losses. Finally, the optimization flowchart is specified with its objective function.

Wake effect model (Katic Jensen model)

The Katic Jensen model [9] [12] is used to calculate the wake effect between turbines. This model was developed by Jensen in 1983 and further improved by Katic et al in 1986. The calculation of the wake effect is not based on an optimization algorithm aiming at positioning the turbines in the optimal location. The calculation approach has as input the fixed positions of the turbines and after that the calculation of the wake effect is estimated by the Katic Jensen analytical model to determine the output power of each turbine taking into account the occurrence of different wind speeds which is determined by the wind rose.

Considering two turbines j and k mounted next to each other in Fig.1, the wind speed seen by the turbine j is given by:

$$U_j = U_0(1\text{-deficit}) \tag{1}$$

Where: U_0 is the ambient wind speed, it represents the wind speed seen by the turbine k and the deficit is the speed decrease caused by wake effects which is expressed by the following equation.

$$\text{deficit}=\sqrt{\sum_{k=1}^{n} U_{kj}^2} \quad \text{with } U_{kj}=\frac{1-\sqrt{1-C_{Tk}}}{(1+\frac{\alpha d_{kj}}{R_j})^2} \frac{A_{kj}}{A_j} \tag{2}$$

Where: C_{Tk} is the thrust coefficient of the turbine k at a given wind speed, A_j is the rotor area for turbine j and A_{kj} is the rotor area for the turbine j influenced by the turbine k.
The A_{kj} can be expressed differently depending on the type of wake effect:

- $A_{kj}=\pi R_j^2$ when the turbine j is totally affected by the turbine k.
- $A_{kj}=0$ when there is no wake effect
- $A_{kj}=\frac{1}{2}$ (R_{kw}^2 (2 arccos ($\frac{R_{kw}^2+C_{kj}^2-R_j^2}{2R_{kw}C_{kj}}$)-sin(2arccos($\frac{R_{kw}^2+C_{kj}^2-R_j^2}{2R_{kw}C_{kj}}$)))) + $\frac{1}{2}$ (R_j^2 (2 arccos ($\frac{-R_{kw}^2+C_{kj}^2+R_j^2}{2R_jC_{kj}}$)- sin (2 arccos ($\frac{-R_{kw}^2+C_{kj}^2+R_j^2}{2R_jC_{kj}}$)))) when the turbine j is partially affected by the turbine k.

The wake expansion is expressed by:

$$R_{kw}=R_{k}+\alpha d_{kj} \tag{3}$$

Where: R_k is the turbine rotor radius and α is the wake decay coefficient that can be calculated by the following expression

$$\alpha=\frac{0.5}{\log(\frac{T_{HUB}}{0.0005})} \tag{4}$$

Where, the T_{HUB} is the turbine hub height.

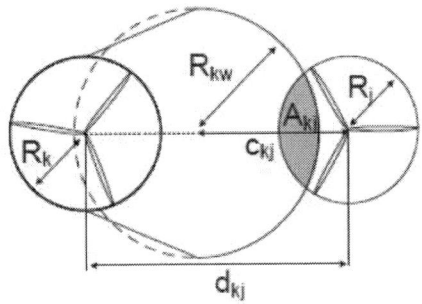

Fig. 1: The partial wake effect caused by the turbine k on the turbine j [5]

The calculation steps of the wake effect are as follows:
Step 1: Inputs: choose the wind direction, coordinates of turbines, hub high and rotor radius of turbines and the thrust coefficient C_T in function of different wind speeds.
Step 2: change of the reference mark related to turbines positions (fixed positions) according to the axis of rotation which is the wind direction,
Step 3: determination of the turbines affected by the wake effect: creation of an influence matrix M which is a square matrix of N dimension where N is the number of turbines. In fact, $M_{ij} = 1$ if there is a wake effect between turbine i and turbine j and $M_{ij} = 0$ if there is no wake effect,
Step 4: exploitation of the matrix M to know each turbine is affected by how many turbines (multiple wake effect),
Step 5: calculate the wake effect for each turbine with the Katic Jensen model.

Load flow calculation

Fig. 2 presents the architecture of a mixed AC/HVDC topology, the turbines are connected in a MVAC 66 kV network and the transport is with a HVDC network under a voltage of +/- 320 kV, therefore, there is a DC offshore substation that contains AC/DC converter. In the onshore part, there is an inverter and a transformer in order to inject the energy transmitted to the terrestrial network.

For the calculation of the AC/DC load flow, the sequential method is used. It consists in evaluating the state variables of the AC and DC systems and iterating them one by one, until algorithm convergence. During this process, the AC energy flow equations and the DC equations are solved separately. The power flow calculation is provided by Mat AC/DC [13] which is a Matlab library that can calculate the mixed AC/DC load flow by integrating the converters modeling and taking into account their control modes.

For a classic power flow calculation, two of four parameters of each node are known previously: V, δ P, Q and the value of the two others can't be known before calculation. Therefore, all the nodes can be divided into PQ nodes, PV nodes and slack buses.

Hybrid AC/DC load flow method is based on separating the AC zones (1 and 2) from DC zone. In each AC zone, one bus placed in the end of the AC network should be chosen as the slack bus to keep the power balanced. The other nodes which are the turbines or the transformers can be PV or PQ buses depending on its controlling strategy. In fact, the turbines are considered to be PQ nodes; P is the power produced by each turbine after wake losses calculation. The turbine is assumed not to produce reactive power (Q=0) i.e. the power factor is equal to 1. Likewise, each transformer block is modeled by two nodes that are at the transformer's input and output. There is no injection of active or reactive power in the input nodes, they represent virtual PQ nodes that facilitates transformer's modelling. The transformer's output nodes are the slack buses because they represent the end of each AC zone. Between these nodes, the transformer will be modelized like a branch with its appropriate parameters (resistance, reactance and susceptance). Besides, each DC network must contain a DC slack bus like a reference node to facilitate the DC load flow calculations.

Furthermore, the AC side of each converter can control the reactive power injection by two control modes:

1. Constant Q: The converter controls the reactive power injected to the AC grid.
2. Constant U_s: The reactive power adapts to keep AC bus voltage U_s constant.

For every HVDC transmission line, one converter must control the active power injection into AC grid (Constant P control) and the other converter must adjust its active power injection to control the DC bus voltage U_{dc} (Constant U_{dc} control).

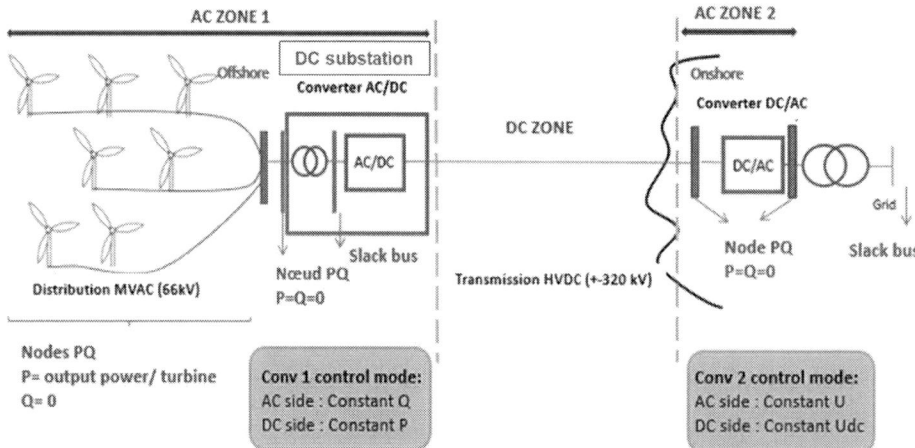

Fig. 2: Wind farm load flow modeling

Optimization framework

Objective function

The LCOE criterion is a reference for the industrialists to assess electrical performance for offshore wind farms. The minimization of the LCOE is the main objective in the optimization framework. This economic function is expressed by the following equation:

$$LCOE = \frac{\frac{CAPEX}{a} + OPEX}{AED} \qquad (5)$$

Where:

a: the annuity factor, $\frac{1-(1+r)^{-N}}{r}$

CAPEX: the annual capital expenditure,

OPEX: the annual operational expenditure,

AED: the annual energy delivered= Annual energy produced (AEP) – the total losses of the architecture including wake losses,

r: the discount rate,

N: number of years of the wind farm exploitation.

Optimization algorithm flowchart

Within the framework of optimization, the electrical topology is built up by steps, like presented in the figure below: the first step is the construction of the MVAC collection network where the turbines are connected. In fact, this phase is based on the clustering concept and that's means gathering the turbines into groups, the genetic algorithm proposes at each iteration the turbine's clustering until the optimal collection network is defined. The output data to be retrieved once the MV network layout is set are: the number of clusters connected to each sub-station, the number of turbines in each cluster and the definition of the MV cables cross section. The second step is the placement of offshores sub-stations. In our case, the maximum possible number of sub-stations that can be installed is 4, the number and location of sub-stations are important aspects that impact the increase of the CAPEX and the total length and dimension of the collection network. The Prim algorithm intervenes to make the connection between the substations and the wind turbines in the nearest clusters. It makes also the connection between the delivery point (terrestrial network) and the nearest substation [14].

The last step presents the construction of the HVDC transmission network, where the number of HV connections is fixed with the definition of the number of the HVDC cables installed in parallel as well as their cross section. The design of the HVDC network layout includes also the connection between substations.

In this work, the cables with different cross sections are used for the purpose of optimization, as long as it reduces the investment costs related to the cables with their installation costs. Designing the optimal electrical architecture according to a specific economic function requests to make the sizing of power components. The power rating of the transformers, the converters and the cables are determined depending on the amount of the power transmitted. In this phase, the constraints relative to cable ampacities and the power rating of transformers and converters are verified.

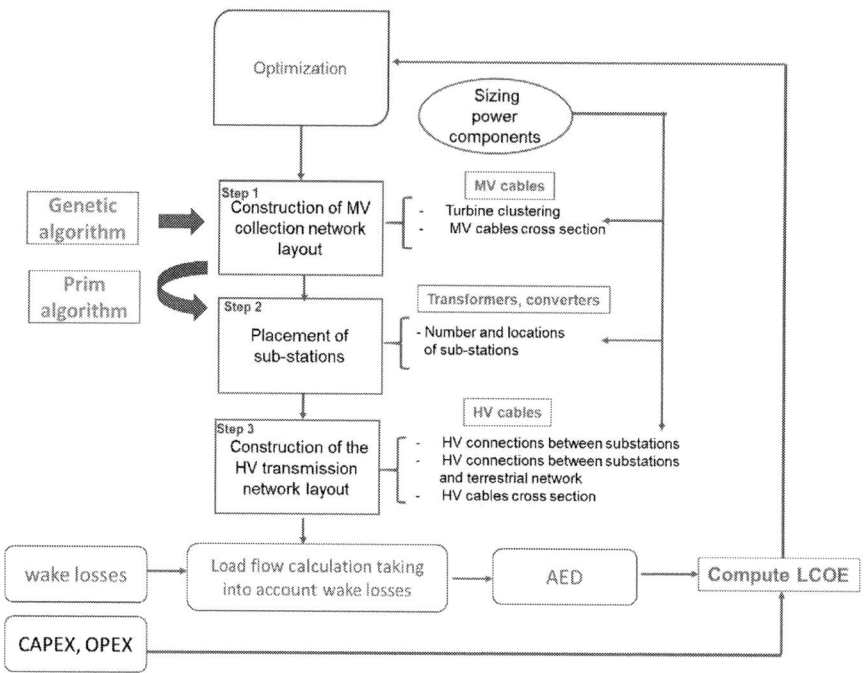

Fig.3: The steps for designing the optimal electrical architecture

Once the 3 steps are done with the right dimensioning of the components, the evaluation of the electrical architecture is done by the LCOE calculation after calculating the CAPEX, OPEX and the annual energy delivered AED which depends on the overall losses including the wake losses. The algorithm continues to run until it reaches an optimal LCOE value.

Results and discussion (case study)

In this section, a case study of a wind farm composed of 40 turbines is presented. Two scenarios are presented for this case study, the first one is the wind farm optimization without the wake losses calculation and the second one is the optimization with consideration of the wake effect impact. Therefore, a comparison between the two scenarios is shown.

For the case study of this wind farm, the following assumptions are set:
- N= 20 years
- r=8%
- OPEX= 50 k€/MW [15]
- Distance of transmission 130 km
- 40 turbines each producing a nominal power 8 MW, hub high = 100 m and rotor radius=150 m
- MVAC network 66 kV, HVDC network +/- 320 kV
- Turbines positions are fixed
- Substations positions are variable and its number vary between 1 and 4

Comparison between optimization with and without wake effect consideration

First scenario (without wake effect)

For the first scenario, the obtained topology after the optimization algorithm is presented in Fig. 4. The topology contains 8 feeders of turbines where the number of turbines vary between 6 and 4 turbines in each cluster and only one DC substation directly linked to the onshore network with the cable HVDC. The length of MVAC cables is 34.30 km and the length of HVDC cables is 131.58 km. The LCOE of this park is 101.48 €/MWh.

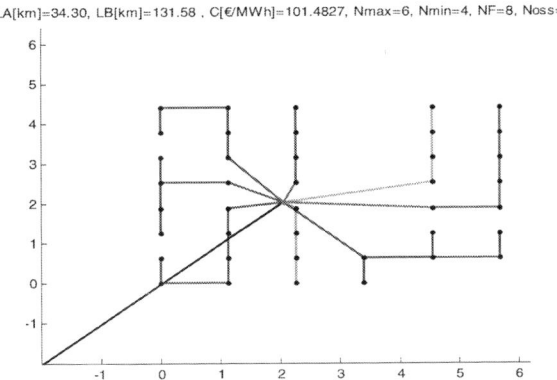

LA[km]=34.30, LB[km]=131.58 , C[€/MWh]=101.4827, Nmax=6, Nmin=4, NF=8, Noss=1

Fig.4: The optimized AC-HVDC topology without wake losses calculation

The following figure summarizes the different components cost with their percentages. In fact, among the most important costs in an AC HVDC topology, there are the converter costs with almost 15% after the turbines and foundations costs that have more than 50% of the total cost. For a distance 130 Km, the HVDC cables also present an important cost due to the utilization of 131.58 km of cables to transport energy to the onshore grid. The total investment cost i.e. CAPEX is 994.17 M€. The total losses of the optimized topology are 22.21 MW that can be divided into three parts, the most significant losses are the converter losses:

- Transmission losses (HV network) = 0.92 MW
- Converters losses = 20.81 MW
- Distribution losses (MV network) = 0.48 MW

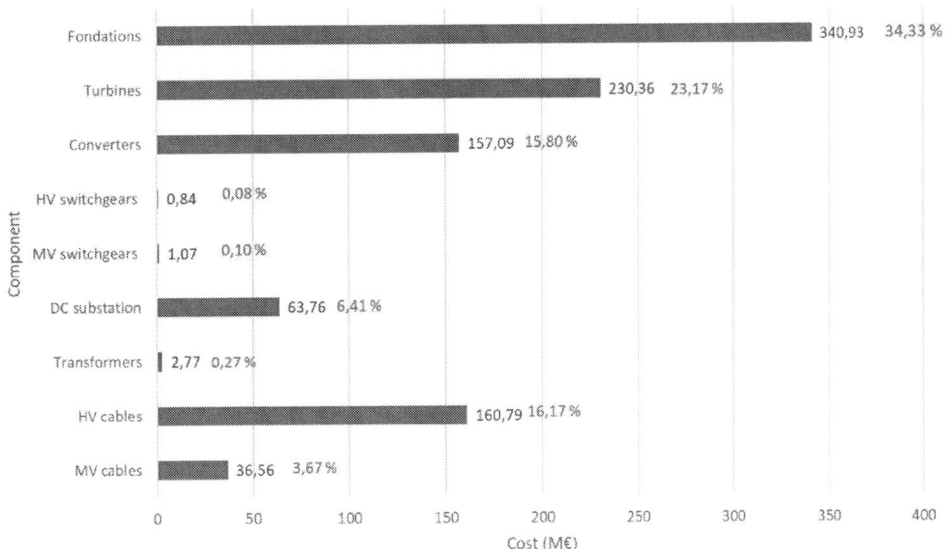

Fig.5: Components cost

Second scenario (with wake effect)

For the second scenario, the CAPEX, OPEX, r and N remain the same as the first scenario but the losses are impacted by the integration of wake losses calculation. The optimized wind farm connection topology is presented in Fig. 6. The topology contains 7 feeders with a transmission distance 130 km. The LCOE evaluated is 139.64 €/MWh.

LA[km]=34.54, LB[km]=130.85 , C[€/MWh]=139.6497, Nmax=6, Nmin=4, NF=7, Noss=1

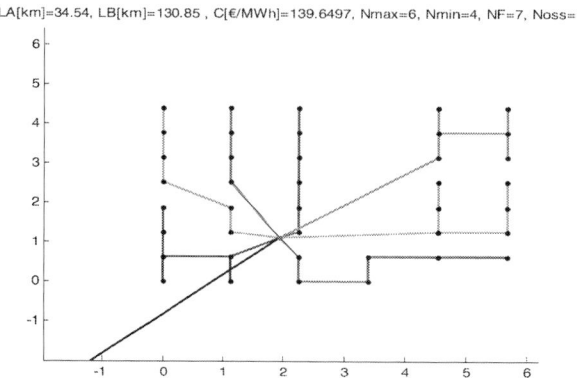

Fig.6: The optimized AC-HVDC topology with wake losses calculation

Comparison between the two scenarios

For the first scenario, the energy yield of the wind farm is obtained by the calculation of the mean power produced by one turbine and then multiplied by the total number of turbines. For the second scenario, the wake effect is considered in the calculation of the total losses so the energy yield is calculated with the exploitation of the wind rose (Fig.7) that presents the estimation of power production of the wind farm for 12 wind directions (0 deg => 330 deg) and 26 wind speeds (0 m/s =>25 m/s). Therefore, the power and the thrust curves in function of wind speeds are exploited to calculate the power produced by each turbine. Then, the energy yield is the sum of energy yields of each turbine.

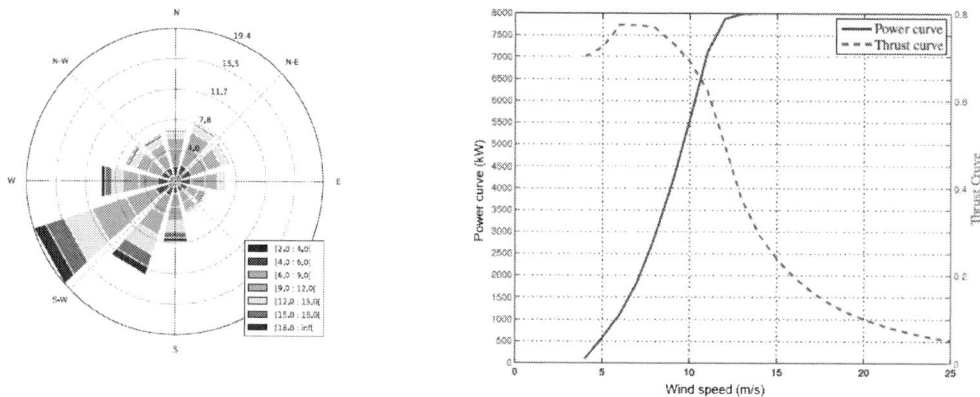

Fig.7: Wind rose, wind turbine power and thrust curves

The calculation results are presented in Table I. It can be noticed that the percentage of wake effect for the energy yield is significant and equals to **22.61%**.

Table I: Parameter comparison between the two scenarios

Parameters	Losses (MW)	Energy yield (MW)	Capacity factor	LCOE (€/MWh)
Without wake effect	22.21	154.11	0.41	101.48
With wake effect	23.48	119.26	0.30	139.64

Results exploitation for scenario 2

Fig.8 shows the distribution of power attenuation for each wind turbine (40 turbines) according to all directions for two wind speeds 7 m/s and 14 m/s. The power production for two wind speeds is different. In fact, with a wind speed 14 m/s, the power of the first 5 turbines in the direction 0˚ can reach 8 MW but it is equal to 1.12 MW with a speed wind 7 m/s. From this graph, one can conclude that the two directions 0˚ and 180˚ are the worst directions where the power is mostly lost (great wake effect). However, the directions 30˚, 150˚, 210˚ and 330˚ are the most favorable directions for power production.

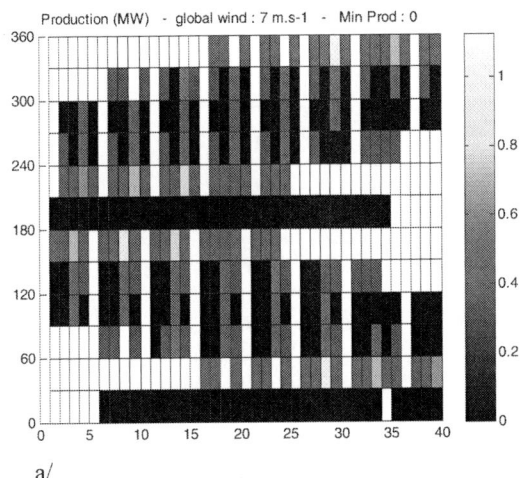
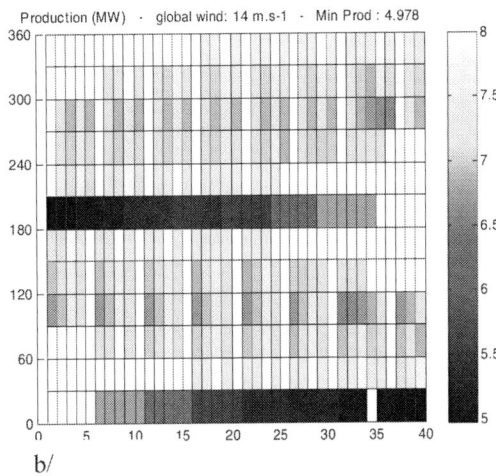

a/   b/

Fig.8: Power attenuation caused by wake effect for all turbines for all directions, a/ wind speed 14 m/s, b/ wind speed 7 m/s

Fig.9 shows the mean power produced for one turbine for all directions and for different wind speeds 7 m/s and 14 m/s. Like mentioned before, with a wind speed 7m/s the mean power per turbine cannot exceed 1 MW but can reach the nominal power 8 MW with a wind speed 14 m/s. This figure validates that the directions 30˚, 150˚, 210˚ and 330˚ are the best ones as long as the rated power can be reached.

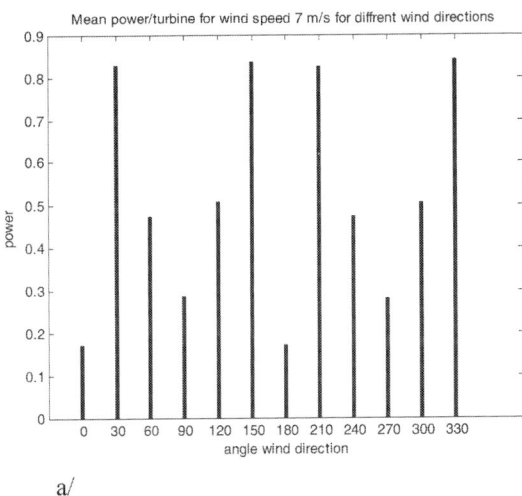
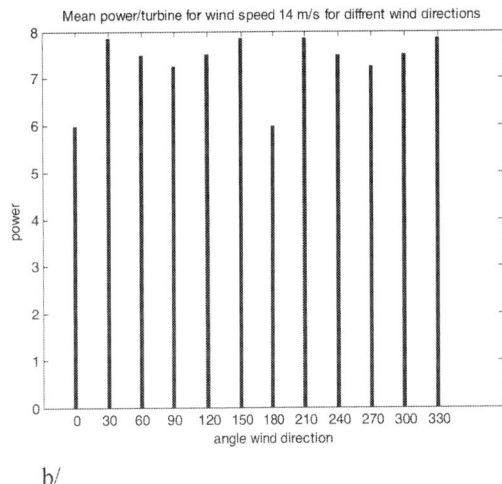

a/   b/

Fig.9: Mean power/ turbine for all directions, a/ wind speed 14 m/s, b/ wind speed 7 m/s

Conclusion

In this paper, an optimization algorithm minimizing the LCOE was done for a mixed AC-HVDC wind farm using the genetic and Prim algorithms. The optimization methodology was studied with the explication of the hybrid AC/DC load flow that takes into consideration the wake losses between turbines. The Katic Jensen model was exploited to calculate the wake effect. A study case of a wind farm composed of 40 turbines is presented with the comparison between park performances with and without wake effect calculation. The wake effect is significant and impacts the total energy yield of the park. Further works can focus on optimizing the location of the turbines and their orientations in relation to the most dominant wind direction (positions and pitch angle optimization).

Acknowledgements

This work was carried out within the framework of the WEAMEC, West Atlantic Marine Energy Community, and with funding from the Pays de la Loire Region.

References

[1] ABB publication, ''It's Time to Connect '', Technical description of HVDC Light® technology, 2008.

[2] Y. K.Wu, C. Y. Lee, C. R. Chen, K. W. Hsu, H. T. Tseng, ''Optimization of the Wind Turbine Layout and Transmission System Planning for a Large-Scale Offshore Wind Farm by AI technology'', Industry Applications Society Annual Meeting (IAS), Industry Applications Society Annual Meeting (IAS), USA, 2012

[3] P. Hou, W. Hu, M. Soltani, Z. Chen ''Optimized Placement of Wind Turbines in Large-Scale Offshore Wind Farm Using Particle Swarm Optimization Algorithm'', IEEE Transactions on Sustainable Energy, Vol 6, Issue 4, 2015

[4] M. S. P.Hou, W.Hu, '' Offshore Wind Farm Layout Design Considering Optimized Power Dispatch Strategy'', IEEE Transactions on Sustainable Energy, Vol 8, Issue 2, pp 638 - 647, April 2017

[5] S.F. Rodrigues, R. T. Pinto, M. Soleimanzadeh, P. A.N. Bosman, P. Bauer, ''Wake losses optimization of offshore wind farms with moveable floating wind turbines'', Energy Conversion and Management, Vol 89, pp 933- 941, January 2015

[6] N. Moskalenko, K. Rudion and A. Orths, '' Study of wake effects for offshore wind farm planning '', IEEE Conference, October 2010.

[7] "Dynamic wake meandering modeling," Riso National Laboratory, Roskilde, Tech. Rep. Riso-R- 1607(EN), 2007.

[8] B. Sanderse, "Aerodynamics of wind turbine wakes - literature review," Energy Research Centre of the Netherlands (ECN), Petten the Netherlands, Tech. Rep, 2009.

[9] N. Jensen, "A note on wind generator interaction," Riso National Laboratory, Roskilde, Denmark, Tech. Rep. Riso-M-2411, 1983

[10] M. Gil, L. Igualada, C. Corchero, O. Gomis-Bellmunt, A. Sumper,'' Hybrid AC-DC Offshore Wind Power Plant Topology: Optimal Design'', IEEE Transactions on Power Systems (Vol: 30, Issue:4), 2015

[11] O. Dahmani, '' Modélisation, optimisation et analyse de fiabilité de topologies électriques AC de parcs éoliens offshore '', University of Nantes, PhD thesis in Electrical Engineering, 2014.

[12] I. Katic, J. Hojstrup, and N. Jensen, "A simple model for cluster efficiency," in EWEC'86. Proceedings.Vol. 1, 1986, pp. 407–410.

[13] J. Beerten, '' Modeling and control of DC grids '', KU Leuven, the PhD in Engineering Science, 2013.

[14] O. Dahmani, S. Bourguet, M. Machmoum, P. Guérin, P. Rhein, L. Jossé, '' Optimization of the Connection Topology of an Offshore Wind Farm Network'', IEEE Systems Journal (Volume: 9, Issue: 4), December, 2015.

[15] B. Associates, '' Offshore wind cost reduction pathways Technology work stream '', prepared for The Crown Estate, 2012.

Small-signal stability analysis of smart grids considering high penetration of power electronics converters and energy markets

Javiera Meneses, Patricio Mendoza-Araya
UNIVERSITY OF CHILE
Electrical Engineering Department, Energy Center
Av. Tupper 2007, Santiago, Chile
E-Mail: jsilvameneses@gmail.com, pmendoza@ing.uchile.cl

Acknowledgements

This work was partially supported by the National Agency for Research and Development (ANID) / Scholarship Program/ MAGISTER BECAS CHILE/ 22200597 and ANID/FONDAP/15110019.

Keywords

Smart grids, Converter machine interactions, Modelling, Renewable energy systems, Stability.

Abstract

Future smart grids are expected to have a high amount of renewable energy sources as well as an advanced metering and communication infrastructure. These communication technologies offer dynamic information that can be sent to the utilities and used by the power system operators. The power system operators drive the scheduling optimization of the generators usually at a different and slower timescale. However, as the exchange of information get closer to real-time, it is possible to turn this into a closed-loop control-based problem, that adjusts the generator and load power outputs constantly as system conditions change. Under this new scenario, it is possible to observe interactions between the markets and the physical power system. This work proposes a methodology to assess the stability of future smart grids, with high penetration of power electronics converters and considering the coupling between the power systems and energy markets. The power system and energy market are modeled to form a feedback system that can be assessed by the Nyquist stability criterion. The simulations are done in the MATLAB/SIMULINK environment using the WSCC 3 machine 9 bus power system.

Introduction

Several advancements in technology have paved the road to enhanced electric power systems, which are usually known as smart grids. These smart grids enable several new features such as the participation of the end-user and the incorporation of distributed generation, among others [1][2]. The penetration of these technologies is usually accompanied by power electronics. The stable operation of power systems with high penetration of power electronics converters is an active topic of research [3][4][5][6], considering novel control strategies for applications in renewable energies [7][8] and storage applied to such systems [9].

Along with several control strategies that have been successfully applied to power systems with high penetration of power electronics converters [10][11], such as the case of microgrids [12], other control strategies have also allowed for the participation of the end-users, for example through demand response programs or demand-side management strategies [13][14]. Some of these strategies operate directly with signals already available in the microgrid or power system, such as droop-based demand response [15]. However, several other strategies depend on signals external to the power system, such as those that depend on the end-user behavior or a price signal, which is the case of real-time pricing [16][17].

Electricity markets are often considered steady in the study of the dynamics of the electrical power system, because of the different timescales that the market (minutes) and the physical power system (seconds) have. However, in the literature, several studies justify the importance of bringing the market closer to the dynamics of the power system. In particular, studies such as [18][19][20] develop dynamic market mechanisms that allow the price signals to be adjusted dynamically in time. Therefore, in this new context, it is most appropriate to consider the dynamics of the electrical system coupled to the market. In this sense, the market signals can be used as input signals in the controls of the electrical power system and, by providing metering information back to the market, they form a closed-loop system that can be studied with conventional feedback systems analysis tools.

When assessing the stability of a smart grid that considers these functionalities, it is necessary to develop a framework that precisely represents the interaction of its components. In particular, the assessment would fall short if it only considered detailed modeling of the power system. For a correct representation of the interactions, a framework that considers the modeling of those external signals (e.g. price signals) is needed. In this sense, the previous work in [21][22] is appealing. However, that framework has not been applied to a power system with penetration of power electronics converters or demand-side management schemes.

In this work, the authors extend this framework to assess the small-signal stability of smart grids, particularly those with high penetration of power electronics converters and end-user participation.

Proposed Methodology

Since the power system models will interact with the external signals, it is convenient to conceptually separate the smart grid model into two subsystems: the power system model and the energy market model.

The power system model traditionally includes a static model for the power grid, usually based on an admittance matrix, dynamic models for generators, and passive loads. For proper representation of a smart grid, an enhanced model should include both machine-based and power-electronics-based generation, energy storage, and active loads. Machine-based generation is vastly studied, and several models, ranging from comprehensive to simplified, are available in the literature [23]. Power-converter-based generation models have also been reported in the literature, ranging from distributed generation in microgrids to large power plants in bulk energy systems. Dynamic or active loads can be modeled in several ways, depending on their actual ability to interact with the system. We are particularly interested in modeling active loads, such as manageable loads, or those with a simple on/off scheme.

The market is composed of three types of participants: producers, consumers, and the system operator. The producers have an associated cost function $C_i(x)$ that can be thought of as the cost of the generator i of produce x units of the resource. The consumers have a value function $V_j(x)$, that can be understood as the value that the consumer j gives to the consumption of x units of electricity. When the market is cleared, it is possible to obtain a market-clearing price, denominated as λ, which is used to determine the utility of each participant. The third participant is the system operator, who is often an independent organization that has the objective of maintaining the balance between the demand and generation in an optimal manner considering the restrictions of the power system. Conventionally, the system operators solve a constrained optimization problem that aims to maximize the aggregated benefits of the producers and consumers.

This investigation uses a dynamic market model, which means that the market is not modeled with a static optimization problem, and dynamic equations are used instead. The dynamic equations represent the behavior of the market participants. However, when a dynamic approach is used, the market operates in a much faster time scale than the conventional static problem. The faster time scale is relevant because the market can be coupled with the physical power system to contribute to the frequency regulation. This coupling between the physical power system and the market forms a closed-loop system that can be modeled and studied with conventional tools of control analysis. The focus in this paper will be the small-signal stability of the closed-loop system, in which we want to study the different interactions of the participants in the coupled system.

Although there are several methods to assess the small-signal stability, such as eigenvalue analysis, the use of the Nyquist stability criterion is appealing in this particular case, since the structure of the model presented in Fig. 1 clearly shows the feedback nature of the coupled system. The details of each block will be explained in the next section.

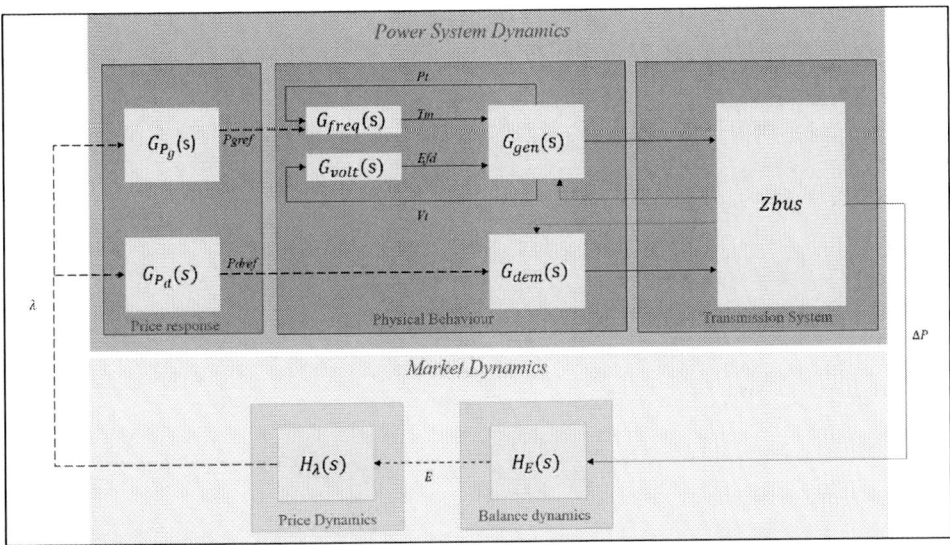

Fig. 1: Modeling of the interconnections between power system and market dynamics.

System Modeling

Considering the full system model presented in Fig. 1, the various control structures that interact in the coupled model are described next. On the one hand, the power system dynamics are represented by synchronous machines, power converters, and the static/dynamic loads models. On the other hand, the dynamics of the market are ruled by first-order differential equations that are derived from a static optimization problem and the behavior of the participants [18].

Power system dynamics

Synchronous Machine and controllers

The synchronous machines are represented by a 3rd order model that does not consider the stator transients and the damping circuits [23]. Then, the machine will be represented by a set of differential-algebraic equations and two controllers, one for the field voltage, denoted as G_{volt} in Fig. 1, and one for the torque of the generator, denoted as G_{freq} in Fig. 1. The field voltage control is a simplified automatic voltage regulator (AVR) shown in Fig. 3.

In this work a simplified speed governor is used, whose block diagram is shown in Fig. 2. Commonly, the power setpoint (P_{ref}) comes from the static power flow solution. In this study, however, this setpoint comes from the solution of the differential equations of the market. Therefore, this is one of the coupling points between the physical power system and the market. The other inputs of the model are the rated frequency of the electrical machine in p.u. (w_{ref}^{pu}) and the actual frequency (w^{pu}). Also, the constant R is the droop of the generator, and T_s, T_3 and T_c are the time constant of the corresponding filters. Finally, the model has a limiter for the active power of the machine, that has the limits P_{max} and P_{min}. The parameters for both the speed governor and AVR are shown in Table I.

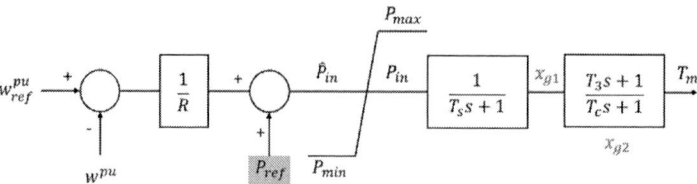

Fig. 2: Speed Governor Model with P_{ref} as an input from the market model.

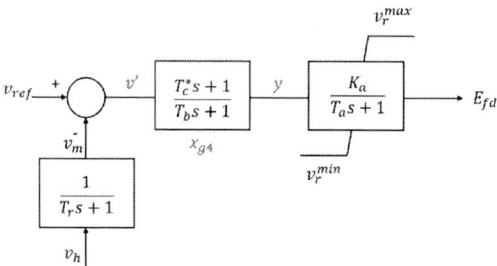

Fig. 3: Automatic Voltage Regulator model.

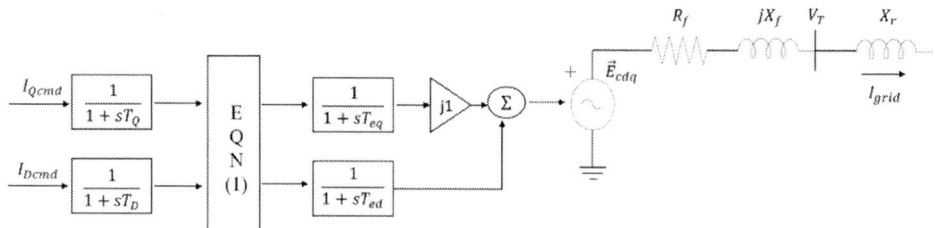

Fig. 4: Power Converter model in dq axis. [10]

Table I: Per unit parameters for speed governor and AVR of synchronous generators (Base 100 MVA).

Governor		AVR	
R	0.05	V_{rmax}	5.000
T_G	0.50	V_{rmin}	-5.000
P_{max}	1.98	K_a	10.000
P_{min}	0.30	T_a	0.005
T_1	1.00	T_c	1.000
T_2	10.00	T_b	10.000
-	-	T_r	0.010

Converter Model

The power-converted-based generation is modeled with the Voltage Source Converter (VSC) model developed in [10] that is used in large scale grid simulations and has the characteristic of being a grid supporting type converter. The converter has a voltage source representation and a coupling inductance (R_f and X_f) that captures the near instantaneous response of the converter interfaced sources. Also, it has both a reactive power controller and a real power controller, making it a good candidate for large penetration of converter studies.

The block diagram of the VSC shown in Fig. 4 includes an inner and outer control loop. The inner loop regulates the instantaneous AC voltage and usually consist of PI controls [11]. The outer loop gives the setpoints to the inner loops (I_{Qcmd} and I_{Dcmd}), where several options of control schemes have been studied. For the purposes of this work, droop control is used. The EQN block correspond to the equations that transform the current from the outer loop to the voltages that feed the coupling inductance and the power grid. In these equations the terms V_{Td} and V_{Tq} are the dq components of the terminal voltage V_T of the converter, and i_d and i_q are the currents that enter the block EQN.

$$E_{cd} = V_{Td} + i_d R_f - i_q X_f$$
$$E_{cq} = V_{Tq} + i_q R_f + i_d X_f$$

(1)

Transmission system and loads

The transmission system is modeled using a matrix representation in MATLAB/SIMULINK. Using the MATPOWER tool in MATLAB it is possible to obtain (i) the admittance matrix of the network and (ii) the power flow solution that is used to calculate the initial conditions of the simulation. In this investigation, the network transients are neglected because the timescales of interest are slower than those of the network. Then, having the admittance matrix of the power system, each generator and load needs to deliver a current to the network to calculate the respective voltage. This can be seen better in Fig. 1. On the one hand, the block called "Physical Behavior" has all the differential algebraic equations of the generators and loads of the power system. On the other hand, the block called G_{dem} is composed by the dynamic model of the loads. Also, at the output of the loads there is a block with initial conditions that is used to mitigate algebraic loops [24]. This loop appears when constant impedance, current and power (ZIP) loads are included in the simulation and two subsystems with only algebraic equations become dependent on each other's outputs. In this case, one subsystem is the algebraic equations of the ZIP loads and the other subsystem is the network algebraic equations.

The loads in this investigation are modeled as constant impedance, however, when they participate in the market process, they behave like a variable power load. This means that the load changes the consumption of power according of the market set-point. The equation that describes the load behavior is shown in (2), where a and b can be chosen to represent the voltage characteristic of the load consumption. In this equation, P_L and Q_L are the active and reactive load power, and P_0 and Q_0 are the initial values for the active and reactive power. In this work both a and b are 2 as the loads are modeled as constant impedance. However, when the loads participate in the market, the active load power P_L has a variable set-point which means that P_0 changes to P_{dref} shown in Fig. 1. The reactive load power Q_0 remains constant when the loads participate in the market.

$$P_L = P_0 \left(\left| \frac{V}{V_0} \right| \right)^a$$
$$Q_L = Q_0 \left(\left| \frac{V}{V_0} \right| \right)^b$$

(2)

Market Dynamics

In this paper, the market is modeled using a dynamic market mechanism that allows the generators and the consumers to participate in a regulation service in real time that reduces the energy imbalance. This role is conventionally fulfilled by the automatic generation control (AGC), but with a real-time market it is reasonable to assume that the market mechanism will perform this regulation service. Also, there is evidence that using a market drive model instead of an AGC will be valuable in future power systems with high penetration of intermittent renewables [25][26]. Then, the real time market has the objective of driving the energy imbalance to zero. To achieve this, the price of the market reflects the degree of energy imbalance. This principle was first shown by [21], and the equations shown here are based on that investigation. However, the market parameters have been adjusted to fit with the initials conditions of the WSCC 3 machines 9 bus power system, and the time constants to best represent a future power system.

The energy market model includes cost functions for producers, benefit functions for consumers and equations related to the system operator. Equation (3) represents the behavior of the producer i, whose decision on whether to increment or reduce the generation (P_{gi}) is a function of the price λ and the marginal cost of the generator $MC_{gi} = b_{gi} + c_{gi}P_{gi}$. The consumers are represented by (4), in which each consumer can modify their demand (P_{dj}) according to the power price and their marginal benefit $MB_{dj} = b_{dj} + c_{dj}P_{dj}$. Then, as the power price λ changes, they can contribute to the energy imbalances. Finally, there is an energy market regulator that ensures that imbalances (E) are signaled back through price signals (λ), as in (5)-(6). The variable P_{ti} in (5) is the terminal power of the generators. The parameters τ_λ and k_E are the price response time constant and the stabilizing gain of the energy imbalance, respectively, and both depend on the design of the market.

$$\tau_{gi}\dot{P}_{gi} = -b_{gi} - c_{gi}P_{gi} + \lambda \tag{3}$$

$$\tau_{dj}\dot{P}_{dj} = b_{dj} + c_{dj}P_{dj} - \lambda \tag{4}$$

$$\dot{E} = \sum_{i=1}^{m} P_{ti} - \sum_{j=1}^{n} P_{dj} \tag{5}$$

$$\tau_\lambda\dot{\lambda} = -k_E E - \lambda \tag{6}$$

Stability Analysis of the coupled system

The WSCC 3 machine 9 bus power system is used to study the coupling between power systems and markets. The market data is selected using as reference the market data shown in [21]. The simulations were performed in the MATLAB/SIMULINK environment. The power system is represented by differential-algebraic equations. To obtain a small signal model, first, the initial conditions of the system are calculated. With the initial conditions and the SIMULINK model, the power system is linearized to obtain the state-space representation, the eigenvalues, and the participation factors of the system. The Simulink model is divided into two subsystems, same ones shown in Fig. 1 as "Market Dynamics" and "Power System Dynamics". The Nyquist plot is obtained after linearizing the two subsystems separately and then forming the open loop system.

The study of the influence of power electronics is done by replacing generators 1 and 3 of the 9-bus power system by power converters. The power converter operates in the system as a grid supporting unit, i.e. the converter can support the local voltage and the frequency of the power system. In this investigation, we consider an associated cost for the power converter operation to make this units dispatchable in real time. Also, in the simulation the loads are considered as fixed demand, however, we add some study cases later in which the demand is participating in the market with a variable power consumption.

The investigation includes 2 study cases. The first study case considers 3 scenarios for the coupled system. The first scenario is the base case that only has the power system dynamics of the 9 bus power systems, i.e. no market interaction. The second scenario couples the market signal (λ) with the physical power system. In the third scenario, two of the generators are replaced by power converters. For the scenarios in which the market is considered, $\tau_\lambda = 18\ s$ and $k_E = 0.48$ are used. Additionally, we considered a second study case, in which several market parameters are tested into a power system with and without penetration of converters. This study case aims to show the effects of the market parameters in a power system with converters replacing conventional units, as well as with flexible demand. In Table II, the values of the market parameters are presented.

Table II: Market Parameters for the second study case.

	Power Price time constant (τ_λ)	Stabilizing Gain (k_E)
Scenario 1	12	0.6
Scenario 2	18	3.0
Scenario 3	18	4.8

Case 1: The effects of the coupled system

The eigenvalues of the 9-bus power system with and without market dynamics are shown in Fig. 5. This figure shows the effect of incorporating the market dynamics into the power system. The poles and zeros in blue color correspond to the power system only, i.e. no market and converter dynamics are included. In this scenario, the results show that the closed loop system is stable because the poles and zeros do not pass to the Right Half Plane (RHP) of the plot. The poles and zeros in green color correspond to the coupled system, i.e. the generators receive the price signals from the market and changes their power references on their governors. In this scenario we can see a critical unstable mode in Fig. 5, almost crossing to the RHP. However, as we can see in Table III this critical mode is still negative on its real part, so the power system coupled with the market is stable. Finally, the poles and zeros in red color correspond to the coupled system when replacing two of the three conventional generators by power converters. In this scenario, two poles than move to the RHP can be observed in Fig. 5, hence, it is possible to identify an unstable configuration. The values of the poles can be observed in the Table III These values are small in comparison of the rest of the poles, but as it will be shown in the next case, it is expected that these poles move farther towards the RHP when the parameters of the market change.

Table III: Modes of interest for the study case 1.

Power System Only	Coupled System	Coupled system+Converters
2,6616e-12 + 0,0000i	-0,00028 + 0,0506i	0,00355 + 0,0815i
-	-0,00028 - 0,0506i	0,00355 - 0,0815i

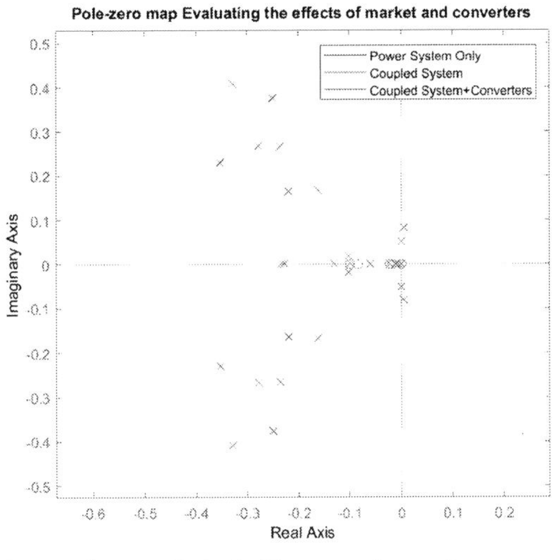

Fig. 5: Pole-zero map for Power System Only (without market and converters), Coupled System (market and physical system) and Coupled System with penetration of converter.

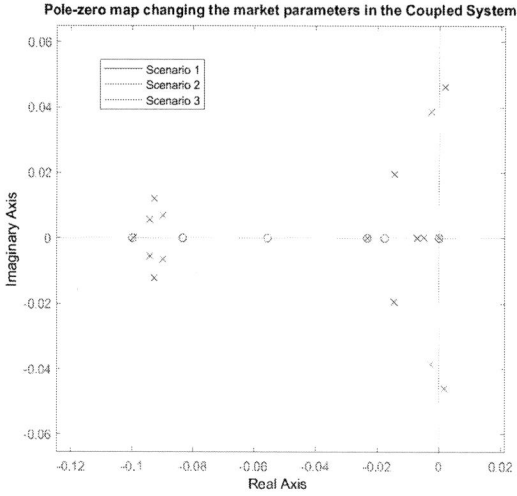

Fig. 6: Pole zero map for the Coupled System without converter, when changing the market parameters.

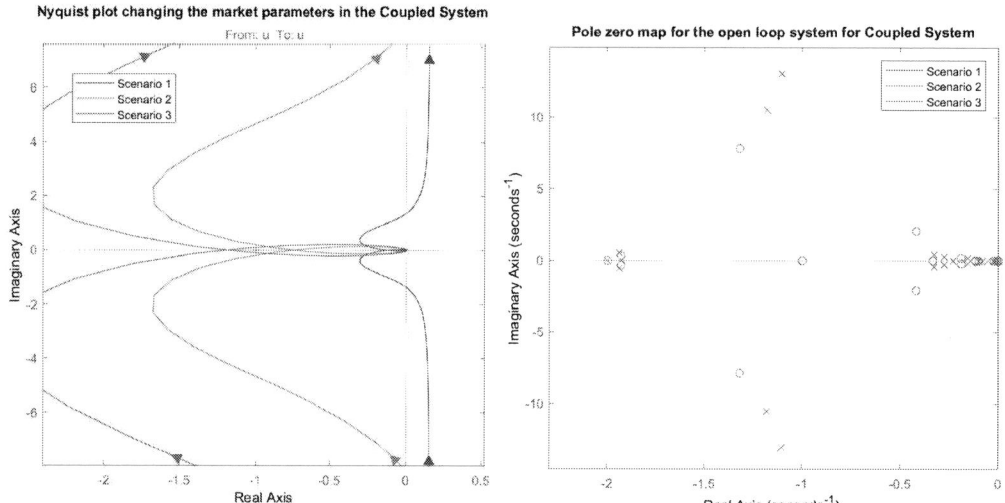

Fig. 7: Nyquist plot and pole zero map (open loop system) for the Coupled system changing the market parameters.

Case 2: The effect of the parameters of the market

In this case, the results presented for three configurations: one configuration is the coupled system without converters. The second configuration is for the coupled system considering the replacement of the generator for converters. The last configuration is the coupled system considering demand response. First, we can observe the pole-zero map of the coupled system in the first configuration shown in Fig. 6. In this plot the scenario 3 makes the system unstable, because of the two poles that move to the RHP. Also, it is possible to say that the poles move to the RHP when we increase either one of the parameters of the market. This can be explained by the ratio (k_E/τ_λ) between the stabilizing gain and the time constant, which represents the sensitivity of the market to energy imbalance. Then, when this ratio increases, the market is more sensitive towards energy imbalance, being this the reason for the coupled system to become unstable.

The Nyquist plot of the coupled power system without converters can be seen in Fig.7. From this plot, it is important to note that the encirclement is not seen because the open loop transfer function has a pole in the origin. Then, the Nyquist plot has infinite values for small frequencies. Nevertheless, we know that the plot closes on infinite to the right of the plot, but the plotting function does not allow this closing to be seen. In addition, the open loop poles-zero map is shown in the Fig. 7, in which we can observe that there are no poles in RHP. This is relevant, because the stability of the closed-loop system depends on both the existence of encirclements of the critical point as well as the RHP poles. Then, these results clearly show the departure of the power system from stable operation when considering the market dynamics, in particular when the parameters of the market increase. In the scenario 1 with the smaller parameters and a smaller ratio between k_E and τ_λ, the Nyquist curve moves away from the critical point -1. But as the market parameters increase, for scenario 2, the Nyquist curve get closer to the critical point until at one combination of parameters, the curve encircles the (-1,0) point and then the closed loop system becomes unstable.

The Nyquist plot for the Coupled System with converters is shown in the Fig. 8. As in the previous case, the poles of the open loop system are not in the RHP of the poles-zero map. Then, every encirclement of the critical point on the Nyquist plot makes the system unstable. Therefore, the system has two scenarios in which it becomes unstable, by judging its Nyquist plot. Unlike the previous case, scenario 2 is unstable when two of the three generators are replaced for power converters. Also, it is clear from Fig. 7 and Fig. 8 that the Nyquist plot with converters lean more towards the left side of the plot. This can be interpreted as a closed loop system that becomes unstable with a smaller ratio of the market parameters, i.e. is more sensitive to the energy imbalance.

Finally, a third configuration is analyzed, in which the loads are no longer fixed, and they participate in the market, changing their power consumption as the power price changes. In Fig. 9 these results are shown. From those plots it is possible to say that the setting of the parameters is critical when demand response is considered as the system become unstable with a smaller ratio of the market parameters when compared to the previous configurations.

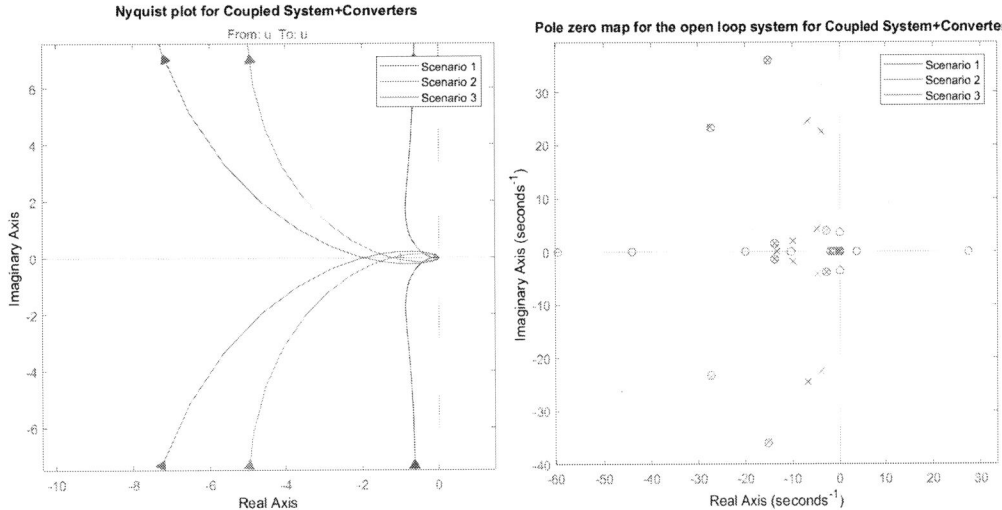

Fig. 8 Nyquist plot and pole-zero map (open loop system) for the Coupled system considering converters.

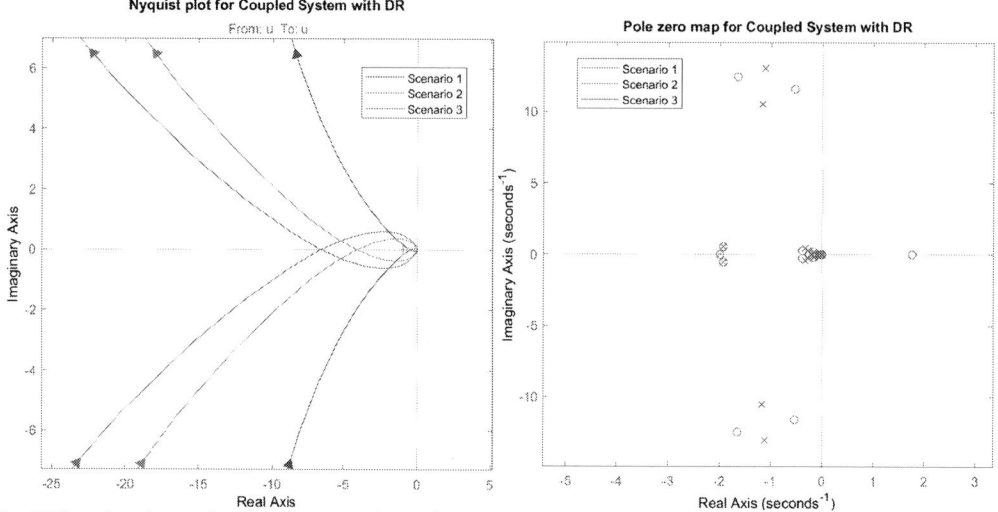

Fig. 9 Nyquist plot and pole-zero map (open loop system) for the Coupled system considering demand response.

Conclusion

The investigation shows that it is possible to have a stable system when the power system and the market are coupled. However, when there are elements of the future power systems such as high penetration of power converters and demand response, the system is more likely to become unstable. This means that, for future systems, special attention must be paid to both market design and power system design. From the small-signal analysis, it could be observed that by including different elements to the coupled system, high and low-frequency modes appear, indicating that different time constants are interacting with each other. The interaction between the different constants is responsible for whether the system is unstable or not. Despite this, it could be shown that the stable operation of the coupled system can be achieved with certain market parameters. The interactions that could arise in future systems can be studied in greater depth by the tools shown in this paper, considering for future work the possibilities that the intelligent network infrastructure will provide.

References

[1] X. Fang, S. Misra, G. Xue, and D. Yang, "Smart grid - The new and improved power grid: A survey," *IEEE Commun. Surv. Tutorials*, vol. 14, no. 4, pp. 944–980, 2012.

[2] M. Liserre, T. Sauter, and J. Hung, "Future Energy Systems," *IEEE Industrial Electronics Magazine*, pp. 18–37, 2010.

[3] B. Kroposki *et al.*, "Achieving a 100% Renewable Grid: Operating Electric Power Systems with Extremely High Levels of Variable Renewable Energy," *IEEE Power Energy Mag.*, vol. 15, no. 2, pp. 61–73, 2017.

[4] D. Ramasubramanian, E. Farantatos, S. Ziaeinejad, and A. Mehrizi-Sani, "Operation paradigm of an all converter interfaced generation bulk power system," *IET Gener. Transm. Distrib.*, vol. 12, no. 19, pp. 4240–4248, 2018.

[5] F. Milano, F. Dorfler, G. Hug, D. J. Hill, and G. Verbič, "Foundations and challenges of low-inertia systems (Invited Paper)," *20th Power Syst. Comput. Conf. PSCC 2018*, pp. 1–25, 2018.

[6] P. G. Thakurta and D. Flynn, "Network studies for a 100% converter-based power system," *J. Eng.*, vol. 2019, no. 18, pp. 5250–5254, 2019.

[7] T. Qoria, T. Prevost, G. Denis, F. Gruson, and F. Colas, "Power converters classification and characterization in power transmission systems," *EPE'19 ICCE Eur. Ital.*, pp. 1–10, 2019.

[8] J. Rocabert, A. Luna, F. Blaabjerg, and P. Rodríguez, "Control of power converters in AC microgrids," *IEEE Trans. Power Electron.*, vol. 27, no. 11, pp. 4734–4749, 2012.

[9] Dong-Jing Lee and Li Wang, "Small-signal stability analysis of an autonomous hybrid renewable energy power generation/energy storage system. Part I. Time-domain simulations," *IEEE Trans. Energy Convers.*, vol. 23, no. 1, pp. 311–320, 2008.

[10] D. Ramasubramanian, Z. Yu, R. Ayyanar, V. Vittal, and J. Undrill, "Converter Model for Representing Converter Interfaced Generation in Large Scale Grid Simulations," *IEEE Trans. Power Syst.*, vol. 32, no. 1, pp. 765–773, 2017.

[11] A. T. Qoria, Q. Cossart, C. Li, and X. Guillaud, "WP3 - Control and Operation of a Grid with 100 % Converter-Based Devices Deliverable 3 . 2 : Local control and simulation tools for large transmission systems," no. 691800, 2020.

[12] Y. S. Kim, E. S. Kim, and S. Il Moon, "Frequency and voltage control strategy of standalone microgrids with high penetration of intermittent renewable generation systems," *IEEE Trans. Power Syst.*, vol. 31, no. 1, pp. 718–728, 2016.

[13] P. Palensky and D. Dietrich, "Demand Side Management: Demand Response, Intelligent Energy Systems, and Smart Loads," *IEEE Trans. Ind. INFORMATICS*, vol. 7, no. 3, pp. 381–388, 2011.

[14] M. H. Albadi, S. Member, and S. Member, "Demand Response in Electricity Markets : An Overview," pp. 1–5, 2007.

[15] D. S. Callaway and I. A. Hiskens, "Achieving controllability of electric loads," *Proc. IEEE*, vol. 99, no. 1, pp. 184–199, 2011.

[16] A. Berger and F. Schweppe, "Real Time Pricing to Assist In Load Frequency Control," vol. 4, no. 3, 1989.

[17] A. J. Conejo, J. M. Morales, S. Member, L. Baringo, and S. Member, "Real-Time Demand Response Model," vol. 1, no. 3, pp. 236–242, 2010.

[18] N. Li, C. Zhao, and L. Chen, "Connecting automatic generation control and economic dispatch from an optimization view," *IEEE Trans. Control Netw. Syst.*, vol. 3, no. 3, pp. 254–264, 2016.

[19] J. Knudsen, J. Hansen, and A. M. Annaswamy, "A Dynamic Market Mechanism for the Integration of Renewables and Demand Response," *IEEE Trans. Control Syst. Technol.*, vol. 24, no. 3, pp. 940–955, 2016.

[20] D. J. Shiltz, S. Baros, M. Cvetkovic, and A. M. Annaswamy, "Integration of Automatic Generation Control and Demand Response via a Dynamic Regulation Market Mechanism," *IEEE Trans. Control Syst. Technol.*, vol. 27, no. 2, pp. 631–646, 2019.

[21] F. L. Alvarado, J. Meng, C. L. Demarco, and W. S. Mota, "Stability analysis of interconnected power systems coupled with market dynamics," *IEEE Trans. Power Syst.*, vol. 16, no. 4, pp. 695–701, 2001.

[22] D. Watts and F. L. Alvarado, "The influence of futures markets on real time price stabilization in electricity markets," *37th Annu. Hawaii Int. Conf. Syst. Sci. 2004. Proc.*, vol. 00, no. C, pp. 1–7, 2004.

[23] P. Kundur, "Power System Stability And Control by Prabha Kundur." 1994.

[24] L. Kunjumuhammed, S. Kuenzel, and B. Pal, "Load modelling," *Simul. Power Syst. with Renewables*, pp. 113–132, 2020.

[25] D. J. Shiltz, M. Cvetković, and A. M. Annaswamy, "An Integrated Dynamic Market Mechanism for Real-Time Markets and Frequency Regulation," *IEEE Trans. Sustain. Energy*, vol. 7, no. 2, pp. 875–885, 2016.

[26] A. T. Al-Awami, "Integrating AGC to Generation Scheduling for Real-Time Operational Optimization," *Arab. J. Sci. Eng.*, vol. 44, no. 8, pp. 7091–7100, 2019.

Component-level Reliability Assessment of a Direct-drive PMSG Wind Power Converter Considering Long-term and short-term thermal cycles

Shuaichen Ye[1,2], Dao Zhou[2], Frede Blaabjerg[2]

[1]School of Aerospace Engineering,
Beijing Institute of Technology, Beijing, 100081, China;
yesc_bit@163.com

[2]Department of Energy Technology,
Aalborg University, Aalborg, 9220, Denmark;
zda@et.aau.dk; fbl@et.aau.dk

Keywords

«wind energy», «reliability», «high voltage power converters», «mission profile», «IGBT».

Abstract

The lifespan of a wind power system is highly influenced by the reliable operation of its power converter. This paper investigates the failure rate and annual consumed damage for a 2 MW direct-drive permanent magnet synchronous generator (PMSG) based power converter in a wind power generation system. The reliability assessment mainly focuses on the component level, namely, diodes and IGBTs, in the machine-side converter (MSC). Annual damages and power cycles for semiconductors are calculated separately under long-term thermal cycles (several minutes to several hours) and short-term thermal cycles (dozens to hundreds of milli-second). A comparison result between different thermal cycles are given and discussed in detail. To ensure an effective lifetime evaluation of the entire converter system, a Monte Carlo method is used to generate the lifetime distributions and entire unreliability functions for power semiconductors. Final B_{10} and B_1 lifetimes can be easily observed from the cumulative distribution functions (CDFs).

Introduction

Recently, the usage of the full-scale power converter in wind power generation systems has been increasing steadily [1, 2]. Compared with the partial-scale power converter, the main advantage of the full-scale power converter lies in its adaptively to the upgraded grid codes [3]. A major category for the full-scale power converter is associated with the permanent-magnet synchronous generator (PMSG), which eliminates the slip rings and supports the grid operation better.

To obtain optimal wind conditions, wind turbines are normally installed in remote areas, such as an offshore wind farm. Due to the high maintenance cost, the lifespan and reliability of the wind power system attract increasing concern from wind turbine manufacturers. According to state-of-the-art researches [4], the power electronic module seems to have the highest failure rate, which takes up 55% among all the components in the wind power system. Moreover, based on the statistics in [5], the thermal stress occupies more than a half proportion in the stressor distribution for a power electronic module. Hence, it is important to investigate of the power module reliability with respect to the thermal profile.

Generally, the power converter reliability is mainly evaluated according to the long-term thermal cycles [6]. As shown in Fig. 1, the wind-speed is regarded as many steady-state values at different sampling points, and the long-term thermal cycles can be calculated during each sampling period. However, although the sampling time is short enough, non-negligible high-frequency junction temperature fluctuations exist in the chip of the power module in the converter. It is introduced by the alternating current through the power converter, which leads to the conduction of power devices with only half of the fundamental period. Therefore, the only calculation of the long-term thermal cycles is not accurate.

A more precise calculation [7], which considers the short-term thermal cycles, should also be taken into account. Although in [8] the thermal cycle for a doubly fed induction generator (DFIG) based power converter reduced to 1 second, it still cannot reflect the damage performance of short-term thermal cycles, whose frequency is usually several to dozens of Hz. Moreover, the final lifetime of the power converter should be generated from both the long-term thermal cycles and the short-term thermal cycles. Moreover, it is still unclear which of these two thermal cycles dominates in the power converter reliability. Thus, this paper aims to compare the reliability of power semiconductors in the converter between using long-term and short-term thermal cycles.

Apart from calculating the accumulated damage for power semiconductors, which is regarded as B_{10} lifetime (a measurement of the time by which ten percent of a population of semiconductors have failed), the overall failure tendency profile for one power component still needs to be generated using a Monte Carlo method. Hence, the power converter reliability can be assessed from a component level and a stricter B_1 lifetime can be subsequently obtained.

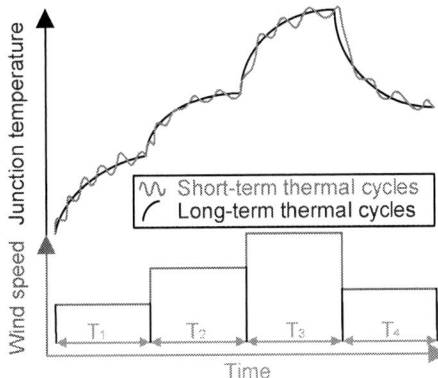

Fig. 1: Concept of long-term and short-term thermal cycles under a random wind profile.

Reliability evaluation of semiconductors with long-term thermal cycles

An overall configuration of PMSG based wind power generation system is shown in Fig. 2. It can be seen that a back-to-back (BTB) power converter consists of a machine-side converter (MSC), a grid-side converter (GSC) and a dc-link capacitor bank. Besides, a filter is applied to eliminate the PWM harmonics and a transformer is used to step up the grid voltage to the transmission level. This paper mainly focuses on the reliability analysis of the MSC and the similar approach could be expanded to the GSC according to the discussions in [9].

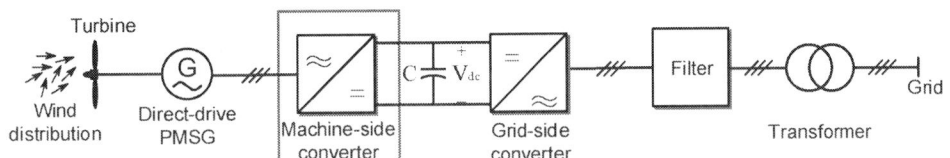

Fig. 2: Overall direct-drive PMSG based wind power generation system.

As a case study, a 2 MW wind turbine [10] and a corresponding direct-drive PMSG system [11] is investigated, whose detailed specifications are listed in Table I and Table II, respectively.

Table I. Parameters for the wind turbine.

Parameters	Values
Rated power (MW)	2.0
Blade radius (m)	41.3
Cut-in wind speed (m/s)	3
Rated wind speed (m/s)	12

Cut-off wind speed (m/s)	25
Optimal tip speed ratio	8.1
Maximum power coefficient	0.41
Range of turbine angular speed (rpm)	6-18

Table II. Parameters for the direct-drive PMSG and the power converter.

Parameters	Values
Rated power (MW)	2.0
Rated stator voltage (V)	477
Rated stator current (A)	3302
Range of stator current frequency (Hz)	2.6-7.8
Maximum rotor angular speed (rpm)	18
Minimum rotor angular speed (rpm)	6
Number of pole pairs	26
Rated rotor flux linkage (Wb)	5.826 (rms)
Stator resistance (mΩ)	0.831
Stator inductance (mH)	1.573
dc-link voltage (V)	1100
Switching frequency (kHz)	2
Rated parameter of power module	1 kA/1.7 kV
Machine-side converter structure	4 modules in parall

The wind speed and the ambient temperature are sampled every 1 hour during one year. An overall flowchart for the power semiconductor B_{10} lifetime in the MSC evaluation is illustrated in Fig. 3.

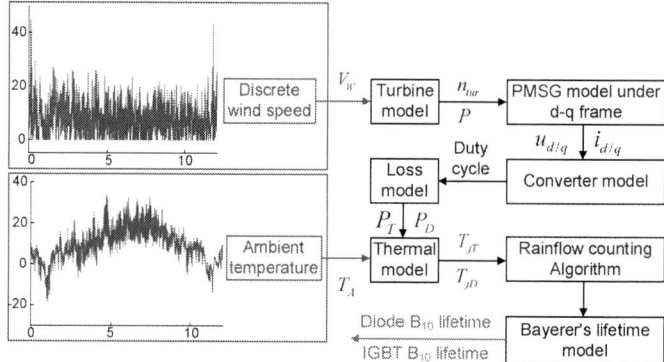

Fig. 3: Flowchart to calculate B_{10} lifetime of IGBT and diode under real-time wind speed and ambient temperature.

In the flowchart, the turbine model can link the relationship between the wind speed to the output power, which is illustrated in Fig. 4(b). Together with the PMSG and the converter models, the voltage and the current of the converter can be calculated at various wind speeds as shown in Fig. 4(c) and Fig. 4(d), respectively. On the basis of calculating the displacement angle, the switching loss and conduction loss of the diode and IGBT can be obtained by loss model according to the following equations [12]:

$$P_{con\_diode} = V_f \left(\frac{I_s}{4} \right) \left(\frac{1}{2\pi} - \frac{1}{8} \frac{U_s}{U_{dc}} \cos\varphi \right) + R_f \left(\frac{I_s}{4} \right)^2 \left(\frac{1}{8} - \frac{1}{3\pi} \frac{U_s}{U_{dc}} \cos\varphi \right) \tag{1}$$

$$P_{con\_IGBT} = V_{ce} \left(\frac{I_s}{4} \right) \left(\frac{1}{2\pi} + \frac{1}{8} \frac{U_s}{U_{dc}} \cos\varphi \right) + R_{ce} \left(\frac{I_s}{4} \right)^2 \left(\frac{1}{8} + \frac{1}{3\pi} \frac{U_s}{U_{dc}} \cos\varphi \right) \tag{2}$$

$$P_{sw\_diode} = f_{sw}\left(\frac{A_{diode}}{2} + \frac{B_{diode}}{\pi}\left(\frac{I_s}{4}\right) + \frac{C_{diode}}{4}\left(\frac{I_s}{4}\right)^2\right)\frac{U_{dc}}{1000} \qquad (3)$$

$$P_{sw\_IGBT} = f_{sw}\left(\frac{A_{IGBT}}{2} + \frac{B_{IGBT}}{\pi}\left(\frac{I_s}{4}\right) + \frac{C_{IGBT}}{4}\left(\frac{I_s}{4}\right)^2\right)\frac{U_{dc}}{1000} \qquad (4)$$

wherein, P_{con} represents conduction loss, P_{sw} represents switching loss, subscript "*diode*" means the calculation result for diode, subscript "*IGBT*" means the calculation result for IGBT, U_s is the stator voltage, I_s is the stator current, f_{sw} is the switching frequency, U_{dc} is the dc-link voltage and φ is the displacement angle between the stator voltage and the stator current, coefficients used in Eq. (1) to Eq. (4) are listed as Table III.

Table III. Coefficients used in the loss model.

Parameters	Values	Parameters	Values
V_f (1 kA/150 °C)	0.66	B_{diode}	0.35
V_{ce} (1 kA/150 °C)	0.67	C_{diode}	-1.20E-4
R_f (1 kA/150 °C)	1.13E-3	A_{IGBT}	32.85
R_{ce} (1 kA/150 °C)	1.64E-3	B_{IGBT}	0.47
A_{diode}	9.14	C_{IGBT}	2.81E-4

The total dissipations of the components (Fig. 4(e)) can be obtained by summing their switching loss and conduction loss. Combined with the real-time ambient temperature, the junction temperature of each components (Fig. 4(f)) are estimated from the thermal models. In this part, only the thermal impedances of power semiconductors and an air-cooling system are considered, moreover, an industrial Foster structure is applied [13], the key thermal specifications are shown in Table IV.

Table IV. Key parameters used in the thermal model.

	Diode		IGBT		Air cooling system	
Thermal resistance (°C/kW)	1st layer	0.48E-3	1st layer	0.30E-3	1st layer	6.60E-3
	2nd layer	3.61E-3	2nd layer	1.60E-3		
	3rd layer	3.46E-2	3rd layer	1.80E-2	2nd layer	1.95E-2
	4th layer	6.47E-3	4th layer	3.10E-3		
Thermal time constant (s)	1st layer	1.80E-4	1st layer	3.00E-4	1st layer	17.93
	2nd layer	8.90E-4	2nd layer	1.30E-3		
	3rd layer	3.00E-2	3rd layer	0.04	2nd layer	5.27
	4th layer	2.00	4th layer	0.40		

Then, the irregular thermal profiles of the IGBT and diode the can be extracted using a Rainflow counting algorithm and the corresponding mean junction temperatures, junction temperature fluctuations and thermal cycle periods are obtained. In Fig. 4(g) and Fig. 4(h), Rainflow counting results for IGBT and diode are presented, respectively. There are 1980 total thermal cycles in both of the IGBT and the diode counting results. However, the mean junction temperatures of the diode are much larger than that of the IGBT, which is mainly caused by the smaller chip size of the diode. Following this, the B_{10} lifetime of diode and IGBT can be calculated using a Bayerer's lifetime model [14], and the annual damage of them can be obtained by accumulating B_{10} lifetime, which is shown in Fig. 4(i). It illustrates that the annual consumed lifetime of diode is 0.52%, which is almost 4 times to the 0.14% of IGBT.

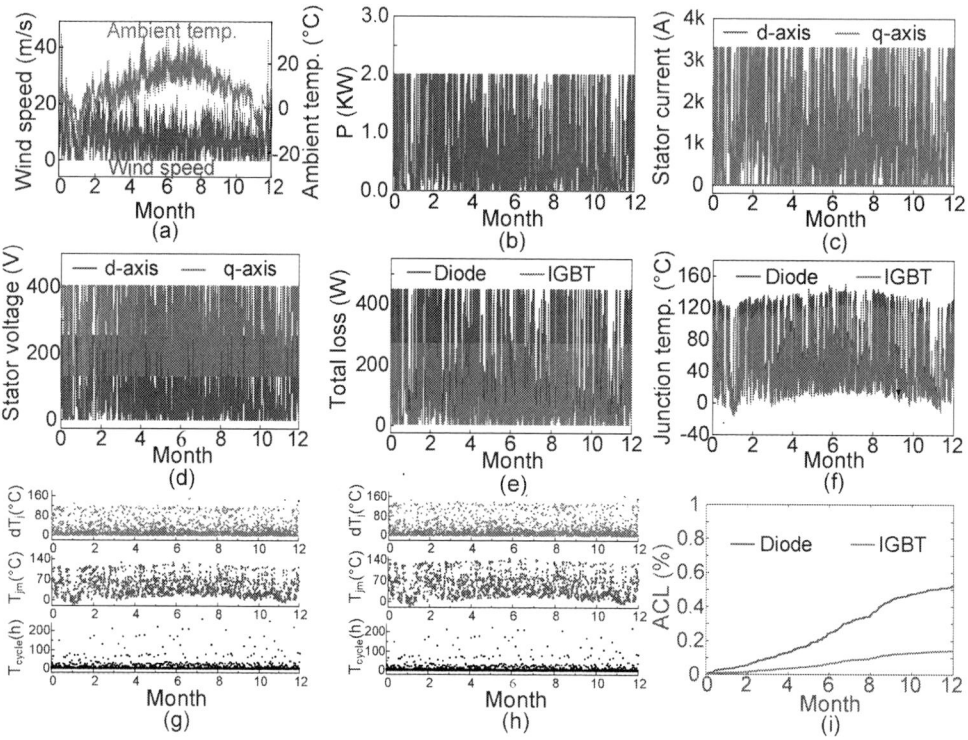

Fig. 4: Results of the long-term thermal cycle based lifetime estimation: (a) Wind speed and ambient temperature; (b) Turbine output power; (c) PMSG stator current in d-q coordinate frame; (d) PMSG stator voltage in d-q coordinate frame; (e) Total loss for diode and IGBT; (f) Junction temperature for diode and IGBT; (g) and (h) Rainflow counting results for IGBT and diode (from top to bottom: junction temperature fluctuation, mean junction temperature and thermal cycle period); (i) Annual consumed lifetime (ACL) for diode and IGBT.

Reliability evaluation of semiconductors with short-term thermal cycles

As indicated in Fig. 1, short-term thermal cycles in thermal profile are mainly related to the frequency of the PMSG stator current. Although a relative short period is selected to sample the real-time wind speed and ambient temperature, the frequency of profile fluctuations caused by short-term thermal cycles is usually several to dozens of Hz, which cannot be reflected by the long-term thermal cycle based calculation. Considering the stator current frequency of the PMSG is decided by the real-time wind speed indirectly, the annual wind speed in Fig. 3 should be sorted by the Weibull distribution. Besides, to obtain the same comparative and subsequent summing conditions, the reliability evaluation under short-term thermal cycle shares a same group of the key parameters with that in the long-term thermal cycle based evaluation. A flowchart of lifetime calculation for the power semiconductors under the Weibull wind speed distribution is shown in Fig. 5.

In the calculation process presented in Fig. 5, the input Weibull wind speed distribution is a statistical result generated from the annual real-time wind speed. In this distribution, the lowest speed is the same as turbine cut-in speed of 3 m/s and the highest speed is the same as the turbine cut-off speed of 25 m/s, from the lowest to the highest speed, a resolution of 1m/s was selected to obtain 23 discrete speed statistical results. By dividing the total wind speed sampling amounts with each speed statistical results, the proportion factor for each speed value can be obtained, which will be used for subsequent accumulation damage calculation. Calculation results for each step in Fig. 5 are show in Fig. 6. In general, the calculation procedure with short-term thermal cycles is similar to that with long-term cycles. Firstly, a relationship between the turbine output power and the input discrete wind speed is established as Fig. 6(a) and (b) using the turbine model, and the discrete stator currents and voltages, which are illustrated in Fig. 6(d) and (e), respectively, for each wind speed can be obtained via the PMSG d-q frame model. Then, the total loss dissipations of diode and IGBT, which include conduction loss of Fig. 6(g) and switching loss of Fig. 6(h), can be calculated from the loss model. Further, thermal profiles of semiconductors, which include mean junction temperature and junction temperature fluctuation, can be illustrated by thermal model. Afterwards, based on a Bayerer's model, total power cycles or B_{10} lifetimes of diode and IGBT under different wind speeds are calculated as Fig. 6(k). Finally, the annual accumulation damage of diode and IGBT can be estimated according to different wind speed distribution mission profiles in Fig. 6(i).

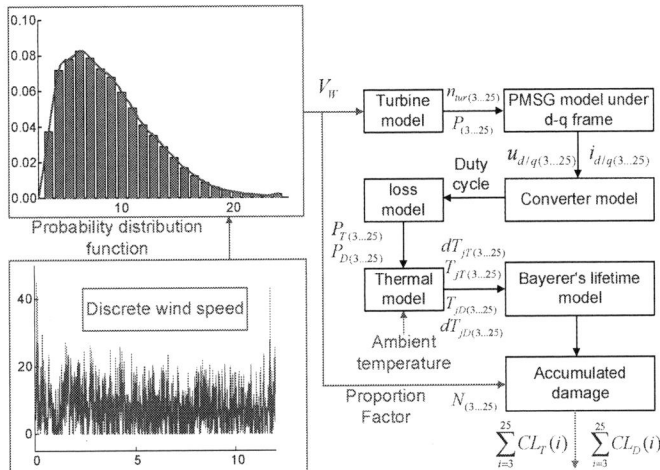

Fig. 5: Flowchart to calculate B_{10} lifetimes and accumulated damages of IGBT and diode under Weibull wind speed distribution.

However, the major difference between the calculations under these two kinds of thermal cycles lies in the obtaining of mean junction temperature and junction temperature fluctuation. For the lifetime calculation under long-term thermal cycles, the mean junction temperature and the junction temperature fluctuation are hard to be estimated directly from the annual discrete junction temperature profile, which leads to employment of a Rainflow counting algorithm. However, for short-term thermal cycle, which assumes the wind speed as a discrete constant value during the 1-year period, the thermal characteristics of the semiconductors with Foster structures can be directly calculated from following formulas [15]:

$$T_{jm\_diode/IGBT} = P_{diode/IGBT}\left(\sum_{i=1}^{m} R_{th\_diode/IGBT}(i) + \sum_{i=1}^{n} R_{th\_c}(i)\right) + T_a \tag{5}$$

$$dT_{j\_diode/IGBT} = 2P_{diode/IGBT}\sum_{i=1}^{m}\left(R_{th\_diode/IGBT}(i)\frac{\left(1-e^{\frac{-t_{on}}{\tau_{diode/IGBT}(i)}}\right)^2}{1-e^{\frac{-1/f_e}{\tau_{diode/IGBT}(i)}}}\right) \tag{6}$$

wherein, T_{jm} represents mean junction temperature, dT_j represents junction temperature fluctuation, subscript "$diode$" means the value for freewheeling diode, subscript "$IGBT$" means the value for IGBT, i is the ith Foster layer, m is the total layers for semiconductors, n is the total layer for air-cooling system, P is the total power loss for each semiconductor, T_a is the constant ambient temperature, which is chosen as the mean value of an annual temperature profile presented in Fig. 4(a), t_{on} is the on-state time for each semiconductor, f_e is the stator current frequency under different wind speeds, and R_{th} and τ denote thermal resistance and thermal time constant, respectively, whose values have been shown in Table IV.

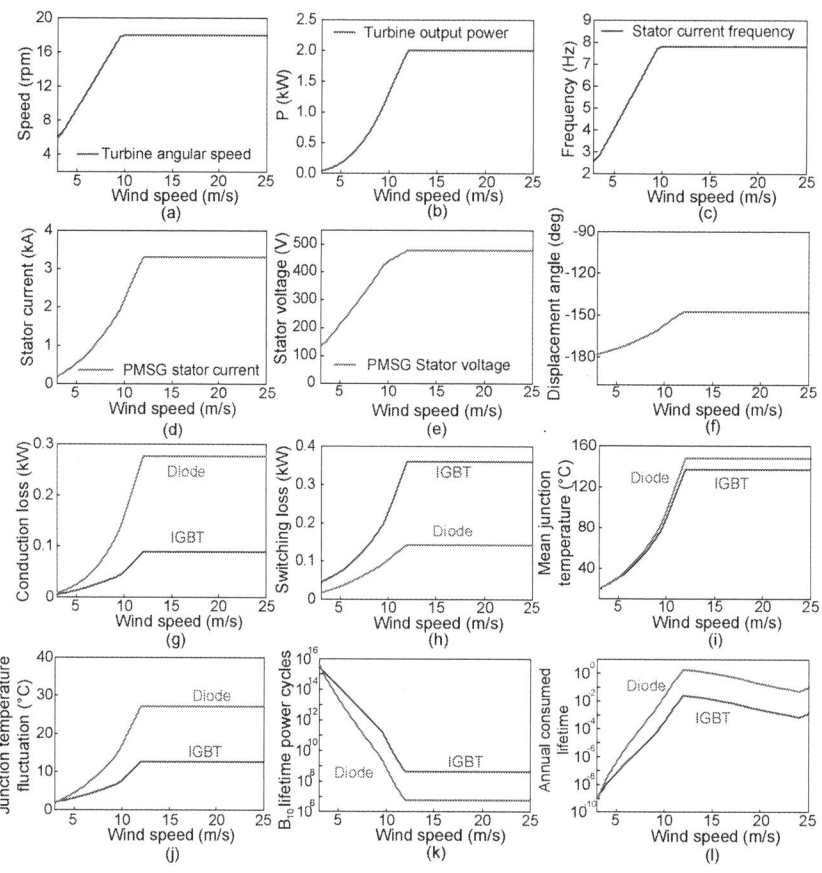

Fig. 6: Calculation results of the short-term thermal cycle based lifetime estimation: (a) Turbine angular speed; (b) Turbine output power; (c) PMSG stator current frequency; (d) PMSG stator current; (e) PMSG stator voltage; (f) Displacement angle; (g) Conduction losses for diode and IGBT; (h) Switching losses for diode and IGBT; (i) Mean junction temperatures for diode and IGBT; (j) Junction temperature fluctuations for diode and IGBT; (k) B_{10} lifetime power cycles for diode and IGBT; (l) Annual consumed lifetimes for diode and IGBT.

Additionally, it is noted that for both of these two evaluations under different thermal cycle circumstances, following assumptions are universally suitable: (1) a unified failure mechanism exists in all Foster layers, which means the soldering cracks are neglected between different layers [13]; (2) the fatigue damage accumulates in a linear manner, which obeys the Miner's rule [16]; (3) parameters in the Bayerer's lifetime model are constant during the B_{10} lifetime calculation.

Comparison between long-term and short-term thermal cycles

Based on the annual consumed lifetime calculation with long-term thermal cycles in Fig. 4 and short-term thermal cycles in Fig. 6, the total consumed lifetimes (TCLs) of the power semiconductors can be compared. It is noted that, for the long-term thermal cycle based evaluation, the TCLs of semiconductors are same as the results obtained in Fig. 4. However, in Fig. 6 (l), the short-term thermal cycle based evaluation contains all the consumed lifetime results under different wind speeds, which should be added after multiplying by their proportions in the annual wind speed profile. The comparative results of the annual TCL for the diode and IGBT with different thermal cycles are shown in Fig. 7.

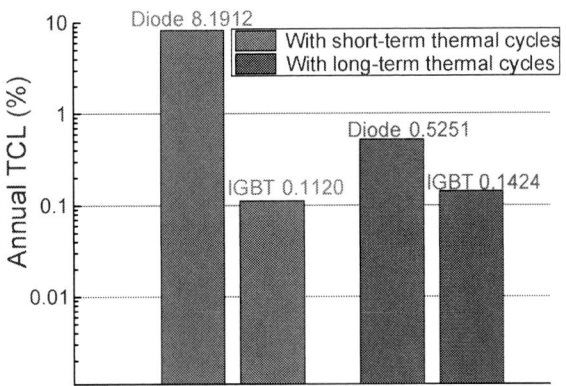

Fig. 7: Annual total consumed lifetimes (TCLs) of diode and IGBT with different thermal cycles.

From the results shown in Fig. 7, following two conclusions can be observed:

(1) No matter under short-term thermal cycles or under long-term thermal cycles, the TCLs for diode are both higher than those of the IGBT, which is mainly caused by a higher mean junction temperature of the diode due to its smaller chip size.

(2) For the TCL of diode, the evaluation result under short-term thermal cycles is much higher than the result obtained under long-term thermal cycles, while for the TCL of IGBT, the TCL with long-term thermal cycles takes a major proportion. Therefore, in the practical design and manufacturing of the power converter, lifetime expectancy for the diode with short-term thermal cycles and the lifetime expectancy for the IGBT with long-term thermal cycles should be given more emphasis.

Component-level reliability distribution for MSC

For the aforementioned calculations in Figs. 4, 6 and 7, only B_{10} lifetimes for one diode or one IGBT are focused, which can only reflect the lifetime when 10% semiconductors fail in an MSC. However, for a complex system with multiple converters in parallel, the B_{10} lifetime is not sufficient to reveal the reliability for the MSC, and a more precise B_1 lifetime or the entire reliability distribution profile are preferred.

To obtain the entire lifetime profile for power components, uncertainties and variations for parameters in the Bayerer's lifetime model should be taken into consideration.

$$N_{f\_diode/IGBT} = A \cdot dT_{j\_diode/IGBT}^{\beta_1} \cdot e^{\left(\frac{\beta_2}{T_{jm\_diode/IGBT} + 273}\right)} \cdot t_{on}^{\beta_1} \tag{7}$$

For Bayerer's lifetime formula in Eq. (7), N_f represents the power cycle, $A, \beta_1, \beta_2, \beta_3$ are coefficients that are fitted by a large amount of test data [17]. Although all these coefficients are statistically fitted and chosen as constants to simplify the calculation, uncertainties for these parameters still exist. It is assumed that all these coefficients are under a standard normal distribution and have 5% variations to their center values; thus, the probability density functions (PDFs) of A to β_3 are shown as Fig. 8.

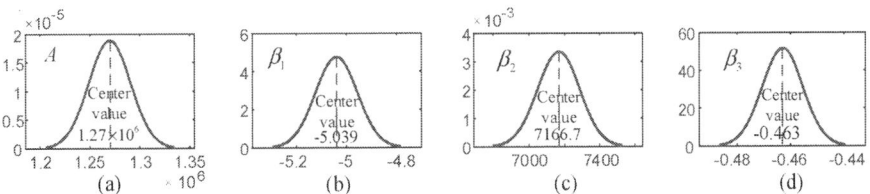

Fig. 8: Normal PDFs for coefficients in Bayerer's lifetime model (a) A, (b) β_1, (c) β_2, (d) β_3.

For the thermal stress related parameters mean junction temperature T_{jm} and junction temperature fluctuation dT_j in Bayerer's lifetime model, variations are induced by the manufacturing process. Normal distributions with 5% variations are also applied to describe temperature related parameters in diode and IGBT.

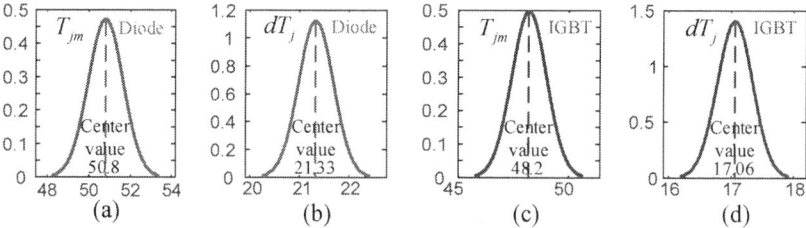

Fig. 9: Normal PDFs for thermal stress related parameters in Bayerer's lifetime model (a) mean junction temperature for diode, (b) junction temperature fluctuation for diode, (c) mean junction temperature for IGBT, (d) junction temperature fluctuation for IGBT.

Taking variations of all parameters into account, a 10000 sampling Monte Carlo method can be utilized to analyze the failure and lifetime distributions for diode and IGBT in the MSC. The annual damages for one IGBT component and one diode component can be obtained by adding their TCLs under long-term thermal cycles and short-term thermal cycles. For the consumed lifetime results under short-term thermal cycles, they are calculated according to different wind speeds, but only the TCLs under the wind speed of 7.17 m/s are used for summing, which equals the average speed in the annual wind speed profile presented in Fig. 4(a). Based on the aforementioned calculation, the TCLs for diode and IGBT under short-term thermal cycles are 4.80E-04 and 2.17E-05, respectively, and under long-term thermal cycles are 5.25E-03 and 1.42E-03, respectively. Therefore, the central TCL values for diode and IGBT can be obtained as 5.731E-03 and 1.445E-03, respectively. Moreover, Weibull distributions are universally used to describe the failure and lifetime (reciprocal of the failure rate) data for power semiconductors, the component-level reliability profiles can be illustrated as Fig. 10.

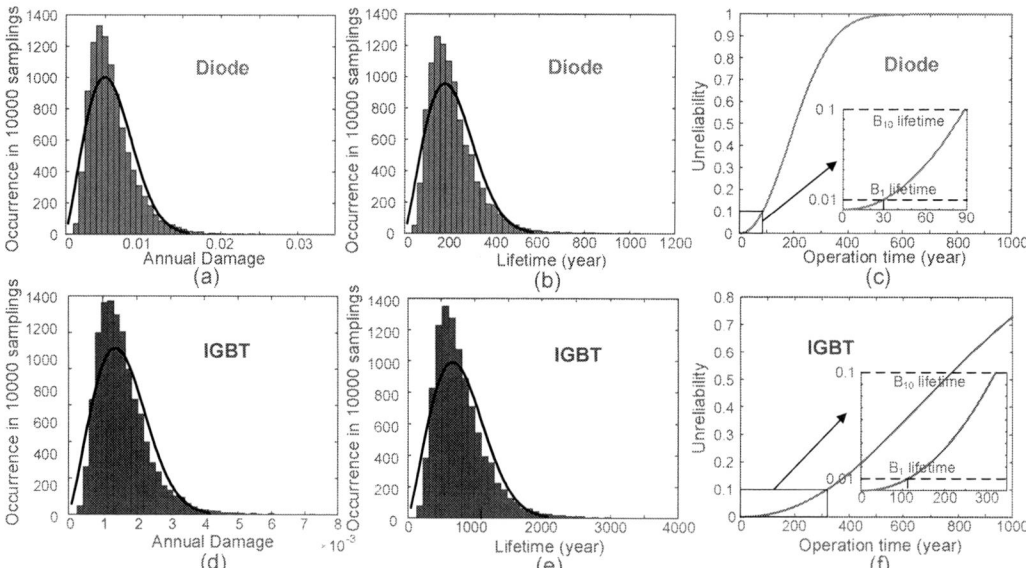

Fig. 10: (a) Annual damage Weibull distribution for diode; (b) lifetime Weibull distribution for diode; (c) unreliability for diode; (d) annual damage Weibull distribution for IGBT; (e) lifetime Weibull distribution for IGBT; (f) unreliability for IGBT.

In Fig. 10(c) and (f), the failure rate cumulative distribution functions (CDFs) for diode and IGBT are obtained by integrating the damage PDFs of diode and IGBT in Fig. 10(a) and (d), respectively. It can be observed that for diode and IGBT, the B_1 lifetime are 30 years and 114 years, respectively, which are highly over the recommendations in [4] for an engineering level.

Conclusions

This paper has presented a reliability comparison for semiconductors between different thermal cycles and evaluated the component-level reliability for a direct-drive PMSG based MSC. Firstly, the TCLs for diode and IGBT under long-term thermal cycles and short-term thermal cycles are obtained separately and compared. Relevant results show the damage under long-term thermal cycles takes a majority proportion for IGBT, while for the diode, short-term thermal cycle damage dominates. Then, based on the Bayerer's lifetime model and the Monte Carlo method, the component-level reliability of MSC is analyzed. It is observed that B_1 lifetime for power components in the selected MSC fulfill the industry requirement.

References

[1] F. Blaabjerg, and K. Ma, "Future on power electronics for wind turbine systems," IEEE Trans. Emerging Sel. Topics Power Electron., vol. 1, no. 3, pp. 139–152, Sep. 2013.

[2] H. Polinder et al., "Trends in wind turbine generator systems," IEEE Trans. Emerging Sel. Topics Power Electron., vol. 1, no. 3, pp. 174–185, Sep. 2013.

[3] M. Tsili, and S. Papathanassiou, "A review of grid code technical requirements for wind farms," IET Renew. Power Gener., vol. 3, no. 3. pp. 308–332, Sep. 2009.

[4] B. Hahn, M. Durstewitz, and K. Rohrig, "Reliability of wind turbines—Experience of 15 years with 1500 WTs," in Proc. of the Euromech Colloquium, pp. 329–332, 2007.

[5] "Handbook for robustness validation of automotive electrical/electronicmodules," Zentralverband Elektrotechnik-und Elektronikindustrie e.V., Frankfurt, Germany Jun. 2008.

[6] D. Zhou and F. Blaabjerg, "Converter-level reliability of wind turbine with low sample rate mission profile," IEEE Trans. on Ind. Appl., vol. 56, no. 3, pp. 2938-2944, May-Jun. 2020.

[7] D. Zhou, F. Blaabjerg, M. Lau, and M. Tonnes, "Comparison of wind power converter reliability with low-speed and medium-speed permanent-magnet synchronous generators," IEEE Trans. on Ind. Electron., vol. 62, no. 10, pp. 6575-6584, Oct. 2015.

[8] G. Zhang, D. Zhou, F. Blaabjerg and J. Yang. "Mission profile resolution effects on lifetime estimation of doubly-fed induction generator power converter," in Proc. of 2017 IEEE SPEC, Puerto Varas, 2017, pp. 1-6.

[9] D. Zhou, F. Blaabjerg, M. Lau, and M. Tonnes, "Optimized reactive power flow of DFIG power converters for better reliability performance considering grid codes," IEEE Trans. Ind. Electron., vol. 62, no. 3, pp. 1552–1562, Mar. 2015.

[10] "Enercon E-82 wind turbine," Enercon, Aurich, Germany. [Online]. Available: www.enercon.de/en-en/62.htm.

[11] J. Chivite-Zabalza et al., "Comparison of power conversion topologies for a multi-megawatt off-shore wind turbine, based on commercial power electronic building blocks," in Proc. of IEEE IECON, pp. 5242–5247, 2013.

[12] ABB Application Notes: Applying IGBTs, Apr. 2009, pp: 22-23.

[13] R. Schnell, M. Bayerer, and S. Geissmann. "Thermal design and temperature ratings of IGBT modules," ASEA Brown Boveri, Zurich, Switzerland, ABB Application Note, 5SYA 2093-00, 2011.

[14] P. Lall, M.N. Islam, M. K. Rahim, and J. C. Suhling, "Prognostics and health management of electronic packaging." . IEEE Trans. Compon. Packag. Technol., vol. 29, no. 3, pp. 666-677, Sep. 2006.

[15] D. Zhou, F. Blaabjerg, M. Lau, and M. Tonnes, "Thermal profile analysis of doubly-fed induction generator based wind power converter with air and liquid cooling methods," in Proc. EPE, 2013, pp. 1–10.

[16] M. A. Miner, "Cumulative damage in fatigue," J. Appl. Mech., vol. 12, pp. A159–A164, 1945.

[17] R. Bayerer, T. Hermann, T. Licht, J. Lutz, and M. Feller, "Model for power cycling lifetime of IGBT modules—various factors influencing lifetime," in Proc. Integr. Power Syst., 2008, pp. 1–6.

A Submodule Implementation for Parallel Conduction of Diodes in Modular Multilevel Converters

Martin GESKE*, Duro BASIC**, Christian KELLER*, Thomas BRÜCKNER***

*GE Power Conversion GmbH, Culemeyerstraße 1, 12277 Berlin, Germany
**GE Power Conversion SAS, 18 Avenue due Québec, 91140 Villebon-sur-Yvette, France
***Universität der Bundeswehr München, Werner Heisenberg-Weg 39, 85579 Neubiberg, Germany
Tel.: +49 / (0) 30 7622 2394*
E-Mail: martin.geske@ge.com, duro.basic@ge.com, christian.b.keller@ge.com
URL: https://www.gepowerconversion.com

Keywords

«Multilevel converters», «Voltage Source Converter (VSC)», «Conduction losses», «HVDC»

Abstract

This paper presents a new double submodule for Modular Multilevel Converters that supports parallel conduction of freewheeling diodes. The new submodule circuit can reduce semiconductor conduction losses through parallel diode conduction and uses a small number of switches.

Introduction

Modular Multilevel Converters (MMC) have become standard for Voltage Source Converters (VSC) used for High-Voltage DC (HVDC) applications [1]. The implemented type of Submodule (SM) is an important factor with regard to the semiconductor losses of the circuit. The Half-Bridge SM achieves low power losses because only one device per submodule is in conduction mode at one time. Alternative SMs, such as the Double-Zero Submodule (DZ-SM) [2] and the Double Submodule (DSM) [3], provide switch states with parallel conduction paths that can reduce conduction losses.

This work presents a new double submodule that is a subset of the DSM, which was introduced by Ilves et al. [3]. This new DSM circuit uses fewer switches compared to the DSM and is called Reduced Full-Bridge DSM (RFB-DSM). The RFB-DSM cannot parallel the SM capacitors like the DSM but still benefits from parallel conduction paths. The new SM is analyzed through system simulations with regard to parallel conduction and related conduction losses.

The Double Submodule

The DSM can parallel both SM capacitors, according to Fig. 1 (a) and (b). These states have been used to analyze the potential reduction of the associated energy storages. Experiments demonstrated a voltage ripple reduction of 18% [3]. The circuitry of the DSM has the same quadrant capability and unipolar voltage characteristic like the HB-SM, as indicated in Fig. 1 (c). The DSM was intended to have the combined power rating of two HB-SMs, whereas the number of devices is doubled [3]. For HB-SMs that use parallel connected semiconductors, the DSM can be implemented with the same number of devices as two respective HB-SMs [3].

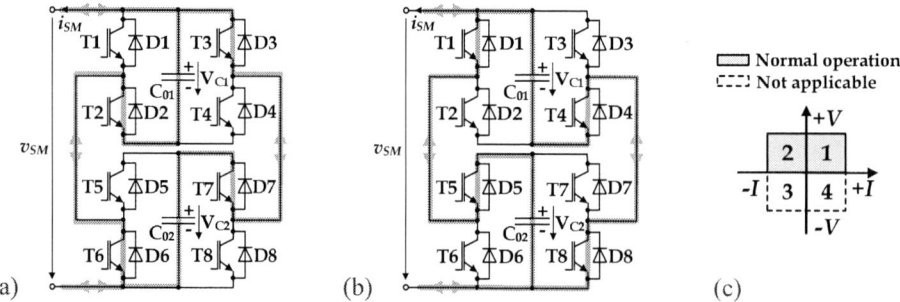

Fig. 1: The double submodule: (a), (b) switch states for parallelization of both SM capacitors, (c) quadrant capability chart with unipolar output voltage characteristic

Table I summarizes possible switch states of the DSM, which differently apply parallel conduction paths between the semiconductor devices. The DSM reduces the voltage ripple of the capacitors through the use of State 4 and 5, which parallel the capacitors during operation.

Table I: Chosen switch states of the DSM, cp. [2]

State	T1	T2	T3	T4	T5	T6	T7	T8	v_{sm}	C_{SM}
1	1	0	1	0	0	1	0	1	0	0
2	0	1	0	1	0	1	0	1	$+V_C$	C_{01}
3	1	0	1	0	1	0	1	0	$+V_C$	C_{02}
4	1	0	0	1	1	0	0	1	$+V_C$	$C_{01}\|C_{02}$
5	0	1	1	0	0	1	1	0	$+V_C$	$C_{01}\|C_{02}$
6	0	1	0	1	1	0	1	0	$+2 \cdot V_C$	$(1/C_{01} + 1/C_{02})^{-1}$

The following analysis relates to the indicated parallel conduction through the freewheeling diodes in Fig. 2 (a) and (b). The parallel paths are envisaged for the reduction of the diode conduction losses apart from the basic intention of the DSM, which considers the same or a similar combined power rating of two HB-SMs. Therefore, the switch states from Table I for the parallel connection of the DSM's capacitors are not applied. Correspondingly, several switch positions of the DSM are not used, which allow to permanently turn-off the switches or to preferably disregard their implementation.

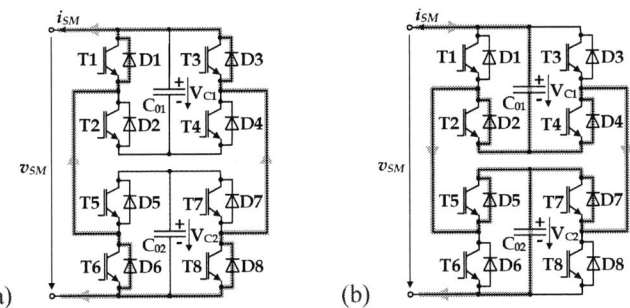

(a)   (b)

Fig. 2: The DSM: (a), (b) switch states for parallel conduction of the freewheeling diodes

The Reduced Full-Bridge Double Submodule

The RFB-DSM presents a subset of the DSM to use the parallel conduction paths that are shown for the DSM in Fig. 2 (a) and (b) but with circuitry that uses fewer switches. Neglecting the States 4 and 5 in Table I of the DSM allows omitting the diagonal switch positions within the Full-Bridges. The tradeoff is that the number of IGBTs has been reduced by 50%, while the SM capacitors cannot be paralleled through respective switch states. The paralleling of the DSM's capacitors can reduce the associated peak-to-peak voltage ripple by about 18% [3], which corresponds to a reduction of the installed capacitance by approximately 10%. The new DSM circuit in Fig. 3 (a) has the same quadrant capability chart as the HB-SM and the DSM that is shown in Fig. 3 (b).

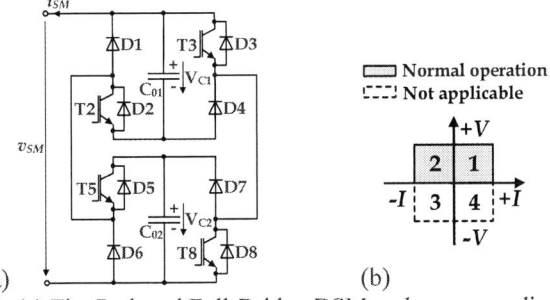

(a)   (b)

Fig. 3: (a) The Reduced Full-Bridge DSM and corresponding (b) quadrant capability chart

Table II summarizes a different composition of switch states that are used for the analysis of the new DSM circuit. Of primary interest are the states of parallel diode conduction and states that are required for the state transitions. For positive current ($i_{SM}>0$), States 1, 3, 5, and 7 provide parallel current paths through the RFB-DSM. States 1, 2, 4, and 6 support the paralleling of the diodes for negative current ($i_{SM}<0$). States 1 to 5 incorporate a dependency of the output voltage v_{SM} on the current direction of i_{SM}, which demands their consideration for the state transitions during operation.

Table II: Chosen switch states of the Reduced Full-Bridge DSM

State	T2	T3	T5	T8	i_{sm}	v_{sm}	Δw_{c1}	Δw_{c2}
1	0	0	0	0	< 0	$+V_{C1}+V_{C2}$	> 0	> 0
	0	0	0	0	> 0	0	0	0
2	0	0	1	0	< 0	$+V_{C1}+V_{C2}$	> 0	> 0
	0	0	1	0	> 0	$+V_{C2}$	0	< 0
3	0	0	0	1	< 0	$+V_{C1}$	> 0	0
	0	0	0	1	> 0	0	0	0
4	1	0	0	0	< 0	$+V_{C1}+V_{C2}$	> 0	> 0
	1	0	0	0	> 0	$+V_{C1}$	< 0	0
5	0	1	0	0	< 0	$+V_{C2}$	0	> 0
	0	1	0	0	> 0	0	0	0
6	1	0	1	0	< 0	$+V_{C1}+V_{C2}$	> 0	> 0
	1	0	1	0	> 0	$+V_{C1}+V_{C2}$	< 0	< 0
7	0	1	0	1	< 0	0	0	0
	0	1	0	1	> 0	0	0	0
8	0	1	1	0	< 0	$+V_{C2}$	> 0	0
	0	1	1	0	> 0	$+V_{C2}$	< 0	0
9	1	0	0	1	< 0	$+V_{C1}$	0	> 0
	1	0	0	1	> 0	$+V_{C1}$	0	< 0

Pulse Width Modulation and Transitions between the Switch States

The RFB-DSM creates three different output voltage levels (see Table II) like the DSM and can be modulated with the Pulse Width Modulation (PWM) techniques that are known from two and three-level converters [4]. Fig. 4 (a) shows the PWM for two-level modulation that was presented in [3], which is here applied for the RFB-DSM to create two output voltage levels through States 6 and 7 from Table II. Furthermore, the new DSM circuit is analyzed for three-level PWM that uses one reference and two carriers in phase disposition [4]. Fig. 4 (b) shows the adapted PWM technique, which has a DC offset for the PWM of the unipolar voltage RFB-DSM. The three-level PWM basically uses the States 6, 7, 8, and 9, which are independent of the current direction.

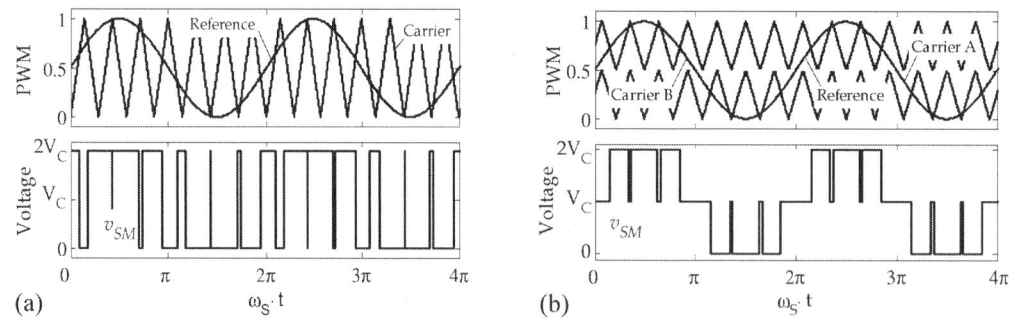

Fig. 4: Modulation of the RFB-DSM: (a) Two-level PWM (top) and switched output-voltage waveform (bottom) (M_{AC}=0.5, M_{DC}=0.5, carrier ratio ω_C/ω_S=7), (b) three-level PWM (top) and switched output-voltage waveform (bottom) (M_{AC}=0.5, M_{DC}=0.5, carrier ratio $\omega_{C,A}/\omega_S=\omega_{C,B}/\omega_S$=7)

The outlined PWM methods are applied under consideration of state transitions that avoid short-circuit conditions for the SM capacitors. Fig. 5 shows an example of the transition from State 6 to 8. This transition requires turning off Switch T2 and turning on Switch T3. If both switching transients are triggered simultaneously and the turn-on process of T3 is faster than the turn-off process of T2, a short circuit of the upper SM capacitor C_{01} is created. Accordingly, dead time requirements and an appropriate choice of the transitioning state are necessary, even though T2 and T3 are not part of the same commutation loop. This demand corresponds to an intermediate transition to State 2, before T3 can be turned on for reaching State 8, which is illustrated through the state machine in Fig. 5 (b). In summary, only two of (any) switches in total may be switched on at any time.

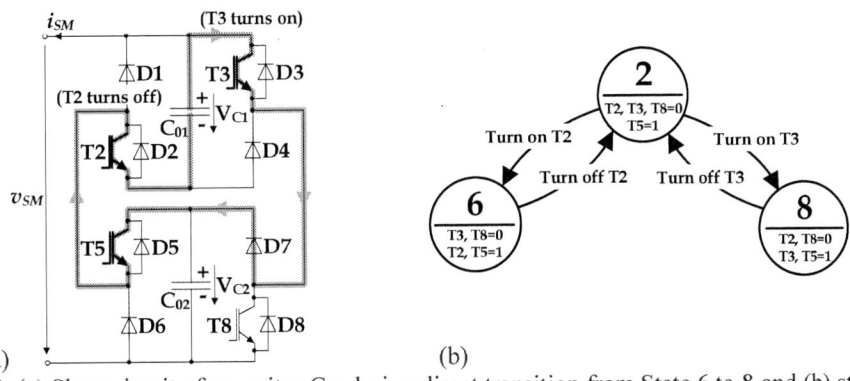

Fig. 5: (a) Short circuit of capacitor C_{01} during direct transition from State 6 to 8 and (b) state-machine representation for the transition between States 6, 2, and 8

Specification of the Analyzed MMC System

The RFB-DSM is analyzed by the use of PLECS simulation software [5]. The investigated MMC topology is the double-star configuration shown in Fig. 6 that connects a three-phase power source with a DC-link. The system parameters of the MMC are summarized in Table III. The model uses the PWM techniques from Fig. 4, and phase-shifted carriers for the modulation of individual RFB-DSMs, which is a known method for MMCs [6].

Table III: Parameters of the MMC system

Parameter	Value
Rated system power S	6.87 MVA
Peak value of phase voltage V_S	5.4 kV
Angular frequency ω_S	314.16 rad/s
DC-link voltage V_{dc}	11.0 kV
Capacitor voltage $\bar{V}_C$	1.1 kV
SM capacitance C_{01}, C_{02}	5 mF
Number of RFB-DSMs per arm N_{SM}	6
Carrier ratio	(7), 21
L_S [mH], R_S [Ω]	3 mH, 0 Ω
L_{arm} [mH], R_{arm} [Ω]	0.25 mH, 15 mΩ
L_{dc} [mH], R_{dc} [mΩ]	3mH, 0.1Ω

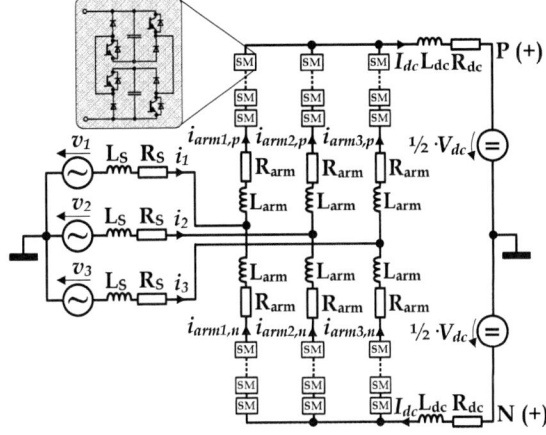

Fig. 6: Simulation model for MMC with RFB-DSMs

Simulation Results

Fig. 7 shows simulated waveforms that have been created for a single RFB-DSM under consideration of the system specification from Table III and the outlined PWM methods in Fig. 4. Here, the reference for the PWM has been assumed to consist of a DC component M_{DC}=0.45 and an AC component M_{AC}=0.45. The two-level (2L) PWM can be considered for MMCs that have a high number of serialized SMs and do not require additional voltage levels. The three-level (3L) modulation applies the intermediate levels per RFB-DSM that are applied through the States 8 and 9. The simulation shows that the 2L PWM creates equal capacitor voltages v_{C1}(2L) and v_{C2}(2L), which is normally not the case due to tolerances of the capacitors that lead to slightly different capacitance values. The 3L PWM requires to balance the voltages v_{C1}(3L) and v_{C2}(3L) during converter operation. This is achieved through a dedicated selection of the redundant States 8 or 9 under considerations of the current direction and the voltages of both capacitors. The current direction of i_{SM} is considered to charge the respective capacitor that has less voltage or to discharge the capacitor that has more voltage than the other capacitor. The same method can be applied for the balancing of the energy storages under consideration of the 2L PWM in case of diverging voltages. Fig. 7 shows a comparison of the upper diode currents of D1 and D3 that are both in on-state for parallel conduction. Here, the 2L modulation yields a proportion of 75% for states with parallel diode conduction paths. The 3L PWM uses the States 8 and 9, which reduces the proportion of parallel paths for current conduction to around 55% of the time.

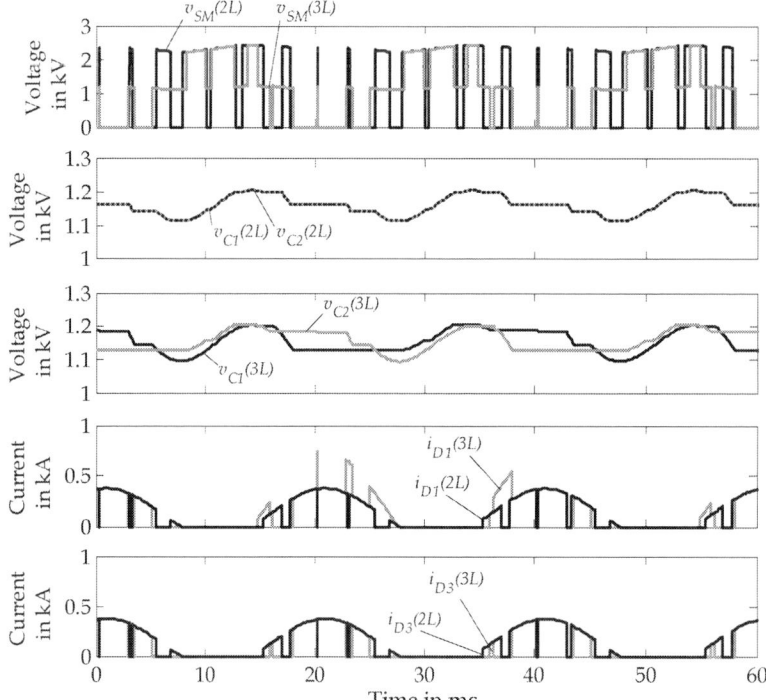

Fig. 7: Simulated waveforms of one RFB-DSM for 2L and 3L PWM M_{AC}=M_{DC}=0.45, carrier ratios: 2L PWM ω_C/ω_S=7, 3L PWM $\omega_{C,A}/\omega_S$=$\omega_{C,B}/\omega_S$=7, cos(φ)= -1)

Fig. 8 shows simulation results for the complete MMC circuit where the 3L PWM from Fig. 4 and the previously outlined capacitor balancing method have been applied. The simulated converter system effectively balances the capacitor voltages during the state transitions to States 8 and 9. The proposed 2L modulation ideally applies States 6 and 7 and uses State 1 only as a transition state. The simulation considers identical DSM capacitors, which is in reality not the case. Accordingly, slightly different capacitors lead to diverging capacitor voltages during operation with the 2L PWM. Then, the voltages of the RFB-DSM capacitors have to be balanced through the States 8 and 9, for example. This aspect causes additional state transitions and switching events, which leads to higher switching losses compared to an ideal modulation with identical capacitors, which is presented here. This undesired effect gets more perturbing when the carrier ratio ω_C/ω_S is below 5 or 3 and where the number of switching events is low or minimized.

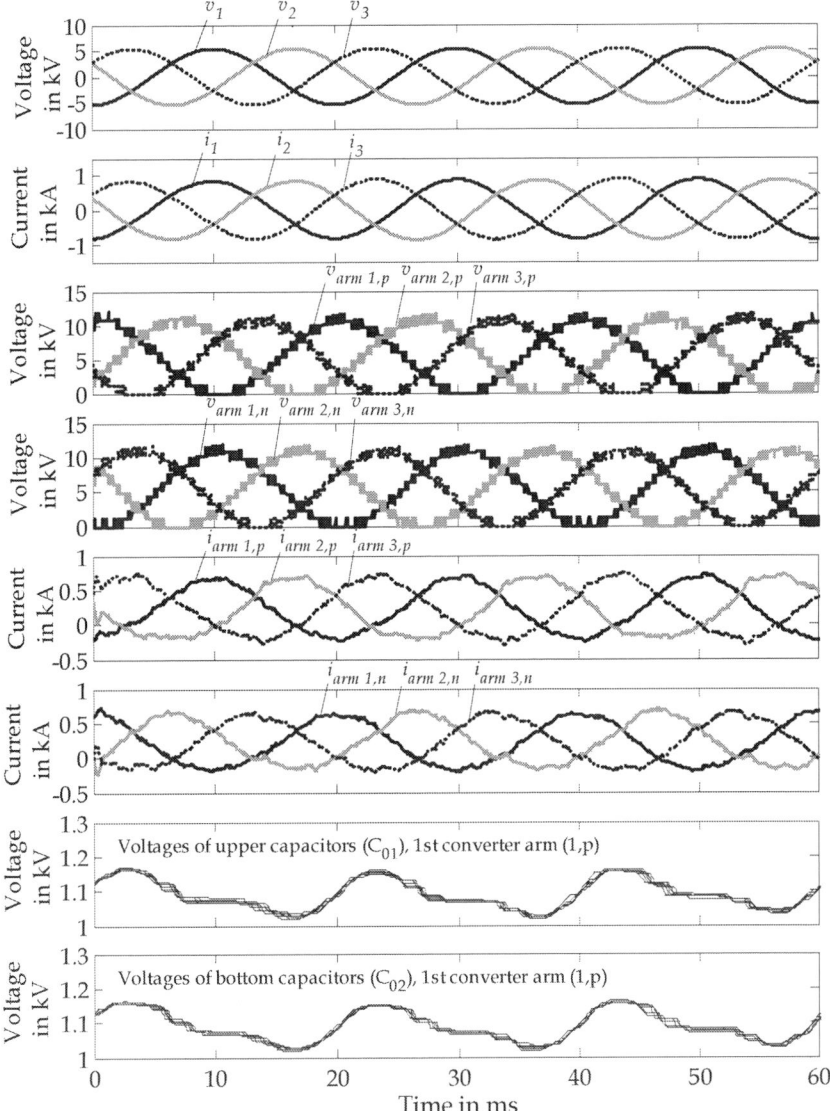

Fig. 8: Simulated waveforms of complete MMC with RFB-DSMs (MMC model from Fig. 6, Table III, 3L PWM with carrier ratio $\omega_{C,A}/\omega_S=\omega_{C,B}/\omega_S=21$, $\cos(\varphi)= -1$)

Analysis of Parallel Diode Conduction

The analysis of parallel conduction paths is done for an IGBT module that was analyzed for an MV MMC in [7]. It is assumed that the single diodes of the RFB-DSM are identical to the diode part of the IGBT module. Following [7], the device characteristics for IGBT and diode are modeled by

$$g(j) = a + b(j)^C,\qquad(1)$$

whose parameters and coefficients for the FZ600R17KE3 are summarized in Table IV. The respective on-state characteristics of the IGBT and the diode are illustrated in Fig. 9 (a).

Table IV: Model of on-state characteristics IGBT module FZ600R17KE3 [7], [8] (T_J=125°C)

g	j	a	b	c
v_{CE}	i_C	0.7 V	10.357 mV	0.79806
v_F	i_F	0.5 V	50.265 mV	0.52041

A generic evaluation of the diode on-state losses is calculated through equation (1). The beneficial effect of parallel paths is shown in Fig. 9 (b), which compares the total diode on-state losses for a single diode versus two diodes in parallel current conduction. For a total current through the diode(s) higher than 300A, the reduction of the associated on-state losses is roughly 20%. The reduction of the on-state losses increases for higher current values because of the on-state characteristic. The parallel conduction of two IGBTs has the same effect but is only feasible with the fully populated DSM.

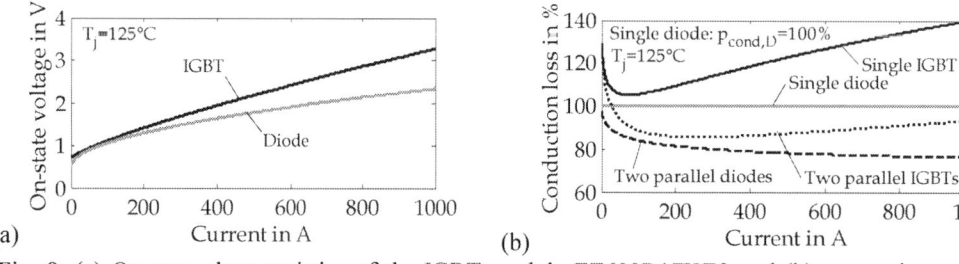

Fig. 9: (a) On-state characteristics of the IGBT module FZ600R17KE3 and (b) comparison of total diode on-state losses for single diode and parallel diode conduction

The models of the IGBT module are further used to compare the previously introduced PWM techniques. The proposed 2L modulation can lower the total conduction losses by approximately 15% for rectifier operation (cos(φ)=-1) compared to the 3L modulation. Accordingly, the positive effect of parallel conduction of two diodes is mostly limited to rectifier operation (cos(φ)<0) and cannot effectively compensate for the IGBT's conduction losses for the inverting operation mode (cos(φ)>0).

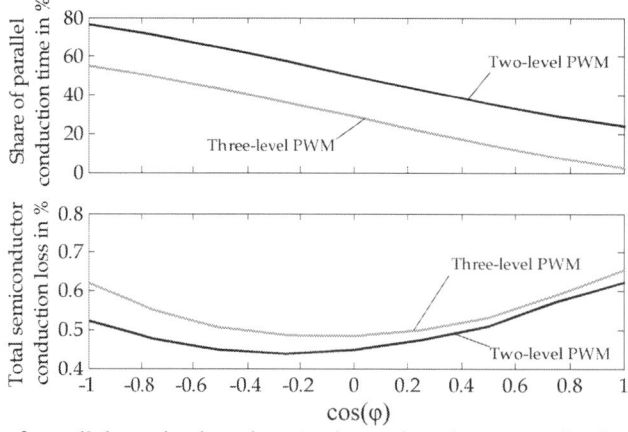

Fig. 10: Share of parallel conduction time (top), semiconductor conduction losses (bottom) of the RFB-DSM versus cos(φ) expressed as a percentage of the rated system power (system from Table III)

The amount of required voltage levels through the converter is important with regard to the proposed 2L and 3L PWM. Both medium- and high-voltage converters benefit from a high amount of voltage levels that enable a low Total Harmonic Distortion (THD) of the electrical output waveforms. In medium-voltage applications, where the number of submodules per arm is rather limited by the ratio between the system voltage and the SM voltage, the 3L PWM would be the preferred implementation. For high-voltage converters, the amount of SMs is typically high enough such that a 2L PWM can be applied for the advantage of an additional reduction of conduction losses. Nevertheless, slight tolerances of the capacitors and controls require balancing the capacitor voltages within the DSMs. Therefore, the intermediate levels have to be used, which gives additional switch state transitions and undesired semiconductor switching loss for the 3L PWM.

Summary

Submodules that provide parallel conduction of power semiconductors are beneficial for the reduction of semiconductor power losses in Modular Multilevel Converters. The Double-Zero Submodule (DZ-SM) and the Double Submodule (DSM) are two variants that use Full-Bridge circuits, which support parallel conduction of IGBTs and diodes by the use of respective switch states. Here, the presented Reduced Full-Bridge Submodule (RFB-DSM) corresponds to a subset of the DSM. The new DSM circuit neglects selected switch positions within the diagonals of the individual Full-Bridge units to apply states with parallel conduction of diodes. Accordingly, the RFB-DSM lowers the number of semiconductors compared to the structure of the DSM but still takes advantage of switch states that connects diodes in parallel. Simulations showed that the states with parallel conduction paths can achieve a share between 20 % ($\cos(\varphi)$=1) to 75 % ($\cos(\varphi)$=-1), which depends on the used PWM technique and the $\cos(\varphi)$. For inverting operation ($\cos(\varphi)$>0), the positive effect of parallel diode conduction is limited because of the increasing time periods of nonparallel IGBT conduction.

The number of IGBTs of the RFB-DSM has been reduced by 50% compared to the DSM, while the SM capacitors cannot be paralleled for the reduction of the installed capacitance. On the other hand, the DSM provides switch states with IGBTs in parallel conduction, which decreases the associated conduction losses in a similar manner.

References

[1] R. Marquardt, "Modular multilevel converters: State of the art and future progress," IEEE Power Electronics Magazine, Vol. 5, Issue 4, 2018.

[2] C. Dahmen and R. Marquardt, "Power losses of advanced MMC submodule topologies using Si- and SiC-semiconductors," European Conference on Power Electronics and Applications, EPE, Genova, Italy, 2019.

[3] K. Ilves, F. Taffner, S. Norrga, A. Antonopoulos, L. Harnefors, and H. P. Nee, "A submodule implementation for parallel connection of capacitors in modular multilevel converters," 15th European Conference on Power Electronics and Applications, EPE, Lille, France, 2013.

[4] D. G. Holmes and T. A. Lipo, "Pulse Width Modulation for Power Converters - Principles and Practice," New York: Wiley, 2003.

[5] Plexim GmbH, "PLECS - The Simulation Platform for Power Electronic Systems," User Manual, Ver. 4.2, Zurich, Switzerland, 2018.

[6] B. McGrath, C. Teixeira, and D. G. Holmes, "Optimized phase disposition (PD) modulation of a modular multilevel converter," IEEE Transactions on Industry Applications, Vol. 53, No. 5, 2017.

[7] S. Rohner, S. Bernet, M. Hiller, and R. Sommer, "Modulation, losses, and semiconductor requirements of modular multilevel converters," IEEE Transactions on Industrial Electronics, Vo. 57, No. 8, 2010.

[8] Infineon Technologies AG, "FZ600R17KE3 - 62mm C-Series module with Trench/Fieldstop IGBT 3 and Emitter Controlled 3 diode," Datasheet, V 3.0, Munich, Germany, 3rd Oct. 2013.

Evaluation of the I_{max}-f_{sw}-dv/dt Trade-off of High Voltage SiC MOSFETs Based on an Analytical Switching Loss Model

Anliang Hu, Jürgen Biela

Laboratory for High Power Electronic Systems (HPE), ETH Zürich, Switzerland

Email: hu@hpe.ee.ethz.ch, jbiela@ethz.ch

Keywords

≪Power semiconductor device≫, ≪Wide bandgap devices≫, ≪Silicon Carbide (SiC)≫, ≪MOSFET≫, ≪Switching losses≫, ≪Device modeling≫, ≪Thermal stress≫

Abstract

Advanced high voltage (3.3-15kV) SiC MOSFETs have been developed for future medium voltage converters over the past decade due to their superior performance. In order to better understand the operation limits and potential of these devices, this paper evaluates the I_{max}-f_{sw}-dv/dt trade-off (maximal current-handling capability at a specific switching frequency and at a defined switching speed) for high voltage SiC MOSFETs based on a proposed linearized analytical switching loss model. There, high voltage SiC MOSFETs manufactured by Cree combined with data from literature for scaling are used as reference.

1 Introduction

High voltage (HV) unipolar SiC MOSFETs (3.3-15kV) are attractive switches for applications in future energy conversion and transmission systems, including railway, HVDC, FACTS, and medium voltage drives [1] - [11]. Due to their high blocking voltage capability, simple two-level topologies can be adopted for medium voltage voltage source converters. Compared to HV bipolar Si devices, these unipolar devices exhibit faster switching speeds, especially the turn-off speed, with high dv/dt rates up to 125V/ns (turn-on) and 70V/ns (turn-off) achieved by a 15kV SiC MOSFET (12kV, 10A) [9]. However, these unipolar devices also have relatively high conduction losses which significantly limit their current-handing capability, since the total losses, i.e. conduction and switching losses, must not exceed the maximum power dissipation capability of the device. In order to increase the maximal current-handling capability I_{max}, the switching losses need to be reduced by either reducing the operating switching frequency f_{sw} or increasing the switching speed dv/dt. However, with higher dv/dt values, the design of EMI filters, magnetics, and motor isolation becomes more difficult and thus the cost increases. As a result, the I_{max}-f_{sw}-dv/dt trade-off must be analyzed to investigate the limit of HV SiC MOSFET applications and to explore how to fully utilize the potential of these devices. So far, in [8] only the I_{max}-f_{sw} trade-off has been discussed.

In order to comprehensively analyze the I_{max}-f_{sw}-dv/dt trade-off of HV SiC MOSFETs, first their characteristic parameters are needed. This requires the parameter scaling of the chosen device's characteristics for different blocking voltages to eliminate the dependency on specific device designs, and to draw a general conclusion. In addition, a scalable, general, and simple switching loss model as a function of the dv/dt is needed to consider the influence of dv/dt, and to evaluate these devices for wide operating voltage ranges (3.3-15kV).

Analytical models enable a fast evaluation for comparison between different semiconductor technologies and provide physical insights into the switching process [12], [13]. Consequently, many studies have attempted to derive accurate analytical switching loss models for power MOSFETs with clamped inductive loads [12] - [19]. If the nonlinearity of the MOSFET's intrinsic capacitances and the parasitic inductances are considered in the model, the dynamic switching behavior can be described by a set of coupled nonlinear differential equations [14]. Therefore, simplifications are required to obtain a closed-form analytical solution for switching transitions. The loss model in [15] provides a simple analytical solution based on the assumption that the drain voltage starts to collapse after the current rise interval is completed during the turn-on transition. However, this model neglects the parasitic inductances. Adopting a piecewise linear approximation, all linearized parasitics are included in [16] without using mathematically complex analytical equations in [12], [13], [17]. Nevertheless, the turn-off transition does not distinguish between the two different scenarios discussed in [18], [19]. To address these problems, this paper proposes a comprehensive analytical switching loss model based on the shifted voltage and current waveform assumption in [15], the piecewise linear approximation in [16], and the separated turn-off scenarios in [18]. The proposed analytical model provides a simple closed-form analytical solution for the switching loss calculation without using iterative process, and the required parameters can be extracted from data sheets.

This paper is organized as follows. In section 2, the analytical switching loss model is derived by a switching transition analysis. Section 3 presents the parameter scaling of state-of-the-art HV SiC MOSFETs and the thermal limit. Based on these parameters, limit, and the proposed switching loss model, the I_{max}-f_{sw}-dv/dt trade-off is analyzed in detail with several figures in section 4. Finally, the conclusions are summarized in section 5.

2 Proposed Analytical Switching Loss Model

In the following, analytical expressions for the proposed switching loss model are derived based on several assumptions and a step-by-step switching transition analysis, where the characteristic switching waveforms are depicted accordingly.

2.1 Assumptions

The proposed linearized analytical switching loss model for SiC MOSFETs assumes a hard-switched boost chopper with a SiC MOSFET and a SiC Schottky diode pair [8], [14] (Fig. 1a), and the linearized MOSFET switching waveforms are depicted in Fig. 1b and 2a. The model assumptions are:

A1) A linearized MOSFET model is considered, which includes three constant intrinsic capacitances (C_{gs}, C_{gd}, C_{ds}) and two lumped parasitic inductances in the commutation loops (L_d, L_s). In saturation mode, a linearized gate-to-source voltage v_{gs} controlled channel current i_{ch} model is assumed

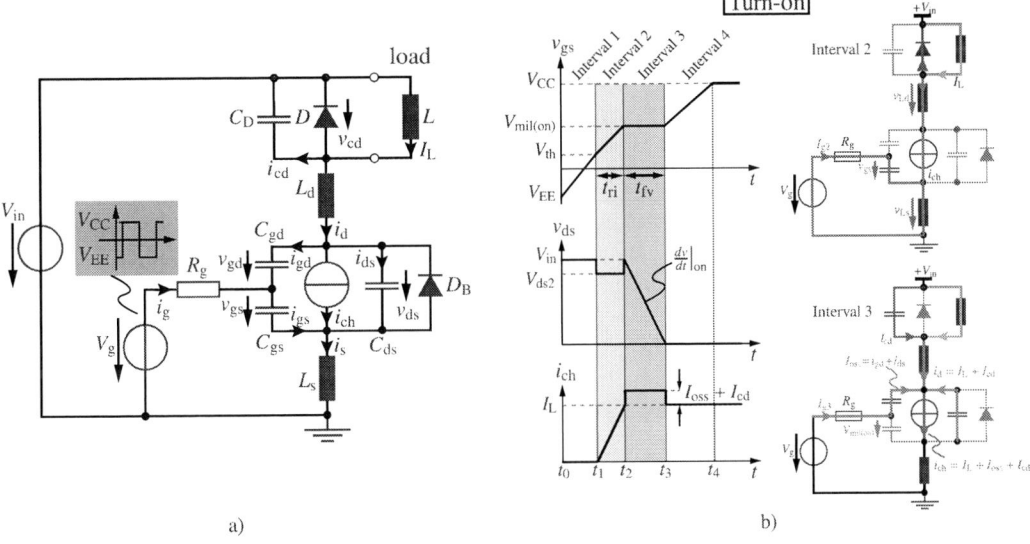

Fig. 1: a) A hard-switched boost chopper with a SiC MOSFET equivalent circuit, parasitics, a SiC Schottky diode and an inductive load. b) Linearized MOSFET switching waveforms and equivalent circuits during the turn-on transition.

with constant transconductance g_m, i.e. $i_{ch} = g_m \cdot (v_{gs} - V_{th})$. The switching waveforms are also linearized except for the ringing period (interval $7'$ in the ZVS turn-off scenario in Fig. 2b) [16]. A constant temperature is assumed so that all MOSFET parameters are constant.

A2) The drain-to-source voltage v_{ds} starts to collapse when $i_{ch} = I_L$ during the turn-on transition (Fig. 1b), and the channel current i_{ch} starts to collapse when $v_{ds} = V_{in}$ during the turn-off transition (Fig. 2).

A3) The DC voltage V_{in} is constant and the inductive load current I_L is assumed to be constant.

A4) An ideal bipolar gate voltage V_g with negligible rise and fall time is assumed.

A5) An ideal SiC Schottky diode without reverse recovery effect is assumed with a junction capacitance C_D (Fig. 1a) and a negligible forward voltage drop.

A6) The charge and discharge of the MOSFET intrinsic capacitances are assumed to be lossless. Therefore, the switching losses are determined by the overlap between i_{ch} and v_{ds}.

2.2 Turn-on Transition

Fig. 1b illustrates the 4 intervals of the turn-on switching transition, which will be discussed in the following. At the beginning of the turn-on transition, the MOSFET is in off-state, therefore the Schottky diode D conducts the full load current I_L and $v_{ds} = V_{in}$.

Interval 1 - Turn-on delay time: At t_0, the gate voltage jumps up from V_{EE} (≤ 0) to V_{CC} based on assumption A4, so the input capacitance $C_{iss} = C_{gs} + C_{gd}$ is charged. The MOSFET remains in the off-state until v_{gs} reaches the threshold voltage V_{th} and the full load current is still conducting via D. No switching losses are generated in this interval.

Interval 2 - Current rise time: At t_1, $v_{gs} = V_{th}$, so that the MOSFET channel starts to conduct the load current. This interval ends when the MOSFET is conducting the full load current and the corresponding gate voltage, called miller voltage (for turn-on), is given by (1) where the actual channel current $I_{ch(on)}$ is solved in interval 3. Assuming a constant gate current I_{g2} during this interval, the current rise time t_{ri} can be derived by (2). According to assumption A1, the constant gate current I_{g2} can be calculated by (3). Furthermore, the constant drain-to-source voltage V_{ds2} of this interval is reduced due to the voltage drops across L_d and L_s, as given by (4). Note that based on the constant current I_{g2} and the constant voltage V_{ds2}, the turn-on current slew rates are equal to each other in interval 2, as denoted by (5).

$$V_{mil(on)} = V_{th} + \frac{I_{ch(on)}}{g_m} \quad (1) \qquad V_{ds2} = V_{in} - (L_d + L_s) \cdot \frac{di_d}{dt} \quad (4)$$

$$t_{ri} = \frac{C_{iss}(V_{mil(on)} - V_{th})}{I_{g2}} \quad (2) \qquad \frac{di_d}{dt} = \frac{di_{ch}}{dt} = \frac{di_s}{dt} = \frac{I_L}{t_{ri}} \quad (5)$$

$$I_{g2} = \frac{1}{R_g} \cdot \left[V_{CC} - 0.5(V_{th} + V_{mil(on)}) - L_s \frac{di_s}{dt} \right] \quad (3)$$

Interval 3 - Voltage fall time: At t_2, the load current has been fully commuted from the diode D to the MOSFET, therefore the MOSFET output capacitance C_{oss} starts to discharge with decreasing v_{ds} and the Schottky diode juction capacitance C_D starts to charge with increasing v_{cd}. The constant gate current I_{g3} given in (6) discharges C_{gd}, so the voltage fall time t_{fv} can be derived by (7). Due to the discharging current I_{oss} of C_{oss} and the charging current I_{cd} of C_D, the channel current is calculated by (8). As a result, $V_{mil(on)}$, t_{ri} and I_{g2} in interval 2 can be solved. Combining (1) and (6)-(8), the turn-on voltage slew rate $\frac{dv}{dt}\big|_{on}$ (absolute value) in interval 3 can be calculated by (9). This interval ends when $v_{ds} = 0$.

$$I_{g3} = \frac{V_{mil(on)} - V_{th}}{R_g} \quad (6) \qquad I_{ch(on)} = I_L + I_{oss} + I_{cd} = I_L + (C_{oss} + C_D) \cdot \frac{dv}{dt}\Big|_{on} \quad (8)$$

$$t_{fv} = \frac{C_{gd}V_{in}}{I_{g3}} \quad (7) \qquad \frac{dv}{dt}\Big|_{on} = \frac{g_m(V_{CC} - V_{th}) - I_L}{C_{oss} + C_D + g_m R_g C_{gd}} \quad (9)$$

Interval 4 - Remaining gate charging time: At t_3, the MOSFET is turned on completely. As the last interval of the turn-on transition, the gate supply continues to charge C_{iss} until $v_{gs} = V_{CC}$. No switching losses are generated in this interval.

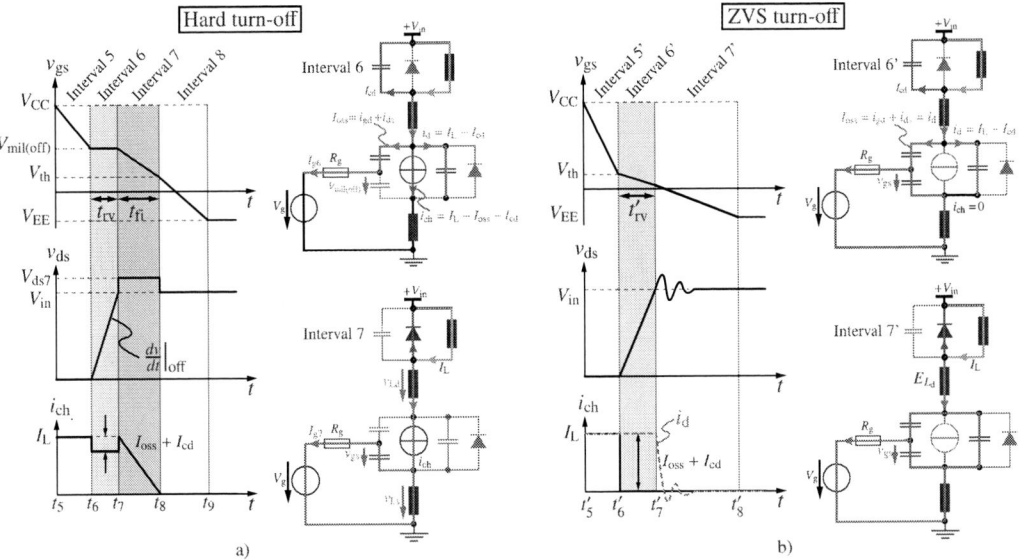

Fig. 2: Linearized MOSFET switching waveforms and equivalent circuits during the turn-off transition including two scenarios, the hard turn-off a) and the ZVS turn-off b).

2.3 Turn-off Transition

At the beginning of the turn-off transition, the MOSFET is in on-state and it conducts the full load current I_L and $v_{ds} = 0$. As discussed for interval 3 during the turn-on transition, the charging/discharging current I_{oss} of C_{oss} and I_{cd} of C_D influences i_{ch}. Therefore, two scenarios of turn-off transition are discussed in the following including the hard turn-off and the ZVS turn-off, as shown in Fig. 2.

2.3.1 Hard turn-off

Interval 5 - Turn-off delay time: At t_5, the gate voltage jumps down from V_{CC} to V_{EE} and C_{iss} is discharged. The MOSFET remains in the on-state because the gate voltage is above V_{th}. This interval ends when the gate voltage reaches the miller voltage (for turn-off) $V_{mil(off)}$, as expressed by (10). Hard turn-off occurs when $V_{mil(off)} > V_{th}$, i.e. $I_{ch(off)} > 0$. No switching losses are generated in this interval.

Interval 6 - Voltage rise time: At t_6, $v_{gs} = V_{mil(off)}$. The MOSFET still conducts the full I_L, because the diode D cannot conduct any current before C_{oss} is fully charged and C_D is fully discharged ($v_{ds} = V_{in}$ and $v_{cd} = 0$). The constant gate current I_{g6} given in (11) discharges C_{gd}, so the voltage rise time t_{rv} can be derived by (12). Due to the charging current I_{oss} of C_{oss} and the discharging current I_{cd} of C_D, the channel current is calculated by (13). Combining (9) with (10)-(13), the turn-off voltage slew rate $\frac{dv}{dt}\big|_{off}$ in interval 6 can be expressed by (14), which is related to $\frac{dv}{dt}\big|_{on}$ in (15). This interval ends when $v_{ds} = V_{in}$.

$$V_{mil(off)} = V_{th} + \frac{I_{ch(off)}}{g_m} \quad (10)$$

$$I_{ch(off)} = I_L - I_{oss} - I_{cd} = I_L - (C_{oss} + C_D) \cdot \frac{dv}{dt}\Big|_{off} \quad (13)$$

$$I_{g6} = \frac{V_{th} - V_{mil(off)}}{R_g} \quad (11)$$

$$\frac{dv}{dt}\Big|_{off} = \frac{g_m(V_{th} - V_{EE}) + I_L}{C_{oss} + C_D + g_m R_g C_{gd}} \quad (14)$$

$$t_{rv} = \frac{C_{gd} V_{in}}{I_{g6}} \quad (12)$$

$$\frac{dv}{dt}\Big|_{off} = \frac{g_m(V_{th} - V_{EE}) + I_L}{g_m(V_{CC} - V_{th}) - I_L} \cdot \frac{dv}{dt}\Big|_{on} \quad (15)$$

Interval 7 - Current fall time: At t_7, $v_{ds} = V_{in}$, so that the load current starts to commute from the MOSFET to the diode D. This interval ends when D is conducting the full I_L. Assuming a constant gate current I_{g7} during this interval, the current fall time t_{fi} can be derived by (16). According to assumption A1, I_{g7} can be calculated by (17). In addition, v_{ds} increases due to the voltage drops across L_d and L_s, as derived by (18). Similar to (5), the turn-off current slew rates (absolute values) are given in (19).

$$t_{fi} = \frac{C_{iss}(V_{mil(off)} - V_{th})}{I_{g7}} \quad (16)$$

$$I_{g7} = \frac{1}{R_g} \cdot \left[V_{EE} - 0.5(V_{th} + V_{mil(off)}) - L_s \frac{di_s}{dt} \right] \quad (17)$$

$$V_{ds7} = V_{in} - (L_d + L_s) \cdot \frac{di_d}{dt} \quad (18)$$

$$\frac{di_d}{dt} = \frac{di_{ch}}{dt} = \frac{di_s}{dt} = \frac{I_L}{t_{fi}} \quad (19)$$

Interval 8 - Remaining gate discharging time: At t_8, the MOSFET is turned off completely. As the last interval of the turn-off transition, the gate supply continues to discharge C_{iss} until $v_{gs} = V_{EE}$. No switching losses are generated in this interval.

2.3.2 ZVS turn-off

As discussed above, ZVS turn-off occurs when $V_{mil(off)} \leq V_{th}$, i.e. $I_{ch(off)} \leq 0$. Combining this equation with (13)-(15), the boundary current I_{ZVS} can be solved by (20).

$$I_{ZVS} = (C_{oss} + C_D) \cdot \frac{g_m(V_{th} - V_{EE}) + I_L}{C_{oss} + C_D + g_m R_g C_{gd}} = (C_{oss} + C_D) \cdot \frac{g_m(V_{th} - V_{EE}) + I_L}{g_m(V_{CC} - V_{th}) - I_L} \cdot \frac{dv}{dt}\Big|_{on} \quad (20)$$

In contrast to hard turn-off, the gate voltage directly drops to the threshold voltage V_{th} at the end of interval $5'$, so that the miller plateau in the gate voltage waveform will not occur in ZVS turn-off. Starting from t_6', the gate circuit loses the control of the channel current i_{ch}, because the MOSFET enters the cutoff region. In interval $6'$, the load current is completely used to charge C_{oss} and discharge C_D. Therefore $i_{ch} = 0$ and no switching losses are generated. The voltage rise time t_{rv}' of this interval is determined by the speed of this process, and the $\frac{dv}{dt}\big|_{off}$ is calculated by $\frac{dv}{dt}\big|_{off,ZVS} = \frac{I_L}{C_{oss}+C_D}$. In interval $7'$, the current and voltage ringing waveforms are depicted in Fig. 2b due to the LC oscillations between the parasitics. Since the losses generated in the gate circuit do not account for the MOSFET switching losses, only the dissipated energy E_{L_d} of L_d contribute to the switching losses. Considering the relatively small L_d and the low operating current level, the complete ZVS turn-off transition is almost lossless.

2.4 Switching Losses

Based on assumption A6 and the aforementioned analysis, Table I summarizes the analytical expressions of the duration and the switching losses of the respective intervals.

Table I: Duration and switching loss equations for the different intervals of the switching transitions.

Transition	Interval	Duration	Switching losses E_{sw}
Turn-on	Interval 2	$t_{ri} = \dfrac{2R_g C_{iss} \cdot A + 2L_s I_L \cdot B}{2(V_{CC} - V_{th}) \cdot B - A}$	$E_2 = 0.5 I_L [V_{in} t_{ri} - I_L(L_d + L_s)]$
	Interval 3	$t_{fv} = \dfrac{V_{in} \cdot B}{g_m(V_{CC} - V_{th}) - I_L}$	$E_3 = \dfrac{0.5 V_{in}^2 I_L \cdot B}{g_m(V_{CC} - V_{th}) - I_L}$
Hard turn-off	Interval 6	$t_{rv} = \dfrac{V_{in} \cdot B}{g_m(V_{th} - V_{EE}) + I_L}$	$E_6 = \dfrac{0.5 V_{in}^2 I_L \cdot B}{g_m(V_{th} - V_{EE}) + I_L}$
	Interval 7	$t_{fi} = \dfrac{2R_g C_{iss} \cdot C + 2L_s I_L \cdot B}{2(V_{th} - V_{EE}) \cdot B + C}$	$E_7 = 0.5 I_L [V_{in} t_{fi} + I_L(L_d + L_s)]$
ZVS turn-off	Interval 6'	$t_{rv}' = \dfrac{V_{in}(C_{oss} + C_D)}{I_L}$	$E_{6'} = 0$
	Interval 7'	-	$E_{7'} = 0.5 L_d I_L^2$

Note: $A = R_g C_{gd} I_L + (C_{oss} + C_D)(V_{CC} - V_{th})$, $B = g_m R_g C_{gd} + C_{oss} + C_D$, $C = R_g C_{gd} I_L - (C_{oss} + C_D)(V_{th} - V_{EE})$

3 HV SiC MOSFET Parameters and Thermal Limit

Table II lists the operating parameters of HV SiC MOSFETs with different maximal blocking voltages V_{bl} and maximal drain currents I_d from Cree, based on an extensive literature review [1] - [11], [20] - [31]. The DC switching voltage $V_{in} = \frac{2}{3}V_{bl}$ is assumed. Because the possible drain current I_d is determined by the cooling systems, the values in the table only give a hint on the current capability of the considered devices. Note that several assumptions have to be made during the parameter scaling process of the considered HV SiC MOSFETs due to the limited data available from literature. In the following, star * marks the scaled or fitted data, while square □ marks the data from literature. The detailed procedures to obtain these parameters are described below:

- $R_{\mathrm{ds(on)}}^{150\,°C}$ is the drain-to-source on-resistance at 150 °C calculated by $R_{\mathrm{ds(on)}} = \frac{R_{\mathrm{ds(on)}}^{A}}{A}$, where $R_{\mathrm{ds(on)}}^{A}$ is the area-specific on-resistance and A is the active die area. For scaling purposes, $A = 32\,\mathrm{mm}^2$ is assumed for all HV (3.3-15kV) SiC MOSFETs according to [1], [4], [5], [11], and the fitted $R_{\mathrm{ds(on)}}^{A}$ values (orange stars in Fig. 3a) are adopted instead of the original ones (red squares in Fig. 3a). The fitting curve (orange dashed line in Fig. 3a) is parallel to the well-known Baliga's figure of merit (FOM) for SiC MOSFETs (black solid line in Fig. 3a), as expressed by $R_{\mathrm{on,ideal}} = \frac{4V_{\mathrm{bl}}^2}{\varepsilon_s \mu_n E_c^3}$ [15].

- $g_{\mathrm{m}}^{150\,°C}$ is the transconductance at 150 °C, which can be estimated by the transfer characteristic I_{ds}-V_{gs} curve based on assumption A1. However, the transconductance can only be extracted from [6] for the 10kV SiC MOSFET die from Cree, while the related information is also missing for the products from other manufacturers/labs. Therefore, TCAD simulations have been performed to estimate $g_{\mathrm{m}}^{150\,°C}$. Since the design parameters of HV SiC MOSFETs are also missing in the literature, 4 groups of geometry parameters are assumed based on Cree's 2nd generation SiC MOSFET [31] ($w_{\mathrm{half-cell}} = 4.55\,\mu\mathrm{m}$, $d_{\mathrm{epi}} = 10\,\mu\mathrm{m}$, $N_{\mathrm{epi}} = 10^{16}\mathrm{cm}^{-3}$), including $w_{\mathrm{half-cell}} = 5\,\mu\mathrm{m}$, $d_{\mathrm{epi}} = 14\,\mu\mathrm{m}$, $N_{\mathrm{epi}} = 10^{16}\mathrm{cm}^{-3}$ (1), $w_{\mathrm{half-cell}} = 6\,\mu\mathrm{m}$, $d_{\mathrm{epi}} = 35\,\mu\mathrm{m}$, $N_{\mathrm{epi}} = 4 \cdot 10^{15}\mathrm{cm}^{-3}$ (2), $w_{\mathrm{half-cell}} = 6.55\,\mu\mathrm{m}$,

Table II: HV SiC MOSFET and thermal parameters.

V_{bl}(kV)	I_{d}(A)	$R_{\mathrm{ds(on)}}^{150\,°C}$(mΩ)	$g_{\mathrm{m}}^{150\,°C}$(S)	C_{iss}(pF)	C_{rss}(pF)	$C_{\mathrm{oss}}+C_{\mathrm{D}}$(pF)	$R_{\mathrm{th,jc}}$(K/W)	P_{D}(W)
15	~10	1800.68 [5]*	2.00*	4458*	17.9*	99 [10]	0.4375 [8]	160
10	~20	800.30 [5]*	2.58 [6]	4561 [7]	14.78 [7]	190 [7]	0.35 [22,23]	200
6.5	~30	338.13 [4]*	3.375*	4634*	12.58*	248.4*	0.3160*	221.5
3.3	~45	87.15 [2]*	5.15*	4700 [2]	10.58 [2]	298 [2,25]	0.296 [21]	236.5
1.2	~100	45.5 [26]	21.7 [26]	2788 [26]	16.14 [26]	463 [26,27]	0.27 [28]	259

Note: [n]*:Fitted based on the data from reference [n]; *:Fitted data

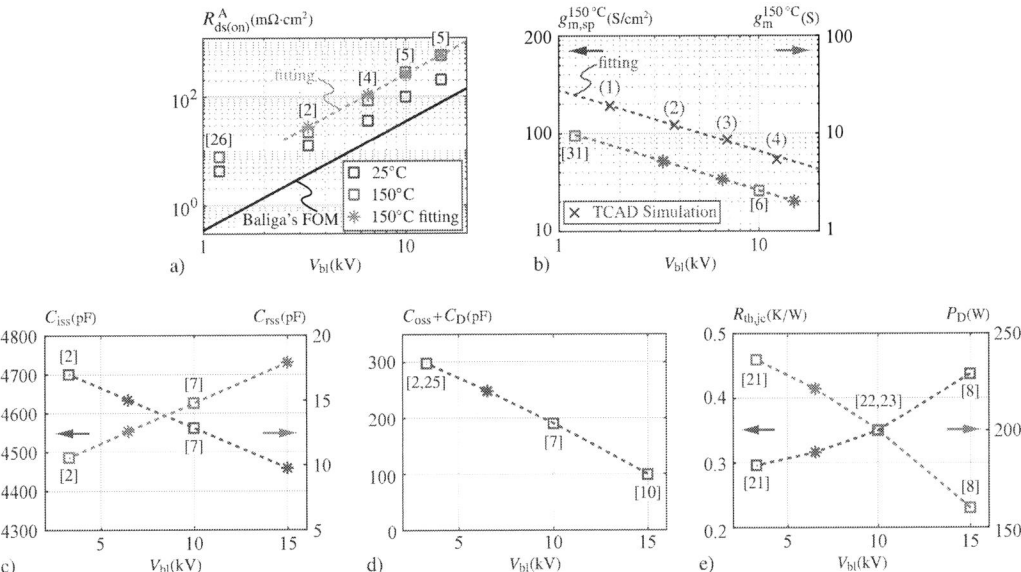

Fig. 3: a) $R_{\mathrm{ds(on)}}^{A}$-V_{bl} scaling, where the fitting curve is parallel to the theoretical SiC limit from Baliga's figure of merit. b) $g_{\mathrm{m,sp}}^{150\,°C}/g_{\mathrm{m}}^{150\,°C}$-$V_{\mathrm{bl}}$ scaling, where g_{m} are fitted based on the fitting curve from TCAD simulation and the g_{m} in [6]. c) $C_{\mathrm{iss}}/C_{\mathrm{rss}}$-$V_{\mathrm{bl}}$ scaling, where C_{rss} is the energy equivalent capacitance at specified V_{in}. d) $(C_{\mathrm{oss}}+C_{\mathrm{D}})$-$V_{\mathrm{bl}}$ scaling, where C_{oss} and C_{D} are energy equivalent capacitances at specified V_{in}. e) $R_{\mathrm{th,jc}}/P_{\mathrm{D}}$-$V_{\mathrm{bl}}$ scaling.

$d_{epi} = 60\,\mu m$, $N_{epi} = 2 \cdot 10^{15} cm^{-3}$ (3), and $w_{half-cell} = 8\,\mu m$, $d_{epi} = 110\,\mu m$, $N_{epi} = 1 \cdot 10^{15} cm^{-3}$ (4). The simulation results for the specific transconductances $g_{m,sp}^{150\,°C}$ are depicted in Fig. 3b, which indicates a linear pattern in the log-log plot. Based on this fitting curve and the transconductance in [6], $g_m^{150\,°C}$ values are fitted respectively in Fig. 3b. The threshold voltage $V_{th}^{150\,°C} = 4V$ at $150\,°C$ based on [3], [4], [6], and the gate supply voltages $V_{CC} = 20V$ and $V_{EE} = -5V$ based on [2], [4], [5].

- The input capacitance C_{iss} is obtained from the C_{iss}-V_{ds} curve. The reverse transfer capacitance C_{rss} is calculated as the energy equivalent capacitance at the specific DC switching voltage V_{in}. However, some C_{rss}-V_{ds} curves are measured only for low voltages, e.g. in [7] (0-600V for the 10kV device). Therefore, the capacitance curves need to be extended to the high voltage region based on the generalized fitting equation $C = \frac{C_0}{(1+V_{ds}/V_b)^r} + C_1$ in [19]. Finally, linear curve fitting is adopted to estimate other C_{iss} and C_{rss} values that are missing in the literature, as shown in Fig. 3c.

- The MOSFET output capacitance C_{oss} and the Schottky diode junction capacitance C_D are calculated similarly to C_{rss}, using the same fitting function to extend C-V_{ds} curves. Since the HV devices in [10] (15kV) and in [7] (10kV) are co-pack modules with a SiC MOSFET chip and an antiparallel SiC Junction Barrier Schottky (JBS) diode chip, $C_{oss} + C_D$ is calculated, which is also in accordance with the switching loss model derived in section 2. For the 3.3kV MOSFET without an antiparallel diode in [2], C_{oss} and C_D are calculated separately, where C_D is based on two 1.7kV Schottky diodes [25] that are connected in series [1]. Finally, data interpolation is adopted to estimate $C_{oss} + C_D$ for the 6.5kV SiC MOSFET and the Schottky diode pair that is missing in the literature, as presented in Fig. 3d.

- The package and module designs for HV SiC MOSFETs [21] - [24] are different from the conventional ones for Si-IGBTs [20]. In order to take full advantage of SiC MOSFETs' characteristics, low thermal resistance and high temperature withstanding power modules has been designed [21]. As a result, the junction-to-case thermal resistances $R_{th,jc}$ are adopted from [8] and [21] - [23] to describe the thermal limit accurately, while the missing $R_{th,jc}$ is interpolated, as illustrated in Fig. 3e. Assuming the maximum junction temperature $T_{j,max} = 150\,°C$ [1] - [11] and the case temperature $T_c = 80\,°C$ with effective cooling, the power dissipation capability P_D can be calculated by $P_D = \frac{T_{j,max} - T_c}{R_{th,jc}}$, as shown in Fig. 3e.

Using similar procedures as mentioned above, all parameters for the commercially available 1.2kV SiC MOSFETs from Cree are extracted from data sheets [26] - [31] without any approximations. The 1.2kV device should be excluded from the scaling of HV devices due to a drastically different device and package design, which could be recognized by the deviated $R_{ds(on)}^A$ value in Fig. 3a, for example. Finally, based on P_D, the thermal limit for the aforementioned MOSFETs can be expressed by (21). The conduction losses P_{cond} are calculated assuming that the devices are switching with a 50% duty cycle [1], [8], and the switching losses P_{sw} can be calculated by the proposed model in Table I, which rewrites (21) as (22) at a defined switching frequency f_{sw} as:

$$P_{loss,tot} = P_{cond} + P_{sw} \leq P_D \qquad (21) \qquad 0.5 \cdot I_L^2 R_{ds(on)}^{150\,°C} + E_{sw} \cdot f_{sw} \leq P_D \qquad (22)$$

4 I_{max}-f_{sw}-dv/dt Trade-off Analysis for HV SiC MOSFETs

Based on the proposed switching loss model in Table I, the device parameters in Table II and the thermal limit in (22), several plots are presented below to analyze the compromise between maximal current-handling capability I_{max} (RMS), feasible switching frequency range f_{sw}, and switching speed dv/dt (i.e. turn-on dv/dt). In order to investigate the dv/dt influence, E_{sw} in Table I is re-written as a function of dv/dt according to (9), so that (22) reveals the relationship between I_{max}, f_{sw}, and dv/dt.

To begin with, Fig. 4 compares I_{max}-f_{sw} curves calculated using the proposed model and the scaled parameters with I_{max}-f_{sw} curves in [8] for the 10kV and the 15kV devices. Considering one I_{max}-f_{sw} curve, the curve bends more in direction A with increasing $R_{ds(on)}$, expands in direction B with increasing P_D, and crosses the f_{sw}-axis to the right in direction C with increasing $C_{oss} + C_D$. As a result, the good matching shown in Fig. 5a indicates that both the scaling of parameters and the thermal limit in section 3 are reasonable, while the proposed switching loss model is also validated with good accuracy. It also explains why the energy-equivalent capacitances are adopted instead of using other capacitance calculation methods in [18], [19], [32].

Fig. 5a illustrates the $I_{\max}$-f_{sw}-dv/dt trade-off for the 10kV device as a 3D boundary surface, where all feasible operating points satisfying the thermal limit in (22) are confined in the region under this surface. $I_{\max}$ is divided by the nominal current I_N as per unit value $I_{\max,pu}$ to show the utilization of the MOSFET's current capability (called "device current utilization" below), and to provide a standardized parameter for comparing HV devices with different V_{bl} fairly. I_N is defined as the maximal DC operating current according to the thermal limit in (22) without switching losses, calculated by $I_N = \sqrt{\frac{2P_D}{R_{ds(on)}^{150\,^\circ C}}}$. In addition, Fig. 5a highlights the boundary current I_{ZVS} in (20) as a cutline, where operating points in the upper region are calculated by the hard turn-off model in section 2.3.1 and the lower region by the ZVS turn-off model in section 2.3.2.

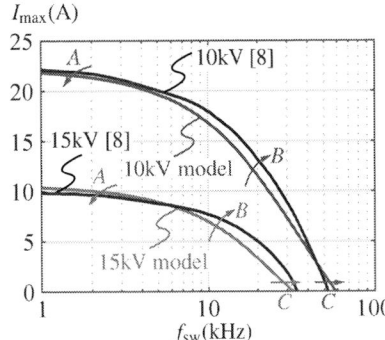

Fig. 4: $I_{\max}$-f_{sw} curves (blue and red) based on the proposed model (dv/dt=67V/ns [8] for the 10kV and 80V/ns [5] for the 15kV device), compared with those (black) in [8].

To further investigate Fig. 5a, the $I_{\max,pu}$-f_{sw} curves given in Fig. 5b are obtained by fixing the dv/dt values, i.e. using planes p_0-p_3 in Fig. 5d to cut the 3D surface in Fig. 5a. Similarly, Fig. 5c is obtained as $I_{\max,pu}$-dv/dt curves by fixing the f_{sw} values. Fig. 5b shows that the device current utilization increases with increasing switching speeds. At 10kHz, the device current utilization grows by 137% if the dv/dt is increased from 10V/ns to 25V/ns. However, only a very limited benefit is obtained by further increasing dv/dt (e.g. 12% growth from 67V/ns to 100V/ns). By focusing on the $I_{\max,pu}$-f_{sw} curve for 25V/ns in Fig. 5b, Fig. 5e results, which depicts two different curve shapes using two turn-off models in Table I. The 10kV device always operates with ZVS turn-off at higher dv/dt, as noted in Fig. 5c, while for lower dv/dt both turn-off scenarios occur. Therefore, two different curve shapes can be observed in Fig. 5b.

Fig. 5c depicts $I_{\max,pu}$-dv/dt curves at fixed switching frequencies, where I_{ZVS} separates the hard turn-off region (grey) and the ZVS turn-off region (white). The device current utilization drops obviously with rising f_{sw}, because the switching losses outweigh the conduction losses. For $f_{sw} \geq 36$kHz, the device

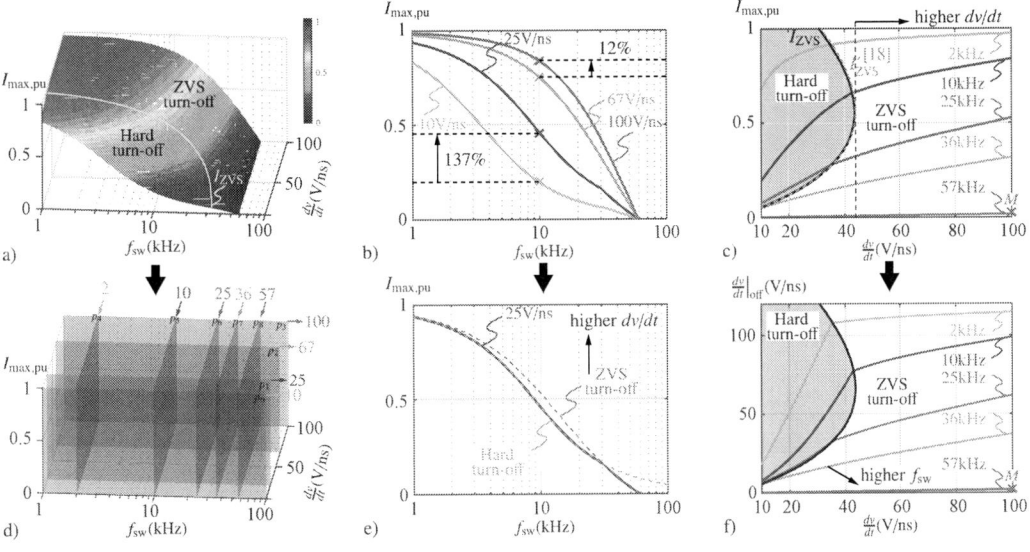

Fig. 5: a) $I_{\max,pu}$-f_{sw}-dv/dt plot for the 10kV device with a highlighted I_{ZVS} cutline. b) $I_{\max,pu}$-f_{sw} curves at fixed dv/dt values. c) $I_{\max,pu}$-dv/dt curves at fixed f_{sw} values, where I_{ZVS} calculated by (20) is depicted and compared to I'_{ZVS} in [18]. d) 4 cutting planes with fixed dv/dt values (p_0-p_3) and 5 cutting planes with fixed f_{sw} values (p_4-p_8). e) $I_{\max,pu}$-f_{sw} curve at dv/dt=25V/ns. f) $dv/dt|_{off}$-dv/dt curves at fixed f_{sw} values.

always operates with ZVS turn-off, as shown in Fig. 5c. In this higher f_{sw} region, the turn-on dv/dt is fast (e.g. 100V/ns at point M), as noted in Fig. 5c and Fig. 5f, but the turn-off dv/dt (2.3V/ns) is strongly limited by the low current level (0.226A) and the slow charging of C_{oss} and C_D. Furthermore, Fig. 5c also indicates that the maximum possible switching frequency is approximately 57kHz, and the useful switching frequency range is within 10kHz assuming a device current utilization above 50%. Finally, Fig. 5c compares the boundary current I_{ZVS} in (20) with the I'_{ZVS} in [18], where the I'_{ZVS} is re-derived in (23) based on the MOSFET and the Schottky diode pair instead of the half-bridge topology in [18]. Although I'_{ZVS} is derived based on different assumptions and procedures from those in section 2, it matches well with the I_{ZVS}, indicating that both assumptions and derivations in section 2 are reasonable.

$$I'_{ZVS} = \frac{V_{in}}{2L_s}\left[-R_g C_{gd} + \sqrt{(R_g C_{gd})^2 - 8(V_{EE} - V_{th})\frac{L_s(C_{oss}+C_D)}{V_{in}}}\right] \tag{23}$$

After analyzing one 10/15kV SiC MOSFET, Fig. 6a compares four state-of-the-art HV SiC MOSFETs to a 1.2kV device CPM2-1200-0025B [26] (Table II). With increasing V_{bl}, the I_{max}-f_{sw}-dv/dt 3D surface moves toward the origin, which demonstrates that devices with higher blocking voltages have more limited operating frequency range, and their current capability is less utilized.

To further investigate Fig. 6a, 2D plots are drawn in Fig. 6b and Fig. 6c. Fig. 6b shows that devices with higher V_{bl} have more limited device current utilization at 10kHz switching frequency and fixed dv/dt values. With increasing dv/dt, the device current utilization improves more significantly for the high V_{bl} devices than for the low V_{bl} devices. By incrementing dv/dt from 10V/ns to 60V/ns, the device current utilization is increased by 27% for the 1.2kV device, which increases monotonically (27%-163%-233%-333%-400%) with increasing V_{bl}! On the other hand, Fig. 6c indicates that the f_{sw} range is more limited for higher V_{bl} devices with the same device current utilization $I_{max,pu} = 0.6$. Although the f_{sw} ranges are wider with increasing dv/dt, they are still strongly limited for the high V_{bl} devices. Note that Fig. 6b and Fig. 6c show similar patterns at other operating points.

Furthermore, Fig. 6b is extended by Fig. 6d showing the comparison of the power loss distribution between conduction losses P_{cond} and switching losses P_{sw}. Fig. 6d illustrates that P_{sw} outweigh P_{cond} at low dv/dt for HV SiC MOSFETs, while P_{cond} gradually outweigh P_{sw} with increasing dv/dt. In addition, P_{cond} always accounts for a larger portion of the total losses for lower V_{bl} devices, compared to higher V_{bl}

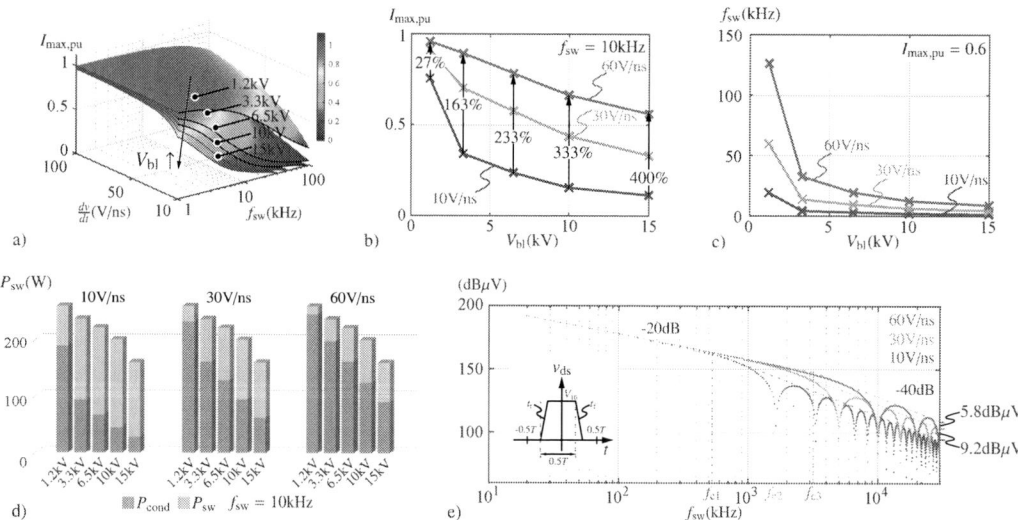

Fig. 6: a) $I_{max,pu}$-f_{sw}-dv/dt comparison between 5 SiC MOSFETs in Table II. The 3D surface edges are outlined for clarity. b) $I_{max,pu}$-V_{bl} comparison at fixed dv/dt values and f_{sw} = 10kHz. c) f_{sw}-V_{bl} comparison at fixed dv/dt values and $I_{max,pu}$ = 0.6. d) Power loss distribution comparison at fixed dv/dt values and f_{sw} = 10kHz. e) Amplitude spectrum of trapezoidal voltage waveforms with different dv/dt values for the 10kV device and f_{sw} = 20kHz. Corner frequencies f_{c1}-f_{c3} and the -20/-40dB asymptotes are highlighted. The trapezoidal voltage waveform is depicted with V_{in} = 6.67kV, assuming identical rise and fall times $t_r = t_f$.

devices, at the same dv/dt. In fact, the channel resistance R_{ch} dominates $R_{ds(on)}$ for low V_{bl} SiC MOSFETs on the market due to processing issues. Therefore, the $R_{ds(on)}$ of low V_{bl} SiC MOSFETs are expected to be further reduced in the future.

To sum up, both Fig. 6b and Fig. 6c show that the current capability can be utilized more and the switching frequency range can be wider with increasing switching speed. In addition, simply increasing dv/dt to 60V/ns can significantly improve the utilization of the current capability and slightly widen the switching frequency range, which is not difficult to achieve in real applications for HV SiC MOSFETs [5]. Nevertheless, it leads to higher EMI in the system, especially in the high frequency range above 1MHz, as presented in Fig. 6e (15dBμV noise increment with dv/dt rising from 10V/ns to 60V/ns), which increases the cost and the design difficulty of EMI filters, magnetics, and motor isolation. In addition, the fast switching speed poses more challenges to the semiconductor package and PCB design, considering the increased parasitics due to larger isolation distances at higher voltages.

The analysis in this section is based on a SiC MOSFET and a Schottky diode pair. If a MOSFET half-bridge without antiparallel Schottky diode is considered, the calculated turn-on switching losses will increase due to the reverse recovery losses from the antiparallel body diode, which will also increase significantly with increasing di/dt and dv/dt. Therefore, the device current utilization and the switching frequency range will be more limited compared to the result presented in this paper. To further analyze this problem, a new switching loss model is required.

5   Conclusion

In this paper, a linearized analytical switching loss model is proposed based on several existing models. A simple closed-form analytical solution is provided for the switching loss calculation without using iterative process, and the required parameters can be extracted from data sheets. Using the presented model, the I_{max}-f_{sw}-dv/dt trade-off of HV SiC MOSFETs is analyzed in detail, based on the parameter scaling of state-of-the-art HV devices from Cree and the thermal limit. The proposed switching loss model, the scaling of parameters and the considered thermal limit are validated by comparing the I_{max}-f_{sw} curve with the curve presented in [8] for a 10kV and a 15kV SiC MOSFET. In addition, the boundary current is verified by the equation derived in [18] in the I_{max}-dv/dt curve.

The device current utilization and the switching frequency range of HV SiC MOSFETs are severely limited at low dv/dt values. With increasing blocking voltage, the current capability is less utilized and the switching frequency range is more limited. Simply by incrementing the switching speed from 10V/ns to 60V/ns, the current capability utilization can be significantly increased by approximately 163%, 233%, 333%, 400% for the considered 3.3kV, 6.5kV, 10kV and 15kV devices at 10kHz switching frequency, which are large improvements over the devices with lower blocking voltages, at the cost of 15dBμV higher EMI levels. However, further increasing the switching speed brings only limited benefits to the increase of the device current utilization. For the considered 10kV SiC MOSFET, this benefit drops from 137% (10V/ns to 25V/ns) to 12% (67V/ns to 100V/ns).

On the other hand, the switching frequency range is always limited for the devices with higher blocking voltages. In order to guarantee a sufficient utilization (around 50%) of the current capability, the switching frequency is limited to approximately 10kHz for the considered 10kV device. In the high frequency range, the SiC MOSFET will always operate with ZVS turn-off, characterized by a fast turn-on and a slow turn-off.

In conclusion, increasing dv/dt to 60V/ns, which is practical in real applications and leads to acceptable EMI levels, can significantly improve the utilization of the current capability and slightly widen the switching frequency range for these HV devices. With the development of these HV SiC MOSFETs, the current utilization is expected to be improved with lower on-resistance in future devices.

Acknowledgements

The authors would like to thank Dr. J. Müting from the Advanced Power Semiconductor (APS) Laboratory at ETH Zürich for providing missing data based on TCAD simulations of high voltage SiC MOSFETs.

References

[1] J. W. Palmour et al., "Silicon Carbide Power MOSFETs: Breakthrough performance from 900 V up to 15 kV," *Int. Symposium on Power Semiconductor Devices IC's (ISPSD)*, pp. 79-82, 2014.

[2] A. Anurag et al., "Static and Dynamic Characterization of a 3.3 kV, 45 A 4H-SiC MOSFET," *Materials Science Forum*, vol. 924, pp. 739-742, 2018.

[3] H. Kono et al., "14.6 mΩcm^2 3.4 kV DIMOSFET on 4H-SiC (000-1)," *Materials Science Forum*, vol. 778-780, 2014, pp. 935–938.

[4] S. Sabri et al., "New Generation 6.5 kV SiC Power MOSFET," in *Workshop on Wide Bandgap Power Devices and Applications (WiPDA)*, Albuquerque, NM, 2017, pp. 246-250.

[5] V. Pala et al., "10 kV and 15 kV Silicon Carbide Power MOSFETs for Next-Generation Energy Conversion and Transmission Systems," in *Energy Conversion Congress and Exposition (ECCE)*, Pittsburgh, PA, 2014, pp. 449-454.

[6] J. Wang et al., "Characterization, Modeling, and Application of 10-kV SiC MOSFET," in *Trans. on Electron Devices*, vol. 55, no. 8, pp. 1798-1806, Aug. 2008.

[7] D. Rothmund, D. Bortis and J. W. Kolar, "Accurate Transient Calorimetric Measurement of Soft-Switching Losses of 10-kV SiC MOSFETs and Diodes," in *Trans. on Power Electronics*, vol. 33, no. 6, pp. 5240-5250, June 2018.

[8] J. B. Casady et al., "New Generation 10kV SiC Power MOSFET and Diodes for Industrial Applications," in Proc. of *Int. Exhibition and Conf. for Power Electronics, Intelligent Motion, Renewable Energy and Energy Management (PCIM)*, Nuremberg, Germany, 2015.

[9] J. Thoma, S. Kolb, C. Salzmann and D. Kranzer, "Characterization of High-Voltage-SiC-Devices with 15 kV Blocking Voltage," in *Int. Power Electronics and Motion Control Conf. (PEMC)*, Varna, 2016, pp. 946-951.

[10] L. Wang, Q. Zhu, W. Yu and A. Q. Huang, "A Study of Dynamic High Voltage Output Charge Measurement for 15 kV SiC MOSFET," *Energy Conversion Congress and Exposition (ECCE)*, Milwaukee, WI, 2016.

[11] A. Q. Huang, Q. Zhu, L. Wang and L. Zhang, "15 kV SiC MOSFET: An Enabling Technology for Medium Voltage Solid State Transformers," in *CPSS Trans. on Power Electronics and Applications*, vol. 2, no. 2, pp. 118-130, 2017.

[12] M. Rodríguez et al., "An Insight into the Switching Process of Power MOSFETs: An Improved Analytical Losses Model," *Trans. on Power Electronics*, vol. 25, no. 6, pp. 1626-1640, June 2010.

[13] Y. Ren, M. Xu, J. Zhou and F. C. Lee, "Analytical Loss Model of Power MOSFET," *Trans. on Power Electronics*, vol. 21, no. 2, pp. 310-319, March 2006.

[14] S. K. Roy and K. Basu, "Analytical Estimation of Turn on Switching Loss of SiC MOSFET and Schottky Diode Pair From Datasheet Parameters," *Trans. on Power Electronics*, vol. 34, no. 9, pp. 9118-9130, Sept. 2019.

[15] B. J. Baliga, *Fundamentals of power semiconductor devices*. New York: Springer Science Business Media, 2008.

[16] K. Peng, S. Eskandari and E. Santi, "Analytical Loss Model for Power Converters with SiC MOSFET and SiC Schottky Diode Pair," in *Energy Conversion Congress and Exposition (ECCE)*, Montreal, QC, 2015, pp. 6153-6160.

[17] D. A. Grant and J. Gowar, *Power MOSFETs: Theory and applications*. New York: Wiley, 1989.

[18] D. Christen and J. Biela, "Analytical Switching Loss Modeling Based on Datasheet Parameters for MOSFETs in a Half-Bridge," in *Trans. on Power Electronics*, vol. 34, no. 4, pp. 3700-3710, April 2019.

[19] X. Wang, Z. Zhao, K. Li, Y. Zhu and K. Chen, "Analytical Methodology for Loss Calculation of SiC MOSFETs," in *Jour. of Emerging and Selected Topics in Power Electronics*, vol. 7, no. 1, pp. 71-83, March 2019.

[20] D. Cottet, U. Drofenik and J. Meyer, "A Systematic Design Approach to Thermal-Electrical Power Electronics Integration," in *Electronics System-Integration Technology Conf.*, Greenwich, 2008, pp. 219-224.

[21] M. Horio et al., "Ultra Compact and High Reliable SiC MOSFET Power Module with 200°C Operating Capability," *Int. Symposium on Power Semiconductor Devices and ICs*, Bruges, 2012, pp. 81-84.

[22] M. Johnson et al., "10 kV SiC Power Module Packaging," *Int. Conf. on Integrated Power Electronics Systems (CIPS)*, Stuttgart, Germany, 2018.

[23] C. DiMarino et al., "Design and Development of a High-Density, High-Speed 10 kV SiC MOSFET Module," *European Conf. on Power Electronics and Applications (EPE)*, Warsaw, 2017, pp. P.1-P.10.

[24] R. Takayanagi et al., "3.3kV All-SiC Module for Electric Distribution Equipment," *Int. Power Electronics Conference (IPEC)*, Niigata, 2018, pp. 3396-3400.

[25] *CPW5-1700-Z050B Datasheet*. Accessed: Rev.-, 2013. [Online]. Available: www.wolfspeed.com

[26] *CPM2-1200-0025B Datasheet*. Accessed: Jan., 2016. [Online]. Available: www.wolfspeed.com

[27] *CPW5-1200-Z050B Datasheet*. Accessed: Rev.A, 2014. [Online]. Available: www.wolfspeed.com

[28] *C2M0025120D Datasheet*. Accessed: Oct., 2015. [Online]. Available: www.wolfspeed.com

[29] *C3M0075120D Datasheet*. Accessed: Feb., 2019. [Online]. Available: www.wolfspeed.com

[30] *C4D10120H Datasheet*. Accessed: Feb., 2018. [Online]. Available: www.wolfspeed.com

[31] *C2M0080120D Datasheet*. Accessed: Sept., 2019. [Online]. Available: www.wolfspeed.com

[32] Laszlo Balogh, "Design and Application Guide for High Speed MOSFET Gate Drive Circuits", Application Note, 2006, Texas Instruments.

Protection Measures for Modular Multilevel Converters in Case of DC Short-Circuit Faults

Martin GESKE*, Duro BASIC**, Roland JAKOB*, Christian KELLER*, Thomas BRÜCKNER***

*GE Power Conversion GmbH, Culemeyerstraße 1, 12277 Berlin, Germany
**GE Power Conversion SAS, 18 Avenue due Québec, 91140 Villebon-sur-Yvette, France
***Universität der Bundeswehr München, Werner Heisenberg-Weg 39, 85579 Neubiberg, Germany
Tel.: +49 / (0) 30 7622 2394*
E-Mail: martin.geske@ge.com, duro.basic@ge.com, roland.jakob@ge.com, christian.b.keller@ge.com
URL: https://www.gepowerconversion.com

Keywords

«Multilevel converters», «Voltage Source Converter (VSC)», «High voltage power converters», «Diode», «Modelling», «Faults», «Fault tolerance», «Protection device»

Abstract

The Modular Multilevel Converter (MMC) technology brings several advantages, although its operation and protection are very different compared to conventional Voltage Source Converters (VSC). Short-circuit faults at the DC-link transmission systems of MMCs that use the Half-Bridge Submodule (HB-SM) are still a major concern due to potentially high fault current magnitudes. The sequence of the pole-to-pole DC short circuit of an HB-MMC is analyzed and modeled through respective equivalent circuits. A simplified calculation is presented that allows estimating the current magnitudes after the DC-fault appearance. Finally, several protection measures for the HB-MMC are described, and a new protection method that uses an AC crowbar is introduced and analyzed.

Introduction

Modular Multilevel Converters (MMC) have become state of the art for demanding high power applications [1]. The implemented type of SM is one key element to limit inflowing fault currents. The HB-SM is a unipolar voltage SM that lacks fault-blocking capability and is subjected to high surge currents passing the lower freewheeling diode in case of a DC short-circuit fault. Alternative SMs such as the Full-Bridge SM or the Clamped Double SM can provide the fault blocking but create higher operational losses [2]. Hence, the implementation of the HB-SM is preferred but requires appropriate protection and system design to avoid severe destruction of the converter. Inflowing surge currents during DC short circuit faults of an HB-MMC potentially rise above the order of 10 kA [3], [4], which typically exceeds the semiconductor limits. The corresponding fault currents can be diverted from the SMs through an additional bypass [5], for example. Furthermore, the impedances of the arm reactors and the phase reactance limit the maximum current magnitudes [3].

This study first provides a detailed analysis of a pole-to-pole DC-side fault of an HB-MMC and introduces simplified calculations of currents for the different fault stages. Therefore, the initial fault stage before the converter detects the fault is analyzed and described. Second, the subsequent fault stages until the fault clearing through the AC circuit breaker are represented by conventional single-phase equivalent circuits. Finally, the known protection measures for HB-MMCs are summarized, and a new protection method that applies an AC crowbar is introduced and investigated.

Pole-to-Pole DC Fault of Half-Bridge MMCs

Faults at the DC side of MMCs are important for DC-transmission systems because of the vulnerability to faults of the transmission link. The pole-to-pole DC-fault sequence of an HB-MMC can be separated into three different stages [6]. The following analysis of the DC-fault sequence compares analytical calculation results with the system simulations performed with PLECS simulation software. The respective simulation model of the HB-MMC in Fig. 1 (a) consists of the double star topology with DC-transmission link. The parameters in Table I refer to an HB-MMC for HVDC whereby the voltage V_S, the power S, and the number of SMs N_{SM} are downscaled by a factor of 10.

Table I: Parameters of the MMC system model

Parameter	MMC Model
Rated system power S [MVA]	72
Peak value of phase voltage V_S [kV]	24
Angular frequency ω_S [rad/s]	314.16
DC-link voltage V_{dc} [kV]	±30
SM capacitor voltage $\bar{V}_C$ [kV]	2.5
SM capacitance C_0 [mF]	4.2
Number of HB-SM per conv. arm N_{SM}	26

Table II: MMC model impedances

Parameter	Value
L_S [mH]	11
R_S [Ω]	0.11
L_{arm} [mH]	1
R_{arm} [Ω]	0.1
L_{dc} [mH]	0.5
R_{dc} [mΩ]	0.1
$V_{T,D0}$ [V] / $r_{T,D}$ [mΩ]	1.55 / 0.75

For the case that the DC side is exposed to a pole-to-pole DC short circuit (cp. Fig. 1 (a)), the DC-link current I_{dc} transiently increases while the converter remains in switch-mode operation between Time Points (1) and (2) in Fig. 1 (b). After detection of the DC fault at Time Point (2), the control system blocks the IGBT firing commands whereby the arm currents $i_{arm1,p}$ to $i_{arm3,n}$ are commutated to the freewheeling diodes D2 within the HB-SMs. Directly thereafter, the converter-arm currents remain in Continuous Conduction Mode (CCM), but the arm currents comprise a decaying DC component, which leads to the transition to Discontinuous Current Conduction (DCM). At Time Point (4), the AC circuit breaker clears the fault. The DC-fault sequence can thus be summarized:

1. Appearance of the DC-side fault (DC-fault time $t_{dc,f(1)}$)
2. DC-fault detection and blocking of IGBT pulses, CCM of converter arms (detection time $t_{det(2)}$)
3. DCM of converter arms (start time of DCM $t_{DCM(3)}$)
4. Opening of AC circuit breaker (opening time of CB $t_{CB(4)}$)

The impedances of the grid, the transformer, and the reactors, the detection time $t_{det(2)}$, and the opening time of the AC circuit breaker $t_{CB(4)}$ are essential for the resulting fault currents. Accordingly, the surge current loading of the lower diode D2 of the HB-SMs depends on multiple system design parameters.

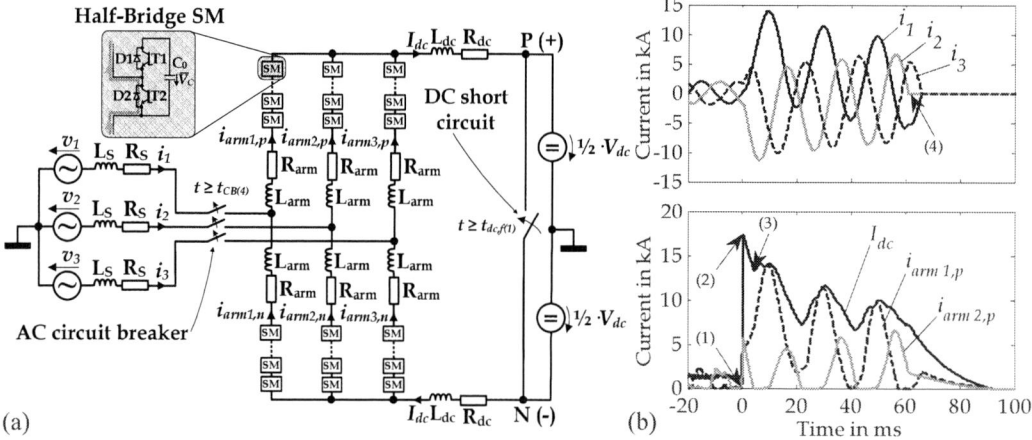

Fig. 1: (a) Converter model for simulation of pole-to-pole DC fault, (b) simulated current waveforms for AC side (top) and DC side (bottom) of HB-MMC ($t_{det((2)}$=500 μs, $t_{CB(4)}$=60 ms, v_l(t=0)=0 V)

The simulated fault scenario in Fig. 1 (b) shows unequal current loading among the converter arms. The phase shift of the AC voltage at the DC-fault instant (Time Point (1)) is crucial for the AC-fault current magnitudes and the distribution of the fault currents among the converter arms. The maximum current magnitude after Time Point (3) occurred when the source voltage v_S passed through zero at the fault instant, which was assumed for the first phase v_l in Fig. 1 (b).

The impedances within the DC-fault path and the time between the DC fault (1) and fault detection (2) define the current magnitude that was above 17 kA in Fig. 1 (b). Before the fault detection in (2), the inserted SMs discharge their capacitors into the DC fault. This time-critical stage can take about 500 μs [7] and requires accurate and fast detection to maintain controllable fault currents.

Stage 1 – Appearance of the DC Fault

During the initial fault stage between (1) and (2) in Fig. 1 (b), the SMs continue a switch-mode operation that creates DC voltage across the fault path and rapidly increases the DC-link current I_{dc}. The first stage equivalent circuit for the DC side shown in Fig. 2 is derived through decoupling from the AC side [8]. It is assumed that the inserted capacitance per converter phase is constant because the converter still modulates the DC voltage V_{dc}, which demands a certain number of SMs. The mean value of the inserted converter-arm capacitance $\bar{C}_{arm}$ is formulated based on the average SM voltage $\bar{V}_C$ and the assumption that each arm modulates half of the DC-link voltage, which yields:

$$\bar{C}_{arm} \approx \frac{2 \cdot \bar{V}_C}{V_{dc}} C_0 . \tag{1}$$

Under the assumption that all converter phases in Fig. 2 (a) consist of identical equivalent parameters, all three phases can be combined and modeled by the RLC circuit shown in Fig. 2 (b).

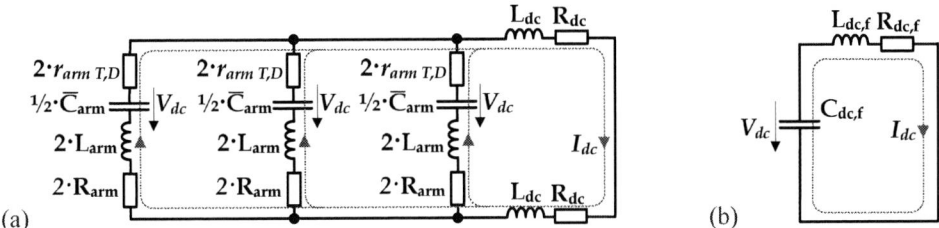

(a)   (b)

Fig. 2: (a) DC side equivalent circuit of the converter during the first fault stage and (b) simplified RLC equivalent circuit (circuits valid up to $t_{det(2)}$)

The simplifications lead to the following equivalent parameters for the RLC circuit in Fig. 2 (b) with

$$C_{dc,f} \approx \frac{3}{2} \bar{C}_{arm} , \tag{2}$$

$$L_{dc,f} = \frac{2}{3} L_{arm} + 2L_{dc} , \tag{3}$$

$$R_{dc,f} \approx \frac{2}{3} \left(R_{arm} + r_{arm\,T,D} \right) + 2R_{dc} . \tag{4}$$

For the period of the first fault stage and given typical parameters, the current rise is calculated as for an under-damped RLC circuit. The time range of the first fault stage is defined by

$$t_{dc,f(1)} < t \le t_{det(2)} : \quad t^* = t - t_{dc,f(1)} . \tag{5}$$

The current I_{dc} is described by

$$I_{dc}(t^*) = \frac{V_{dc} \cdot e^{\frac{-R_{dc,f}}{2 \cdot L_{dc,f}} \cdot t^*}}{L_{dc,f} \cdot \omega_d} sin(\omega_d t^*) \quad \text{with} \quad \omega_d = \sqrt{\omega_0^2 - \left(\frac{R_{dc,f}}{2 \cdot L_{dc,f}} \right)^2} . \tag{6}$$

Fig. 3 compares the results of the calculation with the simulations that used the parameters from Table I and Table II. The detection time was set to 500 μs that covers the delays for communication and fault detection based on the assumption in [7].

The analytical calculation yields a deviation of less than 2% at Point (2) compared to the simulation results. The parasitic resistances given by $R_{dc,f}$ insert little damping and have an insignificant influence on the peak value of I_{dc}. At 500 µs, the calculations yield lower fault currents compared to the simulation. This is because the calculation does not consider the energy balancing of the capacitors or inserting of other SMs during this fault stage. Nevertheless, the proposed method avoids complex modeling for the first stage (cp. [6]) and yields sufficiently accurate values for the considered time range. The calculation through an RLC circuit is well applicable for sufficiently short detection times and comparatively low angular frequencies ω_d, ω_0. The studied case yields a resonance period of the circuit that is more than seven times greater than the detection time of 500 µs.

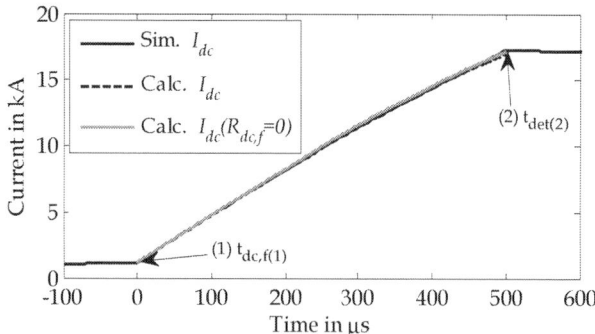

Fig. 3: Comparison of simulation and calculation results during the first stage of the DC fault

Stage 2 – Blocked IGBTs and CCM of Converter Arms

After the detection of the DC fault at Time Point (2), the IGBT pulses are blocked, and the currents are commutated to the diodes. The lower diodes of the HB-SMs bypass the fault current, and these lower diodes are subjected to the high magnitudes of the inflowing AC currents. The AC fault currents are injected through the source power system that is exposed to a three-phase fault. During this fault stage, all converter arms of the HB-MMC are in CCM [6], which results in a complete decoupling of the AC side from the DC side. Thus, two simplified equivalents circuits can be drawn. Fig. 4 (a) shows that the AC power source is subjected to a symmetrical three-phase short circuit that is created through the conducting diodes D*. The HB-MMC operates like a six-pulse bridge rectifier with a short circuit at the DC side [6]. The DC-side equivalent circuit in Fig. 4 (b) demonstrates that the HB-SMs bypass the DC-fault current through the lower diodes D_2*.

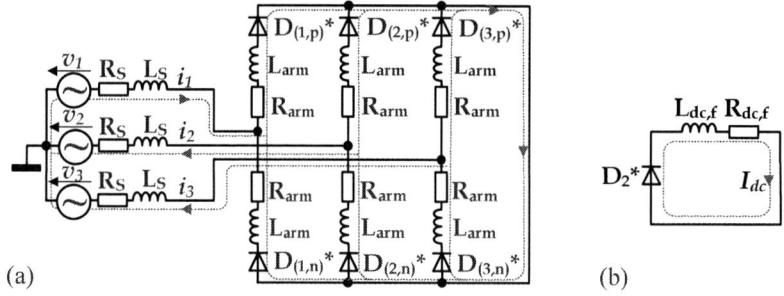

Fig. 4: Stage 2 equivalent circuits of HB-MMC for (a) AC-side and (b) DC-side currents

The equivalent circuit in Fig. 4 (b) for the DC side implies that the DC-link current I_{dc} starts to decay during the second fault stage. The decay of the DC-link current is defined in the time range

$$t_{\det(2)} < t \leq t_{DCM(3)}: \quad t^{**} = t - t_{\det(2)}, \tag{7}$$

$$\text{with} \quad I_{dc}(t^{**}) = I_{dc}(t_{\det(2)}) \cdot e^{\frac{-R_{dc,f}}{L_{dc,f}} t^{**}}. \tag{8}$$

The CCM is further analyzed in the following section for the comparison with the DCM, which is the representative characteristic of stage 3.

Stage 3 – Discontinuous Conduction Mode of Converter Arms

During the third stage, the arm currents are in DCM [6] until the AC circuit breaker disconnects the converter from the AC power source at Time Point (4). Therefore, a simplified calculation approach is introduced for the analysis of inflowing surge currents. The simulated converter-arm currents in Fig. 5 show that the third stage begins at Time Point (3), where $i_{arm2,p}$ first stops to conduct current. The black bars in Fig. 5 (a) indicate the periods at which one, two, and three arms per DC pole are in conduction mode. Accordingly, the short-circuit currents of the AC side flow via the converter arms for around 96% of the analyzed time range. Accordingly, the AC side and the DC side can still be assumed to be decoupled. Here, the surge current pulse is modeled based on an equivalent circuit that is typically used for the short-circuit calculation of AC power systems [9]. The single-phase equivalent circuits in Fig. 5 (b) and (c) incorporate an AC source that feeds an upper and a lower converter arm for CCM and only one converter arm for DCM. The equivalent circuit for CCM Fig. 5 (b) implies that the AC source has short-circuit paths through both star points, represented by the positive and negative DC poles of the converter. The equivalent circuit for DCM in Fig. 5 (c) corresponds to the case that two AC phases build a short circuit through two arms at one DC pole.

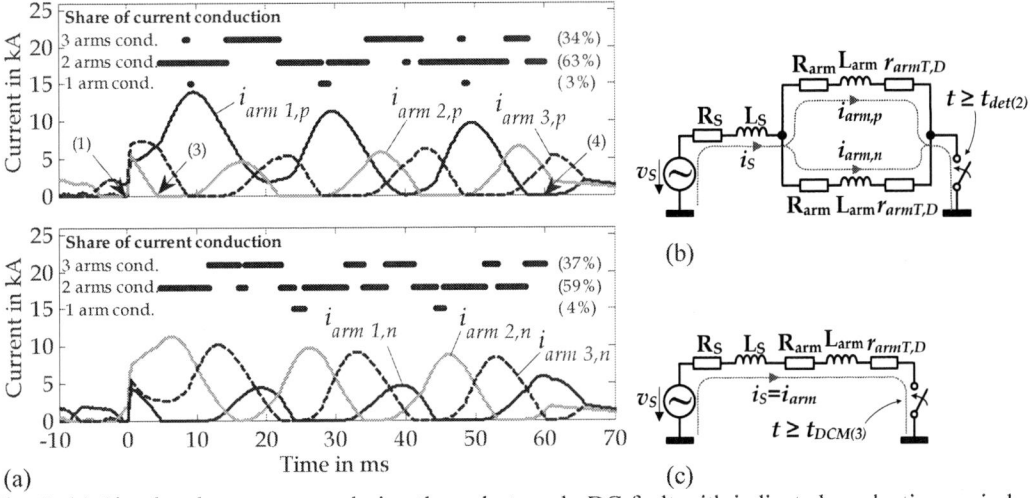

Fig. 5: (a) Simulated arm currents during the pole-to-pole DC fault with indicated conduction periods for one, two, and three arms per DC pole (black bars) and equivalent circuits for the AC-fault current calculation: (b) CCM, (c) DCM

The respective fault currents are further calculated based on the equations for three-phase faults in power systems [9]. The respective equations are adapted for the new equivalent circuits given by

$$i_S(t^*) = \frac{V_S}{\sqrt{R_{ac,f}^2 + X_{ac,f}^2}}\left(sin(\omega_S\, t^* + \varphi) + e^{-\frac{t^*}{\tau}} \cdot sin(\varphi)\right) \tag{9}$$

$$\text{with} \quad \tau = \frac{X_{ac,f}}{\omega_S \cdot R_{ac,f}} = \frac{L_{ac,f}}{R_{ac,f}}. \tag{10}$$

The impedance angle of the grid is not considered here. The calculation covers the maximum current magnitude that occurs if the source voltage v_S passes zero ($\varphi = \pi/2$) during the fault instance. The AC impedances $X_{ac,f}$ and $R_{ac,f}$ are calculated based on the single-phase equivalent circuits and type of conduction mode. The equivalent circuit for CCM in Fig. 5 (b) incorporates the parallel conduction of two converter arms. The equivalent parameters for the CCM are

$$X_{ac,f} = \omega_S L_{ac,f} = \omega_S\left(L_S + \frac{L_{arm}}{2}\right), \tag{11}$$

$$\text{and} \quad R_{ac,f} = R_S + \frac{R_{arm} + r_{armT,D}}{2}. \tag{12}$$

The equivalent parameters for DCM apply for the Fig. 5 (c), which gives the following equations

$$X_{ac,f} = \omega_S L_{ac,f} = \omega_S(L_S + L_{arm}) \tag{13}$$

$$\text{and} \quad R_{ac,f} = R_S + R_{arm} + r_{armT,D} \,. \tag{14}$$

Fig. 6 compares the calculations with the parameters $R_{ac,f}$ and $L_{ac,f}$ ($X_{ac,f}$) for CCM and DCM for the full DC fault sequence. The CCM calculation yields a peak value of i_l and $i_{arm1,p}$ for the first half-wave that deviates by roughly 2%. It is shown that the inflowing surge current pulse is defined by the AC source. The calculation for the DCM parameters is within ±5% of the simulation results.

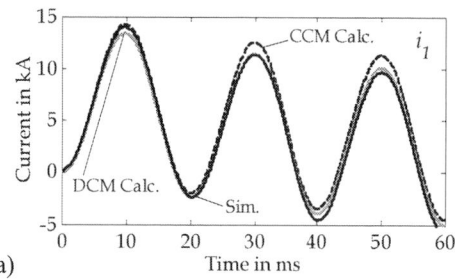 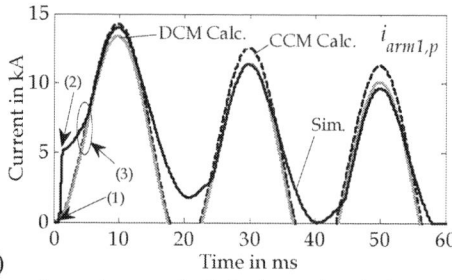

(a)   (b)

Fig. 6: Comparison of simulated and calculated currents for pole-to-pole DC short circuit ($\varphi = \pi/2$): (a) line current i_l and (b) converter-arm current $i_{arm1,p}$

The increasing DC-link current I_{dc} presents the initial fault current just after the fault instance (1) and becomes less important after the AC current exceeds the DC-link current at (3). The third stage of the fault current after (3) is characterized through DCM of the converter arms that is better replicated through the calculation of the DCM equivalent circuit in Fig. 5 (c). The simplified calculation does not consider the complex relations of natural commutation among the converter arms, which would require a more complex modeling. Nevertheless, the worst-case converter-arm peak current can be approximated for the first surge current pulse based on the system impedances.

Protection of Half-Bridge Submodules

The previous analysis demonstrated the high fault currents in conjunction with an HB-MMC that stresses the lower diodes D2 of the SMs until the AC circuit breaker opens. The investigated DC-fault sequence showed converter-arm current magnitudes in the range from 10 kA to 15 kA that can damage the HB-MMC. The following section compares possible protection measures for the HB-MMC and introduces a new protection concept based on an AC crowbar that diverts current from the converter. To prevent damages from the lower diodes, either the surge current loading of the diode has to be reduced, or the surge current capability of the diode has to be increased. Most of the protection methods follow the first approach.

Bypassing the Submodule

A bypass (BP) protection thyristor T3 for the HB-SM is one well-known protection measure for the lower diode [5], which is shown in Fig. 7. The fired thyristor is able to take the majority of the fault current depending on its on-state voltage compared to the diode. Test results in [10] presented a parallel protection device that diverted around 80% from the diode D2, for example.

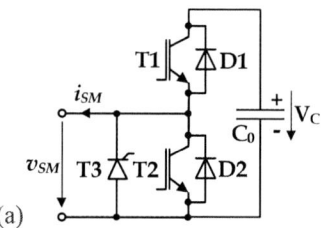 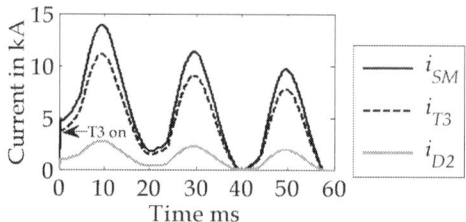

(a)   (b)

Fig. 7: (a) HB-SM with parallel protection thyristor T3 and (b) simulated currents through SM, T3, D2 in case of pole-to-pole DC fault ($i_{T3}=0.80 \cdot i_{SM}$)

Increasing the Surge Current Capability

The parallel protection element T3 of the lower diode shown in Fig. 7 (a) can be omitted if the surge-current capability of the diode D2 is sufficiently high. This can be achieved through paralleling of diodes or by use of press-pack (PP) devices. This packaging technology features double-sided cooling, and the pressure contact omits bond wires, which typically gives a higher surge-current rating compared to wire-bonded diodes. At this point, the diode of the 4.5 kV IGBT module FZ1200R45HL3 [11] is compared to a 4.5 kV single PP diode D1331SH [12]. The wire-bonded diode and the PP diode feature a similar current rating and on-state characteristic but significantly different surge current I_{FSM} and I²t-ratings. Thus, the parallel BP protection through T3 will be considered only in the case with the wire-bonded diode. The I²t-values of the worst-case converter-arm current $i_{arm1,p}$ from the simulation is used for the comparison. The parallel bypass protection through T3 is assumed to take 80% of the fault current 5 ms after the fault instance at Time Point (1). Here, the protection element T3 limits the current magnitude to a maximum value of 8.8 kA, which is below the specified limit of the module based diode of I_{FSM}=10.1 kA for $T_{j,D}$=125°C for a pulse duration of 10 ms. Both, the I_{FSM}-limit and the I²t-limit of 510 kA²s ($T_{j,D}$=125°C) of this diode would be exceeded without BP protection. The overall I²t-value of the fault scenario is 3088 kA²s, which falls within the specification of the PP diode D1331SH of 3920 kA²s for $T_{j,D}$=140°C (I_{FSM}(140°C)=28 kA). It is pointed out that the specified I²t–values relate to one 10-ms pulse, whereas the actual surge current loading as simulated covers a longer period of time, which is less severe if the I²t-limit is not exceeded. The resulting I²t-value of the arm current depends on the opening time of the AC breaker that interrupts the fault at Point (4). A fast opening time of the AC breaker of around 40 ms (cp. [13]) reduces the loading of the diodes.

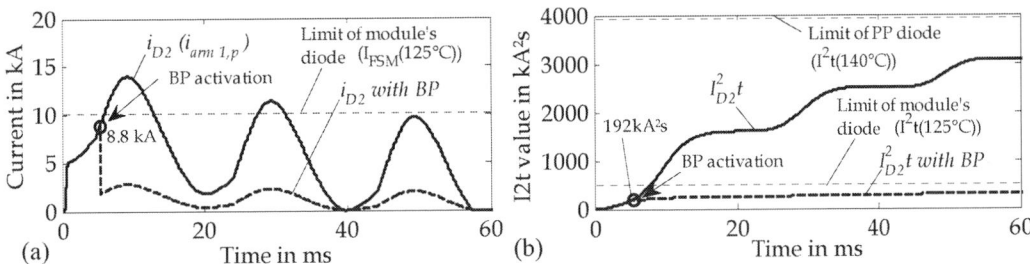

Fig. 8: (a) Simulated diode currents and (b) I²t-values of lower diode D2 of HB-MMC with and without BP protection for DC fault (limits of module FZ1200R45HL3 [11], PP diode D1331SH [12])

Application of DC Breaker

DC breakers directly interrupt DC-side fault currents and separate the fault location from the converter. Hence, a DC breaker can minimize or prevent excessive surge current flowing through the lower diodes of an HB-MMC. Hybrid DC breakers can achieve fault-clearing times in the range from 2 ms to 5 ms, which is much shorter compared to conventional AC breakers that are limited to about 40 ms [13]. Hence, the short fault-clearing time allows reducing the surge-current loading of the lower diodes of an HB-MMC. Fig. 9 shows simulations of a DC breaker for fault clearing during a pole-to-pole DC fault. The initial magnitude of the DC-fault current corresponds to the simulated DC-fault in Fig. 1 (a). The two highest converter arm currents during that time period have been plotted. The simulated opening times of 2 ms and 5 ms limit the maximum converter arm current to 9 kA and 10.3 kA, respectively, which is in the range of the diode specification of the IGBT module $I_{FSM}(T_{j,D}$=125°C)=10.1 kA or the 10-ms pulse. Since the peak surge current capability (I_{FSM}) increases for pulse durations below the typically specified 10 ms, the diodes can be protected in the simulated examples. A corresponding approach for the scaling to different pulse durations can be found in [14]. In summary, a DC breaker with a fault-clearing time in the low millisecond range can protect the freewheeling diodes of the HB-MMC and facilitates the converter protection for DC grid applications.

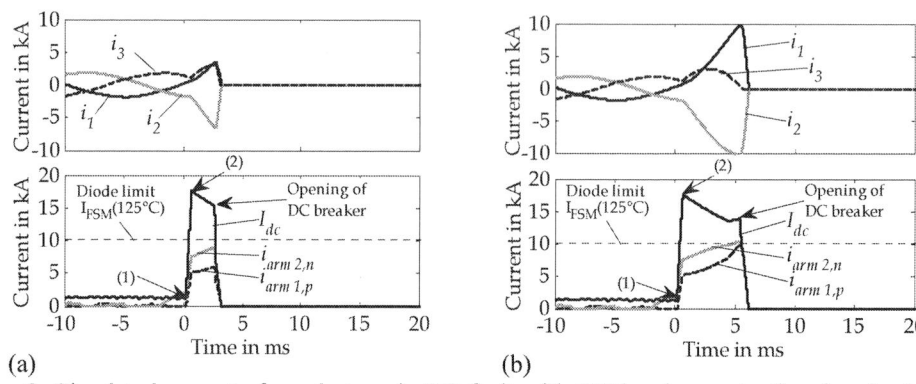

(a)   (b)

Fig. 9: Simulated currents for pole-to-pole DC fault with DC-breaker protection for clearing times: (a) 2 ms, (b) 5 ms (diode limit of module FZ1200R45HL3 [11], HB-MMC model from Fig. 1, Table I)

Application of AC Crowbar

A new type of converter protection can be implemented through an AC crowbar that is activated in case of a converter fault as shown in Fig. 10 (a). The AC crowbar is connected to a tertiary winding of the interface transformer and is used to create a separate fault path that reduces the fault currents flowing into the converter. The single-phase equivalent circuit for CCM after the fault detection shown in Fig. 10 (b) incorporates the path for the activated AC crowbar. The impedances and the magnetic coupling between the three transformer windings define the residual source voltage v_{S*} and the resulting voltage at the converter terminals v_C in case of crowbar activation. The residual voltage v_{S*} is proportional to the fault current i_{S*} that flows into the converter (cp. equation (9)) for the case that the sum of the source impedances remains similar ($L_S \approx L_{S1}+L_{S2}$). Examples for equivalent circuits of three winding transformers are described and analyzed in [15], for example.

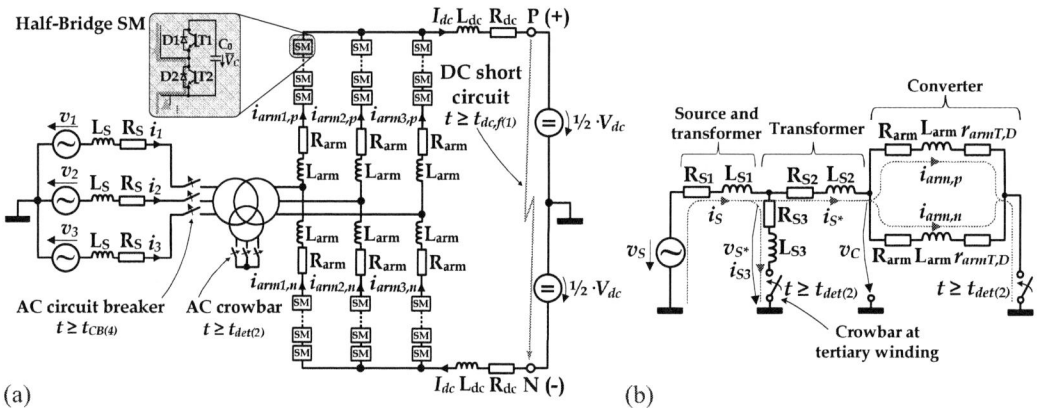

(a)   (b)

Fig. 10: (a) Pole-to-pole DC short-circuit of HB-MMC with AC crowbar and (b) single-phase equivalent circuit (CCM) for DC fault with the crowbar

The influence of the AC crowbar protection on the arm currents is analyzed for three cases that are summarized in Table III. The first case presents the DC short circuit without protection. The second case assumes equal parameters for the AC side with $L_{S3}=L_{S2}=L_{S1}$ and $R_{S3}=R_{S2}=R_{S1}$. The third case considers a much lower impedance of the tertiary winding, which further decreases the voltage v_{S*}.

Table III: Simulation parameters for the AC crowbar ($L_S=L_{S1}+L_{S2}=11$ mH, $R_S=R_{S1}+R_{S2}=0.11$ Ω)

Case	L_{S1}, R_{S1}	L_{S2}, R_{S2}	L_{S3}, R_{S3}
(1) no AC crowbar	5.5 mH, 55 mΩ		-
(2) $L_{S3}=L_{S2}=L_{S1}$	5.5 mH, 55 mΩ		5.5 mH, 55 mΩ
(3) $L_{S3}=L_{S2}/5=L_{S1}/5$	5.5 mH, 55 mΩ		1.1 mH, 11 mΩ

The simulation results in Fig. 11 (a) compare the effect of the AC crowbar depending on the impedance configuration from Table III. The I²t-value of the converter-arm current $i_{arm1,p}$ from Case (2) exceeds the specified I²t-limit of 510 kA²s ($T_{j,D}$=125°C) of the previously considered IGBT module FZ1200R45HL3. Case (3) considers a five times lower impedance of the tertiary winding compared to Case (2) and keeps the I²t-value below the limit of the module as shown in Fig. 11 (a). The positive effect of the AC crowbar on the converter-arm currents during the DC fault causes other unfavorable aspects. The simulated AC currents for i_I in Fig. 11 (b) increase from 14 kA for case (1) to 25 kA for Case (3) because of the reduced impedance. The fired crowbar now constitutes a second short-circuit path for the grid side in parallel to the DC-side short circuits that decreases the resulting impedance.

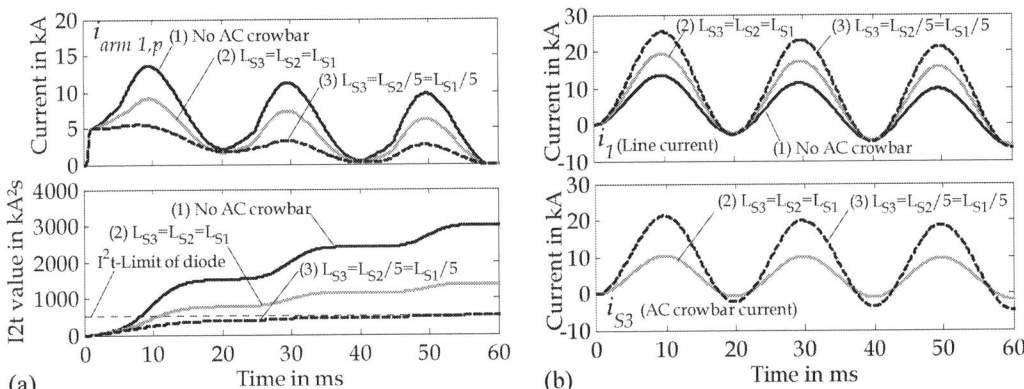

Fig. 11: (a) Simulated converter-arm currents and I²t value, (b) AC currents with and without crowbar (HB-MMC model from Fig. 1, Table I, parameters from Table III, I²t-limit of FZ1200R45HL3 [11])

An AC crowbar can use thyristors with a high surge-current rating, and the voltage rating of the tertiary winding can be scaled accordingly. Therefore, the transformer allows using a lower voltage rating for the crowbar compared to the converter side, which facilitates to optimize the related component efforts. The transformer design gives some degrees of freedom for the coupling between the different windings and the desired set of impedances. Under the assumption, that the tertiary winding has the same voltage rating as the converter winding and the crowbar comprises the same thyristor type like the bypass of the HB-SMs, the effort for thyristors is roughly halved compared to the HB-MMC with bypass protection. This estimate is based on the definition that the crowbar is made of serialized antiparallel thyristor switches in a star configuration (cp. Fig. 10 (a)) and an HB-MMC having the minimum number of SMs, defined by the peak values of the phase voltages of the source V_S and the converter V_C corresponding to half the DC-link voltage (V_S=V_C=V_{dc}/2).

Discussion on the Protection Measures

The protection of the lower diodes of HB-SMs must consider the specification of the chosen device and its packaging. PP diodes with a sufficiently high surge current capability do not necessarily require additional protection. Wire-bonded diodes typically feature a lower surge current capability and rely on supplementary protection circuitry. A fast bypass in parallel to the lower diode is favourable if the protection shall be implemented into the HB-SM. Alternatively, the paralleling of diodes or parallel conduction paths within SMs can be envisaged to decrease the surge current requirement. Protection on the system level can avoid supplementary equipment on the SM level. The opening time of the AC circuit breaker defines the time frame of high current conduction and should be as short as possible. DC breakers are required for DC grids and can support the protection of the SMs when the opening time is sufficiently short. The simulations showed for specific examples that the fault current limitations fall within the specification of relevant 4.5-kV module IGBTs. The total expense of a hybrid DC breaker can amount to two converter arms, which increases for bidirectional current switching capability [16]. The presented AC crowbar protection diverts inflowing AC currents to a tertiary winding that builds a parallel short-circuit path. The design of the transformer is essential for the dimensioning of the respective impedances and the design of the AC crowbar, which defines the corresponding component efforts.

Summary

DC-side short circuits of a Half-Bridge MMC present severe fault conditions that must be safely managed to prevent damages from the converter. A simulated DC-fault sequence has been analyzed, and a simplified calculation was demonstrated that allows the estimation of the fault current magnitudes for a Half-Bridge MMC. The presented calculation omits extensive system simulation and supports fast analysis of the fault current magnitudes based on the system impedances.

Different protection measures for the HB-SMs have been outlined, such as the bypass of the SM, DC breakers, and the increase of the surge current capability of diodes. A press-pack diode with a typically high surge current capability seems to be an attractive solution to withstand severe DC faults, which potentially does not require additional protective equipment. DC breakers with an opening time in the low millisecond range are a promising protection measure, but require a high component effort and are more likely to be implemented for DC grids. Finally, a new protection measure based on an AC crowbar was presented. This protection device establishes a second short-circuit path, which diverts the inflowing AC fault current from the converter. This solution requires an additional tertiary winding for the interface transformer, which is short-circuited by the AC crowbar in cases when the DC fault is detected. Its effectiveness depends on the transformer design and the ratio of the transformer winding impedances. The component effort of thyristors for a crowbar solution is reduced in comparison to the bypass protection of individual HB-SMs of a complete MMC.

References

[1] R. Marquardt, "Modular multilevel converters: State of the art and future progress," IEEE Power Electronics Magazine, Vol. 5, Issue 4, 2018.

[2] A. Nami, L. Wang, and F. Dijkhuizen, "Five level cross connected cell for cascaded converters," Proc. EPE Conf., Lille, France, 2013.

[3] C. Oates, "Modular multilevel converter design for VSC HVDC applications," IEEE Journal of Emerging and Selected Topics in Power Electronics, Vol. 3, No. 2, 2015.

[4] A. Nami, J. Liang, F. Dijkhuizen, and P. Lundberg, "Analysis of modular multilevel converters with DC short circuit fault blocking capability in bipolar HVDC transmission system," Proc. EPE Conf., Geneva, Switzerland, 2015.

[5] S. Cui and S.K. Sul, "A comprehensive DC short-circuit fault ride through strategy of hybrid modular multilevel converters (MMCs) for overhead line transmission," IEEE Trans. Power Electron., Vol. 31, No. 11, 2016.

[6] B. Li, J. He, J. Tian, Y. Feng, and Y. Dong, "DC fault analysis for modular multilevel converter-based system," Journal of Modern Power Systems and Clean Energy, Vol. 5, No. 2, 2017.

[7] R. Marquardt, "Modular multilevel converter topologies with DC-short circuit current limitation," Proc. ECCE Asia, Jehu, South Korea, 2011.

[8] Z. Xu, H. Xiao, L. Xiao, and Z. Zhang, "DC fault analysis and clearance solutions of MMC-HVDC systems," Energies, Vol. 11, No. 4, 2018.

[9] J. C. Das, "Short-Circuits in AC and DC Systems: ANSI, IEEE, and IEC Standards," Power Systems Handbook, CRC Press, Vol. 1, 2017.

[10] T. Xu, P. S. Jones, and C. C. Davidson, "Electrical type tests for the voltage sourced converter valves based on modular multi-level converter," Proc. EPE Conf., Geneva, Switzerland, 2015.

[11] Infineon Technologies AG, "FZ1200R45HL3 - IHM-B Module with Trench/Fieldstop IGBT 3 and Emitter Controlled 3 Diode," Datasheet, V 3.2, Munich, Germany, 19th Feb. 2018.

[12] Infineon Technologies Bipolar GmbH & Co KG, "D1331SH - Fast Hard Drive Diode," Datasheet, V 10.2, Warstein, Germany, 1st Apr. 2015.

[13] W. Leterme and D. van Hertem, "Classification of fault clearing strategies for HVDC grids," CIGRE Symposium, Lund, Sweden, 2015.

[14] T. Hunger, O. Schilling, and F. Wolter, "Numerical and experimental study on surge current limitations of wire-bonded power diodes," Proc. PCIM, Nuremberg, Germany, 2007.

[15] X. Margueron and J. P. Keradec, "Design of equivalent circuits and characterization strategy for n-input coupled inductors," IEEE Trans. Ind. Applicat., Vol. 43, Issue 1, 2007.

[16] Y. Wang and R. Marquardt, "Operation of modular multilevel converter and DC-breaker in large multiterminal-HVDC grids," Proc. PCIM Conference, Nuremberg, Germany, 2013.

Investigation on parallel operation of two MMC-HVDC links in grid forming connected to an existing network

H. Saad, P. Rault, S. Dennetière
RTE
France
hani.saad@rte-france.com
www.rte-france.com

Keywords

« EMT-Type», «grid forming», «HVDC transmission», «Islanding operation», «MMC», «VF control», «VSC».

Abstract

The share of power electronics installations into existing ac power systems is significantly increasing due to the massive penetration of wind power plants and HVDC links. VSC-HVDC link can operate in VF-control (or grid-forming mode) when connected to a weak or islanded network to overcome the PQ-control (or grid-feeding mode) limitation. Up to now, the VF-control represents a challenge due to static and dynamic limits when several HVDC links are in parallel operation. This paper provides a dynamic performance of two HVDC links in grid forming mode connected to an islanded grid. To provide a realistic test case, the islanded grid is based on the Northern France network. AC fault and control instabilities between 2 HVDC links in VF control are analyzed in this work. These set of EMT studies provide an insight on dynamic network behavior when HVDC is operating in VF control.

Introduction

Several power electronics devices as High Voltage Direct Current (HVDC) links and wind farms projects are currently planned or constructed by RTE (French TSO) and worldwide. In the long term, as the share of such power electronics equipment is growing rapidly [1], some synchronous areas might, in the future, be operated with low inertia or even without synchronous generators.

Due to techno-economic reasons, HVDC based on Modular Multilevel Converter (MMC) technology have become the preferable solution for the upcoming HVDC installations. In general, these HVDC links are equipped with two types of controllers [2]: PQ control (or grid-feeding) when HVDC links are connected to a strong network. In this mode of operation, the link does not impose the AC voltage but rather synchronizes with the network voltage and, accordingly, adjusts its generated voltage in magnitude and phase to controls the injected currents into the grid. VF control (or grid-forming) is used to restore blackout scenario or supply islanded network. In this operation mode, the converter synthesizes an instantaneous AC voltage and other network components are synchronized to this voltage waveform. Up to now, the converter in VF control is the main power source of the islanded network and would switch to PQ control as soon as it is coupled to a stronger network supplied by several synchronous machines. In the future, with less synchronous machine and more power electronic devices, several converters in VF control operation would have to work in parallel to supply additional loads or absorbs more power from renewable energy resources. In this context, the Johan Sverdrup Project [3] is expected to be the first application of parallel operation of two HVDC links in grid forming supplying the same offshore passive network.

Currently, power systems are conventionally dominated by synchronous generators which provide their mechanical inertia to the electrical system to absorb the load variations meanwhile the primary frequency control response. This initial response is dictated by the sum of mechanical inertia of synchronous machines that maintains stability of the power system. In addition, during faults and their recovery, synchronous machines, partly thanks to their temporary overcurrent capability, support the AC voltage. With the increase of renewable energy ratio, the power system is expected to evolve from a relatively predictable system, to a fluctuating system dominated by power electronics. For Transmission System Operators (TSOs) which are responsible for the stability of the power system and perform stability studies, the system complexity is increased due to complex power electronic device control & protection, faster dynamic responses, and limited current contribution during faults. To anticipate such situations, in [4] and [5], the increasing penetration of power electronics in the Great Britain power system is studied. However, simulation results are performed in RMS-type software where model accuracies and control structure are limited for such phenomena. In [2], [6] and [7], EMT-type simulations and/or experimental results have been performed to study AC fault performances with power electronics in VF control. These relevant results are performed on a simple network and only one power electronic devices in grid forming mode is considered. This paper intends to provide further realistic results by using an EMT-type software with MMC topology (rather than VSC 2 or 3 level) and the parallel connection of two HVDC links in VF control mode. In addition, the study is conducted using real network data based on the Northern France network area where several HVDC links and wind farm projects are under construction or at planification stage [14].

Issues and challenges associated with parallel operation of MMC-HVDC in grid forming, feeding islanded grid subject to AC faults and control instabilities are the scope of this article. Control instability between the two HVDC links are analysed in frequency domain by performing a frequency scan in time domain simulation to account for control loop's impact as described in [15]. This frequency scan approach provides an insight (in EMT tool) on the root cause of these instabilities and on the parameter sensitivities of controls gain and network configurations.

VF control structure

The general structure of a VF controller is depicted in Figure 1. It contains the Power-Frequency (PF) controller to regulate the active power and the frequency, a Vac controller to regulate the voltage at the point of common connection (PCC), a current limitation controller to limit the current to rated current during transients.

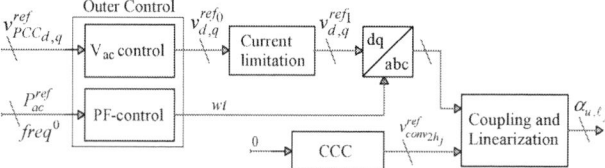

Figure 1: VF control structure

CCC, coupling and linearization

The CCC (circulating current control) eliminates (or reduces) circular currents that are inherent to MMC topology. The "Coupling and Linearization" block allows to couple the reference voltages and to generate references for each arm. For more details see [11] and [12].

PF control

If the HVDC link is the predominant voltage source connected to a passive network with no other devices participating in the energy supply and demand balance, then the link may impose a fixed frequency. However, in practice, the HVDC link may be connected to a network where other HVDC links and/or synchronous generators are present. In this case, a PF droop [Hz/MW] control is necessary to share the load balancing effort. In the literature, several types of controls are proposed for VSC technology and applied to HVDC applications. [8] summarizes the main control structures that can be used. In [9], it was demonstrated that virtual synchronous machine (VSM) and the PF droop control can be interpreted similarly. In this article, the PF droop control presented in Table 1-a is considered. It is similar as the primary frequency control commonly used by synchronous generators in power system. Equation of this control is written as follow:

$$\Delta F = K_P \left(P_{ac}^{ref} - H_{LP}^P(s)P \right) \tag{1}$$

where K_p droop in Hz/MW (or in pu/pu), $freq^0$ nominal reference frequency, ΔF is the frequency deviation and $H_{LP}^P(s)$ low pass filter in order to guaranty the decoupling between inner and outer loops [10].

Table 1 PF-control and Vac-control structure

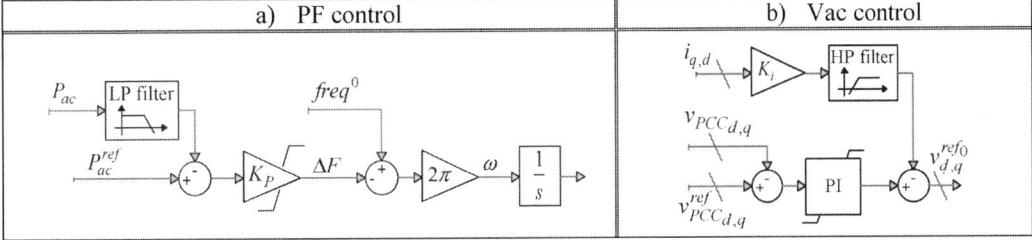

a) PF control	b) Vac control

Vac-control

The AC voltage control in dq frame at the PCC of the converter station is shown in Table 1-b. From [2] and [14], the voltage control is given as follows:

$$v_{d,q}^{ref0} = C_v(s)\left(v_{PCC_{d,q}}^{ref} - v_{PCC_{d,q}} \right) - H_{HP}(s)i_{q,d} \tag{2}$$

where $C_v(s)$ is the transfer function of the PI controller, $H_{HP}(s)$ is a high-pass filter to improve the system damping: $H_{HP}(s) = \dfrac{s}{s+\omega_i}$. The cutoff frequency ($\omega_i$) of this filter must be tuned lower enough to be less sensitive to oscillations from the network.

The feed-forward gain K_i (Table 1-b) provide a voltage compensation (typical value between 0.05-0.3)[6]. The gain K_i impacts the voltage damping and therefore influence the stability of the overall system as shown in the simulations section. v_{dq}^{ref0} is the AC voltage reference for the MMC arms in dq frame, $v_{PCC_{dq}}^{ref}$ and $v_{PCC_{dq}}$ are respectively the voltage operator setpoint and the measured voltage.

Current limitation

During AC fault or transients, current flowing in the converter must be limited to avoid overcurrent on MMC's semi-conductors. There are two main approaches for current limitation: direct voltage reference curtailment by adding a virtual impedance or explicit current control to rated current when it is exceeded. In the following sections these two techniques are presented and compared.

Current limitation by voltage reference curtailment

In this approach, once the overcurrent threshold is detected ("overcurrent detection"), the fault current is limited by adding a virtual impedance in the control. The addition of a virtual impedance in the control is proposed in several articles as [6], [7] and [13] for 2-3 level VSCs topology. In this section, the latter is adapted to the MMC stations. Control structure is shown in Table 2-a. Once the overcurrent has been detected, virtual impedance equation is written in this form:

$$v_{dq}^{ref1} = v_{dq}^{ref0} + Z\left(i_{dq} - I_{dq_{\lim}}^{ref} \right) \tag{3}$$

where v_{dq}^{ref0} is the voltage reference output from Vac-control, v_{dq}^{ref1} is the voltage reference output from the current limiter, Z is the virtual impedance and $I_{dq_{\lim}}^{ref}$ current limit of the converter in dq.

During AC voltage dip, the reference voltage v_{dq}^{ref1} will decrease depending on the value of the virtual impedance Z in order to limit the current. In [7], the analytical approach to deduce the value of the virtual impedance according to the desired current threshold is described.

To illustrate this behavior, a MMC-HVDC link in VF control connected to a resistive load, is considered with such voltage reference curtailment approach. The transmission capacity of the link is 1,000 MW. The dc cable is rated ± 320 kV with 70 km length and is modeled using a frequency dependent model. A MMC 401-level is considered and is modeled using the Model#3 (as defined in [11]). Submodules capacitor stored energy value is $E_{CSM} = 40$ kJ/MVA, arm reactors are equal to 15% and transformer reactor is 18%. The remaining electrical parameters of the converter station can be found in Cigré TB [12]. A solid three-phase fault at the PCC (see Figure 2:) is applied at t=3 sec.

Table 2 Current limitation approaches

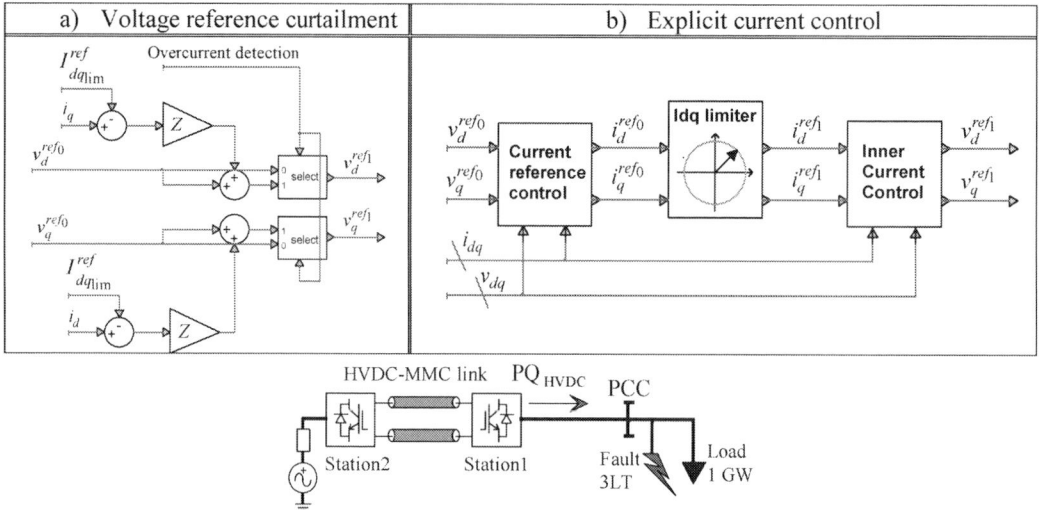

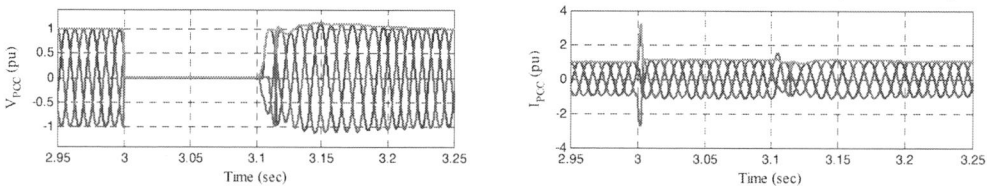

Figure 2: MMC in VF control– solid three-phase fault

Using this current limitation in voltage reference curtailment approach and applying a solid 3-phase fault, instantaneous and in *dq* vector amplitude of voltage/current at PCC results are given in Figure 3.

Figure 3: Tri-phase fault - VF control with current limitation in voltage reference curtailment

Note that during fault, the voltage reference curtailment control manages to limit the fault current (around a set value of $I_{PCCdq} = 1.2$pu). However, at fault ignition, an instantaneous current peak (few ms) around 3 pu is noticed. In practice, this overcurrent spike may damage the semiconductors and/or may lead to temporary block and even trip of the HVDC link. This current spike is mainly related to the activation delay time of the "overcurrent detection" (Table 2-a.). Theoretically, it is, probably possible to reduce this current spike by decreasing the activation delay time but such option may compromise the robustness of the system during transients and would be difficult to achieve with real system because of the acquisition time. Therefore, this phenomenon is inherent to the control structure and cannot be fully avoided. In addition, unlike 2-3 level VSCs topology, MMC stations do not include AC filters (or only small filters), that limit this transient overcurrent on semiconductors. It should be noticed that such transients have a dynamic lower than one period (around 5ms). Therefore, such behavior cannot be reproduced in RMS tools, therefore, EMT tool should be used for such analysis.

Current limitation with explicit current control

In this section, the limitation scheme proposed in [2] for 2-level VSCs is adapted for MMCs [14]. The principle is to obtain a control system including an automatic current control limiting mode once the converter current exceeds the current threshold. The current control system consists of three elements (see Table 2-b) that are described in [14].

When converter current exceeds the maximum limit, the reference currents are limited, and the inner control regulates the current to maintain it at an acceptable value. However, instead of giving a constant current reference, the value of $i_{d,q}^{ref_1}$ is expressed in such a way that in normal operation we have:

$$v_{d,q}^{ref_0} \cong v_{d,q}^{ref_1} \tag{4}$$

In normal operation, as long as $i_{dq}^{ref_0}$ does not exceed the current limits, the block "Idq limiter" does not modify the input:

$$i_{d,q}^{ref_0} = i_{d,q}^{ref_1} \tag{5}$$

However, during transients (i.e. AC fault) if the current is exceeded, block "Idq limiter" will set new $i_{d,q}^{ref_1}$ values and the "inner control control" regulates the current to an acceptable desired value automatically.

The same three-phase fault at the connection point (as in the previous section) is applied with the explicit current control current limiter. The results are given in Figure 4.

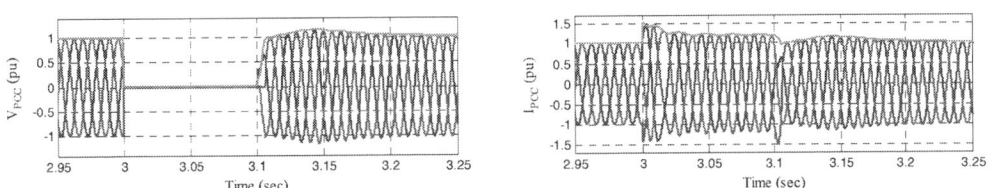

Figure 4: 3-phase fault - VF control with current limitation with explicit current control

AC current is well limited during AC fault. By comparing these results (Figure 4) with those of the voltage reference curtailment approach (Figure 3), one can notice that the instantaneous current peaks at the fault ignition are much lower (1.5 p.u. instead of 3 p.u.). As conclusion, the explicit current control current limiting control is the best candidate because it provides better performances with lower overcurrent spikes and smoother control. The next section will, therefore, considers only the current limitation with explicit current control approach.

Dynamic study using VF control

Islanded grid overview

The following section describes the islanded grid considered in this study. It is based on the 400 and 225 kV of the northern France network area [14]. Figure 5 shows the single line diagram of the considered equivalent AC grid. The high voltage power electronic equipment that are installed are:

- Wind farm with a total capacity of 450 MW and ±150 Mvar
- HVDC 1 and 2: symmetrical monopole HVDC-VSC link with a rated power of 1,000 MW and ± 300 Mvar

The network is modeled and simulated in EMTP-RV software. The wind turbine type 4 is a generic model that includes the detail representation of control and protection system. For such study, the wind farm is aggregated and is operated in grid-feeding mode. For both HVDC links, the same model as in previous section is used with the explicit current controlapproach. VF control parameters are provided in the Appendix. The generator is a synchronous machine (SM) with a rated power of 1.6 GVA. The model includes the exciter, voltage, and speed regulators derived from onsite datasheet. Overhead lines are modeled as a constant distributed parameters and dc cables are represented using frequency dependent models. The two transformers 400/225/20 kV are modeled with their respective magnetization branches derived from the datasheet. In order to study the evolution of the network and the impact of the VF control on system performances, two network configurations are considered by closing and/or opening the switch S3 (see Figure 5):

- Scenario 1: Switch S3 is open, i.e. only HVDC 1 link is connected in the islanded network
- Scenario 2: Switch S3 is closed, the two HVDC links (HVDC1 and 2) are connected in grid forming mode.

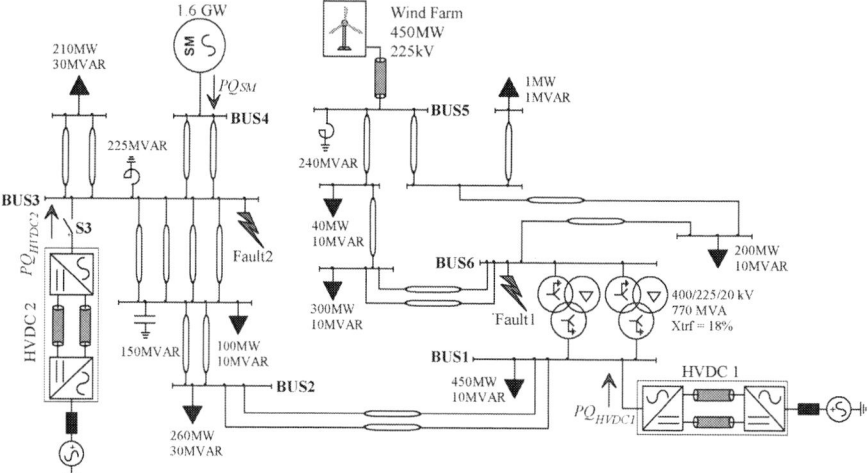

Figure 5: Benchmark network of the Northern France area in EMTP-RV

AC fault performance

The AC fault events simulated in this study corresponds to a solid three-phase faults Fault 1 is located at BUS6 with a duration of 200 ms.

Scenario 1 - One HVDC link connected

Considering a grid development, it is expected that first, one HVDC link is installed before the installation of a second one. Therefore, in this section the behavior of one HVDC link in grid forming is considered (scenario 1). Active/reactive power of HVDC 1 and the generator SM as well as dq voltage/current results are presented in Figure 6. Internal converter variables, as capacitor voltages, $v_d^{ref_1}$ and $i_{d,q}^{ref_1}$ are illustrated in Figure 7.

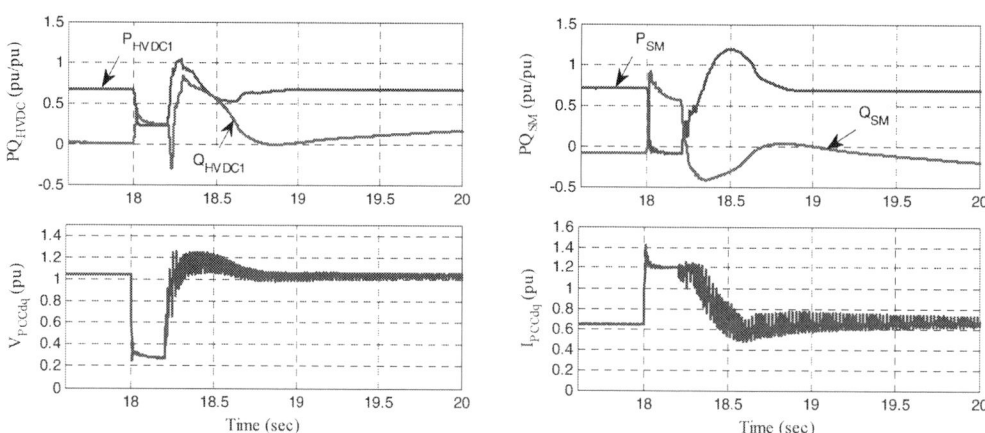

Figure 6: MMC station connected to an islanded network

Note that the system remains stable. The solid tri-phase fault leads to overcurrent peak around $I_{PCCdq} = 1.4$ pu and voltage recovery causes an overvoltage around 1.21 pu. During fault (18 sec $< t <$ 18.2 sec), HVDC active power is reduced and HVDC reactive power is increased to support voltage sag. Recovery time of active and reactive power after fault clearance is around 3 sec because of generator's dynamics. During fault recovery (t>18.2 sec), voltage and current high frequency oscillation are mainly due to the transformer saturation 400/225/20 kV.

Regarding $i_{d,q}$ and the reference $i_{d,q}^{ref1}$, the proper operation of the current loop and the limitation of the overcurrent is noticed. Regarding v_d^{ref0} and v_d^{ref1}, it is clear that as long as the current is in a normal operation, the two variables are equal. However, during faults, the reference voltages v_d^{ref1} is reduced (via the Idq limiter see Table 2-b) to limit the current $i_{d,q}^{ref1}$. Note that the 150 Hz oscillations reported in the dq amplitude vector current/voltage during the voltage recovery are due to transformer magnetization saturation. Such transients will have an impact on system recovery time. Because such transient cannot be reflected in RMS tools, EMT tools should be used to accurately reproduce this phenomena that have an impact on the overall system behavior.

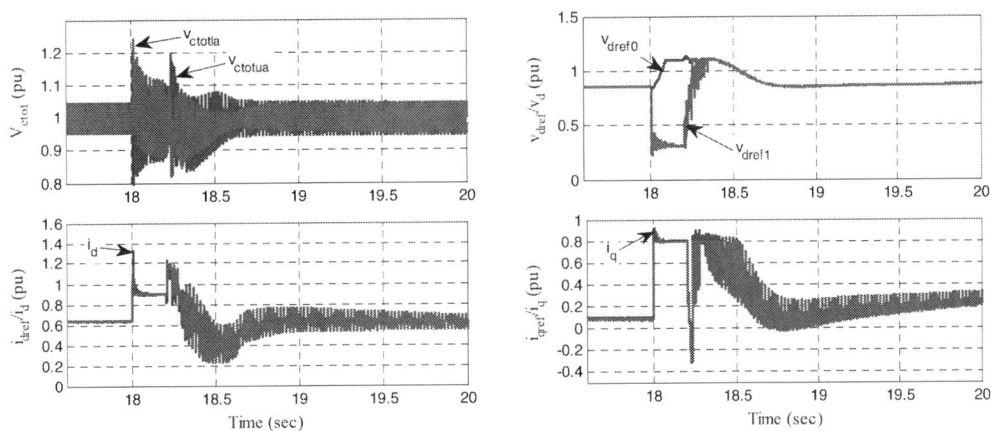

Figure 7: MMC station connected to an islanded network

Comparison between Scenario 1 and 2

In this section, comparison between scenario 1 and scenario 2 is presented. Same tri-phase fault event as in previous section, is simulated and HVDC 1 results are compared in Figure 8 (blue curves for Scenario 1 and green curves for Scenario 2).

Note that the system remains stable. Before fault occurrence (t < 18 sec), in scenario 2, HVDC 1 active power is reduced by half (from 0.67 pu to 0.34 pu) this is because of active power sharing between HVDC 1 and HVDC 2.

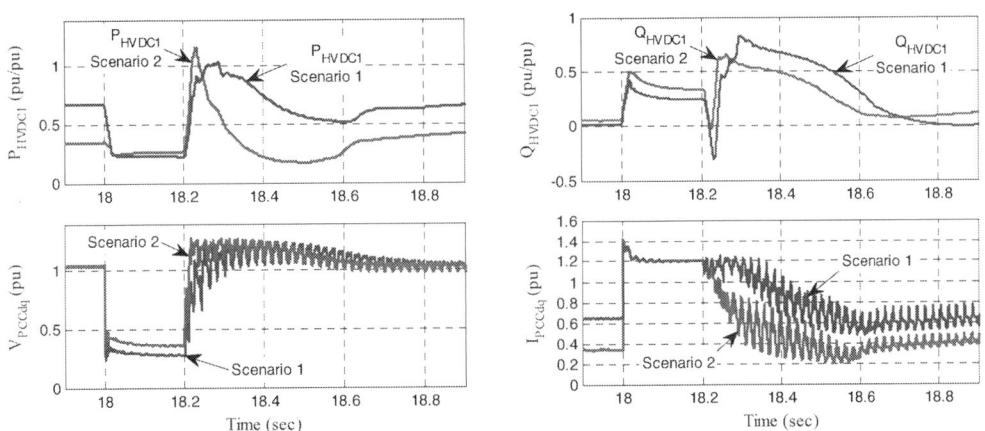

Figure 8: Two HVDC links connected to an islanded network

During fault period (18 sec < t < 18.2 sec), the current limitation strategy provides similar behavior in both scenarios 1/2 and reaches a steady state value around $I_{PCCdq} = 1.2$. This is because AC voltage dips is sufficiently low to activate the maximum current limitation. However, V_{PCCdq} and Q_{HVDC1} have higher value in scenario 2 than in scenario 1. This is due to the presence of the second HVDC 2 in the network (in scenario 2) that provide addition reactive power support.

After fault clearance (t > 18.2 sec), in scenario 2, the presence of a second HVDC link in grid forming improve (and decrease) the fault recovery period (around 2.2 sec for scenario 2 instead of 3 sec in scenario 1). However, during voltage clearance instant, the presence of a second HVDC link (scenario 2) leads to faster transients, higher overvoltage $V_{PCCdq} = 1.28$ pu (respect to Scenario 1 - $V_{PCCdq} = 1.21$ pu) and higher P_{HVDC1} peaks.

Parameter sensitivity study

In previous section, the good performance of both HVDC 1 and 2 (scenario 2) assumes that both HVDC controller settings are properly tuned jointly. However, in real practical situation, it is likely that both HVDC links may be designed by different manufacturers (such as the Johan Sverdrup project [3]) with no coordination between each other. In such a case, parameters would be most certainly tuned differently without knowing the behaviour of the other HVDC links connected to the same network area. If no risk interaction assessment is undertaken during design phase, each HVDC might be tuned independently and without considering the presence of other HVDC links located in the same network area.

This subsection highlights the impact of HVDC link control parameters on the system stability when two HVDC links in grid forming are connected to the same network (scenario 2). Parameter variations on the cutoff frequency of low pass filter $H_{LP}^{P}(s)$ (Table 1-a) and on the current feedforward gain K_i (Table 1-b) is performed. Simulation results P_{HVDC1} when cutoff frequency of low pass filter (5Hz and 10Hz) and the current feedforward gain (0.1 pu and 0.05 pu) are varied and are presented in Table 3-a and Table 3-b respectively. It can be noticed, that when one of these parameter values is increased, unstable oscillations around 15.9 Hz occur. It should be highlighted that the considered parameter values provide stable operation when only one HVDC links (HVDC1 or HVDC2) is connected in the network. Therefore, the observed instabilities are due to the interaction between the two HVDC links.

Table 3: Control instability results: Impact of $H_{LP}^{P}(s)$ and K_i

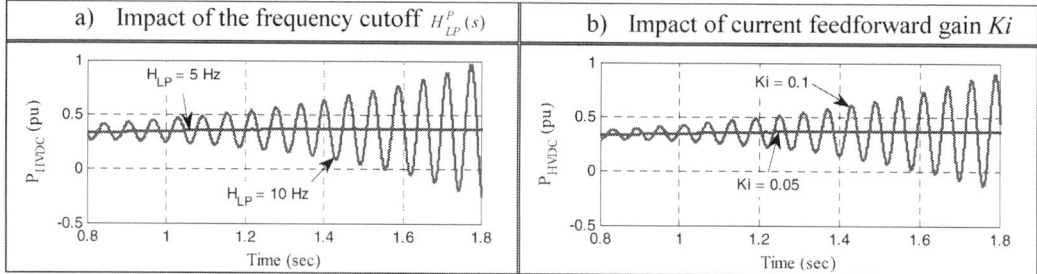

Frequency response analysis – control parameter sensitivities

In order to identify the root cause of these instabilities or undamped oscillations in EMT tool, a frequency domain study is conducted. As described in [15], the frequency response of the EMT closed loop system can be derived by performing a frequency scan during time domain simulation. Same approach is used hereafter, to deduce the frequency response $G(f)$ between HVDC link1's active power and angle.

$$P_{HVDC1}(f)/\delta_{HVDC1}(f) = G(f) \qquad (6)$$

The frequency response results of the initial parameters, the variation of $H_{LP}^{P}(s)$ cutoff frequency and K_i values are plotted in Figure 9.

The pick resonance in $|K|$ is around 15.9 Hz. This frequency value corresponds to the oscillation that are seen in Table 3. Indeed, when $H_{LP}^{P}(s)$ cutoff frequency and K_i increases, it is noticed that the phase margin is decreased, therefore, the system is less damped. Such control parameter variations have an impact mainly on the phase margin and pick amplitude but not on the frequency resonance value.

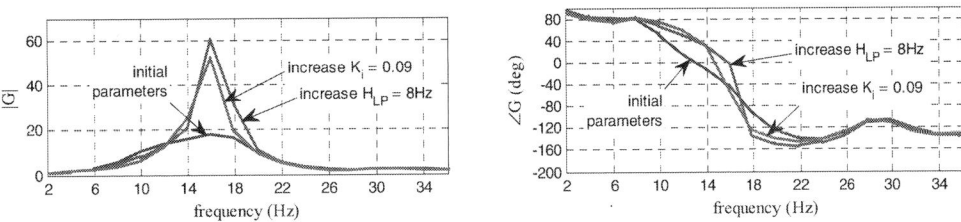

Figure 9: Impact of control parameters - Frequency response

Frequency response analysis - electrical parameter sensitivities

The impact of the electrical parameters poor damped oscillations of the MMC stations and ac grid is analysed in this section. The following electrical parameters are varied: submodule capacitor (E_{CSM}), transformer inductance (L_{trf}) and change in the network topology due to AC line outage in the network (n-1 line outage between BUS1 and BUS2 - Figure 5). Frequency response results are presented in Figure 10.

From Figure 10, the submodule capacitor variation shows a high impact on system resonances. The decreases in submodule capacitor lead to higher pick amplitude and also, decreases the frequency resonances. This shows that internal converter capacitor has an impact on AC system stability. On the other hand, the decrease of L_{trf} leads to lower frequency resonances and higher pick amplitude. This means that when the electrical distance between the two HVDC links decreases, interaction is expected to increase. Similar conclusion can be drawn for the n-1 outage; when the electrical distances between HVDC 1 and HVDC 2 are increased (thus equivalent inductance value increases) frequency resonance and amplitude seems to decrease.

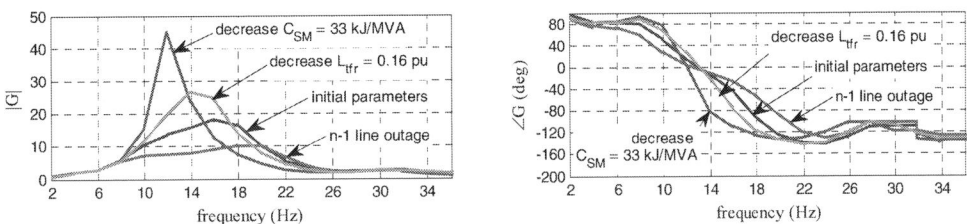

Figure 10: Impact of electrical parameters - Frequency response

Conclusions

This paper investigates the parallel operation of two HVDC links in grid forming connected to an existing network. Two main current limitation approaches (voltage reference curtailment and with explicit current control) are analysed. Current limitation based on explicit current control approach is more appropriate for MMC-HVDC link because it avoid overcurrent spikes at fault instant. In addition, comparison performances when one and two HVDC link in VF control connected in an islanded network is presented. It shows that when two HVDC links are connected and without appropriate control tuning, interaction can lead to oscillations in the system. Finally, control sensitivity study for two parallel HVDC connections is analysed. A frequency domain study on control and circuit parameters sensitivity is proposed. For the considered Northern French area, it shows that the low pass-filter of the PF control should have a low cutoff frequency and the current feedforward gain should be kept low. On the other hand, when VSC capacitors and system inductances decrease, the frequency resonances also decrease. Simulation where conducted in EMT-tool to account for additional realistic phenomena that cannot be reproduced with RMS tools, such as peak fault current behavior in voltage reference curtailment approach and transformer saturation during AC fault performances.

Appendix – HVDC parameters

Inner current control: time response = 2 ms ; Vac-control : time response = 200 ms ; $H_{LP}^{P}(s) = 5Hz$;

$H_{HP}(s) = 11Hz$; $I_{dq_{lim}}^{ref} = 1.2\,pu$; $K_p = 0.05\ pu/pu$; $K_i = 0.05\ pu/pu$

References

[1] B. Franken, G. Andersson, "Analysis of HVDC converters connected to weak AC systems," Power Systems, IEEE Transactions on , vol.5, no.1, pp.235,242, Feb 1990

[2] L. Zhang, "Modeling and Control of VSC-HVDC Links Connected to Weak AC Systems," Ph.D. Thesis, Royal Institute of Technology, Stockholm, Sweden, 2010

[3] K. Sharifabadi, N. Krajisnik, R. Teixeira-Pinto, S. Achenbach, R. Rad, "Parallel operation of multivendor VSC-HVDC schemes feeding a large islanded offshore Oil and Gas grid" CIGRE Conference, Paris, France, Aug. 2018

[4] M. Yu et al., "Effects of swing equation-based inertial response (SEBIR) control on penetration limits of non-synchronous generation in the GB power system," International Conference on Renewable Power Generation (RPG 2015), Beijing, 2015, pp. 1-6

[5] I. Richard, J. Zhu, A. J. Roscoe, M. Yu, A. Dysko, C. D. Booth, and H. Urdal. "Effects of VSM convertor control on penetration limits of non-synchronous generation in the GB power system." In 15th Wind Integration Workshop. 2016

[6] A. D. Paquette and D. M. Divan, "Virtual Impedance Current Limiting for Inverters in Microgrids With Synchronous Generators," in IEEE Transactions on Industry Applications, vol. 51, no. 2, pp. 1630-1638, March-April 2015

[7] F. Salha, F. Colas and X. Guillaud, "Virtual resistance principle for the overcurrent protection of PWM voltage source inverter," 2010 IEEE PES Innovative Smart Grid Technologies Conference Europe (ISGT Europe), Gothenburg, 2010, pp. 1-6

[8] G. Denis, T. Prevost, P. Panciatici, X. Kestelyn, F. Colas and X. Guillaud, "Review on potential strategies for transmission grid operations based on power electronics interfaced voltage sources," 2015 IEEE PES General Meeting, Denver, CO, 2015, pp. 1-5

[9] S. D'Arco and J. A. Suul, "Virtual synchronous machines — Classification of implementations and analysis of equivalence to droop controllers for microgrids," : PowerTech 2013 IEEE Grenoble Conference, Grenoble, 2013, pp. 1-7

[10] J. M. Guerrero, L. G. de Vicuna, J. Matas, M. Castilla, and J. Miret, "A wireless controller to enhance dynamic performance of parallel inverters in distributed generation systems," IEEE Trans. Power Electron., vol. 19, no. 5, pp. 1205–1213, Sep. 2004

[11] H. Saad, S. Dennetière, Mahseredjian, et al, "Modular Multilevel Converter Models for Electromagnetic Transients," *IEEE Transactions on Power Delivery,* vol. 29, no. 3, pp. 1481-1489, June 2014

[12] Cigré, W.G. B4-57." Guide for the Development of Models for HVDC Converters in a HVDC Grid" Cigré Technical Brochure 604, Dec. 2014

[13] J. He and Y. W. Li, "Analysis, Design, and Implementation of Virtual Impedance for Power Electronics Interfaced Distributed Generation," in IEEE Transactions on Industry Applications, vol. 47, no. 6, pp. 2525-2538, Nov.-Dec. 2011

[14] H. Saad, S. Dennetiere, P. Rault, " AC Fault dynamic studies of islanded grid including HVDC links operating in VF-control ", in 14th IET International Conference on AC and DC Power Transmission, 2019

[15] H. Saad, A. Schwob and Y. Vernay, "Study of Resonance Issues Between HVDC Link and Power System Components Using EMT Simulations," 2018 Power Systems Computation Conference (PSCC), Dublin, 2018, pp. 1-8

Modelling and experimental validation of a laboratory-scaled HVDC cable emulator tested in an MMC-based platform

Enric Sánchez-Sánchez[*], Adrià Junyent-Ferré[†], Eduardo Prieto-Araujo[*],
Oriol Gomis-Bellmunt[*] and Tim Green[†]
[*]CITCEA-UPC, [†]Imperial College London
[*]Av. Diagonal, 647 (2nd floor), Barcelona, Spain
Email: enric.sanchez.sanchez@upc.edu - Phone: (+34) 93 401 77 96

Acknowledgments

This work was funded by the EPSRC Centre for Power Electronics Researcher Exchange Scheme, the FI-AGAUR Research Fellowship Program, the FEDER/Ministerio de Ciencia, Innovación y Universidades –Agencia Estatal de Investigación (RTI2018-095429-B-I00), and the Serra Húnter and ICREA Academia programs. The work was also supported by the European Regional Development Fund (ERDF).

Keywords

≪HVDC≫, ≪DC-cable≫, ≪Power transmission≫, ≪Converter control≫.

Abstract

Typical Modular Multilevel Converter (MMC)-based HVDC systems don't include capacitors directly connected to the DC-side in parallel with the HVDC lines. The dynamics of the HVDC line voltage depend on the equivalent capacitance of the HVDC system, which in turn depends on the length of the lines. Recent work in the literature has shown that this can become problematic for HVDC line voltage control in the case of short links. This paper describes the operation of a laboratory-scaled MMC-based HVDC short link platform devised to study this phenomenon. Special emphasis is put on the design and the scaling of a frequency-dependent cable emulator. Experimental results are compared with a benchmark high-voltage simulated model.

Introduction

High-Voltage Direct Current (HVDC) is a key technology for the transmission of electrical power in applications such as offshore wind power transmission and the interconnection of large power systems over long distances. The MMC proposed in [1] is the prevalent technology due to its scalability, controllability and low harmonic distortion. Contrary to two-level VSCs, the energy stored in the MMC is distributed among the capacitors in the submodules inside the converter, and typically there are no additional capacitors directly connected to the DC side [2] in parallel with the HVDC lines. When a compensated modulation is used [3] (ie the measured arm voltages are used to calculate the modulation insertion indexes), it becomes a challenge for the converters to control the DC voltage in short- and medium-length HVDC links. This is due to the low equivalent capacitance, which makes the dynamics ot the voltage be very fast and it can cause stability problems [4].

A crucial aspect is the way the cable is modelled in HVDC system studies. The so-called Wideband Model or Universal Line Model (ULM) [5] is the most commonly accepted model for Electromagnetic Transient (EMT) studies. This model considers the frequency dependency of the cable parameters and it is widely used in computer simulation. However, it is not clear how to physically realise it as a lab-scale model for a low-voltage expeerimental platform. Typically, π-sections have been used in previous works [6]. As the π-section model does not represent the real cable accurately [7], other studies have proposed the use of a scaled real cable with the same voltage drop and wave propagation characteristics as the real high-voltage cable [8].

As discussed in [7], an approximate linear mathematical model with multiple parallel branches can be obtained using the cable data and vector fitting techniques. This model uses time-invariant concentrated parameters and it can be used for small-signal studies. An experimental realisation of a scaled HVDC cable model inspired on this idea is proposed in this paper. Different possibilities are discussed considering the feasibility in terms of volume and cost of the cable model. The validation of the experimental model is performed using a decoupled DC voltage control with the *cross* structure presented in [4]. The tests use the cable model connected between a low-voltage 15 kVA MMC and a controllable DC source.

System description

The system under study consists of a point-to-point MMC-based HVDC link. One of the MMCs controls the DC voltage (master) while the other controls the power transferred by the link (slave). By using the energy-based control from [9], which is a reliable approach in terms of dynamic performance, the internal energy of the converter and the energy of the DC lines can be regulated independently (see Fig. 1a). The first one depends on the capacitance installed in the submodules of the converter, which is linked to the power rating of the converter. The second one depends on the equivalent capacitance of the DC network, which is proportional to the length of the HVDC lines. While the internal MMC capacitance is relatively big, the DC side capacitance can be very small in short and medium length point-to-point links.

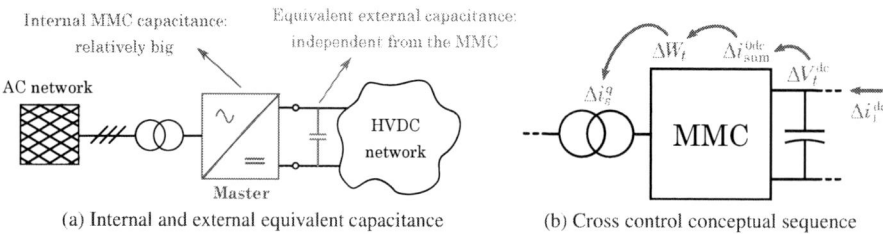

(a) Internal and external equivalent capacitance   (b) Cross control conceptual sequence

Fig. 1: MMC equivalent capacitances and cross control sequence.

Different control structures for a master MMC exist in the literature. In this case, *cross* control has been used: DC voltage (V_t^{dc}) controls DC current (i^{dc}), and total energy (W_t) controls AC current (i_s). This approach provides decoupling between AC and DC sides, as the control sequence affect the different variables in a cascaded way (see Fig. 1b), yielding an appropriate dynamic performance.

Cable model

As mentioned in the Introduction, HVDC cables present frequency-dependent dynamics. Most EMT simulation software packages use the well-known ULM or Wideband model to replicate this effect in fault transient studies. The complexity of this model makes analytical derivations for system-level studies impractical in many cases. A concentrated-parameter linear model to enable small-signal studies would be preferable in many situations; however, wrong fitting of the approximate linear model could lead to wrong conclusions.

Linear models for small-signal analysis

Many studies carried out in the past used the well-known π-section line model to represent HVDC cables in transient studies. For greater accuracy, these used multiple cascaded π sections. However, the work in [7] showed that this model may present unrealistic resonances with low damping. An improvement of the π-section was proposed in [10]. This model improved on the accuracy of the conventional π-section model, but still showed concerning discrepancies with the frequency-dependent model. Consequently, a frequency-dependent model was presented in [7] (Fig. 2a), motivated by the necessity of including the cable model in small-signal studies of power systems with long HVDC lines. This model uses several parallel inductor branches. The number of branches, and also the number of sections (Fig. 2a), determines the order and the accuracy of the model. The method to calculate the parameters of the branches is briefly described next.

Vector fitting fundamentals

Vector fitting [11] of linear systems consists of iteratively relocating the poles of a system until convergence is achieved. The formulation uses simple fractions, therefore avoiding ill-conditioning problems inherent to other approaches based on polynomials. Unstable poles are forced to be stable by moving them to the left hand side of the Laplace plane. Vector fitting is applicable to high order systems and wide frequency bands. The general problem formulation is given as (1), where $f(s)$ is the adjusted function, N is the order of the system, r_i are the residues, a_i are the poles, and d and h are real values. The vector fitting technique aims to match the frequency response of a real system with the frequency response of $f(s)$ (1), by adjusting the parameters r_i, a_i, d and h sequentially [11]. The first step is to obtain the transfer function of the cable model from Fig. 2a. Assuming that the shunt conductance and capacitance do not change substantially in the frequency-domain [7], the system in Fig. 2b is considered, where $l_{z,i}$ and $r_{z,i}$ are the inductance and resistance of branch i, respectively. The equivalent estimated admittance $\hat{Y}(s)$ is derived as (2). Note that $\hat{Y}(s)$ is equivalent to (1) dropping d and h.

$$f(s) = \sum_{i=1}^{N} \frac{r_i}{s - a_i} + d + sh \qquad (1) \qquad \hat{Y}(s) = \frac{I(s)}{V(s)} = \sum_{i=1}^{N} \frac{1}{l_{z,i}s - (-r_{z,i})} \qquad (2)$$

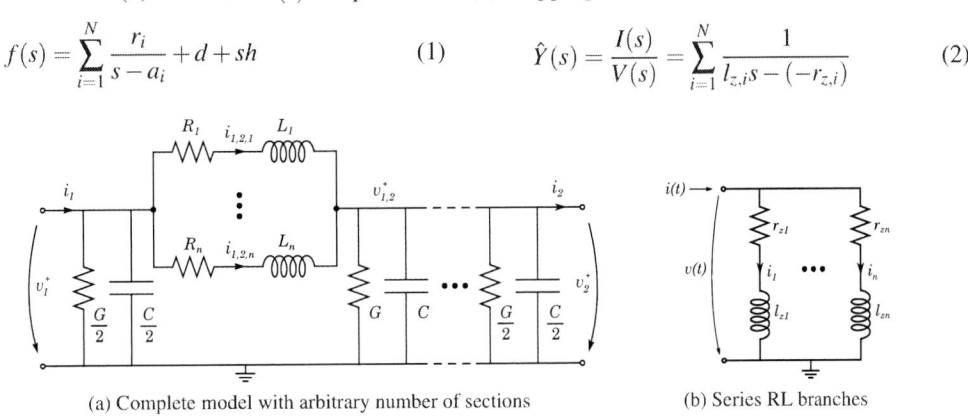

(a) Complete model with arbitrary number of sections (b) Series RL branches

Fig. 2: Parallel branches cable model from [7].

The 320 kV XLPE insulated cable from [12] is used as the reference. The model parameters consider the characteristics of the core conductor, a lead sheath and the steel armor. The vector fitting algorithm iteratively adjusts the parameters from (2) in order to fit the frequency response of the ULM cable model. The algorithm offers different configuration options (number of iterations, initial poles, and frequency weights). In this case, initial real poles logarithmically distributed [11], 10 iterations, and the same weights have been set. The function `vectfit3` [13] returns the poles of $\hat{Y}(s)$, the associated state-space function, and the function `ss2pr` [13] transforms it into a pole-residue model. The branch resistances and inductances can be derived as (3), where Res is the residue and λ the pole associated to branch i.

$$l_{z,i} = \frac{1}{\text{Res}_{z,i}}; \quad r_{z,i} = -\lambda_{z,i} l_{z,i}, \qquad (3)$$

Cable emulator based on the parallel branches model

While prior work in the literature has discussed the trade-offs of replicating the relevant dynamics of 2L-VSCs and MMCs in small-scale platforms, very little has been published about replicating the complex dynamics of HVDC cables in small-scale platforms for converter interaction studies. The approach presented here is inspired by the work on simplified cable simulation described in the previous sections. The implementation is based on the physical realisation of the equivalent circuit models obtained by these methods. The parameters of the branches are calculated using vector fitting using the cable data from [12].

In order to validate the dynamics of the cable model, an open-circuit test of the different models is performed. The effect of changing the number of sections and the number of branches on accuracy of the model is shown in Fig. 3, where a step of 1 p.u. of voltage is applied at t = 0 s. The frequency response of the different models with different number of branches and 10 sections is presented in Fig. 4, comparing it to the ULM cable from the BestPaths project Simulink toolbox [14], for a length of 50 km. It can be seen that:

- A high number of sections makes the model replicate the travelling wave effect and the main frequency components, but causes a high frequency oscillation to appear.
- A high number of branches mitigates the high frequency oscillatory terms.

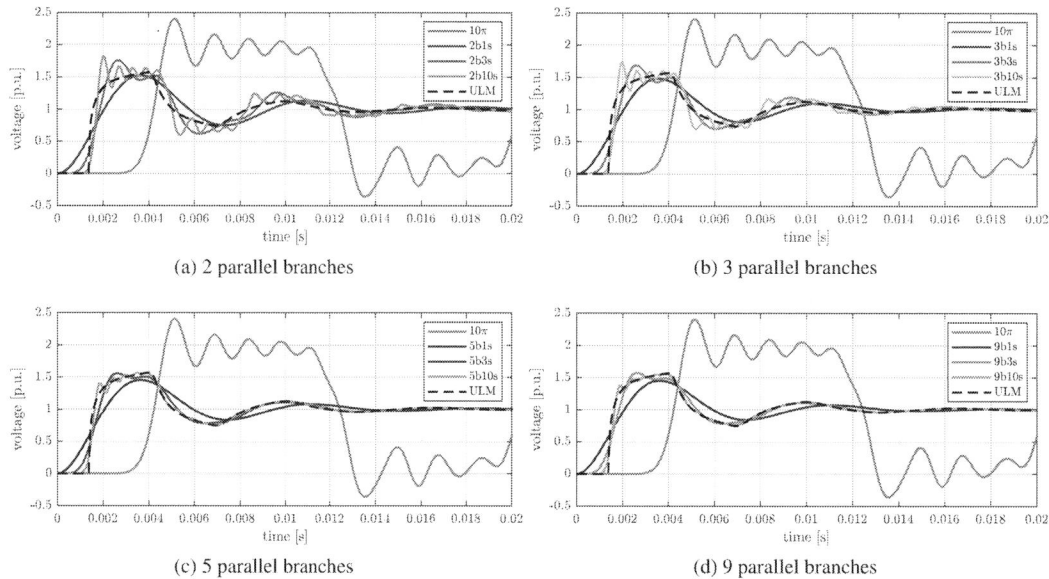

(a) 2 parallel branches

(b) 3 parallel branches

(c) 5 parallel branches

(d) 9 parallel branches

Fig. 3: Cable models transient responses - Voltage step of 1 p.u. in one side, open circuit in the other.

Therefore, ideally it is desirable to include as many sections and branches as possible. However, this is obviously not feasible from an experimental platform point of view, due to hardware volume and cost constraints. A compromise solution is needed as discussed in the next sections.

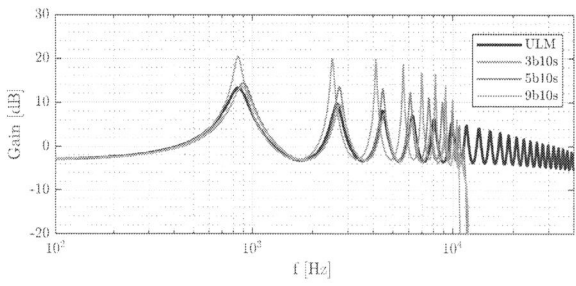

Fig. 4: Cable models frequency response –Different number of branches, 10 sections.

Experimental platform

This section presents the details of a low-voltage experimental platform, specially emphasising on the cable emulator design which aims to emulate the real high-voltage cable dynamics for small-signal stability studies. Testing and validating HVDC systems in laboratory scaled platforms as opposed to using computer simulations has been a controversial topic of discussion among the academic and the power industry communities over the last years. Scaled platforms inevitably miss some details (eg lower efficiency than the real system) [15]. Special attention has to be paid in identifying the limitations of the model, in order for the conclusions drawn to be extrapolated back to the real full-scale system.

The different systems used in the present study are summarised in Fig. 5. Firstly, the original detailed simulation model of an HVDC system is scaled-down. Secondly, a highly detailed simulation model of the experimental platform is used as a previous validation step before carrying out the actual experiments as the third and last step of the process. Simulations are carried out with MATLAB Simulink, and the ULM cable model from [14] is used in the simulated real system, as mentioned before.

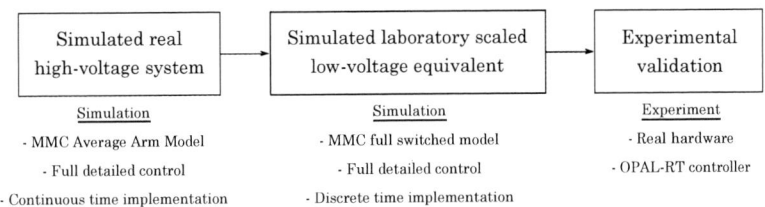

Fig. 5: Overview of the different systems involved in the results validation.

Bearing that in mind, the present study aims to maximise the validity of the results using an accurate MMC low-voltage platform and an improved cable emulator. The test bench (Fig. 6) consists of a master MMC that controls the DC voltage, a DC source that emulates the slave MMC, and a cable emulator between them. In order to have a balanced and harmonic-free AC grid voltage, a programmable AC voltage source is used to emulate a stiff AC system to which the MMC converter is connected.

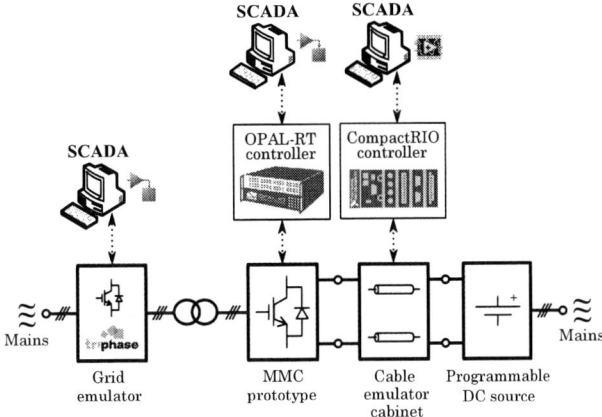

Fig. 6: Control equipment and high level structure of the experimental platform.

MMC prototype

The parameters of the real high-voltage MMC and the scaled low-voltage MMC are summarised in Table I. From the rated DC voltage and the rated power values, the base impedance (DC and AC) of the real and the scaled systems can be calculated as $Z_{dc} = V_N^{dc^2}/S_N$ and $Z_{ac} = U_{ac}^2/S_N$. Note that the energy-to-power ratio of the MMC, H_C, is practically the same for both systems, and that the arm reactor impedance in per-unit is also very similar. This shows a good similarity between the two systems in relative terms. On the other hand, the transformer leakage inductance of the laboratory setup is significantly lower than it would be expected in the large scale system (0.01 pu against 0.2 pu). This is due to the transformer physical characteristics. Although it would be desirable that this value was closer to that of the real system, the inner controllers of the MMC can be designed to provide a similar transient response and compensate for this difference.

Table I: Parameters of the MMC scaled prototype

Parameter	Symbol	Real syst.	Scaled prot.
Rated apparent power	S_N	500 MVA	15 kVA
Rated DC voltage	V_N^{dc}	640 kV	1500 V
Rated line AC voltage (primary)	$U_{ac,p}$	400 kV	340 V
Rated line AC voltage (secondary)	$U_{ac,s}$	320 kV	816 V
Submodules per arm	N_{SM}	400	10
Submodules capacitance	C_{SM}	8 mF	1,1 mF
MMC energy-to-power ratio	H_c	49.2 ms	49.5 ms
Base DC impedance	Z_{dc}	819.2 Ω	150 Ω
Base AC impedance	Z_{ac}	204.8 Ω	40.56 Ω
Arm resistance	R_{arm}	0.0100 p.u.	0.0099 p.u.
Arm inductance	L_{arm}	0.2000 p.u.	0.1743 p.u.
Transf. leakage resistance	R_{arm}	0.0100 p.u.	0.0364 p.u.
Transf. leakage inductance	L_{arm}	0.2000 p.u.	0.0113 p.u.

The cabinets of the AC grid emulator (a 90 kVA converter), the DC power source that is used to emulate the slave MMC (a 15 kVA DC voltage supply), the master MMC prototype (an in-house design from Imperial College London [16]) and the cable emulator are shown in Fig. 7.

(a) AC grid emulator   (b) Master MMC   (c) DC source   (d) Cable emulator

Fig. 7: Pictures of different cabinets of the experimental platform.

HVDC cable emulator

The DC impedance ratio between the real and the scaled systems is used to properly scale the series and shunt parameters of the cable, as indicated in (4). Thus, the per-unit values are equivalent in both the real system and the laboratory prototype, using the respective base impedances.

$$R_{dc}^{lab} = R_{dc}^{real} \frac{Z_{dc}^{lab}}{Z_{dc}^{real}}, \quad L_{dc}^{lab} = L_{dc}^{real} \frac{Z_{dc}^{lab}}{Z_{dc}^{real}}, \quad G_{dc}^{lab} = G_{dc}^{real} \frac{Z_{dc}^{real}}{Z_{dc}^{lab}}, \quad C_{dc}^{lab} = C_{dc}^{real} \frac{Z_{dc}^{real}}{Z_{dc}^{lab}} \quad (4)$$

The design of the parallel branches of the new sections is based on adapting an existing design based on π-sections built for a different purpose. The main difference compared to a design done from scratch is that some of the initial specifications (eg length of emulated cable per model section) were adapted in order to reuse some of the existing material aiming at a reasonable cost and similar physical footprint.

The procedure to determine the number of branches of the cable section is presented in Fig. 8. First, the state-space model is obtained using `vectfit3`, which fits the frequency response of a generic model to a specific frequency response, in this case from the ULM cable data. The algorithm iterates and tries to obtain the best solution based on the root-mean-square error. After fitting the state-space model, the pole-residue model is obtained using `ss2pr`, which gives the branch parameters ($r_{z,i}$ and $l_{z,i}$, $\forall i$). Afterwards, the high-voltage system parameters are scaled-down to the prototype system. The process ends at this stage if the obtained $l_{z,i}$ are feasible in terms of volume and cost. Otherwise, the number of branches is reduced by 1, and the process is started again. The reference value for $l_{z,i}$ is the original series inductance of the existing π-section (5.125 mH), and the following high-level objectives are pursued:

- Length range: each upgraded section should correspond to a few kilometres of emulated cable, in order to reach a relatively long distance when combining all sections in the cabinet.
- Section size: the volume and the weight of an upgraded section must be similar to the original one, in order to avoid making major changes to the cable emulator cabinet.

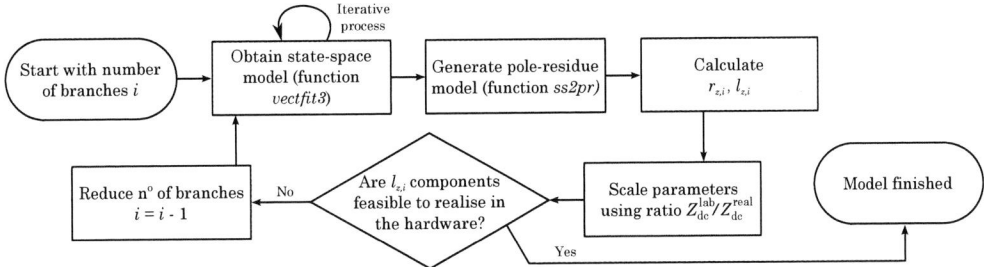

Fig. 8: Flow chart of the cable section design procedure.

The largest inductor of the upgraded section is the largest of the original π-section. It is desirable to use it to model a few kilometres and to keep all other inductors smaller in order to keep the physical footprint small. The parameters obtained for different number of branches (N_b) are shown in Table II. The shunt parameters (independent from the number of branches) are $c = 0.1616\ \mu F/km$ and $g = 0.1015\ \mu S/km$ (real system), and $c = 0.8824 \mu F/km$ and $g = 0.5543 \mu S/km$ (scaled prototype).

Table II: Parameters of the MMC scaled prototype (selected option in bold font)

	N_b	$l_{z,1}$	$l_{z,2}$	$l_{z,3}$	$l_{z,4}$	$l_{z,5}$	$r_{z,1}$	$r_{z,2}$	$r_{z,3}$	$r_{z,4}$	$r_{z,5}$
					Parallel branches						
Real system	5	0.689	0.321	10.2	3.28	30.9	1889	135	511	21	81
	3	0.263	6.51	3.28	-	-	132	242	17	-	-
	2	0.272	2.9	-	-	-	114	16	-	-	-
Scaled prototype	5	0.1262	0.0588	1.8660	0.5999	5,6530	346	25	94	4	15
	3	0.0482	1.1918	0.6009	-	-	24	44	3	-	-
	2	**0.0498**	**0.5303**	-	-	-	**21**	**3**	-	-	-
		[mH/km]					[mΩ/km]				

The largest inductance in the 5-branches model is 5.653 mH. This is approximately the same as the original value and it approximately corresponds to 1 km of cable. This is sufficient reason to not use this model, as it constraints the maximum distance the emulator can reach with the existing number of sections. Regarding the 3-branches model, the largest inductance is 1.1918 mH, and the second largest is 0.6009 mH. This model is more feasible than the previous one, but still requires relatively large additional components, with the corresponding increase in cost and volume. As the difference in accuracy between a 3-branches and a 2-branches model is relatively small (see Fig. 3), the 2-branches option is explored. In this case, the largest inductance is 0.5303 mH. This means that an inductance of 5.125 mH corresponds to 9.66 km of cable, which is reasonable for scalability purposes. Consequently, the the 2-branches model is chosen for the implementation. The rest of the parameters are summarised in Table III.

A few further adjustments are made in order to enable the physical implementation of the circuit. Firstly, parameters are adapted to correspond to the those of the closest commercial components available. In the

Table III: HVDC cable prototype section parameters

Parameter	π-section	New section (ideal)	New section (avail.)	Units
L_1	-	0.481	0.470	[mH/section]
L_2	5.125	5.125	5.125	[mH/section]
R_1	-	203	237	[mH/section]
R_2	13	29	80	[mH/section]
C	3.600	8.524	8.660	[mH/section]
G	-	5.355	-	[mH/section]

case of R_2 this leads to a greater error because its ideal value is lower than the effective series resistance of L_2. Consequently, no discrete resistor is added in series with L_2 in the second branch. Also, G is not included in the circuit given that it has little effect on the dynamics.

Regarding the current rating of the components, it is worth noting that the cable emulator is designed to carry out small-signal experiments; therefore all components are sized to withstand the rated current of the cable with an additional safety margin but they are not suitable to carry out realistic short-circuit studies as the transient current would exceed the saturation point of the inductors.

The original cable emulator cabinet (Fig. 7d) included two full symmetrical monopoles built for a four-terminal DC system. Each of the four cables had 11 π-sections (Fig. 9). At this stage, only two of the four cables were upgraded to the new design because this was the minimum required in order to build a point-to-point DC link. Therefore, the cabinet contains a total of 22 upgraded parallel branches sections, allowing for a maximum length of around 106 km of equivalent cable. The other 22 sections remain as π-sections and are not used in the present study.

Fig. 9: Detailed cable section, corresponding to 9.66 km of cable.

Dynamic performance of the system

The point-to-point system is tested under a DC current disturbance injection, from 0 to rated current in 500 ms. The length of the link is set to be of 48.3 km (ie 5 sections in each pole). The simulation of the real system, using the ULM cable and the Average Arm Model (AAM) of the master MMC, is presented in Fig. 10 (DC current (i_{dc}), DC voltage (v_{dc}), MMC energy (W_t) and submodules voltages (v_{SM})). The slave converter is modelled as a DC source with a first order response of 500 ms. The results of the experimental test in the scaled platform are shown in Fig. 11. The disturbance is applied from the DC source operating in current control mode. Both the simulated real system and the experimental scaled prototype transient performance exhibit similar behaviour. The dynamics are similar in both systems. However, the main difference is found in the DC voltage ripple. This is due to fact that an MMC average model (without switching states) is used in the simulation. A modification of the DC voltage control loop was found to be required to ensure stable operation of the system as described next.

The Bode gain plots of voltage against current for the laboratory scaled cable emulator are presented in Fig. 12. The plots were obtained from the mathematical model of the circuits and they are shown for different number of cable sections (and consequently different cable lengths). The number of resonant peaks is related to the number of cable sections as expected. Peaks appear in the range between 0.7 kHz and 5 kHz regardless of the number of sections under test (ie 1, 3, 5 and 11). This frequency range

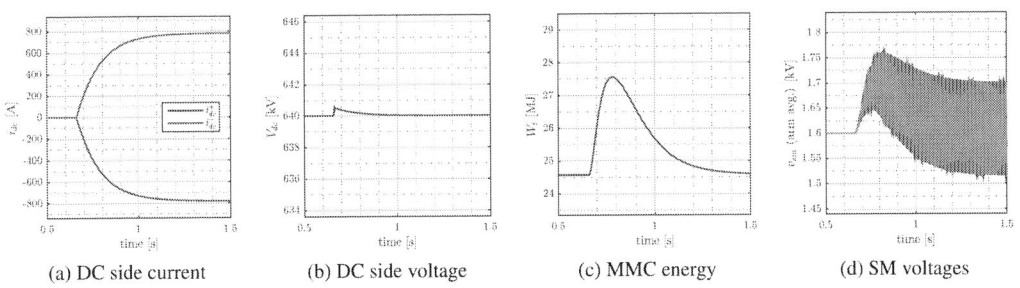

(a) DC side current   (b) DC side voltage   (c) MMC energy   (d) SM voltages

Fig. 10: Dynamic response of the real system –Simulation results.

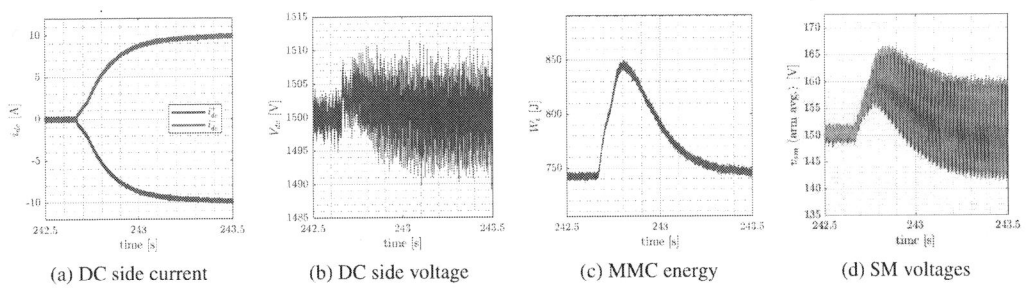

(a) DC side current   (b) DC side voltage   (c) MMC energy   (d) SM voltages

Fig. 11: Dynamic response of the scaled prototype –Experimental results.

is close to the effective switching frequency of the converter and the update frequency of its controller (10 kHz). Therefore, these peaks may be excited by the action of the MMC. Even though 5 kHz is apparently above the bandwidth of the DC voltage controller, the DC voltage measurement is found to cause unstable behaviour in both simulations and experiments. Therefore, a low-pass filter (LPF) is added to the DC voltage measurement to further reduce the gain at the resonant frequencies. This is found to cause negligible effect on the reference tracking capability of the controller. The equation of the LPF is given by (5), where T_s is the sampling frequency (which in this case is of 100 μs) and T_{LPF} is the time constant of the filter, which was set to be of 1 ms. This effectively mitigates the problem described above. The transient response upon by-passing the LPF at time $t = 1.2$ is shown in Fig. 13 for illustrative purposes. Please note that the operation before the disconnection of the filter is perfectly stable.

$$G_{\mathrm{LPF}}(s) = (T_s/T_{\mathrm{LPF}})/(z + T_s/T_{\mathrm{LPF}} - 1) \tag{5}$$

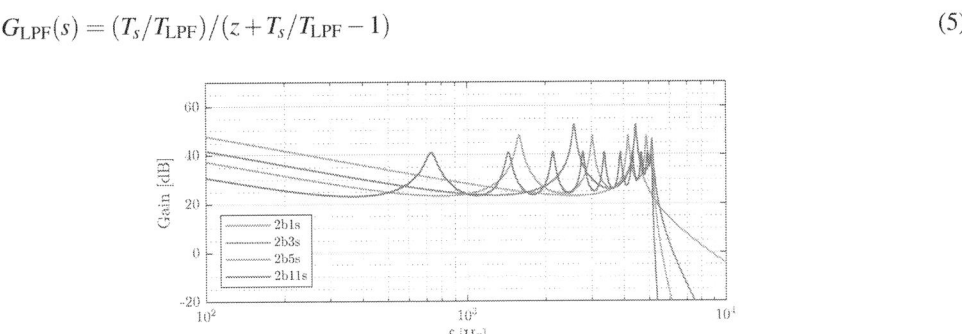

Fig. 12: Cable emulator model Bode gain response (voltage vs. current) –Different number of sections.

Conclusion

The dynamics of an HVDC cable can be modelled using a π-section model with multiple parallel inductor branches. This leads to a more realistic representation than the conventional π-section model. However,

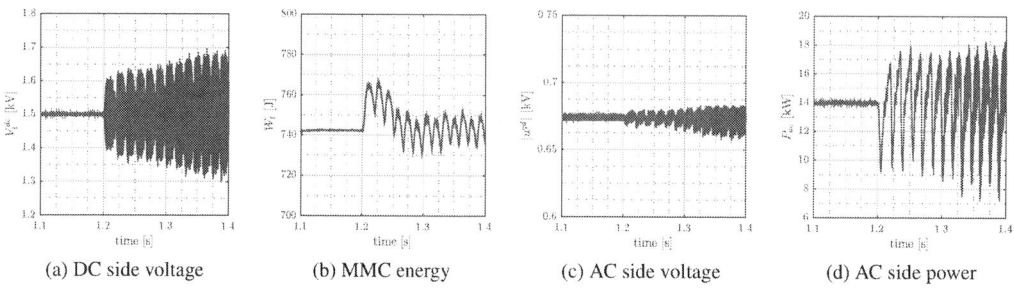

| (a) DC side voltage | (b) MMC energy | (c) AC side voltage | (d) AC side power |

Fig. 13: Experimental platform detailed simulation model –LPF removed at t=1.2 s.

a great number of components may be required to perfectly match the parallel branches model with the frequency-dependent model. While this is less important in a computer simulation, it becomes impractical when implementing the circuit in a laboratory-scaled platform. Our analysis has found that using only two parallel branches leads to a significant improvement over the classic π-section model. However, a series of low-damping peaks may appear in the frequency response of the cable model. These may be excited by the action of the MMC, as they are relatively close to the effective switching frequency of the converter. This can cause instability, which may be prevented by adequate filtering of the DC voltage measurement in the feedback path. The bandwidth of this filter is greater than that of the DC voltage regulator, therefore it may not interfere with the relevant dynamics of the system.

References

[1] Lesnicar, A., and Marquardt, R.: "An innovative modular multilevel converter topology suitable for a wide power range", IEEE Bologna Power Tech Conference, 2003.

[2] Beerten, J., Diaz, G. B., D'Arco, S., and Suul, J. A.: "Comparison of small-signal dynamics in MMC and two-level VSC HVDC transmission schemes", IEEE Int. Energy Conf. (ENERGYCON), 2016.

[3] Bergna-Diaz, G., Suul, J. A., and D'Arco, S.: "Energy-Based State-Space Representation of Modular Multi-level Converters with a Constant Equilibrium Point in Steady-State Operation", IEEE Transactions on Power Electronics, 33(6), 4832–4851, 2018.

[4] Sánchez-Sánchez, E., et al: "Analysis of MMC Energy-based Control Structures for VSC-HVDC Links", IEEE Journal of Emerging and Selected Topics in Power Electronics, 6(3), 1065–1076, 2018.

[5] Morched, A., et al: "A universal model for accurate calculation of electromagnetic transients on overhead lines and underground cables", IEEE Transactions on Power Delivery, 14(3), 1032–1038, 1999.

[6] Egea-Alvarez, A., Bianchi, F., Junyent-Ferre, A., Gross, G., and Gomis-Bellmunt, O.: "Voltage control of multiterminal VSC-HVDC transmission systems for offshore wind power plants: Design and implementation in a scaled platform", IEEE Transactions on Industrial Electronics, 60(6), 2381–2391, 2013.

[7] Beerten, J., D'Arco, S., and Suul, J. A.: "Frequency-dependent cable modelling for small-signal stability analysis of VSC-HVDC systems", IET Generation, Transmission and Distribution, 10(6), 1370–1381, 2016.

[8] Amamra, S. A., Colas, F., Guillaud, X., Rault, P., and Nguefeu, S.: "Laboratory Demonstration of a Multi-terminal VSC-HVDC Power Grid", IEEE Transactions on Power Delivery, 32(5), 2339–2349, 2017.

[9] Prieto-Araujo, E., Junyent-Ferré, A., Collados-Rodríguez, C., Clariana-Colet, G., and Gomis-Bellmunt, O.: "Control design of Modular Multilevel Converters in normal and AC fault conditions for HVDC grids", Electric Power Systems Research, 152, 424–437, 2017.

[10] Akkari, S., Prieto-Araujo, E., Dai, J., Gomis-Bellmunt, O., and Guillaud, X.: "Impact of the DC cable models on the SVD analysis of a Multi-Terminal HVDC system", 19th Pow. Syst. Comp. Conf., 2016.

[11] Gustavsen, B., and Semlyen, A.: "Rational approximation of frequency domain responses by vector fitting", Advances in Mathematics, 14(3), 1052–1061, 1999.

[12] Leterme, W., et al: "A new HVDC grid test system for HVDC grid dynamics and protection studies in EMT-type software", 11th IET International Conference on AC and DC Power Transmission, 2015.

[13] Gustavsen, B.: "The Vector Fitting Website", https://www.sintef.no/projectweb/vectorfitting/

[14] Ugalde-Loo, C. E. et al: "Open access simulation toolbox for the grid connection of offshore wind farms using multi-terminal HVDC networks", 13th IET Int. Conf. on AC and DC Power Transmission, 2017.

[15] Cheah-Mane, M., Adeuyi, O. D., Liang, J., and Jenkins, N.: "A scaling method for a multi-terminal DC experimental test rig", 17th Eur. Conf. on Power Electronics and Appl., EPE-ECCE Europe, 1–9, 2015.

[16] Clemow, P., Judge, P., Chaffey, G., Merlin, M., Luth, T., and Green, T. C.: "Lab-scale experimental multilevel modular HVDC converter with temperature controlled cells", 16th Eur. Conf. on Power Electronics and Appl., EPE-ECCE Europe, 1–10, 2014.

Daisy Chain PN Cell for Multilevel Converter using GaN for High Power Density

Faheem Ahmad*, Asger Bjørn Jørgensen, Szymon Michal Beczkowski, Stig Munk-Nielsen
Department of Energy Technology, Aalborg University
Pontoppidanstræde 101
9220 Aalborg, Denmark
*Email: faah@et.aau.dk
URL: http://www.et.aau.dk

Keywords

≪Multilevel converters≫, ≪Modulation strategy≫, ≪Gallium Nitride (GaN)≫, ≪High power density systems≫.

Abstract

Power semiconductor devices are achieving high switching speed and high breakdown voltage. This improves inverter performance. But, as inverter improves, further challenge of dv/dt noise is generated that needs to be tackled by filter stage. Multilevel inverters can solve this challenge. But there are implementation complexity associated with multilevel topologies like requirement of multiple isolated DC source, complicated charging algorithm, dedicated sensing hardware. This paper presents a switch capacitor type converter topology enabling a DC-AC three level output. Joining multiple iterations of topology in daisy-chained configuration, the converter can achieve voltage gain with multilevel waveform. Requirement of a single DC supply, with inherent charge balancing capability on capacitor, the topology is well suited for low voltage renewable sources like photovoltaic (PV) or fuel cell. The paper presents design of high frequency commutation loop. Utilizing finite element analysis (FEA) tool ANSYS Electronics Desktop (Q3D) to extract PCB parasitics helps in eliminating prototyping cost and time. Designed inverter is then subjected to continuous load test where it shows improving performance with increasing inductive load.

Introduction

Wide bandgap (WBG) based semiconductor technology enables higher switch transition speed, leading to decreased losses in semiconductor devices. To take advantage of higher switching speed, the trend has been towards increasing switching frequency as it reduces output filter requirement. But a fast switching semiconductor device in itself is not going to be enough to achieve high density as it will be limited by passive components and cooling system [1]. In drives application high dv/dt stress on output voltage can lead to motor bearing current and thus reduced lifespan of drive machines [2]. Also, increasing switching frequency will not continuously reduce output filter volume for grid connected inverters [3]. Focus needs to be directed towards different topological solutions as well. For example multilevel inverters can significantly reduce filter volume requirement without pushing dv/dt. In 80s fundamental works were published on multilevel converters [4]. Then in 90s and early 2000 saw the emergence of three most dominant multilevel topology, flying capacitor (FC), neutral point clamped (NPC), and modular multilevel converters (M^2LC) [5], [8]. Still, multilevel converter implementation is concentrated in high power, high voltage applications like high voltage DC (HVDC), medium voltage (MV) motor drives, and offshore wind turbine [6], [7]. To an extent this can be attributed to requirement of complex charge balancing algorithms alongwith additional voltage sensing and digital signal processor (DSP) per sub-module to maintain capacitor charge in multilevel topologies [9]. Furthermore, aforementioned multilevel topologies cannot be directly interfaced to renewable sources like photovoltaic (PV) panel, fuel cell or battery storage applications. An intermediate boost converter is required to boost up low voltage supplied by renewable sources. New versions of multilevel topology are being introduced like modular multilevel series parallel converter (MMSPC) that can interface with low voltage supply but the charge balancing on each sub-module is similar to M²LC [10]. Work done in this paper is based on switched capacitor (SC) topology. For a long time switched capacitor topologies have been used for low power low voltage, regulated DC supply such as in CMOS integrated applications [13]-[14]. Due to high power density, SC have started gaining popularity in power electronics industry [15].

In this paper, a 5-level DC/AC multilevel inverter based on a topology called PN cell is presented. The topology, and its multiple variants is first introduced in [11]. In a recent publication [12], discussion is presented regarding working principle of PN cell. The publication then delves into converter dynamics by developing an averaged model of eight switch variant of the topology. The publication then proceeds with experimental demonstration of a two cell configuration.

The properties of a PN cell topology based multilevel inverter are,

1. Directly interfaces with low voltage DC supply like PV, fuel cell, battery storage systems.
2. Achieves high voltage gain without the use of magnetics.
3. Does not require additional charge balancing algorithm.
4. Produces multilevel waveform with $2n+1$ discrete levels, where n is number of cells.

To illustrate the properties, focus of this paper will be on development of 5-level inverter. The inverter consists of two cell configuration with an input voltage of 40V. Discussion is presented on the importance of reducing resistance of capacitor charge path. In next section three variant of multilevel operation is shown as the topology is capable of producing different modulation schemes. To address the challenge of multiple power supplies for gate drivers and their dependence on cell capacitor, this paper contributes with a novel startup scheme. The startup scheme utilizes eHEMT's reverse conduction for charging the cell capacitor from OFF state. And finally, discussion on hardware development and experimental results are given.

Three level topology - PN cell

In this paper a 6-switch variant of the PN cell topology is utilized. As explained in [12], a single PN cell produces 3-level waveform using operation states termed as P, N, and 0-state. Duty cycle for P-state is d_P, and duty cycle for N-state is d_N. During 0-state (d_0) cell capacitor is connected to the cell input terminal, charging back to input voltage level. The three duty cycles are related as $d_P + d_N + d_0 = 1$. Operation of a single PN cell in these states is shown in Fig. 1.

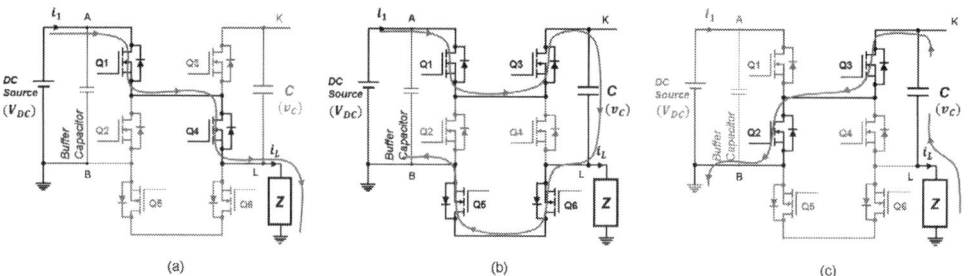

Fig. 1: Single PN cell operation in (a) P-state, (b) 0-state, and (c) N state

Because of inherent charge balancing capability of PN cell, the capacitor (C) maintains its voltage approximately equal to DC link voltage. Therefore output AC voltage across load is expressed by (1),

$$v_{AC} = m \cdot V_{DC} \cdot \sin(\omega t) \tag{1}$$

Where m is modulation index and V_{DC} is cell input voltage. Utilizing the averaged model, [12] has explored the converter cell dynamics. It was shown that the 0-state resistance impacts converter performance. This behavior is apparent when looking at the input terminal current (i_1) in Fig. 1. Input terminal provides load current during P-state and charges capacitor (C) during 0-state. Input current (i_1), thus can be shown to consist of two parts (2).

$$i_1 = d_P \cdot i_L + \frac{(1 - d_P - d_N)(V_{DC} - v_C)}{4R_{DS(on)} + R_{PCB}} \tag{2}$$

where,

$$d_P = \begin{cases} m \cdot \sin(\omega t) & \in & [0, \pi] \\ 0 & \in & (\pi, 2\pi] \end{cases} \qquad d_N = \begin{cases} 0 & \in & [0, \pi] \\ m \cdot \sin(\omega t + \pi) & \in & (\pi, 2\pi] \end{cases} \tag{3}$$

$$i_L = m \frac{V_{DC}}{\sqrt{2}|Z|} \sin(\omega t) \in [0, 2\pi] \quad \text{where,} \quad |Z| >> R_{DS(on)} \tag{4}$$

Substituting (3) - (4) in (2) we obtain.

$$i_1 = \overbrace{m^2 \sin(\omega t) \cdot \frac{V_{DC}}{\sqrt{2}|Z|} \sin(\omega t)}^{\text{load dependent}} + \overbrace{\left(1 - m|\sin(\omega t)|\right) \frac{V_{DC} - v_C}{4R_{DS(on)} + R_{PCB}}}^{\text{capacitor charge}} \tag{5}$$

R_{PCB} is PCB trace parasitic resistance during 0-state. (5) indicates that DC input current to the inverter has two components, one is dependent directly on the load and second component is dependent on capacitor voltage difference from the input voltage. This second component is an additional load to the cells that generate loss. In order to achieve good performance, capacitor charge component needs to be minimized. As can be seen from (5), if the capacitor voltage (v_C) drops much lower than input DC link voltage, ($V_{DC} - v_C$) becomes large invoking higher input current. One key factor in maintaining capacitor voltage close to DC link voltage is RC time constant (τ), formed by $R_{DS(on)} + R_{PCB}$, and C. As the capacitor gets charged during 0-state of switching period. It is important to make sure that even at max. modulation index (m_{max}), RC time constant (τ) is smaller than 0-state time (6).

$$R_{DS(on)} + R_{PCB} < \frac{1 - m_{max}}{k \cdot f_{sw} \cdot C} \quad \text{where,} \quad k \in [3, 5] \tag{6}$$

R_{PCB} is dependent on size of PCB, which is determined by the device footprint and gate driver circuit. Gallium nitride (GaN) based devices exhibit improved performance over silicon counterparts in a smaller package. Lateral GaN eHEMT device (GS61008P) used in this paper is a 100V-90A device and has an on-state resistance of 7mΩ [16].

Modulation strategy in multicell inverter

When more than one cell is daisy chained it creates possibility to obtain multilevel waveform of $2n + 1$ levels where n is number of cells. The topology also has potential to produce standard 3-level waveform. In Fig. 2, two cell configuration in daisy chain is shown to operate in three different states marked as (I), (II), (III). These three operation states are marked accordingly on the different modulation schemes presented in Fig. 2. The flexibility to operate each cells independently gives this topology the potential to produce various modulation schemes. DC link voltage for the simulation shown is 50V.

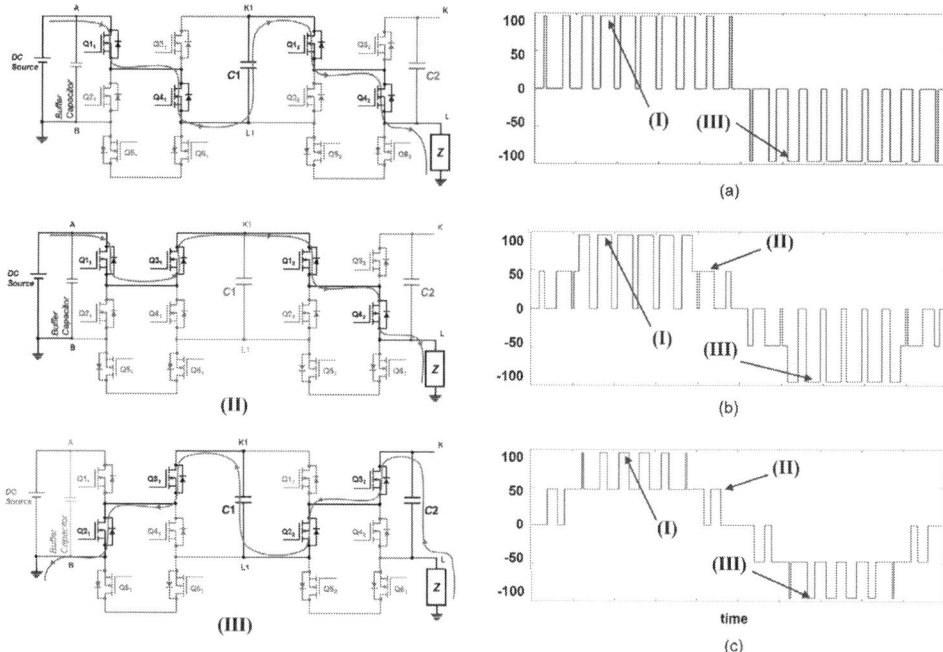

Fig. 2: Output voltage at different modulation strategy (a) 3-level waveform, (b) modified 5-level, and (c) standard 5-level output waveform

Modulation strategy 1 is achieved when all the cells are synchronized in their operation states. Therefore, the capacitors to each cell get charged and discharged together. This modulation is simpler to implement as each cell gets same signal from the controller. Also the modulation is achieved by using a single carrier (triangular) signal compared to a modulating (sinusoidal) signal. With strategy (a) the converter does not provide the advantages of multilevel waveform. Modulation strategy shown in (b), is a modification of strategy (a) where capacitor of each cell are still charged in sync, thus maintaining low ripple current on the capacitors. Finally modulation strategy (c) provides multilevel operation where the variation of voltage at each switching instant is equal to input DC link voltage. Modulation strategy in (c) is achieved using phase disposition of multiple carrier [17]. In the current two cell configuration, two carriers are used which are phase shifted by $\phi = T_{\text{SW}}/2$, where T_{SW} is switching period. Because of phase shift in operation states of the two cells, capacitor charging of the cells are not synchronized unlike the previous two schemes. This leads to higher capacitor ripple current compared to strategy (a) and (b).

Design and Optimization of Power Loop

Power PCB in this paper is designed in conjunction to SPICE simulation platform. SPICE simulation model is used as a digital twin of the actual PCB. For this purpose finite element analysis (FEA) method is employed on each iteration of PCB designed. ANSYS Electronics Desktop (Q3D) is used to extract the parasitic of copper layers which is then fed back into SPICE model to analyze design performance. Necessary changes are made depending on these simulation results [18]. Digital twin methodology accelerates hardware development process by eliminating prototyping cost and time which is a major push by industry. While using this methodology it is important to first identify high frequency commutation loops. Referring to operation states of PN cell. During positive half-cycle of fundamental frequency when inverter is modulating between P-state and 0-state, a high frequency commutation loop is observed on the output side of inverter shown in Fig. 3(a).

During negative half-cycle of fundamental frequency a similar high frequency commutation loop will emerge on the input side of PN cell. Both these loops are identical to power loop present in voltage

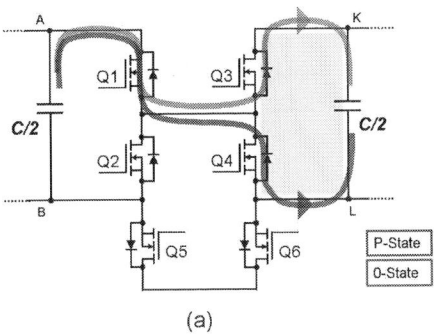
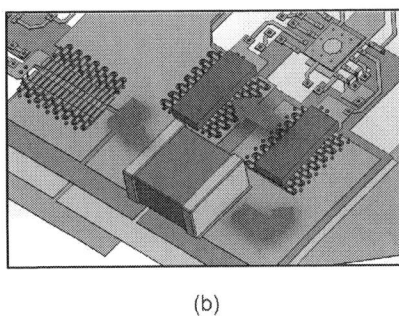

(a)   (b)

Fig. 3: (a) High frequency commutation loop on the output side of PN cell during positive half-cycle of fundamental frequency, and (b) lateral layout design for the commutation loops

source converters (VSCs) between DC-link capacitor and half-bridge. Two most conventional layout for commutation loop are considered - vertical and lateral layout [19]. For vertical layout most optimum performance is achieved when close PCB layers are utilized for return path of loop. Lateral layout is usually contained within either top or bottom PCB layer. In this paper, it was observed that employing multiple internal layers for switch node and paralleled vias results in lateral layout performing better than vertical layout design. A 3-dimensional render of lateral layout is presented in Fig. 3(b).

Using the FEA method, loop inductance for lateral layout is calculated to be 0.42nH. This value consists of the partial self-inductance of individual trace as well as their mutual inductance based on spatial arrangement. Loop inductance value calculated using FEA is then exported as matrix into SPICE simulation. For the GaN eHEMT devices SPICE model is utilized from manufacturer that includes package inductance. The device is kept at 25°C. Gate resistance of 6Ω is used for both high side and low side devices. Using a double pulse test, comparison between simulation (utilizing FEA extracted values) and experimental waveform is provided. Turn ON and turn OFF transients are shown in Fig. 4. Switching transient speed as well as voltage overshoot during turn-off instant for low side HEMT ($Q2$) matches closely between simulation and experimental observation. Although, simulation results oscillations sustain for longer compared to experimental results but results have been successful in validation design methodology utilizing digital twin. Calculating the switch transient speed and voltage overshoot, device is switched at 21A drain current. Turn-off speed of 6ns is measured with dv/dt of 6V/ns. Peak measured voltage of 53V gives an overshoot (ΔV) of 13V which is a 33% overshoot. For turn-on, a speed of 7.5ns is observed which gives a dv/dt of 4.8V/ns.

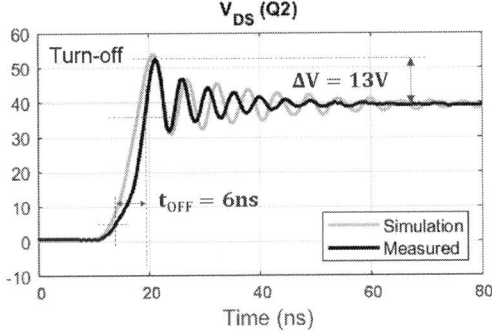
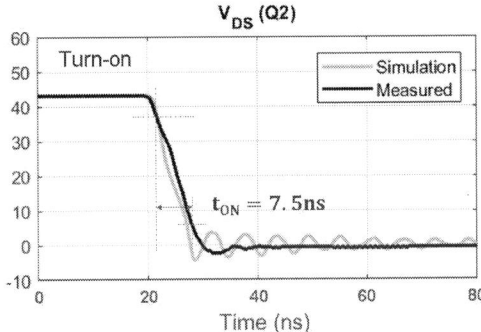

Fig. 4: Voltage surge measurement during double pulse test on low side HEMT in simulation and experimentation

Result

Designed inverter that contains two cells is shown in Fig. 5. Power board contains six switches of each cell, their drivers and output capacitor. Cell 1 has been marked in the figure. A separate PCB is attached to power board, which feeds signal to gate drivers. This second PCB is termed as Controller board. Controller board houses signal isolation circuit as well as a DC-DC circuit to power gate drivers. Although in the current iteration of hardware the controller board is designed to be in perpendicular to the power board for ease of access during debugging. The controller board can be designed to float in parallel to power board, reducing volumetric dimension of the hardware. Fig. 5 shows additional electrolytic capacitors that are added because the multilayer ceramic capacitors (MLCC) were not enough to sustain the capacitor voltage at high load.

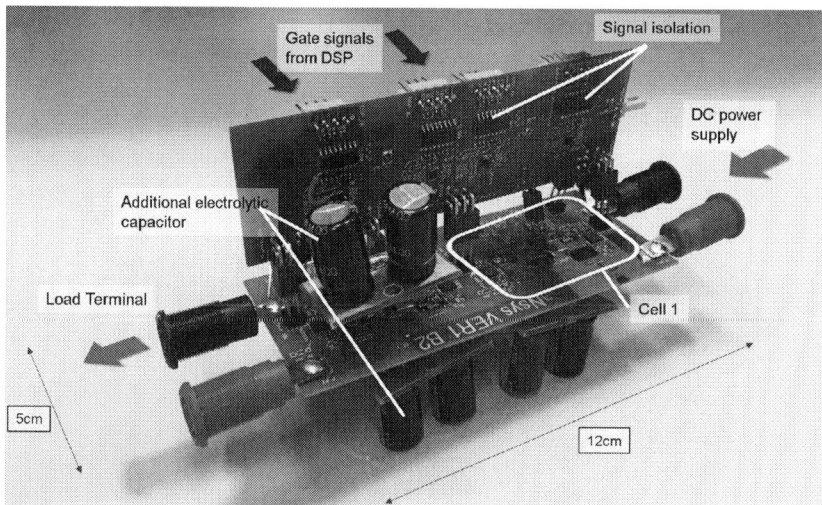

Fig. 5: Power Board and Controller Board fabricated and assembled

Startup sequence

In order to eliminate the requirement of high number of isolated DC-DC regulators or flyback transformer based design to power up gate drivers. An approach is taken where capacitors (C) is used as power source for a non-isolated DC-DC circuit that can convert 40V to regulated 5V supply to be used by gate drivers. But with this interdependent design, challenge arises to charge the capacitors (C) when starting from OFF state. In order to solve this issue, eHEMT's reverse conduction is used. In Fig. 6 a two step process is shown to charge capacitor of cell 1 and cell 2 in a sequential process. At first, when only DC link voltage is available switch $Q1_1$ and $Q5_1$ is supplied with a constant 20% duty cycle to charge up capacitor $C1$ using reverse conduction of $Q3_1$ and $Q6_1$. At this stage, capacitor $C1$ provides power to gate drivers of switch $Q1_2$ and $Q5_2$ of cell 2. And in a similar manner capacitor $C2$ is charged. This two stage concept for the two cell configuration is shown in Fig. 6.

Likewise, if there are more cells, same methodology is capable of charging up capacitors for each cell in a subsequent manner.

Inverter operation under load

At an input DC source voltage of 40V, inverter is tested under load with various power factor. With increasing inductance, power angle is raised from $7° \rightarrow 20° \rightarrow 29°$ at $\approx 20\Omega$ load impedance. DC link voltage is kept at 40V with a modulation index of 0.65 at 15kHz switching frequency. Cell capacitance $C1 = 700\mu F$ and $C2 = 350\mu F$. Input and output waveforms are shown in Fig. 7. RMS value of inverter input current, output voltage and output current for the three cases shown in Fig. 7 is given in Table I. The RMS value is calculated using RMS function in MATLAB on the acquired waveform. There is

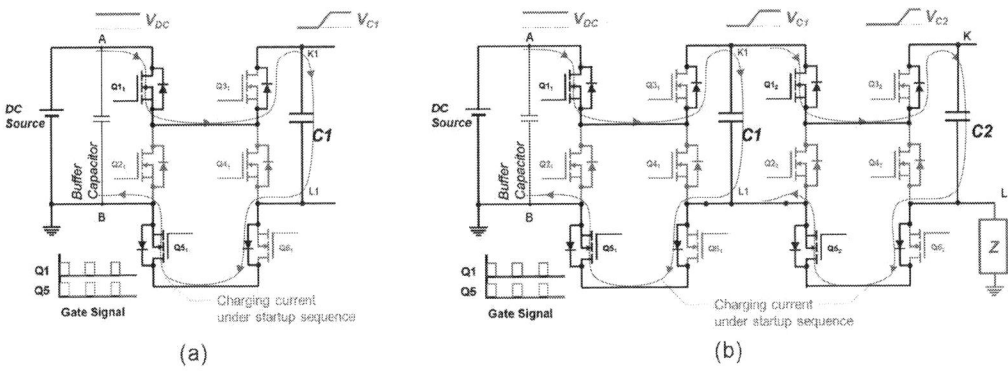

Fig. 6: Startup sequence with (a) capacitor C1 of cell 1 getting charged to DC link voltage, (b) capacitor C2 getting charged

approximately 2W of losses by the gate driver, its power supply, and signal isolation circuit at 15kHz switching frequency. Efficiency given in Table I has been adjusted for this additional loss and thus represents purely the DC to AC power conversion efficiency. The current hardware's power handling is limited by its heat dissipation capability and therefore with proper thermal design power rating can be increased.

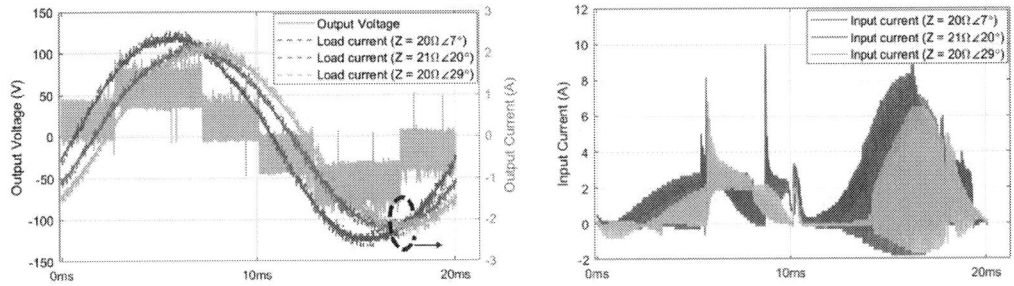

Fig. 7: (a) Inverter output voltage and load current, (b) Input current

Decreasing input current with increasing inductive load shown in Fig. 7(b), can be explained using (5). In [12], experimental demonstration with increased inductive load shows smaller capacitor voltage variation. According to (5), if capacitor voltage variation is decreased, inverter input current decreases and therefore delivers better performance.

Table I: Two cell configuration performance under increasing inductive load

Load Impedance	Input current	Output Voltage	Output Current	Efficiency
$20\Omega \angle 7°$	2.3 A	38 V	1.7 A	71.7 %
$21\Omega \angle 20°$	1.9 A	39 V	1.5 A	79.1 %
$20\Omega \angle 29°$	1.7 A	39 V	1.5 A	88.6 %

The developed hardware is also subjected to purely inductive load. In previous test results, input current during the negative half-cycle of fundamental frequency was highest. In the purely inductive load case however, input current during negative cycle of line frequency is negligible. Output waveform and input current for the purely inductive load is shown in Fig. 8.

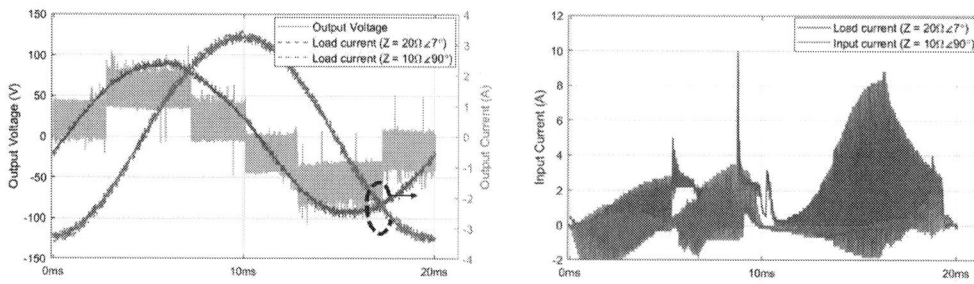

Fig. 8: (a) Inverter output voltage and load current, (b) Input current

Conclusion

This paper has presented PN cell topology based 5-level inverter. This work aimed towards interfacing directly to a low voltage supply, therefore the designed inverter is based on two cell configuration with a 40V DC input voltage. The inverter can be extended to higher number of levels by adding more cells and thus producing higher output voltage that can be interfaced with grid or drive applications. A startup scheme is proposed that utilizes reverse conduction of eHEMT device to charge cell capacitor from complete OFF state. The converter is then subjected to double pulse test. Turn-off time is measured to be 6ns with dv/dt of 6V/ns at 21A drain current. Conformation between simulation and experiment results during voltage transients helps in reducing prototyping cost and time. This digital twin methodology can be used for novel topology or module design where significant amount of literature is not available. Finally, the two cell inverter is operated under varying load conditions.

References

[1] Kolar J. W., Bortis D., and Neumayr D.: The ideal switch is not enough, Proceedings of the 2016 28th International Symposium on Power Semiconductor Devices and ICs (ISPSD).

[2] Kolar J. W., et al.: Advanced SiC/GaN 3-phase PWM inverter systems for VSD applications.

[3] Gurpinar E. and Catellazzi A.: Single-phase T-type inverter performance benchmark using Si IGBTs, SiC MOSFETs, and GaN HEMTs, IEEE Transactions on Power Electronics, vol. 31 no. 10, pp. 7148-7160, October 2016.

[4] Bhagwat P. M. and Stefanovic V. R.: Generalized structure of a multilevel PWM inverter, IEEE Transaction on Industry Applications, vol. 1A-19 no. 6, pp. 1057-1069, Nov. 1983.

[5] Lai J. S. and Peng F. Z.: Multilevel converters - a new breed of power converters, IEEE Transaction on Industry Applications, vol. 32 no. 3, pp. 509-517, May-June 1996.

[6] Shimmyo S., Mochikawa H., and Morishima Y.: Development of high power density converter using flying capacitor multilevel circuits for PV systems, 2016 19th International Conference on Electrical Machines and Systems (ICEMS), Chiba, 2016, pp. 1-6.

[7] Kouro S., Malinowski M., Gopakumar K., Pou J., Franquelo L. G., Wu B., Rodriguez J., Perez M. A., and Leon J. I.: Recent advances and industrial applications of multilevel converters, IEEE Transactions on Industrial Electronics, 2010, vol. 57 no. 8, pp. 2553-2580.

[8] Lesnicar A. and Marquardt R.: An innovative modular multilevel converter topology suitable for a wide power range, 2003 IEEE Bologna Power Tech Conference Proceedings, Bologna, Italy, 2003, pp. 6 vol. 3.

[9] Sztykiel M., Silva R. da, Teodorescu R., Zeni L., Helle L., and Kjaer P. C.: Modular multilevel converter modeling, control and analysis under grid frequency deviations, 2013 15th European Conference on Power Electronics and Applications (EPE), Lille, 2013, pp. 1-11.

[10] Korte C., Specht E., Goetz S. M., and Hiller M.: A control scheme to reduce the current load of integrated bateeries in cascaded multilevel converters, 2019 10th International Conference on Power Electronics and ECCE Asia (ICPE 2019 - ECCE Asia), Busan, Korea (South), 2019, pp. 1-8.

[11] Munk-Nielsen S., Aalborg University: Power circuits for modular multi-level converters (MMC) and modular multi-level converters, WO 2019/201918 A1.

[12] Beczkowski S. M. and Munk-Nielsen S.: Three-level PN Cell for multilevel converters, IET Power Electronics, vol. 13, no. 2, pp. 324-331.

[13] Dickson J. F.: On-chip high-voltage generation in MNOS integrated circuits using an improved voltage multiplier technique, IEEE Journal of Solid-State Circuits, vol. 11, no. 3, pp. 374-378, June 1976.

[14] Arntzen B. and Maksimovic D.: Switched-capacitor DC/DC converters with resonant gate drive, IEEE Transactions on Power Electronics, vol. 13, no. 5, pp. 892-902, Sept. 1998.

[15] Lei Y., et al.: A 2-kw Single-phase sevel-level flying capacitor multilevel inverter with an active energy buffer, IEEE Transactions on Power Electronics, vol. 32, no. 11, pp. 8570-8581, Nov. 2017.

[16] GaN Systems Inc.: Bottom-side cooled 100 V E-mode GaN transistor, Available:https://gansystems.com/wp-content/uploads/2020/04/GS61008P-DS-Rev-200402.pdf, Accessed on: June. 26, 2020.

[17] Konstantinou G. S. and Agelidis V. G.: Performance evaluation of half-bridge cascaded multilevel converters operated with multicarrier sinusoidal PWM techniques, 2009 4th IEEE Conference on Industrial Electronics and Applications, Xi'an, 2009, pp. 3399-3404.

[18] Jørgensen A. B., Munk-Nielsen S., and Uhrenfeldt C.: Overview of digital design and finite-element analysis in modern power electronic packaging, IEEE Transactions on Power Electronics, vol35, no. 10, pp. 10892-10905, Oct. 2020.

[19] Sun B., Jørgensen K. L., Zhang Z., and Andersen M. A. E.: Multi-physic analysis for GaN transistor PCB layout, 2019 IEEE Applied Power Electronics Conference and Exposition (APEC), Anaheim, CA, USA, 2019, pp. 3407-3413.

Grid-frequency Vienna rectifier and isolated current-source DC-DC converters for efficient off-board charging of electric vehicles

Jacek Rabkowski*, Andrei Blinov**, Denys Zinchenko**, Grzegorz Wrona*, Mariusz Zdanowski*

*WARSAW UNIVERSITY OF TECHNOLOGY
Koszykowa 75
Warsaw, Poland
Tel.: +48 / (22) – 234.76.15.
Fax: +48 / (22) – 234.60.23.
E-Mail: jacek.rabkowski@ee.pw.edu.pl
URL: http://www.ee.pw.edu.pl

**TALLINN UNIVERSITY OF TECHNOLOGY
Ehitajate tee 5, Tallinn, Estonia
www.taltech.ee

Acknowledgement

This project was supported in the frame of the ECPE Joint Research Programme and by NPRP12S-0214-190083 from the Qatar National Research Fund (a member of Qatar Foundation).

Keywords

Charging infrastructure for EV´s, converter circuit, emerging topology, multilevel converters, soft-switching, Silicon Carbide (SiC)

Abstract

This paper is focused on an alternative concept of the EV off-board charging system based on three-level Vienna rectifier operating at grid frequency and two soft-switched, isolated current-source DC-DC converters. Operation principles of the system including a low frequency input stage and DC-DC converters are studied, also using circuit simulations. Then, design issues of vital parts are presented. Finally, a 5kVA laboratory model is shown and selected measurements are provided and discussed.

Introduction

One of the key challenges of modern power electronics is fast and efficient charging of batteries in electric vehicles [1]-[3]. On-board chargers are very interesting area due to size and weight limitation; however, off-board chargers are also not trivial issue due to increasing peak power (tens of kW, up to hundreds of kW). Today, most common structure of the off-board charger contains two stages: AC-DC converter operating at unity power factor and one or more isolated DC-DC converters connected via DC link – see a simplified scheme in Fig. 1a. Due to very good performance Vienna rectifier is very often a AC-DC topology of choice when input stage is considered, as for instance was suggested in [4]. Then, three-pole DC link can be merged with cascaded DC-DC converters in series-series or series-parallel configuration (Fig. 1b). Exactly the same power electronic system presented in Fig. 1b may be also applied to reach much higher efficiency by avoiding PWM control of the input AC-DC stage. The concept of a three-level unfolder, which has been proposed back in 1990s [5],[6] and suggested for three-phase inverter [7],[8] may be also utilized in rectifier operation [9]. Three-level rectifier provides variable voltages to the input of the DC-DC converters, which are expected to create parts of the sinusoidal waveforms. Number of similarities to a single-phase PFC concept can be found as DC-DC converters shape input current and control battery current/voltage.

Recently, another solution has been proposed, so called "synergetic control", based on partial PWM operation of the Vienna rectifier has been studied by means of circuit simulations in [10]. The variations of the DC-link voltages are lower and the operating conditions of the DC-DC converters are less critical, however, the concept in general is much closer to a standard two-stage solution. In this paper, the authors focus on the second concept with grid-frequency operating AC-DC stage but instead of complex series-resonant dual active bridges [11],[12], current-source (CS) DC-DC converters are selected. Each CS DC-DC converter (marked as DC-DC1 and DC-DC2 in Fig. 1b) can be controlled to operate in soft-switching mode, but the main advantage is a direct and simple control of the input current. This feature is very important as quality of the grid current (power factor and THD) is influenced mainly by the DC-DC converter performance.

In order to verify this concept, in addition to computer simulations, a laboratory 5 kVA model of the EV charger was designed and first experimental waveforms are provided in this paper. According to gathered experiences most crucial part of the systems are two identical isolated DC-DC converters.

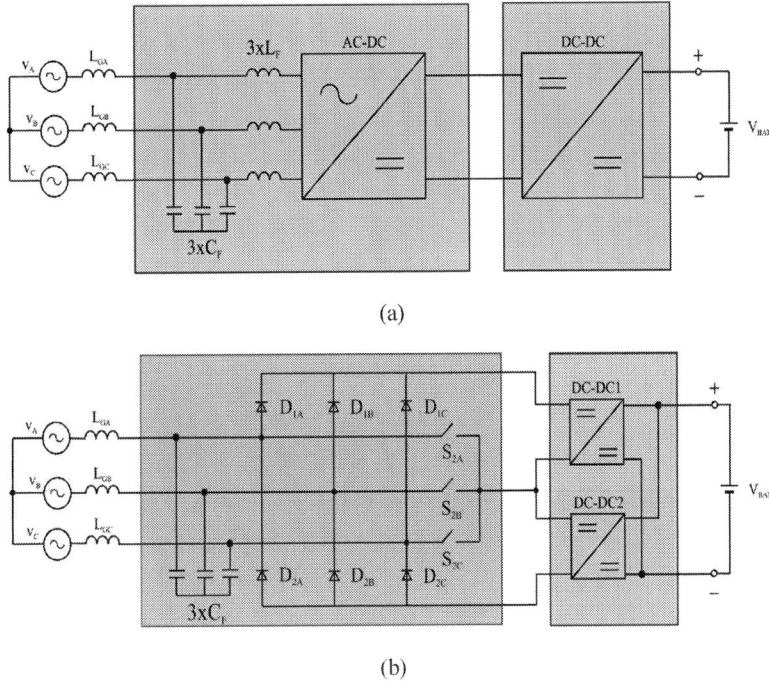

Fig. 1: Typical EV off-board charger structure (a) and Vienna rectifier with two series-parallel DC-DC converters (b)

Operation principles

A concept of three-phase three-level unfolder was developed in [5]-[7] aiming at more efficient three-phase inverter without energy storage. In addition to a low-frequency three-level inverter (usually neutral-point clamped topology), two isolated DC-DC converters were considered as a crucial part. However, recent works shown that this concept may also work in the PV inverter with non-isolated three-level DC-DC stage [13]. Furthermore, this notable method may be easily applied also to three-phase grid-connected inverter [9], [13] or three-phase unidirectional rectifiers [14]. In such a case, instead of a fully-controlled three-level neural-pint clamped a grid-frequency operated Vienna rectifier (VR) seems to be the better choice [14] (Fig. 1b).

The concept can be much more clear when two DC-DC converters shown in Fig. 1b are replaced by two current sources (Fig. 2a). Each phase A, B or C is connected by the rectifier to one of three outputs: "+", "-" by the properly biased diodes or "0" by the bidirectional transistor switches. When current sources track suitable parts of three-phase sinusoidal waveforms and three switches of the VR are properly controlled, three-phase currents at the input side may be sinusoidal and balanced. According to waveforms in Fig. 2b, the current sources i_1 and i_2 follow absolute values of maximum and minimum phase voltages, while the transistor switch connects the middle phase and conducts the difference between i_1 and i_2. In consequence, the system may operate at unity power factor and this topology may be applied as a battery charger with two DC-DC converters connected in series at the input and in parallel at the output (Fig. 1b).

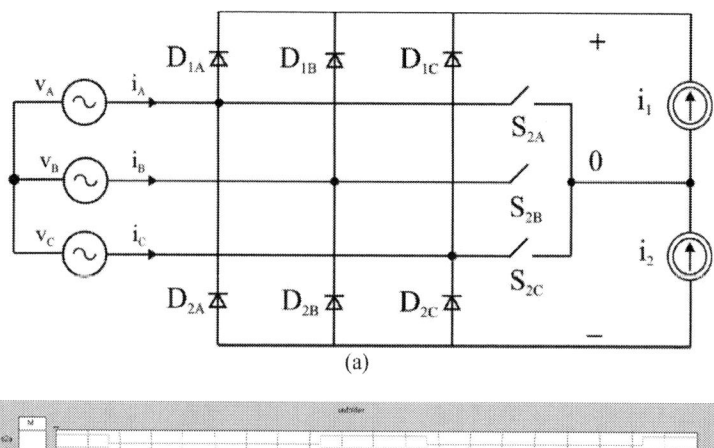

(a)

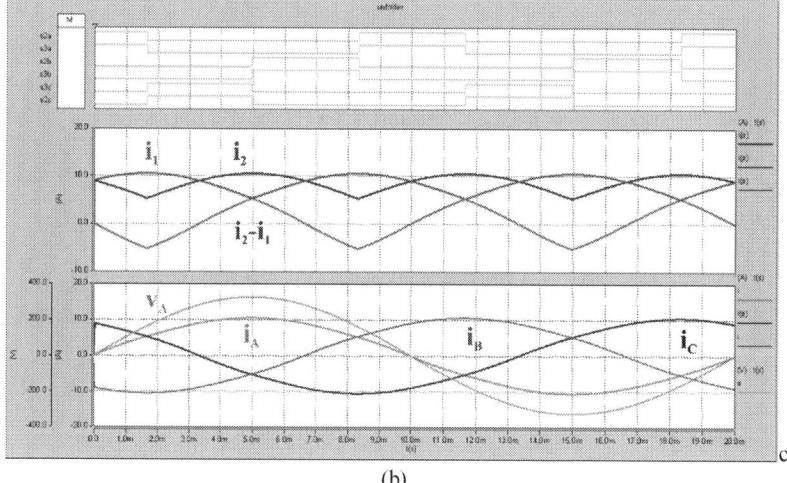

(b)

Fig. 2: The simplified structure of the VR with two current sources (a) and the simulated waveforms displaying control sequence, idealized currents of the DC-DC converters and three-phase input.

Current-source DC-DC converter

Selected topology

Due to specifics of the three-phase unfolder, the input voltage of each DC-DC stage resembles rectified sine with low-frequency harmonic content and amplitude of up to 500 V. Thus, the input voltage ranges from 0 to 500 V with 150 Hz frequency (assuming 3×400V input voltage).

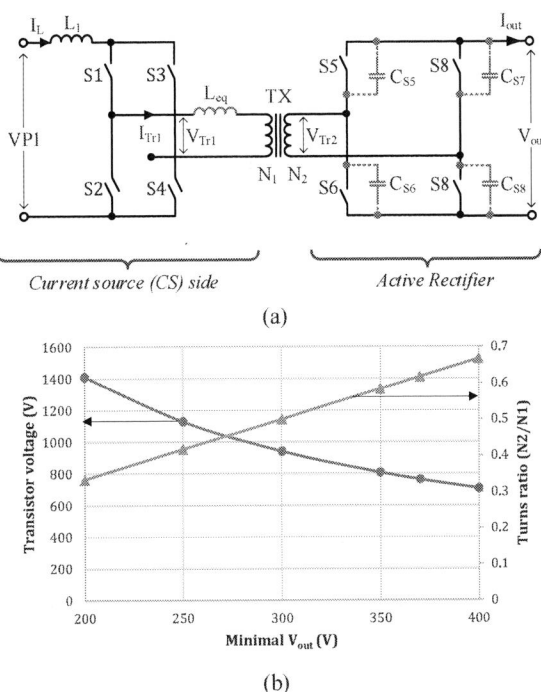

(a)

(b)

Fig. 3: Current-source ASMC topology (a); trade-off between minimal output voltage and voltage stress on the inverter transistors (b).

The required functionality can be obtained using a two-stage converter system, where the first power factor correction (PFC) stage is coupled with the isolated DC-DC stage through high voltage DC-link [15]. Another approach is to apply single-stage isolated converter [16]-[20]. Among the existing topologies, the asymmetric secondary-modulated converter (ASMC) shown in [18],[19] was selected to perform the required input current wave-shaping. Its advantages include soft switching operation of semiconductors, constant switching frequency and relatively low energy circulation. In the general case, it is able to perform PFC using only one independent control variable (phase shift) and relatively simple PI regulator [20], making it an attractive solution when compared to SR-DAB converters with complex multi-mode modulation [11], [12].

Design of the DC-DC converters

The known peculiarity of current-source converters is that the voltage at the output of the inverter stage has to be always higher than the input one. The turns ratio of the transformer has to be determined according to the relation between the peak input voltage and lowest output voltage. Assuming that the minimal boost for the converter primary stage is 10-20% (depends on the transformer leakage inductance and other design trade-offs), the voltage across the transformer primary would be around 550-600 V. The peak voltage stress on the primary switches is then determined by the rectified maximum output voltage (see Fig. 3b).

Assuming 1.2 kV primary devices and steady-state voltage stress of 700-800 V, the turns ratio for 470 V output voltage would be around 1.5:1. This suggests the output voltage range of approximately 360-470 V, which is almost identical to [11],[12].

In order to verify the operation, a simulation model of ASMC converter was created in PSIM11 software. The parameters are chosen in accordance with the specification of the 5 kVA system and are listed in Table I. In the simulation, the components are considered lossless and the influence of the transformer magnetizing inductance is neglected. The PI regulator was used to follow the reference current waveshape. The simulation results are presented in Fig. 4. As observed, the topology can effectively meet the application requirements and provide the necessary functionality under given conditions.

Table I: Parameters of the DC-DC converters

Simulation Parameters	Symbol	A-SMC
Input voltage	$V_{P1,2}$	0-487 V
Amplitude input current	$I_{L1,2}$	10 A
Transformer primary voltage	V_{Trp}	600 V
Transformer turns ratio	N2:N1	1:1.5
Output voltage	V_{out}	400 V
Input inductance	L1,L2	1 mH
Output capacitor	C_{out}	30 μF
Transformer (leakage) inductance	L_{eq}	6 μH
Switching frequency	f_{sw}	75 kHz

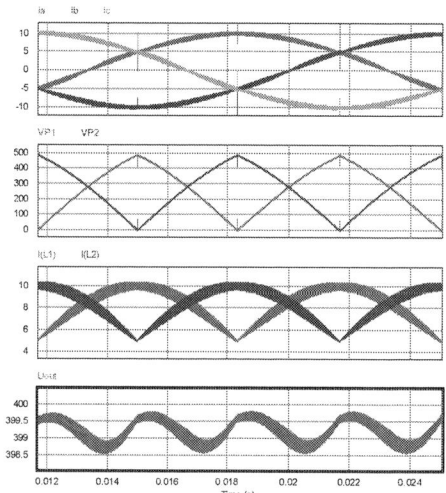

Fig. 4: Simulated waveforms of A-SMC. From top to bottom: line currents; dc-dc stages: input voltages, input currents, output voltage.

Design of the input stage

As first step of the design process, described above operation of the AC-DC stage in three-phase configuration has been simulated. According to the operating principles discussed in details in previous sections, three phase currents are distributed among three phase legs A, B and C of the input stage. At any time, two out of six diodes (D_{1A}, D_{2A}, D_{1B}, D_{2B}, D_{1C} and D_{2C}) are conducting in two phase legs – one in upper arm and one in lower arm, while in the remaining phase leg the middle transistor switch (S_{2A}, S_{2B}, S_{2C}) is active. In consequence, in all three legs the same current waveforms are observed in the mentioned diodes and transistors when single grid period is taken into consideration (Fig. 5). Therefore, when all phase currents are sinusoidal and balanced the same values of the average and RMS currents are measured in all diodes and bidirectional switches – see table 2. In further considerations symbols S_{2x} refers to switches S_{2A}, S_{2B}, or S_{2C} while D_{1x} and D_{2x} will represent diodes D_{1A}, D_{2A}, D_{1B}, D_{2B}, D_{1C} and D_{2C}.

From the values of peak voltages it can be concluded that all selected devices should be rated at 1200 V. Note that in comparison to PWM-operated VR this is a drawback as PWM operated VR requires 600V transistors in middle-point switches. However, it is possible that after further studies of circuit conditions and behaviour of these devices at the 60-degree sector changes, 600 or 650V could be also applied here. Furthermore, low frequency operation of these devices reduces requirements in terms of switching speed and devices optimized in terms of low-voltage drop should be considered.

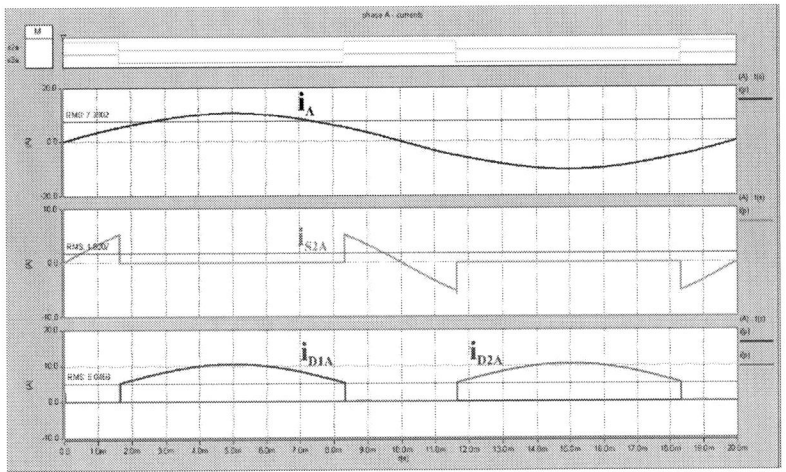

Fig. 5: Single leg of the unfolder during operation at 5 kVA, from the top: control signals of the transistor switches, phase current, currents in the transistor switch S_{2x} and two diodes D_{1x}/D_{4x}.

Table II: Measured currents & voltages in semiconductor devices during three-phase mode for 5 kVA.

Element	Average current [A]	RMS current [A]	Peak voltage [V]
Diodes D_{1x}/D_{4x}	2.86	5.05	562
Switch S_{2x}	-	1.86	488

Taking into account all observations from the simulation study and values in Table II, 1200V devices were selected for the input stage. Analysis of different types of transistors, starting from SiC MOSFETs through Si superjunction MOSFETs and IGBTs ended with IHW15N120R3 – IGBT with low voltage drop and reasonable cost. As devices operate in common-emitter configuration, a low voltage drop of the anti-parallel diode was also important factor. Power loss estimations conducted on the base of datasheet has shown slightly above 1W/device, therefore, low losses in complete VR circuit are anticipated (see Table III). Potentially, even better results can be obtained with reverse blocking IGBTs [21]. Looking at the same crucial aspects: low voltage drop and cost six diodes of the input stage were also selected. As the current of the diodes is much higher, less than 6W of conduction loss per IDP30E120 is expected. Overall, the maximum conduction losses are estimated to be below 31 W and switching losses may be neglected; therefore, the total loss should lower than 0.6% of nominal power. On the base of analysis above a laboratory model of the low-frequency VR was designed and built – see Fig. 6. Six diodes and transistors were mounted on two SK499 heatsinks (Fig. 6a) – natural convection was assumed due to low amount of losses. Three 2.2µF/630V capacitors are applied on DC terminals to create LC filter together with the input inductance of the DC-DC converter modules (Fig. 3a). On the top, the main control with a floating-point DSP was mounted (Fig. 6b) performing voltage measurements and synchronization. This unit sends gate signals to the switches of VR (Fig. 6c) and is additionally responsible for communication to DC-DC submodules.

Table III: Power loss estimations for transistors and diodes (5 kVA).

Devices	Type	Total losses at 5kVA operation [W]
Diodes D_{1x}-D_{4x}	IDP30E120	23.75
Switches S_{2A}-S_{2C}	IHW15N120R3	6.91
All devices (in ref. to the nominal power)		**30.66** (~0.6%)

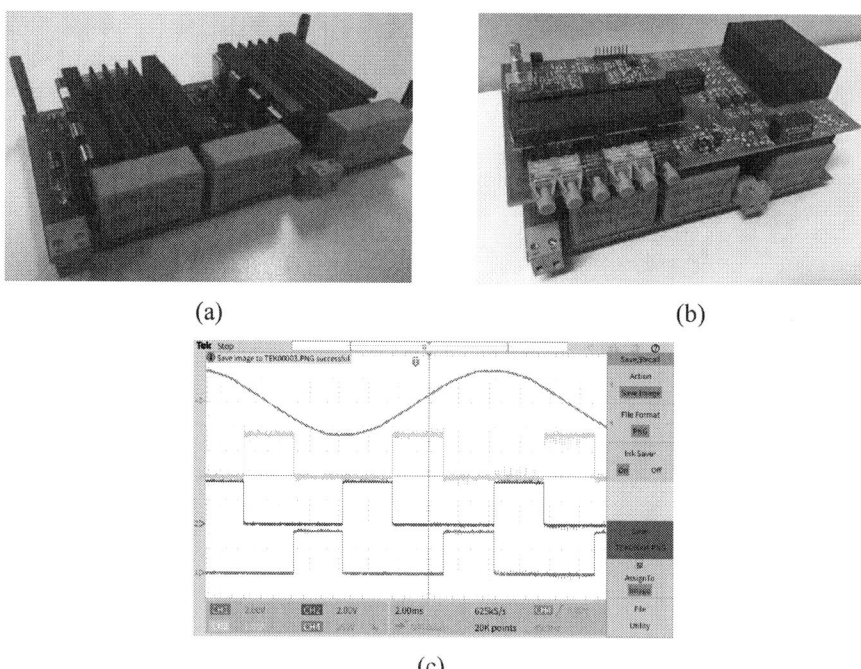

(a)   (b)

(c)

Fig. 6: Photo of the 5 kVA VR power circuit (a) and with the main controller applied on the top (b); control signals versus phase voltage (c).

Experimental verification

In order to verify the proposed charger concept, a scaled prototype with rated power of 5 kVA was assembled. The main components of the converter are listed in Table IV and the photo of the DC-DC converter module is shown in Fig. 7. The experimental waveforms are presented in Fig. 8 and Fig. 9. As observed, the operation is in general agreement with expectations and the DC-DC converters are capable to provide the required input current wave-shaping and deliver output DC voltage with low ripple (Fig. 8d).

Table IV: Main components of the DC-DC converters

Device	Value/Type
Input inductor	1.5 mH
Semiconductors	C2M0160120D
Transformer	n=31:21; L_{eq}=5.6 µH
Output capacitor	150 µF
Microcontroller	TMS320F28335

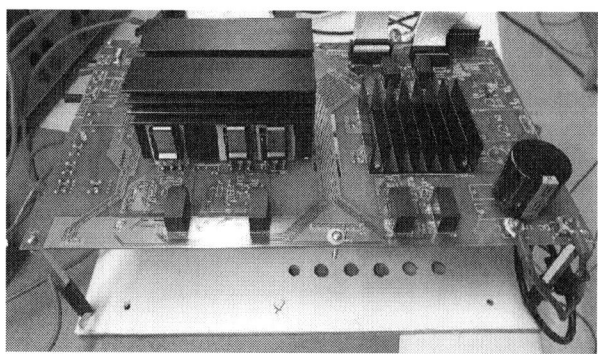

Fig. 7: Photo of the DC-DC converter module

Fig. 9a shows the voltage and current waveforms across transformer at the sine peak. The peak current at the beginning of the active state, which is responsible for soft switching of the current-source transistors, is slightly higher than steady-state current through the transformer. The duration of the duty cycle that determines this current peak value is set to 0.507. The waveforms near sine zero input voltage crossing are shown in Fig. 9b. It can be observed that for the implemented control with single variable the peak current remains the same. However, the energy circulation is not very high since the current in the transformer is at zero during the duration of the shoot-through state. The total efficiency of the system is 94% at maximum input power of 5 kW.

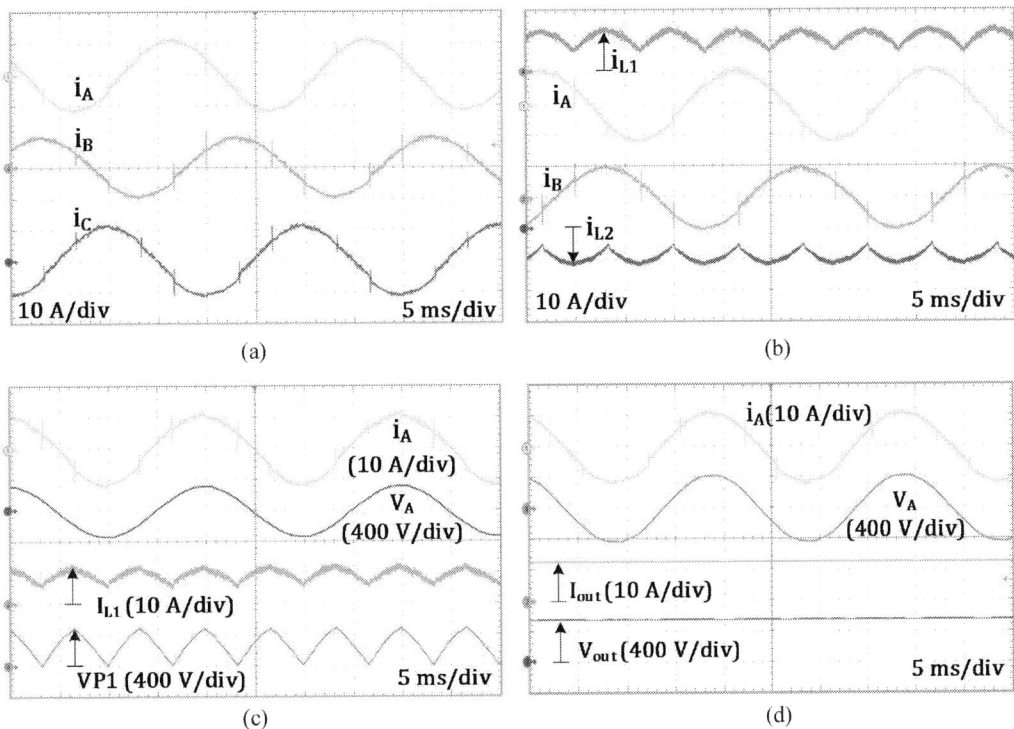

Fig. 8: Experimental results of the experimental system: AC currents (a); AC currents and input currents of DC-DC converters (b); AC currents and input voltage and current of DC-DC module (c); phase voltage and current and output voltage and current of DC-DC module (d).

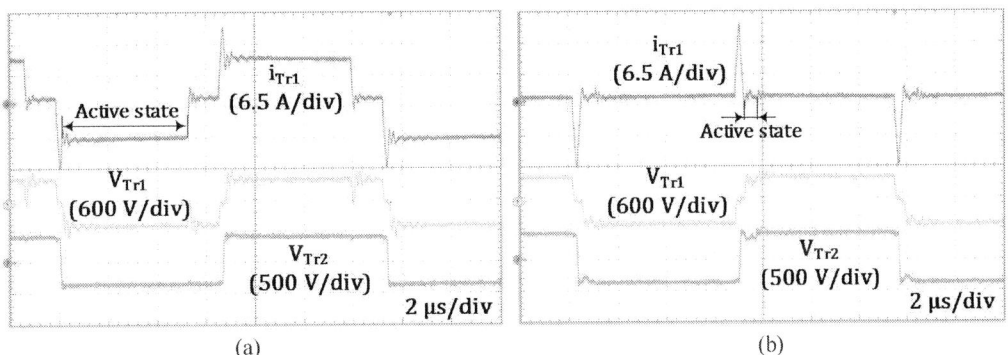

Fig. 9: Experimental results of the experimental system: transformer voltage and current waveforms close to the amplitude value of the input voltage (a), close to the zero crossing of the input voltage (b).

Conclusions

Current paper proposes an EV off-board charger based on the Vienna rectifier unfolder stage and two galvanically isolated current-source DC-DC converters. Thanks to low switching frequency, the unfolder can be designed with designed with low-cost Si IGBTs and diodes. The advantages of the system include simple control of the input current, constant switching frequency and reduced energy circulation. The concept was verified with simulation model and 5 kVA prototype. The experiment showed the capability of the system to source sinusoidal input current form the grid and deliver stable DC voltage at its output terminal.

References

[1] G. E. Sfakianakis, J. Everts and E. A. Lomonova, "Overview of the requirements and implementations of bidirectional isolated AC-DC converters for automotive battery charging applications," *2015 Tenth International Conference on Ecological Vehicles and Renewable Energies (EVER)*, Monte Carlo, 2015, pp. 1-12.

[2] S. S. Williamson, A. K. Rathore and F. Musavi, "Industrial Electronics for Electric Transportation: Current State-of-the-Art and Future Challenges," in *IEEE Transactions on Industrial Electronics*, vol. 62, no. 5, pp. 3021-3032, May 2015.

[3] B. Eckardt, M. Wild, C. Joffe, S. Zeltner, S. Endres and M. Maerz, "Advanced Vehicle Charging Solutions Using SiC and GaN Power Devices," *PCIM Europe 2018; International Exhibition and Conference for Power Electronics, Intelligent Motion, Renewable Energy and Energy Management*, Nuremberg, Germany, 2018, pp. 1-6.

[4] Infineon Technologies AG, Tackling the Challenges of Electric Vehicle Fast Charging, White Paper, Apr. 2019.

[5] K. Oguchi, E. Ikawa and Y. Tsukiori, "A three-phase sine wave inverter system using multiple phase-shifted single-phase resonant inverters," *Conference Record of the 1990 IEEE Industry Applications Society Annual Meeting*, Seattle, WA, USA, 1990, pp. 1125-1131 vol.2.

[6] B. S. Jacobson and E. N. Holmansky, Methods and Apparatus for Three-Phase Inverter with Reduced Energy Storage, US Patent 7,839,023, Nov., 2010.

[7] W. W. Chen, R. Zane and L. Corradini, "Isolated Bidirectional Grid-Tied Three-Phase AC–DC Power Conversion Using Series-Resonant Converter Modules and a Three-Phase Unfolder," in *IEEE Transactions on Power Electronics*, vol. 32, no. 12, pp. 9001-9012, Dec. 2017.

[8] W. Warren Chen, Baljit Riar, Regan Zane, "A three-port series resonant converter for three-phase unfolding inverters", *Control and Modeling for Power Electronics (COMPEL) 2017 IEEE 18th Workshop on*, pp. 1-7, 2017.

[9] W. Chen, B. Riar and R. Zane, "Battery Integrated Modular Multifunction Converter for Grid Energy Storage," *2018 IEEE Energy Conversion Congress and Exposition (ECCE)*, Portland, OR, 2018, pp. 2157-2163.

[10] J. Azurza Anderson, M. Haider, D. Bortis, J. W. Kolar, M. Kasper, G. Deboy, "New Synergetic Control of a 20kW Isolated VIENNA Rectifier Front-End EV Battery Charger", *Control and Modeling for Power Electronics (COMPEL) 2019 20th Workshop on*, pp. 1-8, 2019.

[11] J. Everts, F. Krismer, J. Van den Keybus, J. Driesen and J. W. Kolar, "Optimal ZVS Modulation of Single-Phase Single-Stage Bidirectional DAB AC–DC Converters," in IEEE Transactions on Power Electronics, vol. 29, no. 8, pp. 3954-3970, Aug. 2014.

[12] J. Everts, "Closed-Form Solution for Efficient ZVS Modulation of DAB Converters," in IEEE Transactions on Power Electronics, vol. 32, no. 10, pp. 7561-7576, Oct. 2017.

[13] T. Mannen, P. N. Ha and K. Wada, "Performance Evaluation of a Boost Integrated Three-Phase PV Inverter Operating With Current Unfolding Principle," *2019 21st European Conference on Power Electronics and Applications (EPE '19 ECCE Europe)*, Genova, Italy, 2019, pp. 1-8.

[14] T. B. Soeiro, T. Friedli, J. W. Kolar, "Swiss rectifier - A novel three-phase buck-type PFC topology for Electric Vehicle battery charging", *2012 Twenty-Seventh Annual IEEE Applied Power Electronics Conference and Exposition (APEC)*, pp. 2617-2624, 2012.

[15] B. Singh, B. N. Singh, A. Chandra, K. Al-Haddad, A. Pandey and D. P. Kothari, "A review of single-phase improved power quality AC-DC converters," in IEEE Transactions on Industrial Electronics, vol. 50, no. 5, pp. 962-981, Oct. 2003.

[16] H. Benqassmi, J.-C. Crebier and J.-P. Ferrieux, "Comparison between current-driven resonant converters used for single-stage isolated power factor correction," IEEE Transactions on Industrial Electronics, vol. 47, no. 3, pp. 518-524, 2000.

[17] J. Chen, R. Chen and T. Liang, "Study and Implementation of a Single-Stage Current-Fed Boost PFC Converter with ZCS for High Voltage Applications," in IEEE Transactions on Power Electronics, vol. 23, no. 1, pp. 379-386, Jan. 2008.

[18] A. Blinov, R. Kosenko, A. Chub and D. Vinnikov, "Bidirectional Soft Switching Current Source DC-DC Converter for Residential DC Microgrids," IECON 2018 - 44th Annual Conference of the IEEE Industrial Electronics Society, Washington, DC, 2018, pp. 6059-6064.

[19] A. Blinov, R. Kosenko, D. Vinnikov and L. Parsa, "Bidirectional Isolated Current Source DAB Converter with Extended ZVS/ZCS Range and Reduced Energy Circulation for Storage Applications," in IEEE Transactions on Industrial Electronic

[20] D. Zinchenko, A. Blinov and D. Vinnikov, "Comparison of Isolated Boost Full Bridge Converters for Power Factor Correction Application," 2019 IEEE 60th International Scientific Conference on Power and Electrical Engineering of Riga Technical University (RTUCON), Riga, Latvia, 2019, pp. 1-6.

[21] A. Blinov, O. Korkh, D. Vinnikov and P. Waind, "Characterisation of 1200 V RB-IGBTs with Different Irradiation Levels Under Hard and Soft Switching Conditions," 20th European Conference on Power Electronics and Applications (EPE'18 ECCE Europe), Riga, 2018, pp. 1-10, 2018.

Unidirectional thyristor-based DC-DC converter for HVDC connection of offshore wind farms

Pierre Le Métayer, Piotr Dworakowski, Jose Maneiro
SUPERGRID INSTITUTE
23, Rue Cyprian
69100 Villeurbanne, France
Tel.: +33 7 63 66 19 15
E-Mail: pierre.lemetayer@supergrid-institute.com
URL: https://www.supergrid-institute.com

Acknowledgements

This work was supported by a grant overseen by the French National Research Agency (ANR) as part of the "Investissements d'Avenir" Program (ANE-ITE-002-01)

Keywords

«HVDC», «Converter circuit», « DC collector network», « Wind energy»

Abstract

This paper introduces a unidirectional MVDC-HVDC converter for 'all DC' connected offshore wind farms. This converter combines the low power losses of thyristor based converters and the control abilities of voltage source converters, together with low power electronics component count. A high step-up ratio is achieved by an input parallel output series configuration. The use of medium frequency is discussed and inductance design conditions are given. A control scheme is proposed and validated by simulations. The converter high efficiency is also observed. The presented analysis shows a potential feasibility and benefits of this novel unidirectional thyristor-based DC-DC converter.

Introduction

In a world where the renewable energy transition becomes more and more urgent, the wind energy is one of the most relied upon source. Indeed the wind energy has become the second most important form of installed generation capacity in 2016 [1]. Moreover, the implementation of wind farms in locations remote from the shore allows steadier and stronger wind compared to onshore installations.

Different methods to bring the produced power back to shore have been investigated. For wind farms located far from shore, HVDC connection is required to keep transmission losses low [2]. An attempt to further reduce these losses is the 'all DC' architecture [3-4], using a MVDC collection grid to bring the power from wind turbines to a central platform where the voltage needs to be stepped up to HVDC. Thus, a high step up ratio MVDC-HVDC DC-DC [5-6] converter that can handle with low losses the power generated by a complete wind farm is a key enabling element that is currently missing in this architecture. This converter volume and weight are important parameters, especially considering offshore platform cost [7].

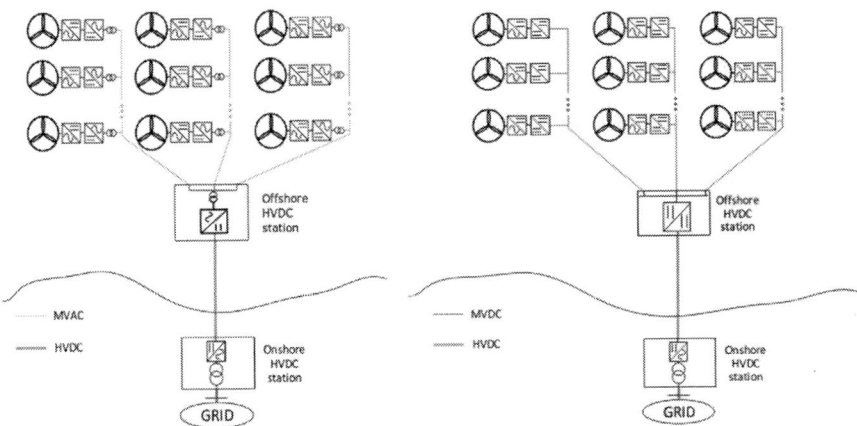

Fig. 1: AC collection offshore wind farm (left), "All DC" offshore wind farm (right)

The power flow of the offshore wind farms being mostly directed towards the shore, the use of unidirectional converters shall be considered offering significant cost reductions. Several step-up DC-DC topologies have been reviewed in [8]. In particular, the unidirectional non-isolated topologies have been considered: chopper-based DC modular converters [9], and the isolated topologies: the phase-shifted full-bridge (PSFB) [10] and the single active bridge (SAB) [11-12]. In order to achieve high voltage stepping ratios [13] while handling important power, converters can be configured in input parallel output series (IPOS) configuration such as presented in [14-16].

The power losses reduction in the transmission systems can be achieved thanks to the use of thyristor valves as in Line Commutated Converter (LCC). A bidirectional thyristor-based converter, the active forced commutated bridge (AFCB), has been proposed in [17-20]. In this topology the thyristor turn-off is forced by a stack of full-bridge chain-link (FB-CL).

One of the main challenges of the "all DC" architecture resides in the use of an isolated DC-DC converter in place of an AC-DC converter and a transformer. The AC solution has one transformation stage less than the isolated DC-DC which makes it difficult for the DC-DC converter to be competitive in terms of costs and power losses. In this article it is proposed to face this challenge by using the AFCB in a SAB configuration, with a diode rectifier on the HVDC port. The AFCB offers low power losses and the SAB configuration offers lower cost compared to a dual active bridge [21]. The proposed topology has only a quarter of the thyristor number compared to its bidirectional counterpart and half the number of FB-CL, foreseeing lower volume, weight and cost. This paper proposes an analysis of the novel topology and controls using the unidirectional AFCB-based step-up DC-DC converter in a 3-cell IPOS configuration for GW range windfarms.

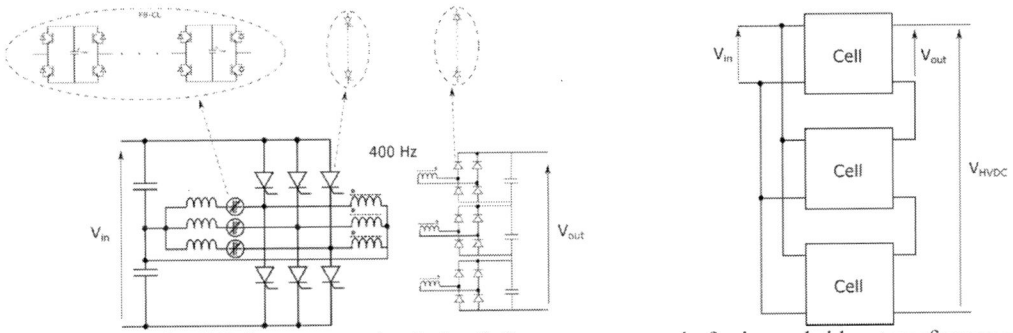

Fig. 2: Proposed DC-DC converter: detailed cell diagram composed of primary bridge, transformer and secondary bridge (left) and 3-cell IPOS configuration (right)

The principle of operation of the converter is described in the first section, followed by a discussion on its component sizing linked to the use of medium frequency. A control structure is proposed in the third section and it is then validated by simulations. An estimation of losses is also given.

Principle of operation

As seen in Fig. 2, the cell primary bridge is composed of a thyristor inverter. Using thyristors for the inverter allows low conduction losses, which is of the outmost importance for a converter of such a power rating. On the other hand, compared to an IGBT-based inverter, the possibility of controlling turn-off is lost. In order to solve this problem, the FB-CL are connected between the inverter phase output and the DC bus middle point. The FB-CL arm is composed of an arm inductor and a stack of IGBT based full-bridge cells allowing positive and negative voltage generation.

The FB-CL arm generates the voltage V_{arm} providing the required negative voltage at thyristor terminals, ultimately forcing its turn-off. As presented in [19] the V_{arm} is defined by:

$$-\frac{V_{in}}{2} - V_{off} \leq V_{arm} \leq \frac{V_{in}}{2} + V_{off} \qquad (1)$$

where V_{in} is the input voltage and V_{off} is the voltage used to force the thyristor valve turn-off. It is defined from the unitary thyristor turn-off voltage $V_{off,th}$ and the number of thyristors n_{th} in the valve:

$$V_{off} = n_{th} \cdot V_{off,th} \qquad (2)$$

The primary and secondary bridge are linked with a 3-phase transformer. The transformer primary winding is Y-connected. At the secondary each transformer winding is connected to a full-bridge diode rectifier. Three cells are connected in IPOS configuration to share the input current and add output voltages, in order to reach the HVDC voltage.

The FB-CL arm and thyristor valve waveforms (for one thyristor leg and FB-CL arm pair) are shown in the Fig. 3.

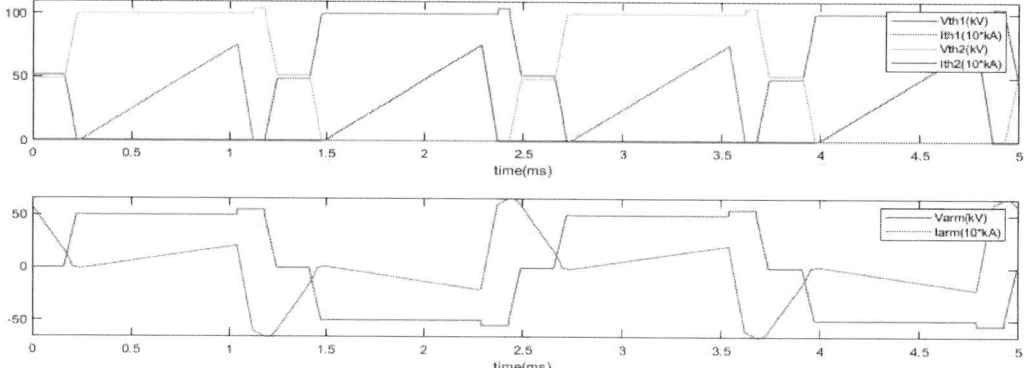

Fig. 3: Thyristor valves waveforms (top), FB-CL arm waveforms (bottom)

It can be seen that when any thyristor in a leg is conducting, it is the FB-CL arm that is conducting the load current. The inverter arms are phase-shifted by 120°. The cells are phase-shifted by 40°, reducing the voltage and current ripples leading to smaller passive components.

Sizing of inductances

The operation at medium frequency allows the size reduction of the passive elements but also requires the use of fast thyristors. Indeed, throughout the turn-off time t_q during which V_{off} must be applied, it is the FB-CL arm that is conducting the load current, bringing down the converter efficiency. The dynamics of the transition between thyristor valve and FB-CL conduction is dictated by the value of the arm inductor L_{arm}, and the shape of the load current by the leakage inductance of the transformer L_{lk}. The value of these inductances must then be chosen carefully in order to ensure a long enough application time of V_{off} on the thyristors.

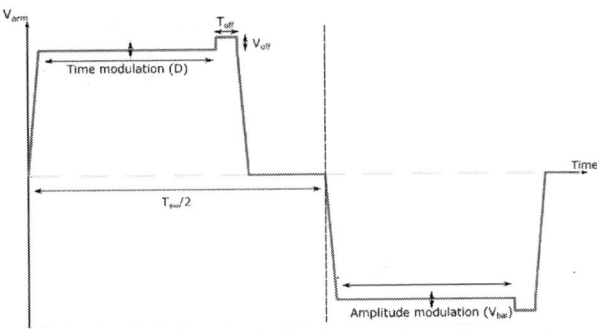

Fig. 4: Definition of quantities on V_{arm} for arm inductance sizing and control

As shown in Fig. 4 the application time of V_{off} by the FB-CL arm t_{off} can be defined as a portion of the half-period ε_{off}:

$$t_{off} = \varepsilon_{off} \cdot \frac{1}{2 \cdot f} \tag{3}$$

where f is the frequency of commutation defined as the inverse of the period T_{sw} presented in Fig. 4. Because of the presence of the arm inductor, the extra voltage V_{off} is not applied on the thyristor arm during t_{off} but during:

$$t_q = t_{off} - t_{Larm} \tag{4}$$

with t_{Larm} following the inductance equation:

$$t_{Larm} = \frac{L_{arm} \cdot I_{ac}}{V_{off}} \tag{5}$$

The peak current I_{ac} in the AC link is obtained from the transformer leakage inductance equation:

$$I_{ac} = \frac{(V_{prim} - \frac{V_{sec}}{n}) \cdot D}{2 \cdot f \cdot L_{lk}} \tag{6}$$

$$t_{Larm} = \frac{L_{arm} \cdot (V_{prim} - \frac{V_{sec}}{n}) \cdot D}{L_{lk} \cdot V_{off} \cdot 2 \cdot f} \tag{7}$$

where V_{prim} is the primary voltage of a cell transformer and V_{sec} is the secondary voltage of the transformer. The transformation ratio n is defined as the ratio of the transformer secondary winding turns over the primary wind turns.

One can notice that t_{Larm} can be reduced by increasing V_{off}. Although it is possible, this would imply an increase in the number of FB-CL and thus an increase of the cost, size and weight of the converter. It is thus recommended to keep V_{off} as the minimum voltage insuring thyristor valve turn-off.

With t_q fixed by the choice of the thyristor t_{off} must fulfill the relation:

$$t_{off} \geq t_q + \frac{L_{arm} \cdot (V_{prim} - \frac{V_{sec}}{n}) \cdot D}{L_{lk} \cdot V_{off} \cdot 2 \cdot f} \tag{8}$$

$$\varepsilon_{off} \cdot \frac{1}{2 \cdot f} \geq t_q + \frac{L_{arm} \cdot (V_{prim} - \frac{V_{sec}}{n}) \cdot D}{L_{lk} \cdot V_{off} \cdot 2 \cdot f} \tag{9}$$

As explained in the control principles section, the time modulation index D varies with respect to the transferred power in order to keep V_{in} to its reference value. Therefore, a condition on the ratio between the FB-CL arm inductance and the transformer leakage inductance can be written for the maximum time modulation index D_{max} corresponding to the maximum power transfer.

$$\frac{L_{arm}}{L_{lk}} \leq \frac{V_{off}(\varepsilon_{off} - 2 \cdot f \cdot t_q)}{(V_{prim} - \frac{V_{sec}}{n}) \cdot D_{max}} \tag{10}$$

The value of ε_{off} being related to the losses in the FB-CL arms, [17] recommends a value not greater than 10%. Since the ratio of inductances can only be positive, ε_{off} must fulfil:

$$0.1 \geq \varepsilon_{off} > 2 \cdot f \cdot t_q \tag{11}$$

The inductance ratio is fixed by (10) in worst case scenario of maximum power transfer. When power transfer is not maximal the time modulation index lowers and ε_{off} can thus be reduced as well while still maintaining (10) true, with limitations given by (11).

Equation (11) also gives an idea of the maximum frequency that this converter can operate at for a given thyristor with a turn-off time t_q.

Control principles

In the context off an offshore wind farm it is considered that the HVDC voltage is imposed by the onshore converter. Thus, the DC-DC converter must regulate its input MVDC voltage. The presence of capacitor-based FB-CL stacks demands the control of their energies. The proposed control is represented in Fig.5 for one phase of one cell in the IPOS configuration. Thyristor valves control signals G_{th1} and G_{th2} and FB-CL arm voltage reference signal are phase-shifted by 120° for following phase of the same cell, and by 40° for the equivalent cell of following cell.

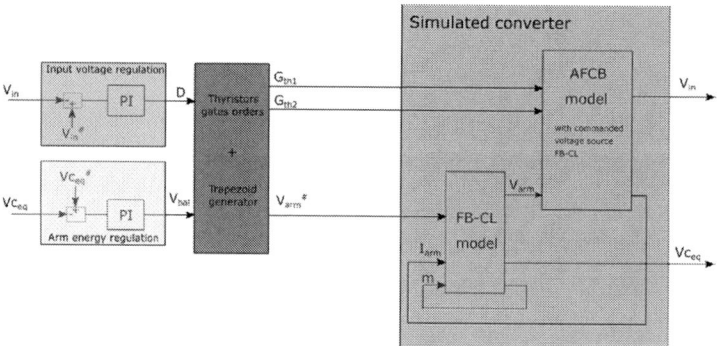

Fig. 5: Proposed control diagram of a cell (only one phase is represented)

The wind turbines connected to the MVDC network are modelled as current sources. These can be considered as an aggregated current source I_{in} flowing through the input capacitors C_1 and C_2 series association C_{in}. In order to control the input voltage, the period of conduction of the thyristors is modulated to charge or discharge the input capacitors.

$$I_{Cin} = I_{in} - I_{conv} \tag{12}$$

$$I_{Cin} = C_{in} \frac{dV_{in}}{dt} \tag{13}$$

Modulation index D of the thyristor conduction time is considered proportional to the current absorbed by the converter I_{conv} and is thus calculated by a PI controller from the error between the sum of both input capacitors voltages and the reference $V_{in}^{\#}$.

The FB-CL arm is composed of many capacitors and the arm energy must be regulated at a certain level chosen by the design. Controlling the energy of the stack amounts to control the voltage V_{ceq} of the stack equivalent capacitor C_{eq}, series association of submodule capacitors.

$$E_{arm} = \frac{1}{2} \cdot C_{eq} \cdot V_{Ceq}^{2} \tag{14}$$

In order to regulate this voltage, the equivalent capacitor is charged or discharged. This is done by modulating V_{arm} amplitude, a higher value than $V_{in}/2$ resulting in a current discharging the equivalent capacitor, and vice versa. The value of the balancing voltage V_{bal} (amplitude modulation) is calculated by a PI controller from the error between the equivalent capacitor voltage and the stack voltage chosen by design $V_{ceq}^{\#}$. The FB-CL arm is modeled by an average model as described in [22]. The equivalent capacitor voltage is calculated from the arm current and the number of currently connected submodules, represented by the utilization index m in figure 5.

In order to keep a balance between both input capacitors voltages V_{c1} and V_{c2}, a correction factor dependent on the ratio between these voltages is added. This factor is applied on the arm voltage during the balancing current generation period. Indeed, without this, charging currents could come from only one of the input capacitor and vice versa. This way, one capacitor would charge up to V_{in} when the other would discharge completely. Thus the arm voltage during the voltage balancing period the positive swings is defined as:

$$V_{arm} = (V_{C1} + V_{bal}) \cdot \frac{V_{C2}}{V_{C1}} \,, if \; V_{C1} > V_{C2}$$

$$\tag{15}$$

$$V_{arm} = (V_{C1} + V_{bal}) \cdot \frac{V_{C1}}{V_{C2}} \,, if \; V_{C2} > V_{C1}$$

And during the negative swings as:

$$V_{arm} = (-V_{C2} - V_{bal}) \cdot \frac{V_{C2}}{V_{C1}} \,, if \; V_{C1} > V_{C2}$$

$$\tag{16}$$

$$V_{arm} = (-V_{C1} - V_{bal}) \cdot \frac{V_{C1}}{V_{C2}} \,, if \; V_{C2} > V_{C1}$$

Simulation Study

In order to validate the correct operation of the proposed circuit, a simulation model of the converter has been built in Matlab-Simulink. Only one of the three IPOS connected cells has been simulated for computational time sake. The ratings of the simulated system are presented in Table I.

Table I: Ratings of the simulated converter

MVDC voltage	V_{MVDC}	100 kV
HVDC voltage	V_{HVDC}	±320 kV
Farm rated power	P	1.2 GW

Cell input voltage	V_{in}	100 kV
Cell output voltage	V_{out}	213 kV
Cell rated power	P_u	400 MW
Switching frequency	f	400 Hz

The frequency of 400 Hz is chosen in regard of existing fast-thyristor and their turn-off time t_q. The considered fast-thyristor [23] has a maximum turn-off time of 60 μs. Following (11), for a chosen ε_{off}=0.07, the maximum frequency of the converter is $0.07/(2*60*10^{-6})$ = 583 Hz. From this frequency a safety margin is taken for thyristor valve turn-off might be different from a single thyristor. The use of such medium frequency enables lower size, weight and thus cost of passive components, especially of FB-CL, since the needed capacitance drops with the increase of frequency, and of transformers [24]. Although, the increased operating frequency of such a high power transformer is a challenge. The selected frequency 400 Hz is relatively high but close to the frequency of 350Hz chosen in [25-26]. Control principles are validated by simulating a change in input power and checking the voltage regulation and FB-CL energy balancing.

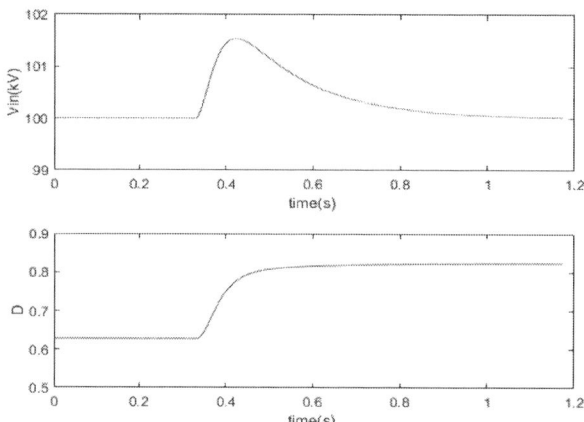

Fig. 7: Input voltage regulation in response to a change from $P_u \cdot 2/3$ to P_u

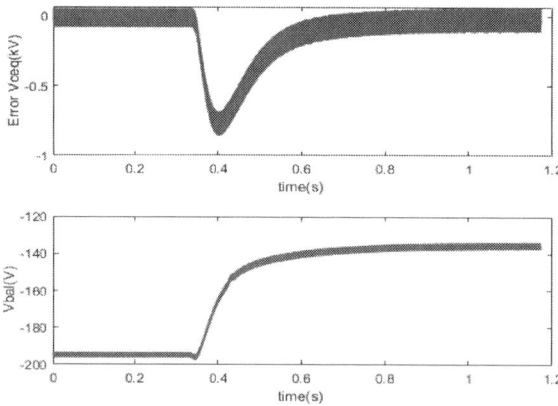

Fig.8: FB-CL arm energy balancing in response to a change from $P_u \cdot 2/3$ to P_u

From (10) the inductance ratio is calculated to be 0.023. The value of the application time of the V_{off} voltage is checked to be superior to the t_q of the thyristor for the worst case, the transfer of the total power P_u. The value measured in simulation is 63 μs.

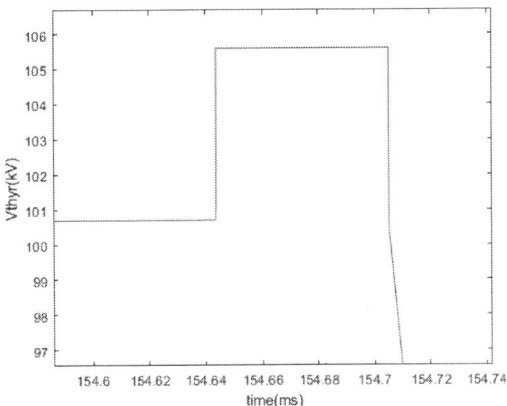

Fig.9: Thyristor valve voltage during V_{off} application time, for power P_u

Simulation of the converter currents through its different parts also enabled an efficiency estimation. Power losses in the FB-CL arms are calculated in regard of the methods given in [27]. Transformer losses estimation are based on the work presented in [28]. From these preliminary calculations an efficiency above 98.5% is expected but a dedicated study is here needed. Such efficiency is higher than those presented in [24] at lower frequency and step-up ratio.

Conclusion

A unidirectional thyristor-based converter has been proposed for the HVDC connection of "all DC" offshore wind farms. The use of thyristors is believed to achieve lower power losses compared to IGBT DC-DC converters. Medium frequency operation is used to enable a reduction of size weight and thus price of the offshore converter. The consequences of the use of medium frequency on certain design parameters have been discussed. A control method has been proposed to regulate input MVDC voltage and internal energy of the converter. For the case study of a 1.2 GW "all DC" offshore wind farm, a simulation of the converter has been done. The simulations enabled to validate the control strategy and the discussed design recommendations. The potential for high efficiency conversion has also been observed. A cost-performance analysis is now needed to rule on the interest of this converter, and globally of the "all DC" offshore wind farm, compared to other proposed solutions.

References

[1] Wind Power, "Wind in power: 2016 European statistics," 2017. [Online]. Available: https://windeurope.org/about-wind/statistics/european/wind-in-power-2016.

[2] K. Meah and S. Ula, "Comparative Evaluation of HVDC and HVAC Transmission Systems," in *2007 IEEE Power Engineering Society General Meeting*, 2007, pp. 1–5.

[3] M. De Prada Gil, J. L. Domínguez-García, F. Díaz-González, M. Aragüés-Peñalba, and O. Gomis-Bellmunt, "Feasibility analysis of offshore wind power plants with DC collection grid," *Renew. Energy*, vol. 78, pp. 467–477, Jun. 2015.

[4] R. Ryndzionek and L. Sienkiewicz, "Evolution of the HVDC Link Connecting Offshore Wind Farms to Onshore Power Systems," *Energies*, vol. 13, no. 8, p. 1914, Jan. 2020.

[5] S. Kenzelmann, A. Rufer, D. Dujic, F. Canales, and Y. R. de Novaes, "Isolated DC/DC Structure Based on Modular Multilevel Converter," *IEEE Trans. Power Electron.*, vol. 30, no. 1, pp. 89–98, Jan. 2015.

[6] S. Cui, J. Hu, M. Stieneker, and R. W. De Doncker, "An Isolated Soft-Switching Hybrid-Source DC-DC Converter for DC Offshore Wind Farms," in *2018 International Power Electronics Conference (IPEC-Niigata 2018 -ECCE Asia)*, 2018, pp. 2484–2489.

[7] National Grid, "Electricity ten year statement Appendix E Technology," 2016. [Online]. Available: https://www.nationalgrideso.com/document/47036/download.

[8] J. D. Páez, D. Frey, J. Maneiro, S. Bacha, and P. Dworakowski, "Overview of DC–DC Converters Dedicated to HVdc Grids," *IEEE Trans. Power Deliv.*, vol. 34, no. 1, pp. 119–128, Feb. 2019.

[9] X. Zhang and T. C. Green, "The Modular Multilevel Converter for High Step-Up Ratio DC–DC Conversion," *IEEE Trans. Ind. Electron.*, vol. 62, no. 8, pp. 4925–4936, Aug. 2015.

[10] N. Hassanzadeh, "Evaluation and comparison of Single-Phase and Three-PhaseFull Bridge topologies for a 50 kW fast charger Station," Master Thesis, Chalmers University of Technology, Department of Energy and Environment Division of Electric Power Engineering, 2018.

[11] K. Park and Z. Chen, "Analysis and design of a parallel-connected single active bridge DC-DC converter for high-power wind farm applications," in *2013 15th European Conference on Power Electronics and Applications (EPE)*, 2013, pp. 1–10.

[12] K. Park and Z. Chen, "Control and dynamic analysis of a parallel-connected single active bridge DC–DC converter for DC-grid wind farm application," *IET Power Electron.*, vol. 8, no. 5, pp. 665–671, 2015.

[13] C. D. Barker, C. C. Davidson, D. Trainer, and R. Whitehouse, "Requirements of DC-DC Converters to facilitate large DC Grids," 2012.

[14] J. You, L. Cheng, B. Fu, and M. Deng, "Analysis and Control of Input-Parallel Output-Series Based Combined DC/DC Converter With Modified Connection in Output Filter Circuit," *IEEE Access*, vol. 7, pp. 58264–58276, 2019.

[15] T. Lagier, P. Ladoux, and P. Dworakowski, "Potential of silicon carbide MOSFETs in the DC/DC converters for future HVDC offshore wind farms," *High Volt.*, vol. 2, no. 4, pp. 233–243, 2017.

[16] P. Wang, L. Zhou, Y. Zhang, J. Li, and M. Sumner, "Input-Parallel Output-Series DC-DC Boost Converter With a Wide Input Voltage Range, For Fuel Cell Vehicles," *IEEE Trans. Veh. Technol.*, vol. 66, no. 9, pp. 7771–7781, Sep. 2017.

[17] P. Li, S. J. Finney, and D. Holliday, "Thyristor based modular multilevel converter with active full-bridge chain-link for forced commutation," in *2016 IEEE 17th Workshop on Control and Modeling for Power Electronics (COMPEL)*, 2016, pp. 1–6.

[18] P. Li, S. J. Finney, and D. Holliday, "Active-Forced-Commutated Bridge Using Hybrid Devices for High Efficiency Voltage Source Converters," *IEEE Trans. Power Electron.*, vol. 32, no. 4, pp. 2485–2489, Apr. 2017.

[19] P. Li, G. P. Adam, S. J. Finney, and D. Holliday, "Operation Analysis of Thyristor-Based Front-to-Front Active-Forced-Commutated Bridge DC Transformer in LCC and VSC Hybrid HVDC Networks," *IEEE J. Emerg. Sel. Top. Power Electron.*, vol. 5, no. 4, pp. 1657–1669, Dec. 2017.

[20] P. Li, S. J. Finney, and D. Holliday, "Grid-connection of active-forced-commutated bridge: Power quality and DC fault protection," in *IECON 2017 - 43rd Annual Conference of the IEEE Industrial Electronics Society*, 2017, pp. 1192–1197.

[21] R. W. De Doncker, D. M. Divan, and M. H. Kheraluwala, "A three-phase soft-switched high power density DC/DC converter for high power applications," in *Conference Record of the 1988 IEEE Industry Applications Society Annual Meeting*, 1988, pp. 796–805 vol.1.

[22] A. Zama, "Modeling and Control of Modular Multilevel Converters (MMCs) for HVDC applications," Université Grenoble Alpes, 2017.

[23] "Fast Thyristor 5STF 23H2040," ABB, Datasheet, Oct. 2014.

[24] J. D. Páez, J. Maneiro, S. Bacha, D. Frey, and P. Dworakowski, "Influence of the operating frequency on DC-DC converters for HVDC grids," in *2019 21st European Conference on Power Electronics and Applications (EPE '19 ECCE Europe)*, 2019, p. P.1-P.10.

[25] D. Jovcic and H. Zhang, "Dual Channel Control With DC Fault Ride Through for MMC-Based, Isolated DC/DC Converter," *IEEE Trans. Power Deliv.*, vol. 32, no. 3, pp. 1574–1582, Jun. 2017.

[26] T. Lüth, M. M. C. Merlin, Tim. C. Green, F. Hassan, and C. D. Barker, "High-Frequency Operation of a DC/AC/DC System for HVDC Applications," *IEEE Trans. Power Electron.*, vol. 29, no. 8, pp. 4107–4115, Aug. 2014.

[27] P. S. Jones and C. C. Davidson, "Calculation of power losses for MMC-based VSC HVDC stations," in *2013 15th European Conference on Power Electronics and Applications (EPE)*, 2013, pp. 1–10.

[28] A. Fouineau, "Méthodologies de Conception de Transformateurs Moyenne Fréquence pour application aux réseaux haute tension et réseaux ferroviaires." These de doctorat, Lyon, 2019.

Inductor Size Evaluation of an Electromagnetic Interference Filter for a Two-Level Power Factor Correction Rectifier Using Different Modulation Techniques

Mohammad Najjar[1], Alireza Kouchaki[2], Morten Nymand[1]
[1]University of Southern Denmark, Odense, Denmark
[2]Danfysik A/S, Copenhagen, Denmark
E-Mail: mohna@mci.sdu.dk, ako@danfysik.dk, mny@mci.sdu.dk
URL: http://power-electronics.sdu.dk

Keywords

«Voltage Source Converter (VSC)», «Pulse Width Modulation (PWM)», «EMC/EMI», «Passive filter», «Power factor correction».

Abstract

Wide band-gap semiconductors exhibit superior performance compared to their silicon counterparts, which has enabled converters to work at higher switching frequencies. However, a higher switching frequency means larger harmonics in the electromagnetic interference (EMI) range which may lead to the need for a larger EMI filter. This paper has essentially two main purposes: to find the impact of switching frequency (swept from 40 kHz to 300 kHz) on the size of EMI filter (especially the inductor part of it) and to evaluate the size using different modulation techniques. This paper shows that the effect of increasing the switching frequency on the size of EMI filter is not linear. The differential mode filter part of EMI filter results differently against the common mode filter part of it when the switching frequency and modulation technique changes. Therefore, depending on the application a suitable modulation and frequency can be picked to result an optimum filter. To support the method of the filter design, a filter is designed and built for a 5kW two-level power factor correction rectifier and the design is supported with EMI measurement on the unit.

Introduction

The advent of wide band-gap semiconductors has given power electronics converters the ability to work at higher switching frequencies. Using a high switching frequency leads to smaller filter size; however, this propagates harmonics with higher content in the electromagnetic interference (EMI) range (>150 kHz). These high frequency harmonics or EMI emissions (i.e. common mode (CM) noise and differential mode (DM) noise) propagate in different paths. The DM noise flows through the phases; however, the CM noise closes its loop via earth and parasitic capacitors.

It is impractical to fulfil the EMI standard for both CM and DM noises using conventional filters such as L or LC filters, because the volume of a single-stage filter would be considerably large [1], [2]. Thus, different higher order filter structures can be employed to handle the DM and the CM noises. Different passive and active EMI filters for various converters with different modulation techniques have been proposed [2]-[9]. However, active filters add more complexity to the system and increase the number of magnetic components [11]-[13].

In [2] and [3], the procedure for designing two stage EMI filters to optimize their size is explained by introducing the ratio between CM and DM inductors found in the first stage of the filter. A review of different passive filters and dampers is done in [4]. The EMI filter design procedure for a 10kW converter with switching frequency of 1MHz is explained in [6]. Analytical design of an *LCL* filter for a power factor correction are carried out in [6]- [10]. In [8], the converter side inductor is designed to reduce the current ripple up to 20 percent of maximum current.

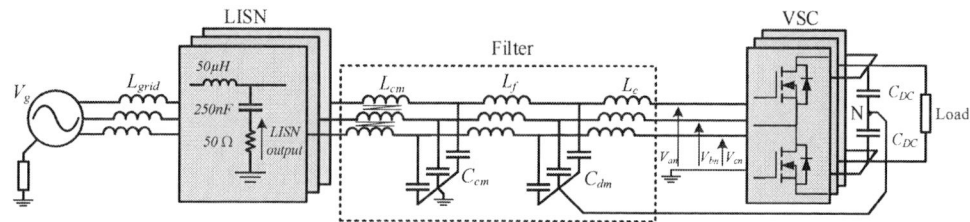

Fig. 1: The structure of the system. VSC: voltage source converter, LISN: line impedance stabilization network, C_{DC}: DC link capacitor, L_c: converter side inductor, C_{dm}: DM capacitor, L_f: grid side inductor, C_{cm}: CM capacitor, L_{cm}: CM inductor, L_{grid}: grid inductor.

Table I: The system characteristics

Parameter	Symbol	Value
Nominal power	P_{nom}	5 kW
AC side voltage frequency	V_g	230 V (rms)
DC link Voltage	V_{dc}	700 V
Grid frequency	f_g	50 Hz

Table II: The CISPR11 Class A EMI Standard.

Frequency range (MHz)	Rated power (≤20 kVA)	
	Quasi-peak dB(µV)	Average dB(µV)
0.15 − 0.50	79	66
0.50 - 5	73	60
5 - 30	73	60

In this work, the effects of different modulation techniques on the value of EMI filter inductors are studied in order to optimize filter design. DM and CM models of the suggested filter are extracted and the relation between the input and the output voltage is calculated. The parameters of the filter are calculated based on the required attenuations for DM and CM noise. The simulation and experimental results confirm the ability of the two-stage filter designed in this paper to fulfil the EMI standard.

System description and filter structure

The configuration of the three-phase power factor correction (PFC) that is under study is shown in Fig.1. Different filter configurations have been proposed to attenuate the DM and the CM noises; however, in this study, a filter with two stages of *LCL* and *LC* [2], [6] is selected to fulfil the considered EMI standard. The reason for this selection is that this configuration has fewer magnetic components than others [6]. Moreover, it can easily provide enough attenuation with just two filter stages.

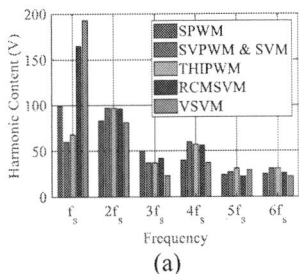 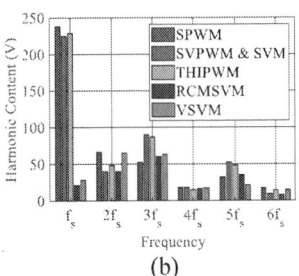

(a)   (b)

Fig. 2: The harmonic content of (a) DM output voltage, (b) CM output voltage.

The system characteristics are listed in Table I. The relevant EMI standard, CISPR11 class A, is shown in Table II [14]. The quasi-peak of DM and CM noises should be less than 79 dB(µV) for the frequency range of 150 kHz up to 500 kHz.

Modulation techniques and harmonics

The following modulation techniques are studied in this paper: sinusoidal pulse width modulation (SPWM), space vector PWM (SVPWM), third harmonic injection PWM (THIPWM), space vector modulation (SVM), reduced common mode space vector modulation (RCMSVM) [15] and virtual space vector modulation (VSVM) [16]. In SVPWM and THIPWM, adding a zero vector to the reference waveforms enables the converter to operate with modulation index up to 1.15. This feature helps the PFC rectifier to work at lower DC link voltage and subsequently decreases the switching losses in the converter. In RCMSVM, instead of using zero vectors, two other active vectors are utilized, which reduces the CM voltage of the converter. To reduce the third-order harmonic component in normal SVM, VSVM has been proposed. By using this modulation technique, the third-order harmonics which flow in the path between star point of DM capacitors and the DC link midpoint (see Fig. 1) is decreased.

The maximum harmonic content of each modulation technique at switching frequency (f_s) and multiples f_s (or at sidebands harmonics) up to 6 f_s for both DM and CM voltage are shown in Fig. 2(a) and 2(b), respectively. These values are obtained based on the Fast Fourier Transform (FFT) algorithm as follows:

$$V_{dm} = V_{an} - V_{cm} \tag{1}$$

$$V_{cm} = \frac{1}{3}(V_{an} + V_{bn} + V_{cn}) \tag{2}$$

where V_{in} ($i=a, b, c$) are the output voltages of the converter with respect to the ground, which can be found in Fig. 1.

For VSCs, the FFT results are not sufficient for designing the filter due to the function of EMI receiver and the effects of side-band harmonics [17]. Nevertheless, these results show the effects of changing switching frequency on the magnitude of the first peak of EMI noise.

Comparing the harmonic contents of both DM and CM output voltages shows that RCMSVM and VSVM have lower CM harmonic content at switching frequency. On the other hand, the DM harmonics for both of these two modulation techniques are higher than the other ones. Meaning, more DM attenuations are required for these two modulation techniques. Compared to other modulation techniques, SPWM has lower harmonic contents for DM at switching frequency multiples above 4 f_s. Comparing the CM harmonic contents of SPWM shows that this modulation technique is more appropriate than others for applications with switching frequencies lower than 50 kHz due to a lower harmonic content above $4f_s$. The reason is that the first major harmonic that appears in the EMI range will be the harmonic with the frequency of $4f_s$ or higher.

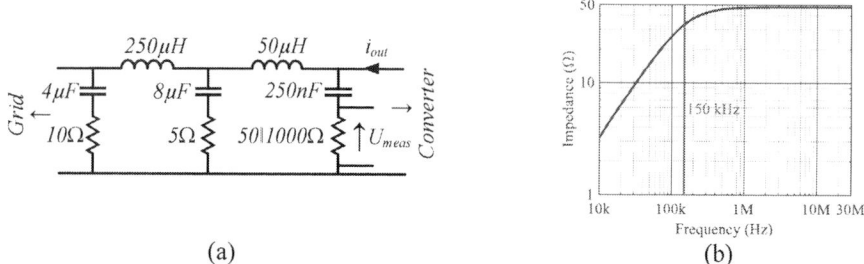

(a)   (b)

Fig. 3: a) LISN equivalent single-phase model. b) impedance frequency behavior of LISN

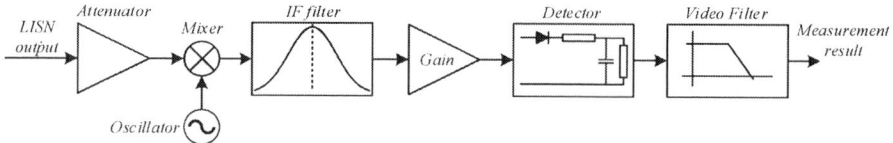

Fig. 4: The harmonic content of (a) DM output voltage, (b) CM output voltage.

Line Impedance Stabilization Network (LISN) and Spectrum Analyzer

A. LISN

The main important features of LISN are: high frequency decoupling between the converter under test and the grid, showing constant impedance at EMI frequency ranges and the ability to repeat the measurements [18]. The equivalent single-phase model of LISN is shown in Fig. 3(a) [19] and the frequency behavior of LISN impedance or U_{meas}/i_{out} in Fig. 3(b) [18]. As can be seen, LISN shows a constant impedance near to 50 ohms for frequencies in the EMI range.

B. EMI Receiver

Fig. 4 shows a simplified block diagram of the EMI receiver. At the first stage, the output signal of LISN is attenuated. The mixer shifts the attenuated signal to an intermediate frequency (IF). Then the IF filter (band-pass filter), also known as a Resolution Band-Width (RBW) filter, is applied to the signal and changes the magnitude of the signal. In this study, a Gaussian function models the IF filter [17]. The bandwidth of IF filter varies according to the frequency range of interest [17]. Moreover, to eliminate the functionality of mixer and oscillator in simulation, the center frequency of IF filter is swept [18]. Finally, the peak, quasi peak and average of the filtered signal are determined.

For both DM and CM noise, the quasi-peak value of the first voltage peak appearing within the EMI range was simulated for switching frequencies between 40 kHz and 300 kHz. These results are shown in Fig. 5(a) and 5(b), respectively.

EMI Filter

The first step in designing a filter is to figure out how much attenuation is needed to fulfill the standard [6]. This can be achieved by using the simulated results from the model EMI receiver as shown in Fig. 5 and calculated as follows:

$$Att_{req} \text{ [dB]} = V_{QP} \text{ [dB}\mu\text{V]} - \text{limit [dB}\mu\text{V]} + \text{ margin [dB]} \tag{3}$$

The simplified equivalent circuits for DM and CM noise are shown in Fig. 6(a) and 6(b), respectively. As mentioned before, by connecting the middle point of the DC link to the star point of the DM capacitors (Fig. 1) a new path for CM noise will be provided. Meaning, with this connection the DM

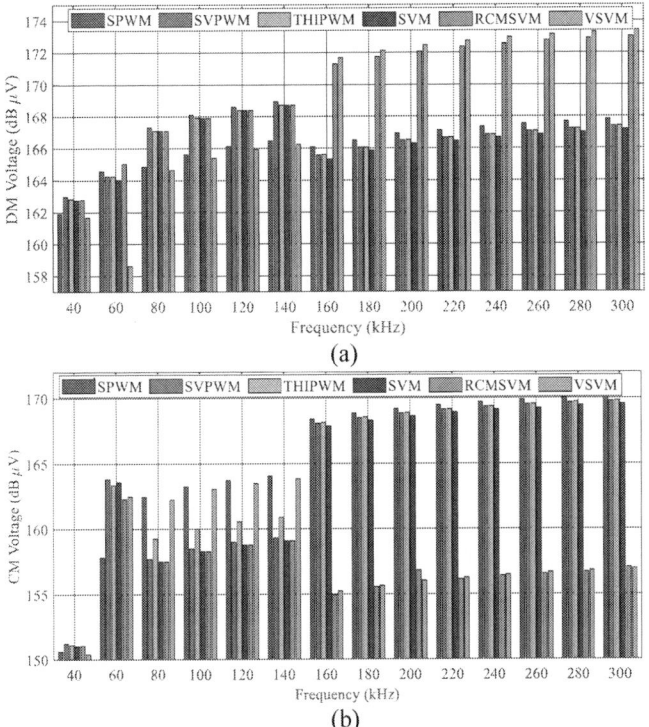

(a)

(b)

Fig. 5: The quasi-peak values of greatest EMI noise in the EMI range based on switching frequency (a) DM output voltage (b) CM output voltage

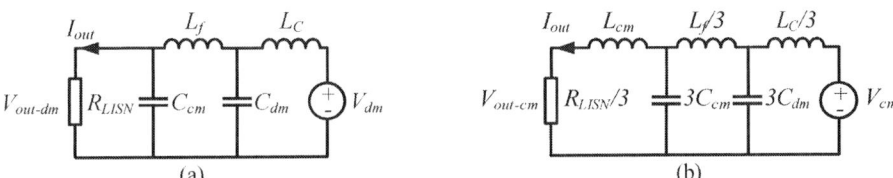

(a)   (b)

Fig. 6: Equivalent model (a). DM model (b). CM model.

capacitors also influence CM noise (Fig. 6(b)) by increasing the order of the characteristic equation of the CM transfer function (between input and output voltages).

To reach a proper attenuation, the relation between the input and the output voltage for both DM and CM must be obtained. Based on the equivalent circuits of Fig. 6, these relations can be calculated as follows:

$$\frac{V_{out-dm}}{V_{dm}} = \frac{R_{lisn}}{as^4 + bs^3 + cs^2 + ds + R_{lisn}}$$

$$a = C_{cm}C_{dm}L_cL_fR_{lisn},\ b = C_{dm}L_cL_f,\ c = C_{cm}L_cR_{lisn} + C_{dm}L_cR_{lisn} + C_{cm}L_fR_{lisn},\ d = L_c + L_f$$

(4)

$$\frac{V_{out-cm}}{V_{cm}} = \frac{R_{lisn}}{as^5 + bs^4 + cs^3 + ds^2 + es + R_{lisn}}$$

$$a = 3C_{cm}C_{dm}L_cL_{cm}L_f,\ b = C_{cm}C_{dm}L_cL_fR_{lisn},\ c = 3C_{cm}L_cL_{cm} + 3C_{dm}L_cL_{cm} + 3C_{cm}L_{cm}L_f + C_{dm}L_cL_f$$

$$d = C_{cm}L_cR_{lisn} + C_{dm}L_cR_{lisn} + C_{cm}L_fR_{lisn},\ e = L_c + 3L_{cm} + L_f$$

(5)

where R_{lisn} is the resistive impedance of line impedance stabilization network (LISN).

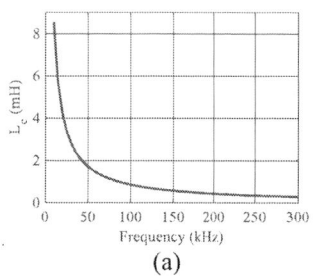 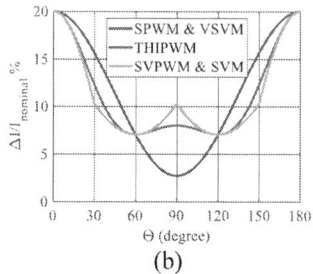

(a)   (b)

Fig. 7: (a) The value of converter side inductor and switching frequency, (b) the degree of current ripple occurrence.

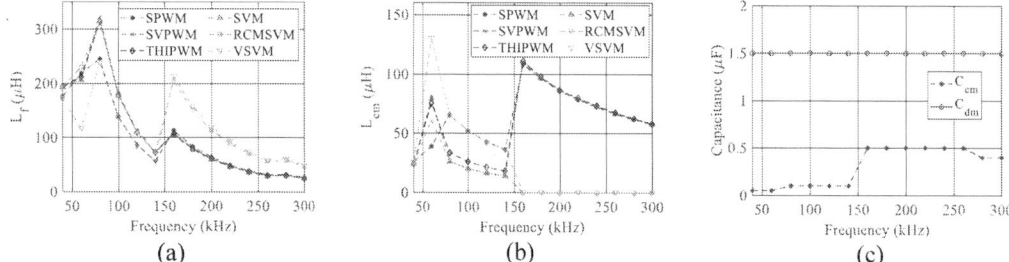

(a)   (b)   (c)

Fig. 8: The filter values for different modulation techniques against switching frequency, (a) grid side inductor, (b) CM inductor, (c) capacitors (C_{dm} & C_{cm}).

A. converter side inductor

The converter side inductor (L_c) is designed based on the maximum acceptable current ripple. Normally, this value is adjusted to limit the current ripple to between 5 and 30 percent of the maximum current peak [8]. L_c can be obtained as follows [8]:

$$\Delta i_{max} = \frac{V_{dc}}{4 \times L \times f_s} \qquad (6)$$

L_c value as a function of the switching frequency is shown in Fig. 7(a) (for keeping the maximum current ripple at 20% of the peak of nominal current). As shown in Fig. 7(b), for 2-level converter, regardless of the modulation technique the maximum current ripple occurs during zero crossing.

B. Grid Side Inductor and Filter Capacitors

The maximum value for filter capacitors is chosen to be lower than 5% of the nominal power. To reach the desired attenuation for the DM noise, C_{dm}, L_f and C_{cm} should be calculated based on equation (4). It is possible to consider two of these variables (for example two capacitors) as having constant values and try to solve the equation based on the next parameter. Normally, in the usual *LCL* filter, the grid-side inductor is selected to reduce the current ripple to 0.2 percent of the maximum current peak to satisfy the grid tie standards for frequencies lower than the EMI range [8]. Thus, the considered value for the grid-side inductor should fulfil both limitations based on maximum acceptable current ripple and on equation (4). By obtaining C_{dm}, L_f and C_{cm}, the value of common mode inductor (L_{cm}) can be calculated based on equation (5). Moreover, the addition of both DM and CM noises should be lower than the specific value defined by standard limitation.

The filter parameters obtained using different modulation techniques are shown in Fig. 8.

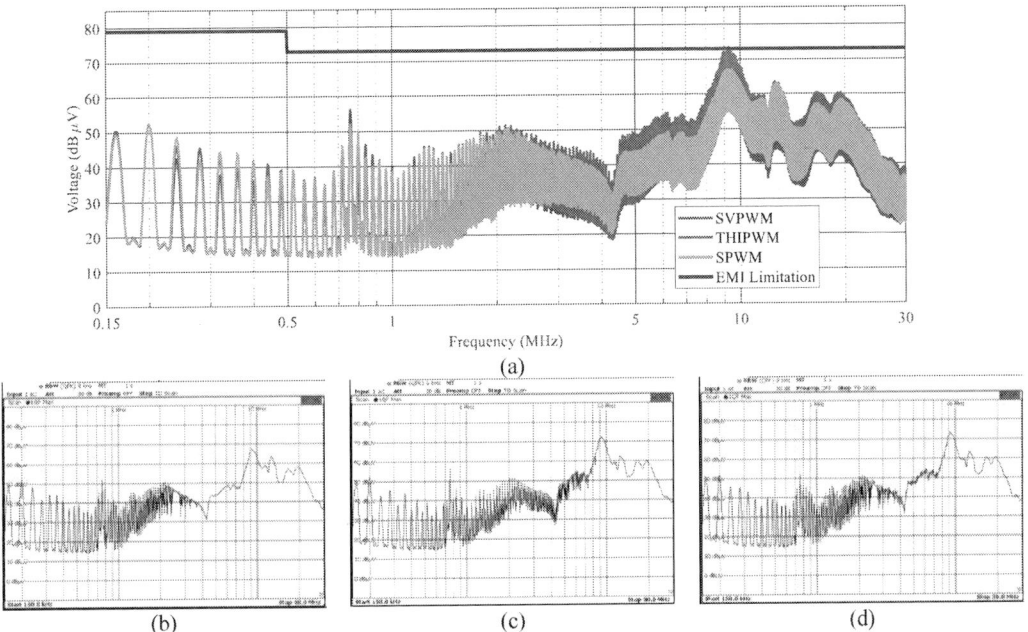

(a)

(b)   (c)   (d)

Fig. 9: The experimental results of the EMI receiver at switching frequency of 40 kHz, (a) SPWM, THIPWM and SVPWM (b) SPWM, (c) THIPWM, (d) SVPWM.

Comparison

To reach a minimal size for EMI filter the values of the inductors should be examined and minimized carefully. Therefore, acceptable values are considered for filter capacitors [1]. As it can be seen in Fig. 5, for all different modulation techniques at the same switching frequency the values of the capacitors were kept constant. Thus, the comparison can be done based on the value of the inductors.

For the switching frequency from 37.51 kHz up to 49.9 kHz the harmonics above $3f_s$ are in the EMI range. Since the EMI filter is designed based on the first peak harmonic falling inside the EMI range ($4f_s$), a smaller L_f can be obtained in this switching frequency range by using SPWM or VSVM.

For a switching frequency from 50 kHz up to 75.49 kHz, the harmonic that determines the size of filter is the third multiple of the switching frequency. In this range, the value of L_f are almost similar for all modulation techniques except VSVM. Since a smaller L_f is required to reach proper attenuation for CM using VSVM, a larger CM inductor is needed.

In the switching frequency range of 75.5 kHz up to 149.9 kHz the EMI filter is designed based on the harmonic of the second multiple of the switching frequency. Here SPWM and VSVM have lower L_f, meanwhile, SVPWM, SVM and RCMSVM require lower CM inductor.

When the switching frequency is above 150 kHz, the value of C_{cm} should be increased to comply with the EMI standard while using reasonable values of inductors. Within this range, RCMSVM and VSVM give higher values for L_f than the other techniques. Since the output CM converter voltage for these two modulation techniques is less than for other techniques, the filter can provide enough attenuation for CM noise without using a CM inductor. In fact, by choosing proper capacitor values and the right modulation technique, it would be possible to choose switching frequencies less than 300 kHz.

Experimental Results

To evaluate the efficiency of the calculated values for EMI filter components, an experiment was performed based on the Fig. 1 and parameters of Table I. The quasi-peak value of the harmonic contents for the considered EMI frequency range (150 kHz up to 30 MHz) is shown is Fig. 9, when the switching frequency is 40 kHz. As it can be seen, the filter can effectively reduce the harmonics content for the tested modulation techniques and thereby satisfy the EMI standard.

Conclusion

In this study, the effects of changing switching frequency from 40 kHz up to 300 kHz on the size of the EMI filter was assessed. It was shown that to comply with the EMI standard for switching frequencies between 150 kHz up to 220kHz (EMI range), the size of filter needs to be increased. The effect of decreasing the size of filter by increasing the switching frequency is not applicable in this range of switching frequencies. Also, a proper modulation technique should be selected to obtain an optimized filter size when the switching frequency is above 150 kHz.

References

[1] Erickson R. W., Maksimovic D., "Fundamentals of Power Electronics", 2nd ed. Norwell, MA, USA: Kluwer, 2001, p. 881.

[2] Boillat D. O., Krismer F., Kolar J. W., "EMI Filter Volume Minimization of a Three-Phase, Three-Level T-Type PWM Converter System," IEEE Trans. Power Electron., vol. 32, no. 4, pp. 2473-2480, April 2017.

[3] Boillat D. O., Kolar J. W., Mu¨hlethaler J., "Volume minimization of the main DM/CM EMI filter stage of a bidirectional three-phase three-level PWM rectifier system," 2013 IEEE Energy Conversion Congress and Exposition, Denver, pp. 2008-2019, 2013.

[4] Beres R. N., Wang X., Liserre M., Blaabjerg F., Bak C. L., "A Review of Passive Power Filters for Three-Phase Grid-Connected Voltage-Source Converters," IEEE j. emerg. sel. top. power electron., vol. 4, no. 1, pp. 54-69, March 2016.

[5] Nussbaumer T., Heldwein M. L., Kolar J. W., "Differential Mode Input Filter Design for a Three-Phase Buck-Type PWM Rectifier Based on Modeling of the EMC Test Receiver," IEEE Trans. Ind. Electron, vol. 53, no. 5, pp. 1649-1661, Oct. 2006.

[6] Hartmann M., Ertl H., Kolar J. W., "EMI Filter Design for a 1 MHz, 10 kW Three-Phase/Level PWM Rectifier," IEEE Trans. Power Electron., vol. 26, no. 4, pp. 1192-1204, April 2011.

[7] Liserre M., Blaabjerg F., Hansen S., "Design and control of an LCL-filter-based three-phase active rectifier," IEEE Trans. Ind. Appl., vol. 41, no. 5, pp. 1281-1291, Sept.-Oct. 2005.

[8] Kouchaki A., Nymand M., "Analytical Design of Passive LCL Filter for Three-Phase Two-Level Power Factor Correction Rectifiers," IEEE Trans. Power Electron., vol. 33, no. 4, pp. 3012-3022, April 2018.

[9] Jalili K., Bernet S., "Design of LCL Filters of Active-Front-End Two-Level Voltage-Source Converters," IEEE Trans. Ind. Electron, vol. 56, no. 5, pp. 1674-1689, May 2009.

[10] Kouchaki A., Nymand M., "LCL filter design for three-phase two-level power factor correction using line impedance stabilization network", IEEE Applied Power Electronics Conference and Exposition (APEC), Long Beach, CA, 2016, pp. 2382-2388, 2016.

[11] Najjar M., Kouchaki A., Nymand M., "An Efficient Active Common Mode Filter: Comparison of Feedback and Feedforward Based Methods for a 20 KW 3-phase Inverter," 2019 IEEE 13th International Conference on Compatibility, Power Electronics and Power Engineering (CPE-POWERENG), Sonderborg, Denmark, 2019.

[12] Najjar M., Kouchaki A., Nymand M., "Evaluation of Active Common Mode Filter Utilization for Size Optimization of a 20 kW Power Factor Correction," 2019 IEEE 13th International Conference on Compatibility, Power Electronics and Power Engineering (CPE-POWERENG), Sonderborg, Denmark, 2019.

[13] D. Xu, C. K. Lee, S. Kiratipongvoot and W. M. Ng, "An Active EMI Choke for Both Common- and Differential-Mode Noise Suppression", IEEE Trans. Ind. Electron., vol. 65, no. 6, pp. 4640-4649, June 2018.

[14] IEC C.I.S.P.R., "Industrial, scientific and medical equipment –Radio-frequency disturbance characteristics – Limits and methods of measurement", 2015.

[15] Hou C., Shih C., Cheng P., Hava A. M., "Common-Mode Voltage Reduction Pulsewidth Modulation Techniques for Three-Phase Grid-Connected Converters," IEEE Trans. Power Electron., vol. 28, no. 4, pp. 1971-1979, April 2013.

[16] Tian K., Wang J., Wu B., Cheng Z., Zargari N. R., "A Virtual Space Vector Modulation Technique for the Reduction of Common-Mode Voltages in Both Magnitude and Third-Order Component," IEEE Trans. Power Electron., vol. 31, no. 1, pp. 839-848, Jan. 2016.

[17] Yang L., Zhao H., Wang S., Zhi Y., "A Technique to Accurately Predict EMI Noise Spectrum in Wide Frequency Ranges Based on the Principles of Spectrum Analyzers," 2018 IEEE Energy Conversion Congress and Exposition (ECCE), Portland, OR, pp. 6410-6417, 2018.

[18] Davari P., Blaabjerg F., Hoene E., Zare F., "Improving 9-150 kHz EMI Performance of Single-Phase PFC Rectifier," CIPS 2018; 10th International Conference on Integrated Power Electronics Systems, Stuttgart, Germany, 2018.

[19] Giezendanner F., Biela J., Kolar J. W., Zudrell-Koch S., "EMI Noise Prediction for Electronic Ballasts," IEEE Trans. Power Electron., vol. 25, no. 8, pp. 2133-2141, Aug. 2010.

Evaluation of MMCs for High-Power Low-Voltage DC-Applications in Combination with the Module LLC-Design

Roland Unruh, Frank Schafmeister, Joachim Böcker
Paderborn University
Warburger Str. 100
Paderborn, Germany
Tel.: +49/ (05251) 60-3492
Fax.: +49/ (05251) 60-3443
E-Mail: unruh@lea.uni-paderborn.de
URL: https://ei.uni-paderborn.de/lea

Acknowledgments

The authors would like to thank the German Research Foundation (DFG) for the funding the research on modular-multilevel converters under the project number 314461654.

Keywords

≪Multilevel converters≫, ≪Resonant converter≫, ≪High voltage power converters≫, ≪ZVS Converters≫, ≪Combination MMC LLC≫

Abstract

In this paper, a full-bridge modular multilevel converter (MMC) and two half-bridge-based MMCs are evaluated for high-current low-voltage e.g. $100 - 400\,\text{V}$ DC-applications such as electrolysis, arc welding or datacenters with DC-power distribution. Usually, modular multilevel converters are used in high-voltage DC-applications (HVDC) in the multiple kV-range, but to meet the needs of a high-current demand at low output voltage levels, the modular converter concept requires adaptations. In the proposed concept, the MMC is used to step-down the three-phase medium-voltage of $10\,\text{kV}$, and provide up to $1\,\text{MW}$ to the load. Therefore, each module is extended by an LLC resonant converter to adapt to the specific electrolyzers DC-voltage range of $142 - 220\,\text{V}$ and to provide galvanic isolation.

The six-arm MMC converter with half-bridge modules can be simplified and optimized by removing three arms, and thus halving the number of modules. In addition, the module voltage ripple and capacitor losses are decreased by 22% and 30% respectively. By rearranging the components of the half-bridge MMC to build a MMC consisting of grid-side full-bridge modules, the voltage ripple is further reduced by 78% and capacitor losses by 64%, while ensuring identical costs and volume for all MMCs.

Finally, the LLC resonant converter is designed for the most efficient full-bridge MMC. The LLC can not operate at resonance with a fixed nominal module voltage of $770\,\text{V}$ because the output voltage is varying between $142 - 220\,\text{V}$. By decreasing the module voltage down to $600\,\text{V}$, additional points of operation can be operated in resonance, and the remaining are closer to resonance. The option to decrease the module voltage down to $600\,\text{V}$, increases the number of required modules per arm from 12 to 15, which requires to balance the losses of the LLCs and the grid-side stages.

Introduction

Applications such as electrolysis or arc welding require high DC-currents and low voltages in the range of $100\,\text{V}...400\,\text{V}$. Same applies for large data centers with DC-power distribution. The state of the art is a medium-voltage transformer to convert the three-phase medium-voltage of $10\,\text{kV}$ to a lower AC-voltage that is rectified for the load. Comparing it to the proposed system, the conventional transformer-based rectifier [1] shows disadvantages in terms of volume, power factor, and modularity [2]. Therefore, a system is desired that has a controllable power factor [3] [4], a low THD [5] [6], and moreover, consists of identical modules to achieve redundancy [7]. So, it should not rely on a single component such as a central HF-transformer [8]. This can be achieved by using the modular multilevel-converter (MMC) [9] shown in Fig. 1a). In this application, an LLC resonant converter is connected to each module capacitance $C_{r,xjn}$, which converts the module voltage to a lower output voltage v_{dcl} shown e.g. in Fig. 1b). The transformers of the LLCs provide the required galvanic isolation, and the bulky grid-frequency transformer is replaced by multiple smaller high-frequency transformers.

The investigated MMCs are similar to a structure being refered to in recent publications as a power-electronic- or solid-state-transformer [10] [11] [12]. All of them use high-frequency transformers to reduce their sizes. However, the investigated MMCs have a dedicated transformer for each module. Furthermore, the proposed MMCs are unidirectional, and use LLCs with a decentralized control [9]. In addition, solid-state-transformers are usually designed to provide a constant output voltage while the investigated MMCs can provide controllable output voltage to control the power of a load such as electrolysis [13] [14] [15].

Three different MMCs are investigated with respect to losses, voltage ripple and module count. This paper will show, that the module number of a three-arm MMC with full-bridge modules Fig. 1b) with e) is halved compared to a half-bridge three-arm MMC Fig. 1b) with d), and even quartered compared to a six-arm MMC Fig. 1a) with d). Additionally, the module voltage ripple and capacitor losses can be decreased by 78% and 64% respectively while all topologies use an identical component count.

Consequently, the LLC is designed for the optimal MMC-topology. Although, the LLC shows the highest efficiency at resonance which can be achieved by reducing the module input voltage at low output voltage, this increases the required number of modules, and also the losses of the grid-side stage. For this reason, a compromise between the number of modules and resonant operation of the LLC is presented which maximizes the overall efficiency.

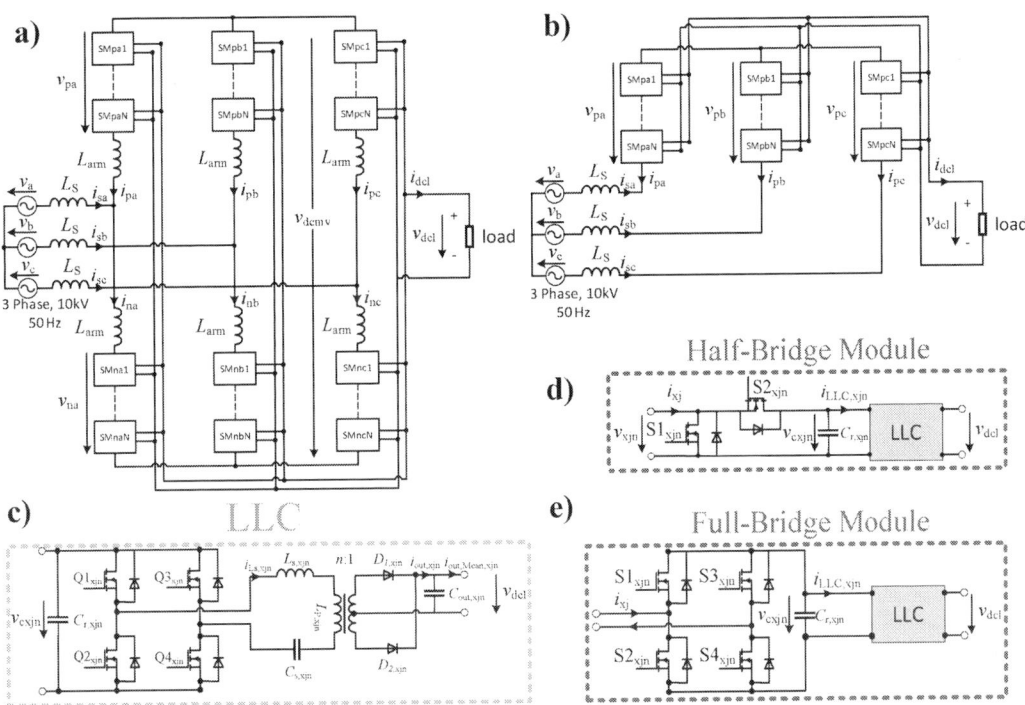

Fig. 1: In a), the structure of an YY-MMC [2] is shown which consists of half-bridge modules d). The Y-MMC [9] in b) can consist of half-bridge d) or full-bridge modules e). Each module employs an LLC c). Both MMCs in a) and b) have the same functionality. Only the full-bridge Y-MMC can charge $C_{r,xjn}$ with a positive and negative arm current i_{xj}, which minimizes its voltage ripple.

Evaluation of Topologies

The topology in Fig. 1a) can provide power to the load by charging the module capacitors $C_{r,xjn}$ with half-bridge modules shown in Fig. 1d), and converting this voltage to the lower load voltage v_{dcl} by using one LLC converter in each module. This topology is called YY-MMC because the modules are arranged in two star connections, while the Y-MMC in Fig. 1b) has only one star connection [9]. All topologies in Fig. 1 can provide power to the load with an operating area as shown in Fig. 5, and a peak power of $\hat{P}_{load} = 1\,\text{MW}$. In steady state the grid currents i_{sa}, i_{sb}, i_{sc} must be sinusoidal and all three MMCs control these currents by adjusting the arm voltages v_{xj}. The YY-MMC is operating to keep $v_{dcmv} = 18\,\text{kV}$ [2] constant as it would be required for HVDC-generation. This allows the usage of the same control structure which balances the module voltages with circular currents [2] [16].

The resulting arm voltages are shown in Fig. 2a). In addition, the voltage within one arm is balanced with conventional voltage balancing [17] for the evaluated MMCs by charging that module with the lowest voltage first. For a comparison of the topologies, Fig. 3b) shows the average module voltage v_{cpjn} of the LLCs within one arm.

Fig. 2a) shows that the upper arm voltages $v_{pj} > 0$ are always positive, but the arm current i_{pj} is sinusoidal and negative for half a grid period which partially discharges the module with $P_{Capacitor} < 0$ [9] as shown in Fig. 3a). As a consequence, the YY-MMC shows the highest voltage ripple and capacitor current stress which results in the highest capacitor losses and required LLC-gain of the investigated MMCs.

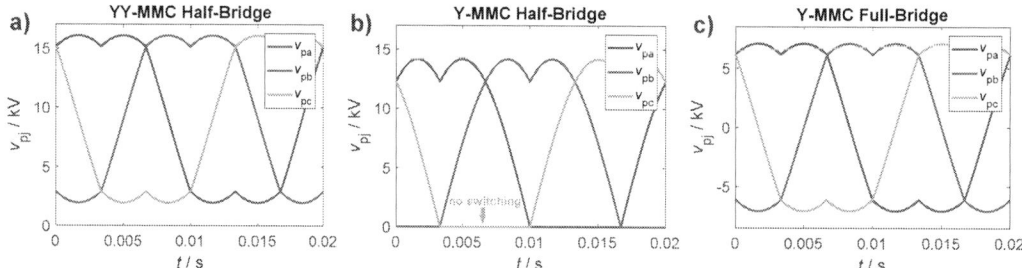

Fig. 2: The YY-MMC a) has always positive arm voltages v_{pj} to achieve a constant $v_{dcmv} = 18\,\text{kV}$ [2]. The Y-MMC b) has always one clamped arm to minimize the switching losses [9] and $\hat{v}_{pj}$. With full-bridge modules c), the maximum required arm voltage is halved from $\hat{v}_{pj} = 14.2\,\text{kV}$ to $\hat{v}_{pj} = 7.1\,\text{kV}$ which reduces the number of required modules by 50% and by 75% compared to the YY-MMC.

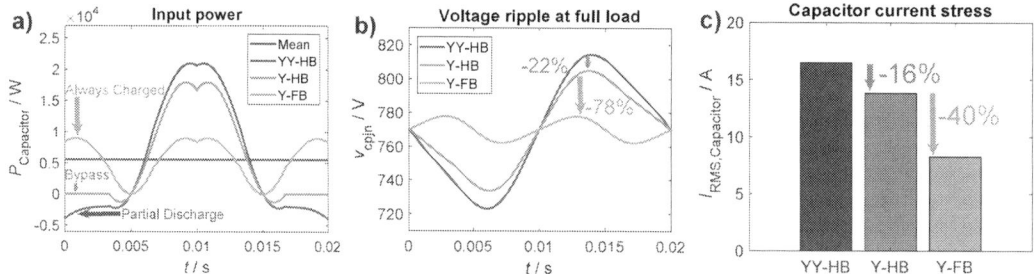

Fig. 3: The six-arm YY-MMC (red) partially discharges the module capacitance $C_{r,xjn}$ during half a grid period resulting in a high voltage ripple and capacitor current stress. The Y-MMC with half-bridges (green) bypasses the negative arm current i_{px} by setting $v_{px} = 0$ which reduces the ripple and current stress. With full-bridges (yellow), $C_{r,xjn}$ is always charged and the ripple and current stress are minimized.

Half-Bridge Y-MMC

The YY-MMC provides a constant v_{dcmv}, which is unnecessary for the proposed application since the LLCs convert the module voltage v_{cxjn} to the load voltage v_{dcl}. Therefore, the three lowers arms can be removed [9] which halves the number of required modules as shown in Fig. 1b). However, the arm voltages v_{pa}, v_{pb}, v_{pc} control the grid currents i_{sa}, i_{sb}, i_{sc}, but the third grid current is determined with Kirchhoff's nodal rule via the two other grid currents. Only two currents have to be controlled explicitly by the three arm voltages. The constraint of keeping v_{dcmv} constant, is not relevant nor possible for the Y-MMC. This offers one degree of freedom for optimization. The voltage ripple of the capacitors and the switching losses are minimized by the introduction of the following equation utilizing this degree of freedom [9]:

$$\min(v_{pa}, v_{pb}, v_{pc}) \overset{!}{=} 0 \tag{1}$$

The resulting arm voltages are shown in Fig. 2b), and the maximum required arm voltage level is $\hat{v}_{px} = 14.2\,kV$. This can be achieved with a nominal module voltage of $v_{cxjn} = 770\,V$ at full load to account for approximately $30\,V$ voltage ripple and $N_{Arm} = 30$ for a sufficient arm voltage reserve, resulting in 90 overall modules instead of 180 for the six-arm YY-MMC.

For the YY-MMC one capacitor with a capacitance of $C = 1.2\,mF$ is selected resulting in $C_{r,xjn} = 1.2\,mF$. The Y-MMC with half-bridge modules (HB-modules) requires half the number of modules, and therefore two of these capacitors are connected in parallel resulting in $C_{r,xjn} = 2.4\,mF$. For a fair comparison, Fig. 3 shows the input power fluctuations and current stress of a single capacitor C since the overall number of capacitors is equal for all three topologies. The peak output power of a single LLC is doubled due to the halved number of modules from $\hat{P}_{LLC} = 5.55\,kW$ to $\hat{P}_{LLC} = 11.1\,kW$. As a result, the reduced voltage ripple and capacitor current stress in combination with fewer LLCs proves the three-arm Y-MMC with half-bridge modules a superior topology in comparison to the six-arm YY-MMC.

Full-Bridge Y-MMC as next Mayor Step

The three-arm Y-MMC can consist of full-bridge modules (FB-modules) to charge the $C_{r,xjn}$ during the whole grid period. In addition, the absolute value of $|\hat{v}_{px}|$ should be minimized to reduce the number of required modules. For this reasons, Eq. (1) is adapted to satisfy both conditions [18]:

$$\max(v_{pa}, v_{pb}, v_{pc}) \overset{!}{=} -\min(v_{pa}, v_{pb}, v_{pc}) \tag{2}$$

The equation Eq. (2) is implemented with the *Voltage Minimizer* shown in Fig. 4. This is similar to a third harmonic injecion [19] [20], and the savings in module count are identical. Consequently, the *Voltage Minimizer* reduces the peak arm voltage by 14% (to $\frac{\sqrt{3}}{2}$), which reduces the required number of modules, as well as the conduction and driver losses per arm by 14%. However, here no tracking of the phase angel is required, since the *Voltage Minimizer* calculates the offset out of the three given arm voltages v_{px}.

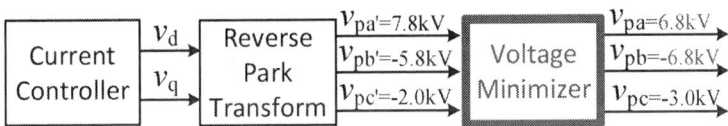

Fig. 4: In this example, the three arm voltages v_{px} are offset by $1.0\,kV$ to satisfy Eq. (2). The half-bridge Y-MMC also uses the *Voltage Minimizer*, but satisfies Eq. (1).

The resulting arm voltages v_{px} are shown in Fig. 2c). The maximum required arm voltage $|\hat{v}_{px}|$ is halved compared to the half-bridge Y-MMC, which again halves the number of modules from 30 to $N_{Arm} = 15$. As a consequence, the module output power is doubled from $\hat{P}_{Mod} = 11.1\,kW$ to $\hat{P}_{Mod} = 22.2\,kW$. This requires to double the output power of a single LLC to $\hat{P}_{LLC} = 22.2\,kW$ or to parallel two LLCs with $\hat{P}_{LLC} = 11.1\,kW$ [21] [22] [23]. The largest advantage of the full-bridge module is the reduction of the voltage ripple by 78% compared to the half-bridge modules and even 83% compared to the six-arm YY-MMC as it is shown in Fig. 3b). In addition, the peak charging current is halved since 4 instead of 2 capacitors with $C = 1.2\,mF$ are connected in parallel for $C_{r,xjn} = 4.8\,mF$. This reduces the RMS-current of one capacitor C by 40%, and therefore the ESR-losses by 64% compared to the half-bridge modules. Consequently, the lifetime of the capacitors is also increased [24]. The capacitor RMS-currents were calculated with the average current model [25] assuming ideal balancing [17]. As a result, the Y-MMC with full-bridge modules is the best of the three investigated MMCs since it has the least voltage ripple, the lowest capacitor losses and requires the least amount of modules.

Trade-Off between LLC Efficiency and Module Count

The LLC is designed for the three-arm Y-MMC with full-bridge modules because this is the most efficient investigated MMC. The operating area of the electrolyser load is shown in Fig. 5. At full load the total power is $\hat{P}_{\text{Load}} = 1\,\text{MW}$, while the load voltage is $v_{\text{dcl}} = 142 - 220\text{V}$. It depends on the load current i_{dcl}, water temperature and gas pressure of the electrolyser [15]. The LLC has the highest efficiency at resonance [26], and at full load resonance can be achieved by selecting the transformer ratio $n = \frac{\hat{v}_{\text{cxjn}}}{\hat{v}_{\text{dcl}}} = \frac{770\,\text{V}}{220\,\text{V}} = 3.5$. By decreasing the module voltage down to $v_{\text{cxjn}} = 500\,\text{V}$, even for the lowest load voltage of $v_{\text{dcl}} = 142\,\text{V}$, the LLCs operate at resonance. However, this would require $N_{\text{Arm}} = 18$ in order to achieve the arm voltages v_{pj} in Fig. 2c) while accounting for a reserve with $\hat{v}_{\text{pj}} = 9\,\text{kV}$. This voltage reserve is selected to account for grid over-voltages of 10 % and a possible damaged module in one arm. In contrast, $N_{\text{Arm}} = 12$ can be selected to minimize the number of modules. As a result, the losses of the grid-side stage are minimized because the grid current i_{sj} causes identical conduction losses in each module within one arm [27]. However, $N_{\text{Arm}} = 12$ would limit the minimum module voltage to $v_{\text{cxjn}} = 750\,\text{V}$ because $\hat{v}_{\text{pj}} = 9\,\text{kV}$ has to be achieved for the chosen arm voltage reserve. As a result, the LLC converter could operate for only a few points at resonance, and the resulting LLC losses are high. In addition, the LLC has to be designed for a relative high gain which further decreases the efficiency even at the few resonance points [28] [29]. This is confirmed by own simulations, which show that a high gain increases the resonance current $i_{\text{Ls,xjn}}$. This increases the conduction, switching and core losses while the peak voltage of the resonance capacitance $C_{\text{s,xjn}}$ is increased as well.

Therefore, a compromise between the efficiency of the LLC converter and the number of modules has to be found to minimize the overall losses and costs. For this reason, $N_{\text{Arm}} = 15$ is selected to achieve resonance in a large area of operation shown in Fig. 5 because the module voltage can be decreased to $v_{\text{cxjn}} = 600\,\text{V}$. The peak power of the LLC converter is $\hat{P}_{\text{LLC}} = 11.1\,\text{kW}$ because two paralleled LLCs are chosen per module, but for demonstration purposes only one is build up in the laboratory.

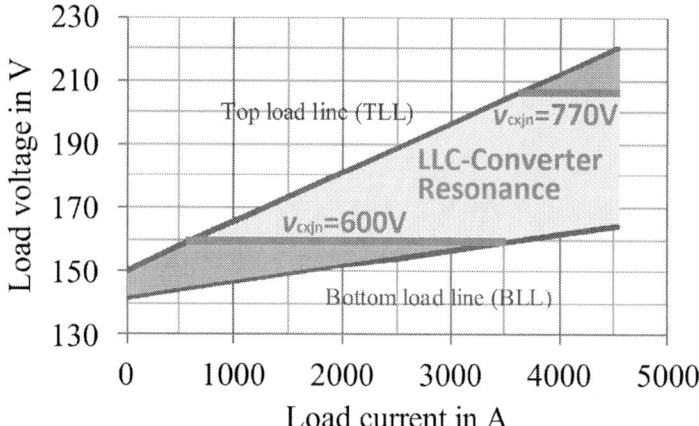

Fig. 5: Operating area of an 1 MW electrolyser [2]. With $N_{\text{Arm}} = 15$, the LLCs operate at resonance for load voltages between $v_{\text{dcl}} = 160\,\text{V}$ and $v_{\text{dcl}} = 206\,\text{V}$ since their input voltage v_{cxjn} is adjusted accordingly. The voltage is not decreased below $v_{\text{cxjn}} = 600\,\text{V}$ to achieve a safe operation of the MMC. Only for $N_{\text{Arm}} = 18$ the LLCs could always operate at resonance.

Components of the LLC Converter

For an efficient design, appropriate values for the components of the LLC have to be determined. The SiC-MOSFETs C3M0075120J of the inverter are selected according to the expected current rating at full load. The rectifier diodes have to withstand a maximum voltage of $2 \cdot \hat{v}_{\text{dcl}} = 440\,\text{V}$ which the selected

650 V SiC-diodes of the type C5D50065D do. Two diodes are connected in parallel for $D_{1,xjn}$ and for $D_{2,xjn}$ to reduce the thermal load of each diode.

The transformer consists of two connected E80/38/20 cores with a center tapped bifilar winding of the secondary side [30] to halve the losses and number of diodes of the rectifier. A litz wire of $600*0.071\,mm$ is chosen for the primary side of the transformer, and $2100*0.071\,mm$ for the secondary side. This results in approximately equal current densities in all wires, and the litz wires have a small diameter to minimize eddy current losses [31] [32].

These wires with 3 layers Mylar insulation are selected since the required isolation of the transformer has to be 10 kV. An own measurement shows that the insulation can withstand at least 1 kV for one minute. This is not sufficient, and for this reason the winding window is only filled partially with Litz wires to leave space for an insulation foil between the primary and secondary side of the transformer. The leakage inductance of the integrated transformer is adjusted to provide the necessary resonance inductance L_s [33]. The resonance capacitance $C_{s,xjn}$ consists of 18 nF foil capacitors B32641B0183J with a permissible current of 2.2 A. The primary current is $\hat{I}_{Ls} = 17.6\,A$ at full load, and therefore at least 8 capacitors have to be connected in parallel. Two strings of capacitors are connected in series to double the voltage rating of the resonance capacitance which results in $C_{s,xjn} \geq 72\,nF$.

An upcoming optimization can influence the parameters $L_{s,xjn}$, $L_{p,xjn}$ and n of the transformer as well as the resonance capacitance $C_{s,xjn}$ to maximize the efficiency.

Optimal Parameters of the LLC Converter

A steady-state model of the LLC resonant converter similar to [29] is implemented to calculate all relevant currents and voltages for different values of $C_{s,xjn}$, $L_{s,xjn}$, $L_{p,xjn}$ and n at all loads. A loss model for each component of the LLC resonant converter is developed, and the losses of the inverter and rectifier are estimated with the given datasheet values. The transformer core losses are calculated with the generalized Steinmetz equation [34], while accounting for the core temperature with a simplified thermal model [35], since the measured core surface temperature reaches 70°C at full load as shown in Fig. 7. However, in the core center the temperature is higher [36], and therefore a sufficient margin is required.

Due to the high switching frequency, the eddy current losses of the litz wires of the transformer are calculated for one point of operation with a FEM-simulation similar to [37]. The result is used to parametrize the equations given in [38]. Higher harmonics of the currents are considered, which also show that resonant operation minimizes the eddy current losses. Performing a FEM-simulation for each point of operation for the investigated parameter space would require a too high computational effort. The losses of the filter capacitors and inductors are also calculated, and different parameters of the LLC are systematically optimized to maximize the average efficiency, while also preventing high losses at full load.

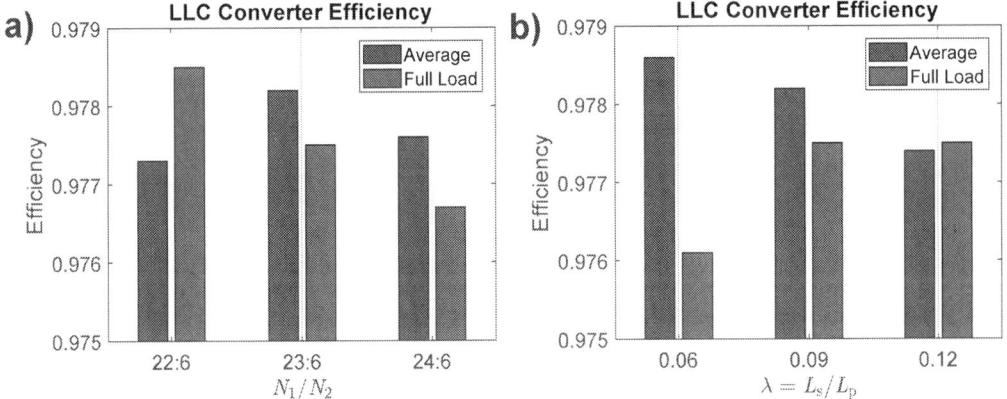

Fig. 6: The maximum average efficiency is achieved for a winding ratio of $\frac{N_1}{N_2} = \frac{23}{6}$ because the LLC converter always operates at resonance or close to it. The transformer is build up with $\lambda = \frac{L_s}{L_p} = 0.09$. Deviations of λ increase either the average losses or the full load losses.

One relevant parameter is the transformer turns ratio n, and at full load its optimal value is $n = \frac{\hat{v}_{cxjn}}{\hat{v}_{dcl}} = \frac{770\,\text{V}}{220\,\text{V}} = 3.5$. This can be achieved with a physical turns ratio of $\frac{N_1}{N_2} = \frac{22}{6}$, while accounting for the integrated leakage inductance [28]. The calculated efficiencies are shown in Fig. 6a). At low load, the module voltage is reduced to $v_{cxjn} = 600\,\text{V}$, and for load voltages $v_{dcl} = \frac{v_{cxjn}}{n} = \frac{600\,\text{V}}{3.5} = 171\,\text{V}$ or above, the LLC converter operates at resonance. However, at lower output voltages v_{dcl}, the switching frequency has to be increased substantially, and the converter operates above resonance, which decreases the average efficiency as shown in Fig. 6a). Therefore, the physical winding ratio is modified to $\frac{N_1}{N_2} = \frac{23}{6}$ to maximize the average efficiency at the expense of the efficiency at full load as shown in Fig. 6.

With the physical winding ratio of $\frac{N_1}{N_2} = \frac{27}{7}$ the same inductances can be achieved. At the full load of $\hat{P}_{LLC} = 11.1\,\text{kW}$, the core losses would be reduced from $\hat{P}_{core} = 40.8\,\text{W}$ to $\hat{P}_{core} = 32.4\,\text{W}$, but the overall winding losses would increase from $\hat{P}_{winding} = 45.8\,\text{W}$ to $\hat{P}_{winding} = 54.9\,\text{W}$. This would result in a too high winding temperature since it is already higher than the core temperature as shown in Fig. 7. Fewer windings increase the core losses and especially the core temperature. Therefore, the physical winding ratio $\frac{N_1}{N_2} = \frac{23}{6}$ is maintained. In addition, the optimal resonance frequency is $f_0 = 70.2\,\text{kHz}$ because smaller switching frequencies result in high core losses, while higher switching frequencies lead to additional eddy current losses.

The ratio of the magnetizing inductance to the resonant inductance $\lambda = \frac{L_s}{L_p}$ is an additional free parameter, and high values of λ are required for a high gain [26]. However, due to the reduction of the module voltage from $v_{cxjn} = 770\,\text{V}$ to $v_{cxjn} = 600\,\text{V}$, the required gain is small. Therefore, small values of λ maximize the average efficiency as shown in Fig. 6b). However, at full load the LLC converter operates below resonance and with $\lambda = 0.06$ with $f_s = 0.69 \cdot f_0$. This results in high conduction losses of the SiC-MOSFETs of the inverter and high winding losses of the transformer. Additionally, with small component tolerances or deviations of the module voltage, full load can not be achieved. Consequently, $\lambda = 0.09$ is selected to maximize the efficiency at full load because the switching frequency $f_s = 0.78 \cdot f_0$ is closer to resonance. A further increase to $\lambda = 0.12$ would reduce the average efficiency, while no benefit at full load is achieved.

The ratio $Z_0 = \sqrt{\frac{L_{s,xjn}}{C_{s,xjn}}} = 18\,\Omega$ is selected because for higher values, the resonance capacitor voltage exceeds the maximum voltage, while a decrease of Z_0 increases the losses [29]. Zero-voltage switching is achieved for all points of operation [39]. The resulting parameters are $L_{s,xjn} = 40.83\,\mu\text{H}$, $L_{p,xjn} = 453.62\,\mu\text{H}$, $n = 3.67$ and $C_{s,xjn} = 126\,\text{nF}$. Consequently, the constraint $C_{s,xjn} \geq 72\,\text{nF}$ is fulfilled, and each capacitor operates below the current, but at its voltage limit. Further details on the design of the resonance converter will be published in [40].

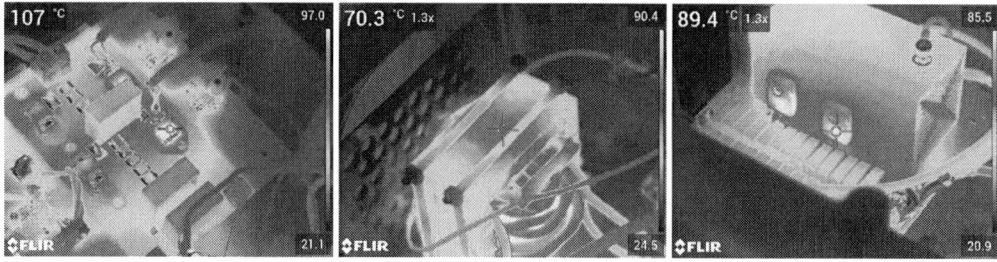

Fig. 7: At full load with $\hat{P}_{LLC} = 11.1\,\text{kW}$, the temperature of the inverter SiC-MOSFETs shows $107°\text{C}$, the core temperature $70°\text{C}$ and the rectifier diodes $89°\text{C}$. The measured overall losses are $P_{Loss} = 280\,\text{W}$ and the resulting efficiency $\eta_{measured} = 97.54\%$. The calculated efficiency is $\eta_{calculated} = 97.75\%$.

Measurements and results

One prototype of the LLC converter is built up in the laboratory with the determined components. In this setup, the component tolerances of $C_{s,xjn}$, $L_{s,xjn}$, $L_{p,xjn}$ and n are approximately 1%. The thermal images of the inverter SiC-MOSFETs, transformer core and rectifier diodes are shown in Fig. 7 at full load of $\hat{P}_{LLC} = 11.1\,\text{kW}$. In addition, Fig. 8 compares the measured efficiency with the calculated efficiency for both load lines. It reveals, that resonant operation is significantly more efficient than non-resonant operation, and that the regulation of the module voltage increases the overall efficiency. As a result, the top load line is operated with an efficiency of $\eta_{\text{measured}} > 98\%$ for load currents i_{dcl} between 30% and 80%. A slow conventional PID controller [41] is used for adjusting the switching frequency, since the electrolysis has no fast load jumps and a higher control speed is not a priority. This even applies if the Power-to-Gas system participates in the (balancing) energy market [42] [43].

One LLC converter in a module can be turned off to operate the remaining LLC converter closer to the peak efficiency [23]. For a load current i_{dcl} of 20%, the efficiency is increased by over 0.5% by operating only one of the two LLCs per module as it can be seen in Fig. 8a). Even for a load current i_{dcl} of 30%, turning off one converter is beneficial, since the efficiency at 60% load current is higher.

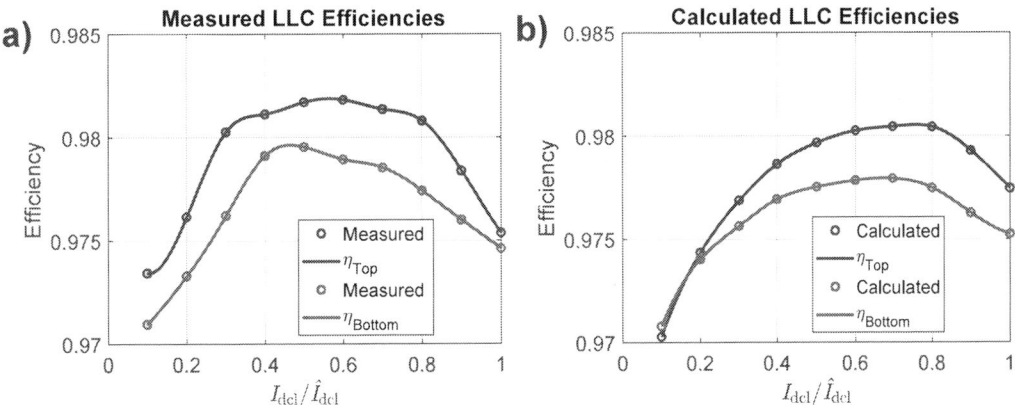

Fig. 8: The measured efficiencies are similar to the calculated efficiencies proving the loss model accurate. However, resonant operation at half load current is more efficient than calculated, while full load is less efficient. At the minimum load current of 10% [15], phase-shifted switching signals are used [44] while the module voltage is decreased to $v_{\text{cxjn}} = 600\text{V}$. The efficiencies η_{Top} and η_{Bottom} refer to the load lines in Fig. 5.

Conclusion

This paper evaluates three different MMC-topologies for high-power low-voltage DC-applications. The optimal Y-MMC has three arms and full-bridge based modules to charge each module's capacitor, which is used to supply the LLC resonant converter. The full-bridge Y-MMC shows a 78% decrease in voltage ripple and a reduction of capacitor losses by 64% compared to the Y-MMC with half-bridge modules because the module capacitor is charged during the whole grid period. In comparison with the YY-MMC, which consists of six arms, the optimal Y-MMC shows an even larger benefit. All MMCs have identical costs and physical volumes because they are a rearrangement of the same components. For this reason, an LLC resonant converter is designed and build for the optimal MMC, which shows a peak efficiency of $\hat{\eta}_{\text{measured}} = 98.17\%$ at resonance at half load. In order to achieve resonance for a large operating range, the module voltage has to be adjusted between 600 V and 770 V. As a trade-off, the number of modules of the Y-MMC has to be increased from $N_{\text{Arm}} = 12$ to $N_{\text{Arm}} = 15$, increasing the efficiency of the LLCs, but also the conduction losses of the grid-side stage by 25%. As a benefit, 50% of operation points of the top load line show an efficiency $\eta_{\text{measured}} > 98\%$ for the LLC resonant converter.

Each module consists of two independently operating LLC resonant converters. A further increase in efficiency can be achieved by interleaving these LLC converters to reduce the output voltage ripple and output capacitor losses. At light load, one LLC can be turned off to operate the remaining one closer to the peak efficiency point.

References

[1] J. Solanki, N. Fröhleke, J. Böcker, A. Averberg, P. Wallmeier, "High-current variable-voltage rectifiers: state of the art topologies", *IET Power Electronics*, vol. 8, 2015.

[2] J. Solanki, "High Power Factor High-Current Variable-Voltage Rectifiers", *Dissertation, University Paderborn, Germany*, 2015.

[3] Feng Dong, B. H. Chowdhury, M. L. Crow, and L. Acar, "Improving voltage stability by reactive power reserve management", *IEEE Transactions on Power Systems*, vol. 20, no. 1, pp. 338–345, 2005.

[4] M. De and S. K. Goswami, "Optimal Reactive Power Procurement With Voltage Stability Consideration in Deregulated Power System", *IEEE Transactions on Power Systems*, vol. 29, no. 5, pp. 2078–2086, 2014.

[5] K. D. McBee and M. G. Simões, "Evaluating the Long-Term Impact of a Continuously Increasing Harmonic Demand on Feeder-Level Voltage Distortion", *IEEE Transactions on Industry Applications*, vol. 50, no. 3, pp. 2142–2149, 2014.

[6] C. Lombard and A. P. J. Rens, "Evaluation of system losses due to harmonics in medium voltage distribution networks", in *IEEE International Energy Conference (ENERGYCON)*, 2016.

[7] G. T. Son, H. Lee, T. S. Nam, Y. Chung, U. Lee, *et al.*, "Design and Control of a Modular Multilevel HVDC Converter With Redundant Power Modules for Noninterruptible Energy Transfer", *IEEE Transactions on Power Delivery*, vol. 27, no. 3, pp. 1611–1619, 2012.

[8] J. Solanki, N. Fröhleke, J. Böcker, and P. Wallmeier, "A modular multilevel converter based high-power high-current power supply", in *IEEE International Conf. on Industrial Technology (ICIT)*, 2013, pp. 444–450.

[9] R. Unruh, F. Schafmeister, N. Froehleke, and J. Boecker, "MMC-Topology for High-Current and Low-Voltage Applications with Minimal Number of Submodules, Reduced Switching and Capacitor Losses", in *PCIM Europe*, 2019.

[10] J. E. Huber and J. W. Kolar, "Solid-State Transformers: On the Origins and Evolution of Key Concepts", *IEEE Industrial Electronics Magazine*, vol. 10, pp. 19–28, 2016.

[11] C. Zhao, S. Lewdeni-Schmid, J. K. Steinke, M. Weiss, T. Chaudhuri, *et al.*, "Design, implementation and performance of a modular power electronic transformer (PET) for railway application", *Proceedings of the 14th European Conf. on Power Electron. and App. (EPE), Birmingham, UK*, 2011.

[12] L. Ferreira Costa, G. De Carne, G. Buticchi, and M. Liserre, "The Smart Transformer: A solid-state transformer tailored to provide ancillary services to the distribution grid", *IEEE Power Electronics Magazine*, vol. 4, no. 2, pp. 56–67, 2017.

[13] T. Grube, L. Doré, A. Hoffrichter, L. E. Hombach, S. Raths, *et al.*, "An option for stranded renewables: electrolytic-hydrogen in future energy systems", *Sustainable Energy Fuels*, vol. 2, pp. 1500–1515, 2018.

[14] S. Clegg and P. Mancarella, "Integrated Modeling and Assessment of the Operational Impact of Power-to-Gas (P2G) on Electrical and Gas Transmission Networks", *IEEE Transactions on Sustainable Energy*, vol. 6, no. 4, pp. 1234–1244, 2015.

[15] Geert Hauke Tjarks, "PEM-Elektrolyse-Systeme zur Anwendung in Power-to-Gas Anlagen", *Dissertation, RWTH Aachen University, Germany*, 2017.

[16] Z. Yan, H. Xue-hao, T. Guang-fu, and H. Zhi-yuan, "A study on MMC model and its current control strategies", in *The 2nd International Symposium on Power Electronics for Distributed Generation Systems*, 2010.

[17] T. Nam, H. Kim, G. Son, Y. Chung, J. Park, *et al.*, "Trade-Off Strategy in Designing Capacitor Voltage Balancing Schemes for Modular Multilevel Converter HVDC", in *Journal of Electrical Engineering Tech.*, 2016.

[18] R. Unruh, F. Schafmeister, N. Froehleke, and J. Boecker, "1-MW Full-Bridge MMC for High-Current Low-Voltage (100V-400V) DC-Applications", in *PCIM Europe*, 2020.

[19] A. Iqbal, E. Levi, M. Jones, and S. N. Vukosavic, "Generalised sinusoidal PWM with harmonic injection for multi-phase VSIs", in *37th IEEE Power Electronics Specialists Conference*, 2006.

[20] J. Jose, G. N. Goyal, and M. V. Aware, "Improved inverter utilisation using third harmonic injection", in *Joint International Conference on Power Electronics, Drives and Energy Systems Power India*, 2010.

[21] H. Figge, T. Grote, N. Froehleke, J. Boecker, and P. Ide, "Paralleling of LLC resonant converters using frequency controlled current balancing", in *IEEE Power Electronics Specialists Conference*, 2008.

[22] H. Figge, T. Grote, F. Schafmeister, N. Fröhleke, and J. Böcker, "Two-phase interleaving configuration of the LLC resonant converter - Analysis and experimental evaluation", in *IECON*, 2013, pp. 1392–1397.

[23] M. Mogorovic and D. Dujic, "Sensitivity Analysis of Medium-Frequency Transformer Designs for Solid-State Transformers", *IEEE Transactions on Power Electronics*, vol. 34, no. 9, pp. 8356–8367, 2019.

[24] H. Wang and F. Blaabjerg, "Reliability of Capacitors for DC-Link Applications in Power Electronic Converters—An Overview", *IEEE Transactions on Industry Applications*, vol. 50, no. 5, pp. 3569–3578, 2014.

[25] M. Evzelman and S. Ben-Yaakov, "Average-Current-Based Conduction Losses Model of Switched Capacitor Converters", *IEEE Trans. on Power Electronics*, vol. 28, no. 7, 2013.

[26] Bing Lu, Wenduo Liu, Yan Liang, F. C. Lee, and J. D. van Wyk, "Optimal design methodology for LLC resonant converter", in *21th Annual IEEE Applied Power Electronics Conf. and Exposition. APEC*, 2006.

[27] S. Rodrigues, A. Papadopoulos, E. Kontos, T. Todorcevic, and P. Bauer, "Steady-State Loss Model of Half-Bridge Modular Multilevel Converters", *IEEE Trans. on Industry Applications*, vol. 52, no. 3, pp. 2415–2425, 2016.

[28] Silvio De Simone, "LLC resonant half-bridge converter design guideline", 2014.

[29] L. Keuck, P. Hosemann, B. Strothmann, and J. Boecker, "A Comparative Study on Si-SJ-MOSFETs vs. GaN-HEMTs Used for LLC-Single-Stage Battery Charger", in *PCIM Europe*, 2017.

[30] J. Jung, "Bifilar Winding of a Center-Tapped Transformer Including Integrated Resonant Inductance for LLC Resonant Converters", *IEEE Transactions on Power Electronics*, vol. 28, pp. 615–620, 2013.

[31] M. Mogorovic and D. Dujic, "100 kW, 10 kHz Medium-Frequency Transformer Design Optimization and Experimental Verification", *IEEE Transactions on Power Electronics*, vol. 34, no. 2, pp. 1696–1708, 2019.

[32] J. Biela, "Wirbelstromverluste in Wicklungen induktiver Bauelemente", in *Skriptum, ETH Zürich*, 2011.

[33] H. Choi, "Analysis and Design of LLC Resonant Converter with Integrated Transformer", in *APEC 07 - Twenty-Second Annual IEEE Applied Power Electronics Conference and Exposition*, 2007, pp. 1630–1635.

[34] Jieli Li, T. Abdallah, and C. R. Sullivan, "Improved calculation of core loss with nonsinusoidal waveforms", in *Conference Record of the 2001 IEEE Industry Applications Conference. 36th IAS Annual Meeting*, vol. 4, 2001, 2203–2210 vol.4.

[35] M. Sippola and R. E. Sepponen, "Accurate prediction of high-frequency power-transformer losses and temperature rise", *IEEE Transactions on Power Electronics*, vol. 17, no. 5, pp. 835–847, 2002.

[36] E. G. teNyenhuis, R. S. Girgis, G. F. Mechler, and Gang Zhou, "Calculation of core hot-spot temperature in power and distribution transformers", *IEEE Trans. on Power Delivery*, vol. 17, no. 4, pp. 991–995, 2002.

[37] L. Keuck, F. Schafmeister, J. Boecker, H. Jungwirth, and M. Schmidhuber, "Computer-Aided Design and Optimization of an Integrated Transformer with Distributed Air Gap and Leakage Path for LLC Resonant Converter", in *PCIM Europe*, 2019.

[38] M.Albach, J. Patz, H. Roßmanith, and A. Stadler, "Optimale Wicklung= optimaler Wirkungsgrad", in *Elektronik power 2010*, 2010, pp. 38–47.

[39] M. Kasper, R. M. Burkart, G. D. Deboy, and J. W. Kolar, "ZVS of Power MOSFETs Revisited", *IEEE Transactions on Power Electronics*, vol. 31, pp. 8063–8067, 2016.

[40] R. Unruh, F. Schafmeister, and J. Boecker, "LLC Design for an MMC Converter with Adaptive Input Voltage for High Efficiency at Light Load", in *Manuscript submitted for publication*, 2020.

[41] F. Kurokawa and K. Murata, "A new fast digital P-I-D control LLC resonant converter", in *2011 International Conference on Electrical Machines and Systems*, 2011.

[42] M. Kuprat, M. Bending, and K. Pfeiffer, "Possible role of power-to-heat and power-to-gas as flexible loads in German medium voltage networks", *Frontiers in Energy*, vol. 11, pp. 135–145, 2017.

[43] Y. Xiao, X. Wang, P. Pinson, and X. Wang, "A Local Energy Market for Electricity and Hydrogen", *IEEE Transactions on Power Systems*, vol. 33, no. 4, pp. 3898–3908, 2018.

[44] J. Kim, C. Kim, J. Kim, J. Lee, and G. Moon, "Analysis on Load-Adaptive Phase-Shift Control for High Efficiency Full-Bridge LLC Resonant Converter Under Light-Load Conditions", *IEEE Transactions on Power Electronics*, vol. 31, pp. 4942–4955, 2016.

Iron Loss Characteristics of MnZn Ferrites under GaN Inverter excitation in the MHz Order

Wilmar Martinez and Camilo Suarez
Department of Electrical Engineering (ESAT)
KU Leuven - EnergyVille
Diepenbeek-Genk, Belgium
wilmar.martinez@kuleuven.be
www.energyville.be

Federico Ibanez
Center of Energy Science and Technology
Skolkovo Institute of Science and Technology
Moscow, Russia
fm.ibanez@skoltech.ru
www.crei.skoltech.ru/cest/

Keywords

«Magnetic Materials», «Gallium Nitride (GaN)», «Ferrites», «Iron Losses», «Harmonic Distortion»

Abstract

More and more a big portion of all electrical power is transported and distributed through a power electronics converter, which comprises invariably magnetic components. In most cases, such components are made with Ferrites which introduce a certain amount of losses and weight/volume to the system. In addition, there is a need for outstanding magnetic components with low-loss and high dense properties able to work under the specifications that novel Wide Bandgap semiconductors demand: high frequency operation, fast transients, compactness, etc. However, there is not a clear ground about how magnetics components operate at high frequencies and under the influence of these semiconductors. Then, magnetic characterization of ferrites under distinct conditions is essential to understand the operation of these components, especially in the MHz order. Harmonic content, introduced by the switching of the fast semiconductors, and fast transients raise higher challenges for an accurate identification of the magnetic properties. Therefore, having a complete understanding of how these harmonic contents affect the operation of the magnetic components could be beneficial for improving the material selection and design procedures themselves. This paper studies the effect of Gallium Nitride inverter conditions (deadtime, fast transients, fundamental frequencies, etc.) on iron loss characterization procedures of Ferrite materials.

I. Introduction

Despite the great efforts of the academic and industrial community towards the development of better magnetic materials and components, still there is the conception that magnetic design and modelling advances are lagging behind the semiconductor advances, especially after the developments of WBG semiconductor devices like Silicon Carbide or Gallium Nitride [1]. Such semiconductor devices have allowed power converters to be operated at higher speeds with outstanding efficiencies allowing, in most cases, weight reduction and compactness in power electronic systems [2], [3]. Size reduction is especially important as the increase of energy composition worldwide goes hand in hand with the demand of power electronic systems and therefore of magnetic materials and wires [4]. Then, depletion of raw materials like Cu, Al, Si, Fe, etc., essential for motors, transformers, and inductors, can be tackled if highly compact magnetic components are utilized in power converters [5].

Nevertheless, it is still said that magnetics are the "bottleneck" of power converter design mainly because of the scaling laws that usually affects the operation of high frequency power electronic systems [6]. This understanding has been changing during the last decade but still there is not a common ground regarding the interaction of magnetic components and semiconductor devices and their effect on each other when they are operating at high switching frequencies. In this context, it is highly important to understand how magnetic components operate under the conditions of fast transients and fast switching that the WBG semiconductor devices generate. Therefore, magnetic expertise can be built up for the improvement of power converters and electric systems.

In order to build such expertise, magnetic characteristics under PWM excitation and especially at high frequencies and fast transients are required. There are effects associated with high dv/dt transients in the order of ~100V/ns, which is common in GaN and SiC, that need to be accounted when iron losses are modelled and calculated [7]. In addition, such fast transients also affect copper losses in magnetic components, mainly caused by skin and proximity effects, which need to be considered as well as the drawbacks from the additional voltage stresses on windings and accelerated insulation breakdown [8]. [9].

In previous studies, the magnetic characteristics of silicon steel materials under the PWM excitation generated by high frequency GaN Inverters were depicted [10]–[12]. Silicon steel laminations were selected because still provide the most cost-efficient solution for transformers, reactors, and some electrical machines [13]. However, silicon steel is not conventionally used in high frequency inductors, transformers, or filters, due to their high loss characteristics, especially in the kHz and MHz area. Ferrites, on the other hand, made of NiZn, MnZn, with additives of Cobalt, Titanium, Strontium, and other metals, have been used for such devices as they offer special characteristics for power converters.

Nevertheless, most of the information that is available from these magnetic materials lacks of high harmonic content effects generated by PWM excitations making deficient the conventional mono-frequency excitation method. This issue becomes of more relevance when the abovementioned fast transients are taken into account. Such transients generate high frequency harmonics and distortions in nonlinear loads making difficult the iron loss assessment, both numerically and experimentally [14].

Iron loss modelling becomes necessary not only for magnetic materials that were conceived and characterized for lower frequencies and slow transients but also for new materials that are intended to be used at high frequency operation, as it is the case of MnZn Ferrites. This iron loss modelling must consider the complex flux-density waveforms with high-frequency components that are generated by WBG semiconductor devices. The characterization procedure, required for constructing a reliable model, is presented in this paper. Several components made with MnZn Ferrite were measured with a high switching frequency inverter constructed with E-mode GaN HEMTs. Two different fundamental frequencies were evaluated, the conventional 50 Hz, used in power systems, and 400Hz, used in electric, more electric aircrafts, and standalone systems [15]–[17]. Then, data was processed and iron losses were simulated using fluxometric techniques.

This paper is divided as follows, first the characterization and measurement procedure to get voltage and current data of the prototype with the selected magnetic material are presented; second, the influence of deadtimes on the iron losses are introduced; then, the harmonic content of the measurements is analysed; finally conclusions are given.

II. Experimental Characterization

a. Characterization Procedure

In order to obtain the iron loss characteristics of certain magnetic material, samples are constructed in ring shapes for better magnetic flux distribution. Then, two windings are wound in the ring and are excited by a GaN Inverter. Fig. 1 shows a schematic of the experimental setup used to obtain the measurements with which iron losses are obtained. The Gallium Nitride inverter is constructed with 650V e-mode GaN HEMTs as it is depicted in Fig. 2. As devices under tests for this characterization procedure, several rings were constructed and wounded. Fig. 3 shows some of them. Different ring sizes were used because with certain sizes is easier to achieve higher frequencies with DC voltages within the limits of the GaN devices (650V). The same happens with extremely low voltages, with certain sizes, high magnetic fluxes are easy to achieve and the measurement procedure becomes compromised as the voltage levels land in the tolerance of the measurement equipment.

For the specific case of this paper, MnZn Ferrites, intended to be used in high frequency applications, were used. Table I shows the test specifications, as well as the ring and the GaN inverter parameters. Table I shows the specific parameters of two of the most used ferrite ring cores for this study.

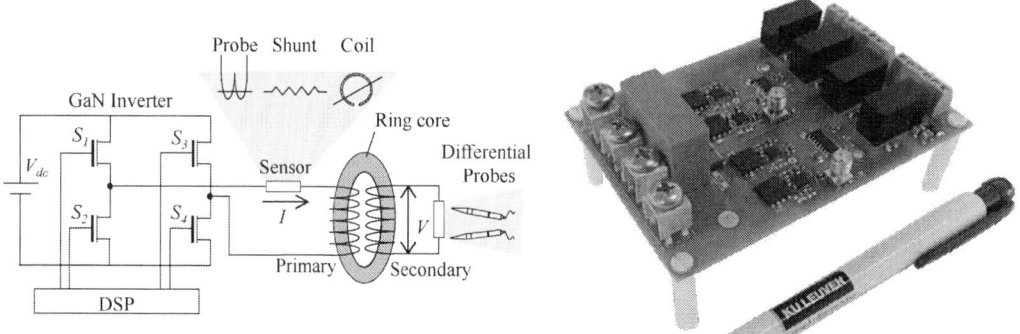

Fig. 1: Measurement circuit   Fig. 2: Inverter setup

Fig. 3: Inverter setup

Table I. Test Setup Specifications

Rings (Ferrite)			GaN Inverter	
Outside diameter d_o [mm]	63	140	Deadtime T_d [ns]	10-300
Inside diameter d_i [mm]	38	103	DC voltage V_{dc} [V]	0-400
Height h [mm]	25	25	Modulation index m	0.2-1
Number of turns of the primary coil N_1	60	171	Fundamental frequency f_o [Hz]	50-400
Number of turns of the secondary coil N_2	60	171	Carrier frequency f_c [kHz]	10-4000

The GaN inverter is controlled with a unipolar modulation programed in a 150MHz DSP. Such inverter is operated with a wide range of switching frequencies between 5kHz and 2MHz. Due to the nature of the unipolar modulation, carrier frequencies in the ring are in a range between 10kHz and 4MHz. Being 4MHz sufficient for most of the power electronic applications where WBG semiconductors are expected to be operated at high frequencies. Moreover, two fundamental frequencies, 50Hz and 400Hz, with different deadtimes between 10 and 300ns are tested.

Voltages in the secondary windings and currents through the first windings in the rings are determined with an oscilloscope capable of getting 10Mpoints per measurement with a maximum sampling frequency of 2.5 GS/s real time. With this scope, the primary-coil current is measured with a wideband current probe (Hall effect-clamp type). Voltage in the secondary coil is measured with a wideband differential voltage probe. Deskewing (synchronization) of the current and voltage probes is conducted before the measurement procedure.

Fig. 4 shows some of the measured voltage and currents at a 400Hz fundamental frequency 100ns deadtime at different conditions of carrier frequency and magnetic flux density. As it is expected, higher DC Voltage at the inverter input is required when higher carrier frequency is required at the same magnetic flux. Moreover, when the magnetic component is working close to saturation, see Fig. 4(c)-(d), current waveforms tend to be deformed requiring higher peak powers.

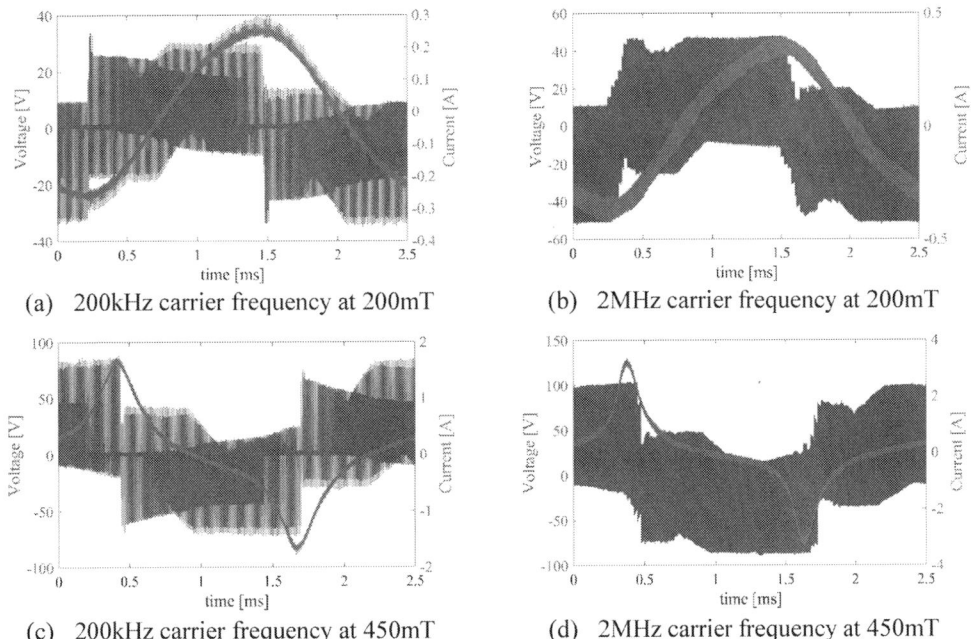

(a) 200kHz carrier frequency at 200mT (b) 2MHz carrier frequency at 200mT

(c) 200kHz carrier frequency at 450mT (d) 2MHz carrier frequency at 450mT

Fig. 4: Voltage and current waveforms of the 400Hz case

b. Post-processing

Once the tests are conducted and data from the current and voltage is obtained, the next step is the mathematic post-processing to find the BH operation and thereby the calculation of iron losses, harmonic content, etc. Fluxometric processing is used to find the BH curves (magnetic flux density B and magnetic field intensity H characteristics) and to calculate the iron loss density of the material under test. The magnetic field intensity H_s on the surface of the iron ring is calculated from the measured primary winding current i as follows:

$$H_s(t) = \frac{N_1}{l_{Fe}} i(t)$$

(1)

where N_1 is the number of turns in the primary winding and l_{Fe} is the length of the flux path, usually calculated as the intermediate path between the inner and outer circles of the ring. Moreover, the average magnetic flux density B_0 in the core is obtained by integrating the back-electromotive force v_2 induced in the secondary winding with N_2 turns as follows:

$$B_0(t) = \frac{1}{N_1 A_{Fe}} \int v_2(t) dt$$

(2)

where A_{Fe} is the cross-sectional area of the ferrite ring. Then, the iron loss density per mass unit (ρ is the material mass density, 4850 kg/m^3 in case of the evaluated ferrite) can be obtained using (3). Full details of the mathematical derivation process can be found in [10].

$$P_{Fe} = \frac{N_1 f}{N_2 A_{Fe} l_{Fe} \rho} \int_0^{1/f_0} i(t) v_2(t) dt \qquad [W/kg] \qquad (3)$$

Finally, as Fig. 4 shows, DC voltage variations are required in order to obtain similar flux densities. However, it is difficult to obtain the exact flux density in all measurements when carrier frequency is swept and more when some parameters like deadtime or modulation index are changed. Nevertheless, in order to have a fair and accurate iron loss estimation, the iron loss measurements should be normalized to a common magnetic flux density B_n, ideally near to the maximum magnetic flux density B_{max} from the measurement set. Therefore, (4) used to normalize the values of iron losses as they are proportional to the square of the magnetic flux density.

$$P_{Fe\_n} = P_{Fe} \frac{B_n^2}{B_{max}^2} \qquad [W/kg] \qquad (4)$$

Fig. 5, 6 and 7 show the reconstruction of two sets of BH loops when the inverter is operating at 400Hz, 100ns deadtime, and a sweep between 10kHz and 2MHz carrier frequency. In the three sets, the comparison between three different maximum flux densities is done (160, 300, and 450mT). In the figures, the effect of the GaN switching and the carrier frequency is evident. The higher the carrier frequency is the more minor loops are present in the BH curve. In addition, it is possible to see the effect of going higher in magnetic flux on the maximum field intensity. Despite the steps of hundreds between the three magnetic fluxes, the generated field intensities have a different order of change. This is more evident in the case of 450mT as it close to the saturation point of the selected material and therefore small changes in voltage will generate great changes in the peak current and therefore on the field intensity.

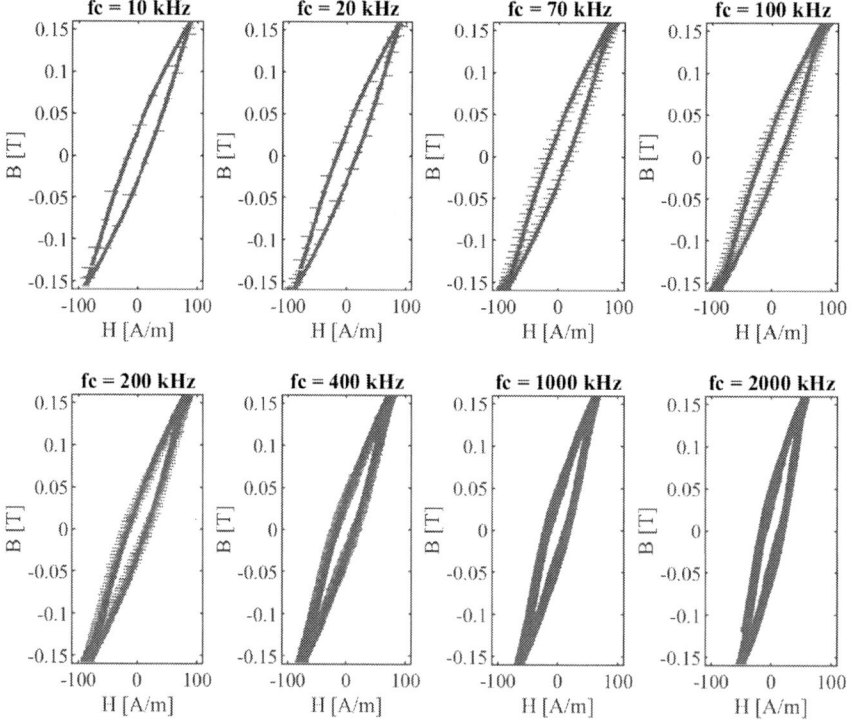

Fig. 5: BH loops at 400Hz fundamental frequency, 100ns deadtime, and 160mT flux density

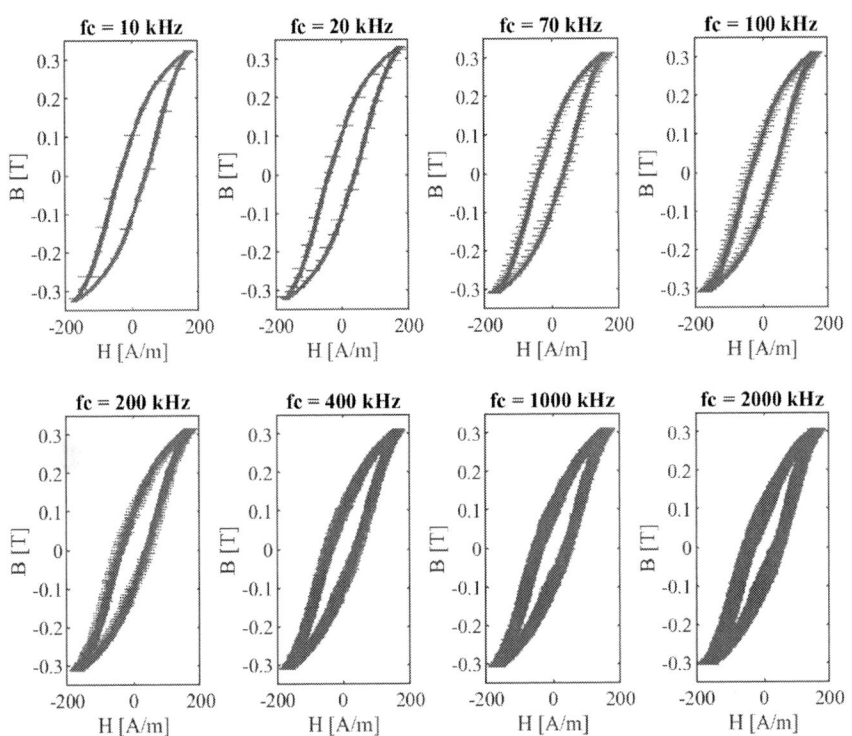

Fig. 6: BH loops at 400Hz fundamental frequency, 100ns deadtime, and 300mT flux density

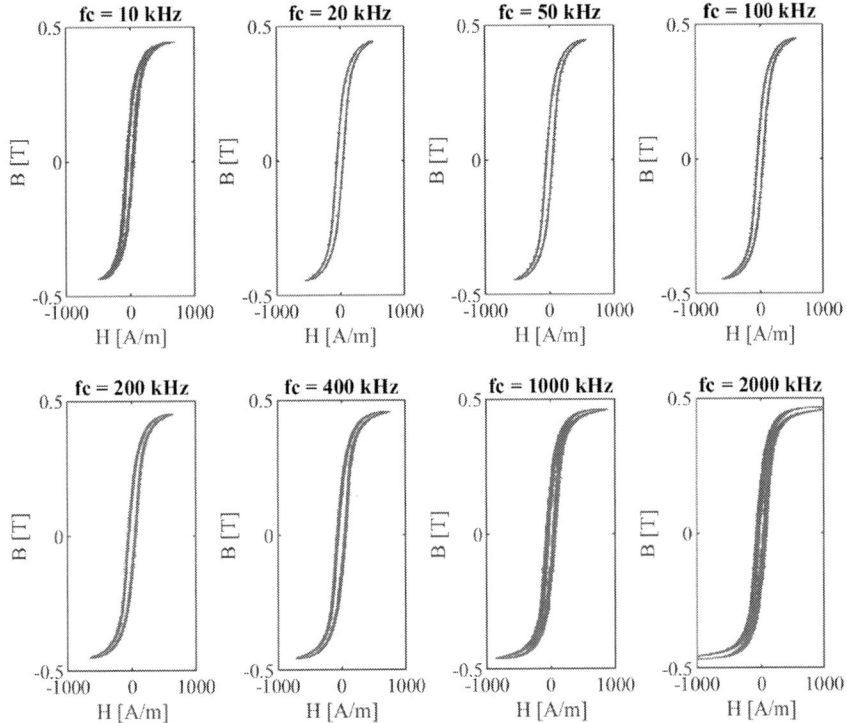

Fig. 7: BH loops at 400Hz fundamental frequency, 100ns deadtime, and 450mT flux density

III. Iron Loss Estimation

After obtaining the BH characteristics in the wide range of frequencies, iron losses were calculated. Fig. 8 shows the comparison of the cases of 50 and 400Hz fundamental frequencies at several flux densities. At a first glance, it is possible to depict that at 400Hz iron losses are higher, however it is worth to mention that usually at higher carrier frequencies smaller components are required. Moreover, when the magnetic component is operating in a zone near saturation, the effect of the carrier or switching frequency becomes more significant. It is the case of the 450mT line in Fig. 8(b). Nevertheless, overall MnZn ferrites offer a stable iron loss operation in the MHz order.

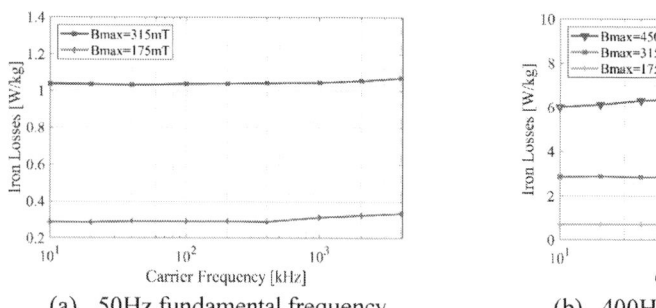

(a) 50Hz fundamental frequency   (b) 400Hz fundamental frequency

Fig. 8: Iron Loss Comparison at two fundamental frequencies and several flux densities

Fig. 9 visualizes the case of 375mT flux density for both 50 and 400 Hz of fundamental frequency. For comparison purposes, Fig. 10 shows the 400Hz evaluation of silicon steel laminations at different flux densities. Such material, tested and evaluated in [11], [12], has a saturation flux density of 1.5T, therefore 450, 670, and 1000mT were measured. It is evident the big difference in the iron loss density of both materials when they are switched at the same switching frequency and fundamental frequency. Here, once again, it is worth to mention, that depending on the application, the amount of material and power conditions of different materials might vary to meet the same requirements.

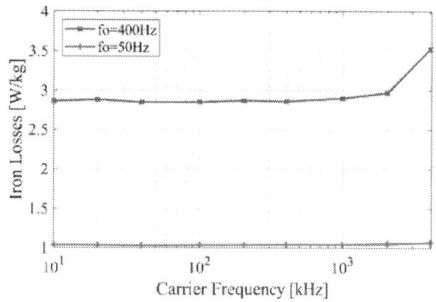 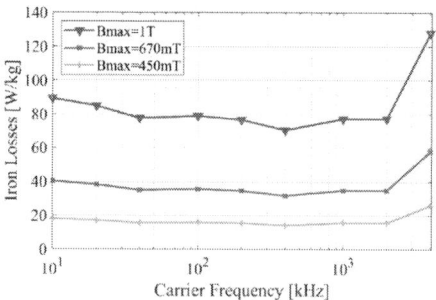

Fig. 9: · Iron loss comparison at two fundamental frequencies at 375mT flux density of MnZn Ferrites

Fig. 10: Iron losses of Silicon Steel laminations at 400Hz fundamental frequency

Finally, an iron loss model for Silicon steel laminations was implemented in MATLAB using Simulink Simscape Power Systems environment[1]. This Simulink model was constructed based on the 50Hz model presented in [10]. With this tool, several analyses for high frequency evaluation of magnetic components under PWM excitation can be done.

---

[1] The 400Hz models are available at https://github.com/whmartinezm/simulink-pwm-inductor-400Hz

IV. Deadtime effect

Another important factor to have in mind is the deadtime effect on the iron losses in ferrites. This is particularly important for converters where half or full bridges are utilized, e.g. DAB, LLC, etc. As Fig. 11(a) shows, when carrier frequency increases, and especially in the MHz order, a constant deadtime will occupy a bigger portion of the duty cycle. Consequently, the effective on-time of the switching process is shorter due to the increase of the carrier frequency. Consequently, higher DC voltage needs to be applied to the inverter input in order to generate the same magnetic flux. As a consequence, iron losses increase due to the change of the field intensity because of the change in the current.

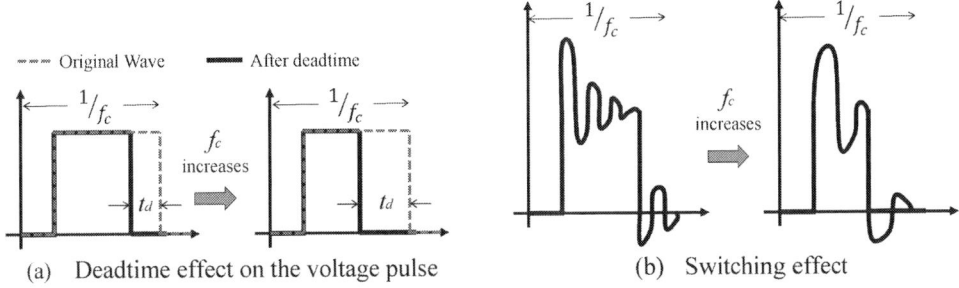

(a) Deadtime effect on the voltage pulse   (b) Switching effect

Fig. 11: High carrier / switching frequency effect on the voltage pulses

Similar situation occurs when fast transients happen during the switching transitions. These transitions generate voltage surges that become significant when carrier frequency is increased, see Fig. 11(b). Consequently, large current surges are generated which are translated into larger field intensities in the minor loops. That is one of the reasons of the wide BH curves at higher carrier frequencies in Fig. 5-7. Fig. 12 shows the comparison of two cases of deadtime (100 and 200 ns) when the ring is excited at 1MHz carrier frequency, 400Hz fundamental frequency, and 450mT magnetic flux.

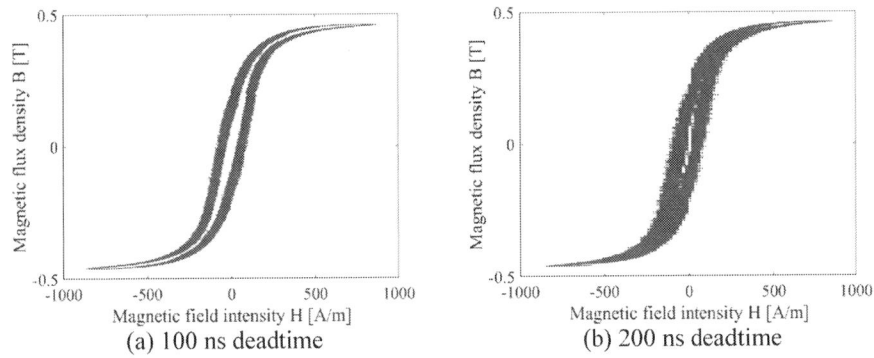

(a) 100 ns deadtime   (b) 200 ns deadtime

Fig. 12: Deadtime effect on BH curves

V. Harmonic content

As aforementioned, fast transients generate an increase in the iron losses of ferrite materials, and therefore the evaluation of the harmonic content of each condition becomes important to see the high frequency effect in terms of noise. Fig. 13 shows the FFT spectrum of four cases of carrier frequency and flux density at 400Hz flux density.

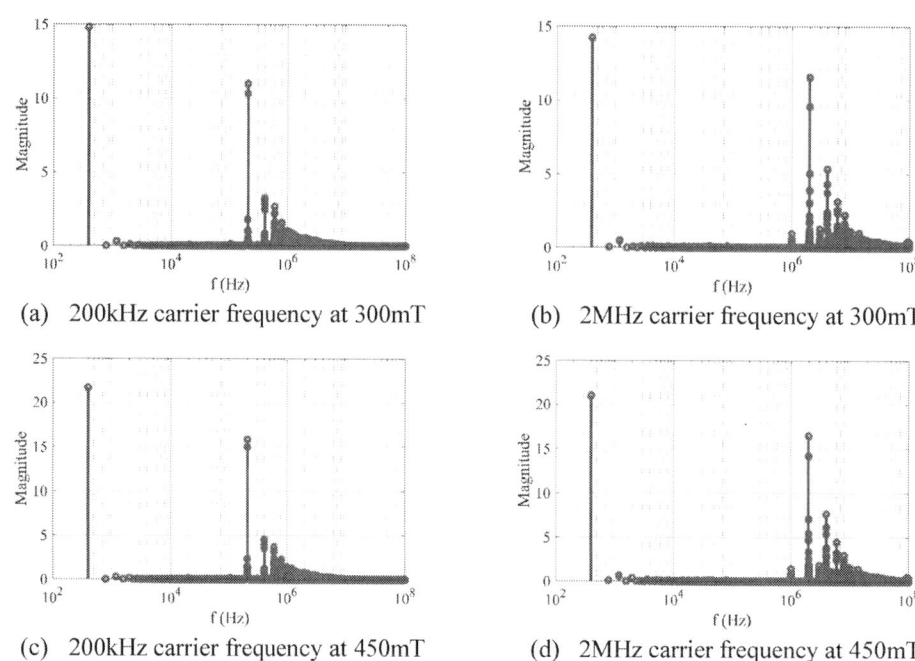

(a) 200kHz carrier frequency at 300mT   (b) 2MHz carrier frequency at 300mT

(c) 200kHz carrier frequency at 450mT   (d) 2MHz carrier frequency at 450mT

Fig. 13: FFT of Voltage waveforms at 100 ns deadtime

As it is shown, magnetic components operating near saturation introduce much higher magnitudes in the harmonics, which is derived from the higher DC voltage required. In the case of higher carrier frequencies, in both flux density cases (Fig. 13(b) and (d)), it is possible to depict the high harmonic content in the MHz range.

VI. Conclusion

Iron losses of MnZn Ferrites under Gallium Nitride inverter excitations in the MHz order were characterized and depicted in this paper. The challenges in measurements and the influence of the inverter and converter conditions on the operation of magnetic components were discussed. Especially, the effect of deadtime and fast transients introduced by WBG semiconductors were analysed. Moreover, the harmonic content introduced by these fast transients and deadtimes was measured.

It was found that the specific switching conditions and derived harmonic content should be considered when magnetic materials are selected and when magnetic components are designed. The effect of deadtime, carrier frequency, and fundamental frequency increases radically the iron loss behaviour of Ferrites. Nevertheless, Ferrites, especially MnZn, showed a good operation when they are operated far from the saturation area. These results open the way to further research for constructing iron loss models and thereby to have a common ground of how magnetic components perform under the conditions that novel semiconductors demand.

References

[1] A. J. Hanson and D. J. Perreault, "Modeling the Magnetic Behavior of N-Winding Components: Approaches for Unshackling Switching Superheroes," *IEEE Power Electron. Mag.*, vol. 7, no. 1, pp. 35–45, Feb. 2020.

[2] N. Yan, J. Hu, J. Wang, D. Dong, and R. Burgos, "Design Analysis for Current-transformer Based High-frequency Auxiliary Power Supply for SiC- based Medium Voltage Converter Systems," pp. 1390–1397, 2020.

[3] G.-S. Seo and H.-P. Le, "An inductor-less hybrid step-down DC-DC converter architecture for future smart power cable," in *2017 IEEE Applied Power Electronics Conference and Exposition (APEC)*, 2017, pp. 247–253.

[4] N. Soltau, D. Eggers, K. Hameyer, and R. W. De Doncker, "Iron Losses in a Medium-Frequency Transformer Operated in a High-Power DC-DC Converter," *IEEE Trans. Magn.*, vol. 50, no. 2, pp. 953–956, Feb. 2014.

[5] A. Klein-Hessling, B. Burkhart, and R. W. De Doncker, "Iron loss redistribution in Switched Reluctance Machines using bidirectional phase currents," in *8th IET International Conference on Power Electronics, Machines and Drives (PEMD 2016)*, 2016, pp. 1–6.

[6] C. R. Sullivan, B. A. Reese, A. L. F. Stein, and P. A. Kyaw, "On size and magnetics: Why small efficient power inductors are rare," in *2016 International Symposium on 3D Power Electronics Integration and Manufacturing (3D-PEIM)*, 2016.

[7] T. Guillod, J. E. Huber, G. Ortiz, A. De, C. M. Franck, and J. W. Kolar, "Characterization of the voltage and electric field stresses in multi-cell solid-state transformers," in *2014 IEEE Energy Conversion Congress and Exposition (ECCE)*, 2014, pp. 4726–4734.

[8] M. Jaritz, A. Hillers, and J. Biela, "General Analytical Model for the Thermal Resistance of Windings Made of Solid or Litz Wire," *IEEE Trans. Power Electron.*, vol. 34, no. 1, pp. 668–684, Jan. 2019.

[9] C. Liu, Z. Zhang, Y. Liu, Y. Si, and Q. Lei, "Smart Self-Driving Multilevel Gate Driver for Fast Switching and Crosstalk Suppression of SiC MOSFETs," *IEEE J. Emerg. Sel. Top. Power Electron.*, vol. 8, no. 1, pp. 442–453, Mar. 2020.

[10] P. Rasilo, W. Martinez, K. Fujisaki, J. Kyyra, and A. Ruderman, "Simulink Model for PWM-Supplied Laminated Magnetic Cores Including Hysteresis, Eddy-Current, and Excess Losses," *IEEE Trans. Power Electron.*, vol. 34, no. 2, pp. 1683–1695, Feb. 2019.

[11] W. Martinez, S. Odawara, and K. Fujisaki, "Iron Loss Characteristics Evaluation Using a High-Frequency GaN Inverter Excitation," *IEEE Trans. Magn.*, vol. 53, no. 11, pp. 1–7, 2017.

[12] W. Martinez, C. Suarez, and W. Lin, "Iron Loss Evaluation of GaN PWM-Supplied Magnetic Cores for MHz Converters in More Electric Aircrafts - 400Hz case," in *29th IEEE International Symposium on Industrial Electronics*, 2020, pp. 1410–1415.

[13] O. de la Barriere, C. Ragusa, C. Appino, and F. Fiorillo, "Prediction of Energy Losses in Soft Magnetic Materials Under Arbitrary Induction Waveforms and DC Bias," *IEEE Trans. Ind. Electron.*, vol. 64, no. 3, pp. 2522–2529, Mar. 2017.

[14] R. V Sabariego, K. Niyomsatian, and J. Gyselinck, "Eddy-Current-Effect Homogenization of Windings in Harmonic-Balance Finite-Element Models Coupled to Nonlinear Circuits," *IEEE Trans. Magn.*, vol. 54, no. 3, pp. 1–4, Mar. 2018.

[15] S. Khalid and B. Dwivedi, "Comparative critical analysis of SAF using soft computing and conventional control techniques for high frequency (400 Hz) aircraft system," in *2013 IEEE 1st International Conference on Condition Assessment Techniques in Electrical Systems, IEEE CATCON 2013 - Proceedings*, 2013, pp. 105–110.

[16] I. Munuswamy, P. W. Wheeler, B. G. Fernandes, and K. Chatterjee, "Electric AC WIPS: All Electric Aircraft," in *2019 Innovations in Power and Advanced Computing Technologies, i-PACT 2019*, 2019.

[17] E. Sener, G. Ertasgin, and D. Zuber, "Design of a 400 Hz Current-Source Single-Phase Converter for Avionic Systems," in *2017 IEEE Vehicle Power and Propulsion Conference, VPPC 2017 - Proceedings*, 2018, vol. 2018-January, pp. 1–6.

Vibration suppression and control parameter design of a sensorless PMSM rotary compressor drive

Tao Li[1], Chaohui Liang[2]

Spintrol Limited

Spintrol Limited, No. 9 Yuexing 1st RD, South Area, Hi-tech Park, Nanshan Dist.

Shenzhen, P. R. China

Tel.: +86 15986747728[1] , +86 13823416665[2]

Fax: +86 0755-26654235

E-Mail: netlitao@126.com[1], chaohui.liang@spintrol.com[2]

URL: http://www.spintrol.com

Keywords

«Adjustable speed drive», «Control of drive», «Active damping», «AC machine»

Abstract

The mechanical vibration and the associated acoustic noise problem of variable frequency rotary compressor systems have been concerned by engineers for years. This paper proposes a scheme to reduce the vibration of a sensorless controlled rotary compressor by using bi-quad filter. The paper presents the overall system control architecture as well as a method to design motor speed adaptation parameters and the bi-quad filter parameters. The impacts of motor parameter variation have been analyzed, too. Both simulation and experiment results show that the bi-quad filter can reduce the mechanical vibration effectively and the way of control parameter selection can guarantee the bandwidth of speed estimation and the stability of the speed control loop. By integrating the result of this paper into an industrial drive control software, automatic control parameter commissioning can be achieved after identifying motor electric parameters, which is a meaningful product feature from an engineering point of view.

Introduction

Variable frequency rotary compressors have been widely used in heating, ventilation, and air conditioning (HVAC) systems for many years due to its high efficiency, simplicity, and low cost. However, rotary compressors' eccentric crankshaft structure leads to some annoying system vibrations [1]. Especially, when the rotation frequency is close to system natural mechanical resonance frequency, the system vibration amplitude will reach its peak value. Sometimes, the vibration and noise become unacceptable according to industrial product standards, it may also cause mechanical failures on the refrigerant pipe and reduce the system lifetime.

To solve this problem, many mechanical improvements have been tried, for example, using twin rotary compressors instead of single rotary ones or adjusting mechanical installations [1][2]. However, compared to these methods, software solutions by adopting active vibration damping algorithms are more favorable since they don't require any change in hardware, which maintains a minimum cost increase. In past decades, active vibration damping techniques have been well developed for high performance field orientation control (FOC) servo drives and many good results have been achieved [3]. But, similar ideas on a speed senseless FOC drive is not so commonly discussed. So far, some effective work has been done in this area, including, [4] proposed a minimum ripple point tracking method to find the best torque compensation value by tuning the amplitude and phase of the compensating torque, in which the convergence rate needs to be tuned by the designer. [5] used a repetitive controller to achieve load torque disturbance rejection, where the weighting factor of the controller needs to be tuned with expertise too. [6] suggested a similar approach while the control coefficient can be automatically adjusted by Fourier transform despite the need of relatively higher computation power. A load torque current compensation with a look-up table method was presented in [7] if an offline tested torque-position table is provided. [8]-[10] use load torque observer to estimate the load disturbance for feedforward compensation, both good dynamic and steady performances had been achieved. However, these methods rely on a relatively more accurate motor inertia parameters which may not always be provided.

This paper proposes a new control structure to reduce vibration. In the following sections, firstly, a motor speed estimation scheme is introduced. Secondly, the architecture of the senseless permanent magnet synchronous motor (PMSM) FOC system based on the bi-quad filter is presented. Thirdly, the selection methods of speed estimation parameters and bi-quad filter parameters are discussed. To further understand the system behavior under various working conditions, the impacts of motor stator resistor variation are also analyzed. Finally, the performance of the proposed design is demonstrated with simulation and experiment results. Moreover, a vibration damping effect comparison between the proposed structure and a widely used resonance damping scheme is made to show the unique character of the HVAC refrigerant compress system.

PMSM model based on MRAS

The following model reference adaptive systems (MRAS) of a PMSM in the synchronous dq frame has been presented in [13][15],

$$U_s = R_s I_s + \dot{\psi}_s + \omega_e J \psi_s \tag{1}$$

$$\psi_s = L I_s + \psi_{pm} \tag{2}$$

where the d-axis is aligned with the rotor permanent flux. $U_s = [u_d \quad u_q]^T$ is the stator voltage vector, $I_s = [i_d \quad i_q]^T$ is the stator current vector, $\psi_s = [\psi_d \quad \psi_q]^T$ is the stator flux, R_s is the stator resistor. $\psi_{pm} = [\psi_{pm} \quad 0]^T$ is the rotor permanent magnet flux vector. ω_e is rotor electrical angular speed. The inductance matrix L and constant matrix J are,

$$L = \begin{bmatrix} L_d & \\ & L_q \end{bmatrix}, \ J = \begin{bmatrix} 0 & -1 \\ 1 & 0 \end{bmatrix}$$

The system takes the physical motor as a reference model, an adjustable model as (3) and (4) is built on an estimated $d'q'$ frame. All estimated variables are marked with ^, and real physical items are marked with '.

$$U_s' = \hat{R}_s \hat{I}_s + \dot{\hat{\psi}}_s + \hat{\omega}_e J \hat{\psi}_s \tag{3}$$

$$\hat{I}_s = L^{-1} (\hat{\psi}_s - \hat{\psi}_{pm}) \tag{4}$$

The rotation angle θ_e of this $d'q'$ frame to the stationary $\alpha\beta$ frame is $\theta_e = \int \hat{\omega}_e dt$

The error between estimated and measured stator current is used for speed adaption.

$$\tilde{I}_s = (I_s' - \hat{I}_s) \tag{5}$$

A speed adaption law is chosen as,

$$\hat{\omega}_e = L_q (-k_p \tilde{i}_q - k_i \int \tilde{i}_q dt) \tag{6}$$

where k_p and k_i are gains of this PI type adaption algorithm.

Vibration suppression using a bi-quad filter

Fig.1 shows the architecture of the proposed system. It consists of an inner current control loop and an outer speed control loop. Some insignificant non-ideal features, such as current feedback delay, inverter delay, and mechanical frictions, are neglected in the transfer function to simplify speed loop analysis [11]. A new speed controller consisting of a PI controller combined with a bi-quad filter on the proportional branch is proposed for mechanical vibration suppression.

EPE'20 ECCE Europe

Assigned jointly to the European Power Electronics and Drives Association & the Institute of Electrical and Electronics Engineers (IEEE)

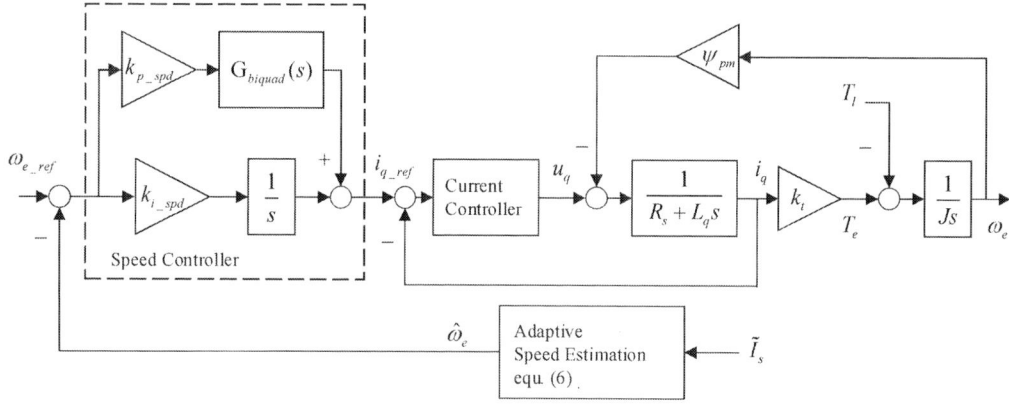

Fig. 1: Block diagram of the sensorless PMSM FOC system with the proposed speed controller

It is shown in [9][10] that, in an HVAC refrigerant system driven by a rotary compressor, an effective way to reduce the vibration is to minimize the compressor's speed variation caused by cyclic pulsive load torque T_l. Based on this, a possible approach is to increase the speed control loop gain to reduce the impact of load disturbance. However, in a sensorless drive system with conventional proportional-integral (PI) type speed controller, due to the limited current control bandwidth and inevitable speed estimation delay, the proportional gain of the speed loop usually cannot be increased as much as in systems with speed and rotor angle sensor, e.g. a servo drive. Therefore, a speed loop controller with a higher gain at a certain frequency rather than the entire frequency range could be an adequate candidate. According to this idea, a bi-quad filter is selected here since it can offer higher gains around its center frequency while maintaining nearly unity gain and a small phase shift at other frequencies [12]. From the engineering point of view, its theory is also easy to understand and simple to be implemented and tuned.

In Fig.1, J is motor inertia, k_t is the motor torque constant, k_{p_spd} and k_{i_spd} are speed PI control gains, the transfer function of a peaking bi-quad filter is designed as,

$$G_{biquad}(s) = \frac{s^2 + k_2 \omega_n s + \omega_n^2}{s^2 + \frac{k_2}{k_1} \omega_n s + \omega_n^2}$$

(7)

where k_1 and k_2 are filter constants, ω_n is the center frequency. The speed loop characters and the selection of these parameters will be discussed in later sections.

Design of control parameters

Speed adaptation parameters

To achieve a stable speed control in the proposed system, the design idea for the speed adaption parameter in (6) is keeping the cut-off frequency of speed estimation sufficiently higher than the mechanical vibration frequency. The dynamic feature of MRAS based speed estimation is discussed in [13] and [14], both these two works have derived transfer function from real speed value ω_e to the estimated one $\hat{\omega}_e$. However, it was found that this transfer function is a high order one, so, it is not easy to properly design the PI gains for the speed adaption in (6). In [15], by subtracting (3) and (4) from the real motor model in the estimated $d'q'$ frame and performing a linearization, the transfer function from stator current error $\tilde{I}_s$ to speed estimation error $\tilde{\omega}_e$ is derived as,

$$F(s) = C(sI - A)^{-1} B$$

(8)

where

$$A = \begin{bmatrix} -R_s L^{-1} - \omega_{e0} L^{-1} JL & \omega_{e0} L^{-1} JLI_{s0} + \omega_{e0} I_{s0} + \omega_{e0} L^{-1} \psi_{pm} \\ 0 & 0 \end{bmatrix}$$

$$B = \begin{bmatrix} JI_{s0} - L^{-1} JLI_{s0} - L^{-1} J\psi_{pm} \\ 1 \end{bmatrix}, \quad C = \begin{bmatrix} 0 & L_q & 0 \end{bmatrix}$$

Variables with subscript '0' are the linearized operation point values. I is the identity matrix. Combining (8) and (6), the open transfer function from speed estimation error to estimated speed $G_{\hat{\omega}_e \tilde{\omega}_e}(s)$ is

$$G_{\hat{\omega}_e \tilde{\omega}_e}(s) = -(k_p + \frac{k_i}{s})F(s) \tag{9}$$

And the speed estimation dynamics is described by

$$G_{\hat{\omega}_e \omega_e}(s) = \frac{-(k_p + \frac{k_i}{s})F(s)}{1 - (k_p + \frac{k_i}{s})F(s)} \tag{10}$$

By expanding (8), the transfer function $F(s)$ is as

$$F(s) = -\frac{1}{s}\left[\frac{b_2 s^2 + b_1 s + b_0}{L_d L_q s^2 + R_s(L_d + L_q)s + R_s^2 + L_d L_q \omega_{e0}^2}\right] \tag{11}$$

Where

$$b_2 = L_d L_q(i_{d0}L_d - i_{d0}L_q + \psi_{pm}), \quad b_1 = R_s(L_q \psi_{pm} - i_{d0}L_q^2 + i_{d0}L_d L_q)$$
$$b_0 = L_d L_q \omega_{e0}^2 (i_{d0}L_d - i_{d0}L_q + \psi_{pm}) + R_s i_{q0} L_q \omega_{e0}(L_d - L_q)$$

It can be seen that $F(s)$ is a multiplication of a second order system and a pure integration item. If an approximation of an average inductance $L_d = L_q \approx L_{avg} = \dfrac{L_d + L_q}{2}$ is assumed for simplicity, $F(s)$ can be further simplified as,

$$F(s) \approx -\frac{1}{s}\left[\frac{L_{avg}\psi_{pm}(L_{avg}s^2 + R_s s + L_{avg}\omega_{e0}^2)}{L_{avg}^2 s^2 + 2R_s L_{avg} s + R_s^2 + L_{avg}^2 \omega_{e0}^2}\right] = -\frac{1}{s}F_1(s) \tag{12}$$

The Bode plot of $F_1(s)$ in Fig.2 indicates that, in low operation frequency range, where $L_{avg}\omega_{e0} \ll R_s$, $F_1(s)$ behaves like a first order system, while, in the high frequency range $L_{avg}\omega_{e0} \gg R_s$, $F_1(s)$ acts as a second order band stop filter with a center frequency around ω_{e0}.

Here, to get a PI parameter design scheme at low frequency range, $F_1(s)$ is reduced to a first order system $F_1'(s)$ by using the Padé approximation method [16].

$$F_1(s) \approx F_1'(s) = \frac{p_0 + p_1 s}{1 + q_1 s} \tag{13}$$

Where

$$p_0 = \frac{L_{avg}^2 \psi_{pm} \omega_{e0}^2}{R_s^2 + L_{avg}^2 \omega_{e0}^2}, \quad p_1 = \frac{L_{avg}\psi_{pm} R_s(2L_{avg}^2 \omega_{e0}^2 - R_s^2)}{L_{avg}^4 \omega_{e0}^4 - R_s^4}, \quad q_1 = \frac{R_s^3 L_{avg} - 3L_{avg}^3 R_s \omega_{e0}^2}{R_s^4 - L_{avg}^4 \omega_{e0}^4}$$

Fig. 3 shows the Bode plot comparison of the approximated transfer function (13) and the original one (12) at 5Hz motor electric angular frequency, which shows a very similar magnitude and phase characteristics between the two, thus it is reasonable to design PI parameter according to (13) in low speed motor operation range, equation (9) can be simplified as,

$$G_{\hat{\omega}_e \tilde{\omega}_e}(s) = -(k_p + \frac{k_i}{s})F(s) \approx -(k_p + \frac{k_i}{s})\frac{1}{s}F_1'(s) = (k_p + \frac{k_i}{s})\left[\frac{1}{s}\left(\frac{p_0 + p_1 s}{1 + q_1 s}\right)\right] \tag{14}$$

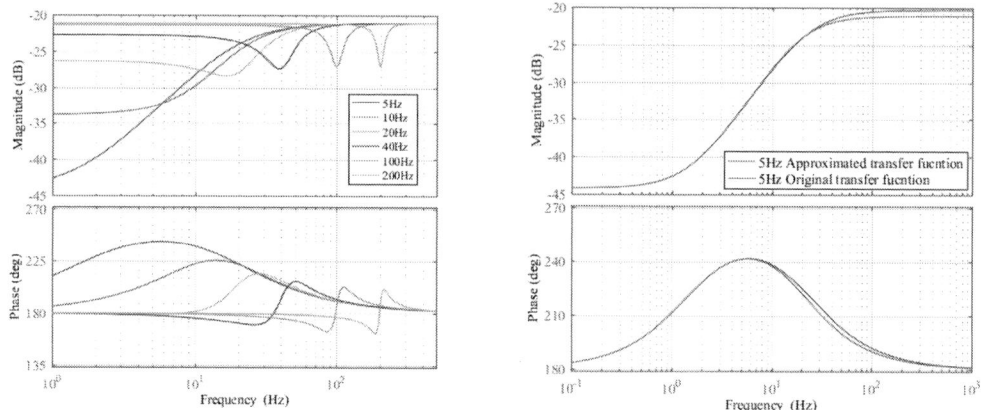

Fig. 2: Bode plot of $F_1(s)$ at different motor speed Fig. 3: Bode plot comparison of $F_1(s)$ and $F_1'(s)$

By choosing $k_p / k_i = q_1$ (14) is now a second order system, If the required speed estimation bandwidth is ω_c, the PI parameters can be selected as,

$$k_{p\_low\_spd} = \frac{\omega_c q_1}{p_1}, \; k_{i\_low\_spd} = \frac{\omega_c}{p_1} \quad (15)$$

As shown in (15), p_0, p_1 and q_1 are functions of motor speed ω_{e0} and motor parameter constants, therefore, after motor parameters is offline identified, k_p and k_i can be pre-calculated offline. Furthermore, from an engineering point of view, considering $L_{avg}\omega_{e0} \ll R_s$, these parameters could be even simplified to,

$$p_1 = \frac{L_{avg}\psi_{pm}}{R_s}, \; q_1 = \frac{L_{avg}}{R_s} \quad (16)$$

At a higher speed range, $L_{avg}\omega_{e0} \gg R_s$, as indicated in Fig. 2, $F_1(s)$ can be seen as second order filter with a center frequency near ω_{e0}, If the PI parameters is to be designed to achieve speed estimation bandwidth $\omega_c < \omega_{e0}$, $F_1(s)$ could be approximated to a pure gain part as $F_1(s) \approx \psi_{pm}$, combining (12), equation (9) is taken as,

$$G_{\tilde{\omega}_e\tilde{\omega}_e}(s) = -(k_p + \frac{k_i}{s})F(s) \approx (k_p + \frac{k_i}{s})\frac{\psi_{pm}}{s} \quad (17)$$

Therefore, equation (10) is now a second order system, to get a bandwidth of higher than ω_c on the speed estimation, the PI parameter can be chosen as

$$k_{p\_high\_spd} = \frac{\omega_c}{\psi_{pm}}, \; k_{i\_high\_spd} = \frac{\omega_c^2}{\psi_{pm}} \quad (18)$$

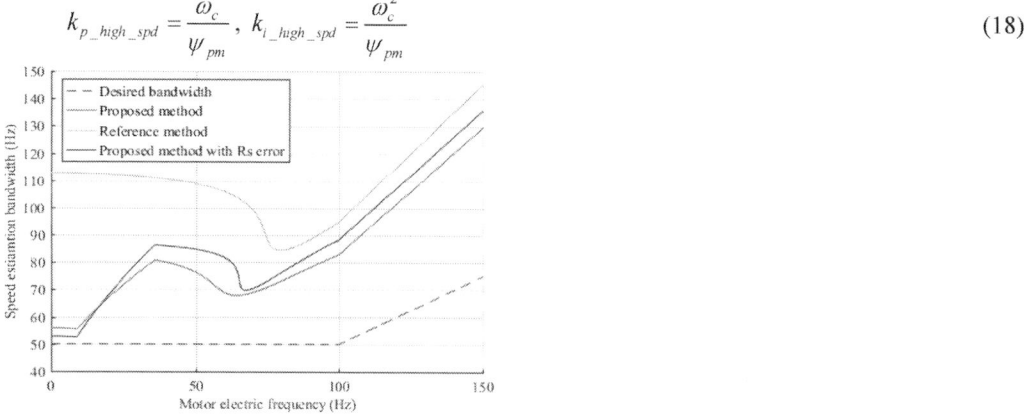

Fig.4: Speed estimation bandwidth from 0.1Hz to 150Hz motor speed.

The design selection in (18) is similar to the parameter design in [15], in which the parameters are selected according to a PLL structure without considering the electrical observer model's dynamics. The proposed method achieves the desired bandwidth by using smaller PI parameters. This provides better noise immunity performance in very low speed operation range operation where the control loop is sensitive to the drive system noises such as current sampling errors and inverter non-linearity features. Apart from low and high speed ω_{e0} operation points, a linear combination of k_p and k_i in (15), (16) and (18) according to the motor speed is used for the mid-speed operation. Fig.4 shows the speed estimation bandwidth of (10) and the result is compared to the reference result of parameter settings in [15]. It can be found that the parameters designed here provide a sufficient speed estimation bandwidth.

In practice, the motor stator resistance $\hat{R}_s$ will change along with environment temperature, inconsistent of mathematic adjustable model's stator resistor and the real motor value will lead to inaccurate estimation. Defining stator resistor error $\tilde{R}_s = R_s - \hat{R}_s$, equation (8), the transfer function from speed estimation error to current estimation error is changed to

$$F_{error} = C(sI - \hat{A})^{-1} B\tilde{\omega}_e - C(sI - \hat{A})^{-1}\tilde{A}\begin{bmatrix} I_s' \\ 0 \end{bmatrix} = \hat{F}(s)\tilde{\omega}_e - D(s) \tag{19}$$

where

$$\hat{A} = \begin{bmatrix} -\hat{R}_s L^{-1} - \omega_{e0}L^{-1}JL & \omega_{e0}L^{-1}JLI_{s0} + \omega_{e0}I_{s0} + \omega_{e0}L^{-1}\psi_{pm} \\ 0 & 0 \end{bmatrix}, \quad \tilde{A} = \begin{bmatrix} -\tilde{R}_s L^{-1} & 0 \\ 0 & 0 \end{bmatrix}$$

Because I_s' is independent to $\tilde{\omega}_e$, (19) is can be seen as a multi-input single output system, therefore, the dynamics from $\tilde{\omega}_e$ to $\tilde{I}_s$ is still decided by $\hat{F}(s)$, the resulting bandwidth is shown in Fig.4. The item $D(s)$ caused by error $\tilde{R}_s$ is taken as a disturbance to speed estimation. Using the same analysis manner as in [15], the speed estimation mechanism in (19) can be represented in Fig.5.

With speed adaption structure shown in Fig.6 and considering (11) and (12), the speed error caused by disturbance is

$$-\frac{G(s)}{1 + G(s)F(s)}D(s) \approx -\frac{s(k_p s + k_i)}{s^2 - (k_p s + k_i)F_1(s)}D(s) \tag{20}$$

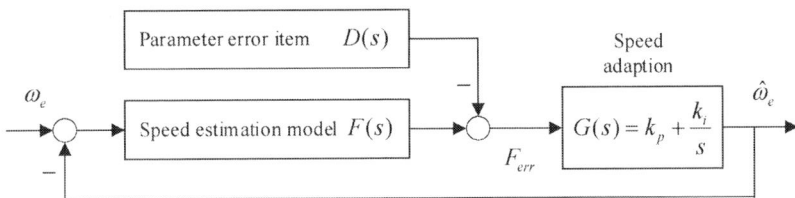

Fig.5: Equivalent mechanism of speed adaption with parameter error

It can be seen that the transfer function $F(s)$ has a pole at the origin in the complex plane. With a PI type $G(s)$, equation (20) has a zero at the origin in the complex plane. This means that the motor resistor error will not cause error in speed estimation at a steady state since the DC gain of (20) is zero.

Bi-quad filter parameters

As presented in (7), there are three parameters in the bi-quad filter to be decided. The center frequency ω_n should be set to the mechanical vibration frequency. The vibration frequency is correlated with rotor mechanical frequency, ω_n is set as $\omega_n = \omega_e / P_n$, P_n is the motor pole pairs. k_1 is related to the filter's peak gain at ω_n, which can be tuned according to vibration phenomenon by an on-site commissioning engineer. k_2 decides the pass band width in the frequency spectrum of the filter.

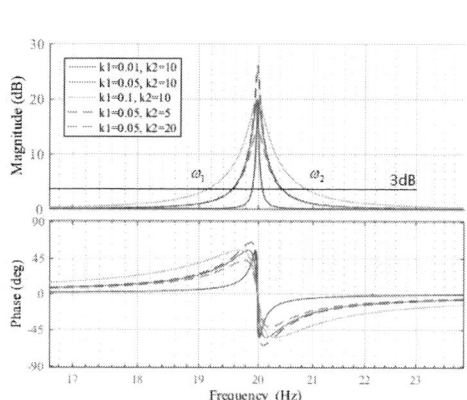

Fig.6: Bode plot of the bi-quad filter

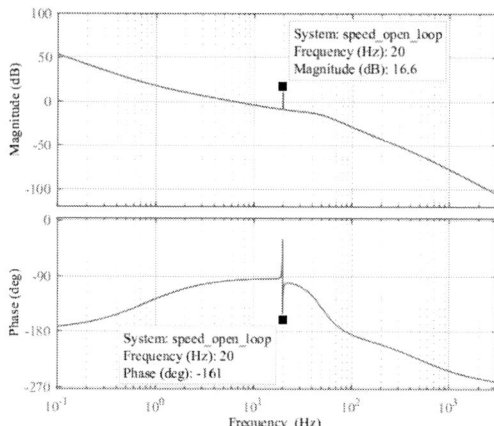

Fig.7: Speed loop gain margin and phase margin

Here, k_2 is designed to changing with k_1 to ensure the upper 3dB pass band frequency ω_2 is lower than the speed estimation bandwidth. The reason for this is that the phase lag of bi-quad filter varies dramatically around ω_n, which might lead to insufficient phase margin in speed loop. The bode plot of a bi-quad filter with different parameter combinations is shown in Fig.6. The relationship of k_1, k_2 and ω_1, ω_2 is

$$\omega_{1,2} = \frac{\omega_n \sqrt{2(2k_2^2 - 2k_1^2 + k_1^2 k_2^2 \pm k_1 \sqrt{(k_2^2 - 2)(4k_2^2 - 2k_1^2 + k_1^2 k_2^2)})}}{2k_2} \tag{21}$$

In this design, ω_2 is set to about $1.05\omega_n$, which makes lower 3dB pass band frequency ω_1 is about $0.95\omega_n$, after k_1 is manually tuned, k_2 is derived by solving (21).

A Bode plot of speed open loop transfer function is shown in Fig.7. The current control loop bandwidth is configurated to 500Hz. The speed estimation feedback bandwidth is set to 50Hz. It shows that at the center frequency of the bi-quad filter, the open loop gain has a peak in magnitude, which can increase the speed controller's ability to resist load disturbance in close loop control. By proper choosing speed control PI parameters, the phase margin at center frequency remains to be adequate.

Simulation and experiment results

The simulation and experiment were carried out on a 220V 1hp three-phase 6-pole PMSM rotary compressor HVAC testing system. Motor parameters are $R_s = 1.511\Omega$, $L_d = 7.7\text{mH}$, $L_q = 18.7\text{mH}$, $\psi_{pm} = 0.0879\text{Wb}$, the rotor inertia is $2.7 \times 10^{-4}\text{kgm}^2$.

System simulation is done on Matlab/Simulink, Fig. 8 shows the motor speed control results. The target motor speed is set to 60Hz electric frequency, which is 20Hz rotor mechanical speed. The load applied at this frequency is 1N.m constant torque with additional 20Hz pulsive load torque varies from 0~1N.m, which simulates the compressing process. The bi-quad filter parameter is $k_1 = 0.1$, $k_2 = 20$, the speed estimation feedback bandwidth ω_c is set above 50Hz at minimum speed. It was found that, under the same PI parameters as the conventional PI controller, the real speed variation is reduced from about 56~64Hz to 59~61Hz by adding the proposed bi-quad filter to the speed loop.

To investigate the impact of stator resistor error, a simulation at low speed is performed, the motor electric speed is set to 20Hz, which is 6.67Hz rotor frequency. Stator resistor is set to 50% less than real motor value. Fig.9 shows the simulation results of electric speed, i_d and i_q estimation. Under this scenario, the speed estimation is still correct, the estimated i_q is converged to its real value. However, an error exists in i_d estimation. This can be explained that, in an MRAS system, as long as the adaption law is designed to be stable under parameter errors, the adjustable model will converge to a certain steady state anyway, from equation (3)~(6), in this two dimensional system, there are three variables to be estimated, if the estimated speed is convergent to real value, then the parameter errors will cause errors on current estimation.

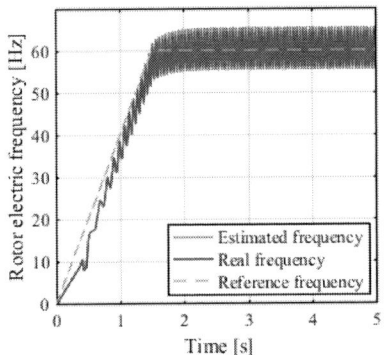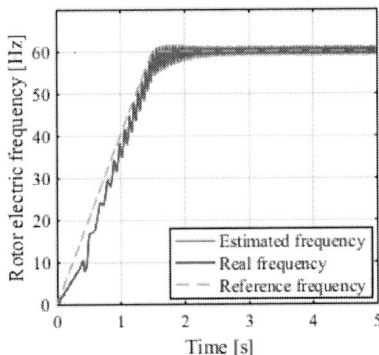

Fig. 8: Speed estimation and control using conventional PI (Left) and the proposed method (Right)

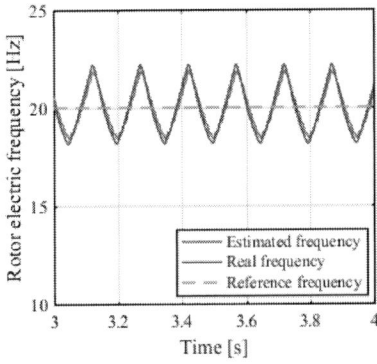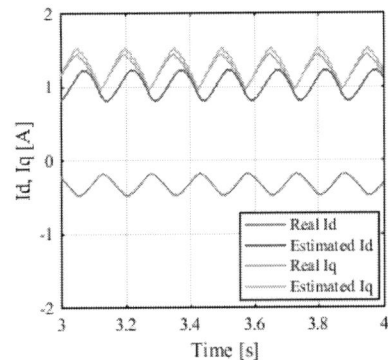

Fig. 9: Speed and current estimation results with 50% R_s error

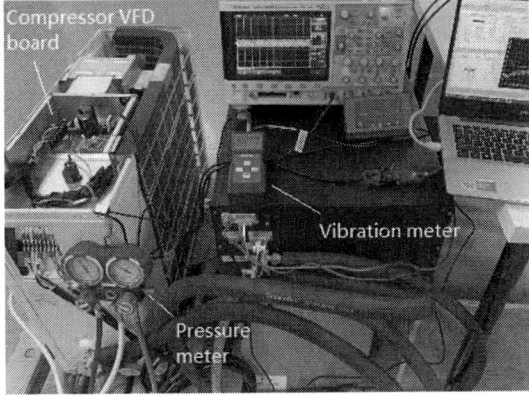

Fig.10: Experiment HVAC compressor drive system

The PMSM compressor is driven by a 220V 1.5kW variable frequency drive (VFD) board using an ARM Cortex M4F processor, on which the control algorithm is implemented by software. All control parameters used in the experiment is calculated based on the proposed design methods. The current controller bandwidth is set to around 500Hz. The software calculation frequency of the speed loop is 2kHz, the calculation frequency of current loop and MRAS motor model is 8kHz.

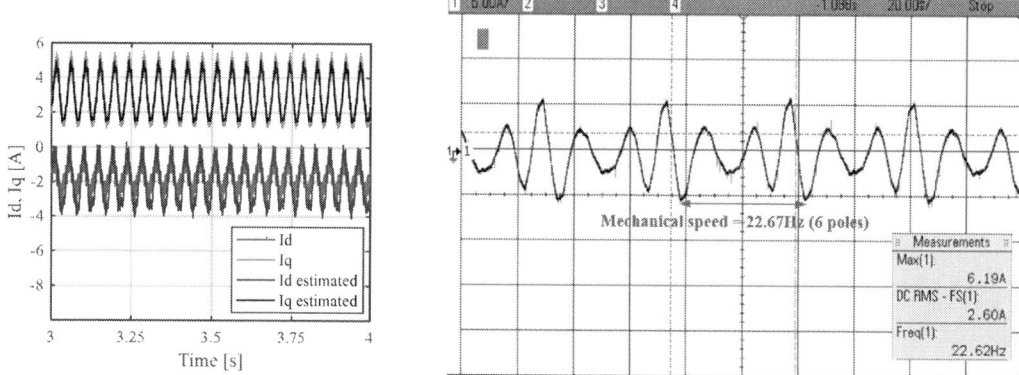

Fig. 11: Experiment current estimation results and motor phase current

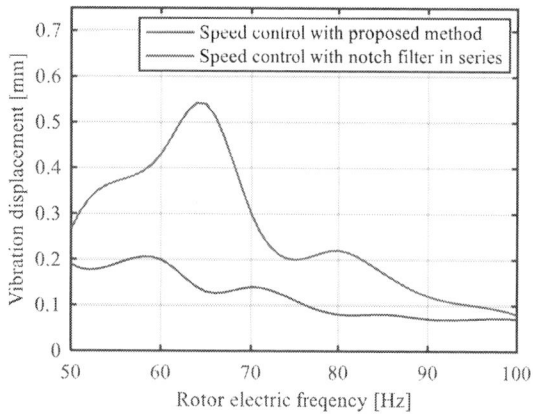

Fig. 12: Compressor mechanical vibration amplitude

Fig.11 shows a valid speed senorless FOC control on the experiment system, current estimation is convergent. When the compressor is running at 68Hz electric frequency, the rotor mechanical frequency is 22.67Hz, which is close to the system's natural vibration frequency, 22.67Hz periodic spikes can be found on motor phase current, this is the current generated by speed controller to resist the load disturbance per round due to refrigerant compressing process. Fig. 12 shows the system vibration suppression effect by using the proposed method, it was compared to a vibration damping method using a notch filter in series connected with conventional PI speed controller, the result shows that the proposed method has a better performance in this HVAC refrigerant compress system.

Conclusion

The simulation and experiment results have demonstrated the effectiveness of the proposed speed control structure. The parameter design method provides a stable speed sensorless FOC control, the vibration amplitude is reduced by the proposed bi-quad filter controller. The analysis has shown that motor stator resistor error does not lead to steady state speed estimation error but has an impact on estimated flux angle accuracy. By integrating these analysis results and approximation methods into software, an automatic parameter commissioning function can be achieved and the on-site engineering tuning effort of the VFD compressor drive product can be greatly reduced.

References

[1] Z. Deng, H. Shen, K. Wang and H. Jin.: Vibration improvement and dynamic balance automatic optimization of rotor compressor, 2019 2nd World Conference on Mechanical Engineering and Intelligent Manufacturing (WCMEIM), Shanghai, China, 2019, pp. 224-228

[2] Lee, J., Ui-Yoon Lee, Jin-Ah Chung and Un-Seop Lee.: Development of a miniature twin rotary compressor, International Compressor Engineering Conference, U. S., 2014, pp. 1350-1~9.

[3] G. Ellis, R.D. Lorenz.: Resonant load control methods for industrial servo drives. Thirty-Fifth IAS Annual Meeting and World Conference on Industrial Applications of Electrical Energy, 8-12 Oct. 2000.

[4] J. Kim, K. Nam.: Speed ripple reduction of pmsm with eccentric load using sinusoidal compensation method, 8th International Conference on Power Electronics - ECCE Asia, pp. 1655-1659, 2011.

[5] H. C. Chen, and C. K. Huang.: Position sensorless BDCM control with repetitive position-dependent load torque, in Proc. of 2010 IEEE/ASME International Conference on Advanced Intelligent Mechatronics, pp.1064-1069, Jul. 2010.

[6] S. Suthep, W. Yankai, I. Yuuma, M. Ishida, K. Yubai and S. Komada.: Frame anti-vibration control for sensorless IPMSM-driven applications, IECON 2016 - 42nd Annual Conference of the IEEE Industrial Electronics Society, Florence, 2016, pp. 2802-2808

[7] Meng Zhang, etc.: A speed fluctuation reduction method for sensorless PMSM-compressor system, ICEMS, Vol.3, 27-29 Sept. 2005.

[8] Z. Zheng, M. Fadel and Y. Li.: High performance PMSM sensorless control with load torque observation, EUROCON 2007 - The International Conference on "Computer as a Tool", Warsaw, 2007, pp. 1851-1855

[9] Z. Youlin, M. Xuetao and X. Min.: Single rotor compressor's low frequency control scheme in air conditioner, 2011 International Conference on Electrical Machines and Systems, Beijing, 2011, pp. 1-5

[10] Cheon-Su Park, SeHwan Kim, Gwi-Geun Park, Jul-Ki Seok.: Active mechanical vibration control of rotary compressors for air-conditioning systems. Journal of Power Electronics, 2012, 12(6), 1003-1010.

[11] J. C. Balda and P. Pillay.: Speed controller design for a vector-controlled permanent magnet synchronous motor drive with parameter variations, Conference Record of the 1990 IEEE Industry Applications Society Annual Meeting, Seattle, WA, USA, 1990, pp. 163-168 vol.1

[12] C. Pradabpet, S. Yimman, W. Hinjit, S. Chivapreecha and K. Dejhan.: Design and implementation of biquad digital filter, 9th Asia-Pacific Conference on Communications (IEEE Cat. No.03EX732), Penang, Malaysia, 2003, pp. 1138-1142 Vol.3.

[13] M. Rashed, P.F.A. MacConnell, A.F. Stronach, P. Acarnley.: Sensorless indirect-rotor field-orientation speed control of a permanent-magnet synchronous motor with stator-resistance estimation, IEEE Trans. Ind. Electron., vol. 54, no. 3, pp. 1664-1675, 2007.

[14] L. Hamefors, H.P. Nee.: A general algorithm for speed and position estimation of AC motors, IEEE Trans. on. Ind. electronics, vol. 47, no. I, pp. 77-83, 2000.

[15] A. Piippo, M. Hinkkanen, J. Luomi.: Analysis of an adaptive observer for sensorless control of interior permanent-magnet synchronous motors, IEEE Trans. Ind. Electron., vol. 55, no. 2, pp. 570-576, 2008.

[16] Xin Luo, Anwen Shen, Renchao Mao.: A double bi-Quad filter with wide-band resonance suppression for servo systems, Journal of Power Electronics, Vol. 15, No. 5, pp. 1409-1420, September 2015.

3D PCB package for GaN inverter leg with low EMC feature

Pawel B. Derkacz*[1][2], *Student Member, IEEE*, Jean-Luc Schanen[1], *Senior Member, IEEE*,
Pierre-Olivier Jeannin[1], Piotr Musznicki[2], Piotr J. Chrzan[2], *Senior Member, IEEE*,
Mickael Petit[3]

[1]Univ. Grenoble Alpes, CNRS, Grenoble INP, G2Elab, F-38000, Grenoble, France
[2]Faculty of Electrical and Control Engineering, Gdansk University of Technology, Gdansk,
Poland
[3]SATIE - CNRS UMR 802, ENS Cachan - CNAM, Cachan, France
*E-Mail: pawel-bogdan.derkacz@grenoble-inp.fr, pawel.derkacz@pg.edu.pl

Acknowledgements

This work has been funded thanks to the French national program "Programme d'investissements
d'Avenir, IRT Nanoelec" ANR-I0-AIRT-05. The authors also want to thanks the CEDMS of IUT
Grenoble for their expertise, advice and capabilities with the PCB assembly and SMD technology.

Keywords

« EMC/EMI », « Gallium Nitride (GaN) », « HEMT », « Wide bandgap devices »

Abstract

This paper presents the adaptation of a 3D integration concept previously used with vertical devices to
lateral GaN devices. This 3D integration allows to reduce loop inductance, to ensure more
symmetrical design with especially limited Common Mode emission, thanks to a low middle point
stray capacitance. This reduction has been achieved by both working on the power layout and
including a specific shield between the devices and the heatsink. The performances of this 3D layout
have been verified in comparison with a more conventional 2D implementation, using both
simulations and measurements.

1. Introduction

During the recent years, an expanse growth of gallium nitride (GaN) transistors manufacturing
technology has been observed. These devices start to be commonly used in modern Power Electronics
applications. The significant increase of the switching speed associated with these new devices [1]
causes several issues such as electromagnetic interference (EMI) non-compliance, voltage overshoot
with oscillations, power drive interactions, etc. One of the biggest challenges in implementing these
devices is an identification and minimization of the origin of these drawbacks. This paper first briefly
reminds the main design rules associated with the layout dedicated to high speed switching (section 2).
Then, the advantages of a 3D integration solution previously used with vertical devices are presented,
and the necessary adaptations for being used with lateral GaN devices introduced in section 3.
A specific attention is paid to the shielding of the leg middle point with respect to ground: both power
layout and heatsink contribute to this stray capacitance is addressed. Finally, in section 4 the
performances of the proposed 3D layout, in comparison with a more conventional 2D implementation
are evaluated.

2. Layout impact on EMC of power converters with GaN devices

The transitions of currents and voltages during switching operations in phase-leg configuration
(further considered as a commutation cell) are well-known sources of electromagnetic disturbances.

Those transients together with existence of capacitive and inductive stray elements generate high frequency currents which are usually separated into differential (DM) and common mode (CM). Fig.1 illustrates the equivalent circuit of a switching cell, including the most impacting stray elements originating the main EMC issues. The inductive nature of the electrical link between the decoupling capacitor and the switching devices is taken into account by the internal stray inductance of the capacitor (ESL - Equivalent Series Inductance), Ls+ and Ls- as well as the two Lc, which are also the parts of the gate circuits. These latter are also accounted using Lg.

The capacitive behavior of the layout is modeled using C+ and C-, between DC bus and ground, and also Cout which represents the stray behavior between the output of the converter and the ground. In addition, other stray capacitances Ciso are important to model the propagation path of Common Mode current through the driver's power supplies. Finally, the stray capacitances of components are also part of the model, since they can generate oscillations with other stray elements of the layout.

The following subsections the impact of all these stray elements on the EMC behavior of the switching cell is detailed.

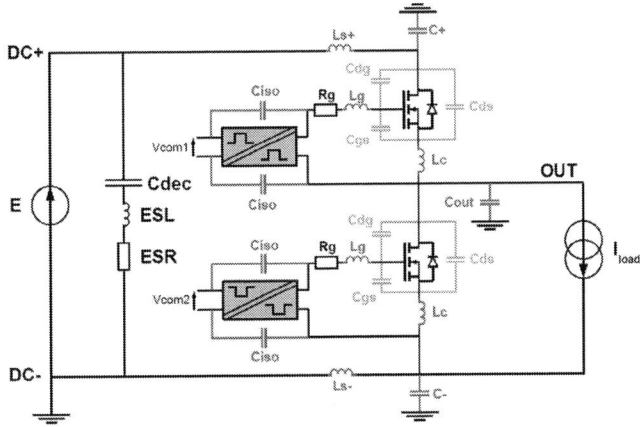

Fig. 1: Electrical scheme of commutation cell with parasitic elements [2]

2.1. Voltage overshoot and ringing

The most visible effect of high-speed switching operation is the voltage overshoot across transistors at turn off. The overshoot is caused by an induced voltage drop across an inductance $L_{switching}$ of the commutation loop ($L_{switching}$ being the sum of inductances: Ls+, Ls-, 2*Lc and ESL according to Fig. 1 [2] also with mutual inductances between them). The high value of di/dt together with $L_{switching}$ can generate overshoot exceeding the voltage rating of the device which may cause its damage. To achieve a very high current commutation speed, the stray inductance of the switching loop has to meet nano – or even subnanohenry range.

Another effect related to the stray inductance $L_{switching}$ are voltage oscillations at transistor turn off. Those oscillations are related to $L_{switching}$ and parasitic output capacitance of a device (Coss). The oscillations cause additional losses and add peaks on a spectrum in the high frequency range. Moreover, the ringing phenomena may change switching conditions for the next commutation if the oscillations are not properly damped. The damping factor describing this process can be calculated by using Eq. 1. According to this equation, the reduction of the stray inductance $L_{switching}$ leads to a significant increase of damping factor [3].

$$\zeta = \frac{R}{2} \cdot \sqrt{\frac{Coss}{L_{switching}}} \tag{1}$$

2.2. Common mode capacitances

The capacitive behavior of the switching cell including the stray capacitances to the ground is shown in Fig. 1. The most impacting capacitance is Cout, since it is submitted to a high dV/dt during transients and generates high CM current. Besides, gate driver insulation and supply stray capacitance Ciso creates an additional path for CM noise. This is not addressed in this paper, but previous work as [4] did focus on this specific aspect.

Moreover, two stray capacitances (C+ and C-) are connected from stable potentials DC+ and DC- to the ground, creating an additional path for CM current and can provide recycling of the current inside switching cell. They should thus be as large as possible. However, C+ and C- should be designed carefully to keep the symmetry. Otherwise high frequency currents flowing in DC bus induce voltage variation, that can create additional common mode current [2].

In summary, the most important design effort is to reduce the capacitance Cout and maintain the symmetry (equal values) of capacitance C+ and C-, which have to be increased to help in recycling CM current.

2.3. Decoupling capacitors and symmetry of DC+ and DC- layout

One of the key aspects for reaching low stray inductance in a power loop is using decoupling capacitors. Their role is to create a low impedance path for high frequency currents circulating in a switching cell. Decoupling capacitors are suppressing high frequency AC signals or voltage spikes at the input of the switching cell and ensure a stable DC voltage at the transistor transients. The selection of decoupling capacitors should be done carefully to ensure a correct value of the capacitance and low ESL for high damping and filtering high frequency transient currents [5].

Moreover, connections of decoupling capacitors and GaN devices (from both sides DC+ and DC-) should be kept symmetrical (the same value of L+ and L- stray inductances). Non-symmetrical design can cause a generation of differential mode current from switching cell to converter's supply and additionally with a lack of symmetry C+, C- capacitances may cause coupling between differential and common mode [6].

2.4. Power and gate circuits coupling

There are two types of coupling between power and gate circuit: inductive and capacitive. Both of them can interact and negatively influence GaN switching performance causing a false turning on or off, or slowing down commutations. The inductive coupling is represented with the common source inductance Lc. A high current variation in a power loop generates the disturbances on the gate circuit and negatively affects the commutation process increasing the switching losses. Thus, above mentioned couplings between power and gate circuits must be minimized, e.g. by designing a Kelvin Source Connection [7]. The inductive coupling also exists due to the mutual inductance between the power and gate part. It should be minimized by designing a small gate loop surface and geometrical separation of power and gate loops. The inductance Lc accounts for both effects, as explained in [8] The capacitive coupling is created through the "Miller capacitance" (Cgd). The voltage variations dVds/dt cause the flow of parasitic current through Cgs and the whole gate circuit which can lead to false turn-on if the inducted voltage drop on the gate impedance is above the threshold value of GaN device [9]. A negative voltage for turning off can significantly improve gate immunity on the Miller effect. However, it can also increase the global losses in case of using large dead times. Therefore, to avoid this disturbance, the gate circuit must be designed with a low value of a stray inductance Lg+Lc.

3. 3D layout for GaN transistors

According to the role of each stray elements underlined in the previous section, designing a power and gate circuit layout suited for high speed devices should meet following requirements:

- Minimizing stray inductances, thanks to reduced loop area
- Reducing mutual inductances between power and gate circuits, using e.g. perpendicular planes
- Reducing Cout and increasing C+ & C-, keeping a perfect symmetry of C+, C-, Ls+, Ls- to avoid mode conversion.

Nowadays, all commercial GaN transistors are produced with a lateral (horizontal) arrangement and adapted for planar modules. This approach corresponds to the classical way of designing power converters, which can be called "2D", where all power components are placed on the one side of PCB for one side cooling system. Using all degrees of freedom in the layout design, this planar design approach prevents to reach previously mentioned guidelines. Furthermore, reaching small loop inductance is easier with vertical implementation, especially for high voltage: the sub-nanohenry performance reached by [10] in a 2D implementation was achieved for a 40V application, using ultra small package.

In this section, we will therefore present how to adapt a promising 3D power layout concept – PCoC: Power Chip on Chip - to lateral GaN devices. At first, the main features of the PCoC concept will be reminded. This implementation will then be adapted to GaN lateral devices. Shielding the middle point of the switching cell is one key feature of the 3D layout, therefore a dedicated section will focus of this aspect.

3.1. Power Chip-on-Chip concept

The solution of "3D" power loop structure (Power Chip-on-Chip, PCoC) can be proposed (Fig. 2) to reduce $L_{switching}$ and decrease EMI generation by keeping symmetrical design and placing components one below the other. This idea has been developed originally for vertical semiconductor devices (Silicon and Silicon Carbide) [11] [2]. The main idea of this concept is an integration of the power devices inside a bus-bar structure. Two distribution planes are placed on the top and bottom of the system, creating large areas of copper connected to the stable potentials DC+ and DC-. These wide conductive sections ensure an extremely small inductance between power ports and switching cell. The decoupling capacitors might be placed vertically between DC+ and DC- very close to the power transistors' pads. In this manner the power loop area is reduced and the stray inductance $L_{switching}$ is reduced to the minimum (Fig. 2). Bus bar idea offers the parasitic couplings minimization and symmetrical arrangement of the copper tracks which affects on EMI reduction.

The other main advantage of the PCoC concept is that the 3D arrangement allows obtaining the middle point of the dies (load connection) surrounded by the DC+ and DC- potentials. Therefore, the DC bus acts as a shield with respect to ground, thus reducing the value of Cout to almost zero, maximizing C+ and C-, and keeping symmetry of all stray elements of the DC bus.

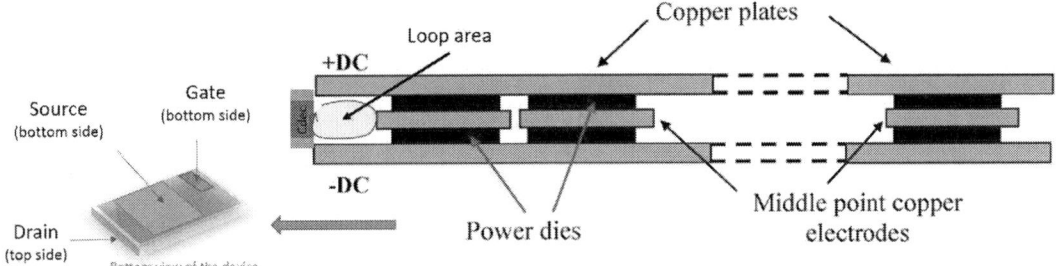

Fig. 2: PCoC concept basis [11] using vertical devices

However, the lateral arrangement of the GaN devices does not allow the simple stacking of the devices to build a switching leg, since all connections are on the same side, the other side being only a thermal pad. Therefore, the concept has to be modified.

3.2. Proposed 3D PCoC design for lateral GaN devices

The proposed conception utilizes only PCB to arrange all connections and provides a high level of integration between discrete components and PCB copper layers (planes/polygons) and tracks. The basic idea from [12] is shown in Fig.3. In this paper, the proposed design focuses on simplicity of manufacturing and assembling the boards. Thus, the multilayer PCB with the minimum number of layers which gives satisfying results (4 layers) was used to implement the PCoC structure.

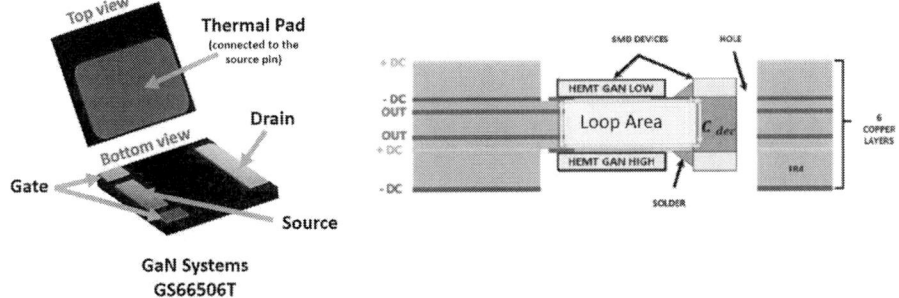

Fig. 3: PCoC concept of [12], which will be used with lateral devices from GaN Systems [13]

In this PCoC concept the thermal path has to be separated from electrical connections since the PCB used for this purpose cannot evacuate easily generated heat. Therefore, external heatsinks are used on both sides of the switching cell as displayed in the Fig. 4. Using other technology as Insulated Metal Substrate (IMS) instead of PCB would allow conducting the heat flux using the same side of the chip, but this is usually only a single layer substrate disabling all necessary connections e.g. for gate circuits. Therefore, we found some specific packages allowing the different use of each sides: one for electrical connection and one thermal pad for the heatsink [13].

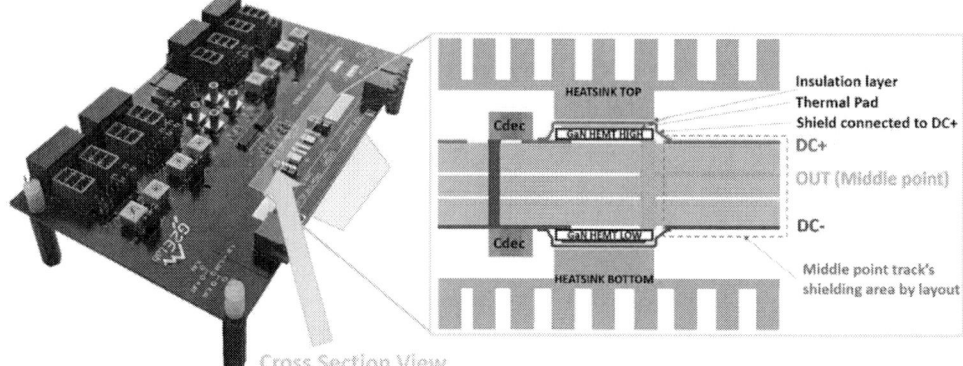

Fig. 4: Prototype of converter with the "3D" PCoC layout and cross view of arranged cooling system

The GaN Systems GS66506T transistors [13] with 650V blocking capability and a continuous drain current 22.5A and top cooling thermal pad are used. Moreover, the decoupling capacitors placement differ from the Fig. 3. They are placed horizontally (on the both sides of PCB) and connected to the bus-bars directly and by dedicated vias. Twelve 100nF 630V SMD capacitors (KEMET C1210C104KBR) in case 1210 were used to decouple two phase-legs of converter (6 capacitors per phase-leg). Those capacitors, together with two EPCOS CeraLink capacitors 500nF 700V, are creating a DC-Link with 2.2µF capacity. All of those components are placed symmetrically on the both sides of PCB.

The exploded 3D view of the power layout is shown in Fig. 5. The GaN switches are marked by arrows, indicating the placement one below the other corresponding to the PCoC idea. The transistors are connected directly to the DC+ and DC- bus bars and the common point (middle point of phase-leg) is created by the vias connection. The converter's phase-leg middle point track is arranged on the inner layer of the PCB, therefore being in the middle of the DC bus, as in the PCoC basic concept. The decoupling capacitors are placed as close as possible to the switches on each side of the PCB (6 capacitors for each phase-leg – 3 on the top and 3 on the bottom). The switching loop is illustrated in Fig. 5: The loop area is as small as possible, since it is contained inside the PCB thickness. In order to reduce more the switching loop, the PCB with a reduced thickness can be considered. Embedded dies technology is another option that could permit also to place GaN components closer one to the other. Regarding gate drive layout design, short connections and perpendicular plane with power layout have been chosen to meet the design criteria of section 2.4.

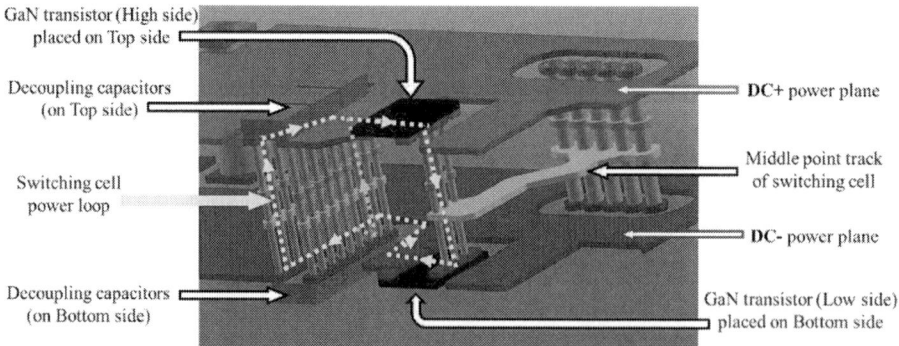

Fig. 5: Exploded 3D view of proposed PCoC layout (thickness of the board in scale 10:1)

3.3. Shielding

Beside symmetry and small switching loop area, the superior advantage of the PCoC concept is the shielding of the middle point of the leg, connected to the output filter of the converter. As displayed in Fig. 5, this middle point is located in the inner layer of the converter, as in the original concept of Fig. 2. Therefore, shielding reduces Cout (middle point to ground capacitance) to almost zero, since it is embedded inside the DC bus. However, there is another stray capacitance to the ground, which is brought by the GaN die itself: the thermal pad of the GaN devices used in our application is actually connected to the source. Therefore, the Top Switch of the inverter leg will also contribute to the Cout value described in Fig. 1, what illustrates the electrical representation in Fig. 6, where Cout has two contributions: Cout_Layout and Cout_Heatsink.

It is worth noting that even without the direct connection of the thermal pad to the source, as in the used dies, a quite large stray capacitance between those potentials would also exist and the issue would remain. In the current situation using package [13], Cout_Heatsink can be roughly evaluated through a parallel plate capacitor formula. The thermal pad area is $5.1*3.1\,mm^2$ and the thickness of the insulator between die and heatsink is approximatively 100µm. Permittivity is considered equal to unity in this first approximation. Therefore, Cout_Heatsink is roughly 1.5pF.

Reducing this value without degrading the thermal path is not possible. Therefore, it has been decided to insert a conductive layer between the die and the heatsink, insulated from both thermal path and heatsink, and to connect this layer to the DC+ potential. This results in capturing all Common Mode currents generated through this stray capacitance and recycle it internally inside the converter through a stable potential. The principle is illustrated in Fig. 7. It has been supposed that the thermal path is constant, therefore the insulation layers are only 50µm, resulting in twice the initial capacitance.

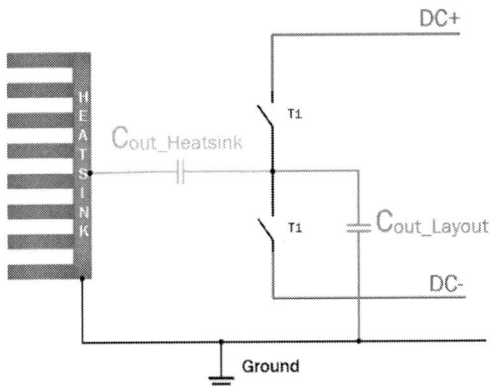 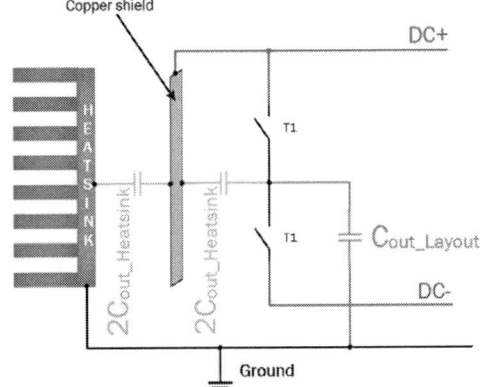

Fig. 6: Different contributions to Cout (middle point to ground stray capacitance): power layout and Top Switch

Fig. 7: Removing the impact of Cout_Heatsink using a shielding layer connected to the DC plus potential

To validate the principle of this shielding approach a simple circuit simulation has been carried out, without the effect of the power layout (Cout_Layout). A simple switching cell has been connected to a LISN (Line Impedance Stabilization Network) in order to obtain the Common Mode current. Fig. 8 shows the comparison in the initial case where the conducting layer is not connected (thus Cout_Heatsink = 1.5pF) with the case where it is connected to the DC+ potential. The decrease of CM current on the whole frequency range is impressive, between 20 and 50dBs.

To verify this result, an experimental validation has also been carried out with the converter built according to the geometry of Fig. 4. To focus on the heatsink shielding only and masking all other stray imperfection an emulation of the phenomena with high values of lumped capacitors (2.2nF) has been used (Fig. 9 right). Again, the CM reduction is huge when the shield is connected to DC+ potential, as displayed in Fig. 9 left. Some resonance around 20MHz results from the effect of the stray inductance of the lumped capacitors.

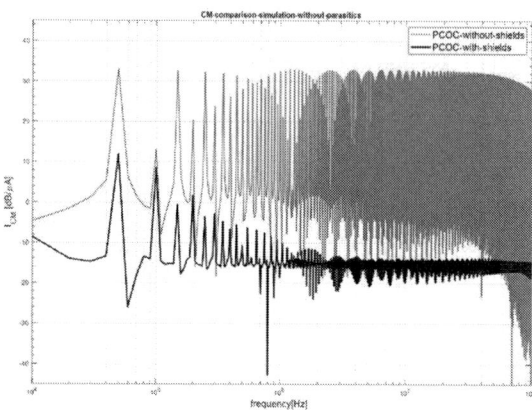

Fig. 8: Effect of heatsink shield connection to the DC+ on CM current– No other stray elements

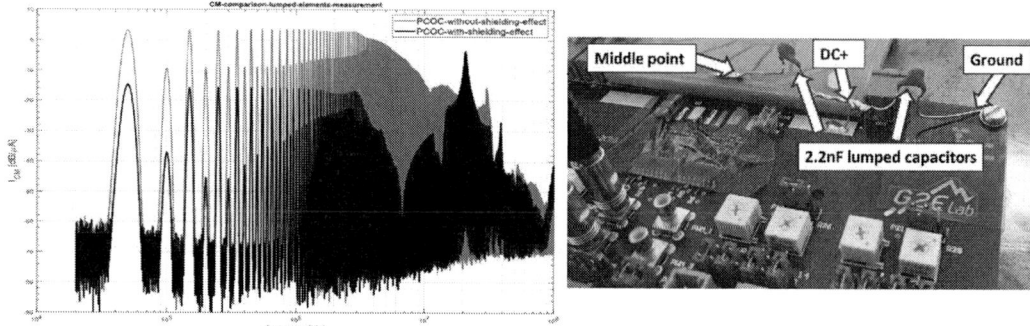

Fig. 9: Emulation of heatsink shield effect with high values lumped capacitors to avoid the impact of other stray elements

Finally, a full electrical model of the 3D power layout has been obtained using Ansys Q3D Extractor [14]. This model of the power layout has been completed with the stray capacitance of the Top switch device to the heatsink, according to Fig.7. In order to compare two extreme cases, two different power layouts have been modeled: one with the middle track embedded in the busbar, as in Fig. 5, and the shield of the heatsink connected to the DC+ potential (best case) and the other one without the heatsink shield and with a power layout of the DC bus not shielding the middle track (worst case). The comparison between these two extreme cases shows the large impact of the contributions of both Cout_Heatsink and Cout_Layout. The reduction of only 10dBs with both shielding effects will be discussed in section 4: it is due to a non-perfect overlap between DC bus and middle point track, which results in a non-zero value for Cout_Layout.

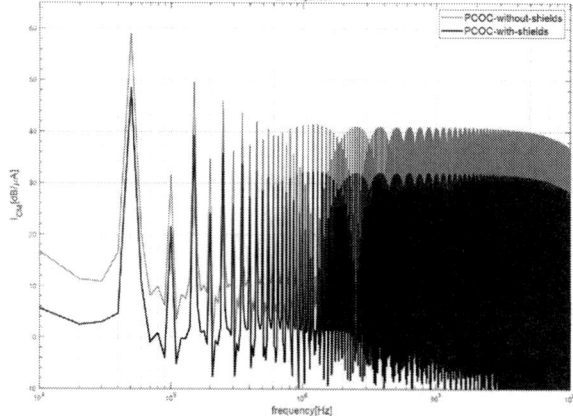

Fig. 10: Effect of a full shielding (both heatsink shield and middle track embedded in the DC bus) in comparison with no shield (no heatsink shield and middle track outside the DC bus) on CM

4. Performance evaluation of PCoC-GaN Layout

With the proposed adaptation and efforts put on the heatsink shielding, the PCoC-GaN can be realized and tested. To assess the EMC performances of the switching leg, two different versions have been built: one 3D corresponding to Fig. 5 and another one using a more conventional 2D layout illustrated in Fig. 11. Both PCBs look quite similar as shown by Fig. 12, except that in the 2D version all 6 decoupling capacitors and 2 power devices are on the same side. Both prototypes include two switching legs in order to realize two inverters in future tests.

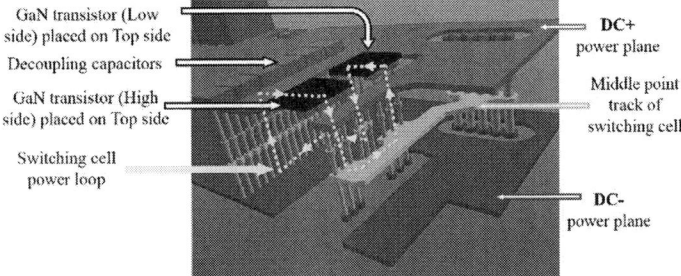

Fig. 11: Exploded 3D view of a conventional 2D layout (thickness of the board in scale 10:1)

Fig. 12: 2 prototypes realized: one PCoC-GaN and one 2D realization

The first test was to measure the performances in terms of voltage overshoot. Using the simulations from Q3D the stray inductance of the 2D and 3D layouts has been compared. They have been computed at 100MHz, which is approximatively the equivalent frequency of the transitions measured in the actual converters (Fig. 13). Results are given in Table 1. The voltage overshoot of the 2D layout is actually larger than the 3D layout, what confirms the stray inductance evaluation with simulation.

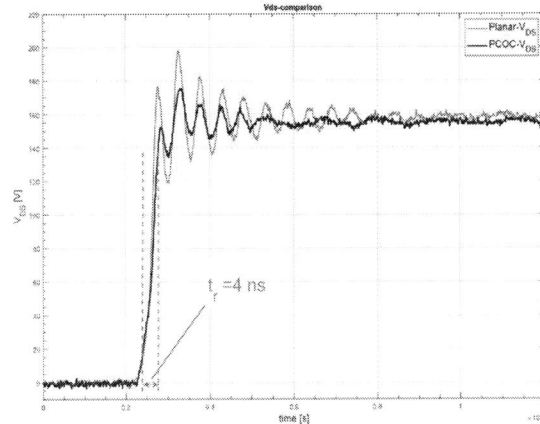

Fig. 13: Voltage overshoot comparison between 2D layout and 3D Layout

Table I: Stray inductance of the switching cell. Comparison between 2D and 3D layouts at 100MHz.

Layout	2D	3D
Stray inductance @100MHz	7.0nH	5.2nH

The second important feature of the switching cell to be verified is obviously the EMC performances. A first simulation based on the Q3D models of the two layouts (including the shielding of heatsink for both layouts, since phenomena are similar in the 2D layout) allows comparing both switching cells in terms of CM generation (Fig. 14). The emission is about 6dBs lower for the PCoC-GaN.

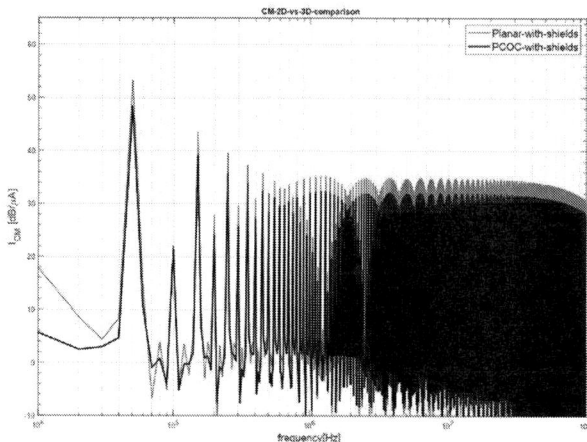

Fig. 14: Comparison in terms of CM current between 2D and 3D layouts based on simulations

After investigation, it appears that this difference can be attributed to a better shielding of the middle track in the 3D case, in comparison with the 2D layout (Fig. 15), due to the higher number of devices on the same level in the 2D layout, which occupies a larger area.

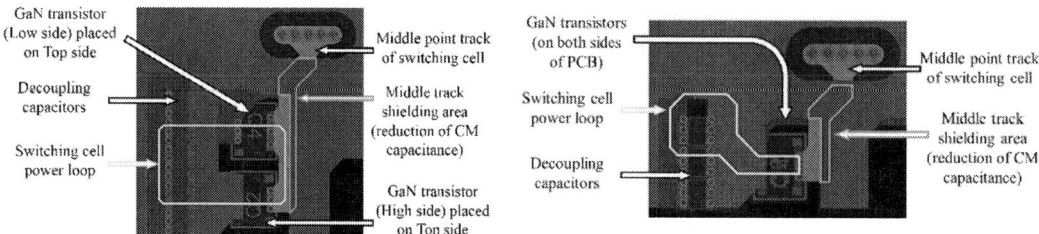

Fig. 15: Middle track embedded in the DC bus: comparison between 2D (left) and 3D (right) layouts. The 3D layout is better shielded (higher surface of middle track embedded in the DC bus). Note also the illustration of switching loop area.

Finally, the comparison of EMI in terms of DM current has also been checked (Fig. 16) and is a bit better for the PCoC-GaN. This can be due to the more symmetrical design even if further investigations have to be carried out regarding this matter.

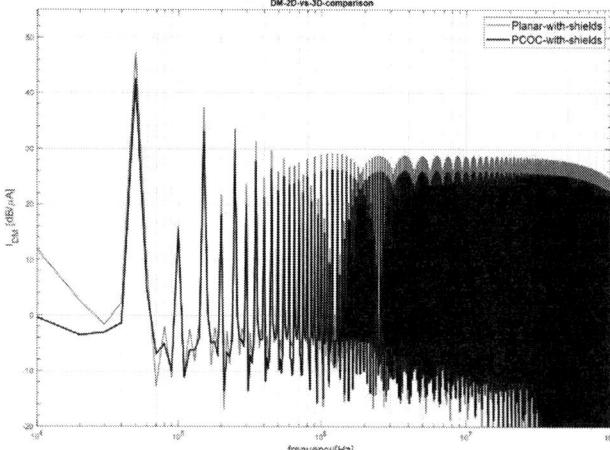

Fig. 16: Comparison in terms of DM current between 2D and 3D layouts based on simulations.

Unfortunately, the two prototypes failed during EMC measurements carried out in a hurry due to reduced amount of time available in the Lab before finalizing the paper (only one half day per week allowed during COVID time for this research action in the EMC Lab). Therefore, no experimental results are available to confirm the simulations.

Conclusion

With dramatically increasing switching speeds of wide bandgap devices, the careful design of power layout becomes a key step to build a converter operating properly. Reduced voltage overshoot and low EMI generation require fulfilling design rules as sub-nanohenry switching cell inductance, symmetrical design of the DC bus in terms of both stray inductance and capacitance, and reduced capacitance of the middle point of the leg with respect to the ground. To achieve these goals a 3D layout (Power Chip on Chip) concept previously introduced has been adapted to GaN devices, which are lateral and not vertical as SiC or Si dies. Multilayer PCB with the minimum number of layers (4 layers) was used to implement the PCoC structure with minimum cost.

A specific attention has been paid to the shielding of the middle track by embedding it inside the DC bus. Another key point has been identified regarding the Common Mode generation: the thermal pad of the GaN devices used is electrically connected to the source, thus increasing the middle point capacitance through the stray capacitance of the top switch with the heatsink. A specific shielding solution has been proposed and validated with both simulations and measurements in a specific test case. Both 3D and conventional 2D power layouts need this solution to reduce CM current generation. Finally, the 3D power layout exhibits lower loop inductance than the conventional 2D layout, thus lower voltage overshoot. Unfortunately, only power waveform has been recorded, since the two prototypes failed during EMC tests. However superior performances of the 3D layout are expected when looking at simulation results.

References

[1] E. A. Jones, F. F. Wang and D. Costinett, "Review of Commercial GaN Power Devices and GaN-Based Converter Design Challenges," in IEEE Journal of Emerging and Selected Topics in Power Electronics, vol. 4, no. 3, pp. 707-719, Sept. 2016, doi: 10.1109/JESTPE.2016.2582685.

[2] G. Regnat, P. Jeannin, D. Frey, J. Ewanchuk, S. V. Mollov and J. Ferrieux, "Optimized Power Modules for Silicon Carbide mosfet," in IEEE Transactions on Industry Applications, vol. 54, no. 2, pp. 1634-1644, March-April 2018, doi: 10.1109/TIA.2017.2784802.

[3] J. Schanen and P. Jeannin, "Integration solutions for clean and safe switching of high speed devices," CIPS 2018; 10th International Conference on Integrated Power Electronics Systems, Stuttgart, Germany, 2018.

[4] Luciano F. S. Alves ; Van-Sang Nguyen ; Pierre Lefranc ; Jean-Christophe Crebier ; Pierre-Olivier Jeannin ; Benoit Sarrazin, " A Cascaded Gate Driver Architecture to Increase the Switching Speed of Power Devices in Series Connection", IEEE Journal of Emerging and Selected Topics in Power Electronics, Year: 2020 | Early Access Article | Publisher: IEEE

[5] Q. Liu, S. Wang, A. C. Baisden, F. Wang and D. Boroyevich, "EMI Suppression in Voltage Source Converters by Utilizing dc-link Decoupling Capacitors," in IEEE Transactions on Power Electronics, vol. 22, no. 4, pp. 1417-1428, July 2007, doi: 10.1109/TPEL.2007.900593.

[6] A. Domurat-Linde and E. Hoene, "Analysis and Reduction of Radiated EMI of Power Modules," 2012 7th International Conference on Integrated Power Electronics Systems (CIPS), Nuremberg, 2012, pp. 1-6.

[7] GaN Systems: How to Drive GaN Enhancement Mode Power Switching Transistors, 2014, n° GN001 Rev. 2014-10-21

[8] M.Akhbari, JL.Schanen, P.Leturcq, O.Berraies, "Accurate modelling of commutation cell for losses calculation and EMC performance prediction in power converter", EPE'99 Lausanne, Sept.1999

[9] K. Murata and K. Harada, "Analysis of a self turn-on phenomenon on the synchronous rectifier in a DC-DC converter," The 25th International Telecommunications Energy Conference, 2003. INTELEC '03., Yokohama, Japan, 2003, pp. 199-204.

[10] D. Reusch and J. Strydom, "Understanding the Effect of PCB Layout on Circuit Performance in a High-Frequency Gallium-Nitride-Based Point of Load Converter," in IEEE Transactions on Power Electronics, vol. 29, no. 4, pp. 2008-2015, April 2014, doi: 10.1109/TPEL.2013.2266103.

[11] E. Vagnon, P. Jeannin, J. Crebier and Y. Avenas, "A Bus-Bar-Like Power Module Based on Three-Dimensional Power-Chip-on-Chip Hybrid Integration," in IEEE Transactions on Industry Applications, vol. 46, no. 5, pp. 2046-2055, Sept.-Oct. 2010, doi: 10.1109/TIA.2010.2057401.

[12] C. Fita, P. Jeannin, P. Lefranc, E. Clavel and J. Delaine, "A novel 3D structure for synchronous buck converter based on nitride Gallium transistors," 2016 IEEE Energy Conversion Congress and Exposition (ECCE), Milwaukee, WI, 2016, pp. 1-7, doi: 10.1109/ECCE.2016.7854725.

[13] GaN Systems, "GS66506T Top-side cooled 650 V E-mode GaN transistor ", GS66506T Datasheet, 2020 [Rev 200402].

[14] https://www.ansys.com/products/electronics/ansys-q3d-extractor, accessed July 21, 2020

Estimation of the winding losses of Medium Frequency Transformers with Litz wire using an equivalent permeability and conductivity method

Mohammad Kharezy
RISE RESEARCH INSTITUES OF SWEDEN
Borås, Sweden
E-Mail: mohammad.kharezy@ri.se
https://www.ri.se/en/

Morteza Eslamian
UNIVERSITY OF ZANJAN
Zanjan, Iran
E-Mail: eslamian@znu.ac.ir
http://www.znu.ac.ir/en/

Torbjörn Thiringer
CHALMERS UNIVERSITY OF TECHNOLOGY
Gothenburg, Sweden
E-Mail: torbjorn.thiringer@chalmers.se
https://www.chalmers.se/en/

Acknowledgements

This project has been funded by the Swedish Energy Agency and with in-kind contribution of Rise Research Institutes of Sweden.

Keywords

«Wind Energy», «DC-DC power converters», «Transformer windings», «Eddy currents», «Finite element analysis».

Abstract

To achieve the highest efficiency of a Dual Active Bridge converter, it is crucial to accurately calculate the winding losses of the Medium Frequency Transformer (MFT) situated inside it. In this article, an effective numerical method for calculation of the copper losses in MFTs with rectangular-shaped windings made up of Litz wire, is utilized and practically verified.

Introduction

In a DC energy collection or a DC energy distribution system, DC/DC converters are used to enable the energy collection, transmission, and distribution and one application where the DC technology has a considerable potential is in the future integration of offshore wind farms. These converters offer the benefit for reduced technical complexity, and accordingly a possibility for reduced life cycle cost. One of the popular topologies used for DC/DC converters is the Dual Active Bridge (DAB) topology (see Fig. 1). The DAB has a key component, a Medium Frequency Transformer (MFT), if this has a high power-density it results in a considerable reduction in weight and volume of the voltage conversion system [1].

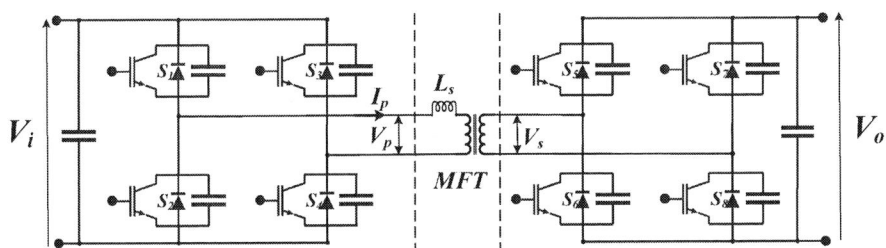

Fig. 1: The Dual Active Bridge (DAB) topology having an MFT in its center

Replacing a normal grid integration with a DC collection system can be claimed to be effective only if it can be shown that the efficiency of the converter is high enough. Since a high loss can adversely affect the converter efficiency, it would thus be of great importance to accurately calculate the winding losses of the MFT at high frequency in the design stage.

At high frequencies, the losses in the copper MFT winding conductors will drastically rise because of the skin effect in the conductors, and the proximity effect due to the adjacent conductors. The winding losses are calculated as the linear superposition of the losses caused by different harmonic contents of the applied AC current [1]. The voltage and current waveforms of DAB converter can be observed in Fig. 2.

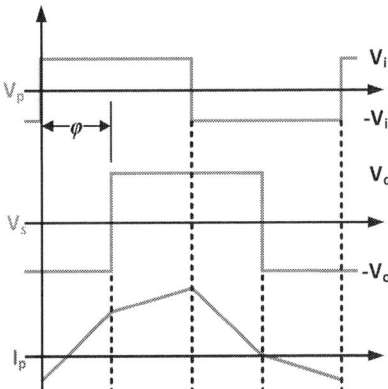

Fig. 2: Voltages and current wave shapes of an MFT placed in the center of a DAB

In order to minimize the extra losses caused by high frequency current harmonics, special conductors should be used, namely foil or Litz conductors. These conductors minimize the proximity and skin effects. Copper foils are suitable alternatives for higher frequencies, but they need a great care during the production and termination. With the introduction of Litz wires (bunches of individually insulated conductors), higher flexibility in the production and achieving lower losses are gained. The use of Litz wires in MFTs is very beneficial and that is why in this activity the determination of Litz winding losses is studied.

Having access to a set of precise formulas and methods, for the calculation of winding losses with an acceptable accuracy, is of high importance. Wide frequency range, possibility of free selection of number of layers, and number of turns per layer are among the features that should be considered during the development of new methods. Previous works [2-3] propose formulas to calculate the AC resistance of the conductors under high frequency switching conditions. These proposed analytical methods for calculation of winding losses are based on empirical relations originally developed for foil windings, which have been modified by researchers to be applicable for Litz wires. However, these methods are approximate, and their application is time consuming, and requiring many simplifying assumptions to solve the problem.

In this article, a numerical method based on FEM simulations is presented, which take advantage of the equivalent permeability and conductivity of stranded copper conductors. The method is applied to the model of a 5 kHz prototype MFT transformer built from Litz wire, and different simulations, with and without usage of the equivalent model, are performed. Winding losses are calculated at different frequencies. Subsequently, the winding losses of the prototype MFT are measured at different frequencies and compared with the results of simulations. The results of simulations have an acceptable conformity with the presented measurement results. This means that the method can be used reliably for calculation of winding losses of the MFTs avoiding time consuming FEM solutions where the current density distribution inside every individual conductor strand of the Litz wire is taken into account.

Equivalent permeability and conductivity method

One method of calculating eddy current losses including skin and proximity effect losses is to make a finite element model, in which each turn in a winding is explicitly represented. By modeling each turn, the distribution of current density within each turn due to these effects, can be accurately determined. However, modeling every wire individually can be computationally expensive. An alternative approach is to replace a wound region composed by many wires with a region with proper equivalent complex-valued material properties.

Continuum representations of wound coils allow proximity and skin effect losses to be represented in numerical models without explicitly modeling each turn in the coil. Moreau et al. [4] described the use of a complex-valued magnetic permeability for the representation of transformer windings with rectangular conductors, presenting closed-form expressions for frequency-dependent permeability. Gyselinck and Dular [5] presented a numerical method for obtaining equivalent properties of a round-wire winding with hexagonal packing. Dowell [6] replaced a winding composed of round wires with an "equivalent" foil winding that admits an analytical solution for proximity losses. Meeker [7] derived approximate closed-form expressions for the equivalent conductivity and permeability of regions filled with hexagonally packed round wires, allowing proximity and skin effects to be included in 2D AC field computations. In this work the method developed in [7] is used for calculation of eddy current losses of MFT windings made by Litz wires.

The effective material properties of the wound region, such as the one shown in Fig. 3, can be expressed as [7]

$$\mu_{eff} = (1-c)\mu_0 + c\mu_{fd} \tag{1}$$

$$\sigma_{eff} = \frac{\sigma fill}{\dfrac{\mu_0}{\mu_{fd}} + \dfrac{(1-c)}{c}j\Omega - \dfrac{1}{3}\dfrac{\mu_{eff}}{c\mu_0}j\Omega} \tag{2}$$

where the parameter c, representing the fill factor of the equivalent foil geometry, is defined as

$$c = \sqrt{\frac{2\sqrt{3}}{\pi}fill} \tag{3}$$

fill is copper fill factor and the non-dimensional frequency Ω is defined as

$$\Omega = \frac{\sqrt{3}\pi c \omega \sigma \mu_0 R^2}{8} \tag{4}$$

and the frequency-dependent permeability of a single equivalent foil is

$$\mu_{fd} = \frac{\mu_0 \tanh\sqrt{j\Omega}}{\sqrt{j\Omega}} \tag{5}$$

Fig. 3. Hexagonally packed Litz wire

Simulation Results

The equivalent material representation method is implemented for calculation of eddy current losses of a 50 kW, 1 / 3 kV shell type MFT [1] (see Fig. 6a). For this purpose, the equivalent complex permeability and conductivity of each Litz wire is obtained from (1) and (2) respectively. Later, the equivalent parameters are used for simulating Litz wires as continuum regions instead of individual strands. A Litz wire with the dimensions of 3.8 mm×2.5 mm is used. The diameter of each individual strand of totally 181 strands of each wire is 0.2 mm representing 61.51% copper fill factor of the equivalent wire region. The primary winding consists of three parallel Litz wires in the axial direction. It is coiled as (7+7+4) turns to have a total number of 18 turns. The secondary winding is coiled as 3 layers of (22+22+10) turns, having a total of 54 number of turns (see Fig. 4).

The finite element simulations were performed in FEMM, a freely available magnetics finite element solver which uses approximate closed-form expressions for the equivalent conductivity and permeability of regions filled with hexagonally packed round wires, allowing proximity and skin effects to be included with ease in 2D AC field computations.

The distribution of flux and current densities with equivalent material representation for 5 kHz is shown in Fig. 4. Solutions were performed over a range of frequencies between 0 and 15 kHz. The value of losses and increase of losses in relation to the DC losses are illustrated in Table I showing an increase of losses equal to 5.73% at 5 kHz, the fundamental frequency, and a value of 51.55% at 15 kHz, the third harmonic frequency of the current waveform in the MFT.

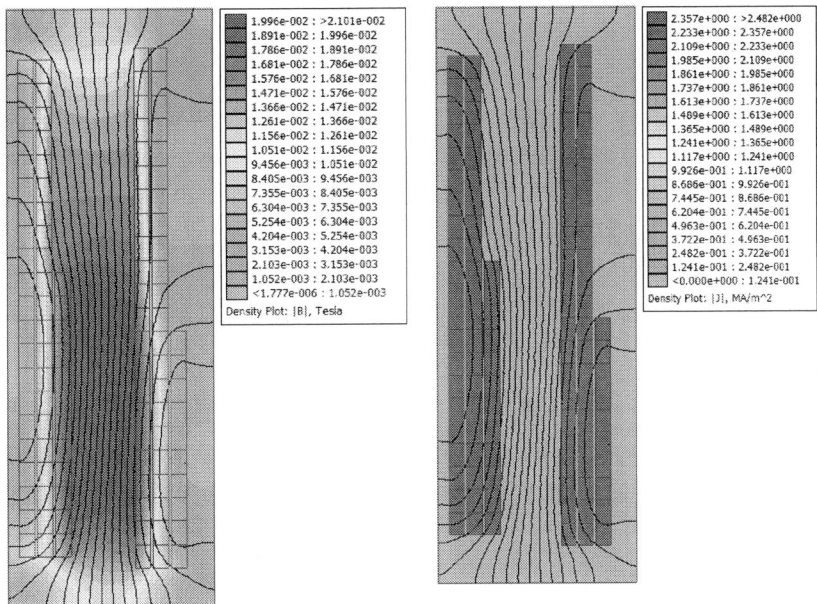

a) Flux density distribution   b) Current density distribution

Fig. 4. winding losses analysis of the MFT with equivalent material representation for 5 kHz

Table I. Results of loss analysis with equivalent material representation method

Frequency (kHz)	Losses (W)	Increase of Losses to DC losses (%)
0	54.43	0
1	54.56	0.23
5	57.55	5.73
10	66.91	22.92
12	72.39	33.00
15	82.49	51.55

The validity of the method was explored by a comparison of finite element solutions for an explicit model and the equivalent material properties model of a wound coil. The explicit model implies calculation of eddy current losses of a Litz wire accounting for the skin and proximity losses inside individual conductor strands. The model with explicit windings needs a fine mesh inside the wires in order to adequately model skin and proximity effects at high frequencies. Since an analysis of a problem with all wires modeled with explicit Litz strands is not possible due to the huge amount of the required mesh at high frequencies, only one wire is modeled explicitly at each time and the total losses occur inside its copper strands is extracted and compared with the one that is obtained from equivalent material representation method. Each wire consists of 186 round copper strands wound in alternating layers of 16 and 15. The change of the current distribution inside the individual strands of a Litz wire at different frequencies is shown in Fig. 5.

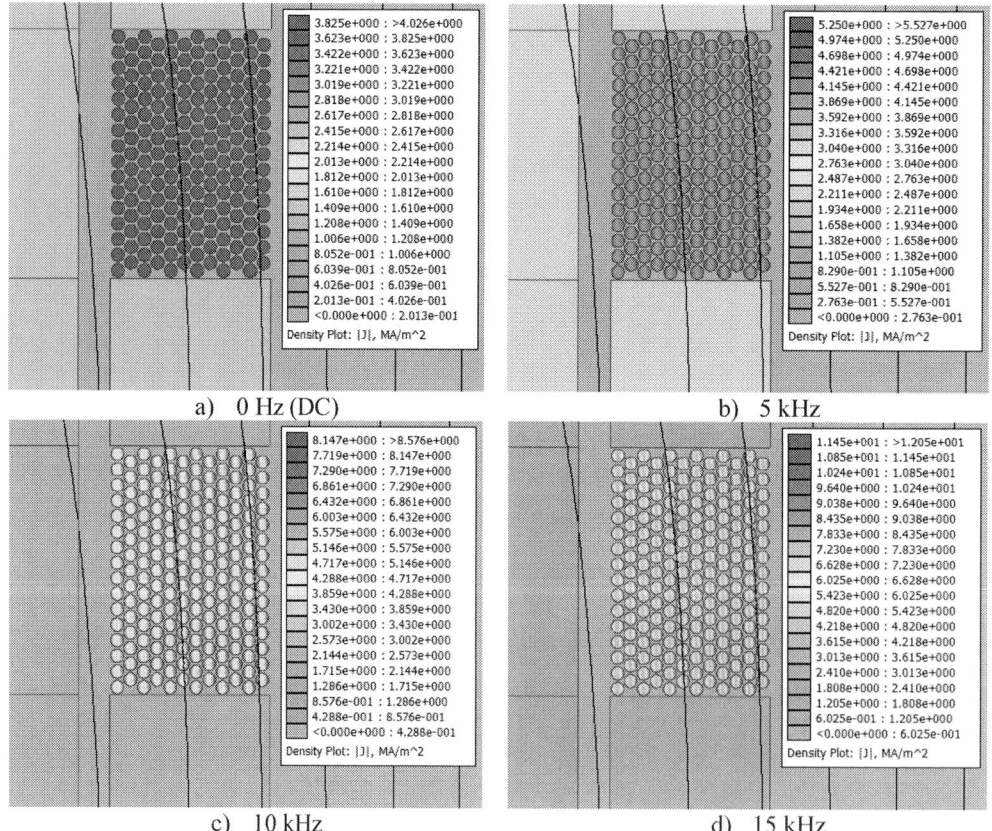

Fig. 5. Current distribution inside individual strands of a Litz wire at different frequencies

Table II. Comparison of Litz wire losses (consisting of 181 strands) from explicit and equivalent material models

Frequency (kHz)	0	5	10	15
Exact Method (W)	0.827109	0.951974	1.32653	1.95066
Equivalent Material Representation (W)	0.827067	0.951871	1.32624	1.95008

The results of the loss analysis for a Litz wire is shown in Table II demonstrating that the equivalent material model provides good agreement with the explicit model over the whole frequency range from 0 up to 15 kHz.

Experimental verification

To verify the calculation method which uses the FEM analysis with the equivalent material properties, measurements have been performed on the prototype MFT. The winding loss measurement setup is demonstrated in Fig. 6b. The voltage, current and the power parameters were registered using a precision Yokogawa WT3000 power analyzer. The supply source used was a California Instrument 4500LX, 5 kHz, PWM converter and a Chroma 61605 AC, 0-1000 Hz power supply. To measure the total losses in the primary and secondary windings, the rated current of the transformer shall be applied to the windings. At the secondary side, when the primary side is short-circuited, the voltage is raised until the rated current passes simultaneously from both windings. The losses are measured as fast as possible to capture the losses at room temperature. At 5 kHz and a supplying voltage of 186.2V (6% of nominal voltage of secondary side), the no-load losses in the transformer core is estimated as 1.5 W and subtracted from the displayed losses. A summary of the calculated and measured losses is presented in Table III which demonstrates the maximum deviation of 0.75 % at 5 kHz.

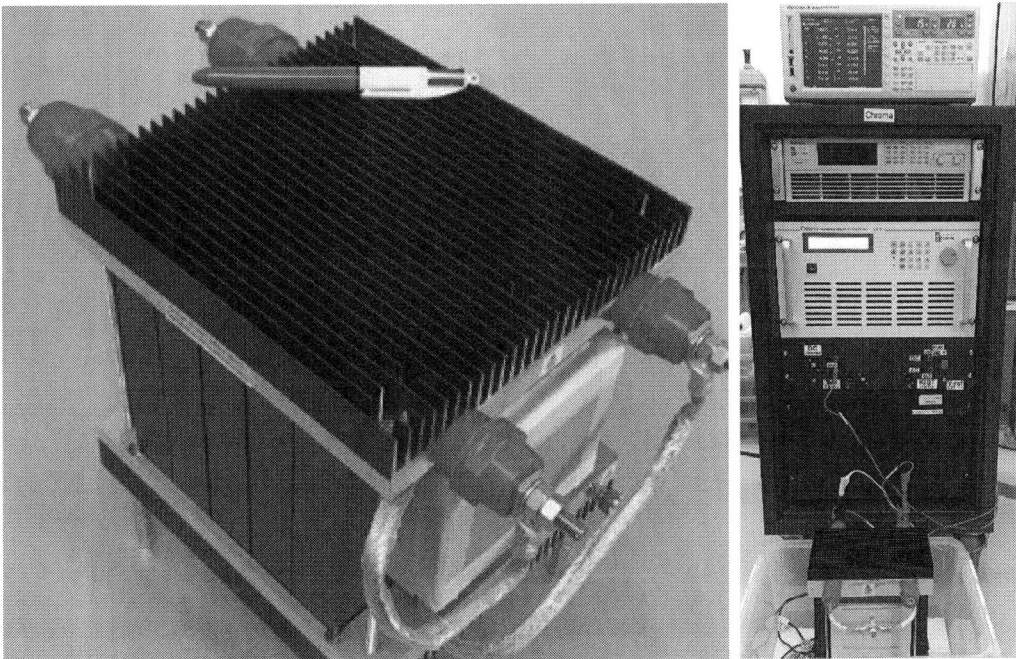

a) Ferrite core MFT   b) Measurement setup

Fig. 6: The prototype MFT and windings loss measurement setup

Table III: Comparison of the measured losses at the rated current & 20°C with the calculated losses using FEM analysis with the equivalent material properties

Frequency (kHz)	Measured (W)	Calculated (W)	Difference (%)
0	54.44	54.43	-0.02
1	54.75	54.56	-0.35
5	57.98	57.55	-0.75

Conclusion

In this article, an effective numerical method is utilized for calculation of eddy current losses in Medium Frequency Transformer (MFT) windings made by Litz wires. The numerical method is based on the equivalent permeability and conductivity of stranded copper conductors. The method is applied to model a prototype 5 kHz MFT, and different simulations are performed. The winding losses of the prototype MFT are then measured at different frequencies and comparison with the simulation results proves the applicability of the method. The proposed method can be used by the designers of MFTs as an effective and quick procedure to easily estimate the winding losses of MFTs where Litz conductors are used.

References

[1] M. A. Bahmani, T. Thiringer, and M. Kharezy, "Design Methodology and Optimization of a Medium-Frequency Transformer for High-Power DC-DC Applications," IEEE Transactions on Industry Applications, vol. 52, no. 5, pp. 4225-4233, 2016

[2] C. R. Sullivan, "Computationally efficient winding loss calculation with multiple windings, arbitrary waveforms, and two-dimensional or three-dimensional field geometry," IEEE Transactions on Power Electronics, vol. 16, no. 1, pp. 142-150, 2001

[3] F. Tourkhani and P. Viarouge, "Accurate analytical model of winding losses in round Litz wire windings," IEEE Transactions on Magnetics, vol. 37, no. 1, pp. 538-543, 2001

[4] O. Moreau, L. Popiel, and J. L. Pages, "Proximity losses computation with a 2D complex permeability modelling," IEEE Transactions on Magnetics, vol. 34, pp.3616-3619, Sept. 1998

[5] J. Gyselinck and P. Dular, "Frequency-domain homogenization of bundles of wires in 2-D magneto dynamic FE calculations," IEEE Transactions on Magnetics., vol. 41, pp. 1416-1419, May 2005

[6] P. L. Dowell, "Effects of eddy currents in transformer windings, "Proceedings of the Institution of Electrical Engineers, vol. 113, no. 8, pp. 1387-1394, 1966

[7] Meeker, D.C., "Continuum Representation of Wound Coils via an Equivalent Foil Approach," available at: www.femm.info/wiki/ProximityLoss

Improvement of Driving Efficiency of PMSM by using Modified Trapezoidal Modulating Signal

Kento Betto , Satoshi Joryo , Toshimitsu Morizane
Osaka Institute of Technology / University
Osaka Institue of Technology / 16-1, 5 chome , Omiya , Asahi-ku , Osaka
Osaka, Japan
Tel.: +81 / 6 - 6954 - 4228.
Fax: +81 / 6 - 6957 - 2133.
E-Mail: withfikento@gmail.com
URL: https://www.oit.ac.jp/index2.html

Keywords

«Electric vehicle», «Permanent magnet motor», «Modulation strategy», «Switching losses», «Harmonics».

Abstract

This paper focuses on improving the efficiency of PMSM for electric vehicles. As a solution, this paper proposes a modulation method to be able to achieve three advantages. This control method provides three advantages, and two of which have been verified by conventional research: torque ripple reduction and increased torque. Therefore, this time, in addition to these two advantages, it was verified whether the third advantage, reducing inverter switching losses, is possible.

Introduction

In recent years, global warming is an international issue. One of the reasons for global warming is exhaust emitted from gas-powered automobiles. Alternative vehicles such as PHEV, HV, and EV are actively developed all over the world as a solution. Among them, EVs have many problems to be solved such as lack of charging infrastructure, short traveling distance, and high prices. These challenges make it difficult for this technology to spread. Additionally, EVs take from tens of minutes to several hours to charge their batteries, and the problem of short traveling distance turn out to be serious since fuel cannot be refilled immediately as with gas-powered automobiles. Generally, a PMSM capable of high-efficiency driving is used for a motor for EVs. And it is possible that the short traveling distance can be solved by improving the control efficiency of the PMSM and developing a high-capacity battery.

The purpose of this research is to improve the efficiency of PMSM control by using Modified Trapezoidal Modulating Signal as a control method to solve the problem of short traveling distance.

Advantage of Modified Trapezoidal Modulating Signal

In the control system of EVs, V/F control and vector control are generally used, and PWM control is performed by an inverter. In this research, using vector control makes it possible to drive with high efficiency. In this control system, Sinusoidal Modulation (SM) is generally used, but in this research, Modified Trapezoidal Modulation (MTM) is applied. The waveform of the MTM is shown in Fig.1.

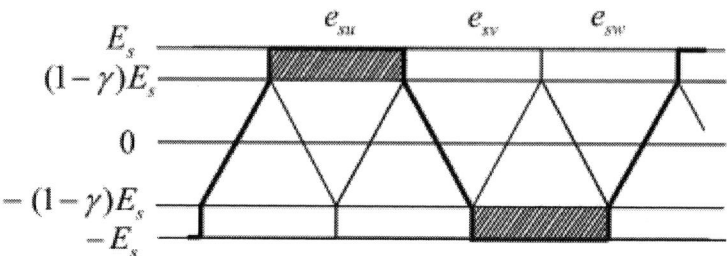

Fig.1: Modified Trapezoidal Waveform Diagram

A modified Trapezoidal Wave is a waveform with two 120-degree-wide flat parts and formed by superimposing a square wave on a trapezoidal wave, and the superposition ratio is defined as γ value. Additionally, e_{su}, e_{sv}, and e_{sw}, are three-phase voltage command values inputted to the inverter. Using the MTM Signal brings about three benefits.

1. Increased torque
    Fig.2 shows a measured value of the amplitude if the γ value of the MTM is altered and the amplitude of the SM.[2]

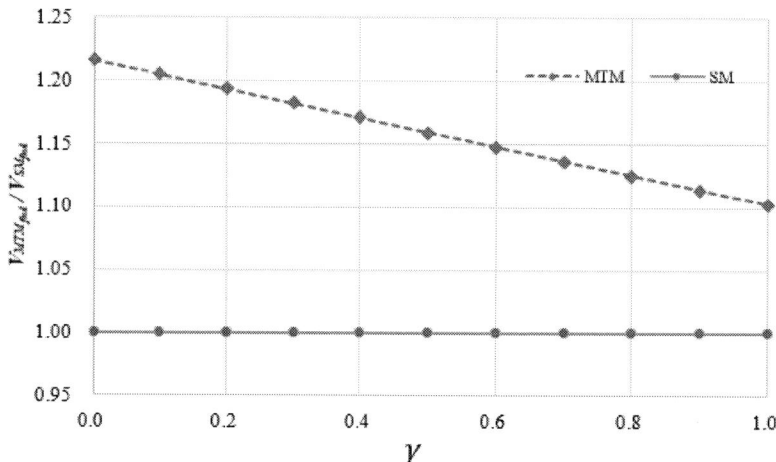

Fig.2: The height of the fundamental amplitude when the γ value is altered

As can be seen in Fig.2, the amplitude of MTM is higher than the amplitude of the SM at any γ value. Therefore, it is possible to get higher torque with MTM.

2. Reduced Torque ripple
    In cases when a block-like wave such as MTM is utilized, the influence of harmonics can't be ignored. As one of that problem, torque ripple occurs in PMSM is considered. In particular, the main component of torque ripple is the 6th harmonic torque. At this time, the 6th harmonic torque generated on the PMSM, which has a non-sinusoidal magnetic flux density, can be calculated by the equation (1). [1]

$$\tau_6 = \frac{3}{2}\frac{1}{\omega_m}\{E_1(I_7 - I_5) + \cdots\} \tag{1}$$

Where, ω_m is the mechanical angular velocity of the rotor [rad/s], E_1 is fundamental voltage [V], I_5 is the 5th harmonic current [A], I_7 is the 7th harmonic current [A]. Further, it is known that the amplitude of the harmonic current when the γ value of the MTM is altered is shown in Fig.3.[2]

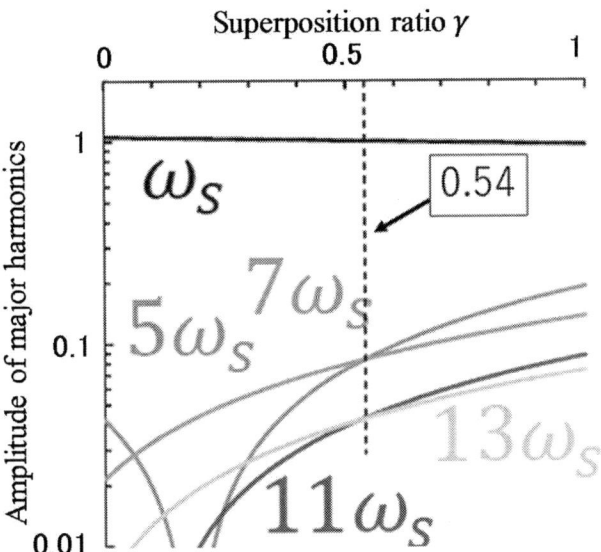

Fig.3: Amplitude of harmonic current compared to fundamental current at each γ value

ω_s is amplitude of the fundamental current. As seen from Equation (1), when the 5th harmonic current is the same as the 7th harmonic current, the 6th harmonic torque is able to be set to zero. Moreover, referring to fig.3, the 5th harmonic current and the 7th harmonic current are theoretically equal when the γ value is 0.54. Therefore, torque ripple can be reduced by choosing an appropriate γ value.

3. Reduced inverter switching losses

Fig.4 shows a comparison of a Modified Trapezoidal Wave and carrier wave under inverter control. As explained in Fig.1, Modified Trapezoidal Wave has two 120-degree-wide flat parts. Therefore, as shown in Fig.4, in a state where the amplitude of carrier wave and Modified Trapezoidal Wave are equal, switching is not performed. Thus, the switching loss can be reduced.

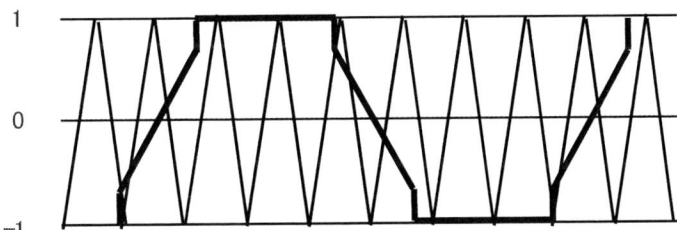

Fig.4: MTM Signal compared to carrier wave

Experimental configuration

In conventional research, it is known that the γ value at which the 6th harmonic torque is smaller than that of SM exists on the PMSM, which has a non-sinusoidal magnetic flux density. Fig.5 shows the height of the 6th harmonic torque when comparing the MTM and SM while changing the γ value. [2] Furthermore, Fig.6 shows the back EMF waveform of PMSM, which has a non-sinusoidal magnetic flux density, and Fig.7 shows the result of FFT for Fig.7. [1][2]

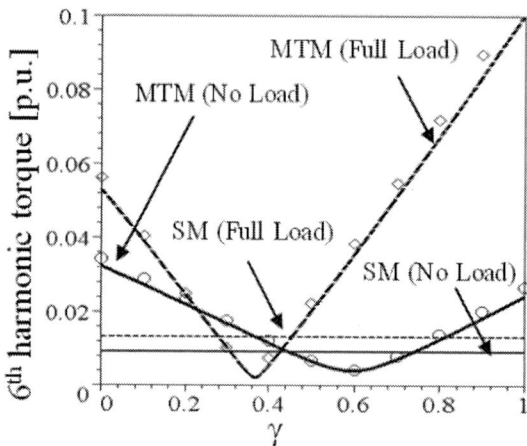

Fig.5: 6th harmonic torque vs. γ

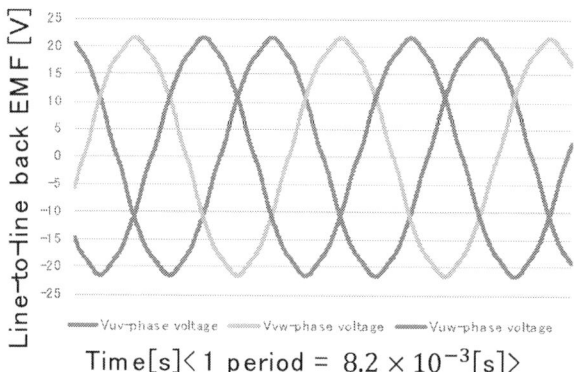

Time[s]⟨1 period = 8.2×10^{-3}[s]⟩

Fig.6: Back EMF waveform of PMSM obtained by experiment

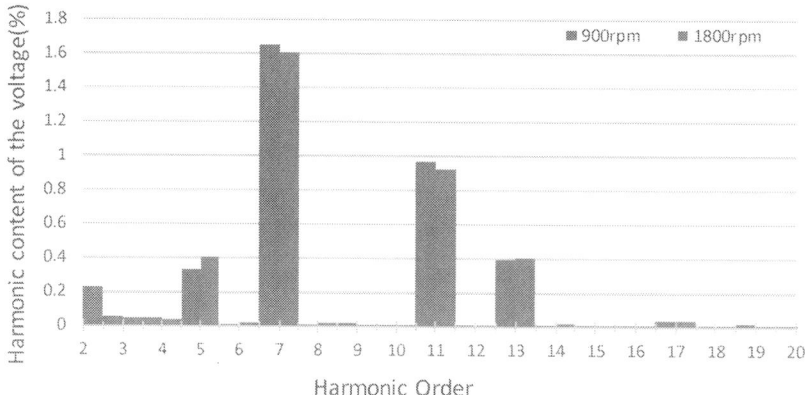

Fig.7: FFT of the back EMF

As can be seen from Fig.6 and Fig.7, 5th and 7th harmonics are high in PMSM, which has a non-sinusoidal magnetic flux density. Additionally, Fig.8 shows a comparison of the 6th harmonic torque in the MTM, SM, Sinusoidal Over Modulation (Over Mod), and one-pulse control (One Pulse). [2] A plurality of the γ values are measured. From Fig.8, it is known that when γ = 0.54, the 6th harmonic torque generated in the MTM is lower than that in the SM. Moreover, it can be seen that the

6th harmonic torque of MTM is lower than the sinusoidal over modulation. From these results, it is possible to confirm an increased torque and reduced torque ripple, which are advantages of the MTM. Therefore, this study focuses on reducing the switching losses of inverter.

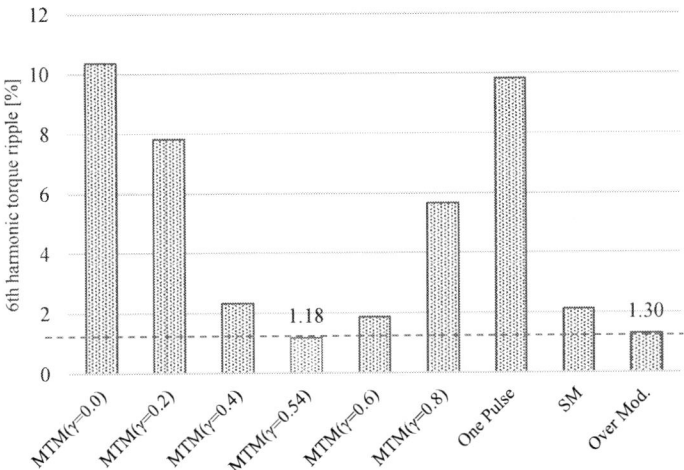

Fig.8: Experimental comparison of 6th harmonic torque ratio by γ

As previously mentioned among the benefits of the MTM, the switching losses of the inverter can be reduced by MTM. And the efficiency improvement of PMSM control can be performed. However, when PWM control, which is a conventional method, is performed by an inverter, switching is performed even in the flat part of a Modified Trapezoidal Wave since the modulation rate (M) becomes 1 or less. Fig.9 shows the PWM control circuit diagram used so far. As explained in Fig.4, in a state where the amplitude of carrier wave and Modified Trapezoidal Wave are equal, switching is not performed. Therefore, the MTM by PAM (Pulse Amplitude Modulation) inverter is proposed as an efficiency improvement method of PMSM. Fig.10 shows a control block diagram of PMSM, which use PAM inverter and buck-boost converter. And Fig.11 shows the output voltage characteristics of PWM control ($V_{dc} = V_{bat}$). Furthermore, Fig.12 shows the output voltage characteristics of PAM control (M=1). The system of fig.10 is suitable for EV because it can perform PMSM drive and regeneration. Furthermore, by performing PAM control, it is possible to create the command value as the modulation rate does not change.

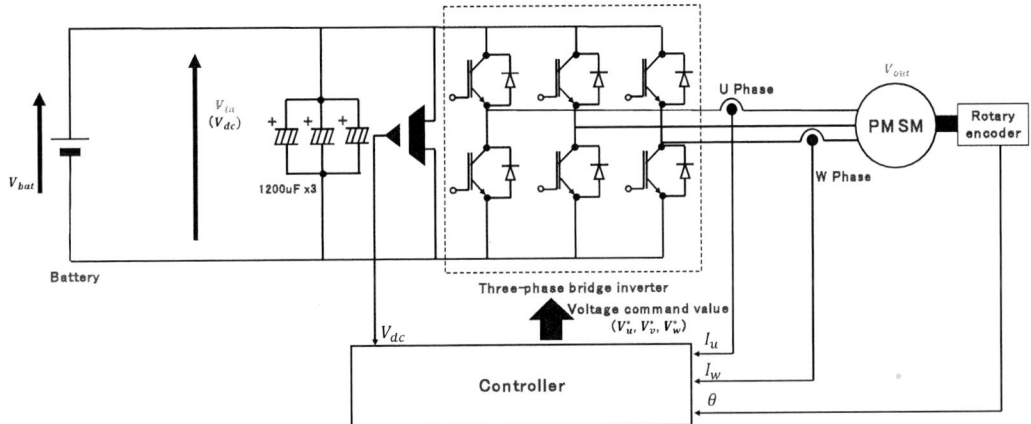

Fig.9: PWM control circuit diagram

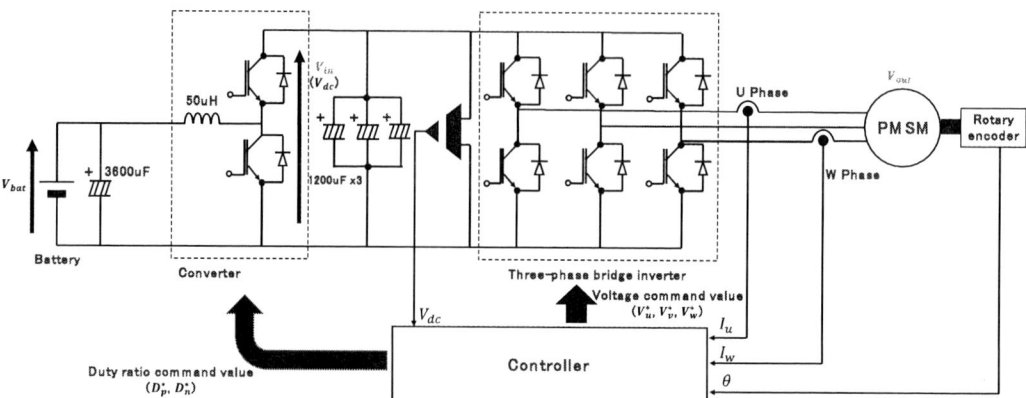

Fig.10: control block diagram of PMSM using buck-boost converter

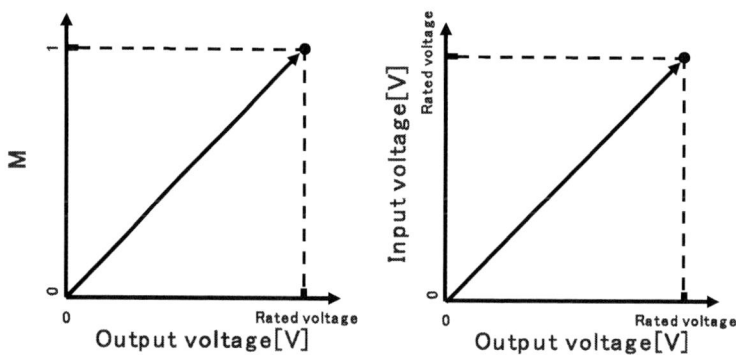

Fig.11: Output voltage characteristics of PWM control (left)
Fig.12: Output voltage characteristics of PAM control (right)

However, this paper aims to compare the efficiency when using PAM control and when using PWM control. And it is a little difficult to control with a circuit that uses a buck-boost converter. Then, in this paper, a control circuit using a buck converter is used. Fig 13 shows a control diagram of PMSM, which use buck converter.

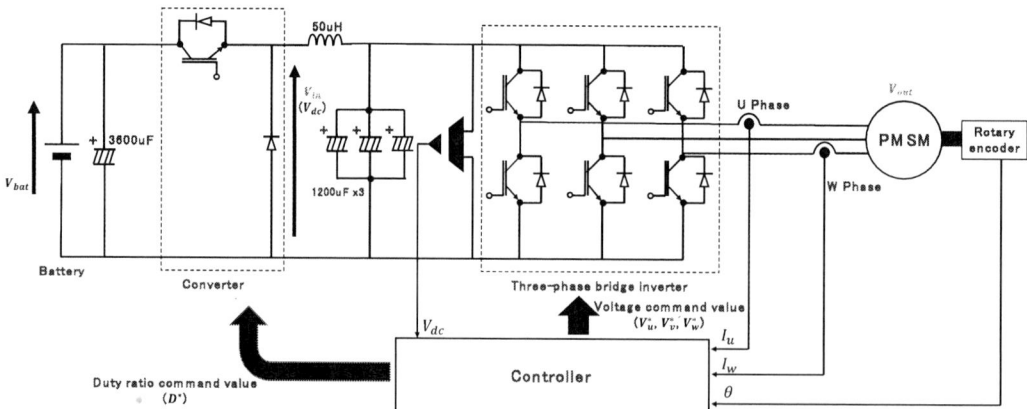

Fig.13: control block diagram of PMSM using buck converter

In the control system of Fig.13, vector control capable of high efficiency is utilized. Then, the three-phase currents I_u, I_w and rotor mechanical angular θ of PMSM are detected by sensors. The current I_v

is calculated using the currents I_u and I_w. In the process of vector control, dq axis voltage command values v_d and v_q are obtained. Then the voltage command value V_{dc}^* of the DC link voltage can be calculated from these v_d and v_q. By performing PAM control with a converter to obtain this V_{dc}^*, the modulation rate is able to be maintained at one.

In this study, Simulation (PSIM) was executed to verify if the PMSM can be driven while maintaining a γ value and that the modulation rate is 1, even if the velocity of PMSM is accelerated with the control system in Fig.13. Fig.14 shows theoretical waveform of Modified trapezoidal Wave and input voltage (V_{in}). The waveforms at low speed and high speed are shown because the PMSM is accelerated in this simulation. Table.1 shows the value of γ, V_{in}, and electrical frequency(freq) at each speed. γ does not change. The V_{in} becomes equal to the maximum voltage of the command value voltage of the Modified Trapezoidal Wave because the ratio of the input voltage and the output voltage becomes equal by PAM control. The electrical frequency increases in direct proportion to the rotation speed of PMSM.

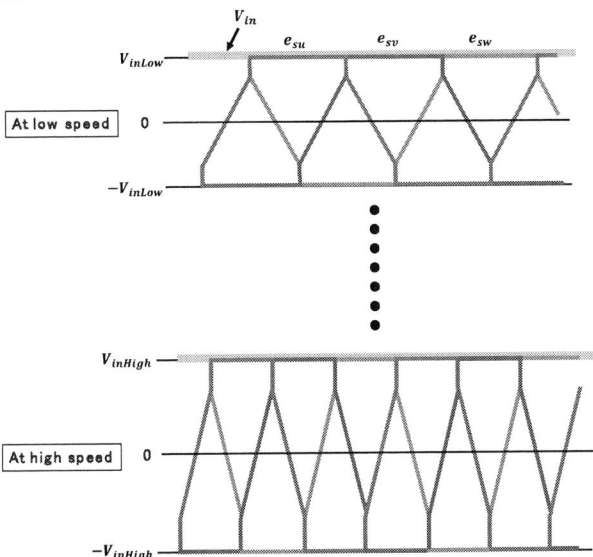

Fig.14: Theoretical waveform of Modified Trapezoidal Wave and input voltage

Table.1: the value of γ, V_{in}, and freq at each speed.

	At low speed		At high speed
γ	γ_{Low}	=	γ_{High}
V_{in}	V_{inLow}	<	V_{inHigh}
freq	$freq_{Low}$	<	$freq_{High}$

Simulating result

Fig.15 shows the simulation result of speed feedback control using the circuit in Fig13. And speed command value is the speed profile that the velocity of PMSM is accelerated for a certain time and then reach a certain speed. Fig.16 shows the U, V, and W-phase voltage command value V_u, V_v, V_w and input voltage V_{in}. The Power-supply voltage (V_{bat}) is set to 20[V], and the γ is set to 0.54 that can make the 6th harmonic torque zero.

from Fig.15, it was verified that the speed control was performed without any problem. From Fig.16, since the amplitude of the input voltage is equal to the amplitude of the voltage command value of the inverter, it is verified that PAM control is performed.

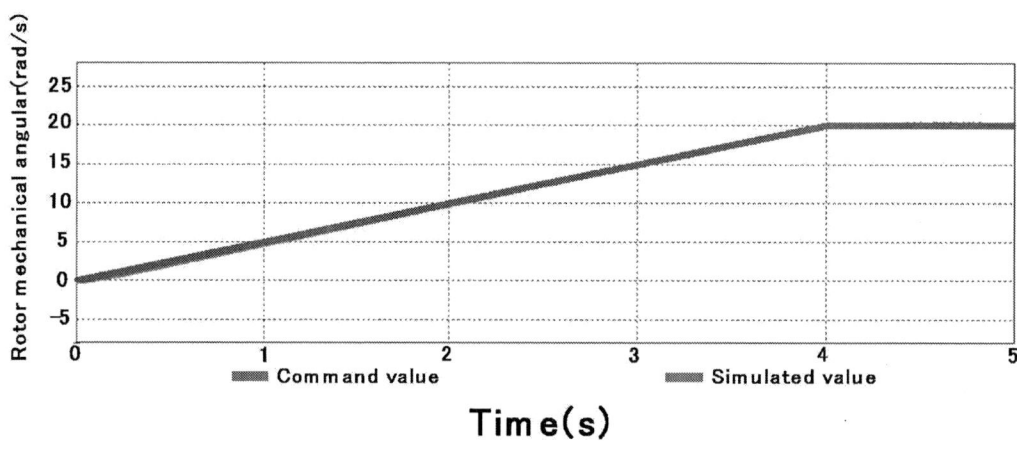

Fig15: Speed command value and simulated value with PAM control

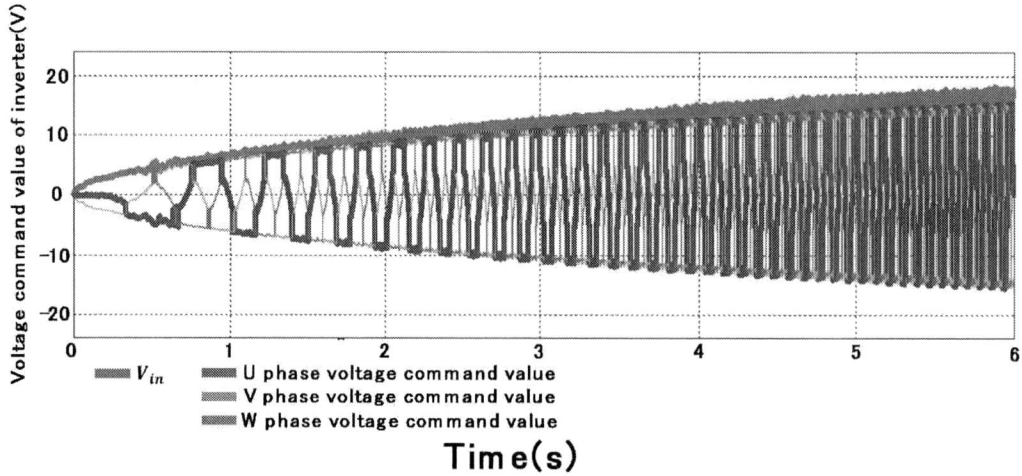

Fig.16: Command voltage and input voltage (V_{in}) with PAM control

However, from Fig.16, it is confirmed that the command values are oscillating. Then, MTM was performed by PWM control using the circuit in Fig.9. Fig.17 shows the simulation result of speed feedback control using the circuit in Fig13. Fig.18 shows the U, V, and W-phase voltage command value V_u, V_v, V_w and input voltage V_{in}. Moreover, Fig.19 and Fig.20 shows the enlarged view of MTM waveform at low speed and high speed. From Fig.19 and Fig.20, It can be seen that oscillation also occurs in the waveform. And oscillation is more noticeable at low speed. Therefore, it is necessary to redesign the circuit of the vector control system using MTM.

Improvement of Driving Efficiency of PMSM by using Modified Trapezoidal
Modulating Signal

BETTO Kento

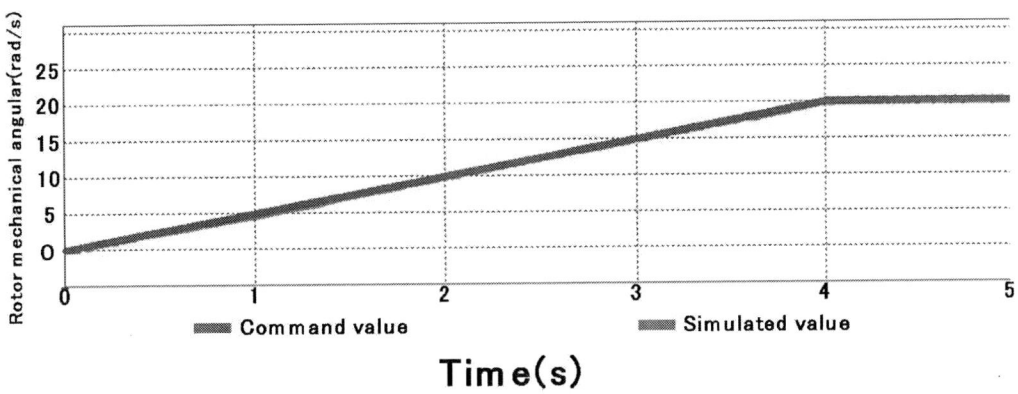

Fig.17: Speed command value and simulated value with PWM control

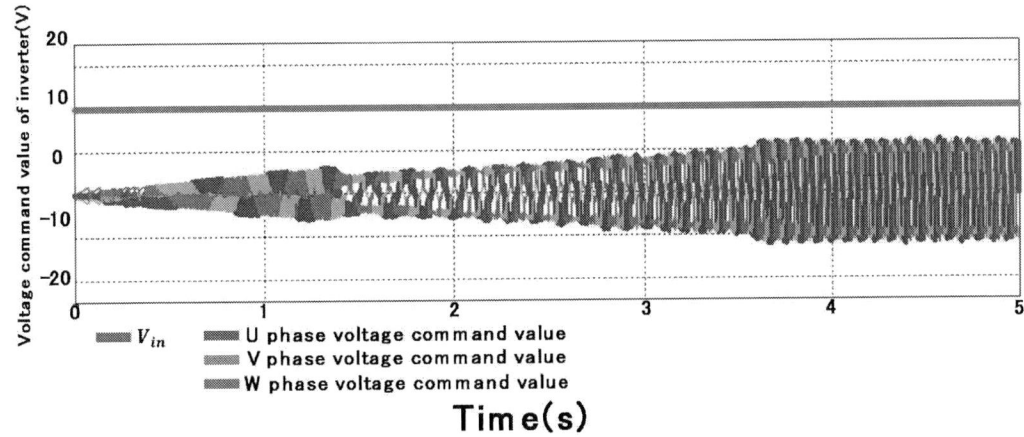

Fig.18: Command voltage and input Voltage (V_{in}) with PWM control

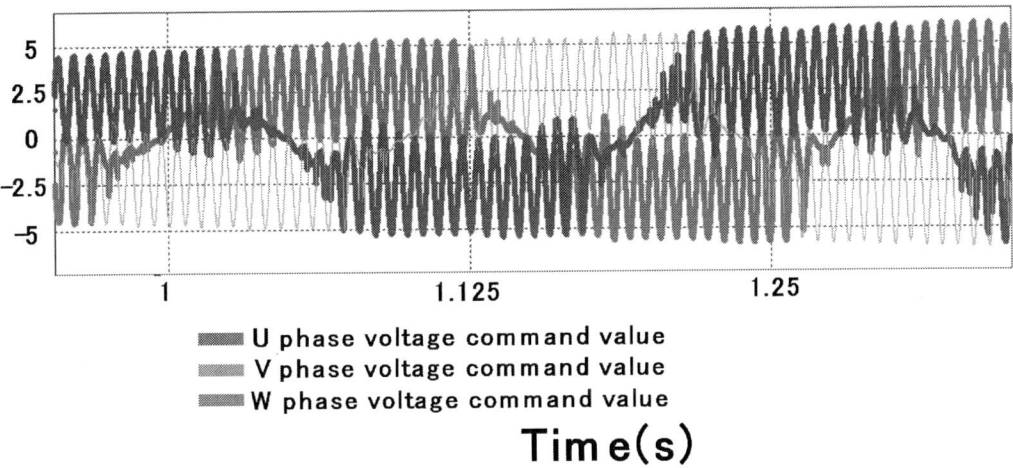

Fig.19: Enlarged view of MTM waveform with PWM control at low speed

EPE'20 ECCE Europe
Assigned jointly to the European Power Electronics and Drives Association & the Institute of Electrical and Electronics Engineers (IEEE)

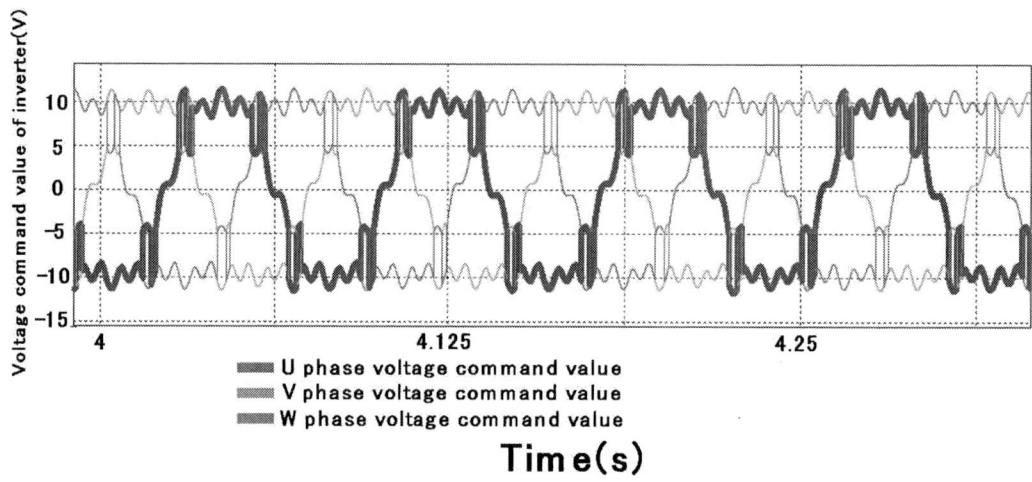

Fig.20: Enlarged view of MTM waveform with PWM control at high speed

Conclusion

In this paper, the switching losses reduction of the inverter is studied. In order to avoid switching at the flat part of the MTM, PAM control was proposed and controllability was verified. As a result, the MTM which has a desired γ value could be applied as the command value while controlling PMSM according to the speed command value. Therefore, it was verified that the switching losses of an inverter are able to be reduced in addition to the torque ripple reduction and torque increase.

However, it is necessary to improve the vector control system of MTM because V_{in} and command voltage value are oscillated. In future research, the efficiency of the circuit in Fig.13 will be measured with an experimental unit.

References

[1]. H. Yonezawa, K. Taniguchi, T. Morizane and N. Kimura, "Modified trapezoidal Modulating Signal suitable for PM Synchronous Motor Drives," *IEEJ Trans. IA*, vol. 125, no. 1, pp. 46-53, 2005. (in Japanese)

[2]. S Joryo, K Tatsumi, T Morizane, K Taniguchi, H Omori, N Kimura "Study on New Modulation Technique to Improve the Performance of PMSM" in Proc. EPE 2019 -ECCE Europe, DS2h, #126, 2019

[3]. K. Taniguchi and T. Morizane, "Characteristics of PAM Inverter System for Electric Vehicle," in *IEE-Japan Industry Applications Society Conference*, 2016. (in Japanese)

[4]. K. Taniguchi, "PWM Power Converter System", Kyoritsu Shuppan, 2007, p. 112. (in Japanese)

[5]. S Joryo, K Tatsumi, T Morizane, K Taniguchi, N Kimura, H Omori "Study of Torque ripple reduction and Torque boost by Modified Trapezoidal Modulation" in Proc. IPEC 2018 -ECCE Asia-, 22P14-4, 2018.

Design and Control of a Virtual DC-Link for a full GaN-based Single Phase Converter with High Power Density

Yugandhara H. Wankhede, Leon Fauth, Jens Friebe
Institute for Drive Systems and Power Electronics (IAL)
Leibniz Universität Hannover
Hannover, Germany
Email: Yugandhara.Wankhede@ial.uni-hannover.de
URL: https://www.ial.uni-hannover.de/de/

Acknowledgments

This work was supported by the Deutsche Forschungsgemeinschaft (German Research Foundation, DFG) through Germany's Excellence Strategy-EXC 2163/1-Sustainable and Energy Efficient Aviation under Grant 390881007 and also by the Ministry of Science and Culture of Lower Saxony and the Volkswagen Foundation.

Keywords

≪PFC≫, ≪Ripple port≫, ≪Single phase ac to dc converter≫, ≪GaN≫, ≪energy storage≫.

Abstract

This paper presents a design and control approach of a virtual DC-Link for a full GaN-based single phase converter with high power density. Low frequency power ripple on the DC bus is an inherent problem for single phase PWM converter. The topology consists of a totem pole PFC with an added virtual DC-Link, which forces this low frequency power ripple on the DC output side to divert to an auxiliary capacitor. It results in significantly reduced required capacitance and e.g. enables the use of film capacitors instead of electrolytic capacitors. The proposed control for the topology is reduced to a minimum amount of sensor requirements while still maintaining the full functionality without bulky DC-link capacitors. Simulation and and experimental results up to 3 kW are provided to validate the performance of the proposed method.

Introduction

To fulfil the demand of different energy utilization in power industries, different forms of power converters are needed. AC to DC converters are needed to exchange power with the AC distribution system. In single phase AC to DC converters, instantaneous power which is twice the AC frequency is flowing between the AC side and the DC side. If the ripple power is not filtered properly on the DC side, it will affect the performance or even the function itself [1][2]. To filter this low frequency ripple the simplest and most common solution is the use of DC link capacitors. A large capacitance is needed on the output side as most of the energy is stored in this capacitor which limits the choice to electrolytic capacitors and increases the volume of the converter. Also, these capacitors can lead to lifetime or temperature limitations. Hence it is important to reduce this DC link capacitor size to achieve better power density of converters with stable output DC side voltage with neglectable low frequency fluctuations [3].

Various active power decoupling methods have been proposed to meet the requirement of high power density and to filter the double line frequency [4, 5, 6, 7]. Active power decoupling ripple port circuits consist of power switches and energy storage elements. The basic approach is to bypass the ripple energy from the DC link, so that both energy storage components and DC link capacitor can be small in size and weight. Control design plays a vital role to improve reliability and efficiency of the ripple port circuit [8].

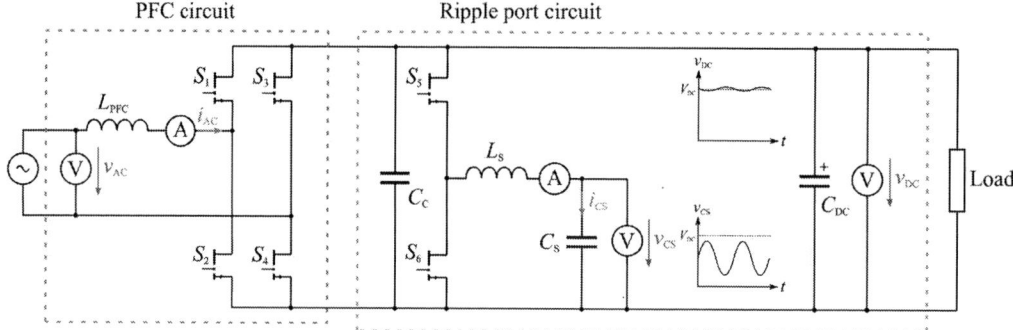

Fig. 1: A single phase PFC rectifier with the ripple port circuit

The proposed ripple energy storage topology contains a PFC circuit and a ripple port circuit (bi-directional buck converter with energy storage component) in parallel with the load as shown in Fig. 1. This is one of the most preferable topology because of its high power density [9]. There is no voltage higher than the DC bus ($V_{cs} < V_{DC}$) in this system as buck type ripple port circuit is adopted for power decoupling [10]. L_s and C_s are energy transfer and energy storage components respectively. Typically in this application, capacitors as energy storing elements are to be preffered against inductors because of smaller size and cost. The DC link capacitor C_{DC} is still needed in parallel at the output to realize a small stabilization of the DC-link voltage due to the high frequency switching of the PFC and the virtual DC-link and also to suppress any voltage slew which could influence the control of the virtual DC-Link [9]. This paper presents a control algorithm for both the PFC circuit and ripple port circuit. Initially power ripple for both the circuit is analyzed through simulation and then ripple port circuit is verified by experimental result.

Capacitor selection

In this type of application, the main challenge is the selection of the capacitors as there are many design criteria like efficiency, size, cost, functionality and lifetime. The auxiliary capacitors act as an energy buffer for the low frequency energy fluctuations. The ripple current, voltage, operating range and the equivalent series resistance (ESR) are acting as rated parameters in the selection of this capacitor [11]. Size reduction of the capacitor is desirable, but it has to be balanced against reliability issues because the capacitor size reduction decreases the surface area for heat removal and increases its temperature during operation. Traditionally, electrolytic capacitors are used in these applications because of their low cost and high energy storage density. The major drawbacks of electrolytic capacitors can be lifetime, large ESR, high leakage current, low insulation resistance and low ripple current rating. So, considering lifelong system requirements and need of improved system performance leads to other capacitor technologies, such as film and ceramic capacitors. Ceramic capacitors have a smaller size, a wider frequency range and higher operating temperature range. The major drawback is that they are relatively costly compared to their energy storage capability. Film capacitors are having a lower energy density than electrolytic capacitors, so there is a trade-off between lifespan and density [12]. At higher voltage, ripple current capability of film capacitors is higher as compared to ceramic capacitors. Comparing the types of capacitors, Film capacitors comes as a better choice for high ripple current applications when careful analysis of performance and reliability is considered. Extremely important is the proper selection of film capacitor to achieve best voltage and current carrying capability [13]. In the design, film capacitors are selected for the capacitors C_s. In practical circuits, the ripple current that the capacitor must handle without overheating by dissipation in the ESR is often the overriding factor. The current can be so high that for a given voltage, a minimum physical size of capacitor is required to achieve low ESR, high dissipation and long lifetime.

Table I: Specifications comparison

Part	LLS2G181MELB	C4AQLEW6210A3BK
Manufacturer	Nichicon	KEMET
Type	Electrolytic capacitors	Film capacitors
Value	180 μF	210 μF
Voltage rating	400VDC	500VDC
Size (mm)	30×25 (D×L)	57.5×45×65 (L×W×H)
Volume (cm^3)	17.67	168.18
Ripple Current	1.5A	34A

In the design, Film or Electrolytic capacitors can be used as a DC link capacitor in ripple port circuit. Table I gives a specification comparison of both the types of capacitors, by searching capacitors in market which can fulfil the requirements of the before mentioned application. As per specifications comparison from Table I, it clearly shows that the volume of the film capacitor is 90% higher than the volume of the electrolytic capacitor. So from the size point of view the electrolytic capacitors are a good choice for the DC link capacitor C_{DC}. Considering ripple current handling capacity, film capacitors are suitable as energy storage capacitor C_s.

Power flow and ripple port circuit analysis

The boost inductor plays a critical role in the function of the boost PFC and to provide a good power factor. Based on the current ripple in the boost inductor its value is selected with respect to the conduction and mainly the switching losses of the power semiconductor half-bridge. The input boost inductor value is calculated as per equation (1) [14]. V_{DC} is generally the output DC voltage, in this case the DC link voltage, f_s is the switching frequency and ΔI_{LPFC} is the maximum inductor ripple current allowed during operation with respect to the inductor losses.

$$L = \frac{V_{DC}}{4 \cdot f_s \cdot \Delta I_{LPFC}} \tag{1}$$

The output capacitor for the PFC, respectively the DC link capacitor, is selected to meet low frequency ripple requirements as per equation (2) [14]. P_o is the output power, f_{line} is input line frequency and ΔV_{DC} is output voltage ripple.

$$C = \frac{P_o}{2 \cdot \pi \cdot \Delta V_{DC} \cdot f_{line} \cdot V_{DC}} \tag{2}$$

The input voltage and current are assumed be sinusoidal with unity power factor in Fig. 1 and are expressed as

$$V_{AC} = \sqrt{2} \, V_{AC} \sin(\omega t) \tag{3}$$

$$I_{AC} = \sqrt{2} \, I_{AC} \sin(\omega t) \tag{4}$$

Where V_{AC} and I_{AC} are root mean square values of voltage and current, and ω is the angular frequency of source. The instantaneous input power calculated as in equation (5) and it consists two parts inductor power and rectifier power.

$$P_{in} = V_{AC} I_{AC} (1 - \cos(2\omega t)) \tag{5}$$

The instantaneous power of the inductor can be expressed with

$$P_{LPFC} = I_{AC} \cdot L_{PFC} \frac{dI_{AC}}{dt} = \omega L_{PFC} I_{AC}^2 \sin(2\omega t) \tag{6}$$

The single phase PFC rectifier output power can be calculated as

$$P_{rec} = P_{in} - P_{LPFC} \tag{7}$$

$$= V_{AC} I_{AC}(1 - \cos(2\omega t)) - \omega L_{PFC} I_{AC}^2 \sin(2\omega t) \tag{8}$$

$$= V_{AC} I_{AC} - I_{AC} \sqrt{V_{AC}^2 + I_{AC}^2 L_{PFC}^2 \omega^2} \sin(2\omega t + \theta) \tag{9}$$

Where $\theta = \arctan(V_{AC}/\omega L_{PFC} I_{AC})$. The rectifier output power consists of constant and ripple components. Ripple power is time varying with twice the grid frequency and its magnitude depends on input voltage and current. This double line frequency needs to be filtered at the output side by the DC-link capacitor.

$$P_o = V_{AC} I_{AC} \tag{10}$$

$$P_{rip} = -I_{AC} \sqrt{V_{AC}^2 + I_{AC}^2 L_{PFC}^2 \omega^2} \sin(2\omega t + \theta) \tag{11}$$

$$P_r = -I_{AC} \sqrt{V_{AC}^2 + I_{AC}^2 L_{PFC}^2 \omega^2} \tag{12}$$

A bidirectional converter is connected in parallel to the DC link of PFC circuit to reduce the DC link capacitance as shown in Fig. 1. This auxiliary ripple port circuit consists of a capacitor C_s, an inductor L_s and converter leg. The voltage and current of the ripple port circuit need to be controlled such that its power is equal to the ripple power flowing into the DC link. The ripple power is equal to the instantaneous power of the energy storage capacitor C_s as

$$\frac{1}{2} C_s \frac{dV_{cs}^2}{dt} = P_r \sin(2\omega t + \theta) \tag{13}$$

By rearranging equation (13), Voltage across C_s [15] is calculated as

$$V_{cs} = \sqrt{\frac{P_r}{C_s \cdot \omega}(K - \sin(2\omega t + \theta))} \tag{14}$$

Where K is constant (K≥1). The current of C_s [15] is calculated as

$$i_{cs} = \frac{P_r \sin(2\omega t + \theta)}{\sqrt{\frac{P_r}{C_s \cdot \omega}(K - \sin(2\omega t + \theta))}} \tag{15}$$

From equation (3), the auxiliary capacitor can be calculated as

$$C_s = \frac{P_r(K + 1)}{V_{cs}^2 \cdot \omega} \tag{16}$$

If K=1, it represents the storage of maximum ripple energy in the capacitor, which means the total charge and discharge. Auxiliary capacitor charging voltage varies from zero to its maximum value and discharging voltage varies from maximum to zero value. Also it is important to control the average inductor current to the calculated current as given in equation (15). Voltage feedforward is implemented for ripple port circuit by using equation (14).

Control analysis

The idea of the proposed control strategy is the strong separation of the control for the PFC and for the ripple port. Therefore, the ripple port control can also be used directly in another single phase application with just information of the power values of the input and output power to be exchanged. Active PFC helps the AC/DC converter to meet harmonic standard requirements without the need of bulky input filter. Two different control schemes are designed for the PFC stage and the ripple port. The control scheme for the PFC stage is to regulate output DC voltage to a fixed voltage and this is achievable with a voltage feed-forward network. There is another current loop control for the PFC stage to regulate the input current. The input current can be controlled to be sinusoidal at a unity power factor with the PFC circuit. This PFC circuit reduces the harmonic current flow distortion in the supply current and creates a current waveform close to a fundamental sine wave, which helps to achieve a high power factor.

The control scheme for the PFC

The dedicated control loop is provided for the PFC as shown in Fig. 2. I_{ref} is the required ideal current which should flow in the input inductor and should also be in phase with the AC input voltage, while I_{LPFC} is the actual inductor current. Low pass filtering is used to reduce the noise in the inductor current. The voltage feed-forward control improves the voltage stabilization and can easily be employed. The PFC stage is driven in Totem Pole mode, therefore S_3 and S_4 are driven based on the sign of the grid voltage. The output signal from the voltage control loop is used to determine the duty cycle of the PWM signal for the converter.

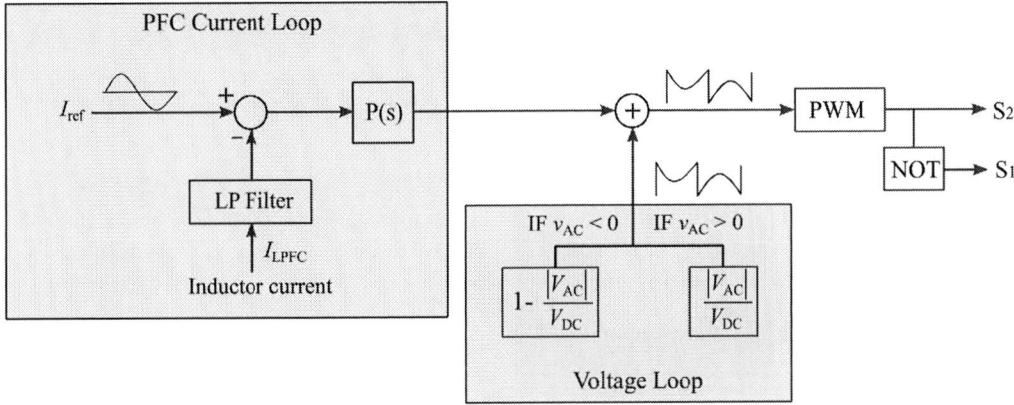

Fig. 2: Boost PFC regulator control loop modelling

The control scheme for the ripple port

The control loop represented in Fig. 3 is for the Ripple port current loop. Here the goal is to reduce the DC link voltage ripple. The current loop for the ripple port regulates the current flowing through the inductor L_s, which corresponds to the ripple power. Input instantaneous power has a double line frequency oscillation which is also delivered to the output. Therefore, a voltage ripple is observed at the output DC bus. While the input instantaneous power is not equal to the instantaneous output power, the ripple port circuit has to compensate the difference between which is an AC power ripple. The ripple power, P_{ripple} calculated in (17), should be absorbed by the ripple port circuit. As the DC link voltage V_{DC} is known, the ripple port current I_{LS^*} can be calculated as per equation (18). This gives the nominal reference ripple port current.

$$\overbrace{V_{AC}\sin(\omega t) \cdot I_{ref}\sin(\omega t)}^{P_{in}(t)} - \overbrace{V_{DC}(t) \cdot I_{DC}(t)}^{P_o(t)} = \overbrace{V_{DC}(t) \cdot I_{LS}(t)}^{P_{ripple}(t)} \tag{17}$$

Rearranging equation (17)

$$I_{LS^*} = \frac{P_{in}(t) - P_o(t)}{V_{DC}} \qquad (18)$$

The actual current through L_s can be gained by measuring the ripple port inductor current I_{LS} and scaling it with the ripple port voltage V_{cs} and the DC link voltage V_{DC} to minimize error differences. The current delta is fed to a P-controller. Then a duty cycle is derived for the charging and discharging phases. When the output DC power $P_o(t)$ is less than input power $P_{in}(t)$, the energy storage capacitor C_s stores the excess energy and the voltage across the capacitor increases. Inversely, when $P_o(t)$ is greater than input power $P_{in}(t)$, then capacitors deliver energy to the output which reduces voltage across C_s.

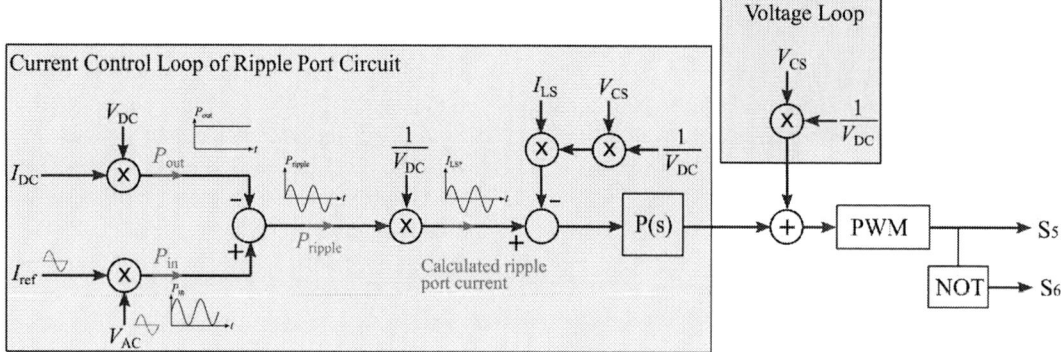

Fig. 3: Current Control loop for the ripple port circuit

Simulation results

A simulation circuit is built in MATLAB Simulink. In this section simulation results of a conventional PFC circuit compared with ripple port integrated PFC circuit are discussed. The power ripple on the DC bus is analysed for the converter with and without ripple port circuit based on simulation.

Table II: Design Parameter of Single Phase Rectifier Circuit

Parameter	Value
Input AC voltage V_{AC}	230 V_{rms}, 50 Hz
Output DC voltage V_{DC}	350 VDC
Input inductor	121 µH
Converter switching frequency	400 kHz
L_s	32 µH
Active capacitor C_s	360 µF
DC link capacitor C_{DC}	180 µF
Output Power	3 kW

Fig. 4(a) shows the results for a single phase PFC rectifier without ripple port circuit with a DC link capacitor of 5 mF. Input voltage and input current are in phase. Here double line frequency observed at the DC link voltage is 3V (0.8% of the DC voltage), but with an DC link capacitor value of 5 mF.

Fig. 4(b) shows the simulation results for the single phase PFC rectifier with ripple port circuit and reduced DC link capacitor size. In this configuration, the energy storage capacitance C_s value is 360 µF and the DC link capacitor C_{DC} value is reduced to 180 µF. The ripple port circuit is active. Most of the ripple energy is observed across the energy storage capacitor C_s. The voltage range V_{CS} is observed across C_s and it is in between 140V to 280V. Output DC-link voltage ripple V_{DC} is within a 2% limit.

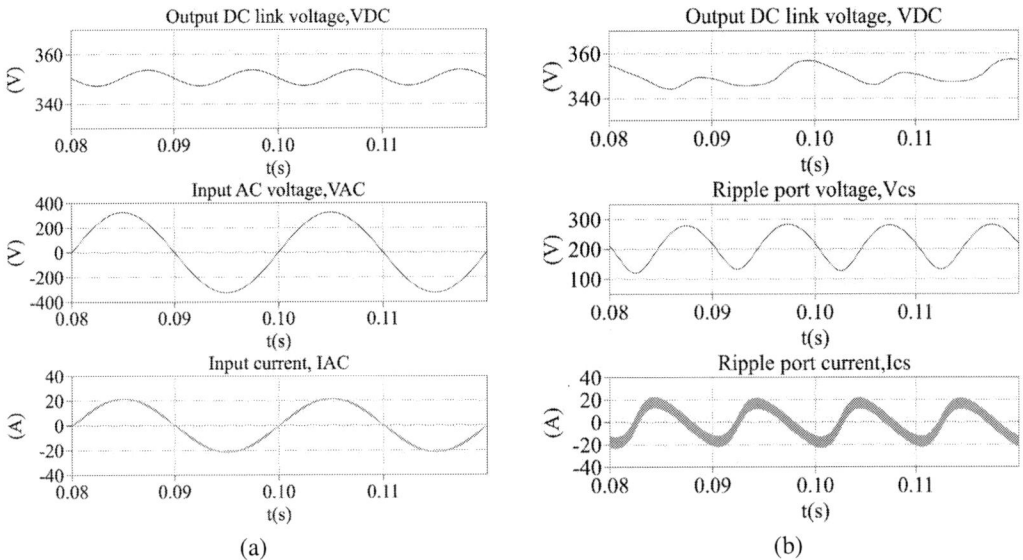

Fig. 4: Time domain results of single phase PFC rectifier after simulation settling time (a) with output DC capacitor value as 5 mF, without ripple port (b) with output DC capacitor value as 0.18 mF, with ripple port

Design of prototype and experimental results

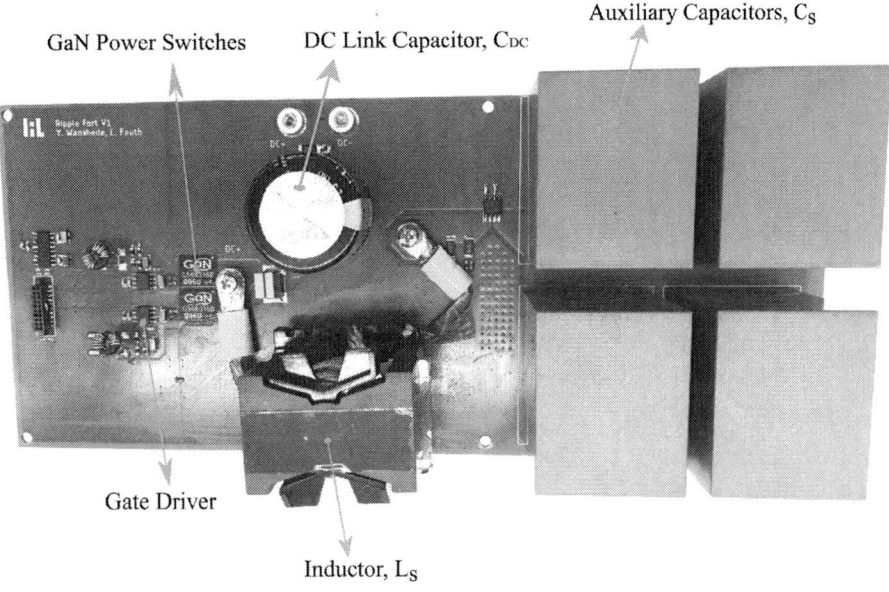

Fig. 5: Hardware prototype

A 3 kW ripple port circuit prototype has been built to verify the functionality of the proposed ripple port scheme. A four-layer PCB-board is designed as shown in Fig. 5. The dimensions of the ripple port circuit hardware prototype are 10 cm×23.5 cm×5 cm with a power density of 2.5 kW/L. As the energy storage capacitor, four 90 μF film capacitors are selected to filter the double line frequency ripple and an inductor with 32 μH is selected. For the power switches, GaN Systems GS66516B are used. External heatsink without optimization of volume 454.90 cm^3 is used for cooling. A 32 bit STM32F microcontroller was used as a controller to generate a PWM signal to drive the converter as per calculated duty cycle using voltage feed forward for the ripple port circuit.

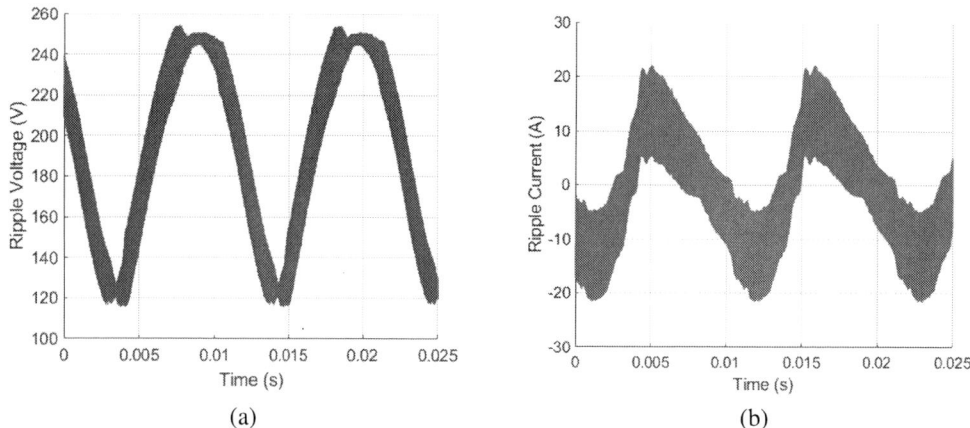

(a)   (b)

Fig. 6: Experimental Results of Ripple port circuit (a) Ripple voltage across auxillary capacitor (b) Ripple current

The ripple port circuit has been tested for 350VDC supply and 200 kHz switching frequency. The ripple port circuit voltage and current are shown in Fig. 6(a) and (b) respectively. The observed peak value of the voltage across the capacitor C_s is 250V and the auxiliary capacitor current is controlled to be the sinusoidal as per simulation , the peak value of the current is approximately 20A.

Fig. 7 shows image taken by an infrared camera to measure the maximum temperature of the switches. The temperature of the GaN power switches reaches up to 102 °C at 3 kW.

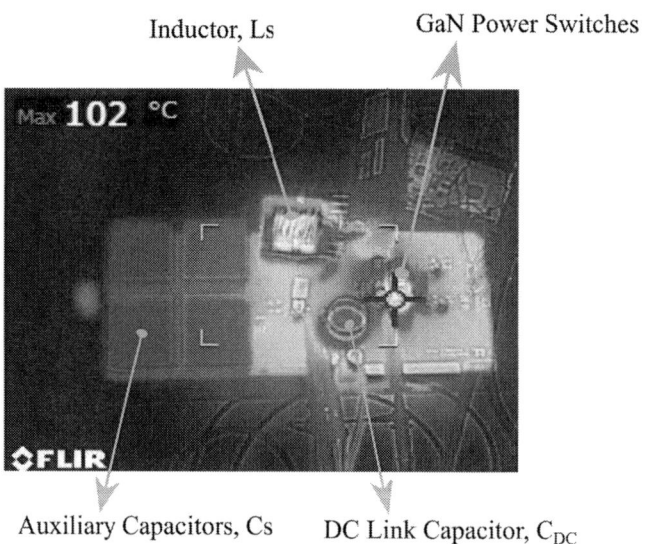

Fig. 7: Thermal measurement of power switches

Efficiency and volume evaluation

To evaluate the efficiency of the ripple port circuit, the power losses of the circuit are observed over the input power range of 250 W to 3 kW. Efficiency performance of the circuit is also verified by changing the value of ON and OFF switching gate resistors as shown in Fig. 8. The efficiency of the circuit was approximately 96.5% to 99.16% over the measured power range.

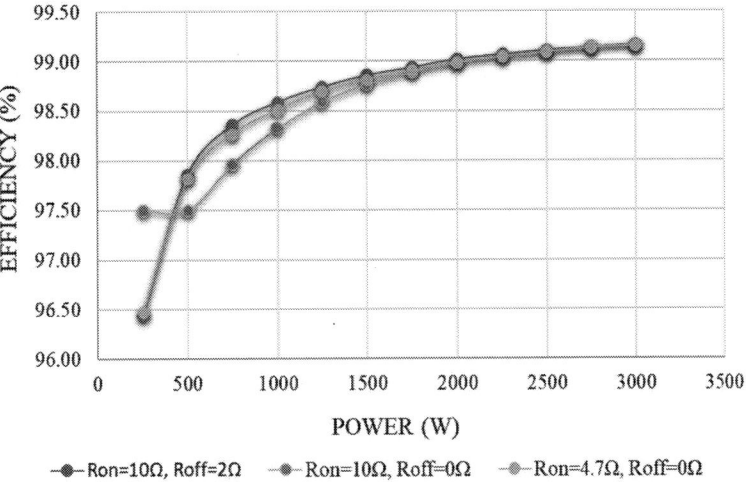

Fig. 8: Efficiency comparison for different gate resistors upto 3 kW at 200 kHz

Fig. 9 shows the volume comparison of the conventional PFC circuit DC link capacitors and ripple port circuit capacitors which includes auxiliary capacitors and the DC link capacitor. In the proposed design, the DC link output capacitor value and size are decreased approximately by 96%. As per simulation result in above section, around 5 mF capacitor value is needed as output DC capacitor in the conventional PFC circuit, while in the proposed converter with the ripple port circuit this value is reduced to 180 μF. The volume of electrolytic capacitor is 17.675 cm^3 and auxiliary capacitors volume is 294 cm^3. Based on this the total capacitor volume in the ripple port circuit design is 311.675 cm^3, while in the conventional PFC circuit the volume of the DC link capacitors is 490.83 cm^3 . Additionally, Fig. 9 shows the volume of the ripple port circuit where the volume of the switches, capacitors (ripple capacitors + DC link capacitor), inductor for ripple port circuit are taken into consideration. Because of the decrease in the volume of the DC-link capacitor, the volume of ripple circuit is reduced in proposed converter.

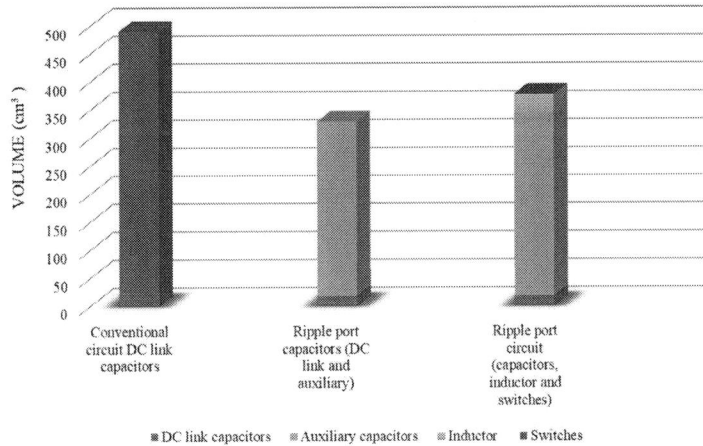

Fig. 9: Volume comparison of conventional PFC circuit DC link capacitors with proposed ripple port circuit and its capacitors

Conclusion

This paper presents a virtual DC link for a single phase converter which is effective for suppressing the double line frequency ripple on the output with reduction of overall capacitor value and size. The control strategy is straightforward and depends on parameters V_{DC}, V_{CS} and I_{ref} current. To decouple the low frequency ripple, the current in the ripple port circuit is regulated. Based on simulation, it is observed that the voltage fluctuation at the DC link is reduced and less than 2% ripple voltage can be reached with an 180 μF DC link capacitor (around 4% of the conventional DC link capacitor value) at a power of 3 kW. Hence, the large electrolytic capacitors can be replaced by film capacitors. The experimental results show that the voltage ripple across the energy storage capacitor is held within the required voltage range depending on the DC link voltage. An efficiency of over 99% was reached at high power values and also a comparison of the impact of different gate resistors is given. The proposed topology shows a good power density and will be optimized for higher power density and even higher efficiency.

References

[1] M. Su, P. Pan, X. Long, Y. Sun, and J. Yang, "An active power-decoupling method for single-phase AC-DC converters," IEEE Trans. Ind. Informat., vol. 10, no. 1, pp. 461–468, Feb. 2014.

[2] H. Li, K. Zhang, H. Zhao, S. Fan, and J. Xiong, "Active power decoupling for high-power single-phase PWM rectifiers," IEEE Trans. Power Electron., vol. 28, no. 3, pp. 1308–1319, Mar. 2013.

[3] B. Tian, S. Harb, and R. S. Balog, "Ripple-port integrated PFC rectifier with fast dynamic response," in Proc. IEEE Midwest Symp. Circuits Syst., College Station, TX, USA, 2014, pp. 781–784.

[4] K. Raggl, T. Nussbaumer, G. Doerig, J. Biela, and J. W. Kolar, "Comprehensive design and optimization of a high-power-density single-phase boost PFC," IEEE Trans. Industrial Electron., vol. 56, no. 7, pp. 2574–2587, Jul. 2009.

[5] B. Mahdavikhah, S. M. Ahsanuzzaman, and A. Prodic, "A hardware efficient programmable two-band controller for PFC rectifiers with ripple cancellation circuits," in Industrial Electronics Society, IECON 2013 - 39th Annual Conference of the IEEE, 2013, pp. 3240-3245.

[6] S. Xu, L. Chang, R. Shao, and Shao, "Evolution of single-phase power converter topologies underlining power decoupling," Chinese Journal of Electrical Engineering, vol. 2, no. June, 2016.

[7] Ruxi Wang, Fred Wang, Puqi Ning, Rixin Lai, Rolando Burgos, Dushan Boroyevich. "Study of Energy Storage Capacitor Reduction for Single Phase PWM Rectifier" IEEE Applied Power Electronics Conference 2008.

[8] Z. Qin, Y. Tang, P. C. Loh, and F. Blaabjerg, "Benchmark of AC and DC active power decoupling circuits for second-order harmonic mitigation in kW-scale single-phase inverters," in Proc. of IEEE Energy Convers. Congr. Expo. (ECCE), 2015, pp. 2514–2521.

[9] R.X.Wang et al.,"A high power density single-phase PWM rectifier with active ripple energy storage," IEEE Trans. Power Electron., vol. 26, no. 5, pp. 1430–1443, May 2011.

[10] Qiu, M.; Wang, P.; Bi, H.; Wang, Z., "Active Power Decoupling Design of a Single-Phase AC-DC Converter," Electronics 2019, 8, 841.

[11] C. B. Barth, I. Moon, Y. Lei, S. Qin and R. Pilwa-Podgurski, "Experimental evaluation of capacitors for power buffering in single phase power converters," in Proc. Energy Convers. Congr. Expo, Sep.2015,pp.6269– 6276.

[12] H. Wang, H. Wang, G. Zhu, and F. Blaabjerg, "An overview of capacitive dc links - topology derivation and scalability analysis," IEEE Trans. Power Electron., pp. 1–1, May. 2019.

[13] H. Wang and F. Blaabjerg, "Reliability of capacitors for DC-link applications in power electronic converters—An overview," IEEE Trans. Ind. Appl., vol. 50, no. 5, pp. 3569–3578, Sep./Oct. 2014.

[14] K. Raggl, T. Nussbaumer, G. Doerig, J. Biela, and J. W. Kolar, "Comprehensive design and optimization of a high-power-density single-phase boost PFC," IEEE Trans. Industrial Electron., vol. 56, no. 7, pp. 2574–2587, Jul. 2009.

[15] H. V. Nguyen and D.-C. Lee, "Reducing the dc-link capacitance: A bridgeless PFC boost rectifier that reduces the second-order power ripple at the dc output," IEEE Trans. Ind. Appl. Mag., vol. 24, no. 2, pp. 2–13, Mar. 2018.

Using Both the Circulating Currents and the Common-Mode Voltage for the Branch Energy Control of Modular Multilevel Converters

Rebecca Dierks, Jakub Kucka, and Axel Mertens
LEIBNIZ UNIVERSITY HANNOVER
Institute for Drive Systems and Power Electronics
Hanover, Germany
Email: rebecca.dierks@ial.uni-hannover.de
URL: https://www.ial.uni-hannover.de

Acknowledgments

This research was funded by the German Research Foundation (DFG) – Project 254417319.

Keywords

≪Converter control≫, ≪Multilevel converters≫, ≪Variable-speed drives≫, ≪Low-voltage ride-through≫

Abstract

Many degrees of freedom are available for the branch energy control of modular multilevel converters. These degrees of freedom comprise different components of the common-mode voltage and the internal circulating currents. If all degrees of freedom are used simultaneously, undesired cross-couplings occur in the branch energy control. In the conventional branch energy control, either the internal circulating currents or the common-mode voltage are used to avoid the coupling terms. The paper presents a novel solution that uses both the circulating currents and the common-mode voltage while omitting the couplings. Thereby, the branch currents and thus the losses in the branches can be reduced without affecting the control dynamics. The proposed control approach is validated through simulations and experimentally on a downscaled converter prototype. Furthermore, the improved performance is demonstrated by means of a comparison to the conventional control. In general, the proposed control approach is expected to be especially beneficial for drive systems operated below rated speed and for a low-voltage ride-through of grid-tied converters.

Introduction

The Modular Multilevel Converter (MMC) proposed by Marquardt in 2001 has been used especially for constant operating frequencies in grid applications [1] [2]. The main advantage of the MMC is the scalability to higher voltages and it is therefore suitable for applications such as the HVDC transmission. By connecting several modules in series, a large number of voltage levels can be achieved and thus an output voltage with low harmonics is generated. A disadvantage of the MMC is the complex structure, resulting in a complicated control system. The first publications dealing with this topic, use an open-loop control without internal current control [3]. However, this approach leads to a dominant second harmonic of the circulating current, which causes additional current stress [4]. In order to be able to reduce the losses, a closed-loop control is used for the internal circulating currents [5]. It is possible to control the energies and currents as a whole system [6]. However, due to the large capacitance in the modules required to buffer the branch energy variation, the time constants of the current and energy differ significantly from each other. Therefore, it is beneficial to use a cascaded control that treats the branch energy control and the current control individually since it reduces the complexity of the control system.

First investigations of MMCs in variable-speed drive applications were done by Hagiwara *et al.* in the year 2010 for a quadratic load torque [7]. The problematic operating range was at output frequencies close to 0 Hz, since at such operating points, the energy variation approaches infinity, which makes a variable-speed operation challenging. By the introduction of the Low-Frequency Mode [8], high branch energies can be reduced during the low speeds. This approach is based on the injection of a higher frequency component in the common-mode voltage. Therefore, the MMC can additionally be used as a variable-speed drive converter.

Considering the literature on the cascaded control of MMCs, there are several methods to control the energy between the branches (branch energy control), whereby the current control is almost always based on the same principles. Various publications, e.g. [9, 10], follow the approach to use the common-mode voltage and the circulating currents as degrees of freedom in the energy control at a variable-speed drive. Whereby the common-mode voltage is used for low output frequencies and the circulating currents at frequencies close to the rated frequency.

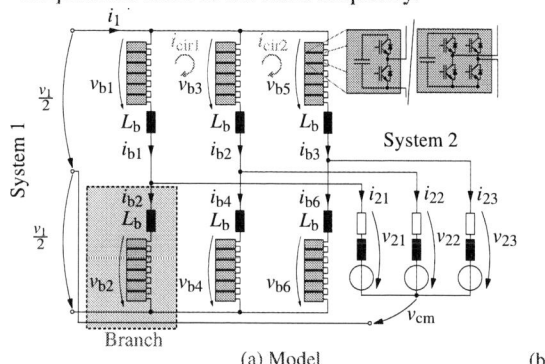
(a) Model

(b) Test bench with nine modules per branch. ($V_1 = 220$ V)

Fig. 1: Three-phase MMC for driving a machine.

In the publications of Karwatzki *et al.* [11, 12], the resulting power equations of the branch energy control are analysed for the case when both the circulating currents and the common-mode voltage are used simultaneously as degrees of freedom. This leads to non-linear cross-couplings between the components of the common-mode voltage and the circulating currents. Furthermore, no approach has been found so far that enables a utilization of all degrees of freedom at the same time without causing a non-linear coupled behaviour between the degrees of freedom.

The basic idea of this paper is to find a solution for the branch energy control that is capable of utilizing all degrees of freedom at the same time while considering the aforementioned coupling between these degrees of freedom. The purpose of this paper is to investigate the effects of a decoupled branch energy control system with regard to a reduced current loading of the modules. This is expected to decrease the losses during the transients

Conventional Control Structure of the MMC

In this paper, a dc-ac configuration of the MMC (Fig. 1(a)) for a variable-speed drive is assumed. System 1 represents the DC link and System 2 is the connected three-phase system. Fig. 1(a) shows the definitions of the circulating currents i_{cir1} and i_{cir2} with the corresponding sign conventions. In the following, lowercase values are assumed to be time-dependent and the bold letters represents vectors or matrices. Generally, the control of the MMC is designed as a cascaded control according to [13]. As shown in Fig. 2, the inner control loop is the current control which is enclosed by the energy control. The balancing of the energies in each branch is based on a two-step modulation, which has been implemented as described in [14]. However, this inner branch balancing is typically negligible for the branch energy performance and is therefore not further discussed in this paper. The current control can be described by a state-space representation with a quadratic input matrix $\mathbf{B}'^{-1}$, since six branch voltages are set by six independent degrees of freedom. The state vector

$$\mathbf{x}' = \begin{bmatrix} \mathbf{x}^{\text{T}} & v_{\text{cm}} \end{bmatrix}^{\text{T}} = \begin{bmatrix} i_{\text{cir1}} & i_{\text{cir2}} & i_1 & i_{2\alpha} & i_{2\beta} & v_{\text{cm}} \end{bmatrix}^{\text{T}} \tag{1}$$

with the addition of the common-mode voltage v_{cm} represents the degrees of freedom that are adjusted during current control and contains the circulating currents i_{cir1} and i_{cir2}, the current of System 1 i_1 and the current of System 2 in alpha-beta coordinates $i_{2\alpha}$ and $i_{2\beta}$. With the exception of the common-mode voltage v_{cm}, all components of the state vector are controlled in a closed loop. The circulating currents (i_{cir1}, i_{cir2}) and the currents of System 2 ($i_{2\alpha}$, $i_{2\beta}$) are controlled by proportional-resonant (PR) controllers. The current of System 1 (i_1) is controlled by a PI controller. Due to the fact that the detailed description of the current control is out of the scope of this paper, the state-space representations with the corresponding matrices can be taken from [13].

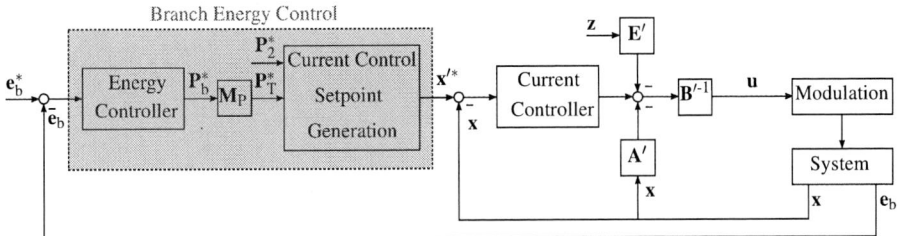

Fig. 2: Block diagram of the cascaded control of the MMC

The branch energy control represents the outer control loop and maintains the mean capacitor voltage at a predetermined level, whereby the setpoints for the current control except of i_2 are used as degrees of freedom (Fig. 2). For the derivation of the branch energy control, it is assumed that the branch currents are ideally set by the current controllers and the voltage drops over the branch inductances are considered negligibly small compared to the branch voltages. Hence, the branch voltages and branch currents can be assumed as

$$
\mathbf{v_b} =
\begin{bmatrix} v_{b1} \\ v_{b2} \\ v_{b3} \\ v_{b4} \\ v_{b5} \\ v_{b6} \end{bmatrix}
=
\begin{bmatrix}
\frac{1}{2} & -1 & 0 & 0 & -1 \\
\frac{1}{2} & 1 & 0 & 0 & 1 \\
\frac{1}{2} & 0 & -1 & 0 & -1 \\
\frac{1}{2} & 0 & 1 & 0 & 1 \\
\frac{1}{2} & 0 & 0 & -1 & -1 \\
\frac{1}{2} & 0 & 0 & 1 & 1
\end{bmatrix}
\cdot
\begin{bmatrix} v_1 \\ v_{21} \\ v_{22} \\ v_{23} \\ v_{cm} \end{bmatrix}
\text{ and } \mathbf{i_b} =
\begin{bmatrix} i_{b1} \\ i_{b2} \\ i_{b3} \\ i_{b4} \\ i_{b5} \\ i_{b6} \end{bmatrix}
=
\begin{bmatrix}
\frac{1}{3} & \frac{1}{2} & 0 & 0 & \frac{4}{3} & \frac{2}{3} \\
\frac{1}{3} & -\frac{1}{2} & 0 & 0 & \frac{4}{3} & \frac{2}{3} \\
\frac{1}{3} & 0 & \frac{1}{2} & 0 & -\frac{2}{3} & \frac{2}{3} \\
\frac{1}{3} & 0 & -\frac{1}{2} & 0 & -\frac{2}{3} & \frac{2}{3} \\
\frac{1}{3} & 0 & 0 & \frac{1}{2} & -\frac{2}{3} & -\frac{4}{3} \\
\frac{1}{3} & 0 & 0 & -\frac{1}{2} & -\frac{2}{3} & -\frac{4}{3}
\end{bmatrix}
\cdot
\begin{bmatrix} i_1 \\ i_{21} \\ i_{22} \\ i_{23} \\ i_{cir1} \\ i_{cir2} \end{bmatrix}. \quad (2)
$$

If energies in branch inductances are neglected, the branch energies can be expressed by

$$
\mathbf{e_b} = \begin{bmatrix} e_{b1} & \cdots & e_{b6} \end{bmatrix}^T = \begin{bmatrix} \int_0^t p_{b1} \ dt + e_{b1}(0) & \cdots & \int_0^t p_{b6} \ dt + e_{b6}(0) \end{bmatrix}^T . \quad (3)
$$

The branch power

$$
\mathbf{p_b} = \begin{bmatrix} v_{b1} \cdot i_{b1} & \cdots & v_{b6} \cdot i_{b6} \end{bmatrix}^T = \mathbf{P_b} + \tilde{\mathbf{p}}_b \text{ with } \mathbf{P_b} = \begin{bmatrix} P_{b1} & P_{b2} & P_{b3} & P_{b4} & P_{b5} & P_{b6} \end{bmatrix}^T, \quad (4)
$$

results from the product of the corresponding branch voltage and branch current, which leads to the mean value $\mathbf{P_b}$ and the time-dependent alternating component $\tilde{\mathbf{p}}_b$. As described in [10], the direct components $\mathbf{P_b}$ are employed to control the energy between the branches. A chain of second order bandstop filters is implemented in front of the energy controllers to filter out the frequencies at the single and double fundamental frequency to control only the low-frequency part of the branch energies and to obtain only the direct components $\mathbf{P_b}$ behind the energy controllers.

The components of the common-mode voltage and the circulating currents have the advantage of not influencing the system variables ($i_1, i_{2\alpha}, i_{2\beta}$), but still being able to transfer branch energy between the branches. For this reason, those are used to control the branch energy as degrees of freedom. Based on [13], the common-mode voltage

$$
v_{cm} = V_{cm0} + \sqrt{2} \cdot V_{cm\alpha} \cdot \cos(\omega_2 \cdot t - \varphi) + \sqrt{2} \cdot V_{cm\beta} \cdot \sin(\omega_2 \cdot t - \varphi) + \sqrt{2} \cdot V_{cmx} \cdot \cos(\omega_x \cdot t) \quad (5)
$$

contains a component V_{cm0}, which oscillates with the frequency of System 1 ($\omega_1 = 0$ for DC grid) and additionally components oscillating with the frequency of System 2. The frequency $\omega_x \notin \{\omega_1, \omega_2\}$ introduced for the operation of the MMC at low speeds using the Low-Frequency Mode [8] is mentioned in the fundamental equations of the branch energy control for the sake of completeness. In the same manner as the common-mode voltage, the two circulating currents

$$\begin{bmatrix} i_{cir1} \\ i_{cir2} \end{bmatrix} = \begin{bmatrix} I_{cir10} \\ I_{cir20} \end{bmatrix} + \sqrt{2} \cdot \begin{bmatrix} I_{cir1\alpha} \\ I_{cir2\alpha} \end{bmatrix} \cdot \cos(\omega_2 \cdot t) + \sqrt{2} \cdot \begin{bmatrix} I_{cir1\beta} \\ I_{cir2\beta} \end{bmatrix} \cdot \sin(\omega_2 \cdot t) + \sqrt{2} \cdot \begin{bmatrix} I_{cir1x} \\ I_{cir2x} \end{bmatrix} \cdot \cos(\omega_x \cdot t) \quad (6)$$

from Fig. 1(a) are also split into the system components and a component oscillating with the frequency ω_x. System 1 is assumed to be an ideal direct voltage source V_1 and System 2 is represented by a symmetrical three-phase grid:

$$\mathbf{v}_2 = \begin{bmatrix} v_{21} \\ v_{22} \\ v_{23} \end{bmatrix} = \sqrt{2} \cdot V_2 \cdot \begin{bmatrix} \cos(\omega_2 \cdot t) \\ \cos(\omega_2 \cdot t - \frac{2\pi}{3}) \\ \cos(\omega_2 \cdot t - \frac{4\pi}{3}) \end{bmatrix} \text{ and } \mathbf{i}_2 = \begin{bmatrix} i_{21} \\ i_{22} \\ i_{23} \end{bmatrix} = \sqrt{2} \cdot I_2 \cdot \begin{bmatrix} \cos(\omega_2 \cdot t - \varphi) \\ \cos(\omega_2 \cdot t - \frac{2\pi}{3} - \varphi) \\ \cos(\omega_2 \cdot t - \frac{4\pi}{3} - \varphi) \end{bmatrix} . \quad (7)$$

The system variables, the common-mode voltage and the circulating currents are inserted in the branch power equation (4) with the equations of branch current and branch voltages (2). The difference between input and output power can be replaced in the resulting branch power equations by $P_{diff} = \Delta P = P_1 - P_2 = V_1 \cdot I_1 - 3 \cdot V_2 \cdot I_2 \cdot \cos(\varphi)$. The Fourier decomposition of the branch power and elimination of the time-dependent parts result in the mean power

$$\mathbf{P}_b = \overbrace{\begin{bmatrix} f_1(\mathbf{x}_e, \mathbf{x}_{in}) & f_2(\mathbf{x}_e, \mathbf{x}_{in}) & \cdots & f_6(\mathbf{x}_e, \mathbf{x}_{in}) \end{bmatrix}^T}^{\mathbf{f}(\mathbf{x}_e, \mathbf{x}_{in})}$$
$$= \mathbf{f}(\underbrace{I_{cir10}, I_{cir20}, I_{cir1\alpha}, I_{cir1\beta}, I_{cir2\alpha}, I_{cir2\beta}, I_{cir1x}, I_{cir2x}, V_{cm0}, V_{cm\alpha}, V_{cm\beta}, V_{cmx}}_{\mathbf{x}_e}, \underbrace{V_1, I_1, V_2, I_2, \Delta P}_{\mathbf{x}_{in}}) . \quad (8)$$

The function $\mathbf{f}(\mathbf{x}_e, \mathbf{x}_{in})$ shows the dependence of the branch power on the components $\mathbf{x}_e$, which represent the degrees of freedom of the branch energy control, and the components of the system variables $\mathbf{x}_{in}$. In this function $\mathbf{f}(\mathbf{x}_e, \mathbf{x}_{in})$, the degrees of freedom and the system variables form several power terms, which are listed in Table I. The table distinguishes between two types of power terms: Terms that do not contain a coupling of degrees of freedom (Group I to Group IV), and terms that contain a coupling between the common-mode voltage and the circulating currents. Hence, the function $\mathbf{f}(\mathbf{x}_e, \mathbf{x}_{in})$ can also be divided in a coupling $\mathbf{f}_{coupl}$ and a decoupled term $\mathbf{f}_{decoupl}$ with

$$\mathbf{P}_b = \mathbf{M}_T \cdot \overbrace{\underbrace{\begin{bmatrix} \mathbf{P}_{cir} & \mathbf{P}_x & \mathbf{P}_{cm} & P_{diff} \end{bmatrix}^T}_{\mathbf{P}_T}}^{\mathbf{f}_{decoupl}} + \mathbf{f}_{coupl}(\mathbf{P}_{coupl}) . \quad (9)$$

The power components with the corresponding power vectors of (9) can be taken from Table I. The matrix $\mathbf{M}_T$ is not quadratic and not invertible and contains the power terms of Group I to IV. The complete matrices and correlations can be found in [11], [12] and [13]. The problematic coupling terms $\mathbf{P}_{coupl}$ and the corresponding vector $\mathbf{f}_{coupl}(\mathbf{P}_{coupl})$ for the upper branches are demonstrated in the appendix. To generate the setpoints for the branch current control, the equation system (9) must be converted into the form $\mathbf{P}_T = \mathbf{M}_P \cdot \mathbf{P}_b$ with the generalized inverse $\mathbf{M}_P = \mathbf{M}_T^+ = (\mathbf{M}_T^T \cdot \mathbf{M}_T)^{-1} \cdot \mathbf{M}_T^T$ [15]. In order to avoid the couplings of the equation (9), the common-mode voltage is negligible when the MMC is operated close to the rated frequency, because in this area the common-mode voltage is insignificant compared to the voltage of the three-phase system (System 2) with a relatively large modulation index. The relationship

$$\mathbf{P}_{T,N} = \begin{bmatrix} \mathbf{P}_{cir} & P_{diff} \end{bmatrix}^T = \underbrace{\begin{bmatrix} \frac{1}{2} & \frac{1}{2} & -\frac{1}{2} & -\frac{1}{2} & 0 & 0 \\ 0 & 0 & \frac{1}{2} & \frac{1}{2} & -\frac{1}{2} & -\frac{1}{2} \\ -\frac{3}{8} & \frac{3}{8} & -\frac{1}{8} & \frac{1}{8} & \frac{1}{8} & -\frac{1}{8} \\ -\frac{1}{\sqrt{3}\cdot 8} & \frac{1}{\sqrt{3}\cdot 8} & \frac{5}{\sqrt{3}\cdot 8} & -\frac{5}{\sqrt{3}\cdot 8} & -\frac{1}{\sqrt{3}\cdot 8} & \frac{1}{\sqrt{3}\cdot 8} \\ 0 & 0 & \frac{1}{4} & -\frac{1}{4} & -\frac{1}{4} & \frac{1}{4} \\ \frac{1}{\sqrt{3}\cdot 4} & -\frac{1}{\sqrt{3}\cdot 4} & -\frac{1}{\sqrt{3}\cdot 2} & \frac{1}{\sqrt{3}\cdot 2} & -\frac{1}{\sqrt{3}\cdot 2} & \frac{1}{\sqrt{3}\cdot 2} \\ 1 & 1 & 1 & 1 & 1 & 1 \end{bmatrix}}_{\mathbf{M}_{P,N}} \cdot \mathbf{P}_b \quad (10)$$

between $\mathbf{P}_b$ and the power components consisting of the degrees of freedom from Table I is represented by the matrix $\mathbf{M}_{P,N}$. The power vector for low speeds

$$
\mathbf{P}_{T,LF} = \begin{bmatrix} \mathbf{P}_x & \mathbf{P}_{cm} & P_{diff} \end{bmatrix}^T = \underbrace{\begin{bmatrix} -\frac{1}{4} & \frac{1}{4} & \frac{1}{4} & -\frac{1}{4} & 0 & 0 \\ 0 & 0 & -\frac{1}{4} & \frac{1}{4} & \frac{1}{4} & -\frac{1}{4} \\ -\frac{1}{2} & \frac{1}{2} & -\frac{1}{2} & \frac{1}{2} & -\frac{1}{2} & \frac{1}{2} \\ -\frac{2}{3} & -\frac{2}{3} & \frac{1}{3} & \frac{1}{3} & \frac{1}{3} & \frac{1}{3} \\ 0 & 0 & -\frac{1}{\sqrt{3}} & -\frac{1}{\sqrt{3}} & \frac{1}{\sqrt{3}} & \frac{1}{\sqrt{3}} \\ 1 & 1 & 1 & 1 & 1 & 1 \end{bmatrix}}_{\mathbf{M}_{P,LF}} \cdot \mathbf{P}_b \tag{11}
$$

consists of the power vectors $\mathbf{P}_x$, $\mathbf{P}_{cm}$ and P_{diff} described in Table I with the common-mode voltage components.

Approximately between one third and two thirds of the rated frequency, the control is switched from operation at low speed using $\mathbf{M}_{P,LF}$ to operation at nominal speed $\mathbf{M}_{P,N}$ in order to reduce losses due to the increasing branch currents of the low-frequency energy control. However, at drive frequencies f_2 between 0 Hz and nominal frequency, it is beneficial to use both the circulating currents and the common-mode voltage in the control system to reduce the current stress in the branches during transient states of the branch energy control. In addition, by limiting the branch current, the energy variation in the branches, which is buffered by the module capacitors, can also be reduced.

Table I: Branch power components dependent on the degrees of freedom

Group	Group I	Group II	Group III	Group IV	Couplings
Power Vector	P_{diff}	$\mathbf{P}_{cir}$	$\mathbf{P}_{cm}$	$\mathbf{P}_x$	$\mathbf{P}_{coupl}$
Constant Power Components	ΔP	$I_{cir10} \cdot V_1, I_{cir20} \cdot V_1,$ $I_{cir1\alpha} \cdot V_2, I_{cir1\beta} \cdot V_2$ $I_{cir2\alpha} \cdot V_2, I_{cir2\beta} \cdot V_2$	$V_{cm0} \cdot I_1,$ $V_{cm\alpha} \cdot I_2,$ $V_{cm\beta} \cdot I_2$	$I_{cir1x} \cdot V_{cmx},$ $I_{cir2x} \cdot V_{cmx},$	$V_{cm0} \cdot I_{cir10}, V_{cm0} \cdot I_{cir20}, V_{cm\alpha} \cdot I_{cir1\alpha}, V_{cm\alpha} \cdot I_{cir1\beta},$ $V_{cm\alpha} \cdot I_{cir2\alpha}, V_{cm\alpha} \cdot I_{cir2\beta}, V_{cm\beta} \cdot I_{cir1\alpha}, V_{cm\beta} \cdot I_{cir1\beta},$ $V_{cm\beta} \cdot I_{cir2\alpha}, V_{cm\beta} \cdot I_{cir2\beta}$

Implementation of the Decoupled Branch Energy Control

This section explains how the decoupling of circulating currents and common-mode voltages can be realized in the branch energy control. In order to limit the losses of the MMC, the largest part of the power $\mathbf{P}_b$ is to be controlled via the common-mode voltage. But the common-mode voltage is limited by the system voltages and the maximum feasible branch voltage, so that the rest of the power is set by the circulating currents. In the first step, the mean value of (8) with the non-linear couplings of the common-mode voltage and the circulating current is examined. For the following considerations, the additional components with ω_x are neglected in order to obtain simpler analytical functions. The dependence of the branch power on the remaining degrees of freedom $\mathbf{P}_b = \mathbf{f}_{decoupl}(\mathbf{P}_{cir}, \mathbf{P}_{cm}, P_{diff}) + \mathbf{f}_{coupl}(\mathbf{P}_{coupl})$ is depicted in Table I.

To keep the circulating currents and thus the branch currents as low as possible, an attempt is made to set the largest part of the power via the common-mode voltages. For this purpose, the matrix $\mathbf{M}_{P,LF}$ of equation (11) for branch energy control at low speeds is used to calculate the power component that can be controlled via the common-mode voltage. On the basis of the resulting power components

$$
\mathbf{P}_{cm} = \begin{bmatrix} P_{cm0} \\ P_{cm\alpha} \\ P_{cm\beta} \end{bmatrix} = \begin{bmatrix} -\frac{1}{2} & \frac{1}{2} & -\frac{1}{2} & \frac{1}{2} & -\frac{1}{2} & \frac{1}{2} \\ -\frac{2}{3} & -\frac{2}{3} & \frac{1}{3} & \frac{1}{3} & \frac{1}{3} & \frac{1}{3} \\ 0 & 0 & -\frac{1}{\sqrt{3}} & -\frac{1}{\sqrt{3}} & \frac{1}{\sqrt{3}} & \frac{1}{\sqrt{3}} \end{bmatrix} \cdot \mathbf{P}_b \ , \tag{12}
$$

the setpoints for the common-mode voltage can be calculated for the given system variables as

$$
\mathbf{V}_{cm} = \begin{bmatrix} V_{cm0} & V_{cm\alpha} & V_{cm\beta} \end{bmatrix}^T = \begin{bmatrix} P_{cm0} \cdot \frac{1}{I_1} & P_{cm\alpha} \cdot \frac{1}{I_2} & P_{cm\beta} \cdot \frac{1}{I_2} \end{bmatrix}^T \ . \tag{13}
$$

The vector $\mathbf{V}_{cm}$ is assumed to be constant within each sampling step. Thus, the circulating currents can be set as a function of the feedforward-controlled components of the common-mode voltages $\mathbf{V}_{cm}$ with

$$\begin{bmatrix} \mathbf{I}_{cir} & \Delta P \end{bmatrix}^{\mathrm{T}} = \begin{bmatrix} I_{cir10} & I_{cir20} & I_{cir1\alpha} & I_{cir1\beta} & I_{cir2\alpha} & I_{cir2\beta} & \Delta P \end{bmatrix}^{\mathrm{T}} = \mathbf{g}_{opt}(\mathbf{P}_b, \mathbf{V}_{cm}, \mathbf{x}_{in}) \ . \tag{14}$$

The system of (14) is under-determined. Thus, one component of $\mathbf{I}_{cir}$ or ΔP can be set freely as an optimization parameter (here: $I_{opt} = I_{cir2\beta}$). In order to make use of this further degree of freedom, the RMS value of the circulating current $I_{cir2\beta}$ is selected in such a manner that the quadratic sum of the branch currents becomes minimal like introduced in [13]. For this purpose, the optimal solution I_{opt} of the RMS value $I_{cir2\beta}$ is calculated by solving the optimization problem

$$\frac{\partial}{\partial I_{cir2\beta}} \left(\sum_{k=1}^{6} \left[i_{bk\alpha}^2 + i_{bk\beta}^2 + i_{bk0}^2 \right] \right) \overset{!}{=} \mathbf{0} \ . \tag{15}$$

With the least squares algorithm, the squared components of the branch current with different frequencies can be summed, as well as the α- and β- components, because they are orthogonal to each other. In this case, the branch currents of (2) are splitted into α-, β- and a component with zero frequency

$$i_{bk\alpha} = i_{bk} \big(\underbrace{i_1 = 0, \mathbf{i}_2 = \mathbf{0}}_{\mathbf{i}_{in} = \mathbf{0}}, i_{cir1} = I_{cir1\alpha}, i_{cir2} = I_{cir2\alpha} \big),$$

$$i_{bk\beta} = i_{bk} \big(\mathbf{i}_{in} = \mathbf{0}, i_{cir1} = I_{cir1\beta}, i_{cir2} = I_{cir2\beta} \big), \quad i_{bk0} = i_{bk} \big(\mathbf{i}_{in} = \mathbf{0}, i_{cir1} = I_{cir10}, i_{cir2} = I_{cir20} \big) \ . \tag{16}$$

In the next step, $I_{opt} = I_{cir2\beta}$ is inserted into (14) to calculate the setpoints of the other RMS values and ΔP. Fig. 3 presents the block diagram of the decoupled branch energy control with the optimization algorithm for limiting the current stress.

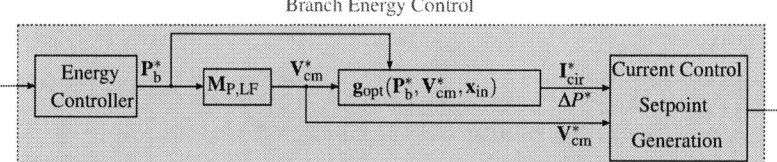

Fig. 3: Decoupled branch energy control block replacing the branch energy control block in Fig. 2

Validation of the Decoupled Branch Energy Control

In the following section, the derived decoupled branch energy control is first validated by simulation using a converter model of the MMC implemented in MathWorks Simulink, with Plexim Plecs toolbox for the physical part of the model. Afterwards, the experimental validation on the test bench (Fig. 1(b)) is presented.

Simulation

In order to validate the proposed decoupled branch energy control, the control structure of Fig. 3 is implemented in Simulink and compared with the conventional control (Fig. 2) that uses the matrix $\mathbf{M}_P = \mathbf{M}_{P,N}$ (10). Consequently, the branch energies of the conventional control are only controlled by the circulating currents. Fig. 4 shows the comparison of both control approaches.

The setpoint of $\hat{v}_{cm}$ in the decoupled control is limited by a certain maximum. If it is assumed that the voltage of System 2 increases proportionally with the output frequency and the current remains constant over all operating points, the achievable maximum of the common-mode voltage can be expressed as

$$\hat{v}_{cm,max} = \underbrace{(1 - M) \cdot \frac{V_1}{2} - \frac{|v_{Lb,set}|}{2}}_{\geq 0} = \max \left(V_{cm0} + \sqrt{\hat{v}_{cm\alpha}^2 + \hat{v}_{cm\beta}^2} \right) \ , \tag{17}$$

which is discussed in detail in the thesis [16]. Thereby, the modulation index $M = \frac{2 \cdot \hat{v}_2}{V_1}$ and $|v_{\text{Lb,set}}|$ are used. The latter represents the minimum required voltage over the branch inductances that is sufficient to control the currents. In the operating point presented in Fig. 4, the amplitude of the voltage of System 2 is $\hat{v}_2 = 1.76\,\text{kV}$, which leads with the given parameters from the caption of Fig. 4 to a maximum common-mode voltage $\hat{v}_{\text{cm,max}} = 4.37\,\text{kV}$.

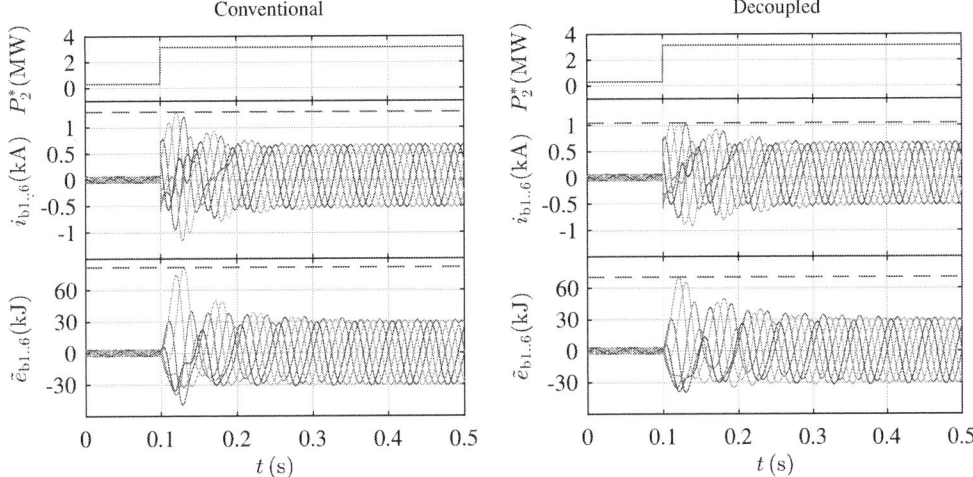

Fig. 4: **Simulations:** Comparison of the conventional (left) to the proposed decoupled branch energy control (right) at the frequency of System 2 $f_2 = 20\,\text{Hz}$ and a setpoint power step from $P_2^* = 0.317\,\text{MW}$ to $3.17\,\text{MW}$. Converter parameters: $V_1 = 13.2\,\text{kV}$, $\hat{v}_2 = 1.76\,\text{kV}$, $\hat{i}_2 = 1.2\,\text{kA}$, $|v_{\text{Lb,set}}| = 900\,\text{V}$, $M = 0.27$, $e_b^* = 287\,\text{kJ}$, $L_b = 350\,\mu\text{H}$, module capacitor $C_m = 15.9\,\text{mF}$. Load: $L_2 = 160\,\mu\text{H}$.

In the simulative investigations, the power setpoint P_2^* is implemented as a step function from 10% to 100% of the full power $P_2 = 3 \cdot I_2 \cdot V_2 \cdot \cos(\varphi) = 3.17\,\text{MW}$ to analyse the performance of the respective control. The branch energy is displayed as controller error with $\tilde{e}_b = e_b - e_b^*$. To compare the two control schemes regarding the maximum branch currents and the branch energy variations, the maximum overshoot r is introduced with

$$r_x = \frac{\max_{k=1..6} |x_k| - \max_{k=1..6} |x_{k,\text{stat}}|}{\max_{k=1..6} |x_{k,\text{stat}}|} \cdot 100\% \quad \text{with} \quad x = \{i_b, \tilde{e}_b\} \ . \tag{18}$$

Thereby, the maximum in the transient state $\max_{k=1..6} |i_{bk}|$ is divided by the maximum current in the stationary state $\hat{i}_{bk,\text{stat}}$. Based on the equation, the maximum overshoots for the conventional and the decoupled control can be calculated with $r_{ib,\text{conv}} = 90\%$ and $r_{ib,\text{decoupl}} = 53\%$. The maximum value of the transient state is highlighted in each case by a dashed line in Fig. 4.

Because the branch energy variation determines the required module capacitance, the maximum overshoot of the branch energy is classified as a relevant quantity to compare the necessary module capacitor volume for variable operating points of different control approaches. At this operating point, the maximum overshoots are $r_{\tilde{e}b,\text{conv}} = 167\%$ and $r_{\tilde{e}b,\text{decoupl}} = 130\%$. These results confirm a reduction of the overshoots of the decoupled control compared to the conventional control. In general, there is no difference in the settling time or control time between the conventional and the decoupled branch energy control, since same parameters of the branch-energy PI controllers were applied.

Based on (17), a higher maximum common-mode voltage should provide a further reduction in branch currents. In order to verify this correlation, Fig. 5 shows the comparison of the maximum overshoot of the branch currents and the branch energies for both control systems over the modulation index. As before, to investigate the dynamic process of the controllers, the power setpoint of System 2 is stepped from 10% to 100% of the power P_2.

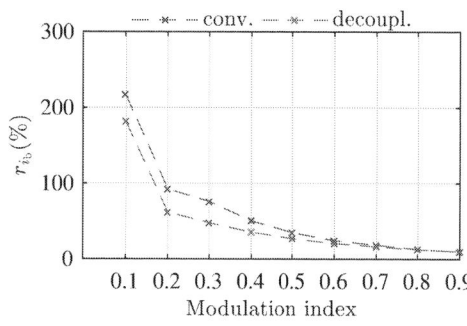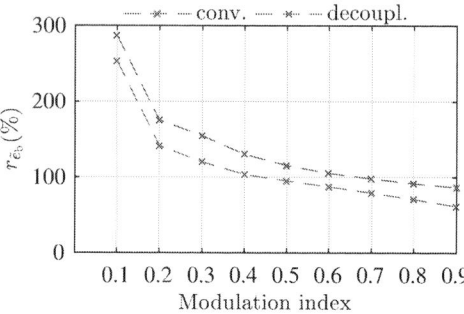

Fig. 5: **Simulations**: Comparison of the maximum overshoot of the conventional (conv.) with the proposed decoupled (decoupl.) branch energy control of the branch current (left) and the branch energy (right) for different modulation indices. Setpoint power step from $P_2^* = 0.317\,\text{MW}$ to $3.17\,\text{MW}$. Converter parameters: $V_1 = 13.2\,\text{kV}$, $\hat{v}_2 = 1.76\,\text{kV}$, $\hat{i}_2 = 1.2\,\text{kA}$, $|v_{\text{Lb,set}}| = 900\,\text{V}$, $e_b^* = 287\,\text{kJ}$, $L_b = 350\,\mu\text{H}$, module capacitor $C_m = 15.9\,\text{mF}$. Load: $L_2 = 160\,\mu\text{H}$.

The simulation results of Fig. 5 show that for both control approaches the maximum overshoots r_{ib} and $r_{\text{ēb}}$ decrease for a higher modulation index. It can be seen that the difference between the control approaches becomes larger with a decreasing modulation index. Especially when considering r_{ib}, the decoupled control is advantageous for a lower modulation index with regard to a reduction of the branch current. This can be explained on the basis of equation (17). With decreasing modulation index, the maximum possible common-mode voltage increases. Thus, the energy between the branches is mainly controlled by the common-mode voltage, which allows the components of the circulating currents to be reduced. When considering the branch energy, the decoupled control is for every operating point the better choice since the value of the maximum overshoot is always lower than with the conventional control. Therefore, the decoupled branch energy control is also very suitable for drive and low-voltage ride-through applications, because especially at low speeds the modulation index is low as well. Consequently, a relatively high common-mode voltage is injected and the branch currents are further reduced during the transient state.

Experiment

The experimental validation is realized with a passive R-L load on a downscaled test bench (Fig. 1(b)). Compared to the simulations, both the voltage and current values are scaled down by a factor of 60. The voltage of System 2 is feedforward-controlled so that the current controllers for $i_{2\alpha}$ and $i_{2\beta}$ are omitted. During the experiment a step of the voltage of System 2 from 10 V to 30 V is initiated. Fig. 6 shows the measured output currents and the branch energies and branch currents for an operating point at the frequency of 20 Hz. Apart from the load resistance, the test bench uses the same converter parameters as the simulation which are listed in the caption of Fig. 6.

In comparison to the conventional control, both the branch currents and the branch energy of the decoupled control are reduced. The effect is not as pronounced as in the simulation, because the energy controllers are set slower to avoid current peaks in the branches. However, the results are sufficient to validate the decoupled control and thus a reduction of the branch currents on the test bench. For a more detailed investigation of the two control approaches, the measured values for the current control ($i_1, i_{\text{cir1}}, i_{\text{cir2}}$) and the common-mode voltage setpoints are displayed. The small oscillations of the common-mode voltage and the circulating currents even in the steady state are caused by measurement noises and the second order bandstop filter in front of the energy control (mentioned in the description of the conventional control) that does not completely filter out the harmonics with the single and double fundamental frequency (here: 20 Hz). In the waveforms of the decoupled control, the oscillations of the common-mode voltage become visible in the branch energy. If necessary, these oscillations can be further attenuated by limiting the possible common-mode voltage or by narrowing the bandwidth of the second order bandstop filter. Nevertheless, it can be seen that the injection of the common-mode voltage in the decoupled control decreases the circulating currents and thus the branch currents.

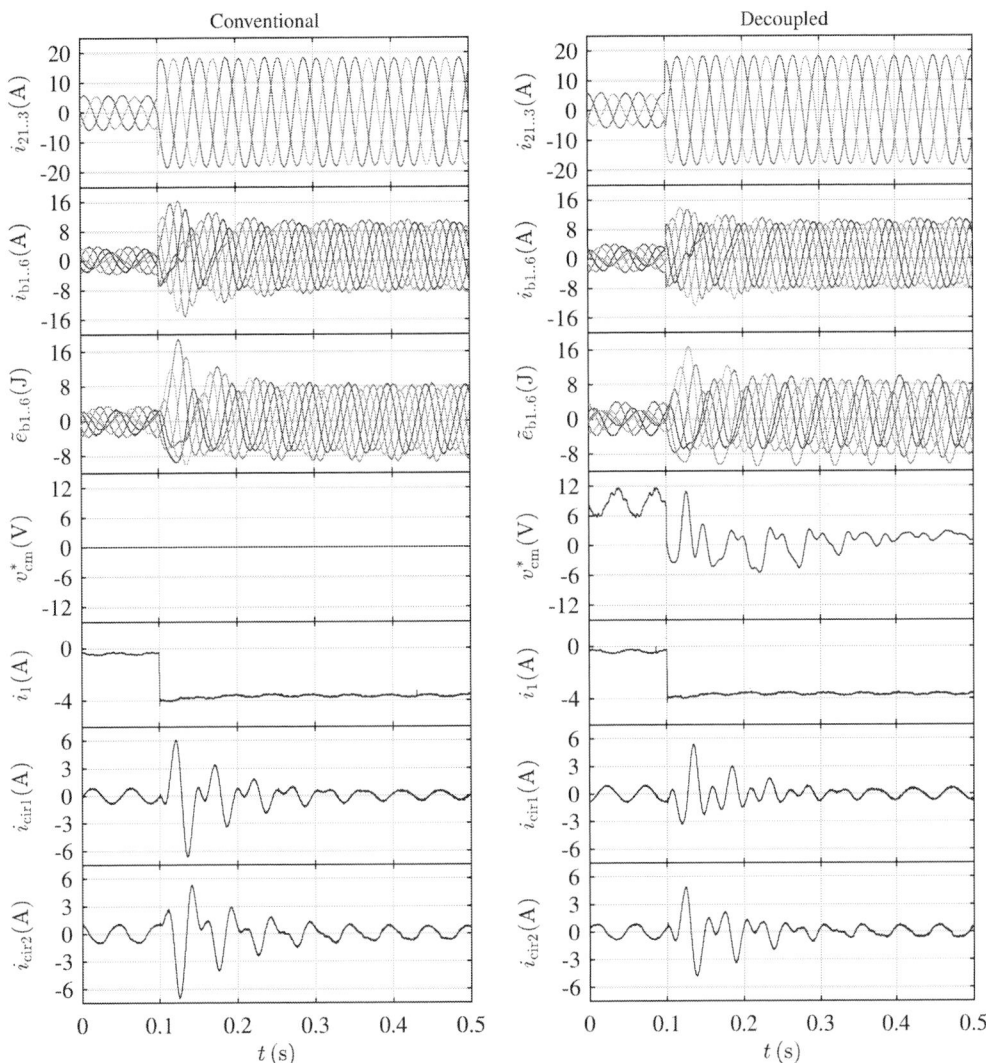

Fig. 6: **Experiments** with a passive load: Comparison of the conventional branch energy control with the proposed decoupled branch energy control at the frequency of System 2 $f_2 = 20\,\text{Hz}$ and a setpoint voltage step from $\hat{v}_2^* = 10\,\text{V}$ to $30\,\text{V}$, $M = 0.27$ (steady state). The waveforms were captured by the logging function of the control system. Converter parameters: $V_1 = 220\,\text{V}$, $e_b^* = 56.1\,\text{J}$, $L_b = 350\,\mu\text{H}$, $C_m = 15.9\,\text{mF}$. Passive load: $R_2 = 1.5\,\Omega$, $L_2 = 160\,\mu\text{H}$.

Conclusions and Outlook

Based on the analytical derivation using a simplified MMC model, the paper presents a way how to utilize all available degrees of freedom in the branch energy control without a coupling in the power components. With the proposed decoupling approach the branch energy is largely controlled using the common-mode voltage, and the circulating currents are adjusted with the aid of an optimization algorithm, so that the branch currents and thus the branch losses are as small as possible. The simulation and the experimental results validate the reduction of the required branch currents and energies during the transient state in the decoupled control. This is expected to reduce the losses and depending on the application even the required capacitance. Investigations of the paper at different modulation indeces reveal that the beneficial effect of the decoupled control is increased at a lower modulation index compared to the conventional control.

Appendix

Mean Branch Power for the Upper Branches as a Function of the Degrees of Freedom

$$P_{b1} = +\frac{1}{6} \cdot \Delta P + \frac{2}{3} \cdot I_{cir10} \cdot V_1 + \frac{1}{3} \cdot I_{cir20} \cdot V_1 - \frac{4}{3} \cdot I_{cir1\alpha} \cdot V_2 - \frac{2}{3} \cdot I_{cir2\alpha} \cdot V_2 - \frac{1}{3} \cdot V_{cm0} \cdot I_1 - \frac{1}{2} \cdot V_{cm\alpha} \cdot I_2$$

$$- \frac{4}{3} \cdot V_{cm0} \cdot I_{cir10} - \frac{2}{3} \cdot V_{cm0} \cdot I_{cir20} - \frac{4}{3} \cdot c_\varphi \cdot V_{cm\alpha} \cdot I_{cir1\alpha} - \frac{4}{3} \cdot s_\varphi \cdot V_{cm\alpha} \cdot I_{cir1\beta} - \frac{2}{3} \cdot c_\varphi \cdot V_{cm\alpha} \cdot I_{cir2\alpha}$$

$$- \frac{2}{3} \cdot s_\varphi \cdot V_{cm\alpha} \cdot I_{cir2\beta} + \frac{4}{3} \cdot s_\varphi \cdot V_{cm\beta} \cdot I_{cir1\alpha} - \frac{4}{3} \cdot c_\varphi \cdot V_{cm\beta} \cdot I_{cir1\beta} + \frac{2}{3} \cdot s_\varphi \cdot V_{cm\beta} \cdot I_{cir2\alpha} - \frac{2}{3} \cdot c_\varphi \cdot V_{cm\beta} \cdot I_{cir2\beta}$$

$$P_{b3} = +\frac{1}{6} \cdot \Delta P - \frac{1}{3} \cdot I_{cir10} \cdot V_1 + \frac{1}{3} \cdot I_{cir20} \cdot V_1 - \frac{1}{3} \cdot I_{cir1\alpha} \cdot V_2 + \frac{\sqrt{3}}{3} \cdot I_{cir1\beta} \cdot V_2 + \frac{1}{3} \cdot I_{cir2\alpha} \cdot V_2 - \frac{\sqrt{3}}{3} \cdot I_{cir2\beta} \cdot V_2$$

$$- \frac{1}{3} \cdot V_{cm0} \cdot I_1 + \frac{1}{4} \cdot V_{cm\alpha} \cdot I_2 - \frac{\sqrt{3}}{4} \cdot V_{cm\beta} \cdot I_2 + \frac{2}{3} \cdot V_{cm0} \cdot I_{cir10} - \frac{2}{3} \cdot V_{cm0} \cdot I_{cir20} + \frac{2}{3} \cdot c_\varphi \cdot V_{cm\alpha} \cdot I_{cir1\alpha}$$

$$+ \frac{2}{3} \cdot s_\varphi \cdot V_{cm\alpha} \cdot I_{cir1\beta} - \frac{2}{3} \cdot c_\varphi \cdot V_{cm\alpha} \cdot I_{cir2\alpha} - \frac{2}{3} \cdot s_\varphi \cdot V_{cm\alpha} \cdot I_{cir2\beta} - \frac{2}{3} \cdot s_\varphi \cdot V_{cm\beta} \cdot I_{cir1\alpha}$$

$$+ \frac{2}{3} \cdot c_\varphi \cdot V_{cm\beta} \cdot I_{cir1\beta} + \frac{2}{3} \cdot s_\varphi \cdot V_{cm\beta} \cdot I_{cir2\alpha} - \frac{2}{3} \cdot c_\varphi \cdot V_{cm\beta} \cdot I_{cir2\beta}$$

$$P_{b5} = +\frac{1}{6} \cdot \Delta P - \frac{1}{3} \cdot I_{cir10} \cdot V_1 - \frac{2}{3} \cdot I_{cir20} \cdot V_1 - \frac{1}{3} \cdot I_{cir1\alpha} \cdot V_2 - \frac{\sqrt{3}}{3} \cdot I_{cir1\beta} \cdot V_2 - \frac{2}{3} \cdot I_{cir2\alpha} \cdot V_2 - \frac{2 \cdot \sqrt{3}}{3} \cdot I_{cir2\beta} \cdot V_2$$

$$+ \frac{1}{4} \cdot V_{cm\alpha} \cdot I_2 - \frac{1}{3} \cdot V_{cm0} \cdot I_1 + \frac{\sqrt{3}}{4} \cdot V_{cm\beta} \cdot I_2 + \frac{2}{3} \cdot V_{cm0} \cdot I_{cir10} + \frac{4}{3} \cdot V_{cm0} \cdot I_{cir20} + \frac{2}{3} \cdot c_\varphi \cdot V_{cm\alpha} \cdot I_{cir1\alpha}$$

$$+ \frac{2}{3} \cdot s_\varphi \cdot V_{cm\alpha} \cdot I_{cir1\beta} + \frac{2}{3} \cdot c_\varphi \cdot V_{cm\alpha} \cdot I_{cir2\alpha} + \frac{2}{3} \cdot s_\varphi \cdot V_{cm\alpha} \cdot I_{cir2\beta} - \frac{2}{3} \cdot s_\varphi \cdot V_{cm\beta} \cdot I_{cir1\alpha} + \frac{2}{3} \cdot c_\varphi \cdot V_{cm\beta} \cdot I_{cir1\beta}$$

$$- \frac{2}{3} \cdot s_\varphi \cdot V_{cm\beta} \cdot I_{cir2\alpha} + \frac{2}{3} \cdot c_\varphi \cdot V_{cm\beta} \cdot I_{cir2\beta} \qquad \text{with} \qquad \cos\varphi = c_\varphi \text{ and } \sin\varphi = s_\varphi$$

References

[1] R. Marquardt et al., "Modulares Stromrichterkonzept für Netzkupplungsanwendungen," *ETG-Symposium, Bad Nauheim, Ger.*, vol. 114, 2002.

[2] A. Lesnicar and R. Marquardt, "An innovative modular multilevel converter topology suitable for a wide power range," *2003 IEEE Bol. PowerTech - Conf. Proc.*, vol. 3, 2003.

[3] M. Saeedifard and R. Iravani, "Dynamic performance of a modular multilevel back-to-back HVDC system," *IEEE Trans. Power Deliv.*, vol. 25, no. 4, 2010.

[4] K. Ilves et al., "Steady-state analysis of interaction between harmonic components of arm and line quantities of modular multilevel converters," *IEEE Trans. Power Electron.*, vol. 27, 2012.

[5] Z. Li et al., "An inner current suppressing method for modular multilevel converters," *IEEE Trans. Power Electron.*, vol. 28, no. 11, pp. 4873–4879, 2013.

[6] S. P. Engel and R. W. De Doncker, "Control of the modular multi-level converter for minimized cell capacitance," *Proc. 2011 14th Eur. Conf. Power Electron. Appl. EPE 2011*, 2011.

[7] M. Hagiwara et al., "A Medium-Voltage Motor Drive With a Modular," *IEEE Trans. Power Electron.*, vol. 25, no. 7, 2010.

[8] A. J. Korn et al., "Low output frequency operation of the Modular Multi-Level Converter," in *2010 IEEE Energy Convers. Congr. Expo.* IEEE, 2010.

[9] J. J. Jung et al., "Control strategy for improved dynamic performance of variable-speed drives with modular multilevel converter," *IEEE J. Emerg. Sel. Top. Power Electron.*, vol. 3, 2015.

[10] J. Kolb et al., "Cascaded control system of the modular multilevel converter for feeding variable-speed drives," *IEEE Trans. Power Electron.*, 2015.

[11] D. Karwatzki and A. Mertens, "Generalized Control Approach for a Class of Modular Multilevel Converter Topologies," *IEEE Trans. Power Electron.*, vol. 33, 2018.

[12] D. Karwatzki, "Analyse und Regelung einer Klasse von modularen Multilevelumrichter-Topologien," Ph.D. dissertation, 2017.

[13] D. Karwatzki et al., "Branch energy balancing with a generalised control concept for modular multilevel topologies - Using the example of the modular multilevel converter," *EPE 2016 ECCE Eur.*, 2016.

[14] J. Kucka, "Quasi-Two-Level PWM Operation for Modular Multilevel Converters : Implementation, Analysis, and Application to Medium-Voltage Motor Drives," Ph.D. dissertation, 2019.

[15] A. Ben-Israel and T. Greville, *Generalized Inverses : Theory and Applications.* Springer, 2003.

[16] J. Kolb, "Optimale Betriebsführung des Modularen Multilevel-Umrichters als Antriebsumrichter für Drehstrommaschinen," Ph.D. dissertation, 2014.

Analytical harmonic current model for a Permanent Magnet Assisted Synchronous Reluctance Motor (PMa-SynRM) fed by PWM inverter

Jessica Neumann[1,2], Carole Hénaux[1], Maurice Fadel[1], Etienne Founier[2], Dany Prieto[2], Mathias Tientcheu Yamdeu[2]

[1]LAPLACE, University of Toulouse
Toulouse, France
E-Mail: neumann@laplace.univ-tlse.fr
henaux@laplace.univ-tlse.fr
fadel@laplace.univ-tlse.fr
URL: http:// www.laplace.univ-tlse.fr

[2]LEROY SOMER - NIDEC
Angoulême, France
E-Mail: etienne.fournier@mail.nidec.com
dany.prieto@mail.nidec.com
mathias.tientcheuyamdeu@mail.nidec.com
URL: https://acim.nidec.com/motors/leroy-somer

Keywords

« Converter machine interactions », « Electrical machine », « Pulse Width Modulation (PWM) », « Harmonics ».

Abstract

This paper presents an analytical model to compute the current harmonics of a PMa-SynRM fed by PWM inverter. Voltage signal expressions are deduced and the machine parameters are provided by an analytical nonlinear model of the machine. The current signal was compared with a nonlinear Simulink model, giving satisfying results.

Introduction

Permanent magnet synchronous motors are used in many applications due to their high performance and torque density. Among the different types of synchronous motors, the PMa-SynRM combines these advantages with those of the reluctant motor that are linked to cost reduction, since a large part of the motor torque is generated by its saliency, and therefore the amount of permanent magnets can be reduced, or less powerful and cheaper ones can be used [1]. Another characteristic of the PMa-SynRM is that it enables sensorless control during starting and low speed operation [2].

Usually PMa-SynRMs have rotors with a complex geometry and therefore FE (Finite Elements) softwares are often used for modeling, but in this case, the simulation time is very significant. Analytical models have been developed, the model presented in [3] takes into account the saturation of the magnetic material, details of rotor geometry and represents the machine adequately. But, as synchronous motor is usually powered by an inverter, it is also necessary to model its effects on the motor. Harmonics created by the inverter affect the current supply signals and generate additional losses in the iron, magnets and copper of the motor. These losses can be significant as shown in [4] and must be also taken into account at the moment of the design of the system.

This paper proposes a complete analytical model for a PMa-SynRM fed by PWM inverter. Firstly, the studied system will be briefly presented as well as some principles of the SVPWM technique that are important for the development of the equations. Secondly, the improved dq model of the machine proposed by [5] will be presented, in this model saturation and cross magnetization effects are considered which allow a better representation of harmonics. And then analytical expressions of the current harmonics are deduced from the improved dq model and analytical expressions of the voltage signal.

The analytical model is validated by means of the comparison of currents curves with a numerical model in Simulink, where the machine is represented by a flux cartography to take into account the nonlinearity of the steel sheet.

2. SVPWM technique

The analyzed system constitutes of a PMa-SynRM fed by a two-stage inverter with six bidirectional switches (six associations of IGBT transistors and diodes).

The control of the switches is done by the Space Vector Pulse Width Modulation (SVPWM) technique, with the control signals of the switches at the bottom (S'u, S'v and S'w) being complementary to those of the top (Su, Sv and Sw) of the converter. The motor is a PMa-SynRM with U-shaped flux barriers filled with NdFeB permanent magnets.

The six switches of the inverter generate 8 different voltage combinations, called switch states. These voltage values in the Clark reference frame become six vectors (v_1, v_2, ..., v_6) separated by 60 degrees and two null vectors (v_0 and v_7) [6], as shown in Fig. 1.a. The region between two adjacent vectors is called sector and is numerated from i=1 to 6.

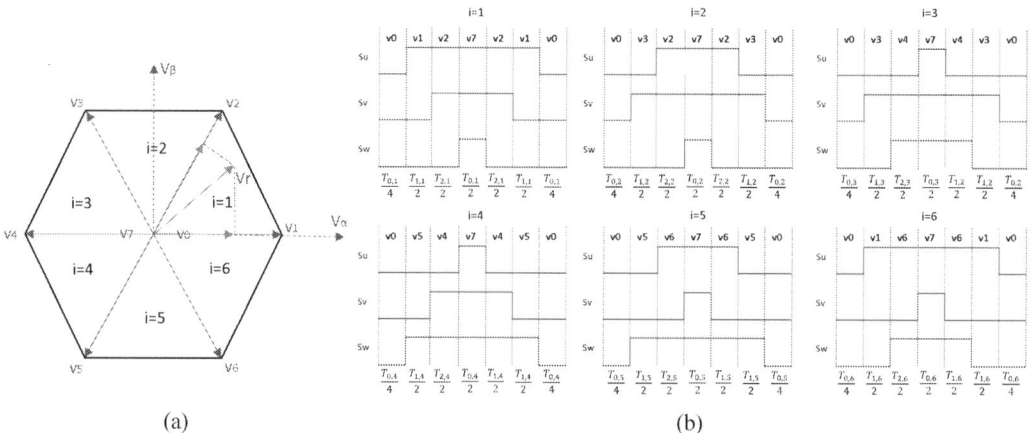

Fig. 1. SVPWM schematic: (a) represents the switch states and sectors and (b) presents the control signal profile

The main idea of SVPWM method is to determine the application time of each switch state in a PWM period (T_{PWM}) according to the control signal profile (Fig. 1.b) to reach the desired output voltage (V_r). The time of application of each state (T_1, T_2, T_3, T_4, T_5, T_6) is known and is listed in Table I for each sector i as a function of V_r.

Table I: Time applications of switch states

$T_1 = \left(\sqrt{\dfrac{3}{2}}V_{r\alpha} - \dfrac{1}{\sqrt{2}}V_{r\beta}\right)\dfrac{T_{PWM}}{U_{DC}}$ $T_2 = \sqrt{2}V_{r\beta}\dfrac{T_{PWM}}{U_{DC}}$	$T_2 = \left(\sqrt{\dfrac{3}{2}}V_{r\alpha} + \dfrac{1}{\sqrt{2}}V_{r\beta}\right)\dfrac{T_{PWM}}{U_{DC}}$ $T_3 = \left(-\sqrt{\dfrac{3}{2}}V_{r\alpha} + \dfrac{1}{\sqrt{2}}V_{r\beta}\right)\dfrac{T_{PWM}}{U_{DC}}$	$T_3 = \sqrt{2}V_{r\beta}\dfrac{T_{PWM}}{U_{DC}}$ $T_4 = \left(-\sqrt{\dfrac{3}{2}}V_{r\alpha} - \dfrac{1}{\sqrt{2}}V_{r\beta}\right)\dfrac{T_{PWM}}{U_{DC}}$
i=1	*i=2*	*i=3*
$T_4 = \left(-\sqrt{\dfrac{3}{2}}V_{r\alpha} + \dfrac{1}{\sqrt{2}}V_{r\beta}\right)\dfrac{T_{PWM}}{U_{DC}}$ $T_5 = -\sqrt{2}V_{r\beta}\dfrac{T_{PWM}}{U_{DC}}$	$T_5 = \left(-\sqrt{\dfrac{3}{2}}V_{r\alpha} - \dfrac{1}{\sqrt{2}}V_{r\beta}\right)\dfrac{T_{PWM}}{U_{DC}}$ $T_6 = \left(\sqrt{\dfrac{3}{2}}V_{r\alpha} - \dfrac{1}{\sqrt{2}}V_{r\beta}\right)\dfrac{T_{PWM}}{U_{DC}}$	$T_6 = -\sqrt{2}V_{r\beta}\dfrac{T_{PWM}}{U_{DC}}$ $T_1 = \left(\sqrt{\dfrac{3}{2}}V_{r\alpha} + \dfrac{1}{\sqrt{2}}V_{r\beta}\right)\dfrac{T_{PWM}}{U_{DC}}$
i=4	*i=5*	*i=6*

3. Improved dq model and motor's magnetic parameters determination

The analytical model of PWM harmonics will be obtained through the representation of the machine in a dq reference frame. The classic dq model of the machine in the Park referential is described as in (1).

$$v_d = Ri_d + L_d \frac{di_d}{dt} - \omega_s L_q i_q$$
$$v_q = Ri_q + L_q \frac{di_q}{dt} + \omega_s(\lambda_d^{pm} + L_d i_d) \tag{1}$$

In this model, the permanent magnet's magnetic flux presents only one component oriented according to the d-axis and it is computed by means of a simulation at no-load condition. In terms of inductances, only the self-apparent inductances are considered and they are determined as in (2). The magnetic flux generated by I_d is assumed to be the difference between the no–load magnet flux and the total flux in the d-axis.

$$L_d = \frac{\lambda_d - \lambda_d^{pm}}{I_d}$$
$$L_q = \frac{\lambda_q}{I_q} \tag{2}$$

With the saturation of the machine's magnetic material, all these assumptions are no more valid. Cross magnetization effects need to be taken into account and the inductances have to be calculated at each operation point. Reference [5] proposes an improved dq model described in (3) and based on [7] and [8], in which cross magnetization is considered through the q-axis component of the magnet flux (λ_q^{pm}) and also through the crossing apparent inductances (M_{qd} and M_{dq}).

$$v_d = Ri_d + L'_d \frac{di_d}{dt} + M'_{dq} \frac{di_q}{dt} - \omega_s(L_q i_q + M_{qd} i_d + \lambda_q^{pm})$$
$$v_q = Ri_q + L'_q \frac{di_q}{dt} + M'_{qd} \frac{di_d}{dt} + \omega_s(\lambda_d^{pm} + L_d i_d + M_{dq} i_q) \tag{3}$$

The parameters L'_d, L'_q, M'_{dq} and M'_{qd} associated with the dynamic term di/dt represent the incremental inductances and are defined as in (4), these parameters differ significantly from the apparent inductances as saturation increases [9].

$$L'_d = \frac{\Delta\lambda_d}{\Delta I_d}\bigg|_{I_q=cst} \qquad L'_q = \frac{\Delta\lambda_q}{\Delta I_q}\bigg|_{I_d=cst}$$

$$M'_{dq} = \frac{\Delta\lambda_d}{\Delta I_q}\bigg|_{I_d=cst} \qquad M'_{qd} = \frac{\Delta\lambda_q}{\Delta I_d}\bigg|_{I_q=cst} \tag{4}$$

Reference [5] presents step by step the methodology to calculate these parameters using a non-linear analytical model of the machine in which the frozen permeability method (FPM) [10] was implemented. This technique consists in fixing the magnetic state (permeability) of the machine to the operating point, and then proceed to the computation of the parameters.

4. Analytical harmonic current model

This chapter presents the equations of the analytical harmonic model and its methodology. To do so, some assumptions are done:

1. The machine is connected in a star configuration
2. The back electromotive force (Back EMF) is considered sinusoidal
3. At t=0s, the voltage of the phase U reaches its peak: the signal is even
4. The electrical period of the fundamental of the voltage signal is considered a multiple (N) of 6 of the switch period as shown in equation (5)

$$T = 6 \times N \times T_{PWM} \tag{5}$$

The expressions of harmonic currents are obtained step by step from the construction of different equations.

Initially, fully analytical expressions are deduced for the three-phase voltage signal as a function of the desired voltage V_r, switch period (T_{PWM}), sector (i) and DC voltage (U_{DC}). These expressions are then represented in the dq reference frame.

Secondly, the improved dq model of the machine presented previously in equation (3) is solved in order to find the current expressions (I_d and I_q). Finally, the inverse Park transform is applied to obtain the current in the three-phase system. Machine's parameters such as coil resistance, apparent inductances, incremental inductances and magnetic flux of permanent magnet are output data of the motor analytical model that uses the FPM.

4.1 Voltage expressions

The three-phase voltage expressions are determined by means of the Fourier series according to equation (6). For the sake of simplicity, only the expressions involving the phase U will be shown. For the phases V and W, the reasoning is the same just by subtracting 120 and 240 degrees in the phase of the signal.

$$V_u(t) = c_0 + \sum_{n=1}^{\infty} a_n \cos n\omega t + b_n \sin n\omega t \tag{6}$$

Since the voltage signal has no DC component and is even, the coefficients c_0 and b_n are zero. Therefore a_n can be determined as in equation (7).

$$a_n = \frac{2}{T} * \sum_{i=1}^{6} \left(\int_{\frac{(i-1)T}{6}}^{\frac{iT}{6}} V_u(t).\cos\left(n\frac{2\pi}{T}t\right).dt \right) \tag{7}$$

For a sector i, remembering the assumption number 4, the integral has to be calculated N times for each T_{PWM} interval term. By generalizing and developing the expression (7) according to the information of Table I, a_n can be analytically expressed by the equation (8).

$$a_n = \frac{U_{DC}}{3\pi n} * \sum_{i=1}^{6} \sum_{j=1}^{N} \left(J_i \left(\sin\left(n\frac{2\pi}{T}\left(\frac{(i-1)T}{6} + (j-1)T_{PMW} + \frac{T_{0,i}}{4} + \frac{T_{1,i}}{2} \right)\right) - \sin\left(n\frac{2\pi}{T}\left(\frac{(i-1)T}{6} + (j-1)T_{PWM} + \frac{T_{0,i}}{4}\right)\right) \right. \right.$$

$$K_i \sin\left(n\frac{2\pi}{T}\left(\frac{(i-1)T}{6} + (j-1)T_{PWM} + \frac{3T_{0,i}}{4} + T_{1,i} + T_{2,i}\right)\right) - \sin\left(n\frac{2\pi}{T}\left(\frac{(i-1)T}{6} + (j-1)T_{PWM} + \frac{3T_{0,i}}{4} + \frac{T_{1,i}}{2} + T_{2,i}\right)\right) \right) +$$

$$K_i \left(\sin\left(n\frac{2\pi}{T}\left(\frac{(i-1)T}{6} + (j-1)T_{PWM} + \frac{T_{0,i}}{4} + \frac{T_{1,i}}{2} + \frac{T_{2,i}}{2}\right)\right) - \sin\left(n\frac{2\pi}{T}\left(\frac{(i-1)T}{6} + (j-1)T_{PWM} + \frac{T_{0,i}}{4} + \frac{T_{1,i}}{2}\right)\right) \right) +$$

$$\left. \left. J_i \sin\left(n\frac{2\pi}{T}\left(\frac{(i-1)T}{6} + (j-1)T_{PWM} + \frac{3T_{0,i}}{4} + \frac{T_{1,i}}{2} + T_{2,i}\right)\right) - \sin\left(n\frac{2\pi}{T}\left(\frac{(i-1)T}{6} + (j-1)T_{PWM} + \frac{3T_{0,i}}{4} + \frac{T_{1,i}}{2} + \frac{T_{1,i}}{2}\right)\right) \right) \right) \quad (8)$$

Where the coefficients J_i and K_i are functions of the sector i as shown in (9).

$$J(i) = \sqrt{3}\cos\left(\frac{\pi}{3}\left(i - \frac{1}{2}\right)\right) + \cos\left(\frac{2\pi}{3}\left(i - \frac{1}{2}\right)\right)$$

$$K(i) = -\sqrt{3}\cos\left(\frac{\pi}{3}\left(i - \frac{7}{2}\right)\right) + \cos\left(\frac{2\pi}{3}\left(i - \frac{7}{2}\right)\right) \quad (9)$$

$T_{0,i}$ represents expressions for application times for the null vectors, $T_{1,i}$ corresponds to the application time of the vectors v_1, v_3 and v_5 in their respective sectors and $T_{2,i}$ represents the application time of the vectors v_2, v_4 and v_6 in their respective sectors. These application times can be expressed as in (10).

$$T_1(i) = \frac{T_{SW}}{U_{DC}} * \left(\left(\frac{1 + (-1)^{i+1}}{2}\right) * \left(2\sqrt{6}V_\alpha\left((j-1)T_{SW} + \frac{T_{SW}}{2}\right) - \sqrt{2}V_\beta\left((j-1)T_{SW} + \frac{T_{SW}}{2}\right)\right) \right.$$

$$\left. + \left(\frac{1 + (-1)^{i}}{2}\right) * \left(\sqrt{2}V_\beta\left((j-1)T_{SW} + \frac{T_{SW}}{2}\right)\right) \right)$$

$$T_2(i) = \frac{T_{SW}}{U_{DC}} * \left(\left(\frac{1 + (-1)^{i}}{2}\right) * \left(2\sqrt{6}V_\alpha\left((j-1)T_{SW} + \frac{T_{SW}}{2}\right) - \sqrt{2}V_\beta\left((j-1)T_{SW} + \frac{T_{SW}}{2}\right)\right) \right. \quad (10)$$

$$\left. + \left(\frac{1 + (-1)^{i+1}}{2}\right) * \left(\sqrt{2}V_\beta\left((j-1)T_{SW} + \frac{T_{SW}}{2}\right)\right) \right)$$

$$T_0(i) = T_{SW} - T_1(i) - T_2(i)$$

Knowing the Fourier coefficient a_n, the equations (6) and (7) are completed. It is therefore necessary to apply the Park transform to the V_u, V_v and V_w expressions to obtain the V_d et V_q as shown in (11).

$$\begin{bmatrix} V_d \\ V_q \end{bmatrix} = k \begin{bmatrix} \cos\theta & \cos\left(\theta - \frac{2\pi}{3}\right) & \cos\left(\theta - \frac{4\pi}{3}\right) \\ -\sin\theta & -\sin\left(\theta - \frac{2\pi}{3}\right) & -\sin\left(\theta - \frac{4\pi}{3}\right) \end{bmatrix} \begin{bmatrix} V_u \\ V_v \\ V_w \end{bmatrix} = \begin{bmatrix} V_{d0} + \sum_{n=2}^{+\infty} V_{n\,d\gamma}\cos(3(n-1)\omega t) + V_{n\,d\sigma}\sin(3(n-1)\omega t) \\ V_{q0} + \sum_{n=2}^{+\infty} V_{n\,q\gamma}\cos(3(n-1)\omega t) + V_{n\,q\sigma}\sin(3(n-1)\omega t) \end{bmatrix} \quad (11)$$

Where $\theta = \theta_0 + \omega t$, and θ_0 and ω are the initial position and speed (electrical values) of the rotor, k is the coefficient of the Park transform, and V_{d0}, V_{q0}, $V_{nd\gamma}$, $V_{nd\sigma}$, $V_{nq\gamma}$, $V_{nq\sigma}$ depend on the Fourier coefficients.

4.2 System of equations solution

To solve the system of equations (3) for the currents I_d and I_q, it is assumed that each voltage harmonic generates a harmonic of the same rank in the current, this method is called *dq Harmonic Balance* and was introduced by [11] and reproduced by [12]. Therefore, the solution will have a format as shown in (12).

$$
\begin{bmatrix} I_d \\ I_q \end{bmatrix} = \begin{bmatrix} I_{d0} + \displaystyle\sum_{n=2}^{+\infty} I_{n\,d\gamma}\,cos(3(n-1)\omega t) + I_{n\,d\sigma}\,sin(3(n-1)\omega t) \\[2mm] I_{q0} + \displaystyle\sum_{n=2}^{+\infty} I_{n\,q\gamma}\,cos(3(n-1)\omega t) + I_{n\,q\sigma}\,sin(3(n-1)\omega t) \end{bmatrix}
\tag{12}
$$

Replacing the expressions (11) and (12) in (3), the problem can be divided in two systems, one for the continuous dq components (13) and another for the harmonics (14). Therefore, expressions for I_{d0}, I_{q0}, $I_{n\,d\sigma}$, $I_{n\,d\gamma}$, $I_{n\,q\sigma}$, $I_{n\,q\gamma}$ were determined by solving these two linear systems shown below.

$$
\begin{bmatrix} V_{d0} + \omega\varphi_{q\,pm} \\ V_{q0} - \omega\varphi_{d\,pm} \end{bmatrix} = \begin{bmatrix} R - \omega L_{qd} & -\omega L_q \\ \omega L_d & R + \omega L_{dq} \end{bmatrix} \begin{bmatrix} I_{d0} \\ I_{q0} \end{bmatrix}
\tag{13}
$$

$$
\begin{bmatrix} V_{nd\gamma} \\ V_{nd\sigma} \\ V_{nq\gamma} \\ V_{nq\sigma} \end{bmatrix} = \begin{bmatrix} R - \omega L_{qd} & 3(n-1)\omega L'_d & -\omega L_q & 3(n-1)\omega L'_{dq} \\ -3(n-1)\omega L'_d & R - \omega L_{qd} & -3(n-1)\omega L'_{dq} & -\omega L_q \\ \omega L_d & 3(n-1)\omega L'_{qd} & R + \omega L_{dq} & 3(n-1)\omega L'_q \\ -3(n-1)\omega L'_{qd} & \omega L_d & -3(n-1)\omega L'_q & R + \omega L_{dq} \end{bmatrix} \begin{bmatrix} I_{nd\gamma} \\ I_{nd\sigma} \\ I_{nq\gamma} \\ I_{nq\sigma} \end{bmatrix}
\tag{14}
$$

Finally, the inverse Park transform is applied to I_d and I_q expressions in order to determine the three-phase currents.

5. Validation of analytical model

5.1 Numeric model

A numerical model was developed in Simulink/Matlab for means of comparison, its general scheme is shown in Fig.2. The SVPWM technique and the inverter were implemented by Simulink block functions.

To represent the machine in a more precise way in Simulink/Matlab, the method presented in [12] was implemented for the PMa-SynRM. This method consists of carrying out FE simulations with different levels of current components I_d and I_q and retaining the values of the coils magnetic flux components λ_d and λ_q. So, two maps of the flux components as a function of the current components $\lambda_d(I_d,I_q)$ and $\lambda_q(I_d,I_q)$ are built and illustrated in Fig. 3.a. These first cartographies are then inverted to obtain current maps as a function of the flux $I_d(\lambda_d, \lambda_q)$ and $I_q(\lambda_d, \lambda_q)$ as shown in Fig. 3.b, which are implemented in the Simulink by means of a lookup table.

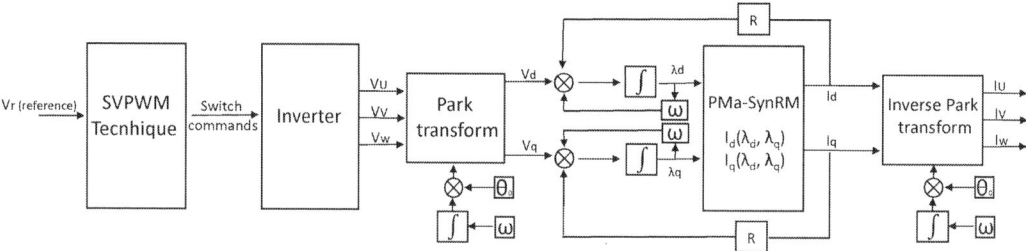

Fig. 2: Scheme model for comparison

The inputs of the motor block λ_d and λ_q are obtained by integrating the voltage supply components as shown in Fig. 3 and equation (15). By means of the cartographies, the corresponding d,q currents are determined and the inverse Park transform is applied in order to obtain the three-phase currents signals with the harmonic components due to the PWM supply.

$$\lambda_d = \int v_d - Ri_d + \omega\lambda_q \ dt$$

$$\lambda_q = \int v_q - Ri_q - \omega\lambda_q \ dt \tag{15}$$

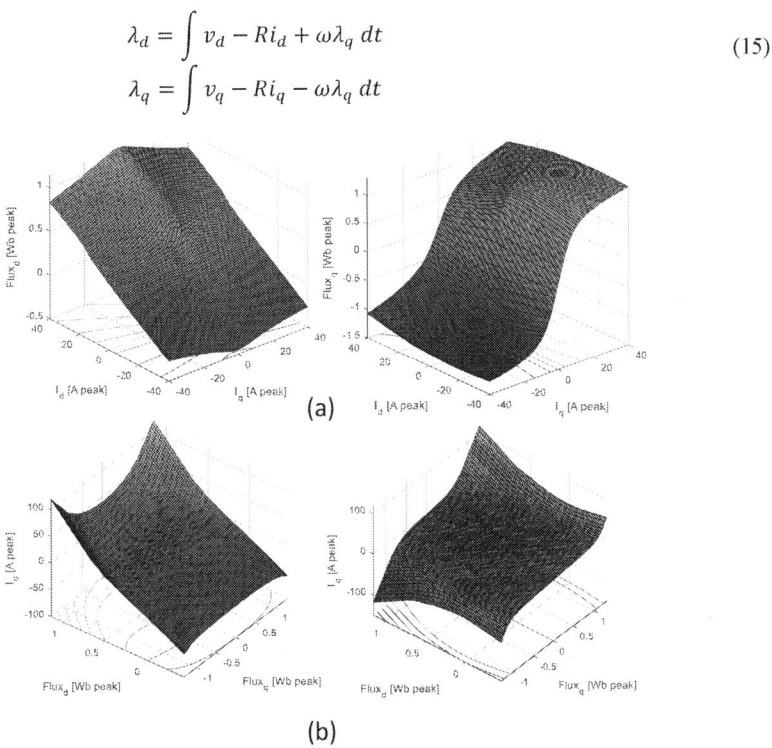

(a)

(b)

Fig. 3: Cartography: (a) λ_d (I_d,I_q) and λ_q (I_d,I_q) (b) I_d (λ_d,λ_q) and I_q (λ_d,λ_q)

5.2 Comparison of results

The analytical model was compared to the Simulink model by employing a fundamental and switching frequency of 50Hz and 3kHz respectively, DC bus voltage of 540V, reference voltage of 100V$_{rms}$ and the sum of the equation (6) goes from 1 to 1000. The motor parameters present in the improved dq model are listed in table II.

Table II: Motor parameters

R (Ω)	L_d (mH)	L_q (mH)	M_{dq} (mH)	M_{qd} (mH)	L'_d (mH)	L'_q (mH)	M'_{dq} (mH)	M'_{qd} (mH)	λ_d^{pm} (Wb)	λ_q^{pm} (Wb)
0.1	8.71	39.1	1.69	1.71	11.9	4.73	-1.5	-1.2	0.51	-0.13

In relation to the Simulink model, the PWM parameters are the same and the flux cartography covers an operation from -40 to 40A. Fig. 4a and Fig. 4b compare the voltage and three-phase current of the two models, zoom of the signals are shown in the same figure.

The curves of the two models overlap, meaning that the analytical expressions found for the harmonics generated by the inverter are correct. The peaks in the voltage curve in the analytic model are due to the Gibbs phenomenon.

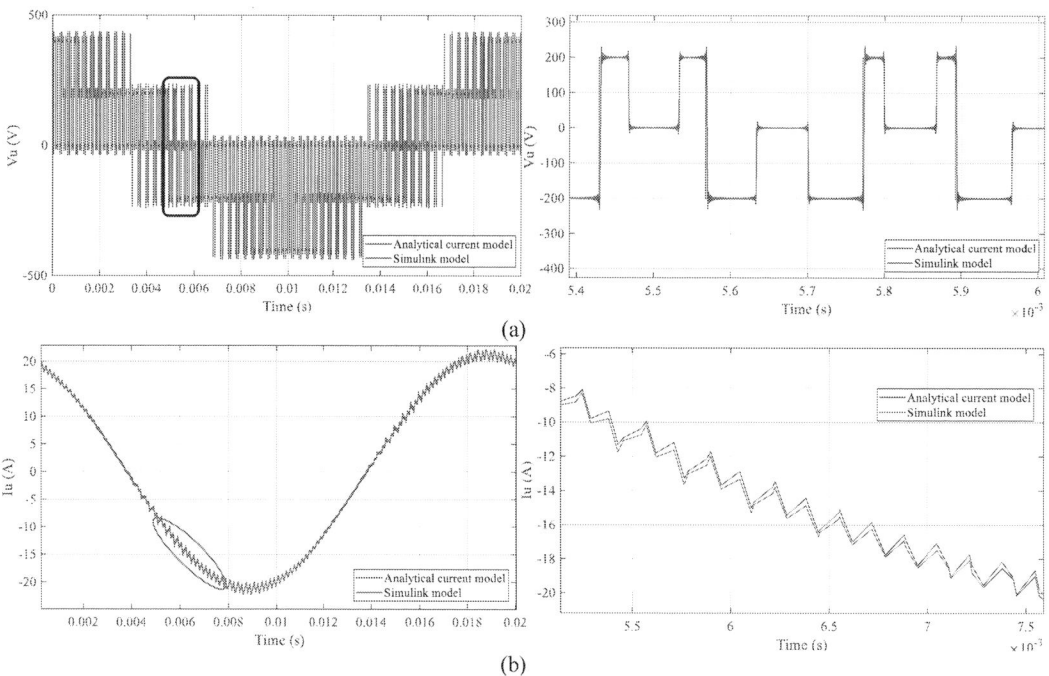

Fig. 4: Comparison between (a) voltage and (b) current curves from Simulink and Analytical models. On the left, one period of the waveform, on the right, the highlighted parts

To analyze the current harmonics, the Fourier transform is applied to the current signal obtained by the two models. The PWM harmonics are around the multiples of the sampling frequency, the more relevant ones are presented in Fig. 5.

The first spectrum shows the harmonics of the current signal from Fig. 4.b. The next ones, Fig.5.b and Fig.5.c, correspond to operating points with higher current levels of 22.5 and 30 A_{rms}. According to this analysis, the model remains representative regardless the current level, even under saturation conditions, bearing a constant average error of 17% for all tested current levels. The error does not increase with the saturation due to the improved dq model incorporated into the analytical expressions, which is able to reproduce the correct behavior of the motor.

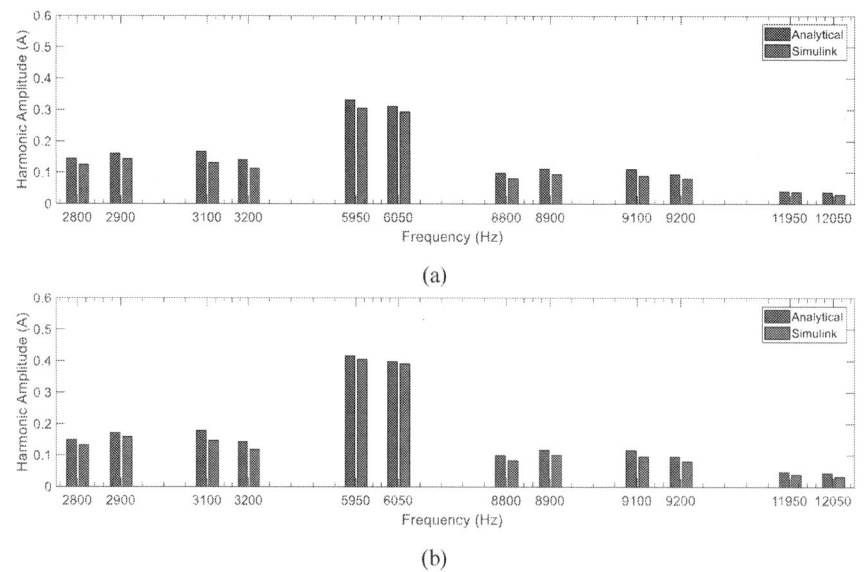

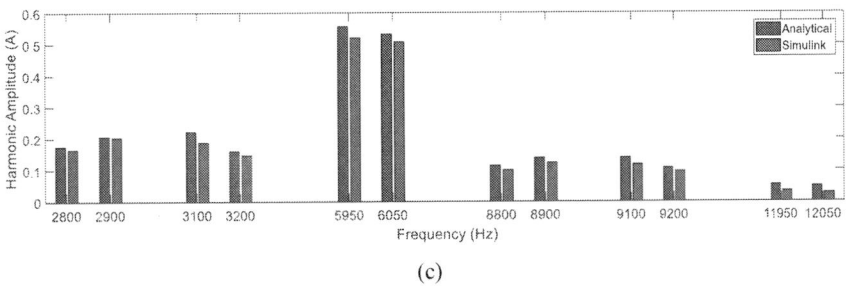

(c)

Fig. 5: PWM current harmonics obtained by analytical and Simulink models at the operating point of (a) 15 A_{rms} (b) 22.5 A_{rms} and (c) 30A_{rms}

The harmonics obtained by the analytic model were also evaluated by employing the Total Harmonic Distortion (THD) of the current curves. The THD is calculated according to its definition in equation (16) and the results are presented in table III. The largest difference between the two models corresponds to 12 % for 15A_{rms}.

$$THD = \frac{\sqrt{\sum(i_{k_{rms}}^2)}}{i_{1_{rms}}} \qquad (16)$$

Table III: THD of current signal

Current level (A_{rms})	Analytical model	Simulink model	Error
15	0,028	0,025	12%
22,5	0,022	0,021	6%
30	0,021	0,020	7%

6. Conclusion

The analytical equations and their assumptions were presented by introducing the development of the analytical voltage expressions in three phase and dq reference frame, the improved dq model and, in sequence, the current harmonics.

The analytical model was validated by means of comparison with a Simulink model of the THD and the harmonics amplitude of the current signal, providing satisfying results and bearing an "instantaneous" computational time, i.e. it does not need iterative steps and hence it is largely faster. This gain in speed is interesting not only in a simulation perspective, but also to optimization applications that usually need to perform a high number of simulations and thus require a large computation time.

The model proved also to be precise for different current levels, therefore robust to the effects of saturation, thanks to the representativeness given to the machine parameters in the dq model incorporated in the equations.

7.References

[1] S. S. Reddy Bonthu, M. Z. Islam, and S. Choi, "Performance Review of Permanent Magnet assisted Synchronous Reluctance Traction Motor Designs," in 2018 IEEE Energy Conversion Congress and Exposition (ECCE), 2018.

[2] E. Armando, P. Guglielmi, G. Pellegrino, M. Pastorelli, and A. Vagati, "Accurate Modeling and Performance Analysis of IPM-PMASR Motors," IEEE Transactions on Industry Applications, vol. 45, no. 1, pp. 123–130, 2009.

[3] D. Prieto, P. Dessante, J.-C. Vannier, X. Jannot, and J. Saint-Michel, "Analytical model for a saturated Permanent Magnet Assisted Synchronous Reluctance Motor," in 2014 International Conference on Electrical Machines (ICEM), 2014.

[4] O. Payza, Y. Demir, and M. Aydin, "Investigation of Losses for a Concentrated Winding High-Speed Permanent Magnet-Assisted Synchronous Reluctance Motor for Washing Machine Application," IEEE Transactions on Magnetics, vol. 54, no. 11, pp. 1–5, Nov. 2018.

[5] J. Neumann, C. Henaux, M. Fadel, D. Prieto, E. Fournier, and M. T. Yamdeu, "Improved dq model and analytical parameters determination of a Permanent Magnet Assisted Synchronous Reluctance Motor (PMa-SynRM) under saturation using frozen permeability method," in 2020 International Conference on Electrical Machines (ICEM), 2020.

[6] D. G. Holmes and T. A. Lipo, Pulse Width Modulation for Power Converters. New York: IEEE/Wiley-Interscience, 2003

[7] A. Pouramin, R. Dutta, M. F. Rahman, J. E. Fletcher, and D. Xiao, "A preliminary study of the effect of saturation and cross-magnetization on the inductances of a fractional-slot concentrated-wound interior PM synchronous machine," in 2015 IEEE 11th International Conference on Power Electronics and Drive Systems, 2015, doi: 10.1109/peds.2015.7203522.

[8] S. Zarate, G. Almandoz, G. Ugalde, J. Poza, and A. J. Escalada, "Extended DQ model of a Permanent Magnet Synchronous Machine by including magnetic saturation and torque ripple effects," in 2017 IEEE International Workshop of Electronics, Control, Measurement, Signals and their Application to Mechatronics (ECMSM), 2017, doi: 10.1109/ecmsm.2017.7945881.

[9] B. Stumberger, G. Stumberger, D. Dolinar, A. Hamler, and M. Trlep, "Evaluation of saturation and cross-magnetization effects in interior permanent-magnet synchronous motor," IEEE Transactions on Industry Applications, vol. 39, no. 5, pp. 1264–1271, Sep. 2003, doi: 10.1109/tia.2003.816538.

[10] X. Chen, V. I. Patel, J. Wang, P. Lazari, L. Chen, and P. Lombard, "Reluctance Torque Evaluation for Interior Permanent Magnet Machines Using Frozen Permeability," in 7th IET International Conference on Power Electronics, Machines and Drives (PEMD 2014), 2014

[11] T. A. Lipo, "Performance Calculations of a Reluctance Motor Drive by dq Harmonic Balance," IEEE Transactions on Industry Applications, vol. IA-15, no. 1, pp. 25–35, Jan. 1979.

[12] X. Jannot, J.-C. Vannier, M. Gabsi, C. Marchand, J. Saint-Michel, and D. Sadarnac, "Steady state performance computation of a synchronous machine using harmonic resolution," in 2010 IEEE International Symposium on Industrial Electronics, 2010.

[13] X. Chen, J. Wang, B. Sen, P. Lazari, and T. Sun, "A High-Fidelity and Computationally Efficient Model for Interior Permanent-Magnet Machines Considering the Magnetic Saturation, Spatial Harmonics, and Iron Loss Effect," IEEE Transactions on Industrial Electronics, vol. 62, no. 7, pp. 4044–4055, Jul. 2015.

Generalized Small-Signal Averaged Switch Model Analysis of a WBG-based Interleaved DC/DC Buck Converter for Electric Vehicle Drivetrains

Sajib Chakraborty[a,b], Dai-Duong Tran[a,b], Joeri Van Mierlo[a,b], Omar Hegazy[a,b*]

[a] Vrije Universiteit Brussel (VUB), Pleinlaan 2, 1050 Brussels, Belgium

[b] Flanders Make, 3001 Heverlee, Belgium

* Corresponding author: omar.hegazy@vub.be

Acknowledgments

This project (HiFi-Elements) has received funding from the European Union's Horizon 2020 research and innovation program under Grant Agreement no. 769935. The authors acknowledge all the project partners involved in the HiFi-Elements project. The authors also acknowledge Flanders make for the support to this research group.

Keywords

«Interleaved Buck converter», «Dual-loop control», «Electric vehicles», «Generalized small-signal», «SiC».

Abstract

To achieve a high-performance index and accurate controllability of power electronics converters (PEC), generalized small-signal analysis of the closed-loop power electronics converters is a key aspect. This article presents a detailed generalized small-signal averaged switch model (GSSASM) of a Wide Bandgap (WBG)-based Interleaved DC/DC Buck converter (IBC) for electric vehicle (EV) drivetrains to better understand the circuit characteristics, performance, stability and control systems. The derived GSSASM considers the power electronics device (e.g., switch & diode) and passive components (inductor, capacitor and internal resistance of input source), which can result in an accurate mathematical model representing the real-time (RT) system. The proposed model can be utilized for any number of phases in the IBC systems. Furthermore, a field-programmable gate array-based (FPGA) programming board of dSPACE MicroLabBox, is used to validate the proper control compensators towards high dynamic performance in the Real-Time Workshop (RTW)-system. Finally, the performance of the proposed mathematical model is verified with a 30-kW SiC-based IBC prototype in both transient and steady-state conditions, respectively.

Introduction

Interleaved DC/DC Buck converter (IBC) has become popular for vehicular drivetrains, photovoltaic microgrids, power distribution systems, aerospace power systems and charger applications [1]–[4]. These power electronics (PE) converters consider as highly non-linear systems due to the presence of at least one non-linear component, i.e., a switch (IGBT/MOSFET) or a diode [5]. These PE DC/DC converters, regulate the DC output voltage/current to stabilize source and load variations using a feedback control circuit. The performance of PEC highly depends on controller stability during steady-state and transient conditions [6]. In literature, the state-space averaging (SSA) technique is used to model the PE DC/DC converters, which requires trisome or matrix manipulation due to the presence of large numbers of parasitic components and passive components [7]. Moreover, the Fourier signal representation of the SSA comprises higher-order harmonics that influence the controller response during transient [8]. Furthermore, Lunze transformation is proposed in [9], [10], for PE converter's controller modelling. This method becomes more difficult to investigate when the number of active phases in the DC/DC converter is increased.

Thus, in this paper, the GSSASM formulates an averaged mathematical model of the non-linear network, which is utilized to design k-factor-based type-II/type-III controller. Hence, this controller is attained fast-dynamic response during source and load variation of the converter plant [11]. Furthermore, in the GSSASM, the zero-order harmonic is the main effective value; hence, the controller response is seamless at the system during transient periods [8].

In this paper, the controller design is based on a stepwise redesign process. The entire transfer function (plant and feedback transfer function) obtained from the Laplace domain (s-domain) is transformed into discrete domain(z-domain) for a real-time (RT) FPGA implementation dSPACE MicroLabBox. Finally, the interleaved buck converter using a k-factor-based type-II controller is experimentally validated with a 30kW SiC-based prototype in both transient and steady-state conditions.

Averaged Switch Modelling Technique (ASMT)

In the ASMT, the non-linear discrete components (switches and diodes) of the PE converters are replaced by average current and voltage sources respectively to emulate the average high-frequency performances. In contrast, linear analog components (e.g., inductor and capacitor with parasitic) have remained unchanged. The average actual switch model network of 3-phase interleaved DC/DC buck converter is shown in Figure 1, where the global unit represents all parameters. Linear small-signal perturbation equations (1) and (2) are obtained at the operating points by expanding Taylor's series, while higher-order terms are ignored.

$$I_{s1} + i_{s1} = (\hat{d} + D)(I_{L1} + i_L) = \hat{d}I_{L1} + \hat{d}i_L + DI_{L1} + Di_L \tag{1}$$

$$V_{LD1} + v_{LD1} = (\hat{d} + D)(V_{SD1} + v_{SD1}) = \hat{d}V_{SD1} + \hat{d}v_{SD1} + DV_{SD1} + Dv_{SD1} \tag{2}$$

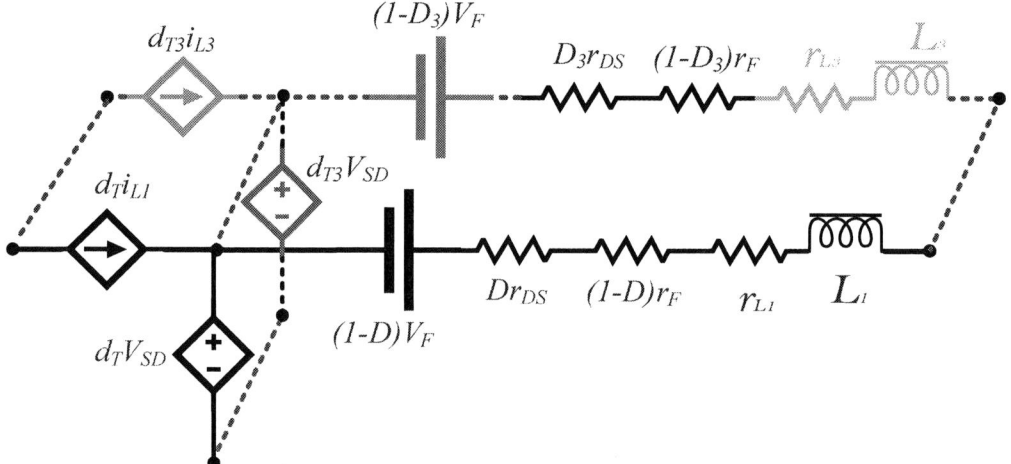

Figure 1. Averaged switch network modelling of the Interleaved DC/DC buck converter.

Generalized Small-Signal Averaged Switch Model Analysis (GSSASM)

The generalized small-signal model of the proposed IBC in continuous current mode (CCM) is shown in Figure 2 with two independent sources $\hat{V}_{in}(s)$ and $d(s)$, where $d(s)=0$ is to formulate audio-susceptibility transfer function. In this study, all phase parameters are considered unified for simple calculation as: $L_1=L_2=L_3=L$, $r_{L1}=r_{L2}=r_{L3}=r_L$, $I_{L1}=I_{L2}=I_{L3}=I_L$.

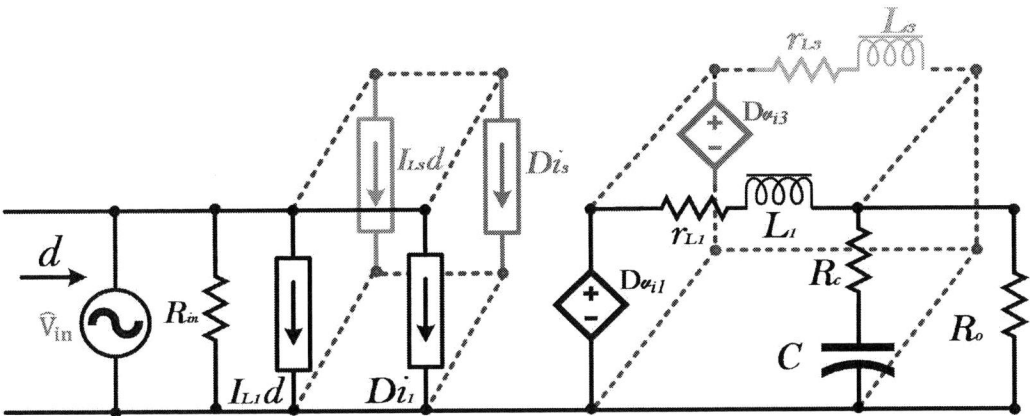

Figure 2. The generalized small-signal model of Interleaved bidirectional DC/DC buck converter for determining $v_o(s)/d(s)$ and $i_L(s)/d(s)$ transfer function.

The GSSASM transfer functions of duty cycle to output voltage in CCM are formulated as (3) to (5):

Transfer Function: $G_{vd} = \dfrac{\hat{\vartheta}_o(s)}{\hat{d}(s)} = G_{dv}\dfrac{(s+\omega_{zv})}{s^2+2\varepsilon\omega_n s+\omega_n^2}$ (3)

Gain: $G_{dv} = \dfrac{\vartheta_o(s).R_o.R_c}{(n.m).D(R_o+R_c).L}$, where n=number of phase and m=parallel switch/phase (4)

ESR Zero: $\omega_{zv} = \dfrac{1}{C.R_c}$; (5)

The SSASM transfer functions of duty cycle to inductor current in CCM are formulated as (6) to (8):

Transfer Function: $G_{id} = \dfrac{\hat{i}_L(s)}{\hat{d}(s)} = G_{di}\dfrac{(s+\omega_{zi})}{s^2+2\varepsilon\omega_n s+\omega_n^2}$ (6)

Gain: $G_{di} = \dfrac{\vartheta_o}{DL}$ (7)

ESR Zero: $\omega_{zi} = \dfrac{1}{(R_o+R_c)}$ (8)

The natural frequency, damping ratio, quality factor and poles of the transfer functions are shown in (9) to (12):

Angular damped Natural frequency: $\omega_n = \sqrt{\dfrac{r_L+R_o}{LC(R_o+R_c)}}$ (9)

Damping Ratio: $\varepsilon = \dfrac{C(r_L R_o+r_L R_c+R_o R_c)}{2\sqrt{LC(R_o+R_c)(r_L+R_o)}}$ (10)

Quality Factor: $Q = \dfrac{1}{2\varepsilon}$ (11)

Poles: $p_1, p_2 = -\varepsilon\omega_n \pm j\omega_n\sqrt{1-\varepsilon^2}$ (12)

Cascaded Dual-loop Controller Design

Figure 3 shows the cascaded control configuration of the IBC. The fast-inner loops are used to regulate the average inductor current, I_{L1}, I_{L2} and I_{L3}. The continuous duty ratios from the controllers are then sent to the PWM modules for the generation of the switching signals for six power semiconductor modules. A phase shift of $1/3f_{SW}$ is added between adjacent PWMs to implement the interleaved switching method. The output voltage is controlled using the slower outer loop using a type-II lead-lag controller. The advantage of the dual loop is that it decouples the two loops and requires only a single lead-lag controller per loop [12].

The loop gains for expressing the outer voltage control loop and inner current control loop are formulated as (13) and (14):

$$T_i(s) = H_i(s).G_{di}(s) \tag{13}$$

$$T_v(s) = \frac{H_V(s).H_i(s).G_{vd}(s)}{1 + T_i(s)} \tag{14}$$

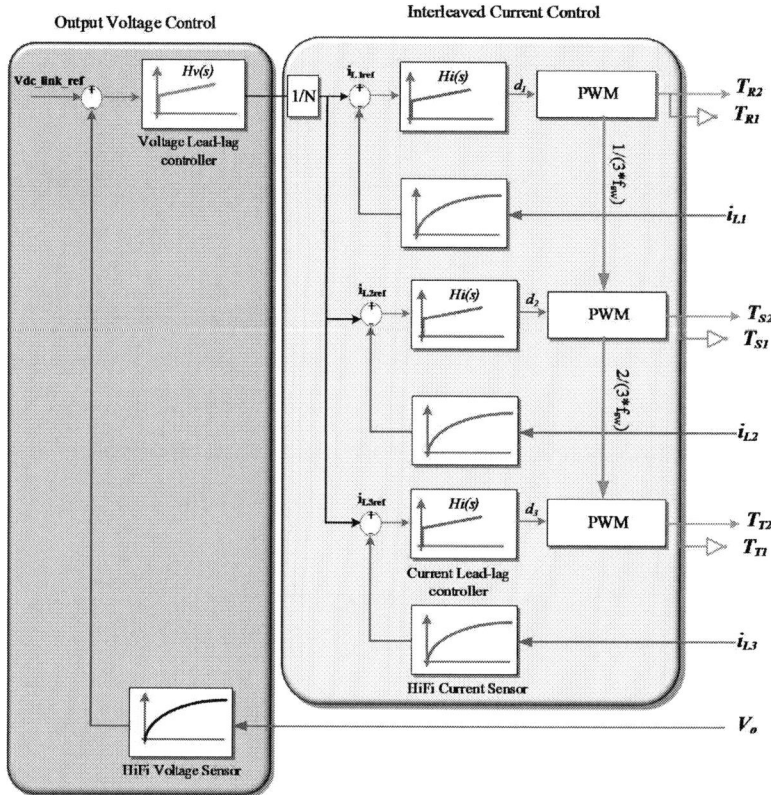

Figure 3. Cascaded dual-loop controller design of the IBC using interleaving technique.

Type-II Controller Design Based on 'k-factor' Technique

k-factor is a mathematical concept illustrated in [11] for stability analysis and synthesis of a control loop quickly and efficiently by solving a few straight-forward algebraic equations. The k-factor is defined as the ratio of the pole frequency to the zero frequency, can be used to derive equations that express the locations of the zeros and poles as a function of phase boost. Given the desired phase margin (Φ_M), the phase shift (Φ_S) of the converter and the cross-over frequency (f_C) of the converter, the following equations (15)- (20) show how to utilize the k-factor for a type-II controller:

Phase boost: $\Phi_{B\_II} = \Phi_M - (\Phi_s - 360^0) - 90^0$ $\tag{15}$

k-factor: $K = (tan \frac{\Phi_{B\_II}}{2} + 45^0)$ $\tag{16}$

Pole frequency: $f_p = f_C \times K$ (17)

Zero frequency: $f_z = \dfrac{f_C}{K}$ (18)

Frequency margin: $f_M = f_p - f_z$ (19)

Controller Gain: $K_C = \left| \dfrac{1}{T(s)} \right|_{f=f_C}$ (20)

The frequency margin is the region where the converter gain will remain constant. The zero and poles can be incorporated into the transfer function of the plant and are given in (21) and (22).

Compensated Type-II current-loop controller: $Gc_{iL}(s) = \dfrac{s+5351}{0.002s^2+133.5s}$ (21)

Compensated Type-II voltage-loop controller: $Gc_{Vd}(s) = \dfrac{s+1942}{0.00021s^2+0.3891s}$ (22)

The bode plot in Figure 4 indicates that the faster current control loop becomes stable with a phase margin of 45^0 and a gain margin of 0.5dB, while the slower outer voltage loop stables with a phase margin of 45^0 and a gain margin of 20.7dB.

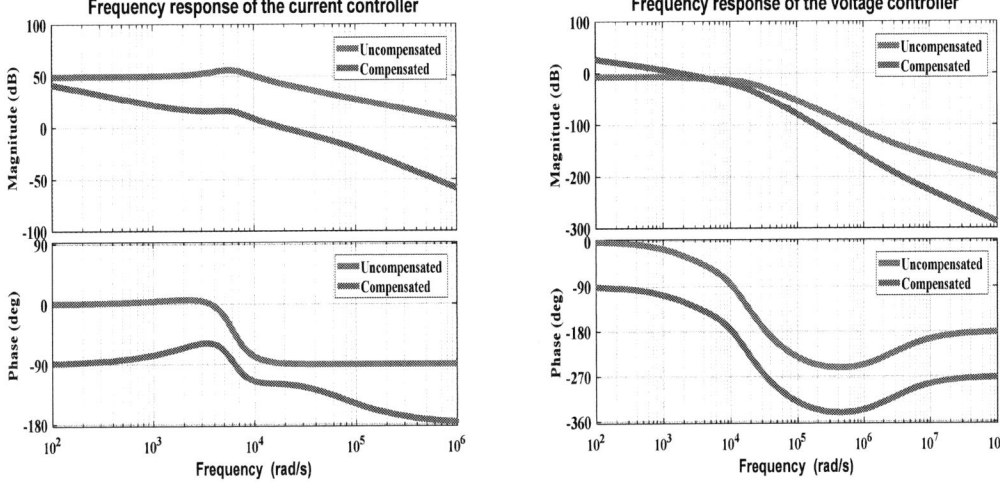

Figure 4. The frequency response of the type-II current and voltage loop controllers at a duty ratio of 50%.

To develop the discrete controller of the IBC, the compensated Type-II voltage control loop $Gc_{vd}(s)$ and current control loop $Gc_{iL}(s)$ are discretized by utilizing Zero-Order-Hold (ZOH) technique. The discrete controller is designed in the z-domain using the time-frequency response method. It offers good agreement among the continuous and discretized transfer function. The discrete-time transfer functions $Gc_{vd}(z)$ and $Gc_{iL}(z)$ comprise the ZOH and computational delay e^{-sT_d} are shown as equations (23)-(24).

Discrete voltage loop controller: $Gc_{vd}(z) = Z\left\{ \dfrac{1-e^{-sT_s}}{s} \cdot e^{-sT_d} \cdot Gc_{vd}(s) \right\}$ (23)

Discrete current loop controller: $Gc_{iL}(z) = Z\left\{ \dfrac{1-e^{-sT_s}}{s} \cdot e^{-sT_d} \cdot Gc_{iL}(s) \right\}$ (24)

Simulation and Experimental Results

The High-Fidelity (HiFi) model mentioned in [5], is used to test the IBC controller's behaviour in steady-state and transient conditions. Figure 5 depicts the simulation results of the IBC controller during load power and source voltage variation. The simulation performances of the IBC due to load and source variations validate the mathematical GSSASM model.

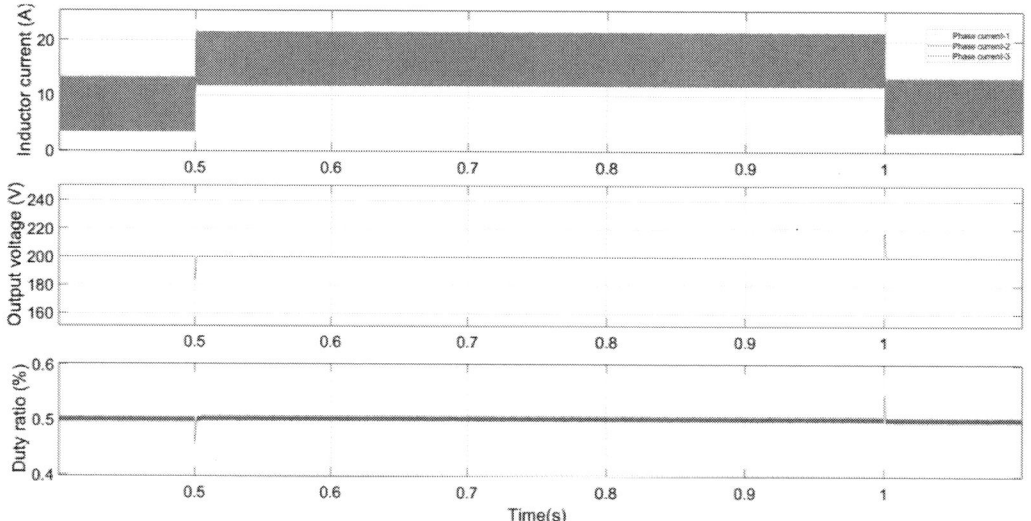

Figure 5. The simulation waveforms of the dynamic response of the IBC during load power variations from 5kW $\rightarrow$ 10kW $\rightarrow$ 5kW at V_{BAT}=395V, V_0=200V.

For the RT dual-loop discrete controller implementation, a dSPACE MicroLabBox based Xilinx® 7 series FPGA programming board is used. The current controller's cutoff frequency is remained at 3kHz, while the voltage controller's cutoff frequency is maintained at 300 Hz. The interleaved inductor's current response of the IBC is shown in Figure 6 (b) at V_{BAT}=395V, V_0=200V & load power 22kW, while Figure 6 (a) shows the converter prototype. The interleaved inductor current response of the RT system validates the susceptibility of controller design at 60kHz switching frequency.

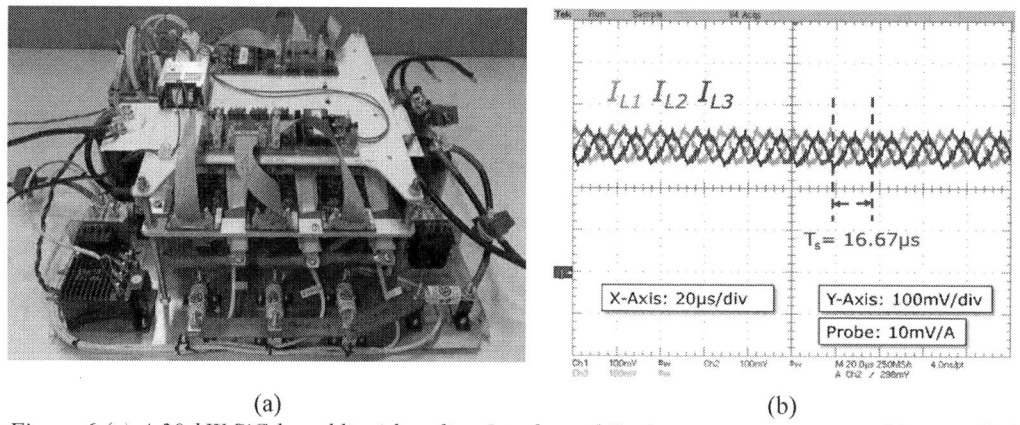

(a)   (b)

Figure 6.(a) A 30-kW SiC-based liquid cooling Interleaved Buck converter prototype, (b) controlled interleaved inductor current at fsw=60kHz and P_o=22kW, the phase difference between $I_{L1,}$ I_{L2} and I_{L3} is 120⁰.

Figure 7 depicts the transient response of the interleaved inductor's current and output voltage at V_{BAT}=395V, V_0=200V & load power changes from 5kW $\rightarrow$ 10kW. The experimental results show a high degree of accuracy with the GSSAM mathematical model expectations, as illustrated in Table 1.

Hence, from the output response, it can be verified that the perfect constant output impedance of the IBC is achieved utilizing the compensators (23) and (24).

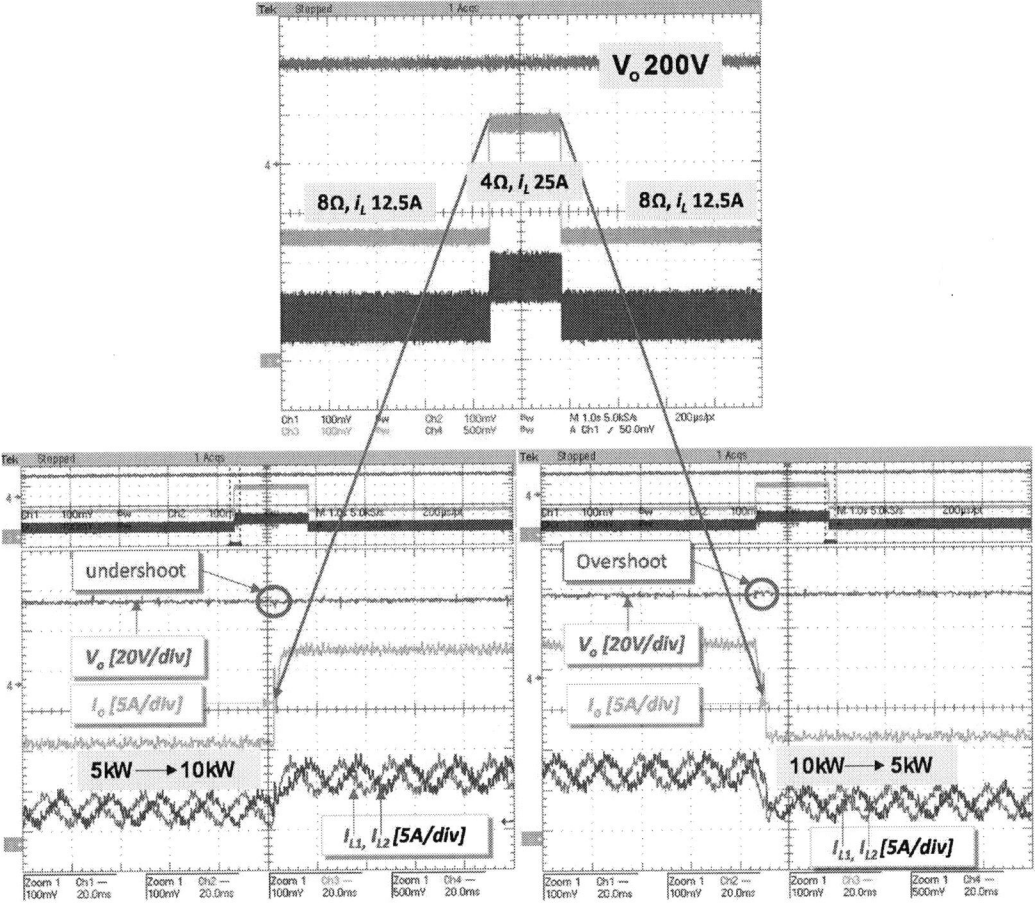

Figure 7. Experimental waveforms of SiC-based Interleaved Buck converter under load variation from $5kW \rightarrow 10kW \rightarrow 5kW$ at V_{BAT}=395V, V_0=200V, where the top figure is shown the entire dynamic load variation, and bottom figures are zoomed at the load transitions period. Here pink color represents the output voltage (200V), green color represents total output current, while red and blue represent interleaved inductor's current of phase1 and phase2 (120^0 phase difference).

The closed-loop performances of the output voltage responses of the k-factor based controller are measured in Table 1 in terms of undershoot/overshoot (in the percentage of V_o) and settling time. A high degree of accuracy has been achieved between simulation and experimental results.

Table 1. Comparative closed-loop performance of the IBC for output voltage response of 'k-factor' based controller.

Specification	V_{in}=395V, V_o=200V, P_o=5kW $\rightarrow$10kW $\rightarrow$5kW		
	'k-factor' controller		Deviation between Sim. & Exp. result
	Sim.	Exp.	
Maximum undershoot	7.5% (15V)	6.5% (13V)	3V
Maximum overshoot	8% (16V)	5.5% (11V)	4V
Maximum settling time	3ms	2ms	1ms

Conclusion

A GSSASM modeling approach for formulating the output voltage-to-duty cycle and inductor current-to-duty cycle transfer functions of the IBC to design a fast, robust and accurate control model has been proposed in this paper. In this paper, stability analysis is accomplished for interleaved inductor current bandwidth and output DC-link voltage concerning input voltage change and dynamic load variation, which is translated into the controlled duty ratio. The proposed mathematical model is verified through input voltage change and dynamic load variation using the HiFi MATLAB Simulink® model and is experimentally validated using a 30-kW SiC-based prototype. Both simulation and experimental results demonstrate a high accuracy at the same controller bandwidth in both steady-state and transient conditions, respectively. Both the simulation and experimental results depict that during 100% load change, the output DC-link voltage is varied within ±10% voltage ripple, and the maximum setting time is less than 5ms. From these obtained results, it can be concluded that as the IBC is a highly non-linear system, the GSSASM is the most appropriate technique to design robust control systems.

References

[1] S. Chakraborty, H. Vu, M. M. Hasan, D. Tran, M. El Baghdadi, and O. Hegazy, "DC-DC Converter Topologies for Electric Vehicles, Plug-in Hybrid Electric Vehicles and Fast Charging Stations: State of the Art and Future Trends," *Energies*, vol. 12, no. 8, p. 1569, Apr. 2019.

[2] M. D. Kankam and M. E. Elbuluk, "A survey of power electronics applications in aerospace technologies," *Proc. Intersoc. Energy Convers. Eng. Conf.*, vol. 1, no. November, pp. 147–153, 2001.

[3] F. Liccardo, P. Marino, G. Torre, and M. Triggianese, "Interleaved dc-dc Converters for Photovoltaic Modules," in *2007 International Conference on Clean Electrical Power*, 2007, pp. 201–207.

[4] S. Cuoghi, R. Mandrioli, L. Ntogramatzidis, and G. Gabriele, "Multileg Interleaved Buck Converter for EV Charging: Discrete-Time Model and Direct Control Design," *Energies*, vol. 13, no. 2, p. 466, Jan. 2020.

[5] S. Chakraborty *et al.*, "Scalable Modelling Approach & Robust Hardware-in-the-Loop Testing of an Optimized Interleaved Bidirectional HV DC/DC Converter for Electric Vehicle Drivetrains," *IEEE Access*, vol. 8, pp. 115515–115536, 2020.

[6] W. Hasan, S. Chakraborty, S. M. Salim Reza, K. Salim, and M. Razzak, "Improvement of Systems Response of a PID Controller in Underdamped Condition," *Int. J. Innov. Appl. Stud.*, vol. 12, no. 4, pp. 864–873, 2015.

[7] A. Skandarnezhad, A. Rahmati, A. Abrishamifar, and A. Kalteh, "Small-Signal Transfer-Function Extraction of a Lossy Buck Converter Using ASM Technique," vol. 2, no. 7, pp. 1760–1763, 2014.

[8] P. Azer and A. Emadi, "Generalized State Space Average Model for Multi-Phase Interleaved Buck, Boost, and Buck-Boost DC-DC Converters: Transient, Steady-state and Switching Dynamics," *IEEE Access*, vol. PP, pp. 1–1, 2020.

[9] H. Shin, J. Park, S. Chung, H. Lee, and T. A. Lipo, "Generalised steady-state analysis of multiphase interleaved boost converter with coupled inductors," *IEE Proc. - Electr. Power Appl.*, vol. 152, no. 3, pp. 584–594, May 2005.

[10] H. Shin, E. Jang, J. Park, H. Lee, and T. A. Lipo, "Small-signal analysis of multiphase interleaved boost converter with coupled inductors," *IEE Proc. - Electr. Power Appl.*, vol. 152, no. 5, pp. 1161–1170, 2005.

[11] D. D. Tran, G. Thomas, M. El Baghdadi, J. Van Mierlo, and O. Hegazy, "Design and Implementation of FPGA-based Digital Controllers for SiC Multiport Converter in Electric Vehicle Drivetrains," *2019 21st Eur. Conf. Power Electron. Appl. (EPE'19 ECCE-Europe), Genova*, pp. 1–7, 2019.

[12] O. Hegazy, R. Barrero, J. Van Mierlo, P. Lataire, N. Omar, and T. Coosemans, "An Advanced Power Electronics Interface for Electric Vehicles Applications," *IEEE Trans. Power Electron.*, vol. 28, no. 12, pp. 5508–5521, Dec. 2013.

Adaptive Predictive-DPC for LCL-Filtered Grid Connected VSC with Reduced Number of Sensors

Hosein Gholami-Khesht, Pooya Davari, Frede Blaabjerg

Department of Energy Technology, Aalborg University, Aalborg, Denmark

hgk@et.aau.dk, pda@et.aau.dk, fbl@et.aau.dk

Acknowledgements

The work is supported by the Reliable Power Electronic-Based Power System (REPEPS) project at the Department of Energy Technology, Aalborg University as a part of the Villum Investigator Program funded by the Villum Foundation.

Keywords

«Voltage source converter (VSC)», «Direct power control», «Model-based predictive control (MPC) », «Adaptive control».

Abstract

An adaptive predictive direct power control (P-DPC) method using a Luenberger observer is proposed in conjunction with conventional P-DPC, which reduces the system sensitivity to parameter mismatches and control delays. Moreover, the observer facilitates sensorless operation of the proposed method. The performance of the proposed adaptive P-DPC is confirmed under various simulation and experimental tests.

Introduction

The control of renewable energy-based power generation systems has attracted considerable attention in recent years. Major control goals are the bidirectional and decoupled active and reactive power control as well as sinusoidal current generation [1]-[15]. Basically, power control methods can be divided into two main categories: indirect and direct power control methods [1]-[14]. The proportional-integral (PI) [4]-[5] and the proportional-resonant (PR) [6]-[7] current controllers are the most popular indirect power control methods. The necessity of using a phase-locked loop (PLL), sensitivity to power grid conditions, and inappropriateness for digital implementation are the main disadvantages of these methods. Switching table based direct power control (ST-DPC) [8]-[10] and predictive based direct power control (P-DPC) [11]-[14] are two traditionally used direct power control methods.

Recently methods based on the predictive control theory have become more popular owing to the various advantages they offer [1]-[3], [11]-[14]. A simple structure, simplicity in digital implementation, coincident to the discrete nature of power electronic converters, and superior dynamic and steady-state performance are the most interesting features of the predictive based control methods. In spite of these advantages, sensitivity to control system delays and parameter mismatches may degrade their performance and even cause instability. These problems can be amplified in systems with higher-order filters, while such systems have many advantages, and they are increasingly used day by day [2]-[4], [9]-[10]. Moreover, in predictive control methods, it is necessary to measure all system states to minimize the steady-state errors, and to do better disturbance rejection [2]-[3]. However, measuring all signals with related sensors increases system cost, volume, and weight. As discussed in [2]-[3], the MPC based direct and indirect power control of an LCL-filtered grid-connected voltage source converter (VSC) needs four current and voltage sensors, which may worsen the system reliability due to additional physical components, which might be subject to damages and thereby failures.

To overcome these problems, an adaptive predictive-DPC (P-DPC) is proposed in this paper, which combines the P-DPC with an identification technique. So far, various identification techniques are applied in VSC based power applications, where the most important are the Luenberger observer [16]-[17], the sliding mode observer [18], the Kalman filter [19], the neural network [20], the steepest descent [21], and recursive least square estimator [22]. The proposed method utilizes a Luenberger Observer (LO) in the P-DPC structure. The LO is a closed-loop estimator that uses the state-space model of the system to predict the states of the system from the measured inputs and outputs based on the minimization of the difference between the measured and the estimated outputs. Besides high accuracy and reliability, and noise immunity of determined signals by LO, it helps to compensate the control delays, and reduces the sensitivity to system uncertainties by estimating states and disturbance inputs one sampling period ahead. Moreover, contrary to the conventional predictive control methods [2]-[3], which need many sensors, the LO reduces the number of sensors in the proposed control structure by replacing the grid current sensors. It is also worth noting that the proposed P-DPC does not need a PLL, which prevents interaction between different control loops and PLL, and consequently improves the system stability.

In summary, proposing an augmented state-space model of grid-connected VSC with an LCL output filter, incorporating the LO in the P-DPC structure, and investigating the feasibility of practical implementation of the proposed adaptive P-DPC are the most significant contributions of this work.

In the following sections, firstly, the dynamics of a grid-connected three-phase VSC with LCL output filter are described in section II. Section III presents the proposed P-DPC, which includes four main subsections, proposed adaptive-predictive direct power control, LO design, complex power reference calculation, and active damping. After that, to evaluate the performance of the proposed sensorless P-DPC and the conventional one, simulation and experimental results are presented in section IV. Finally, conclusions are drawn in Section V.

System Modeling

Basic model of the grid-connected three-phase VSC

The single line diagram of the grid-connected three-phase VSC, which is studied in this paper is shown in Fig. 1. The system of Fig. 1 includes the power source, the LCL-type filter, and the three-phase VSC. Using the space vector theory, the converter model can be readily presented as:

$$
\begin{cases}
L_f \dfrac{di_f}{dt} = v_c - v_{inv} \\[2mm]
C_f \dfrac{dv_c}{dt} = i_g - i_f
\end{cases}
\tag{1}
$$

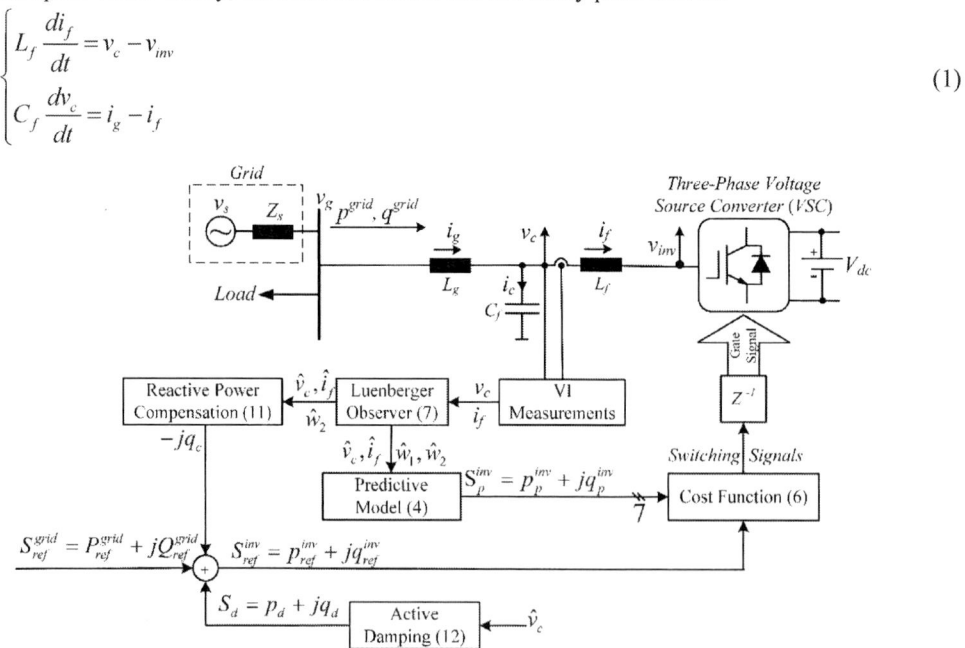

Fig. 1: Proposed adaptive predictive-DPC

where, v_{inv}, v_c, i_f, i_g and i_c are the converter output voltage, capacitor voltage, inverter current, grid current, and capacitor current, respectively. Also, C_f, L_f and L_g are filter capacitance, inverter-side and grid-side filter inductances.

Augmented state-space model

Generally, the perfect implementation of a model-based predictive control method highly depends on the accuracy of the system model and the parameters' values. However, in practice, the accuracy is subjected to un-modeled dynamics, parameter uncertainties, and external disturbances. In this section, to tackle the problem, an augmented discrete state-space model is used, which includes all system parameters uncertainties and unmodeled dynamics, as follows:

$$\frac{dx(t)}{dt} = Ax(t) + Bu(t) + Dw(t)$$

$$x = \begin{bmatrix} i_f \\ v_c \end{bmatrix}, u = \begin{bmatrix} v_{inv} \end{bmatrix}, w = \begin{bmatrix} w_1 \\ w_2 \end{bmatrix} = \begin{bmatrix} \Delta L_f \dfrac{di_f}{dt} + n_1 \\ i_g + \Delta C_f \dfrac{dv_o}{dt} + n_2 \end{bmatrix},$$

$$A = \begin{bmatrix} 0 & \dfrac{1}{L_f} \\ \dfrac{-1}{C_f} & 0 \end{bmatrix}, B = \begin{bmatrix} \dfrac{-1}{L_f} \\ 0 \end{bmatrix}, D = \begin{bmatrix} \dfrac{1}{L_f} & 0 \\ 0 & \dfrac{1}{C_f} \end{bmatrix}. \tag{2}$$

where, parameters with and without "Δ" denote the deviation from nominal values and the nominal values, respectively. Also, n_1 and n_2 represent unstructured uncertainties due to un-modeled dynamics and modeling errors. As seen in (2), in the augmented model, all uncertainties caused by parameter mismatches, grid current disturbances, and other unstructured uncertainties are lumped as a disturbance input $w(k)$.

The practical implementation of the predictive control algorithm is based on the discrete state-space model of the plant dynamics. Discretizing (2) with the sampling period T_S, yields the following discrete state-space equations:

$$\begin{cases} x(k+1) = A_d x(k) + B_d u(k) + D_d w(k) \\ A_d = e^{AT_S} = L^{-1}[(sI - A)^{-1}] \approx I + AT_S \\ B_d = \int_0^{T_S} e^{A_d(T_S - \tau)} B d\tau \approx BT_S \\ D_d = \int_0^{T_S} e^{A_d(T_S - \tau)} D d\tau \approx DT_S \end{cases} \tag{3}$$

Proposed Predictive Direct Power Control (P-DPC)

Proposed adaptive-predictive direct power control

As shown in Fig. 1, the proposed power control contains four main parts: the LO, the model predictive control (MPC), reactive power compensation, and also active damping [4]. The LO provides the predicted states and input disturbances and facilitates a perfect implementation of the MPC by delay compensation of the control system. Based on the augmented model and the predicted variables, the MPC calculates the apparent power for all possible converter voltage vectors and then compares them with the power references. Consequently, the converter voltage vector $(u(k+1))$ that minimizes the error between the predicted and the reference powers is selected and applied at the start of the next sampling period.

To compensate for the inherent delay of digital implementation, the proposed adaptive P-DPC is based on the two-step ahead predicted powers that can be interpreted as:

$$S_p^{inv}(k+2) = p_p^{inv}(k+2) + jq_p^{inv}(k+2) = \hat{v}_c(k+2)\hat{i}_f(k+2)^* \tag{4}$$

where, $S_p^{inv}(k+2)$, $\hat{v}_c(k+2)$ and $\hat{i}_f(k+2)$ are complex power at the beginning of the $(k+2)$th period, predicted capacitor voltage and inverter current vectors, respectively. Predicted voltage and current vectors are computed from (3) as follows:

$$\hat{x}(k+2) = \begin{bmatrix} \hat{v}_c(k+2) & \hat{i}_f(k+2) \end{bmatrix}^T = A_d\hat{x}(k+1) + B_d u(k+1) + D_d\hat{w}(k+1) \tag{5}$$

In the above equations, $\hat{x}(k+1)$ and $\hat{w}(k+1)$ are estimated signals via the LO that is described in the following subsection.

Finally, by substituting (4) into the cost function of (6), it is evaluated for all possible converter voltage vectors, and then the optimal voltage vector that minimizes the cost function is saved and applied at the beginning of the next sampling period.

$$\begin{cases} g = \left| S_{ref}^{inv}(k+1) - S_p^{inv}(k+2) \right|^2 + \lambda_{sw} g_{sw}^{\ 2} \\ g_{sw} = \left| SW_a(k+1) - SW_a(k) \right| + \left| SW_b(k+1) - SW_b(k) \right| + \left| SW_c(k+1) - SW_c(k) \right| \end{cases} \tag{6}$$

In (6), $S_{ref}^{inv}(k+1)$ is the reference complex power at the inverter side. Moreover, g_{sw} includes switching efforts to control switching frequency and losses. Also, SW_i ($i = a$, b, and c) and λ_{sw} are gating signals and weighting factor, respectively [15].

Luenberger observer design

As mentioned in the previous subsection, to evaluate the cost function and accordingly to achieve the optimal converter voltage, the disturbance input and the one sample ahead predicted states are needed. To obtain these quantities, the use of a full-order Luenberger observer is proposed in this subsection. Neglecting the disturbance changes in each sampling period ($w(k+1) \approx w(k)$), the system discrete state-space equation can be rewritten as:

$$\begin{cases} x(k+1) = A_{ob}x(k) + B_{ob}u(k) \\ x(k) = C_{ob}x(k) \end{cases}$$

$$x = \begin{bmatrix} i_f \\ v_c \\ w_1 \\ w_2 \end{bmatrix}, u = \begin{bmatrix} v_{inv} \end{bmatrix}, C_{ob} = \begin{bmatrix} 1 & 0 & 0 & 0 \\ 0 & 1 & 0 & 0 \end{bmatrix}, A_{ob} = \begin{bmatrix} 1 & \dfrac{T_S}{L_1} & \dfrac{T_S}{L_1} & 0 \\ \dfrac{-T_S}{C} & 1 & 0 & \dfrac{T_S}{C} \\ 0 & 0 & 1 & 0 \\ 0 & 0 & 0 & 1 \end{bmatrix}, B_{ob} = \begin{bmatrix} \dfrac{-T_S}{L_1} \\ 0 \\ 0 \\ 0 \end{bmatrix} \tag{7}$$

Based on the above equation, a full-order Luenberger observer can be constructed as follows:

$$\hat{x}(k+1) = A_{ob}\hat{x}(k) + B_{ob}u(k) + GC_{ob}(x(k) - \hat{x}(k)) \tag{8}$$

where, symbol "^" denotes the estimated values, and G is the observer gain. The observer gain is chosen such that the estimation error dynamics of (9) is asymptotically stable, and eigenvalues of the observer are placed in the desired locations, as well [16]-[17].

$$\begin{cases} e(k+1) = (A_{ob} - GC_{ob})e(k) \\ e(k) = x(k) - \hat{x}(k) \end{cases} \tag{9}$$

It is worth to notice that the gain matrix of the observer can be calculated to satisfy the mentioned requirements because the observability matrix of the system ($\begin{bmatrix} C_{ob} & C_{ob}A_{ob} \end{bmatrix}^T$) has full column rank.

Complex power reference calculation

In the proposed power controller, the complex power of the inverter side of the LCL-type filter is controlled, while the grid side complex power tracking is desired in most applications. So, if the grid side power control is required, the inverter side power reference must be simply modified to compensate for the capacitor reactive power, as follows:

$$S_{ref}^{inv}(k+1) = S_{ref}^{grid}(k+1) - jq_c \tag{10}$$

where, $S_{ref}^{inv}(k+1)$, $S_{ref}^{grid}(k+1)$ and q_c are the complex power references of the inverter side and the grid side and the reactive power generated by the capacitor of the LCL-filter, respectively. Based on (1) and (3), the reactive power of the capacitor is calculated as (11).

$$\begin{cases} q_c = v_c(k+1)i_c(k+1)^* \\ i_c(k+1) = w_2(k+1) - i_f(k+1) \end{cases} \tag{11}$$

Active damping

To damp the current oscillation caused by the LCL filter resonance, an active or passive damping method must be used. Work in [9] proposed an effective active damping method that is also employed in this work. Based on this method, the resonance component of the capacitor voltage is extracted to calculate damping components. After that, these derived damping components are added to the power control loop to damp LCL filter resonance.

The damping component can be calculated as:

$$S_d(k+1) = p_d(k+1) + jq_d(k+1) = v_{c,1}(k+1)i_d^*(k+1), \left(i_d = k_d \tilde{v}_c\right) \tag{12}$$

where, $v_{c,1}$ is the fundamental component of the capacitor voltage and $\tilde{v}_c$ is the resonance component of the capacitor voltage. Also, k_d is damping factor. The large value of k_d provides better LCL filter resonance damping, however, it can cause low order harmonic component in the output current, and also output power deviation from the reference one.

Simulation and Experimental Results and Discussion

To evaluate the performance of the proposed adaptive P-DPC under different conditions, a Simulink test bench and laboratory prototype have been prepared, as shown in Fig 2. The experimental setup includes a three-phase 5-kW PWM-VSC, which is supplied from a constant DC voltage and connects to a grid simulator at the AC side. Moreover, the proposed control method is realized in a DS1007 dSPACE system. The parameters of the simulated and experimental systems are the same and given in Table. I.

Table I: System parameters

Nominal power	5 [kVA]
Nominal phase voltage	230 [V]
Grid frequency	50 [Hz]
Inverter-side inductor (L_f)	3 [mH]
Grid-side inductor (L_g)	3 [mH]
Filter capacitor (C_f)	20 [µF]
DC-link voltage (V_{dc})	720 [V]
Sampling period (T_S)	25 [µsec]
Control parameters	
k_d	0.1
λ_{sw}	1000
Observer eigenvalues	[0.01,0.01,0.99,0.05]

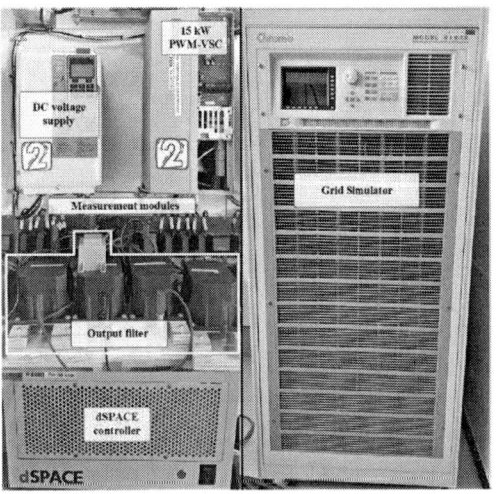

Fig. 2: Laboratory setup used to verify the effectiveness of proposed P-DPC

In the first study, the steady-state performance of the proposed method is presented in Fig. 3. Highly sinusoidal grid currents and regulated injected active and reactive powers with minimum distortions are seem. It is worth note that the average switching frequency of the proposed control method is 3.4 kHz, which is acceptable in point of the switching losses. Although the switching frequency is approximately low, however the total harmonic distortion (THD) of the grid current is 3.8%, which well satisfies the international standard, such as IEEE 519.

To compare performance of the proposed sensorless P-DPC with the conventional P-DPC, the performance of the conventional one is also shown in Fig. 4. It is worth to remark that the conventional P-DPC uses additional sensors for grid-side currents. It can be concluded from Figs. 3 and 4 that both control methods have the same performance.

The transient performance of the proposed adaptive P-DPC and conventional one under consecutive reference power changes is presented in Fig. 5, which shows the performance of both control methods to track the reference powers rapidly. It is worth to point out that this excellent operation is achieved without using any grid current and voltage sensors, and all previous results are obtained using the LO algorithm in the control structure. In Fig. 6, the excellent performance of the LO to estimate the system states and disturbance inputs is shown.

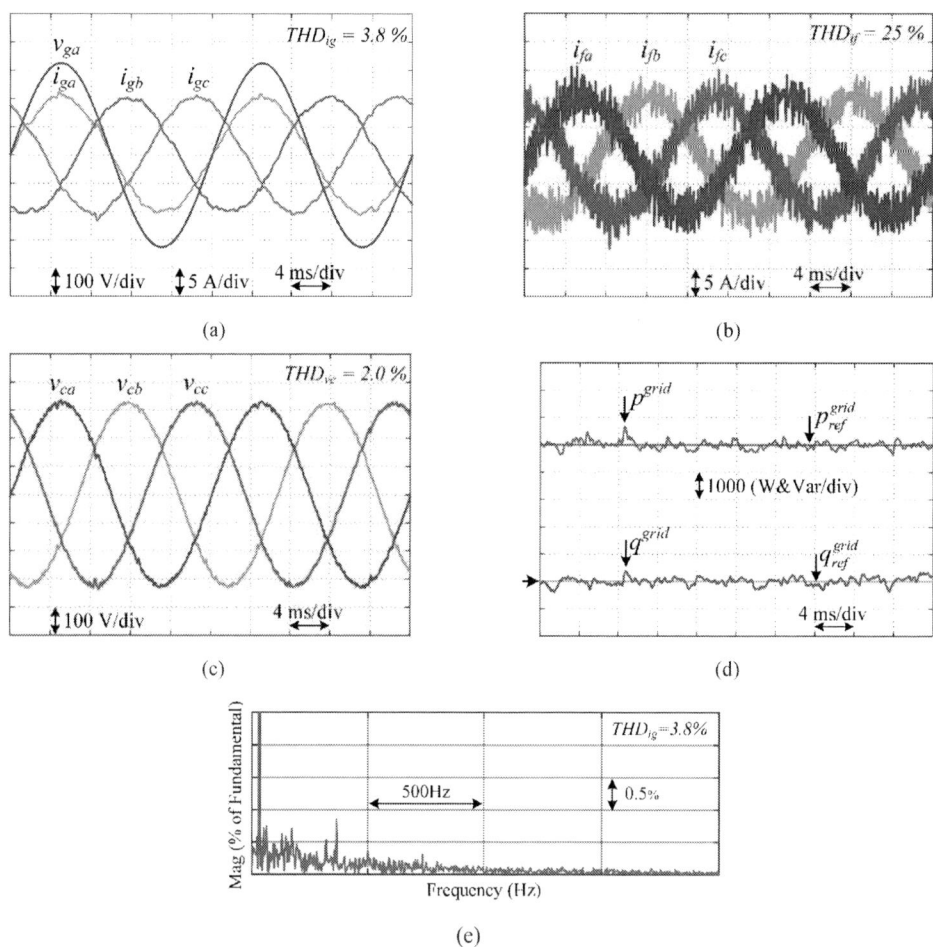

Fig. 3: Simulation results showing steady-state performance of the proposed adaptive P-DPC ($P = 5$ kW, $Q = 0$ Var), (a) grid voltage and currents, (b) inverter currents, (c) capacitor voltages, (d) grid active and reactive powers, (e) (d) harmonic spectrum of the grid current.

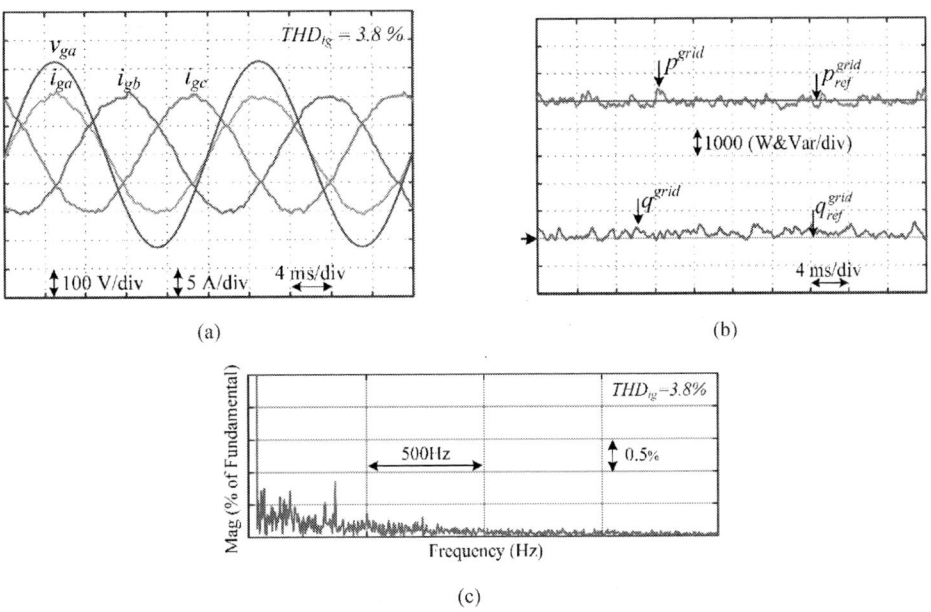

(a)

(b)

(c)

Fig. 4: Simulation results showing steady-state performance of the conventional P-DPC ($P = 5$ kW, $Q = 0$ Var), (a) grid voltage and currents, (b) grid active and reactive powers, (c) capacitor voltages, (d) harmonic spectrum of the grid current.

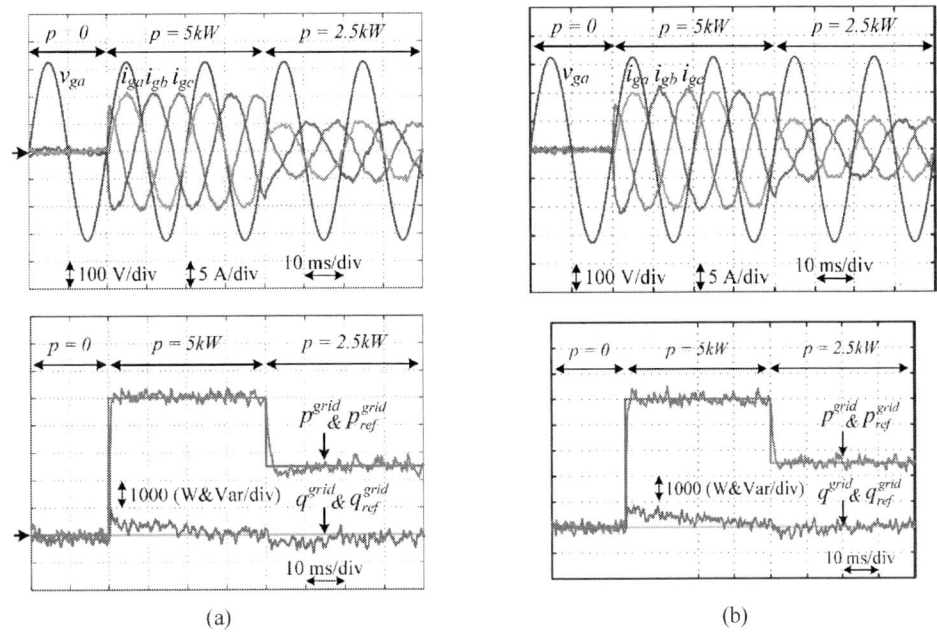

(a)

(b)

Fig. 5: Simulation results showing dynamic performance in response to reference power step changes ($Q = 0$ Var), (a) proposed adaptive P-DPC, (b) conventional P-DPC.

To validate the feasibility of the proposed adaptive P-DPC, experiments are carried out on a laboratory setup. Fig. 7 depicts the steady-state waveforms of the grid currents under pure sinusoidal grid voltages. This figure confirms the high capability of the proposed control method to achieve the sinusoidal current generation and perfect active and reactive power regulation. Fig. 8 shows the dynamic response of the proposed control methods under the step changes in the reference active power as in the simulations.

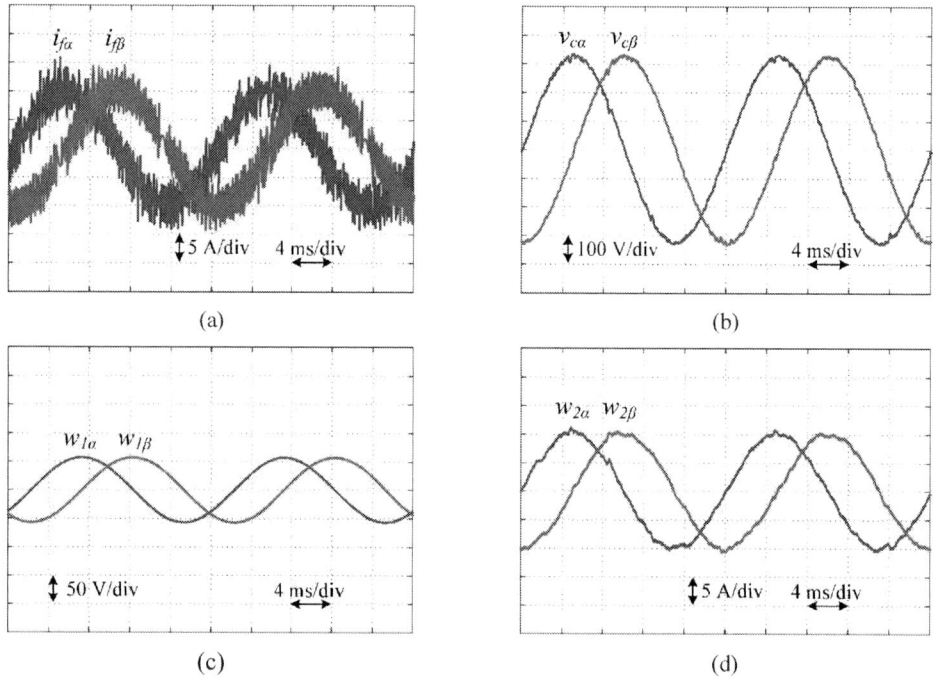

Fig. 6: Simulation results showing outputs of the proposed Luenberger observer (a) estimated inverter currents, (b) estimated capacitor voltages, (c) and (d) estimated disturbance inputs.

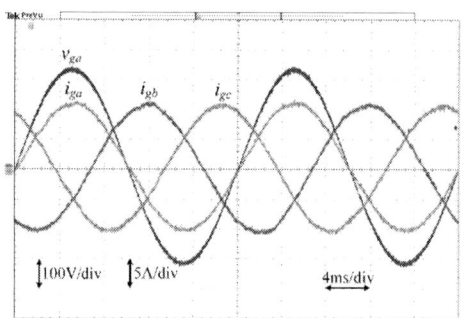

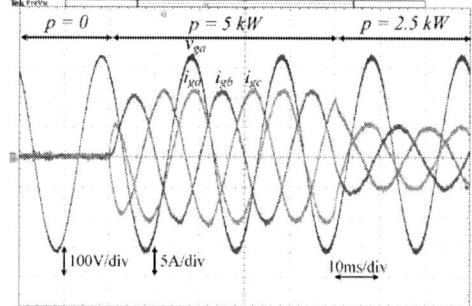

Fig. 7: Experimental results showing steady-state performance of the proposed adaptive P-DPC (P = 5 kW, Q = 0 Var).

Fig. 8: Experimental results showing dynamic performance of the proposed adaptive P-DPC in response to reference power step changes (Q = 0 Var)

Conclusion

In recent years, due to the availability of powerful digital signal controllers (DSCs), more attentions are paid to predictive based control methods. A special application for the predictive methods is the renewable energy-based distributed power generation systems, where bidirectional power flow control is required. However, these methods are susceptible to system control delays and parameter variations and need too many current and voltage sensors. In this study, an improved P-DPC is proposed, which utilizes a LO to improve the system performance by compensating for delays and parameter drifts and uncertainties. Also, the number of sensors is successfully reduced by replacing some of them with estimation already provided by the LO. The performance of the proposed adaptive P-DPC is evaluated under different operation conditions. Simulation and experimental results confirm the superior performance of the proposed method under various conditions.

References

[1] T. Dragicevic, C. Zheng, J. Rodriguez and F. Blaabjerg, "Robust quasi-predictive control of LCL-Filtered Grid Converters," *IEEE Trans. Power Electron.*, vol. 35, no. 2, pp. 1934-1946, Feb. 2020.

[2] P. Falkowski and A. Sikorski, "Finite control set model predictive control for grid-connected AC-DC converters with LCL filter," *IEEE Trans. Ind. Electron.*, vol. 65, no. 4, pp. 2844–2852, 2018.

[3] X. Zhang, L. Tan, J. Xian, H. Zhang, Z. Ma and J. Kang, "Direct grid-side current model predictive control for grid-connected inverter with LCL filter," *IET Power Electron.*, vol. 11, no. 15, pp. 2450-2460, 18 12 2018.

[4] S. Zhou et al., "An improved design of current controller for LCL-type grid-connected converter to reduce negative effect of PLL in weak grid," *IEEE J. Emerg. Sel. Top. Power Electron.*, vol. 6, no. 2, pp. 648-663, June 2018.

[5] C. A. Busada, S. G. Jorge and J. A. Solsona, "A synchronous reference frame PI current controller with deadbeat response," *IEEE Trans. Power Electron.*, vol. 35, no. 3, pp. 3097-3105, March 2020.

[6] H. Wu and X. Wang, "Virtual-flux-based passivation of current control for grid-connected VSCs," *IEEE Trans. Power Electron.*, doi: 10.1109/TPEL.2020.2997876.

[7] H. Gholami-Khesht, M. Monfared, and S. Golestan, "Low computational burden grid voltage estimation for grid connected voltage source converter-based power applications," *IET Power Electron.*, vol. 8, no. 5, pp. 656-664, May. 2015.

[8] Y. Zhang, W. Xie, Z. Piao and C. Hu, "Performance improvement of direct power control of PWM rectifier with simple calculation," *IEEE Trans. Power Electron.*, vol. 28, no. 7, pp. 3428-3437, Jul. 2013.

[9] L. A. Serpa, S. Ponnaluri, P. M. Barbosa and J. W. Kolar, "A modified direct power control strategy allowing the connection of three-phase inverters to the grid through LCL filters," *IEEE Trans. Ind. Appl.*, vol. 43, no. 5, pp. 1388-1400, Sept. 2007.

[10] L. A. Serpa, J. W. Kolar, S. Ponnaluri and P. M. Barbosa, "A modified direct power control strategy allowing the connection of three-phase inverter to the grid through LCL filters," *Industry Applications Conference*, vol. 1, pp. 565-571, 2005.

[11] H. Li, M. Lin, M. Yin, J. Ai and W. Le, "Three-vector-based low-complexity model predictive direct power control strategy for PWM rectifier without voltage sensors," *IEEE J. Emerg. Sel. Top. Power Electron.*, vol. 7, no. 1, pp. 240-251, March 2019.

[12] X. Shi, J. Zhu, L. Li and D. Dah-Chuan LU, "Low-complexity dual-vector-based predictive control of three-phase PWM rectifiers without duty-cycle optimization," *IEEE Access*, vol. 8, pp. 77049-77059, 2020.

[13] A. M. Bozorgi, H. Gholami-Khesht, M. Farasat, S. Mehraeen and M. Monfared, "Model predictive direct power control of three-phase grid-connected converters with fuzzy-based duty cycle modulation," *IEEE Trans. Ind. Appl.*, vol. 54, no. 5, pp. 4875-4885, Sept.-Oct. 2018.

[14] S. Yan, J. Chen, T. Yang and S. Y. Hui, "Improving the performance of direct power control using duty cycle optimization," *IEEE Trans. Power Electron.*, vol. 34, no. 9, pp. 9213-9223, Sept. 2019.

[15] P. Karamanakos and T. Geyer, "Guidelines for the design of finite control set model predictive controllers," *IEEE Trans. Power Electron.*, vol. 35, no. 7, pp. 7434–7450, Jul. 2020.

[16] V. Miskovic, V. Blasko, T. M. Jahns, A. H. C. C. Smith, and C. Romenesko, "Observer-based active damping of LCL resonance in grid-connected voltage source converters," *IEEE Trans. Ind. Appl.*, vol. 50, no. 6, pp. 3977–3985, Nov. 2014.

[17] H. Gholami-Khesht and M. Monfared, "Deadbeat direct power control for grid-connected inverters using a

full-order observer," *Electric Power and Energy Conversion Systems* (EPECS), pp. 1-5, Sharjah, 2015.

[18] Y. A. R. I. Mohamed and E. F. El-Saadany, "Robust high bandwidth discrete-time predictive current control with predictive internal model - A unified approach for voltage-source PWM converters," *IEEE Trans. Power Electron.*, vol. 23, no. 1, pp. 126–136, Jan. 2008.

[19] N. Hoffmann and F. W. Fuchs, "Minimal invasive equivalent grid impedance estimation in inductive–resistive power networks using extended Kalman filter," *IEEE Trans. Power Electron.*, vol. 29, no. 2, pp. 631–641, Feb. 2014.

[20] Y. A.-R. I. R. I. Mohamed and E. F. El-Saadany, "Adaptive discrete-time grid-voltage sensorless interfacing scheme for grid-connected DG-inverters based on neural-network identification and deadbeat current regulation," *IEEE Trans. Power Electron.*, vol. 23, no. 1, pp. 308–321, Jan. 2008.

[21] Y. A. R. Ibrahim Mohamed et al., "An improved deadbeat current control scheme with a novel adaptive self-tuning load model for a three-phase PWM voltage-source inverter," *IEEE Trans. Ind. Electron.*, vol. 54, no. 2, pp. 747–759, Apr. 2007.

[22] Q. Liu and K. Hameyer, "High-performance adaptive torque control for an IPMSM with real-time MTPA operation," *IEEE Trans. Energy Convers.*, vol. 32, no. 2, pp. 571–581, Jun. 2017.

FPGA Implementation of Modified Space Vector Modulation (SVM) for High-Frequency Hybrid Active Neutral-Point-Clamped (NPC) Power Factor Correction Rectifier

Mohammad Najjar[1], Alireza Kouchaki[2], Morten Nymand[1]
[1]University of Southern Denmark, Odense, Denmark
[2]Danfysik A/S, Copenhagen, Denmark
E-Mail: mohna@mci.sdu.dk, ako@danfysik.dk, mny@mci.sdu.dk
URL: http://power-electronics.sdu.dk

Keywords

«Active Front-End», «Power factor correction», «Voltage Source Converter (VSC)», «Pulse Width Modulation (PWM)», «Digital control».

Abstract

This paper presents a model-based implementation of modified space vector modulation (SVM) for a three-level three-phase active neutral-point clamped (ANPC) rectifier. To optimize and reduce the number of high frequency active switches, this paper uses a hybrid modulation technique in which only two switches out of 6 active switches are switching at high frequency and the rest are working at fundamental frequency (50 Hz). However, using space vector-based hybrid modulation for ANPC introduces several limitations on selecting the output vectors such as zero-vector that is generated by connecting all three phases to the positive DC rail. To deal with this issue, a modified hybrid SVM is utilized to adjust the voltages of the DC link. A prototype is made using Gallium-nitride (GaN) FETs and the modified SVM is implemented in a field programmable gate array (FPGA). To show that the modified SVM can control the DC link voltages, an experimental comparison is carried out between a carrier-based modulation and the modified SVM.

Introduction

Multilevel converters are widely used in power electronic applications due to superior characteristics [1]-[2]. Compared with conventional two-level converters, multilevel topologies have better voltage and current THD performance and the lower voltage across semiconductors [2]-[6]. Among multilevel converters, the neutral-point-clamped (NPC) is widely utilized in industry. However, the unequal loss distribution among switches limits the maximum output power of this structure. To deal with this issue, active NPC (ANPC) was proposed, in which the clamping diodes are replaced with active switches. This approach provides more redundancy to make a zero-voltage level at the output waveform of the converter [7].

The main disadvantage of NPC converters is the voltage variation of the middle point of DC-link or neutral-point potential of the converter. This variation increases harmonics, damages switching devices due to the overstress, and reduces output power quality [7]. Thus, this variation should be controlled. Utilizing modified space vector modulations to control the voltage of the middle point of DC-link is proposed in [8]-[14] to reduce the NPC variations.

Changing the switching frequency of the converter is another solution to reduce the losses of the converter. Unlike ANPC, in hybrid ANPC [15]-[21], some of the switches are switching at the fundamental frequency (grid frequency). This action reduces the switching losses of the converter; however, it introduces some limitations, such as reducing the number of redundancies to provide zero output voltage.

A solution to reduce the size of passive components of converters is increasing the switching frequency. This approach can be obtained by utilizing the characteristics of wide band-gap semiconductors, such as shorter switching time and lower switching losses. Thus, these switches can be utilized in multilevel converters to improve their performance.

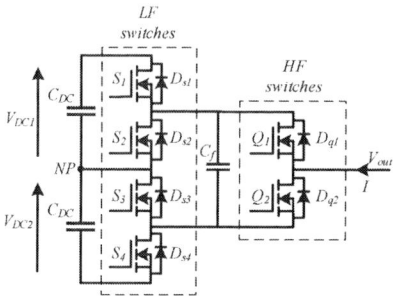

Fig. 1: The structure of Active NPC.

Table I: The switching states of 3-level hybrid ANPC

| Output States | Switching Sequence | | | | | | Output Voltage |
	S_1	S_2	S_3	S_4	Q_1	Q_2	$(V_{out}-V_{NP})$
N	0	1	0	1	0	1	$-V_{DC}/2$
O-	0	1	0	1	1	0	0
O+	1	0	1	0	0	1	0
P	1	0	1	0	1	0	$V_{DC}/2$

In this paper, a modified SVM is utilized for hybrid ANPC. By using GaN FETs, the switching frequency of the converter is increased to 140 kHz. Thus, for the practical implementation of modified SVM with a complicated switching pattern, a Field Programmable Gate Array (FPGA) is utilized. Finally, the experimental results show the performance of the proposed method to control the NP voltage variations.

Active Neutral Point Clamped Converter (ANPC)

Fig. 1 shows the general structure of a 3-level ANPC. Six switches are employed in the structure of ANPC. Two different hybrid modulations have been proposed for ANPC. In one modulation, 4 of 6 switches are modulating with high frequency. Another modulation which is used in this study, two switches are switching with high frequency (HF switches) and the rest are switching at the fundamental frequency of 50 Hz (LF switches). The switching states of the hybrid ANPC are shown in Table I. A decoupling capacitor (C_f) is utilized to reduce the parasitic elements in the switching path of high frequency switches.

Modified Space Vector Modulation

The general space vector diagram (SVD) for 3-level converters is shown in Fig. 2(a). The SVD is divided into six sectors and each sector has four subsectors. There are 27 space vectors: 6 large, 6 medium, 12 small and 3 zero vectors. Each of these vectors has different impacts on the voltages of the DC link capacitors. Large and zero vectors do not have any influence on the voltage of the DC link middle point (NP). The reason is that the current does not close its path through NP or through the DC link capacitors. However, small and medium vectors can change the voltage of NP. This effect for medium vectors is more significant. In this case, the current of one phase flows through the middle point. Small vectors are divided into two groups of positive and negative vectors. One phase is connected to the middle point by the positive vectors; however, the negative ones connect two phases to the middle point. The impact of these vectors on the middle point voltage for the PFC applications are shown in Table II [8],[12]. There are two constraints for hybrid ANPC:

- Two of zero vectors (PPP & NNN) cannot be used.
- For subsector 1 and 3, all the vectors cannot be employed.

These two constraints are related to the hybrid switching of ANPC. Since four of the switches are switching with the fundamental frequency, it is not possible to use all vectors at all subsectors. For

Table II: The impact of small and medium vectors

Positive small vectors	$V_{dc1}-V_{dc2}$	Negative small vectors	$V_{dc1}-V_{dc2}$	Medium vectors	$V_{dc1}-V_{dc2}$
ONN	↓	POO	↑	PON	↓
PPO	↓	OON	↑	OPN	↓
NON	↓	OPO	↑	NPO	↓
OPP	↓	NOO	↑	NOP	↓
NNO	↓	OOP	↑	ONP	↓
POP	↓	ONO	↑	PNO	↓

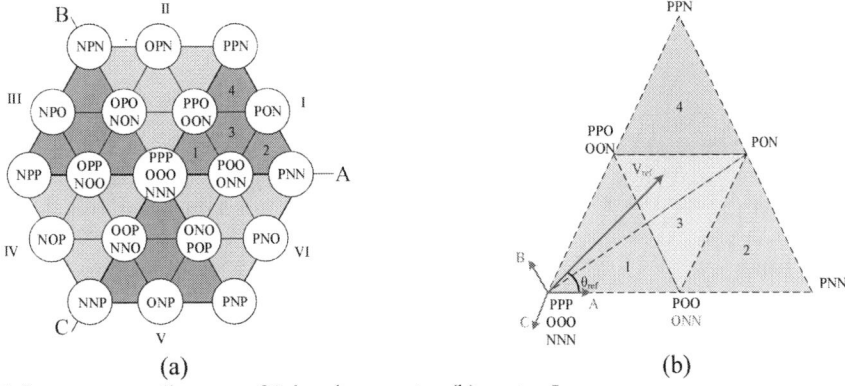

(a)   (b)

Fig. 2: (a) Space-vector diagram of 3-level converter (b) sector I.

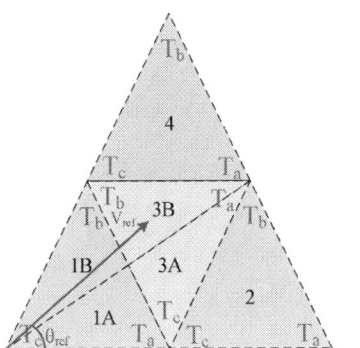

Fig. 3: The modified sector for hybrid ANPC.

example, when one of the phases is switching in the positive half cycle, S_1 and S_3 are conducting and the state of these switches should not change during half cycle (50 Hz switching frequency). Thus, it cannot provide a negative voltage during the positive half cycle. Consequently, the vectors PPP and NNN cannot be produced by hybrid ANPC. Similarly, for subsectors 1 and 3, both pair of small vectors cannot be used at the same time. Fig. 2(b) shows sector I with a reference vector which lies in subsector 3. Considering the position of the reference vector (phase A and B: positive half cycle, phase C: negative half cycle), the small vector of ONN cannot be used. The reason is that at this the position, the output voltage of phase B can be positive or zero. Thus, this vector cannot be applied.

To deal with these constraints, a modified SVD is utilized. As can be seen in Fig. 3, in the modified SVD, subsectors 1 and 3 are divided into two subsectors of 1A, 1B, 3A and 3B, respectively. Still, the main idea of the nearest three vectors (NTV) is utilized for these subsectors. However, for each of them, different vectors are applied with consideration of the middle point voltage.

Table III: The vectors time periods

Subsector	1	2	3	4
condition	$m_1<1$ $m_2<1$ $m_1+m_2<1$	$m_1>1$	$m_1<1$ $m_2<1$ $m_1+m_2>1$	$m_2>1$
T_a	$m_1 \times T_{sw}$	$(m_1-1) \times T_{sw}$	$T_{sw}-T_b-T_c$	$m_1 \times T_{sw}$
T_b	$m_2 \times T_{sw}$	$m_2 \times T_{sw}$	$(1-m_1) \times T_{sw}$	$(m_2-1) \times T_{sw}$
T_c	$T_{sw}-T_a-T_b$	$T_{sw}-T_a-T_b$	$(1-m_2) \times T_{sw}$	$T_{sw}-T_a-T_b$

By considering two following equations, the switching time of each vector for different subsectors can be calculated as follows:

$$m_1 = \frac{V_{ref}}{V_{DC}} 2\sqrt{3} \sin(\frac{\pi}{3}-\theta_{ref}) \tag{1}$$

$$m_2 = \frac{V_{ref}}{V_{DC}} 2\sqrt{3} \sin(\theta_{ref}) \tag{2}$$

which V_{ref} is the magnitude of the reference voltage vector, θ_{ref} is the position of reference vector, and V_{DC} is the voltage of the DC link ($V_{DC1}+V_{DC2}$). Table III shows all the time period for the different vectors, as shown in Fig. 3.

To control the voltage of the middle point, the voltages of the DC link capacitors are measured and the difference between them is calculated and utilized to apply proper vectors. The update rate of this sensing is equal to switching frequency or at the start of the carrier frequency [1].

In the proposed control strategy considering the limitation of hybrid ANPC, only positive or negative vectors are used. For each subsector, there is just one degree of freedom to choose between positive or negative vector. Thus, based on the condition of voltages of the DC link capacitors, the positive or the negative vector is applied. Moreover, if during the switching period, the middle point voltage changes, the new vectors are not applied until the next switching period. This prevents increasing the switching frequency and consequently increasing the losses. The utilized vectors for all sectors are listed in Table IV.

Digital Implementation

The dSPACE's Micro-Lab Box ds1202 FPGA is used to implement and to model modified SVM. For this implementation, different software such as MATLAB/Simulink, dSPACE's ControlDesk, and Xilinx System Generator are used. The control system is modeled in a microprocessor; meanwhile, the FPGA is used for modified SVM implementation. Fig. 4 shows the block diagram of the implemented SVM. The modulation index and angular frequency (ωt) are the input signals for FPGA from microprocessor.

In this study, the frequency of microprocessor is 35 kHz, while the frequency of FPGA is set to 100 MHz. Thus, the update rate of the angular frequency is the same as the microprocessor frequency. To increase the resolution of ωt at FPGA, the value $\pi 10^{-6}$ ($2\pi \times f_{grid}/f_{FPGA}$) can be added to previous value of ωt, until ωt is updated by the microprocessor. Based on the value of ωt, the sector and the phase condition can be measured. For example, if the value of ωt is between 0 and $\pi/6$, then the converter is switching at sector 1. The output voltage of phase A is in the positive half cycle, while phase B and C are working in the negative half cycle. Thus, the states of LF switches can be determined.

The trigonometric values for eq.(1) and eq.(2) can be determined by utilizing a lookup table. The output of lookup table and the value of modulation index are utilized to calculate the time period by eq.(1) and (2). As it was mentioned earlier, to prevent increasing switching frequency the time periods and the voltages of the DC link capacitors are updated at the beginning of each switching cycle.

To reach the switching frequency of 140 kHz, the counter counts to 357 and when reaches to this value it starts to count down to 0 [19]. The output signal of counter is utilized as the carrier. It should be noted that the value of T_S in Table II is set to 357. The function of duty cycle block in Fig. 4 is to calculate the duty cycle of each phase, by using the time periods of Table II and the condition of the voltages of the DC link capacitors. Then, the output of the duty cycle block is compared with the carrier. The output of

Table IV: Utilized DC-link Voltage Balancing Sequence

S¹	S²	DC link Condition	Switching Sequence	S¹	S²	DC link Condition	Switching Sequence
1	1A	$V_{dc1}>V_{dc2}$	OOO-OON-ONN-OON-OOO	2	1A	$V_{dc1}>V_{dc2}$	OPO-OOO-OON-OOO-OPO
		$V_{dc2}>V_{dc1}$	POO-OOO-OON-OOO-POO			$V_{dc2}>V_{dc1}$	PPO-OPO-OOO-OPO-PPO
	1B	$V_{dc1}>V_{dc2}$	POO-OOO-OON-OOO-POO		1B	$V_{dc1}>V_{dc2}$	OOO-OON-NON-OON-OOO
		$V_{dc2}>V_{dc1}$	PPO-POO-OOO-POO-PPO			$V_{dc2}>V_{dc1}$	OPO-OOO-OON-OOO-OPO
	3A	$V_{dc1}>V_{dc2}$	PON-OON-ONN-OON-PON		3A	$V_{dc1}>V_{dc2}$	OPO-OPN-OON-OPN-OPO
		$V_{dc2}>V_{dc1}$	POO-PON-OON-PON-POO			$V_{dc2}>V_{dc1}$	PPO-OPO-OPN-OPO-PPO
	3B	$V_{dc1}>V_{dc2}$	POO-PON-OON-PON-POO		3B	$V_{dc1}>V_{dc2}$	OPN-OON-NON-OON-OPN
		$V_{dc2}>V_{dc1}$	PPO-POO-POO-POO-PPO			$V_{dc2}>V_{dc1}$	OPO-OON-OON-OON-OPO
	2	$V_{dc1}>V_{dc2}$	PON-PNN-ONN-PNN-PON		2	$V_{dc1}>V_{dc2}$	PPN-OPN-OON-OPN-PPN
		$V_{dc2}>V_{dc1}$	POO-PON-PNN-PON-POO			$V_{dc2}>V_{dc1}$	PPO-PPN-OPN-PPN-PPO
	4	$V_{dc1}>V_{dc2}$	PPN-PON-OON-PON-PPN		4	$V_{dc1}>V_{dc2}$	OPN-NPN-NON-NPN-OPN
		$V_{dc2}>V_{dc1}$	PPO-PPN-PON-PPN-PPO			$V_{dc2}>V_{dc1}$	OPO-OPN-NPN-OPN-OPO
3	1A	$V_{dc1}>V_{dc2}$	OOO-NOO-NON-NOO-OOO	4	1A	$V_{dc1}>V_{dc2}$	OOP-OOO-NOO-OOO-OOP
		$V_{dc2}>V_{dc1}$	OPO-OOO-NOO-OOO-OPO			$V_{dc2}>V_{dc1}$	OPP-OOP-OOO-OOP-OPP
	1B	$V_{dc1}>V_{dc2}$	OPO-OOO-NOO-OOO-OPO		1B	$V_{dc1}>V_{dc2}$	OOO-NOO-NNO-NOO-OOO
		$V_{dc2}>V_{dc1}$	OPP-OPO-OOO-OPO-OPP			$V_{dc2}>V_{dc1}$	OOP-OOO-NOO-OOO-OOP
	3A	$V_{dc1}>V_{dc2}$	NPO-NOO-NON-NOO-NPO		3A	$V_{dc1}>V_{dc2}$	OOP-NOP-NOO-NOP-OOP
		$V_{dc2}>V_{dc1}$	OPO-NPO-NOO-NPO-OPO			$V_{dc2}>V_{dc1}$	OPP-OOP-NOP-OOP-OPP
	3B	$V_{dc1}>V_{dc2}$	OPO-NPO-NOO-NPO-OPO		3B	$V_{dc1}>V_{dc2}$	NOP-NOO-NNO-NOO-NOP
		$V_{dc2}>V_{dc1}$	POP-OOP-ONP-OOP-POP			$V_{dc2}>V_{dc1}$	OOP-NOP-NOO-NOP-OOP
	2	$V_{dc1}>V_{dc2}$	NPO-NPN-NON-NPN-NPO		2	$V_{dc1}>V_{dc2}$	NPP-NOP-NOO-NOP-NPP
		$V_{dc2}>V_{dc1}$	OPO-NPO-NPN-NPO-OPO			$V_{dc2}>V_{dc1}$	OPP-NPP-NOP-NPP-OPP
	4	$V_{dc1}>V_{dc2}$	NPP-NPO-NOO-NPO-NPP		4	$V_{dc1}>V_{dc2}$	NOP-NNP-NNO-NNP-NOP
		$V_{dc2}>V_{dc1}$	OPP-NPP-NPO-NPP-OPP			$V_{dc2}>V_{dc1}$	OOP-NOP-NNP-NOP-OOP
5	1A	$V_{dc1}>V_{dc2}$	OOO-ONO-NNO-ONO-OOO	6	1A	$V_{dc1}>V_{dc2}$	POO-OOO-ONO-OOO-POO
		$V_{dc2}>V_{dc1}$	OOP-OOO-ONO-OO-OOP			$V_{dc2}>V_{dc1}$	POP-POO-OOO-POO-POP
	1B	$V_{dc1}>V_{dc2}$	OOP-OOO-ONO-OOO-OOP		1B	$V_{dc1}>V_{dc2}$	OOO-ONO-ONN-ONO-OOO
		$V_{dc2}>V_{dc1}$	POP-OOP-ONO-OOP-POP			$V_{dc2}>V_{dc1}$	POO-OOO-ONO-OOO-POO
	3A	$V_{dc1}>V_{dc2}$	ONP-ONO-NNP--NO-ONP		3A	$V_{dc1}>V_{dc2}$	POO-PNO-ONO-PNO-POO
		$V_{dc2}>V_{dc1}$	OOP-ONP-ONO-ONP-OOP			$V_{dc2}>V_{dc1}$	POP-POO-PNO-POO-POP
	3B	$V_{dc1}>V_{dc2}$	OOP-ONP-ONO-ONP-OOP		3B	$V_{dc1}>V_{dc2}$	PNO-ONO-ONN-ONO-PNO
		$V_{dc2}>V_{dc1}$	POP-OOP-ONP-OOP-POP			$V_{dc2}>V_{dc1}$	POO-PNO-ONO-PNO-POO
	2	$V_{dc1}>V_{dc2}$	ONP-NNP-NNO-NNP-ONP		2	$V_{dc1}>V_{dc2}$	PNP-PNO-ONO-PNO-PNP
		$V_{dc2}>V_{dc1}$	OOP-ONP-NNP-ONP-OOP			$V_{dc2}>V_{dc1}$	POP-PNP-PNO-PNP-POP
	4	$V_{dc1}>V_{dc2}$	PNP-ONP-ONO-ONP-PNP		4	$V_{dc1}>V_{dc2}$	PNO-PNN-ONN-PNN-PNO
		$V_{dc2}>V_{dc1}$	POP-PNP-ONP-PNP-POP			$V_{dc2}>V_{dc1}$	POO-PNO-PNN-PNO-POO

S¹: Sector, S²: Subsector.

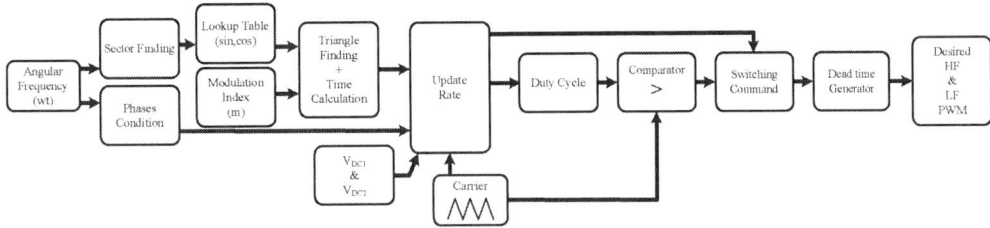

Fig. 4: The block diagram of implemented SVM with FPGA

this comparator is utilized to produce the switching commands for all phases. Fig. 5 shows an example, which one of the switching sequences from Table IV is implemented. As it can be seen, just the time period of T_a is used to produce the duty cycle of phase A; however, for phase C both T_a and T_c are utilized. Moreover, in this sequence phase B does not have any switching transition. Finally, the proper dead time and dead zone using presented method in [21] is employed on switching signals.

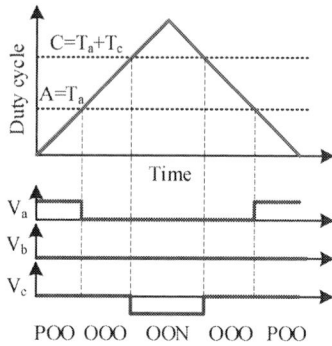

Fig. 5: the implemented switching sequence by using carrier

Table V: The system specification

Parameter	symbol	value
Grid voltage	V_g	65 (rms)
DC link Voltage	V_{DC}	200 V
Grid frequency	f_g	50 Hz
Converter side inductor	L_c	500 µH
Grid side inductor	L_f	140 µH
Common mode inductor	L_{cm}	300 µH
Differential mode capacitor	C_{dm}	2 µF
Common mode capacitor	C_{cm}	1 µF

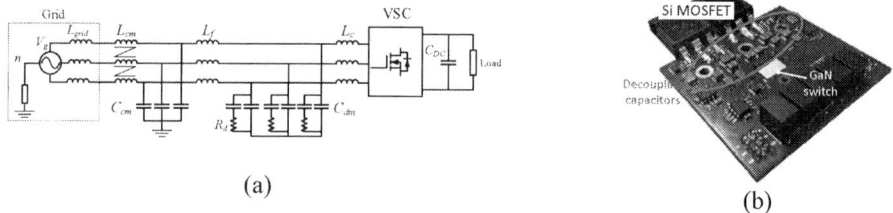

(a)

(b)

Fig. 6: Test setup, (a) built phase leg, (b) The structure of the system under test.

Experimental Results

To show the ability of the modified SVM to control the voltage of DC link capacitors, an experiment is done. In this experiment, a most conventional carrier-based modulation (sinusoidal pulse width modulation (SPWM)) and the modified SVM are compared from different perspectives, especially sharing of the dc link voltage.

Fig 6(a) shows the structure of the system under the test. The converter is connected to the grid by two stage of an *LCL* and an *LC* filters to attenuate both differential and common mode noises. The value of the system parameters under test are listed in Table V. Fig. 6(b) shows one of the switching board, which four Si MOSFET and two gallium-nitride (GaN) are utilized [19].

The experimental results for SPWM and modified SVM are shown in Fig. 7. The voltages of the DC link capacitors using SPWM are shown in Fig 7(a). A low frequency voltage ripple of 150 Hz with the magnitude of 2.5% of V_{DC} is visible on the dc link voltage [22], [23] (the AC part of the DC link voltages are shown separately in Fig. 7(a)). By employing the modified SVM, the low frequency voltage ripple has been canceled out and the voltages of the DC link capacitors are fixed at the same level as it can be seen in Fig. 7(b).

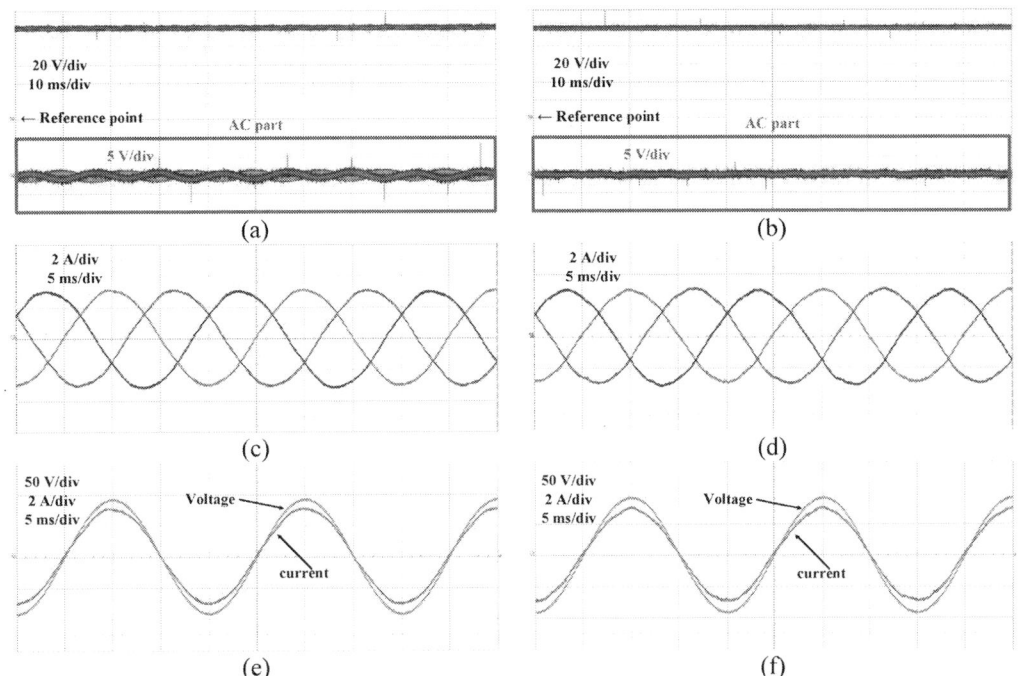

Fig. 7: Experimental results, (a) voltages of DC link capacitors with SPWM, (b) voltages of the DC link capacitors with SVM, (c) three-phase current with SPWM, (d) three-phase current with SVM, (e) phase voltage and current with SPWM, (f) phase voltage and current with SVM.

The three-phase output current for both considered modulation techniques are shown in Fig. 7(c) and 7(d), respectively. A low frequency distortion is observed in the output current using modified SVM. This is the low frequency harmonics that are introduced for instance on phase A during the zero crossing of phase B and C. This mostly contributes to harmonic 5. It needs to be added that the switching frequency of each leg is not the same in the case of modified SVM and based on the condition of the voltages of the DC link capacitors, the switching frequency can vary.

Conclusion

This paper has presented a model-based implementation of a modified SVM for a 3-level active neutral point clamp (ANPC) rectifier. To reduce the losses of the converter, both Silicon (Si) power MOSFET and Gallium-nitride (GaN)-FETs are utilized in the converter. A hybrid modulation technique has been utilized to employ the advantage of each switch technology. It means that the low frequency switches (Si-MOSFETs) are switching at fundamental frequency and the high frequency switches (GaN-FETs) are modulating at higher switching frequency. The low frequency voltage ripple of the middle point of DC-link is the main issue of NPC converters. In addition, hybrid modulation techniques introduce some limitations about using different output voltage vectors. Considering these limitations, a modified SVM has been utilized to control the voltage variation of the middle point of the DC link. To implement the modified SVM, the FPGA of a dSPACE device has been utilized. Finally, the experimental results confirmed the ability of modified SVM about the control of middle point of DC link voltage compared to carrier-based modulation of SPWM.

References

[1] Rodriguez J., Bernet S., Wu B., Pontt J. O., Kouro S., "Multilevel voltage-source-converter topologies for industrial medium-voltage drives," IEEE Trans. Ind. Electron, Vol. 54, No. 6, pp. 2930-2945, Dec. 2007.

[2] Peng F. Z., Lai J.-S., McKeever J. W., VanCoevering J., "A multilevel voltage-source inverter with separate DC sources for static VAr generation," IEEE Trans. Ind. Appl., Vol. 32, No. 5, pp. 1130-1138, Sep./Oct. 1996.

[3] Najjar M., Moeini A., Bakhshizadeh M. K., Blaabjerg F., Farhangi S., "Optimal Selective Harmonic Mitigation Technique on Variable DC Link Cascaded H-Bridge Converter to Meet Power Quality Standards," IEEE Trans. Emerg. Sel. Topics Power Electron., vol. 4, no. 3, pp. 1107-1116, Sept. 2016.

[4] Najjar M., Farhangi S., Iman-Eini H., "A Method to Control the Interphase Power Controller with Common DC Bus", Electric Power Components and Systems, 45:18, 1996-2006, 2017.

[5] Heydari R., Dragicevic T., Blaabjerg F., "High-Bandwidth Secondary Voltage and Frequency Control of VSC-Based AC Microgrid," IEEE Trans. Power Electron., vol. 34, no. 11, pp. 11320-11331, Nov. 2019.

[6] Najafi P., Houshmand Viki A., Shahparasti M., Seyedalipour S.S., Pouresmaeil E., "A Novel Space Vector Modulation Scheme for a 10-Switch Converter". Energies, 13, 1855, 2020.

[7] Barbosa P., Steimer P., Steinke J., Winkelnkemper M., Celanovic N.: Active-neutral-point-clamped (ANPC)multilevel converter technology, Power Electronics and Applications, 2005 European Conference on, Dresden, 2005.

[8] Choudhury A., Pillay P., Williamson S. S., "DC-Bus Voltage Balancing Algorithm for Three-Level Neutral-Point-Clamped (NPC) Traction Inverter Drive with Modified Virtual Space Vector," IEEE Trans. Ind Appl, vol. 52, no. 5, pp. 3958-3967, Sept.-Oct. 2016.

[9] Busquets-Monge S., Bordonau J., Boroyevich D., Somavilla S., "The nearest three virtual space vector PWM - a modulation for the comprehensive neutral-point balancing in the three-level NPC inverter," IEEE Trans. Power Electron., vol. 2, no. 1, pp. 11-15, March 2004.

[10] Le Q. A., Lee D., "Reduction of Common-Mode Voltages for Five-Level Active NPC Inverters by the Space-Vector Modulation Technique," IEEE Trans. Ind Appl, vol. 53, no. 2, pp. 1289-1299, March-April 2017.

[11] Hu C., Yu X., Holmes D. G., Shen W., Wang Q., Luo F., Liu N., "An Improved Virtual Space Vector Modulation Scheme for Three-Level Active Neutral-Point-Clamped Inverter," IEEE Trans. Power Electron., vol. 32, no. 10, pp. 7419-7434, Oct. 2017.

[12] Li C., Lu R., Li C., Li W., Gu X., Fang Y., Ma H., He X., "Space Vector Modulation for SiC and Si Hybrid ANPC Converter in Medium-Voltage High-Speed Drive System," IEEE Trans. Power Electron., vol. 35, no. 4, pp. 3390-3401, April 2020.

[13] Sadeghi A., Mohamadian M., Shahparasti M., Fatemi A., "A new switching algorithm for voltage balancing of a three-level NPC in DTC drive of a three-phase IM." 2013 Twenty-Eighth Annual IEEE Applied Power Electronics Conference and Exposition (APEC), Long Beach, CA, pp. 489-495, 2013.

[14] Najafi P., Hooshmand Viki A., Shahparasti M., "Novel space vector-based control scheme with dc-link voltage balancing capability for 10 switch converter in bipolar hybrid microgrid", Sustain. Energy, Grids Networks., 20, 2019.

[15] Kouchaki A., Kapino G., Nymand M., "Design of a High Frequency 3-Phase 3-Level Hybrid Active-NPC Inverter," 2018 20th European Conference on Power Electronics and Applications (EPE'18 ECCE Europe), pp. 1-10, Riga, 2018.

[16] Ma L., Kerekes T., Rodriguez P., Jin X., Teodorescu R., Liserre M., "A New PWM Strategy for Grid-Connected Half-Bridge Active NPC Converters with Losses Distribution Balancing Mechanism," IEEE Trans. Power Electron, vol. 30, no. 9, pp. 5331-5340, Sept. 2015

[17] Floricau D., Floricau E., Gateau G., "Three-level active NPC converter: PWM strategies and loss distribution", 34th Annual Conference of IEEE Industrial Electronics, Orlando, FL, pp. 3333-3338, 2008.

[18] Wang K., Xu L., Zheng Z., Li Y., "Capacitor Voltage Balancing of a Five-Level ANPC Converter Using Phase-Shifted PWM," IEEE Tans. Power Electroncis., vol. 30, no. 3, pp. 1147-1156, March 2015.

[19] Kapino G., Kouchaki A., Nielsen J., Nymand M., "Simple digital model-based design and implementation of carrier-based PWM for high frequency hybrid 3-L active NPC inverter," IEEE 12th International Conference on Compatibility, Power Electronics and Power Engineering (CPE-POWERENG 2018), Doha, 2018.

[20] Kouchaki A., Nymand M., "Filter design for active neutral point clamped voltage source converter using high frequency GaN-FETs," IEEE 12th International Conference on Power Electronics and Drive Systems (PEDS), pp. 349-354, Honolulu, HI, 2017.

[21] Najjar M., Nymand M., Kouchaki A., "Mitigation Zero-crossing Distortion of Active Neutral-Point-Clamped Rectifier with Improved Hybrid PWM Technique", ISIE2020.

[22] Celanovic N., Boroyevich D., "A comprehensive study of neutral-point voltage balancing problem in three-level neutral-point-clamped voltage source PWM inverters," IEEE Trans. Power Electron., vol. 15, no. 2, pp. 242-249, March 2000.

[23] Pou J., Pindado R., Boroyevich D., Rodriguez P., "Evaluation of the low-frequency neutral-point voltage oscillations in the three-level inverter," IEEE Trans. Ind. Electron, vol. 52, no. 6, pp. 1582-1588, Dec. 2005.

Enhanced Flux Control Including a Closed Loop Voltage Controller to Optimize the Voltage Usage and the Torque Computation for a 48V IPMSM

Felix Bertele, Ulrich Ammann, Christoph Cheshire, Tobias Röser
Faculty of Mechatronics and Electrical Engineering
University of Applied Sciences Esslingen
Robert-Bosch-Str. 1
Göppingen 73037, Germany
Tel.: +49(0)7161.679-1277
{felix.bertele; ulrich.ammann; christoph.cheshire; tobias.roeser}@hs-esslingen.de

Keywords

«Electrical Machines», «Automotive Power Electronic and Drives», «Permanent Magnet Synchronous Machines», «Control of Electrical Drives», «Flux Model», «Optimal Control», «Motor Drives and Electrical Machines»

Abstract

For an efficient operation of a low voltage PMSM an optimized voltage usage is very important. Because of the relation between the low voltage and the high currents in this type of machine, a large voltage reserve is needed to compensate the influence of parameter mismatches and to guarantee a stable current control. As the power is limited by the low voltage in this type of hybrid drive systems, optimizing the voltage usage is also required to maximize the power and the torque availability. This paper describes a closed loop flux control to maximize the voltage usage. The controller feedback is used to estimate and maximize the available torque for each operating point.

I. Introduction

Low voltage hybrid drive systems are commonly used to electrify the drivetrain in passenger cars. Due to the low voltage, no additional safety requirements for voltages higher than 60V have to be considered. However, the car manufacturer can take advantage of the features provided by an electrical machine inside the car's drivetrain. The PMSM is used as a starter for the combustion engine as well as a powerful generator compared to its 12V predecessor. The operating point of the combustion engine can be shifted to run more efficiently and a significant part of the braking energy can be recuperated.

I.1. Flux Based Open Loop Torque Control

A widespread open loop torque control method for an interior permanent magnet synchronous machine (IPMSM) is using look up tables created from a flux measurement ([1], [2] and [3]). These look up tables are used to compute the current reference for the closed loop dq-current controller as shown in Fig. 1. The current reference computation usually has to deal with three input values: The available flux ψ_{max}, which is a result of the available voltage U_{dc} and the speed ω_{el}. The available absolute current I_{max} variates with the internal temperatures and the torque reference M_{ref} is commanded by the user. The open loop torque control generates current references based on a characterization measurement. This characterization is usually done for one machine sample and copied to many others; a parameter variation will result in a mismatch of the resulting torque and voltage. Exceeding the available voltage could destabilize the system, by causing instable or wrong i_{dq} values. One solution in that case is to provide a voltage reserve.

As this reserve reduces the available torque and the efficiency, it should be as small as possible. Therefore the voltage command output of the current controller can be used as a feedback for a superimposed voltage controller, as it is described in ([3], [4] and [5]). The voltage controller affects the available voltage U_{dc} or the available flux ψ_{max} inside the open loop torque control. This results in a reduction of the induced voltages $u_{dq,ind}$ of the IPMSM.

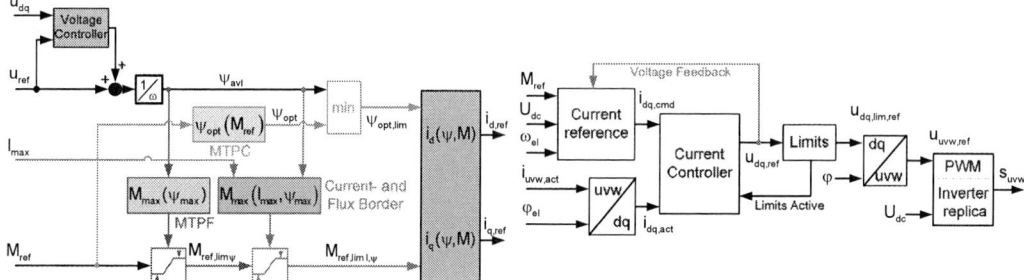

Fig. 1: Current reference computation   Fig. 2: Closed loop current controller

This contribution describes how to optimize this superimposed voltage control by observing the absolute flux. With the design suggested in this contribution it is possible to get a smooth transition at the edge between the base speed range, where the voltage is not at the limit and the flux weakening range.

I.2. Voltage Usage and Torque Accuracy

As the available voltage is very small in this drive system, a high current is required to produce a significant torque. These high currents and the related temperature changes causes some significant variation in the internal parameters. In this contribution, a superimposed closed loop voltage control is used to force the voltage usage to its maximum. The goal here is to get a stable core control system without the necessity of reserving voltage. As the PMSM is used in a hybrid drive system, it is important to control and estimate the torque very accurately. The superimposed system also needs a precise prediction of the maximum and minimum available torque related to the actual operating point. To maximize the voltage usage with a closed loop control and guarantee acceptable torque accuracy and prediction, it is necessary to identify the major reasons for a voltage variation.

- Stator resistance R_{cu}: The high current causes fast temperature changes. This results in an inaccurate measurement related to the position and the time response of the used temperature sensor.
- Permanent magnet flux ψ_{PM}: The permanent magnet flux varies with the temperature of the rotor. Sensor based measurement is not common. However, it can be considered by observing the induced voltage.
- Variation between different machine samples: The characterization data is measured only for a few samples. Related to a variation in the production process, the internal physics of each drive is of course slightly different.

For the application used, it is important to get an accurate torque response as well as a precise prediction of the available torque related to the actual operation point. To improve on this aspect, the observed disturbance of the flux observer is used.

II. Flux Controller

II.1. Closed Loop Structure

The flux controller is designed as an absolute flux observer, observing the difference between the optimum limited flux $\psi_{opt,lim}$ and the actual flux ψ_{act}. The optimum Flux $\psi_{opt,lim}$ is the flux which is commanded by the current reference computation, ψ_{act} is the actual flux calculated from the voltage feedback given by the current controller output considering the ohmic voltage drop $u_{R_{cu}}$ and the speed ω_{el}. The reference Flux ψ_{ref} is the flux the system is commanded to work at in the flux weakening area, calculated in (II.1). m_{idx} represents the commanded modulation index for the PWM. The ohmic voltage drop is calculated based on the reference for current i_{dq} and torque $M_{ref,lim}$. Calculating the ohmic voltage drop $u_{R_{cu}}$ based on the u_{dq} voltage equation results in (II.2).

$$\psi_{ref} = \frac{1}{\omega_{el}}\left(\frac{U_{dc}}{2}\cdot m_{idx} - u_{Rcu}\right) \quad (II.1) \qquad u_{Rcu} = \sqrt{\left(R_{cu}\left|i_{dq}\right|\right)^2 + \frac{4p}{3}\omega_{el}M_{act}} \quad (II.2)$$

Assigned jointly to the European Power Electronics and Drives Association & the Institute of Electrical and Electronics Engineers (IEEE)

II.2. Absolute Flux Observer

The controller structure and the scheme of the used absolute flux observer is shown in Fig. 3. The behaviour is given by the transfer function of the current controller. The relations between the resulting current values and the fluxes is represented by two look-up tables. The discrepancy between the real controller and the model used for the observer is calculated with a PI-controller as shown in Fig. 4. The observed disturbance is interpreted as ohmic resistance or winding temperature deviation.

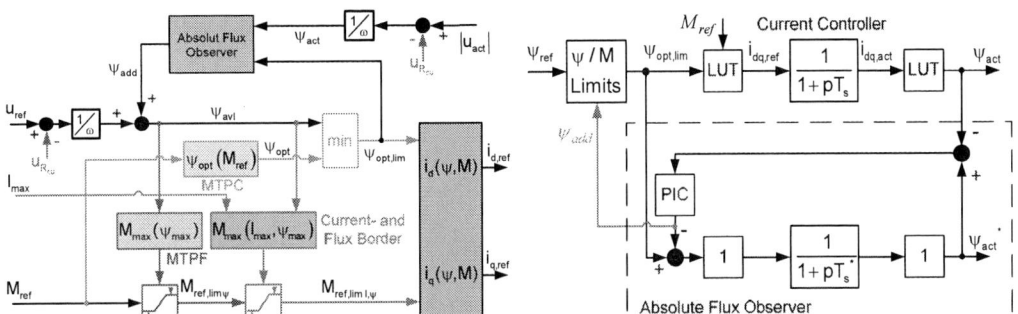

Fig. 3: Flux Controller Closed Loop Structure   Fig. 4: Flux Observer

II.3. Voltage Feedback

The goal is to get a system that works at the voltage limit. Using the limited voltage values for the observer would make it impossible to distinguish if the current controller works at the voltage limits or already exceeds it.

By using the unlimited values, it is possible to respond to an exceeding of the voltage limit. In this case, the flux observer is able to reduce the voltage by reducing the commanded absolute flux ψ_{ref}, to guarantee a stable and accurate settling of the current controller.

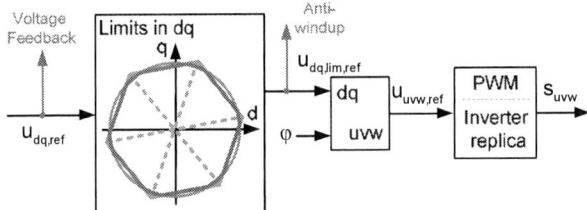

Fig. 5: Limitation for the voltage command $u_{dq,cmd}$

II.4. Inverter Replica and Voltage Limitation

To guarantee a stable and accurate behaviour of the current controller a limitation of the voltage command is necessary. It suppresses commands the PWM unit cannot handle to avoid wrong switching states. For the voltage control scheme, the unlimited output $u_{dq,ref}$ of the current controller is used. It is important to avoid continuous exceeding of the voltage limit, caused by a too large voltage command. To calculate the possible maximum, the physics of the inverter hardware used has to be considered. To improve the torque accuracy of the system, observed internal values are used. Therefore, a closely matching relation between the reference voltages $u_{dq,ref}$ and the actual voltages appearing at the IPMSM is needed. The correction is achieved by the PWM unit by including a inverter replica. The inverter replica considers and corrects the impacts of switching dead time as well as conduction losses of transistors and diodes.

II.5. Transition between Base Speed and Flux Weakening Area

One of the challenges for a stable operation for any design of such a superimposed voltage control for a PMSM is dealing with the transition between the base speed and the flux weakening area. In the base speed area, the current reference calculation operates on the MTPC line shown in Fig. 6. Here the voltage controller should not force the voltage to its maximum, because this would lead to inefficient operation points far to the right of the MTPC line. In the flux weakening area the system operates at the limit, where the voltage controller should force the voltage to its maximum.

In Fig. 6, the flux limitation that represents the transition into the flux weakening area is given by circles of constant flux around the zero flux point. These circles become smaller for a higher speed ω_{el} and a lower voltage U_{dc}. A transition between an operation on the MTPC line and the flux weakening area could be caused by a change in speed, represented by in- or decreasing flux limitation circles. Another reason for a transition is a step in the commanded torque. To deal with this transition some additional limitations and an anti-wind up construct, as shown in ([4], [8]), are necessary.

The use of the solution suggested in this contribution leads to a smooth transition without this additional construction. Hereby the stability around this critical operation points is improved significantly. The flux observer observes the difference between the optimum limited flux $\psi_{opt,lim}$ calculated and commanded by the current reference computation and the actual flux ψ_{act} calculated from the voltage command output as shown in Fig. 3. In the base speed area the optimum flux $\psi_{opt,lim}$ is used as flux limitation for the commanded torque on the MTPC border.

In the flux, weakening area the maximum available flux ψ_{ref} calculated from the available voltage U_{dc} limits the flux for the current reference calculation. Therefore, the observer always deals with the absolute flux that should appear in the PMSM supposing ideal condition.

The output flux of the observer ψ_{ref} is the error between the flux reference $\psi_{opt,lim}$ of the current reference computation and the ψ_{act}. This observed difference is only added to the maximum available flux ψ_{ref}. This is why the flux observer has no impact to the current reference selection when the system operates in the base speed area. The flux observer can only extend or reduce the area limited by the available flux ψ_{ref}. Fig. 7 shows the same MTPC, MTPF, and maximum current borderlines as Fig. 6. Here borders are drawn within a torque M_{ref} over ψ_{ref} grid. In Fig. 7, two different operation points represented by the flux reference ψ_{ref} should show the behaviour of the voltage controller.

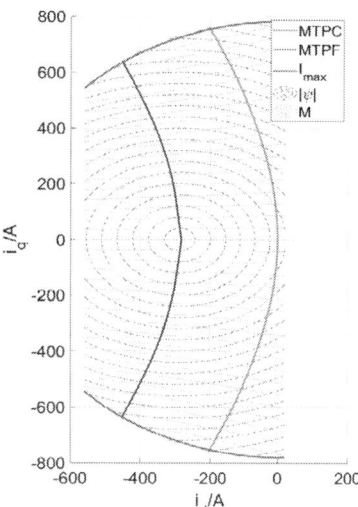

Fig. 6: MTPC / MTPF limitation

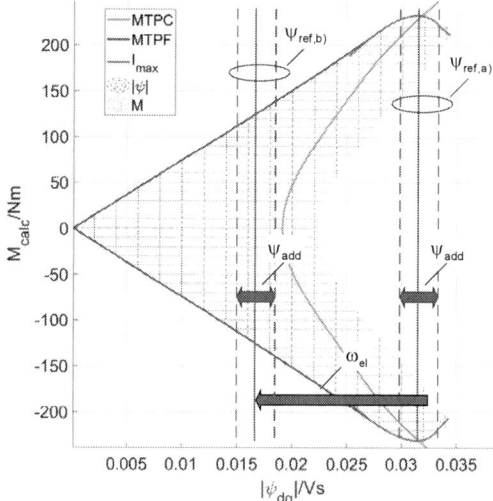

Fig. 7: Limitation over $\left|\psi_{dq}\right|$ and M_{ref}

a) Base speed area: The flux setting is limited by the MTPC. The output of the flux observer has no direct impact on the current reference computation. The observed additional flux will only extend or reduce the area within the system operating on the MTPC line.

b) Flux weakening area: The flux selection is limited directly by the maximum available flux ψ_{ref}. It is directly affected by the additional flux output of the flux observer.

III. Available Torque Calculation and Limitations

For the interaction with the other components of the drive train, a feedback value for the expected maximum and minimum torque related to the actual operation point is needed, as shown in Fig. 8. As the drive operates at its torque limit, the maximum available torque $M_{max,avl}$ is used as limitation for the internal torque reference.

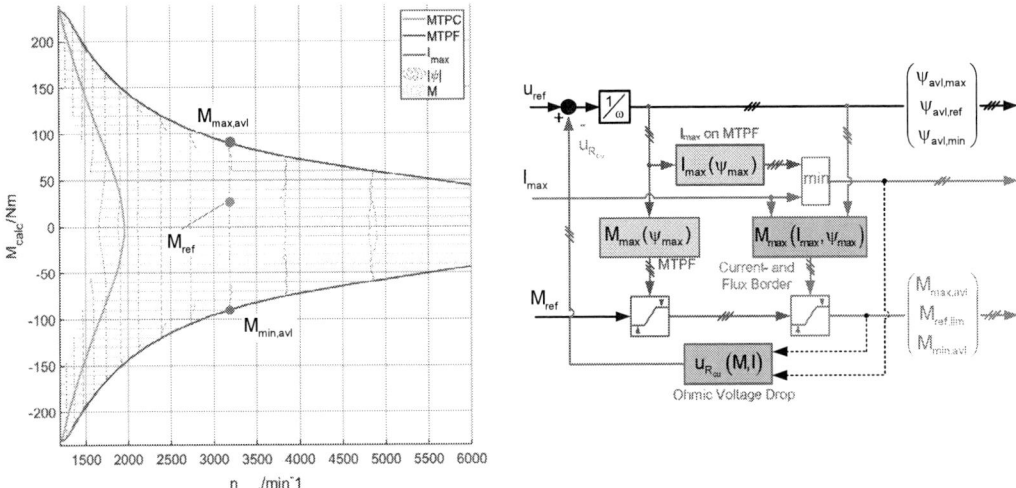

Fig. 8: Limitation over n_{cmd} and M_{ref} Fig. 9: Available Torque Computation

III.1. Available Torque Calculation

As the current and the ohmic voltage drop varies for the three operation points with the torque commands (M_{ref}, $M_{min,avl}$, $M_{max,avl}$), the available fluxes ($\psi_{avl,ref}$, $\psi_{avl,min}$, $\psi_{avl,max}$) are different. To consider this, the whole limitation structure is calculated for these three torque values.

The impact of the ohmic voltage drop requires one iteration as shown in Fig. 9. The value used for the stator resistance can be calculated from the observer disturbance or from the measured stator temperature.

The observed flux mismatch ψ_{add} introduced in Fig. 3 is added to all three available flux values. With this structure, the system can also handle the impact of a mismatch in the absolute flux to the torque limitation. This is necessary to minimize the torque deviation, appearing as the difference between the torque command to the open loop torque control and the real torque on the shaft of the PMSM. In case this observed flux mismatch is caused by an incorrect ohmic resistance, this is not exact but as a flux mismatch can be caused by other impacts, this is an expectable estimation. For the system used in this contribution, the observed value results in an improvement of the accuracy especially for high current amplitudes.

III.2. Torque Limitation

When the ipmsm is commanded to maximum available torque, it is necessary that the torque reference matches with the calculated available maximum. A deviation between these values is interpreted as malfunction by the superposed controller structure.

For the used controller structure, the additional flux calculated by the flux observer is added, as well to the flux path for the optimum limited flux calculation $\psi_{opt,lim}$, as to the available flux for the torque limitation on the current and flux border and the flux border given by the MTPF line.

IV. Results

IV.1. Dynamic Behaviour and Stability

IV.1.1. Torque Step

The dynamic behaviour was evaluated with measurements of internal parameters and values, by commanding steps in different input values. The measurements were done for an IPMSM with the following parameters.

$$U_{dc} = 48V \qquad M_{max} = 200Nm \qquad I_{max} = 750A \qquad n_{max} = 7000\,min^{-1} \qquad p = 12$$

The test performed was a step response for a change in modulation index m_{idx} from 0.9 to 1.2, while the IPMSM was operating in the flux weakening area, with $M_{ref} = 30\,Nm$ and $n_{act} = 1500\,min^{-1}$. Fig. 10 shows the behaviour of the internal flux calculation. With the commanded step in the modulation index m_{idx}, matters the fluxes to increase. As the IPMSM operates at the voltage limit, the actual flux ψ_{act} and the flux reference match. In Fig. 13, the dq-current and voltage values are monitored ($i_{dq,act}$, $u_{dq,act}$). The commanded step mainly effects the d-current and voltage.

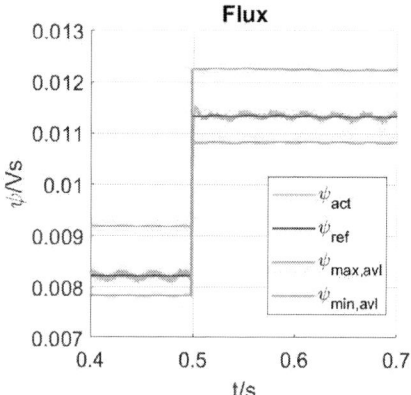

Fig. 10: Internal flux values

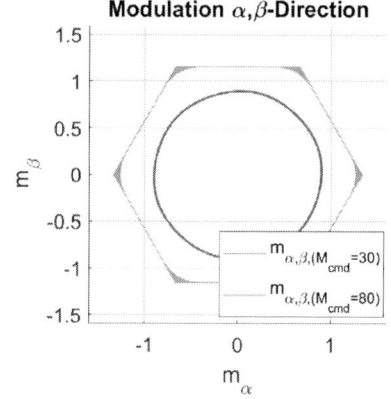

Fig. 11: modulation index over $\alpha - \beta$

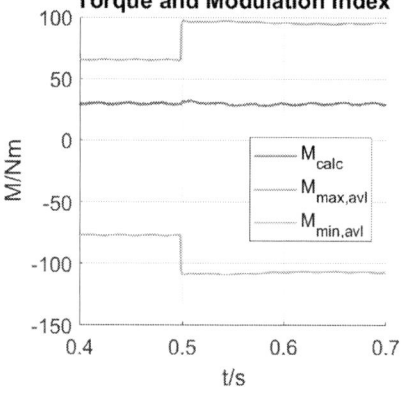

Fig. 12: Torque calculation

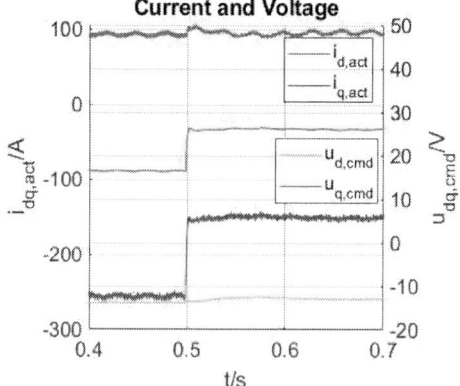

Fig. 13: Current and Voltage

Fig. 11 shows the modulation index vectors $m_{idx,\alpha\beta}$ for the $u_{dq,act}$-voltage after the reverse dq-transformation. The red chart shows vector points before the step. Here the voltage is smaller than the limitation for the linear voltage range. After the step in modulation index m_{idx}, the over-modulation range of the inverter is used, which results in a hexagonal shape for the modulation index locus.

Fig. 12 shows the calculated torque values. The commanded step has almost no impact on the calculated actual torque M_{act}. The limits for the maximum and minimum available torque in generator and motor mode have increased as more voltage and herby more flux is available.

Changing the modulation index might be necessary as the over modulation area must not be used in the lower speed range, where the harmonics harshness has an unacceptable impact on the ac and dc currents. The possibility to reserve some voltage might also be useful, in case a more dynamic step response following a change in the torque command is required.

IV.1.2. Speed Ramp

To demonstrate the stability of the system for the flux observer structure, speed up ramping with different speed slopes dn/dt and torque commands M_{ref} are performed for the full speed range.

$$M_{ref} = -80\,\text{Nm} \qquad n_{min} = 100\,\text{min}^{-1} \qquad n_{max} = 7000\,\text{min}^{-1} \qquad m_{idx} = 1.21 \qquad dn/dt = 4000\,\text{min}^{-1}/s$$

The diagrams show the same values as the diagrams in the previous Figures. For this test, these values have been plotted over the speed n, to represent the behaviour of a fast speed up slope dn/dt. Within the base speed area, the actual flux does not follow its reference. By reaching the voltage limit, the d-current i_d starts to decrease until the systems operating point has reached the MTPF line.

On the MTPF line the value of i_d is nearly constant and the q-current i_q is decreasing. The torque values are still very stable. When the torque reference M_{ref} matches its limitation $M_{max,avl}$, the impact of the flux observer could be seen. This behaviour was forced by using an incorrect value for the ohmic stator resistance R_{cu} for this measurement. At around $1000\,\text{min}^{-1}$, the voltage limit has been reached, and the flux controller forces the actual flux and hereby the output voltage $|u_{dq}|$, to the maximum available value.

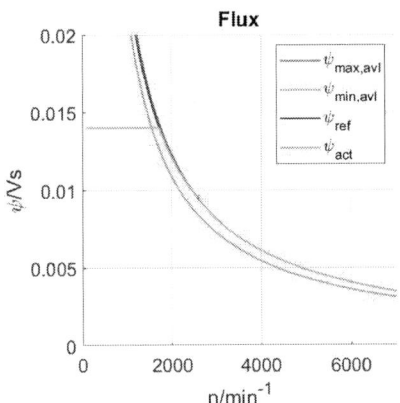

Fig. 14: Internal flux values / Speed ramp

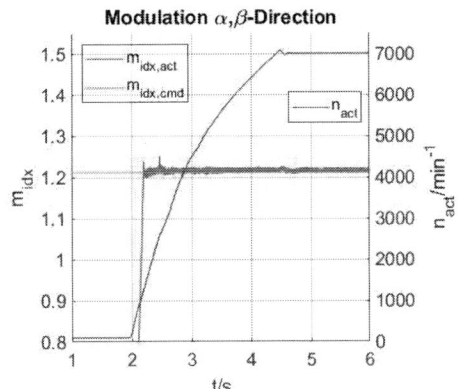

Fig. 15: modulation index over $\alpha - \beta$ / Speed ramp

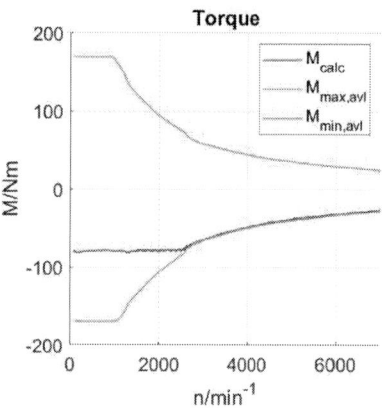

Fig. 16: Torque calculation / Speed ramp

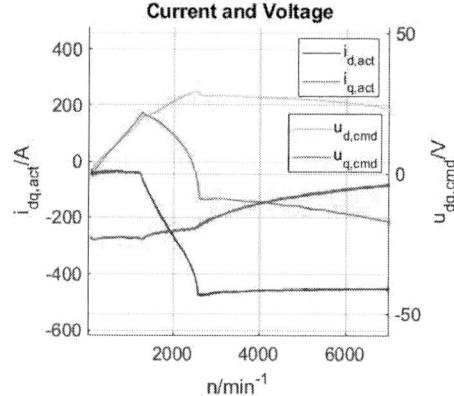

Fig. 17: Current and Voltage / Speed ramp

IV.2. Voltage Usage and Performance Test

A performance test for the IPMSM is done to evaluate the core control system and to generate data required for the system integration, the power and efficiency estimation. For this test, an array of points on a torque over speed map was commanded. The resulting torque, current and voltage values were measured with a power analyser and a torque shaft.

$$M_{min} = -180 \text{Nm} \qquad M_{max} = 170 \text{Nm} \qquad n_{max} = 7000 \text{min}^{-1} \qquad \text{Measured Points} = 3212$$

The performance test results can also be used to evaluate the behaviour of the flux observer. Fig. 18 shows the resulting fundamental voltage across the ac terminals $\hat{u}_{h01}$ with the flux observer turned off. To guarantee a stable settling of the current controller, the modulation index m_{idx} and hereby the voltage $\hat{u}_{h01}$ was reduced. As the system does not operate at the voltage limit in the base speed range, some points of this area were deleted to get a more detailed scaling for the voltage colour chart. Obviously the voltage fluctuates between $\hat{u}_{h01} = 11 \text{V}$ and $\hat{u}_{h01} = 25 \text{V}$, especially in the generating area for very high speed where a misalignment of the absolute flux has the most significant impact, as it is affected by an incorrect voltage amplitude.

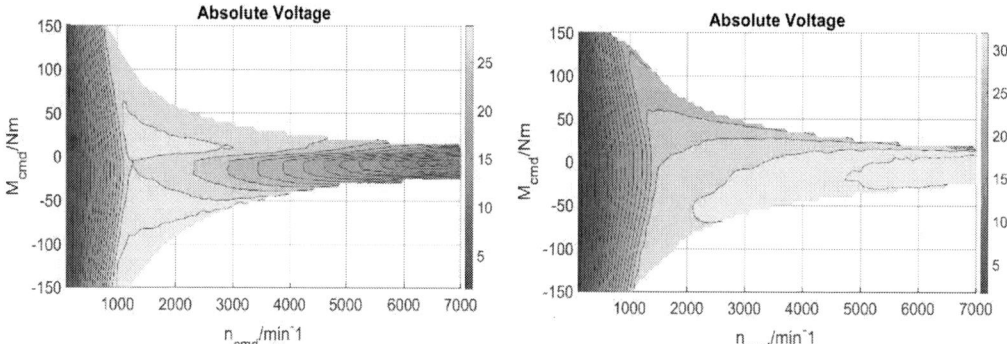

Fig. 18: Voltage usage / Observer deactivated   Fig. 19: Voltage usage / Observer activated

Fig. 19 contains the same values of a performance test with the flux observer turned on. There is still same variation in the voltage amplitude $\hat{u}_{h01}$, but the improvement is significant. The deviation is also caused by a variation of the available voltage of the inverter hardware used. The major problem is the necessary switching dead time, as it causes a nearly constant drop in the available voltage.

II. Conclusions

With the flux observer structure recommended, it is possible to maximize the voltage usage. The observer can handle the transition between the base speed range and the flux weakening area in a very stable way, as there are no additional functionalities like anti-windup or on-off-switching necessary. The observer is using the unlimited voltage command as feedback, this is why it can react to the current controller, when it exceeds the limit and it is possible to operate at the maximum available voltage. The observed flux is also used to adjust the torque limitation. This is needed to reach the required torque accuracy for this hybrid drive system.

The observed flux is considered as the impact of a deviation between real ohmic resistance and the value used in the core control system. This is why it does not affect the torque reference itself. The flux limitation due to the available flux varies with the observed mismatch. The structure considers the variation of the available maximum and minimum torque, related to the available flux and voltage. The behaviour of the system was evaluated with several tests and measurements for the entire torque and speed range. Measurements for commanded steps in different values could demonstrate the stability and settling behaviour of the system. A performance test shows the improved voltage usage for many static operation points.

III.   References

[1]   M. Meyer and J. Böcker, "Optimum control for interior permanent magnet synchronous motors (ipmsm) in constant torque and flux weakening range," (EPE-PEMC), 2006.

[2]   W. Peters, O. Wallscheid, and J. Böcker, "A precise open-loop torque control for an interior permanent magnet synchronous motor (ipmsm) considering iron losses," (IECON), 2012

[3]   W. Peters, T. Huber and J. Böcker, "Control realization for an interior permanent magnet synchronous motor (ipmsm) in automotive drive trains," (PCIM Europe), 2011.

[4]   T. Huber, W. Peters and J. Böcker Voltage controller for flux weakening operation of interior permanent magnet synchronous motor in automotive traction applications (IEMDC), 2015

[5]   T.S. Kwon S.K. Sul "Novel Antiwindup of a Current Regulator of a Surface-Mounted Permanent-Magnet Motor for Flux-Weakening Control " IEEE Transactions on Industry Applications vol. 42 no. 5

[6]   N.V. Olarescu M. Weinmann S. Zeh S. Musuroi "Novel Flux Weakening Control Algorithm for PMSMS " (POWERENG), 2009

[7]   S. Bolognani, S. Calligaro , R. Petrella "Optimal voltage feed-back flux-weakening control of IPMSM" (IECON), 2011

[8]   S. Kim, Y. Lee, W. Choi, M. Kwak, Y. Lee, J. Seok "Maximum voltage utilization of IPMSMs using modulating voltage scalability for wide flux weakening applications" (ECCE), 2012

[9]   Wei Cong ; Xu Huihui ; Zhao Feng ; Zhang Jian "Precise torque control in flux weakening operation based on induction machines system" (ICEMS), 2016

[10]   F. Fernandez-Bernal, A. Garcia-Cerrada, R. Faure: Determination of Parameters in Interior Permanent Magnet Synchronous Motors with Iron Losses Without Torque Measurement, IEEE Transactions on Industry Applications 2001, Vol. 37, No.5.

[11]   M. Ganchev, C. Kral, T. Wolbank: Compensation of Speed Dependence in Sensorless Rotor Temperature Estimation for Permanent-Magnet Synchronous Motor, IEEE IEEE Transactions on Industrial Electronics 2013, VOL. 49, NO. 6

[12]   F. Bertele, U. Ammann, C. Cheshire, M. Neuburger, S. Piriienko and T. Roeser: Interpretation of Measured IPMSM Flux Tables for Parameter Identification, *EPE, 2019*

[13]   O. Wallscheid, A. Specht, J. Böcker: Observing the Permanent-Magnet Temperature of Synchronous Motors Based on Electrical Fundamental Wave Model Quantities, IEEE Transactions on Industrial Electronics 2017, Vol. 64, No. 5.

[14]   W. Peters, Wirkungsgradoptimale Regelung von permanenterregten Synchronmotoren in automobilen Traktionsanwendungen unter Berücksichtigung der magnetischen Sättigung. PhD thesis, Universität Paderborn, Germany, 2015.

Extended Boost PV inverter topology for the reduction of common-mode leakage current in three-phase applications

Georgios I. Orfanoudakis
HELLENIC MEDITERRANEAN
UNIVERSITY (HMU), GREECE
Department of Electrical Engineering
Estauromenos, 71004
Heraklion, Crete, Greece
Tel: +30 2810 379712
E–mail: gorfas@hmu.gr

Eftychios Koutroulis
TECHNICAL UNIVERSITY OF CRETE
(TUC), GREECE
School of Electrical & Computer
Engineering, University Campus
Chania, Greece
Tel.: +30 28210 37233
E–mail: efkout@electronics.tuc.gr

Michael A. Yuratich
TSL TECHNOLOGY LTD
One Ropley Business Park
Ropley, UK
Tel.: +44 / (0) – 1962 772020
E–mail: mike.yuratich@tsltechnology.com
URL: www.tsltechnology.com

Suleiman M. Sharkh
UNIVERSITY OF SOUTHAMPTON, UK
Mechatronics Research Group
Faculty of Engineering and Physical Sciences
Southampton, SO17 1BJ, UK
Tel.: +44 / (0) – 2380 592339
E–mail: s.m.sharkh@soton.ac.uk

Keywords

«Common-mode current», «Ground leakage current», «Boost inverter», «Step-up inverter», «Transformerless», «PV inverter»

Abstract

Photovoltaic (PV) inverters must not generate high levels of common-mode (CM) ground leakage currents, which may flow due to fast variations in their CM voltage. Moreover, a highly desirable feature for PV inverters is their voltage step-up capability, which enables operation with lower DC input voltage levels, thus extending energy harvesting over wider solar irradiance and temperature ranges. This paper presents a new three-phase Boost PV inverter topology and its modulation strategy, designed based on a new concept for suppressing CM leakage currents in three-phase PV inverters. In contrast to past-proposed three-phase step-up topologies, the proposed PV inverter uses a modified Boost converter instead of a transformerless inverter, as well as a modulation strategy generating high-quality PWM voltage. Due to their combined characteristics, the variation of the CM voltage generated by the proposed PV inverter is minimized, which radically improves its CM current suppressing capability. The structure and operating principle of the proposed topology are presented, and its effectiveness with respect to CM current suppression is demonstrated through simulation results in MATLAB/Simulink.

1. Introduction

Common-mode (CM) voltage generated by DC/AC inverters operating with Pulse Width Modulation (PWM) raises concerns in grid-connected photovoltaic (PV) applications, since the ground leakage currents that appear due to CM voltage variations cause deterioration of the PV cells and safety hazards [1, 2]. Typical methods for suppressing CM leakage currents include CM chokes and modified DC/AC inverter modulation strategies, which reduce CM voltage variation. These methods, however, are not effective at suppressing leakage currents in systems with significant capacitive coupling to ground, such as PV systems. The solution initially applied to such systems was the use of isolation/step-up transformers which completely prevent the circulation of ground leakage currents. Nevertheless, due to the increased cost, size and weight of isolation transformers, other solutions were sought, which lead to the development of transformerless PV inverter topologies. These topologies have found great acceptance in the field of PV integration, particularly for single-phase PV inverters [1 – 3]. The single-phase transformerless PV inverter concepts, however, do not perform equally well with respect to CM current suppression when applied to three-phase PV inverters, as the latter still generate a significantly varying CM voltage [4 – 11].

In addition to CM current suppression, a highly desirable feature for PV inverters is their voltage step-up capability that enables operation with lower DC input voltage levels, thus extending energy harvesting over wider solar irradiance and temperature ranges. Several single-phase step-up inverter topologies have been proposed in the literature, operating based on a number of different concepts regarding the waveform of their DC-link voltage [12]. Again, only a few of these concepts are applicable to three-phase inverters.

This paper proposes a three-phase Boost PV inverter topology and its modulation strategy operating based on a new concept for suppressing CM leakage currents in three-phase PV inverters. Due to this feature, the variation of the CM voltage generated by the proposed PV inverter is minimized, which radically improves the CM current suppressing capability. The structure and operating principle of the proposed topology is presented, and its CM current suppression capability is demonstrated through simulation results in MATLAB/Simulink.

2. Background and Literature review

The CM voltage of a three-phase PV inverter is defined as the average of the three output terminal voltages (v_a, v_b, v_c) with respect to a reference point in the inverter circuit, such as the negative DC-link terminal:

$$v_{CM} = \frac{v_a + v_b + v_c}{3} \tag{1}$$

Fast variation of the PV inverter CM voltage gives rise to CM currents, due to the existence of parasitic capacitances between the solar cells of the PV array and the ground. Therefore, CM currents may flow through the (typically) grounded electric grid neutral, as illustrated in Figure 1 [1 – 3]. In order to prevent the generation of CM leakage currents, the CM voltage has to be kept as constant as possible. However, this is not practically possible for conventional three-phase PV inverter topologies, as the generation of PWM voltages by the power circuit of the inverter requires transitions between different inverter states (vectors) during each switching period. For a three-phase two-level PV inverter, the CM voltage, V_{CM}, generated in each inverter state is presented in Figure 2. When using conventional modulation methods, such as the Sinusoidal PWM (SPWM), the Third Harmonic Injection PWM (THIPWM), or Space Vector Modulation (SVM), the value of V_{CM} is altered between different values during each switching period.

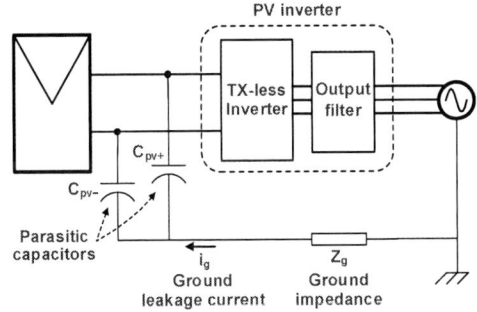

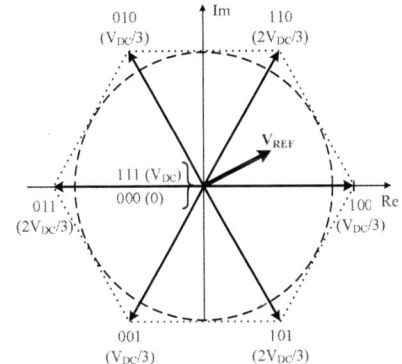

Fig. 1: Ground leakage current circulation path. Fig. 2: V_{CM} for each three-phase PV inverter state.

For example, when the inverter operates with the reference voltage vector, V_{REF}, that is illustrated in Figure 2, V_{CM} successively takes the values of $V_{DC} \times$ 0, 1/3, 2/3, 1, 2/3, 1/3, 0. Because these transitions are performed at a switching frequency of several kHz, at which the impedance of parasitic capacitors is low, they give rise to large CM leakage currents.

In order to reduce these currents, different approaches have been adopted in the literature. First, three-phase PV inverter modulation techniques that reduce the number of CM voltage changes or their effect on the generation of CM currents have been devised. Examples of such techniques can be found in [4 – 7]. Yet, these modulation techniques on their own cannot eliminate the variation of the CM voltage, as the PV inverter inevitably has to apply different vectors to generate the requested output voltage. As a second step towards the enhancement of CM current suppression, modifications to the basic two-level inverter topology have been proposed. Examples of resulting topologies, normally derived from single-phase transformerless PV inverter concepts, can be found in [8 – 11]. Multilevel PV inverter-based topologies have been proposed, as well, which however exhibit requirements for separate/isolated or high PV voltages [15 – 17].

Regarding the application of single-phase step-up inverter concepts to three-phase step-up PV inverters, depending on the form of their DC-link voltage (according to a classification of single-phase step-up topologies presented in [12]), the offered possibilities are the following:

A) Constant DC-link voltage topologies: These normally consist of a voltage step-up stage followed by a transformerless H-bridge inverter. The H-bridge is extended or modified to generate a certain level of CM voltage or to isolate the grid from the DC link during states that generate undesirable CM voltages. For single-phase H-bridge inverters, both approaches are very effective with respect to reducing the CM leakage currents and have led to the creation of commercially successful PV inverter structures such as the H5 topology by SMA and others [18 – 22]. However, for three-phase PV inverters, such a stabilisation of the CM voltage cannot be achieved due to the higher number of switching vectors and the resulting CM voltage levels. Topologies applying this concept to three-phase inverters achieve a reduction but not elimination of the CM voltage variation. Moreover, such topologies may operate based on modified modulation techniques, which typically deteriorate the quality of the PV inverter output voltage.

B) Pseudo-DC-link voltage topologies: These consist of a DC-DC converter, controlled to generate a rectified sinewave voltage, and an H-bridge which simply unfolds it to a sine-wave (i.e. changes the polarity of every other half cycle). This approach cannot be effectively applied to three-phase PV inverters, as three different DC links would be required.

C) Pulsating DC-link voltage topologies: They are Z-source type topologies, which offer several advantages such as high efficiency and high voltage step-up capability, but exhibit high CM currents in single-phase applications. When applied in three-phase PV inverters, they require sophisticated modulation techniques to achieve a reduction of the ground leakage current [23, 24].

D) Integrated DC-link voltage topologies: These topologies do not include a distinct DC link stage; they integrate it in the overall converter circuit. Only a few members of this category can be extended to form three-phase PV inverters. One of them is the half-bridge Aalborg inverter [13, 14], which guarantees minimal CM current, due to the permanent connection of one of its output terminals to the grid neutral. However, since two isolated PV arrays are required for the operation of the single-phase topology, six such PV structures would be required for its three-phase equivalent.

As explained in the following sections, the proposed PV inverter topology is based on the conventional two-stage (step-up, two-level inverter) architecture and operates with a constant DC-link voltage, thus belonging to the first category of three-phase step-up PV inverters. Unlike other members of the same category, it uses a modified Boost converter instead of a transformerless inverter and a modulation strategy generating high-quality PWM voltage.

3. Proposed Boost PV inverter topology

The topology of the proposed PV inverter is illustrated in Figure 3. In this schematic, the DC-link capacitor C2 and the DC/AC inverter block on the right form a conventional two-level inverter configuration. The rest of the components are added to achieve voltage step-up and CM current suppression, as analysed next.

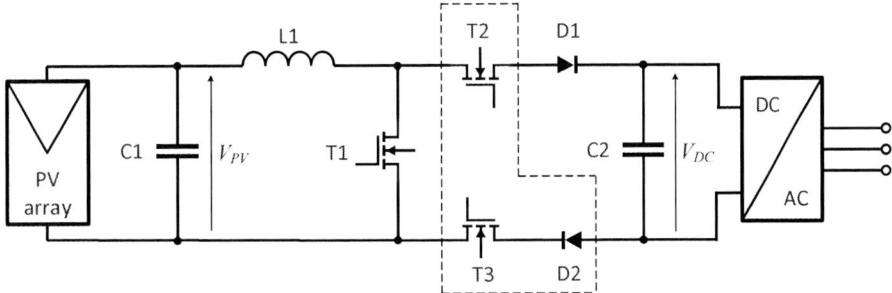

Fig. 3: Proposed Boost PV inverter topology.

The concept of the proposed topology was to use a two-stage DC-link structure, where the first stage (C1):
- charges the second stage (C2) during a set of inverter states that result into the same level of CM voltage, and
- becomes isolated from the second stage (C2) during all other inverter states.

It can be observed that the added configuration (indicated by the dashed line) has characteristics in common with a Boost converter, normally formed only by L1, T1, D1, and C2. Based on that, an extended Boost converter with an additional diode, D2, and switching elements T2-T3 at the positive/negative DC rails is formed. Although these two switching elements are turned ON and OFF simultaneously, thus the second switch does not offer anything to the charging procedure, it is included in the PV inverter circuit in order to block ground leakage currents that can flow through any of the positive and negative rails. Similarly, although diode D2 is not required in a Boost converter circuit, it has to be added to prevent leakage currents from flowing through the anti-parallel diodes of T2-T3.

4. PWM strategy for leakage current suppression and DC-link voltage control

As already stated, the switching elements T2-T3 of the added configuration are turned ON (simultaneously) when the PV inverter outputs a vector belonging in a certain set of vectors that result in the same value of V_{CM}. This set will be herein named as $\{V\}$ and could be one of 0H(igh) – 3H(igh), listed in Table I. T2-T3 are switched off during all inverter states not belonging in $\{V\}$, thus isolating the PV array from C2 (see Figure 3). As explained in Section 2, this operating principle ensures minimal leakage current generation.

Table I: Sets of switching vectors and resulting values of V_{CM}

$\{V\}$ symbol	$\{V\}$ including	V_{CM}
0H	000	0
1H	100, 010, 001	$1/3 \times V_{DC}$
2H	110, 011, 101	$2/3 \times V_{DC}$
3H	111	V_{DC}

1. SVM

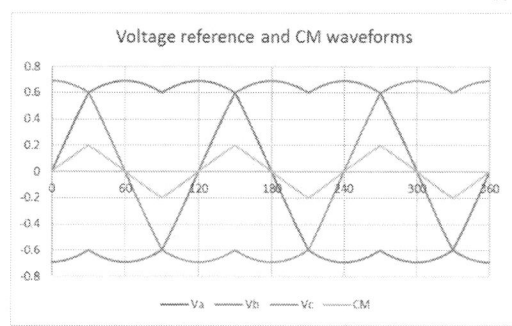
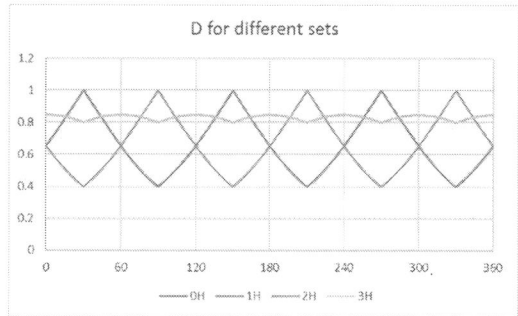

2. Proposed modulation strategy

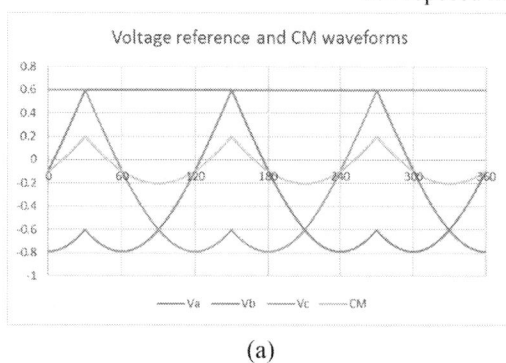
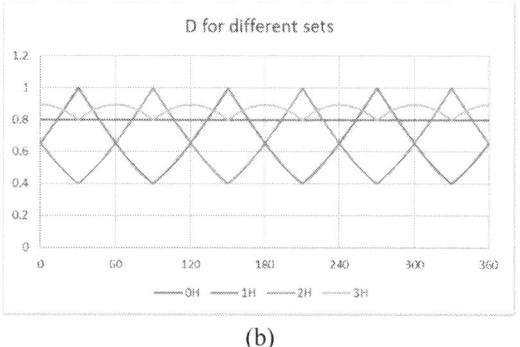

(a)   (b)

Fig. 4: (a) Inverter reference voltages and common-mode signal, and (b) Boost converter duty cycle for each switching vector set, for the classical SVM and the proposed modulation strategies, respectively, with $M = 0.8$.

Since the extended Boost PV inverter proposed in this paper is designed to normally operate in the continuous-conduction mode (CCM), the current of L1 will have to flow through T1 for the time periods that T2-T3 are turned OFF. The duty cycle, D, of the Boost converter stage will therefore be equal to the sum of the duty cycles of the PV inverter states *not* belonging in {V}. Figure 4.1(b) illustrates the variation of D over one fundamental cycle for all possible selections of {V}, assuming that the inverter is modulated using SVM and the modulation index, M, is equal to 0.8 (with a maximum of $2/\sqrt{3}$).

It can be observed that if {V} is selected to be 1H or 2H, significant duty cycle variations occur at 3 times the fundamental frequency. These variations cause undesirable voltage fluctuations on capacitor C2, thus should be avoided. Consequently, {V} is selected to be either 0H or 3H, which results in equal and low duty cycle variations. The selection of 0H or 3H, which corresponds to T1 being turned ON at all times except when the zero vector 000 or 111 is enabled, respectively, also offers another benefit: it provides inherent negative feedback to the DC-link voltage control loop, since the duty cycles of the non-zero vectors, thus also D, decrease as the DC-link voltage increases. The above selection therefore prevents the DC-link voltage from oscillating, increasing excessively or collapsing.

Furthermore, the modulation strategy can be modified to eliminate the duty cycle variations, as follows. The common-mode signal $v_{CM,ref}$ added to the three normalised sinusoidal inverter reference voltages, $v_{a,b,c,ref}$, is set equal to:

$$v_{CM,ref} = 1 - max(v_{a,ref}, v_{b,ref}, v_{c,ref}) - C \tag{2}$$

where C is an offset that can be adjusted to vary the value of the Boost converter duty cycle, D.

The two variables relate to each other through:

$$D = 1 - C/2 \qquad (3)$$

Figure 4.2(b) illustrates the effect of adopting the proposed modulation strategy, with C set to 0.4. It can then be observed that for 0H being the selected {V} set, the value of D is constant and equal to 0.8.

It should be noted that, due to known limitations in the operation of a Boost converter operating with high duty cycle (e.g. high inductor current, low efficiency), the value of C cannot practically be lower than $0.2 - 0.3$, which results in D being in the range of $0.85 - 0.9$. The fact that C cannot be set to zero, results in none of the three voltage reference waveforms being fixed to ± 1 for significant portions of the fundamental period, which happens with discontinuous PWM strategies. Moreover, the proposed strategy does not alter the ordinary sequences of the inverter switching states, which always include both zero vectors (000 and 111). As a result, the proposed strategy retains the high output voltage quality of continuous PWM strategies, which are typically used.

The maximum allowed value for C is determined by the need to avoid inverter over-modulation. Over-modulation will occur if any of the three voltage reference waveforms exceeds ± 1 during the fundamental cycle. Assuming that the selected {V} set is H0, it can be observed in Figure 4.2(a) that the maximum value of these waveforms is equal to $(1 - C)$, thus can never exceed $+1$. Their minimum value is given by the following formula:

$$v_{ref,min} = 1 - C - \sqrt{3}\,M \qquad (4)$$

In order to ensure that the above value does not drop below -1, C must be restricted as follows:

$$C \leq 2 - \sqrt{3}\,M \qquad (5)$$

Equivalently, based on (3), it holds that:

$$D \geq \frac{\sqrt{3}}{2}\,M \qquad (6)$$

Equations (5) and (6) pose a constraint on the maximum acceptable value of modulation index, M, for the PV inverter of the proposed topology. Namely, it can be seen that the proposed PV inverter cannot be operated with M approaching $2/\sqrt{3}$, which corresponds to the limit of the linear modulation region for common modulation strategies (e.g. THIPWM, SVM). Assuming that the highest acceptable value for D is close to 0.9, the maximum value for M is approximately equal to 1 (as for SPWM). In practice, this means that the inverter will have to operate with a higher DC-link voltage level than normal, by at least 15%.

DC-link voltage control is achieved by varying C and therefore D. Increasing C causes a reduction in D, according to (3), which, in turn, reduces the DC-link voltage. Operation of an inverter with a low DC-link voltage (i.e. not much higher than the value required to avoid over-modulation) and a high modulation index, is generally desirable, since it reduces the inverter switching losses and the voltage stress on its components. Nevertheless, according to (5) and (6), C and D have to remain low and high, respectively, if the proposed PV inverter operates with a high modulation index, which leads to a high PV output voltage, V_{PV}, step-up by the Boost converter stage according to:

$$V_{DC} = \frac{V_{PV}}{1 - D} \qquad (7)$$

In order to keep the DC-link voltage as low as possible, C has to be set to its maximum and D to its minimum by using (5) and (6) as equalities, in which the value of M is determined by the magnitude of the reference output voltage. Note that (7) holds for an ideal Boost converter and is adopted to facilitate the theoretical analysis of the proposed approach, while it represents a worst-case condition with respect to a DC-link voltage increment.

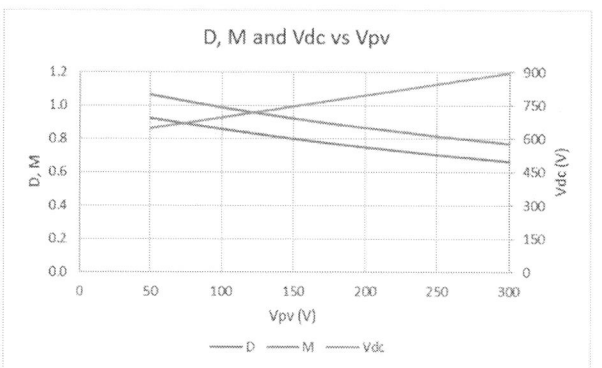

Fig. 5: Variation of D, M (primary axis) and V_{DC} (secondary axis) as a function of V_{PV}, when operating the proposed PV inverter based on the proposed PWM control strategy.

Figure 5 illustrates an example of the proposed PV inverter operation based on the PWM control strategy described above. The figure plots D, M (primary axis) and V_{DC} (secondary axis) as a function of V_{PV}. It is assumed that the PV inverter reference RMS voltage is equal to 420V, which was selected as a typical value for a three-phase inverter connected to a 400V grid through an LCL filter, also accounting for voltage drops across the inverter semiconductors and voltage loss due to dead time. The DC-link voltage required to provide this output voltage when using a conventional modulation strategy is approximately equal to 600V. However, when applying the proposed approach, the resulting DC-link voltage can be increased significantly as the output voltage of the PV array increases. In order to limit the undesirable effects mentioned above on the inverter switching losses and component voltage stress, the PV array must preferably be selected/connected so that V_{PV} is in the range of $100 - 150V$.

It should be noted that the proposed three-phase PV inverter topology and modulation strategy retain the ability of conventional Boost PV inverters with respect to Maximum Power Point Tracking (MPPT). Although the values of D and M are related through (6), fast perturbations over a limited range can be performed independently on each control variable without departing from the operating principle described above. Namely, the PV inverter reference voltages (and thus the value of M) can be varied at a multi-kHz rate as dictated by the PV inverter current control algorithm, to inject a sinusoidal current and the desired amount of real/reactive power to the grid. These variations, however, do not affect the value of D, due to the fact that they are cancelled out by the injected common-mode signal defined in (2). The value of D can then be adjusted according to (3), by varying the offset C in the area allowed by (5).

5. Simulation results

In this section, MATLAB/Simulink simulation results are presented, illustrating the operation of the proposed topology as a 5kVA three-phase step-up PV inverter. The PV array is simulated as shown in Figure 6, as a DC source with voltage $V_{PV} = 100V$ and parasitic capacitances to ground split equally between its positive and negative terminals, with $C_{par+} = C_{par-} = 250nF$ (giving a total of 100nF per PV array kW). The values of L1 and C2 were set to 400µH and 1mF, respectively, so that the Boost stage operates in CCM. The inverter is switched at 10kHz and its three outputs, marked with connection ports 1 – 3, are connected to the three-phase 400V/50Hz grid through an LCL filter with $L_{f1} = L_{f2} = 5mH$, which corresponds to $X_L = 5\%$ p.u. and $C_{delta} = 1µF$ (filter and grid not shown). The grid is assumed to have a phase inductance $L_{grid} = 0.5mH$. The ground resistance R_{ground}, through which the leakage current returns to the grounded neutral of the grid, is assumed to have a value of 1Ω. Figure 6 also presents a simple implementation of the PWM strategy proposed in Section 3.2, according to which T2-T3 are kept turned ON during 0H, that is, for the periods when none of the upper PV inverter modules is in the ON state.

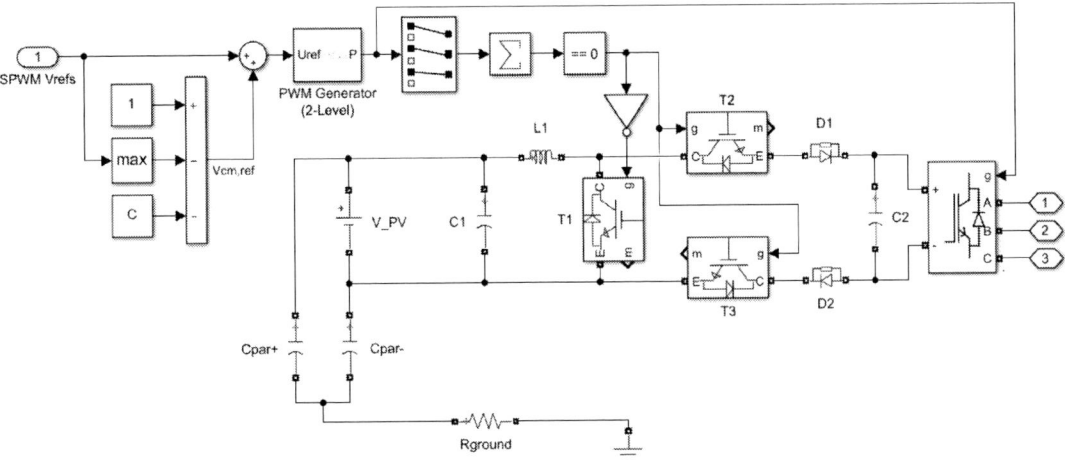

Fig. 6: Subsystem of the MATLAB/Simulink simulation model illustrating the derivation of $v_{CM,ref}$ for the proposed inverter modulation strategy and the method for controlling the Boost converter (T1-T2-T3).

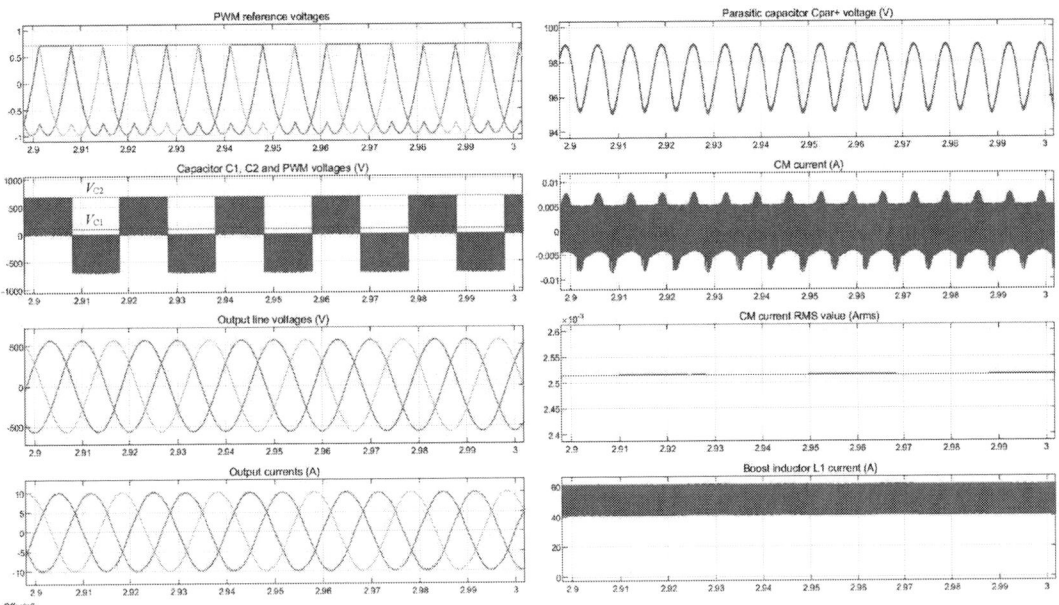

Fig. 7: Simulation results illustrating the operation of the proposed Boost PV inverter topology.

Figure 7 illustrates the resulting set of waveforms (in columns named A – B and rows numbered 1 – 4), which demonstrate the operation of the proposed topology and its effect in suppressing the CM leakage current. The three PWM reference signals (A1) are modified according to the proposed modulation strategy and have the form presented in Figure 4.2(a). The Boost converter operates with $D = 0.86$, which has been adjusted by injecting an offset $C = 0.28$ to the reference signals, so that their negative peaks approach the value of -1 (i.e. the limit of the linear modulation range). The actual value of the modulation index under these conditions is $M \approx 0.98$, while the resulting DC-link voltage is $V_{DC} \approx 685\text{V}$ (A2), which are in agreement with the values shown in Figure 5 for $V_{PV} = 100\text{V}$. Moreover, the modulation strategy can be seen not to adversely affect the quality of the output voltage (A3) and current (A4) waveforms. The effectiveness of the proposed topology in suppressing CM leakage currents is demonstrated by the fact that the parasitic capacitor C_{par+} voltage (B1) exhibits minimal high-frequency variations, while the CM current peak/RMS values (B2, B3) are 8mApeak / 2.52mArms. To provide a comparison, the CM current values given by a model with a conventional Boost converter and the same operating parameters, using typical SVM for the inverter and controlling the Boost converter independently with $D = 0.86$, are 1.5Apeak / 0.75Arms. Finally, the Boost inductor current (B4) can be seen to have no low-frequency variations, which is due to the constant value of D (over a fundamental cycle) resulting from the application of the proposed modulation strategy.

6. Discussion

As illustrated in the previous sections, the proposed three-phase PV inverter topology and modulation strategy achieve the aims of CM current suppression and high output voltage-current quality, as opposed to other alternatives discussed in Section 2. The proposed topology does not require more than one isolated PV array, as multilevel PV inverters commonly do, but reduced efficiency is expected compared to those topologies, mainly due to the added components (T2, T3, D2) and the higher than normal DC-link voltage level. Another limitation of the proposed topology is the operation in a restricted PV output voltage range, required to avoid excessive DC-link voltages. Finally, modifications that are beyond the scope of this paper would be required to enable the proposed topology to encompass more Boost converters performing MPPT for independent PV arrays, as is commonly desirable for three-phase PV inverters. The same, however, is likely to hold for other three-phase topologies derived from single-phase step-up PV inverter concepts summarised in Section 2.

7. Conclusion

This paper presented a new three-phase Boost inverter topology and a modulation strategy adapted for it, to offer PV voltage step-up and CM current suppression. The structure and operation of the topology were analysed to illustrate that they support the fundamental concepts for CM current suppression and derive the limits of the proposed approach. Simulations in MATLAB-Simulink were performed to prove the validity of the analysis and provide estimates for CM current generation. The limitations of the proposed topology were also discussed, despite which it remains a promising candidate among the currently available alternatives for three-phase step-up PV inverter applications.

References

[1] Z. Özkan and A. M. Hava, "A survey and extension of high efficiency grid connected transformerless solar inverters with focus on leakage current characteristics," *IEEE Energy Conversion Congress and Exposition (ECCE)*, Raleigh, NC, 2012, pp. 3453-3460.

[2] H. F. Xiao, and S. J. Xie, "Leakage Current Analytical Model and Application in Single-Phase Transformerless Photovoltaic Grid-Connected Inverter," *IEEE Transactions on Electromagnetic Compatibility*, vol. 52, no. 4, pp. 902-913, 2010.

[3] D. Zografos, E. Koutroulis, Y. Yang and F. Blaabjerg, "Minimization of leakage ground current in transformerless single-phase full-bridge photovoltaic inverters," *17th European Conference on Power Electronics and Applications (EPE'15 ECCE)*, Geneva, 2015, pp. 1-10.

[4] T. K. S. Freddy, N. A. Rahim, W. Hew and H. S. Che, "Modulation Techniques to Reduce Leakage Current in Three-Phase Transformerless H7 Photovoltaic Inverter," in *IEEE Transactions on Industrial Electronics*, vol. 62, no. 1, pp. 322-331, Jan. 2015.

[5] X. Guo, R. He, J. Jian, Z. Lu, X. Sun and J. M. Guerrero, "Leakage Current Elimination of Four-Leg Inverter for Transformerless Three-Phase PV Systems," in *IEEE Transactions on Power Electronics*, vol. 31, no. 3, pp. 1841-1846, March 2016.

[6] Y. Sang, F. He, L. Yuan, Z. Zhao, S. Wei and T. Lu, "High efficient common-mode current suppression SVM method for three-phase three-level transformer-less photovoltaic inverters," *IEEE Energy Conversion Congress and Exposition (ECCE)*, Montreal, QC, 2015, pp. 6871-6876.

[7] J. C. Giacomini, L. Michels, H. Pinheiro and C. Rech, "Active Damping Scheme for Leakage Current Reduction in Transformerless Three-Phase Grid-Connected PV Inverters," in *IEEE Transactions on Power Electronics*, vol. 33, no. 5, pp. 3988-3999, May 2018.

[8] X. Guo, D. Xu and B. Wu, "New control strategy for DCM-232 three-phase PV inverter with constant common mode voltage and anti-islanding capability," *2014 IEEE Energy Conversion Congress and Exposition (ECCE)*, Pittsburgh, PA, 2014, pp. 5613-5617.

[9] L. Concari, D. Barater, C. Concar and G. Buticchi, "A novel three-phase inverter for common-mode voltage reduction in electric drives," *IEEE Energy Conversion Congress and Exposition (ECCE)*, Montreal, QC, 2015, pp. 2980-2987.

[10] L. Concari, D. Barater, G. Buticchi, C. Concari and M. Liserre, "H8 Inverter for Common-Mode Voltage Reduction in Electric Drives," in *IEEE Transactions on Industry Applications*, vol. 52, no. 5, pp. 4010-4019, Sept.-Oct. 2016.

[11] D. Ronanki, P. H. Sang, V. Sood and S. S. Williamson, "Comparative assessment of three-phase transformerless grid-connected solar inverters," *2017 IEEE International Conference on Industrial Technology (ICIT)*, Toronto, ON, 2017, pp. 66-71.

[12] W. Liu, K. Niazi, T. Kerekes and Y. Yang, "A Review on Transformerless Step-Up Single-Phase Inverters with Different DC-Link Voltage for Photovoltaic Applications," in *Energies*, MDPI, Open Access Journal, vol. 12 (19), pp. 1-17, September 2019.

[13] W. Wu, J. Ji and F. Blaabjerg, "Aalborg Inverter - A New Type of "Buck in Buck, Boost in Boost" Grid-Tied Inverter," in *IEEE Transactions on Power Electronics*, vol. 30, no. 9, pp. 4784-4793, Sept. 2015.

[14] H. Wang, W. Wu, H. S. Chung and F. Blaabjerg, "Coupled-Inductor-Based Aalborg Inverter With Input DC Energy Regulation," in *IEEE Transactions on Industrial Electronics*, vol. 65, no. 5, pp. 3826-3836, May 2018.

[15] Y. Wang and F. Wang, "Novel Three-Phase Three-Level-Stacked Neutral Point Clamped Grid-Tied Solar Inverter With a Split Phase Controller," in *IEEE Transactions on Power Electronics*, vol. 28, no. 6, pp. 2856-2866, June 2013.

[16] K. Kim, S. Lee, W. Cha and B-H Kwon, "Three-level three-phase transformerless inverter with low leakage current for photovoltaic power conditioning system," in *Solar Energy*, vol. 142, pp. 243-252, Jan. 2017.

[17] X. Guo, J. Zhou, R. He, X. Jia and C. A. Rojas, "Leakage Current Attenuation of a Three-Phase Cascaded Inverter for Transformerless Grid-Connected PV Systems," in *IEEE Transactions on Industrial Electronics*, vol. 65, no. 1, pp. 676-686, Jan. 2018.

[18] P. Knaup, "Inverter," International Patent Application, Publication Number: WO 2007/048420 A1, Issued May 3, 2007.

[19] M. Victor, F. Greizer, S. Bremicker and U. Hübler, "Method of Converting a Direct Current Voltage from a Source of Direct Current Voltage, More Specifically from a Photovoltaic Source of Direct Current Voltage, into a Alternating Current Voltage," US Patent 7,411,802 B2, Aug. 12, 2008.

[20] S. Heribert, S. Christoph and K. Jurgen, "Inverter for transforming a DC voltage into an AC current or an AC voltage". Europe Patent 1 369 985 A2, 2003.

[21] H. F. Xiao, S. J. Xie, C. Yang, and R. H. Huang, "An Optimized Transformerless Photovoltaic Grid-Connected Inverter," *IEEE Transactions on Industrial Electronics*, vol. 58, no. 5, pp. 1887-1895, 2011.

[22] T. Kerekes, R. Teodorescu, P. Rodriguez, G. Vazquez and E. Aldabas, "A New High-Efficiency Single-Phase Transformerless PV Inverter Topology," *IEEE Transactions on Industrial Electronics*, vol. 58, no. 1, pp. 184-191, Jan. 2011.

[23] N. Noroozi and M. R. Zolghadri, "Three-Phase Quasi-Z-Source Inverter With Constant Common-Mode Voltage for Photovoltaic Application," in *IEEE Transactions on Industrial Electronics*, vol. 65, no. 6, pp. 4790-4798, June 2018.

[24] N. Noroozi, M. Yaghoubi and M. R. Zolghadri, "A Modulation Method for Leakage Current Reduction in a Three-Phase Grid-Tie Quasi-Z-Source Inverter," in *IEEE Transactions on Power Electronics*, vol. 34, no. 6, pp. 5439-5450, June 2019.

A Robust Control Design to Real-Time Conditions and Modelling of a Microgrid

Iréna Horvatic[1,2], Delphine Riu[1], Moataz Elsied[2] et Sébastien Benjamin[2]
[1]Univ. Grenoble Alpes, CNRS, Grenoble INP, G2ELAB
21 Avenue des Martyrs,
38000 Grenoble, FRANCE
[2]SAFT / TOTAL
111 Boulevard Alfred Daney
33074 Bordeaux, FRANCE
E-Mail: irena.horvatic@g2elab.grenoble-inp.fr
[1]URL: https://g2elab.grenoble-inp.fr/

Acknowledgements

The authors would like to acknowledge the people from the BMM team at SAFT for their help and for having provided the data on the battery.

Keywords

« Microgrid», « Robust control», « Modelling».

Abstract

The current energy transition is resulting in the development of microgrids to support massive implementations of renewable electricity generation systems. This leads to a strong constraint on the management and sizing of these networks in order to optimize their performance and make it more reliable as regard a high level of uncertainties related to real-time operations. This paper focuses on the development of an H_∞ robust control strategy applied to an island microgrid for its primary frequency control. An analysis is carried out for unmodelled dynamics uncertainties that could originate from the dynamics of the employed sensors, in order to determine the sizing margin of the system with the aim of guarantying its stability and dynamic performance. The results of the robustness analysis will help us to know what is necessary to integrate from the design phase in order to guarantee the most optimal and robust control during the real-time implementation of the control algorithms.

Introduction

Microgrids become more and more important in strategies to access electricity worldwide, particularly in remote areas [1]. It is therefore crucial to ensure robustness, reliability and durability of these systems, especially when a high level of renewable electrical energy is introduced. Here, we focus on a case study with a standalone microgrid for "Les Saintes" islands in Guadeloupe, supplied by a photovoltaic farm, a diesel generator (DG) and an Energy Storage System (ESS) composed with Li-ion batteries.

In microgrids, it is crucial to ensure the balance between the consumed and the produced power under all operating conditions [1,2]. Indeed, these isolated grids are very sensitive to the slightest variation of the load or the source, which induces a fluctuation in the grid frequency that may exceed the imposed standards (denoted as Grid Codes). The ESS, thanks to its fast dynamics and the control of its power electronics converter, can participate in the primary frequency control and thus limit the grid frequency variation induced by disturbance of the load power.

Furthermore, the limited short-circuit power of the grid and the presence of renewable energy sources having a high variability rate, such as photovoltaic production in an island environment, introduce strong constraints and a lot of uncertainties on the power management and sizing of the grid [1]. In many cases, the theoretical optimality of energy management is lost in its practical implementation. These may be related to "unmodelled dynamics uncertainties" linked to performance and implementation of real components. This is not taken into account in most conventional management strategies, but induces significant differences between the modelled behaviour and the response of the real system [3,4].

The system's sensors play a crucial part in the implementation of the control. They serve to guarantee system properties such as disturbance rejection [3,5]. However, when designing the corresponding control, the actual dynamic behaviour of the sensor is not well known and therefore not taken into account. This neglected or not taken into account unmodelled dynamics uncertainty can affect measurements and potentially result in the loss of stability or desired performance of the system. This article focuses on the primary frequency control of an island microgrid confronted by small signal disturbances related to variations of the photovoltaic production or the load. We have designed an H_∞ robust control which is well indicated for problems of disturbance rejection for uncertain systems [1,2,3]. This control is widely used for the frequency control of isolated microgrids [1]. The design of such controller also involves ensuring stability and robust performance against dynamics uncertainties. Furthermore, an associated method of robustness analysis, denoted as "μ-analysis", will be used to analyse the robust stability (RS) and robust performance (RP) of the system with respect to the studied dynamics uncertainty. Finally, we will see how a modification of the chosen performance criteria of the controller impacts the RS and RP of the system confronted with dynamics uncertainties. The article is organized as follows: the first paragraph presents the studied microgrid, the scientific problem and the objectives of the control. The modelling and detailed design of the H_∞ control are introduced in the second paragraph. The third paragraph will detail the modelling of the studied dynamics uncertainties, the analysis of the robustness of the control with respect to these uncertainties, as well as the analysis of the corresponding robustness on variations of certain performance criteria of the control. Finally, the conclusion of the article will highlight the main ideas and perspectives of this work. All the simulations in this article have been validated with the Matlab©/Simulink© software.

Presentation of the studied system

The Microgrid

The studied microgrid, shown in Fig. 1, is powered by a diesel generator, a field of photovoltaic (PV) panels and a pack of lithium-ion batteries, representing the dynamic response of dispersed or centralized storage means. They are connected in parallel to a Point of Common Connection (PCC) and supply a common load. The PV panels and the storage system are interfaced with power electronics converters: a Maximum Power Point Tracker (MPPT) chopper and an inverter for the PV generator, and an inverter that allows bidirectional power flow for the storage system.

The base power of our microgrid and the moment of inertia are 50 kVA and 50 kWs respectively.

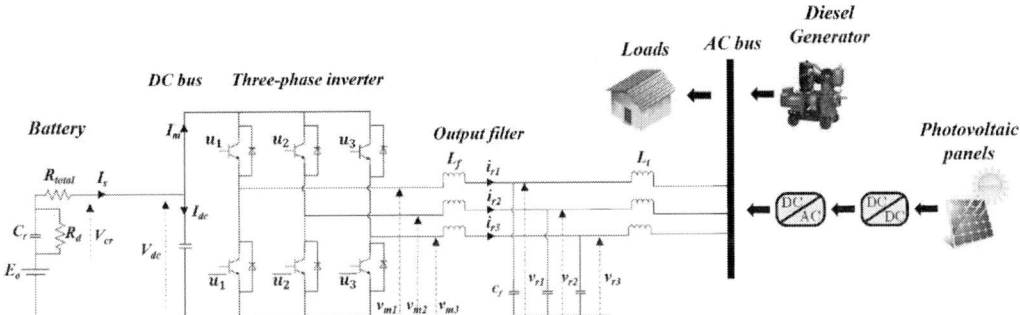

Fig. 1: Studied microgrid

Formulation of the problem

The choice of the H_∞ control for frequency regulation is well adapted for systems with high uncertainties and constraints on performance objectives. The H_∞ can deal with many control objectives such as disturbance rejection or robust stabilization of uncertain systems [1,3,6]. Moreover, the obtained solution is always optimal with respect to the defined criteria, which also means that if no solution to the control objectives is found, then no solution exists at all [6]. Finally, it is possible to combine the H_∞ control with the robustness analysis for unmodelled dynamics uncertainties. Such a control thus offers an ease of design and real-time implementation, which is particularly relevant for a designer of microgrids or highly constrained systems that are very sensitive to uncertainties [7]. Here, we focus on the frequency control to stabilise the imposed value independently of the load perturbations, and thus mainly deal with the problem of perturbation rejection. For this issue, the system performance objectives can be specified in terms of requirements on sensitivity functions [3,4,8]. In the time domain, system performance can be evaluated in terms of the overshoot, response time, damping and steady-state error [6].

Control objectives

In the studied microgrid, the output variable for the control strategy is its frequency (f_{grid}). The objective of the control is to improve the time response of this frequency in case of disturbances from the load or the PV production.

The ESS and the diesel generator are associated to participate in the primary frequency control. The role of the battery is to guarantee fast stabilization of the frequency following a load variation, i.e. to improve dynamic performance such as the overshoot, step response time and steady-state static error. A template is defined for a frequency variation of the microgrid (f_{grid}) in response to a load disturbance on the basis of its time response (see Fig. 2).

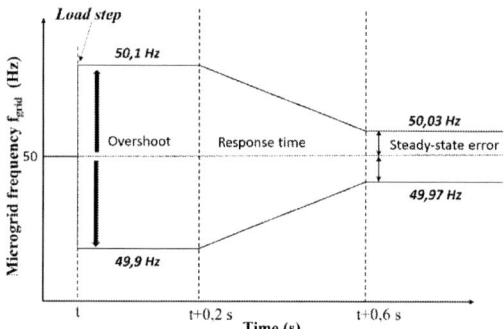

Fig. 2: Time-domain frequency variation performance template

Modelling, Design and Analysis of the H_∞ control

Modelling the system

There are different approaches to model systems with power electronics converters, such as the topological model [9,10,11] or the average model [9,11,12]. In this article, the energy storage system will be modelled using the variables averaged over the scale of the switching period of the switches. This nonlinear average model is linearized around the operation point by assuming that the battery output voltage remains constant. Finally, the linearized model will be described in per-unit variables, in order to improve the numerical conditioning of the system and to facilitate the optimization of the H_∞ controller. We obtain the following equations, where the underlined symbols indicate per-unit quantities:

$$\frac{1}{\omega_{ref}}\frac{d\Delta V_{cr}}{dt} = -\frac{\Delta V_{cr}}{\underline{C_r}*\underline{R_d}} - \frac{\Delta I_s}{\underline{C_r}}, \tag{1}$$

$$\frac{1}{\omega_{ref}}\frac{d\Delta V_{dc}}{dt} = \frac{\Delta I_s}{\underline{C_{dc}}} - \frac{1}{\sqrt{3}*\underline{C_{dc}}}\left(\beta_{de}\,\underline{\Delta I_{rd}} + \underline{I_{rde}}\Delta\beta_d + \beta_{qe}\,\underline{\Delta I_{rq}} + \underline{I_{rqe}}\Delta\beta_q\right), \tag{2}$$

$$\frac{1}{\omega_{ref}}\frac{d\Delta I_{rd}}{dt} = \frac{\sqrt{3}}{L_f}\beta_{de}\underline{\Delta V_{dc}} + \frac{\sqrt{3}}{L_f}V_{dce}\underline{\Delta \beta_d} + \omega_{gride}\underline{\Delta I_{rq}} + I_{rqe}\underline{\Delta \omega_{grid}}, \tag{3}$$

$$\frac{1}{\omega_{ref}}\frac{d\Delta I_{rq}}{dt} = \frac{\sqrt{3}}{L_f}\beta_{qe}\underline{\Delta V_{dc}} + \frac{\sqrt{3}}{L_f}V_{dce}\underline{\Delta \beta_q} - \omega_{gride}\underline{\Delta I_{rd}} - I_{rde}\underline{\Delta \omega_{grid}}, \tag{4}$$

where the status variables are: the battery voltage ΔV_{cr}, the DC bus voltage ΔV_{dc}, and the inverter output currents ΔI_{rd} and ΔI_{rq}, respectively along the d and q axis of the Park's coordinates for the grid voltage. The β_d and β_q are the average values of the inverter switching functions, ω_{ref} is the reference frequency for the transformation into per-unit variables, and ω_{grid} is the grid frequency. The modelling of the diesel generator is based on the Refs. [1,6]. Its power variation ΔP_{diesel}, which participates in primary frequency control, is dependent on the time constant T_{diesel}, the droop value s_{diesel} and the frequency Δf_{grid} [1, 6]:

$$\frac{d\Delta P_{diesel}}{dt} = \frac{1}{T_{diesel}}\underline{\Delta P_{diesel}} - \frac{1}{T_{diesel}\,s_{diesel}}\underline{\Delta f_{grid}}. \tag{5}$$

Following Refs. [1, 6, 13], the frequency dynamics of the microgrid are thus described by

$$\frac{d\Delta f_{grid}}{dt} = \frac{1}{2H}\left(\underline{\Delta P_{bat}} + \underline{\Delta P_{diesel}} + \underline{\Delta P_{PV}} - \underline{\Delta P_{load}}\right) - \frac{D_{load}}{2H}\underline{\Delta f_{grid}}, \tag{6}$$

where H denotes the equivalent inertia constant and D_{load} the load damping constant of the microgrid. We first consider the battery as a perfect source of voltage directly connected to the DC bus of the inverter. The H_∞ control then considers the voltages $\underline{\Delta V_{dc}}$ and $\underline{\Delta V_{cr}}$ as time-invariant parameters, and the state system is thus simplified. The current variation ΔI_{rd} is controlled by a fast current loop, which follows the variation of the reference current generated by the H_∞ control; the output of the controller is then equivalent to $\underline{\Delta I_{rd}} \equiv \Delta I_{rd}{}^{ref}$. Within these hypotheses, where we also neglect the terms $\frac{\omega_{base}}{\sqrt{3}*C_{dc}}I_{rde}\Delta\beta_d$, $\frac{\omega_{base}}{\sqrt{3}*C_{dc}}I_{rqe}\Delta\beta_q$ et $\underline{\Delta I_{rq}} \equiv \Delta I_{rq}{}^{ref} = 0$, the equations of state (1)-(6) are reduced to the following linear system (in per-unit variables) :

$$\begin{cases}\underline{\Delta \dot{x}} = A\underline{\Delta x} + B_1\underline{\Delta w} + B_2\underline{\Delta u} \\ \underline{\Delta y} = C\underline{\Delta x} + D_1\underline{\Delta w} + D_2\underline{\Delta u}\end{cases}, \tag{7}$$

where $\underline{\Delta x} = [\Delta P_{diesel}\ \Delta f_{grid}]^T$ is the state vector, $\underline{\Delta u} = \Delta I_{rd}{}^{ref}$ the input control, $\underline{\Delta w} = -\underline{\Delta P_{PV}} + \underline{\Delta P_{charge}}$ the input disturbance and $\underline{\Delta y} = \underline{\Delta f_{grid}}$ the measured output.

Control and performance architecture

The proposed design of the robust H_∞ control is based on a configuration adapted to the studied microgrid [3,4,7]. We consider that the main disturbance is associated to a variation of the load power (denote ΔP_{load}). For such a model, we chose to treat the mixed sensitivity S/KS [3,4,8]. This allows us to take into account some practical constraints, such as bandwidth limitation, when implementing the controller [6].

When designing the H_∞ controller, we should define which outputs will be optimized in order to obtain the desired performance. The variables to be optimized for our microgrid are chosen to be the frequency Δf_{grid} and the current of the inverter ΔI_{rd}. Modelling of the system can be cast in the P-K form (Fig. 3), where P denotes the system (plant) and K the control [3,4]. The vector $\underline{\Delta z}$ represents the outputs to be optimized.

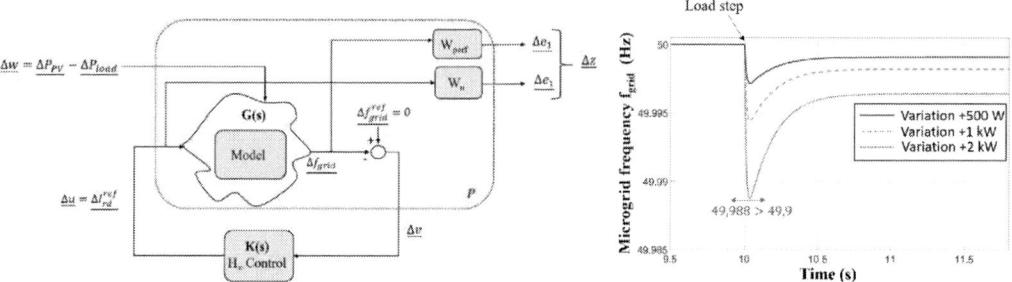

Fig. 3: P-K form for H$_\infty$ control calculation

Fig. 4: Grid frequency response for different load steps

The desired dynamics performance is modelled by the two weighting functions defined in the frequency domain ($s = j\omega$): $W_{perf}(s)$ defines the performance objectives for disturbance rejection, i.e., it determines the performance constraints at the Δf_{grid} output. $W_u(s)$ describes the constraints of practical implementation of control inputs $\Delta I_{rd}{}^{ref}$ taking into account the real dynamic of the actuator, i.e. the inverter. They are expressed as the transfer functions [3,4,8]:

$$W_{perf}(s) = \frac{s/M_s + \omega_b}{s + \omega_b A_\varepsilon}, \quad W_u(s) = \frac{s + {}^{\omega_{bc}}/M_s}{A_u s + \omega_{bc}}. \tag{8}$$

The parameters of the $W_{perf}(s)$ function are chosen according to the desired performance in the time domain (Fig. 2). The high-frequency gain M_s limits the overshoot during a load step, the cut-off frequency ω_b reflects the desired response time, and the low frequency gain A_ε characterizes the steady-state static error. For example, in order to ensure a response time of approximately $t_r = 0.6$ s, we impose a pulse frequency of $\omega_b = 30$ rad/s. In order not to exceed the nominal value of 50 Hz by more than 0.1 Hz we impose $Ms = 0.24$. The definition of these weighting functions is a trade-off between a fast response and a good level of robustness of the system.

After defining the model of our system and the weighting functions, the H$_\infty$ control is determined using the "hinfsyn" function of the Matlab© software. The obtained result corresponds to the minimization of the norm with respect to the sub-optimal value γ [1,6]:

$$\left\| \begin{matrix} W_{perf}S \\ W_u KS \end{matrix} \right\| < \gamma, \tag{9}$$

where S and KS are the system sensitivity functions

$$S = \frac{\Delta f_{grid}}{\underline{\Delta P_{charge}} - \underline{\Delta P_{PV}}}, \quad KS = \frac{\Delta I_{rd}{}^{ref}}{\underline{\Delta P_{charge}} - \underline{\Delta P_{PV}}}, \tag{10}$$

For thus obtained controller, it is important to check its actual performance and robustness, and to evaluate its qualities during *larger* variations of the operating point, in order to evaluate the small-signal operating limits. Fig. 4 shows the real-time frequency response to small load variations ranging from +500 W to +2 kW for a given PV operating point. It can be seen that the frequency remains well within the defined template, so there is a good margin of robustness. More results are presented in [1].

Robustness analysis

After obtaining the controller that responds to the different requirements *and* remains stable, it is important to know whether it will be robust to uncertainties. The design of the control does not take into account the various uncertainties that may arise from the lack of knowledge of some physical characteristics of the system (modelling uncertainties represented by the vague contour in Fig. 3) and that may potentially make it less efficient or even unstable. In this paper, a robustness analysis will be performed using the "μ-analysis" method [3,4,8]. The principle of this method is to compute the structured singular value μ_Δ of the system with the H$_\infty$ controller, in order to analyse the RS and RP and check whether the closed-loop system will remain stable and performing for given uncertainties [3,4,8]. This tool is fully integrated in the H$_\infty$ design methodology, which makes it all the more interesting for a power system designer. However, before using the calculation algorithms of μ-analysis, it is first necessary to define and model the uncertainty to be studied.

In this article, we will focus on the analysis of robustness on uncertainties associated to sensors, especially the delays they may induce.

Unmodelled uncertainty related to sensor delays

General method for modelling dynamic uncertainty

A good model should be simple enough to facilitate design, but also complex enough to give the engineer confidence that the control architecture based on this model will perform equally well in simulations and in real-time operation. Uncertainty is defined as the difference between the chosen mathematical model and the actual model [4]. A control is then robust if it is insensitive to these uncertainties.

Some uncertainties are very complex to model because of the unknown behaviour of different real components used to measure, control and supervise the system. In addition, these components are not ideal, so they can sometimes affect other devices, the power quality of the grid and its dynamic performance. These uncertainties constitute "unmodelled dynamics uncertainty", for which the mathematical models are unknown [3].

Among the various causes of uncertainties in real electrical system, this article will focus on those associated with the sensors used to measure the inverter current of the system. These sensors are essential to ensure observability, disturbance rejection and fault detection [5]. However, they may have imperfections such as delays, offsets or dynamic behaviour, which in turn introduce uncertainties on the control inputs.

In this article, we propose to analyse the impact of imperfections due to an unknown delay on the state variables. We model the dynamics of the unknown delay with the classical transfer function:

$$f_{delay}(s) = e^{-\theta s}, \tag{11}$$

where $\theta \epsilon [\theta_{min} \; \theta_{max}]$ denotes the delay (in seconds) limited to a chosen or given interval. The new disturbed system is then defined as:

$$G_p(s) = f_{delay}(s) * G(s), \tag{12}$$

where G represents the previously defined open loop system without uncertainties, and G_p the new system that takes into account the delay.

However, as the delay transfer function is irrational, it is difficult to perform a robustness analysis for such a system [14,15]. In general, uncertainties are expressed as a transfer function that includes the set of "worst-cases" of all possible uncertainties [3,16]. The uncertainty of the system is expressed in a direct multiplicative form, shown in Fig. 5(a), where ω_I is a stable weighting function representing the dynamics of the uncertainty (i.e., the distribution of the maximum uncertainty amplitude as a function of frequency) and Δ_I is the uncertainty itself, which will be modelled by a stable, linear, bounded, time-invariant, random transfer function [3, 15, 16]:

$$G_p(j\omega) = G(j\omega)\big(1 + \omega_I(j\omega)\Delta_I(j\omega)\big); \; |\Delta_I(j\omega)| \le 1. \tag{13}$$

To define the uncertainty weighting function ω_I, it is necessary to calculate the relative error l_I between the nominal system G and the disturbed system G_p [3,15,16],

$$|\omega_I| \geq l_I(\omega) = \max_{G_p}\left(\left|\frac{G_p(j\omega) - G(j\omega)}{G(j\omega)}\right|\right), \tag{14}$$

ω_I corresponds to the envelope of all relative errors for all possible disturbed systems. Initially, the uncertainty weighting function can be expressed as [3,16]:

$$\omega_{I1} = \frac{\tau s + r_0}{(^\tau/_{r_\infty})s + 1}, \tag{15}$$

where r_0 represents the relative uncertainty in steady state, $1/\tau$ represents approximately the frequency when the relative value of the uncertainty reaches 100% (i.e., when the gain is equal to 1 in absolute value) and r_∞ corresponds to the gain of the weighting function at high frequency, here $r_\infty = 2$. However, this simple transfer function does not cover all cases of uncertainty (see Fig. 5(b)). A (multiplicative) corrective transfer function is then often added in order to come as close as possible to the maximum envelope of the function l_I. Some of the parameters of this term are found empirically in order to better cover all uncertainties [14]:

$$\omega_I = \omega_{I1} * \frac{\left(\frac{\theta_{max}}{2.07}\right)^2 s^2 + \frac{\theta_{max}}{2.07} * 1.6s + 1}{s^2 + \frac{\theta_{max}}{2.07} * 1.2s + 1}. \tag{16}$$

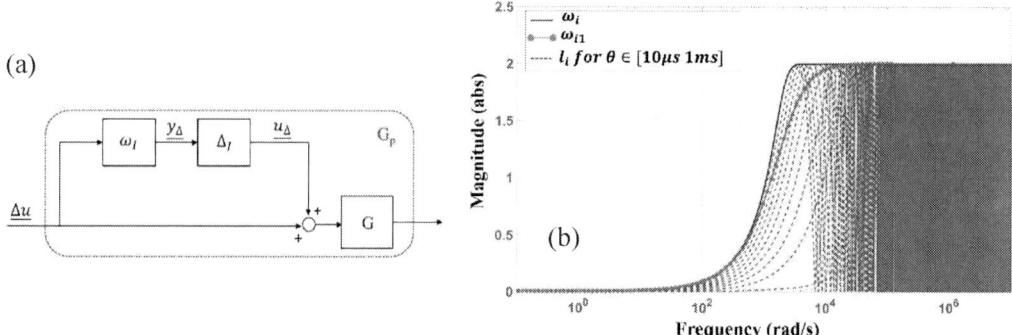

(a)

(b)

Fig. 5: (a) Multiplicative form of the uncertain system G_p and (b) Plot of ω_{I1}, ω_I and l_I for θ varying from 10 μs to 1 ms

Fig. 5(b) shows the uncertainty weighting function ω_I that covers perfectly the maximum of all possible relative errors for a delay interval θ from 10 μs up to 1 ms. For each different range of variation of the delays, it will be necessary to *recalculate* a new function ω_I and thus the new disturbed system G_p. Although this step is not very penalizing from a methodological point of view, it makes the modelling quite demanding.

Robust stability and performance analysis for varied delays

After modelling the delay uncertainty, the system must be modelled in the N-Δ form, shown in Fig. 6, in order to be able to perform further analysis [3,4,17]. The block denoted by P includes the system G and all weighting functions W_{perf}, W_u and ω_I. K represents the initial H$_\infty$ controller and Δ expresses the diagonal block matrix in which Δ_I represents the delay uncertainty. Δ_f represents fictitious uncertainties to account for the uncertainty arising from the performance of the controller [3,17]. The block N is linked to P and K by a fractional linear transformation [3,4], a mathematical modelling step that allows the robustness properties of the system to be characterized in terms of stability and performance via the calculation of the quantity μ.

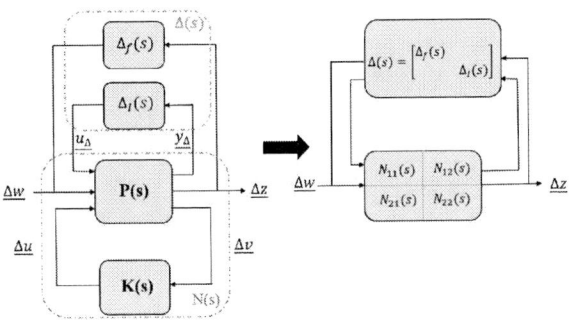

Fig. 6: Restructuring the system into the N-Δ form

After structuring the system, the calculation of the structured singular value $\mu_\Delta(M)$ allows to evaluate the nominal stability (NS), the nominal performance (NP), the robust stability (RS) and the robust performance (RP) of the corresponding closed-loop system (denoted by M) [3,17]:

$$\mu_\Delta(M) = (\min_\Delta(\overline{\sigma}(\Delta)|\det(I - M\Delta))^{-1}, \tag{17}$$

where Δ represents the diagonal block matrix of uncertainties (see Fig. 6) and $\overline{\sigma}$ the maximum singular value. In order to perform a NS, NP, RS and RP analysis, the following conditions must be respected [3,17]:

Nonimal Stability (NS) $\Leftrightarrow$ N(jω) is internally stable, $\tag{18}$

Nominal Performance (NP) $\Leftrightarrow \mu_{\Delta_f}(N_{22}(j\omega)) < 1, \forall\omega$ and NS, $\tag{19}$

Robust Stability (RS) $\Leftrightarrow \mu_{\Delta_l}(N_{11}(j\omega)) < 1, \forall\omega$ and NS, $\tag{20}$

Robust Performance (RP) $\Leftrightarrow \mu_\Delta(N(j\omega)) < 1, \forall\omega$ and NS. $\tag{21}$

When the system satisfies both nominal stability and nominal performance, it would be interesting to check if the closed-loop system will be robustly stable when confronted with various delays. We would also like to know if this system will be robust in performance (for a given performance level), that is to say whether it will be possible to respect the specified dynamic performances in the presence of sensors uncertainties. For this purpose, it is necessary to plot the singular values as a function of the frequency in order to check if the conditions defined above are respected.
The analysis is thus carried out for 3 different delay intervals: from 100 ms to 200 ms, from 1 ms to 10 ms and from 10 μs to 1 ms. The results are plotted on Fig. 7:

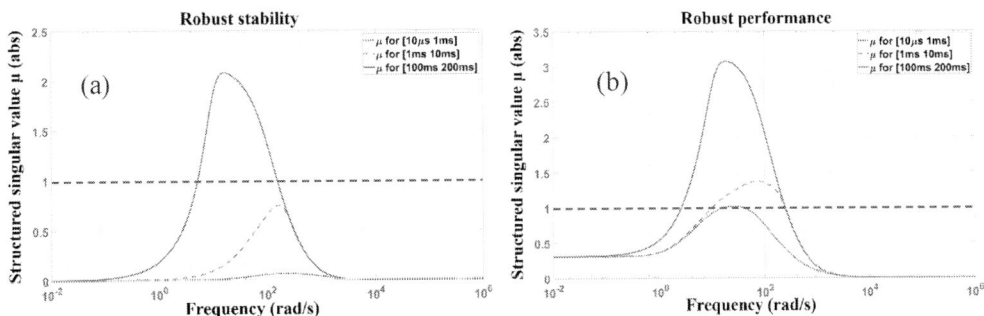

Fig. 7: (a) RS analysis (b) RP analysis

For the RS analysis, Fig. 7(a) shows that for the highest delays the singular value is much greater than 1, which means that the closed loop system is not robust in stability. Thus, when facing this uncertainty, there are significant risks of instability associated with a small disturbance at the operating point. On the other hand, for lower delays, ranging from 10µs to 10ms, the system is very robust in stability.

For the RP analysis, in Fig. 7(b), we can see that for a delay interval of 100-200 ms the singular value is high above 1. For delays ranging from 1 ms to 10 ms, the singular value is still clearly above 1, and even for the shortest delays, 10 µs to 1 ms, it is still slightly above 1. This means that for a given performance, the closed loop system is not robust in performance for all the 3 delay intervals. The desired performance will therefore not be satisfied, which corroborates the results often observed during the implementation of the control and makes it all the more interesting to look at these uncertainties.

Such analysis is then really useful for designers in order to anticipate future problems associated to real-time implementations of controllers and, by the way, find some solutions before the implementation.

Adaptation of performance specifications

The above discussed calculations have shown that some sensor delay intervals induce robustness in stability but not in performance, whatever is the uncertainty associated to the delay. Then, we aim to see if by modifying the dynamics specifications, i.e. some parameters of the weighting functions, we can get both RS and PR for given delays.

We choose to modify one of the parameters of $W_{perf}(s)$, i.e. our requirements on the output network frequency. The high-frequency gain M_s was tested for delays from 1 ms to 10 ms and from 100 ms to 200 ms, the sets of values that are relevant in practice. In order to see the influence of M_s on the robustness in stability and performance of the closed-loop system, we plot the peak values of μ as a function of M_s. This allows us to see if it is possible to find a compromise between system dynamics and robustness. The obtained results are as follows:

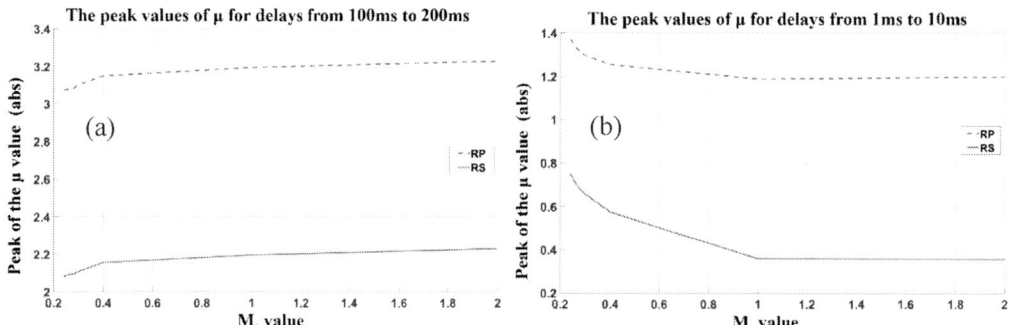

Fig. 8: (a) Peaks of μ value for RS and RP analysis as a function of M_s for delays from 100 ms to 200 ms (a) and from 1 ms to 10 ms (b)

For delays between 100 ms and 200 ms (see Fig. 8(a)), µ is only slightly increasing with a high variation of the coefficient M_s (from 0.24 to 2). Even if RS and RP are somewhat influenced by M_s, this parameter obviously cannot allow guaranteeing the robustness of the system for such large delays. However, for shorter delays comprised between 1 ms and 10 ms, Fig. 8(b) shows that the choice of M_s can have a great influence on RS and RP. This approach can propose a design-making process enabling the designer to ensure the performance of the system in real-time conditions. In this example, the RS is guaranteed (with μ varying from 0.75 to 0.36), but not the RP, as μ is always above 1. Other studies can be similarly done for different parameters of the weighting functions.

Conclusion

This study involves controlling an energy storage system to ensure the primary frequency control of an island/standalone microgrid. An H_∞ robust control of the frequency is investigated, taking into account the uncertainties related to the implementation in real-time conditions. We consider what appears to be the most relevant uncertain behaviour that may come from the real sensors present in the system, which is the delay in measurements. A robustness analysis proves the importance of taking this uncertainty into account in order to guarantee the desired dynamics performance. The main contribution of this work is to provide a framework for designers to implement optimal controls for energy management systems, thus reducing the risk of performance degradation in a real-time implementation.

One of the perspectives is to use the "μ-synthesis", which is another method of designing the controller that ensures its robustness by integrating the modelling uncertainty right from the beginning.

References

[1] Q. L. Lam, Advanced Control of Microgrids for Frequency and Voltage Stability: Robust Control Co-Design and Real-Time Validation, PhD thesis, University Grenoble Alpes, France, 2018.

[2] A. M. Bouzid et al.: A robust control strategy for parallel-connected distributed generation using real-time simulation, in 2016 IEEE 7th International Symposium on Power Electronics for Distributed Generation Systems (PEDG).

[3] S. Skogestad and I. Postlethwaite. Multivariable Feedback Control: Analysis and Design, Second edition. John Wiley & Sons, USA, 2005.

[4] K. Zhou and J. C. Doyle, Essentials of Robust Control, Prentice Hall, 1997.

[5] .M Dion, Christian Commault, and Do Hieu Trinh: Sensor Classification for the Disturbance Rejection by Measurement Feedback Problem, IFAC Proceedings Vol. 41, no 2, pp. 11299-11303 (2008).

[6] Q. L. Lam et al.: Multi-variable H-infinity robust control applied to primary frequency regulation in microgrids with large integration of photovoltaic energy source, in 2015 IEEE International Conference on Industrial Technology (ICIT), pp. 2921-2928 (2015).

[7] A. M. Bouzid et al.: A Survey on Control of Electric Power Distributed Generation Systems for Microgrid Applications, Renewable and Sustainable Energy Reviews, Vol. 44, pp. 751-766 (2015).

[8] K. Zhou, J. C. Doyle and K. Glover, Robust and Optimal Control, Prentice Hall, 1995.

[9] S. Bacha, I. Munteanu and A.I. Bratcu: Power Electronic Converters Modeling and Control: with Case Studies, Springer, 2014, p480.

[10] J. G. Kassakian, M. F. Schlecht and G. C. Verghese, Principles of Power Electronics, Addison-Wesley, 1991.

[11] D. Maksimovic, A. M. Stankovic, V. J. Thottuvelil and G. C. Verghese: Modeling and Simulation of Power Electronic Converters, Proc. IEEE, Vol. 89, no. 6, pp. 898–912 (2001).

[12] S. R. Sanders, J. M. Noworolski, X. Z. Liu and G. C. Verghese: Generalized Averaging Method for Power Conversion Circuits, IEEE Trans. Power Electron., Vol. 6, no. 2, pp. 251–259 (1991).

[13] P. Kundur, Power System Stability and Control, McGraw-Hill, 1994.

[14] P. Lundström, Studies on Robust Multivariable Distillation Control, University of Trondheim, the Norwegian Institute of Technology, July 1994

[15] Z.Q. Wang, P. Lundström, et S. Skogestad: Representation of uncertain time delays in the H∞ framework, International Journal of Control Vol. 59, no 3, pp. 627-638 (1994).

[16] R. Matušů, B. Şenol, C. Yeroğlu: Linear systems with unstructured multiplicative uncertainty: Modeling and robust stability analysis, PLoS ONE 12(7): e0181078 (2017).

[17] M. Sautreuil, N. Retière, D. Riu and O. Sename: A Generic Method for Robust Performance Analysis of Aircraft DC Power Systems, in Proc. 34th Annu. Conf. IEEE Ind. Electron., IECON 2008, pp. 49–54. (2008).

Design of modular low-profile frequency converter for multi-motor manipulators

Tomas Glasberger, Zdenek Kehl, Tomas Kosan, Jan Molnar
UNIVERSITY OF WEST BOHEMIA/REGIONAL INNOVATION CENTRE FOR
ELECTRICAL ENGINEERING
Univerzitni 8
Pilsen/Czech Republic
+420377634442
tglasber@rice.zcu.cz, kehlz@rice.zcu.cz, kosan@rice.zcu.cz, jmolnar@rice.zcu.cz
www.rice.zcu.cz

Acknowledgements

This research has been supported by the Ministry of Education, Youth and Sports of the Czech Republic under the project OP VVV Electrical Engineering Technologies with High-Level of Embedded Intelligence CZ.02.1.01/0.0/0.0/18_069/0009855

Keywords

«Converter control»«Vector control»«Electrical drive»«Semiconductor device»«Regulators»«Field Programmable Gate Array (FPGA)»

Abstract

A new modular low-profile indirect frequency converter is introduced. If fulfills several criterions as mechanical dimensions to fit standard 1U height 19" rack enclosures, energy recuperation, high power factor, high switching frequency and air cooling and high computational performance and connectivity of control system dedicated for special manipulators particularly.

Introduction

Indirect frequency converters can be designed in two different variants. First of them is multi-part conception where every component of an indirect frequency converter is solved as a stand-alone part. The input filter, input rectifier, dc-link circuit, output converter and output filter are designed separately. Control system is also a separate unit. This conception is usually heavy and bulky with poor configuration. In case of industry multi-motor drive applications these frequency converters are usually interconnected via the dc-link. Diode rectifier is usually used as the input part and individual output converters supply individual motors. This conception allows energy sharing via the dc-link but recuperation to the power grid is not possible. [1]

The second conception of the indirect frequency converter contains all important parts in one unit including the control system. This conception can be modular, configurable and scalable using relatively simple parallel connection of several converters. The mechanical dimensions are usually smaller in comparison with the first conception. In the proposed one-unit conception of the indirect frequency converter a voltage source active rectifier (VSAR, AFE) at the input is advantageously used. The active rectifier supports energy recuperation to the power grid and also its transfer among several converters connected in parallel connection. [2]

The proposed conception of the modular low-profile indirect frequency converter is designed for multi-motor drives of special manipulators. The drives of these manipulators are based on linear permanent magnet synchronous motors (LPMSM). The key features of proposed converter are low dimensions, air cooling for simple installation at a customer side, scalability, simple enhancement and modularity including a sophisticated control system with important peripheral interfaces and significant computational power for implementation of complex control algorithms. The proposed converter also represents an universal frequency converter unit. Its first industrial utilization is planned as a supply unit in a special six degrees of freedom motion platform/manipulator [3].

Converter mechanical design and interface

The mechanical conception of modular low-profile frequency converter is designed to fit into a standard 19" rack mount. All components are designed with respect to 1U (44,45 mm) rack mounted enclosure. It brings some advantages: simple mechanical connection and replacement, modularity, configurability and scalability due parallel installation of several converters. A special heatsink for air cooling system was designed. This heatsink is based on two-part construction which are coupled together after electric components are installed on the power board. The shape of the heatsink and its location in rack enclosure are designed for optimal air forced cooling by two fans in back panel. The 3D models which were created during development and final conception of completed modular low-profile frequency converter are shown in Fig. 1.

The hardware interface of the modular low-profile frequency converter is designed respecting needs of industrial manipulators needs. It includes four insulated binary inputs with 24 V logic. There are three 24 V outputs with output current up to 1 A. There is an insulated CAN bus connector and insulated digital synchronization signal respectively on the front panel. The frequency converter disposes with an universal input interface for connection of a position or speed sensor depending on a motor type controlled. The universal input is solved like an add-on module which converts signals from sensor to a main control module logic. In addition, this universal sensor input creates insulation barrier between sensor and converter to avoid ground-loops. It supports both analog (sin-cos) and digital (IRC, ARC) sensor types as it has connection to both analog and digital pins of module.

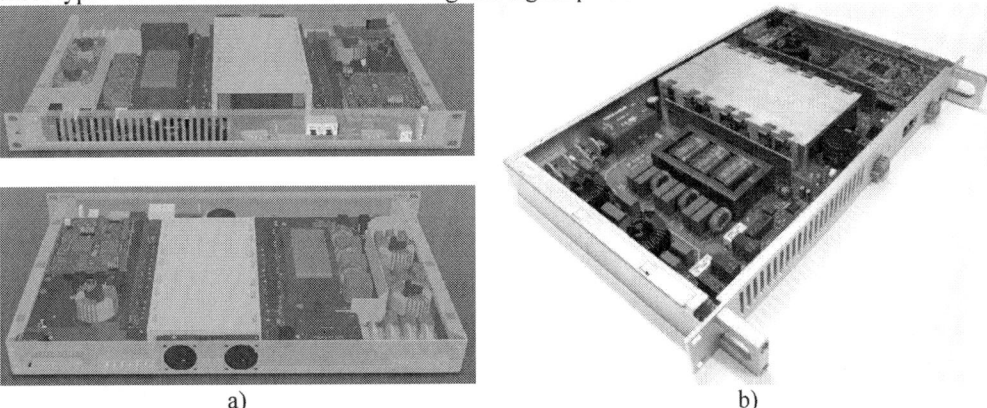

a)   b)

Fig. 1: a) 3D models of proposed converter unit, b) assembled proposed converter unit

Power board design

The input part of the power board topology consists of an input LCL filter, EMI filter and AFE. The input part is designed for connecting to the 3x400 V_{RMS} power grid. The output part of the power board topology consists of dc-link, output voltage source converter and EMI filter. This topology is capable of: 1) stand-alone function, 2) recuperation of energy to the power grid.

The topologies of AFE rectifier and output voltage source inverter (VSI) are the same and both topologies are based on two level voltage source converters (2L-VSC). It brings some advantages: there can be used same power switches, PCB and drivers can be similar for input and output converter and both topologies can be controlled using similar control strategy. These factors can simplify series production of proposed frequency converter.

The power switches used in input and output part of the converter are based on silicon carbide (SiC) technology. Using SiC technology the switching frequency reaches up to 50 kHz. The nominal power of the proposed modular unit is of 2 kW. The nominal power and maximum switching frequency are suppressed by limited cooling possibility caused by the low height of the converter. Power stage is designed to be able to source current upto 5 A$_{RMS}$. The maximum dc-link voltage is 700V. The completed power board is shown in Fig. 2.

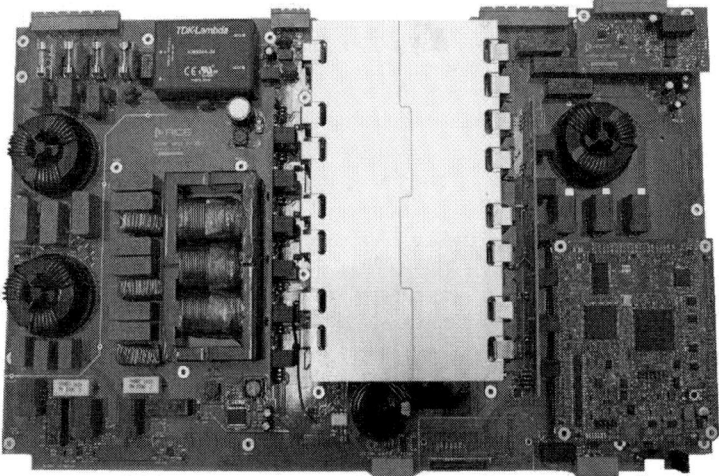

Fig. 2: Detail of the power board

Control module design

The modular low-profile indirect frequency converter is controlled by a single board control module called Rice Universal Microcontroller Module (RUMM). The control module PCB is based on Texas instruments microcontroller TMS320F28377S from latest generation C2000 with combination with Cyclone III FPGA. The RUMM is designed to cover wide range applications. This control unit disposes with important peripheral interfaces and significant computational power for implementation of complex control algorithms. [4] [5] [6] [7]

The control module board contains three pin header connectors. The first on serves for sharing analog signals, another one for digital signals and the last one for optional external EtherCat or similar module. The external module can be connected via SPI or a 16-bit parallel bus. The FPGA Cyclone III is connected with the microprocessor via a 16-bit data parallel bus and a 9-bit address parallel bus. The basic FPGA design contain low level PWM outputs, 18 logical I/O user configurable I/O pins on a side header of module and decoding protocol for data from speed and position sensors. The RUMM block schematics and RUMM module board are shown in Fig. 3.

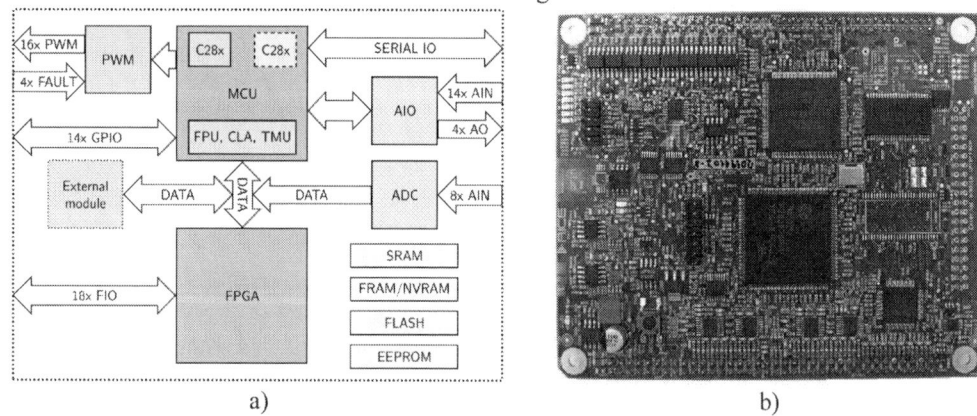

a)   b)

Fig. 3: a) RUMM block structure, b) RUMM PCB module

Control structures

The control structures of input active rectifier (AFE) and output voltage source converter (2L-VSC) are based on cascaded PI controller topology (vector/FOC control). The block scheme of input active rectifier control algorithm is shown in Fig. 4. For synchronization of control algorithm with grid voltages a PLL circuit is used. The output of the PLL represents electrical angle of the grid voltage vector ϑ_e. This angle is used for transformation of measured grid currents to the dq coordinate system. The controller R_{Uc} ensures balancing of DC-link voltage to the required value which is given by input parameter U_{CW}. The output signal of controller R_{Uc} is current component I_{sqw} which is connected to the controller R_{Isq} like required input. This current component represents active power which is necessary for balancing of DC-link voltage. The second current component I_{sdw} is responsible of reactive power. The value of I_{sdw} is set to zero for ensuring zero phase shift between grid currents and grid voltages.

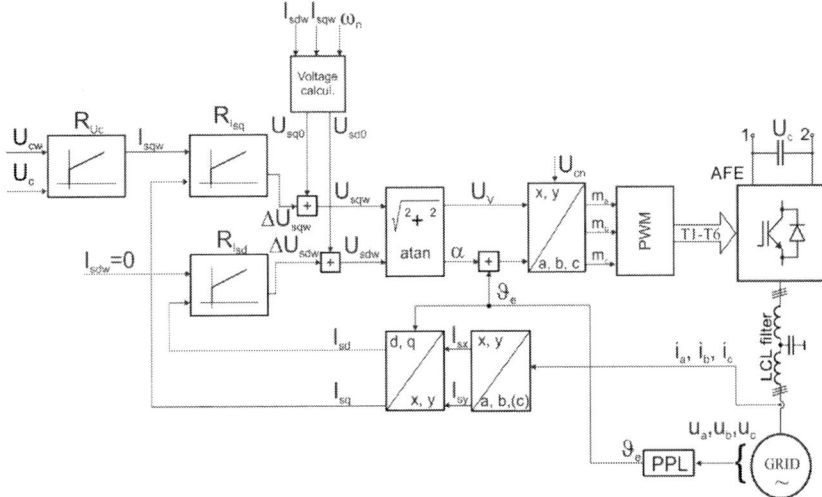

Fig. 4: Control structure of AFE vector control algorithm

The block scheme of the linear permanent magnet synchronous motor (LPMSM) control structure is shown in Fig. 5. It is necessary to measure at least two stator currents i_a, i_b and i_c respectively, and rotor position ϑ_m in this control algorithm. The measured currents are transformed to the motor dq coordinate system. Demanded speed ω_w represents the input quantity of speed controller R_ω. The current component Isqw is the output signal of R_ω controller. The I_{sqw} signal is proportional to the required force. The current component I_{sdw} represents additional magnetic flux. This component allows the flux weakening operation mode which is not assumed in the mentioned application of the converter for the linear motors, therefore this component is set to zero. The position controller can be connected as a superior controller above the speed controller in this application.

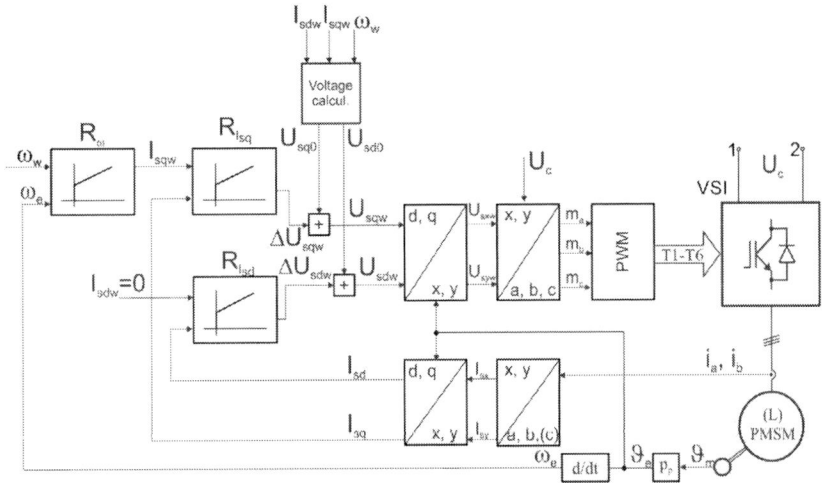

Fig. 5: Control structure of LPMSM vector control algorithm

Experimental results

Temperature progress during loading test is shown in Fig. 6. The frequency converter was loaded by 2.2 kW. The electrical conception of this frequency converter is designed to the maximum heatsink temperature 55°C. The temperature was measured using implemented NTC temperature sensor which is located close to one power switch on the heatsink. The curve of temperature rises up to 55 °C. After that, the fans are activated and after few minutes the temperature is stabilized approximately at 52 °C. The steady state currents corresponding with the power test of the converter can be seen in Fig. 7. The grid current (green) and the output inverter current (blue) are shown in this figure during steady loading test.

A transient in the inverter circuit with motor speed reversal is shown in Fig. 8. The blue waveform represents the rectifier input current in scale 100 mV/A. The cyan curve represents required current component I_{sqw} with scale of 3 A/div. The green waveform represents dc-link voltage with 100V/div scale. The curve with magenta color represents stator electric frequency with scale of 23 Hz/div.

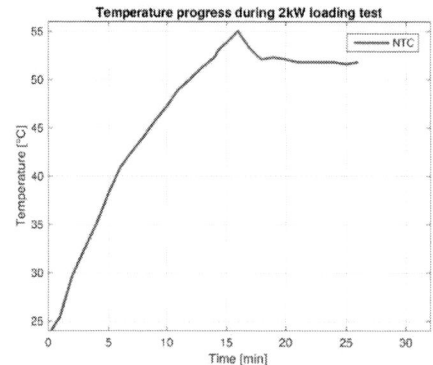

Fig. 6: Temperature progress during 2kW loading test.

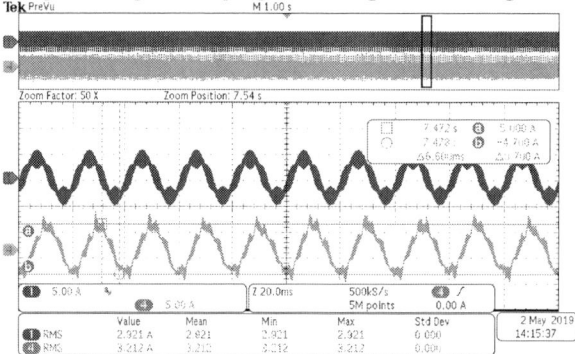

Fig. 7: Experiment: Grid current (green), output rectifier current (blue) during loading test.

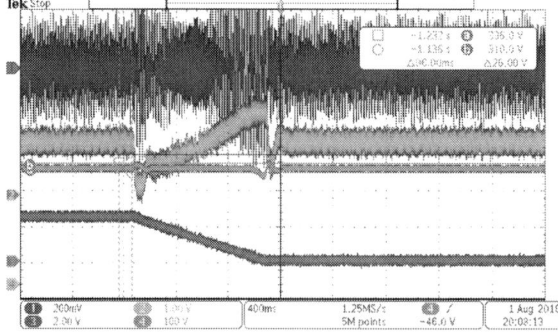

Fig. 8: Experiment: Speed reversal of a motor. Input rectifier current (blue), torque/force current component (cyan), dc-link voltage (green), motor speed (magenta).

Conclusion

The new modular low-profile design of indirect frequency converter has been described in this paper. The hardware conception is designed to fit standard 1U height 19" rack enclosures with air cooling. Input and output part of proposed frequency converter is based on 2L-VSC topology. The rated power of the converter is 2 kW with 700 V maximum in the dc link, supplying from 3x400 V/50 Hz power grid. Switching frequency for introducing experiments is 25 kHz with possibility of its enhancement up to 50 kHz. The converter enables full energy recuperation mode. A universal control module with TMS320F28377 (S, D) is introduced. It fulfills criterion as sufficient peripherals and connectivity features as well as significant computational performance to control special manipulators.

References

[1] J. Li, T. Tang, T. Wang and G. Yao, "Modeling and simulation for common dc bus multi-motor drive systems based on activity cycle diagrams," 2010 IEEE International Symposium on Industrial Electronics, Bari, 2010, pp. 250-255.

[2] D. Kumar, F. Zare and A. Ghosh, "DC Microgrid Technology: System Architectures, AC Grid Interfaces, Grounding Schemes, Power Quality, Communication Networks, Applications, and Standardizations Aspects," in IEEE Access, vol. 5, pp. 12230-12256, 2017.

[3] E. Thndel, "Modelling and simulation of a 6dof motionplatform with permanent magnet linear actuators for testing in wind tunnel," in Modelling and Simulation 2014 - European Simulation and Modelling Conference, ESM 2014, 2014, pp. 403–408, cited By :1. [Online]. Available: www.scopus.com

[4] H. Zhang and Y. Zhao, "Vector Decoupling Controlled PWM Rcetifier for Wind Power Grid-Connected Inverter," 2009 International Conference on Energy and Environment Technology, Guilin, Guangxi, 2009, pp. 373-376.

[5] H. Qi, Y. Wu and Y. Bi, "The main parameters design based on three-phase voltage source PWM rectifier of voltage oriented control," 2014 International Conference on Information Science, Electronics and Electrical Engineering, Sapporo, 2014, pp. 10-13.

[6] Y. Wang, Y. Che and K. W. E. Cheng, "Research on control strategy for Three-Phase PWM Voltage Source Rectifier," 2009 3rd International Conference on Power Electronics Systems and Applications (PESA), Hong Kong, 2009, pp. 1-5.

[7] F. Blaabjerg, R. Teodorescu, M. Liserre and A. V. Timbus, "Overview of Control and Grid Synchronization for Distributed Power Generation Systems," in IEEE Transactions on Industrial Electronics, vol. 53, no. 5, pp. 1398-1409, Oct. 2006.

[8] R. Peña-Alzola, M. Liserre, F. Blaabjerg and T. Kerekes, "Self-commissioning notch filter for active damping in three phase LCL-filter based grid converters," 2013 15th European Conference on Power Electronics and Applications (EPE), Lille, 2013, pp. 1-9.

[9] Z. Yin, Y. Gu, C. Du and F. Gao, "Research On Back-Stepping Control Of Permanent Magnet Linear Synchronous Motor Based On Extended State Observer," 2018 IEEE International Power Electronics and Application Conference and Exposition (PEAC), Shenzhen, 2018, pp. 1-5.

[10] B. Štěpán, K. Tomáš, Š. Václav and P. Zdeněk, "Anti-windup compensation of LQG for single-phase converter with LCL filter," 2017 19th European Conference on Power Electronics and Applications (EPE'17 ECCE Europe), Warsaw, 2017, pp. P.1-P.8.

[11] Fei Liu, Xiaoming Zha, Yan Zhou and Shanxu Duan, "Design and research on parameter of LCL filter in three-phase grid-connected inverter," 2009 IEEE 6th International Power Electronics and Motion Control Conference, Wuhan, 2009, pp. 2174-2177.

[12] M. Liserre, A. Dell'Aquila and F. Blaabjerg, "Stability improvements of an LCL-filter based three-phase active rectifier," 2002 IEEE 33rd Annual IEEE Power Electronics Specialists Conference. Proceedings (Cat. No.02CH37289), Cairns, Qld., Australia, 2002, pp. 1195-1201 vol.3.

Study of the Control of a New AC Voltage Stabilizer using Linear Controller with Reference Frame Transformation

Bunthern KIM, Etienne BOULAUD, Emile BOISAUBERT, Sokchea AM, Phok CHRIN
INSTITUTE OF TECHNOLOGY OF CAMBODIA
Phnom Penh, Cambodia
Tel.: +855 – 77512157
E-Mail: kimbunthern@itc.edu.kh
URL: http://www.itc.edu.kh/en/

Keywords

«Voltage sag compensators», «Converter control», «Converter circuit», «AC/AC converter», «Regulators».

Abstract

This paper explains the control of an automatic voltage regulator (AVR) based on series voltage compensation using a transformer and an AC-AC converter. The paper explores the control of the output of AVR using linear controller using reference transformation technique. Simulation and experimental test has been made to verify the viability of using the proposed control technique for voltage regulation.

Introduction

Computers, information and control equipment employed for various applications are very sensitive to power line voltage variations. To avoid the effect of the random voltage variations, including transients, sags, swells, and surges, the Automatic Voltage Regulator (AVR) is required to avoid undesired effects on the sensitive loads. The conventional voltage regulator usually employed servomotor to change the number of turns of the tap-changing transformer. However, this technique has many drawbacks such as large size, slower dynamic response, and mechanical attrition. A faster response time can be archived with the thyristor-based switches tapped transformer. This technique is still not fast enough since the thyristor switches can be turned on only once per AC cycle. Also, the voltage adjustment is not smooth and has to be done with discrete step changes.

Several topologies of power converters are developed to regulate the voltage from the unregulated source [1]-[6]. A new AC/AC single phase converter (ACVS), in which the power electronic switching devices carry lower current due to integration of an autotransformer with an AC chopper is introduced in [7]. To address the problems of nonlinear and discontinuous relationship within the system, a solution is proposed by applying transformation of the converter model from a fixed reference into a rotating reference frame aligned with the output voltage vector. This paper address further the problematic in the implementation of the control system based on the LQR controller presented in [7]. It proposes the use of a simple all-pass filter to simplify the model of the system by reference frame transformation. The stability issue associated to the proposed technique has been addressed using anti-windup.

Operation of the AC voltage stabilizer

The circuit diagram of the proposed ACVS is shown in Fig. 1. The voltage stabilizer mainly consists of a transformer connected in an auto-transformer form. The primary winding of the transformer is connected to the input power supply through four power transistors (S1-S4). The switch S0 is selected so that the secondary winding voltage will be added to (Step-up Mode) or subtracted from (Step-down Mode) the input voltage. The switching operation of the converter is explained in Table I.

Study of the Control of a New AC Voltage Stabilizer using linear controller with reference frame transformation.

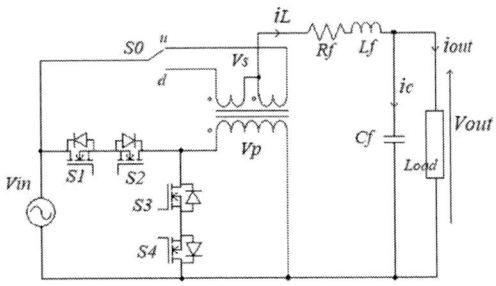

Fig. 1: The proposed AC voltage stabilizer. [7]

Table I: Switching operation.

	Step-up Mode	Step-down Mode
S1, S2	$D(t)$	$D(t)$
S3, S4	$\overline{D}(t)$	$\overline{D}(t)$
S0	u	d

The bidirectional switches S1-S2 are controlled by the PWM signal. The switches S3-S4, controlled by complementary PWM signal, are used to provide a complete path for the current to return from the primary winding. It can be noted that S3-S4 can also be controlled at the line frequency based sign of the input voltage. For instance, if the input voltage is positive, S3 is turned off while S4 is turned on. S4 and the anti-parallel diode of S3 provide the return path for the inductance current.

Operation of the regulator

The relationship between the output and input voltages is explained in [7] as:

$$V_{out}(t) = V_{in}(t)\left[1 \pm \frac{D(t)}{k}\right] - \left[L_f C_f \frac{d^2 V_{out}(t)}{dt^2} + L_f \frac{d i_{out}(t)}{dt} + R_f C_f \frac{d V_{out}(t)}{dt} + R_f i_{out}(t)\right] \quad (1)$$

Where $D(t)$ is the duty cycle of the PWM signal. To eliminate the non-linear form and simply the model, a transformation from the stationary reference frame (a-b) into a rotating reference frame (d-q) is applied. To make the transformation, we assumed a fictive system running in parallel along with the actual one with all the sinusoidal signals shifted to 90 electrical degrees. The parameter of both systems is identical. If we use the index 'a' for the actual system whereas all the variables of the fictive one are marked by index 'b', the transformation to a d-q orthogonal reference frame by using the relation:

$$\begin{bmatrix} V_d \\ V_q \end{bmatrix} = \begin{bmatrix} cos\theta & sin\theta \\ -sin\theta & cos\theta \end{bmatrix}\begin{bmatrix} V_a \\ V_b \end{bmatrix} \quad (2)$$

θ is the electrical angle (the angle between the d-q and a-b frames). The model of the new system in d-q reference frame can be formulated in a state-space form in (4). k is the ratio of primary to secondary turns of the transformer. It has plus sign in the step-up mode and negative sign in the step-down mode. Aligning the d-q reference frame with the output voltage vector, we obtain the following property:
$V_{outd} = |V_{outdq}| = |V_{outab}| = |V_{out}|; V_{outq} = 0; \dot{V}_{outq} = 0$ (3). $|V_{out}|$ is the magnitude of the actual single phase output voltage.

$$\dot{X} = AX + Bu + W \quad (4)$$

$$X = \begin{bmatrix} V_{outd} & V_{outq} & i_{Ld} & i_{Lq} \end{bmatrix}^T ; u = D(t);$$

$$A = \begin{bmatrix} 0 & \omega & 1/C_f & 0 \\ -\omega & 0 & 0 & 1/C_f \\ \frac{-1}{L_f} & 0 & -\frac{R_f}{L_f} & \omega \\ 0 & \frac{-1}{L_f} & -\omega & -\frac{R_f}{L_f} \end{bmatrix}; B = \begin{bmatrix} 0 \\ 0 \\ \pm\frac{V_{inq}}{kL_f} \\ \pm\frac{V_{inq}}{kL_f} \end{bmatrix}; W = \begin{bmatrix} \frac{-1}{C_f} & 0 \\ 0 & \frac{-1}{C_f} \\ 0 & 0 \\ 0 & 0 \end{bmatrix}\begin{bmatrix} i_{outd} \\ i_{outq} \end{bmatrix} + \begin{bmatrix} 0 & 0 \\ 0 & 0 \\ \frac{1}{L_f} & 0 \\ 0 & \frac{1}{L_f} \end{bmatrix}\begin{bmatrix} V_{ind} \\ V_{ind} \end{bmatrix}$$

To obtain a zero steady state control, an additional state variable as an integral of the voltage magnitude error is introduced:

$$v_i = \int_0^t (|V_{out}| - |V_{out}|^*)dt \qquad (5)$$

Where $|V_{out}|^*$ is the rated value of the output voltage. The new state-space model can be written as: $\dot{X}_2 = A_2 X_2 + B_2 u_2 + W_2$ (6). Where $X_2 = [\ |V_{out}|\ \ i_{Ld}\ \ v_i\ ]^T$; $u_2 = |V_{in}|\,D(t)$

$$A_2 = \begin{bmatrix} 0 & \frac{1}{C_f} & 0 \\ C_f\omega^2 - \frac{1}{L_f} & -\frac{R_f}{L_f} & 0 \\ 1 & 0 & 0 \end{bmatrix}; B_2 = \begin{bmatrix} 0 \\ \pm\frac{1}{kL_f} \\ 0 \end{bmatrix}; W_2 = \begin{bmatrix} \frac{-i_{outd}}{C_f} \\ \omega i_{outq} + \frac{|V_{in}|}{L_f} \\ -|V_{out}|^* \end{bmatrix};$$

Control strategy

A Linear-Quadratic Regulator (LQR) for voltage control was employed in [7]. The optimal control law is expressed as: $u_2 = -KX_2$. Where K is the feedback gain matrix obtained by solving the LQR optimization using A_2, B_2, Q and R. The desired duty cycle ratio can be computed as: $D(t) = u_2/|V_{in}|$.

The values of K in step-up and step-down modes are the same with the opposite sign. To obtain the corresponding variable in the d-q reference frame, it is required to calculate the fictive 'b' component of each variable. This orthogonal imaginary component is usually obtained by phase shifting the measured real signals by a quarter of the fundamental period [8]. This approach is relatively simple and tends to deteriorate the dynamic response, which becomes slower and oscillatory due to the introduction the delay in the system. In this paper, we propose using an all-pass filter to obtain a 90 degree shifted version of a variable. The transfer function of the filter is defined as: $H(s) = \frac{1-\tau s}{1+\tau s}$ (7). Where $\tau = \frac{1}{\omega} = \frac{1}{2\pi f}$; $f = 50Hz$ is the electrical frequency. The output response of the filter to a sinusoidal input is shown in Fig. 2. The filter only introduces a 90 degree phase shifted output at around the frequency of the electrical system.

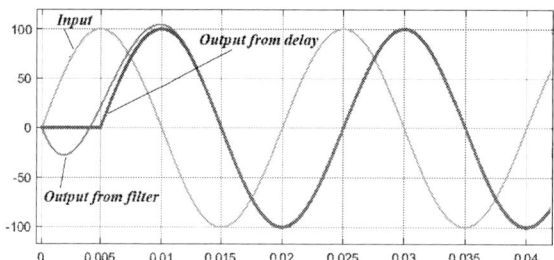

Fig. 2: The output response of an all-pass filter and of a delay of a quarter of the fundamental period.

Once the shifted signal is defined, the variables in d-q reference frame are obtained using the relation (2). Since the sinus and cosinus of θ can be calculated as:

$$cos\theta = V_a/|V_{ab}| \ ; \ sin\theta = V_b/|V_{ab}| \qquad (8)$$

We obtain: $\begin{bmatrix} V_d \\ V_q \end{bmatrix} = \begin{bmatrix} V_a/|V_{ab}| & V_b/|V_{ab}| \\ -V_b/|V_{ab}| & V_a/|V_{ab}| \end{bmatrix} \begin{bmatrix} V_a \\ V_b \end{bmatrix}$ (9) $\rightarrow V_d = |V_{ab}| = \sqrt{V_a{}^2 + V_b{}^2}$; $V_q = 0$; (10)

The block diagram representing the proposed control strategy is depicted in Fig. 3. The first block is the 'All-pass filter' which is used to create the 90° shifted signals of the measured output voltage, measured filter inductor current, and the measured input voltage. The 'a-b to d-q' block generates the corresponding variables in the d-q reference frame using the equation (10). The variables V_{outd} and i_{Ld} are used to find the optimal control input u_2 by multiplying to K.

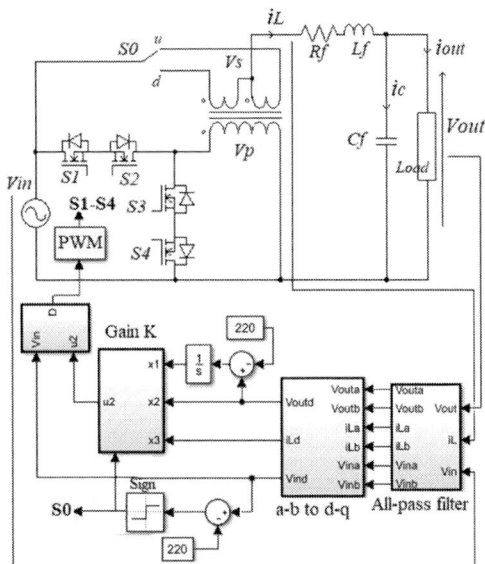

Fig. 3: Block diagram of the proposed control system of the AC voltage regulator.

Since the integral action is introduced in one of the state of the system, a windup phenomenon may happen when the duty cycle command exceed the limit (between 0 and 1). The saturation may happen when the input voltage is too far from the reference value, or too close. It's a real issue as the system is not able to stabilize the output voltage quickly enough in the event of an input transient. For this reason, we propose adding an anti-windup feedback to compensate the effect of the integral effect.

Simulation Study

The model of the voltage regulator is implemented in Matlab/Simulink to test the validity of the proposed control system. The parameters of the components of the system are as followed: $L_f = 3.9$ mH, $C_f = 1$ uF, $R_f = 0.01$ Ω, $k = 4$. The PWM switching frequency is 20 kHz. The LQR gain matrix K is obtained by first guessing the weighting matrix Q and R. If $Q = \text{diag}\{1\ \ 1\ \ 100\}$, $R = 1$, the gain matrix is K = [0.123 61.99 10.00]. The simulated responses of the output voltage to input step change are shown in Fig. 4. It can be seen that the system doesn't work since the measured output voltage is not tuning to the desired voltage level.

To get better performance, the general observation is to choose a large value on the 3rd component of K which corresponds to the state v_i of the system. With K = [0.1 0.1 10000], the responses of the output to input transient without anti-windup are shown in Fig. 5. It can be observed that the control system works correctly for resistive load. For inductive load, the system may become unstable for some interval of time. When this happens, the value of duty cycle is mostly saturated or equal to its maximum or minimum value.

To address the saturation issue, the anti-windup is added to the control system. One way to setup an anti-windup is depicted in Fig. 6. In this configuration, the difference between the duty cycle output of the controller and the actual duty cycle value is multiplied by a factor (Kaw). Then the result of the multiplication is fed to the input of the integrator through a negative feedback loop. Using this setup, the instability problem in inductive load can be solved. It also helps reducing the overshoot of the system response in all cases. The simulated responses of the output voltage for resistive and inductive loads are shown in Fig.7. In these results, the anti-windup is added to the control design (Kaw = 1000). In this case, the regulation works without instability problem. The output transient and the overshot are reduced. The waveforms of the output voltage responses for resistive and inductive loads are shown in Fig. 8.

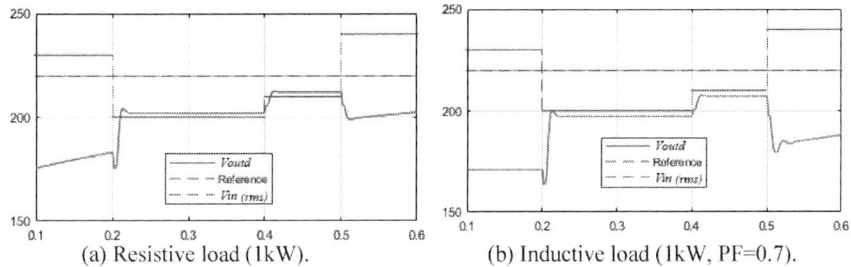

(a) Resistive load (1kW).   (b) Inductive load (1kW, PF=0.7).

Fig. 4: The output voltage response to step change of input voltage. (K = [0.123 62 10])

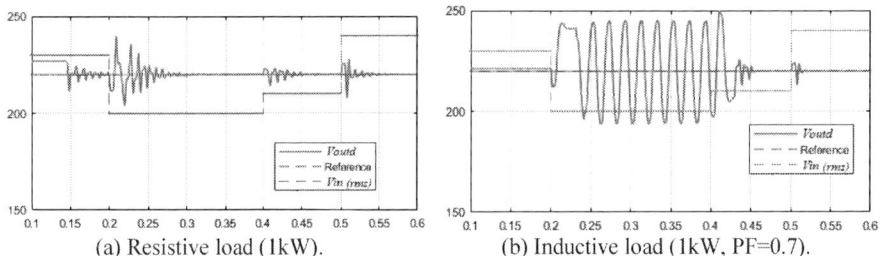

(a) Resistive load (1kW).   (b) Inductive load (1kW, PF=0.7).

Fig. 5: The output voltage response to step change of input voltage. (K = [0.1 0.1 10000], without anti-windup)

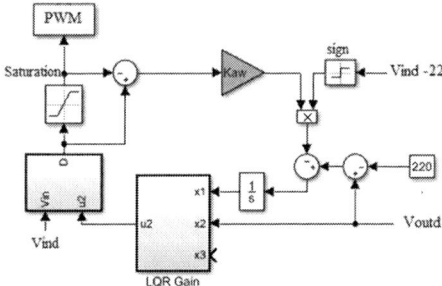

Fig. 6: Adding the anti-windup to the control loop.

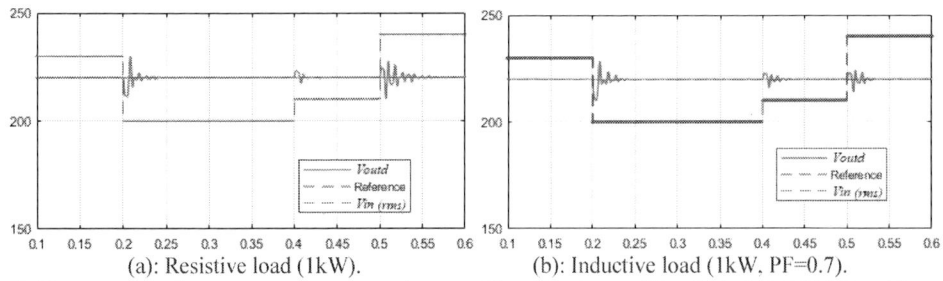

(a): Resistive load (1kW).   (b): Inductive load (1kW, PF=0.7).

Fig. 7: The output voltage response to step change of input voltage. (K = [0.1 0.1 10000], with anti-windup)

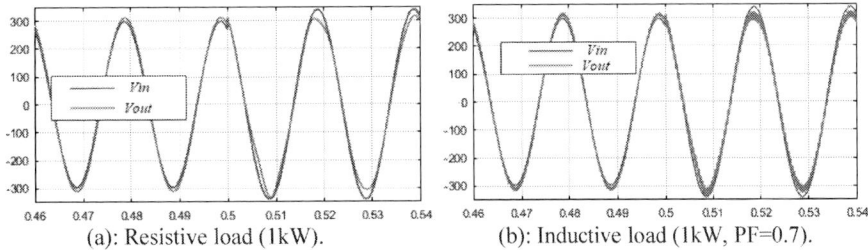

| (a): Resistive load (1kW). | (b): Inductive load (1kW, PF=0.7). |

Fig. 8: Waveform of input and output voltages.

The result of the output voltage response using a quarter-of-the-period delay method is shown in Fig. 9. Different value of the gain matrix K is found for the stability of the system. The result indicates a longer response and higher ripple in comparison to the proposed method.

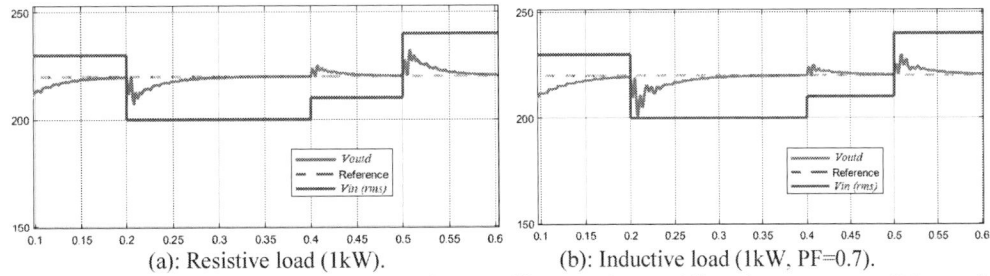

| (a): Resistive load (1kW). | (b): Inductive load (1kW, PF=0.7). |

Fig. 9: The output voltage response to step change of input voltage while using a quarter-of-the-period delay method. (K = [1.4 209.8 316], with anti-windup)

Experimental Result

In order to validate the simulation results obtained, a prototype was developed in laboratory, whose photograph is shown in Fig. 10. The converter was developed according to the diagram of Fig. 3 and the specifications used in the simulation study. The controller board is the TI's F28379D LaunchPad development kit which can be programmed using the Simulink model and Matlab's embedded coder.

Fig. 10: Prototype of the voltage regulator.

For convenience, the switch S0 was not studied in the test prototype. For this reason, the input voltage was only allowed to vary below the reference value in step-up mode, and above the reference value in step-down mode. The output power was around 500W. The step responses of the output voltage are shown in Fig. 11 for resistive load. The waveform of the output voltage responses is shown in Fig. 12.

The experimental results differed significantly from the simulation result. In both cases, the performance was quite similar. The ripple of the output voltage ($Voutd$) might be caused by the noise introduced to the measured voltages and current.

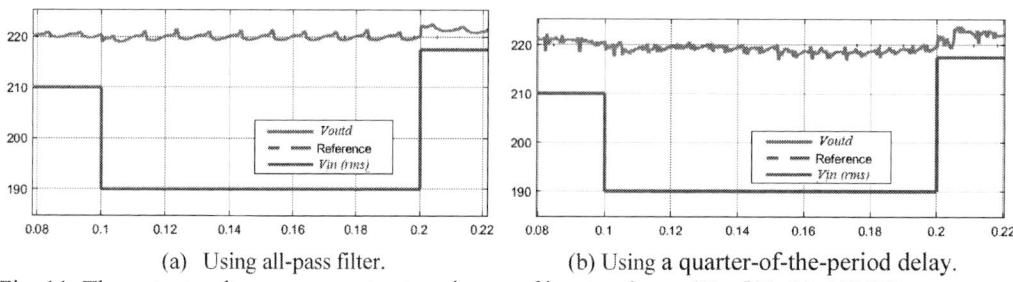

(a) Using all-pass filter.   (b) Using a quarter-of-the-period delay.

Fig. 11: The output voltage response to step change of input voltage. (K = [80 10 10000])

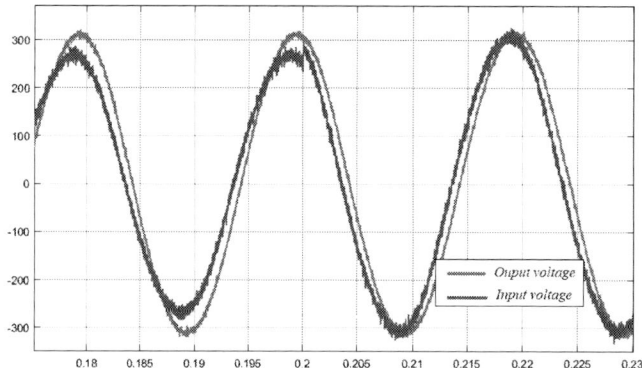

Fig. 12: Waveform of input and output voltages from experimentation (using all-pass filter).

Conclusion

The paper present the study of the control of an AC voltage regulator based on the transformation of electrical signals into synchronous reference frame. It introduces a simple way to implement an LQR controller by using all-pass filters for the transformation. The related stability issues associated to the proposed technique has been addressed using anti-windup. The result of this study has been tested in simulation study and test prototype. However, the experimental results require further investigation in order to validate the finding in the simulation study.

References

[1] Z. Jie, Z. Yunping, Y. Weifu, L. Lei, and L. Fen, "Research on AC chopper power module with module parallel control," in Proc. IEEE Applied Power Electronics Conference and Exposition, 2008, pp. 1324– 1327.
[2] B.-H. Kwon, B.D. Min, and J.H. Kim, "Novel topologies of AC choppers," IEE Proceedings-Electric Power Applications, vol. 143, no. 4, pp. 323-330, 1996.
[3] J.H. Kim, B.D. Min, B.H. Kwon, and S.C. Won, "A PWM buck-boost AC chopper solving the commutation problem," IEEE Trans Industrial Electronics, vol. 45, no. 5, pp. 832–835, 1998.
[4] N. Abd El-Latif Ahmed, K. Amei, and M. Sakui, "Improved circuit of AC choppers for single-phase systems," in Proc. Power Conversion Conference, 1997, vol. 2, pp. 907–912.
[5] F. Peng, L. Chen, and F. Zhang, "Simple topologies of PWM AC-AC converters," IEEE Power Electronics Letter, vol. 1, no. 1, pp. 10–13, 2003.
[6] A. Khoei and S. Yuvarajan, "Single-phase AC-AC converters using power MOSFETs," IEEE Trans Industrial Electronics, vol. 35, no. 3, pp. 442–443, 1988.
[7] H. Liu, J. Wang and O. Kiselychnyk, "Mathematical Modeling and Control of a Cost Effective AC Voltage Stabilizer," in *IEEE Transactions on Power Electronics*, vol. 31, no. 11, pp. 8007-8016, Nov. 2016.
[8] B. Bahrani, A. Rufer, S. Kenzelmann and L. A. C. Lopes, "Vector Control of Single-Phase Voltage-Source Converters Based on Fictive-Axis Emulation," in IEEE Transactions on Industry Applications, vol. 47, no. 2, pp. 831-840, March-April 2011.

Hybrid Energy Storage System for MVDC-Grids

Florian Mahr*, Johann Jaeger
FRIEDRICH-ALEXANDER-UNIVERSITY OF ERLANGEN-NUREMBERG (FAU)
INSTITUTE OF ELECTRICAL ENERGY SYSTEMS

Stefan Henninger
FLUENCE ENERGY GmbH

Hubert Rubenbauer
SIEMENS AG

Cauerstrasse 4, Haus 1
91058 Erlangen, Germany
*Phone: +49 (9131) 85-29532
*Email: florian.mahr@fau.de
URL: http://www.ees.eei.fau.de

Keywords

≪Batteries≫, ≪DC-grid≫, ≪energy storage≫, ≪fuel cell system≫, ≪multilevel converters≫, ≪supercapacitor≫

Abstract

A hybrid energy storage system (HESS) to integrate different energy storage (ES) devices is presented. In this way, ES-devices with complementary physical properties can be used in medium-voltage direct current (MVDC) grids. The aggregation and integration of different ES-technologies is indispensable for operating MVDC distribution grids based on renewable energy sources (RES). State-of-the-art ES-devices are interfaced to alternating current (AC) grids by multiple power electronic converter stages and bulky transformers. Considering the independent operation of MVDC microgrids, an intermediate interface, that aggregates low voltage (LV) ES-devices, is crucial. The proposed HESS topology offers scalability and expendability for the direct integration of ES-devices and avoids lossy AC/DC conversion stages. Besides, great flexibility concerning the installation location inside the MVDC-grid is provided.

Introduction

Decentralized renewable energy sources (RES) combined with energy storage systems (ESS) are essential for future power supply. Typically, RES and ESS, representing direct current (DC) components, are operated at low voltage (LV) or medium voltage (MV) level. Nowadays, they are integrated in alternating current (AC) distribution grids via multiple power electronic converter stages and transformers. Since the penetration rate of RES and ESS in electrical power systems will increase massively, existing grid structures should be reconsidered in this context [1]. Interconnecting DC-components by DC-grids, especially on MV-level, enables new power supply structures and offers numerous advantages over AC-grids [2]. Nevertheless, the direct integration of RES and especially ESS via power electronic interfaces into MVDC-grids is a non-trivial task.

Corresponding to AC-grids, power balance inside DC-grids has to be ensured on different time scales at different power levels. Supercapacitors can provide high power for short time intervals, whereas lithium-ion batteries can cover the mid-term time range (seconds to minutes) for medium power demand. A high amount of energy over a long-time interval can be provided by hydrogen energy carriers via fuel cells. Consequently, it is essential to use power electronic converter interfaces, that aggregate these different storage technologies to combine their physical advantages.

In the literature, modular multilevel converter (MMC) topologies are commonly used to stack low or medium voltage components and use them for high voltage AC/DC applications [3], [4]. In [5] and [6] the integration of energy storage (ES) devices into a multiport AC/DC MMC is presented, but the integration into DC-grids is not investigated. In [7], modular multilevel DC/DC converters for power electronic AC/AC transformers without integrating ES-devices are presented. The use of different MMC topologies as DC/DC converter without multiport functionality is proposed in [8]. Investigations on using exclusively supercapacitors in one storage arm for regenerative breaking applications can be found in [9]. A multiport modular multilevel DC/DC converter for RES and ESS is proposed in [10], but the operation of ES-devices and inner-arm balancing is still an open topic.

In this paper, a hybrid energy storage system (HESS) based on serial connected energy storage modules (ES-modules) inside modular energy storage arms (ES-arms) is presented. An arbitrary number of parallel ES-arms allows to integrate different storage technologies directly into MVDC-grids. In this way, a multiport interface for storage devices with complementary physical properties providing multi-directional power transfer possibilities, is established. Furthermore, energy balancing issues inside this multi-phase system are resolved by using an advanced control algorithm.

The following advantages for energy storage integration into MVDC-grids are gained:

- various grid services provided by different storage technologies
- easy scalability and expandability (maximum power, stored energy, voltage level)
- flexibility of installation location inside the DC-grid
- independent operation from AC-grids or AC/DC interconnection points

This paper is structured as follows: The topology and the operating principle of the HESS is introduced first. Subsequently, the design of the control algorithm is presented. Theoretical considerations are validated by simulations in MATLAB/Simulink afterwards and a summary concludes the paper.

Topology and operating principle

In this section the topology and the operating principle of the HESS is presented. Therefore, the functionality of ES-modules as basic elements and their interaction inside the entire system is analyzed. These theoretical considerations include the derivation of a state space model of the HESS, a mathematical description of ES-arms and the discussion of operational constraints. In order to ensure inner-arm energy balancing in all operating modes, a method to inject AC circulating current components is shown.

Equivalent circuit of the HESS

The equivalent circuit of the HESS consisting of two ES-arms connected to a DC-grid is shown in Fig. 1.

The HESS represents a circuit configuration that consists of an arbitrary number of parallel ES-arms, which are connected to the positive and negative pole of a DC-system. Each ES-arm consists of a number of serial connected ES-modules and an arm inductor. The modules work as interfaces between the ES-arms and the energy storage devices (ES-devices). The following indices are used: Index y represents the number of the ES-arm and index z the number of the ES-module inside one ES-arm.

Components and functionality of ES-modules

An ES-module consists of an ES-arm converter that connects the electrolytic module capacitor to the ES-arm. Considering a full-bridge topology, the output voltage of the ES-module v_{yz} corresponds to the positive or negative value of the capacitor voltage v_{Cyz} or $0\,\text{V}$.

An ES-device converter is used to connect the ES-device to the electrolytic ES-module capacitor. In this way, different storage devices with different operating voltages and specifications can be integrated. In Fig. 2, a half-bridge converter, representing a convenient bidirectional converter topology, is used. Since

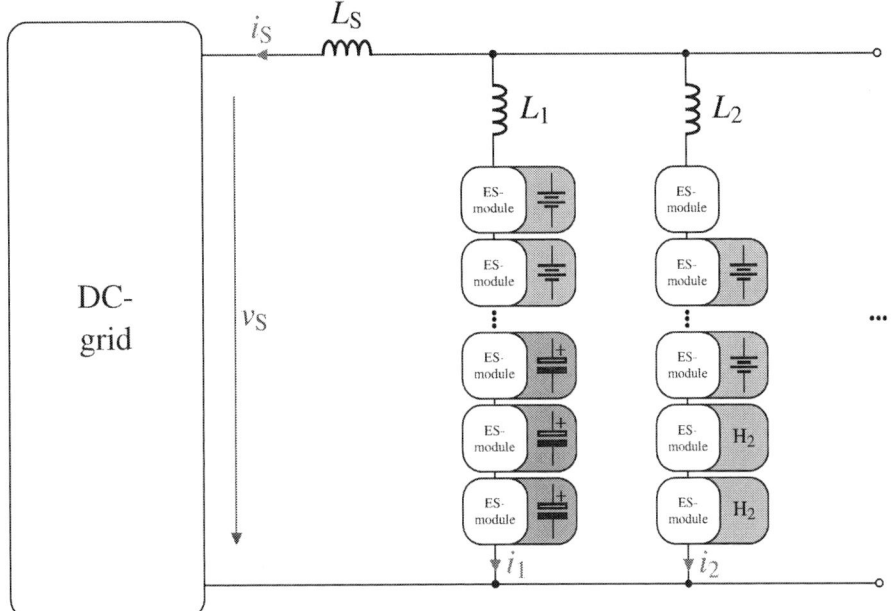

Fig. 1: Topology of the hybrid energy storage system (HESS) with two ES-arms

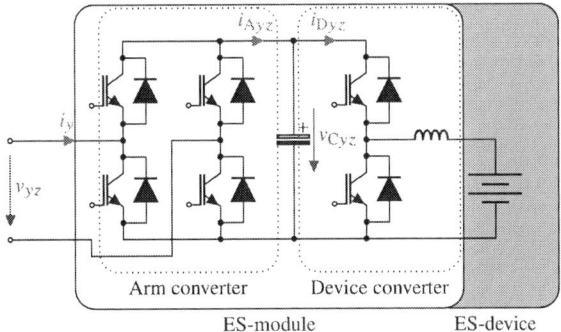

Fig. 2: Components and equivalent circuit of an ES-module connected to a lithium-ion battery

also other converter topologies can be applied, the most appropriate topology for the application can be chosen.

For simplifying isolation coordination, the use of an electrically isolated ES-device converter topology is proposed. The dual active bridge (DAB), for example, provides high efficiency, requires small filter components and causes low switching losses [11].

State space model of the HESS

Regarding short-time mean values of v_{yz}, the serial connection of ES-modules can be modelled as a controllable voltage source from the DC-grid point of view. The simplified equivalent circuit of the HESS consisting of two ES-arms is shown in Fig. 3.

For decoupling the differential equations, the sum current i_S and the circulating current i_Δ are introduced as components of the arm currents i_y, see (1).

$$\begin{pmatrix} i_1 \\ i_2 \end{pmatrix} = \begin{pmatrix} -\frac{1}{2} & 1 \\ -\frac{1}{2} & -1 \end{pmatrix} \cdot \begin{pmatrix} i_S \\ i_\Delta \end{pmatrix} \tag{1}$$

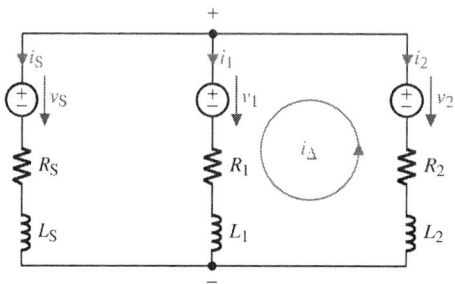

Fig. 3: Equivalent circuit of the hybrid energy storage system (HESS) with two ES-arms

The sum current represents the current flowing between the HESS and the DC-grid. The circulating current flows between the two ES-arms and is used for inter-arm energy balancing. Since it does not appear at the DC-grid, the choice of its frequency components is identified as a degree of freedom.

The ES-arm voltages v_y are transformed according to (2) in an analog way.

$$\begin{pmatrix} v_1 \\ v_2 \end{pmatrix} = \begin{pmatrix} \frac{1}{2} & -1 \\ \frac{1}{2} & 1 \end{pmatrix} \cdot \begin{pmatrix} v_\Sigma \\ v_\Delta \end{pmatrix} \tag{2}$$

The state space model of the HESS is given in (3), where inductors and resistors are considered to be equal inside each ES-arm: $R_1 = R_2 = R_y$ and $L_1 = L_2 = L_y$.

$$\frac{d}{dt} \begin{pmatrix} i_S \\ i_\Delta \end{pmatrix} = \begin{pmatrix} -\frac{R_y + 2R_S}{L_y + 2L_S} & 0 \\ 0 & -\frac{R_y}{L_y} \end{pmatrix} \begin{pmatrix} i_S \\ i_\Delta \end{pmatrix} + \begin{pmatrix} \frac{1}{L_y + 2L_S} & 0 \\ 0 & \frac{1}{L_y} \end{pmatrix} \begin{pmatrix} v_\Sigma \\ v_\Delta \end{pmatrix} + \begin{pmatrix} -\frac{2}{L_y + 2L_S} \\ 0 \end{pmatrix} v_S \tag{3}$$

Mathematical description of ES-arms

The controllable ES-arm voltage v_y equals the sum of the output voltages of the ES-modules v_{yz} inside the ES-arm, see (4).

$$v_y = \sum_{z=1}^{n} v_{yz} \tag{4}$$

The output voltages v_{yz} are formed by the ES-module arm converter out of the capacitor voltages v_{Cyz}. Assuming that all capacitor voltages inside the ES-arm are equal to $v_{Cyz,ref}$ and constant over time, the constraint for ES-arm voltages (5) with n representing the number of ES-modules inside the ES-arm, has to be fulfilled to enable both power flow directions between the HESS and the DC-grid. To this reason, it is a basic prerequisite for the operation of the HESS to keep v_{Cyz} within a defined range around $v_{Cyz,ref}$.

$$n \cdot v_{Cyz,ref} > v_S \tag{5}$$

The active power on the left side of the ES-module capacitor (index A: arm) can be calculated via (6).

$$P_{Ayz} = v_{Cyz,ref} \bar{i}_{Ayz} \tag{6}$$

Neglecting switching and conduction losses of the ES-device converter, the active power transferred to the energy storage device P_{Dyz} (index D: device) can be obtained by (7).

$$P_{Dyz} = v_{Cyz,ref} \bar{i}_{Dyz} \tag{7}$$

Thus, the mean capacitor voltage is constant over a given period of time, if (8) is valid.

$$\bar{i}_{Ayz} = \bar{i}_{Dyz} \tag{8}$$

Operational constraints for inner-arm balancing

The ES-module capacitor current i_{Cyz} is calculated out of the ES-module arm current i_{Ayz} and the ES-module device current i_{Dyz}, see (9).

$$i_{Cyz} = i_{Ayz} - i_{Dyz} \tag{9}$$

For inner-arm balancing via a sorting algorithm, see [12] and [13], it is necessary to open ways to charge and discharge the capacitor in all operating modes and allow both positive and negative values of i_{Cyz}. Since i_{Dyz} is determined by the active power of the ES-device, i_{Ayz} has to be manipulated in order to make inner-arm balancing possible. The current i_{Ayz} is linked to the arm current i_y by the switching function s_{Ayz} of the ES-arm converter, see (10).

$$i_{Ayz} = s_{Ayz} \cdot i_y \tag{10}$$

Assuming a full-bridge topology as ES-arm converter, the switching function can take the following values: $s_{Ayz} \in \{-1, 0, 1\}$.

Subsequent, two different operating modes are described, in which inner-arm balancing is not possible inherently at all times. First it is supposed, that the capacitor is expected to be charged ($s_{Ayz} \cdot i_y > i_{Dyz}$) and the ES-device is being charged ($i_{Dyz} > 0$) at the same time. If $i_{Dyz} > i_y > 0$ is valid, it is not possible to charge the capacitor and to ensure inner-arm balancing. Analog problems occur, if the capacitor is expected to be discharged, the ES-device is being discharged at the same time and $0 > i_y > i_{Dyz}$ is valid. In this paper it is proposed to gain time intervals, in which $i_y > i_{Dyz}$ or respectively $i_{Dyz} > i_y$ is valid by injecting AC current components to i_y. During these time intervals, inner-arm balancing is now possible.

Now it is assumed, that the capacitor is expected to be charged while the ES-device is being discharged. If $i_{Dyz} > i_y$ is valid, inner-arm balancing is not possible. Analog problems occur, if the capacitor is expected to be discharged, the ES-device is being charged at the same time and $i_y > i_{Dyz}$ is valid. In these cases, the sign of i_y has to be switched for certain time intervals. This can also be achieved via the AC current injection.

Injection of AC circulating current components

As explained in the previous section, it is essential for ensuring inner-arm balancing in all operation modes to inject an AC-component $\tilde{i}_y$ with an arbitrary frequency to the arm current i_y according to (11).

$$i_y = \bar{i}_y + \tilde{i}_y \tag{11}$$

In order to inject both DC- and AC-components in the arm currents, the arm voltages are controlled to track a mixed reference signal, see (12).

$$v_y = \bar{v}_y + \tilde{v}_y \tag{12}$$

Since this AC-component should not appear as a part of i_S, $\tilde{i}_y$ is injected as a component of i_Δ. Regarding (1) and (2), the following equations are derived.

$$\tilde{i}_1 = -\tilde{i}_2 \tag{13}$$

$$\tilde{v}_1 = -\tilde{v}_2 \tag{14}$$

Design of the control algorithm

The implemented control algorithm, which enables the operation of the HESS and defines the fundamental operation modes, is described in this section and visualized via signal flow diagrams. The controller of the HESS is separated into two controller levels.

Control of the mean ES-module capacitor voltages

The aim of upper control level is to make sure, that the mean value of all capacitor voltages of the HESS is kept within a defined range around its nominal value and energy is balanced between the ES-arms.

To reach the former control target, a cascaded control structure using two PI-controllers is implemented, see Fig. 4.

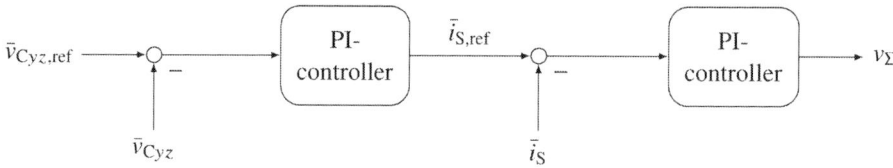

Fig. 4: Control loop for adjusting the mean voltages of the ES-module capacitors

Based on the difference between the mean value of all capacitor voltages of the HESS $\bar{v}_{Cyz}$ and its reference value $\bar{v}_{Cyz,ref}$, the outer PI-controller defines the reference value of the DC-grid current $\bar{i}_{S,ref}$. In this way, the power flow between the HESS and the DC-grid is adjusted to keep the capacitor voltages within a certain range. The current $\bar{i}_S$ is finally controlled by the sum voltage v_Σ. The parameters of this PI-controller are chosen based on the first row of (3).

The power demand on the HESS is sent to the ES-device converters, which adjust the power of each ES-device individually. In this way, different storage devices inside the HESS can provide different and appropriate services to the DC-grid.

Energy balance between the ES-arms and activation of AC circulating current components

The upper control level is complemented by the controller of the energy balance between the ES-arms. Therefore, the difference of the sums of the capacitor voltages of both ES-arms $v_{C\Delta}$ is introduced, see (15).

$$v_{C\Delta} = \sum_{z=1}^{n} v_{C1z} - \sum_{z=1}^{n} v_{C2z} \tag{15}$$

Again, a cascaded control structure is implemented, see the upper path of Fig. 5. The outer PI-controller calculates the reference value of the mean value of the circulating current $\bar{i}_{\Delta,ref}$, while the inner PI-controller adjusts $\bar{i}_\Delta$ by manipulating $\bar{v}_\Delta$ based on the second row of (3). In this way, the DC-component of the circulating current is defined.

The AC-component of the circulating current $\tilde{v}_\Delta$, defined by the amplitude $\hat{v}_\Delta$ and the arbitrary frequency f_c representing a design parameter, is calculated according to the lower path of Fig. 5. Here, the inner-arm voltage unbalance is used to generate $\hat{v}_\Delta$ via a hysteresis-based droop-controller. The slope and the limits of the implemented droop-controller are shown in Fig. 6.

Energy balance inside one ES-arm

The task of the lower control level is to balance energy inside all ES-modules of one ES-arm and ensure equal capacitor voltages inside one arm. Inner-arm energy balance is reached by implementing a sorting algorithm, e.g. a bubble sort-algorithm. This algorithm arranges ES-modules of one ES-arm virtually depending on their capacitor voltage. Depending on the sign of the capacitor current i_{Cyz}, the appropriate values of the switching function of the ES-module arm converter s_{Ayz} are chosen.

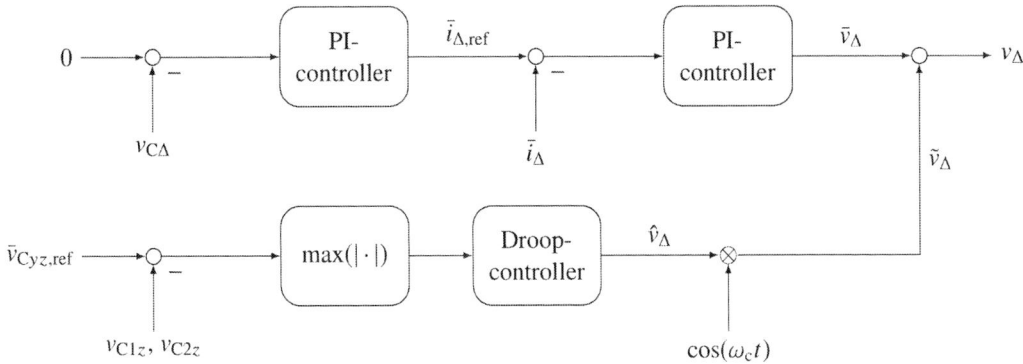

Fig. 5: Control loop for balancing energy between the ES-arms

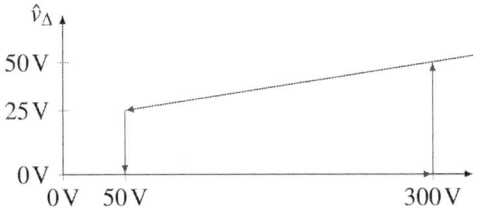

Fig. 6: Droop-controller with hysteresis for AC-current injection

Gate signal generating and signal processing

In Fig. 7 the signal processing steps around the HESS are shown.

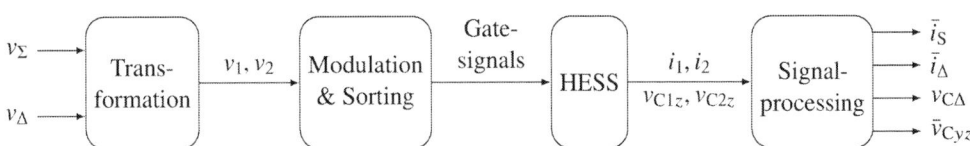

Fig. 7: Generation of gate signals and transformation of variables

First, the controller output signals v_Σ and v_Δ are transformed to the arm voltages v_1 and v_2 via (2). The gate signals for the power semiconductor devices of the ES-arm converters are generated by a modulation and sorting stage ensuring inner-arm balance.

Inside the HESS, the arm currents and all capacitor voltages are measured. Further transformation and filtering steps are needed to derive the mean sum current $\bar{i}_S$, the mean circulating current $\bar{i}_\Delta$, the difference of the sums of the capacitor voltages of both ES-arms $v_{C\Delta}$ and the capacitor voltages $\bar{v}_{Cyz}$. These variables are crucial for the use of the HESS control algorithms.

Simulation results

In order to validate the proposed HESS topology and the control algorithms, time domain (EMT) simulations in MATLAB/Simulink are performed. The results of two different test scenarios are presented in this section. The simulation parameters are listed in Table I.

Hybrid symmetrical storage operation

In this scenario, a hybrid energy storage operation with two ES-arms is investigated. The first ES-arm is exclusively equipped with supercapacitors. This means, that each one of the 25 ES-modules is connected to a supercapacitor. The second ES-arm is exclusively equipped with lithium-ion batteries.

Table I: Simulation parameters

Variable	Value	Description
m	2	number of parallel connected ES-arms inside the HESS
n	25	number of serial connected ES-modules inside one ES-arm
T_S	400 Hz	sample time of the controller
v_S	30 kV	DC-grid voltage
v_{Cyz}^*	1.8 kV	reference value of the ES-module capacitor voltages
C_{yz}	7.5 mF	capacity of the ES-module capacitors
R_S	0.5 Ω	resistance of the DC-grid
L_S	8 mH	inductance of the DC-grid
R_y	0.25 Ω	resistance of the ES-arms
L_y	4 mH	inductance of the ES-arms

The simulation results are shown in Fig. 8. Fig. 8a shows the reference trajectories of the active power of each ES-arm for this operation mode. From $t = 2\,\mathrm{s} - 8\,\mathrm{s}$ all supercapacitors are charged at high power and discharged from $t = 14\,\mathrm{s} - 20\,\mathrm{s}$ on. The batteries are charged from $t = 6\,\mathrm{s}$ on. The sum current and the ES-arm currents are shown in Fig. 8b. Fig. 8c and Fig. 8d show the capacitor voltages of both ES-arms, which are balanced in the desired range around 1800 V.

It is identified, that a safe operation is ensured and the correct functionality of the HESS is reached.

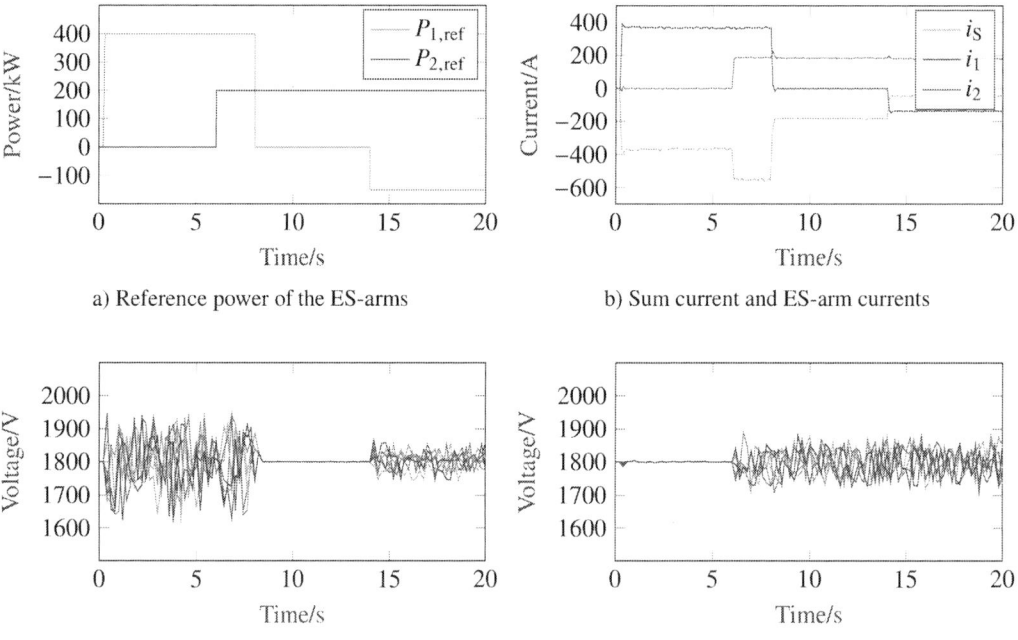

a) Reference power of the ES-arms

b) Sum current and ES-arm currents

c) Selection of module capacitor voltages of ES-arm 1

d) Selection of module capacitor voltages of ES-arm 2

Fig. 8: Simulation results of the hybrid symmetrical storage operation

Inner ES-arm asymmetrical storage operation

In contrast to the first scenario, the ES-arms are not completely equipped with storage devices in scenario two. There are 19 installed ES-modules with supercapacitors in the first arm and 16 installed ES-modules with lithium-ion batteries in the second arm. Each one of the other ES-modules is not connected to any storage device. Compared to scenario one, the relation between the arm currents and the ES-device currents are changed.

The simulation results applying an operation mode without AC current injection are shown in Fig. 9. The reference power trajectories of each arm is depicted in Fig. 9a. It is identified, that inner-arm balancing is not possible and the capacitor voltages diverge in this case due to bad conditions of the arm current and the ES-device currents, see Fig. 9b. For practical use, this scenario has to be excluded to avoid hardware damages.

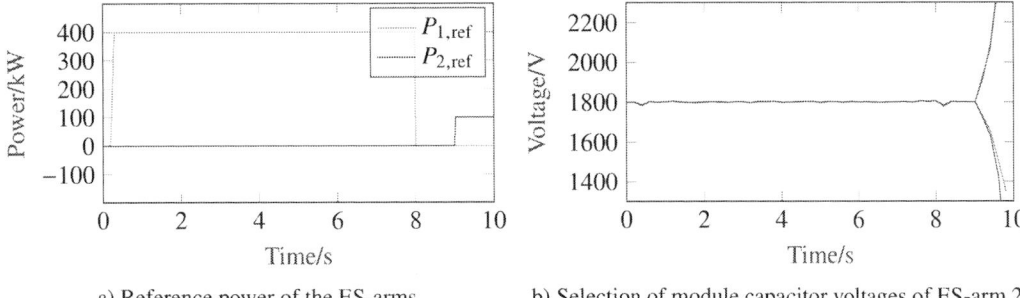

a) Reference power of the ES-arms   b) Selection of module capacitor voltages of ES-arm 2

Fig. 9: Simulation results of the inner ES-arm asymmetrical storage operation without AC-component

For this reason, the injection of an AC-component to the circulating current via a hysteresis-based droop controller (see Fig. 6) is activated. The frequency of the AC-component is chosen to $f_c = 25\,\text{Hz}$. The simulation results are presented in Fig. 10. The power reference trajectories and the sum current and the arm currents are shown in Fig. 10a and Fig. 10b, respectively. In Fig. 10c it is demonstrated, that inner-balancing is possible by injecting an AC current component to the arm currents and the capacitor voltages are kept within the defined range around the reference value of $1800\,\text{V}$. The resulting arm currents are shown in Fig. 10e and Fig. 10f. From Fig. 10d it is seen, that ES-module 11 of ES-arm 2 is charged during time intervals with high values of i_2, e.g. from $t = 11.22\,\text{s} - 11.23\,\text{s}$. Beyond that, this module is discharged during time intervals with low values of i_2, e.g. from $t = 11.24\,\text{s} - 11.25\,\text{s}$. In this way it is confirmed, that by injecting the AC current component indispensable time intervals with appropriate current conditions for inner-arm balancing are gained.

Conclusion

In this paper, a hybrid energy storage system (HESS) to integrate different energy storage (ES) devices into medium-voltage direct current (MVDC) grids is presented. This multiport interface aggregates ES-technologies with complementary physical properties in order to support MVDC-grids. Therefore, individual advantages of each ES-technology can be utilized in an optimal way. The results of EMT-simulations in MATLAB/Simulink demonstrate, that energy balancing between each ES-module and so a safe and reliable operation is ensured under all operation modes by using an advanced control algorithm.

HESS will play a key role in operating and supporting MVDC-grids, which are necessary to acquire power from renewable energy sources (RES). In this way, further steps towards a less-carbon power supply using appropriate grid structures can be initiated.

References

[1] R. de Doncker, "Power Electronic Technologies for Flexible DC Distribution Grids," in 2014 International Power Electronics Conference (IPEC-Hiroshima 2014 - ECCE ASIA). 2014, pp. 736–743.

[2] T. Dragicevic et al., "DC Microgrids—Part II: A Review of Power Architectures, Applications, and Standardization Issues," in IEEE Transactions on Power Electronics. 2016, pp. 3528–3549.

[3] A. Lesnicar et al., "An innovative modular multilevel converter topology suitable for a wide power range," in 2003 IEEE Bologna Power Tech Conference Proceedings. 2003, pp. 272–277.

[4] H. Akagi, "Classification, Terminology, and Application of the Modular Multilevel Cascade Converter (MMCC)," in IEEE Transactions on Power Electronics. 2011, pp. 3119–3130.

[5] M. Lopez et al., "Control strategies for MMC using cells with power transfer capability," in 2015 IEEE Energy Conversion Congress and Exposition (ECCE). 2015, pp. 3570–3577.

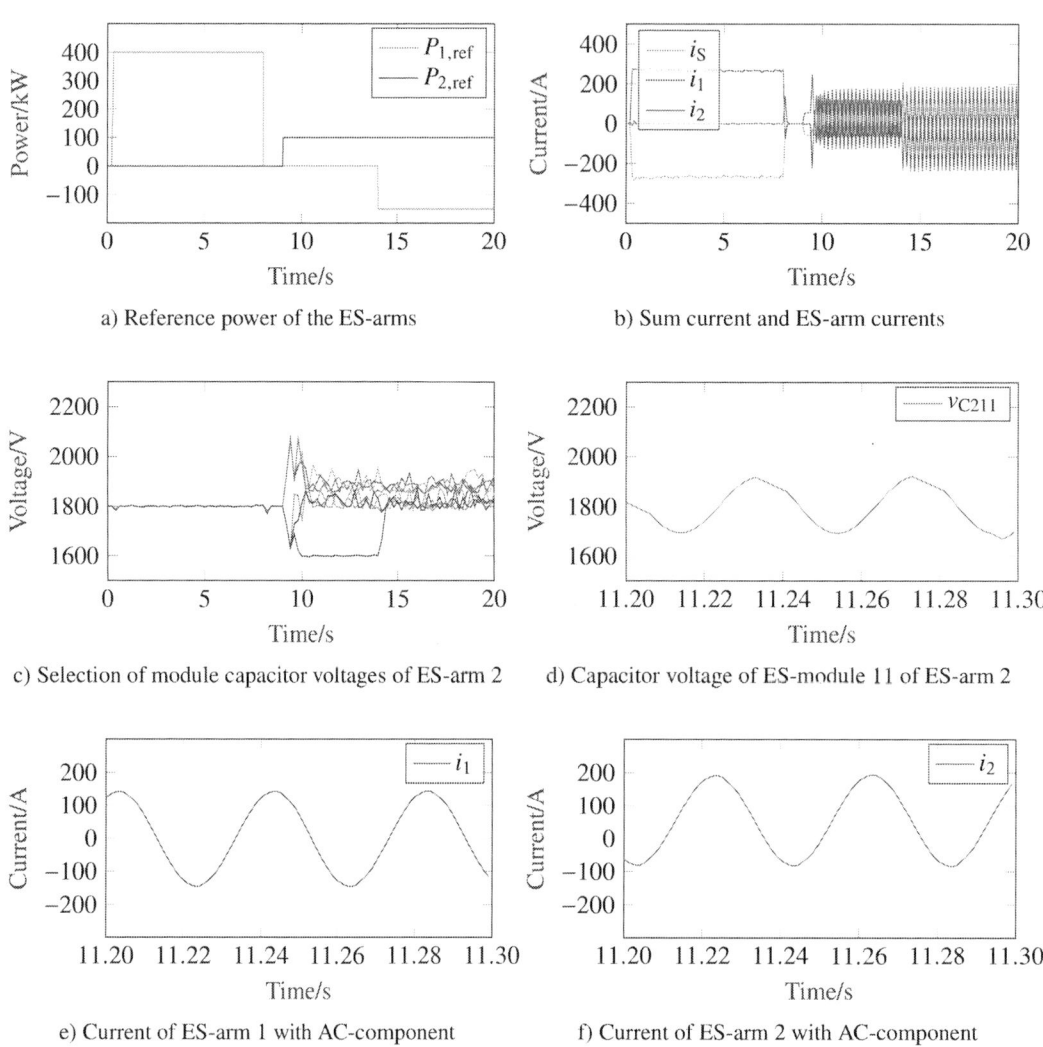

Fig. 10: Simulation results of the inner ES-arm asymmetrical storage operation with AC-component

[6] M. Schroeder et al., "Advanced Energy Flow Control Concept of an MMC for Unrestricted Operation as a Multiport Device," in 2019 IEEE Transactions on Power Electronics. 2019, pp. 11496–11512.

[7] Z. Wang et al., "An Isolated Bidirectional Modular Multilevel DC/DC Converter for Power Electronic Transformer Applications," in Journal of Power Electronics. 2016, pp. 861–871.

[8] J. Ferreira, "The Multilevel Modular DC Converter," in IEEE Transactions on Power Electronics. 2013, pp. 4460–4465.

[9] D. Montesinos-Miracle et al., "Design and Control of a Modular Multilevel DC/DC Converter for Regenerative Applications," in IEEE Transactions on Power Electronics. IEEE, 2013, pp. 3970–3979.

[10] F. Zhang et al., "A multiport modular multilevel DC-DC converter," in 2016 IEEE 7th International Symposium on Power Electronics for Distributed Generation Systems (PEDG). 2016, pp. 1–7.

[11] R. de Doncker et al., "A three-phase soft-switched high-power-density DC/DC converter for high-power applications," in IEEE Transactions on Industry Applications. 1991, pp. 63–73.

[12] D. Siemaszko, "Fast Sorting Method for Balancing Capacitor Voltages in Modular Multilevel Converters," in 2015 IEEE Transactions on Power Electronics. 2015, pp. 463–470.

[13] S. Fan et al., "An Improved Control System for Modular Multilevel Converters with New Modulation Strategy and Voltage Balancing Control," in IEEE Transactions on Power Electronics. 2015, pp. 358–371.

A Combined Model for Optimal Power Flow Applied to MT-HVDC Systems

Fernando Torres[1,3], Javier Muñoz[1], Fredy Muñoz[1] and Claudio Roa[2]

(1)Department of Electrical Engineering, Faculty of Engineering,
Universidad de Talca, Curicó, Chile
(2)Department of Electrical Engineering, Faculty of Engineering,
Universidad de Concepción, Concepción, Chile
(3)Doctorado en Sistemas de Ingeniería, Faculty of Engineering,
Universidad de Talca, Chile
Email: fetorres@utalca.cl

Acknowledgements

This work was funded by Comisión Nacional de Investigación Científica y Tecnológica CONICYT-PFCHA/Doctorado Nacional/2019-21190503 and ANID/FONDECYT/1191028.

Keywords

«Multiterminal DC (MTDC)», «Renewable energy systems», «Power transmission», «Optimal control», «Voltage Source Converter (VSC)»

Abstract

Multi-terminal HVDC grids based on voltage source converters have been considered a good solution to transmit great power bulks coming from different renewable energy sources. Therefore, it is important to generate useful models that contribute to the operational study of such systems. In particular, it is important to make good use of droop schemes, since these are essential for achieving a correct operation. The main contribution of this article is the development of a combined model based on optimal power flow for radial multi-terminal HVDC grids. This model comprises two sub-models. The first sub-model, which considers the droop scheme associated to the active power loop, is established with the objective of minimizing losses over the DC lines. On the other hand, the second sub-model, which considers the droop scheme associated to the reactive power loop, is established with the objective of minimizing the total apparent power in converter stations and, consequently, minimizing their total losses. The combined model allows selecting the parameters and references associated to both droop schemes. The case study presented shows that the use of the combined model results in a correct grid operation, highlighting the contribution to the needs of the AC grid while maintaining the operational restrictions of the entire system. The simulations are obtained using MATLAB/SIMULINK.

Introduction

Multi-terminal HVDC grids based on voltage source converters (VSC) are presented as one of the great options to incorporate renewable energy to power grids, in a massive and secure way [1]. To control these grids, different strategies have been proposed, where droop control is the one presenting the best characteristics. This strategy provides support to the DC grid voltage when stations undergo power variations [2-4]. Additionally, optimal power flow (OPF) based techniques have been developed, which allow modifying droop control parameters, with the objective of optimizing one or more variables in the grid. Among the variables to be optimized we can find, mainly, costs [5-7] and losses.

Other works regarding power flow models associated to the optimization of losses in MT-HVDC grids have studied multi-objective problems that, in addition to decreasing grid losses, seek to improve characteristics associated to power dispatch [8, 9]. On the other hand, some studies indicate when and how to consider losses associated to VSCs and to shunt conductances associated to DC transmission lines [10, 11]. The aforementioned is due to the magnitude of the DC transmission lines losses.

There are also contributions concerning the establishment of converter characteristics to obtain an optimal system operation [12]. Furthermore, there is a study that considers devices such as the interline current flow to improve the control of energy flow and, consequently, reduce the operational costs of the system [13]. It is important to mention that articles [8-13] consider VSC losses to have a quadratic relationship with the AC current, which requires the estimation of some parameters.

The main contribution of this article is developing a combined model, which comprises two sub-models. The first one, based on the droop scheme belonging to the active power loop, allows minimizing losses over the transmission lines. The second one, based on the droop scheme belonging to the reactive power loop, allows minimizing losses in converter stations through their total apparent power. This is an important difference with respect to other models.

The simplicity of the combined model allows establishing reference variables and parameters, associated to the control schemes, with the goal of meeting the needs of the AC grid, while maintaining operational restrictions of lines, converters, and AC and DC buses. It is important to mention that the model is based on steady state conditions obtained from the control schemes, and on circuit equations that govern the behaviour of the DC grid. Furthermore, it is important to mention that the model developed in this article is constructed for radial grids powered by renewable energy sources, which deliver support to AC grids (Fig.1).

This article is structured as follows: Abstract, Introduction, Model Development (considering both sub-model 1, and sub-model 2), Case Study and Conclusions.

Model Development

The modelling in this article is based on the diagram shown in Figure 1 and the diagrams in Figure 2. In Figure 1, a MT-HVDC grid is shown, consisting of n stations connected to n unidirectional power sources. The latter represent the active power coming from different renewable energy sources (RES). Furthermore, each converter station is connected to an AC bus. On the other hand, in Figure 2, the droop schemes belonging to the active and reactive power loops are shown. These schemes are associated with each of the VSCs shown in Figure 1.

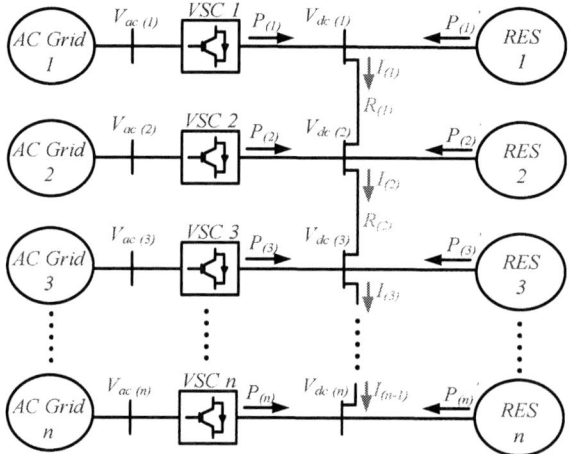

Fig.1: General diagram of the system.

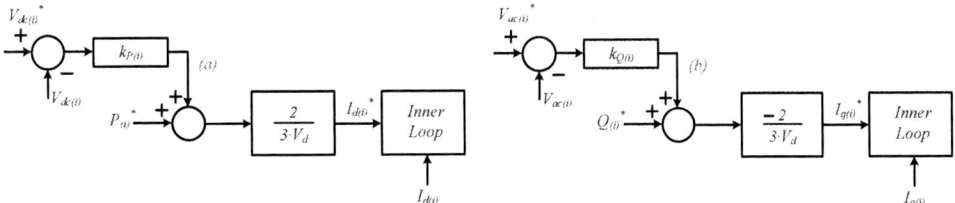

Fig.2: Droop control scheme for the *i-th* station. (a) active power, (b) reactive power.

Definition of Vectors Associated with the Modelling

Starting from Fig.1 and Fig.2 the following vectors can be obtained:

$$V_{ac} = [V_{ac(1)}\ V_{ac(2)} \cdots V_{ac(i)} \cdots V_{ac(n)}] \tag{1}$$

$$V_{ac}^* = [V_{ac(1)}^*\ V_{ac(2)}^* \cdots V_{ac(i)}^* \cdots V_{ac(n)}^*] \tag{2}$$

$$V_{dc} = [V_{dc(1)}\ V_{dc(2)} \cdots V_{dc(i)} \cdots V_{dc(n)}] \tag{3}$$

$$V_{dc}^* = [V_{dc(1)}^*\ V_{dc(2)}^* \cdots V_{dc(i)}^* \cdots V_{dc(n)}^*] \tag{4}$$

$$P = [P_{(1)}\ P_{(2)} \cdots P_{(i)} \cdots P_{(n)}] \tag{5}$$

$$P^* = [P_{(1)}^*\ P_{(2)}^* \cdots P_{(i)}^* \cdots P_{(n)}^*] \tag{6}$$

$$P' = [P'_{(1)}\ P'_{(2)} \cdots P'_{(i)} \cdots P'_{(n)}] \tag{7}$$

$$Q^* = [Q_{(1)}^*\ Q_{(2)}^* \cdots Q_{(i)}^* \cdots Q_{(n)}^*] \tag{8}$$

$$I = [I_{(1)}\ I_{(2)} \cdots I_{(i)} \cdots I_{(n-1)}] \tag{9}$$

$$R = [R_{(1)}\ R_{(2)} \cdots R_{(i)} \cdots R_{(n-1)}] \tag{10}$$

$$k_P = [k_{P(1)}\ k_{P(2)} \cdots k_{P(i)} \cdots k_{P(n)}] \tag{11}$$

$$k_Q = [k_{Q(1)}\ k_{Q(2)} \cdots k_{Q(i)} \cdots k_{Q(n)}] \tag{12}$$

Where V_{ac} is the vector of voltages of each of the AC buses, V_{ac}^* is the vector of voltage references associated to the droop scheme belonging to the reactive power loop of each station, V_{dc} is the vector of voltages of each of the DC buses, V_{dc}^* is the vector of voltage references associated to the droop scheme belonging to the active power loop of each station, P is the vector of DC powers in each of the converter stations, P^* is the vector of active power references associated to the active power loop of each station, P' is the vector of active powers coming from the renewable energy sources, Q^* is the vector of reactive power references associated to the reactive power loop of each station, I is the vector of currents in each of the DC lines, R is the resistance value in each of the DC lines, k_P is the vector of gains associated to the droop scheme belonging to the active power loop of each station, k_Q is the vector of gains associated to the droop scheme belonging to the reactive power loop of each station and n is the number of converter stations. It is important to keep these definitions in mind to understand the development of the model.

Development of Sub-Model 1

In Fig.1, the *p-th* current (I_p) that runs through the *p-th* line is given by:

$$I_p = \sum_{i=1}^{p} \frac{P_{(i)} + P_{(i)}'}{V_{dc(i)}} \tag{13}$$

On the other hand, from Fig.2(a) it is established that, in steady state, the *i-th* DC voltage, in the *i-th* converter, is given by:

$$V_{dc(i)} = \frac{P_{(i)}^* - P_{(i)} + V_{dc(i)}^* \cdot k_{P(i)}}{k_{P(i)}} \tag{14}$$

Furthermore, the Joule losses are determined through:

$$P_T = \sum_{k=1}^{n-1} R_k \cdot I_k^{\ 2} \tag{15}$$

Then, using (13), (14) and (15), the first target function associated to the total Joule losses over DC transmission lines is obtained:

$$\min P_T = \min OF_1 = \sum_{k=1}^{n-1} R_k \cdot \left[\sum_{i=1}^{k} \frac{(P_{(i)} + P_{(i)}') \cdot k_{P(i)}}{P_{(i)}^* - P_{(i)} + V_{dc(i)}^* \cdot k_{P(i)}} \right]^2 \tag{16}$$

On the other hand, each *j-th* current and/or voltage is restricted based on circuit laws and operational restrictions of the grid, i.e.:

$$I_{(j)} = \frac{V_{dc(j)} - V_{dc(j+1)}}{R_{dc(j)}} \tag{17}$$

$$I_{min} \leq I_{(j)} \leq I_{max} \tag{18}$$

$$V_{min\_dc} \leq V_{dc(j)} \leq V_{max\_dc} \tag{19}$$

Furthermore, it is possible to eliminate any *r-th* converter station or *t-th* renewable energy source through the following restrictions:

$$P_{(r)} = 0 \tag{20}$$

$$P_{(t)}' = 0 \tag{21}$$

It is also possible to establish, in each *q-th* renewable energy source, the minimum and maximum energy supplied by it:

$$P_{min}' \leq P_{(q)}' \leq P_{max}' \tag{22}$$

Development of Sub-Model 2

In this case it is desired to minimize the total apparent power associated to the converter stations ($\sum S$).

First, it is established that the total apparent power:

$$S_T = \sum_{i=1}^{n} \sqrt{P_{(i)}^{\ 2} + Q_{(i)}^{\ 2}} \tag{23}$$

On the other hand, from Fig.2(b), in steady state:

$$\left[V_{ac(i)}^* - V_{ac(i)} \right] \cdot k_{Q(i)} + Q_{(i)}^* = Q_{(i)} \tag{24}$$

Then, by combining (24) and (23), the second target function is obtained:

$$\min S_T = \min OF_2 = \sum_{i=1}^{n} \sqrt{P_{(i)}^{\ 2} + [(V_{ac(i)}^* - V_{ac(i)}) \cdot k_{Q(i)} + Q_{(i)}^*]^2} \tag{25}$$

On the other hand, the apparent power of the *j-th* station is restricted according to its capability:

$$\sqrt{P_{(j)}^2 + [(V_{ac(j)}^* - V_{ac(j)}) \cdot k_{Q(j)} + Q_{(j)}^*]^2} \le S_{\max(j)} \tag{26}$$

Furthermore, in order to give physical meaning to the reactive power problem, it is established that in the *j-th* station:

$$k_{(j)} \cdot \left((V_{ac(j)}^* - V_{ac(j)}) \cdot k_{Q(j)} + Q_{(j)}^* \right) \le 0 \tag{27}$$

Where:

$$k_{(j)} = \begin{cases} -1, & V_{ac(j)} < V_{\min\_ac} \\ +1, & V_{ac(j)} > V_{\max\_ac} \\ 0, & V_{\min\_ac} \le V_{ac(j)} \le V_{\max\_ac} \end{cases} \tag{28}$$

A schematic summary of the combined model is presented in Fig.3

Case Study

In this case a study associated to the Chilean electrical grid is considered, which is relevant due to the huge potential of renewable energy to be connected in it [14]. Using this system projected to the year 2050, a hybrid AC-DC grid is generated in order to study the combined model that was developed in the previous sections. The scheme associated with the system under study is shown in Fig.4 Its parameters are presented in Table I, where power, voltage and current are determined considering [15]. Furthermore, the following considerations are made:

1. The input for sub-models 1 and 2, correspond to the DC active power in the converters and the AC voltage in the bus of each converter, respectively (Fig.3). These values are generated randomly for twenty-four conditions (a day of normal operation).
2. AC and DC voltages are restricted to values established by Chilean and European norms, respectively [16,17]. These values are 0.95 and 1.05[pu] for AC voltages and from 0.85 to 1.05[pu] for DC voltages.
3. In sub-model 2, the reference AC voltage is set at a value equal to the nominal value of the grid.
4. In sub-model 1, each renewable energy source is restricted to supplying, at most, a value equal to the sum of the powers set in the converter stations.
5. It is considered that renewable energy sources 3 and 4 do not supply active power.

The results are shown in Fig.5 and Fig.6.

Table I: Parameters of the studied system.

Nominal power of converters	1.85[GVA]
Nominal current of converters	4.5[kA]
Nominal voltage in AC grid	220[kV]
Nominal voltage in DC grid	400[kV]
Resistance of each DC line	0.038[Ω/km]
Crucero - Cumbre distance	569 [km]
Cumbre - Polpaico distance	877 [km]
Polpaico - Charrúa distance	507 [km]
Charrúa - Puerto Montt distance	664 [km]

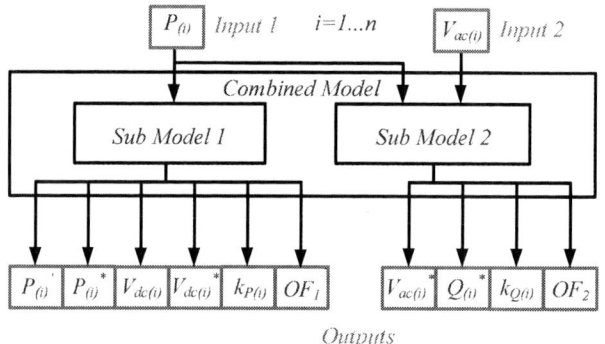

Fig 3: Scheme of the combined model.

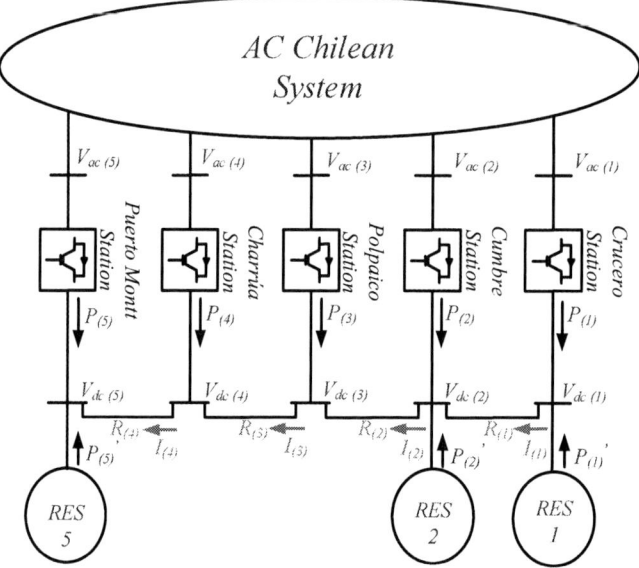

Fig 4: Chilean electric system associated to the case study.

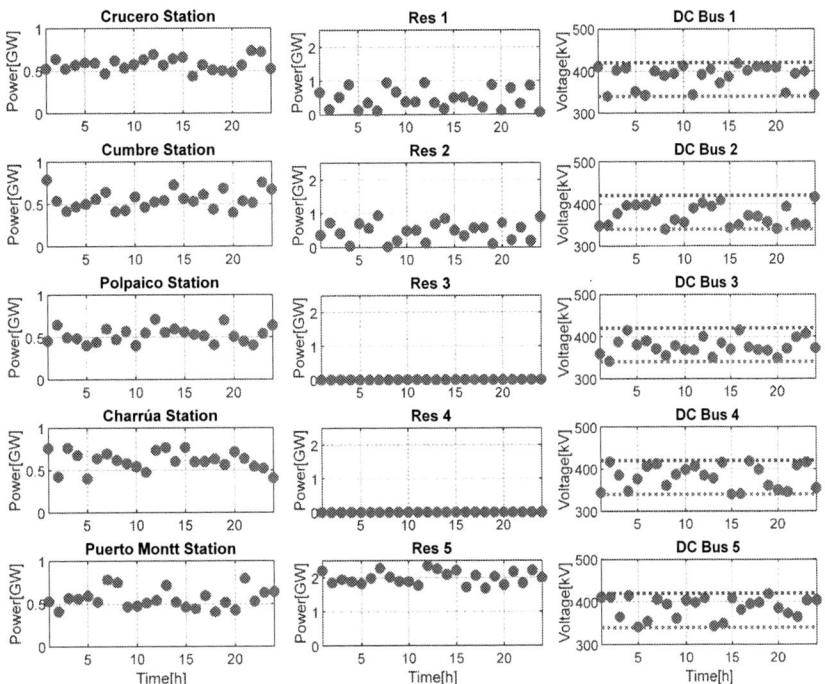

Fig 5: Results associated to sub-model 1.

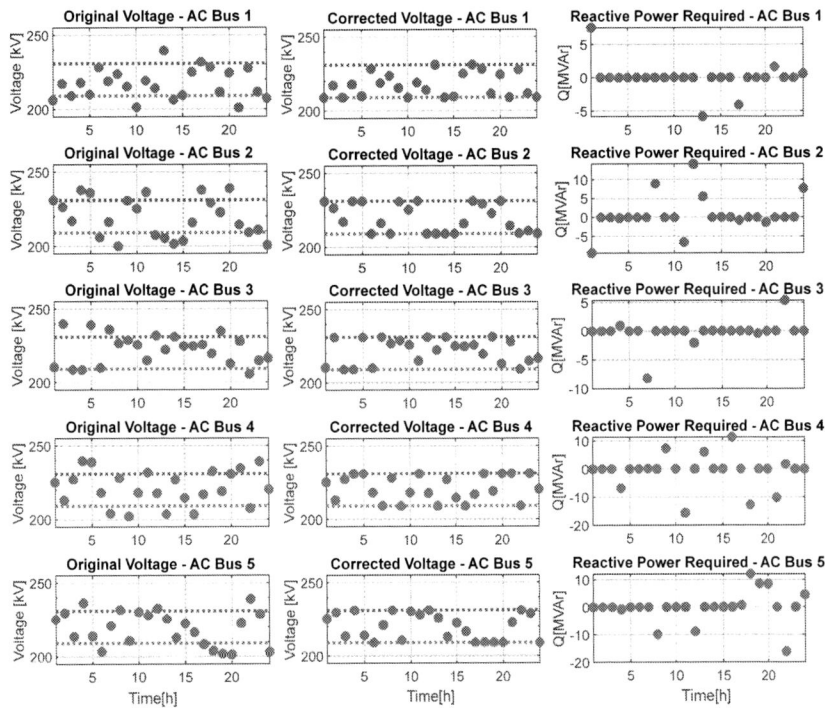

Fig 6: Results associated to sub-model 2.

The first column of Fig.5 (left column) shows the power established in the converters due to the demands of the AC grid. The second and third column show the power coming from renewable energy sources and the DC voltages, respectively. The last two are obtained as output variables from sub-model 1 (Fig.3). It is observed that the renewable energy sources are capable of meeting the power needs of the AC grid, while the DC voltage is maintained within its tolerance band, which was established between 0.85 and 1.05 [pu] (see dotted line). This is achieved with minimal losses over the DC transmission lines.

The first column of Fig.6 (left column) shows the AC voltages established in the AC buses, due to different operational conditions established on the AC grid. The second and third column show the corrected AC voltages and the required reactive power to perform said correction, respectively. In this case, the corrected AC voltage, is established as the voltage closest to the original AC voltage, which establishes its value within the accepted tolerance band. From the aforementioned, the required reactive power is determined as one of the output variables from sub-model 2. Finally, it is observed that, considering the tolerance band of the AC voltage, which is set between 0.95 and 1.05 [pu], (see dotted line) the amount of reactive power needed to fulfil the task is determined, with minimal losses in the stations.

It is important to mention that, the gains associated to the droop control (of active and reactive power), are determined as outputs from the combined model.

Conclusion

In this article a combined model based on optimal power flow for radial grids powered by renewable energy sources was developed, with the goal of (1) satisfying the needs of the AC grid(s) and (2) decreasing total losses of the DC grid (transmission system and converter stations). Furthermore, the operational restrictions associated to the lines, stations, and AC and DC buses are considered.

The main contribution of this article is the development of a combined model that allows, through the DC grid, supplying the needs of the AC grid without exceeding restrictive values associated to variables such as power and voltage in the MT-HVDC grid. This is done using an optimality criterion.

The developed models explicitly considers the references and gains associated to the droop control as decision variables. This allows them to be selected via an optimality criterion.

The results indicate that the system is capable of operating without exceeding restrictive bands associated to design or normative, against every operational condition tested.

References

[1]   L. Guo, Y. Ding, M. Bao, C. Shao, P. Wang, and L. Goel, "Nodal Reliability Evaluation for a VSC-MTDC-Based Hybrid AC/DC Power System," *IEEE Trans. Power Syst.*, vol. 35, no. 3, pp. 2300–2312, May 2020, doi: 10.1109/TPWRS.2019.2951711.

[2]   M. Barnes, J. Y. Chan, and M. Avendano-Mora, "Comparison of Control Strategies for Multiterminal VSC-HVDC Systems for Offshore Wind Farm Integration," in *7th IET International Conference on Power Electronics, Machines and Drives (PEMD 2014)*, Manchester, UK, 2014, p. 4.2.05-4.2.05, doi: 10.1049/cp.2014.0474.

[3]   F. Torres, S. Martinez, C. Roa, and E. Lopez, "Comparison Between Voltage Droop and Voltage Margin Controllers for MTDC Systems," in *2018 IEEE International Conference on Automation/XXIII Congress of the Chilean Association of Automatic Control (ICA-ACCA)*, Concepcion, Oct. 2018, pp. 1–6, doi: 10.1109/ICA-ACCA.2018.8609748.

[4]   S. Wenig, Y. Rink, and T. Leibfried, "Multi-terminal HVDC control strategies applied to the Cigré B4 DC Grid Test System," in *2014 49th International Universities Power Engineering Conference (UPEC)*, Cluj-Napoca, Romania, Sep. 2014, pp. 1–6, doi: 10.1109/UPEC.2014.6934649.

[5]   Y. Li and Y. Li, "Security-Constrained Multi-Objective Optimal Power Flow for a Hybrid AC/VSC-MTDC System With Lasso-Based Contingency Filtering," *IEEE Access*, vol. 8, pp. 6801–6811, 2020, doi:

10.1109/ACCESS.2019.2963372.

[6] K. Meng *et al.*, "Hierarchical SCOPF Considering Wind Energy Integration Through Multiterminal VSC-HVDC Grids," *IEEE Trans. Power Syst.*, vol. 32, no. 6, pp. 4211–4221, Nov. 2017, doi: 10.1109/TPWRS.2017.2679279.

[7] S. Kim, A. Yokoyama, T. Takano, H. Hashimoto, and Y. Izui, "Economic benefit evaluation of multi-terminal VSC HVDC systems with wind farms based on security-constrained optimal power flow," in *2017 IEEE Manchester PowerTech*, Manchester, United Kingdom, Jun. 2017, pp. 1–6, doi: 10.1109/PTC.2017.7981249.

[8] K. Alshammari, H. A. Alsiraji, and R. E. Shatshat, "Optimal Power Flow in Multi -Terminal HVDC Systems," in *2018 IEEE Electrical Power and Energy Conference (EPEC)*, Toronto, ON, Oct. 2018, pp. 1–6, doi: 10.1109/EPEC.2018.8598298.

[9] S. I. Nanou, O. D. Tzortzopoulos, and S. A. Papathanassiou, "Evaluation of an enhanced power dispatch control scheme for multi-terminal HVDC grids using Monte-Carlo simulation," *Electric Power Systems Research*, vol. 140, pp. 925–932, Nov. 2016, doi: 10.1016/j.epsr.2016.04.012.

[10] S. S. H. Yazdi, K. Rouzbehi, J. I. Candela, J. Milimonfared, and P. Rodriguez, "Analysis on impacts of the shunt conductances in multi-terminal HVDC grids optimal power-flow," in *IECON 2017 - 43rd Annual Conference of the IEEE Industrial Electronics Society*, Beijing, Oct. 2017, pp. 121–125, doi: 10.1109/IECON.2017.8216025.

[11] Q. Zhao, J. García-González, O. Gomis-Bellmunt, E. Prieto-Araujo, and F. M. Echavarren, "Impact of converter losses on the optimal power flow solution of hybrid networks based on VSC-MTDC," *Electric Power Systems Research*, vol. 151, pp. 395–403, Oct. 2017, doi: 10.1016/j.epsr.2017.06.004.

[12] D. Kotur and P. Stefanov, "Optimal power flow control in the system with offshore wind power plants connected to the MTDC network," *International Journal of Electrical Power & Energy Systems*, vol. 105, pp. 142–150, Feb. 2019, doi: 10.1016/j.ijepes.2018.08.012.

[13] J. Sau-Bassols, Q. Zhao, J. García-González, E. Prieto-Araujo, and O. Gomis-Bellmunt, "Optimal power flow operation of an interline current flow controller in an hybrid AC/DC meshed grid," *Electric Power Systems Research*, vol. 177, p. 105935, Dec. 2019, doi: 10.1016/j.epsr.2019.105935.

[14] Christian Santana, "Renewable Energies in Chile: Eolic, solar and hydroelectric potential from Arica to Chiloé", Chilean Ministry of Energy and German Cooperation Deustche Zusammenarbeit, 2014.

[15] P. Lundberg, A. Gustafsson, M. Jeroense, "Recent advancements in HVDC VSC systems", Cigré, 2015.

[16] "Technical Standard for Safety and Service Quality", Chilean Ministry of Energy, 2018.

[17] "Establishing a network code on requirements for grid connection of high voltage direct current systems and direct current-connected power park modules", Commission Regulation (EU) 2016/1447, 2016.

Characterization of lithium ion supercapacitors

Zeyang Geng[1], Felix Mannerhagen[2], Torbjörn Thiringer[1]

[1]Department of Electrical Engineering, Division of Electric Power Engineering
Chalmers University of Technology, SE-412 96, Gothenburg, Sweden
E-Mail: zeyang.geng@chalmers.se, torbjorn.thiringer@chalmers.se
[2]Department of Electrical Engineering, Division of Electricity
Uppsala University, SE-751 21, Uppsala, Sweden
E-Mail: felix.mannerhagen@angstrom.uu.se

Acknowledgements

The authors would like to thank Energimyndigheten for the financing of this work.

Keywords

«Batteries», «Supercapacitor», «Device characterisation», «Measurement»

Abstract

A hybrid Li-ion supercapacitor combines a traditional supercapacitor electrode with a Li-ion electrode and thus is expected to offer a high performance in terms of both power density and energy density. In this paper, lithium ion supercapacitors with three sizes, 40 F, 100 F and 270 F, are investigated. Different test methods including cycling at different C-rates and temperatures, pulse tests and electrochemical impedance spectroscopy are used to evaluate the energy and power performance. It is found that this new approach reaches a higher energy density without losing much power density compared with traditional supercapacitors. However a power optimized lithium ion battery shows a strong competitiveness over a lithium ion supercapacitor.

Introduction

Energy storage devices for various electrical applications are continuously improving their performances. Supercapacitors have traditionally been prevailing when considering the power density and the lifetime [1][2]. A commercially available 'mature' conventional supercapacitor [3] has been reported to have a usable power density of 5.8-6.8 kW/kg with an energy capacity of 4.1-6.0 Wh/kg. Its high power density makes supercapacitor an appealing option instead of using lithium ion batteries in many applications. At the same time, there has been a continuous improvement of the power density for lithium ion batteries which make Li-ion batteries strong competitors [4], especially lithium titanate batteries, which has LTO (lithium titanate oxide) as the negative electrode material instead of graphite. It has been reported that a 3 Ah-class LTO cell has a 10 s output power density of 3.6 kW/kg [5].

Although supercapacitors have superior performances regarding the power density, its low energy density is the main limitation, as shown in the Ragone plot in Fig. 1. Therefore, in many applications the supercapacitors need to be used together with batteries [7][8]. Instead of combining supercapacitors and batteries at a system level, another solution is to combine a conventional supercapacitor electrode with a battery electrode in the component level to form a hybrid Li-ion supercapacitor (LISC) [9][10]. Such a device would then be able to offer another balance between the power density and energy density. So far in literature, the available experiment data for LISCs is limited to lab scaled devices and there is very little data reported for LISCs that are commercially available.

It would thus be very valuable to investigate the energy density of a Li-ion supercapacitor towards conventional lithium ion batteries, as well as the power density towards a mature supercapacitor. The purpose of this paper is to characterize a new hybrid Li-ion supercapacitor, i.e. to evaluate its energy and power performance, and furthermore compare its performance in relation to batteries with high power density, as well as ordinary supercapacitors.

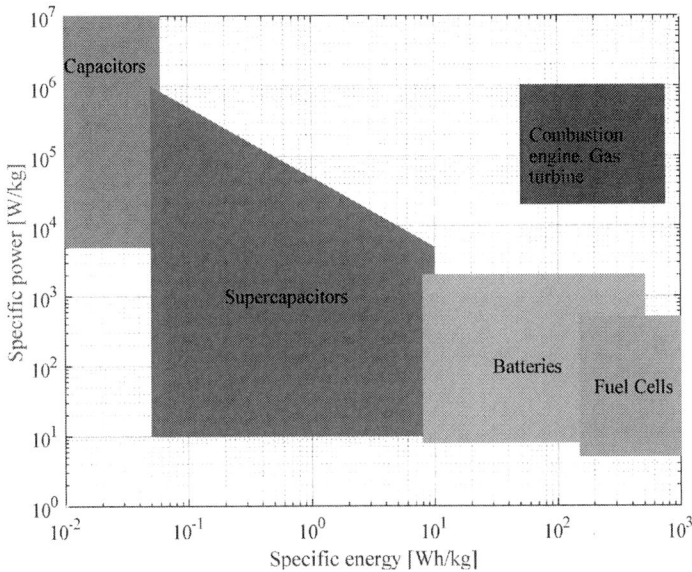

Fig. 1: A simplified Ragone plot (energy versus power density) for energy storage devices. Reprinted with permission from [6]. Copyright (2004) American Chemical Society.

Test object and procedure

The investigated lithium ion supercapacitors with 40 F,100 F and 270 F capacities are presented in Fig. 2. The voltage operation range is between 2.2 V and 3.8 V.

Fig. 2: The investigated lithium ion supercapacitors.

The lithium ion supercapacitors were placed in a climate chamber to have a controlled ambient temperature from -10 °C to 30 °C, with a type K thermocouple to measure the device surface temperature. A high precision battery test equipment GAMRY Reference 3000 was used to perform various tests listed in Table I. A schematic picture of the measurement set-up is presented in Fig. 3.

Table I: Test methods used to characterize the lithium ion supercapacitor.

Test method	State of charge	C-rate	Temperature
Charge and discharge test	0% -100%	50 C, 100 C and 150 C	-10 °C, 10 °C and 30 °C
Electrochemical impedance spectroscopy (EIS) with a range of 1kHz – 10 mHz	20%, 40% and 60% and 80%	Small perturbation signal.	-10 °C, 0 °C, 10 °C, 20 °C and 30 °C
Current pulse test	Every 10%	30 C for 10 seconds	Room temperature

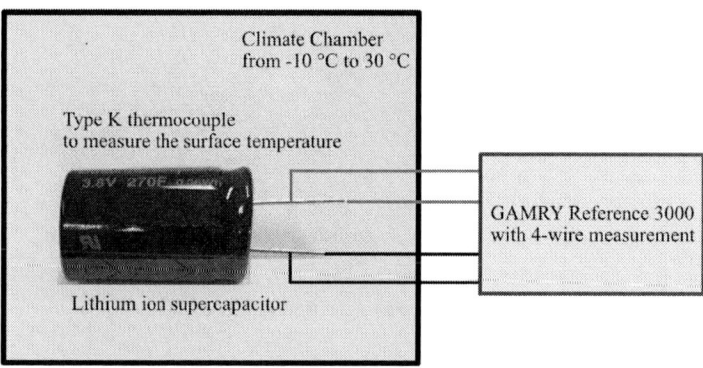

Fig. 3: Measurements set-up for the Li-Ion supercapacitor investigation.

Measurement results

First, the deviation between different specimens of 40 F devices was checked with charge and discharge cycles using 50 C,100 C and 150 C at room temperature. As can be noted from Fig. 4, the deviations between 5 different samples are negligible. Therefore, the following test results are considered representable.

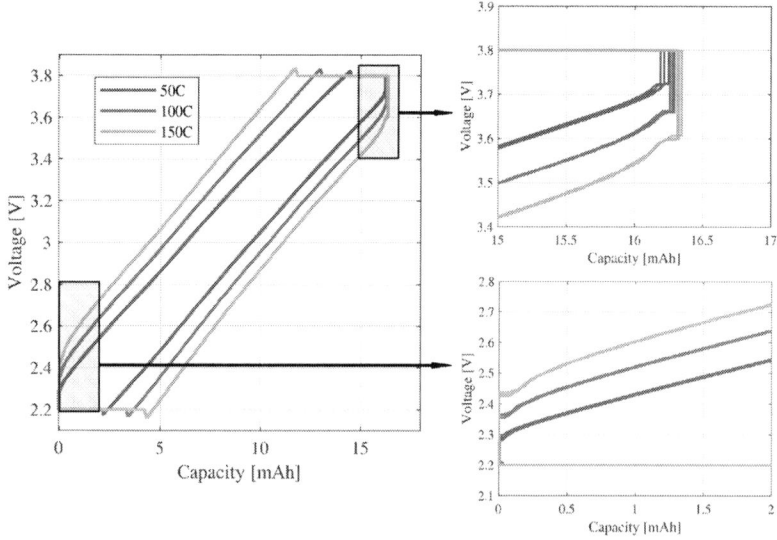

Fig. 4: Quality test of 5 specimens of 40 F lithium ion supercapacitors under 50C, 100C and 150C.

The results of charge and discharge test on a 40 F LISC at various C-rates and temperatures are presented in Fig. 5. During the cycling, constant current and constant voltage (CCCV) procedure is used in both charge and discharge, with 10 mA cutoff current. The open circuit voltage (OCV) is obtained as the average voltage of very slow cycles (1C) tested additionally. It can be observed that the resistance of the object is strongly dependent on the temperature. Furthermore, it can be observed that the charge resistance is higher than the discharge resistance, quite the opposite towards most batteries.

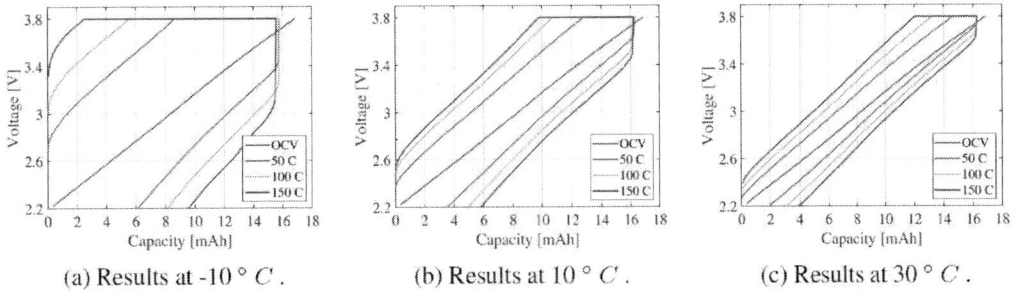

(a) Results at -10 ° C .   (b) Results at 10 ° C .   (c) Results at 30 ° C .

Fig. 5: Charge and discharge test on a 40 F capacitor at various C-rates and temperatures. The blue OCV-line is the average between the voltage for a 1 C charge and discharge sequence.

For electrochemical devices, due to the voltage limitation and the device internal resistance, the usable capacity is less than the actual capacity and the usable capacity is strongly dependent on the C-rate and temperature. In Fig. 6, the usable capacity of the 40 F and 270 F LISC at various C-rates and temperatures is presented. As can be noted from Fig. 6, the large 270 F capacitor has a substantial performance reduction at higher c-rates and lower temperatures regarding the available energy content. Less than 20% of the capacity is reachable at 150 C and -10 °C in Fig. 6(b).

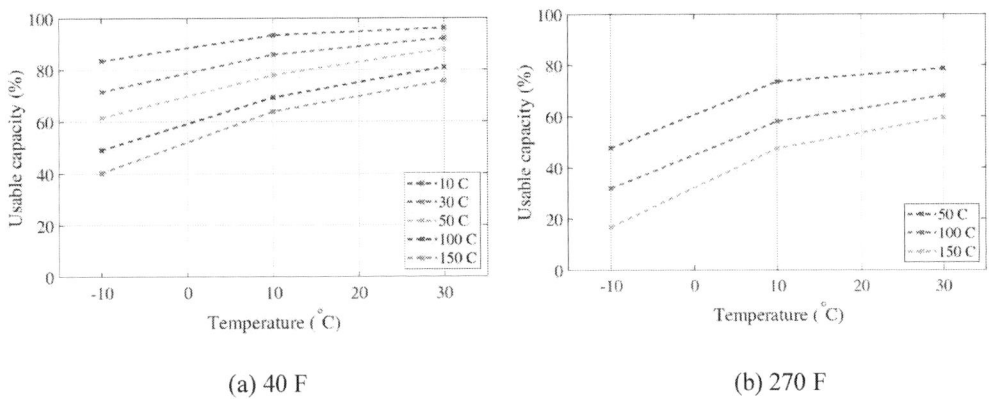

(a) 40 F   (b) 270 F

Fig. 6: Usable capacity as function of current and temperature for the (a) 40 F and (b) 270 F capacitor

The EIS test using a range from 1 kHz to 10 mHz was conducted for various temperatures and state of charge (SOC) levels. One of the examples of the result on a 40 F LISC can be found in Fig. 7. From the results it can be noted that the state of charge level hardly changes the characteristics of the component, both in the case of -10 °C and 30 °C. Whereas there is a stronger dependence on the temperature. In particular the charge transfer resistance (the diameter of the semicircle) is strongly dependent on the temperature, which is around 800 mΩ at -10 °C but only 30 mΩ at 30 °C.

EPE'20 ECCE Europe

Assigned jointly to the European Power Electronics and Drives Association & the Institute of Electrical and Electronics Engineers (IEEE)

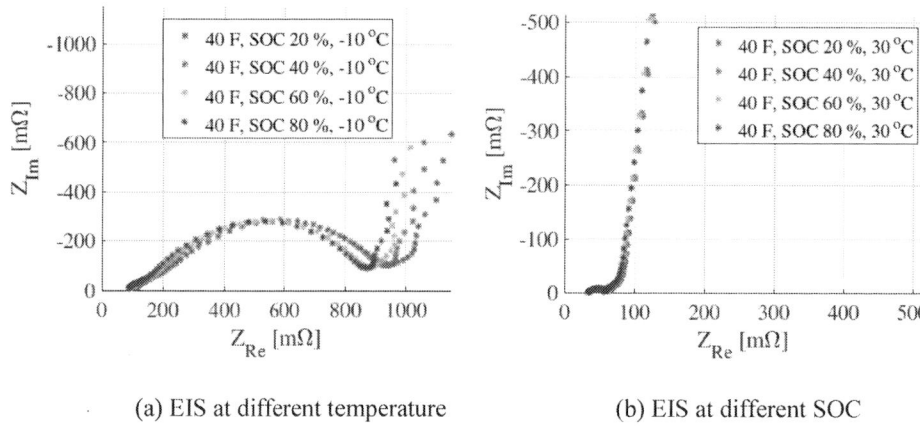

(a) EIS at different temperature   (b) EIS at different SOC

Fig. 7: Results from the EIS test.

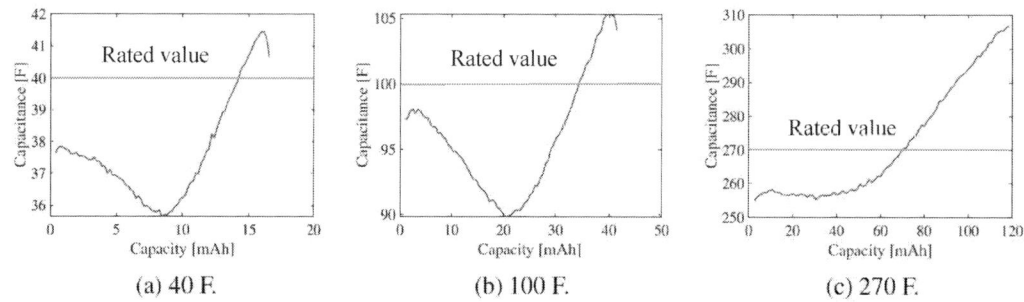

(a) 40 F.   (b) 100 F.   (c) 270 F.

Fig. 8: Capacitance in relation to the capacity.

In Fig. 8 the measured capacitance of the component at different SOC is presented. The increasing capacitance at a higher SOC could be a consequence of the electrochemical reaction on the lithium ion electrode.

Energy density and power density

Finally, the power and energy density were compared with those of power-optimized batteries, the result can be seen in Table II. It can be noted that, especially the 270 F LISC has an energy density of two-three times higher than the traditional supercapacitor, however at the expense of a power density loss of a factor of three. For the 40 F Li-Ion capacitor the energy capacity increase is 25-80 %, with a reduction of about 50 % in usable power density.

So, in applications where more energy capacity is needed, and a reduction in power density can be accepted the Li-ion supercapacitor is a good choice, and obviously there are possibilities regarding the tradeoff between the energy and power density. However, the lithium ion supercapacitor, as well as the traditional supercapacitor, has a substantial way to go to be a strong competitor towards the lithium titanate batteries both regarding power and energy density. In order to be a competitive product there is a need for improving the performance.

Table II: Comparison of power and energy performance

Component	Weight [g]	Energy density [Wh/kg]	Power density [kW/kg] Usable specific power	Power density [kW/kg] Impedance match specific capacity
40 F LISC	7.1	7.5	3.7	7.7
100 F LISC	15.9	8.4	3.2	6.4
270 F LISC	30.8	11.8	2.1	4.55
Maxwell BCAP 650-3000 F [3]	160-510	4.1 - 6.0	5.9-6.9	12-14
LTO Battery 3 Ah [5]	128	64	3.6 (different procedure)	-
LTO Battery 20 Ah [5]	510	90	2.3 (different procedure)	-

Conclusion

In this article, a series of commercially available lithium ion supercapacitors were investigated. The result found was an increase in energy capacity, of varies degrees, depending on the particular specimen investigated. The higher the improvement in performance regarding the energy density, the higher the loss in specific power density performance. Although the devices have a better energy density than usual supercapacitors, it cannot outperform a lithium titanate battery, neither in the power nor in energy density respect. It can be concluded that this product still has some way to go before it becomes a strong competitor to power-optimized batteries.

References

[1] M. C. Argyrou, F. Paterakis, C. Panagi, C. Makarounas, M. Darwish and C. Marouchos, "Supercapacitor application for PV power smoothing," 2018 53rd International Universities Power Engineering Conference (UPEC), Glasgow, 2018, pp. 1-5, doi: 10.1109/UPEC.2018.8541876.

[2] S. Harpool, A. von Jouanne and A. Yokochi, "Supercapacitor performance characterization for renewables applications," 2014 IEEE Conference on Technologies for Sustainability (SusTech), Portland, OR, 2014, pp. 160-164, doi: 10.1109/SusTech.2014.7046237.

[3] 2.7 V 650-3000 F Ultracapacitor cells, Maxwell Technologies, Document number 1015370-En.6

[4] W. Zhuang, S. Lu and H. Lu, "Progress in materials for lithium-ion power batteries," 2014 International Conference on Intelligent Green Building and Smart Grid (IGBSG), Taipei, 2014, pp. 1-2, doi: 10.1109/IGBSG.2014.6835262.

[5] Takami N, Inagaki H, Tatebayashi Y, Saruwatari H, Honda K, Egusa S. High-power and long-life lithium-ion batteries using lithium titanium oxide anode for automotive and stationary power applications. Journal of Power Sources. 2013 Dec 15;244:469-75.

[6] Winter, M., and Brodd, R.J.: 'What Are Batteries, Fuel Cells, and Supercapacitors?', Chemical Reviews, 2004, 104, (10), pp. 4245-4270.

[7] N. K. Medora and A. Kusko, "Battery management for hybrid electric vehicles using supercapacitors as a supplementary energy storage system," Intelec 2012, Scottsdale, AZ, 2012, pp. 1-8, doi: 10.1109/INTLEC.2012.6374473.

[8] I. Ben Amira, A. Guermazi and A. Lahyani, "Lithium-ion Battery/Supercapacitors Combination in Backup Systems," 2018 15th International Multi-Conference on Systems, Signals & Devices (SSD), Hammamet, 2018, pp. 1117-1121, doi: 10.1109/SSD.2018.8570567.

[9] Zhang M, Cheng J, Zhang L, Li Y, Chen M, Chen Y, Shen Z. Activated Carbon by One-Step Calcination of Deoxygenated Agar for High Voltage Lithium Ion Supercapacitor. ACS Sustainable Chemistry & Engineering. 2020 Feb 17;8(9):3637-43.

[10] Polozhentseva YA, Karushev MP, Rumyantsev AM, Chepurnaya IA, Timonov AM. A Lithium-Ion Supercapacitor with a Positive Electrode Based on a Carbon Material Modified by Polymeric Complexes of Nickel with Schiff Bases. TECHNICAL PHYSICS LETTERS. 2020 Feb 1;46(2):196-9.

Grey Wolf Optimizer Based Predictive Torque Control for Electric Vehicle Applications

Ali DJERIOUI[a][b], Azeddine HOUARI[a], Mohamed MACHMOUM[a], Malek GHANES[c], Tedjani MESBAHI[d], Mohamed Fouad BENKHORIS[a]

(a) IREENA Laboratory, University of Nantes, Saint-Nazaire, France
(b) LGE, Laboratoire de Génie Electrique, Mohamed Boudiaf University of Msila, BP 166 Ichbilia, Msila, Algeria
(c) Ecole Centrale de Nantes, LS2N UMR CNRS 6004, 44321 Nantes, France
(d) Department of Electrical Engineering, Strasbourg, France

Keywords

« Predictive Torque Control (PTC », « low speed operation, Electric Vehicles (EVs) », « Grey Wolf Optimization (GWO)», « Supercapacitor », « Fuel cell ».

Abstract

In this paper, an improved Predictive Torque Control (PTC) of a PMSM based on Grey Wolf Optimizer (GWO) is developed for smooth torque operation in electric vehicle applications (EVs). The fuel cell presents the main source and it is complemented with a quick power source (supercapacitors). The embedded Grew Wolf optimizer is used to solve the torque tracking tasks with minimal oscillations at low speed operation of PMSM drive. The new PTC algorithm can successfully ensure smooth time evolution of the torque and the speed. The design methodology is detailed and the provided simulation results show that the proposed PTC-GWO can be implemented in simulink, offering high performance in both steady and transient states of the PMSM drives even at low speed range.

Introduction

Nowadays, permanent magnet synchronous motor (PMSM) is perceived as a competitive candidate for high-performance automotive applications. This attention is partly due to its many attributes like high power density, large torque to inertia ratio, and high efficiency [1]. Today there are basically two types of instantaneous control of PMSM drives: field-oriented contro (FOC) and direct torque control (DTC). Compared with FOC, the DTC has numerous benefits such as lower parameters dependency, simpler digital implementation, and faster dynamic torque response [2]. In the case of classical DTC, the appropriate voltage space vectors are selected according to the errors of flux linkage and torque without need to coordinate transformation and current controller. In this scheme, the use of a switching table and two hysteresis controllers is efficient to carry out the DTC. Although DTC is getting more and more competitive compared to FOC approach, it also has some drawbacks, such as variable switching frequency, torque and flux ripples and high sampling frequency [3].

So, the main challenge that PMSM drives still face is how to reduce effectively their torque ripple, particularly at low speed operation. That is why many research efforts have been carried out to reduce the torque ripples with varying degrees of success. Therefore, the way of minimizing torque ripple becomes the main research subject in DTC of PMSM [4] In this context and thanks to its advantages like low torque ripple, constant switching frequency and excellent DC bus utilization, the DTC associated to voltage modulator (DTCPWM) is widely recognized as valuable solution because it can produce more precise and suitable torque and stator flux [5].

The model predictive torque control (MPTC) with pulse width modulation has been introduced as a serious rival to DTCPWM [5]. In this approach, the future behavior of the drive is predicted using a discrete-time PMSM model. The prediction results are evaluated using a predefined cost function and the reference voltage vector in accordance to the lowest cost would be applied to the machine via pulse width modulator.

A discrete time state-space model based MPTC is described in [2] for PMSMs, to achieve accurate torque control. the developed prediction scheme uses incremental changes in the stator's flux and current, together with voltage vectors.

All the aforementioned predictive approaches have in common an optimization problem resulting from the PTC formulation, which must be somehow solved. According to how the optimization problem is solved: on-line or off-line, PTC can be classified as implicit or explicit. In contrast to the explicit PTC, in the implicit one, the optimization problem must be solved in real-time working conditions which involves the use of fast numerical solvers. In [15], a cost function based optimization algorithm is presented as an efficient algorithm to solve the PTC optimization problem in real-time on embedded hardware. In [16], an on-line PTC method for a PMSM motor was proposed and an embedded quadratic programming solver is used to solve the predictive control problem. In order to confirm the feasibility of the proposed control strategy, its performance was evaluated in processorin-the-loop experiments.

In spite of the effective performance of MPTC methods, they are based on system model knowledge, and, thereby should be sensitive to large model parameters variation. Indeed, in vehicular applications, drives systems are subject to unknown disturbances, e.g., time-varying load, friction forces, and particularly effects of un-

---

EPE'20 ECCE Europe

Assigned jointly to the European Power Electronics and Drives Association & the Institute of Electrical and Electronics Engineers (IEEE)

modeled phenomena at low speed operation. In this way, the present paper aims to present an efficient torque-controller based on a recent optimization method without needing to know the system parameters.

The proposed PTC control approach is based on grey wolf optimizer (GWO) which is also motivated by the usefulness of GW approach for locating the minimum objective function with attracting proprieties like fast convergence and few adjustment parameters. GWO is a newly developed heuristic algorithm to handle nonlinear optimization problems [6].

It simulates the social hierarchy and hunting behavior of grey wolves in nature. As its name implies, the GWO mimics the major steps of grey wolves hunting like, seeking for prey, encircling, and attacking. The algorithm moves the wolves group toward prey by updating location vector, which is an average of best locations of the group. As investigated in [6], this algorithm presents several advantages compared to others heuristic algorithms such as Particle Swarm Optimization (PSO) in terms of low computing complexity, high solution accuracy, convergence independence of initial conditions and its ability to deal with local minima. These advantages are reflected in several works in electrical power systems where optimization is needed [7].

This paper is organized as follows. In section II and III, a description of the studied system of the hybrid vehicle and a brief overview on torque ripple sources of some un-modeled phenomena at low speed operation are receptively presented. Section IV introduces the proposed torque controller. For this end, the cost function considering flux and torque errors is defined and the GW algorithm design methodology is presented. Then, Section V presents simulation results and investigates how the objective function parameters are selected. Finally, section VI summarizes the major contributions of this work.

II. SYSTEM DESCRIPTION

Fig. 1 illustrates a typical circuit representation of an electrical vehicle wherein fuel cells (energy source) and supercapacitors (power source) are connected to the common DC bus via DC/DC converters. While the energy source ensures the steady state operation, the power source is connected to the DC-link through bidirectional DC/DC converter to supply the peak power demand and to keep the DC-bus voltage constant.

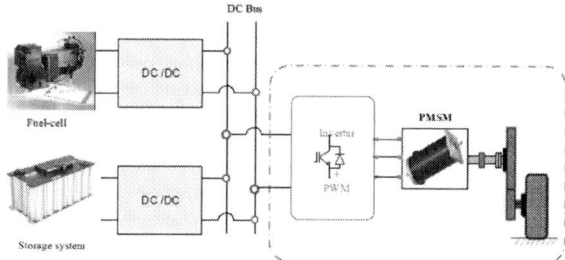

Fig .1. Typical power topology of an electrical vehicle based on PMSM drive

The motor drive includes a three phase inverter and PMSM and a drive train. In this work, the study is focused in the drive side and the dc-link is assumed as a constant voltage source 80 V. The drive train is emulated by a DC generator.

The electrical model of the PMSM with smooth poles expressed in the synchronous Park reference frame is given below:

$$\mathbf{u_s} = R_s \mathbf{i_s} + \dot{\boldsymbol{\psi}}_\mathbf{s} + w_E J \boldsymbol{\psi}_\mathbf{s} \tag{1}$$

where $\mathbf{u_s} = \begin{bmatrix} u_d & u_q \end{bmatrix}^T$ is the stator voltages, $i_s = \begin{bmatrix} i_d & i_q \end{bmatrix}^T$ is the stator current, $\boldsymbol{\psi}_\mathbf{s} = \begin{bmatrix} \psi_d & \psi_q \end{bmatrix}^T$ is the flux vector of stator, R_s is the stator per-phase resistance, $w = p\ \Omega$ is the rotor mechanical speed, p is the number of the pole pairs Ω the electrical angular speed of the a rotor

The stator flux can be formulated as follows:

$$\boldsymbol{\psi}_\mathbf{s} = L_s \mathbf{i_s} + \boldsymbol{\psi}_\mathbf{pm} \tag{2}$$

where $\psi_{pm} = \begin{bmatrix} \psi_{pm} & 0 \end{bmatrix}^T$ is the magnet's flux and L_s is the stator inductances.

The motion equation is expressed as follows:

$$\frac{d}{dt}\omega_e = \frac{p}{J}(\Gamma_e - \Gamma_L - \frac{F}{p}\omega_e) \tag{3}$$

In the above equation Γ_L is the load torques, F is the frictional coefficient and J is the moment of inertia
The electromagnetic torque equation is given by

$$\tag{4}$$

$$\Gamma_e = p \ \psi_{Pm} \ i_q$$

Model of Fuel cell modeling

Fuel cell is used to convert the chemical energy of reactant into electricity. this paper choose PEMFC as the first option of fuel cell. Nernst voltage En can be expressed as follows. The deducing process of En in detail is based on [10].

$$E_n = \begin{cases} 1.299 + (T-298)\dfrac{-44.43}{2F} + \dfrac{RT}{2F} In(P_{H_2}P_{O_2}^{1/2}) \quad when\, T \le 100^o C \\[4mm] 1.299 + (T-298)\dfrac{-44.43}{2F} + \dfrac{RT}{2F} In(\dfrac{P_{H_2}P_{O_2}^{1/2}}{P_{H_2O}}) \quad when\, T > 100^o C \end{cases} \tag{5}$$

Where R is gas constant (8.3145 J/mol K);. T is temperature of operation (Kelvin); F is Faraday constant (96,485 A s/mol); PH2 is partial pressure of H2 (atm); and PO2 is partial pressure of O2 (atm)

Supercapacitor is modeled as a simple RC circuit by the following equations:

$$U_{SC} = \frac{Q_T}{C_{SC}} - R_{SC} i_{SC} \tag{6}$$

where R_{SC} is the internal resistance and C_{SC} is the cell capacity. Therefore, cell terminal voltage depends on the demand current and SC state of charge.

IV. PROPOSED DIRECT TORQUE CONTROLLER

The aim of this section is to describe the design methodology of the proposed control approach based on GWO. The block diagram of the proposed torque controller strategy is shown in Fig.2. In this scheme, a maximum Torque Per Ampere is used to calculate the reference flux. Then proposed PTC-GWO controller ensures the desired tracking performances of both torque and flux with smooth time evolution.

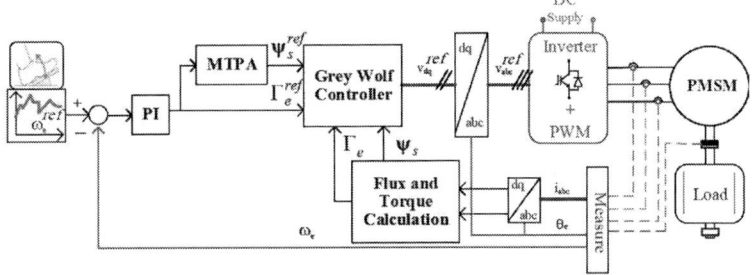

Fig. 3. Schematic of the proposed GWO-PTC controller.

It was found that the stator flux trajectory is necessary to achieve Maximum Torque Per Ampere (MTPA) [8].

B. GWO BASED PTC CONTROL

As discussed in the introduction, the use of GWO is of a growing interest because of its advantages in solving optimization problems with attracting properties like fast convergence, robustness and few adjustment parameters [9]. Herein, the key idea is to take advantage of fast optimization process of the GWO in order to reduce the discrepancies between the estimated torque and flux and their respective references. In this case, GWO algorithms based PTC act as controllers to generate the optimal voltage vector components via fitness functions minimization. For this end, fitness functions considering flux and torque errors are defined.

$$\mathbf{J} = \begin{bmatrix} -K_1 & K_2 T_s & | & 0 & 0 \\ 0 & 0 & | & K_3 + K_4 T_s & K_4 \end{bmatrix} \varepsilon$$

where $\mathbf{J}=[J_d \ J_q]$ is the objective vector, the first fitness function J_d is characterized by ψ_s and only one control input ud while the second fitness function J_q is characterized by Γ_e, it is concerned about minimizing the Γ_{rip} and only one control input uq. $K_1;K_2; K_3;K_4$ are constant gains, T_s is the sample time.

$\varepsilon = [\varepsilon_l[k] \quad \varepsilon_l[k-1] \quad \varepsilon_l[k] \quad \varepsilon_l[k-1]]^T$, the actual values ($X_{1,2}[k]$) of the tracking error are given by:

$$\varepsilon_1[k] = \psi_s^{ref}[k] - \psi_s[k]$$

$$\varepsilon_2[k] = \Gamma_e^{ref}[k] - \Gamma_e[k]$$

Besed on the GW theory, the construction mechanism of the GW based control algorithm involves three steps: hunting solutions evaluation, candidate input control calculation, and control reference selection [9].

IV. ENERGY MANAGEMENT STRATEGY

The aim of this part is to detail the design method of the filter-based strategy. As its name implies, the filter-based strategy synthesizes the reference using simple low-pass filter whose bandwidth is set with respect to the dynamic of the physical system. This method is commonly used for its simplicity and efficiency [10]. In this study, the Filter based EMS calculates the FC current reference that correspond to the steady state load current and presents lower dynamic to variations. The SC controller ensures the regulation of the DC-bus voltage in such a way to maintain the desired value and ensure fast response load solicitation. The relationship between the FC current reference and the load current is expressed in (10).

$$i_{FC\_REF} = \frac{1}{(0.5s+1)(2s+1)} * \frac{i_{load}}{1-d} \tag{10}$$

where, d is the FC converter duty cycle, and i_{load} is the load current.

V. SIMULATION RESULTS

In this section, the objective function parameters are selected through simulations under Matlab/Simulink software. The proposed selection criterion is based on torque ripple reduction at steady state operation when the system is subject to non-sinusoidal flux density distribution.

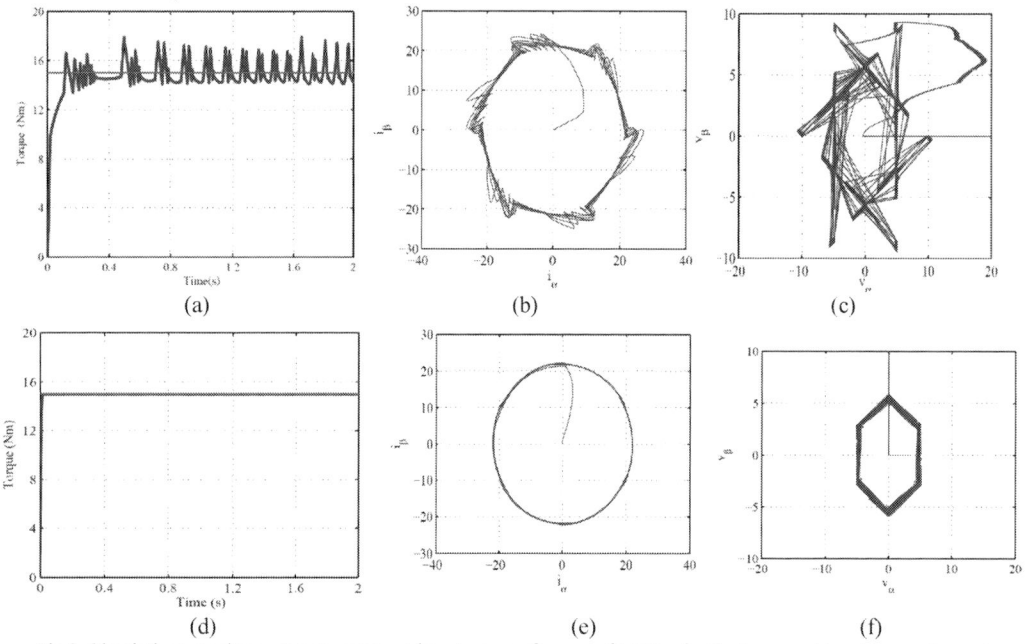

Fig.3. Simulations results: system response under a torque reference of 15 Nm for the two considered parameters sets, with: (a) and (d) torque ; (b) and (e) trajectories of the stator currents in Concordia coordinates; (c) and (f) trajectories of the input control voltages in Concordia coordinates.

The corresponding torque harmonics appear as the sixth, twelfth and other multiples of the sixth harmonics [25], and can be expressed as:

$$\Gamma_e = \sum_{i=1}^{i=n} A_{6i} \cos(6i\theta_E) \tag{11}$$

where A_{6i} is the magnitude of multiple sixth harmonics. The 6th, 12th, 18th and 24th flux harmonics are induced in the torque and are chosen equal respectively to 7%, 3%, 2% and 1% of the rated torque.

Fig. 3 shows the system response under a torque reference of 15 Nm. In this test the torque response, the respective Concordia components of the stator currents, and the input control voltages are highlighted for the two considered parameters sets. Figs. 3 (a) and (d) illustrate that the torque response when the best objective function parameters are applied (Fig.3 (d)) follows its reference trajectory with a smoother torque evolution.

Figs. 3 (b) and (e) illustrate the corresponding trajectories of the stator currents in Concordia coordinates for the two considered parameters sets. With the best objective function parameters, the currents follow circular trajectories which mean that the resulted abc current waveforms are sinusoidal. To investigate the input control behavior under the same test, Figs. 3 (c) and (f) show the respective trajectories of the input control voltages in Concordia coordinates. From this result, it can be seen that the GW algorithm with the best objective function parameters allows generating minimal input control trajectories and it can be noted that the control efforts are well distributed over the six sector of the PWM.

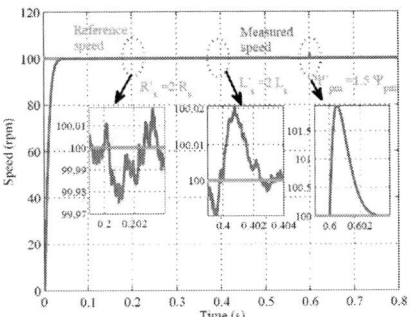

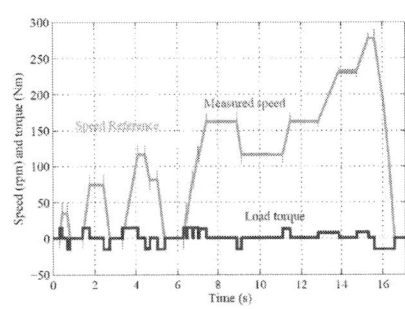

Fig. 4. Speed reponse with parameters variations.

Fig. 5. Motor speed and absolute load torque over the proposed cycle.

Fig. 4 shows the speed response in time domain at 100 rpm under parametric variations. The figure is illustarted in three sequences, respectively: [0.2 to 0.4s] the stator resistance increase of 100%, between [0.4 to 0.6s] the stator inductance increase of 100% and between [0.6 to 0.8s] the stator flux increase of 50%. It is demonstrated that GWO-PTC approache can endure large parametric variations and at the same time maintain good vehicle performance. it can be noticed that the GW control design do not depend on model parameters knowledge (free-model controller). To evaluate the electric vehicle dynamic performances, the speed and Load torque are shown in Fig 5. For that purpose, the speed reference changes during the proposed cycle under torque change.

Conclusion

This paper has proposed Grey Wolf Optimizer based Predictive Torque Control (GWO-PTC) for smooth torque operation in electric vehicle applications. The key idea is to take advantage of the fast optimization process of the GWO to design an efficient predictive torque controller without needing to know the system parameters. The mathematical principle of the proposed GWO-PTC controller is detailed, and the objective function parameters selection is presented in the simulation part. The analysis of the system performance under low speed operation shows that the proposed GWO-PTC ensures smooth-time evolutions of both speed and torque.

ACKNOWLEDGMENT

This work was supported by the International WISE RFI Electronique program, funded by FEDER.

References

[1] C. Xia, B. Ji, and Y. Yan, "Smooth speed control for low-speed synchronous motor using proportional-integral-resonant controller," IEEE Transactions on Industrial Electronics, vol. 62, no. 4, pp. 2123–2134, 2015.

[2] H. Zhu, X. Xiao, and Y. Li, "Torque ripple reduction of the torque predictive control scheme for permanent-magnet synchronous motors," IEEE Transactions on Industrial Electronics, vol. 59, no. 2, pp. 871–877, 2012.

[3] Y. Cho, K.-B. K. Y. Lee, S. Song, and J. Bon, "Synchronous Motor With Concentrated Winding," IEEE Transactions on Industrial Electronics, vol. 45, no. 3, pp. 235–248, 2010.

[4] S. Chai, L. Wang, and E. Rogers, "A cascade MPC control structure for a PMSM with speed ripple minimization," IEEE Transactions on Industrial Electronics, vol. 60, no. 8, pp. 2978–2987, 2013.

[5] Y. Cho, K. B. Lee, J. H. Song, and Y. I. Lee, "Torque-ripple minimization and fast dynamic scheme for torque predictive control of permanentmagnet synchronous motors," IEEE Transactions on Power Electronics, vol. 30, no. 4, pp. 2182–2190, 2015.

[6] S. Mirjalili, S. Mohammad, and A. Lewis, "Advances in Engineering Software Grey Wolf Optimizer," Advances in Engineering Software, vol. 69, pp. 46–61, 2014.

[7] S. Sharma, S. Bhattacharjee, and A. Bhattacharya, "Grey wolf optimisation for optimal sizing of battery energy storage device to minimise operation cost of microgrid," IET Generation Transmission and Distribution, vol. 10, pp. 625–637, 2016.

[8] Y. Zhang and Z. Jianguo, "Direct torque control of Permanent-magnet synchronous motor with reduced torque ripple and commutation frequency," IEEE Transactions on Power Electronics, vol. 26, no. 1, pp. 235–248, 2011.

[9] A. Djerioui, A. Houari, M. Ait-Ahmed, M.F Benkhoris, A. Chouder and M. Machmoum " Grey Wolf based control for speed ripple reduction at low speed operation of PMSM drives" ISA Trans., vol. PP, pp. 1–8, 2018.

[10] A Djerioui, A Houari, S Zeghlache, A Saim, MF Benkhoris, T Mesbahi, M. Machmoum "Energy management strategy of Supercapacitor/Fuel Cell energy storage devices for vehicle applications," International Journal of Hydrogen Energy, vol. 44, pp. 23416-23428, 2019

Operation principle and perspective performances of Metal Oxide Vacuum Field Effect Transistor - MOVFET

Davide Patti

STMICROELECTRONICS, Catania, Italy

G. Busatto, G. Golluccio, D. Marciano, A.Sanseverino, F.Velardi

DIEI University of Cassino and Southern Lazio
Via G. Di Biasio 43
Cassino (FR), Italy
URL: https://www.unicas.it/diei/

Keywords

«Device modeling», «Power semiconductor device».

Abstract

In this paper we present the operation principle of a Metal Oxide Vacuum Field Effect Transistor (MOVFET). The idea is to exploit a technology proposed in the literature to realize nano devices with low threshold voltage for power applications.

The principle of operation of the MOVFET is based on the emission of electrons from a cold cathode in a vacuum in such a way to obtain a pentode-like characteristics with a very low on resistance. This last feature is particularly interesting for the application as switches in power electronics applications. Moreover, the absence of a gate oxide which is always present in traditional MOSFETs suggests that this device can have better reliability features even in rad hard applications.

A numerical simulator is used to evaluate the impact of the technological parameters on the static characteristics of the device and the perspective performances in power electronics applications.

Introduction

The performances of silicon power devices for high frequency application have reached the silicon theoretical limit [1]. To achieve further improvements, it is necessary to move towards Wide Bandgap (WB) materials which have critical electric field and mobility higher than silicon. Nowadays, power electronic switches with very good static and switching performance [2] have been realized in materials like SiC [3] and GaN [4]. The performances of these devices are however limited by the fact that the carriers move in the solid matter with limitations in the carrier mobility which are not found if conduction occurs in a vacuum. In addition, WB devices still suffer from a poor tolerance to cosmic rays [5] which can limit their reliability in many terrestrial and spatial applications. Indeed phenomena like SEB, SEGR and gate damages are also observed in SiC power MOSFET [6 - 9] and GaN power HEMT [10].

In this work, we propose to use the conventional silicon technology to realize power Field Effect Transistors where the carriers' transport takes place in a vacuum. The absence in the proposed devices of a gate oxide make them potentially less sensitive to damages induced by ionizing radiation and more suited to be used in rad hard applications than the WB counterparts.

In the literature, there is great interest in devices that exploits the transport in a vacuum where the velocity of the electrons is orders of magnitude larger than in semiconductors. The possibility to realize two or three terminal field emitting devices, to be integrated in a conventional CMOS technology, is shown in [11] and [12]. Other authors, with the help of numerical finite element simulations, have studied the impact of geometrical and technological parameters on the behavior of nanoscale vacuum channel field effect transistors [13, 14]. All the proposed structures exhibit triode-like characteristics and are not suitable to be used as electronic switches in power applications.

The objective of this paper is to present a new vacuum cold emitter normally off power device with pentode like characteristics exhibiting very good performances in terms of on resistance and blocking voltage capabilities, which could represent a very promising alternative to power MOSFETs particularly in high voltage power electronics applications. The device can be manufactured using consolidated silicon technologies, which were previously used to construct a cold emitter triode [15 -17].

Simulations are used to study the effects of the technological choices on the device perspective performances, which appear to be very promising in power electronics applications.

Principle of operation and simulation tools

The radial section of the elementary cell of the proposed structure is reported in Fig.1 together with its main geometric parameters. The cell includes a cylindrical vacuum cavity surrounded by a thin layer of insulating material. The n+ silicon bottom layer (anode) collects the electrons emitted by the two metal tips, realized on the top of the device (cathode) and electrically connected to the ground. The cold emission of electrons is due to the large electric field induced by the positive voltage applied to the n+ polysilicon gate surrounding the cavity. The electric field at the cathode is enhanced by the pointed shape of the metal tips. A proper choice of the dielectric characteristics of the insulating material between cavity and gate can be used to further increase electric field and electron emission. When the electric field exceeds the value of about 10^7 V/cm at the tip surface, the Fowler-Nordheim emission of electrons [18] becomes significant and establishes a current between the cathode and the positively biased anode. By varying the gate voltage, it is possible to control the electric field at the cathode and therefore the anode current.

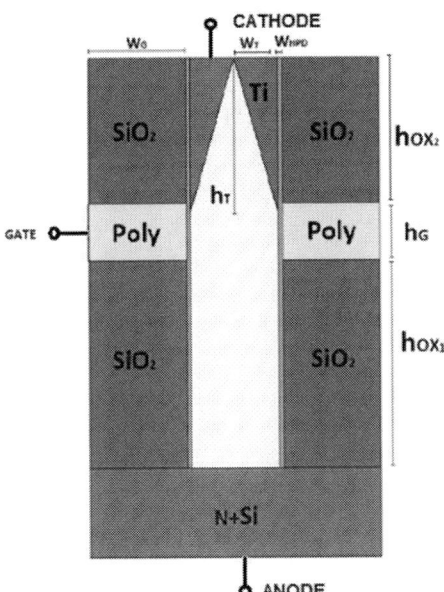

Fig. 1: Radial section of the cylindrical elementary cell of the proposed device

TCAD Sentaurus [19] simulations were used to analyze the behavior of the proposed device. Native TCAD Fowler-Nordheim model is not suitable to describe electron emission from a metal to vacuum because this model was developed to describe the tunnel current between oxide and silicon in digital devices. To solve the problem we developed a generalized model applicable to any interface and in particular to take into account the height of the potential barrier between the metal and the vacuum. In addition, the model includes the reduction of the tunneling barrier due to image forces [20].

2D simulation of the structure in Fig. 1 were carried out using the geometric and technological parameters of the previously presented cell [15] summarized in Table I corresponding to the cell diameter of 0.45 µm and the height of about 2 µm. The structure was biased with positive voltages Vgk and Vak applied to the gate and anode terminals, respectively. The cathode was grounded.

Table I: Main geometric parameters used in simulation

Parameter	Nominal Value [nm]	Description
w_T	225	Tip Width
h_T	700	Tip Height
w_{HPD}	100	Dielectric Width
$w_{OX1} = w_{OX2}$	500	Oxide Width
h_{OX1}	1000	Oxide 1 Height
h_{OX2}	700	Oxide 2 Height
$w_G = w_{OX1} = w_{OX2}$	500	Gate and Oxide Width
h_G	280	Gate Height
h_T	25	Bezier Tip Parameter

The zooms around the metal tips of Fig. 2 a) and b) show the distribution of electric field modulus for Vak=20 V and for two different values of the gate voltages, Vgk=2 and 4 V, respectively. As expected, in both cases the largest values of the electric field are observed in proximity of the tip. However, for Vgk=2 V (Fig. 2a) the simulated maximum electric field of $3.3 \; 10^6$ V/cm is not sufficient to cause the electron emission and therefore no significant anode current is expected. If the gate voltage reaches the threshold voltage of 4 V (Fig. 2b), the electric field at the tip surface becomes $\approx 10^7$ V/cm which can be considered as the threshold conditions for the electron emission to originate the anode current.

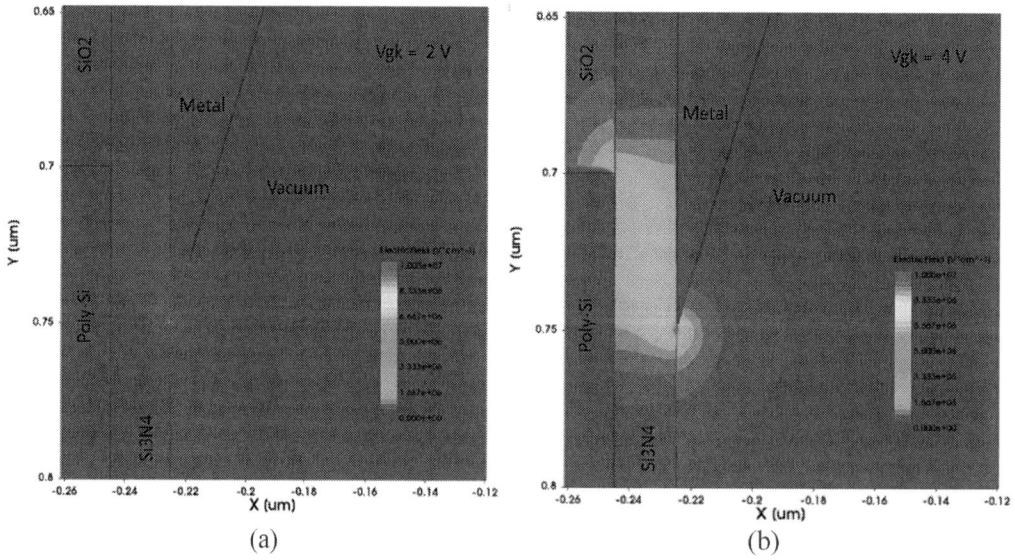

Fig. 2: Detail of the electric field inside the structure @ Vak = 20 V and a) Vgk = 2 V, b) Vgk = 4 V.

Fig. 3 shows simulated anode current produced by a single cell, with thickness and width both of 0.45 µm, as a function of the anode voltage for different values of the gate voltage. Thanks to the large distance between anode and tip (1 µm), the potential applied to the anode only weakly influences the electric field at the tip resulting in a very small Early effect on the output characteristics. Considering the double tips, we can conclude from Fig. 3 that a current of about 100 nA per cell can be achieved for Vgk = 16 V. If we assume a realistic cell pitch of 1µm, a cell density of about 10^8 cells per cm^2 can be obtained so that a current density of 10 A/cm^2 can be expected. It is very important to note that this current density can be obtained also for large distance between anode and gate. In fact the voltage at the anode only marginally influences the field at the tips and therefore the value of the anode current. At the same time, a large distance between anode and cathode allows the device to achieve high blocking voltage capabilities as discussed in the next section.

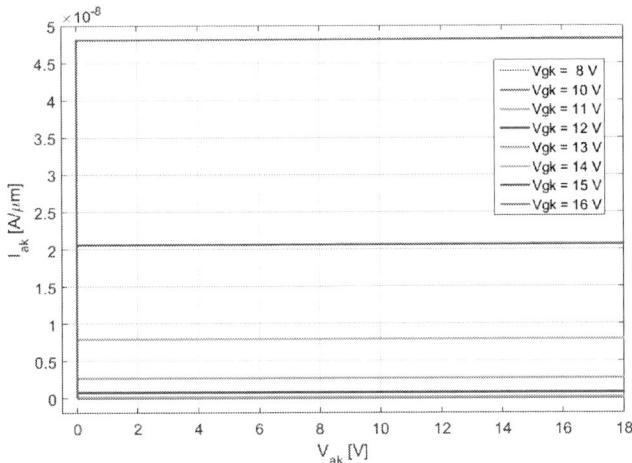

Fig. 3: Simulated output characteristics of MOVFET for different values of the gate bias.

Fabrication and technological improvements

The results presented in the previous section were obtained considering the technologies used in the standard microelectronics processes. In [15] the fabrication process of the structure of Fig.2 has been described in detail. In particular, the paper shows the realization of an array of 20 cells on a silicon substrate of type n. Although the structure was affected by some critical issues, SEM images and numerical simulations indicated a way to develop a more accurate design. In [16] the anode current flowing through the structure was measured and was shown to be an emission current.

Significant improvements of the static characteristics of the device can be obtained acting on the main technologic parameters used to manufacture the MOVFET. Numerical simulations have also been developed to explore these aspect possibilities.

The first parameters taken into consideration were the thickness and the dielectric constant of the material that surrounds the cavity. Fig. 4 and Fig. 5 report the device trans-characteristics, obtained at V_{ak}=20V, for different values of the dielectric thickness (SiO$_2$ in the standard case) and for materials having different dielectric constants, respectively.

The reduction of the thickness from 80 nm (yellow curve) to 50 nm (magenta curve) almost leads to a doubling of the trans-conductance. Moreover, the change of the dielectric from Si$_3$NO$_4$ (k=7), to HfO$_2$ (k=25) and TiO$_2$ (k=80) implies an increase of the I_{ak} at V_{gk}=20V by more than 40%.

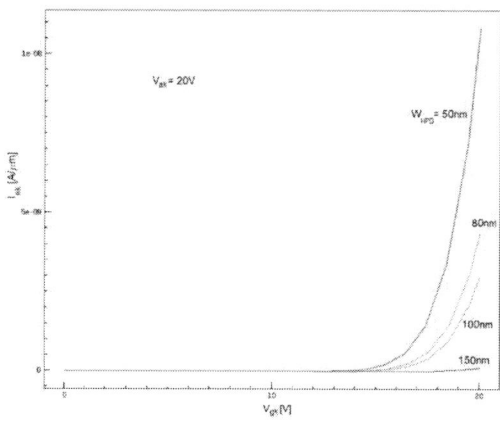

Fig. 4: Simulated output characteristics of MOVFET for different values of the oxide thickness.

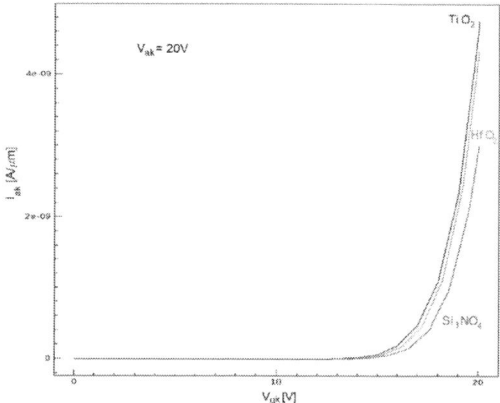

Fig. 5: Simulated output characteristics of MOVFET for dielectric materials with different dielectric constant.

Another parameter, which plays a very important role in improving the static performances of MOVFET, is the work function of the material used for the cathode. Fig. 6 reports the simulated MOVFET trans-characteristics obtained for metal tips made with different materials, namely Mg, Al, Ti, W, and Cu, whose work functions are summarized in Table II. The results show a strong reduction of the gate threshold voltage and a strong increase of the trans-conductance. The current at V_{ak} = 20V and V_{gk} = 20 V becomes larger than 1 $\mu A/\mu m$. If we assume a realistic cell pitch of 1 μm and a cell density of about 10^8 cells /cm² we can expect current density larger than 100 A/cm² which is much better than Silicon Carbide power MOSFET with comparable characteristics.

Table II: Work functions of the tip materials used in the simulation

Metal	Work Function Φ
Mg	3.66 eV
Al	4.10 eV
Ti	4.33 eV
W	4.32 eV
Cu	4.70 eV

It is useful to underline that in this device the distance between cathode and anode plays a negligible role on the flow rate of the current whose intensity is mainly fixed by the potential applied to the gate that determines the electric field at the tip. This means that very high currents can also be reached for devices that have a large blocking voltage, ideally limited only by the value of the electric field in vacuum. This electric field, in principle, is infinite and, in practice, it is limited only by technological aspects, like the minimum pressure attainable in the volume, the nature of the residual gas present in it or the quality of the dielectric materials. Critical Electric field in the order of 2 MV/cm, comparable to that of SiC, can be experimentally obtained in a vacuum [21]. If we consider a realistic cavity aspect ratio of 1/10, a blocking voltage of about 2kV is expected for an anode to cathode distance of 10µm. For this device, the on resistance is expected to be much smaller than a SiC power MOSFET with comparable characteristics thanks to the better current conduction in the vacuum.

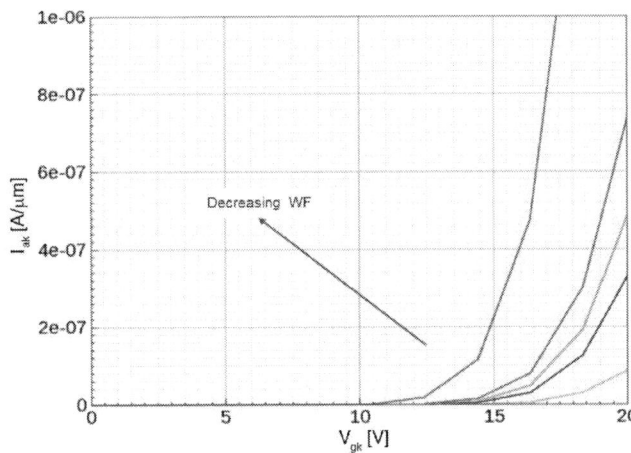

Fig.6 Simulated output characteristics of MOVFET for metal materials of the emitting cathode with different work functions.

Conclusion

We have presented a new power device exploiting the electron emission from a cold cathode and the current conduction in a vacuum. The device can be realized using the standard silicon technology. It is compact and can be easily integrated in a standard microelectronic process. The device exhibits pentode like characteristics. An appropriate choice of geometrical and technological parameters guarantees threshold voltages of few Volts, very high trans-conductance and very good current capability in particular for high blocking voltage devices. 2D simulation result allow us to expect for the MOVFET better performance than those of SiC power MOSFETs.

The expected very good performances make MOVFET a very promising alternative to the other semiconductor devices in power applications.

References

[1] B.J. Baliga, "Power Semiconductor Device Figure of Merit for High-Frequency Applications, IEEE Electron device Letters, Vol.10, NO.10 October 1989.

[2] J. L. Hudgins, G. S. Simin, E. Santi, and M. A. Khan, "An Assessment of Wide Bandgap Semiconductors for Power Devices," IEEE Trans. on Power Electronics, vol. 18, no. 3, MAY 2003, pp. 907-914

[3] M. Östling, R. Ghandi, C. Zetterling, "SiC power devices – present status, applications and future perspective", Proc. ISPSD 2011, pp. 10-15.

[4] X. Ding, Y. Zhou and J. Cheng, "A review of gallium nitride power device and its applications in motor drive," in CES Transactions on Electrical Machines and Systems, vol. 3, no. 1, pp. 54-64, March 2019, doi: 10.30941/CESTEMS.2019.00008.

[5] H.R. Zeller, "Cosmic ray induced failures in high power semiconductor devices," Microelectronics Reliability, vol. 37, Issue 10, pp. 1711-1718, 1997.

[6] S. Kuboyama, C. Kamezawa, Y. Satoh, T. Hirao and H. Ohyama, "Single-Event Burnout of Silicon Carbide Schottky Barrier Diodes Caused by High Energy Protons," in IEEE Transactions on Nuclear Science, vol. 54, no. 6, pp. 2379-2383, Dec. 2007.

[7] A. F. Witulski et al., "Single-Event Burnout Mechanisms in SiC Power MOSFETs," in IEEE Transactions on Nuclear Science, vol. 65, no. 8, pp. 1951-1955, Aug. 2018, doi: 10.1109/TNS.2018.2849405.

[8] C. Abbate, G. Busatto, D. Tedesco, A. Sanseverino, F. Velardi and J. Wyss, "Gate Damages Induced in SiC Power MOSFETs During Heavy-Ion Irradiation—Part I," in IEEE Transactions on Electron Devices, vol. 66, no. 10, pp. 4235-4242, Oct. 2019, doi: 10.1109/TED.2019.2931081.

[9] C. Abbate, G. Busatto, D. Tedesco, A. Sanseverino, F. Velardi and J. Wyss, "Gate Damages Induced in SiC Power MOSFETs During Heavy-Ion Irradiation—Part II," in IEEE Transactions on Electron Devices, vol. 66, no. 10, pp. 4243-4250, Oct. 2019, doi: 10.1109/TED.2019.2931078.

[10] C. Abbate, G. Busatto, F. Iannuzzo, S. Mattiazzo, A. Sanseverino L. Silvestrin, D. Tedesco, F. Velardi: "Experimental Study of Single Event Effects Induced by Heavy Ion Irradiation in Enhancement Mode GaN Power HEMT" Microelectronics Reliability 2015, 55(9-10), pp. 1496-1500.

[11] Jin-Woo Han, Jae Sub Oh, and M. Meyyappan, "Cofabrication of Vacuum Field Emission Transistor (VFET) and MOSFET", IEEE Transaction on Nanotechnology, Vol.13, NO.

[12] William M. Jones, Daniil Lukin, and Axel Scherer, "Practical nanoscale field emission devices for integrated circuits", Applied Physics Letters 110, 263101 (2017).

[13] Jungsik Kim, Jiwon Kim and Hyeongwan Oh, M. Meyyappan and Jin-Woo Han, Jeong-Soo Lee, "Design guidelines for nanoscale vacuum field emission transistor", Journal of Vacuum Science & Technology B, Nanotechnology and Microlectronics: Materials, Processing, Measurement, and Phenomena 34, 042201 (2016).

[14] Jiwon Kim and Hyeongwan Oh, Jungsik Kim, Rock-Hyun Baek, Jin-Woo Han and M. Meyyappan, Jeong-Soo Lee, "Work function consideration in vacum field emission transistor design", Journal of Vacuum Science & Technology B, Nanotechnology and Microlectronics: Materials, Processing, Measurement, and Phenomena 34, 042201 (2016).

[15] S. Pennisi, G. Castorina, Davide Patti, "Dovetail Tip: A New Approach for Low-Threshold Vacuum Nanoelectronics", IEEE Transaction on Electron Devices, Vol. 62, Issue 12, Pagg 4293-4300, 2015

[16] D.Patti, G. Castorina, S.Pennisi, "2-V Turn-on Voltage Field-Emitting Vacuum Nanoelecronic Device, 2016 IEEE International Conference on Electronics, Circuits and Systems (ICECS).

[17] D.Patti, S.Pennisi, S. Lombardo and G. Nicotra, "Toward a Nanofabricated Vacuum Cold-Emitting Triode", 14th International Conference on Synthesis, Modeling, Analysis and Simulation Methods and Applications to Circuit Design (SPAC), Pgg. 1-4, 2017.

[18] Springer, "Field Emission Electronics", N. Egorov, E. Sheshin, ISBN: 978-3-319-56560-6, 2017.

[19] Synopsys™ Sentaurus TCAD 2016 Version.

[20] G. Binnig, N.Garcia, H.Rohrer, J. M. Soler, and F. Flores, "Electron-metal-surface interaction potential with vacuum tunneling: Observation of the image force", Phys. Rev. B 30, 4816, 15 October 1984.

[21] Y. Saito, " Breakdown phenomena in vacuum," in Proceedings of the 1992 Linear Accelerator Conference, Ottawa, Ontario, Canada (1992), p. 575.

Improved methodology for predicting correlated color temperature in mixed LED lighting sources

Thais E. Bolzan, Bruno F. Almeida, Renan R. Duarte, Vitor C. Bender, Rafael A. Pinto
FEDERAL UNIVERSITY OF SANTA MARIA - UFSM
Avenida Roraima 1000, 97105-900, Camobi
Santa Maria, Brazil
Tel.: +55 / (55) – 981297611.
E-Mail: thaisbolzan@gedre.ufsm.br

Acknowledgements

The authors gratefully acknowledge Eletro Zagonel Ltda for the support during the development of this work. This study was financed in part by the Coordenação de Aperfeiçoamento de Pessoal de Nível Superior - Brasil (CAPES/PROEX) - Finance Code 001 and by Conselho Nacional de Desenvolvimento Científico e Tecnológico - CNPq proc 425794/2018-0.

Keywords

«Device characterization», «Device modeling», «Lighting», «Modelling», «Thermal design».

Abstract

This paper proposes an improved methodology for predicting the correlated color temperature (CCT) of a system comprised of one or more LEDs by using just data obtained from the datasheet. Experimental results obtained for two LEDs of distinct CCTs (6500K and 4000K) validate the proposed methodology.

Introduction

Light can affect human behavior in several ways. The human body has a natural 24-hour light-dark cycle called circadian cycle. This cycle is regulated primarily by the incident light on the retina. Depending on the exposure to light, hormone production (such as cortisol and melatonin) can be affected, resulting in changes on the activity-rest pattern of a person [1].

Artificial lighting can, in a negative way, interfere with the circadian cycle of a person, especially considering how much time humans spent indoors nowadays. Previous studies show that be exposed just to solar light during the day helps to re-synchronize the circadian cycle [2]. To prevent cycle shifts, it is desirable to be exposed, even indoors, to light that can change its CCT and luminous flux throughout the day to match the sun characteristics. Furthermore, it is known that a high CCT induces alertness, while a low CCT induces relaxation [3].

When artificial lighting is considered, it is interesting to have a lamp capable of changing its CCT and luminous flux based on the hour of the day or on the situation needed (more alertness, per example). Light-emitting diodes (LEDs) are widely used in general lighting. Some of its characteristics such as high efficacy, long lifespan and more eco-friendly structure are highly desirable in this application [4]. Moreover, LEDs are available in a broad range of CCT values and power ratings.

However, during operation, variations in LED junction temperature can induce notable shifts in its color characteristics [5]. Hence, the CCT vary in terms of current and junction temperature of the LED [6]. The relationship between forward voltage, forward current, optical parameters and operating temperature should be considered when using LEDs. This ensures that system's characteristics match the desirable values during its operation. To do so, it is necessary to obtain a mathematical model which comprises all these factors together and allow for the estimation of the lamp CCT and luminous flux under different operation points. This is especially true when the lamp is made of multiple LEDs, often with different optical characteristics (e.g. different CCTs).

There are already in the literature methods to estimate the correlated color temperature of LEDs [7], [8]. These approaches are, however, based on experimental data. Two key factors in obtaining experimental data from LEDs make these methods unsuited for certain applications: First, specialized (and usually expensive) equipment is mandatory to get accurate CCT and chromaticity values from any given LED. In addition, these values cannot be obtained without having the LED. This makes impossible to speed up the design process as one cannot predict accurately if the chosen LED is suitable for the application.

In this paper is proposed a simple and improved methodology to estimate the CCT of mixed LEDs using just data obtained from the device's datasheet. In doing so, the design process of LED-based lighting systems can be optimized, reducing both development costs and time to market. The methodology is based on the interaction among different parameters that affect the LED performance, such as electric parameters (current, voltage and power), thermal parameters (junction and heat sink temperature) and optical parameters (luminous flux, correlated color temperature, chromaticity coordinates).

The rest of this paper is organized as follows: the proposed methodology is presented in detail in the next section. After, experimental results obtained using two white LEDs with different correlated color temperatures (4000K and 6500K) are presented to validate the methodology. The last section presents the conclusions of this work.

Methodology to predict the correlated color temperature

The main goal of this work is to obtain an estimation of the final CCT for a lamp comprised of one or more LEDs using just the data provided by the manufacturer in the device's datasheet. The first step is to obtain the simplified LED model, comprised of an ideal diode in series with an independent voltage source V_o and a parasitic resistance R_s [9]. By modeling the $I_f(V_f)$ characteristic of the LED (forward current I_f, forward voltage V_f), the parameters of this simplified model can be defined, resulting in (1) [10].

$$V_f = V_o + R_s \cdot I_f \tag{1}$$

In order to account for the changes in the LED's forward voltage due to changes in its junction temperature T_j, most datasheets provide the temperature coefficient k_v, that correlates these two parameters. Assuming a system where the LED is mounted in a heatsink to better distribute the heat loss and keep the device under safe temperature conditions, the junction temperature of the LED can be expressed by (2) [11].

$$T_j = T_{hs} + R_{jc} \cdot V_f \cdot I_f \cdot k_h \tag{2}$$

In this equation, T_{hs} is the heatsink temperature, R_{jc} is the thermal resistance between the LED junction and its case and k_h represents the amount of power applied to the LED that is converted into heat instead of light. This coefficient normally has a value between 65% to 85% and can be found in the LED datasheet as the inverse of radiant efficiency [12].

Later, by applying a linear regression in the data from the luminous flux Φ versus current I_f plot $\Phi(I_f)$ and luminous flux versus junction temperature T_j plot, $\Phi(T_j)$, it is possible to achieve an equation that relates the luminous flux with respect of current and temperature variations, as shown in (3) [11].

$$\Phi = \left[c_0 + c_1 \cdot \left(T_a + \left(R_{jc} + n \cdot R_{hs} \right) \cdot k_h \cdot I_f \cdot \frac{V_o + R_s \cdot I_f + k_v \cdot (T_a - T_o)}{1 - I_f \cdot k_h \cdot k_v \cdot \left(R_{jc} + n \cdot R_{hs} \right)} \right) \right] \cdot \left[d_0 + d_1 \cdot I_f \right] \cdot n \cdot \Phi_{nom} \tag{3}$$

The linear and angular coefficients, c_0 and c_1, respectively, are obtained from the $\Phi(T_j)$ plot while the coefficients d_0 and d_1 are obtained from the $\Phi(I_f)$ plot. n is the number of LEDs of the same type in the lamp (considering a lamp made with m different types of LEDs), Φ_{nom} is the nominal luminous flux of one LED, T_a is the ambient temperature and R_{hs} is the heatsink thermal resistance.

With the chromaticity coordinates y and x of the LED, given in the datasheet, it is possible to evaluate its CCT curve using the McCamy's formula [13]. Plotting the CCT with respect of y variations, a linear behavior can be observed. By the use of a linear regression, the coefficients e_0 and e_1 can be determined. Equation (4) relates variations in the CCT with variations in the y chromaticity coordinate.

$$CCT = e_0 + e_1 \cdot y \tag{4}$$

Later, using the plots of the chromaticity coordinate Δy with respect to junction temperature variations, $\Delta y(T_j)$, and forward current variations, $\Delta y(I_f)$, the coefficients a_0, b_0, a_1 and b_1 (from $\Delta y(T_j)$ and $\Delta y(I_f)$, respectively) can be determined. Afterwards, using these coefficients, the chromaticity coordinate y can be evaluate using (5) with respect of current and temperature variations. Moreover, by combining (4) and (5), one can obtain the CCT of the LED for a given current.

$$
\begin{aligned}
y = &\left[a_0 + a_1 \cdot \left(T_a + \left(R_{jc} + n \cdot R_{hs} \right) \cdot k_h \cdot I_f \cdot \frac{V_o + R_s \cdot I_f + k_v \cdot (T_a - T_o)}{1 - I_f \cdot k_h \cdot k_v \cdot \left(R_{jc} + n \cdot R_{hs} \right)} \right) \right] \\
&\cdot \left[b_0 + b_1 \cdot I_f \right] \cdot y_{nom}
\end{aligned}
\tag{5}
$$

For any given light source, there is a relation between chromaticity coordinates (x, y, z) and tristimulus values (X, Y, Z), as shown in (6) [14].

$$
\begin{aligned}
x &= \frac{X}{X + Y + Z} \\
y &= \frac{Y}{X + Y + Z} \\
z &= \frac{Z}{X + Y + Z}
\end{aligned}
\tag{6}
$$

When a mix of m different light sources are considered, the equivalent tristimulus is the sum of the respective sources, as shown in (7).

$$
\begin{aligned}
X_{mix} &= X_1 + X_2 + \ldots + X_m \\
Y_{mix} &= Y_1 + Y_2 + \ldots + Y_m \\
Z_{mix} &= Z_1 + Z_2 + \ldots + Z_m
\end{aligned}
\tag{7}
$$

By considering this mix, the y chromaticity coordinate given in (6) can be expanded, as shown in (8).

$$
y_{mix} = \frac{Y_1 + Y_2 + \ldots + Y_m}{\dfrac{Y_1}{y_1} + \dfrac{Y_2}{y_2} + \ldots + \dfrac{Y_m}{y_m}}
\tag{8}
$$

From (4), it is known that the chromaticity coordinate y is proportional to the CCT. The tristimulus Y represents luminance, that is proportional to luminous flux [7]. Therefore, equation (8) can be rewritten as (9) and the improved model is obtained. This equation can be used to estimate the final CCT of a mix of LEDs considering their current and temperature variations.

$$
CCT_{mix} = \frac{\Phi_1 + \Phi_2 + \ldots + \Phi_m}{\dfrac{\Phi_1}{CCT_1} + \dfrac{\Phi_2}{CCT_2} + \ldots + \dfrac{\Phi_m}{CCT_m}}
\tag{9}
$$

Figure 1 presents a flowchart summarizing the proposed methodology.

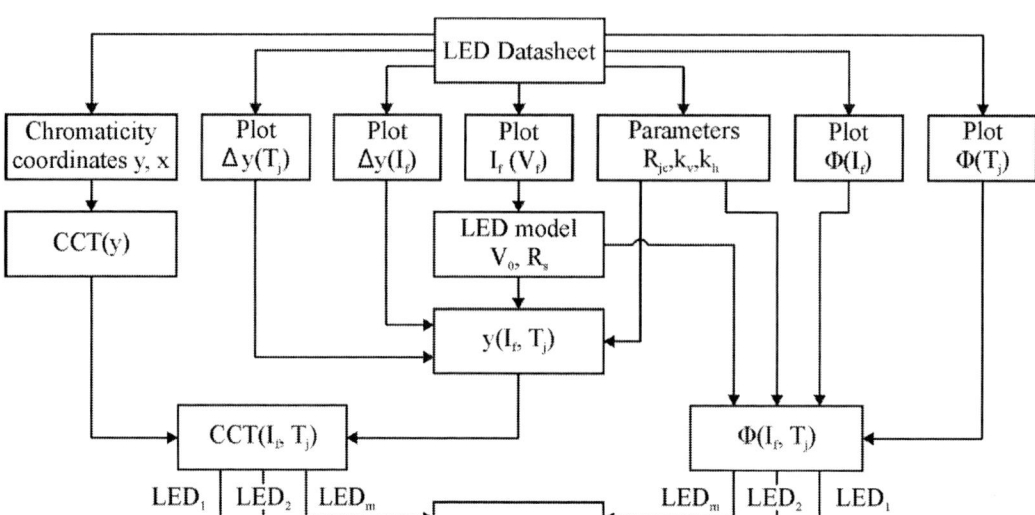

Fig. 1: Summary of the methodology proposed

Experimental validation

To validate the methodology, experimental results were obtained for a lamp made with two LED models from CREE [15]. The system consists of seven cool-white LEDs with a CCT of 6500K (JK3030AWT-00-0000-000B0HL265E) and seven warm-white LEDs with a CCT of 4000K (JK3030AWT-00-0000-000B0HL240E), thus, $n = 7$ and $m = 2$. The LEDs are attached to the same heatsink, as shown in Figure 2. The parameters of the LEDs used in the experiment are shown in Table I.

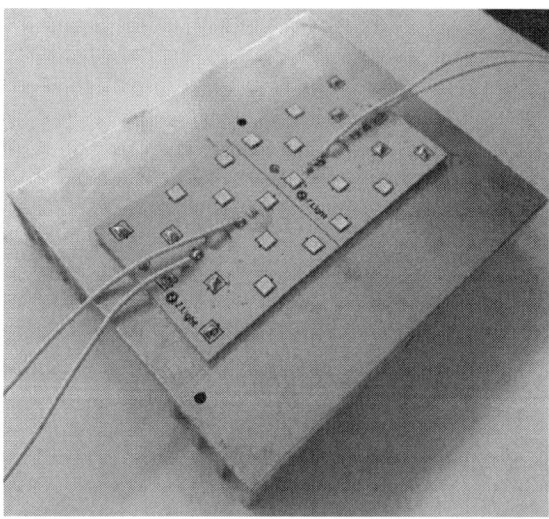

Fig. 2: LEDs assembled on a heatsink

Table I: LEDs Parameters from manufacturer datasheet [15]

Parameters	Value	
Correlated color temperature CCT (K)	4000	6500
Forward voltage V_o (V)	5.2863	
Series resistance R_s (Ω)	4.7246	
Temperature coefficient k_v (V/°C)	-0.0018	
Reference temperature by datasheet T_o (°C)	25	
Nominal current (A)	0.15	
Proportion of power converted into heat k_h	0.75	
LED junction to case thermal resistance R_{jc} (°C/W)	11	
Heatsink thermal resistance R_{hs} (°C/W)	2.92	
Number of LEDs n	7	7
Linear coefficient c_0 from $\Phi(T_j)$	1.064	
Angular coefficient c_1 from $\Phi(T_j)$	-0.002	
Linear coefficient d_0 from $\Phi(I_f)$	0	
Angular coefficient d_1 from $\Phi(I_f)$	6.65	
Nominal luminous flux Φ_{nom} (lm)	124	
Linear coefficient e_0 from $CCT(y)$	6899.3	18582
Angular coefficient e_1 from $CCT(y)$	-7652	-36704
Linear coefficient a_0 from $\Delta y(T_j)$	1.0169	1.0193
Angular coefficient a_1 from $\Delta y(T_j)$	-0.0002	-0.0002
Linear coefficient b_0 from $\Delta y(I_f)$	1.0165	1.0189
Angular coefficient b_1 from $\Delta y(I_f)$	-0.1085	-0.1237
Center point y_{nom}	0.3665	0.3214

A diagram of the test setup is shown in Figure 3. This setup consists of an integrating sphere with spectrophotocolorimeter to measure optical parameters of the system, one thermocouple attached to the heatsink, two current sources and a circuit to change the LEDs average current using pulse width modulation (PWM).

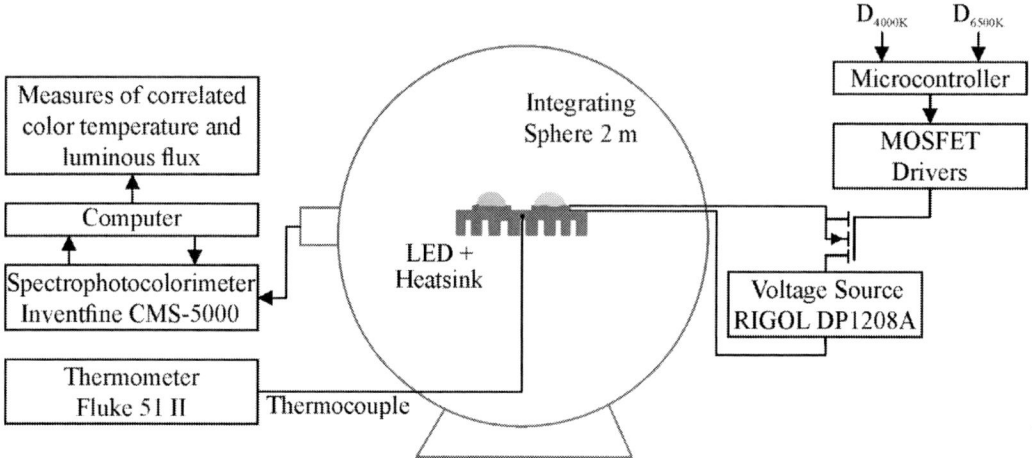

Fig. 3: Setup used for the experimental tests

All tests were made with an ambient temperature T_a = 25°C. Measurements were taken after thermal steady-state conditions were reached. Junction temperatures were estimated using (2).

The current sources were set to a peak value of 150 mA and the values were measured with PWM duty cycles of 25%, 50%, 75% and 100% (average current of 25%, 50%, 75% and 100%, respectively). The currents of warm and cool white LEDs were changed individually. The results were obtained with one LED current fixed and the other varying, for all combinations, totalizing 16 different points.

Test results are presented along with the estimated values. Figures 4 and 5 present results for junction temperature in the 6500K and 4000K LEDs, respectively. The heatsink temperature was measured with the thermocouple, and then the junction temperature were estimated using (2). The average error is 2.49% with a maximum error of 6.31%.

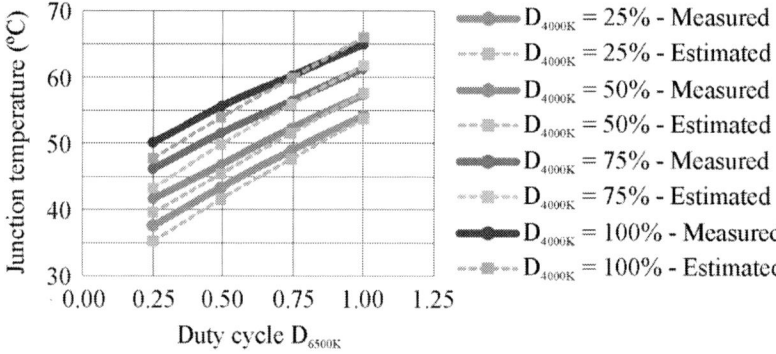

Fig. 4: Comparison of measured and estimated values of junction temperature in the 6500K LED

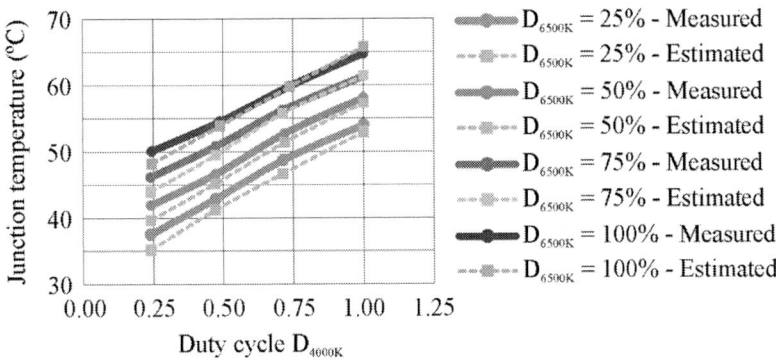

Fig. 5: Comparison of measured and estimated values of junction temperature in the 4000K LED

Figure 6 presents results for luminous flux. The flux of each LED was estimated using (3), and then the values were added. As can be seen, estimated values have good correlation with the experimental data. The average error is 2.28% with a maximum error of 5.5%.

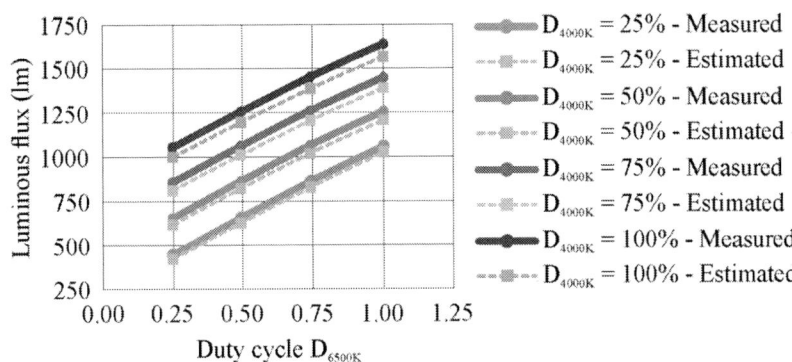

Fig. 6: Comparison of measured and estimated values of luminous flux

In Figure 7, the comparison for CCT is shown. CCT values were estimated using (9). The average error is 0.82%, with a maximum error of 2.08%. In this case, the maximum error is 109K. A ±200K deviation in the CCT is considered non-perceivable by human eyes [7]. The comparison between the measured (meas.) and estimated (est.) in numerical values is shown in Table II. The error (err.) was calculated using (10).

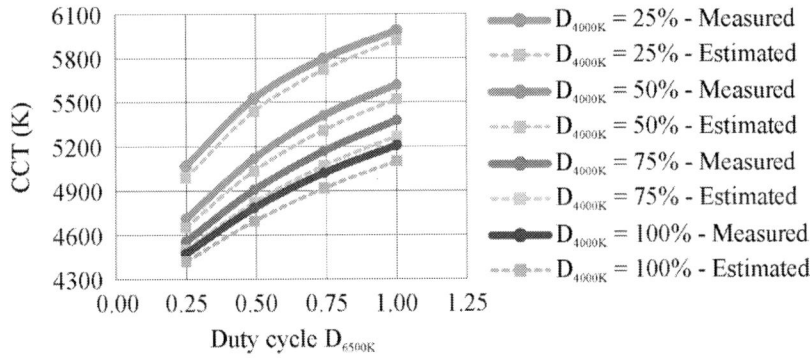

Fig. 7: Comparison of measured and estimated values of correlated color temperature

Table II: Comparison of measured and estimated values

LED	Duty cycle (%)	Junction temperature			Luminous flux			Correlated color temperature		
		Meas. (°C)	Est. (°C)	Err. (%)	Meas. (lm)	Est. (lm)	Err. (%)	Meas. (K)	Est. (K)	Err. (%)
4000K	24	37.5	35.2	6.27	444.2	420.5	5.34	5066	4989	1.52
6500K	25	37.6	35.2	6.24						
4000K	24	41.9	39.7	5.31	656.2	626.8	4.48	5526	5438	1.58
6500K	49	43.3	41.6	4.03						
4000K	24	46.1	43.9	4.78	860.9	828.4	3.77	5794	5719	1.30
6500K	74	48.9	47.6	2.58						
4000K	24	50.0	48.2	3.79	1058.9	1024.5	3.25	5986	5920	1.10
6500K	100	54.2	53.7	0.94						
4000K	47	42.9	41.2	3.78	653.8	617.87	5.50	4711	4651	1.26
6500K	25	41.7	39.6	4.97						
4000K	47	46.6	45.2	3.01	863	821.98	4.75	5120	5035	1.66
6500K	49	46.7	45.4	2.95						
4000K	48	50.8	49.5	2.55	1064.9	1019.6	4.25	5410	5310	1.85
6500K	74	52.3	51.5	1.54						
4000K	49	54.5	53.8	1.22	1256.3	1207.6	3.87	5618	5520	1.74
6500K	100	57.4	57.6	-0.45						
4000K	71	48.7	46.6	4.15	856.1	811.90	5.16	4556	4500	1.22
6500K	25	46.1	43.2	6.31						
4000K	72	52.7	51.4	2.49	1062	1010.5	4.85	4905	4821	1.71
6500K	49	51.5	49.8	3.39						
4000K	71	56.2	55.7	0.90	1261.5	1204.6	4.51	5168	5070	1.90
6500K	74	56.4	55.9	0.85						
4000K	74	59.8	59.7	0.14	1450.4	1390.1	4.16	5378	5269	2.02
6500K	100	61.4	61.8	-0.64						

4000K	100	54.0	52.9	1.93	1053.5	999.13	5.16	4474	4421	1.20
6500K	25	50.1	47.8	4.72						
4000K	100	58.1	57.2	1.41	1255.8	1195.1	4.83	4783	4693	1.88
6500K	49	55.6	53.9	3.04						
4000K	100	61.3	61.4	-0.10	1453.1	1386.2	4.60	5020	4916	2.08
6500K	74	60.2	59.9	0.59						
4000K	100	64.7	65.7	-1.54	1639.2	1567.2	4.39	5208	5101	2.06
6500K	100	65.0	66.0	-1.56						

$$Error = \frac{Value_{measured} - Value_{estimated}}{Value_{measured}} \cdot 100\% \tag{10}$$

Conclusion

A simple and improved methodology to estimate the CCT of white LEDs using just data obtained from the device's datasheet has been developed. This methodology is based on the interaction among electrical, thermal and optical parameters. Even that some parameters (thermal resistances, coefficients k_h and k_v) have dynamic response and could vary across the production lot, they were considered fixed in this methodology, aiming the simplification of the model.

To validate the proposed methodology, a lamp made with 14 LEDs with two different CCTs (7 6500K and 7 4000K) was built and sixteen different operation points were studied. Theoretical results were compared with the experimental data. These results show good correlation between estimated and measured values, with a maximum error in the CCT of 2.08% (109K), demonstrating the feasibility of the proposed methodology.

References

[1] M. G. Figueiro, R. Nagare, and L. L. A. Price, "Non-visual effects of light: How to use light to promote circadian entrainment and elicit alertness," Light. Res. Technol., vol. 50, no. 1, pp. 38–62, Jan. 2018.

[2] K. P. Wright, A. W. McHill, B. R. Birks, B. R. Griffin, T. Rusterholz, and E. D. Chinoy, "Entrainment of the human circadian clock to the natural light-dark cycle," Curr. Biol., vol. 23, no. 16, pp. 1554–1558, Aug. 2013.

[3] P. R. Mills, S. C. Tomkins, and L. J. M. Schlangen, "The effect of high correlated colour temperature office lighting on employee wellbeing and work performance," J. Circadian Rhythms, vol. 5, 2007.

[4] E. F. Schubert, Light-Emitting Diodes, Second Edi. Cambridge University Press, 2006.

[5] K. H. Loo, Y. M. Lai, S. C. Tan, and C. K. Tse, "Stationary and adaptive color-shift reduction methods based on the bilevel driving technique for phosphor-converted white LEDs," IEEE Trans. Power Electron., vol. 26, no. 7, pp. 1943–1953, 2011.

[6] N. Ohta and A. R. Robertson, Colorimetry: Fundamentals and Applications. John Wiley & Sons, 2005.

[7] H. T. Chen, S. C. Tan, and S. Y. R. Hui, "Nonlinear Dimming and Correlated Color Temperature Control of Bicolor White LED Systems," IEEE Trans. Power Electron., vol. 30, no. 12, pp. 6934–6947, Dec. 2015.

[8] H. Chen and S. Y. Hui, "Dynamic prediction of correlated color temperature and color rendering index of phosphor-coated white light-emitting diodes," IEEE Trans. Ind. Electron., vol. 61, no. 2, pp. 784–797, 2014.

[9] R. L. Lin and Y. F. Chen, "Equivalent circuit model of light-emitting-diode for system analyses of lighting drivers," in Conference Record - IAS Annual Meeting (IEEE Industry Applications Society), 2009.

[10] R. W. Erickson, Fundamentals of power electronics, 2nd ed. Norwell: MA Kluwer Academic Publishers, 2000.

[11] V. C. Bender, O. Iaronka, W. D. Vizzotto, M. A. D. Costa, R. N. Do Prado, and T. B. Marchesan, "Design methodology for light-emitting diode systems by considering an electrothermal model," IEEE Trans. Electron Devices, vol. 60, no. 11, pp. 3799–3806, 2013.

[12] G. Farkas et al., "Electric and thermal transient effects in high power optical devices," in Annual IEEE Semiconductor Thermal Measurement and Management Symposium, 2004, vol. 20, pp. 168–176.

[13] C. S. McCamy, "Correlated color temperature as an explicit function of chromaticity coordinates," Color Res. Appl., vol. 17, no. 2, pp. 142–144, Apr. 1992.

[14] D. Malacara, Color Vision and Colorimetry: Theory and Applications. SPIE Press, 2011.

[15] I. Cree, "Product Family Data Sheet Cree J Series 3030 LEDs." 2017.

DC microgrid concept for mine environment

Jooa Pursiainen[a], Jenni Rekola[b], Raimo Juntunen[b], Mikko Valtee[b], Pasi Peltoniemi[a]
[a] LUT University, Yliopistonkatu, Lappeenranta, Finland
[b] Sandvik Mining and Rock Technology, Pihtisulunkatu, Tampere, Finland
E-Mail: pasi.peltoniemi@lut.fi
URL: http://www.lut.fi/en

Keywords

«Distribution of electrical energy», «Electrical drive», «Energy storage», «Microgrid», «Smart grids»,

Abstract

The use of electric vehicles and inverter-controlled ventilation will increase in the future in underground mines. DC microgrids are analyzed during the last years to be used e.g. in solar and wind farms, ships and datacenters. The possibility to use DC microgrid in a mine production area is analyzed in this study. The steady-state calculations are done to choose the suitable DC voltage level, cable type and power level of converters. After that, the dynamic calculations are done by using simulation model to analyze the stability of the system. Based on the calculations, energy storage is added to the grid and the optimal location of the energy storage is analyzed.

Introduction

The use of electric mining machines will increase in the future due to more strict environmental regulation and due to need of lowering of diesel exhaust gases in the underground mines [1-4]. The ventilation is one of the largest electric energy consumers in the underground mines and it can be reduced to be only half of the ventilation power required with the diesel engines if the electric mining machines are used. Moreover, direct on-line (DOL) motor drives of ventilation and pumps are replaced by power electronic converter fed drives to minimize the energy consumption of mines. These inverters, as well as batteries of electric vehicles, are supplied by DC. During the last years, the use of DC microgrids is analyzed e.g. in ships, wind and solar farms, ports and datacenters to achieve multiple benefits. Therefore, the opportunity to supply the whole mine production area by using DC instead of conventional AC is studied [5].

DC-distribution offers several advantages over AC-distribution. The frequency does not need to be controlled and reactive power does not occur. Therefore, the control of DC microgrid is easier due to lack of stability issues and the power losses are lower or cables with smaller diameter can be used due to lack of reactive power losses and elevated voltage level [6-8]. Moreover, the power losses will reduce due to decreased amount of AC/DC/AC transformations. With AC distribution, the frequency and the voltage need to be controlled to fulfill the requirements set for the power quality, whereas in DC power distribution only the voltage level needs to be adjusted [6-8]. The maximum DC and AC low voltage amplitudes are defined in standard, e.g. in Europe, the maximum AC voltage 1000 V but for DC 1500V. Therefore, higher voltage can be used by using cost-efficient low voltage equipment, but still higher power can be transferred due to lower current amplitude as illustrated in Fig. 1 [9].

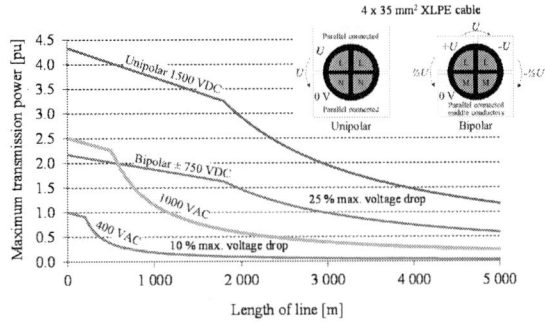

Fig. 1. Transmission capacity with unipolar and bipolar DC network structures. Also shown the transmission capacity of 1 kV and 0.4 kV AC systems [9].

The use of renewable energy is also analyzed in the mines where all the electric power is produced by using diesel-generators now. The first installations are already done e.g. in Australia and Africa [10-15]. Therefore, the strict environmental regulation can be fulfilled and, also the varying price of diesel does not affect to the profitability of mine. These wind and solar farms would be also able to use DC. However, in this very first study, the usability of DC in one production area of underground mine is studied and the whole mine electric grid is not included. An example underground mine and production area is shown in Fig. 2.

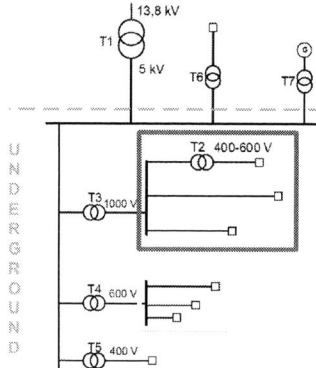

Fig.2. Illustration of underground mine electric grid. Example of production area supply is enclosed in the figure (red).

Typical power quality issue in mines are deep voltage drops due to start-up of high power DOL induction motors of crushers and ventilation fans. The DOL induction motors do not exist in DC grids when all the motors are controlled by power electronic converters hence such power quality issues are avoided. The other typical power quality issue is harmonics caused by high power diode- and thyristor rectifiers of e.g. DC motor of shaft. The harmonics are not an issue in the DC grid and those can be filtered out in point of common coupling (PCC) of DC grid by proper control and LCL-filter design of the active rectifier.

DC grid enables new possibilities to improve the reliability of the mine grid. The economical loss due to even short blackouts in the mine operation are huge and therefore, the reliability of the system is crucial [16-17]. By using DC grid, the short blackouts in the AC supply do not affect the operation of the DC grid in case large capacitance is connected in the DC grid, which operates as an energy storage. If batteries are connected to DC grid, it can operate as an island mode for longer time or the batteries can support the weak grid, so called battery buffering. The start-up of high-power loads can be supported by battery buffer and therefore, the amplitude of the voltage drop can be reduced significantly. On the other side, the faults of the DC grid have no effect to the operation of AC grid.

The frequency of AC grid can be supported by battery buffer. The distribution or transmission company can pay to the mine based on the frequency support capability. Similar grid support capabilities are done already for example to ports (start-up of high-power cranes) and to grid stability control in PCC of high-power wind and solar power farms [18].

In this paper, the usability of DC microgrid in one production area of underground mine is studied. The required DC grid structure, required cable types and diameters are calculated by using steady-state calculations. The required power electronic converters and their power level are specified. The mine loads are highly variable and therefore, also the stability of the DC microgrid is analyzed by creating Matlab/Simulink model to enable dynamic calculations. The real mine operating area and measured mining machine load curves are used in this study.

Loads of the mine

The usability of DC grid in underground mine is studied by using Northparkes E48 block cave type mine production area as an example. The mine is located in New South Wales, Australia. The layout of the mine production area is shown in Fig. 3. The loads of underground mine production area in block cave-type mine are typically ore crusher, underground loaders, ventilation, pumps and other loads, such as lighting. All the loads are three-phase motor loads except lighting. Crusher is usually the largest load of the mine production area and it is almost constantly running [19]. In this case study, the power of the crusher is supposed to be 315 kW and the duty cycle is 80 %.

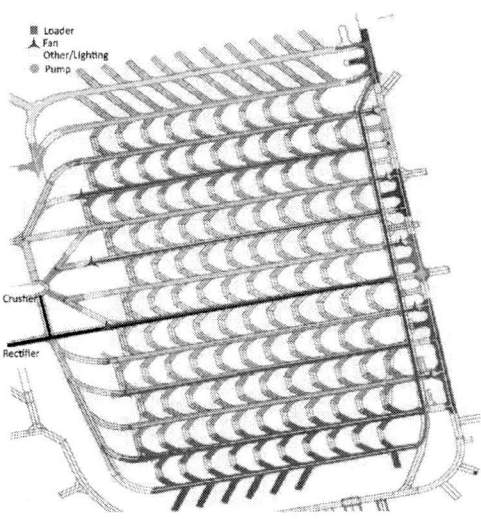

Fig.3. Mine production area analyzed in case study

Underground loaders are used to carry the ore from draw points to the crusher. The electric vehicles are supposed to be used in this study. The power consumption of underground loaders is highly varying, including acceleration and deceleration, due to short driving distance, typically 500m at maximum. The average power level of the loaders is 56 kW and the maximum power 150 kW in the analyzed operating cycle [20]. Eight loaders are supposed to operate in the analyzed production area. Fig. 4 shows the measured power cycle of the loader.

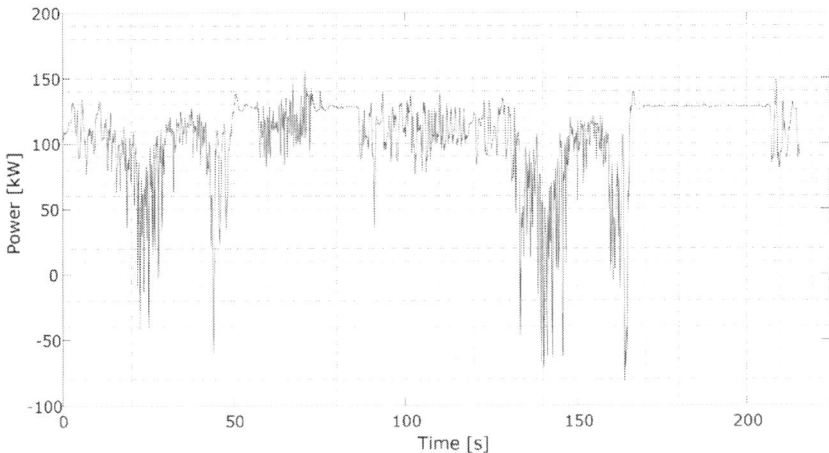

Fig 4: Power consumption measurement of the loader.

Rest of the loads in the production area are supposed to be constant-type loads. The power of the ventilation is 230 kW and the power of the dewatering pump 22 kW. Therefore, the total power consumption of the mine production area is approximately 1500 kW.

DC grid structure and requirements in underground mine

Nowadays, typically 1000 V, 600 V and 400 V low voltage AC power distribution is used underground mines (Fig. 3). To maximize the power transmission, the maximum allowed low voltage DC voltage is used in this study due to high power loads in the underground mine production area. The Australian standards limit the maximum low DC voltage to be 1500 VDC at maximum similarly than in Europe [21-22].

The maximum load in the area is 1500 kW. DC microgrid topology can be either bipolar or unipolar. In the bipolar structure there are three wires; plus, minus, and neutral (i.e. ±750 V and 0 V). In the unipolar structure, there are only two wires, plus and minus. Fig 5. illustrates these two topologies when the DC microgrid is bipolar but the loads can be connected to 750 VDC but also on unipolar way by using 1500 VDC [9]. Multiple grounding methods of DC grid can be used based on the regulation of the country or application, the isolated IT-system is analyzed in this study.

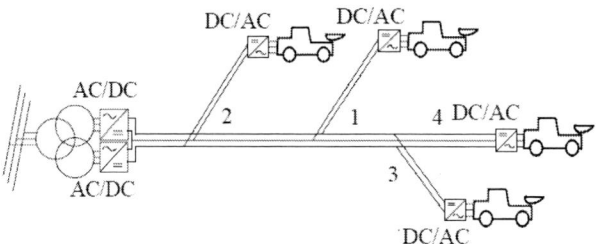

Fig. 5. Bipolar grid with loads connected unipolar way to plus and neutral (1), minus and neutral (2), plus and minus (3) and in bipolar way (4). Modified from [9].

The mine environment differs from DC microgrids designed for single households, group of households, office buildings etc. and other LVDC grid concepts in a way that individual loads can be large and heavily varying. The DC microgrid of ports and mines would be quite similar [23]. DC microgrid would be beneficial especially in the mines, where the layout and infrastructure of the production area stays constant for years or even tens of years such as block caves do. In the production areas of block caves, the distances between the draw points and the crusher are quite limited i.e. also the DC microgrid would be realized for the limited area. In this study, the example production area of 200 m*300 m is analyzed. The underground mine grids are typically radial to minimize the costs. Therefore, radial grid structure is supposed to be used also in this study.

The required cable diameters in the DC microgrid are defined based on the maximum voltage drop of the cables. The specific power quality standards for mine grids do not exist. However, it can be supposed, that the standards defined for distribution grids are applied also to mine grid. According to those standards, the maximum voltage drop is allowed to be e.g. in Europe, 10 % of the nominal voltage and in Australia 230V +10/-6% by 1.7.2020 (Australian Standard AS60038), and after that, 230V +6/-2% (AS 61000.3.100) [24-26]. In addition to that, it should be noticed, that the minimum DC voltage amplitude needs to be defined based on the operation voltage of the loads connected to DC microgrid. In this study, the power electronic converter fed loads require at least 700 VDC voltage to operate.

To be able to justify the design, the voltage drops with corresponding loads need to be calculated. In the bipolar DC microgrid, the voltage drop needs to be calculated for both poles separately. The currents flow opposite directions and can be depicted as

$$I_{\text{pos}} = \frac{P_{\text{pos}}}{U_{\text{pos}}}, \tag{1}$$

$$I_{\text{neg}} = -\frac{P_{\text{neg}}}{U_{\text{neg}}}, \tag{2}$$

where P_{pos} and P_{neg} are loads connected to each pole and U_{pos} and U_{neg} are the voltages of these poles. Both voltages are considered positive here. The voltage drops in both wires can be calculated as

$$\Delta U_{\text{pos}} = I_{\text{pos}} r_{\text{pos}} l + \left(I_{\text{pos}} + I_{\text{neg}}\right) r_{\text{neut}} l, \tag{3}$$

$$\Delta U_{\text{neg}} = I_{\text{neg}} r_{\text{neg}} l - \left(I_{\text{pos}} + I_{\text{neg}}\right) r_{\text{neut}} l, \tag{4}$$

where r_{pos}, r_{neg} and r_{neut} are the resistances (Ω/km) of positive, negative and neutral cables, respectively and l is the length of the cable [10]. Voltages at the end of the cables are then

$$U_{\text{end,pos}} = U_{\text{in,pos}} - \Delta U_{\text{pos}}, \tag{5}$$

$$U_{\text{end,neg}} = U_{\text{in,neg}} - \Delta U_{\text{neg}}. \qquad (6)$$

when calculating the voltage drop, input voltage U_{in} is first used in (1) and (2). As this is not the actual voltage, the load is connected due to the voltage drop, so the voltage amplitude in the end of the cable, U_{end}, is calculated by (3) and (4), and it needs to be used as a new U in (1) and (2). With iterating through (1) to (6) the real end voltages are obtained.

Required power electronic converters

The rectifier is needed to produce DC grid. The passive, i.e. diode-rectifiers or active rectifiers using thyristors or power semiconductors (IGBT, IGCT or GTO) could be used. In this study, some of the loads connected to the grid, needs at least 700 VDC to operate properly. Therefore, the constant, at least 700 VDC voltage level must be maintained by rectifiers. The DC voltage cannot be controlled by using diode rectifiers. On the other hand, thyristor rectifier produces huge amount of low frequency harmonics to AC grid, and therefore the use of it is not recommended. The active rectifier is needed to be able to adjust DC voltage level and to minimize the harmonics produced to AC grid. Moreover, the DC/AC converters are needed in front of each load.

The bipolar DC grid is supplied by two active rectifiers connected in series. A special three winding transformer is needed between the AC grid and the rectifiers. Large capacitors, 25 mF, are added to DC side after the rectifiers to smooth DC voltage variations. The space is very limited in underground mines and therefore, all these components might be liquid-cooled depending on the ambient temperature of the mine [27].

The suitability of DC microgrid in underground mine

The suitability of the DC microgrid for the aforementioned underground mine production area is studied. The active rectifier producing ±750 V DC is used to guarantee the constant DC voltage also during voltage amplitude variations in the AC grid. In the DC microgrid the loads are connected evenly to both poles with unipolar connections. The distribution of the loads in the bipolar grid is shown in Fig. 6. Loaders are distributed on unipolar lines to the tunnels with the furthest being 240 m from the bipolar grid. Ventilation fans are separated to two unipolar groups of five fans each, with the furthest being 330 m from the bipolar grid as shown in Fig. 6.

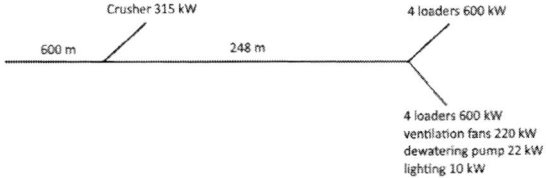

Fig 6: Load distribution in bipolar DC grid

The voltage drop and maximum current of the bipolar line from the rectifier to the connection points of the unipolar loads is calculated to define the required cable diameter or number of parallel-connected cables. The calculation results showed that cable Type 209.1 3x300 would be suitable [28]. The voltage of each load is presented in Table I. The maximum current is 1254 A in the bipolar grid and thermal limit for the cable is 1584A and 1350A when the ambient temperature and the installation method are considered.

According to calculations, the bipolar line needs to be constructed with three cables, one for each wire to withstand the maximum current of 1254 A in positive pole. However, underground loaders need only two parallel cables to withstand the maximum current of 826 A in the negative pole. For other unipolar loads, one cable was enough. Table 1 shows the load voltages, calculated with the Type 209.1 3x300 cable. The π-model of the cable is used in the dynamic calculations.

Table I: Load voltages with Type 209.1 3x300

Pump	Fans	Crusher	Loaders
728 V	728 V	733 V	724 V

The required cable diameters and amount of parallel-connected cables was determined by using steady-state calculations. However, the loads are highly varying and therefore, also dynamic analysis is needed to analyze the stability of the DC microgrid. Therefore, the analyzed DC microgrid has been modeled by using Matlab/Simulink. The dynamic simulation model is verified by comparing it to the results of steady-state calculations.

The DC microgrid is supposed to be supplied by 20 kV, 50 Hz AC grid. The nominal power of the transformer is 2 MVA, to accommodate the theoretical maximum power of a bit over 1.7 MW. Fig. 7 shows the simulated DC voltages of the bipolar DC grid. The voltage decreases temporarily below the minimum limit, 700 VDC.

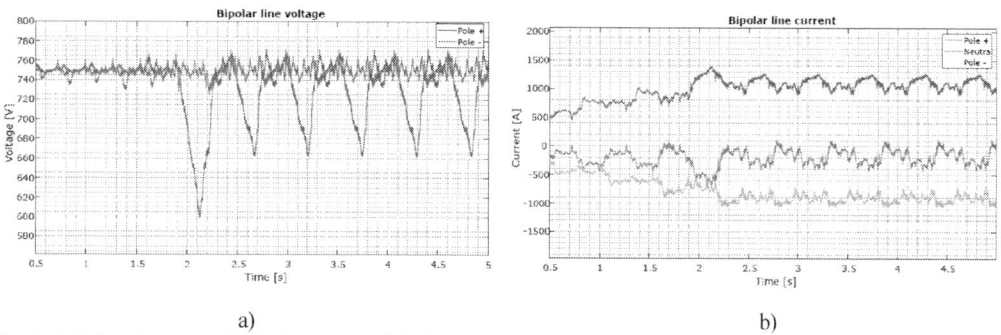

a)   b)

Fig 7: a) DC voltages and b) DC currents of the bipolar grid.

Battery buffer

To manage with the voltage drop during high load situations, an energy storage is introduced to the DC grid. The optimal location of the energy storage is analyzed by adding the storage at first to the beginning of the grid, and then to the end of the grid. The lithium-ion battery is chosen as an energy storage. The nominal voltage of the battery was selected to 1,1 kV and capacity of 900 Ah. Hence, the batteries could supply the nominal load for one hour in case of total blackout. The maximum current drawn from the battery is 900A (i.e.1C).

To control the battery storage unit, the droop control principle is used. The nominal voltage U_n, is 750V and the battery is charged when the voltage is above $1,01*U_n$ (758V DC) and discharged when the voltage is below $0,99*U_n$ (743 VDC). The voltage limits were determined so that the batteries would be charged and discharged during the normal cycle. Fig 8a presents the DC voltages of the bipolar grid when the energy storage is connected in the beginning of the DC grid. When using an energy storage, the voltage level can be maintained above 700 VDC. The maximum power drawn from the energy storage was 317 kW and 182 kW continuously.

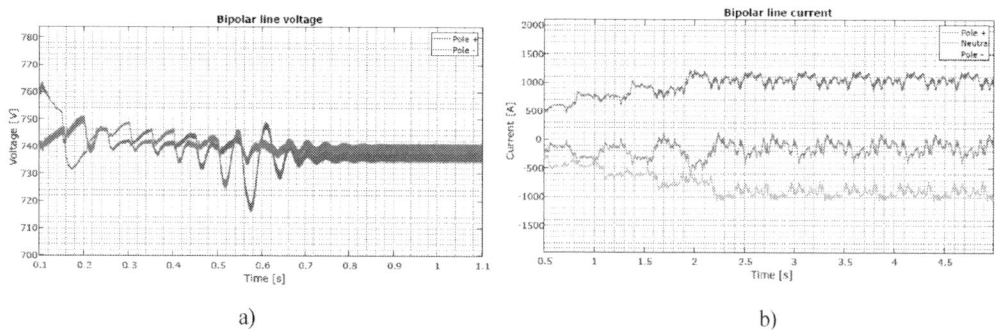

a)   b)

Fig 8: a) DC voltages and b) currents of the bipolar grid when energy storage is connected to the beginning of the grid

The maximum current of the grid is 1506A, as shown in Fig. 8b, which is 8 % higher than the allowed maximum current. Currents in Fig. 8a and 9a are measured in the beginning of the DC grid.

Fig. 9a shows the voltages when the energy storages are located at the end of the grid. The voltages are always above 700 VDC but the positive pole DC voltage starts to oscillate. Large oscillations are caused by the crusher, which is the largest load in the grid. The maximum current is 1198A i.e. below the maximum limit as shown in Fig. 9b.

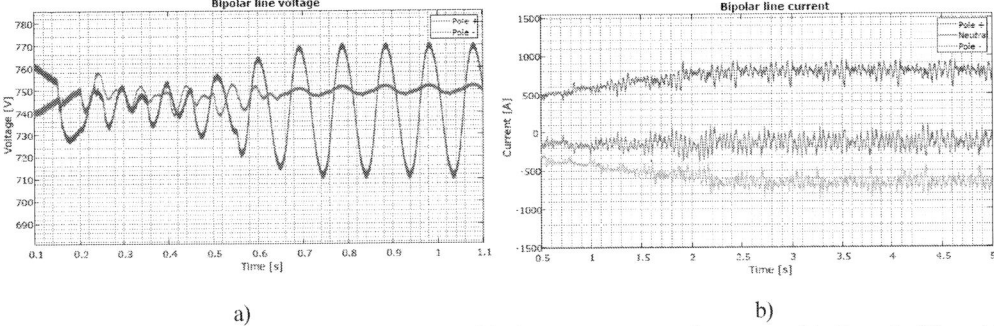

a)   b)

Fig 9: a) DC voltages and b) currents of the bipolar grid when energy storage is connected in the end of the grid

To limit the voltage oscillations, the droop control parameters are adjusted. The battery is charged when the voltage is above $1,1*U_n$ (825 VDC) and discharged when the voltage is below $0,9*U_n$ (675 VDC). The voltages and currents with new control parameters are shown in Fig. 10. The voltage oscillation disappears, voltages are always above 700 VDC and the maximum current is 1210A i.e. below the maximum limit.

According to the results, the energy storage can be used to improve the stability of the grid. The droop control threshold parameters of the energy storage need to be carefully chosen to proof the system stability [29-31]. The maximum current of the DC grid increases to be too high if the energy storage is connected to the beginning of the grid, where also the largest load occurs. Instead, while the energy storage is connected at the end of the grid, the crusher is supplied from two directions, both from AC grid and from the energy storage, and therefore the current stay below maximum limit.

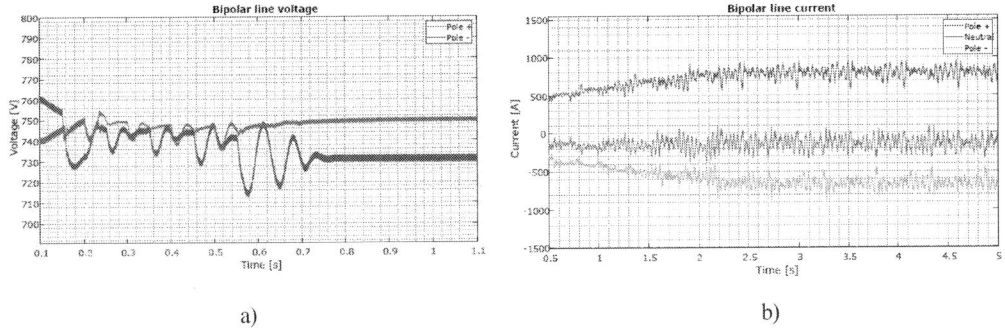

a)   b)

Fig 10: a) DC voltages and b) currents of the bipolar grid when energy storage is connected in the end of the grid and droop control is adjusted.

Battery charging

The batteries of the underground mining vehicles can be charged on multiple ways: fast or slow charging or battery swapping. In this study, the effect of fast charging to stability of DC mine grid is analyzed. Eight battery charging stations, 350 kVA each and four stations per pole of bipolar grid, are added to the grid. It is supposed that other loads, except loaders, are used during charging. The beginning of the charging cycle is the most demanding phase from the grid point-of-view. The energy storage is again added to the beginning or to the end of the grid.

At first, the energy storage is added to the beginning of the DC grid. Fig. 11a shows the DC voltages, which are highly fluctuating, and the voltage decreases below 700 VDC. The voltage fluctuation would be able to decrease by tuning the droop control parameters, but also the maximum currents are 1718A in positive pole and 1507A in the negative pole of the grid, which are higher than the allowed maximum current of the cable as shown in Fig. 11b.

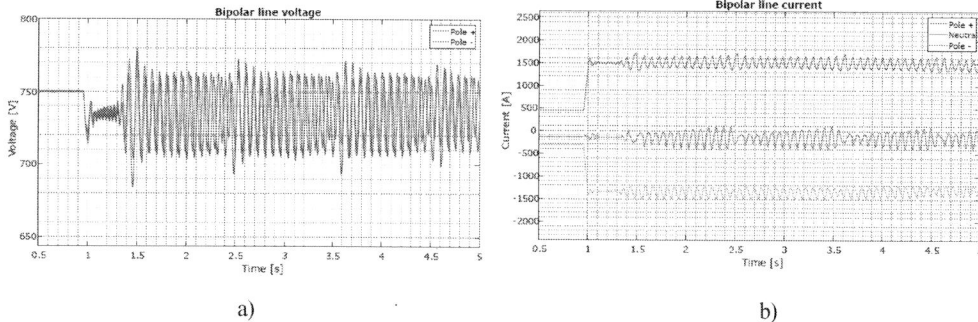

a)   b)

Fig 11: a) DC voltages and b) currents of the bipolar grid when energy storage is connected to the beginning of the grid

Therefore, the energy storage is moved back to the DC grid. In this case, voltages are constant and always above 700 VDC as shown in Fig. 12a. Moreover, the maximum current is 1372A, which is lower than in the previous case but still close to the maximum value as presented in Fig. 12b. The maximum current value can be to limited by smart task management, i.e. all the chargers are not used at the same time with full power. According to these analyses, the battery capacity should be approximately 300-350 kW.

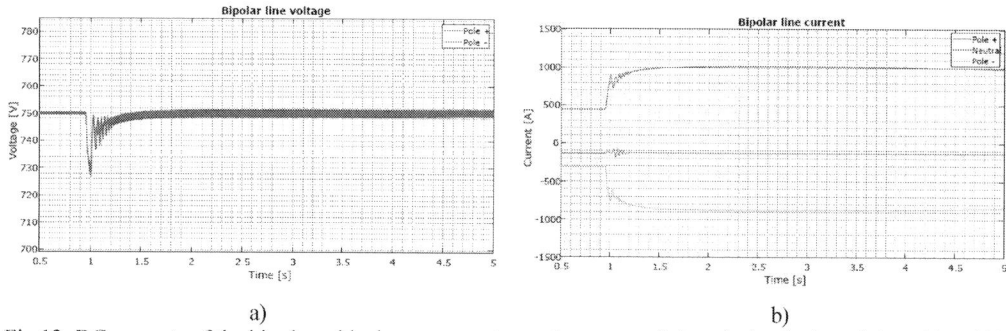

a)   b)

Fig 12: DC currents of the bipolar grid when energy storage is connected a) to the beginning of the grid and b) in the end of the grid and the battery charging stations are added

Conclusion

The use of electric vehicles and other inverter-controlled loads will increase in the future in underground mines due to strict environmental regulations and need to minimize costs of ventilation. DC microgrids have been stiúdied during the last years when applied e.g. in solar and wind farms, ships, ports and datacenters. The possibility to use DC microgrid in mine production area is analyzed in this study by using the real underground mine production area as an example as well as the measured load curves of electric mining vehicles. According to this study, the maximum allowed LVDC voltage level would be suitable to supply the underground mine production area by using realistic cable diameters. However, the voltage drop due to highly varying loads is sometimes too deep and therefore, additional energy storage could be used to improve the voltage quality of the DC network.

The added energy storage removes the voltage fluctuation and enables the constant DC voltage. However, the proper control parameters of the droop control need to be carefully chosen and the parameters vary based on the location of the energy storage. Two possible locations of the energy storage are compared: the storage is located in the beginning or in the end of the grid. The largest load of the DC grid, crusher, is located in the beginning of the grid. Therefore, if both energy storage and the largest load are located in the beginning of the grid, the current increases to be higher than the maximum allowed value of the used cable. However, when the energy storage is located in the end of the grid, the current stays below the maximum limit because in this case, the largest load is supplied from two directions, from AC grid and from energy storage.

Instead of energy storage, the size of the supply transformer, rectifier as well as cable diameter could also be increased to improve the stability of the DC grid. However, the size of the components used in the underground mine should be minimized because the space for the components is expensive. In addition, energy storage can be used to proof the continuation of the mining operation in spite of short blackouts or voltage drops in the AC grid.

The other option would be to use smart task management, i.e. loads are controlled to limit the peak power consumption. All the loads are controlled by inverters and these can be controlled through centralized control room. In addition to that, the batteries of electric mining vehicles can be used to improve the grid stability by efficient task management and proper inverter control.

The optimal energy storage size needs to be analyzed in more detail in the future. The energy storage is expensive device hence required capacity should be carefully chosen. In addition to that, the proper control parameters need to be found. According to this study, simple droop control is accurate enough in most of the situations but also the usability of more sophisticated control methods can be applied as well. The fault-ride-through (FRT) capability of the mine grid can be improved by using energy storage and smart task management. Depending on the reliability and power quality of AC grid, huge productivity increase would be achieved by energy storage, because the voltage of the DC grid can be kept constant also during the short blackouts or voltage drops in the AC grid.

References

[1] EU. "Electrification of mining lowers costs reduces emissions and improves health and safety" 2019, Available: https://www.ey.com/en_lu/news/2019/07/electrification-of-mining-lowers-costs-reduces-emissions-and-improves-health-and-safety

[2] J. Paraszczak, E. Svedlund, K. Fytas, M. Laflamme, "Electrification of loaders and trucks- a step towards more sustainable underground mining", International conference on renewable energies and power quality ICREQ'14, 2014.

[3] Global mining guidelines group: GMG recommended practices for battery electric vehicles in underground mining- 2nd edition, 2018

[4] Tracking the trends 2019- The top 10 issues transforming the future of mining, Deloitte report, available: https://www2.deloitte.com/content/dam/Deloitte/au/Documents/energy-resources/deloitte-au-er-tracking-the-trends-2019-210119.pdf

[5] M. G. Jahromi, G. Mirzaeva, S. D. Mitchell and D. Gay: "DC power vs AC power for mobile mining equipment," 2014 IEEE Industry Application Society Annual Meeting, Vancouver, BC, 2014, pp. 1-8.

[6] J. Rekola, "Factors Affecting Efficiency of LVDC Distribution Network – Power Electronics Perspective", PhD thesis, Tampere University of Technology. 2015.

[7] P. Nuutinen, "Power Electronic Converters in Low-Voltage Direct Current Distribution – Analysis and Implementation", PhD thesis, Lappeenranta University of Technology, 2015.

[8] Salonen P. Kaipia T. Nuutinen P. Peltoniemi P. and Partanen J.: "An LVDC distribution system concept" 2008 Nordic Workshop on Power and Industrial Electronics (NORPIE)

[9] Kaipia, T. et al.," Low-Voltage Direct Current (LVDC) Power Distribution for Public Utility Networks" Tutorial, EPE'14 ECCE Europe, Lappeenranta, Finland, 2014

[10] Kalgoorlie Miner. "Gold Fields opens Australia's biggest hybrid renewable microgrid at Agnew mine near Leinster" 2019, Available: https://www.kalminer.com.au/news/kalgoorlie-miner/gold-fields-opens-australias-biggest-hybrid-renewable-microgrid-at-agnew-mine-near-leinster-ng-b881387711z.amp

[11] Wärtsilä to provide full energy storage solutions for one of the largest power hybrid projects at an off grid mine in Mali, available https://www.wartsila.com/media/news/02-12-2019-wartsila-to-provide-full-energy-storage-solution-for-one-of-the-largest-power-hybrid-projects-at-an-off-grid-mine-in-mali-2590089 the same information here https://www.greencarcongress.com/2019/12/20191203-wartsila.html

[12] Renewables in Mining: Rethink, Reconsider, Replay, More than just a cost play, renewables offer a distinct competitive advantage, Deloitte report, available: https://www2.deloitte.com/content/dam/Deloitte/global/Documents/Energy-and-Resources/gx-renewables-in-mining-final-report-for-web.pdf

[13] ABB- Microgrids for mines, report, available: https://www.nsgm.gov.in/sites/default/files/Microgrid%20for%20Off%20Grid%20Mining%20business%20case%20rev01.pdf

[14] Storage as a service to power Australian gold mine, available: https://www.smart-energy.com/industry-sectors/energy-grid-management/storage-as-a-service-to-power-australian-gold-mine/

[15] Gold Fields opens Australia's biggest hybrid renewable microgrid at Agnew mine near Leinster, available: https://www.kalminer.com.au/news/kalgoorlie-miner/gold-fields-opens-australias-biggest-hybrid-renewable-microgrid-at-agnew-mine-near-leinster-ng-b881387711z.amp

[16] https://im-mining.com/2020/01/07/vale-looks-smart-meters-power-cost-ghg-emission-reductions/

[17] Smart Energy International. Available: https://www.smart-energy.com/industry-sectors/energy-grid-management/mining-industry-leader-believes-meters-can-help-manage-emissions/

[18] L. Peltonen, P. Järventausta, S. Repo, T. Rauhala, "Distributed small loads as fast frequency response: Impact on system performance", IEEE Texas Power and Energy Conference (TPEC), February 2020

[19] Sandvik. CG810i Gyratory Crusher, Available: https://www.rocktechnology.sandvik/en/products/stationary-crushers-and-screens/stationary-gyratory-crushers/cg810-gyratory-crusher/

[20] Sandvik. LH514E Electric Loader, Available: https://www.rocktechnology.sandvik/globalassets/products/underground-loaders-and-trucks/pdf/lh514e-specification-sheet-english.pdf

[21] AS/NZS 3000:2018, Electrical installations, Standards Australia, p. 608

[22] LVD (2014). Low voltage directive 2014/35/EU. European Commission. Available online: https://ec.europa.eu/growth/single-market/european-standards/harmonised-standards/low-voltage_en

[23] Kalmar FastCharge Straddle Carrier, Available: https://www.kalmarglobal.com/4ad2ab/globalassets/equipment/straddle-carriers/straddle-carrier-datasheet-2.pdf

[24] AS60038-2012, Standard voltages, Standards Australia

[25] AS 61000.3.100-2011, Electromagnetic compatibility (EMC) limits – Steady state voltage limits in public electricity systems, Standards Australia, p. 25

[26] EN50160 (1994). Voltage characteristics of electricity supplied by public distribution systems. CENELEC, Belgium.

[27] Karppanen, J., Kaipia, T., Nuutinen, P., Lana, A., Peltoniemi, P., Pinomaa, A., Mattsson, A., Partanen, J., Cho, J., and Kim, J. "Effect of Voltage Level Selection on Earthing and Protection of LVDC Distribution Systems" in 11th IET International Conference on AC and DC Power Transmission, 2015, pp. 1–8.

[28] Prysmian Group. Cable Guide for the Mining Industry, Available https://www.prysmiancable.com.au/documents/2018-mining-cables-guide.pdf/

[29] R. K. Chauhan, K. Chauhan, J. M. Guerrero "Controller design and voltage stability analysis of grid connected DC microgrid", Journal of renewable and sustainable energy, 10(3), May 2018.

[30] L. Meng, Q. Shafiee, G. Ferrari-Trecate, H. Karimi, D. Fulwani, L. Xiaonan, J. M. Guerrero, "Review on control of DC microgrids and multiple microgrid clusters", IEEE Journal of Emerging and Selected Topics in Power Electronics, 2017.

[31] Jung, T., Gwon, G., Kim, C., Han, J., Oh, Y., and Noh, C: "Voltage Regulation Method for Voltage Drop Compensation and Unbalance Reduction in Bipolar Low-Voltage DC Distribution System" 2018 IEEE Transactions on Power Delivery 33(1) pp. 141-149

A Comparison of Two-Stage Inverter and Quasi-Z-Source Inverter for Hybrid Energy Storage Applications

V. Castiglia[1], R. Miceli[1], F. Blaabjerg[2], Y. Yang[2]

[1]Department of Engineering,
Palermo University
Palermo, Italy
E-Mail: vincenzo.castiglia @unipa.it,
rosario.miceli@unipa.it

[2]Department of Energy Technology,
Aalborg University
Aalborg, Denmark
E-Mail: fbl@et.aau.dk,
yoy@et.aau.dk

Acknowledgements

This work was realized with the contribution of: PON R&I 2015-2020 PROpulsione e Sistemi IBridi per velivoli ad ala fissa e rotante PROSIB, CUP no: B66C18000290005; PRIN 2017, Advanced powertrains and systems for full electric aircrafts, prot. no.: 2017MS9F49; RPLab (Rapid Prototyping Laboratory - University of Palermo); project REACTION (first and euRopEAn siC eighT Inches pilOt liNe), co-funded by the ECSEL Joint Undertaking under grant agreement No 783158; PON R&I 2014-2020 - AIM (Attraction and International Mobility), project AIM1851228-1; SDES (Sustainable Development and Energy Savings) Laboratory UNINETLAB of University of Palermo, Laboratory of Electrical APplications LEAP - of University of Palermo.

Keywords

«Z-source converter», «Ultra capacitors», «Batteries», «Battery Management Systems (BMS)», «Energy storage »

Abstract

Batteries are the primary energy sources in many applications such as Electrical Vehicles (EV) and storage facilities. It is known that batteries are high-energy-density and low-power density devices. The hybridization of the storage systems can potentially overcome the batteries drawback, using a high-power-density device such as ultracapacitors (UCs). Many researchers have been focusing on power converters optimal structure for the Hybrid Energy Storage System (HESS). Among them, the Quasi-Z-Source Inverter (qZSI) is a promising solution to simplify the integration of batteries and UCs. In this paper, a comparison of the conventional inverter and the qZSI in HESS is carried out in terms of voltage stresses on the switches, voltage requirements for batteries and UCs, passive component sizing, output voltage THD and system efficiency.

Introduction

Hybrid Energy Storage Systems (HESS) have recently attracted the attention of researchers, for their capacity to greatly improve the storage systems performances. The HESS usually merge a high-energy-density storage, like batteries, and a high-power-density storage, such as ultracapacitors (UCs). HESS can be classified into two categories: passive HESS and active HESS [1]-[6].

The passive HESS couples the energy storages directly, without using additional converters. This solution has the advantage of the easiness of implementation, but it has also a large restriction in managing the power flow of the two sources. Typical passive HESS topologies are represented in Fig. 1 . In Fig. 1 (a) the battery pack and the UC pack are directly connected at the input of the DC/AC inverter [7]. The power sharing depends only on the natural response of the two sources [8]. In Fig. 1 (b), the battery pack and the UC pack are directly connected to the DC/DC converter, which is then connected to the DC/AC inverter. However, in this case, the power management is greatly restricted.

The active HESS couples the energy storage through additional converters. Typical active HESS topologies [9] are shown in Fig. 2. In Fig. 2 (a), the battery pack is connected at the input of the DC/DC converter, while the UC pack is connected at the DC-link. In this way, the power flow management capabilities are increased, keeping meanwhile the design simple [10]. In the system represented in Fig. 2 (b), an additional DC/DC converter between the UC bank and the DC/AC inverter is used. This solution increases the power flow management but yields more complex design and implementation. In a similar way, in the system represented Fig. 2 (c), an additional converter is used, connecting two sources in parallel to the DC-link with the same advantages and disadvantages of the solution shown in Fig. 2 (b).

Substituting the conventional two-stage inverter with a qZSI, a single-stage conversion can be achieved, thus reducing the complexity and cost of the system, and increasing the efficiency and reliability because of the smaller number of switches. In addition, the UC pack can be integrated into the Z-source network to obtain a HESS. A possible qZSI-HESS topology is exemplified in Fig. 3.

The aim of this paper is to compare the usage of conventional inverter and qZSI in HESS. In particular, the topology shown in Fig. 2 (a) and the topology presented in Fig. 3 have been compared. The comparison considers the voltage stresses on the switches, the voltage requirements for the battery pack and the UC pack, the passive component requirements for each solution, the output voltage Total Harmonic Distortion (THD) and the system efficiency. The chosen application is a grid-connected energy storage, with the conventional inverter and qZSI connected to the grid via an intermediate inductive filter. Both solutions are controlled by a Proportional-Integral (PI) regulator, with the DC-side control regulating the battery current and the AC-side control regulating the grid current. The UC pack will automatically supply the difference to match the power balance.

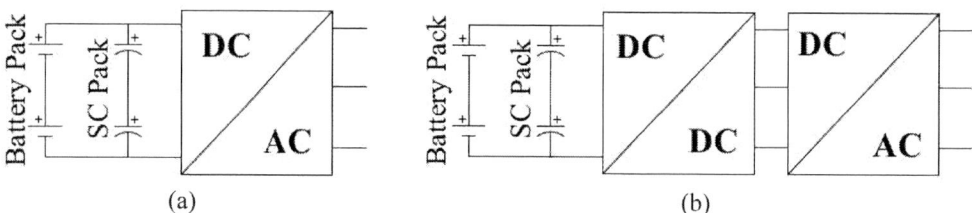

Fig. 1: Passive HESS topologies.

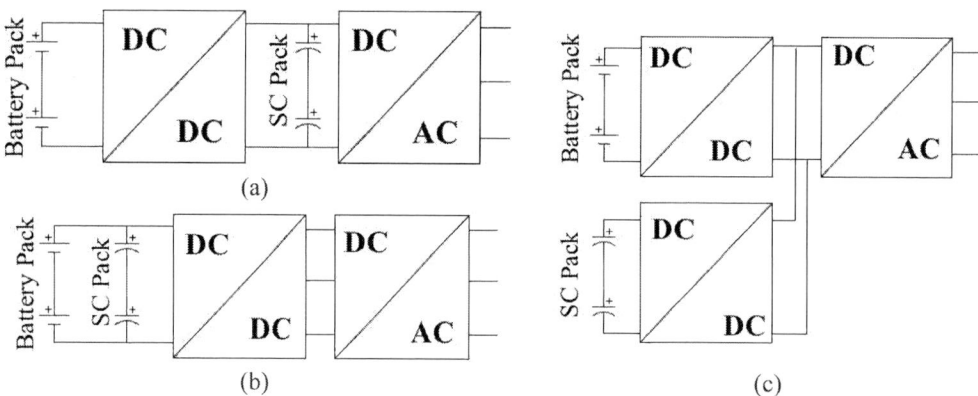

Fig. 2: Active HESS topologies.

Conventional two-stage inverter

As mentioned in the introduction, the chosen topologies for the comparison are the ones shown in Fig.2 (a), consisting of a boost converter and a three-phase voltage source inverter (two-stage inverter), and in Fig. 3, consisting of a qZSI. The grid voltage and frequency, the filter inductance and parasitic resistance and the rated and maximum pulse power are summarized in Table I.

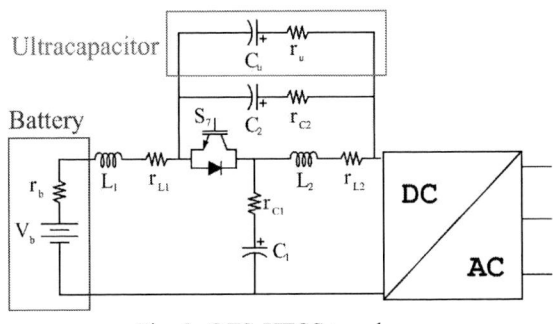

Fig. 3: QZS-HESS topology.

Table I: General parameters of the output stage.

Name	Symbol	Value
Grid peak voltage	$\hat{V}_g$	$230\sqrt{2}$
Grid frequency	f_1	50 Hz
Filter inductance	L_f	12 mH
Filter ESR	R_f	1 Ω
Rated power	P_r	5 kW
Pulse power	P_p	10 kW

Steady-State Analysis

For a three-phase inverter, the first harmonic will have an amplitude $\hat{V}_{a(1)}$ given by:

$$\hat{V}_{a(1)} = \frac{m}{2}\hat{V}_{pn} \tag{1}$$

where m is the modulation index and $\hat{V}_{pn}$ is the peak DC-link voltage. For the boost converter, the relationship between the input and output voltages is

$$\hat{V}_{pn} = \frac{1}{1-D}V_{in} = BV_{in} \tag{2}$$

where V_{in} is the input voltage, D is the duty cycle and B is the boost factor:

$$B = \frac{1}{1-D}. \tag{3}$$

Substituting (2) in (1), (4) is deduced, which represents the peak ac output voltage.

$$\hat{V}_{a(1)} = \frac{m}{2(1-D)}V_{in} = G\frac{V_{in}}{2} \tag{4}$$

where G is the overall voltage gain:

$$G = mB. \tag{5}$$

In order to operate in the linear zone, the maximum modulation index will be

$$m_{max} = 1 \tag{6}$$

while the boost converter duty cycle D can vary in the range:

$$0 \leq D < 1. \tag{7}$$

The voltage stress is defined as the ratio between the DC-link voltage and the input voltage at the maximum modulation index. The maximum voltage gain can be obtained from (6) and (7) as

$$G_{max} = \frac{m_{max}}{1-D} \rightarrow D = \frac{G-1}{G} \tag{8}$$

and the minimum voltage stress can be obtained from (2) and (8) as

$$\frac{\hat{V}_{pn}}{V_{in}} = \frac{1}{1-\frac{G-1}{G}} = G. \tag{9}$$

The input peak voltage ratio can also be defined, by considering (4) and rearranging it as

$$\frac{V_{in}}{\hat{V}_{a(1)}} = \frac{2(1-D)}{1}. \tag{10}$$

Parameters Design

In order to design the inductance, it is to be considered the inductor voltage during the ON state of the boost converter:

$$V_L = V_{in} = L_1 \frac{\Delta i_L}{D} f_{sw}. \tag{11}$$

The current ripple can be set at 30 % of the inductor current I_L at the rated power P_r.

$$V_{in} I_L = P_r \rightarrow I_L = \frac{P_r}{V_{in}} \tag{12}$$

$$\Delta i_L = 0.3 \cdot I_L \tag{13}$$

Accordingly, the minimum inductance can be expressed as:

$$L_1 \geq \frac{V_{in}^2 D}{0.3 \cdot P_r f_{sw}}. \tag{14}$$

In order to design the capacitance, it is considered the capacitor current during the ON state of the boost converter:

$$I_C = C_1 \frac{\Delta v_C}{D} f_{sw}. \tag{15}$$

In this case, all the load current is supplied by the capacitor:

$$I_C = I_{pn} = \frac{P_r}{V_{pn}}. \tag{16}$$

The voltage ripple can be set at 0.3 % of the rated DC-link voltage V_{pn} as:

$$\Delta v_c = 0.003 \cdot V_{pn}. \tag{17}$$

Substituting (2),(16) and (17) in (15), the minimum capacitance value can be calculated as:

$$C \geq \frac{P_r D (1 - D)^2}{0.003 \cdot V_{in}^2 f_{sw_{DC}}}. \tag{18}$$

Quasi-Z-Source Inverter

Steady-State Analysis

The peak output voltage for a qZSI can be obtained according to Eq. (1). The DC-link voltage $\hat{V}_{pn}$ in the qZSI topology, by neglecting all the power losses, is expressed as

$$\widehat{V_{pn}} = \frac{1}{1 - 2D} V_{in} = B_q V_{in} \tag{19}$$

where B_q is the qZSI boost factor:

$$B_q = \frac{1}{1 - 2D}. \tag{20}$$

The peak output voltage is deduced by using (1) and (19):

$$\hat{V}_{a(1)} = \frac{m}{(1 - 2D)} \frac{V_{in}}{2} = G \frac{V_{in}}{2} \tag{21}$$

where G is the overall voltage gain. The qZSI duty cycle D varies in the range:

$$0 \leq D < 0.5. \tag{22}$$

The voltage stresses are then analyzed under different modulation strategies. Based on the modulation technique used to control the qZSI, the modulation index m and the duty cycle D have different relationship and limits.

Simple Boost Control (SBC)

For the *Simple Boost Control* (SBC) [11], the relationship between the modulation index and the duty cycle is:

$$m_{max} \leq 1 - D. \tag{23}$$

Thus, the maximum voltage gain can be expressed as

$$G_{max} = \frac{m_{max}}{1 - 2D} \rightarrow D = \frac{G - 1}{2G - 1} \tag{24}$$

and the minimum voltage stress can be obtained from (19) and (24):

$$\frac{\hat{V}_{pn}}{V_{in}} = \frac{1}{1 - 2\frac{G - 1}{2G - 1}} = 2G - 1. \tag{25}$$

Considering (21) and (24), the input peak voltage ratio can be obtained as:

$$\frac{V_{in}}{\hat{V}_{a(1)}} = 2\frac{(1 - 2D)}{(1 - D)}. \tag{26}$$

Maximum Constant Boost Control (MCBC)

For the *Maximum Constant Boost Control* (MCBC) [12], the relationship between the modulation index and the duty cycle is

$$m_{max} \leq \frac{(2 - 2D)}{\sqrt{3}} \tag{27}$$

and the maximum voltage gain is:

$$G_{max} = \frac{1}{1 - 2D}\frac{(2 - 2D)}{\sqrt{3}} \rightarrow D = \frac{\sqrt{3}G - 2}{2\sqrt{3}G - 2}. \tag{28}$$

Substituting (28) in (19), the minimum voltage stress can be expressed as:

$$\frac{\hat{V}_{pn}}{V_{in}} = \frac{1}{1 - 2\frac{\sqrt{3}G - 2}{2\sqrt{3}G - 2}} = \sqrt{3}G - 1. \tag{29}$$

The input peak voltage ratio can be expressed by considering (21) and (27) as:

$$\frac{V_{in}}{\hat{V}_{a(1)}} = 2\sqrt{3}\frac{(1 - 2D)}{(2 - 2D)}. \tag{30}$$

Maximum Boost Control (MBC)

For the *Maximum Boost Control* (MBC) [13], the relationship between the modulation index and the duty cycle is

$$m_{max} \leq \frac{2\pi - 2\pi\overline{D}}{3\sqrt{3}} \tag{31}$$

and the maximum voltage gain is:

$$G_{max} = \frac{1}{1 - 2D}\frac{2\pi - 2\pi\overline{D}}{3\sqrt{3}} \rightarrow D = \frac{3\sqrt{3}G - 2\pi}{6\sqrt{3}G - 2\pi}. \tag{32}$$

Substituting (32) in (19), the minimum voltage stress can be obtained as:

$$\frac{\hat{V}_{pn}}{V_{in}} = \frac{1}{1 - 2\dfrac{3\sqrt{3}G - 2\pi}{6\sqrt{3}G - 2\pi}}. \tag{33}$$

Finally, the input peak voltage ratio for the MBC is:

$$\frac{V_{in}}{\hat{V}_{a(1)}} = 6\sqrt{3}\frac{(1 - 2D)}{2\pi - 2\pi\overline{D}}. \tag{34}$$

Fig. 4 shows the voltage stresses on the inverter switches versus the voltage gain. The plot highlights that for a voltage gain up to 1.5, the MCBC and MBC show a lower voltage stress. Fig. 5 shows the input peak voltage ratio versus the duty cycle. The two-stage converter characteristic was normalized to $D/2$ in order to directly compare it with the qZS characteristics. A lower ratio means that, to achieve the same output voltage, a lower input voltage is needed. It can be noticed that, if the SBC is used, the ratio will always be higher than the two-stage converter one. Using the MCBC, the ratio is lower up to a duty cycle of 0.12 and using the MBC, the ratio is lower up to a duty cycle of 0.18.

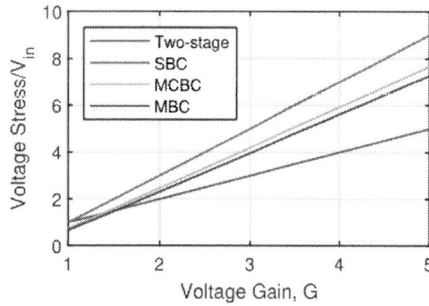

Fig. 4: Voltage stress versus voltage gain.

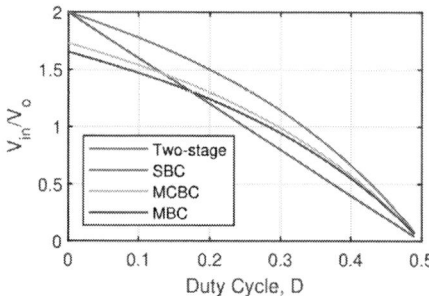

Fig. 5: Input peak voltage ratio versus duty cycle.

Parameters Design

The modulation strategies interfere with the Passive component design.

Simple Boost Control (SBC)

The inductance value for the Z-network can be calculated considering that, during the shoot-through duration, the voltage applied to the inductor is:

$$V_L = V_{in} - \left(-V_{C_2}\right). \tag{35}$$

The steady-state capacitor voltage V_{C_2}, as mentioned in [11], can be calculated as:

$$V_{C_2} = \frac{D}{1 - 2D}V_{in}. \tag{36}$$

Substituting (36) in (35) gives:

$$V_L = \frac{1 - D}{1 - 2D}V_{in}. \tag{37}$$

Accordingly, the minimum inductance can be calculated as:

$$L_1 \geq \frac{D - D^2}{1 - 2D}\frac{V_{in}^2}{0.3P_r f_{sw_{DC}}}. \tag{38}$$

The capacitor value for the Z-source network can be obtained by considering Eq. (18), but noticing that in this case, during the shoot-though, two capacitors are in series:

$$C_1 = C_2 = C \geq 2 \frac{P_r D (1 - D)^2}{0.003 \cdot V_{in}^2 f_{sw_{DC}}}. \tag{39}$$

Maximum Constant Boost Control (MCBC)

If the MCBC technique is used, the capacitor can be chosen in the same way using (39). The inductor current will have a small low frequency component that will increase the ripple. The inductor can be chosen considering (38) and increasing the calculated value by 30%.

Maximum Boost Control (MBC)

To obtain the maximum voltage boost, the MBC technique can be used. In this case, the shoot-though duty cycle D varies at a frequency equal to six times the output frequency. The ripple in the duty cycle will result in ripples in the inductor current and in the capacitor voltage. Indicating the output frequency with f_1 the inductor and capacitor values, as reported in [13], can be chosen as:

$$L \geq \frac{(\sqrt{3}/2 - 3/4) m V_{in}^2}{12 (3\sqrt{3}m - \pi)(1 - 2D) f_1 P_r}, \tag{40}$$

$$C_1 = C_2 = C \geq 2 \frac{P_r D (1 - D)^2}{0.003 \cdot V_{in}^2 6 f_1}. \tag{41}$$

Table II shows all the system parameters for both the 2-stage inverter and the qZSI with the three different modulation schemes. The two-stage converter has the minimum input voltage and the higher UCs voltage. Using the qZSI, the minimum battery voltage will be slightly higher, while the UC voltage can be greatly reduced. Thus, simpler balancing circuits for the UC pack can be used. It should be noted that for the qZSI, the switching frequency of the Z-source network $f_{sw_{DC}}$ is fixed at two times the inverter switching frequency $f_{sw_{AC}}$. Thus, the boost switching frequency for the 2-stage converter is chosen accordingly. It can be noticed that to achieve the same ripple voltage and current percentage, the qZSI needs larger capacitance and inductance. When using the MBC, due to the low frequency component, the values are greatly increased. Fig. 6 shows the simulated system efficiency under different loading conditions. The switching and conduction losses of the IGBTs have been calculated in PLECS, using the IGBT model of IKQ50N120CT2. It can be noticed that, for low load conditions, the qZSI-SBC and qZSI-MCBC have higher efficiency than the two-stage converter. The qZSI with SBC has always lower efficiency. These results are also in accordance to the voltage stresses analysis of Fig. 4, which shows that for a low voltage gain, the qZSI achieves lower stresses. To increase the qZSI efficiency for higher loads, novel modulation strategies that improve the DC utilization and decrease the voltage stresses should be considered [14]-[16].

Table II: System parameters in different cases.

Parameter	Unit	2-Stage	Quasi Z-Source		
			SBC	MCBC	MBC
V_{in}	V	260	370	320	310
V_C	V	650	280	240	230
$f_{sw_{DC}}$	kHz	20	$2f_{sw_{AC}}$	$2f_{sw_{AC}}$	$2f_{sw_{AC}}$
$f_{sw_{AC}}$	kHz	10	10	10	10
D	-	0.6	0.3	0.3	0.3
m	-	1	0.7	0.8083	0.8464
L	mH	1.4	2.4	2.3	7.2
C	μF	120	180	240	2000

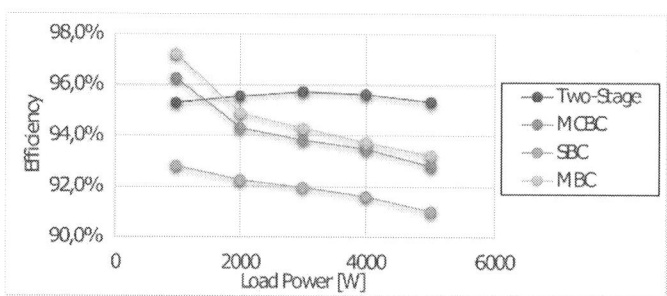

Fig. 6: Voltage stress versus voltage gain.

Fig. 7 shows the simulated inductor current ripple and DC-link voltage ripple in different cases. For the qZSI, the DC-link voltage is the sum of the capacitors voltage V_{C_1} and V_{C_2}. The red dashed line in the sub-figures of Fig. 7 represent the maximum desired ripple, while the blue curves are the actual voltage and current.

Simulation Results

To evaluate the performances of both systems in a grid-connected application, a closed-loop control of the battery current (dc-side control) and of the grid current (ac-side control) was implemented. The dc-side control acts on the shoot-though duty cycle D to regulate the battery current. The ac-side control acts on the modulation index m to regulate the grid current. The same references for the battery power and the grid current were used in all the simulations. shows the simulation results. The power and grid current trends are the same for all the simulations, while the UCs voltage is different. In particular, in the two-stage converter, the total output voltage of the boost converter is applied to the UCs as shown in Fig. 8 (c), while in the qZSI only a fraction of the output voltage is applied to the UCs as shown in Fig. 8 (d). Hence, the rated voltage of the UCs can be significantly reduced by using fewer components connected in series.

In the simulation, the battery power is set to $P_{batt} = 5\ kW$ The grid current reference is set to zero from $t = 0$ to $t = 0.2$ s and the battery power is used to recharge the UCs, as shown in Fig. 8 (a). From $t = 0.2\ s$ to $t = 0.4\ s$ the grid current reference is set to $i_d = 10\ A$, and all the power goes from the battery to the grid keeping the UCs voltage constant. From $t = 0.4$ s to $t = 0.6\ s$, the grid current reference is set to $i_d = 20\ A$, the additional power is provided by the UCs, which starts to discharge, as it can be seen in Fig. 8.

Fig. 9 shows the three-phase output current at steady state, the THD and the fast the phase current Fourier transform for all the cases. The magnitudes are expressed in percent of the fundamental component. It can be observed in Fig. 9 that the output current is fully regulated, with comparable THD. The Fourier analysis shows a similar spectrum for all the simulated schemes.

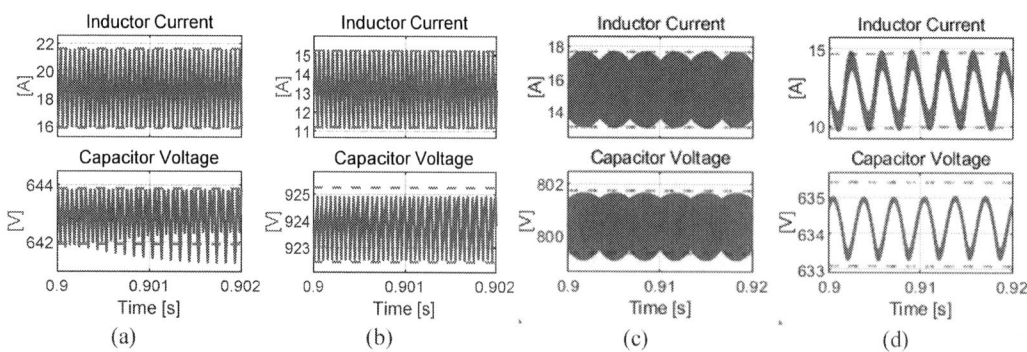

Fig. 7: Current and voltage ripple for (a) the two-stage inverter and qZSI with (b) SBC, (c) MCBC and (d) MBC.

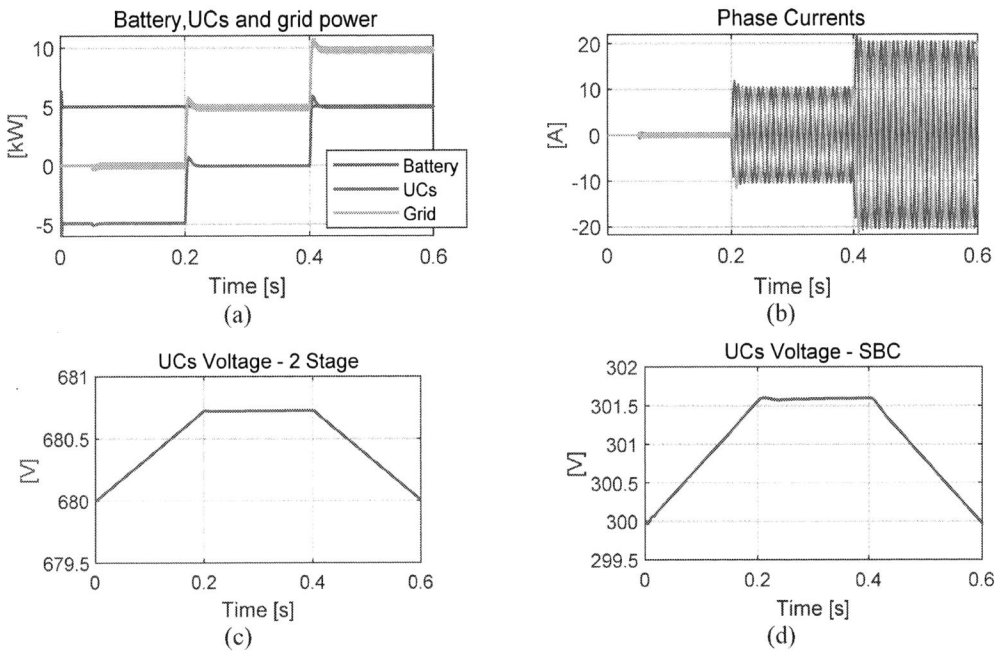

Fig. 8: Simulation results: (a) shows the power trends of the battery (blue), UCs (red) and grid (orange), (b) shows the grid current, (c) shows the UCs voltage for the two-stage converter and (d) shows the UCs voltage for the qZSI with SBC modulation.

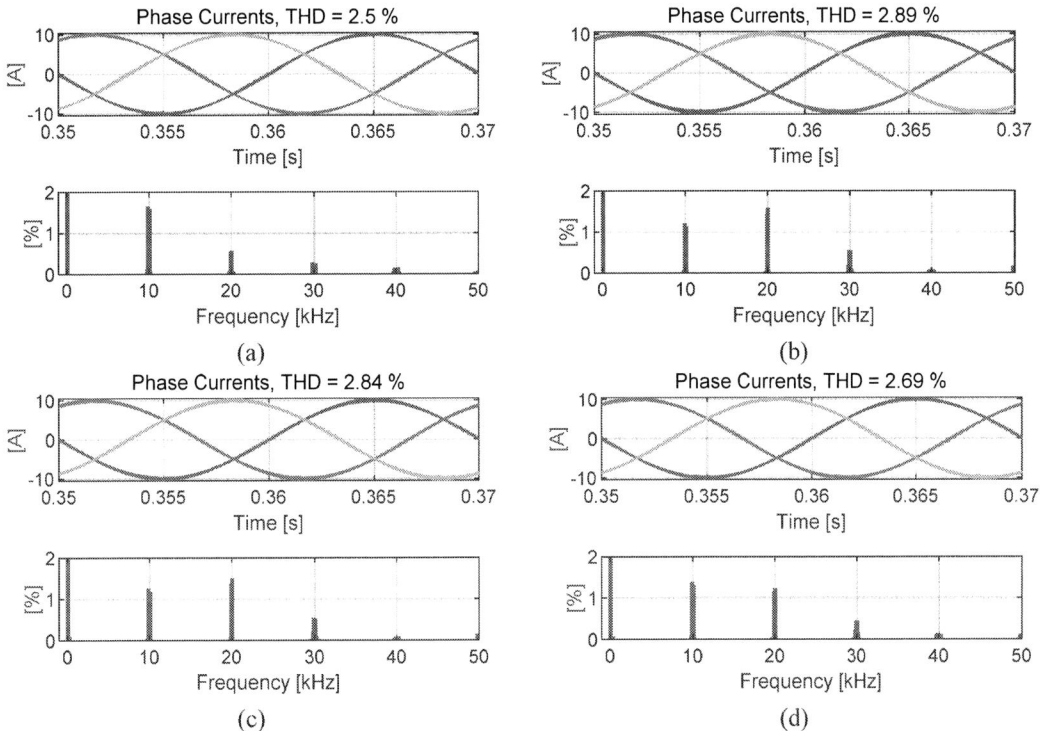

Fig. 9: Grid current at steady state and its fast Fourier transform analysis for (a) the two-stage converter and qZSI with (b) SBC, (c) MCBC and (d) MBC.

Conclusion

In this paper, a comparison between the two-stage converter and the qZSI for the HESS has been carried out. The comparison considered the voltage stresses on the switches, the voltage requirements for the battery pack and the UC pack, the passive component requirements for each solution, the output voltage THD and the system efficiency. For a low voltage gain, the voltage stresses across the switches are lower in the qZSI than in the classical two-stage converter, while for a higher voltage gain, the trend is opposite. The higher voltage stresses cause also lower efficiency in the qZSI for heavy load conditions. The UCs voltage rating can be greatly reduced using the qZSI, while the minimum battery voltage is lower for the two-stage converter. In terms of the output current THD and harmonic contents, the compared systems show almost the same results.

References

[1] J. Li, S. Zhou, and Y. Han, Advances in Battery Manufacturing, Services, and Management Systems. Wiley, Oct. 2016.

[2] M. Choi, S. Kim, and S. Seo, "Energy management optimization in a battery/supercapacitor hybrid energy storage system," IEEE Transactions on Smart Grid, vol. 3, no. 1, pp. 463–472, Mar. 2012.

[3] A. Khaligh and Z. Li, "Battery, ultracapacitor, fuel cell, and hybrid energy storage systems for electric, hybrid electric, fuel cell, and plugin hybrid electric vehicles: State of the art," IEEE Transactions on Vehicular Technology, vol. 59, no. 6, pp. 2806–2814, Jul. 2010.

[4] S. M. Lukic, S. G. Wirasingha, F. Rodriguez, J. Cao, and A. Emadi, "Power management of an ultracapacitor/battery hybrid energy storage system in an hev," in 2006 IEEE Vehicle Power and Propulsion Conference, Sep. 2006, pp. 1–6.

[5] K.V. Singh, H.O. Bansal and D. Singh, "A comprehensive review on hybrid electric vehicles: architectures and components", J. Mod. Transport. 27, 77–107, 2019.

[6] D.-D. Tran, M. Vafaeipour, M. El Baghdadi, R. Barrero, J. V. Mierlo, O. H. , "Thorough state-of-the-art analysis of electric and hybrid vehicle powertrains: Topologies and integrated energy management strategies", Renewable and Sustainable Energy Reviews, Volume 119, 2020.

[7] H.-W. He, R. Xiong, and Y.-H. Chang, "Dynamic modeling and simulation on a hybrid power system for electric vehicle applications," Energies, vol. 3, Nov. 2010.

[8] S. Pay and Y. Baghzouz, "Effectiveness of battery-supercapacitor combination in electric vehicles," in 2003 IEEE Bologna Power Tech Conference Proceedings, vol. 3, Jun. 2003, pp. 1–6.

[9] S. M. Lukic, J. Cao, R. C. Bansal, F. Rodriguez, and A. Emadi, "Energy storage systems for automotive applications," IEEE Transactions on Industrial Electronics, vol. 55, no. 6, pp. 2258–2267, Jun. 2008.

[10] M. Ortuzar, J. Moreno, and J. Dixon, "Ultracapacitor-based auxiliary energy system for an electric vehicle: Implementation and evaluation," IEEE Transactions on Industrial Electronics, vol. 54, no. 4, pp. 2147–2156, Aug. 2007.

[11] F. Z. Peng, "Z-Source Inverters," Wiley Encyclopedia of Electrical and Electronics Engineering, pp. 1–11, 1999.

[12] M. Shen, J. Wang, A. Joseph, F. Z. Peng, L. M. Tolbert, and D. J. Adams, "Maximum constant boost control of the Z-source inverter," in Conference Record of the 2004 IEEE Industry Applications Conference, 2004. 39th IAS Annual Meeting.. vol. 1, Oct. 2004, p. 147.

[13] F. Z. Peng, M. Shen, and Z. Qian, "Maximum boost control of the Z-source inverter," IEEE Transactions on Power Electronics, vol. 20, no. 4, pp. 833–838, Jul. 2005.

[14] R. Miceli, G. Schettino, F. Viola, F. Blaabjerg and Y. Yang, "Performance Evaluation of a Three- Phase Five-Level Quasi-Z-Source Cascaded H-Bridge for Grid-Connected Applications," 2018 IEEE International Power Electronics and Application Conference and Exposition (PEAC), Shenzhen, 2018, pp. 1-6.

[15] R. Miceli, G. Schettino, F. Viola, F. Blaabjerg and Y. Yang, "Modified Modulation Techniques for Quasi-Z-Source Cascaded H-Bridge Inverters," IECON 2018 - 44th Annual Conference of the IEEE Industrial Electronics Society, Washington, DC, 2018, pp. 3743-3748.

[16] A. Abdelhakim, P. Davari, F. Blaabjerg and P. Mattavelli, "Switching Loss Reduction in the Three-Phase Quasi-Z-Source Inverters Utilizing Modified Space Vector Modulation Strategies," in IEEE Transactions on Power Electronics, vol. 33, no. 5, pp. 4045-4060, May 2018.

State Estimation for Medium and Low Voltage Distribution Grids Based on Near Real-time Grid Measurements and Delayed Smart Meters Data

Mohammad RAYATI, Thomas PIDANCIER, Mauro CARPITA, Mokhtar BOZORG
UNIVERSITY OF APPLIED SCIENCES OF WESTERN SWITZERLAND (HES-SO)
Route Cheseaux 1, Case Postale 521, 1401 Yverdon-les-Bains, Switzerland
e-mail : mokhtar.bozorg@heig-vd.ch
URL: http://iese.heig-vd.ch/

Acknowledgements

This research project is financially supported by the Swiss Innovation Agency Innosuisse and is part of the Swiss competence Center for Energy Research SCCER FURIES. The authors would also like to acknowledge the technical support of "Services Industriels de Genève - SIG".

Keywords

«Estimation Technique», «Distribution of electrical energy», «Measurement», «Smart grids».

Abstract

In this paper, first, a distribution system state estimation (DSSE) algorithm based on Distflow model is presented. The model is designed for radial low voltage (LV) distribution grids considering traversal components such as the cable capacitances. Next, to obtain the pseudo-measurements required for the DSSE algorithm, we developed and compared three intraday nodal load forecast (INLF) methods that take; i) near real-time stream of data from the grid measurement devices (MDs), i.e., voltage and current magnitudes at limited number of lines/nodes, and ii) the batch data set from smart meters (SMs) available for previous days. The accuracy associated with the INLF methods is quantified in terms of statistical characteristics such as average of symmetric mean absolute percentage error (sMAPE) over all nodes of the distribution grid. Afterwards, the impact of this accuracy on the DSSE results is investigated using real data available for a distribution grid in city of Geneva, Switzerland.

1 Introduction

The recent increase of uncontrolled renewable production connected to the distribution network raised the need for better network monitoring and control solutions. Regulators encourage the distribution system operators (DSOs) to improve the network observability by installing low voltage (LV) grid measurement devices (MDs) as well as smart-meters (SMs) on the end-user clients' side. For instance, in Switzerland at least 80% of the DSOs' clients should be equipped with SMs by the end of 2027 [1]. The medium voltage (MV) and LV distribution networks are very large in terms of the number of lines/nodes. Therefore, the DSOs are looking for monitoring and supervision systems that require a limited number of affordable MDs [2]. In addition, due to the privacy issues, the measurements coming from MDs installed on the grid and those from the SMs at customer level are not available with the same sampling time and recuperation delay. In particular, in Switzerland the SMs data (i.e., the active/ reactive power consumptions of end-user clients at every 15-minutes interval) are not available in real-time and can only be recuperated once every 24 hours. Hence, the distribution system state estimation (DSSE) algorithms and intraday nodal load forecast (INLF) methods are required to improve the quality of network supervision with a limited number of MDs.

A number of DSSE algorithms are proposed in the literature. These algorithms differ in terms of required measurement data, network model, load flow equations, and estimation technique. A comprehensive review of the state of the art is presented in [3]. Most of the algorithms use the voltage angles (i.e., phasor measurement data), which are not always available with high level of accuracy. Indeed, the distribution grid lines are short; hence, the difference between voltage angles of two neighboring nodes can be as small as the MDs accuracy level. In [4], a DSSE algorithm is presented relying on the solution of Distflow equations [5] to estimate the magnitudes of voltages and line currents. This approach is appropriate for LV distribution grids since it does not require any voltage angle measurements or estimations. However, the shunt components (e.g., cable capacitance) of the lines are not taken into account. In [6], the authors have enhanced the DSSE algorithm proposed in [4] considering the shunt components of the network and using a PI model of the lines.

A DSSE algorithm estimates the state of the grid using: i) the voltage magnitudes and active/reactive power flow measurements (e.g., 1-hour average values) coming from the MDs installed at specified nodes/lines; and ii) the load pseudo-measurements obtained from the intraday forecast of active/reactive power injection/consumption at every node (i.e., INLF).

For benchmarking purposes, we developed the following three INLF methods to quantify the impacts of pseudo-measurements error and eventually the error of the DSSE:

- Seasonal auto-regressive integrated moving average (SARIMA) method using SMs data,
- Random forests method using SMs data,
- Random forests method using SMs and MDs data.

The SMs data includes historic time-series of aggregated active/reactive power consumption profiles at each node of the grid from the previous days (excluding intraday profiles that are not available due to privacy and legal issues). The MDs data includes historic and intraday time-series of voltage magnitude, active/reactive power measurements at specified nodes/lines where MDs are installed.

The rest of this paper is organized as follows: The enhanced DSSE algorithm is briefly presented in Section 2. The INLF forecast methods are discussed in Section 3. Afterwards, a case study is developed and described in Section 4 in order to demonstrate the effectiveness of the DESS algorithm and compare different INLF methods. The numerical results are presented and discussed in Section 5. Finally, summary and conclusions are given in Section 6.

2 Distribution System State Estimation

The DSSE algorithm used in this paper is proposed by the authors in [6]. Indeed, it is an improvement of the DSEE algorithm based on Distflow equation that is originally proposed in [3], in which it does not include the capacitance of the lines (a factor that can be important in LV networks). In order to define and formulate the DSSE algorithm, the following assumptions are considered:

- The network configuration is radial; it is modelled by load-buses and a slack-bus.
- Load pseudo-measures are available at every nodes of the network. Their values can be negative (injection), null (connection node), or positive (consumption). At each node, they consist of the aggregation of the end-user clients connected to that node.
- The MD, installed at a node, gives the voltage magnitude and the active/reactive power flow in the lines/transformers connected to that node.
- The lines and transformers parameters (i.e., rated values and impedances) are known.

Fig. 1 illustrates the hypotheses on the topology and the available measurements. Fig. 2 depicts the PI model of the lines that is used to formulate the Distflow equations in the presence of transversal components. The variables are computed in two steps: first, the power flow in the lines from the bottom-up is computed using the backward-Distflow equation (1); then, the square voltage magnitude of the nodes from the slack bus to the end-nodes is derived using the forward-Distflow Equation (2).

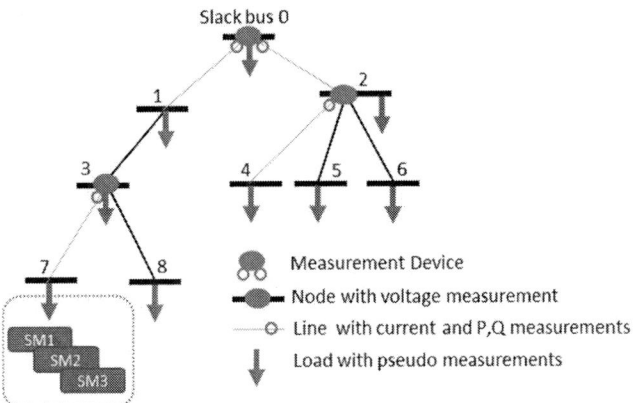

Fig. 1: Network topology example where buildings consumptions (presented by SMs) are aggregated at each node of the network.

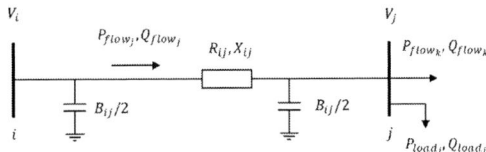

Fig. 2: PI model of a line and notations used in the Distflow.

$$
\begin{cases}
P_j^{flow} = P_j^{load} + \sum_k P_k^{flow} + R_j \dfrac{\left(P_j^{load} + \sum_k P_k^{flow}\right)^2 + \left(Q_j^{load} + \sum_k Q_k^{flow} + Q_{j_B}\right)^2}{V_j^2}, \\[4mm]
Q_j^{flow} = Q_j^{load} + \sum_k Q_k^{flow} + Q_{j_B} + X_j \dfrac{\left(P_j^{load} + \sum_k P_k^{flow}\right)^2 + \left(Q_j^{load} + \sum_k Q_k^{flow} + Q_{j_B}\right)^2}{V_j^2},
\end{cases}
\tag{1}
$$

$$
V_j^2 = V_i^2 - 2\left(R_j P_j^{flow} + X_j Q_j^{flow}\right) + \left(R_j^2 + X_j^2\right)\frac{P_j^{flow\,2} + Q_j^{flow\,2}}{V_i^2}.
\tag{2}
$$

As indicated in Fig. 2, P_j^{flow} and Q_j^{flow} are active/reactive power flow in line j (which is upstream to node j). P_j^{load} and Q_j^{load} are active/reactive power load at node j. V_j is the square of voltage magnitude at node j. R_j and X_j, are resistance and reactance of line j, respectively. $Q_{j_B} = -B_{\Sigma\,j}V_j^2$, where the term $B_{\Sigma\,j}$ is the sum of the shunt components (susceptances) of all the lines connected to node j.

Let us take a radial network with a slack-bus, L lines and L load-buses nodes. At each time step, the DSSE algorithm uses the weighted-least-square (WLS) estimation by iteratively solving (3).

$$
z = h(x) + e,
\tag{3}
$$

where x is the state vector, z is the measurement vector, $h(x)$ is the measurement function that can be derived by (1)-(2), and e is the measurement noise. Using (1) and (2), we can define the DSSE vector $x = \left[P_{load}\ Q_{load}\ V_0^2\right]^T$, where $P_{load} = [P1 \dots P_L]^T$ and $Q_{load} = [Q1 \dots Q_L]^T$ are the active/reactive power load of all load-buses, respectively. Finally, V_0^2 is the square of voltage magnitude at slack-bus. The measurement vector z is composed of the following elements:

- From the MDs installed on the grid:
  - Voltage magnitudes at every measured node $V^2_{n_{mes}}$,
  - Power flow in the measured lines ($P_{flow_{mes}}, Q_{flow_{mes}}$).
- Load pseudo-measurement at every node ($P_{load_{mes}}, Q_{load_{mes}}$).

As formulated in [6], it is possible to solve the DSSE Equation (3) in an iterative manner with the help of (4).

$$x_{(k+1)} = x_{(k)} + (H^t W H)^{-1} H^t W r_{(k)}, \tag{4}$$

where k is the iteration index, H is the jacobian matrix of h, W is the inverse of the variance matrix regarding all the measurements, and $r_{(k)} = z - h(x_{(k)})$ is the error between the measurement and the results of DSSE at iteration k.

3 Intraday Load Forecasting Methods

As remarked in the introduction, the accuracy of DSSE highly depends on the quality of generated pseudo-measurements. Note that unlike transmission systems that enjoys high number of actual measurements [7], the distribution systems observability is undermined with a limited number of SMs and in-field MDs. In addition, the data of SMs for each day is not available in real-time and for implementing the DSSE; thus, we need to generate the pseudo-measurements based on the nodal loads profiles forecasts.

In this section, three INLF methods for generating pseudo-measurements are explained:

- SARIMA method using SMs data – Method A,
- Random forests method using SMs data – Method B,
- Random forests method using SMs and MDs data – Method C.

A: Seasonal auto-regressive integrated moving average method using SMs data

Time series models have been commonly used in a broad range of forecasting applications [8]. The SARIMA method is one of the time series models capturing serial correlation among observations both within and across the seasons.

Here, we use a separate SARIMA model for each node. To configure the SARIMA model of each node, we need the selection of the hyper-parameters for both the trend and seasonal elements of the time-series. There are the following seven elements that should be tuned: (p) trend auto-regression order, (d) trend difference order, (q) trend moving average order, (P) seasonal auto-regression order, (D) seasonal difference order, (Q) seasonal moving average order, and (m) the number of time steps for a single seasonal period. To tune these seven parameters, autocorrelation function (ACF) and partial autocorrelation function (PACF) are used/evaluated [9]. To keep the model as simple as possible and since more confident we can argue that the relationship of the recent data and the forecast is linear, we select the hyper-parameters of all SARIMA models of all nodes as $p = 1$, $q = 1$, $d = 1$, $P = 1$, $Q = 1$, $D = 1$, and $m = 24\ hours$. The performance of our selected hyper-parameters is supported by comparing the Akaike information criteria (AIC) of these models with higher order ones and testing the required time of calculating these models [9].

The final SARIMA model resulted from the above parameter selection of node n is as (4).

$$\left(1 - \phi_{n,24}.B^{24}\right).\left(1 - \phi_{n,1}.B\right).(1 - B).Y_{n,t} = \theta_{n,0} + \left(1 - \theta_{n,1}\right).\left(1 - \theta_{n,24}.B^{24}\right).\epsilon_{n,t}, \tag{4}$$

where $\phi_{n,24}$, $\phi_{n,1}$, $\theta_{n,0}$, $\theta_{n,1}$, $\theta_{n,24}$ are the parameters of the model, B is the backward shift operator, $Y_{n,t}$ is the load of node n at time t (which can be active power P_{load} or reactive power Q_{load}), and $\epsilon_{n,t}$ denotes the noise of node n load at time t. In this paper, we use the function of "SARIMAX" from the package "statsmodels" in "Python" to train/test these models.

It is worth mentioning that an SARIMA model can generally give us a good result when there are no big random changes in the time-series [10]. However, since it is not the case for the nodal loads in LV distribution networks [11], we expect that the generated pseudo-measurements by this method do not have good accuracy.

B: Random forests method using SMs data

Machine-learning methods have attracted scientific attentions for generating the pseudo-measurements that are used in DSSE [12]. One of the big challenges of using these machine-learning methods is that they need huge amounts of data and some methods are computationally expensive. In this paper, to solve this challenge, the random forests method is adopted, which has acceptable computational performance [12].

Random forests method is an ensemble learning method that can be used for both the classification and the regression purposes [6]. Random forests are the collocation of multiple decision trees known as forests. Each tree depends on an independent random sample and in the regression problem, the average of all the trees outputs is considered as the result.

In this study, we use a separate random forests model for each node n. To train each model, we need to define the input patterns $\vec{X}_n$ and output (forecast) one $\vec{Y}_n$. The forecast pattern $\vec{Y}_n$ is the vector of nodal load, i.e., $\left[Y_{n,1} \; Y_{n,2} \; ... \; Y_{n,t} \right]^{\mathrm{T}}$. For the input pattern, we use the SMs data of the previous day; thus, $\vec{X}_n$ is

$$\begin{bmatrix} Y_{n,1-24} & \cdots & Y_{n,1-1} \\ \vdots & \ddots & \vdots \\ Y_{n,t-24} & \cdots & Y_{n,t-1} \end{bmatrix}, \tag{5}$$

where $Y_{n,t}$ is the load of node n at time t (which can be active power P_{load} or reactive power Q_{load}).

Before training the random forests model for each node, we normalize both the input and output patterns. In addition, the outliers are detected and deleted from the time-series to prevent the over-fitting. We use the function "RandomForestRegressor" in the package "sklearn" in "Python" to fit/validate these models.

C: Random forests method using SMs and MDs data

In similar to the explained method B, in method C we use random forests regression models for prediction of nodal loads and generating pseudo-measurements. The only difference is that we modify the input vector $\vec{X}_n$ by adding the data of last 24 hours of the associated MDs (in which these measurements are relevant features to the load data). Thus, the models inputs are twofold: the time-series of grid measurements and the long-term time-series of SMs measurements.

Each MD connects to a feeder including a number of nodes with SMs. For those nodes we modify $\vec{X}_n$ as

$$\begin{bmatrix} Y_{n,1-24} & \cdots & Y_{n,1-1}X_{j,1-1} & \cdots & Y_{j,1-24} \\ \vdots & \ddots & \vdots & \ddots & \vdots \\ Y_{n,t-24} & \cdots & Y_{n,t-1}X_{j,t-1} & \cdots & X_{j,t-24} \end{bmatrix}, \tag{6}$$

where $X_{j,t}$ is the measurement of associated MD j at time t (which measures active power $P_{flow_{mes}}$ or reactive power $Q_{flow_{mes}}$).

It is worth mentioning that before training these models, we normalize both the input and output patterns. In addition, the outliers are detected and deleted from the time-series to prevent the over-fitting. Finally, we use the function "RandomForestRegressor" in the package "sklearn" in "Python" to fit/validate these models.

4 Test case description

In the following, a MV/LV distribution network is used to discuss the performance of different methods and highlight the characteristics of the proposed DSSE. The network and data for the case study were made available by the DSO of the city of Geneva, Switzerland. The network has 52 nodes along two feeders (Fig. 6-a). The branches connecting nodes 2 to 0 and nodes 2 to 5 are MV/LV transformer, while the others are cables. In addition, the locations of in-field MDs are depicted with red line in Fig. 6-a.

In summary, this case study network is in a residential area and has the following characteristics:

- 52 nodes, 49 lines, and 2 MV/LV transformer.
- 58 smart-meters installed (over 89 meters, so 65% of the consumer are covered):
    - Active and reactive power are available with an average over 15 minutes periods.
- 31 conventional meters, in which the annual active power data of these meters is available.
- Average monthly active power consumption of 43200 kWh.
- 2 points for MD installation:
    - 1 in the MV/LV station.
    - 1 at the start of a LV feeder.
    - Active and reactive power, three-phase voltage magnitudes, and current magnitudes are available with an average over 10 minutes periods.

In the numerical results, we compare the accuracy of generated pseudo-measurements by three explained methods A, B, and C in Section 3. In addition, for evaluating the performance of DSSE, we compare the results for the following four cases.

- Case 1: DSSE without using MDs data while the pseudo-measurements are generated by method A, i.e., SARIMA method using SMs data.
- Case 2: DSSE by using MDs data while the pseudo-measurements are generated by method A, i.e., SARIMA method using SMs data.
- Case 3: DSSE by using MDs data while the pseudo-measurements are generated by method B, i.e., random forests method using SMs data.
- Case 4: DSSE by using MDs data while the pseudo-measurements are generated by method C, i.e., random forests method using SMs and MDs data.

Here, the 15 minutes data of SMs and the 10 minutes data of MDs are acquired during 79 days from 17/01/2020 to 04/04/2020. The data of the first 71 days is used for training the models and the data of last 8 days is used for testing. In addition, the measured data of SMs and MDs is resampled into 1 hour and the numerical results of hourly forecasting and hourly DSSE are given in the following.

5 Numerical results

First, we evaluate the performance of the INLF methods for generating the pseudo-measurement as presented in section 3. It is observed that generally, Method C (i.e., random forests method using SMs data and MDs data) outperforms the other two methods. In Fig. 3, the predicted total active power and reactive power of all SMs in two mentioned feeders using method C are compared with the actual data of the last 8 days (testing data). One can observe that the sum of generated pseudo-measurements by method C is close to the actual measured data of active/reactive power.

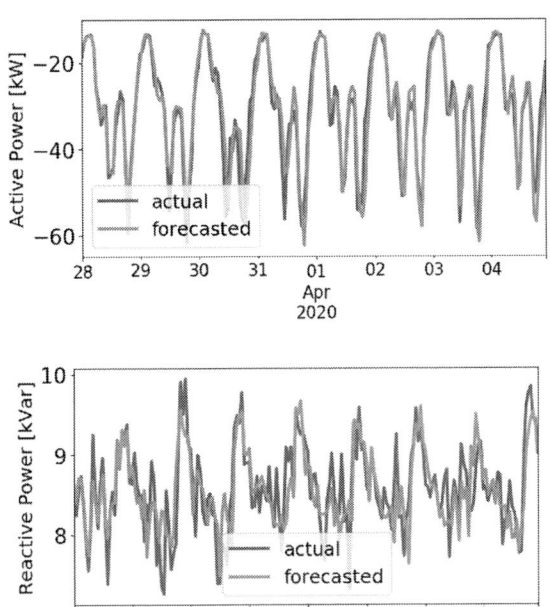

Fig. 3: Actual and forecasted total active/reactive power of all SMs using Method C.

In this study, the symmetric mean absolute percentage error (sMAPE) is used instead of mean absolute percentage error (MAPE) criteria for comparing different methods of INLF to avoid the asymmetry of the MAPE. The sMAPE is introduced by Makridakis [13] and it is more reliable than MAPE. The sMAPE for each time-series data is formulated as follows:

$$sMAPE = \frac{1}{T} \cdot \sum_{t=1}^{T} \frac{|forecasted(t) - actual\ data(t)|}{(|forecasted(t)| + |actual\ data(t)|)/2} \times 100\%. \tag{7}$$

In order to assess the performance of explained methods A, B, C in Section 3, the index of sMAPE is calculated for forecasting the active/reactive power of the nodes with SMs (16 nodes in this case study), separately. For the total active power, sMAPEs of methods A, B, and C are 14.46%, 8.8%, and 8.9%, respectively. In addition, for the reactive power, the sMAPEs of methods A, B, and C are 3.8%, 4.4%, and 3.1%, respectively. However, the sMAPEs of INLF are much higher. The box diagram of sMAPE to forecast the active/reactive power of nodal loads is depicted in Fig. 4. The averages of sMAPEs are also shown in Fig. 4. One can see that the methods B and C (random forest) outperform the method A (SARIMA) in both forecasting the active and reactive power. In addition, the worst case of sMAPEs for both active and reactive power are reduced considerably while using methods B and C instead of method A. Finally, using the MDs in forecasting (method C) has a little advantage in prediction of the nodal active power (less than 1% in average). It can be argued that each MD measures the aggregated active power of the nodes with SMs in that feeder; hence, the MDs data does not reflect the characteristics of each node profile separately. This can be improved by adding further number of MDs. In addition, for the reactive power, the measurements of MDs are even different from the aggregated reactive power of nodes with SMs since it depends on the network impedances. Thus, adding the data of MDs does not have benefit for INLF of reactive power.

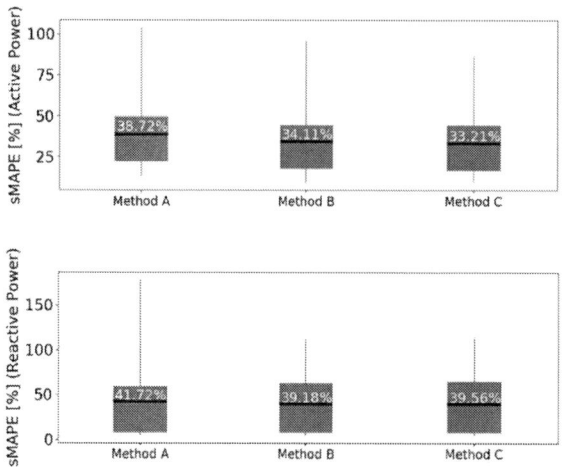

Fig. 4: Comparison of sMAPEs between different methods of A, B, and C.

The results of all four cases, explained in Section 4, are derived and compared. For comparison, the estimation error is defined as below:

$$Estimation\ Error = \frac{1}{T}.\sum_{t=1}^{T} |estimated(t) - actual\ data(t)|. \qquad (8)$$

As one can see in Fig. 5, the voltage estimation error is reduced just by adding the data of MDs into the DSSE. However, for having lower current estimation error, using the random forests method is advantageous since it is conceived to have a better model of pseudo-measurements. In addition, the current estimation error of case 4 (where the data of MDs is also used in generating the pseudo-measurements) is slightly lower compared to the case 3 (where the data of SMs is just used in generating the pseudo-measurements).

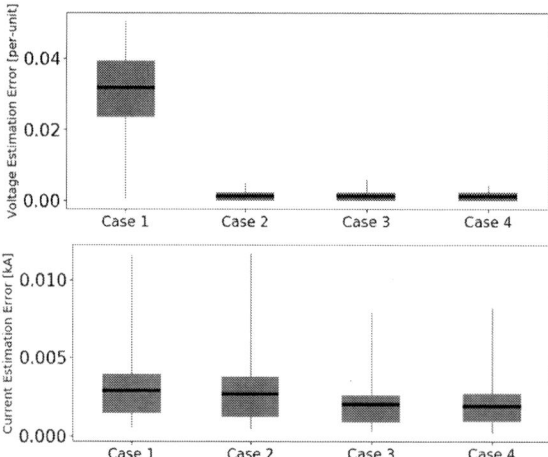

Fig. 5: Comparison of voltage and current estimation errors between different cases of 1-4.

To evaluate the geographical observability improvement of the test case study network using cases 2-4 instead of case 1, we define the improvement index for both the voltage and current estimations of each node and branch as follows:

$$Improvement\ Index\ (Case\ j\ versus\ Case\ 1)$$
$$= \frac{Estimation\ Error\ (Case\ j) - Estimation\ Error\ (Case\ 1)}{Estimation\ Error\ (Case\ 1)} \times 100\%. \quad (9)$$

In Fig. 6-a, the locations of MDs are depicted. The improvement index of cases 2-4 versus case 1 are shown in Fig. 6-b, 6-c, and 6-d, respectively. The voltage estimations of all nodes improve more than 20% in all cases 2-4. Therefore, we conclude that the voltage estimations of all nodes can be better off just by adding MDs data into the DSSE algorithm. On the other hand, the current estimations of distant branches from the MDs locations do not improve in cases of 2-4. However, by comparing the cases 3 and 2, one can see that the current estimation of near branches improves considerably by adopting random forests method for pseudo-measurements generation. In addition, by adding the data of considered MDs in forecasting procedure and using the random forests method (case 4), the current estimations at some branches (such as 10-15, 15-19, 19-21, 21-23, 23-25, 25-29, 29-32, and 32-35) will improve. Therefore, while the advantage of case 4 compared to case 3 was not clear in Fig. 5, the performance of case 4 (using MDs data in forecasting procedure) in the DSSE current results of number of branches can be observed in Fig. 6-d.

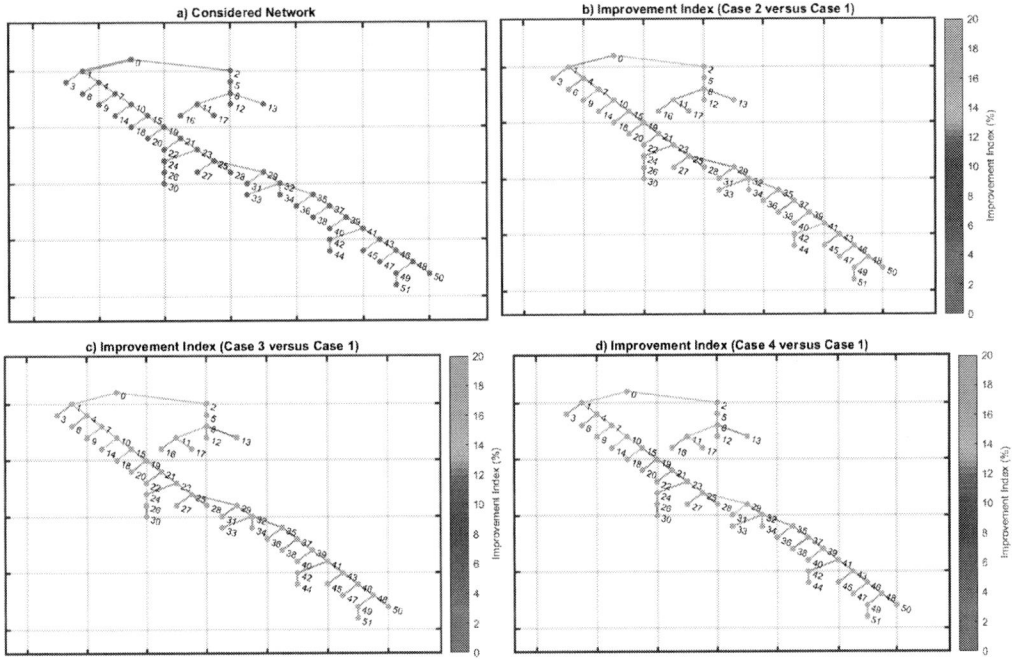

Fig. 6 a) Test case study network with MDs at red nodes/branches, b) Comparing the DSSEs of cases 2 and 1, c) Comparing the DSSEs of cases 3 and 1, and d) Comparing the DSSEs of cases 4 and 1. In figures b), c), and d), the improvement index for estimation of voltage and currents are demonstrated by color bar at each node and branch, respectively.

6 Summary and conclusions

In this paper, a distribution system state estimation (DSSE) algorithm based on Distflow model is presented. The DSSE algorithm takes the following inputs; i) near real-time data from the measurement devices (i.e., voltage magnitudes and active/reactive power flow measurements), and ii) the load pseudo-measurements obtained from an intraday nodal load forecast (INLF). The impact of forecast accuracy and available data on the DSSE results is investigated using the real data available for a distribution grid in city of Geneva, Switzerland.

It is observed that the error of INLF (30-40%) which is much higher than the error of total load forecast (4-9%). Overall, the INLF methods B and C (random forest) outperform the method A (SARIMA) in both forecasting the active and reactive power. In addition, the worst case of sMAPEs for both active and reactive power are reduced considerably while using methods B and C instead of method A. However, using MDs data in INLF (method C) only slightly improved the results (less than 1% in average), since the number of MDs are limited and the MDs data does not reflect the characteristics of each node profile separately. This can be improved by adding further number of MDs.

With reference to the numerical results, the estimation error is reduced just by adding the data of MDs into the DSSE. However, for having lower current estimation error, using the random forests method is advantageous since it is conceived to have a better model of pseudo-measurements. In addition, the current estimation error of case 4 (where the data of MDs is also used in generating the pseudo-measurements) is slightly lower compared to the case 3 (where the data of SMs is just used in generating the pseudo-measurements).

References

[1] Le Conseil fédéral Suisse., *Ordonnance sur l'approvisionnement en électricité (OApEl), section 4a Introduite par le ch. I de l'O du 1er nov. 2017, en vigueur depuis le 1er janv. 2018*. 2017.

[2] M. Carpita, A. Dassatti, M. Bozorg, J. Jaton, S. Reynaud and O. A. Mousavi, "Low Voltage Grid Monitoring and Control Enhancement: The GridEye Solution," 2019 International Conference on Clean Electrical Power (ICCEP), Otranto, Italy, 2019, pp. 94-99.

[3] A. Primadianto and C.-N. Lu, "A Review on Distribution System State Estimation," *IEEE Trans. Power Syst.*, vol. 32, no. 5, pp. 3875–3883, Sep. 2017.

[4] H. A. R. Florez, E. M. Carreno, M. J. Rider, and J. R. S. Mantovani, "Distflow based state estimation for power distribution networks," *Energy Syst.*, Jan. 2018.

[5] M. E. Baran and F. F. Wu, "Network reconfiguration in distribution systems for loss reduction and load balancing," *IEEE Trans. Power Deliv.*, vol. 4, no. 2, pp. 1401–1407, Apr. 1989.

[6] P. Paruta, T. Pidancier, M. Bozorg, and M. Carpita, "Greedy Placement of Measurement Devices on Distribution Grids based on Enhanced Distflow State Estimation", arXiv:2007.15050

[7] Abur, Ali, and Antonio Gomez Exposito. Power system state estimation: theory and implementation. CRC press, 2004.

[8] Ahn, Byung-Hoon, Hoe-Ryeon Choi, and Hong-Chul Lee. "Regional Long-term/Mid-term Load Forecasting using SARIMA in South Korea." Journal of the Korea Academia-Industrial cooperation Society 16, no. 12 (2015): 8576-8584.

[9] Contreras, Javier, Rosario Espinola, Francisco J. Nogales, and Antonio J. Conejo. "ARIMA models to predict next-day electricity prices." IEEE transactions on power systems 18, no. 3 (2003): 1014-1020.

[10] Hor, Ching-Lai, Simon J. Watson, and Shanti Majithia. "Daily load forecasting and maximum demand estimation using ARIMA and GARCH." In 2006 International Conference on Probabilistic Methods Applied to Power Systems, pp. 1-6. IEEE, 2006.

[11] Dehghanpour, Kaveh, Zhaoyu Wang, Jianhui Wang, Yuxuan Yuan, and Fankun Bu. "A survey on state estimation techniques and challenges in smart distribution systems." IEEE Transactions on Smart Grid 10, no. 2 (2018): 2312-2322.

[12] Dudek, Grzegorz. "Short-term load forecasting using random forests." In Intelligent Systems' 2014, pp. 821-828. Springer, Cham, 2015.

[13] Makridakis, Spyros. "Accuracy measures: theoretical and practical concerns." International journal of forecasting 9, no. 4 (1993): 527-529

Ground Fault Active Compensation in Emulated Distribution Grid of 10 kV

Tomáš Komrska[1] Antonín Glac[1] Jakub Talla[1] Bohumil Skala[1] Jan Štěpánek[1] Luboš Streit[1]
Zdeněk Peroutka[1]
[1]Regional Innovation Centre for Electrical Engineering (RICE)
University of West Bohemia
Pilsen, Czech Republic
E-Mail: komrska@rice.zcu.cz

Acknowledgements

This research has been supported by the Ministry of Education, Youth and Sports of the Czech Republic under the project OP VVV Electrical Engineering Technologies with High-Level of Embedded Intelligence CZ.02.1.01/0.0/0.0/18_069/0009855.

Keywords

Power semiconductor device, Distribution of electrical energy, Faults, High voltage power converters.

Abstract

MV distribution grids are often operated with arc suppression coils compensating dominating reactive fault current component when single-phase ground fault occurs. The paper deals with an active compensation in emulated distribution grid of 10 kV performed by a proposed three-phase active fault compensator, based on power semiconductor converter. It enables a complete fault current compensation without residual components and to address resonance-related problems.

Introduction

Distribution power grids require more and more attention, especially due to increasing power consumption, causing several problems in both overhead and cable lines. Many power grids are operated close to the limit of their transmission capacity. Contemporary, safety and reliability of the power delivery are a priority.

In case of medium-voltage distribution power grids, single-phase ground faults are the most frequent faults; they occur several times a week in some power systems, especially in bushes or wooded area. In most cases, they are caused by a tree branch touching a power line during a short period (usually less than 5 s) [1-3].

The ineffectively grounded systems, unlike their effectively grounded counterparts, enable to compensate or limit the fault current and, in most cases, without interruption of the power delivery. Especially the resonant grounded grids with arc suppression coils become popular due to their ability of effective compensation [1, 4, 5]. However, this solution still suffers from resonance-related phenomena, such as overvoltage causing dangerous states and insulation stress, voltage asymmetry during non-faulty conditions, harmonic components and others.

Power semiconductor converters penetrate more and more to the power systems, especially as FACTS, active power filters and HVDC transductors. Active compensation based on power semiconductor converters provides promising solution for ground fault compensation including residual current components, which are not possible to be compensated by traditional passive means such as arc suppression coils.

A solution combining a traditional arc suppression coil (ASC) and a low-power active source is presented in [6, 7]. The reactive component of the fault current is compensated by the passive ASC whilst the active component by the low-power active source of the ground fault neutralizer which is connected to auxiliary winding of the ACS. Thus, this power source (not specified in [6, 7]) is a single-phase solution which must be supplied by an external (auxiliary) power source. Moreover, the ASC still suffers by the resonance-related phenomena.

Our promising solution [8] is based on three controlled current sources connected to phase conductors (see Fig. 1), generating compensation current. This solution can operate without an ASC and it does not require connection to the neutral point of a distribution transformer (possible installation outside the power substation). It ensures compensation of both the active and the reactive component of the fault current, high dynamics and the power is directly supplied by the MV power grid via phase conductors (high power available, no auxiliary power supply needed).

Active compensation in emulated distribution power grid of 10 kV

A prototype of the promising solution based on three active current sources has been connected to an emulated distribution power grid of 10 kV, built in our hall laboratory, as shown in Fig. 1.

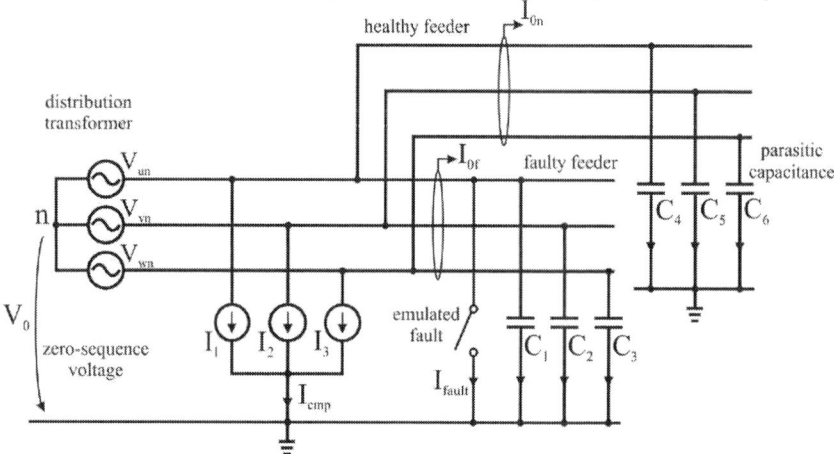

Fig. 1: Emulated distribution power grid of 10 kV with active compensator, parasitic capacitances and emulated single-phase ground fault

The whole emulated grid is supplied by a power source of 10 kV (phase-phase voltage) emulating a distribution transformer of a power substation. The power supply delivers power to two emulated MV feeders, both loaded by their parasitic capacitances and leakage resistance.

In case of wide spread distribution grids with ineffectively grounded neutral, an arc suppression coil of high impedance is connected between transformer center (n) and the ground point. Unlike arc suppression coils, the active ground fault compensator is connected to the phase conductors, in parallel to the power supply. In general, it can be implemented by three controlled active current sources which are connected to a star and the resulting compensation current is generated to the ground point (see Fig. 1). In our case, the active compensator consists of a three-phase transformer of 22/0,4 kV and of a 3Φ power semiconductor converter of 400 V_{ac}, 1.35 MVA (see Fig. 2).

As in real distribution grids, the zero-sequence current of each feeder is measured by current sensors. A manual high voltage power switch is used to emulate single-phase to ground faults. It is possible to test all types of faults, i.e. solid (no resistance), resistive (28 Ω) and arcing faults.

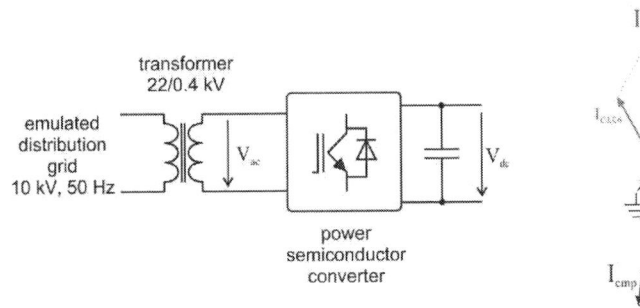
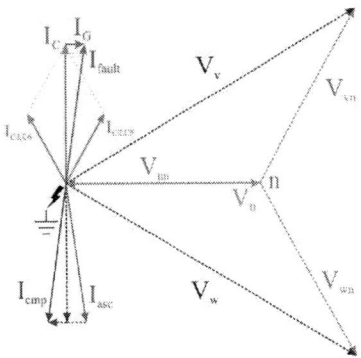

Fig. 2: Active ground fault compensator built by transformer 22/0,4 kV and power semiconductor converter of 400 V_{ac}, 1,35 MVA

Fig. 3: Vector diagram for a solid (0 Ω) single-phase ground fault in phase u.

Single-phase ground faults

Single-phase ground faults cause extreme asymmetry of the line to ground voltage, while the triangle of the line-line voltage, which is critical for the power delivery, remains unchanged and stable. Thus, the end customer usually does not register short ground faults in the grid. However, they cause significant line to ground voltage drop of the faulty phase, while the untouched phases reach the line-line voltage, causing dangerous states and insulation stress.

Fig. 3 illustrates the vector diagram for the solid (0 Ω) single-phase ground fault in phase u. In this case, the phase u is directly connected to the potential of the ground point and its line to ground voltage is almost zero ($V_u \rightarrow 0$). Phase v and w reaches the line-line voltage and their vectors V_v, V_w make an angle of 60 degrees. In the same time, the supply transformer voltage of the faulty phase (V_{un}) becomes a source of the zero-sequence component (V_0) in the system. This voltage V_0 is in opposite to the transformer voltage V_{un} and it causes the zero-sequence currents in the grid, drawn mainly by the parasitic line to ground capacitance (reactive component) and by the leakage resistance as well (active component).

In Fig. 3, the currents I_{C1}, I_{C4} are neglected ($V_u \rightarrow 0$) and the total capacitance current I_C is a sum of I_{C2}, I_{C3}, I_{C5}, and I_{C6}, caused by voltage V_v and V_w. The resulting I_C leads the zero-sequence voltage V_0 by 90 degrees and it builds the dominant component of the fault current (I_{fault}) when uncompensated. This reactive component can be compensated by passive compensation devices such as arc suppression coils. However, on the other hand, a complete compensation is usually not possible, since the coils are often slightly under tuned (by approx. -6%) due to the resonance-related phenomena discussed above.

Active fault compensation

Besides the dominant reactive component, the active (minor) component of the fault current has to be taken into account. It is caused by leakage resistance of the feeders and the resulting current vector I_G is aligned with the zero-sequence voltage vector V_0 (see Fig. 3). Hence, this component cannot be compensated by passive means like ASC, but an active current source is needed to prevent phenomena related to this residual fault current.

Unlike traditional passive means, the proposed active compensator based on the power semiconductor converter enables to generate besides the inductive current also the desirable active component addressing the residual current problem, and thus, it enables the full compensation of the fault current. This is also illustrated in Fig. 3. Whereas the resistance of an ASC even increases the residual fault current (see vector I$_{asc}$ in Fig. 3), it can be seen that the compensation current vector I$_{cmp}$ lags the voltage V_0 by an angle > 90°. Moreover, the complete reactive current component can be compensated without resonance-related issues related to traditional arc suppression coils.

Experimental results

Experimental results have been taken on the compensator prototype in our high-voltage hall laboratory. An emulated distribution power grid with two feeders has been built. Each feeder supplies high voltage capacitors of 3x 3µF and its zero-sequence current component is measured by a current sensor. A distribution transformer is emulated by 3Φ power supply with phase-phase voltage of 10.5 kV/50 Hz. In case of solid single-phase ground faults (0 Ω), the zero-sequence voltage V_0 of 6 kV causes a fault current of approx. 34 A which is compensated by the proposed active compensator. As shown below, the active source enables to significantly speed up the compensation. Fig. 4 and 5 show experimental results of continuous (Fig. 4) and arcing (Fig. 5) resistive ground faults emulated by manual high voltage switch and power resistor of 28 Ω. Fig. 6 illustrates the emulated distribution grid of 10 kV in the hall laboratory with the manually emulated ground faults.

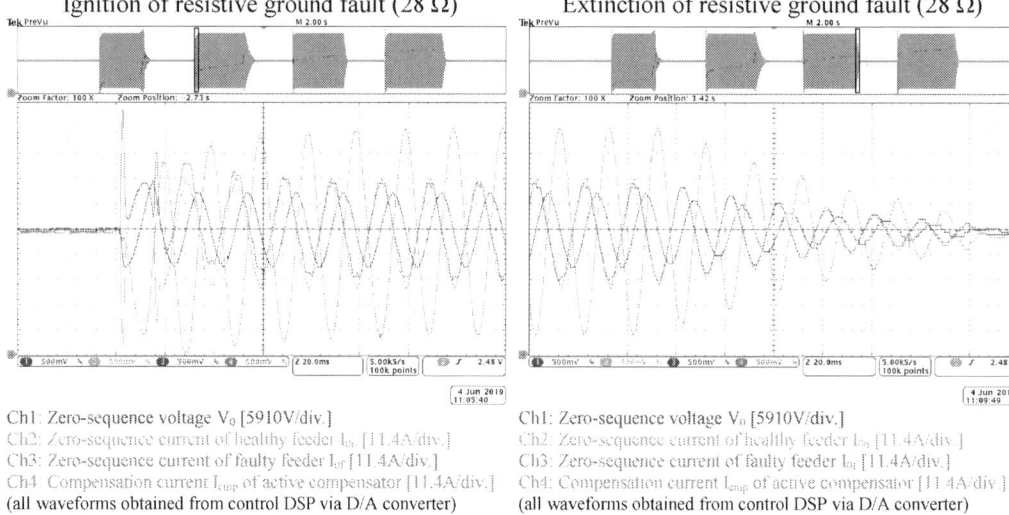

Ch1: Zero-sequence voltage V_0 [5910V/div.]
Ch2: Zero-sequence current of healthy feeder I_{0h} [11.4A/div.]
Ch3: Zero-sequence current of faulty feeder I_{0f} [11.4A/div.]
Ch4: Compensation current I_{cmp} of active compensator [11.4A/div.]
(all waveforms obtained from control DSP via D/A converter)

Ch1: Zero-sequence voltage V_0 [5910V/div.]
Ch2: Zero-sequence current of healthy feeder I_{0h} [11.4A/div.]
Ch3: Zero-sequence current of faulty feeder I_{0f} [11.4A/div.]
Ch4: Compensation current I_{cmp} of active compensator [11.4A/div.]
(all waveforms obtained from control DSP via D/A converter)

Fig. 4: Experimental results - single-phase resistive continuous ground fault compensated by active compensator

As shown in Fig. 4 (top), the resistive ground fault is emulated repeatedly. In the same time, dynamics of the active compensator is very high; the full compensation current is reached within a single period (20 ms). When uncompensated, the current of the faulty feeder is opposite to the current of the healthy one, as can be seen in Fig 4 (left), at the very beginning of the fault.

After a short transient, the currents of both the faulty and the healthy feeder are aligned, i.e. the fault is compensated and the faulty feeder behaves like the healthy one (capacitive load, the current leads the zero-sequence voltage by approx. 90 degrees). Thus, no current flows via the fault connection. After fault extinction (Fig. 4 right), both the zero-sequence voltage as well as the compensation current drops to zero in some few periods.

Fig. 5 shows challenging compensation of arcing ground fault, requiring high dynamics of the compensator.

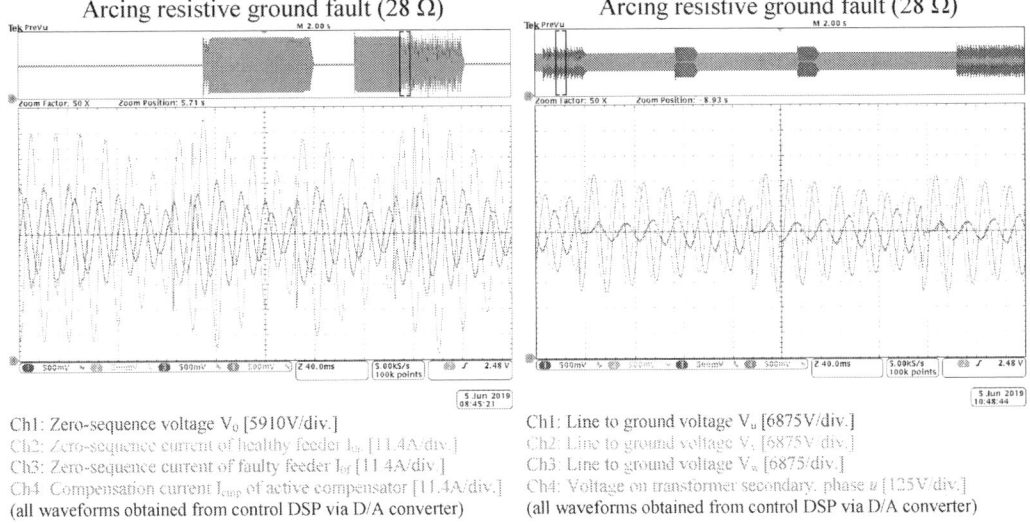

Ch1: Zero-sequence voltage V_o [5910V/div.]
Ch2: Zero-sequence current of healthy feeder I_{oh} [11.4A/div.]
Ch3: Zero-sequence current of faulty feeder I_{of} [11.4A/div.]
Ch4: Compensation current I_{comp} of active compensator [11.4A/div.]
(all waveforms obtained from control DSP via D/A converter)

Ch1: Line to ground voltage V_u [6875V/div.]
Ch2: Line to ground voltage V_v [6875V/div.]
Ch3: Line to ground voltage V_w [6875V/div.]
Ch4: Voltage on transformer secondary, phase u [125V/div.]
(all waveforms obtained from control DSP via D/A converter)

Fig. 5: Experimental results - single-phase resistive arcing ground fault compensated by active compensator

As illustrated in Fig. 5, compensation of the arcing ground fault is a challenging problem requiring high dynamics of the active current source. It can be seen that in the instant of a new arc ignition, the currents of the faulty and the healthy feeder are opposite for a very short period (approx. 2-3 ms). Thanks to the competitive dynamic behavior, the fault compensation is achieved immediately and the arc is extinct until a new arc ignition occurs.

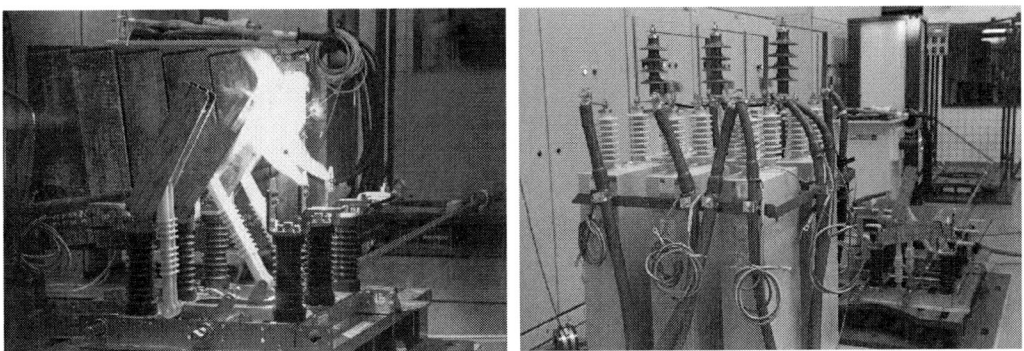

Fig. 6: Experimental measurements in emulated distribution power grid of 10 kV: a. emulated arcing ground fault on manual high voltage switch, b. high voltage capacitors emulating parasitic capacitance of feeder

Conclusion

Active compensation of ground faults in isolated and ineffectively grounded distribution power grids is a promising solution which outperforms traditional passive means. An active ground fault compensator based on a power semiconductor converter has been proposed and built. In order to verify theoretical assumptions, an emulated distribution grid of 10 kV with parasitic capacities has been built and experimental tests have been performed using a manually emulated single-phase ground faults. Experimental results have proven that the technology provides high dynamics and enables complete compensation of the fault current; besides reactive current component, the leakage resistance current is also compensated. In addition, unlike traditional arc suppression coils, the compensator does not suffer from undesired resonance-relevant phenomena and time-consuming tuning procedure.

References

[1]   L. Fickert, G. Achleitner, E. Schmautzer, C. Obkircher, and C. Raunig, "Resonant grounded grids - Quo vadis!?," *CIRED 2009 - 20th International Conference and Exhibition on Electricity Distribution - Part 1*, pp. 1-4, 2009, doi: 10.1049/cp.2009.0672.

[2]   A. M. Dán and D. Raisz, "Towards a more reliable operation of compensated networks in case of single phase to ground faults," *Proceedings of 14th International Conference on Harmonics and Quality of Power - ICHQP 2010*, pp. 1-4, 2010, doi: 10.1109/ICHQP.2010.5625344.

[3]   P. Smirnov, T. Ellinger, and S. Kharitonov, "Compensation of the high-frequency ground fault current components in medium voltage grids," *2019 21st European Conference on Power Electronics and Applications (EPE '19 ECCE Europe)*, pp. P.1-P.9, 2019, doi: 10.23919/EPE.2019.8915047.

[4]   G. Parise, F. M. Gatta, and S. Lauria, "Common grounding system," *IEEE Systems Technical Conference on Industrial and Commercial Power 2005.*, pp. 184-190, 2005, doi: 10.1109/ICPS.2005.1436374.

[5]   Y. Xiuyong and X. Liye, "Faulty Feeder Identification Method in Resonant Grounding Distribution Network," *2019 IEEE Innovative Smart Grid Technologies - Asia (ISGT Asia)*, pp. 391-395, 2019, doi: 10.1109/ISGT-Asia.2019.8881282.

[6]   K. M. Winter, "The RCC Ground Fault Neutralizer — A novel scheme for fast earth-fault protection," *CIRED 2005 - 18th International Conference and Exhibition on Electricity Distribution*, pp. 1-4, 2005.

[7]   K. Winter, "The RCC Ground Fault Neutralizer–a Novel Scheme for Pre-and Post-Fault Protection," presented at the AUPEC, MELBOURNE, AUSTRALIA, 10-13 DECEMBER 2006, 2006.

[8]   Z. Peroutka and I. Matuljak, "The apparatus compensating ground currents connected to phase conductors of a distribution system," Patent No. EP2599180A1 (WO2012013166A1), 2013.

Modeling of a Power Transformer including Higher Order Resonances

Lukas Reißenweber, Alexander Stadler
COBURG UNIVERSITY OF APPLIED SCIENCES AND ARTS
Friedrich-Streib-Str. 2
Coburg, Germany
E-Mail: lukas.reissenweber@hs-coburg.de
URL: http://www.coburg-university.de

Keywords

«Passive component», «Transformer», «Modelling», «Simulation», «Impedance measurement»

Abstract

A calculation method for the transient analysis of transformers is presented. The model is based on a detailed lumped element circuit, including frequency dependent material parameters. Using the example of a 300 W power transformer, it is demonstrated that the model shows good agreement to measurements of the complex impedance up to 100 MHz. Thereby, any load can be considered on the output side. Furthermore, the model allows a precise calculation of the transient current waveshape as well as voltage distribution for any input signal. This is demonstrated by practical measurements at 300 kHz switching frequency.

Introduction

Power transformers are used in a wide range of applications. Increasing switching frequencies enable a reduction of the component size (volume) and its cost. To realize higher switching frequencies and to minimize the losses of semiconductors, the rise and fall times have to be reduced in praxis. At the same time, the trend is towards higher voltage levels. This poses new challenges for inductive components. The reduced component size leads to smaller insulation distances which have to withstand the occurring electric field strengths. When applying high dv/dt values, the transient voltage distribution in the windings is determined by the parasitic stray capacitances. Nonlinear effects can occur [1-5] which lead to higher stress in the insulation and in worst case to a breakdown of the insulation. Depending on the rise time and the waveform of the voltage signal, partial discharges can occur [6], [7]. A calculation method based on a lumped element circuit [8-14] is presented, which enables to calculate the transient voltage distribution in the winding body. Therefore, the complex impedance up to the high MHz range is modeled by simple matrix calculus. The transient voltage distribution can be calculated by transferring the applied input voltage into frequency-domain (FFT). The result can then be traced back to time-domain.

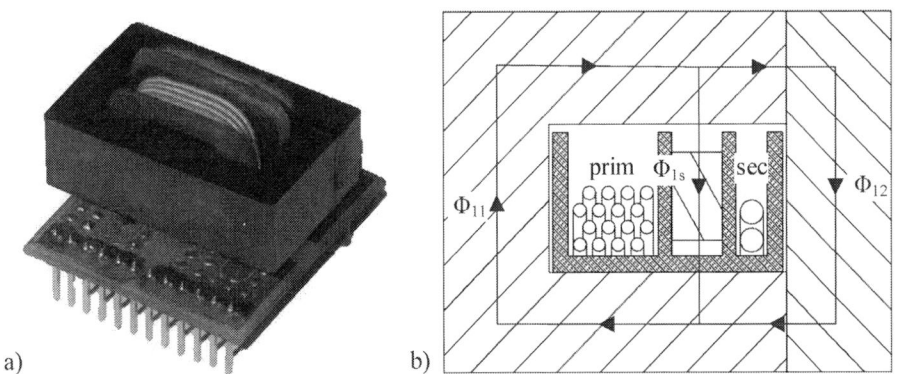

Fig. 1: a) Prototype of the power transformer b) cross-section with magnetic flux-paths

Power Transformer

Fig. 1 a) shows the investigated 300 W power transformer which is based on N87 ferrite core E36/18/11 and yoke I36/6/11 with N = 16 primary and N = 2 secondary windings. The switching frequency of the application is 300 kHz. To minimize the skin and proximity losses, insulated HF litz wire is used for the primary (180 x 0.05 mm) and the secondary (540 x 0.05 mm) as well. To obtain the leakage inductance required for the application, a ferrite element is inserted between the primary and secondary winding as shown in fig. 1 b). This leads to a small interwinding capacitance.

The lumped element circuit for transient analysis

The transient modeling of the transformer is based on a lumped element circuit (PI model), including all magnetic and capacitive couplings. On the one hand, a suitable number of elements for the description of the component has to be selected in order to obtain a sufficient accuracy of the calculated quantities. On the other hand, the calculation effort has to be kept as low as possible. From previous work [13, 14] it is known, that a description of each single turn by a node at the beginning and a node at the end provides valid results. Fig. 2 depicts the lumped elements for a transformer with three primary and three secondary windings. R_{pi}, R_{si} represent the copper losses and the core losses of the selfinductances L_{pi}, L_{si}. M_{ppij}, M_{ssij} denote the mutual inductances between the turns of a particular winding. The related magnetic losses R_{ppij}, R_{ssij} are not shown in the figure. C_{pi}, C_{si} are the capacitances between one and the following turn. The dielectric losses are modeled by the resistors R_{Cpi}, R_{Csi}. The capacitive coupling of each node of a side C_{pij}, C_{sij} and related dielectric losses R_{Cpij}, R_{Csij} e.g. node p1 and node p3 are not shown here. On the secondary side any load can be considered which is modeled by the complex impedance Z_{load}

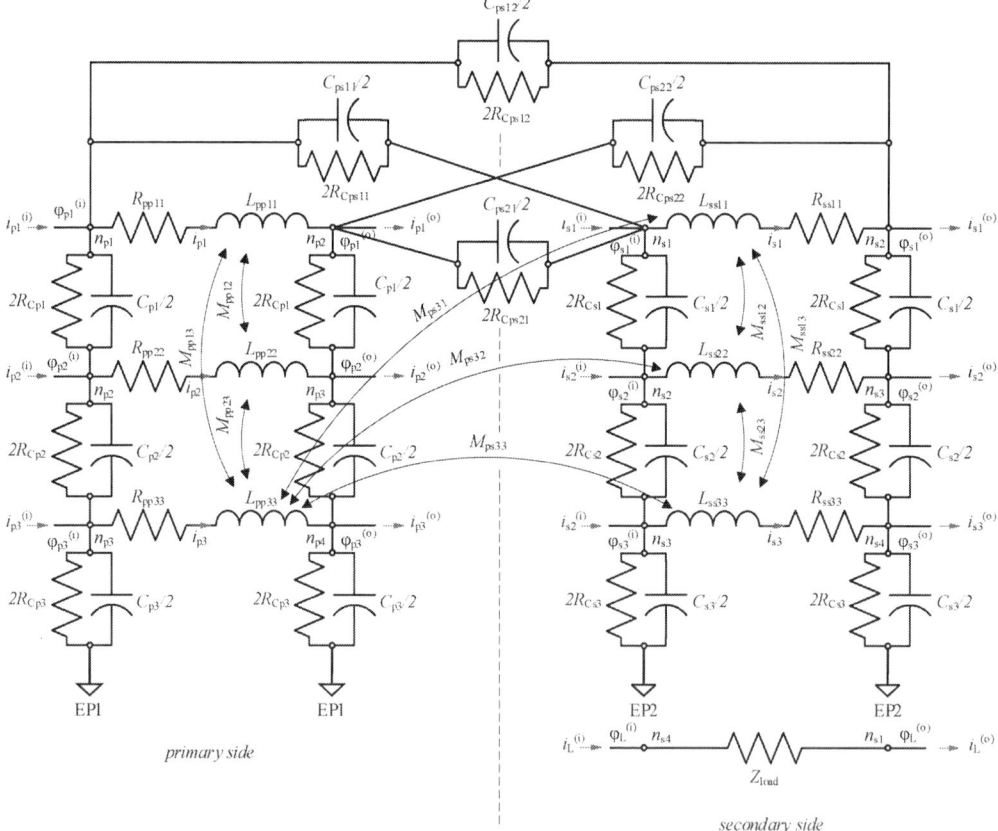

Fig. 2: Lumped element circuit for a transformer with three primary windings and three secondary windings (main elements with indicated coupling inductances and capacitances)

The primary and secondary side are connected by capacitive and inductive coupling. The capacitive coupling is described by the elements C_{psij}. In the picture, this is illustrated by the example of the coupling of the nodes p1, p2 and s1, s2. The dielectric losses R_{psij} are in parallel to the capacitances. The magnetic coupling by the mutual inductances M_{psij} is shown for the selfinductance L_{pp33}. The core losses R_{psij} are not depicted. The input signal is connected between node p1 and p4, the complex load between node s1 and s4. The system is solved by simple matrix calculation using the currents of the nodes and the voltages between the nodes which are shown in table 1, where matrices are underscored. The correlation of equal voltages and equal currents corresponds to the node points in the network. The capacitance matrix $\underline{K}$ describes the capacitive coupling and is used for calculation. The shown partial capacitances $\underline{C}$ can be calculated by the matrix $\underline{K}$, however, they are not necessary for the modeling.

Table I: Equations

	Left nodes	Intermediate nodes	Right nodes
Transformer elements	$\underline{I} = \underline{I}^{(i)} - \dfrac{\underline{R}_K - j\dfrac{1}{\omega \underline{K}}}{-j\dfrac{2\underline{R}_K}{\omega \underline{K}}}\underline{\varphi}^{(i)}$	$\underline{\varphi}^{(i)} - \underline{\varphi}^{(o)}$ $= (\underline{R} + j\omega\underline{L})\underline{I}$	$\underline{I} = \underline{I}^{(o)} - \dfrac{\underline{R}_K - j\dfrac{1}{\omega \underline{K}}}{-j\dfrac{2\underline{R}_K}{\omega \underline{K}}}\underline{\varphi}^{(o)}$
Load	$I_L = I_L^{(i)}$	$\varphi_L^{(i)} - \varphi_L^{(o)} = Z_L I_L$	$I_L^{(o)} = I_L$

Calculation of the lumped elements

For the determination of the lumped element matrices, analytical calculations and 3D finite element method (FEM) simulations are performed. The copper losses, the inductances with the associated core losses and the capacitances with the associated dielectric losses are considered.

Copper losses

The frequency-dependent resistance of each winding is calculated by its power loss [15]:

$$R_{AC} = \frac{P_{v,\text{winding}}}{I_{\text{rms}}^2} \tag{1}$$

The power loss includes ohmic (rms), skin and proximity losses:

$$P_{v,\text{winding}} = P_{\text{rms}} + P_{\text{skin}} + P_{\text{prox}} \tag{2}$$

The ohmic and skin losses depend on the DC resistance and the skin factor F_s:

$$P_{\text{rms,skin}} = I_{\text{rms}}^2 R_0 F_s \qquad \text{with} \qquad F_s = \frac{1}{2}Re\left\{\frac{\alpha a_z I_0(\alpha a_z)}{I_1(\alpha a_z)}\right\} \tag{3}$$

Where α is the skin constant and a_z is the radius of a strand. I_0 and I_1 are the modified bessel functions zeroth and first order. The proximity losses are calculated by the square of the peak value of the external magnetic field, the number of strands N_s and the proximity factor Ds:

$$P_{\text{prox}} = \frac{l}{\kappa} N_s \widehat{H}_{\text{ex}}^2 D_s \qquad \text{with} \qquad D_s = 2\pi Re\left\{\frac{\alpha a_z I_1(\alpha a_z)}{I_0(\alpha a_z)}\right\} \tag{4}$$

In order to calculate the external magnetic field strength for each winding, a magnetostatic field simulation is performed. The square value of the magnitude of the magnetic field in each winding cross-section is integrated and normalized to its volume. For a transformer with N turns (turns of primary and secondary side), the result is a Nx1 matrix. These values will be added to the main diagonal of the core loss matrix.

Inductance and core loss matrix

For the inductance and core loss matrix, it is necessary to perform frequency depending magnetic simulations to consider the increase of the leakage inductance due to the decreasing permeability in the higher frequency range and the associated effect on the core losses. Therefore, the real and imaginary parts of the complex permeability are taken from the datasheet and fitted for the simulated frequency range. For the lower frequencies, the curves are extrapolated linearly. For the higher frequencies, modified Debye-functions are used. For the modeling of the complex permittivity of insulation materials the Debye-functions are common [16]. The real part of the complex permeability at higher frequencies is fitted by:

$$\mu'(\omega) = \mu_\infty + \frac{\mu'_s - \mu_\infty}{1 + \omega^m \tau'^m} \qquad \text{with} \qquad \tau' = \frac{1}{2\pi f'} \tag{5}$$

Where μ_∞ is the complex permeability at infinite, so it is set to 1. μ'_s describes the starting value for the real part of the permeability in the case of N87 it is 2100. The exponent $m = 12$ and the frequency $f' = 1.8$ MHz are chosen so that the so modified function fits the datasheet values. The imaginary part of the complex impedance follows:

$$\mu''(\omega) = \frac{\mu''_s - \mu_\infty}{1 + \omega^n \tau''^n} \, \omega^o \tau''^o \qquad \text{with} \qquad \tau'' = \frac{1}{2\pi f''} \tag{6}$$

For the imaginary part of the complex permeability μ''_s is set to 4000. The exponents $n = 4.5$, $o = 3$ and the frequency $f'' = 1.4$ MHz lead to a good approximation of the datasheet values at higher frequencies. Fig. 3 depicts the fitted curves of the real part and the imaginary of the permeability, the used modified Debye-functions and the measurement points from datasheet.

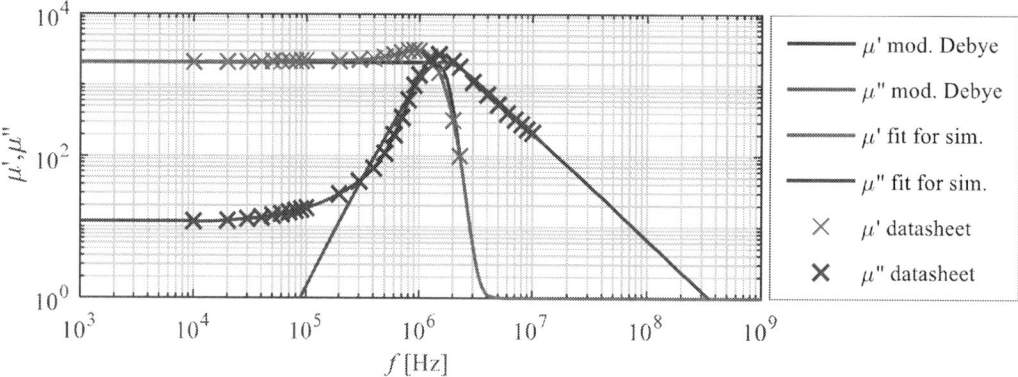

Fig. 3: Real and imaginary parts of the complex permeability of N87

The inductance matrix [15, 17] is calculated by the complex magnetic flux:

$$\begin{bmatrix} \text{real}(\underline{\Phi}_1) \\ \text{real}(\underline{\Phi}_2) \\ \text{real}(\underline{\Phi}_3) \end{bmatrix} = \begin{bmatrix} L_{11} & M_{12} & M_{13} \\ M_{21} & L_{22} & M_{23} \\ M_{31} & M_{32} & L_{33} \end{bmatrix} \cdot \begin{bmatrix} I_1 \\ I_2 \\ I_3 \end{bmatrix} \tag{7}$$

The core loss matrix is taken form the imaginary part of the magnetic flux:

$$\begin{bmatrix} \text{imag}(\underline{\Phi}_1) \\ \text{imag}(\underline{\Phi}_2) \\ \text{imag}(\underline{\Phi}_3) \end{bmatrix} = \begin{bmatrix} R_{\text{core}11} & R_{\text{core}12} & R_{\text{core}13} \\ R_{\text{core}21} & R_{\text{core}22} & R_{\text{core}23} \\ R_{\text{core}31} & R_{\text{core}32} & R_{\text{core}33} \end{bmatrix} \cdot \begin{bmatrix} I_1 \\ I_2 \\ I_3 \end{bmatrix} \tag{8}$$

In fig. 4, several simulated coupling coefficients k are shown as a function of the frequency. All coefficients decrease above 3 MHz. This leads to an increase of stray inductances.

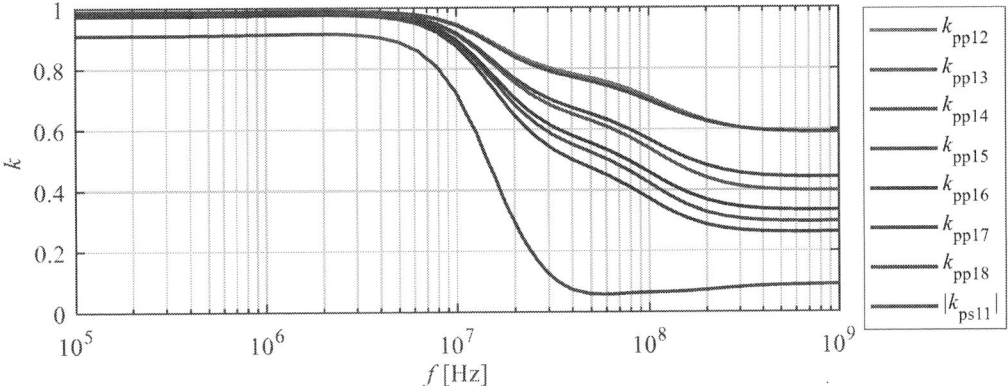

Fig. 4: Several coupling coefficients k as a function of the frequency

The exact knowledge of the leakage inductances is necessary, since the inductive behavior of a transformer at smaller loads and in case of short circuit depends completely on the leakage inductances. Regarding the increasing stray inductance at higher frequencies, the simulations show that the coupling coefficient k between the mutual inductance M_{pp12} and the selfinductances L_{pp11} and L_{pp22} decreases by 28.7 % from 0.986 at 100 Hz to 0.703 at 100 MHz. This can also be seen in the coupling of the primary and secondary turns. The ratio between the mutual inductance M_{ps11} and the selfinductance L_{pp11} and L_{ss11} decreases by 92.5 % from -0.849 at 100 Hz to -0.064 at 100 MHz.

Capacitance and dielectric loss matrix

The capacitance matrix [18-22] is calculated via electrostatic simulation:

$$\begin{bmatrix} Q_1 \\ Q_2 \\ Q_3 \end{bmatrix} = \begin{bmatrix} K_{11} & K_{12} & K_{13} \\ K_{21} & K_{22} & K_{23} \\ K_{31} & K_{32} & K_{33} \end{bmatrix} \cdot \begin{bmatrix} \varphi_1 \\ \varphi_2 \\ \varphi_3 \end{bmatrix} \tag{9}$$

For the dielectric loss matrix, the capacitance matrix is multiplied by dielectric loss tangent $\tan\delta_K$:

$$\underline{R}_K = \underline{K} \cdot \tan\delta_K \tag{10}$$

The exact modelling of the litz wire in the FEM simulation leads to a high computational effort. Therefore, the litz wire is replaced by a solid wire with an equivalent outer diameter. The rel. permittivity of the insulation is set to 3 and a $\tan\delta_K$ of $25 \cdot 10^{-3}$ is used.

Complex impedance with different loads

For the validation of the model, measurements and simulations are compared against each other for different loads. The complex impedance is analyzed for both, primary and secondary feed-in. The measurements are performed in the range of 100 Hz to 100 MHz.

Ohmic loads

Fig. 5 a) shows the absolute value and fig. 5 b) the argument of the complex impedance for primary feed-in and different ohmic loads connected on the secondary side. The curves of simulation and measurement show a good agreement including the damping at the resonance point. For the simulation of the short circuit (sc) at the secondary side, the ohmic resistance was chosen to 1 nΩ. For the open circuit (oc), the resistance value was set to 1 GΩ. Fig. 6 a) shows the absolute value and fig. 6 b) the argument of the complex impedance for secondary feed-in and different ohmic loads connected to the primary side. For the open circuit at the primary side, a second resonance point occurs which is also considered by the simulation.

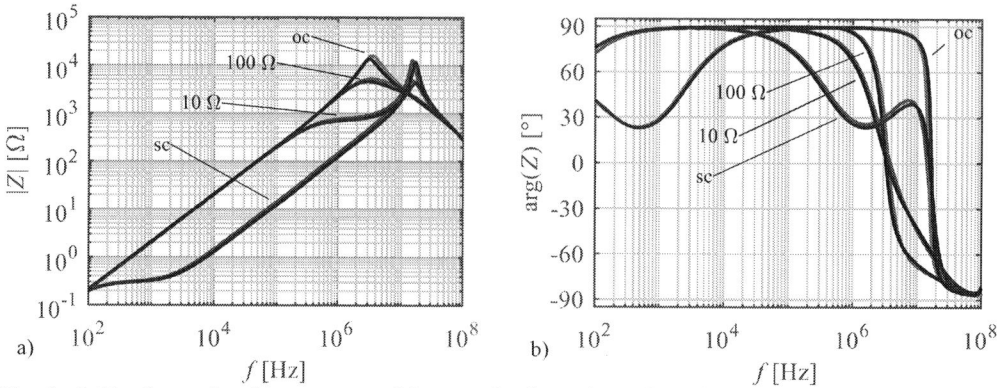

Fig. 5: a) Absolute value, b) argument of the complex impedance for primary feed-in and different ohmic loads

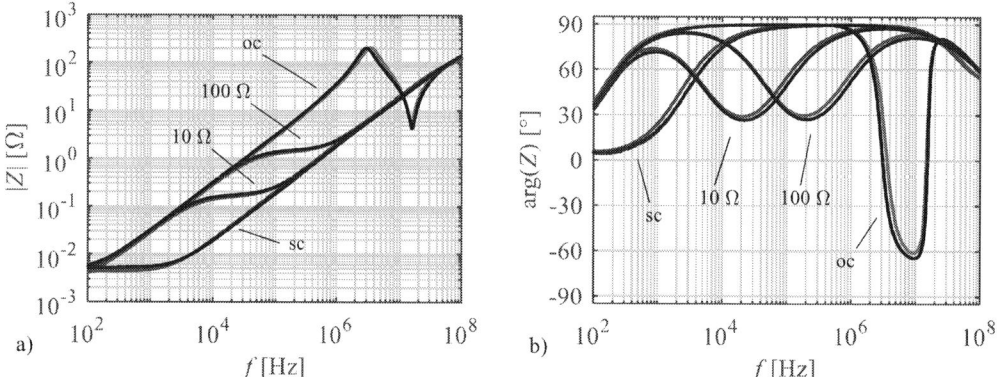

Fig. 6: a) Absolute value, b) argument of the complex impedance for secondary feed-in and different ohmic loads

Capacitive loads

The capacitive load is modeled as a frequency-dependent complex resistance to consider its parasitic elements. The values are taken from measurement. Fig. 7 a) shows the absolute value and fig. 7 b) the argument of the complex impedance for primary feed-in and different complex capacitive loads connected to the secondary side. Higher capacitance values lead to lower resonance points. The simulation models the first three resonance points. The damping is also considered.

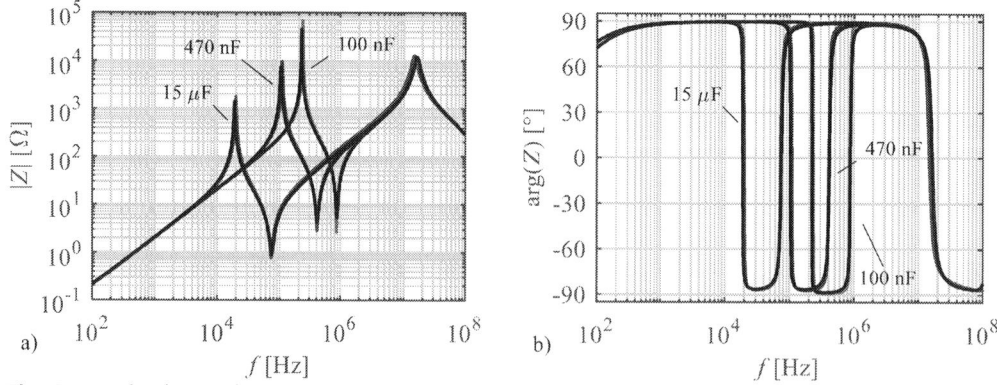

Fig. 7: a) Absolute value, b) argument of the complex impedance for primary feed-in and different capacitive loads

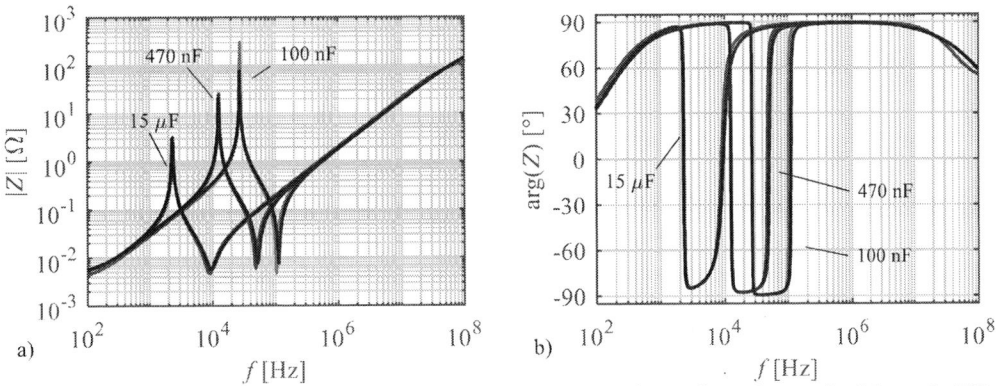

Fig. 8: a) Absolute value, b) argument of the complex impedance for primary feed-in and different capacitive loads

Measurements

The modeled complex impedance can be used to calculate the currents and potentials at each node for any applied voltage signal. A detailed representation of the accuracy for calculating the transient potential distribution at each node with comparison to measurements is shown in [13]. In [14], a transient analysis for rectangular voltage signals with different rise times is presented. At this point, only the global values, the terminal voltages and the winding currents, are examined. The measured square-wave voltage with an amplitude of 250 V at a switching frequency of 300 kHz is depicted in fig. 9 a) The voltage is applied on the primary side. On the secondary side a 50 Ω resistor is connected. For the simulation the measured signal is shortened to one period and transferred into frequency domain by fast Fourier transform (FFT). After the calculation the result is retransferred to time domain. In fig. 9 b) the simulated current is compared to the measurement. The simulation reproduces the measurement well.

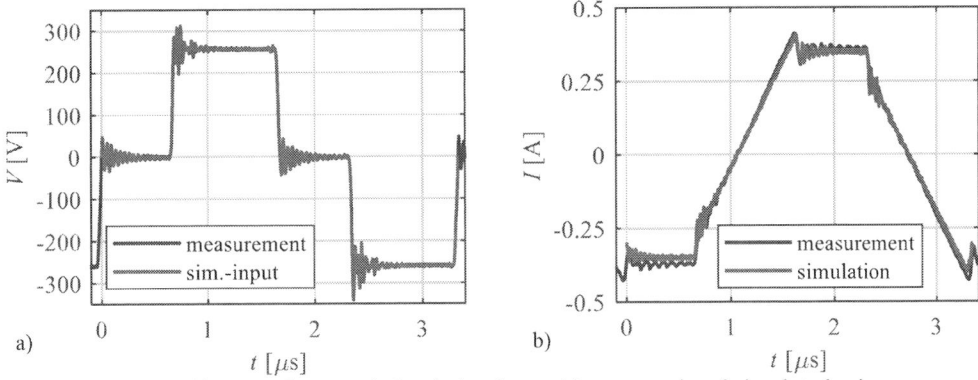

Fig. 9: a) Measured input voltage and simulation input, b) measured und simulated primary current

In Fig. 10 a) the measured and simulated voltage and in fig. 10 b) the measured and simulated current of the secondary side are compared. Overall a good agreement is achieved. Fig. 11 a) depicts a magnification of the measured and simulated voltage and fig. 11 b) of the measured and simulated current. The simulation represents the oscillation ratio in amplitude and frequency.

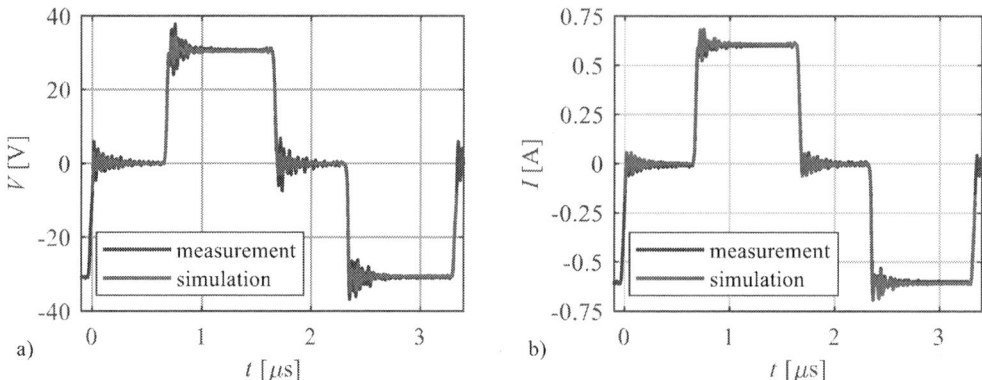

Fig. 10: Measured and simulated a) secondary voltage, b) secondary current

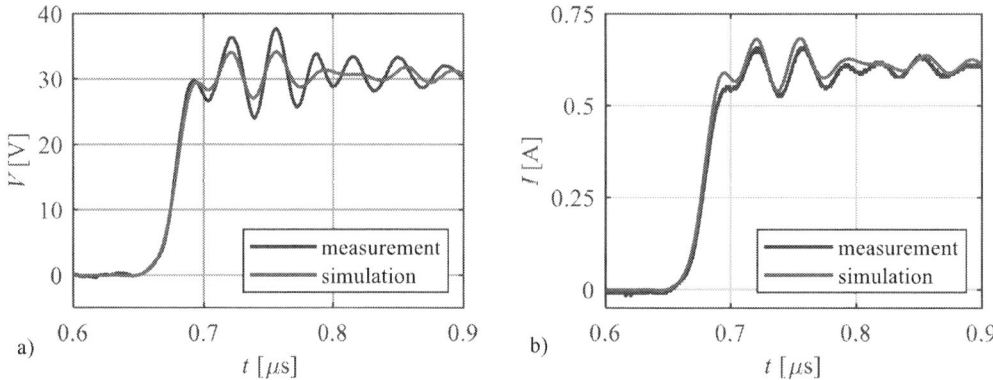

Fig. 11: Magnification of measured and simulated a) secondary voltage, b) secondary current

Conclusion

The presented modeling of a power transformer shows good agreement between the calculated and measured complex impedance up to 100 MHz for arbitrary load. All resonance points are considered as well as the damping at these points. The model parameters are determined in magnetic and electrostatic FEM simulations with datasheet values for the complex permittivity (lower and higher frequencies are extrapolated). By simulating the frequency-dependent magnetic coupling, the two borderline cases transformer open as well as transformer short-circuited at the output can also be modeled. The calculated complex impedance can be used to calculate the voltage distribution and currents at each node. The simulated terminal voltages and winding currents show a good agreement with the measurement for the applied square wave signal of 250 V at 300 kHz.

References

[1] M. Florkowski, J. Furgał. P. Pająk, Analysis of Fast Transient Voltage Distributions in Transformer Windings under Different Insulation Conditions, IEEE Transactions on Dielectrics and Electrical Insulation Vol. 19, No. 6, Dec. 2012, pp. 1991-1998, ISSN: 1070-9878

[2] T. Wang, Z. Wang, Q. Zhang, L. Li, Measurement Method of Transient Overvoltage Distribution in Transformer Windings, Annual Report Conference on Electrical Insulation and Dielectric Phenomena, Oct. 2013, ISBN: 978-1-4799-2597-1

[3] M. Florkowski, J. Furgał, Initial voltage distributions in transformer windings at ultra fast stresses, International Conference on High Voltage Engineering and Application, Oct. 2010, pp. 168-171, ISBN: 978-1-4244-8286-3

[4] P. I. Fergestad, T. Henriksen, Transient Oscillations in Multiwinding Transformers, IEEE Transactions on Power Apparatus and Systems, Vol. 93, No. 2, Mar. 1974, pp. 500-509, ISSN: 0018-9510

[5] P. Gomez, F. P. Espino-Cortes, F. de Leon, Computation of the Dielectric Stresses Produced by PWM Type Waveforms on Medium Voltage Transformer Windings, Annual Conference on Electrical Insulation and Dielectric Phenomena, Oct. 2011, pp. 199-202, ISBN: 978-1-4577-0986-9

[6] P. Wang, N. Yang, C. Zheng, Y. Li, Effect of repetitive impulsive and square wave voltage frequency on partial discharge features, 12th International Conference on the Properties and Applications of Dielectric Materials, May 2018, pp. 152-155, ISBN: 978-1-5386-5788-1

[7] M. Schmidhuber, H. Jungwirth, Partial Discharge of Inductives in a High Frequency Application, PCIM Europe, May 2017, pp. 1274-1277, ISBN: 978-3-8007-4424-4

[8] Ch. Q. Su, Electromagnetic Transients in Transformer and Rotating Machine Windings, IGI Global, 2013, ISBN: 978-1-4666-1921-0

[9] J.P. Bickford, N. Mullineux, J. R. Reed, Computation of Power-Transients, Institution of Electrical Engineers, Revised Edition, 1980, ISBN: 0-906048-35-4

[10] N. R. Watson, J. Arrillaga, Power Systems Electromagnetic Transients Simulation, Institution of Engineering and Technology, 2003, Reprinted, ISBN: 978-0-85296-106-3

[11] A. Greenwood, Electrical Transients in Power Systems, John Wiley & Sons Ltd, Chichester, 1991, ISBN: 0-471-62058-0

[12] Y. Shibuya,S. Fujita, High frequency model of transformer winding, Electrical Engineering in Japan, Vol. 146, No. 3, pp. 8–15, 2004

[13] L. Reissenweber, A. Stadler, Calculation Model for the Transient Voltage Distribution in Inductor Windings Effected by High dv/dt, PCIM Europe, May 2019, pp. 1373-1377, ISBN: 978-3-8007-4938-6

[14] L. Reissenweber, A. Stadler, Extension of a Calculation Model based on the Pi Line Theory for Transient Voltage Distribution in Inductors by Complex Permittivity and Frequency-Depending Complex Permeability, 21st European Conference on Power Electronics and Applications, Sept. 2019, ISBN: 978-9-0758-1531-3

[15] M. Albach, Induktivitäten in der Leistungselektronik, Springer Fachmedien Wiesbaden GmbH, 2017, ISBN: 978-3-658-15080-8

[16] T. Blythe, D. Bloor, Electrical Properties of Polymers, Cambridge University Press, 2005, ISBN-10: 0-521-55219-2

[17] N. R. Watson, J. Arrillaga, Power Systems Electromagnetic Transients Simulation, Institution of Engineering & Technology, 2002, ISBN-10: 3764387742

[18] G. Grandi, M. K. Kazimierczuk, A. Massarini, U. Reggiani, (1996) Stray Capacitances of Single-Layer Air-Core Inductors for High-Frequency Applications, 31st IAS Annual Industry Applications Conference, vol. 3, Oct. 1996, pp. 1384-1388, ISBN: 0-7803-3544-9

[19] J. A. Martinez-Velasco, Power System Transients: Parameter Determination, CRC Press Inc., New Edition, 2009, ISBN-10: 1420065297

[20] J. Biela, J. W. Kolar, (2008) Using Transformer Parasitics for Resonant Converters − A Review of the Calculation of the Stray Capacitance of Transformers IEEE Transactions on Industry Applications, vol. 44, no. 1, Feb. 2008, pp. 223-233

[21] J P. Gómez, F. de León, (2011) Accurate and Efficient Computation of the Inductance Matrix of Transformer Windings for the Simulation of Very Fast Transients IEEE Transactions on Power Delivery, vol. 26, no. 3, 2011, pp. 1423-1431, ISSN: 0093-9994

[22] P. Gomez, F. P. Espino-Cortes, F. de Leon, (2011) Computation of the Dielectric Stresses Produced by PWM Type Waveforms on Medium Voltage Transformer Windings, 2011 Annual Conference on Electrical Insulation and Dielectric Phenomena CEIDP, 2011, pp. 199-202, ISBN: 978-1-4577-0985-2

A Comparison of Two State-Space Models of an Induction Machine Considering Different Sets of Winding Distribution Harmonics

Julien Cordier, Stefan Klass and Ralph Kennel
Chair of Electrical Drive Systems and Power Electronics
Technische Universitaet Muenchen
Arcisstr. 21
Munich, Germany
Phone: +49 (0) 89 289-28415
Fax: +49 (0) 89 289-28336
Email: julien.cordier@tum.de
URL: http://www.eal.ei.tum.de

Acknowledgements

The authors are grateful to the company dSPACE for supporting the research activities presented in this paper.

Keywords

≪induction motor≫, ≪modelling≫, ≪real-time simulation≫, ≪adjustable speed drive≫, ≪harmonics≫

Abstract

The present paper compares two models of a low-power off-the-shelf induction machine. Each of them considers a specific set of winding distribution harmonics. One model takes into account the harmonic orders 1 and 17, while the other includes the orders 1, 5 and 11. These two sets of space harmonics are of particular interest as they give rise to a principal slot harmonic in the stator currents and produce unwanted torque oscillations. The model with space harmonics 1 and 17 has order 6, as does Park's. By contrast, its counterpart with harmonic orders 1, 5 and 11 features two additional states associated with rotor currents and directly linked to the harmonic torque ripple. The potential of each model with respect to drive control applications such as torque harmonic reduction is discussed.

Introduction

Park's modelling approach has been highly popular in drive control applications thanks to its simplicity. However, it does not take into consideration space harmonics which are of importance in applications such as torque harmonic attenuation or sensorless control of three-phase drives. In this context, a model accounting for winding distribution harmonics was derived from the winding function approach (cf. [1]) in [2]. In addition, the general coordinate transformation presented in [3] allows for an optimization of this model, making it promising for real-time applications.

It is well known that different combinations of space harmonics may lead to the same time harmonics in the stator current or torque waveforms (cf. [4, 183]). The method developed in [3] shows that the model order is directly impacted by the space harmonics considered. The present paper aims at discussing possible advantages and drawbacks of additional model states resulting from the choice of space harmonics with respect to control applications.

We will examine this problem by considering the particular case of an off-the-shelf induction machine for which we will derive two models including different sets of winding distribution harmonics. The space harmonics chosen generate a time harmonic in the stator currents in steady state known as principal slot harmonic.

Nomenclature and definitions

Table I lists the symbols used throughout the paper.

Table I: Definition of symbols

Symbol	Meaning
m_s, m_r	Number of stator/rotor electric circuits
N_s, N_r	Number of stator/rotor slots
u, i, ψ	Voltage, current, flux linkage
R, L	Resistance, inductance
ω_r, θ_r	Rotor angular velocity, rotor angle
J_M	Moment of inertia of the rotor
M_M, M_L, C_W	Motor torque, load torque, coefficient of friction
$\mathcal{H}$	Set of space harmonic orders considered in a model ($\mathcal{H} \subset \mathbb{N}$)

Scalar quantities are denoted by means of normal letters, e.g. m_s or ψ_{s0}, while bold fonts are used for matrices, such as $\mathbf{R}$. Vectors are represented using an arrow, for example $\vec{u}$. Subscripts inform about the location of a quantity. For instance, i_{rn} represents the current flowing through rotor circuit number n.

For $m \in \mathbb{N}^\star$, the identity matrix of dimension m is referred to as $\mathbf{I}_m$ and we denote $\mathcal{M}_m(\mathbb{R})$ the set of $m \times m$ matrices with entries in $\mathbb{R}$.

For $(p,q) \in \mathbb{Z}^2$ such that $p \leqslant q$, we denote $[\![p, q]\!] = \{k \in \mathbb{Z} \,|\, p \leqslant k \leqslant q\}$.

We define the following matrices of $\mathcal{M}_2(\mathbb{R})$:

$$\mathbf{J} = \begin{bmatrix} 0 & -1 \\ 1 & 0 \end{bmatrix}; \qquad \mathbf{S} = \begin{bmatrix} 1 & 0 \\ 0 & -1 \end{bmatrix}; \qquad \forall \varphi \in \mathbb{R}, \, \mathbf{T}(\varphi) = \begin{bmatrix} \cos\varphi & -\sin\varphi \\ \sin\varphi & \cos\varphi \end{bmatrix}$$

Modelling methodology

Simplifying assumptions and general approach

The models to be compared in the next sections are derived following the methodology described in [3]. For this reason, we briefly summarize the underlying modelling assumptions here.

The stator windings of the considered induction machine are assumed to be symmetrically distributed in N_s slots. They are modelled as a set of m_s electrical circuits. Depending on the context, a circuit might only consist of a subset of the coils belonging to a winding rather than of the winding itself (cf. [2, 3]). Similarly, a set of m_r electrical circuits is used to model the rotor conductors distributed in N_r slots. In case of squirrel cage machines, we introduce $m_r = N_r$ circuits on the rotor side, each of them being a coil with one turn and a pitch equal to the rotor slot pitch. We presume that all stator circuits (all rotor circuits) have the same resistance R_s (R_r resp.) and the same leakage inductance $L_{\sigma s}$ ($L_{\sigma r}$ resp.).

The current flowing through a stator circuit $m \in [\![0, m_s - 1]\!]$ (a rotor circuit $n \in [\![0, m_r - 1]\!]$) is denoted i_{sm} (i_{rn} resp.). The voltage at the terminals of stator circuit m (rotor circuit n) is u_{sm} (u_{rn} resp.). The respective flux linkages are ψ_{sm} and ψ_{rn}.

We introduce the following vectors:

$$\vec{u}_s = \begin{bmatrix} u_{s0} & \cdots & u_{sm} & \cdots & u_{sm_s-1} \end{bmatrix}^\top \qquad \vec{u}_r = \begin{bmatrix} u_{r0} & \cdots & u_{rn} & \cdots & u_{rm_r-1} \end{bmatrix}^\top$$

$$\vec{i}_s = \begin{bmatrix} i_{s0} & \cdots & i_{sm} & \cdots & i_{sm_s-1} \end{bmatrix}^\top \qquad \vec{i}_r = \begin{bmatrix} i_{r0} & \cdots & i_{rn} & \cdots & i_{rm_r-1} \end{bmatrix}^\top$$

$$\vec{\psi}_s = \begin{bmatrix} \psi_{s0} & \cdots & \psi_{sm} & \cdots & \psi_{sm_s-1} \end{bmatrix}^\top \qquad \vec{\psi}_r = \begin{bmatrix} \psi_{r0} & \cdots & \psi_{rn} & \cdots & \psi_{rm_r-1} \end{bmatrix}^\top$$

In accordance with [2], we assume magnetic linearity and consider magnetic phenomena in a machine cross-section. As described in detail in [5], we assign a conductor distribution and flux density distribution function to each electrical circuit present in the model. Since these functions are 2π-periodic with respect to the air-gap coordinates, they can be represented as Fourier series. In addition, the inductances associated with each space harmonic can be determined efficiently using Parseval's identity [2]. In practice, only a finite set $\mathcal{H}$ of space harmonics can be taken into consideration.

As pointed out in [2, 3, 6], the circuits in the model must be interconnected in order to account for the actual arrangement of electric conductors in the real machine (e.g. a star connection on the stator side). Hence, the currents flowing through them and the voltages at their terminals are not independent variables. We therefore introduce the vectors $\vec{u}'_s$ and $\vec{i}'_s$ ($\vec{u}'_r$ and $\vec{i}'_r$), the components of which represent independent voltages and currents in the network of interconnected stator circuits (rotor circuits resp.). The previous vectors are related by means of stator and rotor interconnection matrices, $\mathbf{C}_s$ and $\mathbf{C}_r$:

$$\vec{i}_s = \mathbf{C}_s \vec{i}'_s; \qquad \vec{i}_r = \mathbf{C}_r \vec{i}'_r; \qquad \vec{u}'_s = \mathbf{C}_s^\top \vec{u}_s; \qquad \vec{u}'_r = \mathbf{C}_r^\top \vec{u}_r \tag{1}$$

Coordinate transformation

The stator windings being considered symmetrically distributed, the electric circuits in the model can be chosen such that the conductor distribution function of stator circuit m is an exact copy of the one of circuit 0 shifted by the angle $2\pi/m_s$. This ensures that the main inductance matrix which links stator currents to stator main fluxes, $\widetilde{\mathbf{L}}_\mathbf{s}$, is *circulant*, i.e. the coefficients in each column of $\widetilde{\mathbf{L}}_\mathbf{s}$ are obtained by shifting the entries in the previous column one position to the bottom with wraparound to the top [3]. The same property holds for the rotor main inductance matrix, $\widetilde{\mathbf{L}}_\mathbf{r}$, as the conductor distribution function of rotor circuit n is the same as the one of rotor circuit 0 shifted by the angle $2\pi/m_r$.

Owing to these considerations, $\widetilde{\mathbf{L}}_\mathbf{s}$ ($\widetilde{\mathbf{L}}_\mathbf{r}$) can be *diagonalized* using a Fourier matrix of order m_s (m_r resp.) [7]. The Fourier matrix of order m_s is defined as follows:

$$\mathcal{W}_{\mathbf{m_s}} = \frac{1}{\sqrt{m_s}} \begin{bmatrix} 1 & \cdots & \cdots & \cdots & 1 \\ 1 & \cdots & W_{m_s}^m & \cdots & W_{m_s}^{m_s-1} \\ \vdots & \vdots & \vdots & \vdots & \vdots \\ 1 & \cdots & W_{m_s}^{mk} & \cdots & W_{m_s}^{(m_s-1)k} \\ \vdots & \vdots & \vdots & \vdots & \vdots \\ 1 & \cdots & W_{m_s}^{m(m_s-1)} & \cdots & W_{m_s}^{(m_s-1)^2} \end{bmatrix} \quad \text{with} \quad W_{m_s} = e^{-j\frac{2\pi}{m_s}} \tag{2}$$

$\mathcal{W}_{\mathbf{m_s}}$ allows us to introduce transformed stator voltage, current and flux linkage vectors, $\underline{u}_s^\natural$, $\underline{i}_s^\natural$ and $\underline{\psi}_s^\natural$. Similarly, the corresponding vectors on the rotor side are obtained using $\mathcal{W}_{\mathbf{m_r}}$:

$$\underline{u}_s^\natural = \mathcal{W}_{\mathbf{m_s}}^{-1} \vec{u}_s \qquad \qquad \underline{i}_s^\natural = \mathcal{W}_{\mathbf{m_s}}^{-1} \vec{i}_s \qquad \qquad \underline{\psi}_s^\natural = \mathcal{W}_{\mathbf{m_s}}^{-1} \vec{\psi}_s$$

$$\underline{u}_r^\natural = \mathcal{W}_{\mathbf{m_r}}^{-1} \vec{u}_r \qquad \qquad \underline{i}_r^\natural = \mathcal{W}_{\mathbf{m_r}}^{-1} \vec{i}_r \qquad \qquad \underline{\psi}_r^\natural = \mathcal{W}_{\mathbf{m_r}}^{-1} \vec{\psi}_r$$

The transformed stator current vector $\underline{i}_s^\natural$ (rotor current vector $\underline{i}_r^\natural$) is found from $\vec{i}'_s$ ($\vec{i}'_r$ resp.) by combining the above definitions with the interconnection relations given in (1):

$$\underline{i}_s^\natural = \mathcal{W}_{\mathbf{m_s}}^{-1} \mathbf{C}_s \vec{i}'_s = \left[\overline{\mathcal{W}_{\mathbf{m_s}}}^\top \mathbf{C}_s \right] \vec{i}'_s = \mathcal{Z}_\mathbf{s} \vec{i}'_s; \qquad \underline{i}_r^\natural = \mathcal{W}_{\mathbf{m_r}}^{-1} \mathbf{C}_r \vec{i}'_r = \left[\overline{\mathcal{W}_{\mathbf{m_r}}}^\top \mathbf{C}_r \right] \vec{i}'_r = \mathcal{Z}_\mathbf{r} \vec{i}'_r \tag{3}$$

In the same manner, the knowledge of the transformed voltage vectors $\underline{u}_s^\natural$ and $\underline{u}_r^\natural$ allows us to work out the expression of $\vec{u}'_s$ as well as $\vec{u}'_r$:

$$\vec{u}'_s = \mathbf{C}_s^\top \mathcal{W}_{\mathbf{m_s}} \underline{u}_s^\natural = \overline{\left[\overline{\mathcal{W}_{\mathbf{m_s}}}^\top \mathbf{C}_s\right]}^\top \underline{u}_s^\natural = \overline{\mathcal{Z}_s}^\top \underline{u}_s^\natural; \qquad \vec{u}'_r = \mathbf{C}_r^\top \mathcal{W}_{\mathbf{m_r}} \underline{u}_r^\natural = \overline{\left[\overline{\mathcal{W}_{\mathbf{m_r}}}^\top \mathbf{C}_r\right]}^\top \underline{u}_r^\natural = \overline{\mathcal{Z}_r}^\top \underline{u}_r^\natural \qquad (4)$$

Eq. (3) and (4) define general coordinate transformations for the electrical quantities of an induction machine.

Investigated machine models

Characteristics of the considered induction machine

We apply the modelling strategy described above to derive two models of the induction machine referred to as IM1 in [2]. Table II provides an overview of the relevant parameters of this off-the-shelf machine.

Table II: Characteristics of the induction machine under consideration

Quantity	Symbol and value (SI)		
Rated power	P_N	= 2.2	kW
Rated torque	M_{MN}	= 7.3	Nm
Number of pole pairs	Z_p	= 1	–
Number of stator slots	N_s	= 18	–
Number of rotor slots	N_r	= 16	–
Number of stator circuits in the model	m_s	= 3	–
Number of rotor circuits in the model	m_r	= $N_r = 16$	–

The machine features three star-connected stator windings, each of them being accounted for in the model using one circuit. The stator interconnection matrix $\mathbf{C}_s$ and the corresponding stator transformation matrix $\mathcal{Z}_s$ are therefore:

$$\mathbf{C}_s = \begin{bmatrix} 1 & 0 & -1 \\ 0 & 1 & -1 \end{bmatrix}^\top; \quad \mathcal{Z}_s = \mathcal{W}_3^{-1} \mathbf{C}_s = -j \begin{bmatrix} 0 & 0 \\ e^{j2\pi/3} & -1 \\ e^{j\pi/3} & 1 \end{bmatrix} \qquad (5)$$

The resulting relations between independent and transformed stator currents and voltages are:

$$\underline{i}_s^\natural = \begin{bmatrix} i_{s,0}^\natural & i_{s,1}^\natural & i_{s,2}^\natural \end{bmatrix}^\top = \mathcal{Z}_s \vec{i}'_s = \mathcal{Z}_s \begin{bmatrix} i'_0 & i'_1 \end{bmatrix}^\top \qquad (6a)$$

$$\vec{u}'_s = \begin{bmatrix} u'_0 & u'_1 \end{bmatrix}^\top = \overline{\mathcal{Z}_s}^\top \underline{u}_s^\natural = \overline{\mathcal{Z}_s}^\top \begin{bmatrix} u_{s,0}^\natural & u_{s,1}^\natural & u_{s,2}^\natural \end{bmatrix}^\top \qquad (6b)$$

The submatrix consisting of the non-zero entries of $\mathcal{Z}_s$ being invertible, the independent currents, i'_0 and i'_1, can be determined from $i_{s,1}^\natural$. Similarly, the transformed voltage component $u_{s,1}^\natural$ can be computed from the impressed voltages u'_0 and u'_1 [5]. These properties are absolutely vital for the implementation of the models to be investigated further on.

Machine model including space harmonics of order 1 and 17

The characteristics of the considered machine make the combination of space harmonic orders 1 and 17 interesting, as this configuration generates current and torque harmonics. A corresponding model was derived in [3] and [5]. We therefore only summarize the main results here.

The transformed stator and rotor inductance matrices, $\mathbf{L}_s^\natural$ and $\mathbf{L}_r^\natural$, involve contributions from the space harmonics of order 1 and 17. As expected, they are diagonal:

$$\mathbf{L}_{\mathbf{s}}^{\natural} = \mathbf{L}_{\sigma s}^{\natural} + \widetilde{\mathbf{L}}_{\mathbf{s}1}^{\natural} + \widetilde{\mathbf{L}}_{\mathbf{s}17}^{\natural} = \text{diag}\left[L_{\sigma s}, \ L_s, \ L_s\right] \qquad \text{where } L_s = L_{\sigma s} + \frac{3}{2}\left[\widetilde{L}_{s,1} + \widetilde{L}_{s,17}\right]$$

$$\mathbf{L}_{\mathbf{r}}^{\natural} = \mathbf{L}_{\sigma r}^{\natural} + \widetilde{\mathbf{L}}_{\mathbf{r}1}^{\natural} + \widetilde{\mathbf{L}}_{\mathbf{r}17}^{\natural} = \text{diag}\left[L_{\sigma r}, \ L_r, \ L_{\sigma r}, \ \dots \ L_{\sigma r}, \ L_r\right] \quad \text{with } L_r = L_{\sigma r} + 8\left[\widetilde{L}_{r,1} + \widetilde{L}_{r,17}\right]$$

The coupling matrix $\widetilde{\mathbf{L}}_{\mathbf{rs}}^{\natural}$ describing the effect of transformed rotor currents on transformed stator flux linkages is a sparse matrix with four non-zero entries:

$$\widetilde{\mathbf{L}}_{\mathbf{rs}}^{\natural} = 2\sqrt{3}\begin{bmatrix} 0 & 0 & 0 & \dots & 0 & 0 \\ 0 & L_{M,1}e^{j\theta'_{r,1}} & 0 & \dots & 0 & L_{M,17}e^{-j\theta'_{r,17}} \\ 0 & L_{M,17}e^{j\theta'_{r,17}} & 0 & \dots & 0 & L_{M,1}e^{-j\theta'_{r,1}} \end{bmatrix}$$

The scalar harmonic inductance coefficients $\widetilde{L}_{s,k}$, $\widetilde{L}_{r,k}$ and $L_{M,k}$ $(k \in \{1, 17\})$ are readily determined from the Fourier coefficients of the stator and rotor conductor distribution functions (cf. [2]).

As demonstrated in [5], the simplified system of voltage equations is:

$$
\begin{bmatrix}
u_{s,1}^{\natural} = R_s i_{s,1}^{\natural} + L_s \dfrac{di_{s,1}^{\natural}}{dt} + 2\sqrt{3}\left[L_{M,1}e^{j\theta'_{r,1}}\dfrac{di_{r,1}^{\natural}}{dt} + L_{M,17}e^{-j\theta'_{r,17}}\dfrac{\overline{di_{r,1}^{\natural}}}{dt}\right] \\[2ex]
\qquad + j\omega_r 2\sqrt{3}\left[L_{M,1}e^{j\theta'_{r,1}}i_{r,1}^{\natural} - 17L_{M,17}e^{-j\theta'_{r,17}}\overline{i_{r,1}^{\natural}}\right] \\[2ex]
u_{r,1}^{\natural} = R_r i_{r,1}^{\natural} + L_r \dfrac{di_{r,1}^{\natural}}{dt} + 2\sqrt{3}\left[L_{M,1}e^{-j\theta'_{r,1}}\dfrac{di_{s,1}^{\natural}}{dt} + L_{M,17}e^{-j\theta'_{r,17}}\dfrac{\overline{di_{s,1}^{\natural}}}{dt}\right] \\[2ex]
\qquad - j\omega_r 2\sqrt{3}\left[L_{M,1}e^{-j\theta'_{r,1}}i_{s,1}^{\natural} + 17L_{M,17}e^{-j\theta'_{r,17}}\overline{i_{s,1}^{\natural}}\right]
\end{bmatrix}
\tag{7}
$$

where for $h \in \mathbb{N}^{\star}$, $\theta'_{r,h} = h\theta_r - \varphi_h$. The phase shift angles φ_h are computed from the Fourier coefficients of the conductor distribution functions of stator and rotor circuits, as described in [2].

The electromechanical torque is given by:

$$M_M = \overline{\underline{i}_s^{\natural}}^{\top}\frac{\partial \widetilde{\mathbf{L}}_{\mathbf{rs}}^{\natural}(\theta_r)}{\partial \theta_r}\underline{i}_r^{\natural} = 4\sqrt{3}L_{M,1}\mathfrak{Re}\left[\overline{i_{s,1}^{\natural}}je^{j\theta'_{r,1}}i_{r,1}^{\natural}\right] + 4\sqrt{3}\cdot 17L_{M,17}\mathfrak{Re}\left[i_{s,1}^{\natural}je^{j\theta'_{r,17}}i_{r,1}^{\natural}\right] \tag{8}$$

The second term in the torque expression arises from the presence of the space harmonic of order 17. It gives rise to additional terms depending on the rotor position which account for the presence of a so-called principal slot harmonic (PSH) in stator currents and produces torque pulsations, as will be discussed later on.

A continuous-time state-space representation with real-valued states is gained from eq. (7) and (8) by splitting the transformed voltage and currents into their real and imaginary parts:

$$u_{s,1}^{\natural} = u_{s,1\alpha} + ju_{s,1\beta} \quad (u_{s,1\alpha}, u_{s,1\beta}) \in \mathbb{R}^2; \qquad u_{r,1}^{\natural} = u_{r,1d} + ju_{r,1q} \quad (u_{r,1d}, u_{r,1q}) \in \mathbb{R}^2$$

$$i_{s,1}^{\natural} = i_{s,1\alpha} + ji_{s,1\beta} \quad (i_{s,1\alpha}, i_{s,1\beta}) \in \mathbb{R}^2; \qquad i_{r,1}^{\natural} = i_{r,1d} + ji_{r,1q} \quad (i_{r,1d}, i_{r,1q}) \in \mathbb{R}^2$$

In practice, the current components $i_{s,1\alpha}$, $i_{s,1\beta}$, $i_{r,1d}$ and $i_{r,1q}$ can be used as state variables alongside the rotor angular velocity ω_r and the rotor angle θ_r [5], leading to a continuous-time state-space representation of the form:

$$\left[\begin{array}{l}\dfrac{d\vec{i}_\circ}{dt} = \mathbf{A}_\circ(\theta_r,\omega_r)\vec{i}_\circ + \mathbf{B}_\circ(\theta_r)\vec{u}_\circ\end{array}\right. \tag{9a}$$

$$\dfrac{d\omega_r}{dt} = \dfrac{1}{J_M}(M_M - C_W\omega_r - M_L) \tag{9b}$$

$$\dfrac{d\theta_r}{dt} = \omega_r \tag{9c}$$

$$M_M = 4\sqrt{3}\begin{bmatrix}i_{s,1\alpha} & i_{s,1\beta}\end{bmatrix}\mathbf{J}[L_{M,1}\mathbf{T}(\theta_r) + 17L_{M,17}\mathbf{ST}(17\theta_r)]\begin{bmatrix}i_{r,1d}\\ i_{r,1q}\end{bmatrix}$$

where $(\mathbf{A}_\circ, \mathbf{B}_\circ) \in \mathcal{M}_4^2(\mathbb{R})$ and

$$\vec{i}_\circ = \begin{bmatrix}i_{s,1\alpha} & i_{s,1\beta} & i_{r,1d} & i_{r,1q}\end{bmatrix}^\top; \qquad \vec{u}_\circ = \begin{bmatrix}u_{s,1\alpha} & u_{s,1\beta} & u_{r,1d} & u_{r,1q}\end{bmatrix}^\top$$

The system (9) has order 6, which is the same as for Park's model. This results in particular from the position of the non-zero entries in the matrix $\widetilde{\mathbf{L}}_{\mathbf{rs}}^\natural$.

Machine model with space harmonics of order 1, 5 and 11

We now focus on an alternative model with space harmonic orders $\mathcal{H} = \{1, 5, 11\}$, for which the stator and rotor main inductance matrices are given by:

$$\mathbf{L}_{\mathbf{s}}^\natural = \mathbf{L}_{\sigma\mathbf{s}}^\natural + \widetilde{\mathbf{L}}_{\mathbf{s1}}^\natural + \widetilde{\mathbf{L}}_{\mathbf{s5}}^\natural + \widetilde{\mathbf{L}}_{\mathbf{s11}}^\natural = \operatorname{diag}\begin{bmatrix}L_{\sigma s}, & L_s, & L_s\end{bmatrix} \quad \text{with} \quad L_s = L_{\sigma s} + \dfrac{3}{2}\left[\widetilde{L}_{s,1} + \widetilde{L}_{s,5} + \widetilde{L}_{s,11}\right]$$

$$\mathbf{L}_{\mathbf{r}}^\natural = \mathbf{L}_{\sigma\mathbf{r}}^\natural + \widetilde{\mathbf{L}}_{\mathbf{r1}}^\natural + \widetilde{\mathbf{L}}_{\mathbf{r5}}^\natural + \widetilde{\mathbf{L}}_{\mathbf{r11}}^\natural$$

$$= \operatorname{diag}\begin{bmatrix}L_{\sigma r}, & L_{r,1}, & L_{\sigma r}, & \dots & L_{\sigma r}, & L_{r,5/11}, & L_{\sigma r}, & \dots & L_{\sigma r}, & L_{r,5/11}, & L_{\sigma r}, & \dots & L_{\sigma r}, & L_{r,1}\end{bmatrix}$$

where $L_{r,1} = L_{\sigma r} + 8\widetilde{L}_{r,1}$ and $L_{r,5/11} = L_{\sigma r} + 8\left[\widetilde{L}_{s,5} + \widetilde{L}_{s,11}\right]$

The entry $L_{r,1}$ in $\mathbf{L}_{\mathbf{r}}^\natural$ is found in row 1 and 15, while the coefficient $L_{r,5/11}$ appears in rows 5 and 11. Owing to the order of the space harmonics considered, the coupling matrix $\widetilde{\mathbf{L}}_{\mathbf{rs}}^\natural$ has now also non-zero entries in columns 5 and 11:

$$\widetilde{\mathbf{L}}_{\mathbf{rs}}^\natural = 2\sqrt{3}\begin{bmatrix}0 & 0 & 0 & \dots & 0 & 0 & 0 & \dots & 0 & 0 & 0 & \dots & 0 & 0\\ 0 & L_{M,1}e^{j\theta'_{r,1}} & 0 & \dots & 0 & L_{M,11}e^{-j\theta'_{r,11}} & 0 & \dots & 0 & L_{M,5}e^{-j\theta'_{r,5}} & 0 & \dots & 0 & 0\\ 0 & 0 & 0 & \dots & 0 & L_{M,5}e^{j\theta'_{r,5}} & 0 & \dots & 0 & L_{M,11}e^{j\theta'_{r,11}} & 0 & \dots & 0 & L_{M,1}e^{-j\theta'_{r,1}}\end{bmatrix}$$

Taking into consideration that for the investigated machine, $\varphi_1 = \varphi_5 = \varphi_{11} = 0$, we obtain the subsequent differential system upon simplification of the voltage equations:

$$\left[\begin{array}{l}u_{s,1}^\natural = R_s i_{s,1}^\natural + L_s \dfrac{di_{s,1}^\natural}{dt} + 2\sqrt{3}\left[L_{M,1}e^{j\theta_r}\dfrac{di_{r,1}^\natural}{dt} + L_{M,11}e^{-j11\theta_r}\dfrac{di_{r,5}^\natural}{dt} + L_{M,5}e^{-j5\theta_r}\dfrac{\overline{di_{r,5}^\natural}}{dt}\right]\end{array}\right.$$

$$\qquad\qquad + j\omega_r 2\sqrt{3}\left[L_{M,1}e^{j\theta_r}i_{r,1}^\natural - 11L_{M,11}e^{-j11\theta_r}i_{r,5}^\natural - 5L_{M,5}e^{-j5\theta_r}\overline{i_{r,5}^\natural}\right]$$

$$u_{r,1}^\natural = R_r i_{r,1}^\natural + L_{r,1}\dfrac{di_{r,1}^\natural}{dt} + 2\sqrt{3}L_{M,1}e^{-j\theta_r}\dfrac{di_{s,1}^\natural}{dt} - j\omega_r 2\sqrt{3}L_{M,1}e^{-j\theta_r}i_{s,1}^\natural \tag{10}$$

$$u_{r,5}^\natural = R_r i_{r,5}^\natural + L_{r,5/11}\dfrac{di_{r,5}^\natural}{dt} + 2\sqrt{3}\left[L_{M,11}e^{j11\theta_r}\dfrac{di_{s,1}^\natural}{dt} + L_{M,5}e^{-j5\theta_r}\dfrac{\overline{di_{s,1}^\natural}}{dt}\right]$$

$$\qquad\qquad + j\omega_r 2\sqrt{3}\left[11L_{M,11}e^{j11\theta_r}i_{s,1}^\natural - 5L_{M,5}e^{-j5\theta_r}\overline{i_{s,1}^\natural}\right]$$

Introducing the subsequent changes of variables

$$i_{r,1}^{\natural} = e^{j\theta_r} i_{r,1}^{\natural}; \qquad u_{r,1}^{\natural} = e^{j\theta_r} u_{r,1}^{\natural}; \qquad i_{r,5}^{\vec{}} = e^{j5\theta_r} i_{r,5}^{\vec{}}; \qquad u_{r,5}^{\vec{}} = e^{j5\theta_r} u_{r,5}^{\natural}$$

enables us to rewrite (10) as follows:

$$
\begin{bmatrix}
u_{s,1}^{\natural} = R_s i_{s,1}^{\natural} + L_s \dfrac{di_{s,1}^{\natural}}{dt} + 2\sqrt{3}\left[L_{M,1}\dfrac{di_{r,1}^{\natural}}{dt} + L_{M,11}e^{-j16\theta_r}\dfrac{di_{r,5}^{\natural}}{dt} + L_{M,5}\dfrac{d\overline{i_{r,5}^{\vec{}}}}{dt} \right] \\[2mm]
\qquad - 16j\omega_r 2\sqrt{3}L_{M,11}e^{-j16\theta_r} i_{r,5}^{\vec{}} \\[2mm]
u_{r,1}^{\natural} = R_r i_{r,1}^{\natural} + L_{r,1}\dfrac{di_{r,1}^{\natural}}{dt} - j\omega_r L_{r,1} i_{r,1}^{\natural} + 2\sqrt{3}L_{M,1}\dfrac{di_{s,1}^{\natural}}{dt} - j\omega_r 2\sqrt{3}L_{M,1}i_{s,1}^{\natural} \\[2mm]
u_{r,5}^{\natural} = R_r i_{r,5}^{\vec{}} + L_{r,5/11}\dfrac{di_{r,5}^{\vec{}}}{dt} - j\omega_r 5 L_{r,5/11} i_{r,5}^{\vec{}} + 2\sqrt{3}\left[L_{M,11}e^{j16\theta_r}\dfrac{di_{s,1}^{\vec{}}}{dt} + L_{M,5}\dfrac{d\overline{i_{s,1}^{\natural}}}{dt} \right] \\[2mm]
\qquad + j\omega_r 2\sqrt{3}\left[11 L_{M,11}e^{j16\theta_r} i_{s,1}^{\vec{}} - 5L_{M,5}\overline{i_{s,1}^{\natural}} \right]
\end{bmatrix}
\tag{11}
$$

We notice the additional equation appearing for $u_{r,5}^{\natural}$ in contrast to the model with $\mathcal{H} = \{1, 17\}$.

The electromechanical torque is given by the relation:

$$M_M = 4\sqrt{3}\left[L_{M,1}\mathfrak{Re}\left[j\overline{i_{s,1}^{\vec{}}} i_{r,1}^{\vec{}} \right] + 5L_{M,5}\mathfrak{Re}\left[j i_{s,1}^{\natural} i_{r,5}^{\vec{}} \right] + 11 L_{M,11}\mathfrak{Re}\left[j i_{s,1}^{\vec{}} e^{j16\theta_r}\overline{i_{r,5}^{\vec{}}} \right] \right] \tag{12}$$

The presence of the harmonic orders 5 and 11 leads to two additional terms in the torque expression compared to Park's model. However, only the last term on the right hand side is dependent on the rotor angle.

We define:

$$u_{r,1}^{\natural} = u_{r,1\alpha} + j u_{r,1\beta} \qquad (u_{r,1\alpha}, u_{r,1\beta}) \in \mathbb{R}^2; \qquad u_{r,5}^{\vec{}} = u_{r,5\alpha} + j u_{r,5\beta} \qquad (u_{r,5\alpha}, u_{r,5\beta}) \in \mathbb{R}^2$$

$$i_{5,1}^{\vec{}} = i_{r,1\alpha} + j i_{r,1\beta} \qquad (i_{r,1\alpha}, i_{r,1\beta}) \in \mathbb{R}^2; \qquad i_{r,5}^{\vec{}} = i_{r,5\alpha} + j i_{r,5\beta} \qquad (i_{r,5\alpha}, i_{r,5\beta}) \in \mathbb{R}^2$$

in order to obtain a state-space representation with real-valued state variables in the form:

$$
\begin{bmatrix}
\dfrac{d\vec{i}}{dt} = \mathbf{\underset{\wedge}{A}}(\theta_r, \omega_r)\underset{\wedge}{\vec{i}} + \mathbf{\underset{\wedge}{B}}(\theta_r)\underset{\wedge}{\vec{u}} & \text{(13a)} \\[3mm]
\dfrac{d\omega_r}{dt} = \dfrac{1}{J_M}(M_M - C_W \omega_r - M_L) & \text{(13b)} \\[3mm]
\dfrac{d\theta_r}{dt} = \omega_r & \text{(13c)} \\[3mm]
M_M = 4\sqrt{3}\begin{bmatrix} i_{s,1\alpha} & i_{s,1\beta} \end{bmatrix}\mathbf{J}\left[L_{M,1}\begin{bmatrix} i_{r,1\alpha} \\ i_{r,1\beta} \end{bmatrix} - 5L_{M,5}\mathbf{S}\begin{bmatrix} i_{r,5\alpha} \\ i_{r,5\beta} \end{bmatrix} - 11 L_{M,11}\mathbf{T}(-16\theta_r)\begin{bmatrix} i_{r,5\alpha} \\ i_{r,5\beta} \end{bmatrix} \right] &
\end{bmatrix}
$$

with $(\mathbf{\underset{\wedge}{A}}, \mathbf{\underset{\wedge}{B}}) \in \mathcal{M}_6^2(\mathbb{R})$ and

$$\underset{\wedge}{\vec{i}} = \begin{bmatrix} i_{s,1\alpha} & i_{s,1\beta} & i_{r,1\alpha} & i_{r,1\beta} & i_{r,5\alpha} & i_{r,5\beta} \end{bmatrix}^{\top}; \qquad \underset{\wedge}{\vec{u}} = \begin{bmatrix} u_{s,1\alpha} & u_{s,1\beta} & u_{r,1\alpha} & u_{r,1\beta} & u_{r,5\alpha} & u_{r,5\beta} \end{bmatrix}^{\top}$$

The components of $\underset{\wedge}{\vec{i}}$ together with ω_r and θ_r constitute the state variables of the model with space harmonic orders 1, 5 and 11. The model order is 8.

The torque expression which appears in (13) is of particular interest as the term depending on the rotor position is independent of the current components $i_{r,1\alpha}$ and $i_{r,1\beta}$. As a result, the current and torque oscillations resulting from the presence of the space harmonic orders 5 and 11 could be cancelled if the states $i_{r,5\alpha}$ and $i_{r,5\beta}$ are maintained to zero. This requires them to be observable and controllable.

The 6×6 matrices $\underset{\sim}{\mathbf{A}}(\theta_r, \omega_r)$ and $\underset{\sim}{\mathbf{B}}(\theta_r)$ can be rewritten as block matrices consisting of 2×2 blocks. It can be shown that the determinant of all blocks of $\underset{\sim}{\mathbf{A}}(\theta_r, \omega_r)$ is independent of θ_r and non-zero for every $\omega_r \in \mathbb{R}$. Similarly, the non-zero blocks of $\underset{\sim}{\mathbf{B}}(\theta_r)$ have a non-zero determinant which does not depend on θ_r. This means that the states $i_{r,5\alpha}$ and $i_{r,5\beta}$ will have an impact on $i_{s,1\alpha}$ and $i_{s,1\beta}$ and vice versa. $i_{s,1\alpha}$ and $i_{s,1\beta}$ representing the stator currents which are usually measured, the controllability and observability of $i_{r,5\alpha}$ and $i_{r,5\beta}$ can reasonably be expected.

Simulation and experimental results

The models with $\mathcal{H} = \{1, 17\}$ and $\mathcal{H} = \{1, 5, 11\}$ have been examined in simulation and experimentally. Extensive investigations into the former may be found in [3] and [5]. Fig. 1 shows simulation results of the stator and rotor currents alongside the electromechanical torque obtained with the two models. They were fed with the same balanced sinusoidal voltages of frequency $f = 50\,\mathrm{Hz}$ and a load torque step of amplitude equal to the machine rated torque applied at $t = 0.5\,\mathrm{s}$. The simulation step size is equal to $10^{-4}\,\mathrm{s}$, in accordance with the sample time used in the experiments. Heun's method was used to determine the value of the model states at each simulation step.

The current and torque waveforms depicted in fig. 1 clearly exhibit a frequency component on top of the fundamental in steady state corresponding to the PSH. However, the two models provide significantly different results with respect to the amplitude of the PSH and the resulting torque pulsations. They are markedly weaker in the configuration with $\mathcal{H} = \{1, 5, 11\}$. This phenomenon is not surprising as the magnitude of the related Fourier coefficients of the conductor distribution functions and therefore the harmonic inductances associated with orders 5 and 11 are smaller.

The models were also implemented in C on a real-time system dSPACE DS1006 to be assessed experimentally. The real-time system was used to control a voltage source inverter feeding the machine under test. A torque-controlled load machine was used to emulate load conditions. During the experiments, the model was fed with the same voltage reference values as the inverter. A torque sensor was used to determine the torque applied on the machine shaft and its signal used as load torque input in the model.

Fig. 2 compares the computed and measured waveforms of stator current i'_{s0} and the ones of the rotor angular velocity for the model with $\mathcal{H} = \{1, 17\}$. The frequency spectra for the computed and measured currents are provided as well. Machine and model were supplied by a set of balanced voltages of frequency $f = 40\,\mathrm{Hz}$. A good agreement regarding the amplitude of the PSH is achieved at no load, while the model tends to overestimate the amplitude of the PSH at rated load. The discrepancies regarding the angular velocity mainly result from the fact that the combination of investigated machine, torque sensor and load machine constitutes a three-mass system, an aspect which was not taken into account in the model. Fig. 3 was obtained under the same operating conditions with the model featuring space harmonic orders 1, 5 and 11. The significantly underestimated amplitude of the PSH is clearly visible under no load and under rated conditions. However, the model parameters could be tuned if the model were to be used for current or torque ripple reduction.

The continous-time state-space equations (9) and (13) were discretized using Heun's method as well as a zero-order hold assumption with third order Taylor approximation. The maximum execution times obtained for the configuration $\mathcal{H} = \{1, 17\}$ were $5.3\,\mu s$ and $3.5\,\mu s$ respectively whereas for the model with $\mathcal{H} = \{1, 5, 11\}$, they were $12.4\,\mu s$ and $7.6\,\mu s$. The increase of computational effort in the latter case follows from the higher model order but does not fundamentally disqualify it for the use in demanding control applications like predictive control schemes.

Conclusion

Two models of an off-the-shelf induction machine considering different sets of winding distribution harmonics were derived using the method presented in [3] and compared. The first model includes space

A Comparison of Two State-Space Models of an Induction Machine Considering
Different Sets of Winding Distribution Harmonics

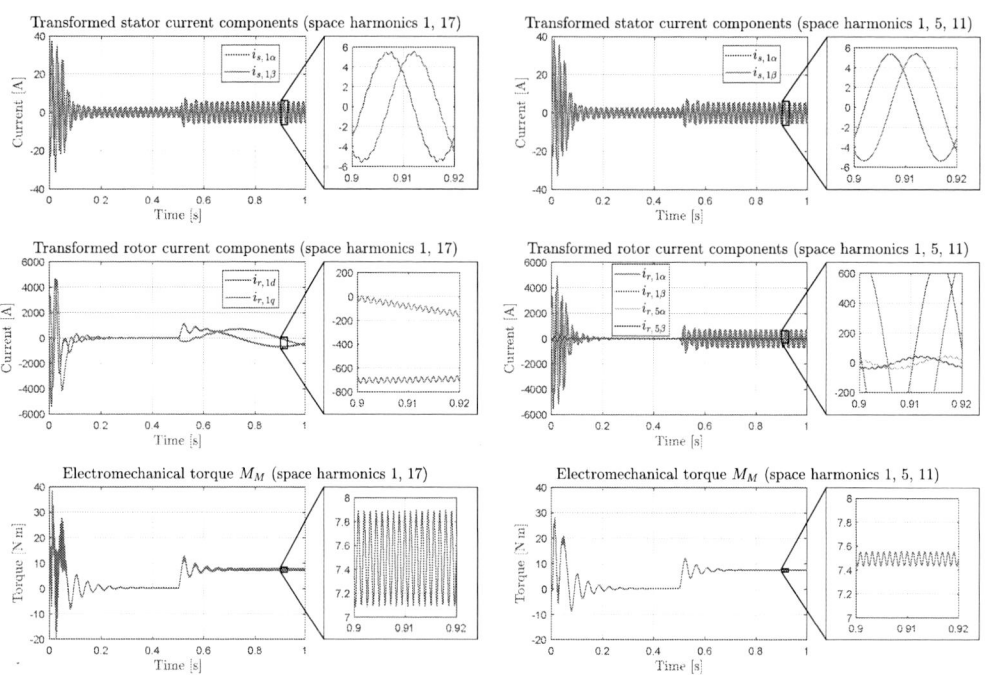

Fig. 1: Comparison of the models with space harmonic orders 1 and 17 (left) and 1, 5 and 11 (right)

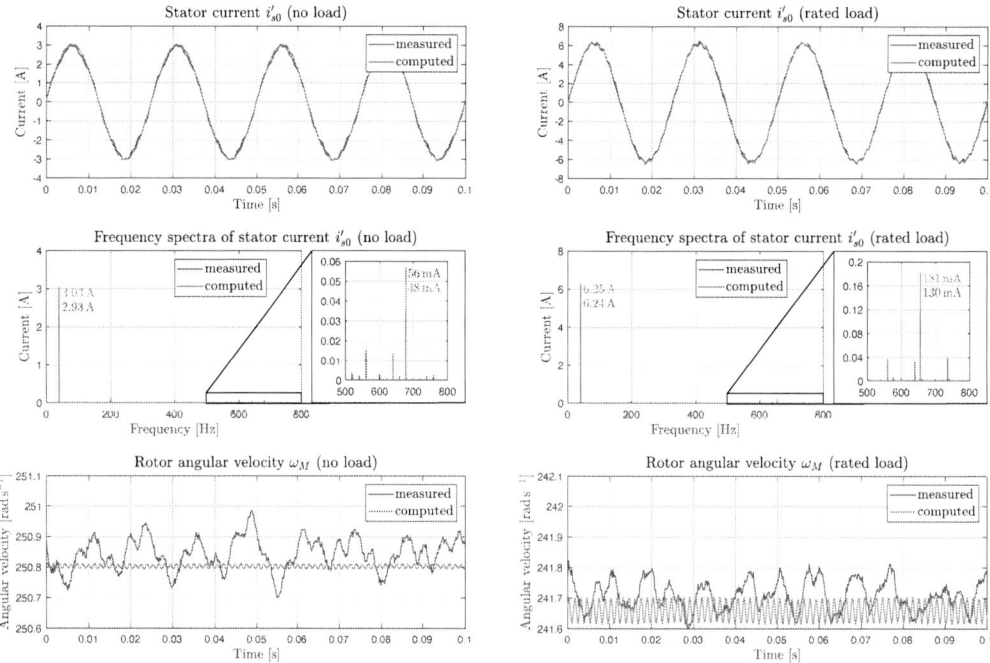

Fig. 2: Comparison of stator current and rotor angular velocity waveforms as well as stator current spectra with experimental results under no load and rated load conditions (Heun's method, model with $\mathcal{H} = \{1, 17\}$)

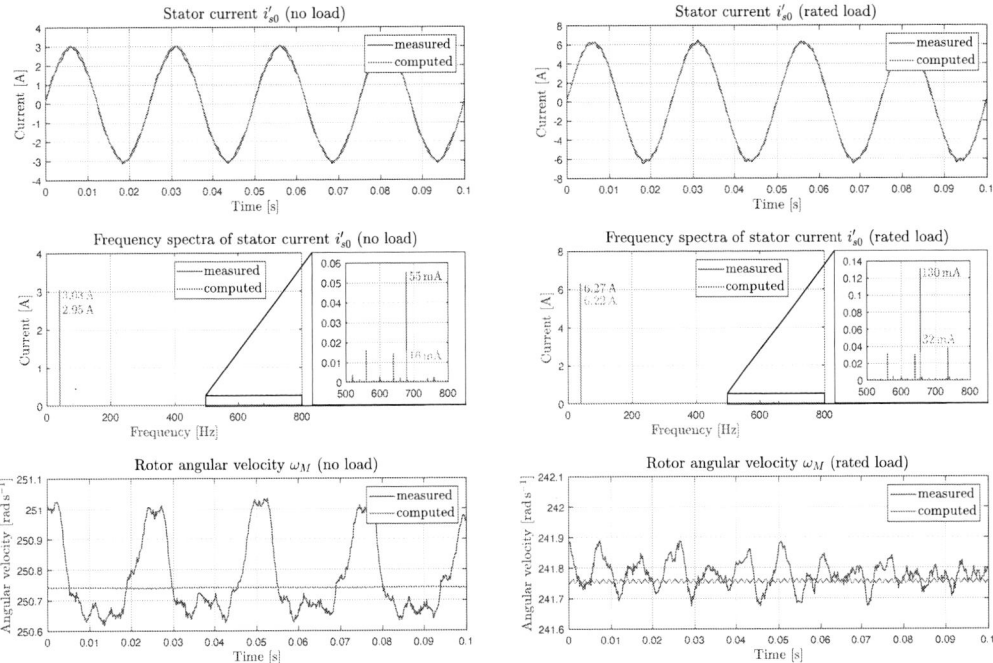

Fig. 3: Comparison of stator current and rotor angular velocity waveforms as well as stator current spectra with experimental results under no load and rated load conditions (Heun's method, model with $\mathcal{H} = \{1, 5, 11\}$)

harmonics 1 and 17 and has order 6. The second takes into account space harmonics 1, 5 and 11, while its order is 8. This higher order results from two additional states on the rotor side and is responsible for an increase of computational burden. However, these extra states being presumably controllable and observable, the model may be of particular interest for torque ripple reduction applications. Further investigations are needed to assess its potential, in particular in combination with a predictive control strategy.

References

[1] J. C. Moreira and T. A. Lipo, "Modeling of saturated ac machines including air gap flux harmonic components," *IEEE Transactions on Industry Applications*, vol. 28, no. 2, pp. 343–349, Mar. 1992.

[2] J. Cordier, S. Klass, and R. Kennel, "A real-time compliant state-space model of induction machines including winding distribution harmonics and winding interconnections," in *2019 21st European Conference on Power Electronics and Applications (EPE'19 ECCE Europe)*. IEEE, Sep. 2019.

[3] ——, "A general coordinate transformation based on Fourier matrices for modelling space harmonics in induction machines," in *2019 IEEE Energy Conversion Congress and Exposition (ECCE)*. IEEE, 2019.

[4] G. Kron, *Equivalent Circuits of Electric Machinery*. John Wiley & Sons, 1951.

[5] J. Cordier, "Modelling space harmonics in induction machines for real-time applications," Ph.D. dissertation, Technische Universitaet Muenchen, Jan. 2020. [Online]. Available: https://mediatum.ub.tum.de/1523443

[6] J. Cordier and R. Kennel, "A simple and efficient state-space model of induction machines with interconnected windings including space harmonics," in *2018 IEEE Energy Conversion Congress and Exposition (ECCE)*. IEEE, 2018, pp. 1595–1602.

[7] J. E. Gentle, *Matrix Algebra - Theory, Computations and Applications in Statistics*. Springer, 2017.

Performance Improvement for Plug-In Reverse Conducting IGBTs through Gate-Voltage Observation

Daniel Lexow, Hans-Günter Eckel

UNIVERSITY OF ROSTOCK
Albert-Einstein Str. 2
18059 Rostock, Germany
Phone: +49 (0) 381-498 7112
Fax: +49 (0) 381-498 7102
Email: daniel.lexow@uni-rostock.de
URL: http://www.iee.uni-rostock.de

Keywords

«IGBT», «switching losses», «optimal control», «power semiconductor device»

Abstract

While driving a plug-in RC-IGBT, the performance outcome can be significantly improved by adapting the turn-OFF process. Hereby, the observation of the characteristic gate-voltage behavior allows the distinction between IGBT and diode turn-OFF. Latter can be easily manipulated in order to utilize the full potential of the inverter locking time.

Introduction

The concept of a *Reverse Conducting* (RC)-IGBT and its different control techniques was widely presented over the past half-decade. Its main feature is the implemented diode function into the IGBT-chip. This provides the quality of a two-way current conduction. The conventional IGBT, which is only able to conduct positive current in forward direction, is reliant on an anti-parallel diode to conduct negative current. This demands two different chip types - diode and IGBT-chips - in one IGBT-module. The RC-IGBT overcomes this standard assembly by only using one type of chips, which are able to handle both current directions. Not only leads this to positive outcomes regarding the reduction of thermal-mechanical stress, but it also entails a major improvement concerning the power density due to a greater diode as well as IGBT chip area[1], [2].

Within the branch of RC-IGBTs, two different ways of controlling the chip are addressed. Firstly, there is the R*everse Conducting Diode Control* (RCDC)-IGBT, which is inevitably dependent on the *"static"* and *"dynamic MOS-control"* techniques. Hereby, the static MOS-control refers to a current direction depending switching behavior. While conducting positive current, the gate needs to be switched ON (V_{ge}= +15 V), while conducting negative current, it has to be turned OFF (V_{ge}= -15 V). The dynamic MOS-control implements a short - some μs - desaturation pulse prior to turning OFF the RCDC in diode conduction mode [3]. For this purpose, the gate is turned ON (V_{ge}= +15 V). This shorts the p-anode under the gate creating the MOS-channel. P-anode efficiency is drastically reduced, and the electron-hole plasma concentration in the device is decreasing. Hence, switching losses are decreased [4]. Both control principles require a complex gate control scheme with an included current direction detection [5], [6], but permit an excellent outcome in terms of electrical performance.

The second type of RC-IGBT is called the *"Plug-In"* (PI-)RC-IGBT. In this regard, the architecture of the RC-chip is highly sophisticated. Therefore, any further gate control techniques can be omitted, and the PI-RC-IGBT can be used by simply replacing the conventional IGBT in the inverter structure. The gate drive unit, as well as the inverter control scheme, do not have to be adapted.

Overall, the power output of the PI-RC-IGBT is consciously reduced on-chip level in order to allow a conventional gate control without the application of further control techniques[7].

Although the PI-RC-IGBT is not reliant on a special gate control scheme, it still reduces its plasma concentration in diode conduction mode while the gate is turned ON at + 15 V [7]. Hence, the device is practically undergoing a very long desaturation pulse as long as it is turned ON and conducting negative current. Deriving from this, it is desirable to immediately start reverse-recovery as long as the plasma in the device is reduced. Therefore, reverse-recovery should directly start when the turn-OFF signal from the inverter control is received. Unfortunately, the beginning of the reverse-recovery is related to the turn-ON of the opposite IGBT, which has to wait for the inverter locking time (t_{lock} ~ 10 µs) to pass until the switching ON signal is forwarded. For this period of time, the plasma concentration of the RC-IGBT in diode mode is building up again. This results from closing the MOS-channel due to V_{ge} = -15 V, enhancing the p-emitter efficiency resulting in an increased plasma concentration and switching losses. All in all, it is preferable to delay the turn-OFF in diode mode as far as possible to the end of the locking time in order to achieve the lowest reverse-recovery losses [4], [7].

Gate-drive unit to manipulate diode turn-OFF

The gate-drive unit presented in Fig. 1, is able to monitor the V_{ge}. This is accomplished by implementing a comparator which is triggered if a certain V_{ge} threshold *("TH_V_{ge}")* is exceeded. The trigger-signal from the comparator *("Comp_V_{ge}")* is forwarded to an FPGA, in which a *Finite-State-Machine* (FSM) processes the information. The output part of the gate-drive unit consists of four gate-states. Two are used as turn-ON (1 and 2), and two are used as turn-OFF (3 and 4) states. Each pair has one slow (large R_g [1 and 3]) and one fast (small R_g [2 and 4]) gate resistor.

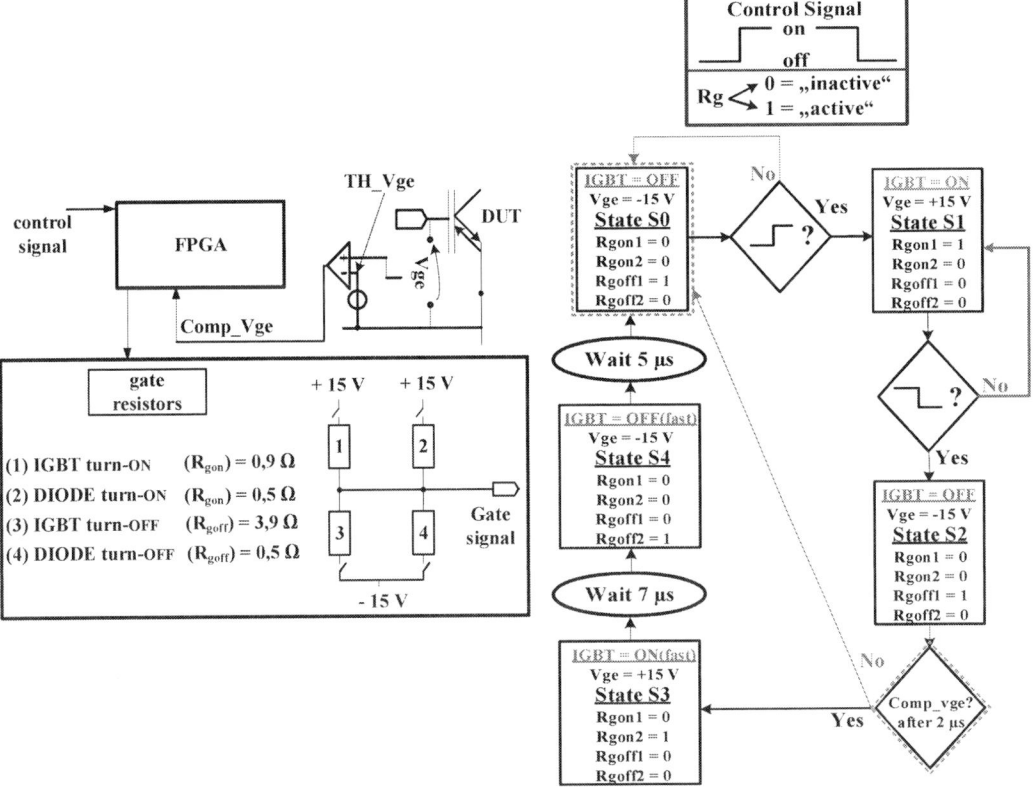

Fig. 1: Gate-drive unit to detect and react to load current zero-crossing

Fig. 2: Finite State Machine with implemented current direction detection

All gate-states are driven by the FSM (Fig. 2). It follows the inverter control signal but uses the sensory input of the comparator to manipulate the turn-OFF process only when the diode conduction mode of the PI-RC-IGBT is recognized. The starting point for the FSM is State S0 (dashed green line), as displayed in figure 2. From here on, the inverter control signal is followed until State S2 is reached. The crucial part of the FSM is the transition from S2 to the next state. Here, two µs after the CTRL turn-OFF signal is received, only V_{ge} is taken into account while deciding whether the RC-IGBT operates in IGBT or diode mode at this exact moment before turn-OFF. If V_{ge} is not triggering the V_{ge} threshold of the comparator, the course of V_{ge} has to show the miller-plateau preventing it from falling beneath TH_V_{ge} (see Fig. 6) Therefore, the FSM decides to perform a regular IGBT turn-OFF with Rgoff = 3,9 Ohm ("3" in Fig. 1). If TH_V_{ge} of the comparator is triggered, the turn-OFF course of the gate-emitter voltage is not showing a miller-plateau. Thus, a diode mode turn-OFF is present (see Fig. 4). As a result, the PI-RC-IGBT is turned ON again with the fast Rgon = 0,5 Ohm ("2" in Fig. 1) From here on, the two counters (state 3; state 4) ensure that the locking time is utilized to its full potential until the IGBT is finally turned OFF. This "observed" switching process is displayed in figure 5.

Test-bench set up for a double-pulse test

All measurements were accomplished using a test-bench set up for a double pulse test, as depicted in Fig. 3. Its main parts are a DC-link capacitor, a load choke, and a half-bridge consisting of two 6.5 kV, 1000A RC-IGBTs. The double pulse test allows elaborating switching losses as well as characteristic current and voltage curves dependent on the DC-link voltage and different adjustable load currents. Firstly, following the pulse pattern, which is also shown in Fig. 3, the Bottom switch (B1) is turned ON, initiating a current flow with a constant di/dt. The adjustment of the specific current test value is made through the turn-ON time, calculated based on the Dc-link voltage and load choke. Now, turn-OFF losses (E_{off}) can be analyzed by turning OFF B1. Furthermore, a current commutation from B1 into the diode of T1 results. The diode conduction time needs to be long enough ($t_{diode} = 400$ µs) to ensure a full electron-hole plasma establishment in the diode. The second turn-ON of B1 causes the current to commutate from the diode T1 into the IGBT B1, provoking the reverse-recovery of the diode. Both processes are measured and evaluated, resulting in turn-ON losses (E_{on}) for the IGBT and the reverse-recovery losses (E_{rec}) of the diode. The locking time was set to $t_{lock} = 10$ µs [7].

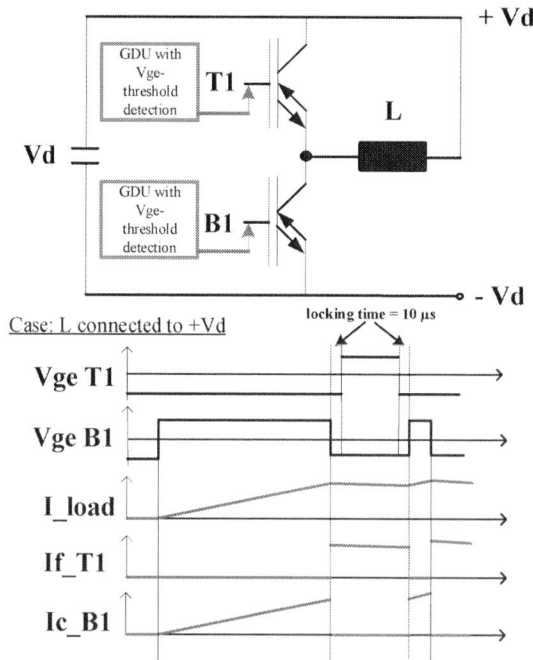

Fig. 3: Test-bench set up in addition to the timing diagram for
double pulse measurements

Experimental Results

All following measurements were accomplished with 6.5 kV (1000A) RC-IGBTs, which are conventional and not Plug-IN since no 6.5 kV PI-RC-IGBTs are commercially available on the market. Nevertheless, the gate-drive concept functionality, as well as the loss reduction potential, can be approved with the conventional RC-IGBTs. The temperature was set to 25 °C.

A) Diode turn-OFF

Measurements were accomplished by following the pulse pattern shown in Fig. 3. In the first part of the test, a conventional GDU was used to show the state-of-the-art switching process. The curves are depicted in Fig. 4. The IGBT turn-OFF in diode conduction mode is performed, and a huge portion of the locking time is wasted due to the unmodified switching behavior caused by the conventional GDU.

Fig. 5 shows the same measurement, as displayed in Fig. 4, but this time the new gate-drive unit utilizing the gate-voltage observation technique is used. It is noticeable that the gate driver recognizes the diode mode turn-OFF process due to falling below TH_V_{ge} two µs after the CTRL turn-OFF signal is received. V_{ge} is immediately turned back ON with the fast R_{gon} ("2" in Fig. 1).

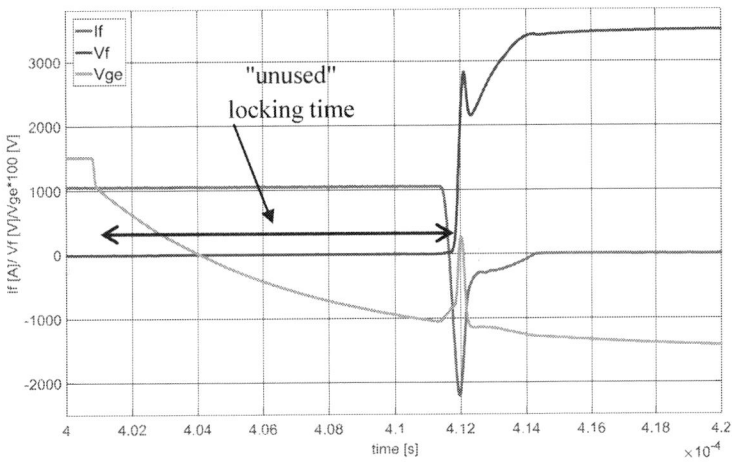

Fig. 4: Regular turn-OFF process in diode conduction mode @ V_d=3600 V; I_c= 1000 A; (conventional GDU)

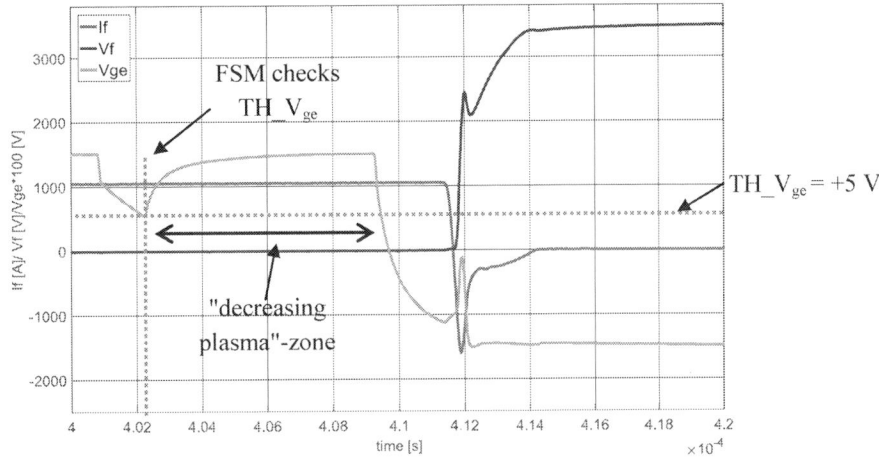

Fig. 5: Turn-OFF process in diode conduction mode @ V_d=3600 V; I_c= 1000 A; T= 25 °C; TH_V_{ge} is triggered

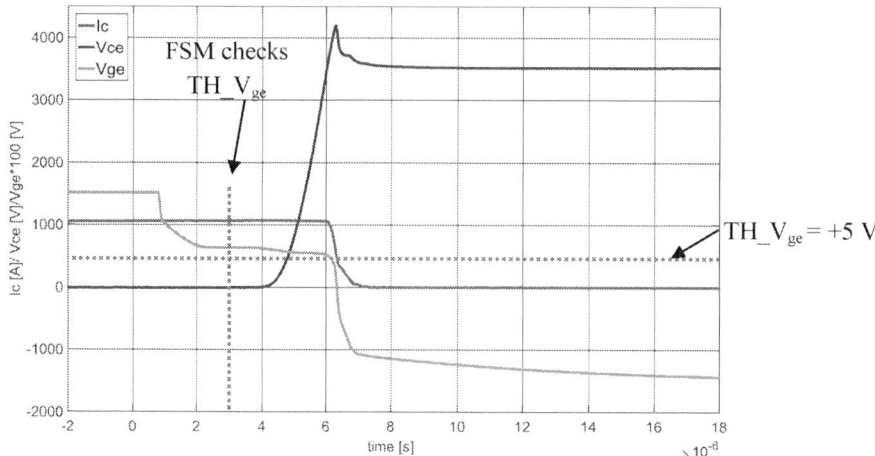

Fig. 6: Turn-OFF process in IGBT conduction mode @ V_d=3600 V; I_c= 1000 A; T= 25 °C; TH_V_{ge} is not triggered

The reduction of the plasma concentration in the device is maintained until the RC-IGBT is turned OFF seven µs later ("decreasing plasma"-zone in Fig. 5) and very close to the turn-ON of the opposite RC-IGBT. Thereby, the lock time between those two switching processes is kept as small as possible to achieve the highest reduction in switching losses [4]. All results concerning the potential switching loss improvement due to the novel gate control technique are displayed in a later chapter.

B) IGBT turn-OFF

In order to validate the function of the novel gate-voltage observation technique, it was examined if the FSM performs a regular switching when a regular turn-OFF in IGBT mode needs to be performed. This is absolutely mandatory to ensure safe inverter operation behavior. Therefore, a conventional turn-OFF process in IGBT mode is depicted in Fig. 6. It can be seen that V_{ge} did not fall below TH_V_{ge} in the given time window due to the miller-plateau. After the waiting time of two µs, Comp_V_{ge} is not triggered, and a regular turn-OFF using the nominal R_{goff} ("1" in Fig. 2) is initiated. The conventional turn-OFF process, which relies on the full ten µs lock time, is accomplished.

C) Current dependency of the miller-plateau

A crucial aspect that needs to be addressed in order to ensure a flawless working behavior of the gate-drive unit, which is using the gate voltage observation technique, is the current dependency of the miller plateau. The level of the miller-plateau is directly dependent on the load current. Therefore, the predetermined gate-voltage threshold (vge_TH) in the FSM of the FPGA needs to be adjusted on this miller plateau fluctuation. Since high collector currents lead to higher gate-emitter voltages of the miller plateau, the lowest miller plateau needs to be found. This is achieved by setting up the load current in the double pulse test as small as possible. This way, the miller plateau minimum can be found, and a V_{ge} value slightly lower than the minimum can be used as the V_{ge}_TH. Therefore, it is ensured that even while conducting the smallest currents, no false diode conduction mode is detected, and each IGBT-mode turn-OFF is properly executed.

Figure 7 shows measurements which are performed with different load currents to observe the behavior of the miller-plateau. Therefore, it is possible to find an optimal value for TH_V_{ge} and to make sure that the gate-drive unit is not falsely operating. Measurements show different collector currents ranging from 1050A to 100 A. Herby, the V_{ge} miller-plateau value for the smallest current (I_c = 100 A), gives 4,9 V after two µs waiting time (see Fig. 7).

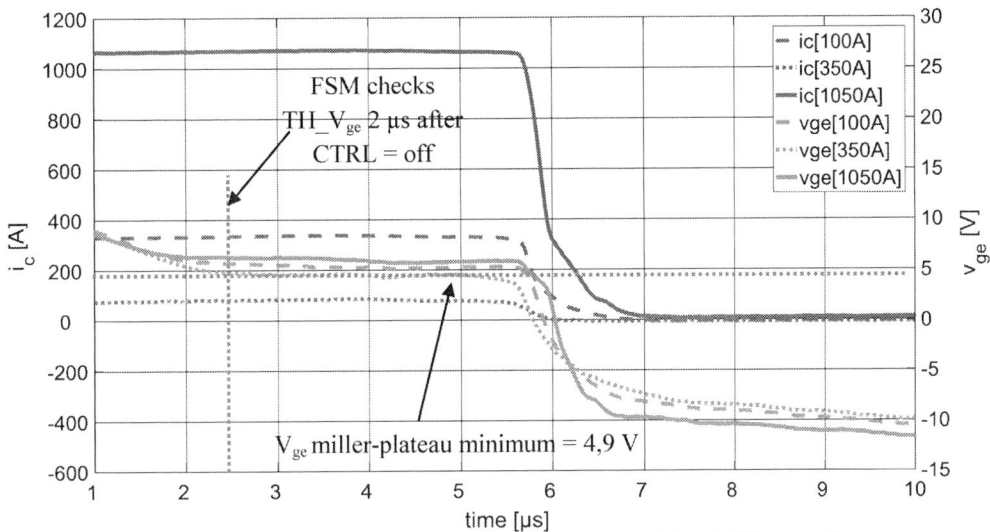

Fig. 7: Turn-OFF process in IGBT conduction mode @ V_d=3600 V; T= 25 °C and different collector currents

In order to provide a safety margin, $V_{ge\_}TH$ would be set to 4 V. Setting it any lower would improve the secure switching behavior, but would also lead to slower reaction times and poorer results for the improved diode conduction mode turn-OFF.

That is caused due to the fact that V_{ge} needs to be safely below the threshold in order to detect the diode conduction mode. For the regular "diode mode" turn-OFF, as seen in Fig. 4, discharging the gate takes some time due to the initially used high R_{goff}. Therefore, it might be necessary to further increase the waiting time (2µs), which reduces the "decreasing plasma"-zone shown in Fig 5. This needs to be verified individually for each Plug-In RC-IGBT type.

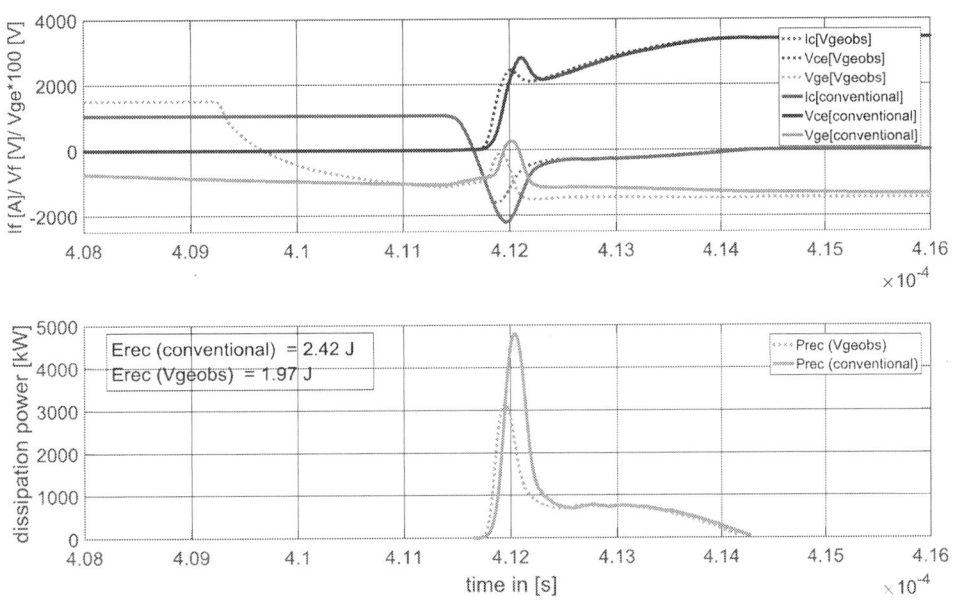

Fig. 8: Comparison of Erec between conventional and novel GDU @ V_d=3600 V; I_c= 1000 A; T= 25 °C;

Outcome of the gate-voltage observation technique for Plug-In RC-IGBTs

To evaluate the outcome of the introduced gate control technique, Fig. 8 compares the conventional diode turn-OFF (Fig. 4) to the manipulated diode turn- OFF (Fig. 5). Herby, in Fig 5, the gate-voltage observation technique is used and shows excellent results in terms of the modified switching behavior. It can be seen that the reverse-recovery current peak is significantly reduced due to the extracted plasma in the device. While the conventional gate control approach (solid lines in Fig. 8) reaches a current peak of If_{max} = -2212 A, the new control approach (dashed lines in Fig. 8) is able to reduce this value to If_{max} = -1606 A. Therefore, a reverse-recovery peak current value reduction of about impressive 27 % can be stated here. Following this, the resulting reverse-recovery losses are also significantly lower with the new gate driver concept. Hereby, the measurements show an E_{rec} of 2,42 J (solid line) for the conventional gate-drive unit. Compared to this value, the novel gate-drive unit is able to achieve about 18 % lower E_{rec}. As indicated by the dashed line in the lower diagram of Fig. 8, total reverse recovery losses add up to the significantly lower value of E_{rec}= 1,97 J).

At this point, it has to be stated again that those measurements were not performed with Plug-In RC-IGBTs due to the lack of access to those new devices. All findings were made with conventional 6.5 kV, 1000A RC-IGBTs. Admittedly, both RC-IGBT types differ in internal structural design but using conventional RC-IGBTs of the same voltage and current class to prove the functionality of the introduced gate diver concept is permissible. The results indicate that by testing real Plug-In RC-IGBTs, highly promising outcomes are expected.

Conclusion

In this paper, a novel approach concerning the turn-OFF process of Plug-In RC-IGBTs to reduce switching losses was demonstrated. With this, a more in-depth insight into the turn-OFF procedure for both current directions within the inverter locking time was given. It was shown, that the new gate-voltage observation technique can significantly reduce reverse-recovery losses (of conventional RC-IGBTS) by utilizing the full potential of the inverter locking time.

References

[1]   D. Werber *et al.*, "6.5kV RCDC: For increased power density in IGBT-modules," *Proc. Int. Symp. Power Semicond. Devices ICs*, pp. 35–38, 2014.

[2]   D. Werber *et al.*, "A 1000A 6.5kV Power Module Enabled by Reverse-Conducting Trench-IGBT-Technology," in *PCIM Asia 2015*.

[3]   R. Hermann, E. U. Krafft, and A. Marz, "Reverse-conducting-IGBTs - A new IGBT technology setting new benchmarks in traction converters," in *2013 15th European Conference on Power Electronics and Applications, EPE 2013*, 2013.

[4]   H. Wiencke, D. Lexow, Q. T. Tran, E. Krafft, and H. G. Eckel, "Plasma dynamic of RC-IGBT during desaturation pulses," in *2016 18th European Conference on Power Electronics and Applications, EPE 2016 ECCE Europe*, 2016.

[5]   D. Lexow, H. Wiencke, and H.-G. Eckel, "Improved Gate-Drive Unit for RC-IGBT to Overcome Load Current Oscillations," in *2018 International Exhibition and Conference for Power Electronics, Intelligent Motion, Renewable Energy and Energy Management (PCIM Europe)*, 2018, pp. 1–9.

[6]   D. Lexow, H. Wiencke, D. Domes, K. Fleisch, and H.-G. Eckel, "Optimized Control Method for Reverse Conduction IGBTs," in *2017 19th European Conference on Power Electronics and Applications, EPE 2017*, 2017, pp. 1–9.

[7]   M. Rahimo, "An Optimized Plug-In BIGT with No Requirements for Gate Con- trol Adaptations n P," no. May, pp. 16–18, 2017.

Differential flatness for smooth transition between grid-connected and standalone mode of three-phase inverter

Abdelhakim Saim [a, b]*, Azeddine Houari[a], Mourad Ait-Ahmed[a], Mohamed Machmoum[a], Josep. M Guerrero[c]

[a] IREENA Laboratory, University of Nantes, France.

[b] LSEI Laboratory, University of Sciences and Technology Houari Boumedien, Algeria.

[c] Center for Research on Microgrids (CROM), Aalborg University, Denmark.

Acknowledgements

This work was supported by the European Fund for Regional Development (FEDER) and RFI Electronique WISE.

Keywords:

«Microgrid», «Voltage Source Inverters (VSI)», «Flatness based Control»

Abstract

In this paper, the use of flatness based control to achieve smooth transition between Grid Connected (GC) and Stand-Alone (SA) is investigated. The idea is to exploit the flatness proprieties to ensure smooth transition between GC and SA modes. The flatness control uses planned flat outputs to calculate the inverter controls that achieve disturbances rejection. The design of the proposed control is detailed and its effectiveness is evaluated.

Introduction

The deployment of microgrid systems based on renewable energy resources is highly encouraged to address both energy consumption increase and gaz emissions reduction. These systems are composed of interconnected distributed generation inverters that operate either in GC or in SA mode. In GC mode, these systems share the produced energy with the main grid and satisfy the local load demand. In case of grid failure, these systems are disconnected from the grid to form a standalone system that satisfies the local load demand with the desired power quality standards [1]. This situation requires advanced control strategies to deal effectively with both operation modes and avoid excessive voltage and current variations during transition [2].

Several control strategies have been proposed in the literature to ensure seamless transition between GC and SA operation modes. These strategies commonly use two distinct control loops, i.e. a voltage loop in SA mode and current loop in GC mode [3]. The transition between GC and SA operation mode results in abrupt changes in current and/or voltage references, which can result in large voltage and current transients. Advanced control strategies suggest the use of parallel methods where both voltage and current loops work simultaneously either in GC or in SA mode [4]. More specifically, in SA mode, the input control takes automatically the output of the voltage loop while the grid current reference is equal to zero, and inversely in the case of GC mode. In the same line, unified control strategies have been proposed where only the output voltage reference is switched [5]. These techniques achieve higher performance under operation mode change. Nevertheless, these methods consider constant voltage references, and this is inappropriate to address the effect of the interconnecting grid impedance. To

---

* Corresponding author e-mail: abdelhakim.saim@univ-nantes.fr / asaim@usthb.dz

address this issues, unified droop based control methods are proposed [6]. These methods allow operating the system in both operation modes with only smooth references changes. In addition, and in order to secure safe connection to the grid, virtual impedance and synchronization algorithms are added.

The transition between GC and SA remains a particular concern with varying control objectives. Therefore, a flatness based control strategy is proposed in this paper to ensure a smooth transition between GC and SA modes. Notice that the flat model is parameterized by the planned trajectories that impose the desired response. To ensure that the flatness based control is able to follow the reference changes, a state to state reference trajectory is calculated using a droop based technique with a virtual impedance algorithm.

1. System description and modeling

Fig. 1 shows the configuration of a three-phase inverter that operates either in GC or in SA mode. An LCL filter connects the inverter to the local load and the grid. The system dynamics can be expressed as follows:

$$\dot{x} = Ax + B \begin{pmatrix} u_d \\ u_q \end{pmatrix} + C \begin{pmatrix} v_{d\_PCC} \\ v_{q\_PCC} \end{pmatrix} \tag{1}$$

With,

$$x = \begin{bmatrix} v_{cd} \\ v_{cq} \\ i_d \\ i_q \\ i_{Ld} \\ i_{Lq} \end{bmatrix}, A = \begin{bmatrix} 0 & \omega & \frac{1}{C} & 0 & -\frac{1}{C} & 0 \\ -\omega & 0 & 0 & \frac{1}{C} & 0 & -\frac{1}{C} \\ -\frac{1}{L_1} & 0 & \frac{-R_1}{L_1} & \omega & 0 & 0 \\ 0 & -\frac{1}{L_1} & -\omega & \frac{-R_1}{L_1} & 0 & 0 \\ \frac{1}{L_2} & 0 & 0 & 0 & \frac{-R_2}{L_2} & \omega \\ 0 & \frac{1}{L_2} & 0 & 0 & -\omega & \frac{-R_2}{L_2} \end{bmatrix}, B = \begin{bmatrix} 0 & 0 \\ 0 & 0 \\ \frac{1}{L_1} & 0 \\ 0 & \frac{1}{L_1} \\ 0 & 0 \\ 0 & 0 \end{bmatrix} \text{ and } C = \begin{bmatrix} 0 & 0 \\ 0 & 0 \\ 0 & 0 \\ 0 & 0 \\ -\frac{1}{L_2} & 0 \\ 0 & -\frac{1}{L_2} \end{bmatrix}$$

where, u_d, u_q: are the dq-axis components of the inverter input voltages. i_d, i_q and v_{cd}, v_{cq} : are the state variables that represent, respectively, the dq-axis components of the filter inductor (L_1) currents and the filter capacitor (C) voltages. i_{Ld}, i_{Lq} : are the dq-axis loads currents (local load current and injected current to the main grid). v_{d_PCC}, v_{q_PCC} are the dq-components of the abc-voltages at the PCC. ω is the angular frequency.

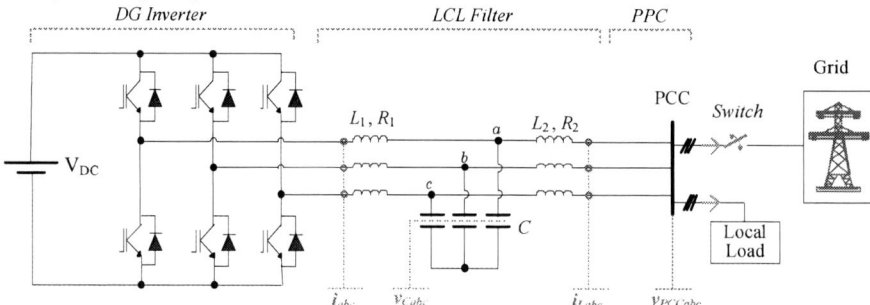

Fig. 1: Circuit diagram of the studied system.

2. Proposed control strategy

The proposed control scheme is shown in Fig. 2.

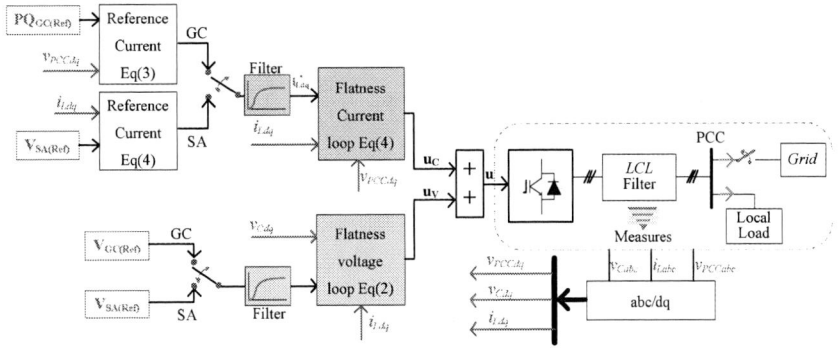

Fig. 2. Block diagram of the proposed strategy.

2.1. Flatness theory

A system is considered to be differentially flat if a set of flat outputs can be found such that all state variables and inputs can be expressed function of these components [7]. More precisely, if the system has a state x ∈ Rn, and an input u ∈ Rm, then the system is differentially flat if an output y ∈ Rm can be found of the form:

$$\begin{cases} y = \phi\left(x, u, \dot{u}, \dots, u^{(l)}\right) \\ x = \varphi\left(y, \dot{y}, \dots, y^{(r)}\right) \\ u = \psi\left(y, \dot{y}, \dots, y^{(r+1)}\right) \end{cases} \tag{2}$$

2.2. Flat model of the system

Voltage loop: The capacitor filter voltages can be defined in function of the flat outputs: $[v_{cd}, v_{cq}]^T = [y_{v_d}, y_{v_q}]^T = \varphi_V(\boldsymbol{y_V})$, while the dq-components of the line current can be defined in function of the flat outputs and their first derivative: $i_{dq} = \varphi_V(y_V, \dot{y}_V)$.

$$\begin{pmatrix} i_d \\ i_q \end{pmatrix} = \begin{pmatrix} \mathcal{C}_d \\ \mathcal{C}_q \end{pmatrix} = \begin{pmatrix} C\dot{y}_{v_d} + i_{Ld} - \omega C y_{v_q} \\ C\dot{y}_{v_q} + i_{Lq} + \omega C y_{v_d} \end{pmatrix} \tag{3}$$

The dq-components of input control can be expressed as:

$$\boldsymbol{u_V} = L_1 \begin{pmatrix} \dot{\mathcal{C}}_d \\ \dot{\mathcal{C}}_q \end{pmatrix} + \begin{bmatrix} R_1 & L_1\omega \\ -L_1\omega & R_1 \end{bmatrix} \begin{pmatrix} y_{Cd} \\ y_{Cq} \end{pmatrix} + \begin{pmatrix} y_{v_d} \\ y_{v_q} \end{pmatrix} \tag{4}$$

with,

$$\begin{pmatrix} \dot{\mathcal{C}}_d \\ \dot{\mathcal{C}}_q \end{pmatrix} = \begin{pmatrix} C\ddot{y}_{v_d} + \dfrac{d}{dt} i_{Ld} - \omega C \dot{y}_{v_q} \\ C\ddot{y}_{v_q} + \dfrac{d}{dt} i_{Lq} + \omega C \dot{y}_{v_d} \end{pmatrix} \tag{5}$$

The input vector can be parameterized as $\boldsymbol{u_V} = \psi_V\left(\boldsymbol{y_V}, \dot{\boldsymbol{y}}_V, \ddot{\boldsymbol{y}}_V\right)$.

Current loop: The proposed flat outputs y_{C_d} and y_{C_q} are related to the dq-components of the filter (L_2) currents.

$$\boldsymbol{y_C} = \begin{pmatrix} y_{C_d} \\ y_{C_q} \end{pmatrix} = \begin{pmatrix} i_{Ld} \\ i_{Lq} \end{pmatrix} \tag{6}$$

Thereby the states i_{Ld} and i_{Lq} can be put function of the flat outputs: $i_{Ldq} = \varphi_C(\boldsymbol{y_C})$.
According to equation (6), the derivation of the flat outputs leads to the expression of a fictious control input u_{C_d} and u_{C_q}. In this mode, the capacitor voltage is imposed by the main grid and the filter influence is neglected.

$$\dot{\mathbf{y}}_C = \begin{pmatrix} \dot{y}_{Cd} \\ \dot{y}_{Cq} \end{pmatrix} = \begin{pmatrix} -\dfrac{R_2}{L_2} y_{C_d} + \omega y_{C_q} + \dfrac{u_{C_d}}{L_2} - \dfrac{v_{d_{PCC}}}{L_2} \\ -\dfrac{R_2}{L_2} y_{C_q} - \omega y_{C_d} + \dfrac{u_{C_q}}{L_2} - \dfrac{v_{q_{PCC}}}{L_2} \end{pmatrix} \tag{7}$$

The control inputs can be expressed as a function of the flat outputs and its first derivative: $\mathbf{u}_C = \psi_C(\mathbf{y}_C, \dot{\mathbf{y}}_C)$.

$$\begin{pmatrix} u_{C_d} \\ u_{C_q} \end{pmatrix} = \begin{pmatrix} L_2 \dot{y}_{Cd} + R_2 y_{C_d} - \omega L_2 y_{C_q} + v_{d_{PCC}} \\ L_2 \dot{y}_{Cq} + R_2 y_{C_{dq}} + \omega L_2 y_{C_d} + v_{q_{PCC}} \end{pmatrix} \tag{8}$$

In summary, the differential flatness conditions are satisfied and the global control input can be expressed as:

$$\mathbf{u} = \mathbf{u}_C + \mathbf{u}_V \tag{9}$$

2.3. Control laws

The control laws calculate the correction terms of the flat model that correspond to the derivative of the flat output components ($\ddot{\mathbf{y}}_V$ and $\dot{\mathbf{y}}_C$) as denoted as follows:

$$\{\ddot{\mathbf{y}}_V = \boldsymbol{\delta}_V, \quad \dot{\mathbf{y}}_C = \boldsymbol{\delta}_C \tag{10}$$

Thus, a first-order and a second-order control laws are associated respectively to the current and voltage errors.

$$\left\{ \boldsymbol{\delta}_C = \dot{\mathbf{y}}_{C(Ref)} + k_{1C}\boldsymbol{\sigma}_C + k_{2C} \int \boldsymbol{\sigma}_C, \quad \boldsymbol{\delta}_V = \ddot{\mathbf{y}}_{V(Ref)} + k_{1V}\dot{\boldsymbol{\sigma}}_V + k_{2V}\boldsymbol{\sigma}_V + k_{3V} \int \boldsymbol{\sigma}_V \right. \tag{11}$$

where, $\boldsymbol{\sigma}_C = (\mathbf{y}_{C(Ref)} - \mathbf{y}_C)$ and $\boldsymbol{\sigma}_V = (\mathbf{y}_{V(Ref)} - \mathbf{y}_V)$. The controller gains (k_{1C}, k_{2C}) and $(k_{1V}, k_{2V}$ and $k_{3V})$ are selected respectively by matching the desired characteristic polynomials.

2.4. Flat outputs trajectories

The flat outputs references depend on the mode operation:

GC mode: The current reference is calculated in respect to the powers participation as follows:

$$\left\{ i_{Ld}^* = \frac{P_L^* v_{d\_PCC} + Q_L^* v_{q\_PCC}}{(v_{d\_PCC}^2 + v_{q\_PCC}^2)}, \quad i_{Lq}^* = \frac{P_L^* v_{q\_PCC} - Q_L^* v_{d\_PCC}}{(v_{d\_PCC}^2 + v_{q\_PCC}^2)} \right. \tag{12}$$

The reference voltage is set equal to the measures $(v_{cd}^* = v_{cd}, \quad v_{cq}^* = v_{cq})$ to not influence the current control.

SA mode: The voltage references are set to fixed voltage magnitude and frequency $(v_{cd}^* = -\sqrt{3}V_{RMS}, \quad v_{cq}^* = 0)$.
The current reference is calculated through the measured load powers.

$$\left\{ i_{Ld}^* = \frac{P_{L2} v_{cd} + Q_{L2} v_{cq}}{(v_{cd}^2 + v_{cq}^2)}, \quad i_{Lq}^* = \frac{P_{L2} v_{cd} - Q_{L2} v_{cq}}{(v_{cd}^2 + v_{cq}^2)} \right. \tag{13}$$

$$\{ P_{L2} = v_{cd} i_{Ld} + v_{cq} i_{Lq}, \quad Q_{L2} = v_{cq} i_{Ld} - v_{cd} i_{Lq} \tag{14}$$

To ensure smooth transition of the flat outputs, second order filters of state to state type, i.e. $F(s) = 1/\left(1 + \dfrac{2\xi}{\omega_F} s + \dfrac{1}{\omega_F^2} s^2\right)$, where ω_F is chosen to set a physical feasible trajectories for the power system.

3. Simulation study

To verify the validity of the proposed control strategy, simulations under Matlab/Simulink are conducted under transient conditions. The system parameters and control gains are summarized in Table 1.

Table I: Power stage and control parameters

Symbol	Value

Vdc	600 V
Vg /F	110 V/50 Hz
L1 / L2	2 / 0.2 mh
R1 / R2	0.1/ 0.16 Ω
Cf	40 µF
Lg / Rg	0.05/0.0016 mh/ Ω
Rload/ Lload/ Cload	50 Ω /0.04 mh /100 µF
Kpv / Kiv	150 /1500- / s-1
Kpc / Kic	250 /2500- / s-1
	20 kHz

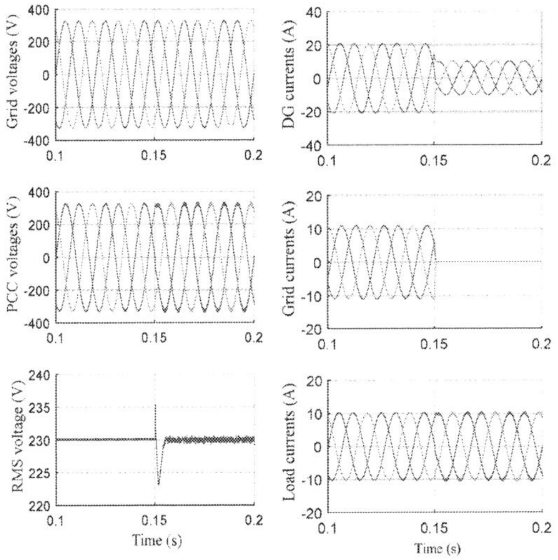

Fig. 3. Transition from GC to SA mode: from top to down (Left): grid voltages, PCC voltages and RMS voltage. (Right): DG currents, grid currents and local load currents.

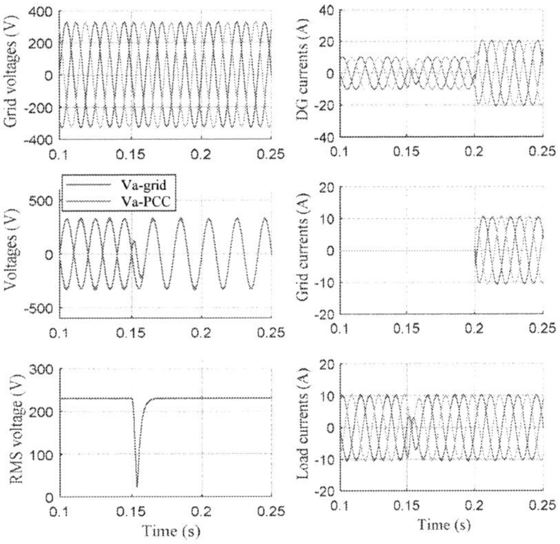

Fig. 4. Transition from SA to GC mode: from top to down (Left): grid voltages, grid side and PCC phase 'a' voltages and the PCC RMS voltage. (Right): DG currents, grid currents and local load currents.

To illustrate the role of the planned trajectory for a smooth transition of the system under disturbances, tests in case of an unintentional islanding and an intentional grid-connection are provided. Fig. 3. shows the system response under an unintentional islanding at t=0.15s. The local load is set to (5-kW, 0 VAR). As it can be appreciated, under GC mode, the inverter supplies the local load and injects power to the grid. It can be noted that disturbances at transition from GC to SA mode is maintained reduced thanks to the proposed control. Indeed, the planned trajectory algorithm provides a smooth reference, and then tracked by the flat outputs. Fig. 4 illustrates that the occurred disturbance on the outer flat outputs y_V at t=0.15 are rejected by the inner loop flat outputs (y_C). Therefore, the effect of the unintentional islanding engenders a low disturbance on the voltage i.e. below 5% of the rated voltage.

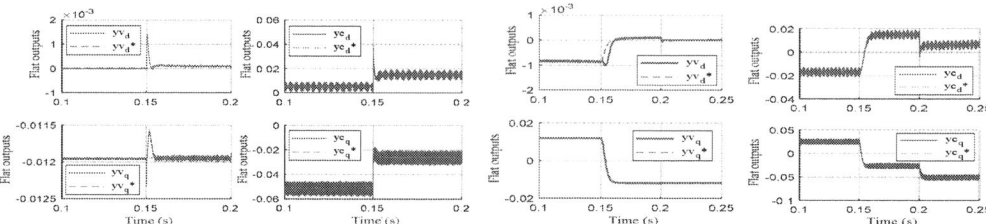

Fig. 5. Transition between GC and SA modes: (Left) outer loop flat outputs and their respective references. (Right) inner loop flat outputs and their respective references.

Fig. 6. Transition from SA to GC mode: (Left) outer loop flat outputs and their respective references. (Right) inner loop flat outputs and their respective references.

Figs.5 and 6 show the system response under an intentional grid connection at t=0.2s. The provided results can be splitted in three steps: (0.1-0.15 s) islanded mode without synchronization; (0.15 to 0.2 s) islanded mode with synchronization and from 0.2 to 0.25 s GC mode. As it can be appreciated in Fig. 5, the abc-voltages of the grid and the DG system are initially shifted by 180° and then synchronized from 0.15s to 0.2s. Note that the synchronization process causes a consequent voltage drop. This aspect can be solved by weighting the synchronization factors, but this engender greater time for the synchronization. When the inverter voltages are synchronized with grid voltages the system is connected to the grid (at t=0.2s). In Fig. 6 it can be noted that the flat outputs follows their reference and not highly disturbed when the reference change. i.e. synchronization and grid connection. As consequence, the synchronization and the grid connection happens without causing any major disturbances on the system.

Conclusion

In this paper, a control scheme based on flatness control is proposed for seamless transfer operation between GC and SA modes of three-phase inverter. The main idea is to generate physically feasible trajectories that the flat model of the system can realize. Hence, the transition effect of a sudden transition between GC and SA modes are rejected and the control performance still relevant. The flatness principle is applied to check the flatness of the studied system and then an appropriate trajectory planning is proposed.

References:

[1]   A. Saim, A. Houari, J. M. Guerrero, A. Djerioui, M. Machmoum, and M. A. Ahmed, "Stability analysis and robust damping of multiresonances in distributed-generation-based islanded microgrids," *IEEE Trans. Ind. Electron.*, vol. 66, no. 11, pp. 8958–8970, Nov. 2019.

[2]   Z. Guo, D. Sha, and X. Liao, "Voltage magnitude and frequency control of three-phase voltage source inverter for seamless transfer," *IET Power Electron.*, vol. 7, no. 1, pp. 200–208, 2014.

[3]   Z. Zeng and W. Shao, "Reconnection of micro-grid from islanded mode to grid-connected mode used sliding Goertzel transform based filter," *IET Renew. Power Gener.*, vol. 11, no. 7, pp. 1041–1048, Jun. 2017.

[4]   G. Lou, W. Gu, J. Wang, J. Wang, and B. Gu, "A Unified Control Scheme Based on a Disturbance Observer for Seamless Transition Operation of Inverter-Interfaced Distributed Generation," *IEEE Trans. Smart Grid*, vol. 9, no. 5, pp. 5444–5454, Sep. 2018.

[5]   Z. Liu, J. Liu, and Y. Zhao, "A unified control strategy for three-phase inverter in distributed generation," *IEEE Trans. Power Electron.*, vol. 29, no. 3, pp. 1176–1191, 2014.

[6]   H. M. Hasanien and M. Matar, "A fuzzy logic controller for autonomous operation of a voltage source converter-based distributed generation system," *IEEE Trans. Smart Grid*, vol. 6, no. 1, pp. 158–165, Jan. 2015.

[7]   A. Djerioui *et al.*, "Flatness-Based Grey Wolf Control for Load Voltage Unbalance Mitigation in Three-Phase Four-Leg Voltage Source Inverters," *IEEE Trans. Ind. Appl.*, vol. 56, no. 2, pp. 1869–1881, Mar. 2020.

Differential Model EMI Filter Analysis for Interleaved Boost PFC Converters Considering Optimal Phase Shifting

Naser Nourani Esfetanaj
Department of Energy
Technology, Aalborg
University
Aalborg, Denmark
nne@et.aau.dk

Yamen Saad
Converdan A/S,
Rødding, Denmark

ys@converdan.com

Omar Ahmed Sakaria, Huai Wang, Pooya Davari
Department of Energy Technology, Aalborg
University
Aalborg, Denmark
soah17@student.aau.dk,
hwa@et.aau.dk,
pda@et.aau.dk

Keywords

« DM EMI Filter analysis », « Interleaved PFC Converters », « phase-shifting ».

Abstract

Interleaved Power Factor Correction (PFC) has become a most popular topology from efficiency and power density point of view over single-switch boost PFC. The dependency of the Differential Model (DM) Electromagnetic Interference (EMI) noise magnitude on input current ripple leads to investigate the influence of the interleaved technique on EMI noise. Hence, this paper provides a comprehensive investigation for the design of DM EMI filter a single-phase interleaved PFC targeting to minimize component size. It is shown how different operation modes (continuous and discontinuous conduction mode) and switching frequency may influence the required filter attenuation and, consequently, the EMI filter size. Furthermore, the impact of the number of interleaved stages and optimal phase shifting on the required filter attenuation is analyzed. Finally, the influence of optimal phase shifting achieve an overall minimum EMI filter corner frequency is discussed. Experimental results from a 2 kW interleaved single-phase boost PFC converter validate the effectiveness of the proposed optimal phase shifting method.

Introduction

Complying harmonic standards and power factor of the input AC power led to the development of boost power factor correction (PFC) circuits to achieve a power factor close to unity. Moreover, using interleaving PFC has many advantages for instance increase power densities, reducing the overall volume of the design, and reduce RMS current in the boost capacitor. Furthermore, the use of the interleaved configuration, which is shown in Fig.1, brings a significant reduction in the switching frequency ripple component due to the ripple cancelation effect [1]. Notably, ensuring sinusoidally shaped input currents in connection with DM EMI input filters, which are limiting high-frequency noise from being transmitted from the converter to the grid [2]. However, high penetration of power electronics converter in the grid causes some challenging EMI issues due to inherent pulse energy conversion characteristics. Thereby, these unintended emissions must be limited to fulfill noise emission standards, such as CISPR 11 for frequencies beyond 150kHz [3]. Whereby, due to the increasing demand of the pulse-width modulated (PWM) converters, some standards are defined under 150 kHz in some application instance, CISPR 14 (Induction hobs) [4], and CISPR 15 (Lighting equipment) [5]. Further, it is accepted that the decreased input ripple current reduces the DM EMI noise magnitude and filter requirement attenuation which makes the DM EMI filter size smaller and corner frequency higher [1]. Furthermore, the EMI filter is one of the effective methods for damping EMI noise emission. From the EMI perspective view, finding the optimal phase shift angles which give the optimal corner frequency is a big challenge. Hence, designing optimal DM EMI filters for interleaved boost PFC applications can be considerably challenging, especially in low-frequency EMI range between 2-150kHz. This paper investigates the effect of phase-shift and number of interleaved stages in a single-phase PFC on DM EMI filter sizing. It is shown that an optimal phase-shift angle can be found depending on the power converter selected switching frequency, number of interleaved stages and its mode of operation which can minimize the EM filter size. Moreover, in order to highlight the importance of phase-shift control

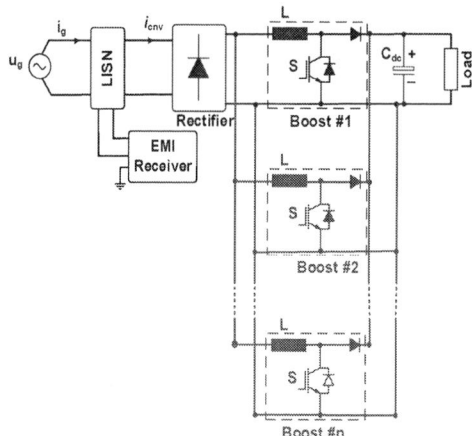

Fig. 1: Topology of Interleaved boost PFC converter with including LISN and EMI receiver.

in different applications, the studies are carried out for both below and above 150 kHz frequency ranges of standard requirements.

This paper is organized as follows. Section II shows the designing process for the two-step EMI filter. Further, it describes the EMI measurement setup according to the CISPR standard, including line impedance stabilizing network (LISN) and EMI receiver. Section III presents to design optimal DM EMI filter for interleaved boost PFC due to finding the optimal filter corner frequencies with considering two-step EMI filters in Band A and B. Subsequently, the benefit of optimal phase shifting in interleaved units will be developed, where it will be shown that the amount of filter attenuation decreasing. Section IV illustrates the experimental results achieved for the two interleaved boost PFC converter by presenting the standard or optimal phase shift. Finally, conclusions are drawn in Section V.

EMI Simulation

DM Filter Design

The EMI filter is used to protect the utility from the high frequency conducted emissions noise, which should comply with EMI standards requirements, more details about the emission standard are discussed before. Therefore, the design of a two-stage filter structure, as shown in Fig. 2, is considered in the following. Further, the primary purpose of the EMI filter is damping emission noise and make it lower than limits to fulfill the standards limit. Furthermore, the selecting filter component is depending on the required filter attenuation Att_{req}, which can be calculated from

$$A_{tt_{req}}(f)[dB] = U_{max}(f)[dB\mu V] - CISPR_{limit}(f)[dB\mu V] + M\arg in[dB] \tag{1}$$

Where, U_{max} is the maximum peak of the spectrum which can measurement based on the PLECS simulation and plotting based on the (4); Att_{req} is the quantity of noise that filters should be damping it. $CISPR_{limit}$ can be found from the standard requirements, which is shown in Fig. 3. Moreover, the filter designing margin is considered 6 dB due to component degradation and EMI parameter tolerance. Hence, two-step EMI filter including inductor and capacitor size can be found as [6]

$$A_{tt_{req}}(f) = \left| ((j2\pi f)^2 . L_{DM} . C_{DM} + 1)^2 + (2\pi f)^2 . L_{DM} . C_{DM} \right| \tag{2}$$

As discussed before, the reduce the input ripple current can be effect on the DM EMI noise magnitude (U_{max}), which will make the DM filter smaller. In addition, the dependency of EMI filter corner frequency to filter component, lead a challenge in selecting filter component size. Hence, lowering the EMI filter component size makes the DM filter corner frequency higher. Whereby, the filter corner frequency can be found from

$$fc = \frac{1}{2.\pi\sqrt{L_{DM}C_{DM}}} \tag{3}$$

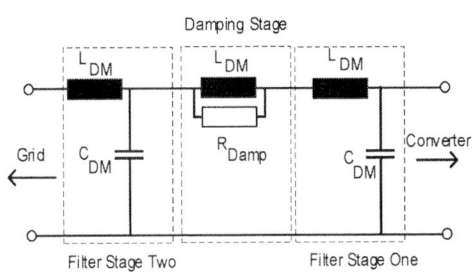

Fig. 2: Two stages DM EMI filter configuration.

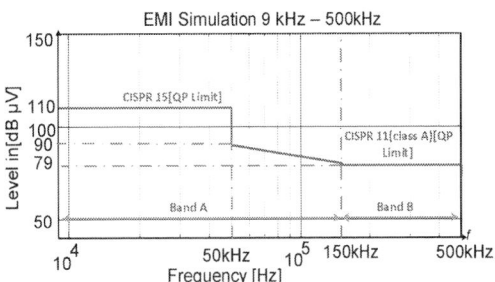

Fig. 3: Considered emission limits following CISPR 15 [5] and CISPR 11 [3] based on QP (Quasi Peak). Band B is only shown up to 500 kHz.

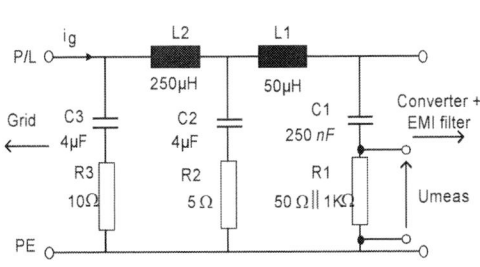

a)

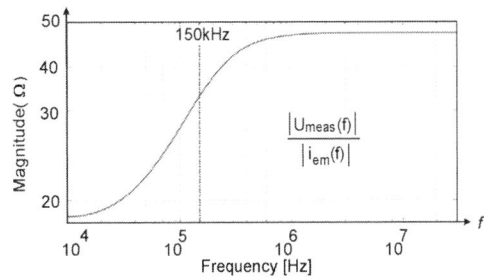

b)

Fig. 4. LISN recommendation from CISPR 16 for band A, (a) per-phase circuit diagram, (b) per-phase DM mode transfer function [6]-[7].

LISN and EMI Receiver

A LISN is specified for EMI tests according to CISPR 16[8], to guarantee the repeatable of the measurements. Moreover, LISN, not only provides the decoupling line from the device under test (DUT) but also provides an interface between the DUT and the test receiver. The structure of the LISN for using in 9kHz-30MHz is shown in Fig 4. Notably, LISN can measure an RMS time-domain voltage (u_{meas}) to define EMI noise based on the (4). Hence, the EMI test receiver used QP detection for detecting EMI peak measurement. Moreover, Bandwidth of the 4th order Butterworth bandpass filter for band A(9-150kHz) is 200Hz, and for band B(150kHz-30MHz) is 9kHz. Finally, with considering of (4), EMI peak measurement [7]-[9] can be estimated, by

$$U_{max}[dB\mu V] = 20\log[1/\mu V \sum_{f=MB-\frac{BW}{2}}^{f=MB+\frac{BW}{2}} u_{meas}(f)\cdot RBW(f)] \tag{4}$$

Investigation of EMI Filter Design

Non-Interleaving Single Unit

In this part, EMI filter design for non-interleaving 1-unit PFC, which is shown with black color in Fig. 1, is studied. The boost-inductor design for continuous conduction mode (CCM) [10]-[11], and discontinuous conduction mode (DCM) [12] operation has already been presented in the literature. Hence, for the sake of simplicity, only the most well-known equations will be specified in this section. Notably, depending on the inductor current, the boost PFC converter can operate in different modes, including CCM and DCM. Hence, designing an EMI filter to fulfill the standard requirements, for both scenarios is different due to different currents ripple. Furthermore, Inductor current for CCM and DCM operations can be calculated respectively from

$$L_{CCM} = \frac{u_o}{4.\Delta i_{L,max}.f_{sw}} \tag{5}$$

$$L_{DCM} = \frac{u_g^2.(1 - \frac{u_g}{u_o})}{4.\Delta i_{L,max}.f_{sw}} \tag{6}$$

Parameter values for single-phase PFC in order to finding filter required attenuation with respect to different frequency switching in CCM and DCM as u_o = 400 V, u_g = 230√2 V, P_{max} = 1000 W, $\Delta i_{L,max}$ = 0.62 A. Hence, inductor sizes are can be calculated from (5) and (6) for the different case studies. So, inductor size for CCM and DCM mode are presented at table I. It is clear that the size of the inductor for DCM mode is lower than CCM mode. Whereby the maximum peak of the spectrum (U_{max}) for different frequency switching can be found based on the PLECS simulation and (4). Therefore, the EMI filter component (L_{DM} C_{DM}) and corner frequency can be achieved from (2) and (3), respectively. In the following, Fig. 5(a) and Fig 5(c) shows the Att_{req} requirement to design a DM two-stage EMI filter based on the (1) in different switching frequency. So, Fig. 5(b) and Fig 5(d) exposes the filter corner frequency (f_c) with considering different switching frequency for two stages DM EMI filter in band A and B based on the (1)-(2). It is clear from Fig. 5(a) and Fig. 5(b), Att_{req} increases significantly at f_{sw} = 50 kHz because the standard limit (CISPR 15) gets more restricted beyond = 50 kHz for band A which is shown from Fig. 3. Furthermore, it is generally known that CCM has the lower current ripple in comparison to DCM modes. Therefore, as a consequence of Fig.5(a) and (d), under CCM operation lower attenuation (Att_{req}) is required. One can see that, due to the first noise peak will be appeared at the switching frequency f_{sw} in band A, the best f_{sw} selection is to be higher than 150 kHz. But if f_{sw} has to be less than 150 kHz,

Table I: Inductor size value for single phase unite in DCM and CCM mode based on the (5) and (6).

	Switching Frequency(kHz)												
	20	25	30	35	37.5	45	50	70	75	140	150	250	500
CCM (mH)	8.06	6.45	5.38	4.61	4.30	3.58	3.23	2.3	2.15	1.15	1.08	0.65	0.32
DCM (µH)	247	197.6	164.7	141.1	131.7	109.8	98.8	70.6	65.8	35.3	32.9	19.7	9.8

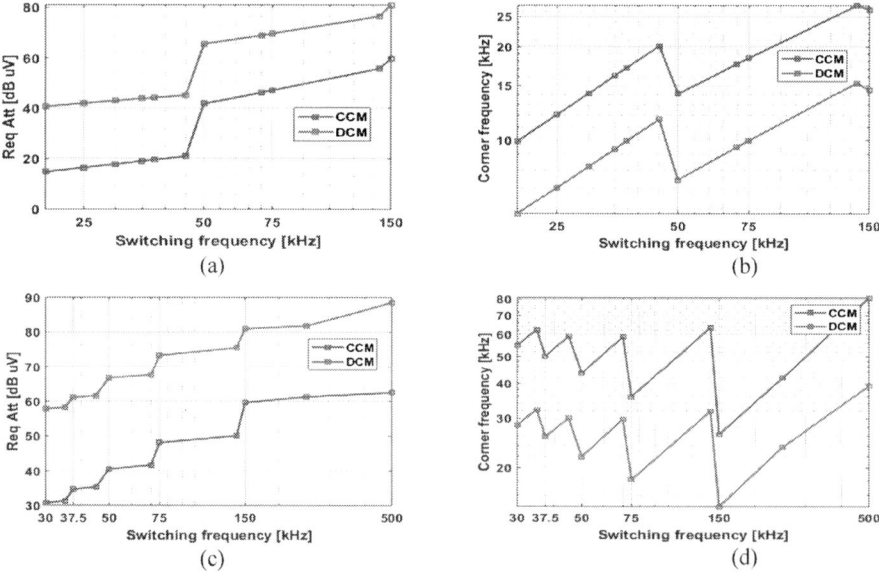

Fig. 5: The relation between the required attenuation and the switching frequency in a) Band A. c) B B base on the (1) for single converter unit. The relation between the two-stage filter corner frequency the switching frequency in b) Band A, d) Band B base on the (1)-(2).

therefore, two or higher stages filter should use to damp attenuation requirement. As it is clear that from the Fig. 5(a), selecting the switching frequency under the 50 kHz is need to low filter requirement attenuation in comparison with selecting above then 50kHz. Moreover, the first noise peak which appears in band B is the kth multiple of the switching frequency f_{sw}. Notably, it clear that from Fig. 5(c) that, the filter corner frequency f_c is significantly decreased at the divisors of $f_{sw} = 150$ kHz (30, 37.5, 50 ...150 kHz). As a result, the main reason for that is choosing the switching frequency to appear first noise peak f_D higher than 150 kHz in-band B. For instance, choosing $f_{sw} = 75$ kHz, the first noise peak appears at 150 kHz which is the second harmonics of $f_{sw} = 75$ kHz. Moreover, if f_{sw} is chosen 70 kHz, the first noise peak appears at $f_{sw} = 210$ kHz which is the third harmonics of $f_{sw} = 70$ kHz. Hence, the first peak above 150kHz for the case with $f_{sw} = 70$ kHz is the lower magnitude and higher frequency in comparison to the second case with $f_{sw} = 75$ kHz. Whereby, it is not efficient to switch at the previously mentioned critical frequencies and use a switching frequency just a bit lower than them, which will increase the filter corner frequency without affecting the boost inductor size.

Interleaving Using Standard Phase-Shift

The interleaved boost PFC and the beneficiary have been previously introduced in the literature [1]. In this section, the interleaving technique to achieve optimal design DM EMI filter has been studied. Hence, up to four interleaved units have been run at different switching frequencies in band A and band B, to achieve the connection between required attenuation with interleaving and optimal phase shift. Notably, the typical phase shift $360°/N$ (N is the number of the interleaved converter) is used here between the interleaved units. As discussed before, the inductor size for interleaved will be $L_1 = NL$ due to ensure constant energy storage. As previously mentioned, in band A, the first noise peak will appear at the switching frequency f_{sw}. By interleaving, the equivalent switching frequency will be $N f_{sw}$, where N is the number of the interleaved units. For example, in two units interleaved at $f_{sw} = 25$ kHz, the equivalent switching frequency will be 50 kHz. Interleaved technique can be the decreased ripple current, and $K_C(d)$ is defined as a cancellation factor, which is the ratio between the input current ripple after interleaving and the inductor current ripple in Non-interleaving one unit:

$$K_c(d) = \frac{N.(d - \frac{m}{N}).(\frac{m+1}{N} - d)}{d.(1-d)} \tag{7}$$

Where N is the number of the interleaved units, $m = floor\ (N.d)$ and, d is the duty cycle. More information about the effect of interleaving on ripple current has been presented on [1]. Fig. 6(a) is shown the connection between the required attenuation and the switching frequency up to four units for CCM mode in Band A. It can be seen in Fig. 6(b) and 6(d) that there is a drop in the filter corner frequency at $f_{sw} = 50$ kHz and $f_{sw} = 25$ kHz for one and two units, respectively. Since the equivalent switching frequency is $N f_{sw}$, then there is no need to use a filter if the switching frequency higher than 75 kHz in two units, 50 kHz in three units, and 37.5 kHz in four units because there will be no peak noise in Band A. The EMI filter component values and corner frequency can be achieved from (2) and (3), respectively. So, Fig. 6(b) shows the filter corner frequency (f_c) with considering different switching frequencies for two stages DM EMI filter in band A based on the (1)-(2). Moreover, Fig. 7 shows the EMI simulation at $f_{sw} = 35$ kHz in non-interleaving (one unit), and in two units interleaved (180° phase shifting). As it is clear that, the first noise peak is appeared at $f_{sw} = 70$ kHz after interleaving while it is at $f_{sw} = 35$ kHz before interleaving. Notably, in two units interleaved, the odd order of the switching frequency harmonics is canceled out, while it does not affect the even harmonics. Hence, according to Fig. 6(d), the filter corner frequency is increasing at a specific switching frequency range (30-37.5kHz, 50-75kHz, and > 150 kHz) in two units interleaved. In addition, for one unit (non-interleaved), the first noise peak beyond 150 kHz will be the odd order of the switching frequency harmonics at those ranges. Hence, interleaving two units will be eliminated odd-order mention noise. As a consequence, then the filter will be designed based on the next even harmonics, which has a higher frequency and lower amplitude, as can be illustrated in Fig. 6(c), where the required attenuation is lower in those ranges. For instance, if $f_{sw} = 35$ kHz, the first noise peak in one unit (non-interleaved) in-band B will be at $f_{sw} = 175$ kHz, which is the 5th harmonic. But by considering interleaving two units with 180° as phase shift, the noise peak will

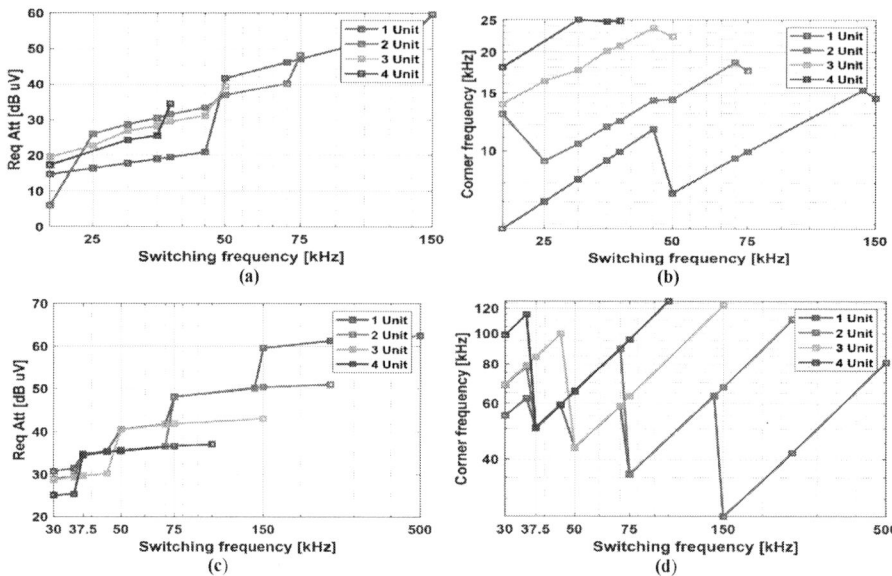

Fig. 6: The relation between the required attenuation and the switching frequency up to four units interleaved in a) Band A. c) Band B. The relation between the 2-stages filter corner frequency and switching frequency up to four units interleaved in b) Band A d) Band B for CCM based on the required attenuation (1)-(2). The typical phase shift is considered 360°/ N.

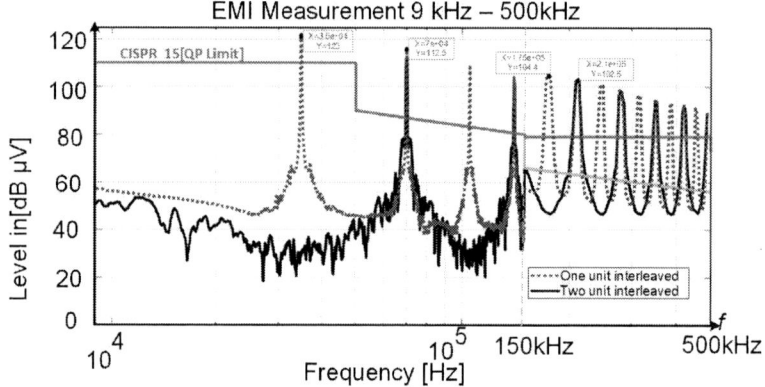

Fig. 7: EMI simulation approach for one and two-unit interleaved CCM at f_{sw} = 35 kHz based on PLECS software (interleaved typical phase shift is equal 180°).

have happened at f_{sw} = 210 kHz (6th harmonic), which is shown in Fig. 7 shows. Therefore, the filter size decreases due to it happened at a high frequency.

Interleaving Using Optimal Phase-Shift

The conventional interleaving with typical phase-shift does not give any beneficiary at some switching frequency ranges like 75-150 kHz in band B, which is clear that form Fig.6(c). Hence, to target the cancellation effect to occur at any order of harmonics, a different phase-shift angle will be presented in this section. Hence, Fig. 8(a) and Fig.8(b) are given an optimal phase-shift effect on required attenuation, and filter corner frequency in different switching frequencies on two unites Interleaved, respectively. As it is visible that the required attenuation and filter corner frequency are the same for one unit non-interleaved and two units with a standard phase shift between 75-150 kHz frequency range. Notably, using optimal phase shifting is can be provided a beneficiary instance decreasing required attenuation

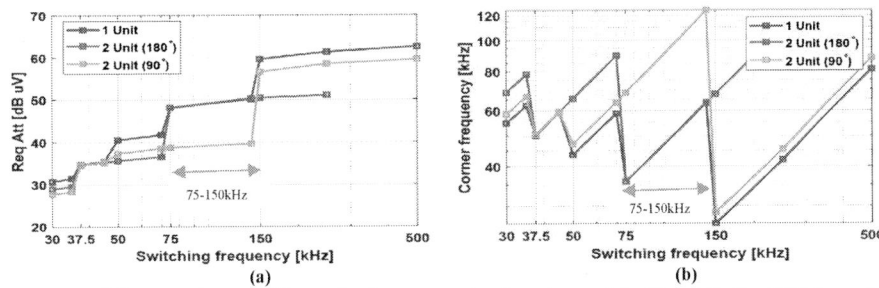

Fig. 8: Choice different phase shift angle for two units interleaved in Band B for CCM mode, a) The relation between the required attenuation with the switching frequency, b) The relation between the filter corner frequency with the switching frequency.

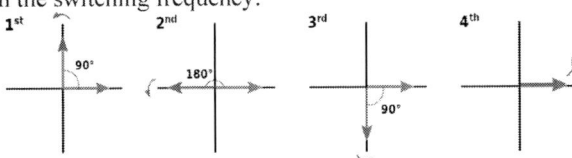

Fig. 9: Noise phasor diagram for two units interleaved with 90° phase shift.

and increasing corner frequency for using Interleaving technique on switching frequency range 75-150kHz. Because of optimal phase shifting can eliminate the second-order harmonics of the switching frequency, which is the first noise will appear above then 150kHz if the switching frequency is choice at the range (75kHz- 150kHz). Notably, Fig. 9 shows the noise phasor diagram for two units interleaved with a 90° phase shift. As can be found on the 2nd, the total noise will be zero due to the noise phase cancelation of two units.

$$\theta = \frac{360^o}{N} \quad \text{if } k \text{ not a multiple of } N \tag{8}$$

$$\theta = \frac{360^o}{\min(\text{floor}(N.k))} \quad \text{if } k \text{ a multiple of } N \tag{9}$$

Where k is the harmonic order of the frequency switching, which will be appeared as a first noise peak in Band B. In addition, the resulting of optimal phase shift angle, found by calculations according to (8) and (9) are summarized in Table II. Whereby, two simulation cases are examined to confirm the optimal phase shift, especially in band B with different phase-shifting (180° and 45°). In the following, Fig. 10 illustrates obtained comparative results with the proposed optimal phase shift with and without EMI filters f_{sw} = 20 kHz. Hence, Table III summarizes the outcomes of two simulation case study, including required attenuation and corner frequency. Notably, a 45° phase shift in comparison of 180° is needed lower filter attenuation in the band B while 180° need to high filter attenuation.

Table II: Optimal phase shift angles at the switching frequency range (30-150kHz) up to four units interleaved based on the (8) and (9).

Frequency	30kHz	37.5kHz	50kHz	75kHz	150kHz
Harm. order:	5th	4th	3th	2th	1th
2 Units	**180°**	45°	**180°**	90°	**180°**
3 Units	**120°**	120°	40°	**120°**	**120°**
4 Units	**90°**	45°	**90°**	90°	**90°**

Table III: Band B EMI filter design based on PLECS software for standard (180°) and optimal phase shift(45°) base on the (8) and (9).

Phase	Δi_L	L(mH)	f_D[kHz]	Attreq[dB]	L_{DM} (µH)	C_{DM} (nF)
180°	2.6	1	33.2	52.3	180	**127**
45°	2.6	1	43	43.5	180	**76**

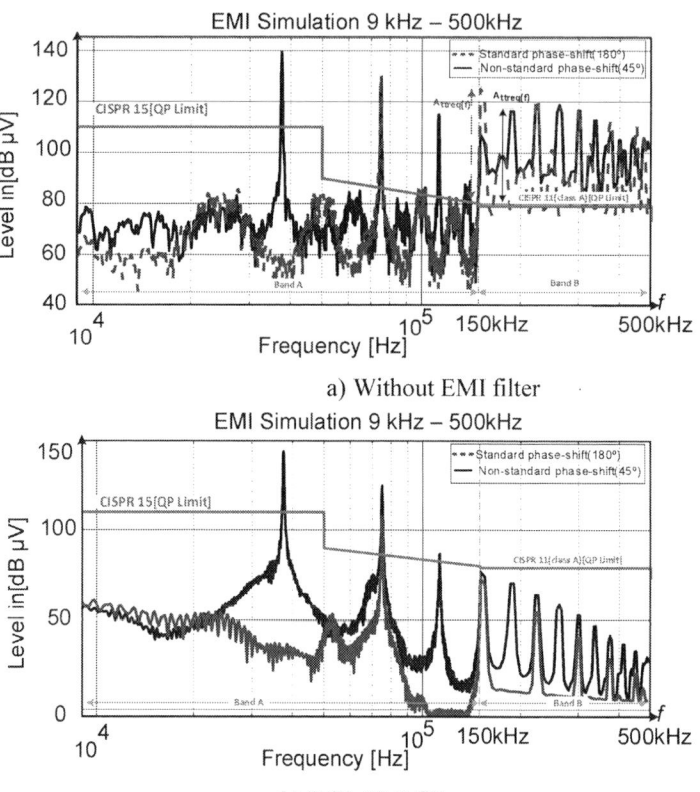

a) Without EMI filter

b) With EMI filter

Fig. 10: EMI simulation approach for two-unit interleaved CCM at f_{sw} = 37.5 kHz based on PLECS software for standard (180°) and optimal phase shift (45°) base on the (8) and (9).

Experimental Results

To evaluate the previously done examinations, the two-unit Interleaved boost PFC rectifier shown in Fig. 1 is considered under CCM, as it is summarized in Table IV. A laboratory setup including EMI receiver and LISN and the two-unit interleaved converter provides simulation verification. Moreover, the simulation model has been run in PLECS software. The sampling frequency of simulation and experimental results is 100 kHz. Whereby, Fig. 11 is measured experimental waveform of two units -

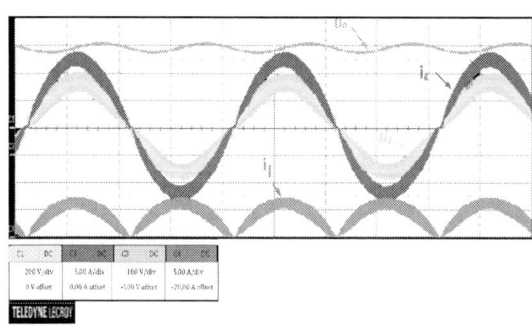

Fig.11 Measured experimental waveform of two units interleaved with Table IV parameters (f_{sw} = 20 kHz) and phase shift 90°

Table IV: Case study specification

Grid Voltage	u_g	230 V
Output Voltage	U_o	400 V
Output Power	P_o	2 kW
Fundamental Frequency	f_o	50 Hz
DC Link Inductor	L	1.8mH
DC-Link Capacitor	C	500 µF
Switching Frequency	f_{sw}	20 kHz
Capacitor Ripple	dV_{dcmax}	20 V
Inductor Ripple	di_{lmax}	20 A
Phase Shift	degree	0°, 90°, 180°

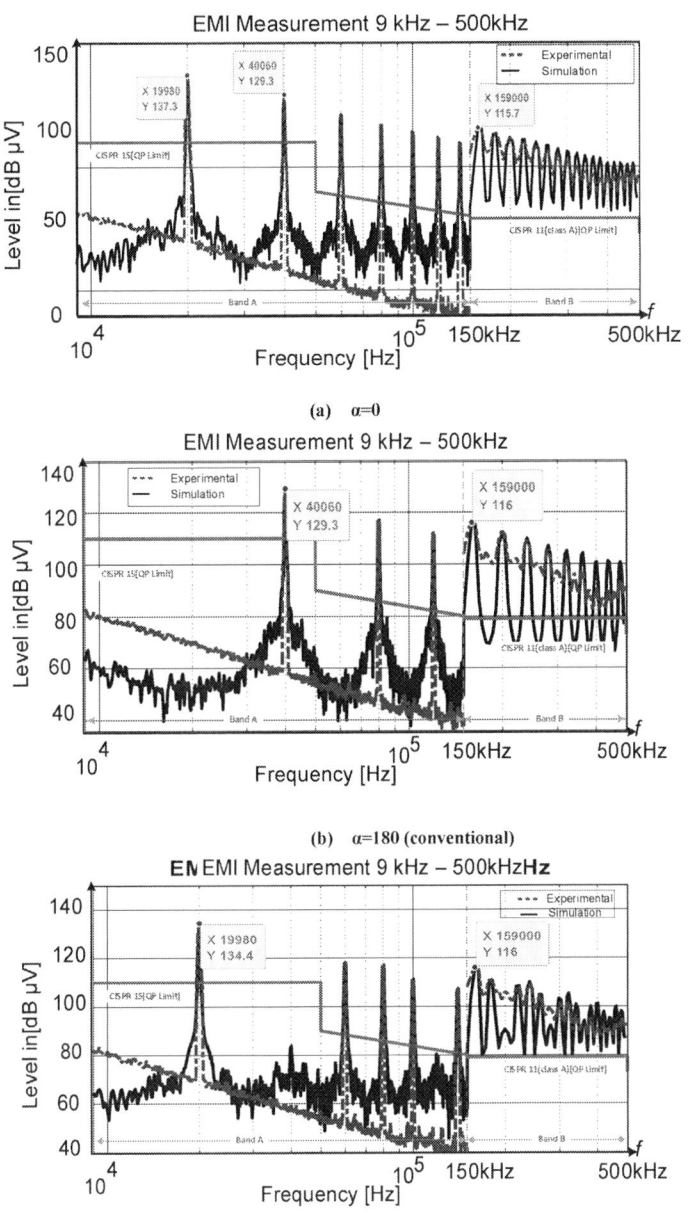

(a) α=0

(b) α=180 (conventional)

(c) α=90 (non-conventional)

Fig. 12: Obtained EMI measurement for 2-unit Interleaved Boost PFC converter, simulation-based PLECS software, and experimental measurement. Test system specification is based on the Table (IV).

interleaved with Table IV parameters (f_{sw} = 20 kHz) and phase shift 90°. Moreover, the first test case is two-unit interleaved with phase shift a = 0° and this the same as using 1-unit (Non-interleaved), but the difference is that total boost inductor will be two times of non- interleaving. Hence, Fig.12(a) illustrates the simulation and experimental results for two-unit interleaved without EMI filter and phase shift α = 0°. In the next step, the second test has a 180-degree phase difference between two Interleave PFC, which call conventional interleave (standard phase shift). It is clear from Fig.12 (b), the experimental outcome is validated simulation by considering the standard phase shift between the units. Notably, the first order of harmonics appears in $2f_{sw}$ in comparison with α = 180° in higher frequency.

Notably, Fig.12(c) is shown the optimal phase shift effects with considering 90° as a phase shift between the units for two-unit interleaved without EMI filter. It is clear that the second-order harmonics is disappeared with considering 90° as a phase shift based on the Fig.9. Hence, it can be used to optimize the filter size in band A. As a consequence, the results are shown the selecting optimal phase shift can be canceled selective harmonics, which is essential in calculating required attenuation in filter designing. Since the noise-emission level is above the standard requirement, which is shown in Fig. 12, designing a proper EMI filter is necessary.

Conclusion

This paper investigates the effect of optimal phase-shift selection on EMI filter optimization for both Band A (9-150 kHz) and B (>150 kHz). The obtained results show that, in band A, the interleaved configuration provides high benefit, which is the possibility of not using the filter if the switching frequency higher than 75 kHz in two units, 50 kHz in three units, and 37.5 kHz in four units. Furthermore, in-band B, it has been seen that from Fig. 8, using the typical phase-shift between the units is not efficient at all the switching frequency ranges. Therefore, different phase-shift is used to achieve a high corner frequency. Hence, a generic formula to find the phase shift angles which give the optimal corner frequency in interleaved boost PFC has been found in band B, which is a consequence of Fig. 10. In addition, this paper shows the beneficiary of the optimal phase shift in band A, to cancel selective harmonics to optimize required attenuation for EMI filter designing. Finally, the experimental outcome is to validate the standard phase shift and optimal phase shift effects for band A.

References

[1] Chuanyun Wang, Ming Xu, Fred C Lee, EMI study for the interleaved multi-channel PFC". In: 2007 IEEE Power Electronics Specialists Conference, IEEE. 2007, pp. 1336–1342.

[2] T. Nussbaumer, K. Raggl and J. W. Kolar, Design Guidelines for Interleaved Single-Phase Boost PFC Circuits, in IEEE Transactions on Industrial Electronics, vol. 56, no. 7, pp. 2559-2573, July 2009.

[3] C.I.S.P.R, Limits and methods of measurement of Radio-frequency disturbance characteristics of Industrial, scientific and medical equipment-Publication 11,2010.

[4] C.I.S.P.R., Electromagnetic compatibility - Requirements for household appliances, electric tools and similar apparatus - Part 1: Emission, vol. 14. 2016, p. IEC Int. Special Committee on Radio Interference

[5] C.I.S.P.R., Limits and methods of measurement of radio disturbance characteristics of electrical lighting and similar equipment Interference, vol. 15, 2015, IEC Int. Special Committee on Radio Interference.

[6] M. Hartmann, H. Ertl, J. W. Kolar, EMI Filter Design for a 1 MHz, 10 kW Three-Phase/Level PWM Rectifier, IEEE Transactions on Power Electronics, vol. 26, No.4, pp.1192-1204, April 2011.

[7] P. Davari, F. Blaabjerg, E. Hoene, F. Zare,: Improving 9-150 kHz EMI Performance of Single-Phase PFC Rectifier, In: CIPS 2018; 10th International Conference on Integrate.

[8] C.I.S.P.R., Specification for Radio Interference Measuring Apparatus and Measurement Methods, Publication 16, 2015, p. IEC Int. Special Committee on Radio Interference.

[9] N. Nourani Esfetanaj; S. Peyghami; H. Wang; P. Davari, Analytical Modeling of 9-150 kHz EMI in Single-Phase PFC Converter, IECON 2019 - 45th Annual.

[10] M. Kazerani, P. D. Ziogas, and G. Joos,: A novel active current wave shaping technique for solid-state input power factor conditioners," IEEE Trans. Ind. Electron., vol. 38, no. 1, pp. 72–78, Feb. 1991.

[11] P. Davari, F. Zare, A. Abdelhakim,: Active Rectifiers and Their Control, Control of Power Electronic Converters and Systems Vol. 2 2018, pp. 3-52.

[12] L. Ping and K. Yong, "Design and performance of an AC/DC voltage source converter," in Proc. IEEE INTELEC, 2000, pp. 419–423.

Modular Hybrid DC Breaker-based Adaptive Auto-Reclosing Method for MMC-HVDC Systems

Hossein Iman-Eini[1], M. Langwasser[2], L. Camurca[2], Marco Liserre[2]

[1]School of ECE, College of Engineering, University of Tehran, Tehran, Iran

[2]Chair of Power Electronics, Christian-Albrechts-Universität zu Kiel, Kiel, Germany

Tel.: +49 (0) 431-880 6101

Fax: +49 (0) 431-880 6103

Email: imaneini@ut.ac.ir, mlan@tf.uni-kiel.de, lc@tf.uni-kiel.de, ml@tf.uni-kiel.de

URL: https://pe.tf.uni-kiel.de

Acknowledgements

This work has been supported by the Alexander von Humboldt Foundation, Germany, and by the German Federal Ministry of Education and Research (BMBF) within the Kopernikus Project ENSURE "New ENergy grid StructURes for the German Energiewende" (03SFK1I0 and 03SFK1I0-2), and by the German Federal Ministry for Economic Affairs and Energy (BMWi) as part of research project RELINK (03ET7562B).

Keywords

«Adaptive auto-reclosing», « Half-Bridge MMC», «HVDC system», «Hybrid DC breaker», « MOV arresters »

Abstract

In this paper, the modular hybrid DC breaker is employed to inject active pulses after fault-current interruption and arc extinction due to a pole-to-pole fault in MMC-HVDC systems with half-bridge submodules. The power electronic branch of hybrid DC breaker is made of n series modules, which allows generating active pulses with a controllable amplitude and pulse width. Each module is a combination of metal oxide varistor (MOV) and parallel IGBTs, which can be bypassed by the IGBTs turn-on. In the proposed method, when the fault current is interrupted and the arc is extinguished, a specific number of modules are bypassed and by this action the operating point of remaining MOVs is pushed from high resistance to temporary overvoltage (or switching surge) mode. At the new operating point, the behavior of MOVs is equivalent to a DC voltage source in series with a low resistance and the voltage difference of MMC DC-link and the equivalent DC source is applied to the line. After a short delay (equal to the desired pulse width), the breaker modules are switched off and the pulse generation is ended. This action is repeated for several times and based on the monitored waveforms on the DC line, the fault type and its characteristics are recognized. In case of temporary fault, the voltage of DC line is gradually increased by sequentially turning the breaker modules on. Finally, the validity of proposed method is verified by simulations and experiments on a scaled down prototype.

Introduction

In the last decade, the modular multilevel converter based high voltage DC (MMC-HVDC) systems have gained more attention in electric power networks. These converters provide a flexible control of active and reactive powers and have a high efficiency. However, handling the dc-side pole-to-pole faults is not an easy task and depending on the MMC type, a fault current interruption method is needed. For example, the Half-Bridge MMC (HB-MMC) is not able to interrupt the DC fault current and extra DC breakers are needed to block the fault current. On the other side, the Full-Bridge MMC (FB-MMC) is able to control the DC side voltage and limit the fault current, but the power losses and cost are much higher than for the HB-MMC [1].

Another challenge in HVDC systems, especially those with overhead lines (OHLs), is to perform reclose/restart function after a temporary fault happened, to improve the power system stability and the power transmission continuity [2]. The OHLs are directly exposed to atmospheric conditions and are more prone to temporary pole-to-pole faults. To address this issue, the fault current should be interrupted in a very short time [3]. Afterwards, a method should be applied to identify the fault type and perform the auto-reclosing function in temporary fault conditions. This topic has recently been investigated in MMC-HVDC systems. In [4], an adaptive auto-reclosing concept has been introduced for single pole-to-ground faults. FB-MMCs are employed in the HVDC system and after a DC fault, all MMCs are forced to control the fault current to zero to interrupt the fault. To identify the end of arc extinction, the grounded converter will inject a sinusoidal test current. In this method, all FB-MMCs must stop active power transmission for several tens of ms and the presented method is dependent on the arc model. In [5], an adaptive auto-reclosing scheme has been proposed for HVDC systems with mixed cell MMCs. This method is applied to a symmetrical bipolar HVDC configuration with a ground return. When a pole-to-ground fault happens, the MMC in healthy pole is employed to inject DC voltage perturbations. The perturbations result in the induced characteristic signals to the faulty line, which has information about the fault property. This method, however, is not suitable for HB-MMCs because the pulse injection will cause ac power perturbations. Also, it is not applicable to monopolar configurations with ground return.

An alternative approach in HVDC systems to interrupt the fault current and to isolate the faulty line is the application of hybrid DC breakers [6-8]. These breakers can selectively isolate the faulty line and provide more flexibility in the operation of multi-terminal DC systems. In [2], the hybrid DC breaker with full-bridge submodules is used to interrupt the fault current and to inject active pulses for DC line fault identification. In fact, after a pole-to-ground fault and current interruption, the breaker remains connected to the MMC station and the MMC synthesizes the nominal dc voltage. To generate the active pulse, a certain number of full bridge cells of the hybrid breaker are bypassed. This breaker, however, needs more IGBTs and the cells capacitors must be discharged in a safe manner during active pulse generation or during auto-reclosing mode. In [9], the modular structure of hybrid DC breaker is employed to perform sequential auto-reclosing in a point-to-point HVDC system. After fault event and current interruption, the breaker waits a fixed time of 200-500 ms, and then sequentially closes the main breaker modules. During this sequence, the fault detection algorithm monitors the line to see whether the fault is permanent or not. This approach has some impact on connected ac grid, if the fault is permanent and requires a fixed time delay before each auto-reclosing attempt, which limits its applicability as fast reclosing and fast power flow recovery solution [5]. Reference [10] takes advantages of the controllability of MMC and hybrid DC breaker to synthesize active pulses. In other words, after fault current interruption and deionization of overhead line, the MMC is controlled to generate a DC voltage less than the rated DC voltage (e.g., kV_d, 0<k<1). Simultaneously, the power electronic branch of the hybrid DC breaker is turned on for a short time to provide a conduction path for the injected pulse. This method, however, needs MMCs with the DC link voltage control capability, which makes the design expensive and complicated. Furthermore, the dynamic response of MMC and control system should be very fast to synthesize the desired active pulses.

In this paper, an HVDC system with HB-MMCs is considered. The modular hybrid DC breakers are considered at the start and end point of each DC line to provide fully selective isolation of the faulty lines. In contrast to [9], the waiting time for auto-reclosing is not a fixed time and is adaptively controlled by applying active short voltage pulses for DC fault identification. With injection of short pulses with low amount of energy, a possible insulation failure and a restrike of the arc is prevented [11]. It is also possible to estimate the location of the fault by injecting active short pulses [10, 12], which is valuable in permanent fault cases. After identifying the fault type, the voltage of DC line is gradually increased by sequentially turning the breaker modules on.

The rest of paper is organized as follows: in section II, the operating principle of modular hybrid DC breaker is explained. Also, the arrester V-I curve is presented and its operating regions for applying the proposed method are discussed. Section III shows simulation results of a MMC-HVDC system with modular hybrid DC breaker and auto-reclosing scheme. In Section IV, experimental results related to a scaled down prototype are given to confirm the performance of the auto-reclosing strategy. Finally, the conclusion part is presented.

Modular hybrid DC breaker and active pulse injection

The structure of an HVDC system based on HB-MMC and a unidirectional modular hybrid DC breaker is shown in Fig.1. Here, for sake of simplicity, a symmetric monopolar HVDC system with one MMC is considered for the study but the idea can be easily extended to other configurations. The topology of modular hybrid DC breaker has already been introduced by ABB [6, 9], but in this paper it is used to inject active pulses after a fault current interruption and to perform the adaptive auto-reclosing function.

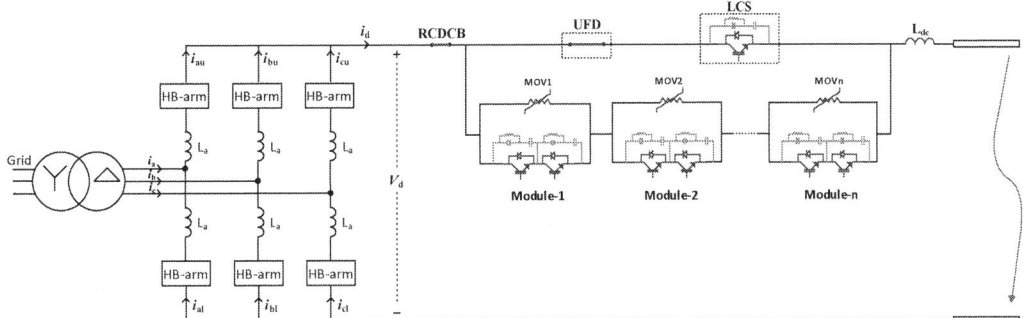

Fig.1. Schematic of a HB-MMC plus a unidirectional modular hybrid DC breaker

Topology of modular hybrid DC breaker

As it is shown in Fig.1, the modular hybrid DC breaker is made of two parallel paths, which are the load current path and the main breaker (power electronic) path. The load current path consists of a series connection of ultrafast mechanical disconnector (UFD) and a semiconductor-based load commutation switch (LCS). The main breaker is made of n series connected modules, where each module has a series connection of IGBTs with parallel RCDs. Parallel to each module, a non-linear resistor, known as metal oxide varistor (MOV) is connected, which is responsible for fault current limiting and energy dissipation after the module turn off.

In normal operation, the UFD and LCS are turned on and conduct the line current i_d. The parallel modules (or their IGBTs) are also on in the parallel breaker but a negligible current is passing through them [13]. When a DC fault current is detected, the LCS is turned off and the line current is commutated to the main breaker. Then, the mechanical switch UFD is turned off at zero current and after the mechanical contacts are fully open (after about 2 ms), the modules are turned off. The current is now passing through the non-linear MOV resistors and a counter voltage greater than V_d is built, e.g. $1.5V_d$, which helps to reduce the fault current to zero.

In the hybrid DC breaker, a residual current disconnecting circuit breaker (RCDCB) is also employed in series to the breaker. When the breaker modules are turned off and the line must be isolated, because the fault is identified to be permanent, the RCDCB disconnects the breaker from MMC and the flow of leakage current in MOVs is stopped.

Arrester VI curve

MOV arresters are highly non-linear resistors and mainly used for protection against temporary overvoltage and switching (or lightning) surges. In Fig.2, a typical arrester VI curve is shown. To be noted that the voltage is in linear scale, while the current is depicted on a logarithmic scale. Based on the definitions given in [14], the VI curve can be divided into four different regions. From left to right, the regions are titled as operating region, temporary overvoltage (TOV) region, switching and surge region. The arrester is normally working in the operating region, where it has very low conduction. Above the knee point of the VI curve, the TOV region is defined, which can extend up to several amperes. In this region, the arrester is rapidly heated due to significant energy dissipation. The next region is the switching surge region, in which the arrester cannot remain more than several milliseconds. The last region is the lightning surge region, where the arresters' conduction is very high.

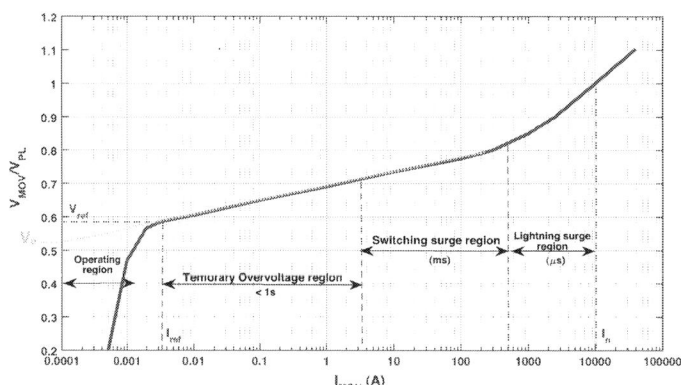

Fig.2. Typical MOV arrester VI characteristic curve and definition of operating regions

Pulse injection principle by the modular hybrid DC breaker

When a pole-to-pole fault occurs and the hybrid dc breaker interrupts the fault, all breaker modules are in off-state. According to [4], after extinction of the so-called *primary arc* at zero current and before auto-reclosing, it is mandatory to wait a certain amount of time, e.g., for several tens of ms. This waiting time allows the temperature of dielectric gas to reduce and then a low energy pulse is applied without the risk of another insulation failure and a restrike of the arc [11]. In this paper, a 50 ms delay is considered and then the first active pulse is applied. To generate the active pulse, some modules of the breaker are turned on for a specified time (here 100 μs) and then switched off again. To study the concept of pulse injection, a simplified model as shown in Fig. 3, can be considered. In this study, the MMC is replaced with a constant DC voltage source V_d and the fault is modeled by R_{sc}, which is a variable resistance in temporary fault cases and a resistance in the range of several ohms, in case of permanent faults [4].

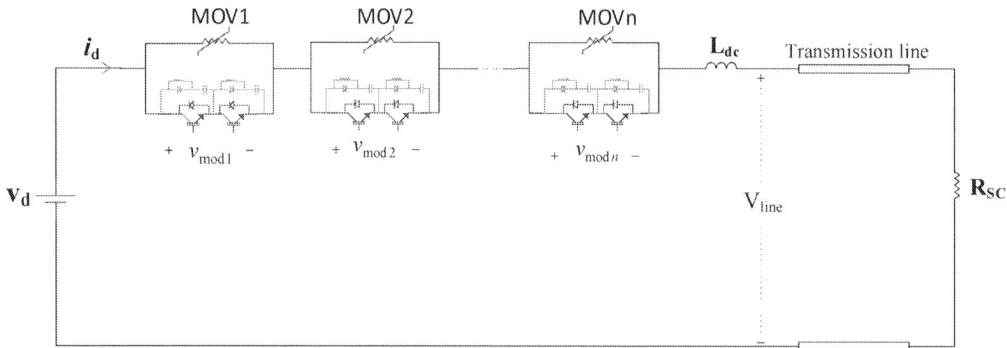

Fig.3. Simplified circuit of the system for the pulse injection study

At first, it is assumed that the R_{SC} has a small value. It is evident that before bypassing some modules, the MMC dc-side voltage V_d is divided equally among series connected MOVs and the voltage of each MOV is equal to V_d/n. From the design point of view, this operating voltage should lie in the range of 0.45 to 0.59 pu or less than V_{ref} in Fig.2. To generate an active pulse, it is assumed that a number m of the MOVs, e.g., 30% of them, are bypassed by turning the module IGBTs on. Bypassing some modules causes the voltage across remaining MOVs to increase and to push them into temporary overvoltage or switching surge regions. By neglecting the dynamic effects of MOVs, one can obtain the equivalent circuit in Fig.4 for the time interval of pulse injection. Here, each MOV is modeled with a DC voltage source and a series resistance, which their values are obtained from the piecewise-linear approximation of MOV VI curve shown in Fig.2. Also, the transmission line is modeled with π equivalent model and the utilized line parameters are R_{line}, L_{line}, and C_{line}.

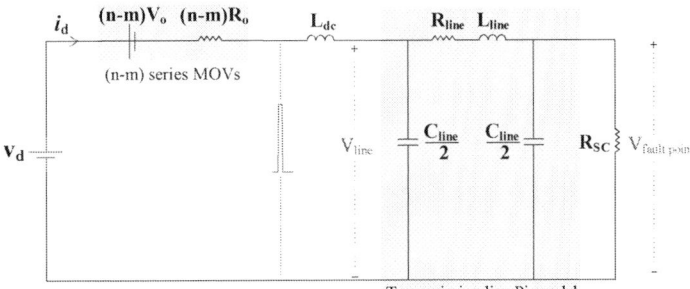

Fig.4. Equivalent circuit of the MMC, hybrid DC breaker, and transmission line during the pulse injection period

As it is seen in Fig.4, after bypassing m modules of the hybrid DC breaker, the voltage difference between the MMC dc-link V_d and the equivalent DC voltage of series MOVs, i.e., $(n-m)V_o$ will generate a voltage pulse with the amplitude of:

$$V_{pulse} = V_d - (n-m)V_o \tag{1}$$

where V_o is extracted from the piecewise-linear approximation of MOV VI curve shown in Fig.2, which is $0.52V_{PL}$ in our case study. By turning the modules off, the MOVs return to their initial condition and show the initial behavior. Hence, by applying this procedure, one can generate a controlled active pulse, where by selecting the number of bypassed modules, the pulse amplitude is controlled, and by adjusting the bypass period, the pulse width is controlled.

After generation of voltage pulse and its injection to the line, a current pulse is developed. Then, by analyzing either the current pulse (and reflected current pulses) or the voltage pulse (and reflected voltage pulses) at the sending-end, one can recognize the fault type and its characteristics. This procedure is further explained in the next subsection.

In brief, to perform the proposed method, three active pulses with a time delay of 15 ms will be applied to the line. If the fault type (after three attempts) is recognized as a permanent one, the auto-reclosing will not be performed and the RCDCB will be opened. However, in case of temporary fault, auto-reclosing is completed by the DC line voltage build-up. In fact, the modular design of hybrid DC breaker allows to increase the dc-side voltage gradually after a temporary fault [9]. To achieve this goal, the breaker modules will be sequentially bypassed with a fixed interval, e.g. 5 ms, and the voltage of DC line will be gradually increased. Afterwards, UFD and LCS are closed and the active power transmission is started.

Fault type recognition with traveling waves

The injected voltage pulse is propagating along the faulted line. The fault is assumed to happen at any arbitrary point x between the sending and the receiving end of the transmission line, forming a physical boundary to the traveling wave. Hence, at the fault point the injected voltage wave will be reflected (going back to the sending end) and refracted (going forward to the receiving end). Being Z_1 the equivalent wave impedance of the transmission line, where the traveling wave first propagates, and Z_2 the equivalent wave impedance of the transmission line, where the traveling wave reflects and refracts, in traveling wave analysis, the reflection coefficient is $\beta = \frac{Z_2 - Z_1}{Z_2 + Z_1}$ and the refraction coefficient $\alpha = \frac{2Z_2}{Z_2 + Z_1}$. In accordance with [10], assuming a permanent fault with fault impedance R_{sc}, and Z_l as the wave impedance of the transmission line, the reflection coefficient at the fault point can be found with $\beta_x = \frac{R_{sc} \| Z_l - Z_l}{R_{sc} \| Z_l + Z_l}$ and the refraction coefficient with $\alpha_x = \frac{2(R_{sc} \| Z_l)}{R_{sc} \| Z_l + Z_l}$. In general, it can be assumed that $R_{sc} \| Z_l < Z_l$ and hence the reflection coefficient at the permanent fault is negative. Since the first detected reflected voltage pulse at the sending end is proportional to $2\beta_x$, for a permanent fault, the sign of the first detected reflected voltage pulse is negative. For the following pulses it is alternating positive, negative, positive.

In the case of a temporary fault, instead, the injected voltage pulse is propagating to the receiving end, where the high impedance leads to $Z_2 \rightarrow \infty$ and hence $\beta_{rec} = 1$. In this case, the first detected reflected pulse is proportional to $2\beta_{rec}$ and hence positive. Since also the sending end DC breaker is opened, the reflection at the sending end is positive and hence all following detected pulses are also positive in sign.

Simulation Part

To verify the presented method in this paper, two different simulations are carried out in PSCAD/EMTDC environment. The studied system is a 150 kV, 150 MW system with the given parameters in Table I. It is also worth noting that the simulations are done at the switch level and in order to increase the simulation speed, only 10 submodules per arm are considered. Also, the main breaker is made of 25 series modules (or MOVs). The transmission line is modeled using a frequency-dependent line model in PSCAD/EMTDC, of which the corresponding lumped parameters are given in Table I.

Table I. Parameters of the system under study

Quantity		Symbol	Value
Nominal Power		P_b	150 MW
Peak of phase voltage		V_m	66 kV
Nominal DC link voltage		V_d	150 kV
Nominal DC line current		I_d	1 kA
Number of HB cells per MMC arm		N_{sm}	10
Grid nominal frequency		f	50 Hz
MMC cell nominal voltage		V_c	15 kV
MMC cell capacitance		C	2 mF
Trip level of fault current		I_f	2 kA
Arm inductance		L_a	10 mH
DC inductance		L_{DC}	50 mH
Fault resistance		R_{sc}	20 Ω
Number of modules in DC breaker		n	25
Number of bypassed modules		m	9
Arrester protection level voltage		V_{PL}	9.5 kV
Line parameters	Line resistance	R_{line}	0.0066 Ω/km
	Line inductance	L_{line}	0.945 mH/km
	Line capacitance	C_{line}	0.0122 μF/km

The first simulation verifies the behavior of proposed approach when the dc-side fault is temporary. In this simulation, a fault occurs at t=0.2s and it is cleared at t=2.04s. The corresponding simulation results, including the DC line current i_d, the MMC dc-side voltage V_d, the voltage across breaker V_{brk}, and the line voltage at the fault point $V_{\text{fault point}}$ are shown in Fig.5(a), respectively. Also, a zoomed view of DC line current i_d and the line voltage at the sending-end V_{line} is shown in Fig.5(b).

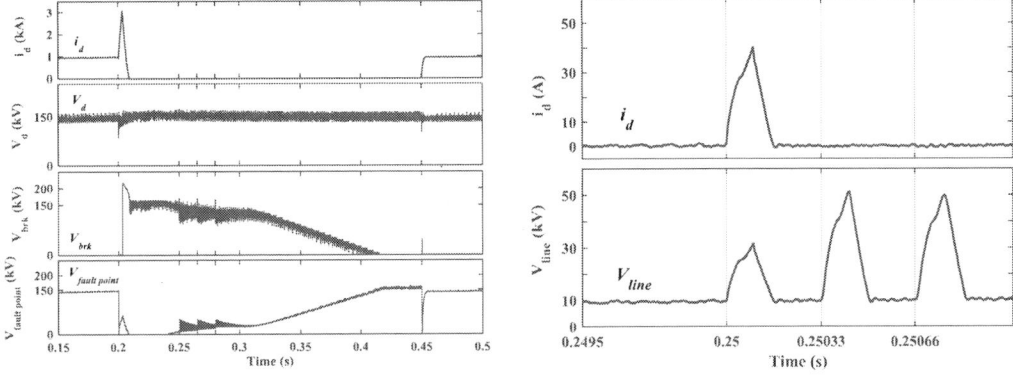

(a) Key waveforms during pulse injection times t=0.25s, t=0.265s, and t=0.28s and subsequent voltage build up

(b) A zoomed view of i_d and V_{line} during first pulse injection

Fig.5. Behavior of proposed method in temporary fault condition

As it is seen in Fig.5(a), after three attempts at time instants t=0.25, 0.265, 0.28s and recognition of fault type from monitoring of line voltage V_{line} (or line current i_d), the breaker modules are sequentially closed and the DC line voltage is gradually built up. Also, the power transmission is restored at t=0.45s. Also, Fig.5(b) shows the idea for recognition of fault type, where there is no reflection of current pulse and the reflections of voltage pulses are greater than zero. In this test, the resistance of fault point has returned to the pre-fault condition and is very high.

The second simulation verifies the system behavior when a permanent fault occurs at t=0.2s. The corresponding key waveforms and a zoomed view of DC line current i_d and the line voltage V_{line} are shown in Fig.6.

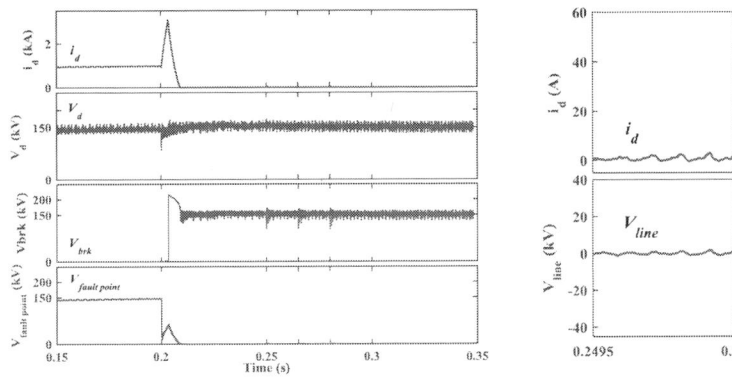 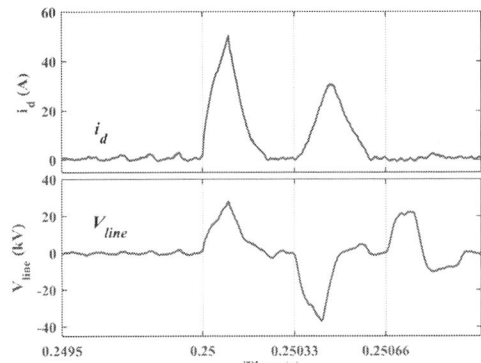

(a) Key waveforms during pulse injection times t=0.25s, t=0.265s, and t=0.28s

(b) A zoomed view of i_d and V_{line} during first pulse injection

Fig.6. Behavior of proposed method in permanent fault condition

As it is seen in Fig.6(a), three voltage pulses are applied at t=0.25, t=0.265, and t=0.28s to distinguish the fault characteristics in the line. By monitoring of either line voltage V_{line} or DC line current i_d (shown in Fig.6(b)), one can recognize the fault type. Here, the permanent fault is identified, due to the negative sign of the first reflected voltage pulse. Moreover, the time difference between the injected pulse and the reflected pulse can be used to estimate the location of fault point. In this simulation, the fault occurs at a distance of 50km and the reflected wave takes about 0.33 ms ($\approx 2 \times 50\text{km}/3 \times 10^5$ km/s) to reach the sending-end. The obtained results in Fig.6(b) confirm the validity of analytical calculations.

Experimental Part

In this part, the proposed method is tested on a laboratory prototype. In this investigation, the MMC has four SMs per arm and the main breaker is made of six series modules. The nominal dc-voltage is 150V, and the nominal dc-current is 5A. Two cases of temporary and permanent fault types are tested and the corresponding waveforms are shown in Fig.7. It is also worth noting that during the pulse injection, two modules (from six available modules) are bypassed. The time step of waveforms is 20ms per cell. Furthermore, the current scale is 10A, and the voltage scale for all channels is 200V. In the following experiment, the fault is modeled with a 2Ω resistance and it is located right after the dc-reactor L_{dc}=5mH.

At the first experiment, a temporary fault condition is examined and a similar procedure is taken to apply three voltage pulses with a time difference of 15ms. Here, to emulate the behavior of arc extinction and end of temporary fault, the fault resistance is eliminated by a solid-state relay after 40ms. Then, the first active pulse is applied with a delay of 10ms. After three pulse injections and monitoring of DC line current (or line voltage) and ensuring about the temporary fault condition, the line voltage is gradually increased and the power transmission is restored.

The second experiment verifies the system behavior when the fault is permanent and the fault resistance does not disappear. Again, three voltage pulses are applied with the explained time plan and the DC line current i_d (or the line voltage V_{line}) is monitored to distinguish the fault characteristics. After recognizing the fault is permanent, the residual current DC circuit breaker should open to disconnect the hybrid DC breaker from the MMC. The operation of this relay, however, is not shown in the following test.

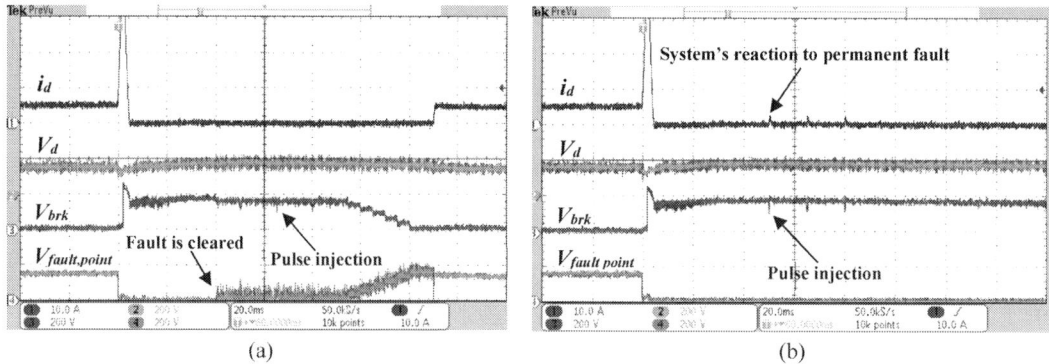

Fig.7. Experimental investigation of the proposed method in a scaled down system, (a) temporary fault condition, (b) permanent fault condition

The obtained results in Fig.7 are in good agreement with the simulation results in Fig.5(a) and Fig.6(a), which confirm the validity and effectiveness of the proposed method.

Conclusions

In this paper, the modular structure of hybrid DC breaker was employed to inject active pulses after fault current interruption in a MMC-HVDC system with half-bridge submodules. This idea was realized by the help of MOV arresters employed in the series modules of hybrid DC breaker. In fact, after operation of hybrid DC breaker and insertion of MOV arresters, a specific number of them are bypassed by turning the modules' IGBTs on. This action changes the operating point of remaining MOVs and push it from high resistance to temporary overvoltage (or switching surge) mode, which is equivalent to insertion of a DC voltage source and a low resistance. Subsequent to this action, the voltage difference of MMC DC link and the equivalent DC source is applied to the line and a voltage pulse is generated. The amplitude of active pulse is controlled by the number of bypassed modules and the pulse width is adjusted by the turn-off time of modules' IGBTs. Then, by monitoring the line current (or line voltage), the fault type and its characteristics are recognized. In case of temporary fault condition, the voltage of DC line is smoothly built up by sequentially turning the breaker modules on. The validity and correctness of the proposed method was confirmed by simulations and experiments on a scaled down prototype.

References

[1] S. Debnath, J. Qin, B. Bahrani, M. Saeedifard, and P. Barbosa, "Operation, control, and applications of the modular multilevel converters: a review," *IEEE Trans. Power Electron.*, vol.30, no. 1, pp. 37-53, Jan. 2015.

[2] G. Song, T. Wang and S. T. H. Kazmi, "DC Line Fault Identification Based on Pulse Injection from Hybrid HVDC Breaker," *IEEE Trans. Power Del.*, vol. 34, no. 1, pp. 271-280, Feb. 2019.

[3] M. Langwasser, G. De Carne, M. Liserre and M. Biskoping, "Fault Current Estimation in Multi-Terminal HVdc Grids Considering MMC Control," *IEEE Transactions on Power Systems*, vol. 34, no. 3, pp. 2179-2189, May 2019.

[4] M. Stumpe, P. Ruffing, P. Wagner, and A. Schnettler, "Adaptive single-pole autoreclosing concept with advanced dc fault current control for full-bridge MMC VSC systems," *IEEE Trans. Power Del.*, vol. 33, no. 1, pp. 321-329, Feb. 2018.

[5] T. Wang, G. Song, and S. T. H. Kazmi, "Adaptive single-pole auto-reclosing scheme for hybrid MMC-HVDC systems," *IEEE Trans. Power Del.*, vol. 34, no. 6, pp. 2194-2203, Jun. 2019.

[6] R. Derakhshanfar, T. Jonsson, U. Steiger, and M. Habert, "Hybrid HVDC breaker – A solution for future HVDC system," in CIGRE 2014, Paris, 24 – 29 Aug. 2014.

[7] M. Langwasser, G. De Carne, M. Liserre and T. Schindler, "Requirement analysis of hybrid direct current breaker in multi-terminal high-voltage direct current grids," *Journal of Engineering*, vol. 2018, no. 15, pp. 1066-1071, 10 2018.

[8] H. Iman-Eini and M. Liserre, "DC fault current blocking with the coordination of half-bridge MMC and the hybrid DC breaker," *IEEE Trans. Ind. Electron.*, vol. 67, no.7, pp. 5503-5514, Jul. 2020.

[9] K Vinothkumar, I. Segerqvist, N. Johannesson, A. Hassanpoor, "Sequential auto-reclosing method for hybrid HVDC breaker in VSC HVDC links," in 2nd Annual Southern Power Electronics Conference (SPEC), Auckland, New Zealand, 2016.

[10] S. Yang, W. Xiang, X. Lu, W. Zou, and J. Wen, "Saizhao Yang et.al. "An adaptive reclosing strategy for MMC-HVDC systems with hybrid DC circuit breakers," *IEEE Trans. Power Del.*, vol. 35, no. 3, pp. 1111-1123, Aug. 2019.

[11] M. Stumpe, R. Puffer, A. Schnettler, and Z. Shi, "Determination of single-pole auto-reclosing restart concept for VSC HVDC with fault current controllability," in the 20th International Symposium on High Voltage Engineering, ISH2017, Buenos Aires, Argentina, 2017.

[12] T. Bi, S. Wang and K. Jia, "Single pole-to-ground fault location method for MMC-HVDC system using active pulse," *IET Gener. Transm. Dis.*, vol. 12, no. 2, pp. 272-278, Dec. 2018.

[13] C. Li, J. Liang, and S. Wang, "Interlink hybrid dc circuit breaker," *IEEE Trans. Ind. Electron.*, vol. 65, no.11, pp. 8677-8686, Nov. 2018.

[14] J. Woodworth. Arrester reference voltage. ArresterFacts-027, 2011. URL http: //www.arresterworks.com.

Multistep MPC of Dual Inverter for Switching Losses Optimization

Martin Votava, Tomas Glasberger, Zdenek Peroutka
UNIVERSITY OF WEST BOHEMIA
REGIONAL INNOVATION CENTRE FOR ELECTRICAL ENGINEERING
Univerzitni 8
Plzen, Czech Republic
Phone: +420 377634149
Email: mvotava@rice.zcu.cz, tglasber@rice.zcu.cz, pero@rice.zcu.cz
URL: http://www.rice.zcu.cz

Acknowledgment

This research has been supported by the Ministry of Education, Youth and Sports of the Czech Republic under the project OP VVV Electrical Engineering Technologies with High-Level of Embedded Intelligence CZ.02.1.01/0.0/0.0/18_069/0009855.

Abstract

A novel control method based on model predictive control (MPC) dedicated to dual inverter (DI) control with minimization of switching losses is introduced. Optimal compromise between switching losses and current tracking error is reached by enlarging of the predicted steps count. The computational burden is reduced by employing the sphere decoding algorithm (SDA).

Keywords

≪Converter control≫, ≪Predictive control≫, ≪Voltage source inverter≫

Introduction

Dual inverters have been employed in a wide range of industrial fields such as adjustable electric drives [1], [2], distributed energy systems with utilization of renewable sources [3], etc. The main advantage of this topology over commonly used topologies is simple dc-link voltage balancing which does not require sophisticated control. Therefore, redundant switching states can be used to employ control which is focused on switching losses reduction.

Nowadays, various controls based on MPC have been developed. MPC control can be modified to control multiple objectives by adding additional terms into a cost function. The disadvantage of MPC is the computational burden is exponentially dependent on the number of total predictions step. The long prediction horizon with a high number of prediction steps leads to a significant improvement in optimality when two terms with significantly different time constants such as load current and switching losses are controlled at the same time. The computational burden is often reduced by employing SDA [4] or preselection. The advantage of SDA over preselection is that the computational burden is reduced without losing the optimality. However, SDA is limited to linear problems. Therefore, the model of the controlled system has to be linearized and SDA employed on the approximated system.

In power electronics, SDA is often utilized for reducing switching frequency [4]. However, switching frequency often does not correspond with switching losses [5]. Therefore, a model of switching losses and linearized cost function, which is utilized by SDA, are introduced in this paper. The objective of the proposed method is to reduce current ripple without increasing switching losses.

Model

The topology of the considered dual inverter with RL load is displayed in Fig. 1. The model of the whole system (voltage-source inverter and load) is now built from models of two individual physical quantities: current and power losses.

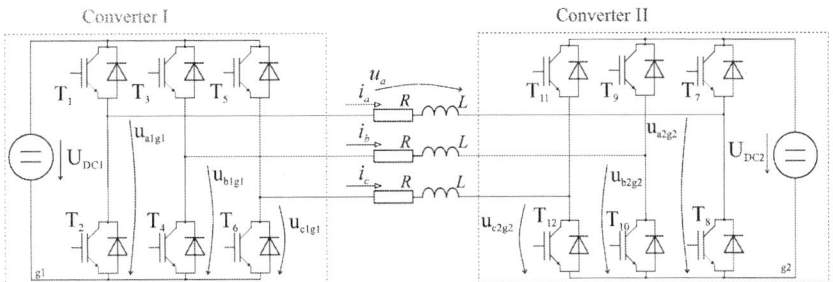

Fig. 1: Configuration of the investigated dual voltage-source inverter.

Model of converter current

The converter current is modeled in the rotational reference frame linked to the demanded current vector described by its magnitude I_m and angle ϕ. For simplification, a three phase RL load without any cross-coupling was chosen. The current submodel is:

$$i_{d,k+1} = i_{d,k} + \frac{u_{d,k}}{L} - \frac{R}{L} i_{d,k} T_s, \qquad i_{q,k+1} = i_{q,k} + \frac{u_{q,k}}{L} - \frac{R}{L} i_{q,k} T_s, \tag{1}$$

where $[i_{d,k}, i_{q,k}]$ are the d- and q-axis components of the current vector in step k, $[u_{d,k}, u_{q,k}]$ are d- and q-axis components of the voltage vector applied in the step k, R is the load resistance, L is the load inductance, and T_s is the sampling period. Here, $u_{d,k}$ and $u_{q,k}$ are obtained by the standard Park/Clarke transformation of phase voltages u_{a1g1}, u_{b1g1}, u_{c1g1} of converter I, and u_{a2g2}, u_{b2g2} and u_{c2g2} of converter II.

Model of switching losses

The switching losses are modeled for twelve elements, where one element is a union of a transistor and the associated diode. This simplification is based on the assumption that the temperatures of both junctions are close to each other. Note that this simplification is also conservative, since the power losses on both junctions contribute to the temperature of the element. Specifically, power losses of the x-th element, $x = 1, \ldots, 12$, in step $k+1$ are computed as

$$P_{x,k+1} = \begin{cases} P_{T_x,k+1} & \text{if } i_{x,k+1} > 0 \\ P_{D_x,k+1} & \text{otherwise} \end{cases} \tag{2}$$

$$P_{T_x,k+1} = \left(\chi_{on,k} \cdot K_{on} + \chi_{off,k} \cdot K_{off} \right) \cdot i_{x,k}, \qquad P_{D_x,k+1} = \chi_{rr,k} \cdot K_{rr} \cdot i_{x,k},$$

where i_x is the x-th element current, $\chi_{on,k}$ indicates if the transistor is switched on in time step k, $\chi_{off,k}$ indicates if the transistor is turned off in time step k, $\chi_{rr,k}$ indicates if the diode is turned off in time step k, K_{on} is the coefficient of the linear approximation of the transistor turn on losses, K_{off} is the coefficient of the transistor turn off losses, and K_{rr} is the coefficient of the diode reverse recovery losses.

Model predictive control

The cost function principle introduced in [4] is adopted over finite horizon N and consists of two main objectives

$$J = \sum_{l=k}^{k+N-1} \left\{ \| i^*(l+1) - i(l+1) \|_2^2 + \lambda_u \| u(l) - u(l-1) \|_2^2 \right\}. \tag{3}$$

The first term penalizes the load current vector deviance from current reference vector $i^* = \left[i_\alpha^*, i_\beta^* \right]^T$. The second term penalizes the change of the switching state represented by the control inputs vector containing the switching state of each phase $u = [u_a, u_b, u_c]^T$. Since there is only one control input per phase, the switching state of a single phases takes the values $\{-1, 0, 1\}$. Here, -1 denotes that the bottom transistor of converter I is turned on and the upper transistor of converter II is turned on; 0 represents two possible switching states producing the same voltage level on the load: both converter I's upper transistor and converter II's upper transistor are switched on or both converter I's bottom transistor and converter II's bottom transistor are switched on; and 1 denotes that the upper transistor of converter I is turned on and the bottom transistor of converter II is turned on. The weight of the second term is modifiable by the weighting coefficient λ_u.

In this paper, the control objective is modified to consider switching losses on each phase $P = [P_a, P_b, P_c]^T = \left[\sum_{x=\{1,2,7,8\}} P_x, \sum_{x=\{3,4,9,10\}} P_x, \sum_{x=\{5,6,11,12\}} P_x\right]^T$

$$J = \sum_{l=k}^{k+N-1} \left\{ \|i^*(l+1) - i(l+1)\|_2^2 + \lambda_u P^T(l+1) P(l+1) \right\}, \qquad (4)$$

According to (2), power losses are dependent on the phase current and the change of the control input between the steps $u(l)$ and $u(l-1)$. Therefore, it is convenient to rewrite the second term $P^T(l+1)P(l+1)$ into the following form:

$$P^T(l+1)P(l+1) = (u(l) - u(l-1))^T K^2 i_{abc}^2(l) (u(l) - u(l-1)), \qquad (5)$$

where $K^2 i_{abc}^2(l) = K^2 \text{diag} \left(i_a^2(l), i_b^2(l), i_c^2(l) \right)$ is a diagonal matrix containing the square value of the phase currents multiplied by the square value of the switching losses coefficient, and $u(l) = (u_a(l), u_b(l), u_c(l))$ is a control input vector containing the switching states in the converter phases corresponding with the order a,b,c. Since there is only one control input per phase, the switching state of a single phase takes the values $\{-1, 0, 1\}$. As the second term is nonlinear, a simple linear approximation based on the assumption that the output frequency (f_{out}) is much smaller than the sampling frequency (f_{sample}) is introduced. For the calculation of power losses, we assume that the phase currents are constant and consider them equal to the current measured in time step k. Then, the cost function takes the following form:

$$J = \sum_{l=k}^{k+N-1} \left\{ \|i^*(l+1) - i(l+1)\|_2^2 + \lambda_u (u(l) - u(l-1))^T K^2 i_{abc}^2(l) (u(l) - u(l-1)) \right\}. \qquad (6)$$

Discrete state space

We consider the following discrete state space

$$x(l+1) = Ax(l) + Bu(l), \qquad y(l+1) = Cx(l+1) \qquad (7)$$

$$A = \begin{bmatrix} 1 - \frac{R}{L}T_s & 0 \\ 0 & 1 - \frac{R}{L}T_s \end{bmatrix} \qquad B = U_{dc}\frac{T_s}{L} \begin{bmatrix} 1 & -\frac{1}{2} & -\frac{1}{2} \\ 0 & \frac{\sqrt{3}}{2} & \frac{-\sqrt{3}}{2} \end{bmatrix} \qquad C = \begin{bmatrix} 1 & 0 \\ 0 & 1 \end{bmatrix},$$

where $U_{dc} = U_{dcI} + U_{dcII}$ is the sum of the dc-links voltage. The presented system has states $x(l) = \left[i_\alpha(l), i_\beta(l) \right]^T$, outputs $y(l) = \left[i_\alpha(l), i_\beta(l) \right]^T$ and reference outputs $y^*(l) = \left[i_\alpha^*(l), i_\beta^*(l) \right]^T$. Substituting $y(l)$ and $y^*(l)$ into (6) yields:

$$J = \sum_{l=k}^{k+N-1} \left\{ \|y^*(l+1) - y(l+1)\|_2^2 + (u(l) - u(l-1)) \lambda_u'(l) (u(l) - u(l-1)) \right\}, \qquad (8)$$

where $\lambda_u'(l) = \lambda_u K^2 i_{abc}^2(l)$ represents the vector of penalization for switching at each phase given by the phase currents.

Control inputs constrains

DI contains six half bridges: two half bridges per phase. Since the discrete state space model has only three control inputs, the control of the single phase half bridges is combined together. The control input of a single phase is restricted into three possible control input levels $U_{av} = \{-1, 0, 1\}$. The reduction of a single phase control input yields the control set shown in fig. 2. Here, -1 indicates that the bottom transistor of converter I and the upper transistor of converter II are turned on; 0 indicates that either the bottom transistor of converter I and the bottom transistor of converter II are turned on, or that the upper transistor of converter I and the upper transistor of converter II are turned on; and 1 indicates that the upper transistor of converter I and the bottom transistor of converter II are turned on. All unspecified transistors are turned off.

Switching losses balancing

The control input level 0 can be produced by two switching states. Therefore, the switching losses balancing among the transistors of individual phases need to be balanced by an additional rule. When the control input level 0 is activated, which switching state will be activated is decided based on which half bridge was switched last time. The switching state which leads to switching of another half bridge is prioritized.

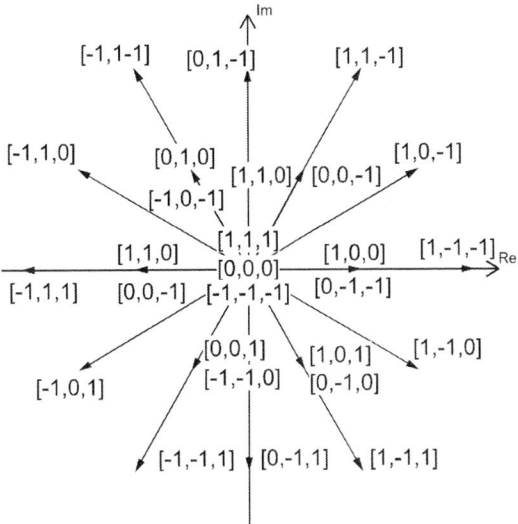

Fig. 2: Available control set.

Integer quadratic form

Vector form

Consider the sequence of output vectors $Y(k) = \left[y^T(k+1), y^T(k+2), ..., y^T(k+N)\right]^T$, the sequence of output reference vectors $Y^*(k) = \left[y^{*T}(k+1), y^{*T}(k+2), ..., y^{*T}(k+N)\right]$, the sequence of input vectors $U(k) = \left[u^T(k), u^T(k+1), ..., u^T(k+N-1)\right]$ and the diagonal matrix of λ_u' sequence $\Lambda_u(k) = \text{diag}\left(\lambda_u'(k), \lambda_u'(k+1), ..., \lambda_u'(k+N-1)\right)$. Substituting $Y(k), Y^*(k)$, $U(k)$ and $\Lambda_u(k)$ into the cost function (8) can be written in the following vector form:

$$J = \|Y^*(k) - Y(k)\|_2^2 + (U(k) - U(k-1))^T \Lambda_u(k)(U(k) - U(k-1)). \tag{9}$$

According to (7), $Y(k)$ is dependent on $x(k)$ and $U(k)$. Substituting these terms into the cost function yields:

$$J = \|Y^*(k) - \Gamma x(k) - \Upsilon U(k)\|_2^2 + (SU(k) - EU(k-1))^T \Lambda_u(k)(SU(k) - EU(k-1)) \tag{10}$$

$$S = \begin{bmatrix} I & 0 & 0 & \dots & 0 \\ -I & I & 0 & \dots & 0 \\ 0 & -I & I & \dots & 0 \\ \vdots & \vdots & \vdots & & 0 \\ 0 & 0 & 0 & -I & I \end{bmatrix} \qquad \Gamma = \begin{bmatrix} CA \\ CA^2 \\ \vdots \\ CA^N \end{bmatrix},$$

$$\Upsilon = \begin{bmatrix} CB & 0 & \dots & 0 \\ CAB & CB & \dots & 0 \\ \vdots & \vdots & & \vdots \\ CA^{N-1} & CA^{N-2} & \dots & CB \end{bmatrix} \qquad E = \begin{bmatrix} -I & 0 & 0 & \dots & 0 \\ 0 & 0 & 0 & \dots & 0 \\ 0 & 0 & 0 & \dots & 0 \\ \vdots & \vdots & \vdots & & 0 \\ 0 & 0 & 0 & 0 & 0 \end{bmatrix}$$

Approximation of switching losses

The analytical solution of (9) is difficult to obtain due to the switching losses dependency on load current. Considering the high ratio between the sample frequency (inverse value of T_s) and the output frequency (f_{out}), we assume constant phase currents for calculation of power losses $\Lambda_u'(k) = \text{diag}\left(\lambda_u'(k), \lambda_u'(k), ..., \lambda_u'(k)\right) \approx \Lambda_u(k)$. Here, both the unconstrained and the constrained solution respectively are possible.

Solution in terms of the unconstrained optimum

The first step for finding the optimal solution in terms of the unconstrained control set $U_{unc}(k)$ is to decompose J from (10) into two terms:

$$J = J_1 + J_2. \tag{11}$$

$$J_1 = 2(\Theta(k))^T U(k) + U(k)^T Q(k) U(k) \tag{12}$$

$$J_2 = \|\Gamma x(k) - Y^*(k)\|_2^2 + (EU(k-1))^T \Lambda_u'(k)(EU(k-1)), \tag{13}$$

where

$$\Theta(k) = \left((\Gamma x(k) - Y^*(k))^T \Upsilon - (EU(k-1))^T \Lambda_u'(k) S \right)^T$$

$$Q(k) = \Upsilon^T \Upsilon + S^T \Lambda_u'(k) S.$$

Here, the second term is independent of $U(k)$ and, therefore, it can be neglected in the unconstrained optimum calculation. The next step is complementing the squares:

$$J_1 = \left(U(k) + Q(k)^{-1} \Theta(k) \right)^T Q(k) \left(U(k) + Q^{-1}\Theta(k) \right) + \text{const}(k). \tag{14}$$

Since J_1 is a quadratic equation, its value is at its minimum over $U(k)$ when

$$U_{unc}(k) = -Q^{-1}\Theta(k). \tag{15}$$

Solutions in terms of the constrained optimum

According to SDA, the cost function of an individual control input vector is calculated as a sum of J_2, $\text{const}(k)$ and the square value of the distance (d) between the control input vector and the solution in terms of the unconstrained optimum

$$J = (U(k) - U_{unc}(k))^T Q(U(k) - U_{unc}(k)) + J_2 + \text{const}(k). \tag{16}$$

The number of calculations is reduced by transforming Q into a lower triangular matrix

$$H^T H = Q. \tag{17}$$

H is calculated using the Cholesky decomposition

$$H = \text{chol}\left(Q^{-1} \right)^{-1}. \tag{18}$$

Substituting H, the cost function can be rewritten as:

$$J = \|HU(k) - \bar{U}_{unc}(k)\|_2^2 + J_2 + \text{const}(k), \tag{19}$$

$$J = d^2 + J_2 + \text{const}(k), \tag{20}$$

where $\bar{U}_{unc}(k) = HU_{unc}(k)$. The solution in terms of the constrained optimum is given as follows:

$$U_{opt}(k) = \arg\min_{U(k)} \|HU(k) - \bar{U}_{unc}(k)\|_2^2. \tag{21}$$

Sphere decoding algorithm

Instead of exhaustive calculation of the cost function of each available control input vector, the sphere decoding algorithm presented in [4] is used to find an optimal control input vector. SDA considers only input control vectors which belong to a sphere radius ρ centered in $\bar{U}_{unc}$

$$\rho^2 \geq \|HU(k) - \bar{U}_{unc}(k)\|_2^2. \tag{22}$$

The advantage of SDA is that H is a triangular matrix and, therefore, (22) can be rewritten

$$\rho^2(k) \geq \left(\bar{U}_1 - H_{(1,1)}U_1\right)^2 +$$
$$\left(\bar{U}_2 - H_{(2,1)}U_1 - H_{(2,2)}U_2\right)^2 +$$
$$\vdots$$
$$\left(\bar{U}_N - H_{(N,1)}U_1 - H_{(N,2)}U_2 \cdots - H_{(N,N)}U_N\right)^2, \tag{23}$$

where $\bar{U}_i$ denotes the i-th element of $\bar{U}_{unc}(k)$, U_i is the i-th element of $U(k)$ and $H_{(i,j)}$ refers to the (i,j)-th entry of H. Therefore, the solution can be found by proceeding in a sequential manner. For details see [4]. In each step, one row of (23) is calculated. Calculation of (23) for a given control input vector stops if the sum of the calculated rows is higher than ρ^2. ρ shrinks every time a control input vector which satisfies condition (23) is found. The initial value of ρ is given by the optimal solution from the previous step

$$\rho^2_{ini}(k) = \|HU_{prev}(k) - \bar{U}_{unc}(k)\|_2^2. \tag{24}$$

Simulations

The proposed control is compared with the MPC described in [4]. The simulation parameters were selected as follows: $L = 5\,mH$, $R = 50\,\Omega$, $N = 5$, $T_s = 25\mu s$, $f_{out} = 50\,Hz$, $\gamma = 0,0525\,mW/A$, $U_{dc} = 600\,V$, and the required amplitude of the load current $I_m = 6\,A$. Coefficients of both algorithms were tuned to have roughly the same value of switching losses: $3.5\,W$. The simulated waveforms are shown for both the MPC [4] and the proposed algorithm in fig. 3. Comparing the algorithms, the proposed MPC has a higher switching number. Switching losses caused by the higher switching number are reduced by prioritizing switching in the phases with the lowest current instant value. Current waveforms of both algorithms were compared by their average single step current tracking error

$$g_{track} = \frac{\sum_{k=1}^{M}\left\{\left(i^*_{\alpha,k} - i_{\alpha,k}\right)^2 + \left(i^*_{\beta,k} - i_{\beta,k}\right)^2\right\}}{M}, \tag{25}$$

where M is the number of simulated steps; and their total harmonic distortion of load current

$$THD_i = \frac{\sqrt{\sum_{l=2}^{50} I^2_{(l)}}}{I_{(1)}}, \tag{26}$$

where $I_{(l)}$ is the amplitude of the fundamental frequency (50 Hz). Both g_{track} and THD_i are, in the case of the proposed MPC algorithm, lower:

Proposed MPC: $g_{track} = 0.2611$, $THD_i = 2.39\%$,
Conventional MPC: $g_{track} = 0.3268$, $THD_i = 4.32\%$.

Further, both algorithms were compared for a wide range of weighting coefficients via the Pareto front of the two objectives. Considering the instant current value in the cost function leads to improvement of current tracking for the same level of switching losses.

The switching pattern obtained by the proposed sphere decoding with the system linearization was compared with the optimal solution calculated by the exhaustive search algorithm without the system linearization. Calculation time for a 100 ms long simulation was reduced from $5565\,s$ to $22.1\,s$. The proposed sphere decoding with the system linearization is able to find an optimal solution in approximately 97% of control steps.

Conclusion

This paper introduces an MPC algorithm with a linearized model dedicated to control of dual converters over a long prediction horizon. The proposed control penalizes both switching losses and current tracking error at the same time. Linearization of the system model allows for employing sphere decoding techniques, which significantly reduces the computational burden of long horizon calculations.

The proper function of the proposed algorithm was tested by an extensive series of simulations. Comparing the performance of the proposed algorithm with the performance of the conventional algorithm, the proposed algorithm has lower switching losses for the same total harmonic distortion of the load current.

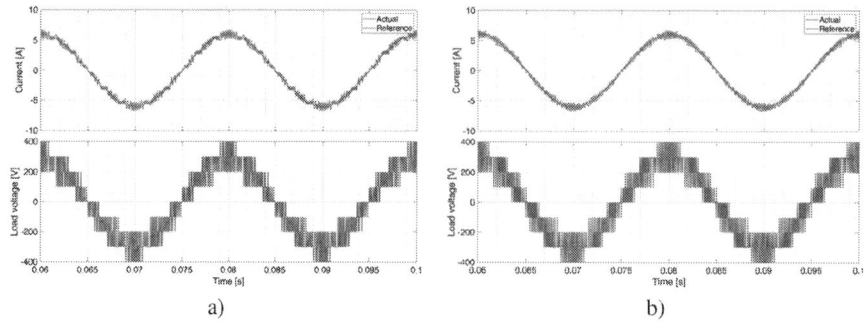

a)   b)

Fig. 3: Simulated steady state for a) MPC [4] and for b) the proposed MPC. Output phase current waveform is displayed in the top plot and phase load voltage is displayed in bottom plot.

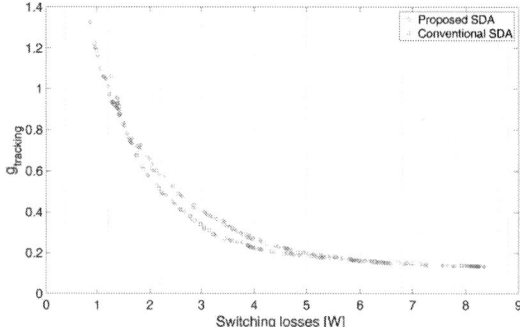

Fig. 4: Pareto front of optimal compromise between the power losses and tracking error.

References

[1] Yongjae Lee and Jung-Ik Ha. Hybrid modulation of dual inverter for open-end permanent magnet synchronous motor. *IEEE Transactions on Power Electronics*, 30(6):3286–3299, 2015.

[2] Luis De Sousa and Boris Bouchez. Combined electric device for powering and charging, September 2 2014. US Patent 8,823,296.

[3] SDG Jayasinghe, D Mahinda Vilathgamuwa, and Udaya K Madawala. Dual inverter based battery energy storage system for grid connected photovoltaic systems. In *IECON 2010-36th Annual Conference on IEEE Industrial Electronics Society*, pages 3275–3280. IEEE, 2010.

[4] Tobias Geyer and Daniel E Quevedo. Multistep finite control set model predictive control for power electronics. *IEEE Transactions on power electronics*, 29(12):6836–6846, 2014.

[5] René Vargas, Ulrich Ammann, and Jose Rodriguez. Predictive approach to increase efficiency and reduce switching losses on matrix converters. *IEEE Transactions on Power Electronics*, 24(4):894–902, 2009.

A High-Efficiency Control of a Double-Input Converter for Renewable Energies and Hybrid Vehicles

Mario Marchesoni, Massimiliano Passalacqua and Luis Vaccaro
UNIVERSITY OF GENOVA
Department of Electrical, Electronics, Telecommunication Engineering and Naval
Architecture (DITEN) - Via all'Opera Pia 11 A
Genova, Italy
Tel : +39 010 3352183
E-Mail: marchesoni@unige.it, massimiliano.passalacqua@edu.unige.it, luis.vaccaro@unige.it
URL : www.diten.unige.it

Keywords:

"Efficiency", "Converter control", "Electric vehicle", "Smart grids"

Abstract

A double-input DC-DC converter is taken into account in this paper. The converter is bidirectional and both currents can be controlled independently. On the one hand it allows to reduce the number of switches, on the other hand, it exhibits a higher efficiency in comparison to a traditional solution with two half-bridge converters in parallel connection. However, the control strategy proposed in the technical literature forces the converter to work in Continuous Conduction Mode at low-load, which leads to a low conversion efficiency. In this paper, a novel control, which exploits Discontinuous Conduction Mode, is presented. Efficiency increase is evaluated with a thermal model in MATLAB/Simulink Plecs environment and the control algorithm is then validated with experimental tests on a converter prototype.

Introduction

DC-DC converters have achieved an increasing role in the last years. Indeed, the spread of renewable energies on the one hand, and of hybrid and electric vehicles on the other, is connected to the use of such a type of converters. As a matter of fact, photovoltaics, wind turbines and storages are often connected to a DC-bus with a DC-DC converter [1-5]. Moreover, DC-DC converters are necessary also in hybrid vehicles, to adapt the battery voltage level to the motor drive voltage level [6-10]. In addition, if storage voltage is not constant (e.g., using supercapacitors), the converter has also to manage the voltage fluctuations [11, 12]. When more DC sources are connected to the same DC bus, the use of a multi-input DC-DC converter could be preferable than the use of various converters [13-18]. In particular, in [19] a new bidirectional double-input DC-DC converter based on a three-switch leg was proposed. In the traditional configuration of a bidirectional DC-DC converter, in which the voltage of one side is always lower than the voltage of the other side, two switches are used (half-bridge converter) and, therefore, four switches are used in the case of two converters in parallel connection. The three-switch leg double input DC-DC converter proposed in [19] allows on the one hand to reduce the number of switches and, on the other hand, to increase the conversion efficiency. Indeed, when current directions are concordant, the efficiency of the three-switch converter is higher than the combined efficiency of two traditional converters in parallel connection [20]. However, the control proposed in [19] forced the converter to work in Continuous Conduction Mode (CCM) also at low-load. In this condition, the currents change direction during the switching period and converter efficiency drops significantly. As a matter of fact, current root-mean-square (RMS) values and peak values remain high, whereas current average values approach zero. Since the losses are connected to RMS and peak values and, on the contrary, the power exchanged by the DC sources is connected to current average values, the converter efficiency in quite low in CCM. In [21], the basis for exploiting Discontinuous Conduction Mode

(DCM) in the three-switch converter was presented. In this paper, a detailed analysis of current paths in all working conditions is carried out, in order to define a proper control. Once the control scheme is defined, the efficiency increase achievable with the new control is evaluated with a converter thermal model in MATLAB/Simulink Plecs environment. Finally, the control is tested on a converter prototype and the current waveforms obtained with simulations are compared with the results obtained with experimental tests.

Converter control strategy

The three-switch converter structure is shown in Fig. 1. One has to note that the sum of E_1 and E_2 should be lower than V_{out}.

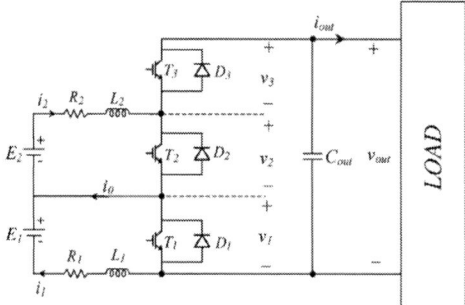

Fig. 1: Converter structure

In a traditional half-bridge converter, DCM is obtained controlling just one switch during the switching period. In this way, while both switches are open, the current can flow only in the free-wheeling diodes and once it reaches zero it cannot invert direction. The same concept can be applied to the three-switch converter considered in this paper. Indeed, one switch has to be always off (i.e., not commanded) during the switching period, in order to achieve DCM at low-load. In particular, one can consider 5 different working conditions:

1.   I1 > 0 , I2 > 0
2.   I1 > 0 , I2 < 0
3.   I1 < 0 , I2 > 0
4.   I1 < 0 , I2 < 0 , $|I_1| < |I_2|$
5.   I1 < 0 , I2 < 0 , $|I_1| > |I_2|$

The current paths in the different working conditions are reported in Fig. 2(a) (condition 1), Fig. 2(b) (condition 2), Fig. 3(a) (condition 3) and Fig. 3(b) (conditions 4 and 5). The corresponding switching states in the different working conditions are reported in Table I. One has to note that in some conditions, different current paths can occur for the same switching state, according to the current values (e.g., in condition 1 while all switches are OFF, currents can flow according to path B, C or D); for this reason the column "$i_{1,2}$ values" shows the current condition. Please note that if the current condition is not specified ("-"), the specific switching state leads just to one current path, regardless the current values. Please note that i_1 and i_2 are instantaneous current values, whereas I1 and I2 are the average current values, calculated over the switching period. T1, T2 and T3 are the three switches, as shown in Fig.1.

From the results shown in Fig. 2, Fig. 3 and Table I, one can obtain the information to design the control scheme. In condition 1, as an example, i_1 increases in paths A and B and decreases in paths C and D, whereas i_2 increases in path A and C and decreases in paths B and D. For this reason, in order to increase I_1, duty cycle of T1 (m_1) should increase; analogously, in order to increase i_2, duty cycle of T2 (m_2) should increase. Similar considerations can be carried out also for the other converter working conditions and the control scheme reported in Fig. 4 can be implemented.

Gains k1 and k2 have to be chosen according to the working condition, as reported in the box on the left in Fig. 4. The outputs of the control ("out1" and "out2") are the duty cycles of the two switches which have to be controlled, whereas the third switch is always off (i.e., duty cycle equal to zero). In the three boxes on the right it is shown which output should be considered for each switch ("out1", "out2" or "0") in the five different working conditions.

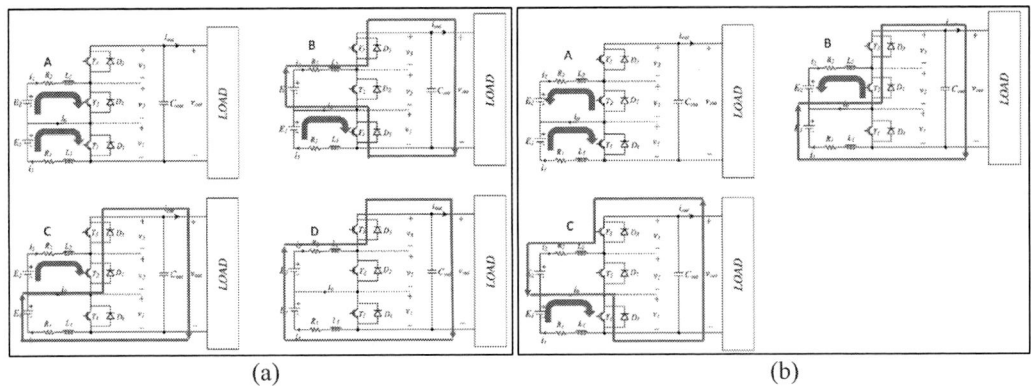

Fig. 2: (a) Current paths in condition 1 (b) Current paths in condition 2.

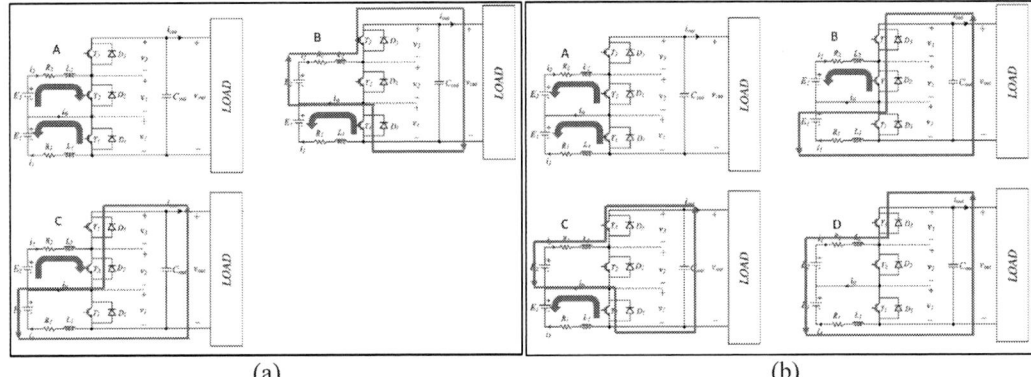

Fig. 3: (a) Current paths in condition 3 (b) Current paths in condition 4 and 5.

Table I: Switch states

Condit.	T1	T2	T3	$i_{1,2}$ values	$i_{1,2}$ paths	Condit.	T1	T2	T3	$i_{1,2}$ values	$i_{1,2}$ paths
1 **Fig. 2a**	ON	ON	OFF	-	A	**4** **Fig. 3b**	OFF	OFF	OFF	-	A
	ON	OFF	OFF	-	B		OFF	OFF	ON	$\lvert i_1 \rvert < \lvert i_2 \rvert$	B
	OFF	OFF	OFF	$\lvert i_1 \rvert > \lvert i_2 \rvert$	B		ON	OFF	ON	-	C
	OFF	ON	OFF	-	C		OFF	OFF	ON	$\lvert i_2 \rvert = \lvert i_1 \rvert$	D
	OFF	OFF	OFF	$\lvert i_1 \rvert < \lvert i_2 \rvert$	C	**5** **Fig. 3b**	OFF	OFF	OFF	-	A
	OFF	OFF	OFF	$\lvert i_2 \rvert = \lvert i_1 \rvert$	D		OFF	ON	ON	-	B
2 **Fig. 2b**	ON	OFF	OFF	-	A		OFF	OFF	ON	$\lvert i_1 \rvert > \lvert i_2 \rvert$	C
	OFF	OFF	OFF	-	B		OFF	OFF	ON	$\lvert i_2 \rvert = \lvert i_1 \rvert$	D
	ON	OFF	ON	-	C						
3 **Fig. 3a**	OFF	ON	OFF	-	A						
	OFF	OFF	OFF	-	B						
	OFF	ON	ON	-	C						

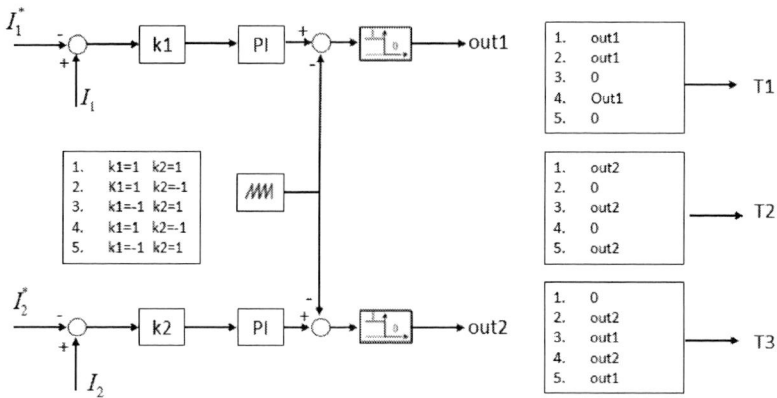

Fig. 4: Control scheme

Simulation results: converter efficiency

The aim of the control shown in the previous Section is to increase converter efficiency at low-load. Indeed, in the control proposed in [19], the converter works in CCM at low-load (i.e., the currents change their signs during the switching period) as shown in Fig. 5(a), whereas with the control shown in the previous Section, current inversion is prevented, as shown in Fig. 5(b). As explained previously, converter losses are related to current maximum and RMS values, whereas the exchanged power is connected to current average values. For these reasons, it is easy to predict that the efficiency will be lower in CCM (Fig. 5(a)) than in DCM (Fig. 5(b)).

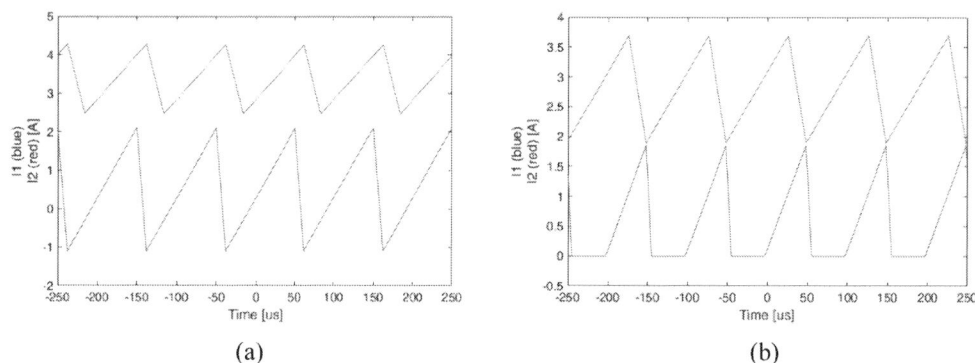

(a)   (b)

Fig. 5: (a) Current waveforms in CCM (b) Current waveforms in DCM

From this point of view, the control proposed in [19] will be referred as CCM whereas the control analyzed in this paper will be referred as DCM. Please note that this definition is valid only at low-load. Indeed, at high-load the converter works in CCM (i.e., the currents do not change direction during the switching period) also with the control proposed in this paper.

Simulations have been carried out considering the same converter parameters as the converter prototype, as shown in Table II, where Fsw is the switching frequency.

Table II: Converter parameters

IGBT	SKM400GA124D	**L2**	950 uH
E1	12 V	**Cout**	4.7 mF
L1	330 uH	**Vout**	100 V
E2	22 V	**Fsw**	10 kHz

One has to note that the converter uses 1200V IGBTs, whereas the simulations and the experimental results have been obtained referring to an output voltage equal to 100 V. 1200V IGBTs were chosen in order to be able of increasing the power level in future works. On the contrary, the aim of this paper is to show the benefits moving from CCM control strategy to DCM control strategy and not to optimize the converter parameters themselves; for this reason, the efficiency obtained with the simulations shown in this paper is far from optimal values, since the converter is working far from the rated working point. The converter thermal model was designed from the data available on the datasheet of the IGBTs. Simulation were performed in the range [-10 A; 10 A]; please note that the DCM limit (i.e., the current average value for which the current reaches null value during the switching period) is 1.6 A for I1 and 0.9 A for I2.

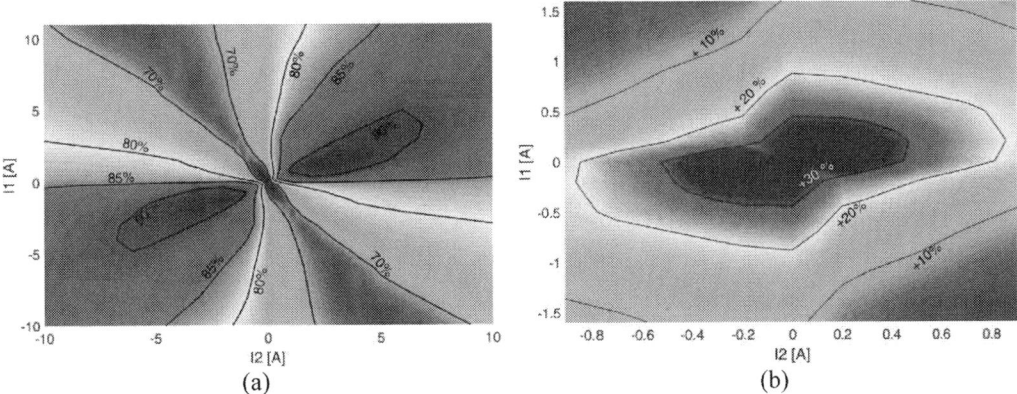

(a)   (b)

Fig. 6: (a) Converter efficiency in DCM [%] (b) Efficiency increase from CCM to DCM (low-load) [%]

Experimental results

A converter prototype has been realized in order to test the proposed algorithm, as shown in Fig. 7. The converter parameters are the same utilized to perform the simulations and shown in Table II. A 12 V battery has been used as voltage source 1 (E_1) and a supercapacitor charged at 22 V has been used as voltage source 2 (E_2), as shown in Fig. 7(a).

The 100 V source has been realized rectifying the output of a variable transformer; a 22 Ω resistance is connected in parallel to the 100 V source. In this way also the 100 V source is bidirectional. With this configuration, the five working conditions could all be tested.

The control has been implemented on Dspace Microlab Box. LEM current and voltage transducers have been used for control feedback to Dspace, whereas two Tektronix hall effect probes have been used to measure the instantaneous current waveforms with the oscilloscope. The current waveforms have been then exported from the oscilloscope and plotted using MATLAB.

Results related to condition 1 (I1=0.5 A and I2=2.8 A) are plotted in Fig. 8, to condition 2 (I1=1.7 A and I2=-0.5 A) in Fig. 9, to condition 3 (I1=-0.4 A and I2=3.9 A) in Fig. 10, to condition 4 (I1=-0.5 A and I2=-3 A) in Fig. 11 and to condition 5 (I1=-3 A and I2=-0.5 A) in Fig. 12. For each condition the simulation results are shown in (a), whereas the experimental results for the same working condition are reported in (b). One can note a great correspondence between the current waveforms obtained with simulations and the ones obtained with experimental tests.

A HIGH-EFFICIENCY CONTROL OF A DOUBLE-INPUT CONVERTER FOR
RENEWABLE ENERGIES AND HYBRID VEHICLES

MARCHESONI Mario

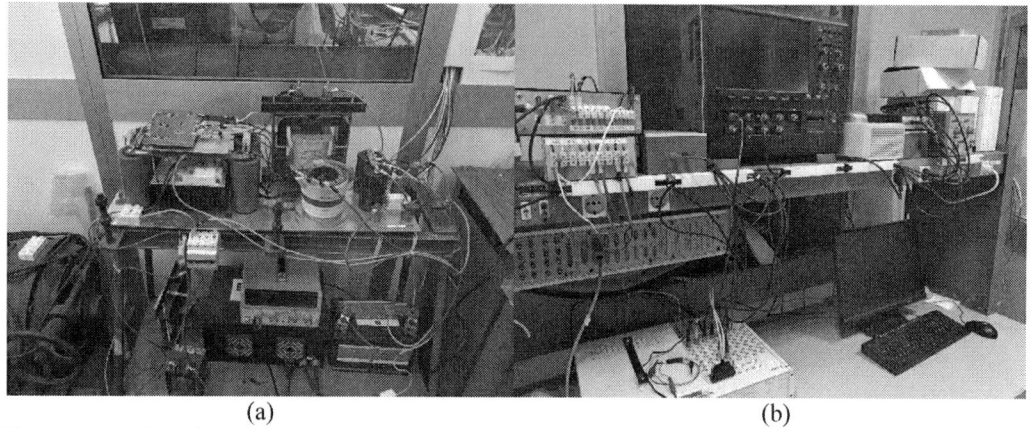

(a)   (b)

Fig. 7: Test bench (a) Converter and storages (b) Dspace Platform, measurement sensors and oscilloscope

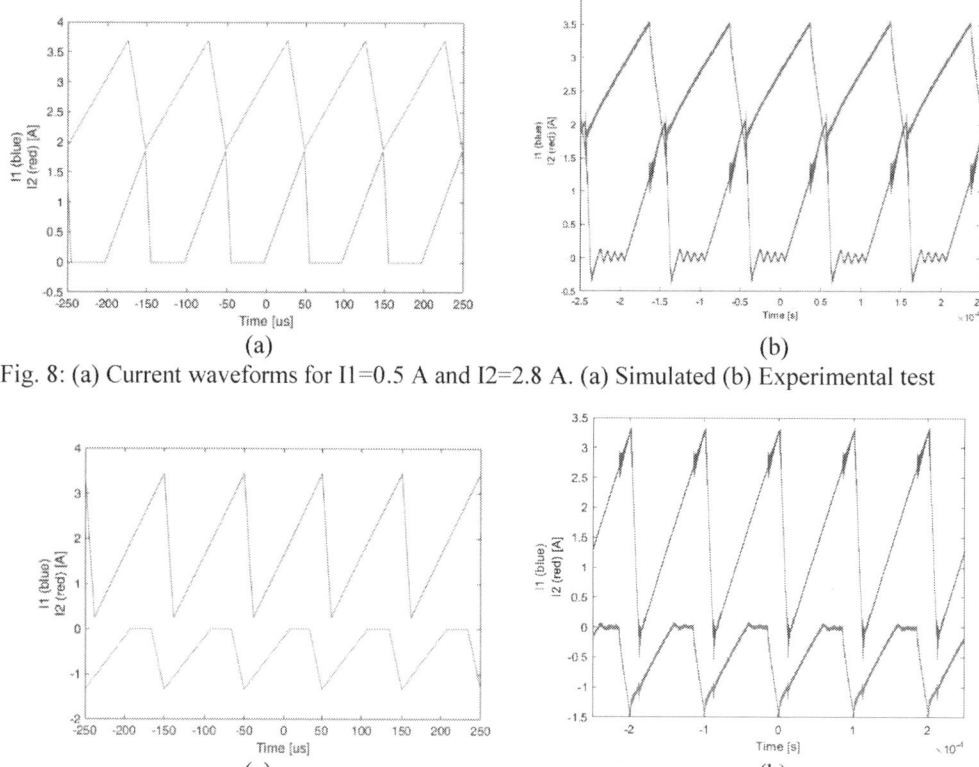

(a)   (b)

Fig. 8: (a) Current waveforms for I1=0.5 A and I2=2.8 A. (a) Simulated (b) Experimental test

(a)   (b)

Fig. 9: (a) Current waveforms for I1=1.7 A and I2=-0.5 A. (a) Simulated (b) Experimental test

EPE'20 ECCE Europe

 Assigned jointly to the European Power Electronics and Drives Association & the Institute of Electrical and Electronics Engineers (IEEE)

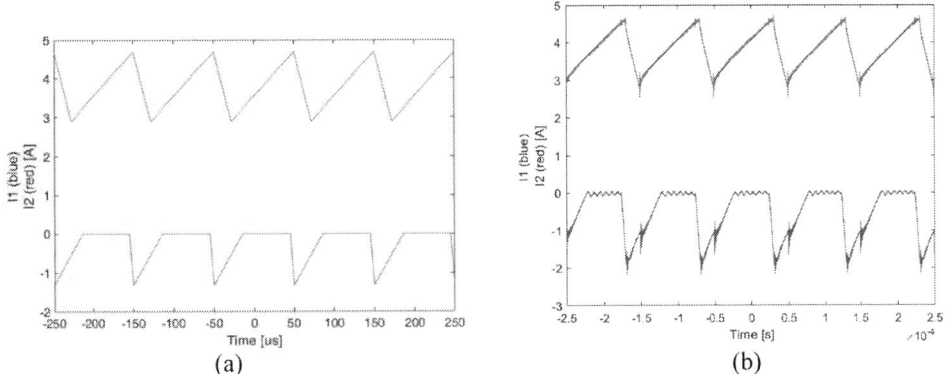

Fig. 10: (a) Current waveforms for I1=-0.4 A and I2=3.9 A. (a) Simulated (b) Experimental test

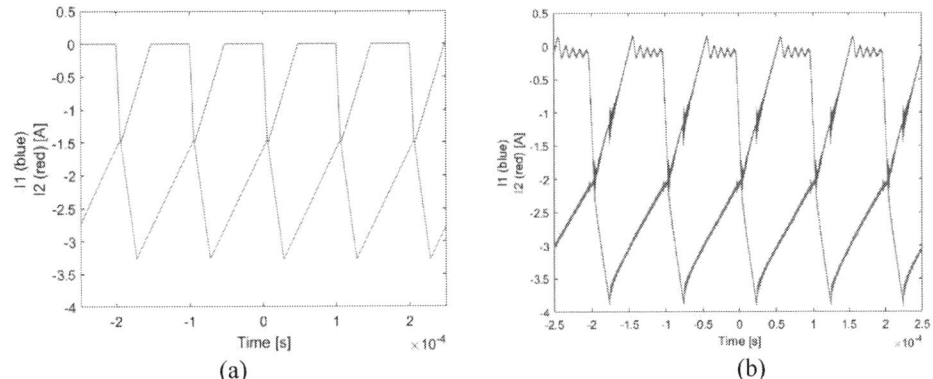

Fig. 11: (a) Current waveforms for I1=-0.5 A and I2=-3 A. (a) Simulated (b) Experimental test

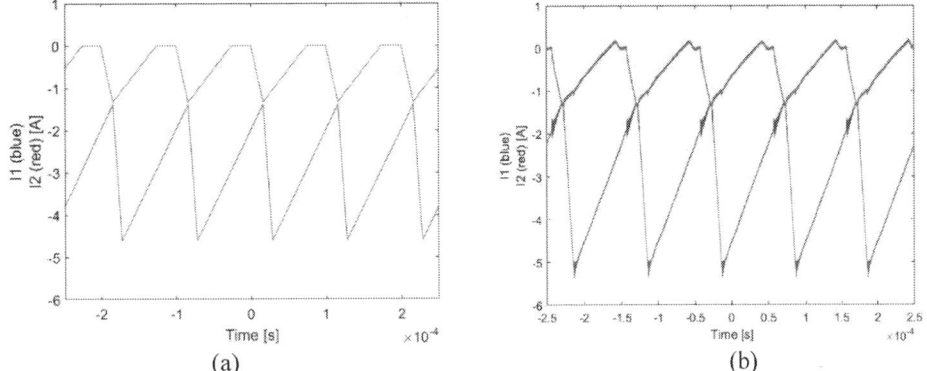

Fig. 12: (a) Current waveforms for I1=-3 A and I2=-0.5 A. (a) Simulated (b) Experimental test

Conclusions

An innovative bidirectional double-input DC-DC converter is taken into account in this paper. Studies in the technical literature have proved that the converter not only allows to reduce the number of switches, but it exhibits also a higher efficiency in comparison with two half-bridge converters in parallel connection. However, the control strategy proposed in the technical literature forced the converter to work in CCM also at low-load; in this way, the currents change direction during the switching period and the converter efficiency drops significantly. In this paper, a new modulation strategy, exploiting DCM at low-load has been analyzed. The benefits in terms of efficiency increase have been evaluated with a converter thermal model in MATLAB/Simulink Plecs environment. The simulations have shown a significant efficiency increase in the low-load region. Subsequently, the proposed control has been validated on a converter prototype. The experimental results have shown a great correspondence between the current waveforms obtained with simulations and the ones obtained with the tests on the prototype.

References

[1] N. Kondrath: Bidirectional DC-DC converter topologies and control strategies for interfacing energy storage systems in microgrids: An overview, in 2017 IEEE International Conference on Smart Energy Grid Engineering (SEGE), 2017, pp. 341-345.

[2] V. Thomas, S. Kumaravel, and S. Ashok: Control of parallel DC-DC converters in a DC microgrid using virtual output impedance method, in 2016 2nd International Conference on Advances in Electrical, Electronics, Information, Communication and Bio-Informatics (AEEICB), 2016, pp. 587-591.

[3] A. K. Rathore, D. R. Patil, and D. Srinivasan: Non-isolated Bidirectional Soft-Switching Current-Fed LCL Resonant DC/DC Converter to Interface Energy Storage in DC Microgrid, IEEE Transactions on Industry Applications, vol. 52, no. 2, pp. 1711-1722, 2016.

[4] K. Nathan, S. Ghosh, Y. Siwakoti, and T. Long: A New DC–DC Converter for Photovoltaic Systems: Coupled-Inductors Combined Cuk-SEPIC Converter, IEEE Transactions on Energy Conversion, vol. 34, no. 1, pp. 191-201, 2019.

[5] R. Suryadevara and L. Parsa: Full-Bridge ZCS-Converter-Based High-Gain Modular DC-DC Converter for PV Integration With Medium-Voltage DC Grids, IEEE Transactions on Energy Conversion, vol. 34, no. 1, pp. 302-312, 2019.

[6] M. Passalacqua, D. Lanzarotto, M. Repetto, and M. Marchesoni: Advantages of Using Supercapacitors and Silicon Carbide on Hybrid Vehicle Series Architecture, vol. 10, no. 7, p. 920, 2017.

[7] J. Hu, Y. Chen, and Z. Yang: Study and simulation of one bi-directional DC/DC converter in hybrid electric vehicle, in 2009 3rd International Conference on Power Electronics Systems and Applications (PESA), 2009, pp. 1-4.

[8] F. Ahmadkhanlou and A. Goodarzi: Hybrid Lithium-ion/Ultracap energy storage systems for plug-in hybrid electric vehicles, in 2011 IEEE Vehicle Power and Propulsion Conference, 2011, pp. 1-7.

[9] A. Bonfiglio, D. Lanzarotto, M. Marchesoni, M. Passalacqua, R. Procopio, and M. Repetto: Electrical-loss analysis of power-split hybrid electric vehicles, Energies, Article vol. 10, no. 12, 2017, Art. no. 2142.

[10] M. Passalacqua, D. Lanzarotto, M. Repetto, L. Vaccaro, A. Bonfiglio, and M. Marchesoni: Fuel Economy and EMS for a Series Hybrid Vehicle Based on Supercapacitor Storage, IEEE Transactions on Power Electronics, vol. 34, no. 10, pp. 9966-9977, 2019.

[11] D. Lanzarotto, M. Marchesoni, M. Passalacqua, A. P. Prato, and M. Repetto: Overview of different hybrid vehicle architectures, IFAC-PapersOnLine, vol. 51, no. 9, pp. 218-222, 2018/01/01/ 2018.

[12] C. Lai, L. Yu-Jen, H. Ming-Hua, and L. Jie-Ting: A newly-designed multiport bidirectional power converter with battery/supercapacitor for hybrid electric/fuel-cell vehicle system, in 2016 IEEE Transportation Electrification Conference and Expo, Asia-Pacific (ITEC Asia-Pacific), 2016, pp. 163-166.

[13] F. Akar, Y. Tavlasoglu, E. Ugur, B. Vural, and I. Aksoy: A Bidirectional Nonisolated Multi-Input DC–DC Converter for Hybrid Energy Storage Systems in Electric Vehicles, IEEE Transactions on Vehicular Technology, vol. 65, no. 10, pp. 7944-7955, 2016.

[14] M. R. Banaei, H. Ardi, R. Alizadeh, and A. Farakhor: Non-isolated multi-input–single-output DC/DC converter for photovoltaic power generation systems, IET Power Electronics, vol. 7, no. 11, pp. 2806-2816, 2014.

[15] S. Dusmez, X. Li, and B. Akin: A New Multiinput Three-Level DC/DC Converter, IEEE Transactions on Power Electronics, vol. 31, no. 2, pp. 1230-1240, 2016.

[16] E. Babaei and O. Abbasi: Structure for multi-input multi-output dc–dc boost converter, IET Power Electronics, vol. 9, no. 1, pp. 9-19, 2016.

[17] F. Kardan, R. Alizadeh, and M. R. Banaei: A New Three Input DC/DC Converter for Hybrid PV/FC/Battery Applications, IEEE Journal of Emerging and Selected Topics in Power Electronics, vol. 5, no. 4, pp. 1771-1778, 2017.

[18] E. Babaei, O. Abbasi, and S. Sakhavati: An overview of different topologies of multi-port dc/dc converters for dc renewable energy source applications, in 2016 13th International Conference on Electrical Engineering/Electronics, Computer, Telecommunications and Information Technology (ECTI-CON), 2016, pp. 1-6.

[19] M. Marchesoni and C. Vacca: New DC–DC Converter for Energy Storage System Interfacing in Fuel Cell Hybrid Electric Vehicles, IEEE Transactions on Power Electronics, vol. 22, no. 1, pp. 301-308, 2007.

[20] M. Marchesoni, M. Passalacqua, and L. Vaccaro:A refined loss evaluation of a three-switch double input DC-DC converter for hybrid vehicle applications, Energies, Article vol. 13, no. 1, 2020, Art. no. 204.

[21] M. Marchesoni, M. Passalacqua, and L. Vaccaro: An Improved Control Strategy for an Innovative DC-DC Converter for Interfacing Energy Storage Systems, in Speedam 2020, 2020.

Dead-Time influence on fast switching pulsed power converters design – A high current application for accelerator's magnets

Ludovic Horrein[1,2], Jean-Marc Cravero[1], Philippe Delarue[2], Alain Bouscayrol[2], Davide Aguglia[1], Carmen Ortega-Perez[1,2]

1: CERN (European Organization for Nuclear Research), Electrical Power Converters group, Geneva, Switzerland

2: Univ. Lille, Arts et Métiers Paris Tech, Centrale Lille, HEI, EA 2697- L2EP

Abstract:

Dead-time in IGBT-based DC-DC power converters is a well-known issue that causes a limitation of the output voltage and the distortion of the output voltage and current. These distortions are more critical when the converter switching frequency increases and in applications requiring high di/dt. This is especially true for power converters used to supply pulsed electro-magnets in particle accelerators where output waveforms with high slew rates are needed. Several solutions exist to compensate the distortions caused by dead-time; however, in fast and precise applications, the compensation method should be analysed in conjunction with design choices, such as the selection of switching frequency, output filter's inductances and topology.

At CERN, hybrid capacitor discharge/switch mode power converters have been developed to supply electro-magnets used for PS beam injection. For standardisation and cost reduction. Because of this design option, dead-time represents 16 % of the switching period thus widely affecting the converter performances. This paper illustrates the effects of dead-time on various design aspects of these hybrid power converters able to produce precise 1ms long current pulses with maximal output current up to 2,5 kA. The converter control loops that have been designed using Energetic Macroscopic Representation approach can partly compensate for the dead-times effects. However, a dead-time compensation technique based on the inductor current sign measurement is nevertheless required to improve the converter performances. It is experimentally demonstrated that this classical compensation method is effective, even in interleaved switching module.

Introduction

At CERN, particle accelerators are used for fundamental research. In particle accelerators, a large collection of magnets are used to control particle beam trajectory and shape. To reduce the converters' size, cost, cooling systems and to save energy, pulsed converters are used when possible [1]. In 2019, CERN accelerators were shut down to perform an important upgrade [2], allowing an increase in the rate of particle collisions per second in the Large Hadron Collider. For that purpose, many power converters have been modified or replaced by new ones with higher performance [3]. In this context, the beam injection scheme in the Proton Synchrotron has been fully reviewed, and a new converter called SIRIUS FP2P2S has been developed. The power converter design, and the associated analog control, have been presented in [4] for two modules in parallel. The adopted solution is a hybrid capacitor discharge/switch mode converter which delivers half-sinusoidal current pulses of 1ms duration with high precision. In order to increase the current magnitude, 4 modules are now used in a series parallel mode. Moreover the control is now implement in a digital controller board and the PWM is considered.

Thousands of power converters are operating at CERN to supply electro-magnets in different particle accelerators. In order to reduce their development, maintenance and operational costs, a modular approach for the converter design has been adopted by creating standard power building blocks. For this application, four standard DC/DC modules are used, they are connected in series/parallel to produce the required output current and voltage. These DC/DC modules are realised using standard IGBT stacks that are not specifically dedicated to fast pulsed applications. In this application, the IGBT modules and associated drivers are used at their maximal switching frequency that is 20 kHz. As a consequence of the use of standard IGBT modules and drivers, the measured turn-on switching time is 3.2 µs and the measured turn-off time is 0.8 µs.

As the IGBT turn-off time is larger than its turn-on time, a delay on the IGBT switching orders must be inserted for the turn-on, in order to avoid a short-circuit of the DC bus. These delays are called dead-time [5] - [7]; they are adjusted at 4 µs in this application. The impact of dead-times have to be considered as regards to the H-bridge switching frequency and if a dead-time of 4 µs generally has low influence with a switching frequency of some kilo Hertz, for the present application, where the dead-time represents 16 % of the half-switching period, the impact is quite significant. There are several approaches (hardware, software) to minimize the effects of dead-times. These solutions have to be considered during the design of the converter so as to select the most appropriate one.

The objective of this paper is to implement the most appropriate method to minimize the effects of dead-times for this specific fast switching pulsed power converter.

The converter specifications, operational requirements and topology are presented in Section 1. Section 2 presents the simulation when dead-times are neglected. A discussion of the different strategies to minimize dead-time effects is given in Section 3. Then, the selected solution is implemented to validate its effectiveness in the fast switching pulsed converter.

1. Specifications and adopted topology

a) Specifications of the power converters

The power converter specifications aim to deliver high current and high precision pulses every 1.2 s. The reference current pulse shape is a half-sinusoidal waveform with a frequency of 500 Hz (1 ms) and an amplitude ranging from 200 A to 2500 A. Based on the load parameters, the converter output voltage range is between -1050 V and +1050 V. In this application, the converter output current is used to produce a magnetic field that will deflect the particle beam so it can be injected in a circular particle accelerator. In order to allow a precise beam trajectory, the magnet current must be controlled with the precision of +/- 3A as regards to its sinusoidal reference.

b) Converters topology

The powering solution is based on a switch mode converter (Fig. 1) developed at CERN [3] that is composed of four DC/DC modules connected in series/parallel. The output of the converter has been modified by the addition of a thyristor that permits to obtain the behaviour of a capacitor discharge converter whose output current can be finely regulated.

The power converter operation is performed in two phases. A first phase that takes place before the current pulse, consists in the charge of the output capacitor bank via the DC/DC converters. The second phase consists in the firing of the thyristor to discharge the capacitor banks and the tracking the magnet current (pulse) by controlling the output voltage using the DC/DC converters.

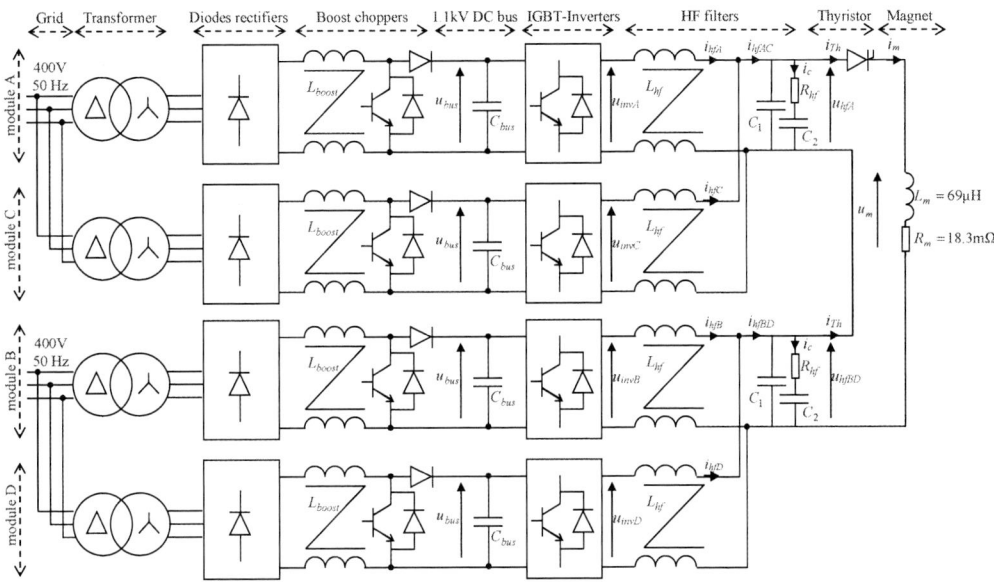

Fig. 1 : Architecture of the SIRIUS FP2P2S converter (with magnet)

Equation (1) gives the reference current that must be tracked. Using the magnet model, the required output voltage u_{hf} is given in (2). By imposing $t=0$, we deduce the initial required voltage to achieve the current reference tracking (3).

$$i_{m\_ref} = \hat{I}_{m\_ref} \sin(2\pi 500 t) \tag{1}$$

$$u_{hf\_ideal} = L_m \frac{d}{dt} i_{m\_ref} + R_m i_{m\_ref} = \hat{I}_{m\_ref}\left(L_m 2\pi 500 \cos(2\pi 500 t) + R_m \sin(2\pi 500 t)\right) \tag{2}$$

$$u_{hf\_0} = \hat{I}_{m\_ref} L_m 2\pi 500 \tag{3}$$

The use of a standard switch mode converter does not allow obtaining such a voltage waveform and a thyristor has to be installed at the output of the converter. Moreover, an output filter is always required at the output of a classic switch mode converter to reduce the voltage ripple due to the PWM operation. This filter limits the bandwidth of the converter and does not allow having a voltage step at the converter output.

2. Study neglecting dead-time

a) Modelling and Control

The modelling of a converter with two parallel modules has been presented in [4]. All the involved subsystems are described using mathematical expressions: inverters (4), HF filter inductors (7), parallel coupling (6) (7), capacitors (9), series coupling (9), thyristor (10), magnet (11).

For the power flow analysis between the interconnected subsystems, the mathematical expressions of the model are organized using the Energetic Macroscopic Representation (EMR) formalism [8][9]. EMR is a graphical description that allows an easy interconnection of models of different subsystems. This interconnection is based on the interaction principle (systemic principle [8]). For a better understanding of the physical power flow, EMR is based on the strict respect of the physical causality [11]. In

EMR formalism, each energetic function is described using specific pictograms. Each pictogram is connected to another using the two action/reaction variables.

$$\begin{cases} u_{invX} = c_{invX} u_{busX} \\ i_{busX} = c_{invX} i_{invX} \end{cases} \tag{4}$$

$$i_{hfX} = \frac{1}{sL_{hf}} \left(u_{invX} - u_{hfY} \right) \tag{5}$$

$$i_{hfAC} = i_{hfA} + i_{hfC} \tag{6}$$

$$i_{hfBD} = i_{hfB} + i_{hfD} \tag{7}$$

$$u_{hfY} = \frac{1 + s\left(R_{hf}C_2\right)}{s^2\left(R_{hf}C_1C_2\right) + s(C_1 + C_2)} \left(i_{hfY} - i_{T_1}\right) \tag{8}$$

$$u_{hf} = u_{hfAC} + u_{hfBD} \tag{9}$$

$$\begin{cases} u_m = u_{hf} & \text{if } T_h \text{ closed} \\ u_m = 0 & \text{if } T_h \text{ opened} \end{cases} \tag{10}$$

$$i_m = \frac{1/R_m}{s\,L_m/R_m + 1} u_m \tag{11}$$

with $X \in \{A, B, C, D\}$ and $Y = AC$ if $X \in \{A, C\}$; $Y = BD$ if $X \in \{B, D\}$ and T_h the thyristor order

Differential relations representing an internal storage energy are described using crossed orange rectangle (HF filter inductors (5), capacitors (8) and magnet (11)) (Fig. 2). To respect the causality principle, output variables of these pictograms correspond to the state variables. Thus, all differential equations are written using integrator Laplace operand ($1/s$).

Coupling relations representing an energy distribution are described using overlapped orange squares (parallel (6) (7) and series (9) connections) (Fig. 2). To respect Kirshhoff's relations, a variable is common to each coupling (voltage for parallel coupling, current for series coupling).

Mono-domain conversion relations representing an energy transformation without energy storage (with or without losses) are described using orange squares (inverter (4) and thyristor (10)) (Fig. 2).

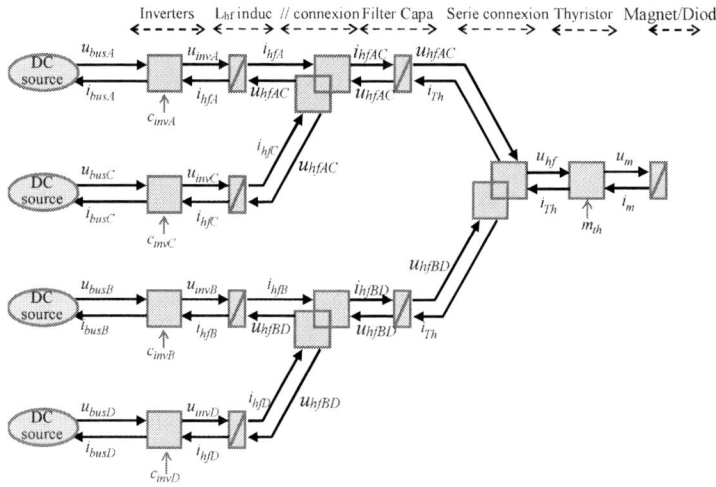

Fig. 2 : Description of the SIRIUS FP2P2S converter with magnet) using EMR

b) Control

The EMR organization allows deducing a control structure with a systematic approach. The tuning path links the tuning inputs (switching orders of the inverters) to the objective (magnet current) (Fig. 3). The control path is the inversion of the tuning path (Fig. 4). A control relation is inserted between each variable of the control path.

If the local inversion concerns an accumulation element, the control relation is a closed-loop controller (HF filter inductors (5)→(12), capacitors (8)→(13) and magnet (11)→(14)).

If the local inversion concerns an energy distribution, the control relation required a distribution criterion to define the energy splitting. In our application, an equal distribution is defined between all modules ($k_{Dp}=k_{Ds}=50\%$) (parallel couplings (6)→(15), (7)→(16), series coupling (9)→(17)).

The control of the inverters is based on the model relation (4)→(18)

The control structure is given in Fig. 5 (EMR) and Fig. 6 (equivalent block diagram).

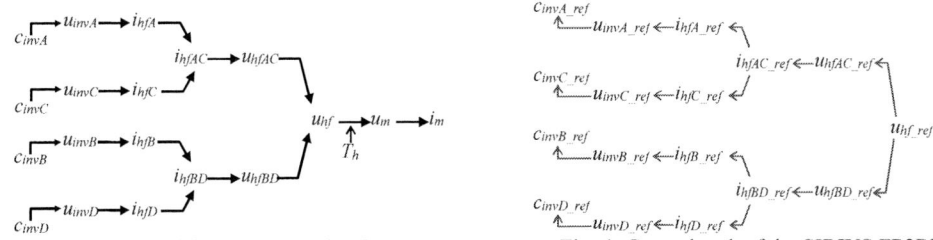

Fig. 3: Tuning path of the SIRIUS FP2P2S converter

Fig. 4: Control path of the SIRIUS FP2P2S converter

$$u_{invX\_ref} = C_{ihf}\left(i_{hfX\_ref} - i_{hfX\_mea}\right) + u_{hfY\_mea} \qquad (12)$$

$$i_{hfY\_ref} = C_{uhf}\left(u_{hfY\_ref} - u_{hfY\_mea}\right) + i_{T_h\_mea} \qquad (13)$$

$$u_{m\_ref} = C_{im}\left(i_{m\_ref} - i_{m\_mea}\right) + u_{hf\_id} \qquad (14)$$

$$\begin{cases} u_{hfAC\_ref} = k_{Ds}u_{hf\_ref} \\ u_{hfBD\_ref} = (1 - k_{Ds})u_{hf\_ref} \end{cases} \qquad (15)$$

$$\begin{cases} i_{hfA\_ref} = k_{Dp1}i_{hfAC\_ref} \\ i_{hfC\_ref} = \left(1 - k_{Dp1}\right)i_{hfAC\_ref} \end{cases} \qquad (16)$$

$$\begin{cases} i_{hfB\_ref} = k_{Dp2}i_{hfBD\_ref} \\ i_{hfD\_ref} = \left(1 - k_{Dp2}\right)i_{hfBD\_ref} \end{cases} \qquad (17)$$

$$c_{invX\_ref} = u_{invX\_ref}/u_{busX} \qquad (18)$$

with $X \in \{A, B, C, D\}$ and $Y = AC$ if $X \in \{A, C\}$; $Y = BD$ if $X \in \{B, D\}$

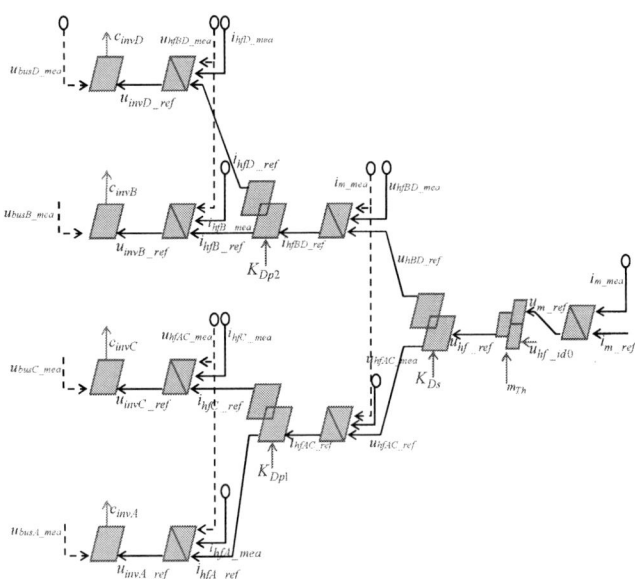

Fig. 5 : Control structure based on EMR inversion of the SIRIUS FP2P2S converter (with magnet) using EMR

This deduced control requires a large number of measurements and closed-loop controllers. To limit the sensors number, the HF filter inductor closed loop controllers are permuted with parallel couplings and merged with the capacitor closed loop controllers (Fig. 7 left part). Using a similar approach, the capacitor closed loop controllers are permuted with the series coupling and merged with the magnet closed-loop controller (Fig. 7 right part). To compensate for the control simplification, the theoretical voltage on the inverter output is deduced considering an equi-distribution of voltage and current between modules:

$$u_{invX\_theo} = \hat{I}_{m\_ref}\left(0.0341\sin(2\pi500t) + 0.4197\cos(2\pi500t)\right) \qquad (19)$$

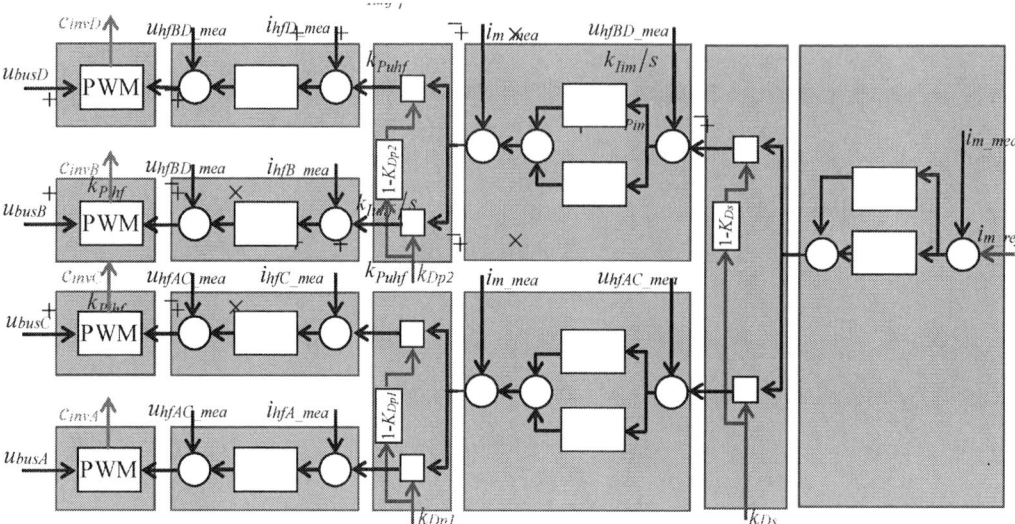

Fig. 6: Details of the EMR based control loops using block diagrams

The different modules of the power converter (series/parallel associations) are assummed to have identical parameter. This is mainly due to an accurate thermal regulation of their cooling systems. This has been confirmed by experimental results. The simplified control structure (Fig. 8) is a first step before the implementation of the complete control scheme.

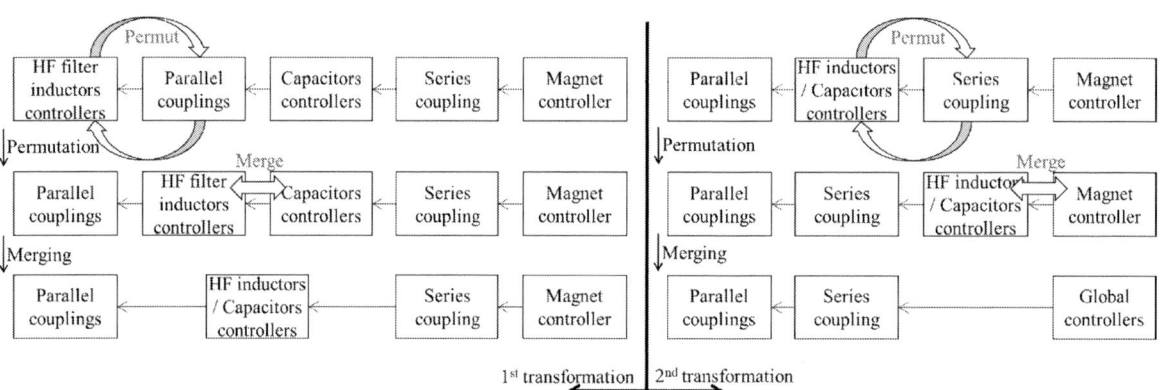

Fig. 7: Control transformations

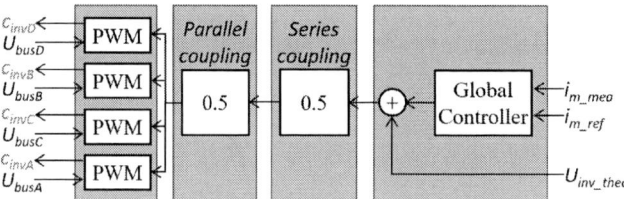

Fig. 8: Details of the simplified control

c) Simulation and experimental comparison

The model and the simplified control system are implemented using Matlab-Simulink. In this section, IGBTs that composed the DC/DC converters H-bridge are supposed to be ideal and no turn-on or turn off delays are implemented in the model.

The power converter is simulated for different reference amplitudes. Fig. 9 shows the results for a 2500 A current pulse reference. When the thyristor is turned on, the output voltage presents a voltage drop that is due to the load connection. This voltage drop produces a small oscillation during the rise of the magnet current. After transient stabilisation, the magnet current is well controlled in respect of the specification.

The control has been implemented on the DSP of the converter (Fig. 10). For the experimental testing, dead-times of 4 μs were inserted to avoid short-circuit of the DC bus. For pulse of magnet current up to 750A, the current in the IGBT of inverters is switched alternatively with a positive and a negative current. Switching with a negative current causes an overvoltage during the dead-time while switching with a positive current causes a voltage deviation during the dead-time. Thus, with alternatively positive and negative switching, the effects of dead-time are offset.

The experimental results show similar current values to the ones from the simulation, between 200 A and 750 A Fig. 11 shows the comparison between simulation and experiment for a pulse of 450 A. The simulation and experimental results are plotted on the same graph for the inverter current (module A), capacitor voltage (module AC) and magnet current. Throughout the duration of the pulse, the switching of the inverter current is done alternatively with a positive and a negative current. If the noise of measurement is removed, the difference between simulation and experimentation is 0.5%. Thus, the simulation model allows a good representation of the real system.

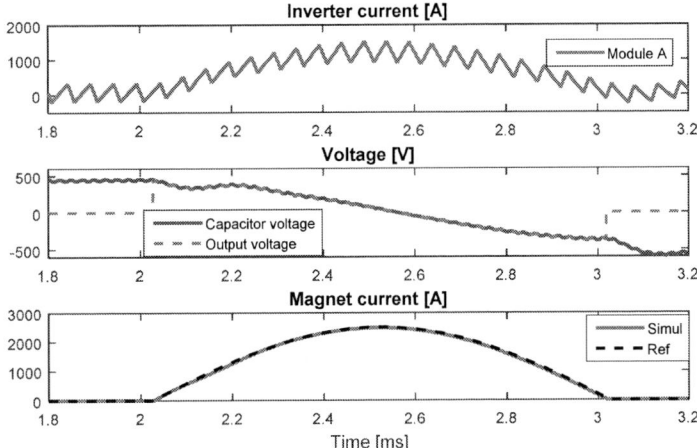

Fig. 9: Simulation results at 2500 A

Fig. 10: SIRIUS FP2P2S Converter

For a current peak higher than 750 A, the switching is done alternatively with positive and negative current when the current is low, and with only positive current when it is high. When the switching is alternatively done with a positive and a negative current, the effects of dead-time are offset. When the switching is only done with a positive current, the dead-time causes undervoltage of:

$$U_{perturbation} = -2U_{bus}\frac{T_{dead-time}}{T_{PWM}} = -176\ V \tag{20}$$

U_{bus} is the DC bus voltage (1100 V), $T_{dead-time}$ is the dead-time duration (4 μs), T_{PWM} is the switching period (50 μs).

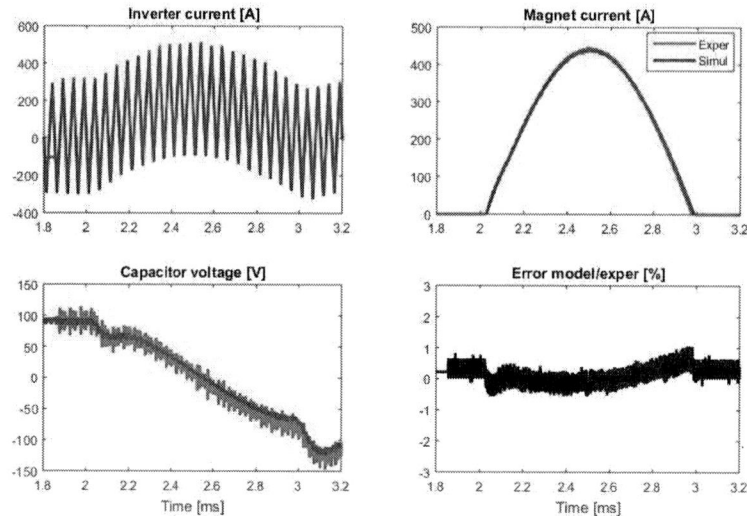

Fig. 11: Comparison simulation – experimentation at 450 A without dead-time compensation

Fig. 12 shows the comparison between simulation and experimental results for a pulse of 900 A. The simulation and experiment results are plotted on the same graph for the inverter current (module A), capacitor voltage (module AC) and magnet current. When the current is low, the switching is done alternatively with a positive and a negative current. During this phase, the difference between simulation and experimental results is small (similar to 450 A). When the current is high, the voltage deviation due to the dead-time is visible on the voltage graph (Fig. 12). This voltage deviation is lower than 176 V due to the control which attempts to compensate for this perturbation. This voltage deviation causes a highly disturbed magnet current. On the inverter current graph, the experimental curve is not only positive (for high value) due to the drop of the magnet current. However, the inverter current cross 0 A during the dead-time and thus has an effect. This is also the reason why the measured drop voltage is lower than 176 V.

The model correctly reproduces the real behaviour of the system when dead-time have no effect. However, it must be improved to take into account the effects of the dead-time.

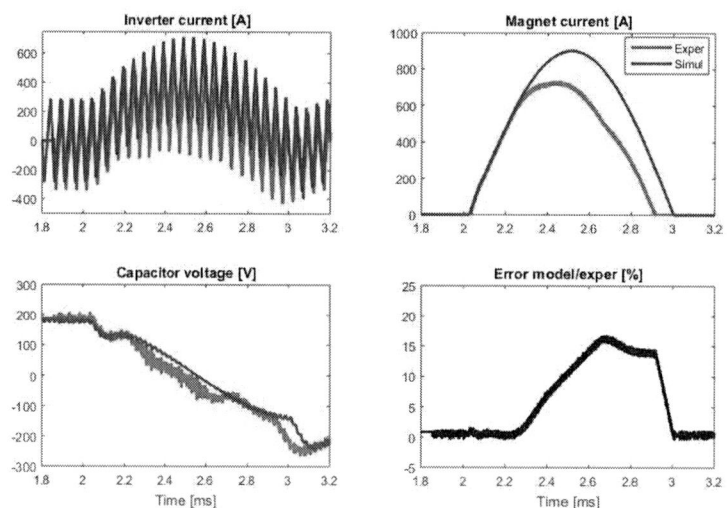

Fig. 12: Comparison simulation – experimentation at 900 A without dead-time compensation

3. Dead-time compensation

a) Dead-time modelling

When a switching order is sent to an IGBT, the effective switching of the IGBT is realized with t_{on} and t_{off} delays. In this application, the PWM operation is implemented using two levels IGBT commands. The effect of dead-time on the IGBT inverter output voltage depends on the sign of the filter inductor L_{hf} current (Fig. 13).

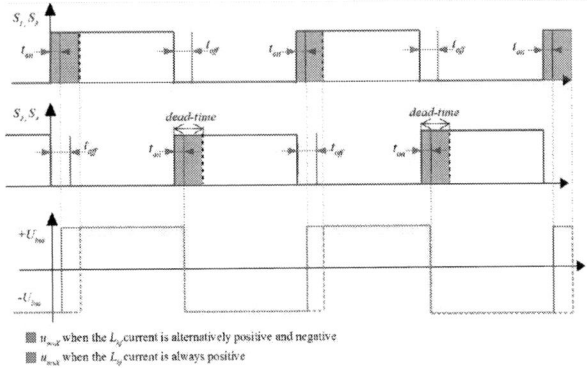

Fig. 13: Effect of dead-times on the IGBT inverter output voltage.

Experimental tests allow to estimate the opening and the closing time of IGBT. The IGBT opening time is measured at 3.2 µs and the closing time is measured at 0.8 µs. Thus, without dead-time, all IGBT are closed during 2.4 µs. To avoid short circuits a dead-time is added to delay the closing order of 4 µs.

The simulation model is modified to insert delays in the switching of the IGBT. The control used in simulation is modified to delay the closing information of the IGBT. The simulation is restarted with Matlab Simulink for a pulse of 900 A as previously. The experiment results and the new simulation results are plotted on the same graph for the inverter current (module A), capacitor voltage (module AC) and magnet current (Fig. 14). When considering the dead-time, the drop voltage, caused by the effects of the dead-time, is correctly reproduced. Now the magnet current obtained in simulation is in accordance with the experiment.

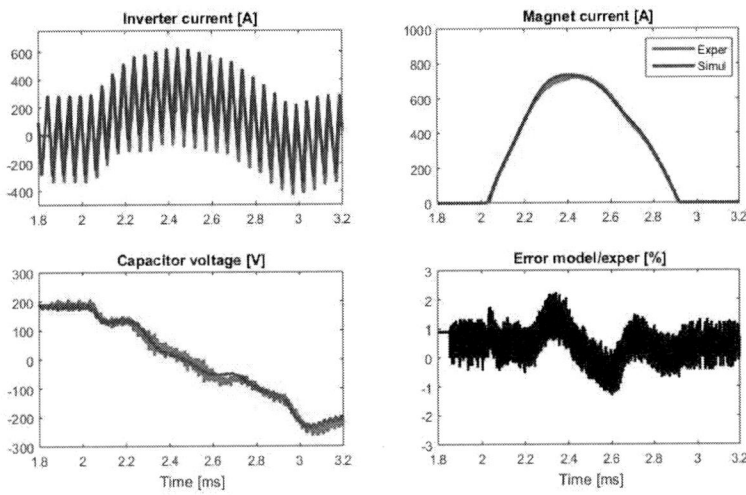

Fig. 14: Comparison simulation – experimentation at 900 A with dead-time compensation

b) Discussion about the different methods to minimize dead-time effects

Dead-time has effects when the modulation ratio is close to 100 % and when the switching is only carried out with a positive (or a negative) current. As presented in the previous section, these effects can have a significant impact. Several hardware and software approaches exist to minimize the dead time effects. Hardware approaches consist in avoiding activation of the dead-time and must be considered during the converter hardware design.

- For hardware example, the size of the output capacitor bank can be increased, making it close to a pure capacitor discharge converter. In this case, the current generated by the discharging capacitor is close to the reference current. Thus, the current produced by the inverters during the pulse is low enough to not activate the dead-time effects. This solution implies that the capacitor bank has to be designed specifically in function of the load. This approach is in contradiction with the wish to standardize the converter.

- Another hardware solution consists in changing the size of the HF inductor, in order to increase the current ripple and avoid switching with only a positive current. This means that the design of the inductor has to support very high current peaks, which are already important with the existing design. Moreover, this solution also leads to an increase of the current peak crossing the IGBT, which, in this case, is not able to support.

- A software solution comprises the compensation of these effects automatically with the close-loop control. Nevertheless, for the proposed application, where the switching operates at 20 kHz and control sampling is at 80 kHz (complete interleaving), the control cannot compensate quickly enough these effects. A modification of the switching frequency is also possible. If the frequency is reduced, the effects of the dead-time are also reduced. This is because the ratio of the dead-time period to the switching period decreases. However, in this case, the control loops cannot work quickly enough to a correct tracking of the fast reference current (half-sinusoidal pulse of 1 ms). Unlike when increasing the switching frequency, which enables the control to work more quickly, but leads to an increase of the dead-time effects due to the increase of the ratio of the dead-time period in the switching period. Thus, it is needed to find a compromise between the ratio of the dead-time period and the switching frequency.

- An alternative software solution consists in adding a non-linear sub-control to compensate the dead-time. This solution requires the measurement of the IGBT current in order to determine if the reference voltage of the inverter must be adapted to dead-time compensation. This solution is common for switched power converters, but, due to the very fast operation of the converter, tests must be done to ensure its viability in the given application.

c) Results with the dead-time compensation

A conventional dead-time compensation strategy [6] [7] has been developed. This strategy defines a compensation voltage which is added to the reference voltage (Fig. 15). This voltage compensation equals the voltage drop due to the dead-time (Fig. 16). When the current of the inverter is positive (current ripple included), this compensation is equal to the maximal voltage drop (20) (Fig. 16). The inverter current is positive when its average value is greater than half the value of the inverter current ripple ($I_{hf\_detect2} = I_{hf\_ripple}/2$).

When the inverter current is small enough to have completely switched (dead-time included) with alternatively positive and negative current, no compensation is necessary (Fig. 16). Switching are done alternatively with a positive and negative current when the average inverter current is lower than:

$$I_{hf\_detect1} = I_{hf\_detect2} - U_{bus}\frac{1}{L_{hf}}T_{death-time} \qquad (21)$$

When the inverter current is close to 0 A to become positive during the dead-time ($I_{hf\_detect1} < I_{hf} < I_{hf\_detect2}$) proportional compensation is imposed (Fig. 16).

The average value of the inverter current is calculated by the average of the four inverter currents. This is possible because all the modules are identical, and thermally regulated, because the voltages and currents are distributed equally between the modules and because the PWM are interleaved with an angle of 90°,

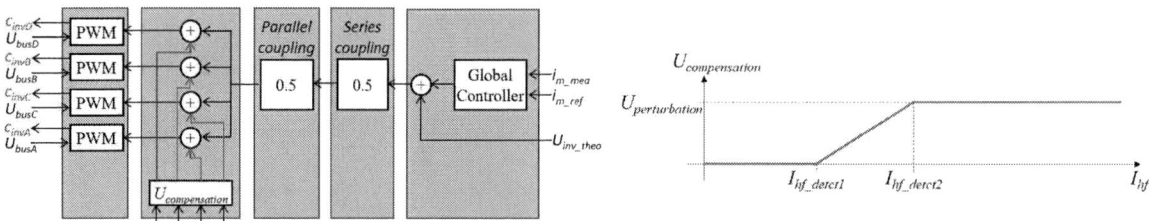

Fig. 15: Control with dead-time compensation   Fig. 16: Dead-time compensation strategy

This strategy is firstly implemented in simulation. Fig. 17 shows the comparison between the simulation results:
- without switching delay, without dead-time and without dead-time compensation (ideal case without dead-time effects),
- with switching delay, with dead-time and with dead-time compensation (real case)

The current obtained with the compensation strategy is very close to the ideal case where the dead-time effects are neglected. Based on the results of Fig. 12, the dead-time produces a current distortion of 17 %. Based on the results of Fig. 17, the dead-time compensation strategy allows to reduce this distortion to 0.3 %

The compensation strategy has been implemented on the DSP controller. Fig. 18.a shows experimental results of a 900 A pulse without and with the compensation of dead-time strategy. By applying this strategy, the pulse is able to reach the right peak with an acceptable precision during the pulse. Thus, the strategy used is as effective as in simulation. Dead-time compensation has also been tested for 2400 A of amplitude with the same effectiveness (Fig. 18.b).

Conclusion

This paper deals with a pulsed power converter where the dead-time represents a significant part of the half-switching period. In a first time, the impact of these dead-time, which generate a discontinue behaviour for current higher than 500 A, is analysed. At 900 A, the dead-time introduced an error of 160 A which correspond to an error of 17 % in comparison to the theoretical case, where the switching times are instantaneous and where no dead-time is required. Several hardware and software design approaches have been discussed and the most appropriate for this fast and specific application has been selected. This solution has been implemented on the pulsed power converter to limit the dead-time effects (detailed in the final paper). With this compensation strategy, the error introduced by the dead-time is of 0.3 %, thus a reduction of 98.2 % of their effects.

This development has been implemented in a simplified control with a unique control loop. An extension of this work will deal with the initial control scheme with 3 closed-loop controls in order to increase the tracking performance and reduce the current error up to +/- 3 A for current reference up to 2.5 kA of magnitude.

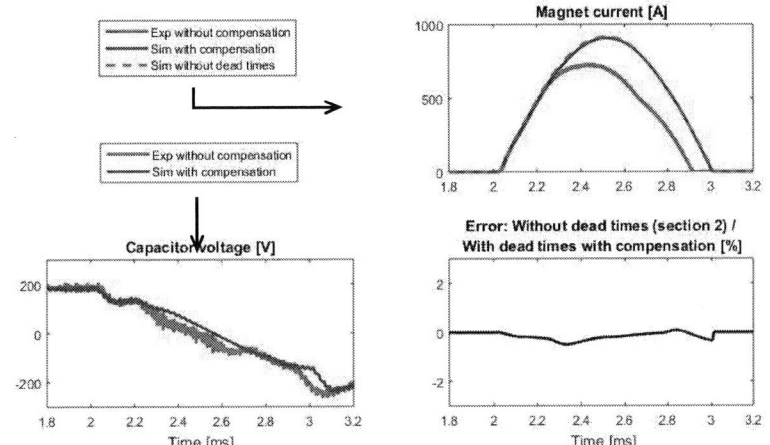

Fig. 17: Comparison of simulation results with and without dead-time compensation at 900 A

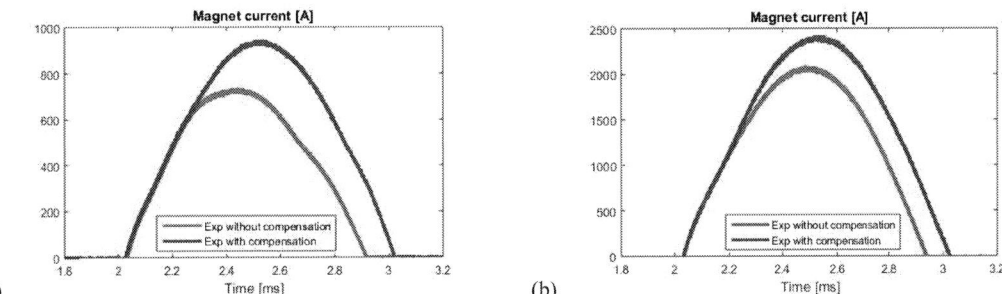

Fig. 18: Experimental results with and without dead-time compensation: at 900 A (a) and 2400 A (b)

References

[1] F. Voelker, "Pulsed capacitor discharge power converter: an introductory overview", *CAS-CERN Accelerator School: Power Converters for Particle Accelerators Conference*, Montreux (Switzerland), March 1990.

[2] CERN website: http://home.cern/topics/high-luminosity-lhc, *website visited in October 2016*

[3] S. Rossini, K. Papastergiou, G. Le Godec, R. Garcia Retegui, S. Maestri, "Power Converter Topologies with Energy Recovery and Grid Power Limitations For Inductive Load Applications", *EPE'15*, Geneva (Switzerland), Oct. 2015

[4] L. Horrein, J. M. Cravero, "Hybrid Capacitor Discharge/Switch-Mode Converter for Pulsed Applications: Topology and Control design", *EPE'17*, Warsaw (Poland), Sep. 2017

[5] S. G. Jeong, M. H. Park, "The analysis and compensation of dead-time effects in PWM inverters", *IEEE Transactions on Industrial Electronics*, vol. 38, pp. 108 – 114, 1991.

[6] D. Leggate, R. J. Kerkman, "Pulse Based Dead-time Compensator for PWM Voltage Inverters", *IEEE Transactions on Industrial Electronics*, vol. 44, pp. 191 – 197, 1997

[7] S. H. Hwang, J. M. Kim, "Dead-time Compensation Method for Voltage-Fed PWM Inverter", *IEEE Transactions on Energy Conversion*, vol. 25, pp. 1-10, 2010

[8] A. Bouscayrol, B. Davat, B. de Fornel, B. François, J. P. Hautier, F. Meibody-Tabar, M. Pietrzak-David, "Multimachine Multiconverter System: application for electromechanical drives", *European Physics Journal - Applied Physics*, vol. 10, no 2, pp. 131-147, May 2000

[9] A. Bouscayrol, J. P. Hautier, B. Lemaire Semail, "Systemic design methodologies for electrical energy systems Analysis, Synthesis and Management", Chapter 3: Graphic formalisms for the control of multi-physical energetic system: COG and EMR, *ISTE and Wiley*, ISBN 978-1-84821-3888-3, 2012

[10] S. Astier, A. Bouscayrol, X. Roboam, "Systemic design methodologies for electrical energy systems Analysis, Synthesis and Management", Chapter 1: Introduction to Systemic design, *ISTE and Wiley*, ISBN 978-1-84821-3888-3, 2012

[11] I. Iwasaki, H. A. Simon "Causality and model abstraction" *Artificial Intelligence*, vol. 67, pp. 143-194, 1994

Dynamic Characterization of a SiC-MOSFET Half Bridge in Hard- and Soft-Switching and Investigation of Current Sensing Technologies

Janine Ebersberger[1,*], Jan-Kaspar Müller[1], Axel Mertens[1,*]

[1]Leibniz University Hannover

Institute for Drive Systems and Power Electronics

Welfengarten 1

Hannover, Germany

[*] Cluster of Excellence SE^{2}A Sustainable and Energy-Efficient Aviation

Braunschweig, Germany

Phone: +49 (0) 511-762 18834

Email: janine.ebersberger@ial.uni-hannover.de

URL: http://www.ial.uni-hannover.de

Acknowledgments

We would like to acknowledge the funding by the Deutsche Forschungsgemeinschaft (DFG, German Research Foundation) under Germany's Excellence Strategy EXC 2163/1 Sustainable and Energy-Efficient Aviation Project ID 390881007.

Keywords

≪Wide bandgap devices≫, ≪Silicon Carbide (SiC)≫, ≪Soft switching≫, ≪Current sensor≫, ≪Double pulse test≫

Abstract

In this paper, the dynamic characterization of a 1.2 kV SiC-MOSFET half bridge module is presented. At first, the switching behaviour of the MOSFET and its body diode is examined with a double pulse test, also targeting the evaluation of the used current sensing technologies. Afterwards, the soft-switching performance is investigated in a three-phase inverter configuration with a sine-wave output filter.

Introduction

Silicon carbide (SiC) semiconductors gain popularity in modern power electronics because of their numerous advantages and growing availability. Due to lower switching energies, the overall switching losses can be reduced for conventionally used switching frequencies in comparison to converters based on silicon semiconductors. Alternatively, the switching frequency can be increased [1]. In order to design an optimized layout, it is crucial to understand the switching behaviour in detail and determine the semiconductor losses as precisely as possible. Therefore, a SiC-MOSFET half bridge module is characterized in this paper by carrying out a double pulse test for the MOSFET itself as well as for its body diode. Other authors have already analyzed the switching behaviour of SiC devices and compared different current sensing technologies. However, mostly commercial Rogowski coils are considered [2, 3]. In this paper, measurements with an in-house designed differential Rogowski coil [4] are performed and compared to measurements with a coaxial shunt in order to point out their main advantages and disadvantages. Following the observations in [5, 6], the soft-switching behaviour of the half bridge in a three-phase inverter configuration connected to an output filter is investigated. Additionally, the proposed approach for calculating the overall switching losses in the inverter is validated with a power analyzer measurement.

Double pulse test (DPT)

Within the framework of the described research, a SiC-MOSFET half bridge module with a nominal on resistance of 23 mΩ [7] is characterized dynamically. It consists of two MOSFETs with their internal body diodes and has maximum ratings of 1200 V and 50 A, respectively. In order to determine the switching losses of the MOSFETs, a double pulse test is executed. Fig. 1 shows the power PCB assembled with the half bridge module and the probes. The measurement equipment can be found in Table I and the equivalent circuit diagram is presented in Fig. 2 including the measurement points. The gate driver is placed on top of the power PCB and sets a fixed gate drive voltage. Thus, the influence of the gate-drive voltages is not further investigated.

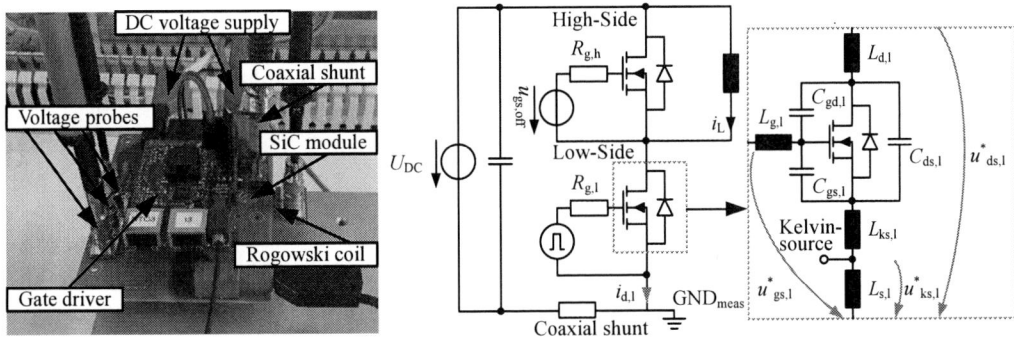

Fig. 1: Double pulse test power PCB   Fig. 2: Equivalent circuit diagram of the DPT setup

Table I: DPT measurement equipment

	Description	**Bandwidth**	**Measured Parameter**
Oscilloscope	LECROY HDO8108A	1 GHz	
Passive probe	LECROY PP023	500 MHz	$u^*_{\mathrm{gs,l}}, u^*_{\mathrm{ks,l}}, u_{\mathrm{Rog1}}, u_{\mathrm{Rog2}}$
Passive probe (HV)	LECROY HVP120	400 MHz	$u^*_{\mathrm{ds,l}}$
Coaxial shunt	T&M RESEARCH PRODUCTS SDN-414-05	2 GHz	$i_{\mathrm{d,l,Shunt}}$
Differential Rogowski coil	in-house design	approx. 37 MHz	$i_{\mathrm{d,l,Rog}}(u_{\mathrm{Rog1}}, u_{\mathrm{Rog2}})$
Current probe	KEYSIGHT N2781B	10 MHz	i_{L}

For the current measurement, two technologies are investigated. On the one hand, a coaxial shunt providing a high bandwidth of up to 2 GHz is used. However, the shunt not only raises the commutation inductance by its self-inductance of 6.7 nH, which is determined by an impedance measurement, but on the other hand, requires a special PCB integration directly affecting the commutation cell. Therefore, a second current measurement with a differential Rogowski coil is considered. The principle is described in [4] as well as its advantages, such as a high dv/dt immunity and a higher bandwidth compared to commercial Rogowski coils. The Rogowski coil is realized as a printed circuit board similar to the one described in [8] and can be placed in between the half bridge module and the power PCB. The Rogowski current sensor consists of two coils with opposed orientations, so that the influence of a parasitic capacitive current is reduced. Thus, two voltages are measured by passive probes and the drain current is calculated by an offline integration. A conventional voltage measurement close to the semiconductor is not possible, as the coaxial shunt sets the reference potential $\mathrm{GND}_{\mathrm{meas}}$ of the oscilloscope. Therefore, the voltage $u^*_{\mathrm{ks,l}}$ is measured additionally and the voltages $u_{\mathrm{ds,l}} = u^*_{\mathrm{ds,l}} - u^*_{\mathrm{ks,l}}$ and $u_{\mathrm{gs,l}} = u^*_{\mathrm{gs,l}} - u^*_{\mathrm{ks,l}}$ are calculated for determining the switching energy.

Results

The switching behaviour is analyzed regarding DC-link voltages from 100 V to 600 V and load currents from 5 A to 50 A with a gate resistance of 1 Ω. Moreover, the DPT is carried out at temperatures from room temperature (approx. 25 °C) to 125 °C. As the switching waveforms are hardly influenced by the temperature raise, the switching behaviour is discussed exemplarily for room temperature in the following. At first, the drain current is measured with the Rogowski coil. Fig. 3 and Fig. 4 show the drain-source voltage and the drain current as well as the corresponding maximum voltage and current slopes depending on different DC-link voltages and load currents. Furthermore, the power and the resulting switching energies are depicted. Increasing the DC-link voltage and/or the load current leads to the rise of the current maximum and the switching energy during turn-on. Moreover, the drain-source voltage slope depends strongly on the load current during turn-off. Different DC-link voltages influence the oscillation frequency, as the output capacity of the MOSFETs depends on the applied voltage, as can be seen in Fig. 4. In Fig. 5 and Fig. 6, the switching energies E_{on} and E_{off} as well as their dependency on the examined parameters are depicted. The load current and the DC-link voltage have a significant impact on the turn-on energy E_{on} whereas there is less influence on the turn-off energy E_{off}. Since the largest losses are caused by turning on the MOSFET, its soft-switching behaviour is investigated in the last section.

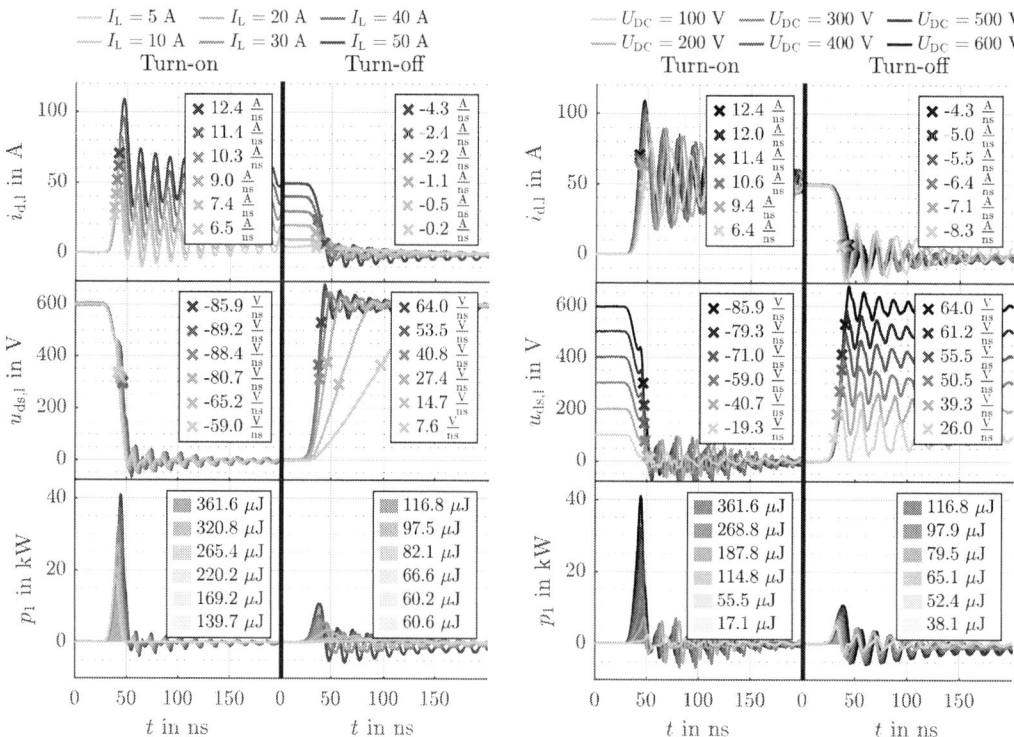

Fig. 3: Double pulse test switching transients of the low-side MOSFET at different load currents I_L, $U_{DC} = 600\,\mathrm{V}$ and $R_{g,l} = 1\,\Omega$

Fig. 4: Double pulse test switching transients of the low-side MOSFET at different DC-link voltages U_{DC}, $I_L = 50\,\mathrm{A}$ and $R_{g,l} = 1\,\Omega$

In order to investigate the switching behaviour of the body diode with the same test setup, the circuit from Fig. 2 is modified in the way that the load inductance is in parallel to the low-side. Now, the low-side driver signal is set low and the high-side is pulsed. In Fig. 7, the switching transients of the body diode are displayed. Typically, the turn-on process of a diode is lossless, as it cannot conduct as long as the blocking voltage is applied. However, in Fig. 7, the drain current and the drain-source voltage are overlapping for several nanoseconds during turn-on. This apparent contradiction is caused by the capacitive current which discharges the body diode's junction capacitance, and consequently no Joulean

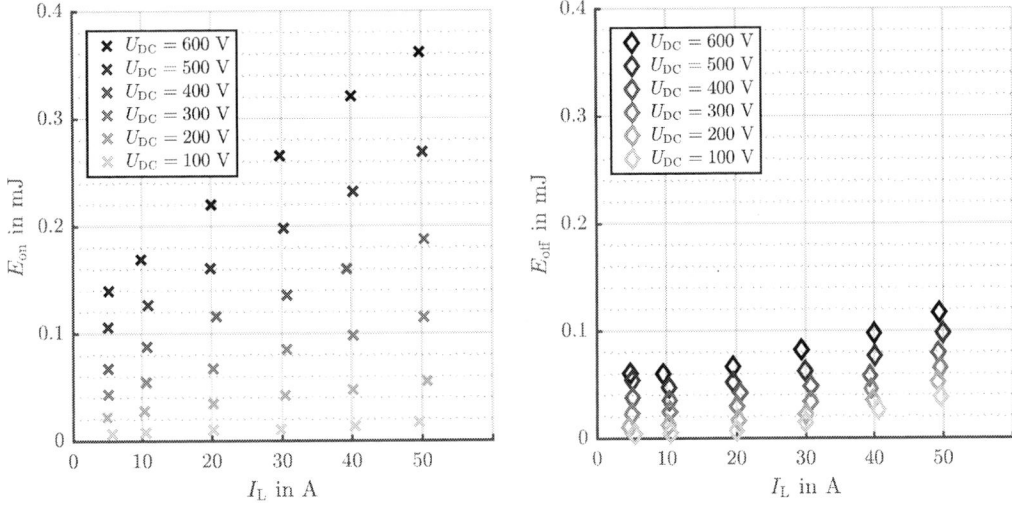

Fig. 5: Turn-on energy E_{on} of the low-side MOS-FET at different load currents and DC-link voltages

Fig. 6: Turn-off energy E_{off} of the low-side MOS-FET at different load currents and DC-link voltages

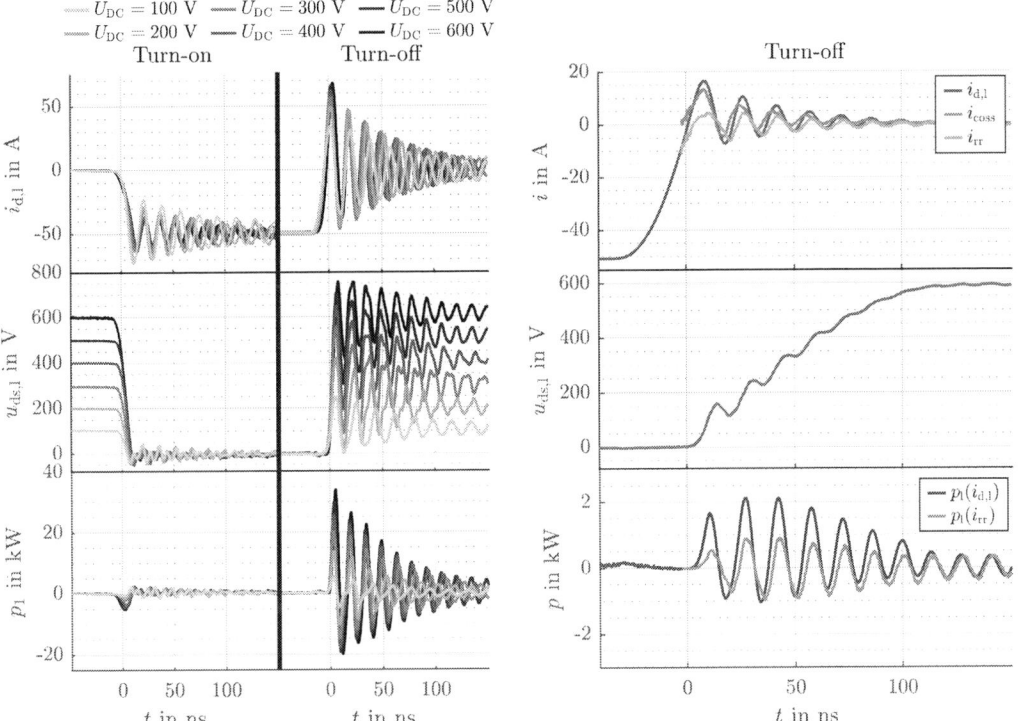

Fig. 7: Double pulse test switching transients of the low-side body diode at different DC-link voltages U_{DC}, $I_L = 50\,A$ and $R_{g,h} = 1\,\Omega$

Fig. 8: Separation of the measured current $i_{d,l}$ in the capacitive current i_{coss} and the reverse-recovery current i_{rr} at $I_L = 50\,A$, $U_{DC} = 600\,V$ and $R_{g,h} = 15\,\Omega$

heat losses are caused. When the junction capacitance is discharged completely, the body diode starts to conduct the load current. This instant can be identified by the drain-source voltage reaching 0 V.

Body diodes usually have a so called snappy turn-off behaviour, meaning that when returning to 0 A the reverse current and its slope are large in combination with a high voltage overshoot and strong oscillations. Such a behaviour can be observed on the right side of Fig. 7. At a DC-link voltage of 600 V, the drain-source voltage and the current almost reach 800 V and 70 A, respectively.

In order to determine the reverse-recovery energy, the switching process is analyzed in detail. Two effects define the body diode's turn-off. First, the bipolar charge of the diode is removed and second, the sudden voltage drop at the high-side MOSFET forces a high dv/dt across the low-side output capacitance. Both effects cause currents which overlap and their sum is measured by the Rogowski coil. The first is called reverse-recovery current and generates losses in the device, whereas the second current charges the output capacitance and does not cause losses. The capacitive current is determined according to $i_{\text{coss}} = C_{\text{oss}}(u_{\text{ds,l}}) \cdot \frac{du_{\text{ds,l}}}{dt}$ and by calculating the voltage change across the capacitance as well as extracting the voltage-dependent value of the output capacitance $C_{\text{oss}}(U_{\text{ds}})$ from the chip's SPICE model. The calculation of i_{coss} is limited by the definition of $C_{\text{oss}}(U_{\text{ds}})$ for $U_{\text{ds}} > 0$ and by voltage slew rates that are too high to be evaluated. This method is applied on a switching process with a gate resistance of $R_{\text{g,h}} = 15\,\Omega$. In Fig. 8, the measured current $i_{\text{d,l}}$, the calculated capacitive current i_{coss} and the resulting reverse-recovery current $i_{\text{rr}} = i_{\text{d,l}} - i_{\text{coss}}$ are depicted for the operating point of 50 A / 600 V. The bottom graph in Fig. 8 shows the resulting power p_{l} calculated on the one hand with the measured current $i_{\text{d,l}}$ and on the other hand with the reverse-recovery current i_{rr}. Since the calculated reverse-recovery current i_{rr} does not show the expected typical waveform, the calculated power $p_{\text{l}}(i_{\text{rr}})$ is not suitable for deriving the reverse-recovery energy.

Comparison of the current sensors

Next, the drain current is measured additionally with the coaxial shunt. Fig. 9 compares the prior measurement without the shunt (measurement (a)) to the measurement with the shunt soldered to the board (measurement (b)), meaning the drain current is measured by the Rogowski coil and the shunt in parallel. Regarding Fig. 9, two aspects can be discussed. First, the drain-source voltages $u_{\text{ds,l}}$ of measurement (a) and (b) show that inducing an additional inductance in the commutation loop results in a higher initial voltage drop during turn-on and a higher voltage overshoot during turn-off, as well as in a lower oscillation frequency. The overall switching energy from measurement (b) is 9.1 % smaller than from measurement (a) (calculated with $i_{\text{d,l,Rog}}$).

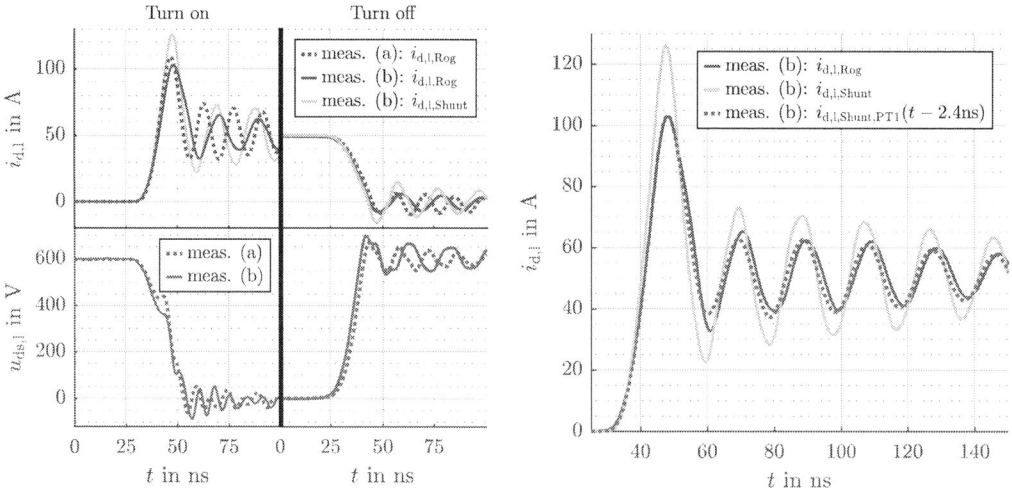

Fig. 9: Comparison of an additional inductance's influence in the commutation loop (meas. (a) vs. meas. (b)) and of the current sensors (meas. (b): $i_{\text{d,l,Rog}}$ vs. $i_{\text{d,l,Shunt}}$)

Fig. 10: Comparison of the drain current measured with the shunt $i_{\text{d,l,Shunt}}$ and the Rogowski coil $i_{\text{d,l,Rog}}$ to the filtered and shifted shunt signal $i_{\text{d,l,Shunt,PT1}}$

Second, the comparison of the two current sensors (measurement (b) $i_{\mathrm{d,l,Rog}}$ and $i_{\mathrm{d,l,Shunt}}$) shows that the coaxial shunt measures a steeper rise/fall and higher maximum/minimum current, respectively. The mentioned observations indicate an insufficient bandwidth of the Rogowski current sensor which is examined next. The switching energy calculated with $i_{\mathrm{d,l,Shunt}}$ is 8.7 % higher than the one calculated with $i_{\mathrm{d,l,Rog}}$ (measurement (b)).

For estimating the bandwidth of the Rogowski coil, the shunt's current signal is filtered by a PT1-element and its cut-off frequency is decreased until both current signals are matched. By applying this method at the operation point of 50 A / 600 V, the equivalent-PT1 bandwidth of the measurement system with the Rogowski coil and the passive probes is determined to 37 MHz. Furthermore, the shunt signal is shifted by −2.4 ns in order to align the filtered shunt current and the current measured by the Rogwoski coil. Fig. 10 compares the currents $i_{\mathrm{d,l,Shunt}}$ and $i_{\mathrm{d,l,Rog}}$ to the filtered and shifted shunt current $i_{\mathrm{d,l,Shunt,PT1}}$. If solely the deskew is applied to the shunt current, the result is a 13.2 % higher switching energy compared to the calculation without deskew.

Analyzing the different switching energies demonstrates the importance of choosing appropriate measurement equipment, as insufficient bandwidth, time misalignment and modification of the commutation circuit influence the measured switching energy severely.

Switching energies of a three-phase inverter

In [9], the necessity of determining the switching energy in a three-phase configuration is stated, as the setup strongly differs between a double pulse test under ideal conditions and a three-phase configuration in an actual application. Therefore, the switching performance of a three-phase inverter connected to a sine-wave filter, as described in [10] and shown in Fig. 11, is investigated.

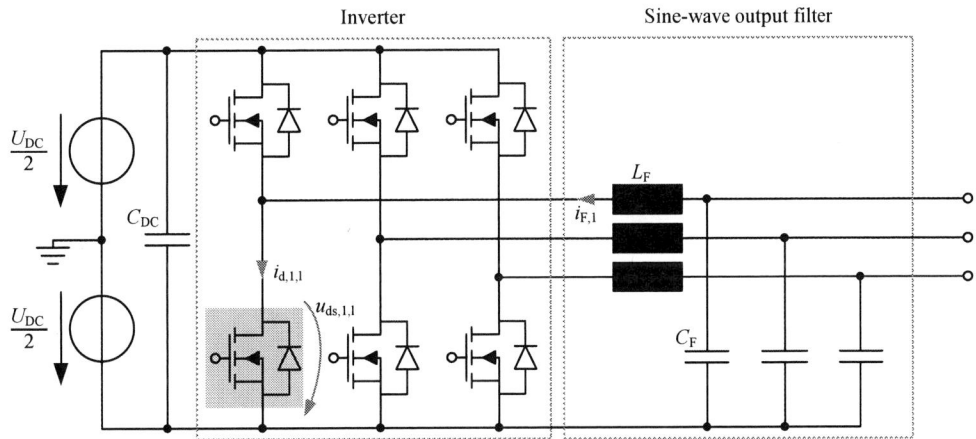

Fig. 11: Equivalent circuit diagram of the three-phase inverter and the sine-wave output filter

The inverter PCB is optimized in terms of the connection between the modules and the DC-link in order to minimize the commutation inductance. Hence, a current measurement with the coaxial shunt is not possible and the drain current is solely measured by the Rogowski current sensor. Due to the filter sizing presented in [10], the switching frequency is 140 kHz. The experiments are carried out at no load. Furthermore, a triangular filter current is achieved by 50 % duty cycle, so that the considered low-side switch (Fig. 11 highlighted in green) turns on at a negative filter current $i_{\mathrm{F,1}}$ (definition see Fig. 11) and turns off at a positive filter current. The DC voltage is varied which also leads to different output currents. The MOSFET's turn-off is displayed on the right side of Fig. 12 and the switching energies are calculated from the resulting power. During turn-on, the MOSFET's body diode conducts first due to the interlock time of 100 ns, and afterwards the current commutates to the channel of the same MOSFET under nearly zero voltage condition. This turn-on process is shown on the left side of Fig. 12, where the junction capacitance of the body diode is discharged by a capacitive current, as described before. According to [6], the net switching energy has to be calculated for determining the power losses in this case, as part of the turn-on energy is fed back into the circuit. Thus, the turn-on energy reduces the consumed energy in the switching cycle and therefore is negative, as shown in the graph at the bottom left of Fig. 12. The single points in Fig. 13 compare the switching energies from Fig. 12 to the fitted turn-off energy of the MOSFET E_{off} and the turn-on energy of the body diode $E_{\text{on,D}}$, which were obtained during the double pulse test (solid lines). A high accuracy of the switching energies is shown, with the highest deviation of 14 % for E_{off} at a DC-link voltage of 600 V. The authors in [9] conclude that the motor load and the cable connecting the motor with the inverter have the most significant influence on the switching losses. As the experiments in this section are carried out at no load, the results are in good agreement with the findings in [9].

Finally, the input power is measured by a ZES ZIMMER LMG671 precision power analyzer and compared to the calculated losses, as presented in Fig. 14.

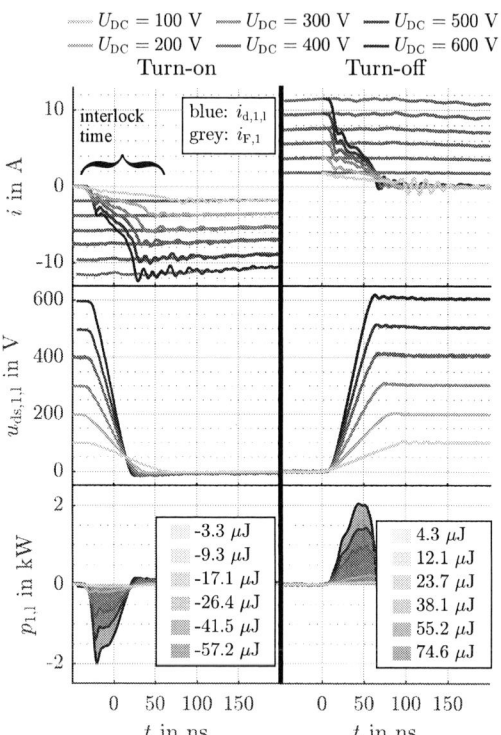

Fig. 12: Switching transients of the SiC-half bridge in a three-phase inverter configuration with triangular output current at different DC-link voltages

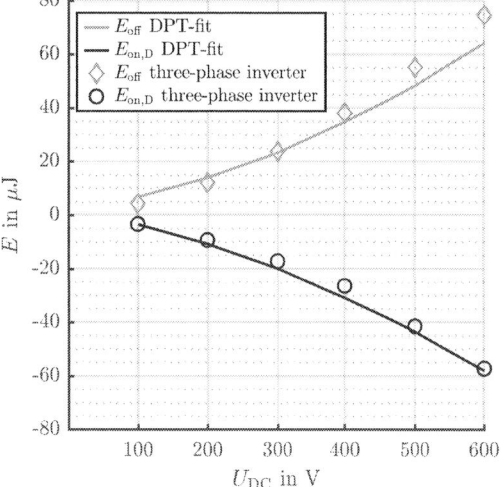

Fig. 13: Comparison of the switching energies between the double pulse test and the three-phase inverter configuration

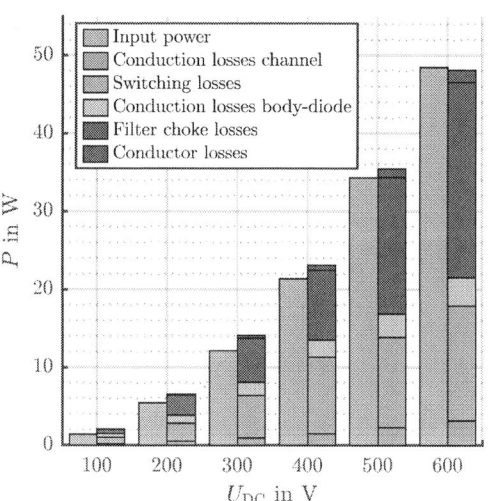

Fig. 14: Comparison of the measured input power and the determined losses in the three-phase inverter

The semiconductor conduction losses are based on information from the datasheet and the SPICE model. By calculating the net switching energy from Fig. 12 and taking the switching frequency f_s into account, the switching losses are identified $P_{v,sw} = f_s \cdot (E_{off} - |E_{on,D}|)$. For determining the losses in the conductors between the inverter and the output filter as well as in the return conductor, the frequency dependent AC losses are neglected and the DC losses are taken into account instead. The simulation software GECKO-MAGNETICS is used for simulating and calculating the losses caused by the filter choke. The overall losses in Fig. 14 are dominated by the switching losses and the losses in the filter choke. Furthermore, the difference between the measured input power and the calculated losses is small, especially at high voltages (0.73 % at a DC-link voltage of 600 V). Thus, the proposed calculation of switching losses under soft-switching conditions is confirmed. The results point out possibilities for reducing switching losses in the application by using a high current ripple which is carried out similarly in [11] for buck converters.

Conclusion

This paper presents the dynamic characterization of a SiC-MOSFET half bridge module. At first, the setup for the double pulse test is described and the investigated current sensing technologies are introduced. The double pulse test is used to investigate the switching behaviour of the MOSFET and its body diode depending on the parameters DC-link voltage and load current. This setup allows for a fair comparison of the measurement quality of the differential Rogowski coil and the coaxial shunt. It is shown that the bandwidth of the Rogowski coil is insufficient regarding current rise times as seen in wide bandgap devices. However, the coil is advantageous towards a coaxial shunt regarding its small invasivity, so that it can be used at a PCB optimized for fast switching, as shown in the last section. Additionally, the soft-switching condition of a three-phase inverter, connected to a sine-wave filter with a triangular filter current, is examined. In this operation mode, only the difference between the hard turn-off energy and the soft turn-on energy causes Joulean heat losses, which is in good agreement with the system losses characterized with a power analyzer.

References

[1] A. Merkert, T. Krone and A. Mertens: Characterization and Scalable Modeling of Power Semiconductors for Optimized Design of Traction Inverters with Si- and SiC-Devices, in IEEE Transactions on Power Electronics, vol. 29, no. 5, pp. 2238-2245, May 2014

[2] Z. Zhang, B. Guo, F. F. Wang, E. A. Jones, L. M. Tolbert and B. J. Blalock: Methodology for Wide Band-Gap Device Dynamic Characterization, in IEEE Transactions on Power Electronics, vol. 32, no. 12, pp. 9307-9318, Dec. 2017

[3] H. Li, S. Beczkowski, S. Munk-Nielsen, K. Lu and Q. Wu: Current measurement method for characterization of fast switching power semiconductors with Silicon Steel Current Transformer, 2015 IEEE Applied Power Electronics Conference and Exposition (APEC), Charlotte, NC, 2015, pp. 2527-2531

[4] S. Hain and M. Bakran: New Rogowski coil design with a high DV/DT immunity and high bandwidth, 2013 15th European Conference on Power Electronics and Applications (EPE), Lille, 2013, pp. 1-10

[5] A. März, T. Bertelshofer, M. Helsper and M. Bakran: Comparison of IGBT and SiC MOSFET in resonant application, 2018 20th European Conference on Power Electronics and Applications (EPE'18 ECCE Europe), Riga, 2018, pp. P.1-P.8

[6] D. Rothmund, D. Bortis and J. W. Kolar: Accurate transient calorimetric measurement of soft-switching losses of 10kV SiC MOSFETs, 2016 IEEE 7th International Symposium on Power Electronics for Distributed Generation Systems (PEDG), Vancouver, BC, 2016, pp. 1-10

[7] Infineon Technologies AG.: Datasheet FF23MR12W1M1_B11 V2.2, 07.2018

[8] A. Merkert, J. Müller and A. Mertens: Component design and implementation of a 60 kW full SiC traction inverter with boost converter, 2016 IEEE Energy Conversion Congress and Exposition (ECCE), Milwaukee, WI, 2016, pp. 1-8

[9] Z. Zhang, F. Wang, L. M. Tolbert, B. J. Blalock and D. J. Costinett: Evaluation of Switching Performance of SiC Devices in PWM Inverter-Fed Induction Motor Drives, in IEEE Transactions on Power Electronics, vol. 30, no. 10, pp. 5701-5711, Oct. 2015

[10] J. Müller, T. Manthey, D. Han, B. Sarlioglu, J. Friebe and A. Mertens: Output Sine-Wave Filter Design and Characterization for a 10 kW SiC Inverter, 2019 IEEE Energy Conversion Congress and Exposition (ECCE), Baltimore, MD, USA, 2019, pp. 359-366

[11] B. Cougo, H. Schneider and T. Meynard: High Current Ripple for Power Density and Efficiency Improvement in Wide Bandgap Transistor-Based Buck Converters, in IEEE Transactions on Power Electronics, vol. 30, no. 8, pp. 4489-4504, Aug. 2015, doi: 10.1109/TPEL.2014.2360547

Power Supply Design Considerations For 400Hz Aircraft Applications

Bilal Ahmad and Jorma Kyyrä
School of Electrical Engineering and Automation
Aalto University
Espoo , Finland
Email: bilal.3.ahmad@aalto.fi
URL: http://www.aalto.fi

Juha Mäkelä
Vensum Power
Otakaari 5
Espoo , Finland
Email: juha.makela@vensum.com
URL: http://www.vensum.com

Keywords

≪DC power supply≫, ≪Design≫, ≪EMC/EMI≫, ≪Gallium Nitride (GaN)≫, ≪Transformer≫

Abstract

In this study a 5kW power supply for aircraft applications has been designed. Boost type power factor correction topology Delta rectifier complies with military standards of conducted emissions and is selected as a front-end rectifier for 400 Hz mains power supply. For isolation and output Dc voltage regulation a full bridge Dc-Dc converter with 99% efficient high frequency transformer has been designed.

Introduction

To improve the life time and performance of an aircraft a concept known as "More Electric Aircraft (MEA)" has been introduced in early 90's . Futuristic MEA aims to have an electric power as its major source of energy instead of hydraulic, pneumatic and mechanical power for its numerous auxiliary functions [1].This will not only enhance the performance of an aircraft but will also reduce its weight and hence fuel consumption and CO_2 emissions will be reduced [2].

Conventional aircraft has an AC grid of $115V_{RMS}$ with 400Hz fundamental frequency [3]. It utilizes a two stage DC power supply (AC-DC Rectifier followed by a DC-DC converter) to power its numerous auxiliary functions which are rated at 28VDC - 270VDC. However, the existing power converters in an aircraft are bulky and inefficient [4].And to move towards the realization of MEA, it is vital to have high power dense and efficient power conversion solutions.In addition to efficiency and power density, it is also critical for converters to comply with strict military electro-magnetic interference (EMI) standards.[5].

In this study, a design procedure of an efficient and power dense DC power supply for an aircraft has been presented.Fig. 1 shows a typical two-stage isolated power supply.

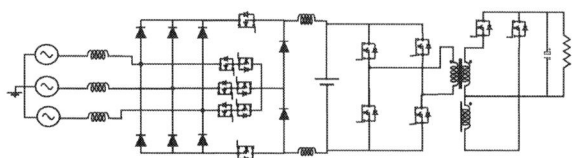

Fig. 1: Isolated DC Power Supply

Main design challenges that have been discussed in this study are as follows:
- Selection of an efficient and power dense appropriate AC-DC rectifier topology that complies with military standards (MIL-STD-461G) for power quality.

- Design optimization of an isolation transformer for DC-DC stage to achieve high efficiency and power density.

This study is done in two stages. First stage includes reviewing and evaluating different three phase PFC topologies. Based on efficiency and compliance with EMI standards an appropriate PFC topology is chosen. Next stage includes selection of an isolated DC-DC converter and implementation of a MATLAB based optimization algorithm for design of isolation high frequency transformer.

Three Phase PFC Topologies

A detailed review of different three phase PFC topologies has been presented in [6]. In our application we require unidirectional flow of power, hence bi-directional three-phase PFC topologies are not considered for this study. A PFC topology can be either Buck or Boost type. Based on number of high-frequency switching devices and its conversion efficiency, one topology from each category has been chosen for further analysis.
- Swiss Rectifier Topology from Buck Type PFC Rectifier Family.
- Delta Rectifier Topology from Boost Type PFC Rectifier Family.

[7] did a comprehensive study on design of Swiss rectifier and has shown that very high conversion efficiency of 99.3% can be achieved. Similarly, [8] presents the design and evaluation of Delta rectifier and has shown that high conversion efficiency ($> 90\%$) can be achieved. However, both these studies do not discuss the compliance of these topologies with military power quality standards (MIL-STD-461G) [9]. Hence, in next sections compliance of both these topologies with MIL-STD-461G along with brief explanation of their operation principle is presented.

Swiss Rectifier

Swiss rectifier shown in Fig. 2 consists of a three-phase bridge to convert 3-phase AC voltage into DC pulsating voltage. Three bi-directional bi-polar switches are utilized for third harmonic injection

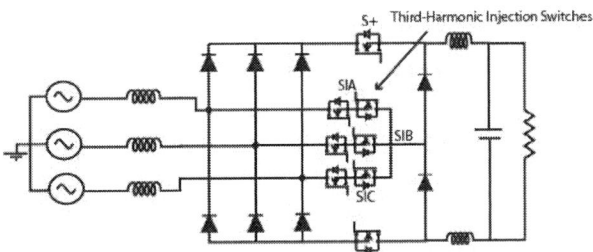

Fig. 2: Swiss Rectifier

to achieve unity power factor. These switches are realized by the common drain connection of two GaN HEMTs [5]. These switches operate at twice the fundamental frequency (i.e. 800Hz) and their switching state depends on the operation sector of three-phase input voltage. To explain their switching pattern symmetrical three-phase waveform is shown in Fig. 3 and it is divided into six sectors. Each

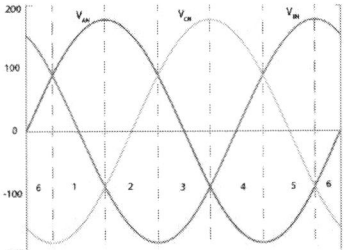

Fig. 3: Three Phase AC Voltage

Bi-directional switch is switched ON when its corresponding phase (SIA $\rightarrow V_{AN}$, SIB $\rightarrow V_{BN}$, SIC

$\rightarrow V_{CN}$) (c.f. Fig. 2) has the medium value. As in sector 1, phase V_{AN} has maximum and phase V_{BN} has minimum value, hence switch SIC is kept ON throughout this sector. Detailed switching pattern of all third harmonic injection switches is presented in [5].

Switches S+ and S- are high frequency switches that realizes two buck-converter for output voltage regulation. The fact that this topology has only two high frequency switches makes it a suitable candidate for high efficiency applications.Equations for duty ratio calculations of high frequency switches (S+,S-) which are responsible for output DC voltage regulation are given in (1) and (2) respectively. Detailed derivation of these equations has been presented in [7].

$$dS+ = \frac{2V_{DC}}{3\hat{V_{PH}}^2}max(V_{AN},V_{BN},V_{CN}) \tag{1}$$

$$dS- = \frac{2V_{DC}}{3\hat{V_{PH}}^2}min(V_{AN},V_{BN},V_{CN}) \tag{2}$$

Delta Rectifier

Delta Rectifier topology shown in Fig. 4 belongs to the Boost type three-phase PFC topologies. Similar to Swiss rectifier (c.f. Fig. 2) it has a three phase bridge to convert AC voltage into pulsating DC voltage. However, unlike in Swiss rectifier, bi-directional bi-polar third harmonic injection switches in Delta Rectifier are switched at high frequency [8].

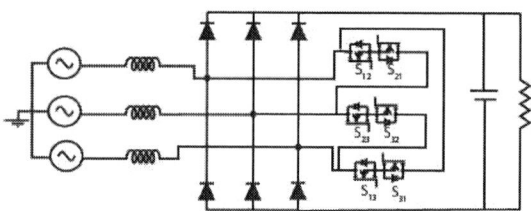

Fig. 4: Delta Rectifier

Since third harmonic injection switches are connected in delta configuration in this topology, switching of one switch will effect the current in its two correspondence phases. Hence, a multiple input multiple output (MIMO) control loop will be required to generate PWM signals for unity power factor and output voltage regulation. However, it has been proven in [10] that the cross-coupling between two phases can be neglected and this rectifier can be controlled by a single input single output (SISO) control loop. Details of control design are not presented in this paper, however it can be found in [10]. Transfer function of the rectifier for design of SISO control is given in (3).

$$i_N = \frac{2V_O}{3L_N s}\delta_1 \tag{3}$$

In (3) i_N is the phase current and δ is the duty ratio of its corresponding bi-directional bi-polar switch. Fig. 5 shows the PWM signals for switch S_{12} and S_{21}. In this figure, sectors are defined based on line-line voltages. Bi-directional switches S_{12} and S_{21} are only switched when their corresponding line-line voltage V_{AB} has either maximum or minimum value. When V_{AB} has maximum value switch S_{12} is switched at high frequency while S_{21} is kept on. Similarly when V_{AB} has minimum value S_{21} is switched at high frequency and S_{12} is kept on. PWM signals for all third-harmonic injection switches are generated by following the same principle.Detailed wave-forms can be found in [10].

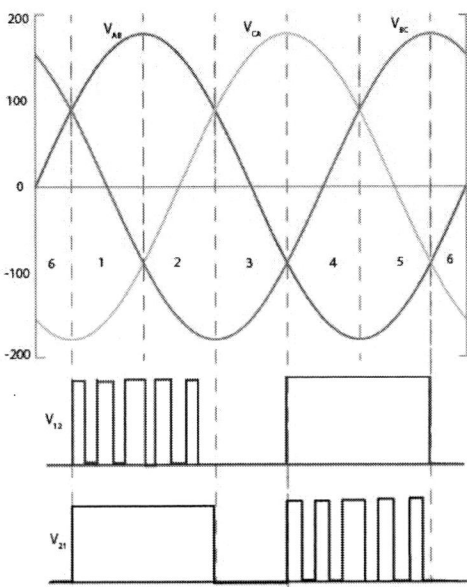

Fig. 5: Clamped PWM Signals

Selection of Rectifier Topology

As we have discussed during intoduction that for selection of the rectifier topology for its application in an aircraft, it is vital to check its compliance with military EMI standards (MIL-STD-461G) [9] . Fig. 6 shows a standard EMI measurement setup for a three-phase system. With the test apparatus shown

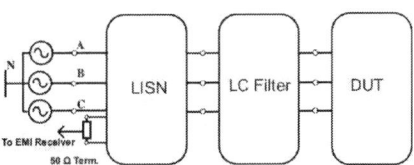

Fig. 6: EMI Measurement Setup

in Fig. 6, high frequency noise component from input current is obtained through the 50 ohm terminal resistance and viewed on EMI receiver. However, for complete compliance with MIL-STD-461G, low frequency emissions (800Hz- 10kHz) also need to be measured and contained below the standard limit. Fig. 7 shows the measurement setup for CE-101 (800-10kHz Noise Component) and CE-102 (10kHz-30MHz) compliance. PLECS and MATLAB based simulations are made for both rectifier topologies

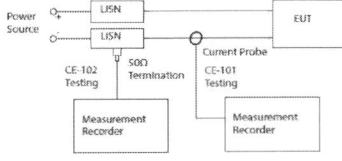

Fig. 7: MIL-STD 461G Measurement Setup

including LISN and filter block. Apart from measuring the conducted emissions, total harmonic distortion (THD) in input current is also measured. FFT is performed in MATLAB on current waveform obtained from PLECS to test compliance with CE-101. For CE-102 compliance testing, voltage waveform is recorded across the LISN and is exported in MATLAB to perform FFT.

Fig. 8 shows the simulation results for Swiss Rectifier. As shown in Fig. 8 (a) noise peaks around frequency range of 8kHz -10 kHz exceed the standard limit. These high noise peaks are introduced because of LISN resonance at 8kHz.

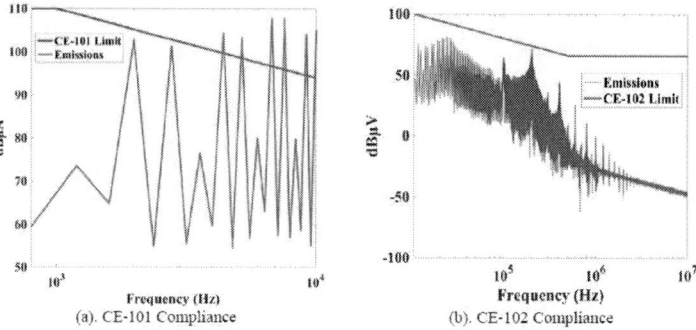

(a). CE-101 Compliance

(b). CE-102 Compliance

Fig. 8: Swiss Rectifier Simulation Results

Fig. 9 shows the simulation results for the Delta rectifier topology and it can be seen that the noise peaks are well under the standard limit. Hence, Delta type rectifier is chosen for this application.

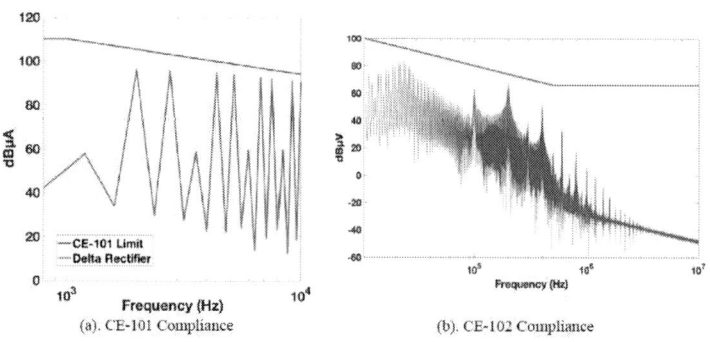

(a). CE-101 Compliance

(b). CE-102 Compliance

Fig. 9: MIL-STD 461G Measurement Setup

Isolated DC-DC Converter

When it comes to high power density DC-DC converter there exists two converter types, resonant and non-resonant converters. Resonant converters such as LLC are typically used to achieve very high switching frequency as well as soft switching. However as is demonstrated in the transformer design parts, high power density high current transformer typically operate much lower switching frequencies as high voltage transformers, and therefore non-resonant converter was selected. Using phase shifted full bridge as in Fig. 10.

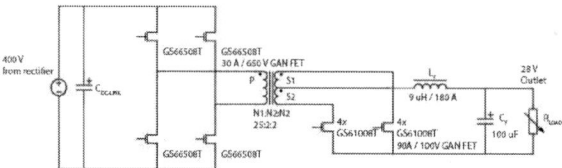

Fig. 10: Isolated DC-DC Converter

Converter can be used in both hard-switched mode and partial soft-switched mode. Allowing focus on magnetic design, and later the control can be improved to reduce switching losses.

98.8 % efficient hard switched converter was shown in [11] and its partial soft-switched counterpart in [12]. This topology allows for quick prototyping and easy implementation of gate driving using minimal computational effort.

Transformer Design

The trend towards higher power densities and higher operation frequencies in DC-DC converters will push the boundaries of regular switch mode power supply (SMPS) transformer design. This section presents an optimization tool for design of highly efficient, high frequency transformers by minimizing the core and winding losses. Appropriate core material, core dimensions, wire thickness, required interleaving and optimal switching frequency for the given high current step down DC-DC converter is evaluated using the proposed algorithm. Transformer design based on this algorithm has better efficiency as compared to one based on conventional empirical formulas. A 2000 W center tapped transformer with output voltage of 28 V and 72 A is tested and verified, input is 400 V and 5 A. Optimal operation frequency is 60 kHz. Overall transformer efficiency is 99%.

Transformer Loss Structure

Transformer losses are divided to two categories, core and winding losses. Winding losses can be then divided in to two factors, the dc-resistance of the windings and high frequency eddy current and proximity effect losses which are a results of the high frequency eddy currents inside the windings causing the effective current carrying cross sectional area of the winding to decrease [13].

Transformer core losses are defined by Steinmetz equation described by Steinmetz [14][15]. Steinmetz equation is described in (4)

$$P_{\mathrm{V}} = k * f^a * B^b \tag{4}$$

Where, P_{V} is the volumetric losses in the material and k, a, b are material dependent empirical coefficients for determining losses, f is frequency and peak flux density is represented by B. Unfortunately, a single three-parameter Steinmetzs equation is inadequate for accurate core loss prediction over a large range of frequency and flux density values [16]as well as describing losses using square waveform excitation with varying duty cycle. To reach an analytical formulation the simplest form of loss table used to obtain single analytical formulation for the optimization algorithm.

For transformer winding the losses are described in terms of average winding current density J [17] and is shown in (5).

$$P_{\mathrm{CU}} = \rho_{\mathrm{CU}} * V_{\mathrm{CU}} * J^2 \tag{5}$$

In addition to average current density, additional losses are explained with the inclusion of eddy current losses in terms of resistance factor:

$$F_{\mathrm{R}} = \frac{R_{\mathrm{AC}}}{R_{\mathrm{DC}}} \tag{6}$$

F_{R} represents all the increased losses due to high frequency effects [18][19].

Total winding losses are:

$$P_{\mathrm{CUTot}} = F_{\mathrm{R}} * P_{\mathrm{CUDC}} \tag{7}$$

Method for calculating resistance factor F_{R} is the Dowells equation. High frequency winding losses are greatly influenced by winding structure and transformer wire thickness as seen in (8). These are parameters in the optimization algorithm.

$$F_R = \frac{R_{AC}}{R_{DC}} = \phi \frac{\sinh 2\phi + \sin 2\phi}{\cosh 2\phi - \cos 2\phi} + \frac{2(m^2 - 1)}{3} \phi \frac{\sinh \phi - \sin \phi}{\cosh \phi + \cos \phi} \tag{8}$$

Where ϕ is a relation between wire thickness and conductor skin depth at a given frequency:

$$\phi = \frac{d}{\Delta f} \tag{9}$$

and m is the number of sheets of winding or number of litz wire turn on top of each other.

Algorithm Logic

Transformer optimization algorithm design flow is explained in Fig. 11.

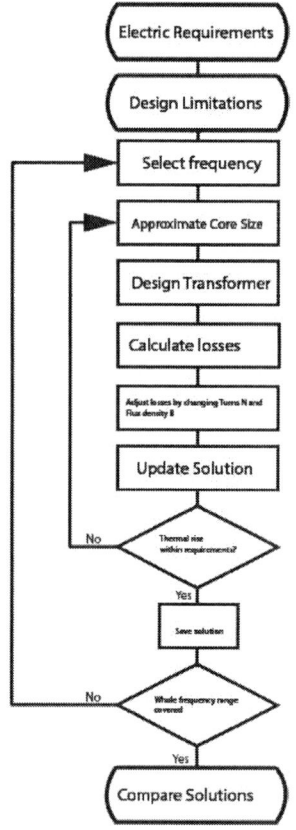

Fig. 11: Transformer Design Flow Chart

After design parameters and limitations are given, the algorithm estimates the initial core size by equaling the losses between winding and core. Then calculating the required surface area of the core to dissipate losses as is done in [17]. This process is iterative and the initial losses and core size might be far from optimal. As the algorithm constantly improves the winding structure (by changing number of turns) to minimize both core and winding losses in each step, the end result is the smallest possible losses for a given transformer size. After single loop, results are stored in a table. The algorithm changes the frequency and size of the transformer. A quality function Q includes weight and efficiency of the transformer and is used to select the best transformer. Example of a quality function would be $Q = \eta/W$ (efficiency / weight).

Reducing Winding Losses

Key to reducing windings losses is the adjustment of high frequency eddy current effects, this is accomplished by having thinner foil and extensive interleaving of the transformer windings [19]. Results from MATLAB optimization algorithm are shown in Fig. 12 and Fig. 13.

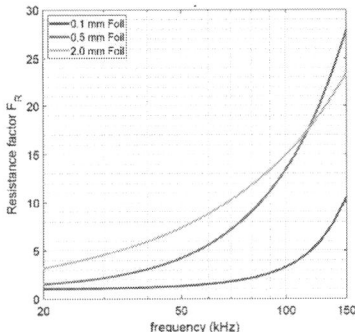

Fig. 12: Resistive Losses Vs Foil Thickness

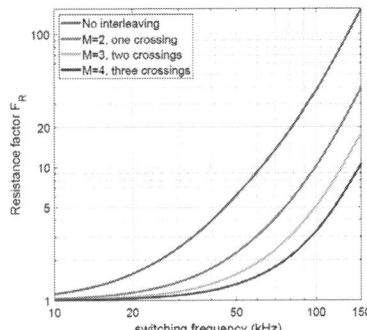

Fig. 13: Resistive Losses Vs Interleaving

Reducing Core Losses

Eddy current losses in the core increase as the frequency is increased in a transformer core this phenomenon is also captured by the curve fitted Steinmetz equation.

Increment in the core losses as a function of frequency is counteracted by the fact that less flux is required for the same terminal voltage when frequency is increased. Comprehensive table of available materials from Ferroxcube, TDK, Mag-Inc and Vitroperm was created, and using material comparison, best material was selected. Fig. 14 shows that each material has an optimal operational switching frequency. One should note that the optimal switching frequency is dependent on the winding as well, for this reason it appears that all material operate best in the 30 kHz to 60 kHz region, but this is a mere artifact of the high current requirement (72 A) in secondary.

Conclusion

In this paper the design methodology of a power supply for aircraft applications has been discussed. This study consists of two phases. During the first phase a three-phase rectifier topology has been selected. Selection criteria of the rectifier topology includes efficiency, power quality and compliance with MIL-STD-461G. It has been shown that with application of Delta rectifier it is possible to comply with aircraft emissions standards. Second phase of the study consists of Dc-Dc converter design for output voltage regulation. This phase emphasized on design of high frequency transformer for obtaining overall

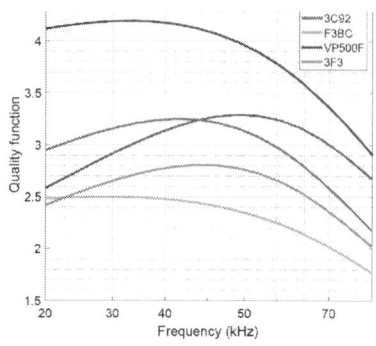

Fig. 14: Magnetic Material Selection

high efficiency and power density. It has been shown that with the application of new nano-crystalline magnetic core, it is possible to obtain very high efficiency and power density. However, these cores are yet not available in EE or EI shapes, which makes the manufacturing process very complicated. Hence, the performance of the converter can be further improved once the nano-crystalline cores are available in more practical shapes.

References

[1] J.A. Rosero, J.A. Ortega, E. Aldabas and L. Romeral, "Moving towards a more electric aircraft," IEEE Aerosp.Electron.Syst.Mag., vol. 22, no. 3, pp. 3-9 2007.

[2] K. Graage, "Energy Supply Unit on Board an Aircraft,", 2001.

[3] P. Wheeler and S. Bozhko, "The more electric aircraft: Technology and challenges." IEEE Electrification Magazine, vol. 2, no. 4, pp. 6-12 2014.

[4] J. He, D. Zhang and D. Torrey, "Recent Advances of Power Electronics Applications in More Electric Aircrafts," in 2018 AIAA/IEEE Electric Aircraft Technologies Symposium (EATS), 2018, pp. 1-8.

[5] B. Ahmad, J. Kyyra, M. Routimo and W. Martinez, "EMI Standard Compliance of Three-Phase Buck Type PFC Rectifier For Application in Aircraft," in IECON 2019-45th Annual Conference of the IEEE Industrial Electronics Society, 2019, pp. 6355-6362.

[6] J. W. Kolar and T. Friedli, "The essence of three-phase PFC rectifier systemsPart I," IEEE Transactions on Power Electronics, vol. 28, (1), pp. 176-198, 2012.

[7] L. Schrittwieser et al, "99.3% Efficient three-phase buck-type all-SiC SWISS Rectifier for DC distribution systems," IEEE Transactions on Power Electronics, vol.34, (1), pp. 126-140, 2019.

[8] M. Hartmann, J. Miniboeck, H. Ertl and J.W. Kolar, "A three-phase delta switch rectifier for use in modern aircraft," IEEE Trans.Ind.Electron., vol. 59, no. 9, pp. 3635-3647 2011.

[9] (Mar 2,). DEPARTMENT OF DEFENSE INTERFACE STANDARD. Available: http://www.interferencetechnology.com/wpcontent/uploads/2015/04/461G.pdf.

[10] M. Hartmann, "Ultra-compact and ultra-efficient three-phase PWM rectifier systems for more electric aircraft," 2011.

[11] R. Ramachandran and M. Nymand, "Experimental demonstration of a 98.8% efficient isolated DCDC GaN converter," IEEE Trans. Ind. Electron., vol. 64, (11), pp. 9104-9113, 2016.

[12] J. Baek, C. Kim, J. Lee, H. Youn and G. Moon, "Gate driving method for synchronous rectifiers in phase-shifted full-bridge converter," 2015 9th International Conference on Power Electronics and ECCE Asia (ICPE-ECCE Asia), Seoul, 2015, pp. 753-758, doi: 10.1109/ICPE.2015.7167867.

[13] P. L. Dowell. Effects of eddy currents in transformer windings. In: Proceedings of the Institution of Electrical Engineers 113.8 (Aug. 1966), pp. 13871394.

[14] Ouyang, Z., Thomsen, O.C. and Andersen, M.A., 2010. Optimal design and tradeoff analysis of planar transformer in high-power DCDC converters. IEEE.

[15] C. P. Steinmetz. Theory of the General Alternating Current Transformer. In: Transactions of the American Institute of Electrical Engineers XII (Jan. 1895), pp. 245256. issn: 0096-3860. doi: 10.1109/T-AIEE.1895.4763861..

[16] W. Chen et al. Predicting Iron Losses in Soft Magnetic Materials Under DC Bias Conditions Based on Steinmetz Premagnetization Graph. In: IEEE Transactions on Magnetics 52.7 (July 2016), pp. 14. issn: 0018-9464. doi: 10.1109/TMAG.2015.2514239.

[17] W. G. Hurley, W. H. Wolfle, and J. G. Breslin. Optimized transformer design: inclusive of high-frequency effects. In: IEEE Transactions on Power Electronics 13.4 (July 1998), pp. 651659. issn: 0885-8993. doi: 10.1109/63. 704133.

[18] R. Ramachandran and M. Nymand. A 98.8 % efficient bidirectional full-bridge isolated dc-dc GaN converter. In: 2016 IEEE Applied Power Electronics Conference and Exposition (APEC). Mar. 2016, pp. 609614. doi: 10.1109/ APEC.2016.7467934.

[19] M. Nymand et al. Reducing ac-winding losses in high-current high-power inductors. In: 2009 35th Annual Conference of IEEE Industrial Electronics. Nov. 2009, pp. 777781. doi: 10.1109/IECON.2009.5415018.

DC Capacitor Voltage Feedback Method for a Peak Voltage Suppression Control with Multiple Leg-Short-Circuits Using SiC-MOSFETs Employed in Power Converters

Tomoyuki Mannen, Takanori Isobe,
University of Tsukuba,
Tsukuba, Japan,
Email: mannen@ieee.org

Keiji Wada
Tokyo Metropolitan University
Tokyo, Japan
Email: kj-wada@tmu.ac.jp

Keywords

≪Feedback control≫, ≪Initial charge≫, ≪Short circuit≫, ≪SiC-MOSFETs≫, ≪Three-phase converter≫

Abstract

This paper proposes control method for initial charge of grid connection converters using leg short-circuits with a dc capacitor voltage feedback. The proposed method makes short-circuits of a fixed time duration based on the dc capacitor voltage. In the experiments, the proposed method exhibits smaller energy consumption in each short-circuits than the control using an adjusted timing and duration. As a result, the proposed method can increase the number of short-circuits without complexity and extend the lifetime of power converters employing the initial charge method.

1 Introduction

Improvement of power semiconductor devices and application of new circuit topology and/or control method suitable for specific application make it possible to reduce passive components in power converters [1]. Recent power converters tend to employ a small film capacitor as the dc capacitor instead of a large electrolytic capacitor and try to reduce its volume and cost [1–3]. In addition, power converters equipped with wide-band-gap (WBG) semiconductor devices, such as SiC and GaN, are being put into practical use. The WBG devices can reduce a loss in the power devices resulting in size reduction of the heat sink [4].

However, the power converters usually do not consist of only a main circuit. For example, grid connection converters have an initial charge circuit for preventing inrush current. A conventional initial charge circuit consists of a resistor and mechanical contactor [5]. These components are designed by the rated power of power converters. Therefore, the initial charge circuit makes it difficult to reduce the size and cost of power converters even though small passive components and heat sink are applied.

When the power converter employs a small dc capacitor, the inrush current becomes small because of a large characteristic impedance of ac inductors and the dc capacitor. However, the dc capacitor voltage is increased to double of the source voltage when the power converter has no initial charge circuit. Since the rated voltage of the dc capacitor is usually designed as 150–160% of the source voltage, overvoltage occurs in the dc capacitor.

Reference [6] proposed an overvoltage suppression method using leg short-circuits in the power converter. The leg short circuit discharges the dc capacitor when the capacitor voltage reaches the set voltage.

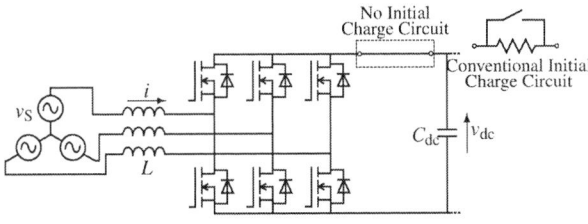

Fig. 1: Circuit diagram of a voltage source converter

Fig. 2: Picture of the experimental setup

Many papers have reported the results of short-circuit test [7–11], which indicate that the short-circuit current is limited by not only parasitic parameters of power devices [12] but also equivalent resistance of power devices. The short circuit causes a large amount of loss and heats the power devices. As a result, long short-circuit may cause destructive damage [13] or repetitive short-circuits causes deterioration of power devices [14]. The overvoltage suppression method using single short-circuit [15] can apply only several hundred times without any damage to the power devices [16] because of its too-long short-circuit time duration. On the other hand, the overvoltage suppression method using multiple short-circuits is applicable more than 10000 times with slight degradation in the power devices [17]. However, it is difficult to increase the number of short-circuit in the multiple short-circuit method because the degree of freedom of the timing and duration of the short-circuit increases, thereby increasing the complexity.

This paper proposes control method of the overvoltage suppression for the initial charge using the multiple short-circuits with a dc capacitor voltage feedback. The proposed method makes short-circuits in each leg of the power converter when the detected dc capacitor voltage reaches the set value so as to suppress the peak voltage. After a several-μs, the proposed method automatically finishes the short-circuit. Therefore, the proposed method can suppress the overvoltage without adjusting the timing and duration of the short-circuit even though the power converter has parameter error and/or variation in its components. In the experimental verification, in spite of no adjustment, the proposed method exhibits lower energy consumption in each short-circuit than the control using the adjusted timing and duration. As a results, this paper reveals that the proposed method makes it possible to increase the number of the short-circuit without complexity.

2 Circuit Configuration

Fig. 1 shows a circuit configuration of a grid-connected converter. The circuit is designed for a rectifier of electrolytic-capacitorless three-phase motor drive system rated at 5 kVA. Fig. 2 is a picture of the experimental setup. The circuit parameters used for experiment are listed in Table I. These circuit parameters are designed enough to operate the power converter at a switching frequency of 20 kHz [1, 2].

The power converter employs SiC-MOSFET (Cree C2M0040120D) and SiC-SBD (Cree C4D10120D) as semiconductor switches. AC inductors L are connected to the ac side of the converter as a switching ripple filter. Only a small film capacitor C_{dc} is installed in the dc side as a dc capacitor. The power converter circuit has no initial charge circuit such as a combination of resistor and mechanical contactor.

3 Overvltage Suppression Method Using Multiple Short-Circuits in Each Leg

When the converter connects to the ac voltage source, the dc capacitor starts to be charged. If the converter has no initial charge circuit, the dc capacitor directly connects to the ac voltage source and ac inductors in series during the initial charge. The charging current is limited by the impedance of the ac

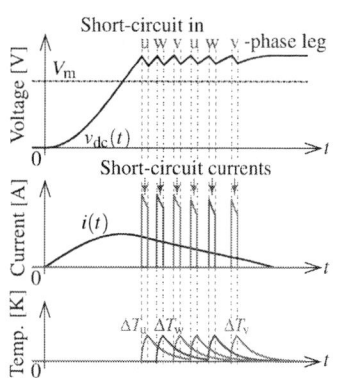

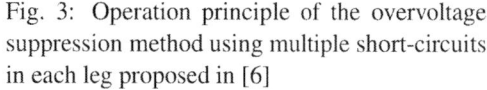

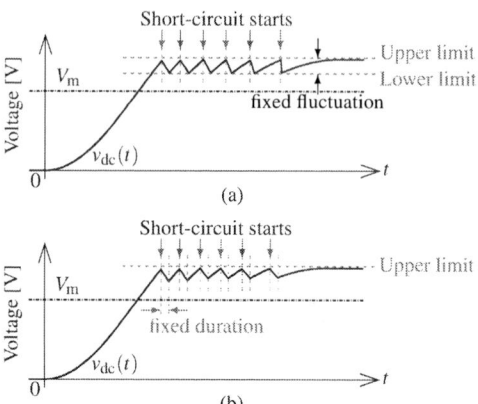

Fig. 3: Operation principle of the overvoltage suppression method using multiple short-circuits in each leg proposed in [6]

Fig. 4: Operation principle of the overvoltage suppression method using multiple short-circuits with a dc capacitor voltage feedback based on (a) a hysteresis control and (b) a fixed time duration.

inductors and dc capacitor, which is not large enough to damage the semiconductor switches due to the small film capacitor. However, the dc capacitor voltage may eventually charge up to twice the ac source voltage without any control.

The overvoltage suppression method proposed in [6] makes several short-circuits on each leg for discharging the dc capacitor by using the semiconductor switches in the power converter. Fig. 3 shows an operation principle of the overvoltage suppression method using multiple short-circuits in each leg. When the dc capacitor voltage reaches to the set voltage, both upper and lower semiconductor switches in one of three legs turns on simultaneously. This short circuit discharges the dc capacitor as if small resistor is connected to the dc capacitor in parallel. After a several μs, the semiconductor switches turn off to finish short circuit duration and the dc capacitor starts to charge again. The overvoltage suppression method repeats charge and discharge by making short-circuits on each leg sequentially. Suitable timing and time-duration of short-circuits makes it possible to suppress the overvoltage during the initial charge duration with almost no damage to the power devices.

Reference [17] has reported that multiple times of short-time short-circuits has smaller damage to semiconductor devices than the long-time short circuit once. The overvoltage suppression method makes almost the same amount of total energy consumption in semiconductor switches during the short-circuits. The energy consumption with one long-time short-circuit is enough large to damage the switches. On the other hand, each short-circuit has small energy consumption in multiple short-circuits because total energy consumption splits into several times. Intervals between the short-circuits also allow the semiconductor switches to cool and prevent from damage. As a results, the multiple short-circuits reduce the maximum junction temperature of the semiconductor switches and the thermal stress on them.

4 Control and Implementation of a Dc Capacitor Voltage Feedback for the Overvoltage Suppression Method

The multiple short circuit method has high flexibility in timing and duration of short-circuits which directly affect the lifetime of semiconductor switches. If the short-circuit duration is long, the lifetime becomes short because of high peak junction temperature. If the short-circuit timing is early, the total energy consumption becomes large. In addition, if the short-circuit timing is late, the peak voltage becomes high.

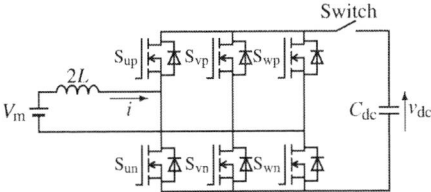

Fig. 5: A circuit diagram used in the experimental verification.

Table I: Circuit parameters of the converter.

Source Voltage	V_S	200 V	
Source Frequency	f_S	50 Hz	
Rated Power	S	5 kVA	
AC Inductor	L	0.43 mH	(1.7%)
DC Capacitor	C_{dc}	50 μF	(0.5 ms)

Usually, grid connection converters are connected to a fixed-voltage ac-source and designed to have fixed ac inductors and dc capacitor. Therefore, these parameters give the start-up waveforms of the grid connection converters. The optimal timing and duration of short-circuits may be derived from the waveforms. However, it is difficult for general controllers to calculate the optimal timing and duration online because the optimization requires large amount of calculation and iteration within a couple of hundred-μs. Therefore, the timing and duration are calculated offline and implemented into the controller. In addition, the circuit parameters include error and variation in the actual system. If the timing and duration are fixed at the set values in the controller, tuning can cover effect of the parameter error but cannot cover the parameter variation.

On the other hand, it is possible for the timing and duration based on a dc capacitor voltage feedback to suppress effect of the parameter error and variation. It is easy for grid connection converters to employ the dc capacitor voltage feedback because they are usually equipped with a voltage sensor for its dc capacitor. There are two ways to achieve the voltage feedback in the overvoltage suppression method: one is upper and lower voltage limit triggers so called hysteresis control, and the other is an upper voltage limit trigger and fixed time-duration.

Fig. 4 shows operation principles of the voltage feedback in the overvoltage suppression method. The hysteresis control makes a leg short-circuit, when the dc capacitor voltage reaches the upper-limit set value, and finishes the short circuit, when the dc capacitor voltage reaches the lower-limit set value. The hysteresis control can minimize the dc capacitor voltage fluctuation and the total energy consumption in the start-up because the dc capacitor voltage with small fluctuation keeps higher than that with large fluctuation. However, it is impossible for the hysteresis control to regulate energy consumption during a short-circuit. The short-circuit duration in the hysteresis control only depends on the dc capacitor voltage. If the initial charge current is large, the time-duration of and energy-consumption in a short-circuit tend to be large. As a result, the hysteresis control may damage the power devices.

On the other hand, the fixed time duration control can regulate the energy consumption during a short-circuit. The fixed time duration control also makes a leg short-circuit, when the dc capacitor voltage reaches the set value. In this control, all the short-circuit time duration is set to a fixed value. The fixed time duration control automatically finishes the short-circuit duration several-μs after the start of the short-circuit. The short-circuit energy depends on the device-voltage and time-duration during the short-circuit. Since the dc-capacitor voltage during the short-circuit is almost constant around the set voltage for the start timing, the short-circuit time duration is a dominant factor of the energy consumption in each short-circuit. Therefore, the set value for the short-circuit time duration in the fixed time duration control can effectively regulate the energy consumption in each short-circuit.

5 Experimental Verification

Fig. 5 shows experimental circuit configuration for the initial charge. In the following experiments, a dc voltage power supply is used instead of a three-phase ac power source so as to assume the initial charge

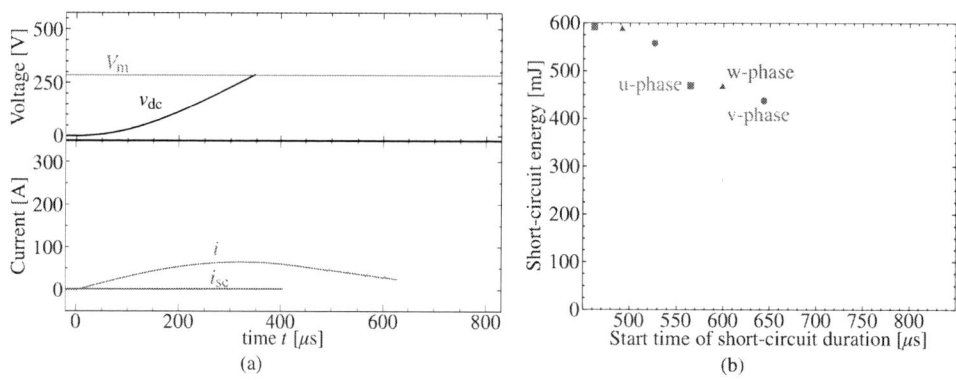

Fig. 6: The experimental results of the initial charge with the overvoltage suppression method using leg short-circuits with controller-implemented adjusted timing and duration of short-circuits: (a) waveforms and (b) short-circuit energy in each leg.

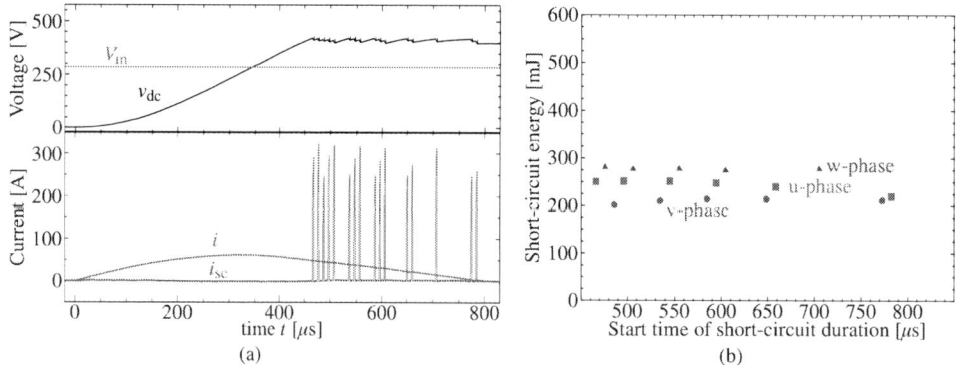

Fig. 7: The experimental results of the initial charge with the overvoltage suppression method using leg short-circuits with the voltage feedback and fixed short-circuit duration of 2.5 μs: (a) waveforms and (b) short-circuit energy in each leg.

starts from the same phase angle. An ac inductor $2L$ is connected to the ac side instead of the three-phase ac inductor L. The voltage source V_m was set to 283 V, the maximum line-to-line voltage in ac 200 V. A switch in the dc side connects the dc capacitor to the power source, as if it works like a circuit breaker in the ac side. The switch turns on at $t = 0$, which connects the power converter and the ac source.

Fig. 6 shows experimental results of the initial charge for the dc capacitor using the overvoltage suppression method with controller-implemented adjusted timing and duration of the short-circuits. Fig. 6(a) is waveforms of the dc capacitor voltage v_{dc}, source current i, and short-circuit current i_{sc}. The timing and duration of the short-circuits were tuned offline so as to suppress the dc capacitor voltage fluctuation and were implemented into the controller in advance. In the first short-circuit of Fig. 6, the peak of the short-circuit current was 334 A and the dc capacitor discharged to 394 V. After that, the dc capacitor voltage increased and decreased alternately. Finally, the dc capacitor voltage fluctuation during the initial charge was 32 V, between 425 V and 393 V. Fig. 6(b) shows the start-timing of and energy consumption in each short-circuit. The energy consumption in the first short-circuits in each leg was higher than 550 mJ and that in the second ones was 450–470 mJ because the first and second short-circuits in each leg was set to 5 μs and 4 μs, respectively.

Figs. 7–10 show experimental results of the initial charge method with the voltage feedback and fixed

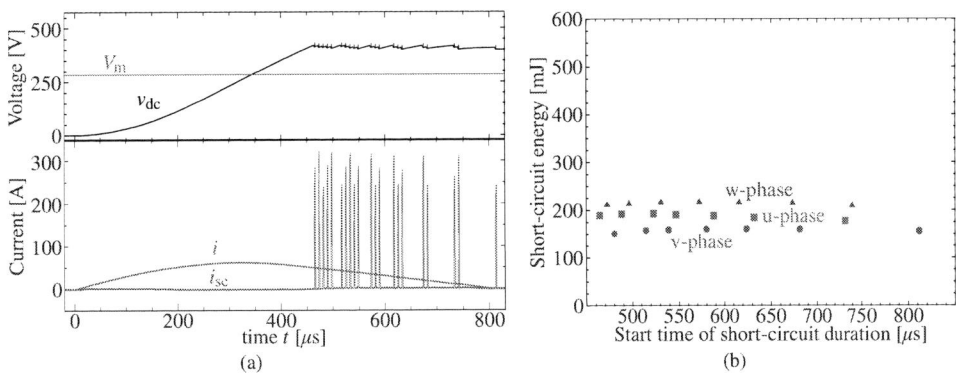

Fig. 8: The experimental results of the initial charge with the overvoltage suppression method using leg short-circuits with the voltage feedback and fixed short-circuit duration of 2.0 μs: (a) waveforms and (b) short-circuit energy in each leg.

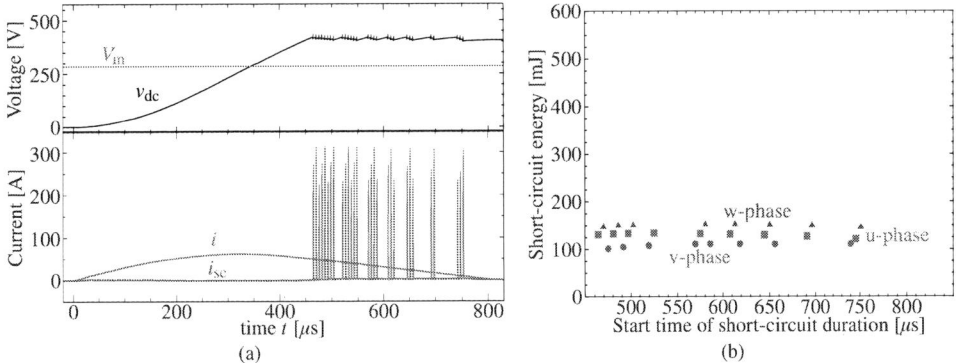

Fig. 9: The experimental results of the initial charge with the overvoltage suppression method using leg short-circuits with the voltage feedback and fixed short-circuit duration of 1.5 μs: (a) waveforms and (b) short-circuit energy in each leg.

short-circuit duration. The short-circuit duration in Figs. 7–10 was set to a fixed value, 2.5 μs, 2.0 μs, 1.5 μs, and 1.0 μs, respectively. Fig. 7(a) is the waveforms under the condition of 2.5 μs short-circuits. The peak of the short-circuit current was 324 A and ripple of the dc capacitor voltage was 30 V, between 428 V and 398 V. Therefore, the dc capacitor voltage in Fig. 7(a) was also suppressed without any tuning. Fig. 7(b) is the start-timing of and energy consumption in each short-circuit. Each short-circuit energy was reduced to 200–280 mJ because the short-circuit duration was reduced to and fixed at 2.5 μ.

The peak of the short-circuit current in Figs. 8(a)–10(a) reduced to 321 A, 310 A, and 281 A, respectively, because the fixed time duration of the short-circuit also decreased. Stray inductance in the grid connection converter makes the short-circuit current increase slowly and the short-circuit duration is not enough to saturate the short-circuit current. On the other hand, the dc capacitor voltage in Figs. 8(a)–10(a) still included a voltage ripple between 400 V and 430 V. In the feedback method, the start time of the short-circuit depends on the sensor for the dc capacitor voltage. Delay in the sensor and/or gate driver makes it difficult to set the optimal timing of the short-circuits. However, the energy consumption during the short-circuit in Figs. 8(b)–10(b) effectively reduced, for example, the energy in Fig. 10(b) was less than 90 mJ. Smaller short circuit energy decreases the device stress and increases the lifetime of the converter.

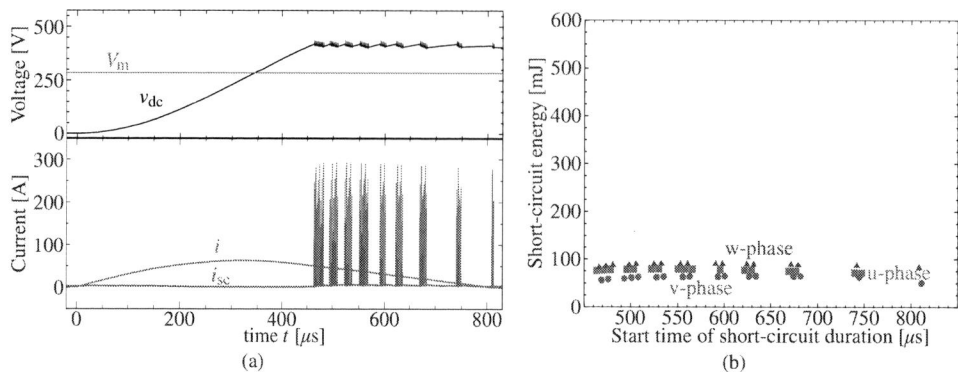

Fig. 10: The experimental results of the initial charge with the overvoltage suppression method using leg short-circuits with the voltage feedback and fixed short-circuit duration of 1.0 μs: (a) waveforms and (b) short-circuit energy in each leg.

Therefore, the set value of the short-circuit duration should be set as low as possible and the number of short-circuit should be as high as possible. The voltage feedback method can easily reduce the short-circuit time duration and increases the number of short-circuits because there is no offline tuning in the feedback method. Therefore, the feedback method is suitable for the initial charge method from a view point of easy implementation and long lifetime.

6 Conclusion

This paper proposed a dc-capacitor voltage feedback method for peak voltage suppression in the initial charge method using leg short-circuits. The proposed method automatically starts the short-circuit in a power converter leg so as to discharge the dc capacitor, when the dc capacitor voltage reaches to the set value. After a fixed time duration, the proposed method also ends the short-circuit automatically. The power consumption in the devices during the short-circuit are regulated by changing the fixed time duration for each short-circuit.

The experimental results revealed that the proposed method with 1 μs short-circuit can reduce the power consumption in each short-circuits to 1/6, even though both of the dc capacitor voltage ripple was 30 V. Therefore, the proposed feedback method has a capability of reduction in the power consumption of each short-circuit and increase the number of the short-circuits compared with the initial charge method using the controller-implemented adjusted timing and duration. The proposed feedback method can extend lifetime of the power converter using the initial charge method using leg short-circuits.

7 Acknowledgment

This paper is based on results obtained from a project supported by Nagamori Foundation.

References

[1] I. Takahashi, H. Haga, "Inverter control method of IPM motor to improve power factor of diode rectifier," *IEEE Proceedings of the Power Conversion Conference-Osaka 2002*, pp. 142–147, 2002.

[2] H. Yoo, S. Sul, "A Novel Approach to Reduce Line Harmonic Current for a Three-phase Diode Rectifier-fed Electrolytic Capacitor-less Inverter," *IEEE Applied Power Electronics Conference and Exposition (APEC) 2009*, pp. 1897–1903, 2009.

[3] T. Shimizu, K. Wada, N. Nakamura, "Flyback-Type Single-Phase Utility Interactive Inverter With Power Pulsation Decoupling on the DC Input for an AC Photovoltaic Module System," *IEEE Trans. Power Electron.*, vol. 21, no. 15, pp. 1264–1272, 2006.

[4] J.W. Kolar, D. Neumayr, D. Bortis, "Google Little Box Reloaded: How to Achieve 200W/in^3 & Beyond? Concepts - Evaluation - Barriers - Future," *IEEE Applied Power Electronics Conference and Exposition (APEC) 2017*, 2017.

[5] A. M. Y. M. Ghias, J. Pou, V. G. Agelidis, M. Ciobotaru, "Initial Capacitor Charging in Grid-Connected Flying Capacitor Multilevel Converters," *IEEE Trans. Power Electron.*, vol. 29, no. 7 pp. 3245–3249, 2014.

[6] T. Mannen, H. Mishima, K. Wada, "Peak Temperature Reduction Method of SiC-MOSFETs Employed in the Initial Charge for the DC Capacitor Using Leg Short-Circuits," *IEEE International Conference on Power Electronics and ECCE Asia (ICPE 2019 - ECCE Asia)*, pp. 1–6, 2019.

[7] A. E. Awwad, S. Dieckerhoff, "Short-circuit evaluation and overcurrent protection for SiC power MOS-FETs," *17th European Conference on Power Electronics and Applications (EPE'15 ECCE-Europe)*, pp. 1–9, 2015.

[8] C. Ionita, M. Nawaz, K. Ilves, F. Iannuzzo, "Short-circuit Ruggedness Assessment of a 1.2 kV/180 A SiC MOSFET Power Module," *IEEE Energy Conversion Congress and Exposition (ECCE) 2017*, pp. 1982–1987, 2017.

[9] F. Boige, F. Richardeau, D. Tremouilles, S. Lefebvre, G. Guibaud, "Investigation on damaged planar-oxide of 1200 V SiC power MOSFETs in non-destructive short-circuit operation," Microelectronics Reliability, vol. 76–77, pp. 500–506, 2017.

[10] P. D. Reigosa, F. Iannuzzo, H. Luo, F. Blaabjerg, "A Short-Circuit Safe Operation Area Identification Criterion for SiC MOSFET Power Modules," *IEEE Trans. Ind. Appl.*, vol. 53, no. 3, pp. 2880–2887, 2017.

[11] L. Ceccarelli, P.D. Reigosa, F. Iannuzzo, F. Blaabjerg, "A survey of SiC power MOSFETs short-circuit robustness and failure mode analysis," Microelectronics Reliability, vol. 76–77, pp. 272–276, 2017.

[12] Y. Liu, Z. Song, S. Yin, J. Peng, H. Jiang, "Analytical and Experimental Validation of Parasitic Components Influence in SiC MOSFET Three-Phase Grid-connected Inverter," Journal of Power Electronics, vol. 19, no. 2, pp. 591–601, 2019.

[13] G. Romano, A. Fayyaz, M. Riccio, L. Maresca, G. Breglio, A. Castellazzi, A. Irace, "A Comprehensive Study of Short-Circuit Ruggedness of Silicon Carbide Power MOSFETs," *IEEE Journal of Emerging and Selected Topics in Power Electronics*, vol. 4, no. 3, pp. 978–987, 2016.

[14] J. Wei, S. Liu, L. Yang, J. Fang, T. Li, S. Li, W. Sun, "Comprehensive Analysis of Electrical Parameters Degradations for SiC Power MOSFETs Under Repetitive Short-Circuit Stress," *IEEE Trans. Electron Devices*, vol. 65, no. 12, pp. 5440–5447, 2018.

[15] T. Mannen, K. Wada, "Control Method for Overvoltage Suppression Across the DC Capacitor in a Grid-Connection Converter Using Leg Short Circuit of Power mosfets during the Initial Charge," *IEEE Trans. Ind. Appl.*, vol. 55, no. 4, pp. 4012–4019, 2019.

[16] T. Mannen, K. Wada, "Operating-waveform analysis based reliability evaluation of power MOSFETs used for a leg short-circuit initial charge method," Microelectronics Reliability, vol. 88–90, pp. 589–592, 2018.

[17] T. Mannen, K. Wada, "Reliability evaluation of power MOSFETs used for an initial charge method using multiple short-circuits in each leg," Microelectronics Reliability, vol. 100–101, no. 113428, 2019.

Investigation of Bond Wire Lift-Off by Analyzing the Controller Output Voltage Harmonics for the Purpose of Condition Monitoring

Firat Yüce, Marc Hiller
Karlsruhe Institute of Technology (KIT)
Elektrotechnisches Institut (ETI) – Power Electronic Systems
Kaiserstr. 12, 76131 Karlsruhe
E-Mail: firat.yuece@kit.edu, marc.hiller@kit.edu
URL: http://www.eti.kit.edu

Keywords

«Condition Monitoring», «Power Semiconductor», «IGBT», «Reliability», «Artificial Intelligence»

Abstract

This paper presents a new approach in the field of condition monitoring of power semiconductors in power converters. The new approach avoids additional sensors and uses only the data that is already available in a power electronic system. This results in advantages such as saving extra costs and eliminating potential sources of failure. In this paper, the aging mechanism of bond wire lift-off is investigated. First, it is shown by simulations that output voltage harmonics of the current controller contain information about the bond wire lift-off aging mechanism. For testing the new approach in practice, real data are recorded in a test bench for a power module with unharmed bond-wires in order to train an algorithm for normal behavior. Then the aging mechanism bond wire lift-off is induced by cutting off bond wires of a power module with the aim of testing the functionality of the algorithm. The results of this investigation show that the proposed approach is able to detect the anomaly in the data set.

Introduction

Power Electronics is a key technology for sustainable energy production and environmentally-friendly mobility. The requirements for power electronics such as power density in energy and drive technology are steadily increasing. In addition to the power density and the energy efficiency, the reliability of power electronic systems is playing an increasingly important role – driven not least by trends such as electromobility, electric aircraft and the growing importance of renewable energies.

One way to increase the reliability of power electronic systems is to use condition monitoring. Condition monitoring has the goal of identifying the condition of a system during operation and thus providing information for necessary repair and maintenance work. This information allows the selection of optimal maintenance intervals in order to reduce the downtimes of a plant.

For this reason, wind turbines are a representative application. 13% of all wind turbine failures can be traced back to the converter, and Power electronics is the second most common cause of a wind turbine failure [1]. Only the pitch system for adjusting the pitch angle of the rotor blades has a higher failure rate [1]. The availability of wind turbines can be increased by knowing when the converter has to be serviced. This can significantly reduce maintenance costs and enable better planning of downtimes, especially for offshore wind turbines that are difficult to access.

In this paper, the specific aging process bond wire lift-off is investigated as an example. Such aging processes at the module level are mainly caused by the different coefficients of thermal expansion (CTE) of the materials [2]. In thermal cycles, due to the different material expansions at the interface, thermo-mechanical forces are generated, which can lead to cracks and fractures over time. For example, the bond wires are made of aluminum, which has a CTE app. ten times higher than that of the silicon chip [8]. The interface between the bond wire and the silicon chip is therefore prone to failure [8].

A major disadvantage of the existing methods of condition monitoring is the need for additional sensors and additional measuring equipment [3][5], which entails additional costs and additional sources of failure. The proposed approach overcomes this disadvantage and operates without additional hardware. This results not only in a significant cost advantage, but also in a higher reliability of the power converter system due to the elimination of additional potential sources of failure. Therefore, the already available operating data is analyzed for certain patterns and characteristics. Additional effort only results from the increased need for computing power in the signal processing of the converter system.

State of the art

Currently, there are two approaches in the field of condition monitoring of power converters, which are reviewed in this chapter. The aging-based modeling approach aims to establish a model for a specific aging mechanism that leads to the failure of the converter. This approach requires detailed knowledge of the aging process and is only able to detect the modeled aging mechanism. For example, increased leakage currents due to humidity can be simulated using thermo-mechanical and electrochemical models [2]. These models are then used during converter operation to estimate the current condition and the remaining lifetime of the power converter.

Another approach deals with the monitoring of sensitive parameters. Sensitive parameters are failure indicators and provide information about the condition of the converter. For instance, the differential resistance, the threshold voltage, and the turn-on and turn-off times of the power semiconductors are sensitive parameters [3][4][5]. This approach is able to detect all aging mechanisms affecting the sensitive parameter. In order to monitor these parameters, additional measuring equipment and sensors are commonly used. For example, in [5] a measuring circuit has been developed which is able to detect the collector-emitter threshold voltage of the transistor and thus detect bond wire lift-off. Since the sensitive parameters are also temperature-dependent, the information about the sensitive parameter alone is not sufficient to determine the condition of the converter. For this reason, additional approaches are being developed to overcome this issue [5].

Table I below summarizes the two approaches that exist in the field of condition monitoring of power converters.

Table I: Summary of previous approaches to condition monitoring of power converters

Aging-based modeling approach	Monitoring of sensitive parameters
Physical modeling of the aging process	Monitoring of the sensitive parameter mostly via direct measuring methods
Requires detailed knowledge of the aging mechanism	Requires sensitive parameters
Usually requires additional sensors	Usually requires additional sensors
Can only detect the modeled aging mechanism	Monitors all aging mechanisms affecting the sensitive parameter

Proposed approach

This chapter explains the proposed approach. In the first subchapter modeling, it is shown by simulations that the information about sensitive parameters already exists in the available operating data of the converter control. This is the basis for condition monitoring without using additional sensors. In the next subchapter, the practical implementation of the approach is explained. Therefore, the test bench is described. Then the data caption and the mathematical processing of the data for the purpose of condition monitoring is presented.

Modeling

The investigations in [4] have shown that the differential resistance R_{on} of an IGBT increases when bond wire lift-off occurs since the current has to flow through a smaller cross-section [4]. In this work, a simulation model is developed that allows changing the sensitive parameter R_{on} with the aim of noticing effects on the measurable variables of a power electronic system.

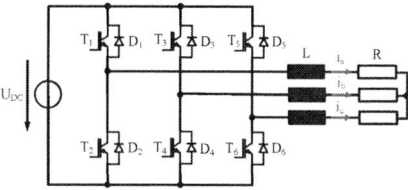

Fig. 1: Three-phase bridge circuit

A three-phase 2-level converter is analyzed in this work as shown in Fig. 1. Among others, it is used in electric vehicles, industrial machines and wind power applications. In a power electronic system, different signals are processed for control purposes. In addition to the DC link voltage U_{DC} and the AC currents, which are the variables to be controlled, the output control variables of the current controller are also recorded. In [6] it has been determined that the output voltages of the current controller contain information about the condition of the converter. Since a controller has to react to system changes, the aging of semiconductors leads to changes of the controller output voltages.

In order to determine to what extent changes in the power semiconductors affect the control voltage, two models are used. The first model represents an ideal converter and serves as a reference model (Fig. 2a). The power semiconductors are modeled as ideal switches S with ideal switching and conduction characteristics. The second model uses parasitic power semiconductors that have non-ideal characteristics (Fig. 2b). Both the conduction and switching characteristics are non-ideal. The properties of the conduction characteristics can be set by the threshold voltage U_f and the differential resistance R_{on}. The characteristics of the blocking behavior is modeled with the leakage current gradient G_{Leak}. The switching behavior is characterized using the turn-on time t_{on} and the turn-off time t_{off}.

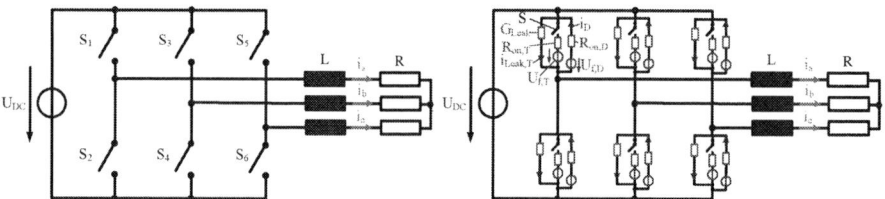

Fig. 2a): Model with ideal power semiconductors   Fig. 2b): Model with non-ideal power semiconductors

In the following, the controller output voltages of both models are compared depending on the semiconductor parameters of the non-ideal model (Fig. 3).

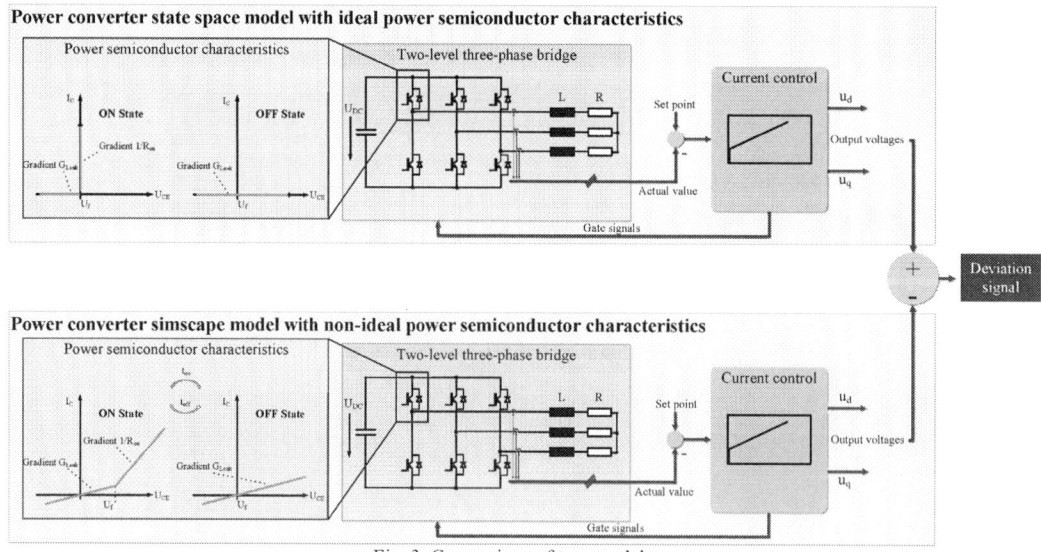

Fig. 3: Comparison of two models

Simulation results

For the simulation, the parameters of Table II are used. It is found that the control voltages of both models differ marginally (Fig. 4).

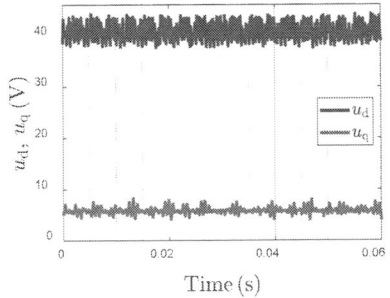

Fig. 4a): Control voltages of ideal model

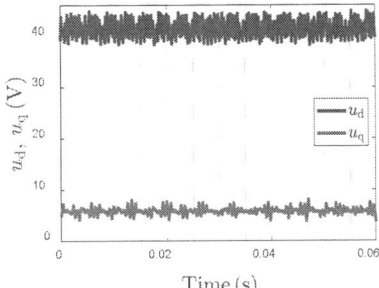

Fig. 4b): Control voltages of non-ideal model

Table II: Simulation parameters

Operating point	Value	Semiconductor properties	Value
Current in d-axis	10 A	Forward voltage IGBT	0,5 V
Current in q-axis	0 A	Differential resistance IGBT	5 mΩ
DC link voltage	200 V	Leakage conductance	0,83 µS
Electrical frequency	50 Hz	Forward voltage Diode	0,6 V
Pulse frequency	16000 Hz	Differential resistance Diode	6 mΩ
Load resistor	4,1 Ω	Switch-on time	0 µs
Load inductance	2 mH	Switch-off time	0 µs

The difference between the control voltages results in deviation signals in both d- and q-axis (Fig. 5). The deviation signals represent the difference between the non-ideal and the ideal converter behavior.

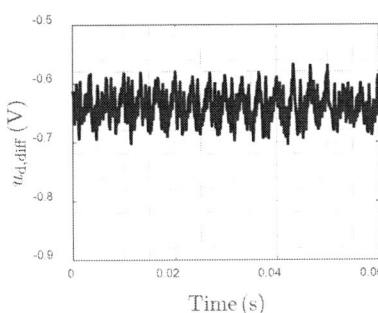

Fig. 5a): Deviation signal in d-axis

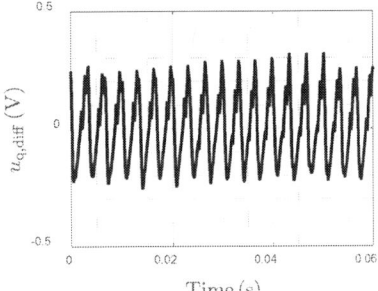

Fig. 5b): Deviation signal in q-axis

The deviation signal changes marginally as the parameters of the non-ideal power semiconductor model change. Here, the simulation of the aging mechanism bond wire lift-off is presented as an example. In Fig. 6, the failure mechanism is simulated by increasing the differential resistance R_{on} of the model with non-ideal power semiconductor characteristics from $R_{on,IGBT} = 5$ mΩ to $R_{on,IGBT} = 15$ mΩ. In [4] it is shown that the values are reasonable in terms of their order of magnitude.

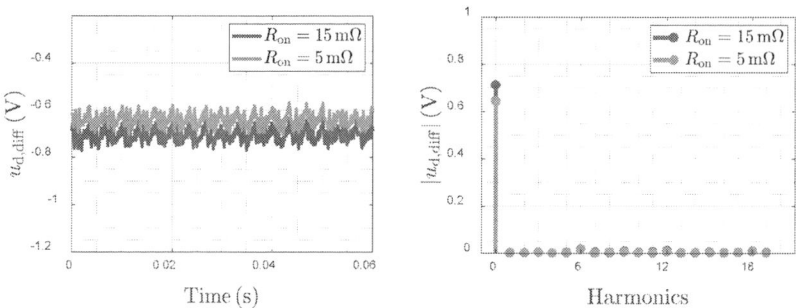

Fig. 6: Simulation result - change of the deviation signal when differential resistance R_{on} increases

Fig. 6 shows that the average of the deviation signal changes while the harmonics remain the same. This pattern is a characteristic feature of this aging mechanism.

Practical implementation

Test bench

In this chapter, the test bench is discussed. The block diagram of the test bench is shown in Fig.7.

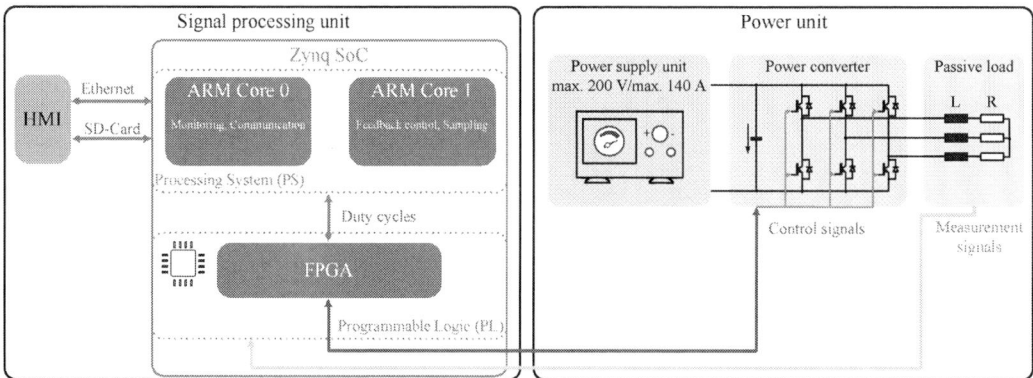

Fig. 7: Block diagram of the test bench

The test bench consists of a signal processing unit and a power unit. The converter converts the DC voltage provided by the power supply into a three-phase AC voltage. Passive components consisting of an ohmic resistor and an inductor are used as load. The power unit is composed of a simple structure in order to minimise the influences on both the input side and the load side. Therefore, it has been decided not to use a motor as load. As a result, only the converter can be investigated without external interference from other system components. This is necessary to gain first experiences with the algorithms for failure detection. In further research work, additional system components will be added to be able to investigate also larger real systems.

The converter is equipped with the Infineon Si-IGBT module FS75R12KT4 and has an output rating of 30 kW. It is equipped with measuring systems for the AC output currents, the heatsink temperature, the DC-link voltage and the output voltages. For the current measurement, CMS3000 family current sensors from SENSITEC with a bandwidth of 2 MHz are applied. Furthermore, the integrated NTC of the Infineon module FS75R12KT4 measures the module temperature.

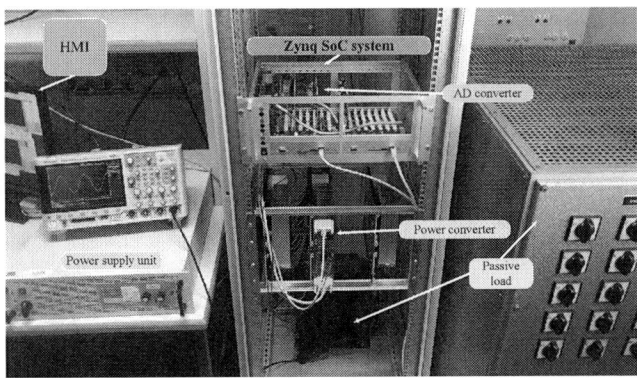

Fig. 8: Test bench

The signal processing unit is a System on Chip (SoC) system Zynq 7000 consisting of a dual core ARM Cortex A9 processor and a Kintex-7 FPGA [7]. For a real-time implementation, the ARM cores are used as an asymmetric multiprocessing system. ARM core 0 functions as a communication system and ARM core 1 as a processing system. On the ARM core 0, a slim real-time operating system FreeRTOS for embedded devices is implemented which is responsible for the data exchange with the Human Machine Interface (HMI).

On ARM core 1, the feedback control algorithm is executed. The program code can be written in C or directly generated with the Matlab/Simulink C-code generation out of the model-based simulation. Due to this structure, the control algorithm for the ARM core 1 can be flexibly reconfigured online. For the data transfer between the FPGA and the ARM core 1 the Xilinx AXI4 stream protocol, part of the Advanced Microcontroller Bus Architecture (AMBA), is used. On the HMI a Monitor Control Tool, which is based on LabView, is used to visualize the received data. The test bench is shown in Fig. 8.

Data caption

Instead of analyzing the sensor data, the controller output variables are considered. The information content in the data recorded by the sensors is much less, as these values are forced to the desired set point by the control system. Instead, the output control variables of the controller are investigated. The controller always reacts to changes of the system. This means that the controller also reacts to aging of the system components. This principle can be used to detect aging mechanisms that have occurred during operation.

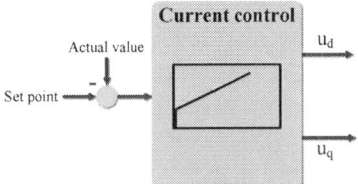

Fig. 9: Current controller

In Fig. 9 the current control system is shown. Since a PI current controller is used, the output voltages u_d and u_q can be divided into a d- and a q-component which represent the d- and q-axis in the rotational reference system. The PI controller combines the properties of a proportional controller with the properties of an integral controller. The proportional controller ensures that the controller reacts quickly to a control deviation. The integral component, on the other hand, allows steady-state accuracy. In this work, the output voltages are also divided into the proportional part and the integral part.

Table III: Measurement parameters

Operating point	Value
Current in d-axis	10 A
Current in q-axis	0 A
DC link voltage	200 V
Electr. frequency	50 Hz
Pulse frequency	16000 Hz
Load resistor	4,1 Ω
Load inductance	2 mH

As an example, the measurement with the operating point of Table III has been performed. Fig. 10 shows the integral component of the current controller output voltage over time. A closer look reveals that the output voltage contains harmonic oscillations. That is why, in addition to the DC component of the output voltages, the harmonics of the fundamental oscillation are also analyzed. For this purpose, a Fast Fourier Transformation (FFT) is performed for each electrical period to transform the voltages from the time domain to the frequency domain.

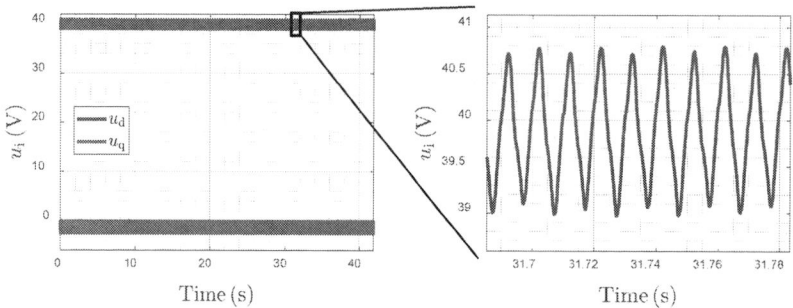

Fig. 10: Measurement result – Integral component of current controller output voltage

The complete data is now visualized in the following diagrams. The data set consists of 2093 data points since this is the number of electrical periods that have been captured. Besides the DC component, the 6th, 12th and 18th harmonics are also considered and visualized in 2-dimensional charts in Table IV.

Table IV: Measurements of controller output voltage harmonics

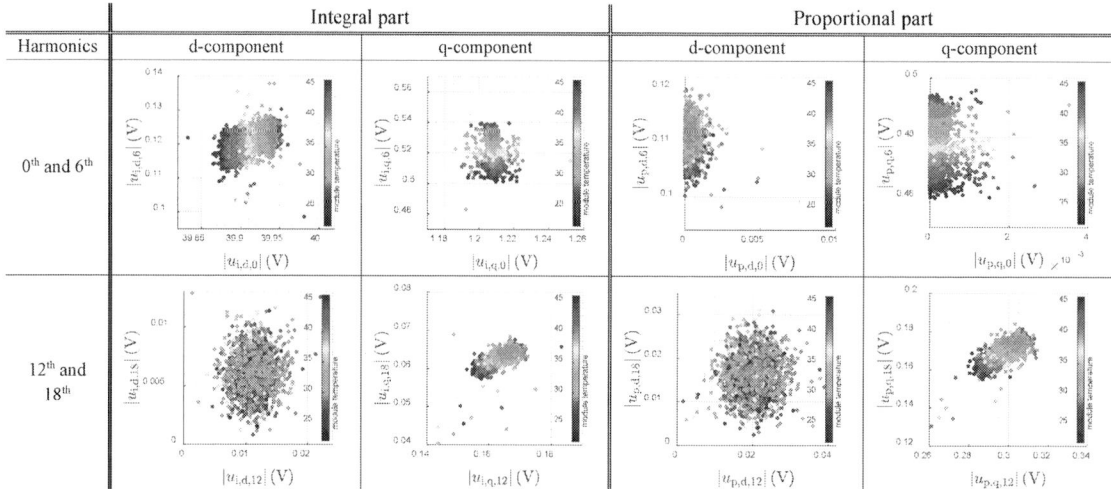

The module reference temperature is measured by using the built-in NTC resistor within the module. It can be seen that the recorded data has a temperature dependence. This corresponds to the temperature dependence of the semiconductor characteristics, which is reflected in the recorded data.

Minimum covariance determinant for anomaly detection

The measured data points contain outliers. To eliminate the outliers, the minimum covariance determinant (MCD) method is used [9]. The method is based on the robust Mahalanobis distance. In a two-dimensional space, all points at equal Mahalanobis distance from a center $\vec{\mu}$ form an ellipse, while in the Euclidean distance it is a circle. The formula of the Mahalanobis distance D_M for the vector $\vec{x}$ is shown in (1).

$$D_M(\vec{x}) = \sqrt{(\vec{x} - \vec{\mu})^T \cdot C^{-1} \cdot (\vec{x} - \vec{\mu})} \tag{1}$$

In the Mahalanobis distance formula a covariance matrix C is included, which represents the spreading in both axes. The Mahalanobis distance has the disadvantage that the mean value and the covariance matrix are distorted by the outliers. The tolerance ellipse becomes more accurate if the robust Mahalanobis distance is calculated instead of the usual Mahalanobis distance. For this purpose, both a robust mean value $\vec{\mu}_R$ and a robust covariance matrix C_R are determined, which are not distorted by the existing outliers. Equation (2) shows the formula for calculating the robust Mahalanobis distance D_{RM}.

$$D_{RM}(\vec{x}) = \sqrt{(\vec{x} - \vec{\mu}_R)^T \cdot C_R^{-1} \cdot (\vec{x} - \vec{\mu}_R)} \qquad (2)$$

In order to find the robust values the method of the minimum covariance determinant is used [9]. This method is based on the mathematical theorem that the determinant of the covariance matrix of a subset H_1 of the overall data set X ($H_1 \subset X$) is always less than or equal to the determinant of the covariance matrix of the overall data set X. This theorem serves as the basis for the algorithm to determine the robust mean and the robust covariance matrix. First, a subset H_1 of the data set X is selected and the determinant $\det(C_1)$ of the covariance matrix C_1 is determined for the subset H_1. For this new subset, the individual data points are now sorted beginning with the smallest Mahalanobis distance. The next step is to determine another subset $H_2 \subset H_1$, which contains only those data points that have a Mahalanobis distance less than $s = \sqrt{\chi^2_{p,0.975}} = 2{,}7162$. This threshold value is based on the chi-square distribution χ^2 and is discussed in [9]. With this new subset H_2, the determinant $\det(C_2)$ of the covariance matrix C_2 is now calculated again. The mathematical theorem described above states that $\det(C_2) \leq \det(C_1)$ applies. This step of determining a new subset is performed until the determinant of the covariance matrix either does not change any more or reaches the value 0. The last determined mean value and the covariance matrix are now considered to be robust. Finally, the ellipse is calculated, which outlines the data points. The general ellipse equation in parameter form is given in (3).

$$\begin{pmatrix} x \\ y \end{pmatrix} = \begin{pmatrix} \mu_{R_x} + a \cos t \cos \alpha - b \sin t \sin \alpha \\ \mu_{R_y} + a \cos t \sin \alpha + b \sin t \cos \alpha \end{pmatrix} \qquad (3)$$

The parameter t has the value range $0 \leq t < 2\pi$, μ_{R_x} and μ_{R_y} represent the centre of the data set and the angle α defines the orientation of the ellipse towards the x-axis.

For the determination of the ellipse, three critical values still have to be identified. First, the angle of the ellipse α towards the x-axis is determined. This is calculated using the largest eigenvector of the covariance matrix. If $\vec{E}_L$ is the eigenvector corresponding to the largest eigenvalue λ_L of the covariance matrix, the slope α of the ellipse is calculated with $\alpha = \arctan \frac{\vec{E}_{L_y}}{\vec{E}_{L_x}}$.

In addition, the deviations a and b in x and y direction are calculated in (4) and (5) using the threshold value $s = \sqrt{\chi^2_{p,0.975}}$ and the largest and smallest eigenvalues λ_L and λ_S of the covariance matrix.

$$a = \sqrt{\chi^2_{p,0.975}} \cdot \sqrt{\lambda_L} \qquad (4)$$

$$b = \sqrt{\chi^2_{p,0.975}} \cdot \sqrt{\lambda_S} \qquad (5)$$

Results

Measurements

Table V shows the measurement results of the DC component of the integral part of the output voltage divided into temperature ranges. In the following, only the DC component of the integral part is considered. It is divided into temperature ranges, which are evaluated separately. The aim of this consideration is to separate the aging impacts from purely temperature-dependent effects. This enables to isolate the aging-dependent effects for the purpose of condition monitoring.

For every temperature range, an ellipse is determined by means of the MCD method described above. On the one hand, the goal of this approach is to detect and eliminate the outliers from the data set, which are located outside of the ellipse. On the other hand, anomalies can be detected if a shift of the ellipse occurs during converter operation. The shift of the ellipse may have several causes. One reason is a change in the environmental conditions. If the environmental conditions stay constant or are also detected and tracked during converter operation, the shift of an ellipse indicates an aging of a component of the power electronics system.

Table V: Measurements of $u_{i,d,0} - u_{i,d,6}$ divided into temperature ranges

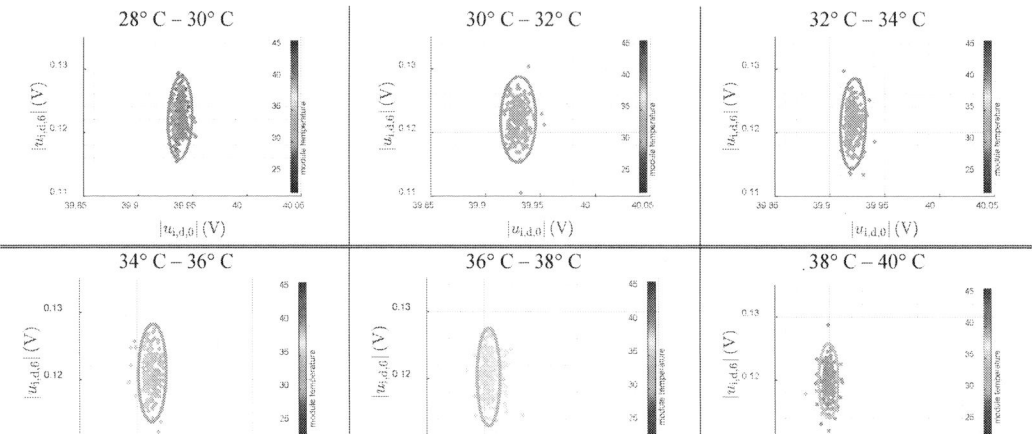

Implementation of an aging mechanism

In order to test the MCD algorithm for anomaly detection, an aging mechanism was intentionally implemented. For this purpose, some of the bond-wires of the IGBT power module EconoPack2 FS75R12KT4_B15 (1200 V/75 A) were cut off. Fig. 11 shows the unharmed and the altered module. First, a series of measurements with the unharmed module was done. The MCD algorithm was used to describe the recorded data sets using an ellipse. Secondly, some bond-wires between the IGBT chip and the diode chip were cut off in order to increase the differential resistance. Here the redundant bond-wires of the semiconductor chips were cut off. Cutting off these bond-wires does not result in a total failure of the power module. Furthermore, the bond wires that are located near the chip, were separated since these bond wires are most exposed to temperature fluctuations making them prone to failure. Then a measurement with the altered module has been performed. These data sets were also trained with the MCD algorithm. For the measurements, the operating point of Table III is used.

Fig. 11a): Module with unharmed bond-wires   Fig. 11b): Module with separated bond-wires

Table VI shows the measurement results of the unharmed module and the modified module. In addition to the data set, the corresponding ellipse is also shown. A clear shift of the ellipse can be observed for each temperature range. It results in an increase in the DC component of the d-component of the integral part. In conclusion, the aging mechanism bond wire lift-off can be detected using the available data. Even if the effect is marginal, its detection is possible with a large number of data points.

Table VI: Measurements of $u_{i,d,0} - u_{i,d,6}$ with unharmed bond wires and modified bond wires

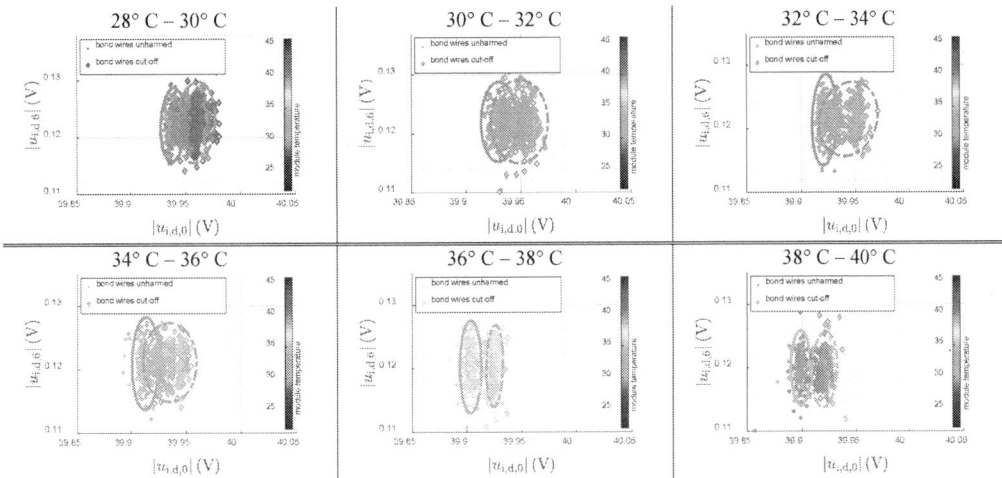

Conclusion

Most existing approaches in the field of condition monitoring of power semiconductors require additional sensors and measuring equipment. The aim of the research project presented in this paper is to develop a new approach that overcomes this disadvantage. The proposed approach replaces the sensors that are typically required for condition monitoring with software-based techniques. In order to obtain information on the condition of the converter from the data that is already available, the data is analyzed more precisely by means of signal analyzes. In simulation, it was shown that there is a relation between the characteristics of a power semiconductor and the harmonics of the output voltages of the current controller. Finally, measurement results confirming the simulation are presented. The aging mechanism bond-wire lift off was investigated. This aging mechanism has the effect that the mean value of the output voltage increases marginally. It was shown on a test bench that the detection is possible with a large number of data points. In the future, further failure mechanisms will be introduced into the semiconductors of the power converter. As a next step, a broad spectrum of modern algorithms from the field of machine learning can be used in order to trace back to the cause of the failure.

References

[1] Fischer K.: Zustandsüberwachung von elektronischen Systemen, Methoden und Technologien, ECPE Workshop, Nuremberg, Germany, 2018

[2] Holzke W., Brunko A., Groke H., Kaminski N., Orlik B.: A condition monitoring system for power semiconductors in wind energy plants, PCIM Europe, Nuremberg, Germany, 2018

[3] Krone T., Hung L., Jung M., Mertens A.: Advanced Condition Monitoring System Based on On-Line Semiconductor Loss Measurements, IEEE Energy Conversion Congress and Exposition, 2016

[4] Eleffendi M., Johnson M.: In-Service Diagnostics for Wire-Bond Lift-off and Solder Fatigue of Power Semiconductor Packages, IEEE Transaction on Power Electronics, September 2017

[5] Choi U., Blaabjerg F., Jorgensen S., Munk-Nielsen S., Rannestad B., Reliability Improvement of Power Converters by Means of Condition Monitoring of IGBT Modules, IEEE Transactions on Power Electronics, Vol. 32, No. 10, October 2017

[6] Richter J., Modellbildung, Parameteridentifikation und Regelung hoch ausgenutzter Synchronmaschinen, KIT Scientific Publishing, 2017

[7] Schwendemann R., Decker S., Hiller M., Braun M.: A Modular Converter- and Signal-Processing Platform for Academic Research in the Field of Power Electronics, IPEC'18 ECCE Asia, Niigata, Japan, 2018

[8] Ciappa M., Selected failure mechanisms of modern power modules, Microelectronics Reliability, 2002

[9] Hubert M., Debruyne M.: Minimum covariance determinant, 2009 John Wiley & Sons, Inc. WIREs Comp Stat 2010 2 36-43, 2010

Frugal Innovation for Sustainable Rural Electrification

Bunthern Kim[1][2], Phok Chrin[1][2], Maria Pietrzak-David[1], Pascal Maussion[1]

(1)　LAPLACE, UNIVERSITE DE TOULOUSE, INP, UPS, CNRS, Toulouse, France,

(2)　INSTITUTE OF TECHNOLOGY OF CAMBODIA, Phnom Penh, Cambodia,

Phone: +33 5 34 32 23 59, +33 5 34 32 23 64, Fax: +33 5 61 63 75 88

bunthern.kim@laplace.univ-tlse.fr, pchrin@itc.edu.kh, Maria.David@laplace.univ-tlse.fr,
Pascal.Maussion@laplace.univ-tlse.fr,

Keywords

«Induction motor», «Generation of electrical energy», «Generator excitation system», «Photovoltaic», «Sustainable system/technology»

Abstract

In this article an original solution is proposed by using wasted electric and electronic equipment (second-life components) to create the new power generation systems for remote rural areas. This frugal innovation for rural electrification guarantees an important support social, educational and economic development goals especially in Southeast Asian countries.

Introduction

The use of second-life components, available at low cost locally, can prove to be a viable solution for the electrification of isolated villages in developing countries. If based on renewable energies, they contribute to the fight against global warming and can promote economic development and education. With local available energies, for example biogas or electricity, better education, reduced time for collecting wood, access to information or entertainment (TV, radio, laptops), improve economic activities, human health and life could be better. Indeed, many electrical and electronic products are often discarded even before their end of life [1], for reasons of fashion, marketing or change of use. This leads to an increase in the consumption of energy and raw materials. The innovative solutions proposed in this paper could offer to a new sustainable economic strategy developing countries.

This study focuses, of course, on the application of WEEE reuse in a stand-alone renewable energy system as a solution for electrification in rural areas in some developing countries. Solar, hydro or wind energy are chosen thanks the outcomes of many international reports such as [2] in a global survey regarding the energy access in the Least Developed Countries. Moreover, these energies have also been selected by many of these countries in their Intended Nationally Determined Contribution [3] for COP21, in 2015.

The different architectures developed in this study are described in next section. They include two parallel energy sources: solar panels and hydraulic generation and energy storage in in lead-acid batteries. Re-use power electronic devices with minimum modifications are used for DC/DC or DC/AC or even AC/DC conversions of energy. Different solutions have been proposed for this objective, but the main idea of this study concerns the research of minimal modifications in the existing products, so as not to increase the environmental impacts and also encourage a large dissemination. The main elements of a Life Cycle Analysis (LCA) of the solar chain are given in [4]. The minimum modifications of one ATX PC Power Supply Units (PSU) are presented and an MPPT is implemented in a Arduino microcontroller. Moreover, this sections also provides some experimental results of several PSU associations in order to increase the supplied power. Finally, simulations and tests on a test bench, with a 1,5kW 3 phase induction motor used as a single phase generator validate the system feasibility.

Global presentation of frugal solution

The global architecture of the studied system is shown in Fig. 1. It consists of two interconnected "solar" and "hydraulic" lines. The solar chain includes photovoltaic (PV) panels and a Personal Computer (PC) power supply (PSU) instead of a commercial solar converter. We opt for a Maximum Power Point Tracking (MPPT) control strategy which has been successfully implemented in an Arduino Due Microcontroller. This microcontroller is very popular because of its low cost, its high availability in the market and its open source software.

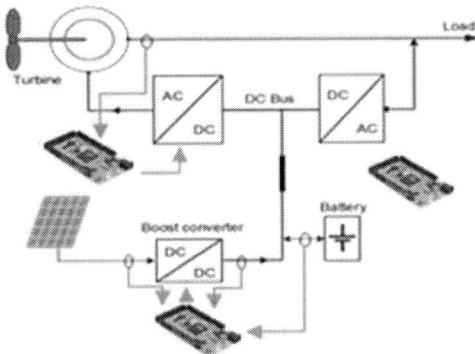

Fig. 1: Rural electrification architecture

The hydraulic line includes a second life squirrel cage induction motor which has been purchased from a local scrap dealer (Fig. 2) and a voltage source inverter (VSI), initially dedicated to supply PCs as emergency network (DC/AC converter). The energy storage will be provided by second-life automotive lead-acid batteries.

Fig. 2: Induction recovery motors in Cambodia (low cost second hand products from China, Taiwan, Malaysia, Thailand...)

Since the original objective of this frugal innovation is the electrification of remote villages in Southeast Asian countries, the functional unit is defined to satisfy the daily energy needs of a rural village over a period of 20 years. The load profile of a typical small village in Thailand [5] was chosen for this case study (Fig. 3).

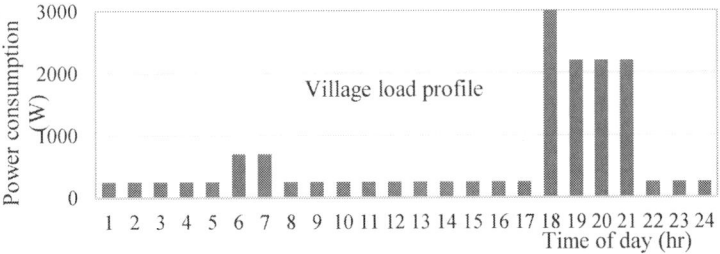

Fig. 3: Load profile of a single village in Chi Angria, Thailand [5], [6]

Solar Line Presentation

Power supply unit description and modification

The ATX PC Power Supply Units (PSU) which replace the charge controller unit are slightly modified in order to first, disable their protection functionalities, second to allow the interface between the microcontroller and the DC/DC converter of the power supply and to reduce their input voltage range (from 280V to 100V as to become more suitable to the output of a small PV generator).

The internal circuit board of HP305-00 is given in Fig. 4. The basic functionality of PSU shown in Fig. 5 is to generate regulated constant DC voltages for computer hardware from the grid (230V AC) and +12V DC as the most powerful output.

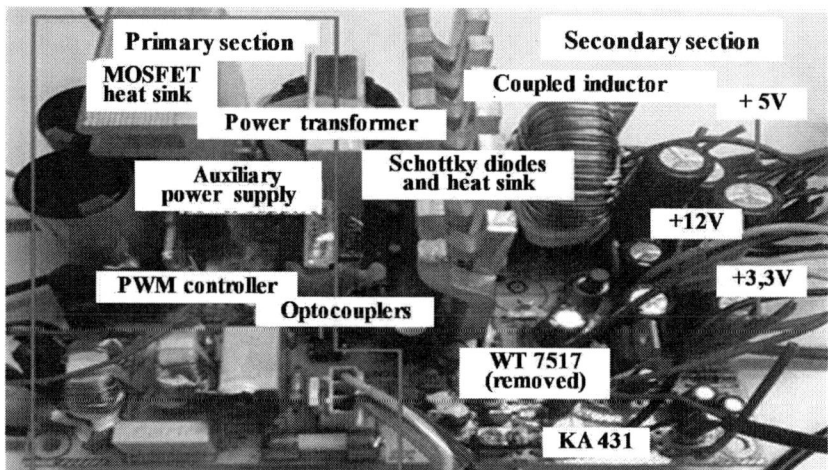

Fig. 4: HP305P-00 internal circuit board.

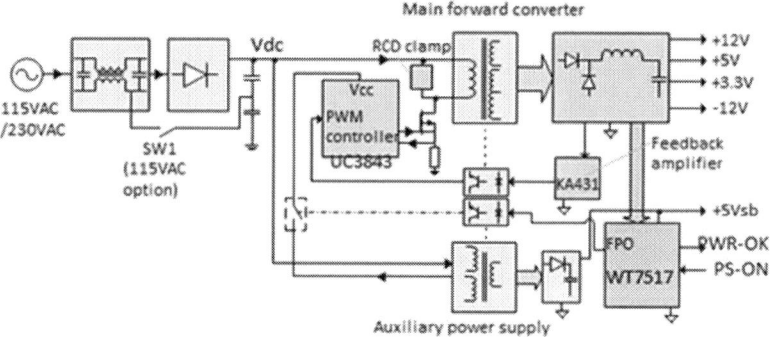

Fig. 5: Functional block diagram of DELL HP305P-00. [7]

This functionality does not fit the requirements of small renewable energy systems which provide low input DC voltage (100V) and need variable output voltage. Figure 5 shows the components which have been modified. Firstly, the protection functionality of WT7517 should be removed to allow the converter to be used independently of the limitation of the standard power supply. In addition, the range of the PSU's input voltage should be reduced by using the primary side center tap pin of the main transformer. Detailed analysis and modifications are shown in figures 6 and 7. The objectives of the feedback circuit modifications are that the output voltage can be changed by the controller through an external input voltage, given by a MPPT microcontroller for example. The different modifications are depicted in Figures 6.a and 6.b.

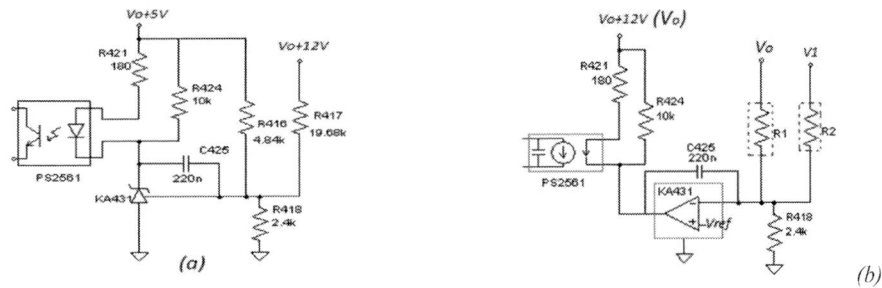

Fig. 6: Initial (a) and modified feedback (b) circuits

Then, the Schottky diodes, resistors, capacitors, inductors of the +5V, +3.3V and -12V circuits are removed and the second Schottky diode of the +12V circuit has to be grounded. After these modifications, shown in Figure 7, only the +12V works while the 5VSB is generated from another part of the converter.

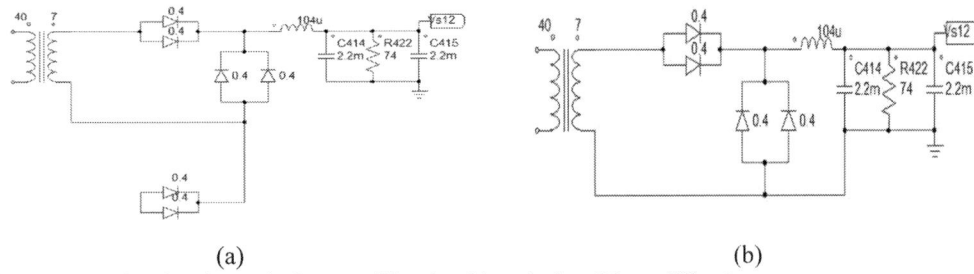

(a)   (b)

Fig. 7: +12V circuit scheme before modification(a) end after (b) modification

A problem on the +12V circuit has to be solved: the maximum output voltage cannot exceed 16V, due to output filter capacitor ratings. The solution is the serial insertion of another capacitor (to increase the voltage) and a parallel connection of two identical legs to come back to the initial value of the capacitance.

The detailed analysis and experimental results are presented in [8]. The value of the resistors R_1 and R_2 in Fig. 6b are changed from 10kΩ to 18kΩ in order to avoid the ground problem and use any PSU without any danger. The relation of V_o and V_1 at steady state can be written by (1) and the output value V_o can be controlled by the injected signal V_1 following the relation (2).

$$\left(1 + \frac{R_{418}}{R_1} + \frac{R_{418}}{R_2}\right) V_{ref} = \frac{R_{418}}{R_1} V_o + \frac{R_{418}}{R_2} V_1 \quad (1) \qquad\qquad V_o = 20 - V_1 \qquad (2)$$

As a consequence of this section and without the need for a new printed circuit board, contrary to [4], this initial AC/DC static converter with regulated fixed output voltage is transformed into a DC/DC converter with variable voltage (or current) controlled. The complete modifications on the PSU are described in [8] and [9].

Control of the PSU for photovoltaic systems

The performance of the modified PSU with MPPT control have been successfully implemented an Arduino Due microcontroller and has been tested for load step changes and sun light intensity modification. The parameters used in this test for control discretised algorithms [11], i.e. Perturb-and-Observe (P&O) and Incremental Conductance (INC) methods are made with the sampling period Ta=0,02s and duty ratio $\Delta\alpha = 0.01$. In this way the validation of satisfactory operation of this system is confirmed (Fig. 8).

For experimental validation, the solar array consists of twelve FVG 10P solar panels (10W). These panels have been connected in two parallel strings of six panels to match the input voltage range of PSU. In this topology, at the maximum power point, the PV output voltage is equal to 105V at Standard Test

Condition (STC). The 12V, 44Ah lead-acid battery has been sized for daily energy storage compliant with one day needs of a typical rural village in such rural areas [6].

Several units must be associated in serial and/or parallel configurations in order to increase the power delivered to the end users. More details could be found in [8], [9].

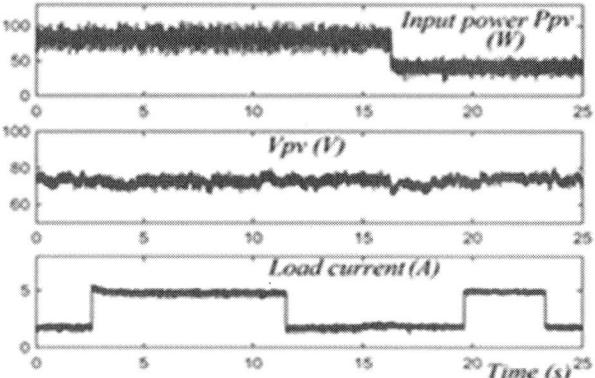

Fig. 8: PSU operation with load changing and light modification incident (INC algorithm).

Electro-Hydraulic Line

System presentation

Several work have been done according to the idea frugal innovative hydroelectric generation system, based on recycled materials, located directly in isolated villages or nearby. The electrical generation is achieved thanks to an old induction machine (IM) purchased at low price from a local scrap dealer. One phase of this IM is excited by a voltage inverter (VSI), initially devoted to PC power supply in an emergency network. As shown in Figure 9, the excited phase of the EIG is connected to the very simple push-pull DC/AC of the UPS circuit to step up the voltage through a transformer.

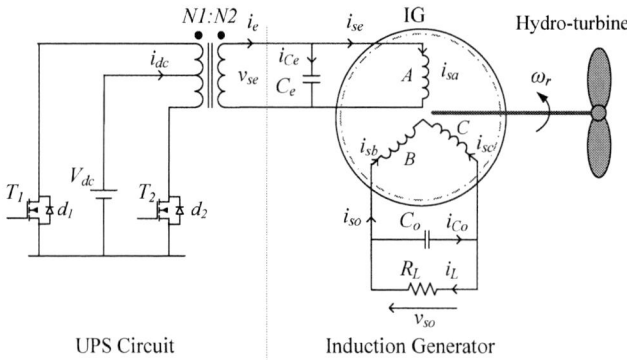

Figure 9: EIG architecture supplied by recovery UPS, transformer and reactive energy compensation capacitor

The equivalent circuit of this transformer is shown in the Figure 10. Its primary winding parameters, resistance R_1 and reactance X_1, are referred to secondary winding, i.e. $R_1'=(N_2/N_1)^2 R_1$, $X_1'=(N_2/N_1)^2 X_1$. So, the equivalent resistance R_{eq} and reactance X_{eq} can be expressed as shown in the Fig. 11: $R_{eq}=R'_1+R_2$ and $X_{eq}=X'_1+X_2$ ($X_{eq}=2\pi f L_{eq}$). The transformer output voltage can be expressed as follows:

$$v_2' = v_2 + R_{eq} i_2' + L_{eq} \frac{d i_2'}{dt} \qquad (3)$$

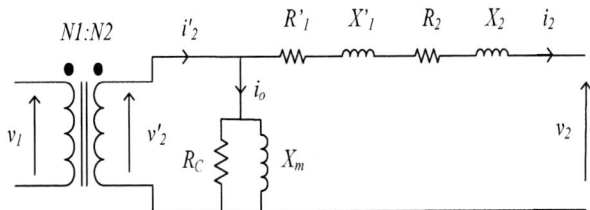

Figure 10: Equivalent transformer circuit

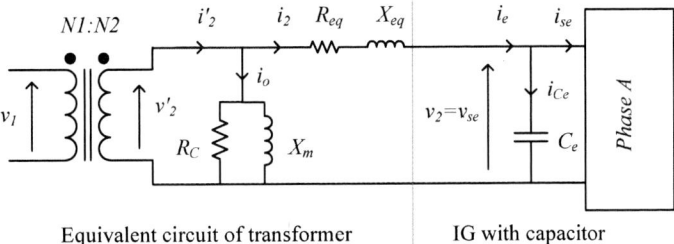

Equivalent circuit of transformer   IG with capacitor

Figure 11: Equivalent circuit of transformer and EIG with capacitor [12]

As shown in the Figure 11, the transformer secondary winding is connected to the excitation phase A of EIG. So, the $v_2'=v_{se}$ and $i_2=i_e$. In fine, the state space equation describing the EIG behavior is formulated as below:

$$\dot{x}_T = [A_T]x_T + [B_T]u_T \tag{4}$$

with $x_T = \begin{bmatrix} i_e & v_{se} & v_{so} & i_{se} & i_{so} & i_{r\alpha} & i_{r\beta} \end{bmatrix}^T$ system state space vector and $u_T = v_2'$ system input, expressed by (3).

The system matrices are expressed as follows:

$$[A_T] = \begin{bmatrix} [A_{T1}] & [A_{T2}] \\ [A_{T3}] & [A_{T4}] \end{bmatrix}, \quad [A_{T1}] = \begin{bmatrix} -R_{eq}/L_{eq} \end{bmatrix}, \quad [A_{T2}] = \begin{bmatrix} -1/L_{eq} & 0 & 0 & 0 & 0 & 0 \end{bmatrix};$$

$$[A_{T3}] = [B_c], \quad [A_{T4}] = [A_c], \quad [B_T] = \begin{bmatrix} 1/L_{eq} \\ 0_{6\times1} \end{bmatrix}$$

The matrices $[A_c]$ and $[B_c]$ are defined from the EIG modeling with capacitors C_e and C_0 only.

$$[A_c] = \begin{bmatrix} A_{c0} & A_{c1} \\ [B] & [A] \end{bmatrix}, \quad [B_c] = [1/C_e \quad 0_{2\times2}]^T \quad \text{with } [A_{c0}] = \begin{bmatrix} 0 & 0 \\ 0 & -1/R_L C_o \end{bmatrix},$$

$$[A_{c1}] = \begin{bmatrix} -1/C_e & 0 & 0 & 0 \\ 0 & -1/C_o & 0 & 0 \end{bmatrix}$$

The matrices $[A]$ and $[B]$ were defined in our previous study concerning the EIG modeling [12]. Finally, the dynamic system matrix $[A_T]$ corresponds to the 7th order system written as follows:

$$[A_T] = \begin{bmatrix} -\dfrac{R_{eq}}{L_{eq}} & -\dfrac{1}{L_{eq}} & 0 & 0 & 0 & 0 & 0 \\[2mm] \dfrac{1}{C_e} & 0 & 0 & -\dfrac{1}{C_e} & 0 & 0 & 0 \\[2mm] 0 & 0 & -\dfrac{1}{C_o R_L} & 0 & -\dfrac{1}{C_o} & 0 & 0 \\[2mm] 0 & -\dfrac{L_{rr}}{L_1} & 0 & \dfrac{R_s L_{rr}}{L_1} & -\dfrac{\sqrt{3}L_{ms}^2 \omega_r}{L_1} & -\dfrac{R_r L_{ms}}{L_1} & \dfrac{L_{ms}\omega_r L_{rr}}{6L_1} \\[2mm] 0 & 0 & -\dfrac{L_{rr}}{L_b} & \dfrac{\sqrt{3}L_{ms}^2\omega_r}{L_2} & \dfrac{2R_s L_{rr}}{L_2} & -\dfrac{\sqrt{3}L_{ms}\omega_r L_{rr}}{6L_2} & -\dfrac{\sqrt{3}R_r L_{ms}}{L_2} \\[2mm] 0 & \dfrac{L_{ms}}{L_1} & 0 & -\dfrac{R_s L_{ms}}{L_1} & a_{t(5,4)} & \dfrac{R_r(L_{ls}+L_{ms})}{L_1} & a_{t(5,6)} \\[2mm] 0 & 0 & \dfrac{\sqrt{3}L_{ms}}{L_b} & -\dfrac{2L_{ss}L_{ms}\omega_r}{L_2} & -\dfrac{2\sqrt{3}L_{ms}R_s}{L_2} & a_{t(6,5)} & \dfrac{2R_r L_{ss}}{L_2} \end{bmatrix} \quad (5)$$

With $a_{t(5,4)} = \dfrac{\sqrt{3}(L_{ls}+L_{ms})L_{ms}\omega_r}{L_1}$, $a_{t(5,6)} = \dfrac{6\omega_r L_{s1}-7L_{ms}^2\omega_r}{6L_1}$, $a_{t(6,5)} = \dfrac{7L_{ms}^2\omega_r-4\omega_r L_{s2}}{2L_2}$, ω_r – rotation speed

L_{ls} - leakage inductance of stator winding, L_{ms} - maximal value of magnetizing inductance of stator

winding

As this system is non stationary, its dynamics using pole locus evolutions and frequency responses were analyzed according to the parameter variations to find most preponderant of them modifying the system behavior [13]. This analysis helps to design the voltage control which will be discussed with details in final version of this article.

Output voltage control of the EIG

The output voltage of the push pull inverter is controlled by its duty ratio thanks to a feedback loop, given below. The output voltage of EIG depends on its excitation voltage v_{se} and its rotation speed. In this application of power generation output voltage and frequency must be always kept constant even in any disturbance condition. Moreover, in hydropower applications rotation speed variations are not easy to manage, except in high quality hydropower stations. Consequently, speed and load current variations will be considered as system disturbances. Our experimental set up of an EIG 1.5kW and disposal UPS are shown in Fig. 12.

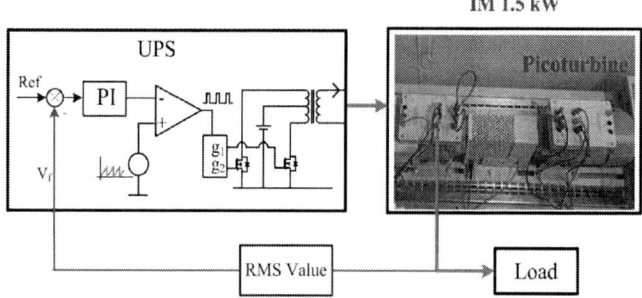

Fig. 12: Experimental setup of the 1,5 kW EIG + reused UPS, f_{sw}=50Hz, R_L=240Ω, Ce=20µF, C_o=10µF

Two identical 3-phase IM are connected on the same shaft axis. As usual and only for development purpose during lab work, a torque sensor is placed between two machines. One IM is considered as the pico-turbine and runs as a motor in order to drive the other machine operating as an induction generator. The small Single Phase Induction Generator (SPIG) has a nominal output power of 500W (on one phase) since the nominal power of the 3-phase machine is 1,5kW.

First PI controller is tested for the closed loop systems. EIG associated with the push-pull inverter of the UPS is described by a transfer function [12],[13] of a high degree (7th order). In this study, the well-known and robust Ziegler-Nichols tuning method is applied [14] to determine the controller gains. It is important to justify this choice in the framework of the re-used elements, where no data is available, nor data sheets.

Experimental results

Fig. 13 illustrates the experimental results of the 1.5 kW EIG supplied by the re-used UPS. For output voltage stability test, the feedback was tested under load resistance variations from 240Ω and 127Ω (output powers from 200W to 380W). Four variables were recorded and compared to simulation results.

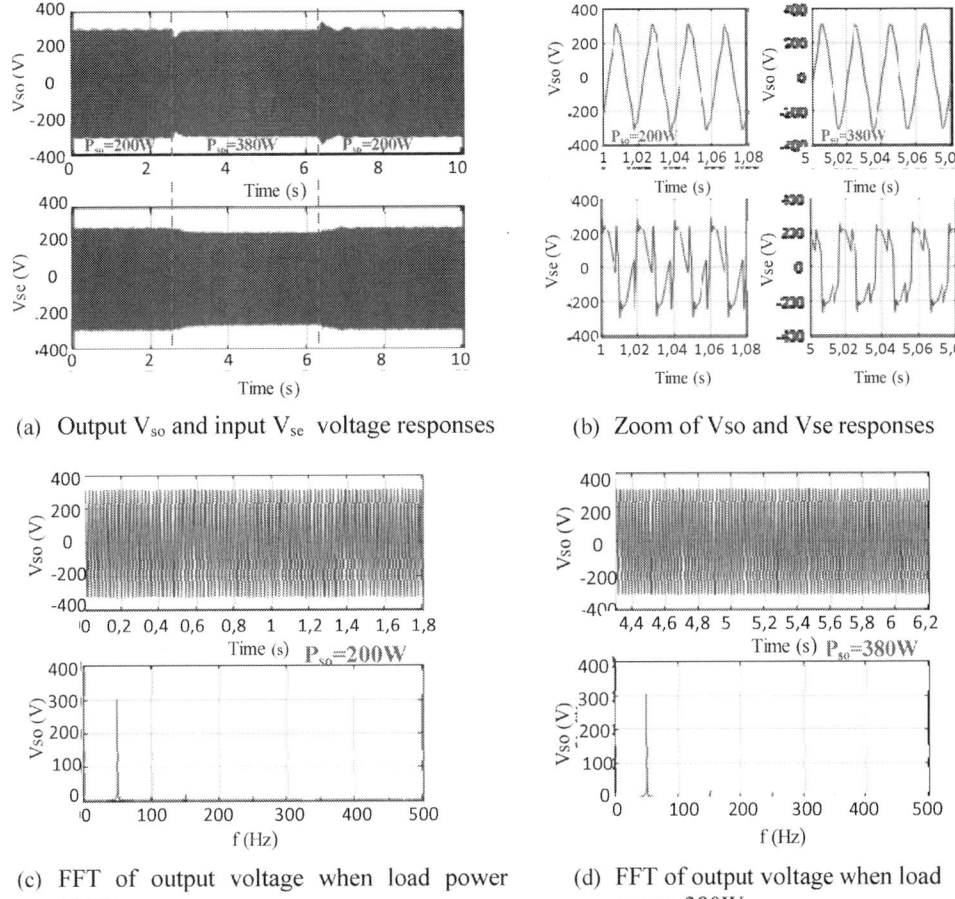

(a) Output V_{so} and input V_{se} voltage responses

(b) Zoom of Vso and Vse responses

(c) FFT of output voltage when load power 200W

(d) FFT of output voltage when load power 380W

Fig. 13: Experimental results with PI control: voltage response of the 1.5 kW IG with reuse UPS during step load change (F_{sw}=50Hz, R_L=240Ω- 127Ω -240Ω, Ce=20μF, C$_o$=10μF)

The output voltage raises/drops about 55V (peak voltage in Fig. 13) and gets back to its reference with a settling time around 200ms. System voltage is stable and at the required amplitude after the transient due to load disturbances and its frequency remains constant as the excitation voltage is always 50 Hz, as shown in FFT analysis (Fig. 13 (c) and (d)).

Conclusion

In this paper some innovative and original frugal solutions for rural electrification are presented and tested. The obtained experimental results are very encouraging. They are based on modeling, sizing, characterization and control of re-used materials for rural electrification in developing countries. In final version this study will supplemented by a life cycle analysis that justifies the choice of environmental impacts avoided. The reused components of power electronics and also the EIG, presents very attractive economically suitable solution for rural electrification of developing countries. Low cost of local components and frugal proposed innovative solution for rural electrification allows a creation of new life conditions for local peoples and also new future local jobs concerning all light modifications of reused products.

References

[1] Blenkinsopp, T.; Coles, S.R; Kirwan, K.; 2013. Renewable energy for rural communities in Maharashtra. India, Energy Policy, Volume 60, pp 192-199

[2] The Least Developed Countries Report 2017, Transformational energy access. United Nations Conference on Trade and Development, United Nations Publication (UNCTAD/LDC/2017). http://unctad.org/en/PublicationsLibrary/ldcr2017

[3] Intended Nationally Determined Contribution (INDCs), as communicated by Parties for the COP21. http://www4.unfccc.int/submissions/indc/Submission%20Pages/submissions.aspx

[4] Kim, B.: Contribution to the design and control of a hybrid renewable energy generation system based on reuse of electrical and electronics components for rural electrification in developing countries, 2019, PhD thesis at Toulouse INP

[5] Ketjoy,N. : Photovoltaic Hybrid Systems for Rural Electrification in the Mekong Countries, 2005, PhD Thesis at University of Kassel

[6] Nayak,C.K.; Nayak,M.R.: Optimal size and cost analysis of standalone PV system with battery energy storage using IHSA, International Conference on Signal Processing, Communication, Power and Embedded System, SCOPES, 2016

[7] Rogers,D.; Green, J.; Foster, M.; Stone, D.; Schofield, D.; Abuzed, S.; Buckley, A.: Repurposing of ATX computer power supplies for PV applications in developing countries, International Conference on Renewable Energy Research and Applications, ICRERA, 2013.

[8] Hop Dinh, T.B.; Phan,Q.D.; Mausion, P.: Associations of Second Life of Power Supply Units as Charge Controllers in PV System, IECON 2018, Washington, USA,

[9] Kim,B.; Cenni,H.; Roth,A.; Bun,L.; Pietrzak-David,M.; Maussion,P.: How to reuse PC power supply for renewable energy applications", EPE 2015, Suitzerland

[10] Rogers, D.; Green, J.E.; Foster, M.P.; Stone,D.A.; Schofield, D.; Buckley, A.; Abuzed, S.: ATX power supply derived MPPT converter for cell phone charging applications in the developing world, 7th IET International Conference on Power Electronics, Machines and Drives, PEMD 2014

[11] Sera, D.; Mathe, L.; Kerekes, T.; Viorel Spataru, S.; Teodorescu, R.: On the Perturb-and-Observe and Incremental Conductance MPPT methods for PV Systems, IEEE Journal on Photovoltaics, Vol.3, N° 3, July 2013

[12] Chrin, P.; Maussion, P.; Pietrzak-David, M.; Dagues, B.; Bun, L.: Modeling of 3-phase Induction Machine As Single Phase Generator for Electricity Generation from Renewable Energies in Rural Areas, IEEE International Electric Machines and Drives Conference, IEMDC 2015, USA

[13] Chrin, P.: Contribution to electric energy generation for isolated-rural areas using 2nd life components and renewable energies: modelling and control of an induction generator, 2016 PhD Thesis at Toulouse INP

[14] Ziegler, J.G.; Nichols, N.B.: Optimum settings for automatic controller, Transaction, ASME 1942, vol. 64, pp. 759–768,

A current-modulus derivative-based protection method in a flexible DC grid

Jianquan Liao, Niancheng Zhou, Qianggang Wang
State Key Laboratory of Power Transmission Equipment and System Security and New Technology
Chongqing University, Chongqing, 400044, China.
Corresponding author: jquanliao@cqu.edu.cn

Keywords

«DC grid», «DC fault», «protection», «current modulus», «current derivative».

Abstract

This study develops a rapid protection method for flexible DC grids. Based on the principle of modulus decomposition, common-mode (CM) and differential-mode (DM) circuits of MMC, DC transmission line, and fault transition resistance under different faults are derived, and the synthesized equivalent circuit of DC grids are finally obtained. The method utilizes a single-end derivative of CM and DM currents to construct a protection phase plane that divides different DC fault types into different regions. The proposed method is verified by simulation and experiment. This method can identify and detect faults within 1ms, which can significantly reduce the breaking current of the DC circuit breaker (DCCB).

Introduction

Modular multilevel converter (MMC)-based DC grids have broad application prospect in fields, such as bulk power delivery, island or weak AC system power supply, and interconnection of asynchronous power grids, due to its numerous advantages, such as independent active and reactive power control, power supply for passive networks, no commutation failure, and high power quality [1-3]. The MMC-based DC grid contains numerous power electronic equipment, and the system presents a low-inertia feature due to the zero-frequency characteristics of DC voltage and current [4]. When a DC fault occurs, the fault current can rise to dozens of kA within several milliseconds (ms) in a case of high-voltage direct current (HVDC) transmission [5]. This occurrence poses a threat to the safety of the DC grid and places high demands on the rapidity of DC grid protection. Therefore, the protection of DC grids must recognize the DC fault and eliminate it at a high response speed [6].

Several protection methods have been proposed in previous studies. These methods can be classified into three generic concepts: 1) Implementation of the converter with DC fault self-clearance capability, such as full-bridge MMC. This method requires shutting down the converter station for a long time. In addition, the uncontrollable high current rushing through freewheeling diodes into the DC side may damage the semiconductor devices of the HB-MMC; 2) Implementation of unit protection or double-ended pilot protection with a DC circuit breaker (DCCB). When the transmission line is long, time delay caused by communications and signal errors may restrict the protection's response speed. Therefore, unit protection is suitable for the backup protection of DC grids to overcome the defects of non-unit protection, such as tolerance of considerable transition resistance and impossible protection of full line length; 3) Implementation of a non-unit or single-end protection scheme with a DCCB. The proposed method is a non-unit protection method, which only utilizes the single-end information to protect the DC grid.

To realize a high-speed protection, an effective method for distinguishing internal and external, positive and negative faults is necessary to be studied. In this paper, a single-end information based protection is proposed. The positive and negative pole currents are measured and decomposed into common-mode (CM) and differential-mode (DM) currents. After that, the CM currents are considered as x- coordinate, and the increment of DM currents are viewed as y- coordinate. In this manner, the internal and external positive ground fault (PGF), negative ground fault (NGF), and positive-to-negative fault (PNF) will be divided on a protection phase-plane of different regions. In [7], this "phase-plane" is constituted with the increment of double-end currents of DC transmission line. This method can be realized to recognize DC faults within 1 ms and has a specific ability to resist noise and transition resistance. However, the approach to distinguish the PGF and NGF is not investigated. In addition, the protection action may be extended when the DC transmission line is long. The contributions of this study are as follows:

1) The synthesized CM and DM circuits of single-end MMC based DC grid are derived. Then, the corresponding expressions of CM and DM currents under different faults are obtained. These circuits and illustrations provide a useful perspective to analyze the fault current characteristics and recognize DC faults.

2) A single-end protection method based on the CM and DM components of the protection phase plane is proposed. This method can recognize different faults within 1 ms, and it is less affected by the fault transition resistance or other parameters in the fault loop.

I. Modulus current analysis of DC grid under different fault types

MMC can be equivalent to an RLC series circuit, as shown in Fig. 1. Here, L_m and R_m are the equivalent inductance and resistance of its upper and lower bridge arms. The equivalent capacitance (C_{eq}) of each-phase bridge arm can be obtained according to the conservation of capacitive energy storage, as illustrated in Eq. (1).

$$2N \times \frac{1}{2}C_{sm}U_{sm} = \frac{1}{2}C_{eq}(NU_{sm})^2 \tag{1}$$

where N is the number of submodules input in each phase bridge arm; C_{sm} is submodular capacitance; U_{sm} is submodular capacitor voltage; According to Eq. (1), $C_{eq}=2C_{sm}/N$. For the equivalent RLC series circuit, $R_a=2R_m/3$, $L_a=2L_m/3$ and $C_a=3C_{eq}$.

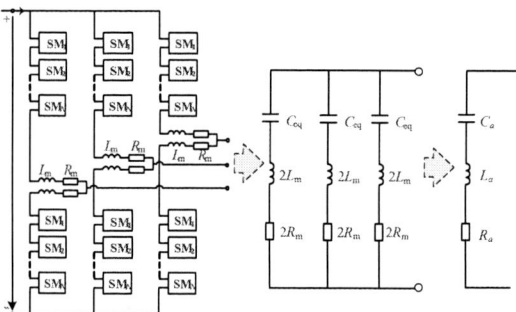

Fig. 1 Equivalent circuit of MMC.

If a PGF occurs, the equivalent circuit of the MMC based DC grid is shown in Fig. 2. This equivalent circuit is challenging to transform into CM and DM circuits due to it involves many elements. Therefore, this circuit is divided into different parts, including converter, DC transmission line, and fault transition resistance. Their CM and DM circuits are respectively solved. Finally, these CM and DM circuits will be synthesized to derive the CM and DM fault current expressions.

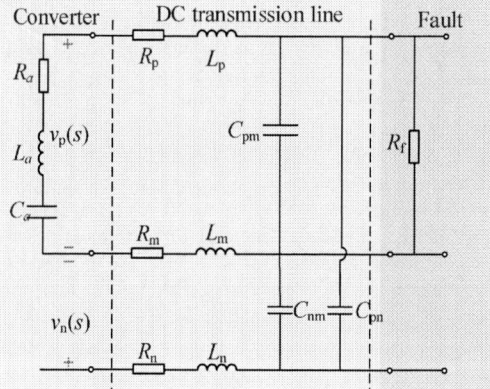

Fig. 2 Equivalent circuit of single-end MMC based DC grid under PGF.

The positive and negative pole voltages and currents can be transformed into CM and DM components via modulus decomposition, which satisfy

$$\begin{bmatrix} X_0 \\ X_1 \end{bmatrix} = A \begin{bmatrix} X_p \\ X_n \end{bmatrix} = \frac{1}{2}\begin{bmatrix} 1 & 1 \\ 1 & -1 \end{bmatrix}\begin{bmatrix} X_p \\ X_n \end{bmatrix} \tag{2}$$

The reverse transformation of Eq. (2) is

$$\begin{bmatrix} X_p \\ X_n \end{bmatrix} = A^{-1}\begin{bmatrix} X_1 \\ X_2 \end{bmatrix} = \begin{bmatrix} 1 & 1 \\ 1 & -1 \end{bmatrix}\begin{bmatrix} X_1 \\ X_2 \end{bmatrix} \tag{3}$$

where X denotes the voltage or current; subscript 0 represents the CM component; index 1 represents the DM component; subscript p is the positive component; subscript n is the negative component.

Through the synthesis of the derived CM and DM circuits, the equivalent circuits are shown in Fig. 3. These circuits are mostly a capacitor-discharging loop, allowing the DC fault current and voltage to be calculated analytically under different fault types. The expressions of CM and DM fault current is derived following Fig. 3, which satisfy

$$
\begin{cases}
i_1(s) = i_0(s) = \dfrac{U_d(s)}{Z_a(s) + s(L_{l1} + L_{l0}) + (R_{l1} + R_{l0} + 2R_f)} \Rightarrow \text{PGF} \\[3mm]
i_1(s) = -i_0(s) = \dfrac{U_d(s)}{Z_a(s) + s(L_{l1} + L_{l0}) + (R_{l1} + R_{l0} + 2R_f)} \Rightarrow \text{NGF} \\[3mm]
i_0(s) = 0, i_1(s) = \dfrac{U_d(s)}{Z_a(s) + sL_{l1} + (R_{l1} + R_f)} \Rightarrow \text{PNF}
\end{cases} \tag{4}
$$

Where PGF represents a positive pole to ground fault, NGF represents a negative pole to ground fault. PNF represents a positive pole to the negative pole fault. i_0 is a common mode (CM) current, i_1 is the differential mode (DM) current. Following Eq. (4), the expression of i_0 and i_1 under PGF are

$$
i_0(t) = i_1(t) = -\frac{1}{\sin\theta} i_{dc}(0) e^{-\frac{t}{\tau}} \sin(\omega_n t - \theta) + \frac{U_{dc}(0)}{R_e} e^{-\frac{t}{\tau}} \sin(\omega_n t) \tag{5}
$$

Where $i_{dc}(0)$, $U_{dc}(0)$ is the initial current and voltage during DC fault, and other parameters satisfy. It can be seen from Eq. (5) that the ratio of i_1 and i_0 is kept as a constant 1 under a PGF, and keep as -1 under an NGF. This ratio is not affected by the fault transition resistance or inductance in the fault loop. Besides, when an external fault occurs, the direction of i_1 is contrary to the internal fault. Moreover, i_0 under external PGF and NGF also have opposite directions. These characteristics could be used to protect the DC gird.

$$
\begin{cases}
\tau = \dfrac{4L_m + 6(L_{sr} + L_{l1} + L_{l0})}{2R_m + 3(R_{l1} + R_{l0} + 2R_f)} \\[4mm]
\omega_n = \sqrt{\dfrac{2N(2L_m + 3(L_{sr} + L_{l1} + L_{l0})) - C_{sm}(2R_m + 3(R_{l1} + R_{l0} + 2R_f))^2}{4C_{sm}(2L_m + 3(L_{sr} + L_{l1} + L_{l0}))^2}} \\[4mm]
\theta = \arctan(\tau\omega) \\[4mm]
R_{eq} = \sqrt{\dfrac{2N(2L_m + 3(L_{sr} + L_{l1} + L_{l0})) - C_{sm}(2R_m + 3(R_{l1} + R_{l0} + 2R_f))^2}{36C_{sm}}}
\end{cases} \tag{6}
$$

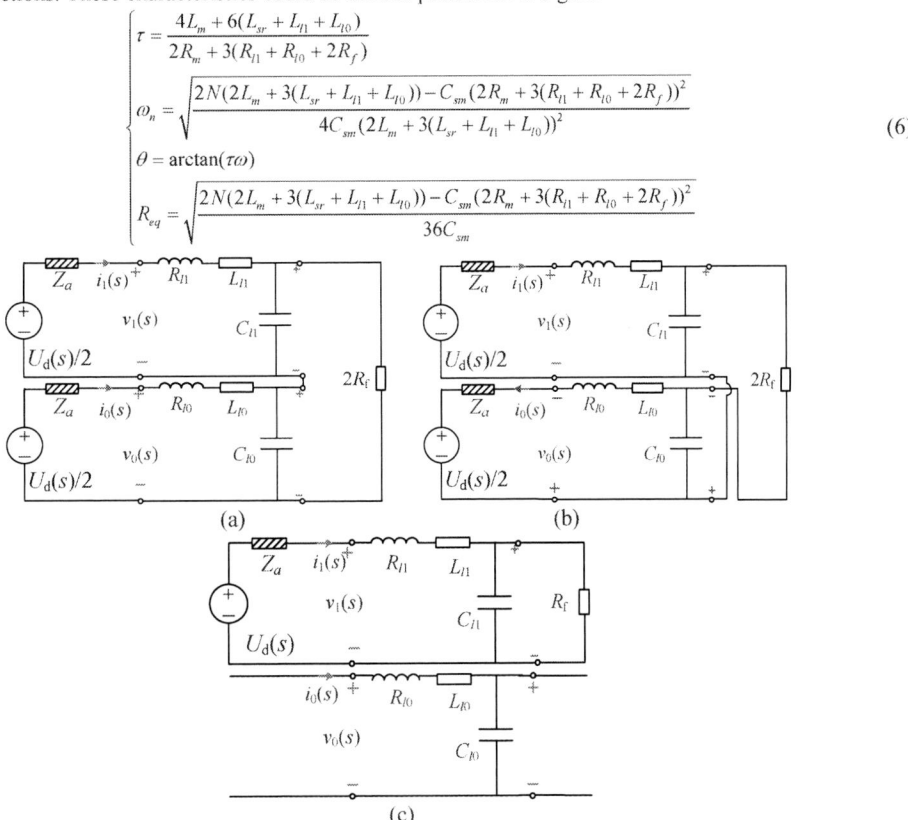

Fig. 3 Synthesized CM and DM circuits under different fault types: (a) PGF, (b) NGF, (c) PNF.

II. Protection method

A) Protection start-up criterion

When a DC fault occurs, the change rate (k) of fault current can reach 3–4 kA/ms under PNF. However, k is close to 0 under regular operation. Therefore, k is taken as the criterion for protection start-up, which is expressed as

$$
k = \frac{1}{n\Delta t}\left(\sum_{j=1}^{N}\left|i_{j+1} - i_j\right|\right) \tag{7}
$$

where i_j is the current data of the jth sampling point; Δt is the sampling interval; n is the number of sampling data points within a sampling data window. When k exceeds the threshold g (set as 0.5 kA/ms in this study), the protection is activated.

B) *Fault recognition method based on a protection phase plane*

On the basis of the synthesized CM and DM circuits and the derived expressions of CM and DM currents, the criterion of internal and external PGF, NGF, and PNF is shown in Eq. (5). Here, k_{set1} is the threshold of an internal fault. The current change rate of external fault is relatively lower than the internal fault. Therefore, its limit is set as k_{set2} rather than k_{set1}. k_{set1} and k_{set2} are satisfied using the proportional relationship. k_{set3} is the threshold of asymmetrical fault, namely When $|i_0|<k_{set3}$, this fault is then recognized as a PNF (symmetrical fault). When an internal PGF occurs, Δi_1 is more significant than k_{set1}, and i_0 is more exceptional than k_{set3}; when an internal NGF occurs, Δi_1 is greater than k_{set1}, and i_0 is smaller than $-k_{set3}$.

$$\begin{cases} (\Delta i_1 > k_{set1}) \cap (i_0 > k_{set3}) \Rightarrow \text{Internal PG F} \\ (\Delta i_1 > k_{set1}) \cap (i_0 < -k_{set3}) \Rightarrow \text{Internal NG F} \\ (\Delta i_1 < -k_{set2}) \cap (i_0 > k_{set3}) \Rightarrow \text{External PG F} \\ (\Delta i_1 < -k_{set2}) \cap (i_0 < -k_{set3}) \Rightarrow \text{External NG F} \\ (\Delta i_1 > k_{set1}) \cap (|i_0| < k_{set3}) \Rightarrow \text{Internal P N F} \\ (\Delta i_1 < -k_{set1}) \cap (|i_0| < k_{set3}) \Rightarrow \text{External P N F} \end{cases} \qquad (8)$$

The corresponding expressions of k_{set1}, k_{set2}, and k_{set3} are defined as

$$\begin{cases} k_{set1} = [\Delta i_1]_{op} = 2(k_{rel}I_N - I_N) \\ k_{set2} = k_e[\Delta i_1]_{op} = k_e k_{set1} \\ k_{set3} = [i_0]_{op} = 5\%I_N \end{cases} \qquad (9)$$

where the subscript op denotes the action value of protection; k_{rel} is the reliability factor, which is set as 1.2 in this paper; k_e is the ratio of internal and external fault current, which is set as 0.25.

If i_0 is taken as the x-coordinate, and Δi_1 is taken as the y-coordinate, then different faults are divided on a protection phase plane. This protection phase plane is shown in Fig. 4. The corresponding boundary conditions are consistent with Eqs. (8) and (9). All DC fault types are included in Fig. 4, and clear boundaries can be observed among different faults. Moreover, the fault recognition method is slightly affected by noise and transition resistance because of the sufficient margin among different limits. When the sampling frequency and fault recognition time is increased, the difference between various faults in this phase plane is significant. Such a large difference is beneficial for fault recognition.

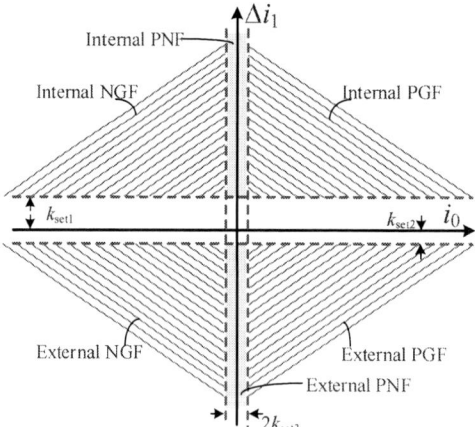

Fig. 4 Protection phase plane.

C) *Protection procedure*

Fig. 5 illustrates the protection procedure. After the condition of protection start-up is satisfied, fault detection is conducted to confirm whether a DC fault has occurred. Fault detection can be implemented using low-voltage terms. If the fault detection condition is satisfied, then the fault recognition component will work. The internal and external faults are then recognized in this link. Afterward, fault pole selection is implemented to distinguish PGF, NGF, and PNF.

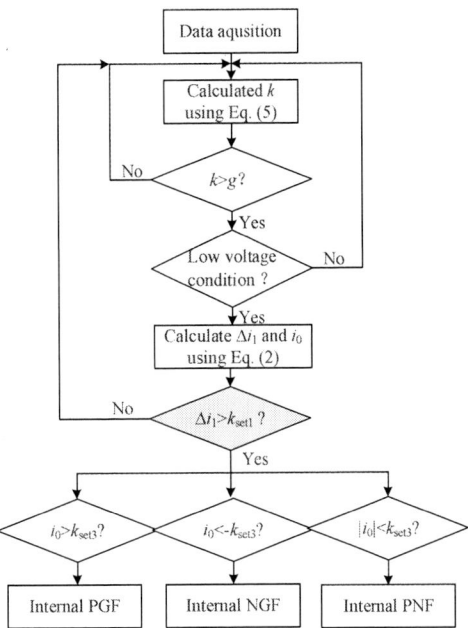

Fig. 5 Protection procedure.

III. Simulation and Experiment Results

Simulation method

A four-end MMC-based DC grid simulation model is established on the basis of PSCAD/EMTDC. The structure of the simulation model is shown in Fig. 6, and the corresponding circuit parameters are listed in Tab. I. Here, L_{dc} represents the smoothing reactor, whose value is 150 mH. L_1–L_4 represent the DC transmission line. The corresponding distance is displayed in Fig. 6. DCCB is placed at both ends of each DC transmission line. F_1 represents internal fault; F_2 and F_3 represent external faults. The blue arrow represents the power flow direction of the transmission system during regular operation.

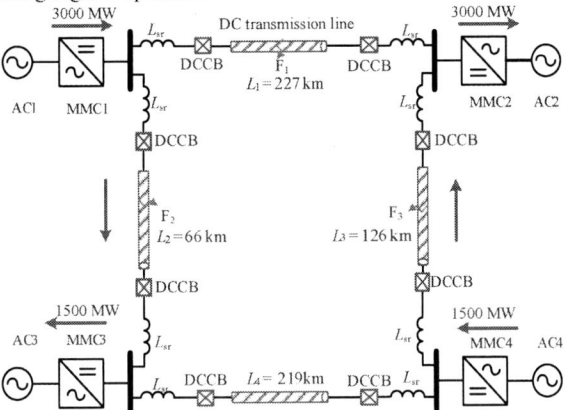

Fig. 6 Structure of the simulation model.

TABLE I

Circuit parameters of simulation

Parameters	MMC1	MMC2	MMC3	MMC4
Rated capacity/MW	3000	3000	1500	1500
Rated DC voltage/kV	±500	±500	±500	±500
Rated AC voltage/kV	220	220	500	500
Ratio of transformer	230/260	525/260	230/260	525/260
Leakage reactance /p.u.	0.15	0.15	0.15	0.15
Bridge arm inductance/mH	44	88	44	88
Submodule capacitance /mF	15	15	7	7
Number of submodule	233	233	233	233
Polarity control cycle /μs	50	50	50	50

The fault types are set as internal PGF, and the fault locations are 100 km away from MMC1. The trajectory in the protection phase plane and the fault pole current under different L_{sr}, R_f and fault current limiters are displayed in Fig. 7. Here, MOA represents a metal oxide arrester. The withstand voltage of MOA is 1500 kV. It can be seen that the recognition results of the proposed method under different L_{sr}, R_f are the same, and The discriminant delay due to parameter changes is significantly reduced. If the fault pole current is used to recognize the fault types (the threshold is set as 4kA), there is an inevitable delay for the protection. Due to the protection method is less affected by the circuit parameters, the fault current limiter can be activated earlier than DCCB, which is beneficial for the fault current inhibition. Besides, the simulation results indicate that the proposed protection method does not depend on the boundary formed by the smoothing reactor, and is more applicable in different DC grids.

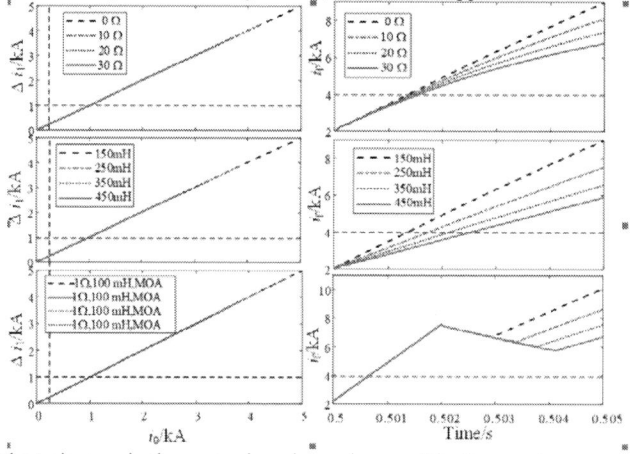

Fig. 7 Comparison of the trajectory in the protection phase plane and fault current.

In order to further eliminate the influence of transition resistance, we further proposed a protection plane based on the differential of i_0 and i_1, namely di_0 and di_1. $di_0=|i_{0j}+ i_{0j+1}|$, $di_1=|i_{1j}- i_{1j+1}|$。 Here, i_{0j} or i_{1j} is the current data of the jth sampling point. The corresponding protection phase plane is shown in Fig. 8. For the internal fault, the threshold is set as 0.5kA/ms. When there are 5 points exceeds the threshold, the recognition result will be sent to the protection system. It can be seen from Fig. 8 that the first point enters into the protection phase plane has the maximum value. This is because, at the beginning of the DC fault, the fault current is small. Therefore, the reverse on the transition resistance is small. Therefore, the differential of i_0 and i_1 has the maximum value.

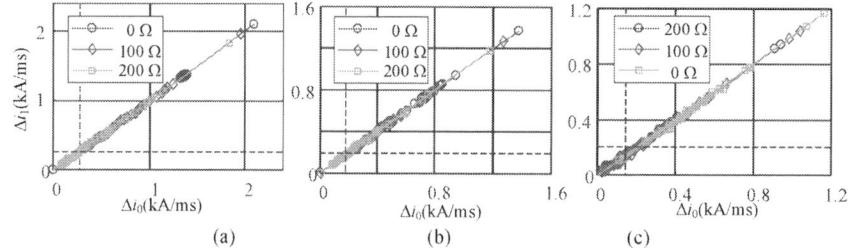

(a)   (b)   (c)

Fig. 8 The protection phase plane of internal PGF under different fault locations. (a) d=0%, (b) d=50%, (c) d=100%.

Experiment method

The experimental platform of a bipolar DC grid is constructed to verify the aforementioned analysis, as depicted in Fig. 9. The parameters are listed in Table II.

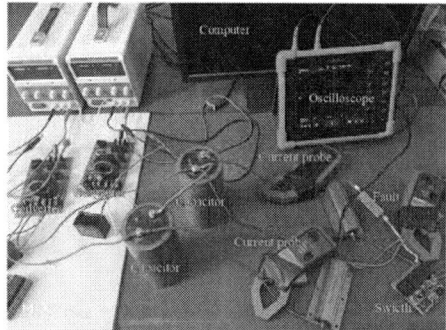

Fig. 9 The structure of experiment platform.

TABLE II
CIRCUIT PARAMETERS OF EXPERIMENT

Parameters	positive pole	negative pole
DC voltage/V	10	10
DC load/Ω	20	20
Capacitor/uF	5600	5600
inductance/mH	5	5
Fault resistance/Ω	5	5
Maximum current/A	5	5
controller	DSP28335	

Fig. 10 (a) illustrates the experiment results under PGF. Under normal operation, the positive and negative currents are both 500mA. When the PGF is turned on, the positive current rises rapidly, and the maximum current can reach 6A. This maximum current is closely related to the capacity of the capacitor. After 2~3 ms, this bipolar system will enter into another steady state, and the stable current is 2 A. This is very similar to HB-MMC. Fig. 10(a) also shows the waveform of i_0 and i_1, which is consistent with previous analysis. The 2ms current data after the fault is used to draw the protection phase plane, which is demonstrated in Fig. 10(b). When the sampling frequency is set as 20 kHz, 40 sampling data will be obtained. According to Eq. (5), k_{set1}=0.2 A, k_{set2}=25 mA. It can be seen that N_1=2, N_2=38. This means that the DC fault will be recognized within 1 ms. Fig. 10(c) illustrates the experiment results under internal NGF. It can be seen that the difference between PGF and NGF is the polarity of i_0. When an NGF occurs, i_0 is negative. Other results are consistent with PGF.

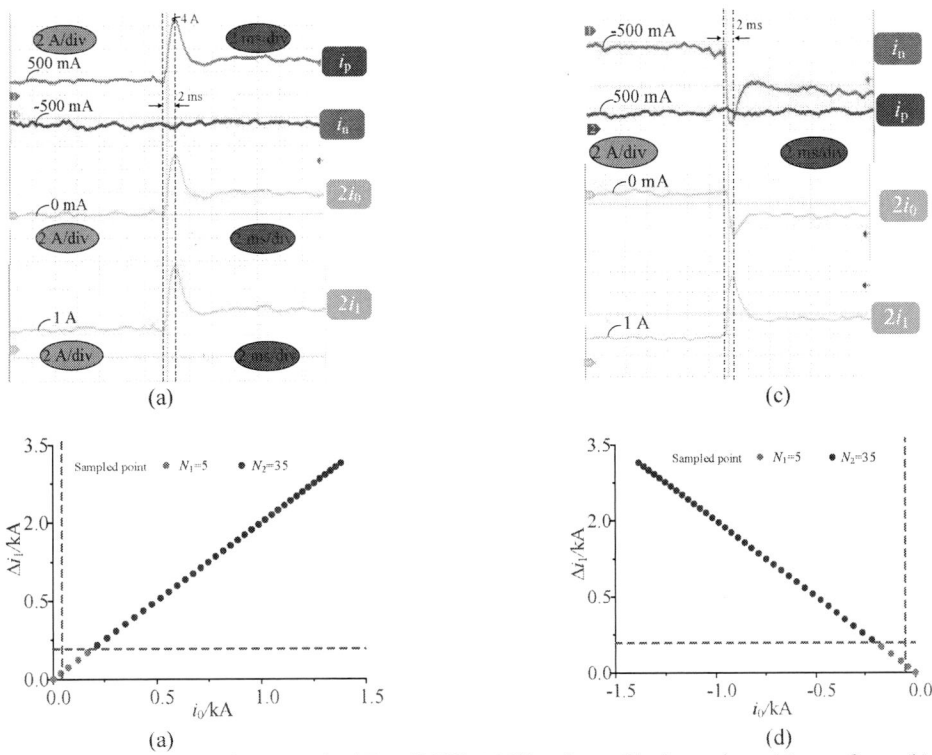

Fig. 10 Experiment results under internal PGF and NGF: (a) Waveform; (b) phase-plane. (a) waveform; (b) phase-plane.

IV. Conclusion

In this paper, a DC grid protection method based on phase-planes of single-end common-mode and differential-mode components is proposed. When an internal fault occurs, its differential-mode current is positive, whereas it is negative for an external fault. When a PG fault occurs, its common-mode current is positive, and it is negative for the NG fault. If the common-mode current is taken as x-coordinate, and the increment of differential-mode current is made as y-coordinate, different faults can be divided into different regions on a phase-plane. This phase plane can be utilized for fault recognition. This protection method is simple and has high reliability. Moreover, it only uses single-end information, so the protection can realize fault recognition within 1~2ms.

References

[1] K. A. Saleh, A. Hooshyar and E. F. El-Saadany, "Hybrid Passive-Overcurrent Relay for Detection of Faults in Low-Voltage DC Grids," in IEEE Transactions on Smart Grid, vol. 8, no. 3, pp. 1129-1138, May 2017.

[2] J. D. Páez, D. Frey, J. Maneiro, S. Bacha and P. Dworakowski, "Overview of DC–DC Converters Dedicated to HVdc Grids," in IEEE Transactions on Power Delivery, vol. 34, no. 1, pp. 119-128, Feb. 2019.

[3] R. Majumder, S. Auddy, B. Berggren, G. Velotto, P. Barupati and T. U. Jonsson, "An Alternative Method to Build DC Switchyard With Hybrid DC Breaker for DC Grid," in IEEE Transactions on Power Delivery, vol. 32, no. 2, pp. 713-722, April 2017.

[4] L. Mackay, E. Vandeventer and L. Ramirez-Elizondo, "Circulating Net Currents in Meshed DC Distribution Grids: A Challenge for Residual Ground Fault Protection," in IEEE Transactions on Power Delivery, vol. 33, no. 2, pp. 1018-1019, April 2018.

[5] Mokhberdoran A , Carvalho A , Silva N , et al. Design and implementation of fast current releasing DC circuit breaker. Electric Power Systems Research, 2017, 151:218-232.

[6] Li S , Zhao C , Xu J . A new topology for current-limiting solid-state HVDC circuit breaker. In: Power Electronics Conference. IEEE, 2017:1-6.

[7] R. Mohanty and A. K. Pradhan, "A Superimposed Current Based Unit Protection Scheme for DC Microgrid," in IEEE Transactions on Smart Grid, vol. 9, no. 4, pp. 3917-3919, July 2018.

Comparative assessment of voltage modulation methods for asymmetric six-phase machines

R. S. Kanchan	Omer Ikram ul Haq	Luca Peretti
CORPORATE RESEARCH	CORPORATE RESEARCH	DEPT. OF ELECTRICAL
ABB AB	ABB AB	POWER AND ENERGY
Västerås, Sweden	Västerås, Sweden	SYSTEMS
Tel.: +46 (0)21-345175	Tel.: +46 (0)21-324249	KTH Royal Institute of
rahul.kanchan@se.abb.com	omer.ikramulhaq@se.abb.com	Technology
		Stockholm, Sweden
		lukap@kth.se

Keywords

«AC machine», «Multiphase drive», «Modulation strategy», «Harmonics», «Electrical drive».

Abstract

In the most common industrial adaptation of electric drives for asymmetric six-phase machines, two independent three-phase converter units are configured to supply two three-phase winding sets. In principle, the converter units may use their own internal clock, acting independently on the two sets of windings with both the voltage pulse-width modulation and the sampling of phase currents. In the case, the machine is controlled with a vector space decomposition approach, this may cause some synchronization issues with voltage references and current sampling.

In this paper, an asymmetric six-phase permanent-magnet machine is controlled with a vector space decomposition architecture has been controlled with individual current controllers in each harmonic plane. The output voltage references from the current controllers is fed to pulse-width modulators. The modulators are selected among the most promising methods for two-level converter-fed six-phase machines from the literature. A comparative analysis is performed through simulation studies, with the objective of comparing the performance of various modulators considering synchronization of voltage reference signals and phase current sampling.

Introduction

A conventional multi-phase power converter consists of a n number of legs equal to the number of machine phases connected to it. While it is common to have a single common star point in odd-phase machines like five- or seven-phase machines, machines with n multiple of three phases could be designed differently. Their windings can be grouped into symmetrical three-phase winding sets, each having an individual star point, and each set shifted by π/n with respect to the a nearby set. The most common machine of this type is the asymmetrical six-phase machine, designed with two sets of three-phase windings with $30°$ phase shift between them. When the sets are star-connected, the machine is also known as a YY30 machine, and it is usually fed by n parallel three-phase converters. A general schematic of the phase winding of a YY30 machine and n three-phase converters arrangement is shown in Fig. 1.

The common modelling approaches used for multi-phase machine analysis and control are the dual three-phase (DTP) modelling approach [1] and the vector space decomposition (VSD) [2]. The DTP approach originate from the modelling of traditional three-phase machines, and its extension to six-phase machines with separated star points. Each set of three-phase windings is represented in its $d-q$ coordinate system with the torque equation representing sum of the torque contributions from each set. The VSD approach, on the contrary, involves transforming the machine equations into a number of orthogonal subspaces by using appropriate transformation matrix, and it considers the machine as a whole object with no parallel of three-phase systems.

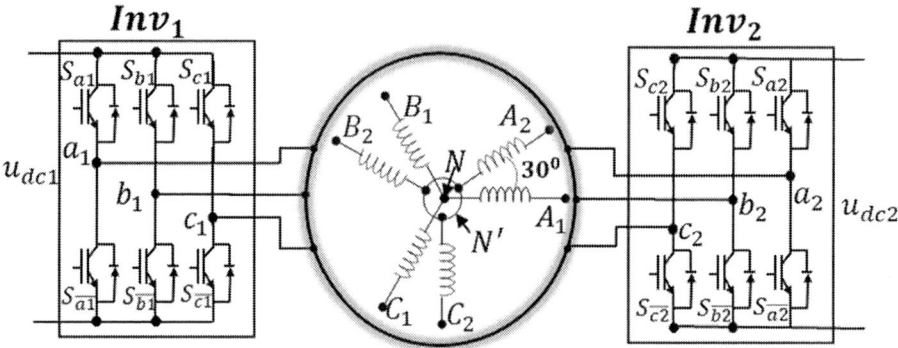

Fig. 1. Electrical connection arrangement for asymmetrical six-phase drive.

A generalized current control structure based on VSD theory is shown in Fig. 2 for six-phase machines. The closed loop control structure is composed of co-ordinate transformations to convert the control variables in different reference frames, current control loops in various subspaces rotating at different speeds, assisted by feed-forward compensation or decoupling blocks per each loop. The control variables appear in orthogonal pairs per each reference frame. Based on the individual switching states $S_{a1}, ..., S_{c2}$ of the inverter legs in Fig. 1, the combined six-phase voltage vector can be defined by a binary number $S_{a1} S_{b1} S_{c1} S_{a2} S_{b2} S_{c2}$ where a total of $2^6 = 64$ permutations are possible. By utilizing the VSD transformation $T_{abc \to \alpha\beta0}$ as described in (1) [2] [3], these inverter voltage vectors can be transformed into three two-dimensional orthogonal subspaces $\alpha\beta_1, \alpha\beta_3, \alpha\beta_5$, each containing (among others) the fundamental, 3rd and 5th harmonic frequency, respectively.

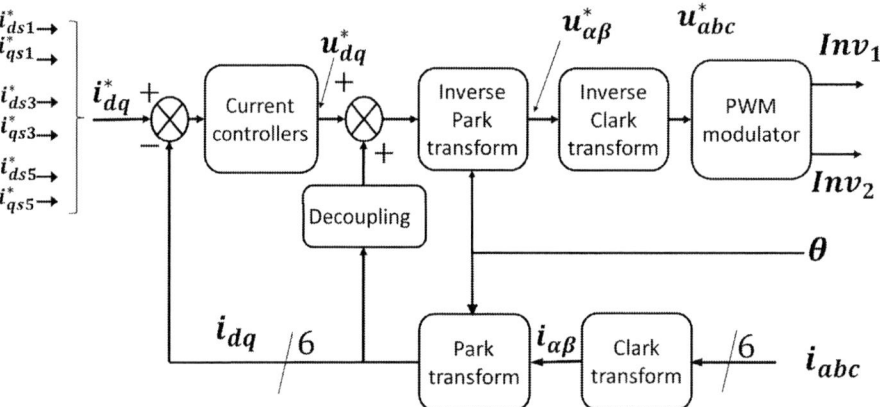

Fig. 2. VSD based current control structure.

$$
\begin{bmatrix} v_{\alpha1} \\ v_{\beta1} \\ v_{\alpha3} \\ v_{\beta3} \\ v_{\alpha5} \\ v_{\beta5} \end{bmatrix} = \underbrace{\begin{bmatrix} 1 & -1/2 & -1/2 & \sqrt{3}/2 & -\sqrt{3}/2 & 0 \\ 0 & \sqrt{3}/2 & -\sqrt{3}/2 & 1/2 & 1/2 & -1 \\ 1 & 1 & 1 & 0 & 0 & 0 \\ 0 & 0 & 0 & 1 & 1 & 1 \\ 1 & -1/2 & -1/2 & -\sqrt{3}/2 & \sqrt{3}/2 & 0 \\ 0 & -\sqrt{3}/2 & \sqrt{3}/2 & 1/2 & 1/2 & -1 \end{bmatrix}}_{T_{abc \to \alpha\beta0}} \begin{bmatrix} v_{a1} \\ v_{b1} \\ v_{c1} \\ v_{a2} \\ v_{b2} \\ v_{c2} \end{bmatrix} \qquad (1)
$$

The voltage space vector orientation of the fundamental and 5th harmonic planes for all possible switching permutations is shown in Fig. 3. If a YY30 machine with isolated star points is considered, no voltage components from the 3rd harmonic subspace are of interest. Moreover, in YY30 machines with sinusoidally-distributed windings, only the fundamental subspace components contribute to torque production, while the voltage component in 5th harmonic subspace should be kept to zero to avoid additional, unnecessary losses. However, in converter-fed machines, the pulse-width modulation nature of the voltage induces harmonic currents, whose magnitude is dependent on the equivalent impedance in the respective harmonic planes (stator resistance and leakage inductance) [2].

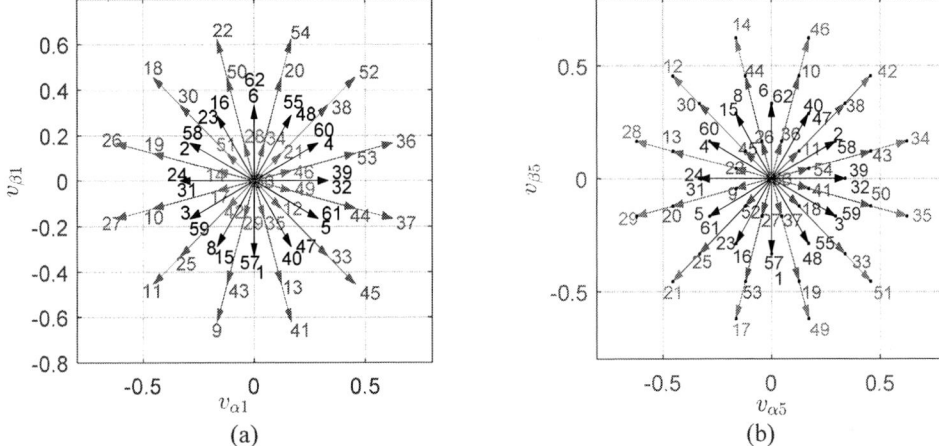

(a)   (b)

Fig. 3. Voltage space vector representation in (a) the $\alpha\beta_1$ subspace, and (b) the $\alpha\beta_5$ subspace.

Voltage modulation in multi-phase drives

Reviews of pulse-width modulation (PWM) methods for multi-phase converters are available [4] [5] [6]. In this section, a more specific perspective on the YY30 case is taken. From the analysis of Fig. 3, a possible classification of the voltage vectors depend upon their magnitude in the $\alpha\beta_1$ subspace. Thus, four main categories can be identified: large (L), large-medium (B), medium (M) and small (S) voltage vectors. The application of L vectors produces the smallest voltage magnitude in the $\alpha\beta_5$ subspace, and vice-versa. Whereas, the magnitude of B and M voltage vectors is the same in both $\alpha\beta_1$ and $\alpha\beta_5$, although the orientation of the vectors is different in the two subspaces.

Unlike the three-phase case, the problem of selecting a suitable voltage vector is multi-planar. The main challenge of a multi-phase modulator connected to a YY30 machine with sinusoidally distributed windings is therefore to produce the output voltage waveforms dictated by $\alpha\beta_1$ voltage references (those producing torque), while at the same time avoiding the production of large current harmonics in $\alpha\beta_5$. However, due to the independence of each current controller in the different subspaces, it is unlikely that the reference voltage vector in $\alpha\beta_1$ and $\alpha\beta_5$ subspaces will occupy the same sectors. Thus, it is very difficult to satisfy the demands of both $\alpha\beta_1$ and $\alpha\beta_5$ at the same time.

This problem has been approached differently throughout the years. Sometimes, in its simplest form, the chosen modulator is a carrier-based PWM (CBPWM) type, where the reference voltage signals are compared with triangular carriers to generate the PWM signals. However, already in [2] there was a proposal of a six-phase space vector PWM (SVPWM), where only the L vectors were used. This section presents an overview of the most promising CBPWM and SVPWM methods which are used in the analysis presented in later sections of this paper.

Carrier-based double zero-sequence injection modulation

A formulation of CBPWM for YY30 machines has been proposed by [7], referred here as "double zero-sequence injection" (DZSI). The DZSI consists of two separate three-phase modulators having their own reference and zero-sequence computation [7]. The voltage reference from the current controllers is first converted into independent three-phase references, where the voltage reference to the second modulator is generated by a 30° vector rotation to the input reference of a six-phase modulator. The reference signals are added with zero-sequence voltages calculated separately for each three-phase modulator, according to (2) where $x = \{1,2\}$.

$$v_{cmx} = -\left(\frac{max(v_{ax}^*, v_{bx}^*, v_{cx}^*) + min(v_{ax}^*, v_{bx}^*, v_{cx}^*)}{2}\right) \tag{2}$$

The carrier comparison directly generates the switching sequences for the inverter legs and no separate dwell time calculations or lookup tables for selection of vectors is required. The schematic of the DZSI is shown in Fig. 4. This is the simplest implementation of a YY30 voltage modulator, and it is also extendable for asymmetric n-phase machines where n is a multiple of three.

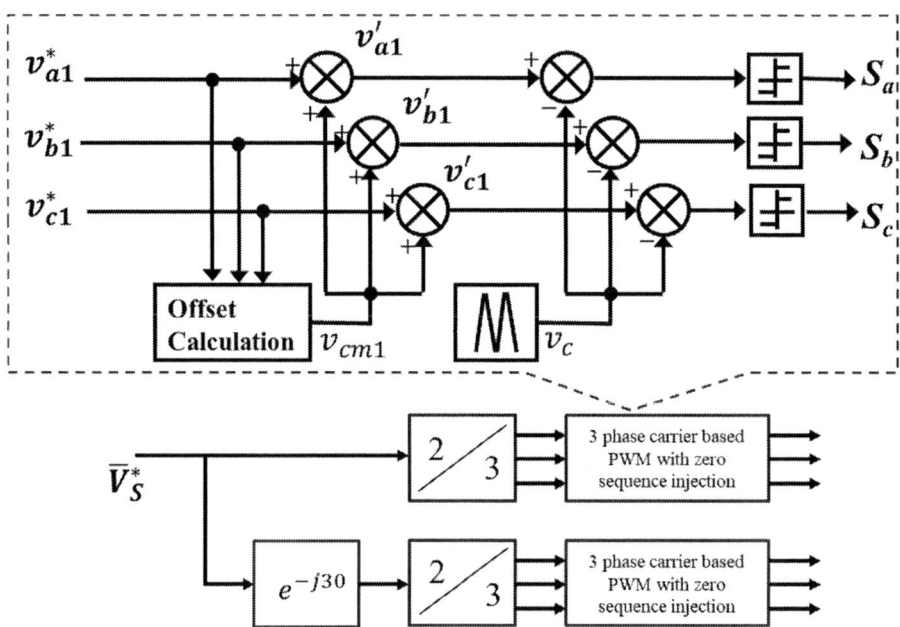

Fig. 4. Schematic of the DZSI modulation method *[7]*.

Space vector PWM for asymmetric six-phase converter

The initial attempt to implement a space vector-based modulation technique was found in [8], where the fundamental frequency plane is divided into 12 sectors separated by the 12 L vectors. The PWM strategy consists of switching two large vectors adjacent to the reference vector and zero vectors. The dwell times are calculated based on the volt-second balance during the switching interval. The PWM scheme only considers the volt-second balance in the $\alpha\beta_1$ subspace, leaving large harmonic currents to flow in the machine due to voltages generated in $\alpha\beta_5$ plane.

The following work [2] proposed a space vector-based modulation method based on the VSD theory (referenced hereafter as 12SSVPWM), aiming to improve the harmonic performance of the method proposed in [8]. Since all L voltage vectors in $\alpha\beta_1$ map into small vectors in the $\alpha\beta_5$ subspace, the harmonics in $\alpha\beta_5$ plane can be controlled to great extent if the SVPWM is formulated using only these L vectors. In particular, the average cumulative volt-second generated by these vectors must equal the volt-second of the reference vectors from the $\alpha\beta_1$ and $\alpha\beta_5$ subspaces, which sets four constraints. An additional constraint comes from the fact that the sum of the dwell times of all vectors must equal switching interval T_s. The minimum number of vectors required to fulfil these constraints is therefore five, as described in (3).

In (3), $v_{\alpha1}^k, v_{\beta1}^k$ and $v_{\alpha5}^k, v_{\beta5}^k$ with $k = 1, ..., 5$ are the projections of the kth vectors on the $\alpha\beta_1$ and $\alpha\beta_5$ planes. The expression (3) can be solved to obtain the dwell times of the voltage vectors but it does not reveal which five vectors shall be selected. There can be many possibilities to pick voltage vectors at random which satisfy (3). In [2], four L vectors adjacent to reference voltage vector in $\alpha\beta_1$ subspace are selected together with one zero vector, as shown in Fig. 5(a).

$$
\begin{bmatrix}
v_{\alpha 1}^1 & v_{\alpha 1}^2 & v_{\alpha 1}^3 & v_{\alpha 1}^4 & v_{\alpha 1}^5 & v_{\alpha 1}^6 \\
v_{\beta 1}^1 & v_{\beta 1}^2 & v_{\beta 1}^3 & v_{\beta 1}^4 & v_{\beta 1}^5 & v_{\beta 1}^6 \\
v_{\alpha 5}^1 & v_{\alpha 5}^2 & v_{\alpha 5}^3 & v_{\alpha 5}^4 & v_{\alpha 5}^5 & v_{\alpha 5}^6 \\
v_{\beta 5}^1 & v_{\beta 5}^2 & v_{\beta 5}^3 & v_{\beta 5}^4 & v_{\beta 5}^5 & v_{\beta 5}^6 \\
1 & 1 & 1 & 1 & 1 & 1
\end{bmatrix}
\begin{bmatrix}
T_1 \\ T_2 \\ T_3 \\ T_4 \\ T_s
\end{bmatrix}
=
\begin{bmatrix}
v_{\alpha 1}^* T_s \\ v_{\beta 1}^* T_s \\ v_{\alpha 5}^* T_s \\ v_{\beta 5}^* T_s \\ T_s
\end{bmatrix}
\tag{3}
$$

This technique needs additional logic for sector identification, and accordingly a lookup table for the selection of the active vectors. The sequence of the active voltage vectors and the selection of zero vectors is done such that minimum switching transitions occur during the switching interval. The zero vectors are split equally to be placed at the beginning and end of the switching interval, resulting into six voltage vectors and five vector transitions in a switching interval. This causes simultaneous switching in two inverter legs at least once in a sampling interval. Note, however, that the 12SSVPWM can be conveniently applied for both situations when sinusoidal output is demanded in the $\alpha\beta_1$ plane (zero reference voltages in $\alpha\beta_5$ plane), but also when harmonic injection in $\alpha\beta_5$ plane is demanded, for example to achieve torque enhancement with non-zero reference voltages in $\alpha\beta_5$ plane. This is achievable only in machines with non-sinusoidal back EMF [5].

24-sector space vector PWM

An improvement of the method in [2] was presented in [9], mainly targeting at reducing the switching frequency of the PWM modulator. The $\alpha\beta_1$ voltage vector plane is now divided into 24 sectors (thus, the method is hereafter called 24SSVPWM), due to introduction of M voltage vectors placed in the middle of the L vectors. The PWM formulation is based on the use of three L voltage vectors and one M voltage vector in a switching period. As an example, Fig. 5(b) shows a case when the reference voltage vector is in sector 1.

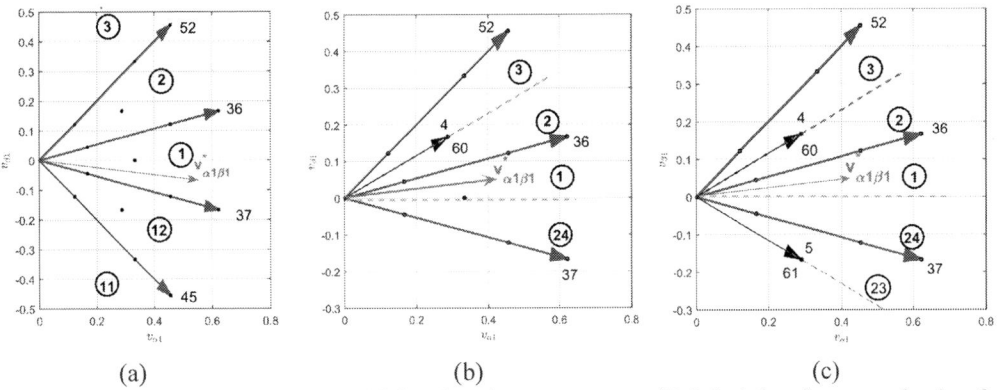

Fig. 5. (a) Example of 12SSVPWM utilizing four large L vectors. (b) Principle of vector selection for 24SSVPWM. (c) Principle of vector selection for 24SSVPWM$_M$.

Voltage vectors 36, 37, 52 and 4 or 60 are used together with zero voltage vectors to generate the PWM sequences. Since there are two switching combinations which can be used to generate the M vector (4 or 60), the selection of correct switching combination is done considering minimal switching transitions while building up the sequence of these voltage vectors. The selection of the zero voltage vectors is also done similarly to avoid double switching during the switching interval. The calculation of dwell times for switching vectors is similar to the 12SSVPWM, while the sector identification requires additional logic to identify 24 sectors instead of 12. The lookup tables now include three L and one M voltage vectors.

It shall be noted that the M voltage vectors also generate equal amplitude voltages in $\alpha\beta_5$ subspace. However, the harmonics flux analysis of the 24SSVPWM has better performance at lower modulation index range compared to 12SSVPWM, depending on the equivalent leakage inductance in the subspaces. The higher the fundamental equivalent leakage inductance as compared to the fifth-harmonic leakage inductance, the better is the performance of the 24SSVPWM over the 12SSVPWM.

Modified 24-sector space vector PWM

The 24SSVPWM described above involves switching of four active vectors along with two zero vectors placed at beginning and at the end of a switching interval, thus involving six switching vectors. This causes at least one simultaneous switching in two inverter legs during a switching interval. A variant of the 24SSVPWM (referred as 24SSVPWM$_M$ henceforth) has been proposed to overcome this drawback [10]. The proposed method uses three L vectors and two M vectors in the $\alpha\beta_1$ subspace for generating the sequence in the switching interval. For example, when the reference vector is in sector 1 as shown in Fig. 5(c), three L vectors 37, 36 and 52, two M vectors 4 or 60 and 5 or 61 are used.

The selection of M vectors and zero vectors is done in such a way that only one inverter leg switching transition is happening during every vector change. This is possible if vectors 60 and 5 are selected together with 56 and 7 for zero vectors. The vector sequence during two consecutive switching cycles will be [56-60-52-36-37-05-07 | 07-05-37-36-52-60-56]. This sequence ensures only one inverter leg switching during vectors changes. The dwell times for different vectors are calculated assuming equal dwell times for zero vectors and small vectors, which simplifies the dwell time calculations further. Following the same approach as in the 12SSVPWM, the dwell time calculations is now represented as in (4).

$$
\begin{bmatrix}
v_{\alpha 1}^1 & v_{\alpha 1}^2 & v_{\alpha 1}^3 & (v_{\alpha 1}^4 + v_{\alpha 1}^5)/2 & v_{\alpha 1}^6 \\
v_{\beta 1}^1 & v_{\beta 1}^2 & v_{\beta 1}^3 & (v_{\beta 1}^4 + v_{\beta 1}^5)/2 & v_{\beta 1}^6 \\
v_{\alpha 5}^1 & v_{\alpha 5}^2 & v_{\alpha 5}^3 & (v_{\alpha 5}^4 + v_{\alpha 5}^5)/2 & v_{\alpha 5}^6 \\
v_{\beta 5}^1 & v_{\beta 5}^2 & v_{\beta 5}^3 & (v_{\beta 5}^4 + v_{\beta 5}^5)/2 & v_{\beta 5}^6 \\
1 & 1 & 1 & 1 & 1
\end{bmatrix}
\begin{bmatrix}
T_1 \\ T_2 \\ T_3 \\ T_4 \\ T_s
\end{bmatrix}
=
\begin{bmatrix}
v_{\alpha 1}^* T_s \\
v_{\beta 1}^* T_s \\
v_{\alpha 5}^* T_s \\
v_{\beta 5}^* T_s \\
T_s
\end{bmatrix}
\tag{4}
$$

In (4), v_x^4 and $v_y^5 (x, y = \alpha 1, \beta 1, \alpha 5, \beta 5)$ are projections of M voltage vectors on the $\alpha 1, \beta 1, \alpha 5, \beta 5$ axis respectively. These voltage vectors are switched for an equal time T_4 during the switching interval. Similarly, zero vectors are also used in beginning and at the end of the switching duration for equal time. With this approach the structure of the SVPWM implementation (particularly, the dwell time calculation) is still retained as the original implementation used for the 12SSVPWM. The selection of zero voltage vectors and sequencing of active voltage vectors is done such that only one inverter leg switching is involved during vector change. The zero voltage vector pairs change after every other sector and thus there will be one instance of simultaneous switching in one three-phase set.

Carrier-based implementation of 24-sector SVPWMs

The 24-sector SVPWM techniques described above overcome many challenges of the 12-sector SVPWM and improve harmonic performance as well as the computational effort. However, the implementation is still very complex and time consuming as compared to carrier-based techniques. Attempts have been made to implement 24−sector PWMs using carrier-based approaches and one such attempt is documented in [11], hereafter referred as 24SSVPWM$_C$. The authors establish an average per-phase voltage expressions for the inverter leg voltages to derive a common-mode signal which can be generated from reference voltages. It turns out that the zero-sequence signal v_{cm} is a simple expression with different combinations of reference signals, depending upon the angle information. This is represented as $v_{mid}/2$ for ωt between $\{0 - 15, 45 - 75, 105 - 120\}°$, $(\sqrt{3}/2 - 1)v_{min}$ for ωt between $\{15 - 45\}°$, and $(\sqrt{3}/2 - 1)v_{max}$ for ωt between $\{75 - 105\}°$, where v_{min}, v_{mid} and v_{max} are the minimum, middle and maximum of the three-phase reference voltages. The resulting v_{cm} is a trapezoidal-shaped waveform with $120°$ symmetry. The carrier waves for each three-phase modulator are inverted after every four sectors, and such inversion for the second three-phase modulator is done with a delay of two sectors [11]. The complete schematic of the 24SSVPWM$_C$ is shown in Fig. 6. In addition to the DZSI, the implementation 24SSVPWM$_C$ involve a sector identification scheme and a carrier adjustment scheme based on the sector information.

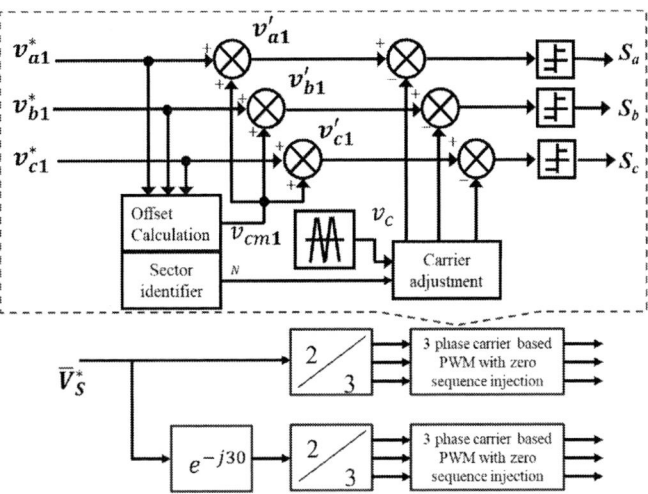

Fig. 6. Schematic of 24SSVPWM$_C$

The carrier-based implementation of 24SSVPWM$_M$ is also demonstrated in [12], in the similar way as done in [11] for the 24SSVPWM. The principle is again the same - determine the zero-sequence offset voltage which, when added to the reference signals, will result into similar patterns as the original 24SSVPWM$_M$. It turns out that such a zero-sequence voltage expression is the well-known common mode voltage expression used in three-phase modulators, similar to the DZSI (2). Thus, the 24SSVPWM$_{MC}$ is exactly like a DZSI for the addition of the offset voltage v_{cm}, supplemented with carrier inversion after every four sectors, which is also similar to 24SSVPWM$_C$ as shown in Fig. 6.

Comparative assessment of the PWM methods

The 24-sector space vector PWMs and their carrier-based counterparts (24SSVPWM, 24SSVPWM$_M$, 24SSVPWM$_C$ and 24SSVPWM$_{MC}$) are implemented in the control architecture of Fig. 2 for comparing their performance with 12SSVPWM and DZSI techniques. The only flux-, and torque-producing components for this control configuration are i_{d1} and i_{q1}, for which the reference is generated by the external control loop, while the reference currents to other current regulators are set to zero. The output voltage from current controllers, through inverse Park and inverse Clarke transformations, is input to the PWM modulators, which generates the switching signals for two three-phase inverters. The internal clock of these inverters is not synchronized, this can produce two samples of three phase voltage references and feedback currents acquired at two different instants by each inverter. This phenomenon is termed as asynchronous sampling in this paper. The objective of this simulation analysis is to compare the performance of the above PWM modulators with respect to different asynchronous conditions. The control scheme described in Fig. 2 can include a phase-shifted sampling of feedback currents and voltage references by time interval T_{shift} for one of the two converter units. Four different scenarios can be thought:

Case A. Synchronous sampling of modulator references and phase currents;
Case B. Synchronous sampling of modulator references only;
Case C. Synchronous sampling of phase currents only;
Case D. Asynchronous sampling of modulator references and phase currents.

The performance of different modulators is verified for the above situations and compared with DZSI technique which is taken as reference modulator.

Case A: Synchronous sampling of currents and references

The phase currents are sampled synchronously and transformed into the $\alpha\beta_1$ and $\alpha\beta_5$ subspaces to obtain the space vectors $i_{\alpha1\beta1}$ and $i_{\alpha5\beta5}$. The vectors are shown in Fig. 7 for the different modulation techniques. The performance of 12SSVPWM is satisfactory and comparable to DZSI modulator, because it makes use of large vectors only which have the smallest magnitude in the $\alpha\beta_5$ subspace. All three modulators can maintain a fundamental sinusoidal current vector, while achieving minimum current in the $\alpha\beta_5$ subspace. The 24SSVPWM and 24SVPWM$_M$ suffer in controlling the harmonic current vector in the $\alpha\beta_5$, mainly due to an incorrect sector identification when the sampled voltage references are used for angle calculation inside the modulators. The performance of carrier-based modulators 24SSVPWM$_C$ and 24SSVPWM$_{MC}$ is similar to the DZSI modulator.

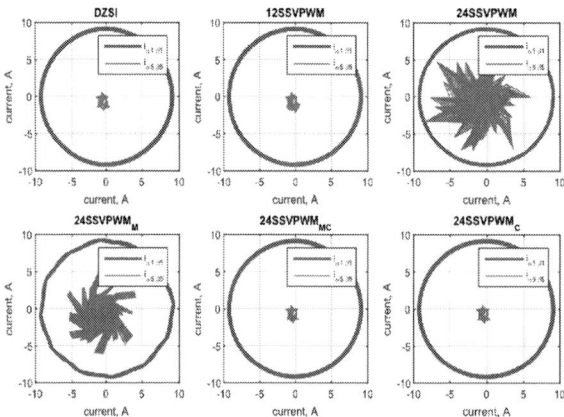

Fig. 7. Current vector in $\alpha\beta_1$ and $\alpha\beta_5$ subspaces for the simulation case A

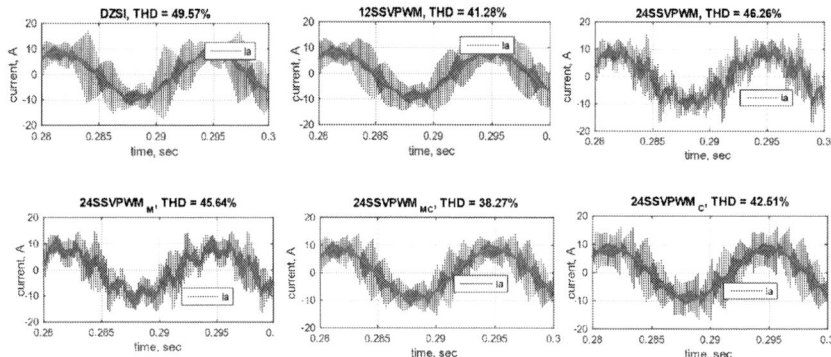

Fig. 8. Machine phase current (actual- blue, sampled- red)

The use of M vectors results in comparatively higher ripple in the $\alpha\beta_5$ subspace for all 24SSVPWM methods. However, the frequency of harmonic currents for 24SSVPWM methods is double than that of the 12SSVPWM. This impacts the total harmonic distortion of the phase currents, as shown in Fig. 8. The lowest THD of 38% is achieved with the 24SSVPWM$_{MC}$ compared to the 49% observed with the DZSI technique.

Case B: Asynchronous sampling of second set of currents

This situation appears when PWM sampling time synchronization is possible but separate hardware for phase current sampling is used. The major effect on asynchronous current sampling is the appearance of low frequency harmonics in the fundamental frequency subspace. The observation is true for all modulators as shown in Fig. 9(a). The current THD for all modulators is higher as compared to the previous case.

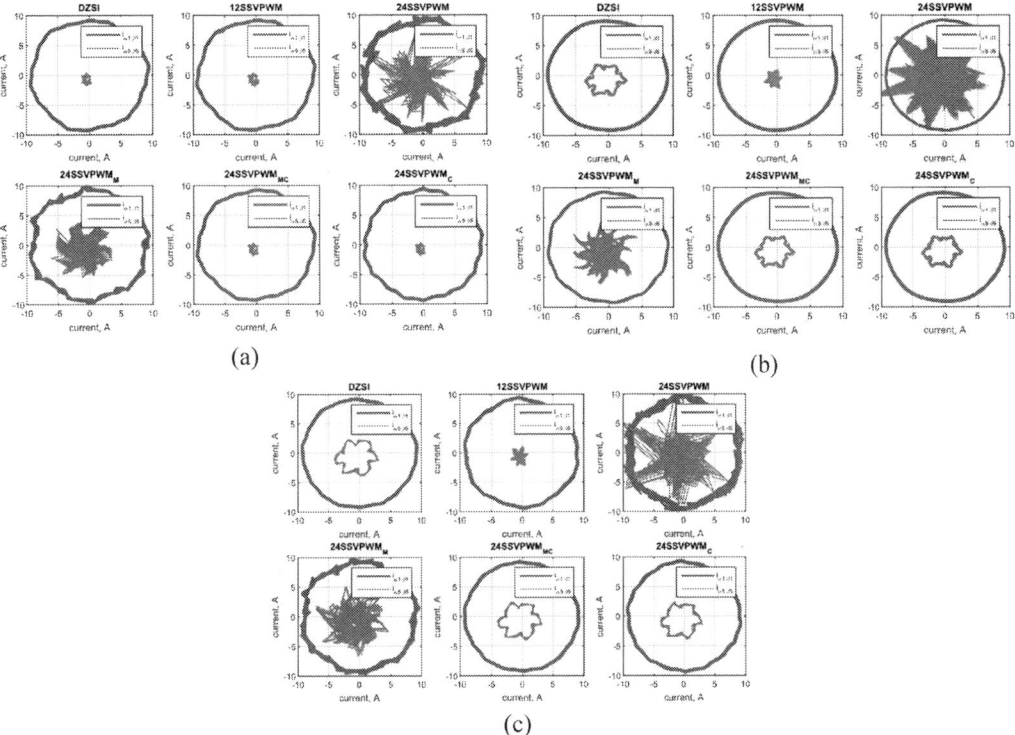

Fig. 9. Current vector in $\alpha\beta_1$ & $\alpha\beta_5$ subspaces for the simulation (a) Case B (b) Case C (c) Case D

Case C: Asynchronous sampling of second set of references

This simulation case corresponds to a synchronous sampling of phase currents through the use of external hardware, with no possibility of synchronization between the two PWMs. As compared to previous case, the major effect on asynchronous sampling of modulator references is the magnitude of $i_{\alpha5\beta5}$ as shown in Fig. 9(b). This observation is true for three modulators: DZSI, 24SSVPWM$_C$ and 24SSVPWM$_{MC}$. Interestingly, 12SSVPWM does not seem to be affected.

Case D: Asynchronous sampling of currents and references

The simulation case corresponds to two individual three-phase converter units with their own control boards and no means of synchronization for sampling and PWM. It is rather difficult to set an exact phase shift between sampling instances, because it can be anything between 0 and T_s. However, by setting a certain shift T_{shift} for the second set of sampled three-phase currents and voltage references, the combined effects of asynchronous sampling from the case B (low-order harmonics in $i_{\alpha1\beta1}$) and case C (magnitude of $i_{\alpha5\beta5}$) can be seen, as shown in Fig. 9(c). The carrier-based modulators, in particular 24SSVPWM$_{MC}$ modulator, is again best in terms of the fundamental current THD.

Conclusion

This paper makes a comparative analysis of six different voltage modulation schemes for asymmetric six-phase (YY30) permanent magnet machine, some of which are of the space-vector modulation type, and others of the carrier-based type. Among them, the performances of 24SSVPWM$_{MC}$ is found to be optimal in terms of THD and switching transitions. This modulator allows only one switching per vector change during a PWM period, and it also results in lowest THD among all modulators for different scenarios of synchronization in phase current sampling and voltage modulator references. The implementation of all the modulators is very similar, with some differences in the logic for sector identification and carrier shifting. The 24-sector SVPWM methods can be equally implemented using simple methods like carrier-based (DZSI) techniques but with improved THD performance.

References

[1] R. H. Nelson and P. Krause, "Induction machine analysis for arbitrary displacement between multiple winding sets.," *IEEE Trans. Power App. Syst,* Vols. PAS-93, no. 3, pp. 841-848, May 1974. [Online]. Available: http://dx.doi.org/10.1109/TPAS.1974.293983.

[2] Y. Zhao and T. A. Lipo, "Space vector PWM control of dual three-phase induction machine using vector space decomposition.," *IEE Trans. Ind. Appl.,* vol. 31, no. 5, pp. 1100-1109, Sep./Oct. 1995 [Online]. Available: http://dx.doi.org/10.1109/28.464525.

[3] A. A. Rockhill and T. A. Lipo, "A generalized transformation methodology for polyphase electric machines and networks," in *Proceedings of the 2015 IEEE International Electric Machines and Drives Conference IEMDC,* Idaho, USA, May 2015.

[4] E. Levi, R. Bojoi, F. Profumo, H. A. Toliyat and S. Williamson, "Multiphase induction motor drives - a technology status review," *IET Electr. Power Appl.,* vol. 1, no. 4, pp. 489-516, Jul. 2007 [Online]. Available: http://dx.doi.org/10.1049/iet-epa:20060342.

[5] E. Levi, "Multiphase electric machines for variable-speed applications," *IEEE Trans. Ind. Electron.,* vol. 55, no. 5, pp. 1893-1909, May 2008 [Online]. Available: http://dx.doi.org/10.1109/TIE.2008.918488.

[6] ——, "Advances in converter control and innovative exploitation of additional degrees of freedom for multiphase machines," *IEEE Trans. Ind. Electron.,* vol. 63, no. 1, pp. 433-448, Jan. 2016. [Online]. Available: http://dx.doi.org/10.1109/TIE.2015.2434999.

[7] R. Bojoi, A. Tenconi, F. Profumo, G. Griva and D. Martinello, "Complete analysis and comparative study of digital modulation techniques for dual three-phase ac motor drives," in *Proceedings of the 33rd Annual IEEE Power Electronics Specialists Conference,* Cairns, Queensland, Australia, Jun. 23-27 2002, pp. 851–857. [Online]. Available: https://doi.org/10.1109/PSEC.2002.1022560.

[8] K. Gopakumar, V. T. Ranganathan and S. R. Bhat, "Split-phase induction motor operation from PWM voltage source inverter," *IEEE Trans. Ind. Appl.,* vol. 29, no. 5, pp. 927-932, Sep./Oct. 1993.

[9] K. Marouani, L. Baghli, D. Hadiouche, A. Kheloui and A. Rezzoug, "A new PWM strategy based on a 24-sector vector space decomposition for a six-phase VSI-fed dual stator induction motor," *IEEE Trans. Ind. Electron.,* vol. 55, no. 5, pp. 1910-1920, May 2008 [Online]. Available: https://doi.org/10.1109/TIE.2008.918486.

[10] W. Kun, Y. Xiaojie, W. Chenchen and Z. Minglei, "An equivalent dual three-phase SVPWM realization of the modified 24-sector SVPWM strategy for asymmetrical dual stator induction machine," in *Proceedings of the IEEE Energy Conversion Congress and Exposition (ECCE),* Milwaukee, WI, USA, Sep. 18-22 2016. [Online]. Available: https://doi.org/10.1109/ECCE.2016.7854842.

[11] P. R. Rakesh and G. Narayanan, "Investigation on zero-sequence signal injection for improved harmonic performance in split phase induction motor drives," *IEEE Trans. Ind. Electron.,* vol. 64, no. 4, pp. 2732-2741, May 2017. [Online]. Available: https://doi.org/10.1109/TIE.2016.2643620.

[12] K. Wang, X. You and C. Wang, "An equivalent carrier-based implementation of a modified 24-sector SVPWM strategy for asymmetrical dual stator induction machines," *Journal of Power Electronics,* vol. 16, no. 4, pp. 1336-1345, Jul. 2016. [Online]. Available: http://dx.doi.org/10.6113/JPE.2016.16.4.1336.

Simulation and Measurement-Based Analysis of Efficiency Improvement of SiC MOSFETs in a Series-Production Ready 300 kW / 400 V Automotive Traction Inverter

A. Nisch, M. Heller,
W. Wondrak
MERCEDES-BENZ AG
Hanns-Klemm Straße 45
D-71034 Böblingen
alexander.nisch@
Daimler.com

A. Bucher, C. Hasenohr,
K. Kefer, B. Lunz,
A. Pawellek, A. Smit
VALEO SIEMENS
eAUTOMOTIVE
GERMANY GMBH,
Frauenauracher Str. 85,
D-91056 Erlangen
alexander.bucher.jv@
valeo-siemens.com

M. Gärtner, N. Twardon,
U. Kirchenberger
STMICROELECTRONICS
Bahnhofstrasse 18,
D-85609 Aschheim-Dornach
manuel.gaertner@st.com

Keywords

Automotive Electronics, Electric Vehicle, Power Converters for EV, Voltage Source Converter (VSC), Silicon Carbide (SiC)

Abstract

As experienced for most power electronic applications in the past, efficiency is also becoming one of the key characteristic for automotive applications, affecting the design of future high-voltage components for electric vehicles. In case of traction inverters, mainly semiconductor losses deteriorate system efficiency and thus, upgrading the switches in hard-switched PWM-controlled 2-level inverters is a direct measure to significantly reduce system losses. For this task, SiC MOSFETs are a promising class of power semiconductors with superior device characteristics. With SiC high power modules becoming available on a broader basis, proper design-in approaches are necessary in order to obtain robust and cost-effective system solutions. Addressing this topic, a new full-SiC power module suited for direct replacement of existing Si-based solutions is described in this paper. A simulation and measurement-based analysis of the efficiency benefits of such a retrofit solution for a 400 V based drivetrain with up to 300 kW of output power is presented. Simulation results predict a loss reduction on inverter level of around 50 % compared to the Si IGBT based solution with respect to steady-state part load operation. The simulation results are verified by drivetrain test bench measurements showing a very high accuracy of the simulation model. Drive cycle simulations indicate an advantage in terms of energy consumption of up to 6.6 % for the drivetrain equipped with SiC MOSFETs over its counterpart equipped with Si IGBTs.

Introduction

Since the introduction of SiC as base material for power semiconductors, novel SiC devices have been promoted as huge step towards more efficient power electronics. SiC MOSFETs offer superior conduction as well as switching losses over their Si-based counterparts [1]. Compared to the Si IGBT, partial load losses and especially switching losses are significantly reduced, making them a technically very interesting solution for traction inverters which are predominantly operated under light load conditions in typical mission profiles [2–6]. Due to these advantages, SiC MOSFETs are a promising option to increase the efficiency of the drivetrain with vehicle energy consumption being one of the key parameters especially for battery electric vehicles [7].

As discussed in [8], [9], 1200 V SiC MOSFETs have been demonstrated to significantly push inverter efficiency to levels unreachable with existing state-of-the-art Si-based semiconductors for upcoming 800 V electric vehicle designs. With new devices with blocking voltages suitable for 400 V vehicle architectures becoming available on a broader scale, the SiC MOSFET also represents a very promising candidate for retrofit solutions for existing Si IGBT drivetrain inverters. Also for these voltage classes, significant improvements in terms of part load efficiency have been reported in [10], [11].

Automotive high voltage inverter power module

Different power modules have been designed over the last years with well-established silicon technologies, which are now the standard solution for electrical vehicles, either hybrid cars or full electric ones. Today IGBT technologies still offer some significant room to improve the vehicle efficiency by optimizing the losses inside the inverter. This is very important to be compliant with future legislations. Wide bandgap technologies are considered accordingly as the foundation of the future power electronics due to their superior behavior. They are strongly optimizing the static on losses, switching losses and are having the possibility to be operated at higher junction temperatures. After 20 years of research and development of SiC components by today MOSFETs designed in this wide bandgap technology have reached a maturity level, to comply with the demanding automotive requirements. Main points are quality and reliability aspects, the supply chain and the manufacturing setup. And for sure also the wafer size has been increased from 2" in the nineties to 6" inch which is the standard SiC wafer size in STMicroelectronics by today. All these points are important to reach a competitive solution, from technical as well as commercial standpoint, to introduce the material in high volumes to the market and not to enter niche applications only. The focus of the last years has been put on optimizing the ST**POWER** SiC MOSFET technology nodes to reach a higher quality level as well as continuously shrinking the R_{DSon} for a given die size, this parameter is specified with the FOM (mΩ x cm^2). The evolution of this parameter is shown in Fig. 1.

As SiC is still in an early phase of its industrial life cycle, compared with Si that has been optimized over more than 50 years, it is very important to optimize the structures and to well understand the failure mechanism of the devices. For this reason, it has been decided by STMicroelectronics to still optimize the planar technology instead of immediately designing in a trench approach. The FOM of the 2nd planar generation is at the same level as known and state of the art trench technologies and the robustness of the structure has been optimized by lessons learned from previous technology steps. This has been achieved beside other taking benefit from automotive series programs, having by today already a huge number of transistors in the field being designed in by early adaptors and being in development programs using SiC MOSFETs in different applications like e.g. on-board charger, DC-DC converter or traction inverter. The next technology node Gen 3, that will reach its automotive maturity in 2020, will be characterized by an even higher robustness with significant lower FIT rates accordingly. The technology will be available in 750 V to support the well-established 400 V nominal DC bus voltage as well as in 1200 V for the upcoming 800 V board net topology.

To bring the SiC material in automotive applications further innovation on packaging level is needed. Here the mainstream solutions are not so easy to define as by today there is a huge number of different solutions for power packaging and modules ongoing to be supplied on the market. On the one hand side optimized modules are under development (ACEPACK^{TM} DRIVE, ACEPACK^{TM} SMIT), that are compatible with available solutions used in the automotive market. This enables a fast replacement of a running Si traction inverter application by a SiC one, without applying major changes to the ECU (Electronic Control Unit) including for example the form factor, EMC measures, gate drive board and cooling concept. This is considered as a very fast "plug and play" concept, that will be described in this paper. On the other hand, more scalable solutions like the STPAK^{TM} have been developed and are available, based on a standard block that can be used in different number of devices per inverter switch. By this concept power can

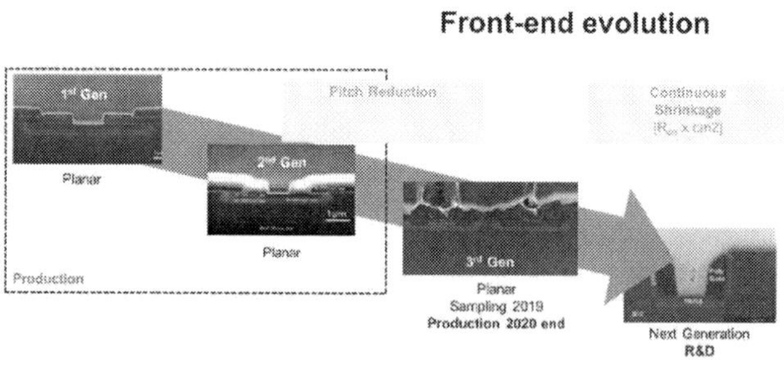

Fig. 1 ST**POWER** SiC MOSFET technology roadmap (Front-end evolution)

be easily scaled by the number of used packages per switch. For sure bringing such a new solution needs a re-design of the complete application including for example the bus bar and the cooler concept. Typical power packages for SiC components are shown in Fig. 2. For the ACEPACK™ Drive dedicated SiC dice have been designed optimized for the layout of the package in 750 V as well as 1200 V breakdown voltage. This paper describes the use and evaluation of a 2nd generation technology in 750 V, as the target was to use a mature and series proven semiconductor material for the comparison with the state of the art IGBT solution that is in use in several applications. Even if it is only the 2nd generation of SiC MOSFETS the typical R_{DSon} of the modules is around 2 mΩ per switch, which is indeed an impressively low value. Next step will be to introduce the technology node gen 3 which will be clearly below 1.5 mΩ. As there is a certain power dissipation switching RMS currents of up to 600 A, as substrate to spread the heat, a powerful Si3N4 substrate is used. To achieve the needed reliability results, the dice are sintered on the substrate. The direct cooled Cu base plate is designed with pins fins to reach the needed cooling capability.

SiC packaging solutions

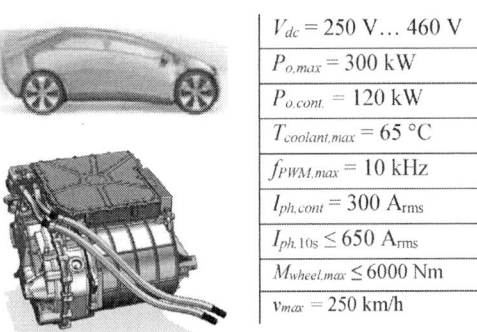

Fig. 2 Different SiC packaging solutions

Electrical characterization and design-in of full SiC power module

For designing in a full SiC B6 power module for a 3-phase 2-level DC/AC traction inverter under automotive constraints, several key aspects of the inverter design have to be re-considered. Most obviously, gate driver as well as EMI filter design need to be investigated in order to cope with these novel semiconductors. On system level, also motor aspects such as winding isolation and potential impacts of higher switching speeds on the performance of the machine need to be investigated in order to fully meet all requirements with respect to electrical performance and lifetime of the drivetrain system. The corresponding key parameters of the drivetrain are summarized in Fig. 3.

From module package point-of-view, additional modifications beyond the Si state-of-the-art with respect to low stray inductances [8], [9] and more reliable interconnection technologies are under discussion [12], [13] for being able to fully utilize the full potential of SiC power semiconductors in the future. However, in order to emphasize the direct efficiency impact of the power semiconductors alone, no additional mechanical and electrical modifications than exchanging the power module and the gate driver of the inverter were implemented for the investigations described in this paper. In order to directly assess the efficiency benefits of SiC MOSFETs compared to state-of-the-art Si IGBTs, an existing 400 V inverter is upgraded with a full SiC retrofit power module. It has to be noted, that under these constraints, the SiC MOSFET's switching performance will be limited by this approach due to the fact that admissible stray inductance is higher in case of using packages that have been originally designed for slow-switching Si IGBT devices. Thus, an increase in switching frequency above today's levels was not further investigated. However, design efforts are drastically reduced by using such a foot-print compatible power module within an existing inverter design.

One of the main aspects of paralleling a high number of power semiconductors in high power module is switching performance [14]. In order to fully utilize the module's available voltage SOA within the final application, a

$V_{dc} = 250$ V … 460 V
$P_{o,max} = 300$ kW
$P_{o,cont.} = 120$ kW
$T_{coolant,max} = 65$ °C
$f_{PWM,max} = 10$ kHz
$I_{ph,cont} = 300$ A$_{rms}$
$I_{ph,10s} \leq 650$ A$_{rms}$
$M_{wheel,max} \leq 6000$ Nm
$v_{max} = 250$ km/h

Fig. 3 Electrical specification for high-end drivetrain variant

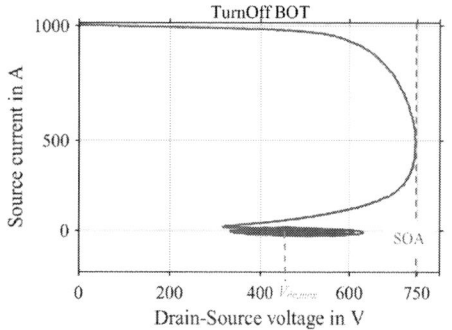

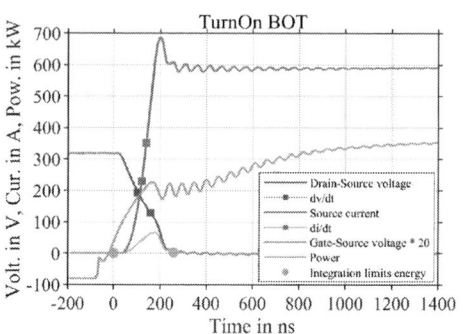

Fig. 4 Turn-off trajectory of BOT switch fully utilizing the voltage SOA in terms of blocking voltage for V_{dc} = 460 V, I_L = 1000 A, T_j = 25 °C

Fig. 5 Turn-on waveforms of BOT switch for V_{dc} = 325 V, I_L = 600 A, T_j = 25 °C

good controllability of the switching slopes is desirable. Using a classical driver architecture, it was possible to fine-tune the turn-off trajectory for maximum turn-off speed as shown in Fig. 4. A corresponding turn-on waveform is depicted in Fig. 5. Typical for SiC MOSFETs, visible ringing is encountered due to linear LC oscillations within the commutation cell, see [15] for a generic modelling approach applicable for MOSFETs in general. On first glance, this ringing is more pronounced than in case of Si IGBTs due to the unipolar characteristic, the higher slew rates and lower output capacitances of the SiC MOSFETs. Nevertheless, exploiting the advantages of improved switching losses inherently makes it necessary to accept higher slew-rates in order to reduce the switching times with both voltage and currents being present across the switches. For valid measurement results regarding the switching energy, the resulting time delay between the current and voltage sensor has to be compensated [8], [16–18].

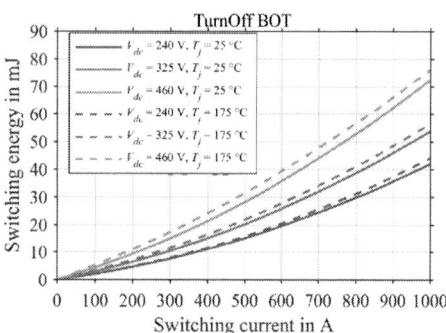

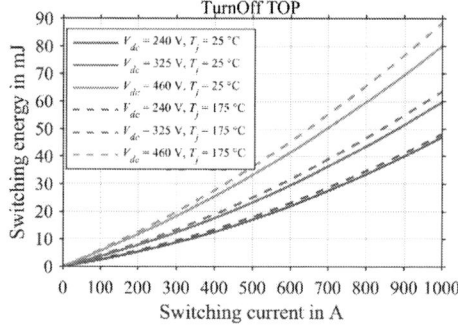

Fig. 6 Turn-off energies of BOT switch

Fig. 7 Turn-off energies of TOP switch

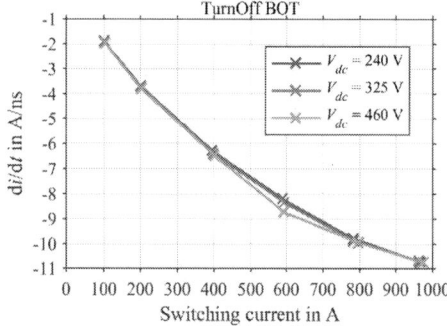

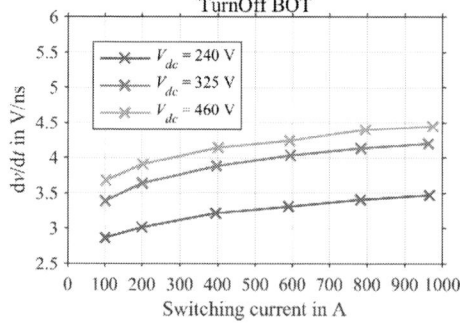

Fig. 8 Turn-Off di/dt slopes evaluated for 40 … 60 % of switched current for T_j = 25 °C, see Fig. 5

Fig. 9 Turn-Off dv/dt slopes evaluated for 40 … 60 % of switched voltage for T_j = 25 °C, see Fig. 5

Under the assumption that all components show linear behaviour for the duration of this oscillation, the sensor delay is adjusted in order to correct the phase delay between the voltage and the reactive current across the switch. With a total output capacitance of $C_{oss} = 3$ nF per switch, a total stray inductance of $L_\sigma = 23$ nH for the complete commutation cell in the high frequency domain is calculated for the observed oscillation frequency of 19 MHz for $V_{dc} = 325$ V. This result is in line with alternative measurements in the time domain correlating the turn-off overshoot and di/dt values. The resulting switching energies for the turn-off transition are compared in Fig. 6 vs. Fig. 7. Switching speed was maximized for both switches independently, resulting in almost identical switching energies, indicating a very symmetric power module layout. As also reported for 1200 V SiC MOSFETs, temperature influence on switching losses is almost negligible [8], [9], compare Fig. 6 and Fig. 7.

However, striving for lower switching losses inevitably leads to higher switching gradients with potential negative consequences regarding EMI noise [19], [20]. The resulting slew-rates for the retrofitted commutation cell are depicted in Fig. 8 and Fig. 9. In terms of di/dt switching speed was maximized to fully utilize the available breakdown voltage of the semiconductor devices and resulting dv/dt values are well within an acceptable range regarding EMI.

Inverter loss simulation and test-bench measurements

With respect to designing a traction inverter for electric vehicle applications, proper simulation tools for predicting the junction temperatures of the semiconductors are mandatory. In order to safely operate the SiC MOSFETs within their thermal SOA, all relevant temperature-dependent losses have to be taken into account in order to guarantee sufficient margin in junction temperature to prevent thermal run-away conditions as discussed in [21]. In order to predict thermal equilibrium for each individual point of operation with low computational efforts, transient simulations are not efficient and steady-state calculations are sufficient. Due to the strong dependence of conduction losses on junction temperature, assuming fixed R_{DSon} values for a pre-determined junction temperature leads to inacceptable inaccuracies whenever large load variations are to be considered.

For this task, a straight-forward simulation model was implemented which calculates the semiconductor losses for different junction temperatures. Depending on the transferable heat across the thermal path into the coolant water which is influenced by flow-rate and inlet temperature, the intersection between generated and removable heat is calculated representing the thermal steady-state as described in [21]. In case of SiC MOSFETs, no influence of the power factor on the loss distribution occurs due to synchronous rectification during reverse conduction conditions. Furthermore, thermal cross-coupling effects are less pronounced than in case of Si IGBT power modules with additional free-wheeling diodes. Especially for peak-load, correct calculation of the steady-state junction temperature is crucial, see Fig. 10.

Assuming typical R_{DSon} characteristics for the power module under electrical worst-case conditions yields the loss distribution per logical switch depicted in Fig. 11. Here, the impact of the package's stray inductance is directly visible with switching losses accounting for a significant amount of total losses despite the unipolar switching characteristics of the SiC MOSFETs.

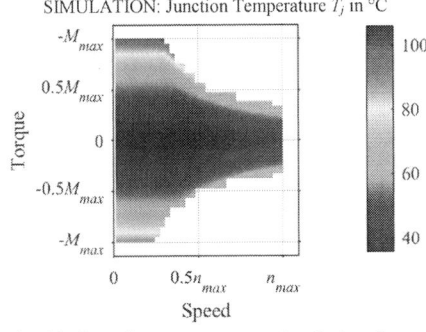

Fig. 10 Junction temperature simulation for $V_{dc} = 325$ V, $f_{PWM} = 10$ kHz, $T_{inlet} = 35$ °C, typical R_{DSon}

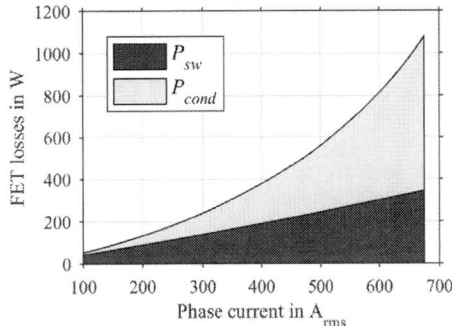

Fig. 11 Simulated loss distribution per switch for $V_{dc} = 460$ V, $f_{PWM} = 10$ kHz, typical R_{DSon}, typical R_{th}, $T_{coolant} = 65$ °C

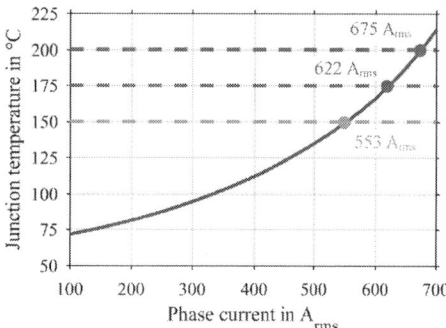 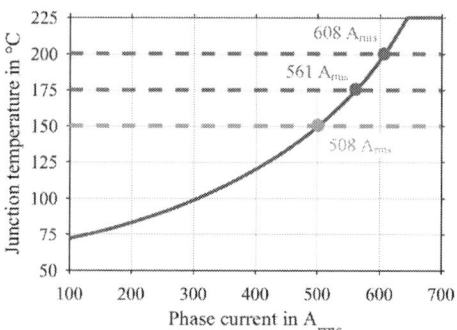

Fig. 12 Ampacity for $V_{dc} = 460$ V, $f_{PWM} = 10$ kHz, typical R_{DSon}, typical R_{th}, $T_{coolant} = 65$ °C for different admissible junction temperatures

Fig. 13 Ampacity for $V_{dc} = 460$ V, $f_{PWM} = 10$ kHz, worst-case R_{DSon}, typical R_{th}, $T_{coolant} = 65$ °C for different admissible junction temperatures.

With SiC representing a wide-band gap semiconductor, high temperature operation is not as directly limited by the substrate itself [22] as it is the case for Si. For reliable operation, packaging aspects become relevant with respect to thermal cycling capabilities of the interconnection technologies that are used. Fig. 12 highlights the impact on admissible junction temperature for the achievable output current of the inverter under typical conditions, see also [22], [23]. Increasing $T_{j,max}$ from 150 °C to 175 °C reflects into 70 A$_{rms}$ of additional phase current that the inverter is able to supply under electrical worst-case conditions. An additional increase in $T_{j,max}$ of 25 K then yields another 50 A$_{rms}$. Of course, higher peak temperatures directly lead to higher thermomechanical stress which is a challenge for classical packaging technologies in case of SiC [12], [13], [24], [25].

However, in case of automotive applications covering peak-load conditions that are rarely encountered in real-world driving scenarios by specifying short-term overload capabilities of the power module seems a promising approach in order to obtain a cost-effective solution. In addition to packaging aspects, also device specifications must be within a reasonable tolerance range in terms of conduction losses since this loss mechanisms dominates peak-load losses. Assuming a worst-case R_{DSon} that is increased by 30 % compared to the typical case results in a severe impact on the inverter's ampacity as illustrated in Fig. 13.

In order to validate the simulation model, first inverter tests were performed using an inductive load. Resulting phase current waveforms are given in Fig. 14 with a clear sinusoidal waveshape as also encountered for motor operation. For nominal conditions, the measured inverter losses are compared to the simulation results in Fig. 15. For part load operation up to 300 A$_{rms}$ excellent agreement between model prediction and measurement was observed. For higher load currents it was found that the early engineering samples outperformed the typical target values and re-measurements of the R_{DSon} indicated

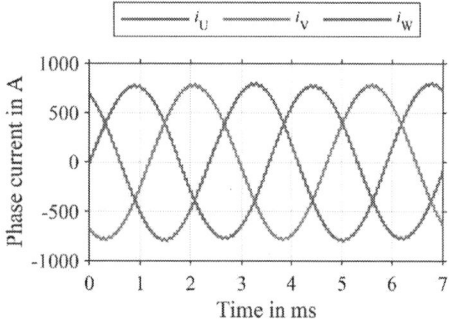 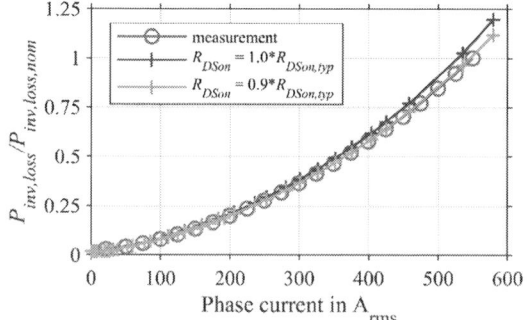

Fig. 14 Measured phase current waveforms for $V_{dc} = 450$ V, $I_{ph} = 550$ A$_{rms}$, $f_{PWM} = 10$ kHz, $f_{AC} = 280$ Hz

Fig. 15 Comparison of measured and simulated inverter losses under inductive load operation for $V_{dc} = 450$ V, $f_{PWM} = 10$ kHz, $f_{AC} = 200$ Hz, $T_{coolant} = 25$ °C

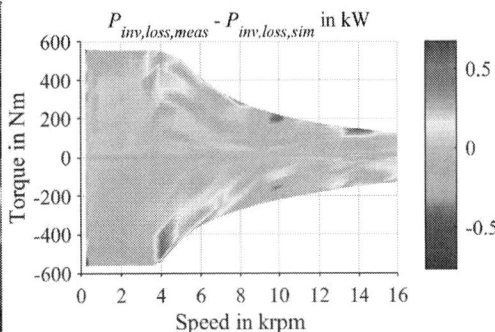

Fig. 16 eAxle prototype under test

Fig. 17 Difference between measured and simulated inverter losses on motor test bench for V_{dc} = 325 V, f_{PWM} = 10 kHz, $T_{coolant}$ = 35 °C

better conduction characteristics than originally assumed based on datasheet characteristics. Taking an effective reduction of the R_{DSon} by 10 % into account, model prediction and measurements showed an even better agreement.

In order to demonstrate the full achievable performance, test on motor test benches have been performed. Actual motor test bench results of the retrofitted drivetrain are given in Fig. 17. It was possible to cover the complete range of operation as defined in Fig. 3. In terms of loss prediction, significant deviations were only encountered in the region of peak output power and at the border of the field-weakening region. These deviations had been most likely caused by unstable measurements of the fundamental frequency on the test bench. Besides these deviations, efficiency projections obtained by the model showed excellent agreement throughout the whole range of motor operation as can be seen in Fig. 18. Here, the SiC specific advantages in terms of part load efficiency for $M < 0.5 M_{max}$ become clearly visible. Peak efficiency values in the sweet spot above $n > 0.4 n_{max}$ amount up to 99.4 %. Increasing the DC voltage to 460 V increases switching losses, affecting especially the part load region with $M < 0.5 M_{max}$, see Fig. 19. Compared to typical peak efficiency values of Si IGBT inverters, a loss reduction by a factor of 2 had been demonstrated by the SiC inverter prototype especially in the part-load domain.

Discussion of efficiency improvement and impacts on vehicle level

Efficiency improvements by SiC MOSFETs in the 800 V battery systems have already been published in [8], [9], [11]. The investigations presented here aim at clarifying the potentials for consumption reduction by simply replacing the power module in an existing 400 V drive train system. For this purpose, a module with Si IGBTs is retrofitted with an identical module with SiC MOSFETs without major adjustments to the EMC filter or the e-motor design. Based on the measured efficiency of the drivetrain

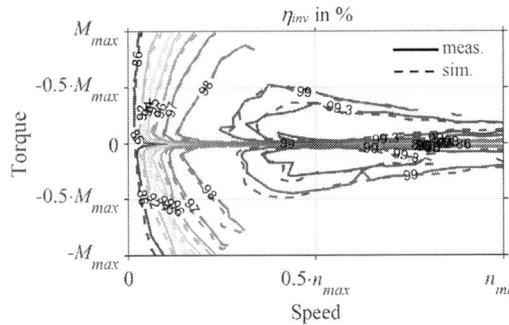

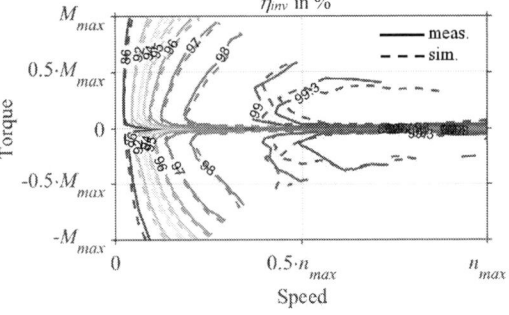

Fig. 18 Comparison of measured and simulated efficiency map for V_{dc} = 325 V, f_{PWM} = 10 kHz, $T_{coolant}$ = 35 °C

Fig. 19 Comparison of measured and simulated efficiency map for V_{dc} = 450 V, f_{PWM} = 10 kHz, $T_{coolant}$ = 35 °C

we calculated the efficiency improvements for a luxury sedan as illustrated in Fig. 20. The vehicle parameters of this luxury sedan being currently under development such as weight, aerodynamic drag, tire characteristics, rolling resistance, cross-sectional area front face and energy consumption for comfort systems are taken into account for several driving cycle simulations.

For both drivetrain variants, the energy consumption values for the driving cycles UDDS, Highway, WLTP and a customer driving cycle were calculated and the consumption advantages were determined as depicted in Fig. 21.

As the results show, a simple update from the Si-IGBT to the SiC MOSFET can achieve consumption reductions in the case of a city cycle of more than 6 % and for cycles with an overland share of more than 3 %. In the case of the Highway profile, the consumption advantage is significantly lower. This is due to the high proportion of working points with high current content, where the conduction losses are the dominant quantity, as illustrated in Fig. 22.

The red dots mark operation points in the cycle, the size of the dots represent the frequency of how often the operation point occurs in the cycle. With switching losses dominating under part load conditions, conduction losses become dominant when ap-

Fig. 20 Illustrative rendering of a full electric luxury sedan

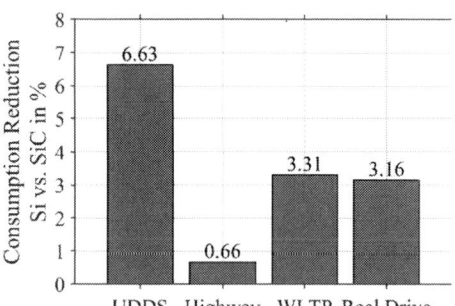

Fig. 21 Reduction of SiC drivetrain energy consumption for different drive cycles

proaching full load conditions as can be seen from the three operation points highlighted in Fig. 22. In order to achieve a further loss reduction with today's semiconductor, the switching losses should be reduced. This can be achieved by higher switching gradients, but limits are given by the stray inductance of the commutation cell. Additionally the effects on EMC behaviour must be evaluated. By using new SiC MOSFET generations with reduced R_{DSon}, the losses can be reduced during cycles with high-load operating points and the cooling system can be optimized.

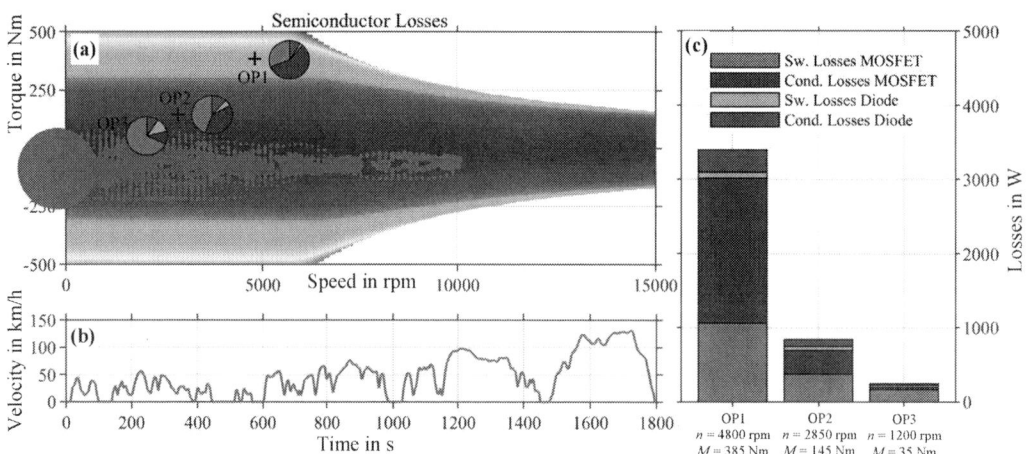

Fig. 22 Distribution of semiconductor losses
    (a) Heat map of MOSFET losses vs. speed and torque; red dots: occurrence of drive cycle operation points
    (b) Corresponding vehicle speed vs. time of drive cycle
    (c) Distribution of switching and conduction losses

Conclusion

After numerous years of discussing the theoretical advantages of SiC power semiconductors, first high power off-the-shelf SiC power modules are ready to be introduced into the field. The proposed paper demonstrates the direct efficiency impact of SiC MOSFETs retrofitted to a 400 V / 300 kW drivetrain unit. A simulation model capable of predicting the junction-temperature dependent semiconductor losses confirms the superior efficiency values achievable by these new devices even in standard automotive IGBT B6 packages. Test-bench measurements on inductive loads as well as on a motor test bench demonstrate the accuracy of the used inverter loss model. A significant impact on total drivetrain efficiency as well as vehicle range is identified, with inverter efficiencies up to 99.4 % which seem to be unreachable for Si IGBTs. In terms of energy consumption, drive cycle simulations based on measured loss maps of the SiC inverter indicate an improvement of up to 6.6 % for drive cycles with significant amount of part-load operation. With the SiC module having the same package outline compared to de-facto standard Si IGBT power modules a straight-forward retrofitting of existing inverter designs becomes feasible with low development effort and manufacturing impact.

References

[1] N. Kaminski, "The ideal chip is not enough: Issues retarding the success of wide band-gap devices," *Japanese Journal of Applied Physics*, vol. 56, 2017.

[2] M. André, "Real-world driving cycles for measuring cars pollutant emissions – Part B: Driving cycles according to vehicle power," Institut National de Recherche sur les Transports et leur Securité, 2006.

[3] M. André, "Real-world driving cycles for measuring cars pollutant emissions – Part A: The ARTEMIS European driving cycles," Institut National de Recherche sur les Transports et leur Securité, 2004.

[4] L. Beaurenaut, "SiC-Based Power Modules Cut Costs for Battery-Powered Vehicles," *Power Electronics Europe*, no. 3, pp. 22–25, 2018.

[5] S. La Mantia, V. Giuffrida, and S. Buonomo, "Benefits and advantages of using SiC," in *Proceedings PCIM*, 2019.

[6] J. Rice and J. Mookken, "Economics of High Efficiency SiC MOSFET based 3-ph Motor Drive," in *Proc. Int. Conf. for Power Electronics, Intelligent Motion Renewable Energy and Energy Managment PCIM Europe*, 2014, pp. 1003–1010.

[7] "Technology Roadmap - Electric and plug-in hybrid electric vehicles," International Energy Agency, Jun. 2011.

[8] A. Bucher, R. Schmidt, R. Werner, M. Leipenat, C. Hasenohr, T. Werner, S. Schmitz, and A. Heitmann, "Design of a full SiC voltage source inverter for electric vehicle applications," in *2016 18th European Conference on Power Electronics and Applications (EPE'16 ECCE Europe)*, 2016, pp. 1–10.

[9] A. Bucher, C. Hasenohr, A. Hoefer, H. Hofmann, B. Lunz, M. Usman, and R. Zaeh, "SiC Inverter 2.0: Compact Design for Ultra High Performance Applications," in *Proc. Int. Conf. on Automotive Power Electronics (APE)*, 2017.

[10] T. Bertelshofer, R. Horff, A. März, and M. M. Bakran, "Comparing 650V and 900V SiC MOSFETs for the application in an automotive inverter," in *2016 18th European Conference on Power Electronics and Applications (EPE'16 ECCE Europe)*, 2016, pp. 1–10.

[11] A. Nisch, C. Kloeffer, J. Weigold, W. Wondrak, C. Schweikert, and L. Beaurenaut, "Effects of a SiC TMOSFET Tractions Inverters on the Electric Vehicle Drivetrain," in *PCIM Europe 2018; International Exhibition and Conference for Power Electronics, Intelligent Motion, Renewable Energy and Energy Management*, 2018, pp. 1–8.

[12] C. Herold, M. Schaefer, F. Sauerland, T. Poller, J. Lutz, and O. Schilling, "Power cycling capability of Modules with SiC-Diodes," in *CIPS 2014; 8th International Conference on Integrated Power Electronics Systems*, 2014, pp. 1–6.

[13] A. Streibel, M. Becker, O. Muehlfeld, B. Hull, S. Sabri, D. Lichtenwalner, and J. B. Casady, "Reliability of SiC MOSFET with Danfoss Bond Buffer Technology in Automotive Traction

Power Modules," in *Proc. International Exhibition and Conference for Power Electronics, Intelligent Motion, Renewable Energy and Energy Management (PCIM Europe)*, 2019.

[14] H. Li, "Parallel Connection of Silicon Carbide MOSFETs for Multichip Power Modules," Department of Energy Technology, Aalborg University, 2015.

[15] D. Kübrich, T. Dürbaum, and A. Bucher, "Investigation of Turn-Off Behaviour under the Assumption of Linear Capacitances," in *Proc. Power Conversion Intelligent Motion Conf. PCIM*, 2006.

[16] K. Ammous, B. Allard, O. Brevet, H. E. Omari, D. Bergogne, D. Ligot, R. Ehlinger, H. Morel, A. Ammous, and F. Sellami, "Error in estimation of power switching losses based on electrical measurements," in *Power Electronics Specialists Conference, 2000. PESC 00. 2000 IEEE 31st Annual*, 2000, vol. 1, pp. 286–291 vol.1.

[17] K. Ammous, H. Morel, and A. Ammous, "Analysis of Power Switching Losses Accounting Probe Modeling," *Instrumentation and Measurement, IEEE Transactions on*, vol. 59, no. 12, pp. 3218–3226, Dec. 2010.

[18] G. Laimer and J. W. Kolar, "Accurate Measurement of the Switching Losses of Ultra High Switching Speed CoolMOS Power Transistor SiC Diode Combination employed in Unity Power Factor PWM Rectfifier Systems," in *Proceedings of the 8th European Power Quality Conference (PCIM)*, 2002, pp. 72–78.

[19] X. Gong and J. A. Ferreira, "Comparison and Reduction of Conducted EMI in SiC JFET and Si IGBT-Based Motor Drives," *Power Electronics, IEEE Transactions on*, vol. 29, no. 4, pp. 1757–1767, Apr. 2014.

[20] C. M. Johnson and P. R. Palmer, "Current measurement using compensated coaxial shunts," *Science, Measurement and Technology, IEE Proceedings -*, vol. 141, no. 6, pp. 471–480, Nov. 1994.

[21] C. Buttay, C. Raynaud, H. Morel, G. Civrac, M.-L. Locatelli, and F. Morel, "Thermal Stability of Silicon Carbide Power Diodes," *Electron Devices, IEEE Transactions on*, vol. 59, no. 3, pp. 761–769, Mar. 2012.

[22] B. T. Wrzcionko, "High Temperature / Power Density / Output Frequency SiC DC-AC Converter System for Hybrid Electric Vehicles," ETH Zürich, 2013.

[23] B. Wrzecionko, J. Biela, and J. W. Kolar, "SiC power semiconductors in HEVs: Influence of junction temperature on power density, chip utilization and efficiency," *Industrial Electronics, 2009. IECON '09. 35th Annual Conference of IEEE*, pp. 3834–3841, Nov. 2009.

[24] T. McNutt, B. McPherson, B. Passmore, Z. Cole, B. Whitaker, A. Barkley, B. Reese, R. Shaw, and A. Lostetter, "High Temperature (> 200 °C), High Frequency Multi-Chip Power Modules," in *Proc. International Conference on Compound Semiconductor Manufacturing Technology*, 2013.

[25] B. Passmore, Z. Cole, B. McPherson, B. Whitaker, D. Martin, A. Barkley, B. Reese, R. Shaw, T. McNutt, K. Olejniczak, and A. Lostetter, "Wide bandgap packaging for next generation power conversion systems," in *Power Electronics for Distributed Generation Systems (PEDG), 2013 4th IEEE International Symposium on*, 2013, pp. 1–5.

Validity of power cycling lifetime models for modules and extension to low temperature swings

Josef Lutz, Christian Schwabe, Guang Zeng, Lukas Hein
Chemnitz University of Technology
Reichenhainerstr. 70, 09126 Chemnitz
Chemnitz, Germany
Tel.: +49 / 371 531 33618.
E-Mail: josef.lutz@etit.tu-chemnitz.de
URL: https://www.tu-chemnitz.de/etit/le/

Acknowledgements

We thank the Siemens AG for providing the modules and support. We thank also for the fruitful discussions and critical reviews, to bring the research in this field to a new level.

Keywords

Power cycling, Lifetime models, Elastic deformation, Plastic deformation

Abstract

Various papers in power electronics contain a part of lifetime estimation depending on power cycles in application. The used model is often the CIPS08 lifetime model published at the conference CIPS 2008. In many applications, a lot of cycles with low temperature swings occur. The used model, however, is only valid for temperature swings above 40 K. For temperature swings < 30 K, there are strong deviations, since some materials are now approaching the elastic region. First experimental power cycling results are gained below 30 K temperature swing. Also an approximation for the reliability of low temperature swing is given in this paper.

Introduction

Power electronic devices are exposed to high thermo-mechanical load in field. During operation, lateral and vertical temperature gradients evolve in the layers of the modules. It leads to expansion and deformation of each layer, especially due to the different coefficients of thermal expansion (CTE) of the involved materials. The resulting mechanical stress causes degradation in the interconnection interfaces, which finally leads to failure and to a limited lifetime in application. Bond wires and soldered interconnections were found to be weak points limiting the lifetime in application, which can be reproduced using the power cycling test.

The first large-scale research on power cycling capability of power module was the LESIT project [1]. It delivered a data base and an equation to calculate the lifetime of the that-time standard power modules: Base plate, soldered substrate with Al_2O_3 ceramics, soldered chip and aluminum wire bonding on top. The LESIT results have been widely used for power system designs in following. In 2008, an improved empirical model for the same type of packages was given with the CIPS08 equation, considering the meanwhile progress in technology, based on a large number of tests and containing more parameters [2]. With the application of power modules in motor drives for hybrid- and electric cars, the interest on power cycling lifetime of power modules is increasing.

The CIPS 2008 Model

Based on a large number of power cycling results of power modules, Bayerer et al. [2] derived the equation:

$$N_f = K \cdot \Delta T_j^{\beta 1} \cdot \exp\left(\frac{\beta 2}{T_{min}}\right) \cdot t_{on}^{\beta 3} \cdot I^{\beta 4} \cdot V^{\beta 5} \cdot D^{\beta 6} \tag{1}$$

N_f is the number of cycles to failure. The first term, the Coffin-Manson-term, describing the dependency on the temperature swing ΔT_j, as well as the second term, the Arrhenius term describing the dependency on the operation temperature are already contained in [1]. As parameter K the value $9.30 \cdot 10^{14}$ is given in [3]. The other parameters ß2 ... ß6 are given in Table I [2].

The parameters for a high-power DAB module were derived in [4] and are discussed in the section Results of this paper.

Table I. Parameters for calculation of power cycling capability according to Eq. (1)

	CIPS08 [2]	DAB module [4]
K	9.3E14	5.65E14
β_1	-4.416	-4.1
β_2	1285	1285
β_3	-0.463	-0.484
β_4	-0.716	-0.716
β_5	-0.761	-0.761
β_6	-0.5	-0.5

Equation (1), which is known as CIPS 08 model, contains additionally the dependence on the heat-up time t_{on} in seconds, the current per bond stitch on the chip I in A, the voltage range of the device V in V/100 (reflecting the impact of the semiconductor die thickness), and the bond wire diameter D in μm. The CIPS 08 model holds for standard power modules with Al_2O_3 substrates, it is not valid for high-power traction modules which are built with the AlN substrate and AlSiC baseplate.

Equation (1) was a result of purely statistical analysis and is not a result of physics-based models [2]. The paper has 328 citations in Scopus, however many of them do not consider the validity range of the parameters, despite it is mentioned: "As this is an empirical approach, the formula is limited to the test data range and cannot be used for extrapolation" [2]. The range for ΔT_j reaches from 43 K to 130 K, the range for ton from 1 s to 15 s. Since at 15 s the temperature profile in the module is assumed to be stationary, for $t_{on} >15$ s the value 15 should be used in Eq. (1). Nevertheless it is more often used outside the defined range (e.g. [5], [6]), these results have to undergo a critical review.

The Coffin-Manson-term in (1) describes crack propagation. It used in mechanical engineering for plastic deformation. It is well known that many materials have a transition from plastic to elastic deformation for low deformation amplitudes. In this point of view, a model from Hartmann et al. for bond wires was published [7] which contains a "cut-off line" for low temperature swings, where no lifetime is consumed by power cycles. The range for ΔTj reaches from 40 K to 75 K in the data used in [7].

Measurement delay time and delay time correction

At turn-off of I_{load} and the measurement of the temperature sensitive electrical parameter $V_j(T)$, there is the delay time t_d between turning-off the load current and the instant of measurement, as shown in Fig. 1. This delay time is necessary for several reasons. First, there can occur some ringing due to unavoidable parasitic inductances in the circuit, as to be seen in Fig. 1. Second, the measurement current source I_{sense} is exposed to a load leap if the voltage at its output changes from on-state voltage at high current to low measurement current. A current source I_{sense} with fast response behavior is necessary. Thirdly, what is less known: If a bipolar device is used, there is a necessary recovery time for the internal charge carriers.

The time delay t_d leads to a systematic measurement error ΔT_d. The cooling-down ΔT_d in this interval lies for Si devices in the range of 2 K for usual power density, and up to 4 K for modules with high power density and advanced cooling systems. For tests executed for the models [1] [2], this measurement error was neglected. However, if the temperature swing range < 30 K is addressed, this is a significant error. A correction of this measurement error is possible with the square-root-t method [8]. It holds under

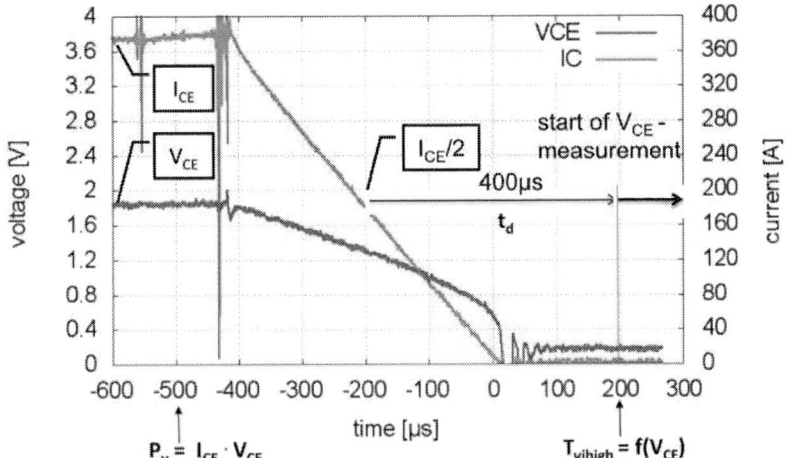

Fig. 1 Detail of measurements at turn-off of the load current with an IGBT. Fig. from [3]

boundary condition of a planar heat source at the surface of a semi-infinitely thick cylinder assuming one-dimensional heat flow. Since the heat source in SiC devices is in a narrow region close to the device surface, that method is found to hold for SiC devices with good accuracy [9]. However, for Si-IGBTs the use of the square-root-t method leads to a significant error, since there the heat generation is across the whole thickness of the device. For IGBTs, the correction can be done with thermal FEM simulation. It is possible to simulate only the layers close to the device, since during the short t_d they are dominating.

First results for short on-time and low temperature swings

A high-current power cycling test-bench for short load pulse duration was described in [10] and first results for a t_{on} time down to 10 ms were published in [4]. For low temperature swings, the short heating time is essential since a large number of cycles is expected. The devices under test were fabricated with copper-substrates DCB or with Al-substrates DAB.

Modules with DAB substrate have reached lower power cycling lifetime as the modules with the DCB substrate. However there are more differences in the packaging technology of these two tested modules. The difference in the substrate is not solely responsible for the difference in the power cycling lifetime. But several results were gained with the DAB based modules. It was found that for DAB the power cycling lifetime does not change anymore, if t_{on} is reduced from 40 ms to 20 ms or even smaller. A saturation of the t_{on}-dependency was found in the very small t_{on} range. The lifetime strongly increased below 30 K, if one approximates the region from plastic to elastic deformation. Results are summarized in Fig. 2.

There is a strong lifetime increase for T_{jmean} 100 K below ΔT 30 K. Three of 6 devices failed, 3 were still within the limits. The N_f is lower than predicted by the hyperbolic model in [7], and the Coffin-Manson exponent ß1 is approximated as -10.1. All failures are due to V_{CE} increase by bond wire lift-off. After the test with more than 500 Mio cycles, reported in [4], several parts of the test bench had to be renewed and the test with the 3 still functioning devices was continued up to $1.17 \cdot 10^9$ cycles (green circles in Fig. 2). No further device failed.

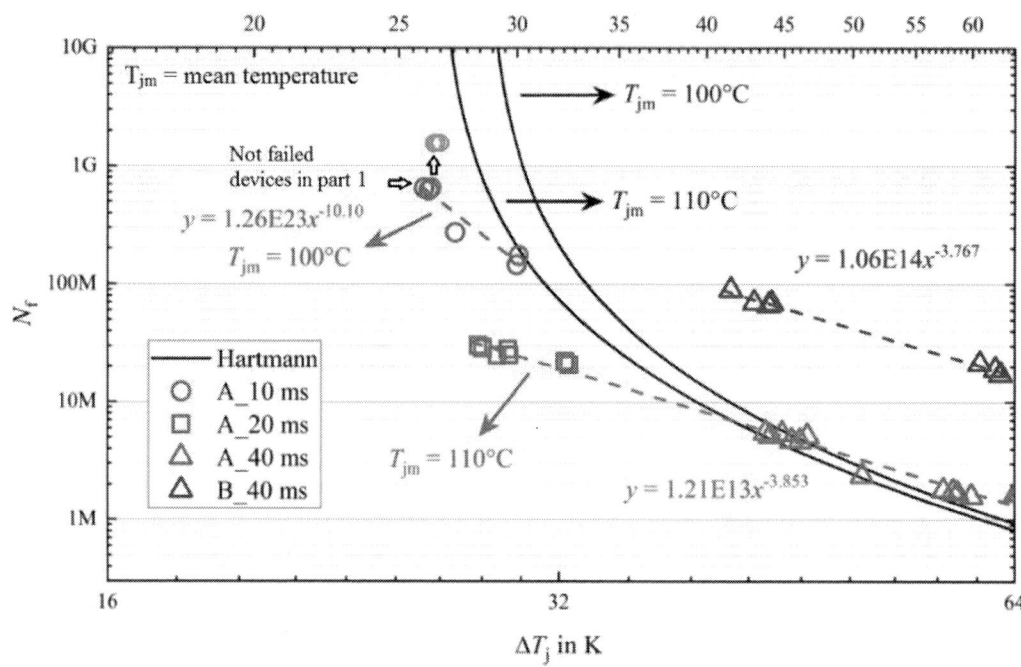

Fig. 2 Power cycling capability at low temperature swings, taken from [4]. Compared are results with DCB (B) and DAB (A) and the model of Hartmann [7], while $T_{jm} = T_{jmin} + \Delta T/2$. Not failed devices in part 1 [4] actualized (green circles)

Table II: Test conditions for continuation ($t_{on} = 0,01$ s, $t_{off} = 0.02$ s, $T_{inlet} = 22$ °C)

	ΔT_j in K	T_{jmin} in °C	P_V in W	I_L in A	ΔT_j in K (corr)	cycles	status
A_020_Sys2_D	23.7	80.4	2554	755	27,3	1,169,199,601	Not EOL
A_022_Sys3_D	23.2	76.6	2475	755	26,7	1,169,199,601	Not EOL
A_026_Sys1_D	26.0	105.4	2580	755	29,6	1,169,199,601	Not EOL

The failed modules (Fig.2) and the not failed were inspected regarding number of lifted-off bond wires and the adhesive force of the remaining bond wires. For the failed modules, 60.3 % were lifted-off, and the adhesive force of the remaining bond wires was in average 0.8 N. For the "Not EOL" modules the result is given in table III. For evaluation a self-constructed bond pull tester was designed (see Fig. 3). It is a force measuring equipment in a rack, to ensure vertical load. A hook is attached in the bond wire loop and pulled upwards. The remaining force of the bond interconnection is measured.

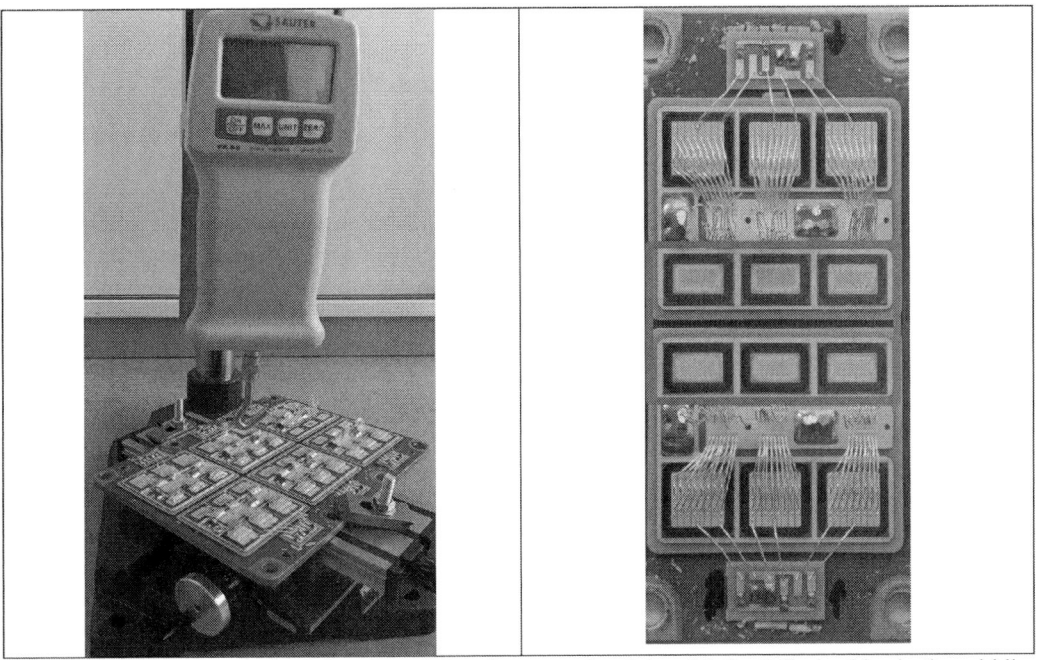

Fig. 3 Bond pull tester on the left and module scheme on the right with the 6 diode chips in the middle

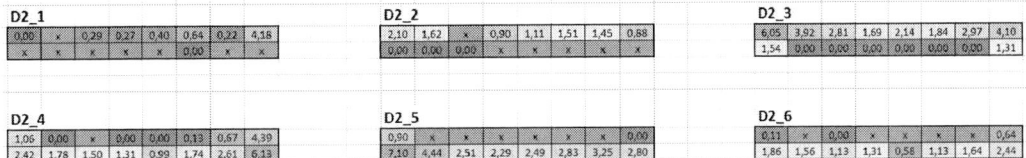

Fig. 4 Example of bond pull forces in N for the 6 diode chips with color scheme from blue → lifted off during test to green → still good contact force

Table III: Evaluation of bond wires and estimation final EOL

	Lifted-off wires in %	Lifetime consumption for lift-off	adhesive force, average in N	Lifetime consumption adhesive force in %	Expected EOL in Gigacycles „worst-case"
A_020_Sys2_D	31.2	51.8	1.7	47.1	**2.26**
A_022_Sys3_D	32.2	53.6	1.1	72.7	**1.61**
A_026_Sys1_D	39.6	65.6	2.2	36.4	**1.79**

The expected lifetime in table III is calculated with the lower value of lifetime consumption from lifted wires and from adhesive force. The lifetime model Eq. (1) holds for standard DCB modules, for the DAB high-power modules a lifetime model was derived in [4]. The equation is the same, some parameters are different, and they are given in table I. They hold for $\Delta T_j > 43$ K.

With focus on the temperature swing the Coffin-Manson relation can be described as:

$$N_F \sim K' \cdot \Delta T_j^{\beta 1'} \tag{2}$$

Considering the data in Fig. 2 and the estimated lifetimes for the remaining devices in table III, the parameter $\beta 1'$ can be approximated with the equation:

$$\beta 1' = e^{-\frac{\Delta T_j - 27{,}1\,K}{2{,}08\,K}} + \beta 1 \tag{3}$$

All other parameters for the DAB modules are used from table I. ß1' approximates to -4.1 for $\Delta T_j > 43K$. It was also considered that the t_{on}-influence saturates for $t_{on} < 0.04$ s [4] and therefore for 0.01 s in the experiment, the value of 0.04 s is used in Eq. (1).

The comparison of the Eq. (1) with the parameters for DAB modules table I and the extension for temperature swings below 30 K is given in Fig. 6. It becomes visible that soon a failure of one to two decades will occur if the original model of CIPS2008 [2] is used and the approximation to the elastic range is not considered.

With equation (3) and the CIPS08 model with the parameters for DAB module (see table I), it is possible to normalize the results and to do statistical analysis. Fig. 5 shows a Weibull plot for the experimental results. The confidence interval with the upper and lower percentile of 95 % of the probability distribution is quite large, due to the small amount of data points. The results are normalized to $T_{jmin} = 90$ °C and $\Delta T = 28$ K for the statistical analysis to allow a comparison between the cycles until failure.

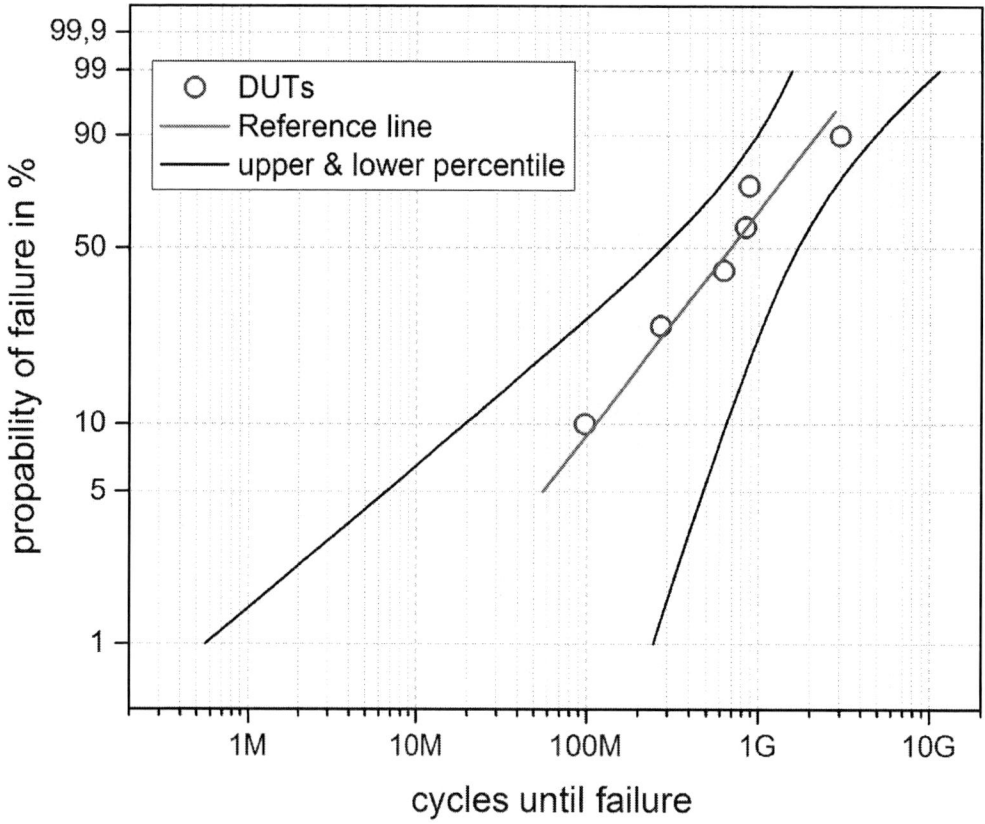

Fig. 5 Weibull-Plot for the experimental results normalized to T_{jmin} 90 °C and $\Delta T = \mathbf{28\ K}$

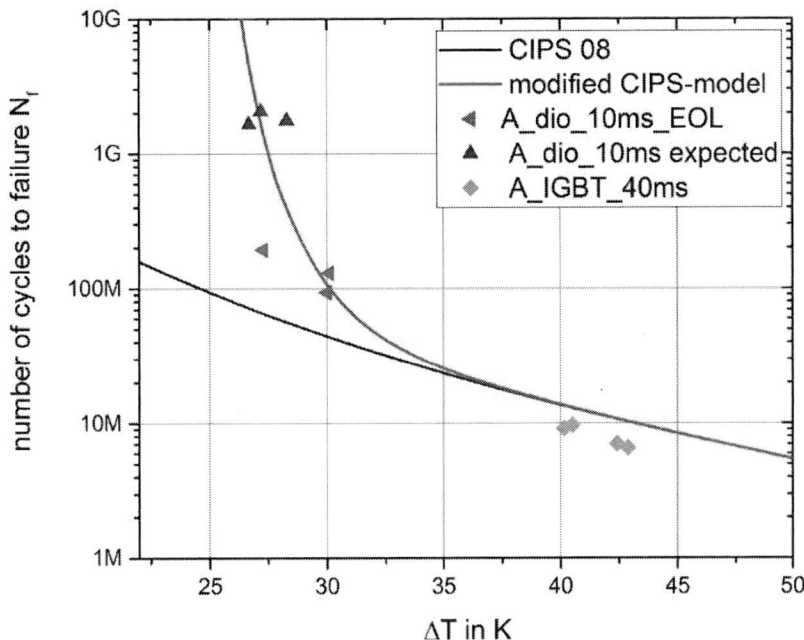

Fig. 6 Comparison of CIPS08 model and exp. data, normalized for T_{jmin} 90 °C, and modified CIPS-Model with Eq. 3 for β1'

Since there are slightly different conditions for the devices, the experimental values in Fig. 6 are normalized to T_{jmin} 90 °C using the Arrhenius-term in Eq. (1) with the data in table I. This allows a better comparison between the results.

Discussion of the results

Fig. 5 and Fig. 6 show that only six experimental values were determined for N_f, three of them are expected values. The data also showed significant scattering. On the other hand, this is to our knowledge the first time that a power cycling test up to more than 10^9 cycles was executed. Therefore even this low amount of data is valuable. All devices failed due to bond wire ageing. The analysis of the three failed modules and the three not failed modules are consistent. Besides, the measured ΔT_j is in the range of 23 K to 26 K, which raises a much higher requirement on the accuracy of the chip temperature measurement compared to the classical power cycling test with ΔT_j in the range of 80 K to 150 K.

Further, the data are in agreement with material science which gives an elastic and no-linear elastic and plastic range for material deformation. This is already contained in the model of Hartmann et al. [7]. However in [7] the approximation is hyperbolic,

$$N_f = K2 \left(\frac{1}{c_\Delta \Delta T_j + c_T T_{j,max} - c_0} \right)^m \qquad (4)$$

with $c_\Delta = 7.6 \cdot 10^{-6}$ K^{-1}, $c_T = 1.7 \cdot 10^{-6}$ K^{-1}, $c_0 = 4.1 \cdot 10^{-4}$, m = 2.4. This model has a pole at

$$c_\Delta \Delta T_j + c_T T_{j,max} = c_0 \qquad (5)$$

at which the lifetime becomes infinite. The approximation (3) presented in this paper is an exponential approach where the lifetime never becomes infinite. This is in agreement with [11] where an exponential approach for the elastic behavior is assumed, where the material never becomes ideal elastic, however approaches to this region exponentially. The results in Fig. 2 already show that the lifetime becomes very high, but not as high as calculated by Eq. (4).

The results are gained with DAB based modules. Fig. 2 shows that DCB based high-power modules reach one decade higher No. of cycles in the range $\Delta T_j > 43$ K. The root cause is supposed to be the found Al-reconstruction in the Al layer of the DAB which increases the temperature swing in the center of the chips. However, since the failures are always the bond wires, we can assume that the approximation (3) can also be used for DCB based high-power modules, however with a different K in Eq. 1, which is given in [4].

For standard DCB-based modules, whose lifetime is described in Eq. (1) for temperature swings above 40 K, we preliminary suggest to use Eq. (3) for extension down to $\Delta T_j > 25$ K. Table 1 is to be considered, and the second factor in (3) must be replaced by -4.416 to be in agreement with the CIPS2008 equation for cycles $\Delta T_j > 43$ K.

We have also to consider that 160 % of rated current had to be used to achieve a significant temperature swing for the short t_{on} of 10 ms. Because the current density is of influence, the results are probably worst-case, and the real lifetime for rated current will be higher. In the next step, a power cycling test with standard DCB modules is running, where the power for heating is generated by an adjustable part of switching losses [12]. Due to the high experimental effort and the expected cycles to failure of above 10^9, it will take some time for the next experimental data.

Conclusion

The model published at CIPS 2008 [2] is only valid for temperature swings above 40 K. For temperature swings < 30 K, the error becomes more than a factor of 10. Using the assumption of exponential approaching the ideal elastic behavior, the factor of underestimation of lifetime increases exponentially if used for temperature swings < 25 K. A first preliminary approximation for low temperature swings is given in this paper. Further work is in progress, however to gain more data is extremely time consuming.

References

[1]   M. Held, P. Jacob, P. Scacco und M. H. Poech, „Fast power cycling test of IGBT modules in traction application," in *Proc. of PEDS*, 1997.

[2]   R. Bayerer, T. Licht, T. Herrmann, J. Lutz und M. Feller, „Model for Power Cycling lifetime of IGBT Modules – various factors influencing lifetime," in *Proceedings of the 5th International Conference on Integrated Power Electronic Systems*, p 37-42, 2008.

[3]   J. Lutz, H. Schlangenotto, U. Scherumann und R. De Doncker, Semiconductor Power devices - Physics, Characteristics, Reliability, 2nd edition: Springer, 2018.

[4]   G. Zeng, R. Alvarez, C. Künzel und J. Lutz, „Power cycling results of high power IGBT modules close to 50 Hz heating," in *Proc. of EPE 2019, p. 1-9*, Genova, 2019.

[5]   H. Wang, K. Ma und F. Blaabjerg, "Design for Reliability of Power Electronic Systems", Montreal : IEEE , 2012.

[6]   Y. Zhang, H. Wang, Z. Wang, Y. Yang und F. Blaabjerg, „Impact of Lifetime Model Selections on the reliability prediction of IGBT modules," in *IEEE Energy Conversion Congress and Exposition (ECCE)*, 2017.

[7]   S. Hartmann und E. Özkol, „Bond wire life time model based on temperature dependent yield strength," in *Proc. PCIM Europe*, Nuremberg, 2012.

[8]   D. Blackburn und F. Oettinger, „Transient Thermal Response Measurement of Power Transistors," in *IEEE Transactions Electrical and Control Instrumentation*, Vol. IECI-22 , pp.134-141.

[9] C. Herold, J. Franke, R. Bhojani, A. Schleicher und J. Lutz, „Requirements in power cycling for precise lifetime estimation," in *Microelectronics reliability*, 2015.

[10] G. Zeng, C. Herold, M. Beier-Möbius, S. Kubera, R. Alvarez und J. Lutz, „High-current power cycling test bench for short load pulse duration and first results," in *Proc. of PCIM Europe*, Nuremberg, 2016.

[11] W. Illg, „Fatigue tests on notched and unnotched sheet specimens of 2014-T3 and 7075-T6 aluminum alloys and of SAE 4130 Steel with special consideration of the life range from 2 to 10`000 cycles," NACA Technical Note 3866, 1956.

[12] P. Seidel, C. Herold, J. Lutz, C. Schwabe und R. Warsitz, „Power Cycling with switching losses," ECPE final report, Nuremberg, 2017.

Roadmap for DC

Prof. Dr. Ir. Pavol BAUER
Delft University of Technology
Electrical Sustainable Energy Department
DC Systems, Energy Conversion & Storage Group
Postbus 5, 2600 AA Delft
The Netherlands

Abstract:

DC grids are considered to be a key technology for the connection, collection and integration of renewable energy resources, for the realization of integrated power systems, for mobile applications (electric ships, aircrafts), for new types of urban and industrial distribution power networks and to bridge and support existing AC systems. Advanced power electronic components, power converters and system protection are enabling DC grids on multiple voltages levels. Especially medium voltage DC grids are expected to play a key role in managing the higher power flows in our future distribution grids. Roadmap for DC and different steps and research at the TUD is presented with focus on DC grids and DC microgrids. Problem of Power Flow control in DC grids is addressed first an Power flow controller introduced. A zonal protection framework where the low voltage dc grid is partitioned according to short-circuit potential and provided degree of protection, and several known protection schemes that ensure selectivity, sensitivity and security will be discussed.

Biography:

Pavol Bauer is currently a full Professor with the Department of Electrical Sustainable Energy Of Delft University of Technology and head of DC Systems, Energy Conversion and Storage group. He was also appomitend as a professor by president of Czech Republic at the Brno University of Technology (2008) and honorary professor at Politehnica University Timisioira in Romania (2018). From 2002 to 2003 he was working partially at KEMA (DNV GL, Arnhem) on different projects related to power electronics applications in power systems. He published over 120 journal and 500 conference papers in his field (with H factor Google scholar 42, Web of Science 27), he is an author or co-author of 8 books, holds 7 international patents and organized several tutorials at the international conferences. He has worked on many projects for industry concerning wind and wave energy, power electronic applications for power systems such as Smarttrafo; HVDC systems, projects for smart cities such as PV charging of electric vehicles, PV and storage integration, contactless charging; and he participated in several Leonardo da Vinci, H2020 and Electric Mobility Europe EU projects as project partner (ELINA, INETELE, E-Pragmatic, Micact, Trolly 2.0, OSCD, Power2Power, Progressus) and coordinator (PEMCWebLab.com-Edipe, SustEner, Eranet DCMICRO).

Virtual EPE'20 ECCE Europe

His main research interest is power electronics for charging of electric vehicles and DC grids. He is a Senior Member of the IEEE ('97), former chairman of Benelux IEEE Joint Industry Applications Society, Power Electronics and Power Engineering Society chapter, chairman of the Power Electronics and Motion Control (PEMC) council, member of the Executive Committee of European Power Electronics Association (EPE) and also member of international steering committee at numerous conferences.

The role of collaborative research to support innovation for clean energy transition

Hubert DE LA GRANDIERE

SuperGrid Institute
23 rue Cyprian, CS 50289
69628 Villeurbanne Cedex
France

Abstract:

Making the transition towards clean energy requires a tremendous transformation of the current energy system. System structure, scale, economics, and energy policy must all be addressed in order to achieve significant change. Within the current sanitary crisis and its resulting impact on the economy, private sector industrial companies are looking to short term returns to boost their recovery rather than investing in long term R&D projects, thus further widening the gap between industrial and academic mind sets. But achieving energy transition requires coordination, collaboration & the creation of internationally recognised standards. How can we bridge the gap between academia and the private sector?

Biography:

Hubert de La Grandière received his two engineering degrees from Ecole Polytechnique and from Ecole nationale Supérieure de Techniques Avancées respectively in 1996 and 1998. He joined Alstom transport in 1998 where he held various positions, from Melbourne tramway project manager in 2000 up to tramway product line director from 2003 to 2005. He moved to Alstom Signalling business end of 2005, where he managed Alstom center of excellence for control electronics based in Lyon, before becoming Signalling products Vice-President in 2007 and eventually Main Line Signalling business Vice-President in 2011. In 2013, He moved to the energy business, heading Alstom Grid Solutions circuit breaker site based in Villeurbanne. In May 2016, He joined SuperGrid Institute as a Chief Executive Officer. SuperGrid Institute is a research, test and consulting centre created in 2014, performing collaborative research in Direct Current as well as medium and high voltage Power Electronics

Thomas Edison vindicated – the resurgence of DC in MV and HV power grids

Colin DAVIDSON
GE Grid Solutions - HVDC Activity
Stafford, UK

Abstract:

From the very earliest days of electrical power transmission, in the 1880s, the advantages of DC (as promoted by Thomas Edison) were already clear but despite this, the "Battle of the Currents" was won by Westinghouse and Tesla's AC solution, mainly because two 19th century inventions, the transformer and the circuit-breaker, were much easier to realise using AC than DC. Nevertheless, the use of DC in certain, niche point-to-point transmission applications never completely went away, with the first electromechanical conversion systems installed in the 1890s and electronic AC/DC conversion starting to appear in the 1930s. Today, HVDC is widely used for point-to-point power transmission applications where very high powers need to be transmitted for long distances, and the first commercial applications of meshed HVDC grids and medium-voltage DC (MVDC) for reinforcement of distribution grids, are starting to appear. With the drive for ever-increasing levels of renewable energy generation, along with drastic changes in consumption patterns as transportation and domestic heating are electrified, much greater use of DC for both transmission and distribution are inevitable. This talk will present a short historical perspective of how the industry got to its present position, a description of the present state of the art and predictions of how the grid will evolve in the coming decades.

Biography:

Colin Davidson is Consulting Engineer - HVDC, at GE Grid Solutions HVDC Activity, whose Centre of Excellence is in Stafford, UK. He joined the company in January 1989, when it was part of GEC, and progressed through the positions of trainee Thyristor Valve Design engineer; Manager, Thyristor Valves; Engineering Director and R&D Director, to his current role. He is a Chartered Engineer and a Fellow of the Institution of Engineering and Technology in the UK, and has served on several IEC standardisation committees for HVDC and FACTS. He has a degree in natural sciences, specialising in physics, from the University of Cambridge.

Virtual EPE'20 ECCE Europe

Integration of Electric Mobility in the French public electricity distribution network

Anne-Sophie COCHELIN
Senior Project Manager
ENEDIS
Electric Mobility Team
Paris, France

Abstract:

As a Distribution System Operator (DSO), Enedis is one of the key players in the development of electric mobility on the French territory.Indeed, charging infrastructures are directly or indirectly connected to the distribution network. In addition, Enedis runs electric vehicles on a daily basis and has the second largest electric fleet in France.

More broadly speaking, Enedis is committed to working alongside industrial and public players. The aim is to develop charging solutions for the various use cases of electric mobility, to identify territories needs, and to facilitate electric vehicle charging control, in order to optimise its cost for users and for the community.

In this presentation, we propose to present some key elements that were published by Enedis in 2019 about grid integration of electric mobility. We will also focus on the aVEnir project, led by Enedis with 11 industrial and academic partners, aimed at controlling in real conditions and in collaboration with users of electric vehicles, questions relating to charging flexibility.

Biography:

Anne-Sophie COCHELIN studied in Ecole Polytechnique (Palaiseau, France) and in McGill University (Montréal, Canada), including an internship at CSIRO (Australia). She started to work at EDF in 2004, in the Research and Development Department, about energy production and climate change / water resource. Then she joined the nuclear engineering teams and worked during 10 years on the « Grand Carénage » program, whose goal is to have the agreement to extend the lifetime of french nuclear power plants beyond 40 years. Since 2019, she is a senior project manager in the Enedis Electric Mobility Team, leading the aVEnir project.

Virtual EPE'20 ECCE Europe

A critical role for R&I for clean energy for the EU green and digital recovery

Hélène CHRAYE

European Commission
Directorate-General for Research and Innovation
Clean Planet – Clean Energy Transition
Brussels, Belgium

Abstract:

The recovery from sanitary and economic crisis will imply massive investments and in depth reforms. Even before, the EU set up extremely ambitious objectives for a Green Deal, making the EU a carbon free economy at latest for 2050, now merged in the challenge of a green and digital recovery for Europe. Energy represents the major share of the GHG emissions and has a critical place in this challenge. The transition to clean energy should rely upon a massive switch towards green energy as well as innovative and breakthrough solution in the demand side, whether on technology, on business processes, on social. Research and innovation will play a critical role for this and as well on supporting private and public investment decisions.

Biography:

French State Civil Engineer by education, Hélène Chraye graduated then in Economics and Public Law at Sciences – Po / Paris.

After a stay in the French Administration to build the Energy Observatory and then on State Aids to the industry, she joined the European Commission and worked successively on various domains of the European Transport policy: Inter-modality, Trans-European Networks, Single Sky, as well as dealing with enlargement and with Mediterranean area.

After a few years as Head of Operations in the EU Delegation to Belarus, Moldova and Ukraine, she joined DG RTD where she built the European Research Council Executive Agency and then managed financially and legally the EU programme NMBP, part of FP7 and H2020.

Since 1st June 2019, she is heading the unit in charge of designing and implementing the European Research policy for Clean Energy Transition within the Directorate Clean Planet.

Virtual EPE'20 ECCE Europe

AUTHOR INDEX

Aarniovuori, Lassi 2829
Abbate, Carmine 2802
Abbosh, Amin 1006
Abdel-Rahim, Naser 352
Abdelhakim, Ahmed 2220
Abdelrahem, Mohamed 900
Abramson, Rose A. 1934
Abusara, Mohammad 471
Aganza-Torres, Alejandro 1813
Aguglia, Davide 3330
Ahmad, Bilal 3348
Ahmad, Faheem 2987
Ahola, Jero 2753
Ait-Ahmed, Mourad 3289
Aizpuru, Iosu 251, 1205
Alam, M. M. 480, 1551
Alam, Muhammad Farhan 416
Alatise, Olayiwola 2241
Alawieh, Hadi 1685
Albach, Manfred 173, 193
Alexandre, Philippe 1905
Ali, Ahmed Ismail M. 1417
Ali, Marwan 1118, 2039
Ali, Mohammad 2743, 2763
Ali, Waqas 871
Alisar, Ibrahim 1205
Alishah, Rasoul Shalchi 460
Alkama, Kouceila 2564
Allard, Bruno 522, 829, 1470, 1874
Almaksour, Khaled 1700
Almeida, Bruno F. 3217
Alonso, C. 919
Alqatamin, Moath 65
Am, Sokchea 3172
Ammann, Ulrich 3137
Amrane, Fayssal 362
Anders, Erik 944
Andrade, Fabio 1400, 1841, 1850
Andresen, Jan 2303, 2545
Anzola, Jon 251
Aoustin, Yannick 1923
Arandia, Nerea 1524
Arazi, M. 2881
Arrizabalaga, Antxon 1205
Arrozy, Juris 1067
Arruti, Asier 251
Artiglia, Melissa 2791
Aríztegui, Raquel González 1515

Asllani, Besar 1279
Avenas, Yvan 1685, 2564
Averbukh, Moshe 2439
Averous, Nurhan Rizqy 153
Azizian, Mohammadreza 2860
Baburajan, Silpa 2573
Bacha, S. 406
Bacha, Seddik 820
Baërd, H. 95
Bagaber, Bakr 2613
Bahman, Amir Sajjad 2704
Bahrani, Behrooz 787
Bai, Wenshuai 667
Baker, Erik 512
Bakran, Mark-M. 644, 686, 1252, 1533, 1831, 1885,
.... 2106
Bakri, R. 2554
Bakri, Reda 2078
Balkowiec, Tomasz 2029
Barazi, Yazan 1057
Barelli, Linda 292
Barg, Sobhi 416
Barwig, Markus 1460
Basic, D. 37, 95
Basic, Duro 2938, 2957
Bauer, Pavol 1224, 1233, 1561, 3422
Bazin, Pascal 2467
Beczkowski, Szymon Michal 2987
Beerten, J. 1551
Belhaouane, Moez 1158, 1756
Bello, Guilherme 2049
Benchaib, A. 745
Benchaib, Abdelkrim 820, 1215
Bender, Vitor C. 3217
Bendfeld, Christian 163
Benjamin, Sébastien 3156
Benkhoris, Mohamed Fouad 3205
Bensebaa, S. 1363
Bentivegna, N. 2293
Benzagmout, Abdelhadi 1905
Beranger, Bruno 2467
Berkani, M. 1363
Bernet, Steffen 124, 1569
Bertele, Felix 3137
Bertilsson, Kent 416, 460
Betto, Kento 3071
Betz, Robert Eric 927
Bevilacqua, Pascal 1279

Beza, Mebtu 969, 1952
Bhajana, V. V. Subrahmanya Kumar 2068
Bidini, Gianni 292
Biela, Juergen 2331, 2583, 2791
Biela, Jürgen 2230, 2409, 2446, 2475, 2513, 2524,
....2684, 2712, 2780, 2946
Bier, Anthony 2638
Bikinga, Wendpanga Fadel 2564
Binder, Andreas 2049
Birou, Camille 332
Bissal, Ara 871
Blaabjerg, F. 3237
Blaabjerg, Frede 1, 460, 810, 927, 2088, 2135,
.................................2220, 2393, 2573, 2888, 2898, 2928, 3119
Blanco, Marcos 1076
Blanquez, Francisco R. 1336
Blaquiere, Jean-Marc 1057
Blinov, Andrei 2996
Blume, Sebastian 2684
Böcker, Jan 2341
Böcker, Joachim 1638, 3024
Boersma, S. 745
Bohlen, Oliver 707
Bohnke, M. 1613
Boige, François 1057
Boisaubert, Emile 3172
Bolzan, Thais E. 3217
Bolzoni, A. 1306
Bombois, X. 745
Bongiorno, Massimo 969, 1952
Borcherding, Holger 608
Boulaud, Etienne 3172
Bourennane, Abdelhakim 1096
Bourguet, Salvy 2907
Bouscayrol, Alain 3330
Boutleux, Emmanuel 433
Boutry, Arthur 2366
Bozorg, Mokhtar 3247
Boškovic, N. 2812, 2820
Branca, Xavier 522
Briff, Pablo 9
Bringezu, Thilo 2780
Brockhage, Torben 1766
Brooks, Michael 2773
Bruyere, Antoine 1756
Brückner, Thomas 2938, 2957
Bründlinger, Roland 2723
Büdel, Johannes 1718
Bucher, A. 3403
Bucher, Alexander 203
Budo, Kohei 1450
Buigues, Garikoitz 85
Burgos, Rolando 2366
Burgos-Mellado, Claudio 1354
Busatto, G. 3210
Busatto, Giovanni 2802
Buttay, Cyril 1106, 2265, 2366
Cacciato, M. 909
Cai, Pei 2135
Camail, Philippe 2000, 2265
Camara, M. B. 2881
Camurca, L. 3305
Cardelli, Ermanno 292
Cárdenas, Roberto 1354
Carnielutti, Fernanda 2851
Caron, Hervé 1700
Carpita, Mauro 3247
Carpiuc, Sabin 962
Cascino, S. 2293
Castellazzi, Alberto 2210
Castellini, Simone 292
Castelltort, Arnaud 1747
Castiglia, V. 3237
Catellani, Stéphane 2638
Cavallaro, D. 2293
Chaiba, Azeddine 362
Chakraborty, Sajib 2320, 3111
Cheaito, Hassan 829
Chen, Linglin 1224, 1233
Chen, Qing 542
Chen, Yu 804
Chen, Zhengxin 637
Cheshire, Christoph 3137
Chevalier, Florian 1895
Chillón-Antón, Cristian 853, 1542
Chiumeo, Riccardo 424
Chraye, Hélène 3427
Chrin, Phok 3172, 3376
Chrzan, Piotr J. 3054
Chédot, L. 183
Chédot, Laurent 433, 1106
Ciupageanu, Dana-Alexandra 292
Clerc, Guy 433, 829
Clerici, Alessio 424
Cochelin, Anne-Sophie 3426
Coelho-Medeiros, Rafael 1479
Colak, Ilknur 871
Colas, F. 1579, 2554
Colas, Frédéric 1158
Colmenero, Manuel 1336
Connaughton, Alexander 1866
Cordier, Julien 3272
Corentin, Darbas 1803
Cornea, Octavian 2192

Costa, François	503
Coujard, Clementine	1270
Cravero, Jean-Marc	3330
Crebier, Jean-Christophe	2010
Da Cunha, Julian	2312
Dabbabi, Asma	2907
Dahmen, Christopher	843
Dai, Jing	820, 1215, 1479
Dakyo, B.	2881
Dang, Ziyue	804
Danzer, Michael A.	707
Darivianakis, Georgios	998
Davari, Pooya	726, 1006, 2573, 3119, 3295
David, Romain	522
Davidson, Colin	2366, 3425
Davoodi, Amirali	2393
De Doncker, Rik W.	153, 163, 1627
De Jaeger, Jean-Claude	1895
De Jódar, Esther	1658
De La Grandiere, Hubert	3424
De Lauretis, Maria	512
De Mora, Pablo Rodriguez	1533
De Oliveira, Eduardo F.	2535
De, Dipankar	2210
De-Preville, Guillaume	9
Defrance, Nicolas	1895
Degrenne, N.	2865
Delamea, R.	2865
Delarue, Philippe	1158, 3330
Delette, G.	1613
Delhommais, Mylène	1737
Delpech, F.	2554
Demidov, Iurii	2419
Denis, Guillaume	1270
Dennetière, S.	2967
Derbey, Alexis	2039
Derkacz, Pawel B.	3054
Despesse, Ghislain	503
Despouys, Olivier	1373
Dessante, Philippe	1858
Devos, Guillaume	1858
Di Gregorio, Francesco	1747
Dieckerhoff, Sibylle	2274, 2341, 2603
Dierks, Rebecca	3091
Dietz, Armin	1316
Dincan, Catalin	2873
Dinkel, Daniel	1297
Dinulovic, Dragan	2773
Djerioui, Ali	3205
Dong, Dong	2366
Doppelbauer, Martin	1589
Douine, Bruno	1373

Doumiati, Moustapha	2251
Drabek, Pavel	2068
Dragicevic, Tomislav	2393, 2898
Driesen, J.	480
Driesen, Johan	27
Drofenik, Uwe	627
Duarte, J.	2812, 2820
Duarte, Jorge L.	1067, 2656
Duarte, Renan R.	3217
Duerbaum, Thomas	203
Dujic, Drazen	1776, 2486
Dürbaum, Thomas	173, 193, 1460
Dworakowski, Piotr	406, 1106, 2000, 2265, 3006
Džonlaga, Bogdan	1479
Ebersberger, Janine	3340
Ebrahimi, Reza	2860
Eckel, Hans-Günter	532, 1666, 2059, 2126, 3282
Eckerle, Richard	765
Ecrabey, Jacques	2467
Egrot, Philippe	1479
Eguia, Pablo	85
Ehlich, Martin	608
El Baghdadi, Mohamed	2320
El Jihad, Hamza	1982
Elizondo, Laura Ramirez	1561
Ellul, Racquel	1025
Elsabrouty, Ibrahim	871
Elsied, Moataz	3156
Elthokaby, Youssuf	352
Endisch, C.	551
Enjeti, Prasad	1390
Erenler, Yeliz	571
Errigo, F.	183
Escobar, Gerardo	1390
Escofficr, Réne	1470
Esfetanaj, Naser Nourani	2860, 3295
Eskandari, Bahman	863
Eslamian, Morteza	3064
Eslampanah, Vahid	2860
Espina, Enrique	1354
Etoz, Burhan	2241
Fadel, Maurice	3101
Fauth, Leon	3081
Fazli, Nastaran	2126
Fehr, Hendrik	1030, 2116
Ferreira, Jan Abraham	892
Ferrieux, Jean-Paul	2039
Finkenzeller, Michael	835
Fischer, Manuel	1168, 1605, 1709
Fogsgaard, Martin Bendix	2258
Foray, Etienne	1874
Forsyth, A. J.	1306

Fort, Jiri	1086
Founier, Etienne	3101
Francois, Bruno	362, 1700, 1756
Frédèric, Poitiers	1803
Frey, D.	406
Frey, David	2695
Freytes, Julián	9
Friebe, Jens	2743, 2763, 3081
Fromme, Christopher	2603
Frost, Damien	223
Fruchier, Olivier	1905
Fu, Siqi	302
Fuchs, Simon	2409, 2513
Fürst, Markus	1885
Galeshi, Soleiman	2695
Gamatié, Abdoulaye	1747
Gandolfi, Chiara	424
Ganjavi, Amir	726, 1006
Gao, Fei	1224, 1233
Gao, Jianbo	1262
Garate, José Ignacio	1524
Garbuio, Lauric	1685
García-Torres, Felix	134
Garnier, Laurent	2467
Gaubert, Jean-Paul	396, 2284
Gauthier, Jean-Yves	590, 736
Gautier, Cyrille	1118, 1858, 2039
Gautier, Maxime	1923
Geng, Zeyang	3198
Gensior, Albrecht	581, 1030, 2116
Gentejohann, Marius	2274
Georges, Didier	820
Gerada, Chris	1944
Gerada, David	1944
Geramirad, Hadiseh	2000
Gerstner, Michael	1316
Geske, Martin	2938, 2957
Geury, Thomas	2320
Geyer, Tobias	2145
Ghamrawi, Ahmad	2284
Ghanes, Malek	3205
Gholami-Khesht, Hosein	3119
Giacomazzo, M.	490
Gierschner, Sidney	2059, 2126
Giewont, William	1016, 1972
Giotakos, Panagiotis I.	2172
Girbau-Llistuella, Francesc	1542
Gireada, Mihaita	2192
Glac, Antonín	3257
Gladen, Marcel	1148
Glasberger, Tomas	3166, 3314
Gleissner, Michael	644, 686
Glushakov, Vasiliy V.	47, 56
Gnärig, Jan Lasse	1030
Golluccio, G.	3210
Golsorkhi, Mohammad S.	2733, 2898
Gomez, Juan S.	1354
Gomis-Bellmunt, Oriol	1542, 2977
Gonzalez, Jose Ortiz	2241
Gonzalez-Torres, J-C.	745
González-Fontderubinat, Paula	1542
Gosses, Kilian	143
Gou, Wanchao	153
Govaerts, G.	480
Gradinger, Thomas B.	627
Grainger, Brandon M.	65
Grbovic, Petar J.	2723
Grecki, Filip	627
Green, Tim C.	276
Green, Tim	2977
Griepentrog, Gerd	1675
Gruson, F.	1579
Gruson, Francois	1158
Gu, Chunyang	1944
Guerrero, Josep. M	3289
Gui, Qiuye	1030
Guichon, Jean Michel	2564
Guillaud, X.	1579
Guillaud, Xavier	1158, 1756, 1952
Guo, Mingzhu	718
Guo, Xuan	317
Gutierrez, A.	919
Gutierrez, Sebastian	598
Gärtner, M.	3403
Götting, Gunther	1289
Hackl, Christoph	900
Hage-Hassan, Maya	1858
Hai, Jie	231
Hallemans, L.	480
Hamid, Muhammad	1289
Hammerer, Horst	2182
Hammes, David	2059
Han, Hua	260, 285, 302, 326
Han, Lubin	370
Han, Weiji	765
Haq, Omer Ikram Ul	3393
Häring, Johannes	644, 686
Harnefors, Lennart	1952
Hartmann, Michael	2258
Harzig, Thibaut	65
Hase, Genki	467, 498
Hasenohr, C.	3403
Hasenohr, Christian	203
Hatori, Kenji	1489

Haug, Martin 2773
He, Maojun 804
He, Yuying 1410, 1435
Hegazy, Omar 2320, 3111
Hein, Lukas 3413
Helle, Lars 2873
Heller, M. 3403
Hénaux, Carole 3101
Henninger, Stefan 3179
Heredero-Peris, Daniel 853, 1542
Herkommer, Christian 1718
Hernandez, Fernando Davalos 598
Herwig, Daniel 1766
Heucke, Sören 2341
Heydari, Rasool 2733, 2898
Hideaki, Yano 1442
Higashihata, Takeshi 1489
Hijazi, A. 183
Hiller, Marc 3366
Hillermeier, Claus 1297
Himker, Niklas 979
Hinkkanen, Marko 2495
Hiraki, Eiji 386
Hirayama, Hiroshi 2376
Hiwatari, Daichi 2376
Hofer, Matthias 2403
Hoffmann, Felix 954
Hofmann, Harald 203
Homann, Michael 307
Hong, Yang 231
Horrein, Ludovic 3330
Horvatic, Iréna 3156
Houari, Azeddine 3205, 3289
Hu, Anliang 2946
Hu, Rui 2594
Huang, Han 1128
Huang, Pin-Yu 2666
Huang, Xingxuan 1972
Huisman, Henk 1067, 1186
Hulea, Dan 2192
Hussain, E. K. 471
Iannuzzo, Francesco 2258, 2704
Ibanez, Federico 598, 3034
Idarreta, Aitor 1205
Idir, Nadir 1793, 1895, 2078
Iman-Eini, Hossein 3305
Ingman, Jonny 563
Inoue, Sadayuki 19
Iraola, Unai 1205
Isaksson, Dan 1016
Ishihara, Hiroki 19
Isobe, Takanori 3358

Itoh, Jun-Ichi 1380, 2200
Jackiewicz, Krzysztof 2029
Jaeger, Johann 3179
Jakob, Roland 2957
Jaritz, Michael 2684
Jasim, Omar 9
Jean-Christophe, Olivier 1803
Jeannin, Pierre-Olivier 3054
Jehle, Andreas 2475
Jelena, Popovic 892
Jeong, Min 2409, 2513
Ji, Shiqi 1972
Jia, Ming 1627
Jiang, Jinhai 637
Jiaqi, Diao 231
Joebges, Philipp 1627
Jonokuchi, Hideki 2376
Jorge, Tenorio 1400
Joryo, Satoshi 3071
Jotwani, Ankit 1138
Joubert, Charles 522
Judge, Paul D. 276
Jun-Ping, He 1599
Junge, Patrick 2613
Juntunen, Raimo 3227
Junyent-Ferré, Adrià 2977
Justin, Elissa Cresenta Anak 2265
Jäppinen, Janne 563
Järvisalo, Heikki 1016
Jørgensen, Asger Bjørn 2987
Kado, Yuichi 2666
Kadri, R. 2554
Kahl, Tino 2603
Kahle, Karsten 1336
Kaiser, Ingmar 1666
Kallfass, Ingmar 1915
Kaminski, Nando 954
Kampen, Dennis 2049, 2153
Kanchan, R. S. 3393
Kang, Yong 797, 804
Karaventzas, Vasilios 2583
Karlsson, Martin 512
Kaszewski, Arkadiusz 2029
Kawabata, Yoshitaka 1177, 1243
Keel, Oliver 2791
Kefer, K. 3403
Kehl, Zdenek 3166
Keller, Christian 2938, 2957
Kennel, Ralph 542, 900, 1262, 2386, 3272
Kersten, Anton 765
Kesbia, Nasreddine 1685
Kestelyn, X. 1579

Ketchedjian, Vasken	1605
Keysan, Ozan	1823
Khanzadeh, Babak	1040
Kharezy, Mohammad	3064
Kikuchi, Naoto	2200
Killeen, Peter	777
Kim, Bunthern	3172, 3376
Kimura, Norihito	105
Kimura, Shota	377
Kindl, Vladimir	1086
Kirchenberger, U.	3403
Kirowitz, Thomas	2403
Kitagawa, Wataru	1425
Kitamura, Taishi	1177
Kiviniemi, Mika	563
Kjaer, Martin Vang	2888
Kjær, Philip	2873
Klass, Stefan	3272
Klier, Samantha	707
Koch, Dominik	880, 1915
Kohlhepp, Benedikt	266, 1460
Kojabadi, Hossein Madadi	2860
Komma, Thomas	835
Komrska, Tomáš	3257
Kondo, Keiichiro	2675
Kone, Lamine	1793
Kopacz, Rafal	2457
Korhonen, Juhamatti	1016
Kosan, Tomas	3166
Kouchaki, Alireza	3015, 3129
Koutroulis, Eftychios	3146
Krall, Felix	1196
Krim, Youssef	1700
Krischan, Klaus	1866
Kroneisl, Michal	1992
Krug, Dietmar	2059
Kucka, Jakub	2020, 3091
Kuder, Manuel	765
Kuebrich, Daniel	203
Kuhlmann, Kai	1718
Kukkola, Jarno	2495
Kumar, Dinesh	726, 1006, 2573
Kuring, Carsten	266
Kusaka, Keisuke	1380, 2200
Kuwana, Kazuki	1425
Kuwata, Akiko	19
Kwasinski, Alexis	2594
Kyyrä, Jorma	3348
Kärkkäinen, Hannu	2829
Kärkkäinen, Tommi J.	563
Kübrich, Daniel	266
Küster, Pierre	571

Labiano, Daniel	1205
Labouré, Eric	1858
Lacarnoy, Alain	654
Lacressonnière, Fabien	332
Ladoux, Philippe	1106, 2265
Lafon, Frederic	1793
Lafoz, Marcos	1076
Lagier, Thomas	1106, 2000, 2265
Lana, Andrey	2419
Langbauer, Thomas	1866
Langmaack, N.	241
Langwasser, M.	3305
Lapassat, N.	37
Larruskain, Marene	85
Lautner, Frank	1831
Lazaroiu, Gheorghe	292
Le Moigne, Philippe	1158, 2078
Le Métayer, Pierre	3006
Le, Hoai Nam	2200
Lee, Seong-Yong	2486
Leedham, Rob	871
Lefebvre, Bruno	1106, 2000, 2366
Lefebvre, S.	1363
Lefebvre, Stéphane	1118
Lehmann, Franziska	944
Lehn, Peter	223
Lembeye, Yves	2010, 2695
Lemmen, Erik	2350, 2358
Leo, Jacopo	75
Leppänen, Joonas	563
Leterme, Willem	276
Letrouvé, Tony	1700
Lexow, Daniel	3282
Li, Boyang	1289
Li, Chi	317
Li, Dingrui	1972
Li, Hui	2704
Li, Jiaqi	9
Li, Jing	1944
Li, Lang	285
Li, Qi	1262
Li, Tao	3044
Li, Weilin	2135
Li, Yongdong	317, 1944
Li, Yu	2386
Li, Zheming	2106
Liang, Chaohui	3044
Liang, Lin	370, 443
Liao, Jianquan	3385
Liao, Yuefeng	1498
Licari, John	1025
Lima, Glauber De Freitas	2010

Lin, Lei	937
Lin-Shi, Xuefang	590, 736
Liserre, Marco	3305
Liu, Cuicui	2505
Liu, Fuxin	1410, 1435
Liu, Libo	1289
Liu, Xudan	804
Liu, Yao	260
Liu, Zhangjie	260, 302, 326
Liukkonen, O.	2630
Llanos, Jacqueline	1354
Llonch-Masachs, Marc	1542
Locment, Fabrice	667, 677
Loisel, Rodica	2907
Loiselay, Florent	1648
Lomonova, E. A.	2812
Lorenz, Malte	2020
Ludois, Daniel C.	777
Lunz, B.	3403
Luo, Fang	370
Luo, Xian	153
Lutz, Josef	3413
Lutze, Marcel	1675
López-Alcolea, Fco. Javier	134
Ma, Yixiao	1435
Mabe, Jon	1524
Machmoum, Mohamed	2907, 3205, 3289
Maekawa, Sari	2838, 2844
Maerz, Martin	1316
Magambo, Jean Sylvio Ngoua Teu	2078
Maharana, Manoj Kumar	2068
Maharana, Suman	2210
Mahr, Florian	3179
Maier, Robert W.	1252, 2106
Mäkelä, Juha	3348
Mallwitz, R.	241, 490
Mallwitz, Regine	213, 1515
Maneiro, Jose	1106, 3006
Mannen, Tomoyuki	3358
Mannerhagen, Felix	3198
Mantellini, Mattia	1076
Mantzanas, Panagiotis	203
Mao, Saijun	892
Marcault, E.	919
Marchesoni, Mario	3321
Marciano, D.	3210
Marciano, Daniele	2802
Margueron, Xavier	2078
Marmolejo, Narciso G.	1589
Marquardt, Rainer	843, 1297
Martin, Christian	1874
Martin, Jérémy	2638

Martinez, Wilmar	598, 3034
Martire, Thierry	1905
Martínez-Gómez, Manuel	1354
März, Martin	143
Mateos, Felix Rodriguez	2583
Mattar, Rita	1118
Mattsson, Aleksi	1016
Maussion, Pascal	3376
Mayorga, John Paul	1658
Mazuela, Mikel	1205
McIntyre, Michael	65
Mehdi, Driss	2284
Meißner, Markus	124, 1569
Mendoza-Araya, Patricio	2917
Meneses, Javiera	2917
Meng, Qingchao	2446
Mercier, Adrien	1858
Mercier, Sylvain	2467
Mertens, Axel	979, 1766, 2020, 2303, 2545, 2613, 2743, 2763, 3091, 3340
Mesbahi, Tedjani	3205
Meynard, Thierry A.	654
Mezrag, Bachir	2564
Mezzetti, Margarita	944
Micallef, Alexander	1025
Miceli, R.	3237
Milas, Nikolaos T.	2172
Miletic, Zoran	2723
Millinger, Jonas	512
Minami, Masataka	467, 498
Mirtchev, Alex V.	2429
Mitani, Kohei	1425
Miyauchi, Tsutomu	377
Mizutani, Hiroto	386
Mohamed, Abdalla Hussein	115
Mohamed, Islam	352
Molina-Martínez, Emilio J.	134
Mollov, S.	2865
Molnar, Jan	3166
Monmasson, Eric	1118, 1823
Montero, E. Rodriguez	2098
Montesinos-Miracle, Daniel	853, 1542
Moraes, Tiago José Dos Santos	1693
Morel, Cristina	2251
Morel, F.	183, 406
Morel, Florent	2000, 2366
Morel, Hervè	1279, 1648
Mori, Osamu	386
Morici, Riccardo	1076
Morizane, Toshimitsu	1047, 3071
Mortimer, Benedict	163
Motegi, Shin-Ichi	1786

Mourouvin, Rayane	820
Mourtzis, Dimitris A.	2172
Muehlbauer, Markus	707
Muetze, Annette	1196
Müller, Jan-Kaspar	2763, 3340
Mumtaz, Muhammad Adnan	1326
Munk-Nielsen, Stig	2987
Muñoz, Fredy	3189
Muñoz, Javier	3189
Muntean, Nicolae	2192
Murata, Ryo	386
Musznicki, Piotr	3054
Nabatirad, Mohammadreza	787
Nada, Kaho	19
Nadh, Greeshma	1619
Nair, Durga S.	1619
Najera, Jorge	1076
Najjar, Mohammad	3015, 3129
Nakagaki, Akito	498
Nakamura, Keiichi	1489
Nakashima, Osamu	2376
Nakatani, Shota	1047
Nakazawa, Yosuke	2675
Narula, Anant	969, 1952
Natori, Kenji	2675
Navarro, Gustavo	1076
Navas, Alex F.	1354
Ndagijimana, Fabien	2010
Nee, Hans-Peter	2220
Neumann, Jessica	3101
Nevaranta, N.	2630
Ngoua-Teu, J-S	1613
Nguyen, Ngac Ky	1693
Nguyen, Van Sang	2638
Nicolas, Ginot	1803
Nie, Cheng	1972
Nie, Qingqing	797
Niemelä, M.	2630
Niemelä, Markku	563, 2829
Nikowitz, Mario	2403
Nisch, A.	3403
Nishida, Yasuyuki	1786
Nitzsche, Maximilian	880, 1605, 1709, 1915
Niu, Liyong	342
Norambuena, Margarita	2851
Nymand, Morten	3015, 3129
Obernolte, Urs	608
Oberschelp, Wolfgang	1727
Odriozola, Kepa	654
Oguma, Kenji	377
Ohta, Takahiro	2666
Okamori, Daichi	1047
Okazaki, Akihiro	2838
Okazaki, Yuhei	1040
Oliveira, Joao	1648
Olivier, Jean-Christophe	2251
Omori, Hideki	1047
Omrane, A.	2554
Ordoño, Ander	1524
Orfanoudakis, Georgios I.	3146
Ortega-Perez, Carmen	3330
Ota, Kenji	1489
Ottaviano, Andrea	292
Oumaziz, Amirouche	1096
Pace, Loris	1895
Páez, Juan	1106
Paez, J. D.	406
Palensky, Peter	810, 1561
Pallier, Joris	829
Palm, Herbert	707
Pan, Xuejiao	1128
Passalacqua, Massimiliano	3321
Patarroyo-Montenegro, Juan F.	1841
Patti, Davide	3210
Paulus, Sebastian	2303, 2545
Pavlicek, Vladimir	1086
Pawellek, A.	3403
Pawellek, Alexander	203
Payman, A.	2881
Pei, Xiaoze	342
Peller, Stefan	266
Pelosi, Dario	292
Peltoniemi, Pasi	3227
Peña, R. A.	183
Pendharkar, Ishan	1346
Peng, Han	797, 804
Peng, Hao	804
Peng, Tao	1498
Penin, Carolina	1905
Peralta, Patricio	75
Perenyi, Christian	65
Peretti, Luca	3393
Pérez-Molina, María José	85
Peric, V.	745
Peroutka, Zdenek	3257, 3314
Perriard, Yves	75
Petit, M.	1363
Petit, Mickael	1118, 3054
Petkovic, Marko	1776
Peyghami, Saeed	810, 2573
Pfeifer, Markus	1675
Pfeiffer, Jonas	571, 2535
Phulpin, Tanguy	618
Pidancier, Thomas	3247

Pietrzak-David, Maria...3376
Pilawa-Podgurski, Robert C. N......................................1934
Pinheiro, Humberto...2851
Pinomaa, Antti...2419
Pinto, Rafael A..3217
Pirsto, Ville...2495
Pitel, Ira..1390
Planson, Dominique...1279, 1648
Plaza, Jesus D. Vasquez...1841
Plissonnicr, Marc..1470
Poebl, Monika..835
Polacek, Libor..1086
Pollet, Benjamin..503
Pommier-Petit, Pascal..829
Pool-Mazun, Erick I..1390
Popuri, Madhuchandra..2068
Pouresmaeil, Edris..863
Pöyhönen, Santeri...2753
Prevost, Thibault..1270
Prieto, Dany..3101
Prieto-Araujo, Eduardo..2977
Pronin, Mikhail V...47, 56
Puls, Simon...608
Pulvirenti, M...909, 2293
Pursiainen, Jooa...3227
Putkonen, A..2630
Pyrhönen, Juha...2829
Pyrhönen, O..2630
Pyrhönen, Olli...2419
Qashlan, Ziyad H. S...571
Qiang, Jin..163
Qoria, T..1579
Qoria, Taoufik...1756
Queval, Loic..1473, 1479
Rabba, Heiko..307
Rabkowski, Jacek...2457, 2996
Radet, Hugo..332
Rahmoun, Yasser..2182
Rahul, Arun S..1619
Rajput, Shailendra...2439
Ramm, Hannes...307
Ramírez-Scarpetta, J. M..1850
Rasmussen, Tonny Wederberg...1138
Rathnayake, Hansika..726, 1006
Rault, P...2967
Raute, R..989, 1506
Rautio, Juuso..563
Ravyts, S...480
Ravyts, Simon...27
Rayati, Mohammad...3247
Razzaghi, Reza..787
Rehlaender, Philipp..1638

Reigosa, Paula Diaz...1346, 2258
Reißenweber, Lukas...3263
Rekola, Jenni..3227
Ren, Chunpin...718
Restrepo, Jose Alex..1850
Retianza, Darian V...1067
Retianza, Darian Verdy...1186
Rezaee, Ali Yahya...460
Richardeau, Frédéric...1057, 1096
Rietmann, Stefan...2712
Rigot, Valentin..618
Risch, Raffael...2230
Riu, Delphine..3156
Roa, Claudio...3189
Robet, Pierre-Philippe...1923
Roboam, Xavier...332
Robyns, Benoit...1700
Rodriguez, Jose...900, 2851
Rodriguez, José Luis...1205
Rokrok, Ebrahim..1756
Roncero-Sánchez, Pedro..134
Roose, T...1551
Röser, Tobias..3137
Roth-Stielow, Jörg.................................880, 1168, 1605, 1709
Rouger, Nicolas..1057
Routimo, Mikko...2495
Rouzbehi, Kumars..863
Ru, Yang...231
Ruan, Xinbo..1410, 1435
Rubenbauer, Hubert...3179
Rufer, Alfred..696
Rute, Erwin..1354
Ruthardt, Johannes.................................1168, 1605, 1709
Saad, H..2967
Saad, Yamen..3295
Saber, Christelle..1118
Sácz, Doris..1354
Sadarnac, Daniel...618
Saeedian, Meysam..863
Saggio, M..2293
Sah, Gyanendra Kumar..532
Sahin, Ilker...1823
Saim, Abdelhakim...3289
Sakai, Kazuto..1442
Sakai, Norikazu..1489
Sakai, Yasuhiro..1489
Sakaria, Omar Ahmed..3295
Sakly, Jihen...618
Sakurazawa, Yoshiki..2675
Saleh, Bassem..2648
Salem, Qusay...1289
Sallot, P..1613

Salvo, L. 909
Sánchez-Sánchez, Enric 2977
Sandelic, Monika 2088
Sandik, Diane -Perle 512
Sangwongwanich, Ariya 1, 2088
Sanseverino, A. 3210
Sanseverino, Annunziata 2802
Santos, Miguel 1076
Sari, A. 183
Sarraute, Emmanuel 1096
Sassatelli, Gilles 1747
Sathik, Mohd. Ali Jagabar 460
Saudemont, Christophe 1700
Savaghebi, Mehdi 2733, 2898
Savarit, Elise 1982
Sawada, Takashi 2623
Sayed, Mahmoud A. 1417
Scarpetta, Jose Miguel Ramirez 1400
Scelba, G. 909
Schafmeister, Frank 1638, 3024
Schanen, Jean-Luc 1685, 2039, 3054
Schiesser, Matthias 962
Schleippmann, Nico 143
Schlesinger, Richard 2524
Schlüter, Michael 2274
Schmidt, Dimitri 1168
Schmitt, Alexander 2182
Schmitt, N. 1363
Schmitt, Stefan 954
Schmitz, Jan 124, 1569
Schobre, Thorben 213, 1515
Schröder, Günter 1727
Schrödl, Manfred 2403
Schulte, Hendrik 1709
Schulz, Matthias 143
Schulz, Nicola 1346
Schulze, Torben A. 307
Schwabe, Christian 3413
Schütt, Michael 532
Sciacca, A. G. 909
Scicluna, K. 989, 1506
Sechilariu, Manuela 667, 677
Segur, R. 745
Seiler, Pascal 2331
Semail, Eric 1693
Sergeant, Peter 27, 115
Shah, Chirag 153
Sharkh, S. M. 471
Sharkh, Suleiman M. 3146
Shi, Guangze 326
Shinoda, Kosei 820, 1215
Shiozaki, Koji 2623

Shirakawa, Tomohide 386
Shousha, Mahmoud 2773
Si-Yuan, Cai 1599
Siala, S. 95
Siala, Sami 1982
Siebke, K. 490
Siemens, Ag 2603
Silventoinen, Pertti 563, 1016
Simola, Aleksi 2753
Singer, Arthur 765
Skala, Bohumil 1086, 3257
Smailus, Erik 1675
Šmídl, Václav 1992
Smit, A. 3403
Snook, Mark 871
Soeiro, Thiago Batista 1224, 1233, 1561
Sokur, Pavel V. 47
Soltau, Nils 1489
Song, Kai 637
Soumaoro, Ousmane 2251
Soupremanien, U. 1613
Sprunck, Sebastian 2535
Stadler, Alexander 3263
Staines, C. Spiteri 989, 1506
Stathis, Spyridon 2684
Staudt, Volker 1148
Stecca, Marco 1561
Steckler, Pierre-Baptiste 736
Stengl, Josef 1086
Stenglein, Erika 173, 193
Štepánek, Jan 3257
Stock, Alexander 1718
Stöckl, Johannes 2723
Stöttner, J. 551
Stotckaia, Anastasiia D. 47, 56
Stras, Andrzej 2029
Streit, Lubeš 3257
Strittmatter, Tobias 1346
Strunk, Robin 979
Su, Guoxing 1410
Su, Mei 260, 285, 302, 326, 1498
Suarez, Camilo 3034
Sugiyama, Kohei 1177
Sun, Jian 1962
Sun, Yao 260, 285, 302, 326, 1498
Svensson, Jan R. 1952
Tadano, Hiroshi 2623
Taheri, Shamsodin 863
Takahara, Takaaki 386
Takahashi, Hirotaka 377
Takahashi, Hiroyuki 1243
Takano, Tomihiro 19

Takeshita, Takaharu 1417, 1425, 1450
Takeuchi, Somi 1243
Talbert, Thierry 1905
Talla, Jakub 3257
Tan, Guoqiang 443
Tanaka, Ami 2844
Tanaka, Miwako 19
Tanaka, Nobuhiko 1489
Tang, Bojin 718
Tang, Houjun 1224, 1233
Tang, Xiaohu 1589
Tannhäuser, Marvin 2603
Tant, Jeroen 27
Tareilus, G. 241
Tarisciotti, Luca 1224, 1233
Tárraga, Sergio 1658
Tatakis, Emmanuel C. 755, 2172, 2429
Taul, Mads Graungaard 927
Tedesco, Davide 2802
Teigelkötter, Johannes 1718
Teirelbar, Ahmed 2648
Tenorio, Jorge 1850
Teramura, Keiko 377
Terbrack, C. 551
Thal, Eckhard 1489
Thiringer, Torbjörn 765, 1040, 3064, 3198
Tibola, Gabriel 2656
Tikhonov, Sergey 1638
Todd, R. 1306
Tolbert, Leon M. 1972
Torres, Alfonso Parreño 134
Torres, Esther 85
Torres, Fernando 3189
Torres, Jorge 1076
Torres, Jose Rueda 810
Touhami, Mustapha 503
Tran, Dai-Duong 2320, 3111
Traoré, Bakou 2251
Tremmel, Werner 2723
Tremouilles, D. 919
Trillaud, Frédéric 1373
Trochimiuk, Przemyslaw 2457
Tröster, Nathan 1168
Tsolaridis, Georgios 2331
Tsoumas, Ioannis 998, 2145, 2163
Turjanica, Pavel 1086
Turki, Faical 307
Twardon, N. 3403
Ufnalski, Bartlomiej 2029
Ulissi, Gabriele 2486
Umetani, Kazuhiro 386
Unruh, Roland 3024

Uwai, Shuto 1177
Vaccaro, Luis 3321
Valtee, Mikko 3227
Van Den Broeck, G. 480, 1551
Van Duivenbode, Jeroen 1186
Van Mierlo, Joeri 2320, 3111
Van Tichelen, P. 480
Vanfretti, L. 745
Vannier, Jean-Claude 1479
Vansompel, Hendrik 115
Vashishtha, Anushruti 1138
Vasquez-Plaza, Jesus D. 1850
Vázquez, Javier 134
Vecchia, Mauricio Dalla 27
Vechiu, Ionel 396
Velardi, F. 3210
Velardi, Francesco 2802
Velazco, Diego 433
Venet, P. 183
Verbelen, Florian 27
Vermeersch, Pierre 1158
Vermulst, Bas 2350, 2358
Viana, Caniggia 223
Videt, Arnaud 1793, 1895
Vienot, Stephane 1793
Vieto, Ignacio 1962
Villarejo, José 1658
Villegas, Carlos 962
Vip, Stephan 2303, 2545
Vitan, Danut 2192
Voborník, Ales 1086
Vogelsberger, M. 2098
Voigt, Matthias 944
Voldoire, Adrien 2039
Vollaire, Christian 2000
Vollmaier, Franz 1866
Vorontsov, Aleksey G. 47, 56
Votava, Martin 3314
Vu, Duc Tan 1693
Wada, Keiji 3358
Wallart, François 433, 736, 1106
Wang, Bo 450
Wang, Dian 677
Wang, Feng 2505
Wang, Fred 1972
Wang, Huai 1, 2888, 3295
Wang, Meiqi 1944
Wang, Qianggang 3385
Wang, Qiwu 1262
Wang, Tianqing 450
Wang, Xuehua 1410, 1435
Wang, Ziyue 443

Wankhede, Yugandhara H.	3081
Wasfi, Amr	2648
Watanabe, Hiroki	1380
Weicker, Martin	2049, 2153
Weimer, Julian	880, 1915
Weinert, Tristan	1727
Weiss, Sébastien	1793
Wernicke, Laurenz	2603
Weyh, Thomas	765
Wickramasinghc, Thilini	1470
Wijnands, Korneel	2350, 2358
Wilk, Andrzej	2265
Will, Frank	944
Winzer, Patrick	2182
Wolbank, T.	2098
Wondrak, W	3403
Wondrak, Wolfgang	644, 686
Wrona, Grzegorz	2457, 2996
Wu, Hailong	1693
Wu, Xiaohua	2135
Wunder, Bernd	143
Xi, Jiawen	342
Xia, Qingping	260
Xiang, Yusheng	1289
Xie, Jian	1289
Xin, Wei	370
Xiong, Weijing	1498
Xu, Chaoqun	718
Xu, Chen	937
Xu, Dianguo	450
Xu, Guo	1498
Xu, Junzhong	1224, 1233
Xu, Lie	1944
Xu, Xinwei	2656
Yakop, Netan	223
Yamada, Shota	1243
Yamashita, Daniela Yassuda	396
Yamazaki, Osamu	2675
Yamdeu, Mathias Tientcheu	3101
Yan, Xiaoxue	443
Yan, Zheng	1326
Yang, Bo	1944
Yang, Guang	637
Yang, Y	3237
Yang, Yongheng	2135, 2393, 2888
Yang, Zhiqing	153, 163
Yano, Takahiro	1177
Yao, Ran	2704
Yao, Wenli	2135
Ye, Shuaichen	2928
Ye, Zichao	1934
Yin, Tianxiang	937

Yu, Qihao	2350, 2358
Yu, Yong	450
Yu, Zhanqing	718
Yuasa, Hiroaki	105
Yüce, Firat	3366
Yuki, Kazuaki	2675
Yuratich, Michael A.	3146
Zacharias, Peter	571, 1813, 2535
Zafar, Talha	2773
Zanetti, E.	2293
Zaoskoufis, Konstantinos	755
Zare, Firuz	726, 1006
Zdanowski, Mariusz	2996
Zehelein, Matthias	880, 1915
Zeller, Valentin	1460
Zeng, Guang	3413
Zeng, Xianwu	342
Zhai, Dongling	718
Zhang, Haibo	1158, 1756
Zhang, He	1944
Zhang, Li	1128
Zhang, Peng	637
Zhang, Xiaokang	590
Zhang, Zhenbin	2386
Zhang, Ziqian	2505
Zhao, Biao	718
Zheng, Zedong	317
Zhou, Dao	2860, 2928
Zhou, Niancheng	3385
Zhu, Chunbo	637
Zhu, Q.	1306
Zhu, Yangming	450
Zhu, Yuanhao	326
Zhuo, Fang	2505
Zi-Fan, Li	1599
Ziegler, Philipp	1168, 1605, 1709
Zinchenko, Denys	2996
Zucuni, Jordan P.	2851
Zuolian, Liu	231

IEEE
445 Hoes Lane
Piscataway, NJ 08854-4141

ISBN 978-1-7281-9807-1